The Rise of Modern China

中國近代史

The Rise of
MODERN CHINA

Fifth Edition

IMMANUEL C. Y. HSÜ

New York   Oxford
OXFORD UNIVERSITY PRESS
1995

Oxford University Press

Oxford   New York
Athens   Auckland   Bangkok   Bombay
Calcutta   Cape Town   Dar es Salaam   Delhi
Florence   Hong Kong   Istanbul   Karachi
Kuala Lumpur   Madras   Madrid   Melbourne
Mexico City   Nairobi   Paris   Singapore
Taipei   Tokyo   Toronto

and associated companies in
Berlin   Ibadan

Published by Oxford University Press, Inc.,
200 Madison Avenue, New York, New York 10016

Oxford is a registered trademark of Oxford University Press

Library of Congress Cataloging-in-Publication Data
Hsü, Immanuel Chung-yueh, 1923–
The rise of modern China / Immanuel C.Y. Hsü. — 5th ed.
p. cm.
ISBN 0-19-508720-8
1. China—History—Ch'ing dynasty, 1644-1912.
2. China—History—20th century. I. Title.
DS754.H74   1995
951'.03—dc20   94-13369
Cover Illustration: Demonstration at Tiananmen Square, 1989.
Reuters/Bettmann

2 4 6 8 9 7 5 3 1
Printed in the United States of America
on acid-free paper

To the Memory of
John King Fairbank

Preface

The T'ien-an-men Square crackdown in June 1989 brought China widespread condemnation, diplomatic ostracism, and international economic and military sanctions. Teng Hsiao-p'ing determined that the following three to five years would be most critical to the survival of Chinese Communism and that China should adopt a low posture of noninvolvement in international affairs while striving to achieve domestic stability. "Action is not as good as inaction," he declared.

But in less than six months all of Eastern Europe was engulfed in popular protests, precipitating the downfall of the Communist regimes in Poland, Hungary, Czechoslovakia, and East Germany. By the end of 1991 the Soviet Union itself had disintegrated. What would be the future of China?

Shaking off the initial shock of these cataclysmic events, the Chinese leadership reached some basic decisions. First and foremost, it would never voluntarily abdicate its monopoly of power as the Soviet leaders had done, and the media were instructed to play up the specter of disorder and chaos (*luan*) in the Soviet Union as a warning to the Chinese people should they think of following the Soviet course. Secondly, economic development must continue and the people's livelihood must improve. Thus, the authorities promised economic prosperity and political stability as long as they were in control.

What followed was an unprecedented economic boom, especially in the south and along the coast. With an amazing annual increase of 10 to 13.5 percent, China's GDP, in terms of "purchasing power parity," grew to

$1.66 trillion in 1991 as estimated by the International Monetary Fund—the third largest in the world after those of the United States and Japan. By A.D. 2020 the Chinese economy could be the largest on earth. The prospect recalls Napoleon's remark: "China? There lies a sleeping giant. Let him sleep. For when he wakes he will move the world."

The rise of semicapitalism in a Leninist dictatorship is unprecedented, and the Chinese model intrigues Western social scientists and some in the former Soviet Union and Eastern Europe as a viable Third Way between Communist central planning and Western democratic freedom and market economy. But the question of the durability of the Chinese model remains.

Conventional wisdom suggests that economic prosperity and market dynamism will inevitably lead to political change, as Margaret Thatcher openly proclaimed.* The Chinese may experience a quiet transformation of their Communist system from within. Since the old ideology has become largely irrelevant, they are looking for ways to reinforce it with a new unifying philosophy that is a mixture of traditional Confucianism, nationalism, patriotism, and a touch of "Great Chinaism" that subtly reflects the Middle Kingdom concept. Continued economic prosperity will give rise to a growing middle class in an emerging civil society, in which the new rich and powerful—the entrepreneurs, the bankers, the investors, the stockholders, and the professionals—will demand greater political participation and the rule of law, popular elections, and a voice in legislation and budgetary decisions. Step-by-step the government will find it necessary to accommodate them, and eventually—perhaps early in the 21st century—a Chinese-style "restrained" democracy may emerge.

In preparing this edition, I have received valuable advice from Drs. Leonard Tourney and Robert H. Smitheram. To them I wish to express my deep gratitude.

Santa Barbara, California   I.C.Y.H.
June 1, 1994

---

* Margaret Thatcher, "The Future Has No Borders," *Free China Review*, Jan. 1993, 30–35.

Preface to the First Edition

This comprehensive history conveys primarily a Chinese view of the evolution of Modern China, reinforced by the fruits of Western and Japanese scholarship which has been considerable in the past three decades. We see here a turbulent era of Chinese history in which domestic and foreign forces interacted to transform the Confucian universal empire into a modern national state. The metamorphosis was labored, slow, and at times painful; an inside view of the process promises to shed light on the contemporary behavior of China.

As the present volume represents the author's long-time study of history, both Western and Chinese, he is deeply grateful to many of his former professors at Harvard who taught him the historical discipline: Drs. John K. Fairbank, L. S. Yang, Edwin O. Reischauer, William L. Langer, and Serge Elisséeff. To Harvard-Yenching Institute he is indebted for four years of fellowship which enabled him to pursue graduate studies in a most stimulating atmosphere. He is also thankful to the numerous authors whose works have directly or indirectly benefitted him in the preparation of this volume. While it is impossible to acknowledge them all here, he should like to mention a few of the most helpful scholars and research centers. Hsiao I-shan's monumental work, *Ch'ing-tai t'ung-shih* (A general history of the Ch'ing period), which first appeared in two volumes in 1927–28 and was enlarged into a five-volume set in 1963, is truly a mine of information, which materially enriched this work. The writings of John K. Fairbank have been a source of inspiration to a generation of students, and under his resourceful directorship the Harvard East Asian Research Center has published several dozen monographs which have significantly elevated the level of scholarship on Modern China. The series of publications by the Institute of Modern History, Academia Sinica, on Taiwan, the Toyo Bunko in Tokyo, and the Historical Research Association on mainland China, have all been useful in various ways. Special thanks are due Dr. S. Y. Teng, University Professor at Indiana University, for his perceptive comments on the manuscript which enabled the author to make numerous corrections and improvements. To his many students over the past decade, the author is indebted for their stimulating questions which alerted him to their needs and frequently opened new vistas for discussion. Thanks are also due Mrs. Alice Kladnik for her typing of the manuscript and to Mr. En-han Lee for preparing the index. Last but not least, the author's wife, Dolores, deserves a special word of appreciation: but for her constant encouragement, moral support, and loving companionship, this work would not have been possible.

Despite all the help and inspiration he has received, the author alone is responsible for all the inadequacies of the book. He submits it to the public with the sincere hope that it will prompt other scholars to make worthier contributions, in the spirit of the Chinese saying "May the brick I have thrown attract jade from others" (*p'ao-chuan yin-yü*).

Santa Barbara, California   I. C. Y. H.
New Year's Day, 1970

Contents

VI The Rise of the Chinese People's Republic

Maps

A Note on the Text

1. All translations from the Chinese, unless otherwise indicated, are by the author.
2. Footnotes, usually considered superfluous in a work of synthesis, are nonetheless given, where essential, to indicate the sources of direct quotations and important statements, and thus facilitate the reader's further study.
3. A reading list appears at the end of each chapter, consisting mostly of easily available Western works. A few important Chinese and Japanese works are included for the advanced student. These reading lists are not intended to be exhaustive bibliographies of the chapters, but merely aids to further reading.
4. All Chinese, Japanese, and Russian dates have been transcribed to correspond with the Western dates according to the Gregorian calendar.
5. Chinese place names are given in accordance with the Chinese Postal Atlas; hence many of them appear in Anglicized forms such as Peking, rather than Pei-ching, and Nanking, rather than Nan-ching.
6. Chinese personal names are usually given in their original order, with the family name preceding the given name, such as Li Hung-chang, Mao Tse-tung. However, where a man is better known under some other form, such as Confucius rather than K'ung-fu-tzu, Chiang Kai-shek rather than Chiang Chung-cheng, T. V. Soong rather than Sung Tzu-wen, the more familiar form is adopted.
7. Transliteration of Chinese names follows the Wade-Giles system, originally devised by the British diplomat-linguist, Thomas F. Wade,

for his Peking *Syllabary* of 1859. Wade became the first professor of Chinese (1888-95) at Cambridge University after his retirement from the foreign service. The system was slightly revised by Herbert A. Giles, another British consular officer and Sinologue, for his *Chinese-English Dictionary* of 1912. The Wade-Giles system is designed to achieve pronunciation in the official Peking dialect (*Mandarin*); since many of the Chinese sounds have no exact counterpart in English, a number of letters used in the system are pronounced in a "peculiar" manner to approximate the Chinese sound. Though not perfect by modern linguistic standards, the Wade-Giles system has been widely followed by the Sinologists for many decades, and it is used here for convenience and consistency. Its main features may be summarized as follows:

Vowels

a is pronounced as in f*a*ther   *ai* as "i" in *i*ce   *ao* as in c*ow*
e as in *e*bb
i as in mach*i*ne
o as in s*aw* or *o*ft
u as in l*u*nar   *ü* as the German *ü*

Consonants

UNASPIRATED	ASPIRATED, AS MARKED BY AN APOSTROPHE
ch as J in *J*oe	*ch'* as in *ch*in
j as in *r*um	*k'* as in *k*ing
k as in *g*ive	*p'* as in *p*an
p as in *b*an	*t'* as in *t*ip
t as in *d*ip	*ts'* or *tz'* as in Bet*sy*
ts or *tz* as *dz*	
ss as in hi*ss*	

8. Japanese names are transliterated according to the Hepburn system in which:

a is pronounced as ä in f*a*r
i as e in m*e*te
u as oo in f*oo*d
e as e in b*e*t
o as o in h*o*me

9. Since 1979 a number of Western journals and newspapers have begun to use the Pinyin system of Romanizing Chinese adopted by the People's Republic of China. Its essential features are as follows, with corresponding letters used in the Wade-Giles system in parentheses:

a (a) as in father	p ('p) as in pan
b (p) as in boy	q (ch') as in chin
c (ts) as in its	r (j) as in run
d (t) as d in dip	s (s, ss, sz) as in sister
e (e) as in uh	t ('t) as in tip
ei (ei) as way	u (u) as in too
g (k) as in go	ü (ü) as the German ü
i (i) as in eat	x (hs) as in she
ie (ie) as in yes	y (y) as in yet
j (ch) as in jeep	z (ts, tz) as in zero
k ('k) as in king	zh (ch) as in jump
o (o) as in saw	

Appendix I at the end of the book consists of Chinese names, places, and terms mentioned in Part VII in both Wade-Giles and Pinyin systems; Appendix II is a conversion table of the two systems.

Major Chronological Periods

The Ch'ing Dynasty (1644-1911)

    Shun-chih   1644-61

    K'ang-hsi   1662-1722

    Yung-cheng   1723-35

    Ch'ien-lung   1736-95

    Chia-ch'ing   1796-1820

    Tao-kuang   1821-50

    Hsien-feng   1851-61

    T'ung-chih   1862-74

    Kuang-hsü   1875-1908

    Hsüan-t'ung   1909-11

The Republic of China (1912-　)

The People's Republic of China (1949-　)

Conversion Tables
of Currencies, Weights, and Measures

CURRENCIES (1600-1814)

1 tael 兩 = 1 Chinese ounce, or 1.208 English ounce, of pure silver
  = £⅓ = 6s. 8d. (6 shillings and 8 pence)
  = U.S. $1.63
  = Spanish $1.57
(In 1894 the value of tael dropped to 3s. 2d., and in 1904, 2s. 10d.)
1£ = 3 taels (Tls.) = Spanish $4
1 Spanish $ = 0.72 tael or 5s.

WEIGHTS

1 picul (shih 石) = 100 catties (chin 斤)
  = 133⅓ lbs.
  = 60.453 kilograms
1 catty (chin) = 16 taels (liang 兩)
  = 1⅓ lbs.
  = 604.53 grams
1 tael (liang) = 1⅓ oz.
  = 37.783 grams
16.8 piculs = 1 long ton
16.54 piculs = 1 metric ton

MEASURES

1 li 里 = ⅓ mile = ½ kilometer
1 ch'ih 尺 = 1 Chinese foot or cubit = 14.1 inches
1 mou = ⅙ acre
15 mou = 1 hectare

The Rise of Modern China

1
A Conceptual Framework
of Modern China

With a recorded history of nearly 4,000 years, Chinese civilization is one of the oldest in the world. Until modern times its development had been largely indigenous, partly because of the independent spirit of the Chinese people and partly because of China's isolation from the other great civilizations. However, with the advent of the Age of Discovery a drastically different situation set in. Portuguese and Spanish explorers and envoys began to arrive in South China via the new sea routes in the 16th century, and in their wake came the traders and missionaries. Shortly afterwards, the Russians marched across Siberia and reached the Manchurian border in the mid-17th century. These events were nothing less than epochal for China, for they broke its age-old isolation and initiated the beginning of direct East-West contact, which, though weak and faltering at first, was to grow to such force in the 19th century as to effect a head-on collision between China and the West. Moreover, when viewed in the context of China's domestic development, the arrival of the Europeans takes on added significance: it coincided with the rise of the Manchus and the establishment of the alien Ch'ing dynasty. These momentous foreign and domestic developments left behind far-reaching consequences which endowed the period that followed with characteristics markedly different from earlier times.

First and foremost, the convergence of Chinese and Western history ended China's seclusion and resulted in its increasing involvement in world affairs, until today what happens in one immediately affects the other. Second, the interplay of foreign and domestic elements gave rise to revolutionary changes in the Chinese political system, economic institutions, social structure, and intellectual attitudes. "Change" thus became a key feature of the period, making it far more complex than ever before. Third, the forcible injection of alien elements into Chinese life—the Westerners from without and the Manchus from within—generated a strong sense of nationalistic-racial consciousness, which was to influence deeply the future course of Chinese history. So distinct is this period from earlier ones that it justifiably forms a separate period of historical investigation.

When Does Modern China Begin?

Although the meeting of Western and Chinese history began in the 16th century, its effect did not become significant until the middle of the 19th century, when the intensified activities of the West led to radical changes in China. Scholars therefore have differed on whether the 16th century or the 19th should be regarded as the beginning of modern China. One influential school, consisting mostly of Western historians and political scientists, Marxist scholars, and many Western-trained Chinese scholars, takes the Opium War of 1839-42 as the point of departure. For the Chinese historians of this school, the war marked the beginning of foreign imperialism in their country, and Chinese history thereafter was largely one of imperialism in China. For the Western historians, the war signified the acceleration of foreign activities which shattered Chinese isolation and ushered in a period of revolutionary changes in China. For Marxist historians, the war was the epitome of the evils of capitalism and imperialism, which plunged the "semifeudal" Chinese state into the abyss of "semi-colonialism."

A second school, consisting mostly of the more traditional Chinese historians[1] (who are beginning to win converts among Western scholars), disputes the propriety of regarding the Opium War as the beginning of a new era. They consider the arrival of European explorers and missionaries during the transitional period from the Ming (1368-1643) to the Ch'ing (1644-1911) as a more justifiable starting point, for internally it was the time of the rise of the Manchus and the establishment of the Ch'ing dynasty, and externally it was the period when Western learning was first

---

1. Such as Hsiao I-shan, *Ch'ing-tai t'ung-shih* (A general history of the Ch'ing period), rev. ed. (Taipei, 1962); Li Shou-k'ung, *Chung-kuo chin-tai shih* (Chinese modern history), (Taipei, 1961); Li Fang-ch'en, *Chung-kuo chin-tai shih* (Chinese modern history), (Taipei, 1960).

introduced into China. They argue that for all its spectacular effects the Western impact in the 19th century was only an extension and intensification of a process already set in motion two and one half centuries earlier, and that the hundred or so years since the Opium War is scarcely enough time to represent the modern period of a 4,000-year-old history. Moreover, to date modern China from around sixteen hundred is to place it more in congruity with the beginning of modern Europe.

Both approaches are valid, and both have certain shortcomings. From the standpoint of effects, the burgeoning of the Western impact in the 19th century certainly was more instrumental in transforming traditional China into modern China than the arrival of the European explorers and missionaries in the 16th and 17th centuries. To be sure, the Jesuits introduced the Western sciences of astronomy, mathematics, geography, cartography, and architecture, but their influence was limited to a small group of Chinese scholars and officials in the ruling circles. They left little imprint on China's political institutions, social structure, and economic systems, which remained much as they were before the coming of the missionaries. From this point of view, the first school would seem to have a strong claim.

Yet no one can fully appreciate the changes in the 19th and 20th centuries without knowing the institutions of the earlier period. The study of the Western impact must be preceded by a knowledge of what that impact was on. Furthermore, in view of the crucial role that the West and Russia played in influencing the destiny of modern China, one should not lose sight of the import of the early contacts, nor the manner of their advance—the Western maritime powers pushed from the south upward and the Russian land power thrust from the north downward, constituting something of a pincer movement pointing at the heart of China, Peking.[2] Indeed, from a historical standpoint the arrival of the western Europeans and the Russians in the 16th and 17th centuries paved the way for the intensified activity of the West in the 19th century. For these reasons, the second school would seem to have a plausible argument.

However, I believe that the two schools can be reconciled through the development of an eclectic approach. Even accrediting the Opium War as a viable point of departure, one still needs to be familiar with the traditional Chinese state and society, which conditioned China's reaction to

2. Immanuel C. Y. Hsü, *China's Entrance into the Family of Nations: The Diplomatic Phase, 1858-1880,* 2nd printing (Cambridge, Mass., 1968), 108. See also T. F. Tsiang, "Chung-kuo yü chin-tai shih-chieh ti ta-pien-chü" (China and the great changes of the modern world), *Tsing-hua hsüeh-pao* (The Tsing-hua Journal), 9:4:783-828 (Oct. 1934); "China and European Expansion," *Politica,* 2:5:1-18 (March 1936).

the foreign challenge of the 19th century. The intrusion of the West can be construed as a kind of catalyst, precipitating traditional China into its modern counterpart. Hence, one can hardly understand the result of the transformation without a fair knowledge of the mother institutions.

Consequently, a general discussion of the major developments in domestic and foreign affairs from 1600 to 1800 will provide the background information necessary for a sound understanding of modern China. During this period, China's political system, social structure, economic institutions, and intellectual atmosphere remained substantially what they had been during the previous 2,000 years. The polity was a dynasty ruled by an imperial family; the economy was basically agrarian and self-sufficient; the society centered around the gentry; and the dominant ideology was Confucianism. With a knowledge of this traditional complexion of China one can more readily appreciate Chinese conduct vis-à-vis the accelerated Western activity in the 19th century. Hence, this eclectic approach preserves the historical completeness of the second school without sacrificing the realistic considerations of the first.

The questions may be asked, why did the convergence of Chinese and Western history not begin before the 16th century, and why did the Western influence not intensify before the 19th? To answer these questions, one must remember that for over two millennia before the last century, the main streams of Chinese and Western civilizations moved in divergent directions. Western civilization originated in Greece, moved westward to Rome, and spread over western Europe and into America, while Chinese civilization developed in the Yellow River valley, moved southward toward the Yangtze River valley, and then to other parts of China. The major currents of the two civilizations thus moved farther and farther apart rather than toward each other. They could not meet until one of them had developed sufficient power and technology, coupled with interest, to reach the other.

To be sure, there were intermittent contacts between the two civilizations before the Age of Discovery. The Han dynasty (202 B.C.-A.D. 220) and its European contemporary, the Roman Empire, knew something of each other. The Chinese addressed the Roman Empire with the honorable title of *Ta-Ch'in,* the Great Ch'in state. The famous general, Pan Ch'ao, active in Central Asia from A.D. 73 to 102, even sent an envoy[3] to look for the Roman Empire, and he reached as far as the Persian Gulf. Other contacts included the importation of Chinese silk to Rome, and the

3. Kan Ying.

arrival in China of Roman jugglers and merchants in A.D. 120 and A.D. 166. During the T'ang period (618-907), Nestorian Christianity and Mohammedanism entered China, and the Arabs were active in Chinese foreign trade during both the T'ang and Sung (960-1279) periods. The Yüan period (1280-1367) witnessed the coming of the Venetian merchants, Maffio and Niccolo Polo and the latter's famous son, Marco, as well as many Franciscan friars. The Ming period (1368-1643) saw the rise of the great maritime expeditions conducted by Cheng Ho, who reached as far as the eastern coast of Africa, and in the 15th century the Chinese technique of blockprinting was transmitted to Europe. Thus, for many centuries before the Age of Discovery, there were sporadic contacts between China and the West, but a direct confrontation of the two had to wait until one of them could make a *sustained* drive to reach the other.

By the time of the Age of Discovery, Europe had developed enough geographical knowledge and technical skill in shipbuilding to be able to reach the East. Portuguese oceanic expeditions carried explorers and empire-builders to Asia, followed immediately by traders and missionaries, bringing with them the new scientific knowledge of the Western world. The beginnings of a more than occasional East-West contact were established, but it was still not powerful enough to bring the two civilizations face to face. Europe had to wait until after the Industrial Revolution to generate power sufficient to make a vigorous and sustained effort to reach China. It is no accident that England, as the cradle of the Industrial Revolution, took the lead in this drive. Clearly, the convergence of Chinese and Western history could not have occurred before the Age of Discovery, and the direct confrontation of the two civilizations could not have occurred before the Industrial Revolution.

The Shaping Forces of Modern China

The key to an understanding of any period of history lies in the discovery of its major shaping forces. In the case of modern China, we discern several powerful forces at work, some overt and some covert. First was of course the government's policies and institutions which to a large extent determined the ebb and flow of the country's fortune. During the Ch'ing period, the overriding consideration of the dynasty was to maintain itself. To win Chinese good will and acceptance, it identified itself with the traditional order, kept the Ming government and social institutions, embraced Neo-Confucianism as a state philosophy, and absorbed the Chinese into its bureaucracy to work alongside the Manchus in a sort of dyarchy. It carried out ruthless literary inquisitions to punish critics of the

alien rule. It strove to preserve the Manchu identity through the establishment of the Imperial Clan Court to supervise the births, education,
and marriages of the Manchu nobles, the prohibition of Manchu-Chinese
intermarriages, and the ban on Chinese immigration to Manchuria. New
government offices were created to suit its particular needs, such as the
Li-fan yüan in 1638 to manage affairs relating to Tibet, Mongolia, and
the Western Region (Sinkiang); the Grand Council in 1729 to centralize
the decision-making process; and the Tsungli Yamen in 1861 to direct foreign relations with the Western powers. Military expeditions were dispatched to the far corners of the empire to quell revolts and expand frontiers, making the dynasty the second largest in Chinese history. After the
mid-19th century, many foreigners were appointed to government service,
thereby enlarging the Manchu-Chinese dyarchy into a Manchu-Chinese-
Western *synarchy*.[4] All these policies and undertakings as well as a host
of others, designed to ensure the permanence of the Ch'ing rule, strongly
influenced and guided the main streams of the country's political life
from 1644 to 1911. After the fall of the Manchu dynasty, the primary
concern of the republican government was to unify the country internally
and abolish the unequal treaties externally. Coming down to the Communist period since 1949, we notice an intense drive toward socialist transformation, rapid industrialization, and big-power status. These overriding
policies of the government in power obviously played a major role in
guiding the destiny of the country and the people and should therefore
earn the unremitting attention of the historian.

Yet we must not ignore the more elusive subterranean currents of history as a shaping force. Indeed, in an autocracy like the Ch'ing where no
legal or loyal opposition was allowed, underground activities at times
played a vital role in the unfolding of history. While many Chinese
joined the Manchu government and tacitly accepted the Ch'ing rule, a
great many more remained in silent opposition. The very fact that the
Ch'ing was an alien dynasty continuously evoked Chinese protest in the
form of secret society activities and nationalistic-racial revolt and revolution. Initially, the anti-Ch'ing sentiment was accompanied by a desire to
revive the Ming—as witness the various Ming loyalist movements, the resistance of Koxinga and his son on Taiwan, and the Revolt of the Three
Feudatories (*San-Fan*). When these movements failed, the "Anti-Ch'ing,

4. John K. Fairbank, "Synarchy under the Treaties" in John K. Fairbank (ed.), *Chinese Thought and Institutions* (Chicago, 1957), 204-31; "The Early Treaty System in the Chinese World Order" in John K. Fairbank (ed.), *The Chinese World Order: Traditional China's Foreign Relations* (Cambridge, Mass., 1968), 257-75.

Revive-Ming" (*Fan-Ch'ing fu-Ming*) idea was covertly preserved and nurtured among the secret societies such as the Heaven and Earth Society (*T'ien-ti hui*) and the White Lotus Sect (*Pai-lien chiao*), waiting for a chance to burst out. It was no coincidence that when central vigilance relaxed toward the end of the Ch'ien-lung reign (1736-95), the White Lotus Rebellion broke out in 1796 and lasted until 1804. After its pacification, the nationalistic-racial protest once again subsided, only to be revived by the Taipings in 1850-64. However, the Taipings kept only the "Anti-Ch'ing" portion of the slogan and dropped the idea of reviving the Ming, since they wanted to create a kingdom of their own. After the failure of the Taipings in 1864, the nationalistic-racial revolution again subsided into secret society activities, giving inspiration to later revolutionaries such as Dr. Sun Yat-sen. By then the scope of revolutionary aspirations had broadened to include a vendetta against foreign imperialism as well. With the downfall of the Manchu dynasty in 1912, the original "Anti-Ch'ing" objective had been realized, and the nationalistic revolution turned against foreign imperialism, with European colonial powers as the chief target in the 1910s and 20s, Japan in the 1930s and first half of the 40s, and Russia since the late 1950s—the last, it may be observed, as much the target of the Chinese Nationalists as of the Communists.

Thus, throughout the three-hundred-odd years of modern China, the thread of nationalistic-racial protest against foreign elements in Chinese life formed a distinct theme of history, now coming to the surface and now going underground. So persistent was this force that a noted historian remarked, perhaps somewhat exaggeratedly, that the history of modern China may be construed as a history of nationalistic revolution.[5]

The third shaping force was the search for a way to survival in the new world that had been forcibly thrust upon China by the West after the mid-19th century. Ironically, Western civilization, so creative and vital elsewhere, proved more destructive than constructive in its immediate confrontation with China. It precipitated the breakdown of the old order without substituting another, leaving the Chinese the difficult task of forging a new order out of the ruins of the old. Burdened by tradition and heritage, and as yet ignorant of the nature of the Western world, the Chinese groped in the dark, looking for a way to live with what the great statesman Li Hung-chang described as "a great change in more than three thousand years of history."[6] They were faced with the agonizing problem

5. Hsiao I-shan, I, 15.
6. Li Hung-chang, *Li Wen-chung-kung ch'üan-chi* (Complete works of Li Hung-chang), (Shanghai, 1921), "Tsou-kao" (Memorials), 19:45. Memorial dated June 20, 1872.

of deciding how much of Old China must be discarded and how much of the modern West must be accepted in order for China to exist and win a respectable place in the community of nations.

The search for a new order involved an extremely hard struggle against the weight of pride, disdain for things foreign, and the inveterate belief that the bountiful Middle Kingdom had nothing to learn from the outlandish barbarians and little to gain from their association. However, after China's second defeat in war and the Anglo-French occupation of Peking in 1860, the more progressive mandarins[7] realized that the Western challenge was inescapable and that China must change if it was to survive. They initiated what was known as the Self-strengthening Movement in the early 1860s, invoking the slogan of the famous scholar Wei Yüan, "learn the superior barbarian technique with which to repel the barbarians." In this spirit, translation bureaus were established, and arsenals and dockyards were created after the Western models, supported by military industries. The movement, which lasted some thirty-five years, was a superficial attempt at modernization; only those aspects having immediate usefulness were adopted, while the more commendable parts of the Western civilization—political systems, economic institutions, philosophy, literature, and the arts—were totally ignored. Even the progressive Chinese at the time believed that China had little to learn from the West except weaponry.

The Self-strengthening Movement was proved totally inadequate by the defeat in the Japanese war in 1895. Realizing the limited scope of their movement, Chinese scholars and officials determined that they must broaden the modernization program to include political reform as well. K'ang Yu-wei, an assertive thinker, and his famous student, Liang Ch'i-ch'ao, urged the emperor to make institutional reform after the fashion of Peter the Great and Emperor Meiji. However, even at this late stage reformers did not advocate complete Westernization but rather the creation of a hybrid polity containing both Chinese and Western elements. The spirit of the movement was "Chinese learning for fundamentals, Western learning for practical application," as described by the famous scholar-official Chang Chih-tung. The result of K'ang's program was the "Hundred-Day" Reform of 1898, which ended abruptly in failure.

Meanwhile, secret revolutionary activities were initiated by Dr. Sun Yat-sen, a Western-trained physician, who believed that China's problems could not be remedied by a partial institutional reform, but only by a complete revolution. Carrying the torch of nationalistic-racial revolution,

7. Such as Prince Kung, Wen-hsiang, Tseng Kuo-fan, Tso Tsung-t'ang, and Li Hung-chang.

he advocated overthrow of the Manchu rule. Operating on the fringes of society, he won support from the secret societies, the lower classes, and the overseas Chinese, but not from the scholars and the gentry, who generally followed K'ang and Liang. But after the Boxer Rebellion of 1900, which severely discredited the dynasty, an increasing number of educated people joined Dr. Sun's cause, and his earlier image of disloyal rebel was transformed into that of patriotic revolutionary. The success of the revolution in 1911 was followed by the establishment of a Western-style republic the next year, and the imperial dynasty was abolished for the first time in 4,000 years.

Although a break was made with the outdated political system, the hand of the past continued to weigh heavily in social habits and intellectual life. The government had had a face-lifting but its spirit remained the same; corruption, warlordism, attempts at reviving monarchism, and disorder were rife. The institution of the republic was not accompanied by the expected peace and order, and Chinese intellectuals became convinced that without a thorough thought-reform no good government and no social improvement were possible. Those who had studied in Japan, Europe, and the United States—men such as Ch'en Tu-hsiu, Ts'ai Yüan-p'ei, and Hu Shih—returned home in the second half of the 1910s to promote a New Cultural Movement and an intellectual revolution, culminating in the May Fourth Movement of 1919. The spirit of the age opposed traditionalism and Confucianism and advocated complete Westernization, "science," and "democracy." In this period of ideological ferment, two main philosophies emerged. John Dewey's pragmatism, espousing an evolutionary approach to social improvement, was introduced by his student Hu Shih, while the Marxist revolutionary approach was propagated by Ch'en Tu-hsiu and Li Ta-chao under the influence of the Bolshevik Revolution.

China had come a long way from a sneering rejection of the West in the early 19th century to the worship of it by 1920. A learned political scientist summarizes the sequence of change in the following words: "First, technologies affecting material existence; then principles concerning state and society; and finally, ideas touching the inner core of intellectual life. The Self-strengthening Movement of the T'ung-chih period, the reform movement of 1898, and the May Fourth movement of 1919 marked the climactic points of these three stages."[8] One could add a fourth stage as following the May Fourth Movement: the period of Contemporary China, which is generally not considered separately from the Age of Modern China.

8. Kung-chuan Hsiao, "The Philosophical Thought of K'ang Yu-wei—An Attempt at a New Synthesis," *Monumenta Serica,* XXI (1962), 129-30.

The dominant theme of Contemporary China has been the struggle between the Nationalists and the Communists for the supreme power of state. The Chinese Communist Party was founded in 1921 amidst the intellectual revolution surrounding the May Fourth Movement. Under the guidance of the Comintern, it began to collaborate with the Nationalist Party in 1923. Dr. Sun, impressed by the success of the Bolshevik Revolution and desirous of Soviet assistance in reorganizing his party and army, was anxious to cooperate with the Soviets and the Chinese Communists. However, his death in 1925 doomed the alliance, and an open split came in 1927. Chiang Kai-shek, the Nationalist military leader, emerged as the new strong man, with a Nationalist government established in Nanking in 1928.

After the split, Mao Tse-tung and Chu Teh formed their own soviets in Kiangsi, virtually independent of the underground central party organization in Shanghai. Chiang launched five campaigns against them, routing them from southeastern China late in 1934. The Communists embarked upon the epic Long March of 25,000 *li*[9] (actually 6,000 miles) to the Northwest, where they re-established themselves. But hardly had the Communist problem been resolved when the Japanese struck in 1937. Facing a common enemy, the Nationalists and the Communists formed a United Front, though neither trusted the other. No sooner had the Japanese war ended in 1945 than civil strife erupted again. Depleted by the long foreign war, threatened by runaway inflation, burdened by the age-old problem of landlordism, and weary of fratricidal fighting, the Nationalists, despite their apparent military superiority and American aid, lost the mainland and took refuge on Taiwan. Mao established the Chinese People's Republic in 1949.

This cursory survey reveals several important stages in the development of modern China: from rejection of the West in the pre-Opium War period to the Self-strengthening Movement of 1861-95, to political reform and revolution of 1898-1912, to intellectual revolution around 1917-23, and finally to the rise of Chinese Communism in 1949. While history seldom moves in a simple linear fashion, the general pattern of development and major landmarks nevertheless serve as useful guideposts in evolving a conceptual framework.

Accompanying the great political changes described above was a fundamental transformation of the economy and the society. Modern industries and enterprises sprang up during the Self-strengthening Movement in the second half of the 19th century, while foreigners operated manufac-

9. *Li* is one-third of a mile.

turing, shipping, banking, and trading firms in the treaty ports under the protection of the "unequal treaties." The juxtaposition of these different types of activity gave rise to a mixed economy, tinted with a semicolonial flavor, which characterized the Chinese economic scene for nearly a century.

Socially, the gentry who dominated Chinese society for centuries began to fade away after the abolition of the civil service examinations in 1905. The traditional social stratification of the four classes—scholar-official, farmer, artisan, and merchant—also crumbled in the face of two rising groups: the compradores and the militarists, who represented the new rich and the new power. Moreover, the influx of the Western ideas of individualism, freedom, and equality of the sexes eroded the Confucian precepts of family loyalty, filial piety, the Three Bonds, and the Five Relations.[10] When the individual rose to assert his status as a member of the state rather than of the family, the kinship society disintegrated. The pace of social changes accelerated after the Communist take-over in 1949, and the most dramatic change of all was perhaps the transformation of the peasant from an inert entity to an activist member of the state.

Modern China represents such a broad spectrum of kaleidoscopic change that I believe it cannot be satisfactorily explained by the restrictive theories of foreign imperialism, Western impact, or capitalistic and feudal exploitation. The dynamics of change suggests that modern Chinese history is not characterized by a passive response to the West, but by an active struggle of the Chinese to meet the foreign and domestic challenges in an effort to regenerate and transform their country from an outdated Confucian universal empire to a modern national state, with a rightful place in the family of nations. This view avoids the pitfall of "foreign causation" in the interpretation of Chinese history and the implication that China merely "reacted."

Toward a New Synthesis

I propose to open the study of modern China with a survey of the "traditional" state and society during the years 1600-1800 as a prerequisite to the discussion of the above-mentioned stages of development. This approach differs notably from several that have been attempted in the past. Pioneer Western scholars of the World War I vintage tended to favor general works on Chinese foreign relations and relied almost exclusively

10. The Three Bonds are those between the ruler and the subjects, father and son, and husband and wife. The Five Relations include two more: between brothers and between friends.

on Western sources; they ignored or treated superficially the internal conditions. Scholars of the next generation shifted their focus from general works to monographic studies, taking China's response to the West as a main theme of investigation. Works of this period strove to see history from the Chinese side, and made much use of Chinese materials along with Western sources. They forged a new trend in research and enriched considerably our knowledge of modern China. More recently, there have been attempts to probe basic Chinese social, economic, and intellectual forces independent of the Western impact, or to examine the changes in terms of the domestic milieu, all of which result in a more sophisticated scholarship.

The study of modern China has also been pursued with great zeal by Communist historians on the mainland, apparently in response to Mao's call that special attention be devoted to the modern period of Chinese history. They approach the subject from the standpoint of dialectical materialism, class struggle, and the changing nature of the society. While no consensus has been reached, there seems to be some tentative agreement among a majority of them on the question of periodization: (1) the period of the invasion of foreign capitalism and of peasant revolution, 1840-64; (2) the period of semicolonialism and semifeudalism, 1864-95; (3) the period of deepening national crises and the emergence of patriotic movements, 1895-1905; and (4) the period of the rise and fall of the bourgeois revolution, 1905-19. These four periods of modern Chinese history are said to constitute "The Age of the Old Democratic Revolution," as opposed to the period of contemporary Chinese history from 1919 to 1949, which constitutes "The Age of the New Democratic Revolution."[11]

The increased sophistication of Chinese, Western, and Japanese scholarship on modern China, and the flurry of activity among the Marxist writers lead one to think that the relatively young field of modern Chinese studies is coming of age. Through the application of new methodologies of the social sciences and the humanities, and the employment of multi-archival and interdisciplinary approaches, a great many sound and perceptive monographs have appeared in several major languages over the past three decades. Productive research and the general elevation of scholarship encourage the appearance of a comprehensive history, which should include the distillation of Chinese, Japanese, Western, and, where applicable, Marxist scholarship. China's deep involvement with the West makes

11. *Chung-kuo chin-tai shih fen-ch'i wen-t'i t'ao-lun chi* (A collection of papers on the question of periodizing modern Chinese history), compiled by the editorial department of *Li-shih yen-chiu* (Historical Research), (Peking, 1957).

it highly desirable, if not imperative, that its modern history be written in such a way as to reflect both the insight of Chinese scholarship and the kind of objectivity that foreign scholars can bring to bear from without. Such a synthesis promises to place modern China in a proper historical perspective.

Further Reading

Banno, Masataka 坂野正高, *Kindai Chūgoku gaikōshi kenkyū* 近代中國外交史研究 (A study of modern Chinese diplomatic history), (Tokyo, 1970).

Chung-kuo chin-tai shih fen-ch'i wen-ti t'ao-lun chi 中國近代史分期問題討論集 (A collection of papers on the question of periodizing modern Chinese history), compiled by the editorial department of *Li-shih yen-chiu* (Historical Research), (Peking, 1957).

Cohen, Paul A., *Discovering History in China: American Historical Writing on the Recent Chinese Past* (New York, 1984).

Dawson, Raymond, *The Chinese Chameleon: An Analysis of European Conceptions of Chinese Civilization* (London, 1967).

Etō, Shinkichi 衛藤瀋吉, *Higashi Ajia seijishi kenkyū* 東アジア政治史研究 (A study of East Asian political history), (Tokyo, 1968).

Fairbank, John K., Edwin O. Reischauer, and Albert M. Craig, *East Asia: The Modern Transformation* (Boston, 1965).

Fan, Wen-lan 范文瀾, *Chung-kuo chin-tai shih* 中國近代史 (Chinese modern history), Vol. I (Peking, 1949).

Feuerwerker, Albert, Rhoads Murphey, and Mary C. Wright (eds.), *Approaches to Modern Chinese History* (Berkeley, 1967), Intro. 1-14.

Ho, Ping-ti, and Tang Tsou (eds.), *China in Crisis*, Vol. I, *China's Heritage and the Communist Political System*, Vol. II, *China's Policies in Asia and America's Alternatives* (Chicago, 1968).

Hsiao, I-shan 蕭一山, *Ch'ing-tai t'ung-shih* 清代通史 (A general history of the Ch'ing period), rev. ed. (Taipei, 1962), I, ch. 1.

Kuo T'ing-i. *Chin tai Chung-kuo shih-shih jib-chih* 近代中國史事日誌 (Chronology of events in modern Chinese history), 2 vols. (Taipei, 1963).

———, *Chung-hua min-kuo shih-shih jih-chih* (Chronology of events in the history of the Republic of China), Vol. I, 1912-1925 (Taipei, 1979).

———, *Chin-tai Chung-kuo shih-kang* (An Outline of modern Chinese history), (Hong Kong, 1979).

Li, Fang-ch'en 李方晨, *Chung-kuo chin-tai shih* 中國近代史 (Chinese modern history), (Taipei, 1960), Intro.

McAleavy, Henry, *The Modern History of China* (New York, 1967).

Morse, H. B., *International Relations of the Chinese Empire* (London, 1910-18), 3 vols.

Nathan, Andrew J. (ed.), *Modern China, 1840-1972: An Introduction to Sources and Research Aids* (Ann Arbor, 1973).

Teng, S. Y., and John K. Fairbank, *China's Response to the West* (Cambridge, Mass., 1954).

Tsiang, T. F., "China and European Expansion," *Politica,* 2:5:1-18 (March 1935).
————, "Chung-kuo yü chin-tai shih-chieh ti ta pien-chu"中國與近代世界的大變局 (China and the great changes of the modern world), *Tsing-huah hsüeh-pao,* 9:4:783-828 (Oct. 1934).

I
The Persistence
of Traditional Institutions
1600-1800

2

The Rise and Splendor
of the Ch'ing Empire

By 1600 a new age was about to dawn in China. For the first time Western European explorers, traders, and missionaries began to arrive by sea in considerable numbers, bringing with them the seeds of a new civilization, while the Russians, having marched across Siberia, were pushing toward the Manchurian border. Domestically, a momentous transition was in the offing. The Ming dynasty, which came into power in 1368, had long since passed its zenith[1] and entered a state of rapid decline, beset with problems usually associated with the end of a "dynastic cycle"—eunuch domination of the court, moral degradation, political corruption, intellectual irresponsibility, high taxes, and famine. In this period of decadence, two peripatetic rebel bands under Chang Hsien-chung and Li Tzu-ch'eng overran the greater part of the country for nearly twenty years (1628-47), generating untold misery and causing widespread unrest. Taking advantage of this dynastic decline and general disorder, an alien frontier tribe in the Northeast, the Manchus, rose to challenge the central power, and ultimately established a new dynasty in China.

The Founding of the Ch'ing Dynasty

Historically, the Manchus[2] were a hardy stock of the nomadic Jurched tribe, living in what is today's Manchuria where they subsisted by hunt-

1. Reached during the Yung-lo period (1403-24).
2. Known in ancient times as Su-shen.

ing and fishing. During the 12th century they founded the Chin (Gold) dynasty (1115-1234), which had threatened the existence of the Southern Sung dynasty (1127-1279). Though conquered by the Mongols in the 13th century, they regained something of their former independence under the Ming (1368-1643) emperors, who divided them into three commanderies: Chien-chou, Hai-hsi, and Yeh-jen. They sent horses, furs, and ginseng[3] as tribute to the Ming court, and received Chinese agricultural products as gifts in return.

Geopolitics played an important part in the future development of the Jurched tribes. They lived in an area north of Korea and east and northeast of Liaotung where the Chinese had already settled, and were thus in a position to observe and profit from the Chinese experience and institutions.[4] Increasingly they came under the Chinese influence in modes of dwelling, eating, and drinking. After the mid-16th century, a growing number of Chinese crossed the frontier and taught the Jurcheds to farm and to build castles and fortresses. The resultant economic and technological advancement considerably altered the character of the once nomadic society, and historical hindsight reveals that the time was propitious for the appearance of a great leader to lift the Jurcheds out of their feudal frontier existence.[5]

As reward for having assisted China in smoothing eastern border hostilities, the head of the Chien-chou commandery was made a frontier officer of the Ming with an imperially bestowed name, Li. This commandery was divided into the Left Branch (*Tso-wei*) and Right Branch (*Yu-wei*), both under the Chinese commander in Liaotung. In 1574 the chieftain of the Left Branch, Giocangga, and his son Taksi, allied themselves with the Chinese commander, Li Ch'eng-liang, in a campaign to chastise the unruly chieftain of the Right Branch. A second campaign was launched in 1582 against the son of the said chieftain, and in the melee that followed Giocangga and Taksi were killed. What followed was an intense internecine struggle within the household of the deceased, and in 1583 Taksi's

3. A kind of root which the Chinese considered an elixir with rejuvenating properties.

4. Franz Michael, *The Origin of Manchu Rule in China* (Baltimore, 1942), 3, 11; Wada Sei, *Tōashi kenkyū-Manshū hen* (Studies on the history of the Far East-Manchurian volume), (Tokyo, 1955), chs. 15-16.

5. Wada Sei, "Some Problems Concerning the Rise of T'ai-tsu, the Founder of Manchu Dynasty," *Memoirs of the Research Department of the Toyo Bunko*, Tokyo, 16:71-73 (1957); David M. Farquhar, "The Origins of the Early Manchu State." Paper read before the 62nd annual meeting of the Pacific Coast Branch, American Historical Association, San Diego, August 28, 1969.

25-*sui*-old son,[6] Nurhaci, emerged victorious and won the right to succeed his father as chieftain.[7]

THE ASCENT OF NURHACI. In his youth Nurhaci (1559-1626) is said to have frequented the house of General Li, the Chinese commander, and developed an interest in the Chinese novels *Romance of the Three Kingdoms (San-kuo yen-i)* and *All Men Are Brothers (Shui-hu chuan)*. Well acquainted with frontier affairs, this ambitious Jurched chieftain was determined to avenge the deaths of his father and grandfather and was equally set on advancing his own cause. But for the moment his innate shrewdness dictated that he recognize his limitations and suppress his hostility toward the Ming court. It was clear to him that the unification of the Jurched tribes must be achieved before any projected attack on China. By two skillfully contrived marriages and a series of successful military campaigns, he rose rapidly in power and status. A Chinese captive, Kung Cheng-liu, became his trusted adviser and handled documents and correspondence for him.[8] During all these years of preparation, Nurhaci showed the greatest fealty toward the Ming court. In fact, in 1590 he personally brought his tribute to Peking, and in 1592-93 offered to lead an army to defend Korea against the Japanese invasion under Toyotomi Hideyoshi. In appreciation the Ming emperor bestowed on him the coveted title of "Dragon-Tiger General," the highest designation granted to a Jurched chief.

As a prospective empire-builder, Nurhaci sought the economic foundations of military conquest, drawing upon his earlier business experience to monopolize the trade in pearls, furs, and ginseng. He succeeded in amassing a considerable fortune, and by 1599 he was sufficiently prepared to start his campaign by attacking his neighbors while befriending the distant tribes. One by one the various Jurched tribes fell to him. His position had become so strong by 1607 that the Mongols conferred on him the title of Kundulen Han (The Respected Emperor); and in 1608, by a formal agreement with the Ming general in Liaotung, he was able to fix the boundary of his domain and forbid any Chinese to cross it. By 1613 Nurhaci had conquered all the Jurched tribes except one, the Yehe of the Hai-hsi commandery, which resisted him with the aid of the Ming troops.

6. *Sui*, the Chinese age, is counted from the day of conception; hence, 25 *sui* is actually 24 years of age in Western calculation.
7. Wada Sei, "Some Problems," 41-50. Nurhaci's surname in Manchu, Aisin Gioro, means the "Golden (Chin) Clan." See Inaba Iwakichi, *Shinchō zenshi* (A history of the Ch'ing dynasty), (Tokyo, 1914), Chinese tr. by Tan T'ao, *Ch'ing-ch'ao ch'üan-shih*, reprinted (Taipei, 1960), ch. 7, p. 71.
8. Wada Sei, *Tōashi kenkyū*, 637-49.

As a step toward creating a new state, in 1599 Nurhaci caused a Jurched alphabet to be created to replace the Mongolian script which had hitherto been used among the Jurched peoples since at least 1444.[9] In 1601 he invented the ingenious military "Banner System," whereby his warriors were organized into four companies (*niru*) of 300 men each, represented by banners of four different colors: yellow, white, blue, and red. By 1615 the number of companies had grown to 200,[10] and four more banners were created, with the same four colors but bordered in red, except for the red banner itself, which had a white fringe. Later, the size of the banner (*gūsa* or *gusai*) grew to 7,500 men, divided into five regiments (*jalan*), each of which comprised five companies (*niru*).

The Eight Banners were more than a simple military organization; they functioned as rudimentary administrative units during the transition from tribal feudalism to military administration and proto-nationhood. Everyone under Nurhaci, except for a few princes, belonged to a banner. Registration, taxation, conscription, and mobilization of the members were all managed by the banner organization. In time of peace the bannermen and their families engaged in farming and craftsmanship, and in time of war each banner contributed a certain number of men who fought under it. With this system Nurhaci organized his people into a war machine, which proved to be most effective. Chinese captured in the early campaigns were made bondservants, who were also organized into companies after the banner fashion, although probably they did not participate in actual fighting.[11] Eight Mongol and Eight Chinese Banners were added in 1634 and 1642 respectively, bringing the total number to twenty-four.

In 1616 Nurhaci boldly announced the establishment of the Chin state, and proclaimed himself "the Heaven-designated" emperor (*T'ien-ming*). Two years later, with the aid of the Qalqa Mongols he was ready to attack China. He enumerated seven grievances against the Ming, including the murder of his father and grandfather, the aid given to the Yehe tribe

---

9. David M. Farquhar, "The Origins of the Manchus' Mongolian Policy" in Fairbank (ed.), *The Chinese World Order*, 203.

10. The figure 400 is given in many accounts, such as the *Great Ch'ing Collected Statutes,* the *Veritable Records* of the Ch'ien-lung Period, and Meng Shen, *Ch'ing-tai shih* (A history of the Ch'ing period), (Taipei, 1960), 21-22. But it has been found unreliable. See Chaoying Fang, "A Technique for Estimating the Numerical Strength of the Early Manchu Military Forces," *Harvard Journal of Asiatic Studies,* 13:195, 208 (1950); a recent study suggests that the banner system was a mixed product of Mongol influence and Manchu hunting tradition, and that the size of the *niru* was not fixed at 300 until 1615. See Farquhar, "The Origins of the Early Manchu State." Cited with permission.

11. For a succinct description of the bondservant system, see Jonathan D. Spence, *Ts'ao Yin and the K'ang-hsi Emperor: Bondservant and Master* (New Haven, 1966), 1-18.

against him, the fact that Chinese subjects were permitted to cross over his border, and that the Ming had sent an inferior envoy to him.[12] This public airing of complaints was in fact a feudal declaration of war. Nurhaci swiftly advanced to the Chinese border and took the important city of Fu-shun. He captured a Chinese scholar, Fan Wen-ch'eng, who was persuaded to change sides and became his and his successors' trusted adviser.

The Ming court sent General Yang Hao and 90,000 soldiers to chastise Nurhaci but met with a disastrous defeat at Sarhu, east of Fu-shun. Having annihilated the main forces of the Ming, Nurhaci, riding the tide of victory, swept into the recalcitrant Yehe tribe and conquered it in September 1619. In the ensuing campaign against the Ming, Nurhaci took the important cities of Liaoyang and Mukden in May 1621. Later, in 1625 Nurhaci moved his capital to Mukden. A year later, at the crest of his success, he led a vehement attack on Ning-yüan. The defending Ming general Yüan Ch'ung-huan, using cannons cast by Jesuit missionaries, for once repelled the invader. In this first major defeat of his life, Nurhaci was wounded but his pride was hurt even more; seven months later he died.

Abahai (1592-1643), eighth son of Nurhaci, carried on his father's unfinished work. He attacked Korea first to safeguard his rear and to force the Koreans to send him annual tribute in silver. He then turned on China. A breakthrough was made in the Great Wall at Hsi-feng-k'ou, and Abahai advanced to Peking, looting and returning to Mukden with rich spoils. There, in 1631, Abahai set up a civil administration with six boards, patterned after the Ming court, thus marking a further constitutional transition from the banner-style military administration to the Chinese-style civil administration. However, the organization of the boards was different—there were no presidents or vice-presidents as in the Ming system. Each board was put under the nominal direction of a Manchu prince (*beile*), who was usually away in the battlefield, leaving the actual administration to three to five assistants (*ch'eng-cheng*), including one Mongol and one Chinese except for the Board of Punishments which kept two Chinese assistants, presumably due to the greater need for experienced Chinese to handle complicated legal matters. Herein lay the origin of the Manchu-Chinese dyarchy, or rather Manchu-Mongol-Chinese synarchy, which characterized the Ch'ing administration for 268 years.[13]

12. Arthur W. Hummel (ed.), *Eminent Chinese of the Ch'ing Period* (Washington, D.C., 1943-44), I, 597.
13. Piero Corradini, "Civil Administration at the Beginning of the Manchu Dynasty," *Oriens Extremus*, 9:2:136-38 (Dec. 1962).

THE MEANING OF MANCHU AND CH'ING. On the advice of his Chinese collaborators, Abahai in 1635 barred the use of "Jurched" and "Chien-chou," in favor of the term "Manchu." On May 14, 1636, he changed the dynastic title "Chin" to "Ch'ing" and proclaimed himself emperor. Abahai apparently wanted to obliterate anything reminiscent of Chinese suzerainty and to obscure the vassal status the tribes had endured under the Ming.

The origin of "Manchu" is of some interest. According to Emperor Ch'ien-lung, Manchu 滿洲 was the Chinese corruption of Man-chu 滿珠, which was the old name of the Jurched state in the beginning.[14] The noted Japanese scholar, Inaba Iwakichi, agreed with this explanation, and suggested further that Man-chu 滿珠 was an honorable and respected designation among the Jurched, Tibetans, and Mongolians.[15] Another interpretation is that Manchu came from a similarly pronounced Buddhist term Man-chu 曼珠, meaning "wonderful luck" 妙吉祥, which appeared in Tibetan Buddhist scriptures that were sent to the Jurched tribes. A fourth interpretation, rather mystical in nature, holds that Manchu is derived from the first character of Nurhaci's honorable designation Man-chu 滿柱 and the second character in Chien-chou 建州, to which the water-radical 氵 is appended to make it 洲. That these characters—Manchu 滿洲 and Ch'ing 清—contain the same water radical is a deliberate creation according to the principles of Yin-yang and Five Elements. The character for the Ming 明 dynasty means "bright," and that for the surname of the imperial family, Chu 朱, means "red." The combined image of "bright" and "red" is "fire," which can melt gold, the character indicated by the Chin 金 dynasty. An inauspicious name such as Chin, then, must be changed. The creation of the terms Manchu and Ch'ing, all written with the water radical, augured the future quenching of the Ming fire.[16]

Political expedience provided yet another motive for the change of dynastic title. Nurhaci had first adopted the title "Chin" in 1616 to arouse nostalgia in the Jurched tribes and to prompt them toward the creation of a new empire like the Chin state of the 12th century. By Abahai's time in the 1630s, this emotional appeal was no longer useful. Many Chinese had come over to join his ranks and the objective of the new regime was to overthrow the Ming dynasty. It was therefore necessary to win Chinese support and avoid offending their sensibilities. The dynastic name Chin could not serve this purpose, for the Chinese associated it with the killing,

14. Hsiao I-shan, I, 49. Professor Kenneth Ch'en of the University of California, Los Angeles, informed me that Manchu was derived from the Sanskrit term, Mañjusri 文殊, meaning "wonderful luck." I am indebted to him for this information.
15. Inaba Iwakichi (Chinese trans.), chapter 18, pp. 58-61.
16. Li Fang-ch'en, *Chung-kuo chin-tai shih* (Chinese modern history), (Taipei, 1960), 16.

pillaging, and invasion of their country by the old Chin state. To expunge this odious connotation, Abahai dropped the name Chin in favor of Ch'ing 清, meaning "pure." The two characters are pronounced nearly alike, but their meanings are vastly different. Furthermore, Ch'ing sounds more like Chinese and could more easily win Chinese acceptance.

Another theory is based on the historical fact that the old Chin state had conquered only the northern half of China, not all of it. The ambitious Abahai could not be satisfied with this imperfect record and so determined to make a fresh start with a new dynastic name.

By assuming the title of emperor and adopting the new dynastic name, Abahai made apparent his intent to overthrow the Ming dynasty. But first he forced the Korean king to break off relations with the Ming court and accept the Ch'ing suzerainty. The *Li-fan yüan* (Court of Colonial Affairs) was then set up (1638) to take charge of Korean and Mongolian affairs.

THE SEIZURE OF PEKING. Abahai now prepared to invade North China. In 1640 he attacked Chinchow with a large force. In defense, the Ming court appointed Hung Ch'eng-ch'ou commander in Liaotung, transferring eight generals, including Wu San-kuei, and 130,000 troops to reinforce the city. Abahai routed more than 50,000 Ming troops and overpowered stubborn resistance. Chinchow fell and Hung was captured in 1642. He was treated with honor, and subsequently capitulated and joined the Ch'ing cause. Abahai now extended his territory to the key pass in the Great Wall, Shanhaikuan (Mountain and Sea Pass), but for the time being, he chose to avoid a direct confrontation with the heavy Ming forces there. He turned instead to Northern Manchuria, and by 1643, he had brought the whole Amur Region under the Ch'ing rule. At this point his health failed, and he died at the age of fifty-one. His ninth son, Fu-lin (1638-61), a boy of six *sui*, was chosen to succeed him, under the regency of Jirgalang, a nephew of Nurhaci, and Dorgan, Nurhaci's fourteenth son.

The Ming court was not only threatened by the rising power of the Manchus but also by devastating internal rebellions. Dorgan had intended to make connections with some of the rebels, but before his plans matured the swift-moving rebel leader Li Tzu-ch'eng (1605?-45), nicknamed the "Dashing King" (*Ch'uang Wang*), advanced to Peking in late April 1644. Wu San-kuei, the brigade-general of Liaotung and commander of large forces at Shanhaikuan, was ordered by the Ming emperor to come to the rescue, but the city fell before his arrival. The rebel leader Li entered Peking on April 25, and the Ming emperor hanged himself on Prospect

Hill, overlooking the Forbidden City. The rebels captured Wu's father, then in Peking, and forced him to urge his son to surrender. Caught between the rebels and the Manchus, Wu decided to invite the latter to join him.

Dorgan, who had encamped near Sanhaikuan to await developments in China, was delighted with the invitation. Manchu troops poured through the opened gates of Shanhaikuan, where Wu welcomed Dorgan in person. As the Ch'ing forces advanced toward Peking, Li burned part of the palaces and the towers of the nine gates of the city. At daybreak on June 4, 1644, he escaped westward before the approaching Ch'ing forces, which entered Peking on June 6.

To win the support and confidence of the Chinese, Dorgan buried the Ming emperor and empress with honors, and paid tribute to those Ming officials who had lost their lives in the turmoil. Piously he declared that the Manchus had come to save the country from the rebels. Forces were sent out under Wu San-kuei and several Manchu generals to exterminate Li, who was killed in June or July of 1645, reputedly by villagers in Hupeh province while he was making a raid for food. The other rebel leader, Chang Hsien-chung, notorious for his reckless killing of millions, was defeated and killed in Szechwan in 1647 by the Ch'ing forces. Thus, the two largest rebellions that had troubled China for nearly two decades were suppressed.

Although the Manchus had announced that they had entered China to avenge the death of the Ming emperor and to save the country from the rebels, it was evident that their motives were not entirely noble. They cleverly defended their occupation of Peking by announcing that they had recovered it from the rebels rather than having seized it from the Ming rulers. In October 1644 the Ch'ing court was moved from Mukden to Peking, marking the beginning of a new dynasty—one which was to last until 1911. The first emperor of the dynasty was Fu-lin. In accordance with the practice which discouraged use of the ruler's personal name, he was better known by his reign title, Shun-chih. The power of state was in the hands of Regent Dorgan, who decided state policies and directed the unfinished task of the conquest of China.

Ming Loyalist Movement

Though the Manchus had established a court in Peking, South China remained in the hands of Ming loyalists. In 1645 these loyalists set up Prince Fu as emperor in Nanking, the subsidiary capital of the Ming dynasty, to continue their resistance. However, he proved to be a feeble ruler, who displayed more interest in the pursuit of pleasure than in state affairs.

Several other uncoordinated resistance movements came into being to continue the Ming cause. One group of loyalists proclaimed Prince Lu their new leader in Shao-hsin, while another group established Prince T'ang in Foochow. The two princes were uncle and nephew, but could not stand one another. In the end both were defeated by the Ch'ing army. Prince T'ang's brother, known as the New Prince T'ang, was then proclaimed emperor in Canton by still another faction of Ming loyalists, but his rule only lasted forty days (1646). As these movements fizzled, a new and more durable loyalist regime under Prince Kuei, a grandson of Emperor Wan-li (1573-1619), came into being in Chao-ch'ing, Kwangtung province. Prince Kuei succeeded in regaining control of seven southern and southwestern provinces by 1648. But ultimately the movement collapsed under powerful attacks from several directions by Chinese collaborators of the Manchus.

While these movements rose and fell in rapid succession, a more sustained resistance was organized along the coast by the Ming loyalist Cheng Ch'eng-kung, better known as Koxinga (1624-62). He was the son of Cheng Chih-lung, a onetime supporter of Prince T'ang, and a Japanese woman of the Tagawa family. Prince T'ang, much taken with young Cheng, had bestowed on him the imperial surname Chu in 1645; hence he was popularly known as the "Lord of the Imperial Surname" (*Kuohsing-yeh*), from which the Dutch derivation "Koxinga" comes. Prince T'ang treated him as an imperial agnate. In early 1646 the prince further favored him with the rank of earl and the title "Field Marshal of the Punitive Expedition" against the Ch'ing. In grateful acknowledgment of the imperial grace, Koxinga pledged lifelong allegiance to the Ming cause. However, late in 1646, his father defected to the Ch'ing, thus providing a way for Ch'ing forces to attack Prince T'ang. In disgust Koxinga gathered a following of several thousand men and took Amoy and Quemoy as his bases, declaring his support of Prince Kuei, who was then fighting the Manchus. Early in 1655 he established seventy-two military stations and six civil bureaus in Fukien, perfecting his military and civil organization, which had a total strength estimated at 100,000 to 170,000 men. Koxinga patronized Ming officials and scholars, and engaged in foreign trade to raise revenue for his movement.

In 1658-59 Koxinga raided Chekiang and Kiangsu from the sea and took the key city of Chinkiang. He could have taken Yangchow and cut off the supply lines of the Ch'ing army, but instead he decided, against his generals' counsel, to advance to Nanking. There he suffered a fatal defeat in September 1659, and saw five hundred of his ships burned. He had to retreat to Amoy to recoup. With this fiasco, prospects for restoring the Ming dimmed. He now found Amoy and Quemoy too restricted as operational bases, and turned his attention to Taiwan, or Formosa ("beautiful" in Portuguese), then under Dutch occupation. Koxinga launched an all-out attack in 1661 with 900 ships and 25,000 marines. The Dutch defenders were overpowered, and on February 1, 1662, a treaty was concluded between Koxinga and Governor Frederick Coyett, ending Dutch rule on Taiwan. With Taiwan as his new base, Koxinga was prepared for a long drawn-out struggle against the Ch'ing, who were virtually helpless to stop him. They could do no more than to have his father and brothers executed (1661), order the coastal inhabitants to move inland by 30 to 50 *li* (1662), and forbid fishing boats and commercial vessels to sail from the coast, thus cutting off Koxinga's sources of supply. Koxinga now represented the only thread of hope for the Ming loyalists. But he died suddenly on June 23, 1662, at the age of 38, reputedly of malaria but possibly by his own hand. The resistance movement was carried on by his son Cheng Ching, but the spirit was no longer the same, and there was continuous internal dissension. In 1683 Taiwan fell to the Ch'ing forces and a year later was made a prefecture of the Fukien province. With the defeat of this last loyalist group, the Ch'ing had completed the conquest of all China.

Dynastic Consolidation and Splendor

THE REIGN OF EMPEROR SHUN-CHIH, 1644-61. Shun-chih became emperor of China on October 30, 1644, at the age of seven *sui*. The power of state was exercised by his uncle Dorgan, who was given the affectionate title "Uncle Prince Regent." In 1645 Dorgan's title was expanded and raised to "Imperial Uncle Prince Regent" and in 1648 or 1649 to the most exalted "Imperial Father Prince Regent." Dorgan was the most powerful man in the country; his word was law. All high policies were made by him and the imperial seals were kept in his residence. Those who memorialized[17] the emperor sent duplicate copies to Dorgan and awaited his reaction. So imposing was his position that he was excused from kowtowing to the emperor during audience.

Dorgan's contribution to the young dynasty cannot be ignored, however. Under his direction, the Ch'ing forces took the provinces of Shensi,

17. A memorial was a minister's report to the emperor.

Honan, and Shantung; and in 1645 Kiangnan, Kiangsi, Hupeh, and part of Chekiang. In 1646 Szechwan and Fukien were added. In civil administration he retained most of the Ming institutions and practices, and welcomed Chinese officials into the government service, even permitting them the privilege of wearing the Ming costumes. He retained the service of the German Jesuit Adam Schall von Bell as director of the Imperial Board of Astronomy. Two of Dorgan's orders, however, greatly irritated Chinese sensibilities: the compulsory pigtail after the Manchu fashion and the encircling of rich Chinese farms for allotment to Manchu princes, nobles, and bannermen.

In reaching the summit of power at a relatively early age, Dorgan in effect halted his own career; he seems to have experienced the frustration of having no higher estate to reach for. He began to indulge himself in pleasure-seeking. Toward the end of 1650, he died suddenly during a hunting trip at Kharahotun near the Great Wall at age 39 *sui*.

When Emperor Shun-chih took over the reins of the government in 1651, he continued the same policy that Dorgan had found successful, of using Chinese assistance in the civil administration. He made diligent efforts to learn Chinese so that he could read Chinese memorials without relying on Manchu translations. He continued the "single-whip taxation system"[18] and improved the accounting system to reduce corruption and abuse. Irregularities were dealt with severely. To forestall secret opposition among the Chinese scholars, literary societies were forbidden. In institutions he made several innovations, among them the Imperial Clan Court and the positions of sub-chancellors of the Grand Secretariat, chancellors of the Hanlin Academy, and readers, expositors, as well as sub-readers and sub-expositors, of the same organization. The Ministry of Imperial Household was abolished in 1653 and in its place were established thirteen departments in the palace, run by the eunuchs, who wielded considerable influence over the young emperor. Although he warned them not to interfere with politics, it was impossible to keep them out of public affairs altogether, and in 1660 these departments were abolished. But his reign was short as he died of smallpox in 1661.

THE REIGN OF EMPEROR K'ANG-HSI, 1662-1722. Emperor Shun-chih was succeeded by his third son Hsüan-yeh, better known by his reign title K'ang-hsi, a boy of eight *sui*. He was selected primarily because he had already survived the dreaded smallpox and was in that measure more nearly certain of longevity. During his minority four regents were ap-

18. Consolidation of all taxes into one general sum; hence the name. For details, see Chapter 3.

pointed: Soni, Suksaha, Ebilun, and Oboi. The last named was the most domineering; and K'ang-hsi, though a mere boy, resented his conduct. In 1667 K'ang-hsi assumed personal rule at the age of thirteen *sui* and obtained the help of Songgotu, the uncle of his empress, to send Oboi to prison on charges of thirty high crimes. The courage, sagacity, and decisiveness with which the emperor disposed of the case were to characterize his long reign of sixty-one years.

K'ang-hsi was extremely energetic and resourceful. He followed a prodigious daily schedule of work. He rose before dawn, listened to a short exposition of Confucian Classics presented by an imperial lecturer, and then beginning at 5 a.m. conducted the daily audiences. However, after October 21, 1682, to accommodate officials not living close to the imperial palace, the audiences were rescheduled to commence at 7 a.m. in the spring and summer and 8 a.m. in the autumn and winter. At the receptions, K'ang-hsi first received memorials from heads of the Boards and Departments and discussed with them the problems involved. Next he received the grand secretaries who advised him on important and pressing issues of state. They were followed by the representatives of the Imperial Household who sought guidance in palace administration. Finally, he gave personal audiences to provincial officials or foreign envoys. When the morning audiences were over, K'ang-hsi read memorials submitted by the Office of Transmission on behalf of junior officials not permitted personal interviews. The rest of the day was occupied with lectures by imperial tutors, solicitations to the empress dowager about her health, practicing calligraphy or writing poems and essays, or studying Western science and mathematics with the Jesuit Fathers in his service. As a result, K'ang-hsi rarely retired before midnight.

As a ruler K'ang-hsi approached the ideal. He was intelligent, understanding, lenient, diligent, conscientious, and attentive to state affairs. He often admonished himself: "One act of negligence may cause sorrow all through the country, and one moment of negligence may result in trouble for hundreds and thousands of generations." His reign was marked by careful performance of official duties and frugal living in the court. It was under K'ang-hsi that the insecure Manchu rule was turned into a stable and prosperous state.

In civil administration K'ang-hsi performed a number of laudable works. Mindful of the suffering of the people, he put an end to the unscrupulous practice which had allowed the Manchus to take over good Chinese farms in desirable locations in exchange for their poor and badly located ones. He was concerned over the flooding of the Yellow and Huai rivers, and made personal inspections of the conservancy work being done there. Six times he toured the southern provinces of Kiangsu and Chekiang; four times he went beyond the passes in the north; and four times he visited the Wu-t'ai Mountain in Shansi. These tours served to acquaint the ruler with the local conditions and strengthen the ties between the central government and the various parts of the country.

K'ang-hsi also appointed a number of his trusted Chinese bondservants to various provincial posts as financial, textile, salt, or judicial commissioners to insure the flow of funds to the Imperial Household and the transmission of confidential information to himself. He issued them secret instructions and they furnished him with information in "secret palace memorials" (*mi-che*) on which the emperor marked his endorsements. In this manner, K'ang-hsi built up a personal bureaucracy as well as an intelligence network.[19]

To demonstrate his benevolence K'ang-hsi reduced the land and grain taxes countless times; during his first forty-four years (1662-1705) of reign he remitted some 90 million taels of taxes, and in 1712 alone, some 33 million. In that latter year, he issued the famous decree that the tax quota was to be based on the population of that year, and that no more taxes would be imposed on the basis of the number born after that year. He cleaned up corruption in government by a relentless application of justice; irregularities in civil service examinations were punished with severity.

K'ang-hsi was a dedicated patron of learning. He was said to be well versed in Chinese classics and philosophical writings. In 1679 he called for a *Po-hsüeh hung ts'u* special examination to recruit fifty learned men to compile the *History of the Ming* (*Ming-shih*). They were given preferred posts in the Hanlin Academy, to the jealousy of those who had risen from the regular examinations, who therefore called the fifty fortunate men "Wild Hanlin." K'ang-hsi's Imperial Study (*Nan shu-fang*) was filled with literary men, artists, and calligraphers. Frequently he gave banquets in honor of famous scholars and artists, and on those occasions they would drink and compose poems freely.

As a result of K'ang-hsi's patronage of learning, several monumental works were compiled; of these, the most famous are the *K'ang-hsi Dictionary*, the important phrase dictionary called *P'ei-wen yün-fu*, the *Complete Works of Chu Hsi*, and a grand encyclopedia called *Ku-chin t'u-shu chi-ch'eng*, which dealt with ancient and modern books and comprised 5,020 volumes (*ts'e*). Many of these works contained a preface written by the emperor himself, and therefore bore the impressive and authoritative mark "Imperial Edition"; but apparently most of the prefaces came from the hands of his learned Chinese scribes.

19. Spence, 14-16, 222-40.

The emperor's widely acknowledged fondness for learning included an extensive interest in the arts and sciences. He had an impressive collection of paintings and calligraphy, and his imperial kilns turned out many beautiful pieces of porcelain which today are priceless. Many Chinese and European artists worked in his palace; the *Jui-i kuan* (Hall of Satisfaction) was said to be filled with artistic Jesuit missionaries who drew, painted, and engraved for the emperor. With the missionaries he studied mathematics and his accomplishments were described by his admirers as remarkable. It was said that K'ang-hsi so loved learning that "his hands were never free from books." But the extent of his accomplishments was probably exaggerated. His "vermilion remarks"[20] on the memorials have been found rather childish in style and poor in calligraphy.[21] Father Matteo Ripa, who served in the court for thirteen years and who engraved a map of China for K'ang-hsi in 1718, remarked in his memoirs: "The emperor supposed himself to be an excellent musician and a still better mathematician, but though he had a taste for the sciences and other acquirements in general, he knew nothing of music and scarcely understood the first elements of mathematics."[22] Nonetheless, K'ang-hsi was a conscientious ruler of unusually broad interests. He regarded learning as the basis of good government, and people's welfare as the root of peace and order. Constantly he scrutinized himself and his administration in the light of these two standards. A noted scholar of Ch'ing history characterized K'ang-hsi's reign of sixty-one years aphoristically: "Diligence in administration, concern for people, and orthodoxy in thought."[23] K'ang-hsi was truly one of the greatest and most admirable of emperors in the history of China. Some have compared him to Louis XIV and Peter the Great.

In military conquest K'ang-hsi completed the unfinished work of his predecessors and laid the foundations of an empire which turned out to be the largest in China since the Mongols. His greatest accomplishment was the suppression of the Rebellion of the Three Feudatories (*San-Fan*). It will be recalled that the Ch'ing conquered China with the help of many Chinese defectors. General Wu San-kuei, who opened the gates at Sanhaikuan to welcome Dorgan and later fought all over the country

20. Marginal or interlinear comments written in vermilion ink.
21. Jonathan Spence, "The Seven Ages of K'ang-hsi (1654-1722)," *The Journal of Asian Studies*, XXVI: 2:206 (Feb. 1967).
22. Matteo Ripa, *Memoirs of Father Ripa, during Thirteen Years' Residence at the Court of Peking in the Service of the Emperor of China,* tr. from the Italian by Fortunato Prandi (London, 1855), 63.
23. Hsiao I-shan, *Ch'ing-tai shih* (A history of the Ch'ing dynasty), (Chungking, 1945), 64.

for the Manchus and drove Prince Kuei into Burma, was rewarded with the title of West-Suppressing Prince (*P'ing-hsi wang*) and given jurisdiction over Yunnan. Shang K'o-hsi and Keng Chi-mao, commanders of Ming forces in Liaotung who had surrendered to the Manchus, were made, respectively, South-Suppressing Prince (*P'ing-nan wang*) in Kwangtung and South-Pacifying Prince (*Ching-nan wang*) in Fukien. These were known as the Three Feudatories. Wu controlled an army of over 100,000 men while the other two maintained forces of 20,000 each. The military forces of the Three Feudatories cost the Ch'ing court some 20 million taels annually by 1667—more than half the total state expenditures—while at the same time they were virtually independent within their own realms.

The Three Feudatories were a source of great irritation to the Ch'ing dynasty. Emperor Shun-chih had put up with them because the new regime had not dared risk a civil war, but when K'ang-hsi came to power, the dynasty had been considerably consolidated and he decided to abolish the feudatories and cut the military power of the three princes.

Wu reacted by open rebellion on December 28, 1673, calling himself generalissimo of all the forces of the country and proclaiming the establishment of a new dynasty, the Chou. He ordered the restoration of Ming style costumes and haircut; his army used white flags and his soldiers wore white caps.[24] He announced his intention of overthrowing the Ch'ing and reviving the Ming. The other two feudatories joined him, and for a time it looked as if the Ch'ing dynasty might be toppled. The Manchu bannermen were unable to suppress them. K'ang-hsi used a number of Chinese generals in the campaign, and in 1681, after eight years of hard fighting, the Three Feudatories were finally put down. Two years later (see previous section) Taiwan under Koxinga's grandson[25] was also taken and turned into a prefecture of the Fukien province.

Once freed from the civil war, K'ang-hsi turned to the two problems posed by the Ölöd Mongols in the Northwest and the Russians in the Northeast. These two problems directly involved one another, for there seemed to be a good possibility that the Ölöd and the Russians might form an alliance against the Ch'ing dynasty. The Russians had conquered Siberia and reached the Amur River; and during the 1640s and 1650s the Cossacks from Siberia had made repeated raids on the Amur Region. In 1666 they founded Albazin (Ya-k'e-sa) as an advance base, threatening the Manchu homeland. Nearly simultaneously with this, during the 1670s, Galdan (1644?-97), khan of the Dzungars (West Mongols), a

24. White, the funeral color, was presumably being worn to mourn the passing of the Ming dynasty.
25. Cheng K'o-shuang.

tribe of the Ölöd, had risen to power and aspired to build a Central Asian empire. He conquered Eastern Turkestan in 1679, invaded Outer Mongolia in 1687, routed the Qalqa (Khalkha, East Mongols), and penetrated as far as the Kerulen River. Thus an alliance with the Russians seemed nearly inevitable.

To prevent this, K'ang-hsi's strategy was first to crush the Russians at Albazin and then offer them a liberal treaty as a kind of sop. His general Pengcun attacked Albazin in 1685, ruthlessly demolishing it. Russian reinforcements were sent in the following year and new fortifications were constructed. A Ch'ing expedition was dispatched to besiege Albazin. Informed that a Russian diplomatic mission under Fedor A. Golovin was on its way, K'ang-hsi, with an eye to winning Russian good will, lifted the siege and prepared to enter negotiations with the envoy.

The result was the signing of the Treaty of Nerchinsk in 1689, China's first agreement with a "Western" power. In the treaty, Russia agreed to destroy its fortresses at Albazin and evacuate its subjects, while China agreed to cede some territory along the undecided frontiers and extend certain trade privileges to Russia. With this diplomatic exchange, K'ang-hsi felt relatively assured of Russian neutrality in his war against Galdan, whom in 1696 he finally defeated at Jao Modo. The following year the Ölöd chieftain died, and K'ang-hsi extended the Ch'ing rule to Outer Mongolia and Hami, clearing the way for his grandson, Emperor Ch'ien-lung, to complete the conquest of Chinese Turkestan in the 1750s.[26]

Galdan's death did not put an end to the Ölöd problem, however. His nephew Cewang Arabdan gradually built up his power and in the early 1710s became a new threat to the Ch'ing. Cewang had married the daughter of Ayüki, chief of the Türgüd tribe which had migrated to Russia in 1630. This marriage made it possible that Cewang and Ayüki might join forces against the Ch'ing. To forestall this possibility and strengthen the Ch'ing tie with the Türgüd—and possibly to persuade the Türgüd tribe to return to China—K'ang-hsi dispatched a mission in 1712 to Ayüki. The mission, under the leadership of Tulisen, traveled through Siberia and reached the Volga in 1714. Tulisen met with Ayüki and presumably accomplished the objectives. He returned with an account of his travels which was entitled *I-yü lu* (Description of a foreign land), probably the first authentic Chinese work written on Russia during the Ch'ing period.

In both internal and external affairs, K'ang-hsi truly had accomplished much. He had developed a stable, frugal, and efficient civil administration, patronized learning, suppressed the Rebellion of the Three Feudatories,

26. For details of K'ang-hsi's war against Galdan and the early Russian-Chinese relations, see Chapter 5.

crushed the resistance movement on Taiwan, established diplomatic relations with Russia, and defeated the Ölöd under Galdan. Dynastic splendor had replaced the earlier insecurity, and the country took on the appearance of an empire. K'ang-hsi closed his reign a contented ruler in 1722.

THE REIGN OF EMPEROR YUNG-CHENG, 1723-35. Yung-cheng ascended the throne at the age of 45 *sui*. By nature he was severe, suspicious, and jealous, but extremely capable and resourceful; these characteristics were clearly exhibited in his administration. Yung-cheng felt that his father's rule, especially during the last years, had been too relaxed. He therefore began his reign by centralizing power in his own hands. Not only did he reject the request of the various princes that they be established as feudal lords with territories of their own, but he went further to deprive them of their military power. Whereas in the early Ch'ing period the emperor had directly controlled only the three superior Manchu banners—yellow, bordered yellow, and white—Yung-cheng took over control of all eight banners.

In civil administration Yung-cheng was indefatigable. He personally read and commented on numerous memorials daily, working late at night to deliberate over state policies. He was probably the hardest-working man in the empire. His control over officials was tight and autocratic; enforcement of law was carried out inexorably and vigorously, and spies were sent all over the empire to inform on officials' dereliction of duty. To prevent secret opposition, associations and cliques among officials and scholars were emphatically prohibited; the emperor himself in 1725 wrote "A Discourse on Parties and Cliques" (*P'eng-tang lun*) to warn the more venturesome against attempts in this direction. In financial matters, he combined the poll tax with the land tax. He also instituted the practice of granting officials liberal stipends called "anti-corruption fund" (*yang-lien chin*, lit., integrity-nourishing allowance) while prohibiting them from charging surtax or practicing irregularities, which, when discovered, were severely punished. Socially, he took the egalitarian step of raising beggars, hereditary servants, and boatmen from the despised status of "mean people" to that of the common people.

Institutionally, he brought in two innovations. One was the practice of depositing the name of the heir-apparent in a sealed box to prevent tampering with the machinery of succession. The box was placed behind a big tablet that was hung in his palace. Copies of the secret designation were locked in other safe places so that the authenticity of the selection could be double-checked upon the death of the reigning emperor. This practice was observed until the last decades of the Ch'ing dynasty. The other institutional contribution was the creation of the Grand Council (*Chün-chi ch'u*) by 1729 to assist the emperor in drafting edicts and offer advice on military and state policy during the campaign against the Ölöd. The Grand Council originally had three members who stayed within the palace grounds so as to be available for service at any time. This small, tightly knit group could reach fast decisions, offer quick counsel, and guard secrets. Because it proved so useful, the Grand Council continued to exist even after peace was restored. It pre-empted the powers of the Great Secretariat, which was reduced to handling only routine matters.[27]

Military and diplomatic matters had not changed much from K'ang-hsi's reign: the twofold threat from the Ölöd and the Russians continued. The Treaty of Nerchinsk with Russia had not settled the boundary between Siberia and Outer Mongolia, and the traffic between the Ölöd leader Cewang Arabdan and the Russians renewed Chinese fears of secret plotting between them. To carry out his father's policy of isolating Mongolia from the Russians, Yung-cheng was anxious to settle all pending issues with Russia by a new agreement. The resultant Treaty of Kiakhta of 1727 secured for China a clear delineation of the Mongolian-Siberian frontier, while Russia gained territorial concessions of nearly 40,000 square miles between the Upper Irtysh and the Sayan Mountains as well as land south and southwest of Lake Baikal. In addition, Russia was granted trade privileges and permitted to establish a religious mission in Peking.

Having settled the Russian problem, Yung-cheng sent an expedition to the Ölöd. During the campaign, however, reports came of complaints made by the Russians over Mongolian border raids: the Mongolian bandits had been looting horses, camels, oxen, and sheep. Wanting Sino-Russian relations to remain unimpaired and desirous of keeping Russia neutral, Yung-cheng dispatched T'o-shih, a vice-president of one of the boards, on a diplomatic mission to Russia in 1729—the first official Chinese mission to a "Western" state. The mission was sent on the pretext of congratulating Peter the Second on his coronation, but upon his arrival in St. Petersburg, T'o-shih learned that the tsar had died and that the new ruler was Anna Ivanovna, a niece of Peter the Great. The Manchu envoy returned to Peking to obtain new credentials and in 1731 set out for Russia again. To St. Petersburg he suggested that if the Chinese attack drove the Ölöd to seek refuge in Russia, the Russian government should extra-

---

27. Alfred K. L. Ho, "The Grand Council in the Ch'ing Dynasty," *The Far Eastern Quarterly*, XI:2:167-82 (Feb. 1952). See also Silas Hsiu-liang Wu, "The Memorial System of the Ch'ing Dynasty (1644-1911)," *Harvard Journal of Asiatic Studies*, 27:30 (1967). For further discussion of the Grand Council, see pp. 48–51.

dite their rulers and nobles but should keep the tribesmen in the country, under control, and prevent them from troubling China; in return China would give Russia part of the land that should be seized from the Ölöd. The Russian government was noncommittal, replying that it would discuss the question of extradition when it arose. Although the mission achieved no concrete results, it soon became clear that Russia, then fighting the Polish Succession War, was in no position to aid the Ölöd. The Ch'ing forces sent to fight the Ölöd were at first defeated by Galdan Cereng, the son of Cewang Arabdan, who had died in 1727, but were able to score a victory in 1732 at Erdeni Tsu, making possible a negotiated peace at no great loss of prestige to China.

Yung-cheng has often been accused of being excessively autocratic and despotic, particularly in his literary inquisitions. A celebrated case involved Lü Liu-liang, who was accused of having written an anti-Manchu book which stressed the differences between the Chinese and the barbarians (i.e. Manchus). Lü was put to the most severe punishment—the "lingering death"—and his son and students were all beheaded. Emperor Yung-cheng himself even wrote a treatise[28] to justify the Manchu rule in China and to warn against the danger of advocating racial revolution by the Chinese.

If K'ang-hsi's reign was characterized by tolerance, leniency, and liberality, Yung-cheng's was marked by strict control, severe punishment, and high efficiency. The spirit of his administration was reflected in the manner of his favorite grand secretary and president of the Board of War, O-erh-t'ai, who was noted for his unbending, high-handed measures. To be sure, the emperor himself declared that a proper balance between strictness and leniency was essential to good government; but he did not take this to mean the tempering of justice with mercy. Rather he meant that a careful weighing of a given situation should elicit a clear picture of whether severity or leniency was called for and to what degree of either the administrator should resort. When strictness was needed, he should be strict; when leniency was warranted, he should be lenient. But there was no profit in mixing the two; the important consideration in each situation was "appropriateness" (*i*).

Yung-cheng could accurately be described as a Legalist statesman. Under him the highest form of absolute monarchy was achieved. All powers of state were concentrated in his hands. His reign was sometimes described as cruel, despotic, and autocratic—the opposite of K'ang-hsi's; but the opposites apparently supplemented each other and made possible the glorious reign of the next emperor.

28. *Ta-i chüeh-mi lu.*

THE REIGN OF EMPEROR CH'IEN-LUNG, 1736-95. Emperor Yung-cheng was succeeded by his fourth son, Hung-li, whose reign title was Ch'ien-lung. As a boy Ch'ien-lung had been a favorite of his grandfather, K'ang-hsi, after whom he sought to model himself. In disposition the two were in fact much alike, both straightforward, open-minded, and rather lenient. When Ch'ien-lung ascended the throne late in 1735 at the age of 25 *sui,* he prayed to heaven that he might be granted a reign almost as long as, but no longer than, his grandfather's, which had been sixty-one years.

Ch'ien-lung was well prepared for the throne, for during his prince-hood he had been given a rigorous indoctrination in the idealized role of the sovereign. When barely ten and a half years old, he was ordered by K'ang-hsi to the Palace School for Princes (*Shang shu-fang*), where ten Chinese and five Manchu tutors eagerly taught him the Confucian ethics and the Manchu military arts. Class hours lasted from dawn till mid- or late afternoon, and the curriculum included the study of the classics, history, literature, philosophy, court manners, filial duties, ritual performances, and later, administrative techniques. Riding and archery were also practiced. Ch'ien-lung was a keen student of history, especially of the chronicles that provided historical models of imperial perfection. His paragon and all-time favorite was the martial and heroic emperor, T'ang T'ai-tsung (A.D. 627-47), whose reign of military splendor and material prosperity, tempered by a studied humility and benevolence, was a source of inspiration to the young prince.[29]

During his years of study, Ch'ien-lung learned that the ideal ruler was one who had the "ability and desire to discover, select, and use ministers of high talent" and to "exhaust their talent in the service of the state." He was also taught to shun favoritism, to beware of clique machination and eunuch dominance, and to treat the lowly "according to their merits" and the superior talent with the respect due a teacher, with ample material rewards.[30]

Thus, when ascending the throne, Ch'ien-lung was thoroughly grounded in the art of emperorship. He was conscientious and responsible, if somewhat pompous and ostentatious. He thought his father's rule too severe just as the latter had thought K'ang-hsi's too indulgent, and he deliberately announced his own preference for the "middle road," i.e. the golden mean. During the early years of his reign he was assisted by ex-

29. Harold L. Kahn, "Some Mid-Ch'ing Views of the Monarchy," *The Journal of Asian Studies,* XXIV:2:230-31 (Feb. 1965).

30. Harold L. Kahn, "The Education of a Prince: The Emperor Learns His Roles" in Albert Feuerwerker, Rhoads Murphey, and Mary C. Wright (eds.), *Approaches to Modern Chinese History* (Berkeley, 1967), 15-44.

perienced statesmen like O-erh-t'ai (1680-1745) and Chang T'ing-yu (1672-1755), whom he inherited from his father. The work begun by his predecessors now reached fruition. The country enjoyed peace and prosperity; the treasury was full; the dynasty glowed with an opulence and affluence it had never known before.

After O-erh-t'ai's death in 1745 and Chang's retirement four years later, Ch'ien-lung came into his own. After the pattern of his grandfather, he made elaborate tours of the country. Six times he journeyed to the south, ostensibly to inspect the water conservancy, but actually to enjoy the wealth and luxury of the southern provinces. Four times he toured the east; five times he went to the west; and numerous times he visited the birthplace of Confucius in Shantung. Wherever he went, elaborate preparations were made to welcome him, and a trend of luxury set in.

Ch'ien-lung considered himself the lord-patron of letters. He revived special examinations for erudite men, of the type that K'ang-hsi had given, and invited famous scholars and recluses to join his government. His own accomplishments in the arts and letters were not particularly impressive, although he boasted of having composed 43,000 poems—a prolificness that defies credulity. A number of them doubtless were accomplished with the assistance of his Chinese secretaries. Moreover, Ch'ien-lung's penchant for displaying his calligraphy and seals on old master paintings raised the question of good taste, and when he jotted down 54 inscriptions on one handscroll and fixed 13 seals on another, he did not endear himself to the artistic world.[31] Nonetheless, Ch'ien-lung exhibited a great interest in the arts and maintained exquisite collections of paintings, calligraphy, porcelain, and cloisonné. His imperial kilns produced some of the most elegant porcelain and cloisonné the world has ever known, with designs that occasionally reveal European influence, for many missionaries taught Western drawing to Chinese court artists and amused the emperor with their accomplishments. For instance, in 1747 Michel Benoist built a Western style fountain for him, and a number of Italian-style buildings were designed by G. Castiglione for the Summer Palace, called the Yüan-ming Yüan, located about five miles to the northwest of Peking.

The greatest literary project sponsored by Ch'ien-lung was the compilation of the *Complete Library of the Four Treasuries* (*Ssu-k'u ch'üan-shu*), which contained more than 36,000 volumes, arranged in the four main categories of the literature: classics (*ching*), history (*shih*), philosophy (*tzu*), and belles-lettres (*chi*). Even the printed catalogue of this

31. Kahn, "The Education of a Prince," 30-31.

monumental library was an impressive work of scholarship: it consisted of succinct comments on 10,230 titles of works. Seven sets of the *Four Treasuries* were made and deposited in different parts of the empire.

Ch'ien-lung's sponsoring of literary projects was to some degree prompted by political motives: it provided the means to control virtually everything that was written and to expunge seditious references to the Manchus. Suspect and heretical views could be discovered and suppressed and their authors brought to account. According to the reports of the Board of War, the destruction of "unacceptable" books took place 24 times between 1774 and 1782, amounting to 13,862 works in 538 titles. Many considered his destruction of books the greatest catastrophe since the book-burning of the First Ch'in Emperor in 213 B.C. Indeed, the imperial control of learning led to more than sixty cases of literary inquisition during the Ch'ien-lung era.

Ch'ien-lung's military record was splendid. He solved once and for all the Ölöd problem, which had troubled the dynasty since its early days. So confident was he, in fact, of defeating the Ölöd that he did not even consider the possibility of Russian intervention. In 1759 the whole of Chinese Turkestan was pacified, and a military occupation followed. A military governor was stationed in Ili, with jurisdiction over the northern and southern sides of the T'ien-shan Mountains. Large columns of troops and a number of imperial agents and councillors were stationed at key points. In 1768 the area known as the Hsi-yü, or Western Region, was renamed Sinkiang, meaning "New Dominion" or "New Territory." The Ch'ing dynasty now joined the ranks of the three great dynasties before it—the Han, the T'ang, and the Yüan—in the distinction of having extended the Chinese rule into the Tarim Basin in the heart of Central Asia.

In addition to the conquest of Sinkiang, Ch'ien-lung was notably successful in a number of lesser campaigns. Enormously proud of these feats, he composed a eulogy of them in 1792 and called it *A Record of Ten Perfect Accomplishments (Shih-ch'üan chi)*. This *Record* included two campaigns against the Dzungars in northern Sinkiang (1755, 1756-57), the pacification of the Moslems in southern Sinkiang (1758-59), the annihilation of the Chin-ch'uan rebels in two campaigns (1747-49, 1771-76), the suppression of a rebellion on Taiwan (1787-88), the subjugation of Burma (1766-70), the subjection of the Annamese to his suzerainty (1788-89), and two separate conquests of the Gurkhas. To give things their proper perspective it must be noted that except for the conquest of Sinkiang, which was a great military accomplishment by any standard, the other victories enumerated in the *Record* were only police actions or localized campaigns which deserved little special recognition. But in the very act of compiling such a record and calling himself the "Old Man of Ten Perfect Accomplishments" (*Shih-ch'üan lao-jen*), Ch'ien-lung demonstrated his self-satisfaction and love of show.

Indeed, Ch'ien-lung had much to be proud of and much to be grateful for. He ruled over an empire that stretched from Outer Mongolia in the north to Kwangtung in the south, and from the coast in the east to Central Asia in the west. Peace and prosperity prevailed within the country, and numerous peripheral states came to pay tribute. Dozens of nations in East, Southeast, and Central Asia acknowledged Chinese suzerainty over them: from Korea in the northeast, to Annam, Burma, and Siam in the south; Bhutan, Nepal, and the Gurkhas in the southwest; and a number of khanates in Central Asia such as Khokand, Bukhara, Burut, Badakshan, Afghanistan, and the Kazaks. Ch'ien-lung proudly presided over this vast empire, which was larger than those of the Han and T'ang and second only to the Mongol Empire of the 13th century. It was a golden era in Ch'ing—and Chinese—history.

But in the very splendor of the dynasty the elements of ultimate ruin were already present. Ch'ien-lung's senility and the degeneration of his judgment had much to do with the decline. At the age of 65, he noticed a 25-year-old handsome imperial bodyguard called Ho-shen (1750-99). Within a year Ho-shen was promoted to vice-president of the Board of Revenue, two months later was made a grand councillor, and a month after that a minister of the Imperial Household—posts usually filled by the most meritorious and venerable officials. In 1777, when barely 27, Ho-shen was given the unusual privilege of riding a horse in the Forbidden City, a favor reserved for the highest officials who were too old to walk. Later he was given control of the Boards of Revenue and of Civil Office, enabling him to control the revenue of the empire and appoint his henchmen to important and lucrative positions. His hold on the senile emperor was further strengthened when in 1790 his son was married to Ch'ien-lung's youngest daughter. Secure of the emperor's favor and approbation, Ho-shen enjoyed complete freedom of action. He was openly corrupt and practiced extortion on a large scale. His satellites in the government followed suit, and his associates in the military service unnecessarily prolonged campaigns so as to have the benefit of additional funds.

The last years of Ch'ien-lung's reign were indeed shameful. Although he retired in 1795 after sixty years of reign, he still ruled behind the scenes as Super Emperor (*T'ai shang-huang*). Not until his death in 1799 was his son, Emperor Chia-ch'ing, able to execute Ho-shen. From the time he caught Ch'ien-lung's attention in 1775 until his death in 1799,

Ho-shen plundered the state and amassed an incredible fortune. His confiscated property was estimated at 800,000,000 taels—nearly one and a half billion dollars!

The influence of Ho-shen spread like a dye. There was corruption at different levels within and without the capital, among civil as well as military personnel. Bannermen developed licentious habits and became totally useless as a military force. The Chinese Green Standard army was beset with irregular practice and lost much of its earlier fighting spirit. Military defense of the frontiers was neglected. The habit of luxury and big spending contributed to moral degradation and a general decline of the dynasty. Ch'ien-lung's six tours to the south cost at least 20 million taels, while the expenditures of his other tours to the east, west and north were not even known. His Ten Perfect Accomplishments were made at the cost of 120 million taels, against an average annual revenue of some 40 million taels. These massive spendings and the general trend toward luxury set the machinery in motion for great financial difficulties in the future.

Thus as Ch'ien-lung's reign drew to a close, China was experiencing the beginning of a dynastic decline. The splendor of past glory remained on the surface, but beneath it the substance of grandeur was gone. It was at this juncture that the Westerners began to intensify their bid to open China to trade and diplomacy, and a new phase of history began.

Further Reading

A-kuei 阿桂 (ed.), *Huang-ch'ao k'ai-kuo fang-lüeh* 皇朝開國方略 (A brief account of the founding of our imperial dynasty), (1887), 6 ts'e.

Ames, Roger T., *The Art of Rulership: A Study in Ancient Chinese Political Thought* (Honolulu, 1983).

Chan, Albert, *The Glory and Fall of the Ming Dynasty* (Okla., 1982).

Chan, Hok-lam, *Legitimation in Imperial China: Discussions Under the Jurchen-Chin Dynasty (1115-1234)*, (Seattle, 1985).

Chou, Ju-hsi, and Claudia Brown, *The Elegant Brush: Chinese Painting Under the Qianlong Emperor, 1735-1795* (Phoenix, 1985).

Corradini, Piero, "Civil Administration at the Beginning of the Manchu Dynasty," *Oriens Extremus*, 9:2:133-38 (Dec. 1962).

Crossley, Pamela Kyle, "*Manzhou yuanli Kao* and the Formalization of the Manchu Heritage," *The Journal of Asian Studies*, 46:4:761-90 (Nov. 1987).

Fang, Chaoying, "A Technique for Estimating the Numerical Strength of the Early Manchu Military Forces," *Harvard Journal of Asiatic Studies*, 13:192-215 (1950).

Feuerwerker, Albert, *State and Society in Eighteenth-Century China: The Ch'ing Empire in Its Glory* (Ann Arbor, 1976).

Fletcher, Joseph, "Ch'ing Inner Asia c. 1800" in John K. Fairbank (ed.), *The Cambridge History of China* (Cambridge, Eng., 1978), Vol. 10, 35-106.

Guy, R. Kent, *The Emperor's Four Treasuries: Scholars and the State in the Late Ch'ien-lung Era* (Cambridge, Mass., 1987).

Hsiao, I-shan 蕭一山, *Ch'ing-tai shih* 清代史 (A history of the Ch'ing dynasty), (Chungking, 1945), chs. 3-4.

————, *Ch'ing-tai t'ung-shih* 清代通史 (A general history of the Ch'ing period), rev. ed. (Taipei, 1962), I, chs. 1-5, 8-21, 26-30; II, chs. 1-4.

Huang, Pei, "Five Major Sources for the Yung-cheng Period, 1723-1735," *The Journal of Asian Studies*, XXVII:4:847-57 (Aug. 1968).

————, *Autocracy at Work: A Study of the Yung-cheng Period, 1723-1735* (Bloomington, 1974).

Hummel, Arthur W., *Eminent Chinese of the Ch'ing Period* (Washington, D.C., 1943-44). Biographies of Nurhaci, Abahai, Dorgan, Shun-chih (Fu-lin), K'ang-hsi (Hsüan-yeh), Yung-cheng (Yin-chen), and Ch'ien-lung (Hung-li).

Inaba, Iwakichi 稲葉岩吉, *Shinchō zenshi* 清朝全史 (A history of the Ch'ing dynasty), (Tokyo, 1914), Chinese tr. by Tan T'ao under the title, *Ch'ing-ch'ao ch'üan-shih* 清朝全史 (Taipei, 1960), chs. 1-2, 7-12, 17-18, 24-32, 39-43, 47-48.

Ishida, Mikinosuke, "A Biographical Study of Giuseppe Castiglione (Lang Shih-ning), A Jesuit Painter in the Court of Peking under the Ch'ing Dynasty," *Memoirs of the Research Department of the Toyo Bunko*, 19:79-121 (Tokyo, 1960).

Kahn, Harold L., "The Education of a Prince: The Emperor Learns His Roles" in Albert Feuerwerker, Rhoads Murphey, and Mary C. Wright (eds.), *Approaches to Modern Chinese History* (Berkeley, 1967), 15-44.

————, "Some Mid-Ch'ing Views of the Monarch," *The Journal of Asian Studies*, XXIV:2:29-43 (Feb. 1965).

————, "The Politics of Filiality: Justification for Imperial Action in Eighteenth Century China," *The Journal of Asian Studies*, XXVI:2:197-203 (Feb. 1967).

————, *Monarchy in the Emperor's Eyes: Image and Reality in the Ch'ien-lung Reign* (Cambridge, Mass., 1971).

K'ang-hsi, *The Sacred Edict, Containing Sixteen Maxims of Emperor Kang-hi*, tr. by the Rev. William Milne, 2nd ed. (Shanghai, 1870).

Kessler, Lawrence D., *K'ang-hsi and the Consolidation of Ch'ing Rule, 1661-1684* (Chicago, 1976).

Lee, Robert H. G., *The Manchurian Frontier in Ch'ing History* (Cambridge, Mass., 1970).

Ma, Feng-ch'en, "Manchu-Chinese Conflicts in Early Ch'ing" in E-tu Zen Sun and John DeFrancis (eds.), *Chinese Social History* (Washington, D.C., 1956), 333-51.

Meng, Shen 孟森, *Ch'ing-tai shih* 清代史 (A history of the Ch'ing period), (Taipei, 1960), chs. 1-3.

Michael, Franz, *The Origin of Manchu Rule in China* (Baltimore, 1942).

Miyazaki, Ichisada 宮崎市定, *Yō-sei-tei, Chūgoku no dokusai kunshu* 雍正帝, 中國の獨裁君主 (The Yung-cheng Emperor, China's autocratic ruler), (Tokyo, 1950).

Naitō, Torajirō 内藤虎次郎, *Shinchōshi tsuron* 清朝史通論 (A survey of the history of the Ch'ing dynasty), (Tokyo, 1944).

Nivison, David S., "Ho-shen and His Accusers: Ideology and Political Behavior in the Eighteenth Century" in David S. Nivison and Arthur F. Wright (eds.), *Confucianism in Action* (Stanford, 1959), 209-43.

Oxnam, Robert B., *Ruling from Horseback: Manchu Politics in the Oboi Regency, 1661-1669* (Chicago, 1974).

Ripa, Matteo, *Memoirs of Father Ripa, during Thirteen Years' Residence at the Court of Peking in the Service of the Emperor of China,* tr. from the Italian by Fortunato Prandi (London, 1855).

Rossabi, Morris, *China and Inner Asia: From 1368 to the Present Day* (New York, 1975).

Sanjdorj, M., *Manchu Chinese Colonial Rule in Northern Mongolia* (New York, 1980).

Shen, Yün 沈雲, *Tai-wan Cheng-shih shih-mo* 台灣鄭氏始末 (A complete account of the Koxinga family on Taiwan), (1836), 6 *chüan.*

Smith, Richard J., *China's Cultural Heritage: The Ch'ing Dynasty, 1644-1912* (Boulder, 1983).

Spence, Jonathan D., *Ts'ao Yin and the K'ang-hsi Emperor: Bondservant and Master* (New Haven, 1966).

————, "The Seven Ages of K'ang-hsi (1654-1722)" *The Journal of Asian Studies,* XXVI:2:205-11 (Feb. 1967).

————, *The Emperor of China: Self-Portrait of K'ang-hsi* (New York, 1974).

Spence, Jonathan D., and John E. Wills, Jr. (eds.), *From Ming to Ch'ing: Conquest, Region, and Continuity in Seventeenth-Century China* (New Haven, 1979).

Sugimura, Yūzō 杉村勇造, *Ken-ryū kotei* 乾隆皇帝 (Emperor Ch'ien-lung), (Tokyo, 1961).

Tao, Jing-shen, *The Jurchen in Twelfth-Century China* (Seattle, 1977).

Tsao, Kai-fu, *The Relationship Between Scholars and Rulers in Imperial China: A Comparison Between China and the West* (Lanham, Md., 1984).

Wada, Sei 和田清, *Tōashi kenkyū—Manshū hen* 東亞史研究 (滿洲篇) (Studies on the history of the Far East—Manchurian volume), (Tokyo, 1955).

————, "Some Problems Concerning the Rise of T'ai-tsu, the Founder of Manchu Dynasty," *Memoirs of the Research Department of the Toyo Bunko* (Tokyo, 1956).

Wakeman, Frederick, Jr., *The Great Enterprise: The Manchu Reconstruction of Imperial Order in Seventeenth Century China,* 2 vols. (Berkeley, 1985).

Waley-Cohen, Joanna, *Exile in Mid-Qing China: Banishment to Xinjiang, 1758–1820* (New Haven, 1991).

Wills, John E., *Embassies and Illusions: Dutch and Portuguese Envoys to K'ang-hsi, 1666-1687* (Cambridge, Mass., 1985).

Wu, Silas Hsiu-liang, "Emperors at Work: The Daily Schedules of the K'ang-hsi and Yung-cheng Emperors, 1661-1735," *The Tsing Hua Journal of Chinese Studies,* New Series, VIII:1-2:210-27 (Aug. 1970).

————, *Passage to Power: K'ang-hsi and His Heir Apparent, 1661-1722* (Cambridge, Mass., 1979).

Yang, Lu-jung 楊陸榮, *San-Fan chi-shih pen-mo* 三藩記事本末 (A complete account of the Three Feudatories), (1717), 4 *chüan.*

3

Political and Economic
Institutions

Political Structure

Though the Ch'ing was an alien dynasty, it accepted the traditional Confucian order and recruited Chinese scholars into the officialdom to work with the Manchus, forming a kind of ethnic dyarchy within a political monarchy. The government was essentially an autocracy. There was no division of power in the Western sense; the emperor was the absolute ruler in every branch of the regime, whether executive, legislative, or judicial. He governed without a prime minister, and was truly qualified to pronounce: "L'état, c'est moi!" This high concentration of power placed far greater demands on the emperor than on any other man in the empire; K'ang-hsi once remarked that though his ministers were free to come and go, he was not. Ch'ing absolutism was doubtless an inheritance from the Ming, as indeed were most of its institutions and practices; only a few additions were made to suit special occasions and needs.

THE EMPEROR AND THE NOBILITY. The emperor, at the summit of the hierarchy, gave to state affairs a personal attention seldom tendered by earlier rulers. He read all the memorials from every corner of the empire—from 50 or 60 to 100 daily—and wrote marginal or interlinear comments on each of them in vermilion ink. In his executive capacity he decided all important state policies, made appointments, conferred titles, approved promotions, demotions, and dismissals, awarded pensions, commanded the army, and ratified treaties with foreign powers. As the supreme legislator he enacted, annulled, and amended laws by decrees and edicts. Judicially,

45

he was the highest court of appeal, granting pardons and reprieves as a mark of favor. Indeed, the highest form of absolute monarchism was reached under the Ch'ing.

The emperor was the religious head, too. He sanctioned the Dalai Lama, the chief Taoist, and the Duke K'ung (direct descendants of Confucius), and offered sacrifices to heaven, earth, Confucius, Buddha, and other focuses of reverence. In times of natural calamities—manifestations of nature's anger—he made offerings to heaven to expiate his sins, which were considered to be the cause.

Finally, in his sponsoring of the compilation and publication of books and encyclopedias, and in his patronage of learning generally, he showed himself the intellectual leader of his people. He ordered the administering of provincial and metropolitan examinations and conducted the palace examinations himself. Often he personally questioned the candidates and decided the ranking of the first ten successful ones; and occasionally he would even lecture to the Imperial College (*Kuo-tzu chien*). All degrees were conferred in his name.

For all his near omnipotence, the emperor was nevertheless subject to some restrictions. The Confucian cult demanded that he be moral, virtuous, and attentive to the needs of his subjects; and it bound him to be respectful on ceremonial occasions and to follow the good precedents of the past, setting a living example for the millions. He should not run counter to traditions and social customs, nor should he ignore the "public opinion" of the literati and the gentry. Except in great emergency he could not call into service officials who were in mourning. When he referred cases to his high ministers—the Six Boards (*Liu-pu*) and the Nine Ministers (*Chiu-ch'ing*)—he was morally bound by their unanimous recommendations. As a member of the imperial household, he could not ignore the imperial family law or take lightly the admonitions and instructions of his ancestors, which were considered sacred and inviolable. To neglect these restraints would be to justify remonstration by the censors or a *coup d'état* or even a rebellion. Rebellion—a corollary of the Mencius idea of the popular right to revolt—was the strongest check on a ruler's conduct. If the emperor exercised his supreme powers conscientiously and at the same time honored these provisions, he could be reasonably sure of his ministers' admiration and the support of his subjects, and could thereby justify his exercising of the fictitious Mandate of Heaven in his role as the Son of Heaven (*T'ien-tzu*) and mediator between man and nature.

The nobility in the Ch'ing system consisted of three categories: the imperial clansmen, the titular nobles, and the bannermen. The clansmen were the direct male descendants of Nurhaci, numbering about 700 from the late 16th century through the end of the 19th. They were governed by an Imperial Clan Court (*Tsung-jen fu*) whose function it was to keep their identity. This Court kept records of them from birth to death, including marriages, appointments, promotions, demotions, dismissals, or whatever. It also operated their schools, conducted separate examinations for literary degrees, tried them for offenses, and in general supervised their activities. The emperor gave them land, official residences, and annual silver and rice allowances. But they were kept more or less isolated: they were not allowed to communicate with the provincial authorities, and as a rule were not appointed to the all-important Grand Secretariat or, later, to the Grand Council. Exceptions were made only rarely during the 17th and 18th centuries, though in the last decades of the dynasty the restriction was relaxed somewhat: Prince Kung, for instance, was appointed to the Grand Council in 1853.

The titular nobility were divided into five classes: duke (*kung*), marquis (*hou*), earl (*po*), viscount (*tzu*), and baron (*nan*). Most of these titles were accorded civil and military officers who had merited such honor. The titular nobility did not function as a class by itself, and as a group had little influence in the society.

The bannermen, the third type of nobility, were also given preferential treatment by the emperor in the form of annual pensions, land for cattle-raising, and allotments of rice and cloth. To preserve their dignity and special status all bannermen were excluded from participation in trade and labor. Offenses committed by the bannermen were tried not by the ordinary civil magistrates but by the Tartar General (Manchu General-in-Chief). A large portion of the bannermen were stationed in Peking and its vicinity, while the rest were assigned garrison duties throughout the country.

CENTRAL GOVERNMENT ORGANIZATIONS. Before 1729 the most important organ in the central government was the Grand Secretariat (*Nei-ko*), which the first Ming emperor had instituted after he abolished the office of prime minister in 1380.[1] The Ch'ing dynasty inherited this institution and appointed four grand secretaries and two associate grand secretaries to comprise it, half of whom were Manchus and half Chinese. They formed an advisory group to the emperor and were the closest equivalent to the old office of prime minister, but there was no leader among them officially, and they could not issue orders directly to the Six Boards or

1. S. Y. Teng, "Ming T'ai-tsu's Destructive and Constructive Work," *Chinese Culture*, VIII:3:20 (Sept. 1967).

provincial governments: only the emperor could do that. The grand secretaries drafted edicts, declarations, and manifestos for the emperor and assisted him in deciding high policies. As they controlled access of memorials and were able to pass judgment on them before submitting them to the emperor, they had the power to influence his decisions. Because they were close to the source of power, they commanded great respect and were regarded as the highest officials in the empire. Only holders of the *chin-shih* degree could be appointed grand secretaries, who enjoyed an indefinite tenure in this office, simply because there existed no higher one for them to be advanced to. The average length of tenure was 8 years and 9 months, but between 1644 and 1773 one grand secretary held office for more than 30 years, and twenty-four lasted more than 10 years.[2] All the grand secretaries held concurrent appointments either as presidents of the Six Boards or in other important offices.

The Grand Secretariat suffered some loss of power during K'ang-hsi's reign, when the emperor came to rely on his own secretaries in the Imperial Study (*Nan shu-fang*) to draft edicts and decrees for him. And by 1729 came the really severe blow: the establishment of the Grand Council (*Chün-chi ch'u*). This new organization pre-empted the Grand Secretariat's role as the closest adviser to the sovereign and usurped most of its original functions, leaving it only routine matters to handle. Grand secretaryships then became merely honorary titles granted to meritorious high civil officials; they did not require attendance to any regular business.

The origin of the Grand Council has been much discussed in recent years. Yet, despite discordance, it appears certain that its establishment was not immediate but gradual, evolving over a number of years. As early as 1726 the decision was made to prepare a campaign against the Dzungars, a tribe of the Ölöd in the Northwest, and in the following year Prince I (Yin-hsiang, a favorite brother of Emperor Yung-cheng) and Grand Secretaries Chang T'ing-yü and Chiang T'ing-hsi were secretly put in charge of military affairs and other related matters. Because of the confidential nature of the assignment, their appointments were not disclosed for more than two years. Therefore, 1729 has often been accepted as the year the Grand Council was established, even though the official seal of the office did not appear until 1732. Recently, a new study strongly suggested 1730 as the probable date of the creation of the Grand Council. The existence of these dates, all products of scholarly research, reflects that, depending on one's viewpoint, the origin of the Grand Council could be any one of the following years: 1726, 1727, 1729, 1730, or 1732.

2. Pao Chao Hsieh, *The Government of China* (*1644-1911*), (Baltimore, 1925), 74-75.

This ambiguity leads to the conclusion that the Grand Council evolved informally and gradually over several years and that *de facto* grand councillors were functioning before the official establishment of the office during the middle of the Yung-cheng period (1723-35).

The Grand Council was created partly as a result of Emperor Yung-cheng's need for a small, tightly knit group of aides to help draft edicts and to offer confidential advice on military and state affairs, and partly as a clever device to bypass the powerful princes, further enhancing the emperor's power and efficaciousness. Thus, the creation of the office constituted a milestone in the development of Ch'ing autocracy.

The grand councillors were generally selected from the grand secretaries, presidents, and vice-presidents of the Six Boards and other officials of the second civil service rank or higher, although occasionally a fourth- or fifth-ranking official might be appointed as a mark of imperial favor. Initially, the number of the councillors was three but increased to ten in 1745; their average number, however, was five or six, divided between Manchus and Chinese. Officially, the councillors were equals, but actually there was always a ranking member (*Ling-pan chun-chi ta-ch'en*), who was usually a Manchu grand secretary in the early period and a Manchu blood prince after the mid-19th century. They were stationed within the palace precincts (as opposed to the grand secretaries, who maintained offices outside the palace) and were therefore in a position to respond immediately to imperial summons. They began their daily work before daybreak, usually between 3 a.m. to 5 a.m., and studied the memorials that had been reviewed by the emperor and bore his vermilion endorsements. Next, audiences with the emperor—between 7 a.m. and 9 a.m., convened individually before 1749 and collectively thereafter—were held. During this two-hour period they discussed the affairs of the state with the ruler, suggested action to be taken regarding memorials not yet endorsed by him, memorized the imperial instructions as best they could, and then retired to draft edicts that came to be known as "court letters" (*t'ing-chi*). After 1749 the chore of drafting was left to the secretaries. After the draft was proofread by the ranking secretary it was presented to the grand councillors for further review. The final version was then presented to the emperor for approval.

As confidential secretaries and advisers to the emperor, the grand councillors conferred with him at least once a day. They offered him counsel on nearly every subject: military affairs, frontier defense, finance, taxation, and diplomacy. They recommended policies, suggested appointments and dismissals, participated in key trials, went on special missions for the ruler, and at times prepared the palace examinations. They read, transmitted, and kept memorials and military archives for the masters, prepared edicts and court letters, and served as the closest advisers to the throne on myriad matters. They followed the emperor wherever he went, even on furloughs, hunting trips, visits—and in such circumstances they were usually received by him after dinner. But powerful as they were, like the grand secretaries they had no authority to issue orders directly to the Six Boards or the provinces, since such power belonged exclusively to the emperor. Because of their sensitive position, they were prohibited from maintaining private liaisons with provincial authorities. In principle, princes and adjutant generals (*Yü-ch'ien ta-ch'en*) were excluded from the Grand Council, lest they become too powerful. But with the appointment of Prince Kung in 1853 the rule was broken; thereafter every ranking councillor until the end of the dynasty was a Manchu prince.

The grand councillors, who were concurrent appointees, drew no salaries except from their original posts. The budget of the Grand Council was therefore a limited 10,500 to 11,000 taels a year.

The grand councillors did not serve for definite terms: Tung Kao served for 39 years (1779-97, 1799-1812, 1814-18) while others served only a few months. Of the 145 grand councillors appointed during the Ch'ing period, 72 were Manchus, 64 Han Chinese, 3 Chinese bannermen, and 6 Mongols. Thus, numerically, the distribution of appointments between Manchus and Chinese seemed quite equitable, and indeed it was a court device to pacify the Chinese. But the numbers hardly suggest equivalence of power. Power was in proportion to the closeness of the councillor's relationship with the emperor, who generally trusted the Manchus more than the Chinese, as indicated by the large majority of special and secret assignments entrusted to the Manchu grand councillors. Nonetheless, the number of Chinese in the Grand Council is indicative of the considerable opportunity afforded them for participation in the core of the central administration.

Under the grand councillors were 32 secretaries—16 Chinese and 16 Manchus—who worked in shifts, half during the day and the other half at night. They performed the regular administrative and secretarial work, and being privy to important state affairs, they were known as the Little Councillors (*Hsiao chün-chi*). Indeed, during the Ch'ing period, thirty-four of them eventually rose to be grand councillors. If especially favored by the emperor, the secretaries could occasionally become more influential than the grand councillors, as in 1898 when Emperor Kuang-hsü appointed four reformists as secretaries in the Grand Council to take charge of the Hundred-Day Reform.

The usefulness of the Grand Council—its efficiency, its ability to reach fast and secret decisions, its enhancement of imperial powers, and its strengthening of the links between the throne and the provinces—justified its continuance after the military campaign against the Dzungars had been concluded. Only in 1735 was it temporarily suspended by Emperor Ch'ien-lung, and when it was revived the next year it permanently replaced the Grand Secretariat as the most important organ in the central government.[3]

Next to the Grand Secretariat and the Grand Council were the Six Boards, which formed the backbone of the central administration. These were the boards of Civil Office (*Li-pu*), Revenue (*Hu-pu*), Rites (*Li-pu*), War (*Ping-pu*), Punishments (*Hsing-pu*), and Public Works (*Kung-pu*). Each Board had two presidents and four vice-presidents, the offices being equally divided between the Manchus and the Chinese, and each board had four bureaus, except for the Board of Revenue which had fourteen bureaus and the Board of Punishments which had eighteen. Conspicuously absent in the central government was the Foreign Office, for the Confucian universal empire traditionally maintained no equal diplomatic relations as understood in the West; it recognized no foreign affairs but only tributary, or barbarian, or trading, affairs.

Of the Six Boards, that of Civil Office headed the list. Appointments, except those of the grand secretaries and grand councillors, were usually made for three years, at the end of which time a nominal examination was held to determine promotion or demotion. The "Law of Avoidance" required that no one be appointed to high positions in his native province and no two members of the same family be allowed to work in the same locality or service, so as to prevent nepotism and the forming of cliques. There were exceptions to these rules, but they were rare.[4] Retirement age was fixed at 55 *sui* in 1757 and raised to 65 in 1768, but this rule was not strictly enforced.

3. Fu Tsung-mou, *Ch'ing-tai Chün-chi ch'u tsu-chih chi chih-chang chih yen-chiu* (A study of the organization and functions of the Grand Council of the Ch'ing period) (Taipei, 1967), 121, 147ff, 166-67, 182-83, 213-16, 239-42, 246, 263, 321, 336; Wu Hsiu-liang, "Ch'ing-tai Chün-chi ch'u chien-chih ti tsai chien-t'ao" (A reappraisal of the establishment of the Grand Council under the Ch'ing), *Ch'ing Documents at National Palace Museum*, Taipei, II:4:21-45 (October 1971); Alfred K. L. Ho, "The Grand Council in the Ch'ing Dynasty," *The Far Eastern Quarterly*, XI:2:167-82 (February 1952); Pei Huang, "Aspects of Ching Autocracy: An Institutional Study, 1644-1735," *Tsing Hua Journal of Chinese Studies*, New Series, VI, Nos. 1-2 (December 1967), 123-25, 132; Thomas A. Metzger, *The Internal Organization of Ch'ing Bureaucracy: Legal, Normative and Communication Aspects* (Cambridge, Mass., 1973), 435-36.

4. In 1865, for instance, Li Hung-chang was appointed governor-general of Liang-kiang, with jurisdiction over Kiangsu, Kiangsi, and Anhwei, the last-named being his native province.

The Board of Revenue was second in rank to that of Civil Office. It had its usual two presidents and four vice-presidents, but in addition it retained a general superintendent, usually a Manchu but occasionally a Chinese. Under this Board's custody, naturally enough, came the collection of taxes and, since the biggest single tax was the land tax, the management of land registration.

The Board of Rites attended, as might be supposed, to rituals. These included obvious things like court ceremonies, state sacrifices, official costumes, wedding and funeral rites; but under the same rubric came, perhaps surprisingly, tributary affairs, education, and the administering of the civil service examinations at the district, provincial, and metropolitan levels. The successful candidates became degree-holders, the *shen-shih,* who formed a privileged class, from which the government selected its officials.

The Board of War concerned itself with military policies and the appointments and dismissals of officers. However, it did not exercise jurisdiction over the Imperial Bodyguard, whose 8,646 officers were directly under the control of the emperor.

An interesting feature of the Board was its control of official communication. It raised horses and supplied them to the relay teams throughout the country which carried messages between the capital and the provinces. High provincial and military authorities were given a certain number of credentials to use the teams. According to the importance of a document that was to be delivered, one or another statutory speed was required of the relay teams in transmitting it. The highest statutory speed was 600 *li* a day ordinarily, although 800 *li* a day was occasionally required of the teams. Other speeds ranged from 500 to 300 *li* a day; and routine communications were forwarded by foot couriers at 100 *li* a day. Thus a routine report from Nanking to Peking (2,300 *li* or 766 miles) was 23 days in transit, and from Canton to Peking it was 56 days.

The Board of Punishments directed matters of law, including punishments, pardons, and confiscation, and in conjunction with the Censorate and the Court of Judicature and Revision (*Ta-li ssu*) it reviewed cases in which the provincial judges had issued the death sentence.

The legal philosophy and practice of the Ch'ing were markedly different from those of the West. The judiciary was not independent of the executive branch—it was merely a part of it. There was no such thing as due process of law or advice of counsel in a trial. The judges were not protected by lifelong tenure of office. A case would more likely be tried on the basis of its moral implications than on any consideration of legality. Lawsuits and litigations were regarded as manifestations of unvirtuous behavior—therefore an appearance in court was a blow to one's social prestige. Only as a last resort—after all persuasion and ethical appeal had failed—would a man take recourse to law.

The Board of Public Works was the lowest in rank of the Six Boards. It controlled the construction and repair of public buildings, purchased and sold properties for the government, and maintained the streets and ditches in the capital. The repair of river dikes, dams, and irrigation systems was an important part of its work.

The Grand Secretariat, the Grand Council, and the Six Boards were the essential organs of the central government, but there were several other important "coordinate" offices.

The Li-fan yüan ranked directly after the Six Boards. Unlike many Ch'ing institutions, this one was not inherited from the Ming dynasty, nor had it any other historical precedent. It was established about 1636 as the Mongolian Office (*Meng-ku ya-men*). As the Ch'ing demesne expanded, however, the office took over the relations with Tibet, Sinkiang, and Russia as well; and two years after its inception its name was changed to the Li-fan yüan. It was headed by a president and a senior and a junior vice-president, and during the Ch'ien-lung period a supernumerary vice-president—usually a Mongol Prince—was added. The presidency and the two regular vice-presidencies were usually held by Manchus but occasionally by Mongol bannermen until the Ch'ien-lung period; thereafter they went to Manchus exclusively. No Chinese were ever appointed.

The Ch'ing Censorate (*Tu-ch'a yüan*) was headed by two senior presidents (Left Grand Censors) and four senior vice-presidents (Associate Left Grand Censors)—offices equally divided between the Manchus and Chinese. The titles of junior president (Right Grand Censor) and junior vice-presidents (Associate Right Grand Censor) were usually conferred concurrently on the governors-general and governors respectively. There were 24 censors for the Six Boards and 56 for the provinces, likewise composed of equal numbers of Manchu and Chinese.

The censors were known as the speech officials (*yen-kuan*) because they supposedly enjoyed freedom of speech: they were allowed to address the emperor on any subject. They served as his "eyes and ears," entrusted with the duty of discovering secret opposition. They could impeach, attack, criticize, or praise any official and any policy, openly or secretly as they saw fit. Although their institutional function was to detect dereliction of duties among the officials and not to concern themselves with policies, through their watchful supervision of the execution of policies and their readiness to impeach or attack the officials in charge, they actually exerted an influence on both the administration of current policies and the formulation of new ones. They considered themselves guardians of the Confucian principle of propriety (*li*), and sometimes their straightforward admonition or remonstration of the emperor cost them their jobs or lives.

Another remarkable feature of the central government was the Hanlin Academy. The Academy's function was primarily literary; its two chancellors—one Manchu and one Chinese—lectured on the classics to the emperor, or recommended lecturers to do so. They prepared edicts and manuscripts for imperial lectures, and officiated at the sacrificial offerings to Confucius. They were assisted by six readers, six expositors, six sub-readers and as many sub-expositors, comprising equal numbers of Chinese and Manchus. In addition there were a number of compilers and correctors.

The Hanlin Academy maintained a magnificent library which contained duplicates of all the books in the Imperial Library and held as well a great stock of memorials and documents. The State Historiographer's Office within the Academy prepared a chronicle of each reign called the veritable records (*shih-lu*), which was not made public until after the death of the reigning emperor. It also collected materials and drafted manuscripts for the biographies of emperors, empresses, nobles, officials, and scholars; but it never wrote out the history of the ruling dynasty: this was a task reserved for the next dynasty.

Membership in the Academy was limited to metropolitan graduates of the highest honors. It was a haven for the bright young talents and was an excellent training center for their political careers. Within a term of three years they could expect to win good appointments and rapid promotion in the official hierarchy. It was not infrequent that a member rose to the highest rung of the official ladder within ten years.

Two organizations were significant in handling the flow of documents. The Office of Transmission (*T'ung-cheng ssu*) received the "routine memorials" (*pen-chang*) from the provinces and was empowered to open them to see whether their delivery had been delayed by the relay teams and whether all the proper forms of elevation and phraseology were observed. Another organ, the Chancery of Memorials (*Tsou-shih ch'u*), received special "palace memorials" (*tsou-che*) from civil and military officials above the rank of 4-a, whether in the capital or in the provinces. In no case was the Chancery allowed to open the "palace memorials," but only the accompanying papers to identify the couriers and to ascertain their masters' qualifications to address the emperor. If all the credentials were in order, the Chancery quickly transmitted the memorials to the chancery eunuchs for presentation to the emperor, who was the first to read the "palace memorials." Frequently he wrote interlinear remarks on the memorials; other times he gave oral instructions to the grand councillors on the drafting of replies. The memorials were then returned to the original senders, who read the imperial comments and returned the memorials to the capital. In this manner, the emperor kept himself well informed on the conditions of the country.

The "routine memorial" system, however, fell increasingly into disuse after the Chia-ch'ing period (1796-1820). In 1901 it was finally abolished and five months later the Office of Transmission went out of existence.

LOCAL ADMINISTRATION. China's local administration—exclusive of special administrative areas such as Mongolia, Manchuria, Sinkiang, Tibet, and Chinghai—was of four kinds: province, circuit, prefecture, and district. There were altogether 18 provinces,[5] 92 circuits, between 177 and 185 prefectures, and about 1,500 districts and departments.

The eighteen regular provinces were put under the control of governors-general and governors (numbering eight and fifteen respectively during the Ch'ien-lung reign, but varying from one reign to the next). Two of the governors-general controlled only one province each—Chihli and Szechwan—but the other six usually had jurisdiction over two or three provinces.[6] The fifteen governors each controlled one province, and in the remaining three provinces, Chihli, Szechwan, and Kansu, the governors-general performed the duties of governor. The ranks for governor-general and governor were 2-a and 2-b respectively.

The court in Peking seemed to derive some sense of security in interposing Manchus with Chinese in these high provincial posts; whenever a Manchu was appointed governor-general, the governors under him were usually Chinese, and vice versa. Taking the Ch'ing period as a whole, the ethnic distribution was fairly even: 57 percent of the governors-general and 48.4 percent of the governors were Manchus, compared with 43 percent and 51.6 percent, respectively, Han Chinese.[7]

Under each governor was a financial, a judicial, and an educational

5. Chihli, Shantung, Shansi, Honan, Kiangsu, Anhwei, Kiangsi, Chekiang, Fukien, Hupeh, Hunan, Shensi, Kansu, Szechwan, Kwangtung, Kwangsi, Yunnan, and Kweichow. Actually, at the beginning of the Ch'ing dynasty there were only 15 provinces. Because of their large size, Emperor K'ang-hsi (1662-1722) split Kiangnan into Kiangsu and Anhwei, Shensi into Shensi and Kansu, and Hukwang into Hunan and Hupeh, making a total of 18 provinces, as stated. Later in 1884 and 1887, Sinkiang and Taiwan were respectively made provinces, but Taiwan was ceded to Japan in 1895. In 1907 Manchuria was turned into three provinces: Fengtien, Kirin, and Heilungkiang, making a total of 22 provinces by the end of the dynasty.
6. These were the governors-general of Liang-Kiang (Kiangsu-Anhwei-Kiangsi), Minche (Fukien-Chekiang), Liang-Kwang (Kwangtung and Kwangsi), Hu-Kwang (Hupeh-Hunan), Shen-Kan (Shensi-Kansu), and Yun-Kwei (Yunnan-Kweichow).
7. Lawrence D. Kessler, "Ethnic Composition of Provincial Leadership during the Ch'ing Dynasty," *The Journal of Asian Studies*, XXVIII:3:496, 500 (May 1969); Hsiao I-shan, I, 533-37; Pao Chao Hsieh, 294.

Ch'ing Central Government

Principal Offices
- Grand Secretariat (*Nei-ko*)
- Grand Council (*Chün-chi ch'u*)
- Six Boards
  - The Board of Civil Office (*Li-pu*)
  - The Board of Revenue (*Hu-pu*)
  - The Board of Rites (*Li-pu*)
  - The Board of War (*Ping-pu*)
  - The Board of Punishments (*Hsing-pu*)
  - The Board of Public Work (*Kung-pu*)

Coordinate Offices
- The Censorate (*Tu-ch'a yüan*)
- The Court of Judicature and Revision (*Ta-li ssu*)
- The Court of Colonial Affairs (*Li-fan yüan*)
- The Hanlin Academy (*Han-lin yüan*)
- The Office of Transmission (*T'ung-cheng ssu*)
- The Imperial College (*Kuo-tzu chien*)
- The Imperial Board of Astronomy (*Ch'in-t'ien chien*)

Imperial Departments
- The Imperial Clan Court (*Tsung-jen fu*)
- The Imperial Household (*Nei-wu fu*)
- The Supervisorate of Imperial Instruction (*Chan-shih fu*)
- The Court of Sacrificial Worship (*T'ai-ch'ang ssu*)
- The Banqueting Court (*Kuang-lu ssu*)
- The Imperial Stud (*T'ai-p'u ssu*)
- The Court of State Ceremonial (*Hung-lu ssu*)
- The Imperial Medical Department (*T'ai-i yüan*)

Adapted from Hsiao I-shan, I, 503, table 1.

commissioner—all appointed by the emperor. The government supplied them with a staff, but usually they maintained their private staffs too. Besides those just listed, there were a number of special commissioners who were placed in charge of salt, grain transport, customs, rivers and waterways, and postroads. A provincial commander-in-chief took charge of military affairs.

Further down the ladder in the provincial administration came the intermediary offices of the circuit intendants and prefects, and at the bottom was the district (*hsien*). Some large districts were bigger than the small states in the United States; the average population of a district was 200,000. The district magistrate collected taxes, settled litigations, and generally maintained peace and order in the locality. He was known as the "father-mother official" (*fu-mu kuan*) because he dealt directly with the people and was supposed to take care of them. Usually, upon assumption of office he entered into some kind of understanding or contractual relationship with a local group well conversant with the affairs of the district. This group functioned as an unofficial local permanent civil service, organizing itself into six "houses" (*fang*): (1) civil and administrative affairs; (2) census and taxation; (3) protocol and ceremonies; (4) militia; (5) crime or constabulary; and (6) public works. Members of this extralegal bureaucracy received no pay from the magistrate but were allowed to collect surcharges in his name. They were required to hand over a prefixed amount of their collections while keeping the rest for themselves. It is this little publicized body which performed the bulk of the day-to-day operation of the magistrate's yamen.[8]

At the same level as the district were some slightly larger administrative units called department (*chou*) and subprefecture (*t'ing*). Some departments were under the direct control of the provincial government, in which case they enjoyed a higher status than the regular department.

SUBADMINISTRATIVE RURAL CONTROL. In each district there were a number of villages (*ts'un* or *chuang*), cities (*ch'eng*), towns (*chen*), countryside settlements (*hsiang*), and rural markets (*shih, chi, ch'ang*). Operation of these rural divisions was left to the local inhabitants rather than to government officials, as the imperial administration stopped at the district level. Imperial control, however, was still made manifest through the development of two neighborhood organization systems called the *pao-chia* and the *li-chia*. The former was established in 1644 to facilitate police control and the latter in 1648 to help with tax collection.

8. K. C. Wu, "Local Government in Imperial China" in his *Why Is America Not Better Informed on Asian Affairs* (Savannah, Georgia, 1968), 8-9.

In the *pao-chia* system every ten households (*hu*) formed a *p'ai*, with a headman called the *p'ai-chang*; every ten *p'ai* constituted a *chia*, for which there was a headman called *chia-chang* (or *chia-t'ou*); and every ten *chia* in turn made up a *pao*, headed by a *pao-chang*. Each *pao* thus consisted of 1,000 households. Each household hung on its door an official placard bearing the names of its members. The *pao-chia* maintained a census, kept track of the movements of individuals, and made periodic recounts of the local population. Members of the *pao-chia* were supposed to watch for and report crimes and criminals in the neighborhoods to the heads of the *pao-chia*, who relayed the information to the district magistrate. Every member of the *pao-chia* was therefore a potential informer on every other member. The fear and suspicion thus created inhibited the villagers from entering into seditious plots with their fellow citizens, and thus reduced the chances of uprisings or revolts. Failure to report crimes and secret plots would bring collective punishment. At the end of the month the head of each *pao* was required to submit a "voluntary bond" (*kan-chieh*) to the district magistrate to assure that everything was well in his neighborhood.

The *li-chia*, often confused with the *pao-chia*, was a totally different system. Every 110 households in the *rural* area constituted a *li*, and the ten households within it which contained the largest number of tax-paying adults were chosen heads of the *li* (*li-chang*). The 100 households that remained were divided into ten *chia*, each with a head. Every three, and after 1656 every five, years a census was taken as the basis of the land and poll taxes. The function of the *li-chia* was to assist in the registration of the local inhabitants, to assess and collect the land (*ti*) and poll (*ting*) imposts, and to help in compiling the Yellow Register,[9] which recorded all taxable individuals in the area.

The nature of the *li-chia*, however, underwent some change after 1712, when Emperor K'ang-hsi froze the *ting* (labor service) quota as it was established in that year and announced that no more new imposts would be made on the population increase thereafter. By 1740 the labor service impost had been merged into the land tax in nearly all the provinces, and the Yellow Register had virtually lost its original usefulness. The *li-chia* compilation of taxable adult records was discontinued in favor of the more general *pao-chia* census registration. In 1772 the practice of quinquennial register of the *ting* was abolished altogether. The main function of the *li-chia* changed from providing materials for the compilation of the Yellow Register to the urging of prompt tax payments. Not infrequently

9. So called because of its yellow covers.

the *li-chia* heads were held responsible for the failure of the villagers to honor their tax obligations.

By using the local inhabitants to control themselves at the subadministrative level, the Ch'ing government exercised an ingenious means of extending imperial control to the very root of the society, at the same time obviating the expenses of local government and the need to appoint officials.

At the very bottom of the rural community was the docile, passive, and hard-working peasant, who worked all year round for a hand-to-mouth subsistence. The peasants were mostly resigned to their fate and accepted the social circumstances as the yoke they had to bear. Nevertheless, if the taxes became too onerous and their living too hard, they would rebel under the leadership of the more venturesome scholars or members of the gentry. The government therefore saw to it that periodic favors were extended to them; these usually took the form of tax remission in years of bumper crops. A good government was one which could provide the peasantry with a passable living and at the same time temper control with a certain degree of latitude.[10]

Economic Institutions

In a predominantly agrarian country such as China, soil and human labor constituted the economic foundations of state. The bulk of the revenue came from the land and poll taxes (*ti-ting*), supplemented by incomes from the salt tax, tea tax, the native customs houses, the commercial license tax, and others. On the whole, revenue did not increase notably during the first century and a half of the dynasty. During the Shun-chih period (1644-61), the annual income was about 28 million taels; during the K'ang-hsi period (1662-1722), about 35 million; during the Yung-cheng period (1723-35), 40 million; and during the Ch'ien-lung period (1736-95), between 43 and 48 million, against an annual expenditure of roughly 35 million (1765). The slow rise in income was largely a result of the modest increase in land cultivation; in 1661 there was a total of 549 million *mou* of arable land and in 1766 it rose to only 741 million.

THE LAND AND TAXATION SYSTEMS. The land tax was by far the greatest of all taxes in the early and middle Ch'ing periods. The basis of its collection was the 1646 edition of the *Fu-i ch'üan-shu* (Complete text of land and labor), in which were spelled out the total amount of cultivated land

10. For details of the *pao-chia* and *li-chia* systems, see Kung-ch'üan Hsiao, *Rural China: Imperial Control in the Nineteenth Century* (Seattle, 1960), chs. 2-4.

in the country, the quotas of land (*ti*) and labor (*ting*) imposts in the various provinces, the number of persons liable to the labor impost, and the quotas of the revenue to be sent to the imperial treasury. The regular land tax came from the people's land, with the rates contingent upon the fertility of the soil and the size of the property. Odd bits of people's land too small to be worth assessment were exempt from taxes, as were the official lands and public lands assigned for sacrificial and educational purposes, and those lands allotted to the imperial household or bannermen. The land measurement registers, called the Fish-scale Registers (*Yü-lin ts'e*)[11] showed the total land area in each locality, and the Yellow Registers (*Huang-ts'e*) gave the total number of taxable adults in the area.

In 1712 K'ang-hsi issued his well-known edict, which was to have far-reaching consequences, that the *ti* and *ting* taxes would be permanently based on the quotas of 1712. This document read in part as follows:

> The empire has enjoyed continued peace for a long time, and the number of households and inhabitants has been multiplying daily. If additional taxes are assessed on the basis of the current population figures, it is really quite improper. For although there is an increase in population, there is no increase in the amount of cultivated land. We deem it fit, therefore, to instruct governors-general and governors of the provinces to take the number of registered *ting* listed in the *current registers,* which number is not to be augmented or diminished, *as the permanent, fixed quota* [for collecting the *ting* imposts]. *All inhabitants born hereafter shall be exempted from* [additional] *imposts.* In taking the census it will be necessary merely to ascertain the actual amount of population increase and report it in separate registers.[12]

This freezing of the *ting* quota rendered the poll tax meaningless, and it was gradually merged into the land tax. In 1716, with K'ang-hsi's approval of the merger in Kwangtung province, the precedent was set, fixing the poll tax at .1064 tael for each tael of the land tax. Other provinces followed suit until by the early 19th century the combination of the two taxes became standard practice throughout the country. As a rule, the poll tax was light where the land tax was heavy, and vice versa. After the merging of the two taxes, the usefulness of the Yellow Registers decreased, whereas that of the Fish-scale Registers increased.

11. So called because of its fish-scale design on the covers, supposedly symbolic of plots of land.
12. Kung-ch'üan Hsiao, 89-90. Italics added.

The method of collection was the "Single-whip system" (first begun in 1581), which consolidated all taxes into one compound sum to be paid twice a year. The summer payment was generally made between the second and fifth months, and the autumn payment between the eighth and eleventh months. These two periods of payment were popularly known as "The Upper Busy (Season)" and "The Lower Busy (Season)."

The following table shows the land-poll tax in money and in grain during different years:[13]

Year	Tax in taels	Tax in shih (133⅓ lbs.) of grain
1661	21,576,006	6,479,465
1753	29,611,201	8,406,422
1812	32,845,474	4,356,382

STATE REVENUE AND EXPENDITURES. In addition to the land-poll tax, a number of other smaller taxes made up a considerable part of the revenue. The surtax called *huo-hao*, or allowance for wastage in silver-melting, was originally collected by district magistrates illegally, but since it could not be stopped in spite of repeated warnings, Emperor Yung-cheng decided that the government should take it over. The rate of *huo-hao* ranged between 4 percent or 5 percent to 20 percent of the regular (i.e. land) tax. From this source the government derived an annual income of about 4.5 million taels during the Ch'ien-lung period. The total revenue of state toward the end of the 18th century may be summarized as follows:[14]

1.	Land-poll tax	approximately 30,000,000	taels
2.	Wastage allowance (*huo-hao*)	4,600,000	"
3.	Miscellaneous surtax on tribute grain	2,000,000	"
4.	Salt tax quota	7,500,000	"
5.	Customs dues	4,000,000	"
6.	Land rents	260,000	"
7.	Tea tax	70,000	"
8.	Tribute grain for Peking	4,000,000	shih

According to these figures, the total income of state should have been around 48 million taels and 4 million *shih* of grain, but since the salt tax quota of 7.5 million taels was often filled only 50 percent to 60 percent, the actual revenue was around 43 or 44 million.

13. Hsiao I-shan, II, 386-87.
14. Hsiao I-shan, II, 432.

Against this income, the largest items of debit were military expenditures and officials' salaries and allowances. The Manchu banners and the Chinese Green Standard Army, totaling more than 200,000 and 600,000 men respectively, represented some 20 million taels of the expenditure. The stipends for the nobles and the salaries of the officials are shown in the following selective scale:[15]

Post	Annual Salary (taels)	Rice stipend (shih)
Prince of the Blood	10,000	5,000
Duke, First Grade	700	350
Earl, " "	610	305
Count, " "	510	255
Viscount, " "	410	205
Baron, " "	310	155
Civil official, Grade 1-a, b	180	90
" " " 2-a, b	155	77.5
" " " 3-a, b	130	65
" " " 4-a, b	105	52.5
" " " 5-a, b	80	40
" " " 6-a, b	60	30
" " " 7-a, b	45	22.5
" " " 8-a, b	40	20
" " " 9-a	33.114	16.557
" " " 9-b	31.5	15.75

It will be noted that the civil officials received very meager salaries indeed. A grand secretary at the rank of 1-a received only 180 taels a year, and a governor-general, whose rank was 2-a, received only 155 taels. To compensate for the low salary scale, officials were given what was euphemistically called "integrity-nourishing allowances" (*yang-lien fei*),[16] which were frequently a hundred times the amount of the regular pay.

A governor-general whose regular salary was 155 taels received an "integrity-nourishing allowance" of 13,000 to 20,000 taels; and in addition he was given some "official expenses." The total "integrity-nourishing allowances" for all civil and military officials amounted to more than four million taels a year, while "official expenses" ran to a quarter of a million. A breakdown of the approximate expenditures of the central and local governments of the year 1765 is shown as follows:

15. Hsiao I-shan, II, 411-16.
16. Sometimes translated as "anticorruption fee" or "honesty-fostering allowance."

Central Government Expenditures:

1. Salaries for princes, nobles, and officials	930,000
2. Military expenses	6,000,000
3. Military expenses in Mukden and Jehol	1,400,000
4. Stipends for Mongolian and Moslem nobles	120,000
5. Office and food expenses for metropolitan officials	110,000
6. Mess allowances for the Grand Secretariat, etc.	18,000
7. "Integrity-nourishing allowances" for the Boards of Civil Office and Rites	15,000
8. Gifts to Mongolian and Korean tribute bearers	10,000
9. Miscellaneous	900,000
	9,503,000

Local Government Expenditures:

1. Military expenses	15,000,000
2. Salaries of officials	1,000,000
3. "Integrity-nourishing allowances"	4,220,000
4. "Expense" allowances	200,000
5. Repairs of rivers and ponds	4,000,000
6. Miscellaneous (post stations, charity, etc.)	1,400,000
	25,820,000

The total expenses of the central government were approximately 9.5 million taels, and those of the local government about 25.8 million, making a grand total of 35 million taels, against a revenue of 43 or 44 million taels. The government during the Ch'ien-lung period enjoyed a surplus of eight or nine million taels annually.[17]

POPULATION. In an agricultural economy land and population cannot be considered separately: the merging of the land and poll taxes was a recognition of this. Every three or five years during the early Ch'ing period, the *li-chia* heads conducted a census of taxable adult males between the ages of 16 and 60 (*sui*); it was on the basis of their investigation that the provincial financial commissioners compiled the Yellow Registers. Taxable individuals attempted to avoid being included in these registers, and for good reason: the poll (*ting*) tax actually consisted of a certain number of compulsory labor services commuted into money payment, which might be as high as 8 or 9 taels a year in some localities. Emperor K'ang-hsi reported in 1712 that during his tours of the country he frequently found only one out of five or six persons in a household

17. Hsiao I-shan, II, 432-35.

who paid taxes, and sometimes in households of nine or ten members only one or two were paying taxes. Provincial authorities did not report the actual population for fear of increases in their tax quotas. K'ang-hsi's 1712 decree that froze the poll tax quotas to the population figure for that year somewhat alleviated the popular fear of registration, but the habit of avoiding the census was deep-rooted. With the increasing merging of the poll tax into the land tax during the Yung-cheng period, the Yellow Registers became more and more superfluous. On the other hand, the *pao-chia* census-taking took on a much greater importance. But the rate of increase in the population was not marked until the Ch'ien-lung period (1736-95). The difference is shown roughly in the following listing:[18]

1660	19,088,000 *ting*
1700	20,411,000
1730	25,480,080
1741 (first year of figures based on *pao-chia* census)	143,411,559 *k'ou* (mouths)
1753	183,678,259
1779	275,042,916
1800	295,273,311
1821	355,540,258
1850	429,913,034

One sees from this that the figures in the early Ch'ing period rose only slightly between 1660 and 1730, but jumped abruptly in 1741 when the census came to be based on the *pao-chia* figures; thereafter the increase was steady. Actually, the figures for 1660-1730 represented neither the entire population nor returns of households nor tax-paying adult males, but were rather "tax-paying units" or *ting*—which were quotas of compulsory labor service commuted into money.[19] The figures for 1741 and thereafter were those of the individuals (*k'ou,* mouths) of every status.

The steady increase since 1741 may be explained in several ways, though the absolute figures themselves may not. For one thing, the merging of the land and poll taxes, whereby land and not person became the basis of taxation, erased much of the fear the people had had about reporting the actual numbers. Another possible reason to account for it is Emperor Ch'ien-lung's natural ostentatiousness and love of grandeur. He

18. Ping-ti Ho, *Studies on the Population of China, 1368-1953* (Cambridge, Mass., 1959), 281-82.
19. Ping-ti Ho, 35. Thus, although the official figure for 1660 was 19 million, the actual population was probably somewhere around 100-150 million.

was fond of large figures as they attested to the prosperity of his reign, and he repeatedly warned the local officials not to conceal the true census figures. People formerly excluded from the census—women, the elderly, children, slaves, and mean persons—now were all included. Hence the sudden increase in 1741.

Chinese historians and demographers tend to believe that the census figures after 1741 were under- rather than over-reported; for a large section of the gentry harbored an inveterate resentment of being made to reveal the size of their households, and misrepresented wherever possible. One modern demographer suggests that the under-reporting of the census between 1741 and 1775 might be as high as 20 percent of the total.[20]

Actually, the growth of population from 275 million in 1779 to 430 million in 1850 represented an increase of 56.3 percent at an annual rate of 0.63 percent, which is much lower than the 2 percent annual increase in many of the rapidly expanding industrialized countries today.[21] The sharp increase from 1741 to 1779 might be ascribable to the generally favorable economic and political conditions and the long period of peace during the Ch'ien-lung era. Moreover, the increase of arable land and the introduction of foreign food products like maize, sweet potatoes, and peanuts from America in the 16th or 17th century also contributed to population increase. One is inclined to the opinion of the well-informed scholar, Wang Ch'ing-yün, that Ch'ing population figures showed less, rather than more, of the total picture.

20. Ping-ti Ho, 46.
21. *Ibid.*, 64.

Further Reading

Backhouse, Edmund, and J. O. P. Bland, *Annals and Memoirs of the Court of Peking from the 16th to the 20th Century* (Boston, 1914).

Baker, Hugh D. R., *Chinese Family and Kinship* (New York, 1979).

Bartlett, Beatrice S., *Monarch and Ministers: The Grand Council in Mid-Ch'ing China, 1723–1820* (Berkeley, 1991).

Bodde, Derk, and Clarence Morris, *Law in Imperial China* (Cambridge, Mass., 1967).

Chang, Chun-shu, and Shelley Hsueh-lun Chang, *Crisis and Transformation in Seventeenth-Century China: Society, Culture, and Modernity in Li Yü's World* (Ann Arbor, 1992).

Chen, Shao-kwan, *The System of Taxation in China in the Tsing Dynasty, 1644–1911* (New York, 1914).

Chi, Ch'ao-ting, *Key Economic Areas in Chinese History, as Revealed in the Development of Public Works for Water Control* (London, 1936).

Ch'ü, T'ung-tsu, *Local Government in China Under the Ch'ing* (Cambridge, Mass., 1962).

Ch'uan, Han-sheng, and Richard A. Kraus, *Mid-Ch'ing Rice Market and Trade: An Essay in Price History* (Cambridge, Mass., 1975).

Fairbank, John K., and S. Y. Teng, *Ch'ing Administration: Three Studies* (Cambridge, Mass., 1960).

Farquhar, David Miller, "The Ch'ing Administration of Mongolia up to the Nineteenth Century" (Ph.D. thesis, Harvard University, 1960).

Fu, Tsung-mou 傅宗懋, *Ch'ing-tai Chün-chi ch'u tsu-chih chi chih-chang chih yen-chiu* 清代軍機處組織及職掌之研究 (A study of the organization and functions of the Grand Council of the Ch'ing period), (Taipei, 1967).

Grantham, Alexander E., *Manchu Monarch: An Interpretation of Chia Ch'ing* (London, 1934).

Hinton, Harold, *The Grain Tribute System of China, 1845-1911* (Cambridge, Mass., 1961).

Ho, Alfred K. L., "The Grand Council in the Ch'ing Dynasty," *The Far Eastern Quarterly*, XI:2:167-82 (Feb. 1952).

Ho, Ping-ti, *Studies on the Population of China, 1368-1953* (Cambridge, Mass., 1959).

Hsiao, I-shan 蕭一山, *Ch'ing-tai t'ung-shih* 清代通史 (A general history of the Ch'ing period), rev. ed. (Taipei, 1962), I, chs. 19-21; II, chs. 7-9.

Hsiao, Kung-chüan, *Rural China: Imperial Control in the Nineteenth Century* (Seattle, 1960).

————, *History of Chinese Political Thought*, Vol. 1 (Princeton, 1978).

Hsieh, Pao Chao, *The Government of China (1644-1911)* (Baltimore, 1925).

Huang, Liu-hung, *A Complete Book Concerning Happiness and Benevolence: Fu-hui ch'üan-shu, A Manual for Local Magistrate in Seventeenth-Century China* (Tucson, 1984). Tr. by Djang Chu.

Huang, Pei, "Aspects of Ch'ing Autocracy: An Institutional Study, 1644-1735," *The Tsing Hua Journal of Chinese Studies*, New Series, VI:1-2:105-48 (Dec. 1967).

Huang, Philip C. C., *The Peasant Economy and Social Change in North China* (Stanford, 1985).

Hucker, Charles O., *The Censorial System of Ming China* (Stanford, 1966).

Kessler, Lawrence D., "Ethnic Composition of Provincial Leadership during the Ch'ing Dynasty," *The Journal of Asian Studies*, XXVIII:3:489-511 (May 1969).

Liang, Fang-chung, *The Single Whip Method of Taxation in China* (Cambridge, Mass., 1961).

Mayers, W. F., *The Chinese Government* (Shanghai, 1897).

Metzger, Thomas A., "The Organizational Capabilities of the Ch'ing State in the Field of Commerce: The Liang-huai Salt Monopoly, 1740-1840," in W. E. Willmott (ed.), *Economic Organization in Chinese Society* (Stanford, 1972), 11-45.

————, *The Internal Organization of Ch'ing Bureaucracy: Legal, Normative and Communications Aspects* (Cambridge, Mass., 1977).

Morse, H. B., *The Trade and Administration of the Chinese Empire* (London, 1908).

Saeki, Tomi 佐伯富, *Shindai ensei no kenkyū* 清代鹽政の研究 (A study of the salt administration of the Ch'ing period), (Kyoto, 1956).

Shih, Min-hsiung 施敏雄, *Ch'ing-tai ssu-chih kung-yeh ti fa-chan* 清代絲織工業 的發展 (The development of the silk-weaving industry during the Ch'ing period), (Taipei, 1968).

Sprenkel, Sybille van der, *Legal Institutions in Manchu China, A Sociological Analysis* (London, 1962).

Sun, E-tu Zen 任以都. "Ch'ing Government and the Mineral Industries before 1800," *The Journal of Asian Studies*, XXVII:4:835-45 (Aug. 1968).

———, "The Board of Revenue in Nineteenth-Century China," *Harvard Journal of Asiatic Studies*, 24:175-228 (1962-63).

———, "Ch'ing-tai k'uang-ch'ang kung-jên," 清代礦廠工人 (Mining labor in the Ch'ing period), *Journal of the Institute of Chinese Studies*, Hong Kong, III:1:13-29 (1970).

Taeuber, Irene B., and Wang Nai-chi, "Population Reports in the Ch'ing Dynasty," *Journal of Asian Studies*, XIX:403-17 (1959-60).

Tai I 戴逸, *Chung-kuo chin-tai shih-kao* 中國近代史稿 (A draft history of Chinese modern history), (Peking, 1961), third printing I, ch. 1.

Tang, Edgar C., "The Censorial Institution in China, 1644-1911" (Ph.D. thesis, Harvard University, 1932).

Torbert, Preston M., *The Ch'ing Imperial Household Department: A Study of Its Organization and Principal Functions, 1662-1796* (Cambridge, Mass., 1978).

Waltner, Ann, *Getting an Heir: Adoption and the Construction of Kinship in Late Imperial China* (Honolulu, 1990).

Wang, Yeh-chien, *An Estimate of Land Tax Collection in China, 1753 and 1908* (Cambridge, Mass., 1973).

———, *Land Taxation in Imperial China, 1750-1911* (Cambridge, Mass., 1974).

Watt, John R., *The District Magistrate in Late Imperial China* (New York, 1972).

Wiens, Mi Chu, *The Origins of Modern Chinese Landlordism* (Taipei, 1976).

Wilkinson, Endymion, *Landlord and Labor in Late Imperial China* (Cambridge, Mass., 1977).

Williams, S. Wells, *The Middle Kingdom* (New York, 1883), 2 vols.

Wu, Silas Hsiu-liang, "The Memorial Systems of the Ch'ing Dynasty (1644-1911)" *Harvard Journal of Asiatic Studies*, 27:7-75 (1967).

———, *Communication and Imperial Control in China: Evolution of the Palace Memorial System, 1693-1735* (Cambridge, Mass., 1970).

——— 吳秀良, "Ch'ing tai Chün-chi ch'u chien-chih ti tsai chien-t'ao," 清代軍機 處建置的再檢討 (A reappraisal of the establishment of the Grand Council under the Ch'ing), *Ku-kung wen-hsien* 故宮文獻 (Ch'ing documents at the National Palace Museum), Taipei, 2:4:21-45 (Oct. 1971).

Yen, Chung-p'ing 嚴中平 (ed.), *Ch'ing-tai Yun-nan t'ung-cheng k'ao* 清代雲南 銅政考 (A study of the copper administration in Yunnan during the Ch'ing period), (Peking, 1957).

Zelin, Madeleine, *The Magistrate's Tael: Rationalizing Fiscal Reform in Eighteenth Century Ch'ing China* (Berkeley, 1984).

———, "The Rights of Tenants in Mid-Qing Sichuan: A Study of Land-Related Lawsuits in the Baxian Archives," *The Journal of Asian Studies*, XLV: 3:499-526 (May 1986).

4
Social and Intellectual Conditions

The Chinese Society

The nature of Chinese society has been an object of keen interest to historians and sociologists. Marxist scholars have pointed to the combined exploitation of the peasantry by the landlords, usurers, and reactionary Manchu overlords, and have damned the Ch'ing society as feudal and bureaucratic. Others have emphasized the gentry as the dominant feature of the society. A more recent view appraises the Chinese society as an archetype of "Oriental despotism," characterized by a centralized monolithic government which kept the peasant masses in line by controlling the large-scale public projects, such as the building of roads and the raising of defensive walls along the frontiers, and especially the vast system of waterworks which the peasants needed for irrigation, flood control, drainage, and canalization.[1]

These three views—that of a feudal bureaucracy, a gentry-based elite, and an "Oriental despotism"—may at first glance appear contradictory, but they really are not, for each stresses an important aspect of Chinese society without canceling out the other two: the Ch'ing state was indeed a despotic autocracy, in which the bureaucrats within the government and the gentry outside of it dominated the political and social worlds; likewise, it was the peasant who paid the greatest part of the taxes to the government, the highest rents to the landlords, and the most exorbitant interest rates to the usurers. Each of these appraisals adumbrates a characteristic of the

1. Karl A. Wittfogel, *Oriental Despotism: A Comparative Study of Total Power* (New Haven, 1957).

state not exclusive of the others; taken together they provide us with a fuller picture of the Chinese society.

THE FAMILY. The basic unit of the Chinese society was the family rather than the individual. The average size of the family was five members, contrary to the common belief that Chinese families were usually large. The notion that several generations lived in one big household is true only in the case of the well-to-do; ordinary families could not afford that luxury. Within the family, respect was claimed and given according to age and sex, the older members enjoying a status superior to the younger ones, and the male members to the female. The family head was the father, who had complete authority over the other members. He decided all family issues, arranged his children's marriages, disciplined the unfilial and disobedient, and could even sell them. Yet for all his authority, he still had to act within the moral code of Confucianism and behave like a father—strict yet benevolent, authoritative yet paternalistic—so that his children would likewise fulfill their expected roles. Status-consciousness led the father to speak respectfully to his own parents and authoritatively to his children. In the same vein, an elder brother conducted himself submissively toward his father but confidently before his younger brother. Indeed, the Chinese family was a laboratory of human relationships.

The status of women was very unlike that of the Western world. The wife was supposed to be obedient to her husband. She had no property rights and enjoyed no economic independence. The widow was generally expected not to remarry, although the husband could take a concubine even if his legal wife was still alive.

THE CLAN. Families of common ancestral lineage that settled in a certain locality formed a clan (*tsu*)—an institution which was strong in the south, less so in central China, and weak in the north. Although structural organization varied, there was usually a clan leader (*tsu-chang*), who was generally an older, prominent member. With the assistance of other members he discharged clan affairs, particularly the management of clan property and the ancestral hall and the reward and punishment of clan members. Clan activities frequently encompassed the following: (1) compiling and updating genealogies (practiced for the most part by the wealthier clans); (2) maintaining ancestor worship, ancestral halls, ritual land, and ancestral graveyards; (3) aiding clan members; (4) educating young clansmen; (5) rewarding meritorious members and punishing unworthy ones; (6) settling disputes; and (7) defense.

Clans maintained rules of behavior (*tsung-kuei*), which frequently echoed the moral teaching of Confucianism. Posted in the ancestral halls or recited on suitable occasions, the moral didactics usually consisted of the following tenets: "Render filial piety to parents; show respect to seniors by the generation-age order; remain in harmony with clan members and the community; teach and discipline sons and grandsons; attend to one's vocation properly; and do not commit what the law forbids." Sons were urged to be filial to their fathers, wives to be dutiful to their husbands, and brothers to be affectionate toward each other. Also, clan rules warned against offensive conduct such as laziness, extravagance, violence, and gambling. Serious transgressions were openly handled in the ancestral hall in the presence of all clansmen, so that the penalty would be a lesson to each member. According to the seriousness of the offense (unfilial conduct and adultery were considered the most reprehensible), one of the following punishments was administered; moral injunction, oral censure, monetary fine or ritual punishment, flogging, forfeiture of clan privileges, expulsion from the clan and expurgation from its genealogy, or even death or order to commit suicide. Although severe corporal and capital punishment were illegal, the state seldom interfered with clan justice.

With their special status and functions, small wonder that the family and the clan have been considered the characteristic institutions of the traditional kinship society of China.[2]

SOCIAL STRATIFICATION. Chinese society was highly stratified. Among the many criteria for social classification, a very common one separated the farmers who constituted some 80 percent of the population, from the other 20 percent of the population, who lived in the urban areas and represented a composite stratum of scholars, gentry, officials, absentee landlords, artisans, merchants, militarists, etc. Another method of classification was to follow the Confucian principle of distinguishing the ruling group from the ruled on the basis of mental as opposed to menial work. "Those who labor with their minds rule others," the philosopher Mencius postulated, "and those who labor with their physical strength are ruled by others." Yet in reality not all brain workers were members of the ruling bureaucracy; at any given time during the Ch'ing period only a very small fraction of the 1.1 million degree-holders held the 27,000 official positions. Strictly speaking, Chinese society has never been a simple, bipolarized

2. Hui-chen Wang Liu, *The Traditional Chinese Clan Rules* (Locust Valley, N.Y., 1959), 5-6, 8, 23, 40-45; Hsiao Kung-chuan, "The Role of the Clan and Kinship Family," in William T. Liu (ed.), *Chinese Society under Communism: A Reader* (New York, 1967), 36, 40.

structure of the ruling and the ruled, but always a multiclass one in which four major "functional orders" coexisted: scholars, farmers, artisans, and tradesmen. Above them were the government bureaucrats, and below them were the "declassed" or "degraded" people,[3] less than a fraction of 1 percent of the population who, until their legal emancipation by Emperor Yung-cheng (1723-35), were denied the rights enjoyed by the common people.

Though stratified, the society was egalitarian in that there was no caste system. Except for the degraded, whose descendants for three generations were barred from the civil service examinations, the ladder of success was available to everyone, regardless of family, birth, or religion. There was in fact considerable movement among the different social groups: powerful or high-status families could fall because of incompetent offspring, while men of humble circumstances rose through their successes in the open competitive government examinations and in receiving official appointments. More than anything else, individual merit based on literary excellence, as evidenced by the successful passing of the examinations, formed the basis of recognition.

It is to be noted that the merchants were at the bottom of the social scale. They included not only the wealthy monopolistic traders but also the small shopkeepers and the clerks and apprentices. Powerful tea and silk merchants who controlled the distribution of these commodities throughout the country were enormously rich. The salt merchants of Yangchow were particularly noted for their great wealth and luxurious living; their estimated aggregate profit in the second half of the 18th century was something like 250 million taels.[4] The various houses of foreign trade were likewise famous for their riches; Howqua of Canton, for instance, built up a fortune of 26 million silver dollars in 1834 which was, according to H. B. Morse, the largest mercantile fortune on earth.[5] But by and large commercial activities were regarded as beneath the dignity of the scholar-gentry, and the pursuit of profit was frowned upon by cor-

3. Ping-ti Ho, *The Ladder of Success in Imperial China: Aspects of Social Mobility, 1368-1911* (New York, 1962), 18. These traditionally included singers, dancers, entertainers, beggars, the "lazy people" of Chekiang, boatmen, hereditary servants, bonded servants, prostitutes, actors, entertainers, and certain types of menial workers in government called "runners."

4. Ping-ti Ho, "The Salt Merchants of Yang-chow: A Study of Commercial Capitalism in Eighteenth-Century China," *Harvard Journal of Asiatic Studies*, 17:149 (1954).

5. H. B. Morse, *The International Relations of the Chinese Empire* (London, 1910), I, 86. Twenty-six million silver dollars was the equivalent of US $52 million. See Frederic Wakeman, Jr., *Strangers at the Gate: Social Disorder in South China, 1839-1861* (Berkeley, 1966), 44.

rect Confucianists. Such an attitude inhibited the growth of business enterprises.

THE GENTRY: THEIR PRIVILEGES AND FUNCTIONS. The gentry, or *shen-shih*—those scholars who had passed the governmental examinations—played a dominant role in the society and enjoyed many unique privileges. Only they, for instance, could attend official ceremonies in the Confucian temples, and usually they led the ancestral rituals that were performed in clans. The gentry were distinguished from the commoners in style of dress and in embellishments. They wore black gowns with blue borders, and decorated their saddles and reins with splendid articles such as fur, brocade, and fancy embroidery. None of the commoners, no matter how rich, were allowed the same privilege. The *sheng-yüan* wore buttons of plain silver on their hats, and the *chü-jen* and *chin-shih* wore plain gold buttons.[6] When a degree-holder rose to an official position of the first rank, his gold button had a flower design added to it, with a ruby on top and a pearl in the middle, and his robe was embroidered with nine pythons.

The gentry were protected against insults from commoners and interference from officials. A commoner who offended a gentry member would be punished more severely than one who had committed the same offense to another commoner. Furthermore, ordinary people were not allowed to involve the gentry as witnesses in lawsuits. If a member of the gentry was involved in such a suit himself, he was not required to appear before the court in person but could send a servant instead. A member of this upper class who had committed a crime presented an awkward problem, for gentry status made one immune to action by the local magistrate. In order to be prosecuted, such a person had first to be stripped of his gentry status. But the district magistrate could not perform this act, because the gentry member was his social equal. The divesting could be performed only by the educational commissioner, and it was he whom the magistrate had to consult before any punishment could be issued. Violation of this rule could bring about the impeachment of the magistrate.

The gentry were exempt from *corvée* labor service, for their station and cultural refinement forbade them to engage in manual labor. They were also excused from paying the poll tax in order that they might devote themselves to the studies that would qualify them for future government examinations and service. When the poll and land taxes were merged in

6. These were the scholars who had passed, the district, provincial, and metropolitan examinations, respectively, upon which their status as "lower" or "upper" gentry depended. The examinations and degrees are discussed in some detail in the next section.

1727, the gentry managed to pay less of the combined assessment than did the commoners. They called their houses "scholar households" (*ju-hu*), or "gentry households" (*shen-hu*), or "big households" (*ta-hu*), as distinguished from the "commoner households" (*min-hu*) or "small households" (*hsiao-hu*), so as to make differentiations in tax payments. The gentry household might pay as little as 2,000 or 3,000 cash (copper coins) for each picul (133⅓ lbs) of rice that was collected as tax, or sometimes no grain tribute at all, whereas the commoner's household had to pay 6,000 or 7,000 cash for each picul.[7] Not infrequently the commoner and the gentry colluded to falsify land registration: the farmer would use the gentry member's name for the farmland and thus pay a lower tax and avoid labor conscription. In times of hardship or poor harvest the gentry often asked for official remission or reduction of taxes in the name of the people; when the request was granted, it was they and not the people who benefited the most.

Privileged as they were, the scholar-gentry were not part of the ruling bureaucracy; they were the intermediary agent between the local magistrate and the people. The magistrate had to rely on them for information and advice on local affairs, and they in turn promoted the welfare and interest of their locality. The magistrate, usually a degree-holder from another province, had no great interest in local affairs and was reluctant to start long-term projects that would not yield results during his short term of office.[8] Such projects therefore customarily fell to the gentry. They financed the construction and repair of public works such as the bridges and ferries. They raised funds for dredging rivers, building dikes and dams, and improving irrigation systems, and contributed to the upkeep of local temples, shrines, and memorial arches. Often they were involved in local charity and welfare cases such as the organizing of rice distribution centers for the poor.

A major function performed by the gentry in the local community was the settlement out of court of civil disputes between individuals and communities, using persuasion and arbitration rather than law. Since appearance in court was detrimental to one's reputation, disputes were more frequently settled privately under the direction of the gentry than in court under the magistrate.

The gentry considered themselves guardians of the cultural heritage. They took it upon themselves to disseminate moral principle and they contributed heavily to the establishment of private academies. Twice a month they expounded K'ang-hsi's Sacred Edict of Sixteen Politico-moral

7. Chung-li Chang, *The Chinese Gentry* (Seattle, 1955), 43.
8. Averaging 1.7 to 4.5 years before 1800, and .9 to 1.7 years afterward; *ibid.,* 53.

Maxims[9] to the villagers in their respective prefectures. They supported the examination system and often contributed money for the construction and repair of the local examination halls. Because the examples of loyal, filial, chaste, and virtuous persons were considered beneficial to public morality, the gentry compiled local gazetteers, which recorded the history of the localities and the biographies of prominent figures of the area.

In times of turmoil and unrest, when government troops were unable to afford protection to the local areas, the gentry would organize militia and at times even lead the fighting themselves. They raised funds to construct or repair the fortresses or city walls, and so strengthened the defense of their localities.

From arbitration of disputes to the sponsorship of public works to the organization of local defense, the gentry performed an indispensable function in their home areas. They served as a link between the government and the people: on the one hand they advised the officials on local affairs, and on the other they promoted the local interests to the officials as the commoners could not do. Because they were the social equals of the magistrates they could communicate freely, without the fear and diffidence of the commoner; thus they enjoyed the best of both worlds. If the magistrates represented the formal power, the gentry represented the informal.

9. Issued by Emperor K'ang-hsi in the eleventh year (1672) of his reign. They were read to students and expounded to the people on the 1st and 15th of each lunar month:

  1. Stress filial piety and brotherly love to exalt human relations.
  2. Be sincere to your kindred to manifest the virtue of harmony.
  3. Maintain peace in your local communities to absolve quarrels and litigations.
  4. Emphasize agriculture and sericulture to insure a full supply of food and clothing.
  5. Promote thrift to save expenditures.
  6. Expand schools to rectify the behavior of scholars.
  7. Reject heterodox doctrines to honor the orthodox learning.
  8. Make known the laws to warn the foolish and obstinate.
  9. Manifest propriety and righteousness to cultivate good customs.
  10. Accept your own calling to the end that the minds of all may be stabilized.
  11. Admonish your children and youngsters against evil-doing.
  12. Eliminate false accusations to preserve the good and innocent.
  13. Refrain from protecting fugitives to avoid collective punishment.
  14. Complete tax payments to dispense with official prompting.
  15. Cooperate with the *pao-chia* neighborhood organizations to forestall burglary and thievery.
  16. Resolve vengeance and animosities to guard your own lives.

The above is my own translation. Cf. three other versions: Rev. William Milne, *The Sacred Edit, Containing Sixteen Maxims of Emperor Kang-hi*, 2nd ed. (Shanghai, 1870); John K. Fairbank, Edwin O. Reischauer, and Albert M. Craig, *East Asia: The Modern Transformation* (Boston, 1965), 85; and Dun J. Li, *The Ageless Chinese: A History* (New York, 1965), 323.

In normal times the power of both derived from the same political order, and hence their interests coincided. But at other times, when their interests collided, the gentry could remonstrate with the officials with impunity, because they were the only pressure group in the locality. And if the worst should come, the gentry might organize an uprising in righteous protest of government oppression. Without question, the gentry were the most important single group in Chinese society. It is not without good reason that China was sometimes described as a "gentry state."

THE GOVERNMENT EXAMINATIONS. After observing the enormously influential stature of the gentry in the society, it is imperative to investigate the procedure by which one became a member of this group. Gentry status was conferred upon one largely as a result of one's winning a literary degree in the civil service examinations. The ability to compose what was called the "eight-legged (i.e. eight-paragraphed) essay" (*pa-ku wen*) was essential to success in the examinations. This essay demonstrated a formal and rigid style of writing, requiring great literary skills but no profound knowledge. The essay opens with two sentences of preliminary remarks, followed by three sentences of introduction and a short paragraph of general discussion. Then comes a short paragraph composed of one to three sentences that deal specifically with the subject matter, followed by two verse-paragraphs—a short one and a long one—composed of phrases of four or six words each, in rhymed couplets. The essay moves gradually into a preliminary closing paragraph, and perorates in a final paragraph of conclusions. The length of this performance ranges between 360 and 720 words. Successful writers had to excel in rhyme and diction as well as in calligraphy and poetical expression. Poor phraseology or slipshod calligraphy reflected an unsound training in fundamentals or even "village mediocrity," which could doom the candidate from the start.

The examinations were conducted on the district, provincial, and metropolitan levels. In order to qualify for the first of these, the preliminary district examination (*t'ung-shih*), the candidate had to present a guarantee of his origin and character from a member of the gentry. This examination was held twice every three years and consisted of three sessions. The first session, conducted by the district magistrate for candidates of his locality, required two "eight-legged essays" on subjects taken from the Confucian Four Books[10] and a short poem of six couplets, with five characters to the line. Many candidates were eliminated in this first session for such things as the wrong use of words, violation of the rules of rhyme, and poor cal-

10. *The Analects, The Book of the Mean, The Book of Menicus,* and *The Great Learning.*

ligraphy; the successful candidates (*t'ung-sheng*) proceeded to the second session. This was given by the prefect or independent department (*chou*) magistrate. A test similar to the first one was administered to make sure that no one had passed the first one by luck. The candidates who passed this hurdle then took the *yüan* examination given by the provincial educational commissioner (whose title was *hsüeh-yüan*, whence the name of the examination). The number of candidates who could emerge successfully from the sessions was prearranged by quotas set by the government. In the entire country, for instance, only 25,089 were allowed for each *yüan* examination. Of this number, Chihli was allotted the highest provincial quota (2,845) and Kweichow the lowest (753.)[11] Only 1 or 2 percent of the candidates passed the *yüan* examination, taking the degree or title of *sheng-yüan* (government student), more commonly called *hsiu-ts'ai* (beautiful talent). With this degree they were admitted to membership among the gentry, but they were only the "lower gentry." Their average age was 24. Given a life expectancy of 57, they enjoyed gentry status for 33 years. At any given time before 1850 there were about 526,869 civil *sheng-yüan* and 212,330 military *sheng-yüan* in the whole country, i.e. a total of 740,000 extant at any one time.[12]

The *sheng-yüan*, or government students, became members of the district college and received allowances from the provincial government with which to prepare themselves for the higher examinations. The local gentry gave them travel subsidies to the provincial capital for the next examinations. These were conducted triennially by a chief examiner and an associate examiner, whom the emperor appointed from among the officials holding the metropolitan degree. By the "law of avoidance" these examiners had to come from other provinces than the one in which the examinations were being held. They were aided by eight to eighteen assistant examiners, appointed by the governor-general or governor from among the provincial officials who held at least the provincial degree, i.e. *chü-jen*. As the government allowed a quota of only 1,400 successful candidates for the whole country, the competition in the provincial examination was very keen.

Like the district examination, the provincial examination consisted of three sessions. The sessions began usually on the ninth day of the eighth lunar month. On the day before, the candidates entered the examination halls; they were locked up in cells for three days to write three essays on subjects chosen from the Confucian Four Books and to compose an eight-couplet poem with five words to each line. Released from their cells on

11. Chung-li Chang, 73, 141-42.
12. *Ibid.*, 97-98.

the tenth of the month, they re-entered them on the eleventh for the second session, this time to compose five essays on the Five Classics.[13] They were let out again on the 13th, only to return once more on the 14th for the third session, in which five more essays were required, these to be on government. They came out of the cells on the 16th, exhausted. The results of the examination were announced in 30 to 45 days.

Extreme care was taken in the examination hall to avoid corruption in its various forms, especially favoritism. All papers written by the candidates were anonymous, and all the assistant examiners were held incommunicado while grading the papers. They recommended the better papers to the chief and associate examiners, who made the final decision. On the day the results were to be announced, the chief examiner, accompanied by the governor or governor-general, conferred the degree of *chü-jen* (employable men) on the successful candidates in the emperor's name, and on the following day the governor-general or governor gave a banquet in honor of the new degree-holders.

The examination papers were then forwarded to the Board of Rites in Peking for scrutiny and safekeeping. Those who failed in the provincial examinations but who otherwise showed high scholastic promise were given the title of *kung-sheng* (*senior licentiates*) and retired to their home towns as local community leaders or teachers, to await the next examination. The successful provincial graduates returned home in glory, for they reflected honor on their families and their districts. These fortunates became members of the "upper gentry"; their average age was 31. They were given traveling expenses by the provincial government so that they might compete in the triennial metropolitan examination in Peking, usually given in the third month of the following year.

The metropolitan examination was also in three sessions. In the first, the candidates wrote four essays—three expository and one critical—on historical subjects; in the second, four expository essays on the classics and one short poem of eight rhymes, five characters per line; and in the third session, an essay on a current political topic.

The results of this examination were known in thirty days. Successful candidates were designated *kung-shih* (presentable scholars) and were qualified to take the palace examination a month and a half later, of which the emperor, with the assistance of fourteen ranking officials, took personal charge. There was only one session to this examination. The candidates wrote an essay of a thousand or more words on current problems. Although the substance of the essay was important, artistic penman-

13. *The Book of Odes, The Book of History, The Book of Changes, The Book of Rites, and The Spring and Autumn Annals.*

ship and an exceptional literary style could catch the attention of the graders and create a good impression at the start. The examiners selected the ten best papers for the emperor, who indicated his preferences and his ranking of the candidates with a vermilion brush. The successful candidates of the palace examination were given the degree of *chin-shih* (advanced scholar). They were divided into three ranks: the first consisted of the three candidates with the highest honors, the second contained roughly the next 30 percent of the candidates, and the third held the remainder. An imperial banquet was given in their honor; a gift of 80 taels was bestowed on the three in the first rank, and 30 taels were given each of the rest. The average age of the *chin-shih* was 34 or 35. The government quota allowed only one out of ten to succeed in the metropolitan examination. From 1644 to 1911, a total of 112 such examinations were held and 26,747 *chin-shih* degrees granted: an average of 238 degrees per examination and about 100 per year.[14]

It is often assumed, with some justification, that only the children of the rich could afford the prolonged years of education necessary to prepare for the examinations. Of course the rich could meet the costs of education more easily; nevertheless, many poorer families managed to produce successful candidates. Recent research has revealed a considerable social spread among the degree-holders during the Ming (1368-1643) and Ch'ing (1644-1911) periods. During the Ming rule, 47.5 percent of the *chin-shih* came from families which had had no degree holders at all for the preceding three generations, and 2.5 percent came from families whose past three generations had produced no one holding a degree higher than the elementary *sheng-yüan;* about 50 percent came from families having holders of higher degrees in the last three generations. During the Ch'ing times, 19.1 percent of the *chin-shih* came from families who were degreeless for three generations back; 18.1 percent came from families that had had one or more *sheng-yüan* but no holders of higher degrees. These figures show that a total of 37.2 percent of the *chin-shih* came from families whose educational background for the preceding three generations had been low or nil, while 62.8 percent came from families with holders of higher degrees or offices in the same three generations.

The provinces which were most prolific were Kiangsu and Chekiang—accounting for 2,920 and 2,808 *chin-shih,* respectively, out of a total of 26,747 granted during the entire Ch'ing period—followed by Chihli (with 2,701), Shantung (2,260), and Kiangsi (1,895). There were approximately 130 *chin-shih* per million people in Chekiang and 93 per million

14. Ping-ti Ho, *The Ladder of Success,* 189.

in Kiangsu. Within the provinces, the most successful prefectures were Hangchow in Chekiang, with 1,004 *chin-shih* during the Ch'ing period, and Soochow in Kiangsu, with 785.[15]

Since success and honor in the Ch'ing system was based so predominantly upon scholarship, an attitude came to prevail in the society that "all activities are unworthy; only learning is lofty." A student spent his entire youth getting ready for the examinations, and it was not unusual for an unlucky one to fail a dozen times or more in the triennial event—thereby virtually forfeiting his entire lifetime. Even the successful ones were marked by exhaustion from intellectual strain. They were stunned into submissiveness and became cautious and meek officials of the court, offering little threat of inciting rebellion. No wonder the ruler could remark with satisfaction: "All the brilliant men of the country have been trapped in my bag!"

The greatest shortcomings of the examination system were its narrow scope and impractical nature. Literary excellence and administrative ability were different matters: proficiency in one did not necessarily imply competence in the other. Conformity to the rigid pattern of the "eight-legged essay" tended to stifle free expression and encourage orthodoxy of thought. What is perhaps most important, the examination system stressed the Confucian values only and rewarded literary and humanistic accomplishments at the expense of science, technology, commerce, and industry.

On the other hand, the system selected men of superior intelligence and common sense for public service. It set up objective and impartial standards for social advancement and reduced the chances for nepotism and other forms of favoritism. It made the society more egalitarian by permitting nearly all its members to rise to the top through individual merit rather than through birth and wealth. It encouraged social flexibility and tended to blur class distinctions. The convergence in government of educated men from all walks of life and all parts of the country created a force of unity. The intellectuals of the country formed an educated bureaucracy which assisted the government rather than criticized it, as opposed to intellectual habits in the West. Put on the scales, the advantages of the examination system probably outweighed the disadvantages.

Though passing the civil service examinations was the regular way to acquire gentry membership, it was not the only way. The academic title of *chien-sheng*—student status in the Imperial College—could also be bought. Occasionally, so could the title *kung-sheng*. The purchasers were usually literate men of means who had failed to win a regular degree or who

15. Ping-ti Ho, 114, 228-29, 247.

wished to attain the coveted gentry status by short cut. The purchasers were "irregular" members of the gentry; they did not enjoy the same prestige as the regular members, and their appointments were usually made at a low level. However, they could regularize their status by passing the provincial and metropolitan examinations. Some intelligent students who also happened to be wealthy bought the *chien-sheng* title to qualify for the provincial examinations, without going through the pains of the district examinations.

It should be mentioned that there was a complete set of military degrees, corresponding to the civil ones, which could be won either by examination or by purchase, but most of the army officers rose from the ranks rather than through the examinations. Their official position gave them the status of gentry, too.

At any one time before 1850 the total number of the gentry was about 1.1 million, of whom 4,000 were civil and military *chin-shih,* and the rest were holders of other degrees and titles. There were, at the same time, only 27,000 official positions to be had in the whole country—20,000 civil and 7,000 military. Regular degree-holders occupied the more important half of the 20,000 civil jobs, while the less important half went to the holders of purchased degrees. Since there were so many more successful candidates than available positions, the greater part of the degree-holders had to stay out of the government. But nearly all the metropolitan graduates and about one-third to one-half of the provincial graduates won appointments, while a small percentage of the *kung-sheng* and *sheng-yüan* were also able to step into official posts.[16] Those who were not part of the bureaucracy became the gentry and community leaders in the society at large.

Intellectual Trends

EARLY CH'ING REACTION AGAINST MING IDEALISM. The early Ch'ing intellectual world was split into two circles. The government officially sponsored the Sung school of Neo-Confucianism as it had been taught in the 11th and 12th centuries by the Ch'eng brothers[17] and Chu Hsi. Chinese officials who served in the government and scholars who wished to do so attached themselves to this "Sung learning" (*Sung-hsüeh*) as to a kind

16. Chung-li Chang, 116-18. The majority of the degree-holders had to wait ten to twenty years before they received appointments. John R. Watt, "Leadership Criteria in Late Imperial China." Paper read before the 62nd annual meeting of the Pacific Coast Branch, American Historical Association, San Diego, August 28, 1969.
17. Ch'eng Hao (1031-85) and Ch'eng I (1032-1107); Chu Hsi (1130-1200).

of state philosophy. On the other hand, there were in the country a great many Ming loyalists who refused to serve the Manchus and who rejected Neo-Confucianism in favor of the so-called "Han learning" (*Han-hsüeh*), by which they hoped to create a new intellectual climate that would help topple the Ch'ing dynasty and restore the Ming.

The Ch'ing rulers used Neo-Confucianism as an instrument to win control over the intellectuals, the habitual ruling class in China, and through them over the people. K'ang-hsi's Sacred Edict of Sixteen Maxims, required reading for all subjects, was full of the Confucian ideas of loyalty, obedience, duty, morality, and propriety. The task of the government, it was reasoned, would be facilitated if everyone followed these maxims and if the scholars and officials could set an example for the rest of the empire. K'ang-hsi was particularly impressed with the Sung philosopher Chu Hsi for his balanced and proper interpretation of the Confucian classics. Every question and every answer on the Four Books and Five Classics in the civil service examinations had to conform to Chu's commentaries, which K'ang-hsi had praised as the "grand synthesis of hundreds and thousands of years of untransmitted learning, capable of opening the minds of fools and children and of establishing the ultimate goal (truth) for a myriad of generations."

The Sung school of Rationalism (*li-hsüeh*), commonly known in the Western world as Neo-Confucianism, was a syncretic philosophy which contained elements of Confucianism, Buddhism, and Taoism, and which provided a system of metaphysics to sanction the old Confucian moral order. The Sung scholars advanced the dual concepts of the rational principle (*li*) and its material manifestation (*ch'i*). According to this school, everything has a rational principle for its being. Thus, a tree or a blade of grass has its own rational principle which makes it what it is. There is only one universal rational principle, although there are many manifestations of it. For instance, the concept of *jen*, often translated as humanity or benevolence, is manifest in filial piety, in affection for one's children, in loyalty to one's ruler, etc. Thus, there is only one *jen* but many manifestations.

Chu Hsi synthesized the Neo-Confucian ideas into a systematized school of philosophy. He called the sum total of all rational principles the "Supreme Ultimate" (*T'ai-chi*).[18] With regard to the dual concepts of *li* and *ch'i*, he said, rational principle was immanent in the material force; the two did not exist separately, or independently of each other, much less in opposition. It is clear that although Chu advanced the dual con-

18. Wing-tsit Chan, "The Evolution of the Neo-Confucian Concept Li as Principle," *Tsing Hua Journal of Chinese Studies*, New Series, IV:2:139-141 (Feb. 1964).

cepts of *li* and *ch'i,* he did not preach the dualistic nature of things, as many assumed he did.

The Sung Neo-Confucianists, though metaphysically oriented, did not lose sight of the practical aspects of Confucianism. They regarded the rational principle as the moral law that must be followed, and declared that an understanding of it could be achieved through the investigation of the nature of things and the study of history and the classics. They continued to stress the cultivation of the self and the management of the family in preparation for service to the state and the world at large. Even in their abstract discourses on the rational principle they emphasized the importance of book-reading and nature-investigating—both pursuits involving substantial effort. Nevertheless, it was the fresh and stimulating metaphysical side of Neo-Confucianism, rather than the practical side, that caught the attention of their disciples.

In time, Neo-Confucianism became rarefied. As the school entered the Ming period (1368-1643), the scholars persisted in the general tendency toward abstract discourse and metaphysical argumentation. The practical aspects of Confucianism were neglected and the classics went untouched. Students drifted in sublime philosophizing of Mind (*hsin*) and Nature (*hsing*) without finding it necessary to read books.

The philosopher Wang Yang-ming (1472-1529) rose to propound his own idealistic school of Neo-Confucianism (*hsin-hsüeh*) as a revolt against Chu Hsi's rationalistic school. Wang was influenced by the meditative Ch'an (Zen) sect of Buddhism and by the Sung philosopher Lu Hsiang-shan (1139-93), who took the view that "the universe is my mind, and my mind is the universe. . . . What permeates the mind, what emanates from it and extends to fill the universe, is nothing but the principle." Wang found very congenial the idea that "the mind is the universe," and he argued further that when the mind was cleared of selfish desires it would be congruent with the principle of nature and hence would know the good from the evil. From this came Wang's theory of the "intuitive knowledge" (*liang-chih,* literally, "good innate knowledge"), which held that all things were complete within oneself; one needed only to search inwardly through meditation to discover the good "intuitive knowledge." Thus, instead of Chu Hsi's "extension of knowledge through the investigation of things," Wang stressed the nurture of intuitive knowledge through meditation and introspection. It should be noted, however, that even Wang did not neglect the importance of practical action. The mind, which is supreme, must "always be doing something," that is, must be actively engaged in human affairs. A sincere innate knowledge of filial piety was not its own end; rather it was desirable insofar as it would lead a man to serve his parents filially. Hence it was important to unite knowledge with action: "Knowledge is the beginning of action," he maintained, "and action the completion of knowledge." The unity of the two could be achieved through self-discipline and self-cultivation.[19]

In spite of his underscoring the importance of action, Wang's philosophy further strengthened rather than weakened the general tendency toward metaphysical discourse. Students packed books away as if reading were anathema and languished in the opiate of aimless abstract dialogue. The latter-day followers of Wang abused his teachings to the point of proclaiming that wine, women, wealth, and emotion were not obstacles to Enlightenment. Uninhibited sexual freedom and unabashed drinking orgies were celebrated as free exercise of the "intuitive knowledge." To maneuver themselves into official positions they took to currying favor with the court eunuchs. When social and moral behavior reached such a nadir, a reaction against that school was bound to arise.

A group of serious scholars at the Tung-lin Academy in Wusih, Kiangsu, attempted to turn the tide of intellectual irresponsibility and moral degradation by shifting attention from the abstract back to the practical, and from individual introspection to engagement in public affairs. In their "moral crusade," they visited an ungenerous attack on political corruption which led, unfortunately, to their political destruction at the hands of the powerful eunuch Wei Chung-hsien; but they succeeded at least in arousing an interest in public affairs among scholars.[20]

The great scholars of the early Ch'ing period reacted vigorously against the Ming intellectuals, whom they held accountable for the moral and social degradation and the ultimate collapse of the dynasty. They called upon scholars to free themselves of the shackles imposed on them by the Sung and Ming learning and to seek truth directly from the old classics. Their advocating of practical studies and serious scholarship had itself a realistic purpose: to create a healthy intellectual atmosphere which would contribute to the overthrow of the Manchus.

Ku Yen-wu (1613-82) of Kunshan, Kiangsu, was the first to launch a powerful attack on the unhealthy Ming intellectual disposition. He censured Wang Yang-ming for gathering disciples to engage in "lofty and esoteric" doctrines of Mind and Nature without extensive study of the classics and history and without showing any concern for the problems of the society. He chided the Wang followers as "unlettered men who bor-

19. Wing-tsit Chan, 142, 213.
20. For a study of the Tung-lin movement, see Charles O. Hucker, "The Tung-lin Movement of the late Ming Period" in Fairbank (ed.) *Chinese Thought and Institutions*, 132-62.

row accepted sayings to hide their ignorance," and ridiculed Ming intellectuals by calling their works "nothing but plagiarism." The theory of "intuitive knowledge" he attacked as the basic cause of confusion and disorder. The Wang school of idealism never recovered from his attack. It should be noted that Ku did not attack Chu Hsi and the Ch'eng brothers personally, for he still had a high regard for their encouragement of bookreading and their exhortation to investigate the nature of things.[21]

Ku traveled widely in North China and studied the practical problems of geography, frontier defense, farming, and trade. Of the dozen or so works to his name, the most famous was the *Record of Daily Knowledge* (*Jih-chih lu*), which was a crystalization of his lifelong devotion in the form of a notebook.

Ku's chief contribution to the Ch'ing school of learning was his construction of a revolutionary research methodology distinguished by three characteristics: (1) *Originality:* In his preface to the *Jih-chih lu,* Ku wrote, "Since my childhood studies, I have always noted down what I perceived . . . If earlier men had said it before me, I omitted it entirely." It is certain that Ku's writings did not contain a single point that was borrowed. (2) *Utility:* Just as Confucius had edited the Six Classics in order that they might keep people out of trouble, Ku decided "not to do any writing unless it had a relation to the actual affairs of the contemporary world as indicated in the Six Classics." His emphasis on utility brought knowledge and society into a closer relationship, in contrast to the Ming habit of abstract discussion divorced from social reality. (3) *Extensive Evidence:* Ku would not write a page without scrutinizing every fact thoroughly and supporting it with solid evidence. Consequently, citations in his work were numerous and extensive, and his writings were sound, broad, and consistent. His study of phonology relied on primary and secondary sources of evidence; only when these were lacking would he fall back on evidence at several removes. His research method approached the standards of modern historical investigation.[22]

For his generally destructive influence upon the Ming intellectual trends and for his constructive development of a new research methodology, Ku was honored as the founder of the Ch'ing school of learning. It was from his initial efforts that a new trend of "sound learning" or "unadorned learning" (*p'u-hsüeh*) and textual criticism evolved.

Another important figure was Wang Fu-chih (1619-92), who made a very perceptive statement on the relation of human desire to the rational

21. Liang Ch'i-ch'ao, *Intellectual Trends in the Ch'ing Period,* tr. from the Chinese by Immanuel C. Y. Hsü (Cambridge, Mass., 1959), 30.
22. Liang Ch'i-ch'ao, 31-32.

principle: "The rational principle of nature resides in human desires; without human desires there can be no discovery of the rational principle of nature." It is from this line of thinking that Tai Chen (1724-77), perhaps the greatest of all Ch'ing scholars and thinkers, later developed his famous doctrine on human feelings and desires.

Wang's views on history and politics seem startlingly refreshing and modern. He denied that supernatural forces, mysticism, fate, or luck could influence the course of history, and boldly advanced the evolutionary and progressive view that history unfolds itself unceasingly in an orderly direction which perforce makes later periods better than earlier ones. Hence there was no point to revive ancient institutions and ways for modern application, inasmuch as each age has its own characteristics and needs. This view of continuous societal advancement contradicted the traditional cyclical theory of history, which asserts that a period of order is always followed by a period of disorder, and vice versa.

Wang has been described as a materialist in that he believed that progress could best be achieved under conditions of economic well-being. He advocated the security of life and satisfaction of basic needs; development of natural resources; and encouragement of domestic and foreign trade. The state should consider the people's welfare as its chief occupation, and it should belong to the people and not to any one hero or any one dynasty, much less a foreign dynasty. Alien domination of China was utterly insufferable and illegitimate; hence the Chinese could justifiably cheat or kill barbarians. Because of their nationalistic and anti-Manchu overtones, Wang's voluminous writings were not published for two hundred years. It was not until the late 19th century that reformers and revolutionaries publicly distributed them.

These early Ch'ing masters—from Ku Yen-wu to Wang Fu-chih—reacted vigorously against the abstract and metaphysical intellectual trends of the Ming, and created a new climate of learning, in which stress was laid on the study of the old classics, textual research based on extensive evidence, and practical application of knowledge to society. All of them exhibited a strong spirit of doubt which led them to scrutinize works which had been accepted for ages, if their authenticity seemed at all in doubt. Their revival of the old classics confronted them with problems of correctly understanding the ancient texts. To clarify the meaning of ancient words and phrases as well as the pronunciation of words so as to reconstruct the rhymes, they devoted themselves to the study of phonology, philology, the semantics of technical terms, and ancient regulations and institutions, as prerequisites to their classical studies. Their investigations led them deeper and deeper into antiquarian textual studies, and paved the way for the emergence of the School of Empirical Research (*K'ao-cheng hsüeh*) in the middle Ch'ing period.

MIDDLE CH'ING EMPIRICAL RESEARCH. The term *k'ao-cheng hsüeh*, literally "search for evidence," has sometimes been translated as "textual criticism." But since this type of research included the examination of artifacts as well as textual study of ancient works it is more appropriate to give the term the broader translation, School of Empirical Research. Scholars of this discipline employed the inductive method of investigation, collecting evidence from a wide range of sources and testing their various hypotheses against it. Their motto was to get at the truth through concrete proof and to hold to no belief without it. This school, which made a tentative beginning in the early Ch'ing period, grew to its full stature during the middle period. Among the many scholars of empirical research, two were particularly outstanding: Hui Tung of the Soochow group and Tai Chen of the Anhwei group.

Hui Tung (1697-1758) came from a family with a scholarly tradition. His approach to scholarship was characterized by extensive reading and a ready acceptance of Han works, which were considered authentic because the Han was not far removed from antiquity. The guiding spirit of the Hui school was, "That which is ancient must be authentic and that which is of the Han must be good." In this vein, Hui Tung made a number of studies of ancient works.[23] His worship of Han scholarship prompted his attempt to elevate the views of the Han masters to the rank of the classics.

Tai Chen (1724-77) of Siuning, Anhwei, was probably the greatest of all Ch'ing scholars. He regarded Hui Tung as a teacher and friend, but his approach to learning was vastly different. Whereas Hui was partial to the Han learning to the exclusion of anything else, Tai would not let himself be bound by any school. Imbued with a strong spirit of doubt, he would not accept any statement—be it from the sage, or his father, or his teacher—without conclusive evidence. He maintained a high degree of objectivity in his work and sought truth by investigating the facts without patronizing any school. His guiding principle in research was "not to be deluded by others and not to be deluded by oneself." Tai wanted to liberate scholars from dependence of all kinds. He respected the methodology of the Han scholars but would not ask anyone to follow it blindly. When he had doubts about a point, he had no peace with himself until

---

23. Including the *Ancient Interpretation of the Nine Classics* (*Chiu-ching ku-i*), *Studies of the Book of Changes According to Han Tradition* (*I Han-hsüeh*), *A Study of the Book of History in Ancient Text* (*Ku-wen Shang-shu k'ao*).

he had repeatedly checked the references and evidence to his satisfaction. By virtue of his ability to make incisive observations and critical judgments in his investigation, he was able to elevate the standards of research to a new height. The thoroughness of his scholarship is illustrated in his statement: "Having an inaccurate understanding of ten [things] is worth less than having a true understanding of one."

Tai's learning was extensive, yet by no means superficial. His areas of specialization were traditional Chinese linguistics, the calendar and mathematics, and waterworks and geography.[24] In late life Tai went beyond the realm of empirical research to evolve a philosophy of his own. He completed a masterwork called *The Elucidation of the Meaning of Words in the Book of Mencius (Meng-tzu tzu-i shu-cheng)*, by which he intended to substitute his own "philosophy of feeling" for the "philosophy of rational principle" which the Ch'eng brothers and Chu Hsi had evolved. He attacked the Sung philosophers for the two sins of adulterating Confucianism with Taoism and Buddhism, and of ignoring desires in favor of rational principle:

> The way of the sages was to see to it that there were no unexpressed feelings in the world; it sought to realize desires so that the world could be governed well. Later scholars did not understand that it is precisely when feelings reach their fullest and most unreserved expression that the "rational principle" is fulfilled. Their so-called "rational principle" was similar to what cruel officials called "law": cruel officials killed men with their "law," and later scholars killed men with their "rational principle."
>
> When the superior man governs the world, he makes it possible for everyone to express his feelings and satisfy his desires, not contradicting "truth" [*tao*] and righteousness. In governing himself, the superior man unifies feelings and desires in truth and righteousness. The evil of suppressing desires is even worse than that of blocking a river; it kills feelings and eradicates intelligence as well as stifles benevolence and righteousness.[25]

Tai's "philosophy of feeling" was undoubtedly influenced by Wang Fu-chih. Tai was very proud of his philosophy and spoke of his *Elucidation*

---

24. Among his numerous accomplishments, the following works were particularly noteworthy: *A Study of Phonetics (Sheng-yün k'ao)*, *A Study of the Language in the Erh-ya (Erh-ya wen-tzu k'ao)*, *An Inquiry into the Origin of Astrology (Yüan-hsiang)*, *A Study of Ancient Calendar (Ku-li k'ao)*, *A Record of Waterworks and Geography (Shui-ti)*, and *Collation of the Commentary on the Water Classics (Chiao Shui-ching chu)*.

25. Liang Ch'i-ch'ao, 59-61.

as his greatest work. But unfortunately most of his students could not understand it and failed to take it seriously. Even though this work had little influence during the middle Ch'ing period, Tai's contributions to research methodology and linguistics, calendar and mathematics, and waterworks, were overwhelming. His investigations far exceeded the scope of the Han learning. Therefore, rather than call it "Han learning," it is both more accurate and more appropriate to refer to Tai's school as the "Ch'ing learning."

The School of Empirical Research reached its zenith during the middle Ch'ing period and dominated the intellectual horizon of the country. Even the court ceased to sponsor the now outmoded Sung learning. Emperor Ch'ien-lung's bureau for the compilation of *The Complete Works of the Four Treasuries* (*Ssu-k'u ch'üan-shu*) was in fact the headquarters of some 300 Han scholars, including Tai Chen. They edited 3,457 works in 79,070 *chüan* (tomes). The synopsis of each entry in the printed catalogue represented the crystallization of the Han school approach to the subject.

Scholars of empirical research re-examined nearly every aspect of the Chinese cultural heritage with thoroughness, objectivity, alertness, and open-mindedness. Because of their solid research and unpretentious style of writing, they described their works as "unadorned learning." The core of their study was still the classics, but they extended their activities into such fields as traditional linguistics, phonology, history, astronomy and mathematics, geography, government institutions, and artifacts. The great corpus of Chinese classical literature since the Han (and even earlier) was put through a rigorous examination, with the result that difficult ancient books could be read and understood, forgeries were exposed, and lost subjects were restored.

Ch'ing scholars often called themselves Han scholars and their learning the Han learning. Doubtless they used the impressive name of Han in order to eclipse the Sung learning. But actually their reverence for the classics and their habit of extensive reading and the writing of commentary were very much in line with the spirit of the Sung scholars. In all fairness, one cannot say that the Ch'ing learning stood diametrically opposed to the Sung; the difference between the two lay chiefly in research methodology rather than in any essential difference in the spirit of learning. Nor can one equate Ch'ing learning with Han learning, for the Ch'ing was much broader in scope than the Han. It has been suggested that Ch'ing learning was established under the Han banner but in the Sung spirit.

Examining the merits and demerits of the School of Empirical Research, one is struck by the radical shift in its objective. Whereas in the early Ch'ing period the masters advocated the pursuit of knowledge for practical use, in the middle period knowledge was pursued for its own sake. The idea of utility was totally cast aside. Of course, such a turn of attitude could be ascribed in great part to the frequent literary inquisitions that were instigated as a result of anti-Manchu writings. Scholars found refuge in pure scholarship and antiquarian research, which was politically safe and intellectually rewarding. Empirical research, which had been a means by which the early Ch'ing scholars hoped to create the intellectual climate that would restore the Ming, became an end in itself during the middle period. When men of talents immersed themselves in ancient textual research and writing exegeses and commentaries, they lost contact with the reality of the society and deprived the country of practical leadership. It became again an intellectual irresponsibility which indirectly fostered the growth of political corruption—precisely the situation that the early Ch'ing scholars were anxious to correct.

On balance, the Ch'ing scholars re-evaluated and reordered the rich Chinese cultural heritage, but they created no new strands of thought or major schools of philosophy. They were diligent interpreters and devoted editors of Chinese culture but they were not its creative builders. Liang Ch'i-ch'ao, observing the step-by-step revival of ancient studies from the Ming to the Sung to the Han and the pre-Han periods, remarked that the Ch'ing learning was "a reverse development of the intellectual trends of the previous two thousand years. It was like peeling a spring bamboo-shoot: the more it is peeled, the closer one gets to the core."[26] Liang compared the revival of antiquity to the European Renaissance. Such a comparison is of course forced, but it is undeniable that access to the treasury of the Chinese cultural heritage was made much easier by the endeavors of the Ch'ing scholars.

26. Liang Ch'i-ch'ao, 14.

Further Reading

Beattie, Hilary J., *Land and Lineage in China: A Study of T'ung-ch'eng County, Anhwei, in the Ming and Ch'ing Dynasties* (Cambridge, Eng., 1978).

Buxbaum, David C. (ed.), *Chinese Family Law and Social Change in Historical and Comparative Perspective* (Seattle, 1977).

Cahill, James, *Chinese Painting* (Geneva, 1977).

———, *The Compelling Image: Nature and Style in Seventeenth-Century Chinese Painting* (Cambridge, Mass., 1982).

Chan, Wing-tsit, "The Evolution of the Neo-Confucian Concept Li as Principle," *Tsing Hua Journal of Chinese Studies*, New Series, IV:2:123-49 (Feb. 1964).

Chang, Chung-li, *The Chinese Gentry* (Seattle, 1955).

———, *The Income of the Chinese Gentry* (Seattle, 1962).

Chang, Te-ch'ang 張德昌, *Ch'ing-chi i-ko chiang-kuan ti sheng-huo* 清季一個京官的生活 (The life of a metropolitan official in the late Ch'ing period), (Hong Kong, 1970).

Cheng, Chung-ying, *Tai Chen's Inquiry into Goodness* (Honolulu, 1971).

Chia, Chih-fang 賈植芳, *Chin-tai Chung-kuo ching-chi she-hui* 近代中國經濟社會(Modern Chinese economy and society), (Shanghai, 1950).

Chin, Ann-ping, and Mansfield Freeman, *Tai Chen on Mencius* (New Haven, 1990).

Ching, Julia, *To Acquire Wisdom: The Way of Wang Yang-ming* (New York, 1976).

Ch'ü, T'ung-tsu, *Law and Society in Traditional China* (Paris and The Hague, 1961).

Cohen, Jerome A., et al. (eds.), *Essays on China's Legal Tradition* (Princeton, 1979).

Cole, James H., *Shaohsing: Competition and Cooperation in Nineteenth-Century China* (Tucson, 1986).

Creel, H. G., *Chinese Thought from Confucius to Mao Tse-tung* (Chicago, 1953).

de Bary, William Theodore, "Chinese Despotism and the Confucian Ideal: A Seventeenth-Century View" in John K. Fairbank (ed.), *Chinese Thought and Institutions* (Chicago, 1957), 163-203.

———, *The Liberal Tradition in China* (Hong Kong, 1983).

———, Wing-tsit Chan, and Burton Watson, *Sources of Chinese Tradition* (New York, 1960), chs. 19-22.

Eastman, Lloyd E., *Family, Field, and Ancestors: Constancy and Change in China's Social and Economic History, 1550-1949* (New York, 1988).

Ebrey, Patricia Buckley, and James L. Watson (eds.), *Kinship Organization in Late Imperial China, 1000-1940* (Berkeley, 1986).

Elman, Benjamin A., "Political, Social, and Cultural Reproduction via Civil Service Examination in Late Imperial China," *The Journal of Asian Studies* 50:1:7-28 (Feb. 1991).

Fei, Hsiao-tung, *China's Gentry* (Chicago, 1953).

———, *Peasant Life in China* (London, 1945).

Freedman, Maurice (ed.), *Family and Kinship in Chinese Society* (Stanford, 1969).

———, *Chinese Lineage and Society* (London, 1966).

———, *The Religion of the Chinese People* (New York, 1975).

Fried, Morton H., *Fabric of Chinese Society* (New York, 1953).

Fung, Yu-lan, *A History of Chinese Philosophy*, tr. by Dark Bodde (Princeton, 1953).

Goodrich, L. Carrington, *The Literary Inquisition of Ch'ien-lung* (Baltimore, 1935).

Ho, Ping-ti, *The Ladder of Success in Imperial China: Aspects of Social Mobility, 1368-1911* (New York, 1962).

Hsiao, Kung-ch'üan, *Rural China: Imperial Control in the Nineteenth Century* (Seattle, 1960).

Lang, Olga, *Chinese Family and Society* (New Haven, 1946).

Liang, Ch'i-ch'ao, *Intellectual Trends in the Ch'ing Period* (*Ch'ing-tai hsüeh-shu kai-lun*), tr. by Immanuel C. Y. Hsü (Cambridge, Mass., 1959), Parts I and II.

Liu, Hui-chen (Wang), *The Traditional Chinese Clan Rules* (Locust Valley, N.Y., 1959).

Mackerras, Colin P., *The Rise of the Peking Opera, 1770-1870: Social Aspects of the Theatre in Manchu China* (Oxford at Clarendon, 1972).

Mann, Susan, *Local Merchants and the Chinese Bureaucracy, 1750-1950* (Stanford, 1987).

Marsh, Robert M., *The Mandarins, the Circulation of Elites in China, 1600-1900* (Glencoe, Ill., 1961).

Miyazaki, Ichisada, *China's Examination Hell: The Civil Service Examinations of Imperial China*, tr. by Conrad Schirokauer (Salem, Mass., 1976).

Moore, Charles A. (ed.), *The Chinese Mind* (Honolulu, 1967).

Nivison, David S., *The Life and Thought of Chang Hsüeh-ch'eng (1738-1801)*, (Stanford, 1966).

Overmyer, Daniel L., *Folk Buddhist Religion: Dissenting Sects in Late Traditional China* (Cambridge, Mass., 1976).

Peterson, Willard J., "The Life of Ku Yen-wu (1613-1682)," Part I, *Harvard Journal of Asiatic Studies*, 28:114-56 (1968).

Qian, Wen-yuan, *The Great Inertia: Scientific Stagnation in Traditional China* (Dover, N.H., 1985).

Rawski, Evelyn S., *Education and Popular Literacy in Ch'ing China* (Ann Arbor, 1978).

Shang, Yen-liu 商衍鎏, *Ch'ing-tai k'o chü k'ao-shih shu-lu* 清代科舉考試述錄 (A study of the civil service examinations system of the Ch'ing period), (Peking, 1958).

Shapiro, Sidney, (tr. and ed.), *Jews in Old China: Studies by Chinese Scholars* (New York, 1984).

Shih, Vincent Yu-chung (tr.), *The Literary Mind and the Carving of Dragons* (Hong Kong, 1982).

Skinner, G. William et al., *Modern Chinese Society: An Analytical Bibliography* (Stanford, 1973), 3 vols.

Strassberg, Richard E., *The World of Kung Shang-jen: A Man of Letters in Early Ch'ing China* (New York, 1983).

Teng, S. Y. 鄧嗣禹, "Wang Fu-chih's Views on History and Historical Writing," *The Journal of Asian Studies*, XXVIII:1:111-23 (Nov. 1968).

——, *Chung-kuo k'ao-shih chih-tu shih* 中國考試制度史 (A history of Chinese examination system), (Taipei, 1967).

Tu, Wei-ming, *Neo-Confucian Thought in Action: Wang Yang-ming's Youth (1472-1509)*, (Berkeley, 1976).

Wang, Teh-chao, *Ch'ing-tai K'o-chu chih-tu yen-chiu* (Studies of the Civil Service Examination System of the Ch'ing Dynasty), (Hong Kong, 1982).

Waley, Arthur, *Yuan Mei: Eighteenth-Century Chinese Poet* (Stanford, 1956).

Weber, Max, *The Religion of China: Confucianism and Taoism*, tr. by Hans H. Gerth (Glencoe, Ill., 1951).

Wittfogel, Karl A., *Oriental Despotism: A Comparative Study of Total Power* (New Haven, 1957).

Wolf, Arthur P. (ed.), *Religion and Ritual in Chinese Society* (Stanford, 1974).

5

Foreign Relations

During the transitional period from late Ming to early Ch'ing, the western Europeans began to arrive in China—and almost simultaneously with them, though independently, the Russians were marching across Siberia toward the Manchurian border. With this unprecedented confrontation of the East and the West, there began a new era in China's relations with the outside world.

The Arrival of the Western Europeans

Europe during the Age of Discovery was suffused with a new spirit of adventure, fed by the lust for empire, by the evangelical zeal to spread Christianity to the heathen (i.e. non-Western) world, and by the mercantile search for the spice trade. Under the patronage of Prince Henry the Navigator (1394-1460), Portuguese captains set out for the vaguely known continent of Africa, and in 1487 Bartholomeo Dias rounded the Cape of Good Hope. Only a few years later, in 1492, Spain, moving in a different direction, sponsored Columbus's voyage which inadvertently led to the discovery of America. Competition between the Portuguese and the Spaniards was so fierce that Pope Alexander VI intervened and issued the famous Bulls of May 3 and 4, 1493 (ratified a year later at the Convention of Tordesillas), dividing the as yet unexplored world between them for future expeditions—Portugal taking Brazil and most of the non-Christian world of the East and Spain taking most of the Americas, the Pacific, the Philippines, and the Moluccas. With this division of sphere of operation, Vasco da Gama reached India in 1498 via the Cape, and the route to the East was thereupon carved out; Portuguese, Spaniard, Dutch, and English, one after another, arrived in Asia.

THE EXPLORERS AND THE TRADERS. In those days little distinction existed between geographical explorers and empire-builders. When Alfonso d'Albuquerque captured Goa in 1510 and Malacca in 1511, he broke the ground for the Portuguese empire in the East and also gained control of the gateway to the spice land of Malaya and the East Indies. At Malacca the Portuguese met Chinese traders, who had come with silk, satins, chinaware, and pearls to exchange for spices, ginger, incense, and gold thread. The Portuguese began to think about reaching China. In 1516 Rafael Perestrello sailed there in a European vessel, made a handsome profit from his trade, and probably established himself as the first Portuguese to appear in China. The Chinese called the Portuguese "Fo-lang-chi," a corruption of the Arabic name for the Europeans, "Feringhi," which in turn was derived from the name "Franks" of the Crusades.

In 1517 Tomé Pires was sent by the king of Portugal as ambassador to the Ming court, and the mayor of Goa, Fernao d'Andrade, was given command of an exploratory mission along the China coast. Together they arrived in Canton in September of that year in eight ships, from which they fired a thunderous gun salute—their first act, and a cause of great alarm to the Chinese, who could not understand the meaning of the fusillade. However, the governor-general[1] received them with some courtesy and allowed them to anchor at St. John's Island (*Shang-ch'uan*).

Large contingents of Portuguese gradually came to settle on St. John's Island, on Lambacao, and on Macao. In 1535 the Portuguese secured, by way of bribery, official Chinese permission to dry their cargoes in Macao, thus winning legal sanction to reside and trade there. They agreed to pay an annual customs dues of 20,000 taels on ships and commodities and an annual 1,000 taels rent, which was reduced in 1582(?) to 500 taels at their repeated request.[2] China did not cede the territory of Macao, but by 1557 the Portuguese appointed officials themselves to govern the area, as though it were a colony. The Ming court did not protest but instead in 1573 constructed a wall along the narrow isthmus of Macao, guarded by soldiers, ostensibly to check the kidnapping of Chinese coolies but actually to keep a close watch over the Portuguese and limit their expansion. This act was tantamount to extending recognition of the Portuguese occupa-

1. Ch'en Chin.
2. Kuo T'ing-i, *Chin-tai Chung-kuo shih* (Modern Chinese history), (Taipei, 1963), I, 117-18.

tion, and with the firmness of their entrenchment thus established in Macao, the Portuguese monopolized China's foreign trade at Canton and strove to exclude other foreigners from sharing the profit.

The fortunes brought back by the Portuguese flotillas from the spice trade in Malacca and the East Indies kindled the jealousy of the Spaniards. Columbus's discovery of America convinced the Spaniards that they could reach the East by sailing west, skirting the American continent via its southernmost tip. In 1519 Ferdinand Magellan, a Portuguese navigator in the service of the Spanish King Charles V, led an expedition of five ships along the eastern coast of South America and emerged into the Pacific. After thirty-three months of navigation he reached the Luzon (later known as the Philippine Islands), completing the first voyage of a European ship from America to the East. Magellan and most of his followers were killed by the natives, and the survivors returned to Spain by way of the Indian Ocean and the Cape of Good Hope in 1522.

With Cortés's conquest of Mexico, the Spaniards had a point from which to embark for Asia. In 1564 Miguel Lopez de Legazpi, sent from Mexico by Philip II, seized the Luzon, which was renamed the Philippine Islands, with Manila as the capital.

Meantime, many Chinese were engaged in lucrative trade in this very area which the Spaniards had taken over. The problem of piracy became increasingly serious. In 1574 a Chinese pirate fleet of 62 armed vessels and 2,000 men under Lin Feng (Limahong) attacked Manila. The Spaniards repulsed them and followed up by burning their fleet, thereby earning the gratitude of the Chinese fleet commander from Fukien, who had been sent to chastise the pirates. The Spaniards seized this opportunity to develop relations with China and invited the Chinese commander to Manila. A Spanish deputation, consisting of two Augustinian friars[3] returned with the Chinese fleet to Fukien in 1575—the first official contact between the two countries. The Chinese authorities treated the delegates well and allowed the Spaniards to trade along the Fukien and Chekiang coast, but not to maintain a settlement such as the Portuguese had done in Macao. From then on trade flourished between Foochow, Amoy, Ch'uan-chow (Zayton) and Manila, Mexico, and Spain.

The medium of exchange was the Mexican dollar, since the Philippines were under the administrative jurisdiction of Mexico. Spanish and Peruvian dollars were also used but to a lesser extent. This was the beginning of the influx of Mexican dollars into Chinese ports. The Chinese called the Spaniards "men of Luzon" since they came from the Luzon;

---

3. Geromine Marin and Martin de Rada.

and often they were known as the "Fo-lang-chi" (Franks), as the Chinese did not distinguish them from the Portuguese.

In 1626 the Spaniards descended upon Keelung, Taiwan, and established a base at Tamsui for trade and the propagation of religion. There they remained until they were expelled in 1642 by the Dutch.

The Dutch arrived in China in 1604, some thirty years after the Spaniards and ninety years behind the Portuguese. The forces behind their vigorous push were their nationalistic sentiments and their spirit of Protestant Reformation. The Dutch had been under Spanish rule but had succeeded in throwing it off in 1581; for this action King Philip II of Spain, who also ruled Portugal, punished them by barring them from the port of Lisbon in 1594, thereby depriving them of their share of the spice trade. The Dutch then determined to reach the East Indies themselves and the needed information on the trade possibilities was provided by J. H. van Linschoten, a Dutchman who had spent many years of service with the Portuguese in the East. In 1595 the merchants of Amsterdam organized a private East India Company to explore the routes to the East, and Cornelius Houtman spearheaded the first voyage to Sumatra and Java in 1596.

In 1602 the Netherlands East India Company was officially established, with authority from the government to maintain troops, colonize overseas territories, declare war, and conclude peace with countries in the East. From the Portuguese they seized Sumatra, Java, and the Moluccas, and won the right to trade with Japan under the Tokugawa Bakufu. In 1619 the great organizer and empire-builder Jan Pieterszoon Coen established the Batavian government on Java, which became the center of Dutch enterprise in the East, covering a vast area from India to Japan.

With the establishment of the Ch'ing dynasty in 1644, the Dutch, who had been barred from Canton by the Portuguese during the Ming era, had their hopes for trade revived. In 1656 two envoys[4] were dispatched from Java to Peking. They accepted the role of tributary envoys, performed the full kowtow to Emperor Shun-chih, and presented gifts as tribute. The Ch'ing court permitted the Dutch to send a tributary mission by way of Canton every eight years. The size of the mission was limited to four ships and 100 men, of whom twenty were allowed to come to Peking.

The Dutch rule on Taiwan, which began in 1642, came to a sudden end in 1662, when Koxinga, the Ming loyalist who had been raiding the China coast from Amoy, descended on the island and drove them out.

4. Pieter de Goyer and Jacon de Keyser.

After the pacification of Taiwan in 1683, the Ch'ing granted the Dutch permission to trade in Kwangtung and Fukien and pay tribute every five years.

The expansive Elizabethan Englishmen naturally would not fall far behind in the race for the spice trade. In 1600 the queen of England granted a charter for fifteen years to "The Governor and Merchants of London trading into the East Indies"—the coterie that was the origin of the British East India Company. A fleet of five ships under James Lancaster and John Davis then set out for Sumatra and Java, marking the beginning of the English commercial empire in the East. In the following years the Company rapidly established a number of agencies, called "factories," in key points of trade. Because of conflicts of interest with the Dutch in the East Indies and with the Portuguese in Macao, the British focused their attention on India.

Just as the Portuguese and the Spaniards were indiscriminately called "Fo-lang-chi" (Franks), the British and the Dutch, so far as the Chinese were concerned, were both the "Red Hair." As a result, predatory activities committed by the Dutch on the high seas caused Chinese resentment to be directed against the Englishmen as well. Added to this, the Portuguese in Macao, who intended to monopolize the Canton trade as long as possible, did their best to denigrate the English. The Chinese therefore had a very poor image of the English from the start.

After the Ch'ing pacification of Taiwan, K'ang-hsi lifted the ban on sea trade; in 1685 customs houses were opened at four places: Canton, Chang-chou (in Fukien), Ningpo, and Yün-t'ai-shan (in Kiangsu). Among the ports Canton was the most prosperous because of its proximity to Southeast Asia, and in 1699 an English factory was established there.

France's attempt to establish its interests in the East was beset with internal dissension. In 1604 King Henri IV granted a charter of fifteen years of monopolistic trade with the East Indies to La Compagnie française des Indes. The company did little, and soon a rival organization—employing mostly Dutchmen for its fleet—came into existence. By 1719 all eastern trade was given to a new organization called Compagnie des Indes. A factory was established at Canton in 1728, but French trade remained insignificant throughout the 18th century.

On the whole, foreign traders in China, who were mostly profit-seeking adventurers and uncouth men of little culture, made a poor show of themselves. Their violent and reckless conduct confirmed the Chinese view of foreigners as barbarians. They were not welcomed but tolerated in China as a mark of favor from the emperor for men from afar. The proud and self-sufficient Chinese refused to admit of a need for foreign products.

Quarantined in a few pockets along the coast, the foreign traders made little constructive impact on the Chinese state and society. Of far greater importance were the missionaries, particularly the Jesuits.

THE MISSIONARY ACTIVITIES. With the discovery of the new sea route to the East, members of the Catholic Church—which had weakened since the Reformation—looked to foreign lands for spreading the faith. The Society of Jesus, founded by St. Ignatius of Loyola in 1540, in particular was fired with the zeal natural to a new order to evangelize the East. St. Francis Xavier, having opened Japan to Catholicism, came to China with the dream of converting the Chinese, only to die in 1552 at the gates of his promised land. In 1573 Alessandro Valignano, an Italian Jesuit with a degree of doctor of civil law, who had served in the court of Pope Paul IV, was made Superior of all the Jesuit missions in the East Indies, which included China and Japan. With forty-one Jesuits he left Lisbon in 1574 and arrived in Macao in 1577. A man of rare intellectual and spiritual gifts, he charted a new course of action in China. Instead of imposing Christianity as a foreign religion and forcing the converts to take up Christian names and a foreign style of dress, as was the habit of missionary activity at the time, he decided that Christianity should function like a leaven, entering China quietly and transforming it from within. "Europeanism" was to be replaced by cultural adaptation. The Jesuits in China were instructed to learn to read, write, and speak Chinese, to "Sinicize" themselves rather than to "Portugalize" the converts.[5] Two Italian priests, Michele Ruggieri and Matteo Ricci, were sent as pioneer missionaries to carry out this policy and to continue the unfinished work of Xavier.

Ruggieri and Ricci settled in Chaoching, Kwangtung, in 1583. To lay the groundwork of an effective apostolate, they changed into Chinese attire, studied the Chinese language, adopted Chinese mannerisms, and learned the moral doctrines of Confucianism.[6] Their first aim was not to win converts but to earn for Christianity an accepted place in Chinese society. They did not work toward the statistical and dubious success of multiplying baptisms each year, but instead concentrated on diffusing Christian ideals and ideas through a widening circle of sympathetic contacts. Through their knowledge of the Chinese language and culture, and their astronomical, mathematical, geographical, and other scientific

5. George H. Dunne, S.J., *Generations of Giants: The Story of the Jesuits in China in the Last Decades of the Ming Dynasty* (Notre Dame, 1962), 17.
6. However, it should be noted that they rejected the Sung school of Neo-Confucianism on the grounds that it was an adulterated form of philosophy which distorted the teachings of Confucius.

achievements, they made friends with the more open-minded Chinese scholars and officials. The Chinese were impressed with European mechanical devices such as clocks and solar quadrants, and with the technique of perspective drawing, as well as European cartography. Ricci, conscious of the Chinese ignorance of cosmography, constructed a map of the world, but unthinkingly placed America on the left side, Europe in the center, and Asia on the right. This was a tactical error, for it challenged the concept of China as the Middle Kingdom, and the map understandably did not win immediate acceptance. Fortunately the mistake was an easy one to set right; with the new knowledge that the earth is round rather than flat—as the Chinese still believed—Ricci revised the map, placing China at the center. The map then won high praises from the Chinese and was widely circulated in the country, earning much credit for the missionaries.

Ruggieri, on his part, had written in Latin a work of apologetics which he called a catechism. With the help of Ricci and a Chinese scholar it was translated into Chinese and published in 1584 under the title *T'ien-chu sheng-chiao shih-lu* (A true account of God and the Sacred Religion). This work, probably the first Christian literature to appear in China, discussed the existence and the attributes of God, the immortality of the soul, the natural law, the sacrament of baptism, etc.

Though they discussed religion with their Chinese visitors with full integrity, Ruggieri and Ricci were careful to present themselves initially as scholars and scientists; conversion naturally was the ultimate end, but they realized that it could not be hurried until the ground had been prepared. Ricci successfully established himself as a learned scholar of Chinese culture, maker of the famous world map, teacher of mathematics, astronomy, and other scientific truths, and only lastly as a missionary of Catholicism. After spending fifteen years in Chaoching and Shaochow and five years in Nanchang and Nanking, and having made many friends among leading Chinese scholars and officials, Ricci went to Peking in 1601, to seek imperial patronage. He presented to the throne many gifts, including a reproduction of the famous painting of the Madonna attributed to St. Luke, a painting of the Blessed Mother with the infant Jesus and John the Baptist, a Roman breviary, a reliquary in the form of a cross, two glass prisms, a spinet, two clocks, and a map of the world. Accompanying the gifts was a memorial in highly polished Chinese drafted by a friend and admirer.[7] In the memorial, beautifully copied out by the best calligrapher available, Ricci stated that, attracted by Chinese

7. Liu Tung-hsing, who held the title of president of the Board of Public Works and president of the Censorate.

civilization to leave his own country, he had spent three years traveling to Kwangtung, had studied the Chinese language and the classics, had spent twenty years in various parts of China; and had now come to the capital to pay his respects and to offer his service. He also wrote that as a religious who had never married he had no family burden and sought no favor; that he was an established scholar in his own country well conversant with astronomy, geography, map-making, calendar, and mathematics; that it would indeed be a great honor if the emperor chose to enlist his service. Ricci and his associates were well treated in the Residence for the Tributary Envoys. They never saw the emperor in person, but that they were allowed to reside in Peking implied imperial sanction of Christian activities in China.

Ricci's strategy was pacific penetration, cultural adaptation, and the avoidance of needless conflicts with Chinese prejudices and suspicions—methods calculated to win good Christians rather than a horde of indifferent baptisms. To his realistic mind, God's grace did not operate in a vacuum but through human instrumentalities. It was not improper, therefore, that he compose the motets for the eunuchs, since to do so was to help the cause of Christianity. With this approach, Ricci quickly won friends and admirers among prominent officials and scholars in Peking, among whom were dignitaries no less than a grand secretary,[8] a president of the Civil Office,[9] and a president of the Board of Rites.[10] The most famous converts were of course (Leo) Li Chih-tsao (d. 1630), a director in the Board of Public Works, and (Paul) Hsü Kuang-ch'i (1562-1633), a member of the Hanlin Academy who later rose to be a grand secretary.

Ricci's fame as a scholar and scientist drew a great number of admirers to his residence to hear him discourse on science, philosophy, and religion. From twenty to a hundred visitors came to him daily, and the street outside his home was always filled with carriages. The last nine years of his life were especially gratifying: he had become so familiar to—and popular with—the Chinese that they ceased to regard him as a foreigner but rather as one of themselves. His fame and popularity came at a price, however; the schedule he was forced to maintain in Peking eventually undermined his health, and in 1610 he died. His work was carried on by generations of dedicated Jesuits.[11]

Catholicism in China enjoyed a boom: in 1640 the total number of

8. Shen Yi-kuan.
9. Li T'ai-tsai.
10. Feng Ch'i.
11. Among whom the most famous were Diego de Pantoia, Sabbathinus de Ursis, Julius Aleni, Adam Schall von Bell, Ferdinand Verbiest, Thomas Pereira, Jean-François Gerbillon, and Michel Benoist.

converts was between 60,000 and 70,000 and by 1651 it had risen to 150,000.[12] In 1642 fifty of the high-ranking ladies in the palaces became Christians, and when Prince Kuei was established emperor to continue the fight against the Manchus, his empress, the crown prince, both dowagers, and several high officials all accepted Catholicism. One of the dowagers—christened Helena—even sent a message to Pope Innocent X in 1650, and another to the general of the Society of Jesus, asking for prayers for the Ming cause.[13]

The Ch'ing dynasty was not vindictive toward the Jesuits for their services to the Ming court; in fact, Emperor Shun-chih, who was religiously inclined, was very much drawn to the missionaries for several years. He was fond of Adam Schall von Bell, whom he appointed court astronomer and to whom he gave numerous favors and honors. In 1653 Adam Schall was given the title "Master of Universal Mysteries" and in 1657 the title "President of the Imperial Chancery." Between 1656 and 1657 the emperor visited him twenty-four times, even celebrating his own birthday in 1657 in Schall's home. The following year Schall was made a mandarin of the first class with the title of Imperial Chamberlain.

A setback came during the early years of K'ang-hsi's reign when the Regent Oboi, sharing the native technicians' resentment of foreigners holding high posts in China, replaced Schall with the Chinese calendar-maker Yang Kuang-hsien as the court astronomer. However, after Emperor K'ang-hsi had begun his personal rule, he dismissed Yang in 1669, and appointed F. Verbiest as court astronomer. Keenly interested in Western science and mathematics, K'ang-hsi often asked the Jesuits to lecture on these subjects. It was a period of triumph for Western learning and for Christianity in China: churches were established in various parts of the country and the number of converts grew steadily. The prospect was indeed promising, but before long there emerged elements of discord from within the Church that would bring ruin.

THE DECLINE OF THE JESUIT INFLUENCE: THE RITES CONTROVERSY. As a matter of strategic policy Ricci and his followers had avoided conflict with Chinese sensibilities and customs as long as these customs did not contradict the basic teachings of the Church. They accepted established Chinese terms to express Christian ideas, related Confucian moral concepts to Christian teachings, refrained from interfering with Chinese rites honoring Confucius and one's ancestors, and allowed the converts to perform the kowtow as a form of civil obeisance. In short, they accepted the prin-

12. Dunne, 212, 314.
13. The Polish Jesuit, Michael Boym, was the courier.

ciple of cultural accommodation and rejected the spirit of "Europeanism" prevalent among many other religious orders since the Age of Discovery.

The "Europeanist" orders, such as the Franciscan and the Dominican, looked upon non-Christian cultures as the work of the devil and tolerance of these cultures as betrayal of Christian principles. Their missionaries, attacked the Jesuits for compromising the integrity of the Roman Catholic faith and for misleading the Christians in China. The Franciscan and Dominican monks would not make concessions to Chinese susceptibilities or to local conditions, but sought to impose upon the converts the doctrine of faith as well as all ecclesiastical laws and customs as observed in Europe and in the Spanish possessions.

In 1634 Francisco Dias, a Dominican, and Francisco de la Madre de Dios, a Franciscan, arrived in China, and were joined three years later by another Franciscan, Gaspar Alexda. They were horrified by what they saw: the painting of Christ and the Twelve Apostles, hung in the Jesuit chapel in Peking, showed the figures with shoes on their feet; and the chapel itself maintained two "altars," one for the Christ and the other for the emperor. They charged the Jesuits with distorting the picture of Christ and the Apostles and of raising the pagan emperor to a rank co-equal with the Savior's. They took no cognizance of Chinese feelings on the subject of bare feet; nor did they care to know that the chapel was a gift of the emperor in honor of Ricci, and that in acknowledgment of the imperial grace the Jesuits had placed a wooden plaque on a table with an inscription equivalent to "Long live the emperor." The Dominicans and the Franciscans attacked the Jesuits for their (1) improper use of Christian terminology, (2) tolerance of questionable Chinese rites honoring ancestors, the recently deceased, and Confucius, (3) refusal to say that Confucius was in hell, and (4) failure to promulgate the church laws and to preach the crucifixion of Christ.

The question of expressing Christian ideas in the Chinese language was a major point of contention. Ricci had chosen to find the closest existing Chinese concepts to express Christian ideas rather than to transliterate Western terms. The Latin word "gratia," for example, was not transliterated as "ke-la-chi-a" but rendered *t'ien-en* (heavenly grace) or *sheng-en* (sacred grace). He equated the Chinese concept of *t'ien* (heaven) with the Christian concept of God; hence *T'ien-chu* (Lord of Heaven) or *Shang-ti* (Lord on High) for God or Lord, *t'ien-shen* for angel, and *ling-hun* for soul.[14]

14. This approach, it might be added, ran into some opposition even within his own order: a few Jesuits themselves, such as Longobardo, would rather have had a transliterated form of the Latin *Deus* for God.

The problem of rites was even more controversial. Ricci and the Jesuits regarded the rites performed before ancestral tablets as expressions of reverence and respect. They permitted the keeping of a tablet in the house with names of the family ancestors inscribed on it surrounded by flowers, candles, and incense. They did not believe that the Chinese supposed their ancestors to dwell in these tablets; ancestral rites were accepted as nothing more than a civil act expressing one's respect and filial piety, with superstition having very little part. Likewise, Ricci considered the burning of incense a social custom without religious connotation. The kowtow before the ancestral tablets and the coffins of the recently deceased, which to many missionaries of other orders was an act of adoration which should be reserved for the Divinity alone, was construed by the Jesuits as a symbol of ceremonial obeisance and civil rites intended to console the afflicted and to manifest sorrow, without religious significance. They saw nothing religious or blasphemous in children kowtowing to their parents or officials to the emperor; the Jesuits themselves performed the kowtow when receiving imperial gifts or emissaries. They also tolerated simple ceremonies in honor of Confucius during the conferring of the *hsiu-ts'ai* degree as an expression of customary obeisance to the Master. Past a certain point, however, the Jesuits drew a line: in matters of solemn ceremonies in honor of Confucius, Chinese converts were not permitted to take part because such occasions took on the appearance of a sacrifice. Solemn rites observed in honor of familial ancestors were permitted only on condition that there be no burning of paper money, no prayers or petition to the dead, and no expression to the effect that the spirits of the dead derived sustenance from the food offerings.[15]

The Jesuits had come to these decisions because they believed that the majority of the Chinese scholar class observed ritual as a part of good citizenship. Prohibition of the rites would make it impossible for them to become Christians and would render the policy of peaceful penetration unworkable: Christianity, instead of becoming a leaven working quietly from within the Chinese society, would be hostile to the Chinese way of life. Accordingly, the papal decree of 1656 allowed the practice of rites under the conditions observed by the Jesuits.

However, the matter of rites was not put to rest; it continued to plague the European intellectual and religious world and became a *cause célèbre*, with leading theologians and philosophers taking part in the debate. More than 262 works were published on the subject, in addition to a hundred or so that remained unpublished. In 1704 the Pope reversed the church's

15. Dunne, 292.

stand and banned the rites, prohibiting the use of *T'ien* or *Shang-ti* for God, while approving the term *T'ien-chu* (Lord of Heaven).

In 1715 Pope Clement XI issued the bull *Ex Illa Die* to reaffirm the anti-rites stand and warn violators of excommunication. It further strained the "Emperor-Pope" relations. K'ang-hsi decided that to avoid further complication all missionaries would be repatriated except those who were scientists and technicians, such as court astronomers. Although the order was not strictly enforced, the status of the missionaries was slipping badly, and during the Yung-cheng period it deteriorated further. The new ruler was unfavorably disposed toward the Jesuits for their support of K'ang-hsi's ninth son against him in the succession struggle. He stated: "China has its religions and the Western world has its religions. Western religions need not propagate in China, just as Chinese religions cannot prevail in the Western world." Prohibition of Christianity was much more strictly enforced under Emperor Yung-cheng.

In 1742 Pope Benedict XIV reiterated the anti-rites stand, and the missionaries in China were put in an extremely difficult, if not impossible, position. Their work and influence fell to a low ebb. And with the dissolution of the Society of Jesus in 1773, the moving spirit of Catholicism in China was gone.

INTRODUCTION OF WESTERN SCIENCE AND TECHNOLOGY. From the late Ming period until the middle Ch'ing, a total of about 500 Jesuits came to China, of whom 80 made substantial contributions to cultural exchange. From them the Chinese learned the Western methods of cannon-casting, calendar-making, cartography, mathematics, astronomy, algebra, geometry, geography, art, architecture, and music. At the same time, the Jesuits introduced Chinese civilization to Europe. It was the initial meeting of China and the West in modern times, and provided China with the chance to modernize itself. Some of the major Jesuit contributions to science and technology were as follows:

1. Cannon-making. From the Dutch the Chinese first learned about the cannon, which they promptly dubbed *Hung-I p'ao* (Cannon of the Red-haired Barbarians). Too proud to adopt this foreign weapon, they were taught a lesson by the Japanese invaders of Korea in 1592-97, who used the cannon. Later, threatened by the rising power of the Manchus, the Ming court swallowed its pride and in 1622 sent for the Jesuits in Macao to cast guns for its army. The ban on Catholicism was, perforce, automatically relaxed. In 1642 Adam Schall was asked to cast guns and teach the technique to the Chinese officials in charge of gun-making. He made twenty probational guns and, having won imperial praises for them,

was commissioned to make 500 more. Schall also wrote a book on the manufacture and operation of guns, balls, mines, and rockets. Under his instruction, (Paul) Hsü Kuang-ch'i, (Leo) Li Chih-tsao, and a few others among the Chinese mastered the technique of gun-making.

2. Calendar-making. Besides the cannon, the Jesuits brought to China new knowledge of astronomy and calendar-making. The two existing Chinese almanacs—the Grand Calendar (*Ta-t'ung*) of Liu Chi based on a Yüan calendar, and the Moslem calendar (*Hui-li*)—were both exposed by Ricci as inaccurate and obsolete. Several of Ricci's associates were well conversant with the technique of calendar-making. In 1629, on recommendation of (Paul) Hsü Kuang-ch'i, the court appointed the Jesuits Longobardo and Terrenz to the Calendrical Bureau (*Li-chü*). When Terrenz died in the spring of 1630, Adam Schall was named as his successor and proved to be even more skilled than his predecessor.

The missionaries made astronomical instruments and directed Chinese officials to translate astronomical and logarithmic tables. During the eclipse of 1643, their calculations proved much more accurate than those of the official astronomer, and consequently the court agreed to accept the Jesuit calendar; but there was secret opposition among the courtiers.

When the Ming dynasty was replaced by the Ch'ing, the first emperor of the new regime, Shun-chih, appointed Schall court astronomer, and the Calendrical Bureau merged with the Imperial Board of Astronomy. The Jesuits holding these positions were given stipends and official residences, and Schall enjoyed the trust and respect of the emperor for several years, as noted before. A calamity befell him, however, when he presented to Emperor K'ang-hsi a 200-year calendar. The antiforeign astronomer Yang Kuang-hsien accused Schall of implying by this act that the dynasty could last only 200 years; using this pretext as a *point d'appui*, Yang went on to charge Schall with errors in his astronomical calculations and with indoctrinating people with false ideas. Oboi, the dictatorial regent during K'ang-hsi's minority, pronounced Schall's action "highly improper" and threw him into prison late in 1664; his life was spared only because the empress dowager interceded. Yang now became the court astronomer and the old calendar was revived, but shortly thereafter he made an erroneous calculation of the solar eclipse. Though released in May 1665, Schall, now old and paralyzed, died a year later. The post of the court astronomer, after Yang's downfall in 1669, went to Verbiest, and from that time until 1838 it remained occupied by foreigners exclusively.[16]

3. Geographic survey and map-making. Under the patronage of K'ang-

16. Yang made a defense of his stand in a treatise called *Pu-te-i* (*I could not do otherwise*) in which he bluntly stated that "he would rather have no good calendar than have foreigners in China" and that "it is exactly because of their excellent instruments

hsi, Joachin Bouvet, a French Jesuit, led a group of missionaries in a geographical exploration of the empire between 1708 and 1715. From the data they gathered, an atlas of China was completed in 1716, with detailed maps of the provinces. K'ang-hsi proudly bestowed on it the title *Huang-yü ch'üan-lan t'u* (A complete map of the Imperial Dynasty). This was the first map of China to be marked with the longitudes and latitudes.

4. Other activities. The Jesuits introduced a number of other elements of Western learning, too. Ricci and (Paul) Hsü Kuang-ch'i translated Euclid's *Elements of Geometry,* and Ricci and (Leo) Li Chih-tsao translated a work on mathematics. Rho and Hsü translated Archimedes' plane and spherical trigonometry; Aleni wrote on geometry, trigonometry, and geography; Terrenz wrote on human physiology; T. Pereira wrote on music; and Schall wrote on the principle of the spectrum and telescope. Aristotelian philosophy and perspective drawing were also introduced.

By the same token, the missionaries transmitted Chinese learning back to Europe. An Italian translation of the Confucian Four Books by Ricci was followed by a Latin version by Ignatius de Costa, Prosper Intorcenta, and Philippus Couplet, published in Paris in 1687. To the Pope in 1682 Couplet presented Jesuit translations of more than 400 Chinese works. For the first time Europe was learning something of the richness of Chinese culture, and great scholars and thinkers such as Spinoza, Leibniz, Goethe, Voltaire, and Adam Smith became admirers of Chinese civilization.[17] During the Age of Enlightenment, the Chinese rational approach to life and the secular government—totally divorced from church—won praise from Voltaire, Holbach, and Diderot. In art, the Rococo movement, which liberated Europe from the stilted Baroque art form of Louis XIV, was also in part a product of Chinese influence. Chinese porcelain was imitated by the Italians, the Dutch, and the Germans. French brocade with Chinese designs became a vogue. The Chinese garden, with its stone bridges, artificial hills, and goldfish, was much admired, and the Garden at Kew of the Duke of Kent was particularly famous for its good Chinese taste.

CHINA'S LOST CHANCE FOR MODERNIZATION. In spite of the samplings of Western civilization which they carried in, the Jesuit had not been the

---

and excellent weapons that they are a potential enemy." (Hsiao I-shan, I, 680-81) The apparently unreasoning petulance of these xenophobic outbursts can be understood in the light of the Spanish conquest of the Philippines and the rapid rise of the Catholic influence in Japan during the early Tokugawa period.

17. David E. Mungello, *Leibniz and Confucianism: The Search for Accord* (Honolulu, 1977).

catalysts for modernization in China. The missionaries represented a thin ray of Western learning which shone feebly among a small coterie of progressive Chinese scholars and officials, but it never penetrated beyond the surface. They caused at best a slight tremor in the otherwise immutable Sinic civilization. Chinese scholars and officials on the whole were too proud of their cultural heritage to admit of the need for foreign learning.

Moreover, the Jesuits who came with this new knowledge of science and technology were basically men of religion rather than men of science. Apart from the several dozen very gifted, most of the missionaries were limited in their capacity as cultural transmitters. Rather than presenting a broad front of European civilization, they merely introduced a few branches of Western science that happened to attract Chinese attention. Even this partial introduction was interrupted when the missionary movement was put to an end in the 18th century. Thus, the feebleness of the Jesuit efforts, the ethnocentric complacency of Confucian intellectuals, and the imperviousness of Chinese culture to outside stimuli inhibited any process of modernization of China at this point.[18]

Ironically, it was precisely after this disruption of Western learning in China that great progress was made in Western political, economic, social, and scientific fields. The American Revolution, the French Revolution, and the great reforms in England set the stage for the rise of modern democracy, while the Industrial Revolution ushered in a new age of technological development, nationalism, expansion, capitalism, and imperialism. The shibboleth of progress permeated the air of Europe. In contrast, Chinese intellectuals still looked to their "golden past" for guidance and absorbed themselves in antiquarian textual research. Europe surged ahead in its search for progress, while China slept in its dream of glory. To jostle China out of its sleep would require efforts far more bombastic and powerful than the Jesuits had been able to provide. Britain, the front-runner in the Industrial Revolution, unhesitatingly accepted the challenge, as we will see in later chapters.

18. The limitations of the missionaries were only too obvious to themselves. Repeated efforts were made by Sabbathinus de Ursis and Nicolo Longobardo to secure the services of famed astronomers and mathematicians, but the General of the Society felt that Europe could not spare such talents. In rejecting the applications of three noted mathematicians for missionary work in China—Gregory St. Vincent (1584-1667), Christopher Scheiner (1575-1650), and John Cysat (1588-1657)—the General stated that "for the greater glory of God and for the good of Society it was preferable for them to stay in Europe, and energetically promote mathematical studies. Thus, they will be able to do by means of their disciples in China what they will not be able to do themselves." For details, see Pasquale M. D'Elia, *Galileo in China* (Cambridge, Mass., 1960), 21-24.

The Russian Advance

At about the same time the Western Europeans were reaching China via the sea route, the Russians were marching toward it across Siberia. China's confrontation with the European world, then, was on two fronts. The sea-faring European explorers and traders from the south and the land-based Russians from the north closed in on the previously impervious empire like a pair of pincers, and China's destiny was never afterwards the same.

THE MARCH ACROSS SIBERIA. The conquest of Siberia was largely the work of Russian explorers, adventurers, hunters, and trappers. By the middle of the 16th century the Russians had reached the Urals and won submission from some of the tribal chieftains such as Ediger (khan of Sibir) and Kuchun Khan, and in 1554 Ivan the Terrible had assumed the title "Lord of all Sibir." In 1558 the rich merchant family of Stroganov secured permission from the tsar to explore beyond the Urals. In 1581 Yermak (Vasili Timofeiev), a onetime brigand and a chieftain in Stroganov's private army, led 800 Cossacks in an eastward drive and reached the Irtysh the following year, taking possession of the town of Sibir, from which the name Siberia was derived. The territory he occupied was offered to the tsar, along with a tribute of furs which he sent to redeem his past misconduct. Awarded a medal and made a hero, Yermak carried on his trek along the Irtysh and the Ob. He was drowned in 1584 but the march went on. In 1587 the town of Tobolsk was established and in 1590 the Russian government settled 3,000 peasant families in West Siberia. In 1604 Tomsk was founded on the Ob as the seat of government for Siberia, and in 1619 Yenisseisk was established. By 1628 the Cossacks had reached the Lena River in East Siberia. In 1632 and 1638 were founded Yakutsk and Okhotsk, respectively, from which a number of expeditions were sent still further east. Reaching Kamchatka and what is today's Bering Strait in 1648, the Russians had completed the march to the Pacific within seventy years of Yermak's drive from the Urals (1581), conquering more than four million square miles. In 1651 Lake Baikal was reached and the town of Irkutsk founded.

From the Siberian tribes the Cossacks learned of the rumored riches of the Amur Region—the "Eldorado of Eastern Asia"—where gold and silver, cotton and silk, cattle and grains, were said to abound. A number of exploratory expeditions were sent into this promising land. In 1658 Pashkov, the governor of Yenisseisk, penetrated to the Shilka, a tributary of the Amur and founded Nerchinsk. In 1666 Nikitor Chernigovskii, a Polish exile, built the fort of Albazin (Ya-k'e-sa or Yacsa), and was appointed governor by the tsar in 1669. Now firmly entrenched in the Amur Region, the Cossacks, known to the Chinese as the *Lo-ch'a,* were determined to push further into Manchuria.

The advance of the Russians coincided with the rise of the Manchus in China. The founders of the Ch'ing dynasty, though troubled and worried about the Russian threat, had to postpone any large-scale punitive action because of their preoccupation with the conquest of China and the consolidation of the dynasty. It was not until the suppression of the "Revolt of the Three Feudatories" (*San-Fan*) in 1681 that Emperor K'ang-hsi was free to deal with the Russian problem.

EARLY DIPLOMATIC MISSIONS TO CHINA. Along with their conquest of Siberia and penetration into the Amur Region, the Russians sent a number of exploratory and diplomatic missions to China. Russian knowledge of China at this point was pitifully limited; some believed that China was neither big nor rich—"completely surrounded by a brick wall, from which it is evident that it is no large place."[19] In 1618 Tsar Mikhail Theodorovich dispatched Ivan Petlin of Tomsk to Peking to seek information about this strange country. He brought no tribute and was refused an audience.[20]

The first Russian ambassador, Feodor Baikov, was sent off in 1654 to find out the best routes to reach China, the distance involved, and the kinds of goods the Chinese had to trade. He was also to look into the Chinese military and economic strength, and to investigate the local agricultural products and precious stones. He brought a letter from the tsar to the Bogdikhan (Great Khan), a designation for the Manchu emperor which the Russians had learned from the Mongols. Baikov was instructed to present the letter to no one but the Bogdikhan himself and during the audience he was not to perform the kowtow. Somewhat amusingly, his instructions went on to state: "he is by no means to kiss the Bogdikhan's foot [as the Chinese do]; but if he is called upon to kiss hands he need not refuse."[21] Because of his adamant refusal to comply with Chinese ceremonies, Baikov was not granted an audience, and his gifts were peremptorily returned. Though a failure as a diplomat, Baikov brought back "valuable" information on China.

In 1675 the flamboyant envoy, Nikolai G. Spathary, whose original family name was Milescu, was sent to find out the routes to China, feel

19. John F. Baddeley, *Russia, Mongolia, China* (London, 1919), II, 67-68.
20. *Ibid.,* 83.
21. Baddeley, 134, 442.

out its reaction toward relations with Russia, and learn something about the type of people dwelling between Siberia and China. Spathary was a man of erudition, prepared to defend the honor of his master without compromise. He would not hand over the tsar's letter to Chinese officials but insisted on its personal delivery to the emperor. He refused to have his gifts designated as tribute. After twenty-six days of contention with the Board of Rites, he finally gave in and performed the kowtow to K'ang-hsi, who then favored him with a dinner in the palace. Spathary was in Peking for three and a half months without accomplishing his mission, but he did learn the important information from the Jesuit, F. Verbiest, that K'ang-hsi would go to war to destroy Albazin and Nerchinsk.[22]

THE TREATY OF NERCHINSK, 1689. With the "Revolt of the Three Feudatories" put down (1681), the Ming loyalist movement on Taiwan suppressed (1683), and the dynastic hold on the country made fast, K'ang-hsi was ready to take on the Russian problem. In 1685, after several years of elaborate preparation, General Pengcun marched from Tsitsihar with 10,000 soldiers, 5,000 sailors, and 200 pieces of artillery. Against these numbers, the 450 Cossack defenders under Aleksei Tolbuzin stood very little chance, and were, predictably, completely overpowered. Forty-five Russians were taken prisoner, and Albazin was reduced to ashes. Tolbuzin, however, managed to escape to Nerchinsk.[23]

After having leveled Albazin, Pengcun returned home; but Tolbuzin, with the help of 336 Cossacks, soon re-established himself on the ruins. New fortifications were erected, and in March 1686, Tolbuzin renewed his raids on the Amur. K'ang-hsi once again dispatched an expedition to Albazin. This time the Russians resisted the Chinese siege for more than a year, but, as before, the sides were hopelessly mismatched. Tolbuzin was killed in action and many of his men died of disease. At last, in mid-1687, when only 66 Cossacks were left and when one more concerted attack would have given Albazin to the Chinese, K'ang-hsi suddenly ordered the siege lifted. His general Sabsu even offered provisions to the starving Cossacks. Ostensibly, the emperor was acting in deference to news from the tsar that a diplomatic mission was on its way, actually he had been looking for a chance to win Russian good will. K'ang-hsi did not want to goad the Russians into an alliance with the still unpacified West-

22. Baddeley, 395-411.
23. Ho Ch'iu-t'ao (ed.), *Shuo-fang pei-sheng* (A manual of northern places), (Peking?, 1881), 6:16b-17.

ern Mongols, the Ölöd. Moreover, China needed a rest after years of internal campaigns expended against the Three Feudatories: it was not in its own interest to prolong hostilities against the Russians.

Nor was Russia in any position to make war: Peter, not yet "the Great," was in his early teens sharing a shaky throne with his invalid brother; the country itself was preoccupied with military affairs in the Baltic; and the treasury was depleted by military expenses and internal economic depression. It was far preferable for Russia to follow a policy of *peaceful penetration* of China through trade contact than one of naked aggression and *territorial expansion* along the Amur. For this reason peace with China was imperative; hence the diplomatic mission. It was dispatched under Fedor A. Golovin, the son of the Tobolsk governor. He was instructed to meet the Chinese at Selinginsk and to try to set the boundary line along the Amur and Bystra, or, failing that, at least along the Amur and the Dzeya. On October 22, 1687, Golovin reached Selinginsk.

The Chinese delegation was led by Prince Songgotu (So-o-t'u) and a number of high dignitaries, and with them as interpreters two Jesuit priests, Jean-François Gerbillon and Thomas Pereira. They left Peking in May 1688, but found the road to Selinginsk blocked because of the Ölöd chieftain Galdan's invasion of the Eastern Mongols, the Qalqa (Khalkha). Nerchinsk was then selected as an alternative site for the diplomatic negotiations. Emperor K'ang-hsi, anxious to win Russian good will and prevent a unity between Galdan and the Russians, had instructed Songgotu that China might grant Nerchinsk to Russia and accept the boundary at the Argun River. At the conference, Golovin proposed to fix the boundary along the Amur, while Songgotu pressed for Russian evacuation of Nerchinsk and Albazin as well as relinquishment of the land beyond Selinginsk. With both sides insisting on their demands, a deadlock resulted. The two Jesuit priests shuttled between the two camps as mediators, while the Chinese delegation threatened to use force.[24] They had 10,000 soldiers and 90 armed vessels as support, while Golovin had only 1,500 troops at his disposal. The Russians gave in at last. The Treaty of Nerchinsk (Ni-pu-ch'u) was signed on September 7, 1689; it was drawn up in five languages: Chinese, Russian, Manchu, Mongolian, and Latin—the Latin version serving as the official text. It contained six articles:

1. The Siberian-Manchurian border would be set along the Argun, would continue along the Amur to the mouth of the Kerbechi, and along the Outer Hsing-an (Stenovoi) Mountains to the sea.

24. For a study of the role of the Jesuits in the treaty negotiations, see Joseph Sebes, S.J., *The Jesuits and the Sino-Russian Treaty of Nerchinsk (1689)* (Rome, 1961).

2. Albazin would be demolished and its Russian residents repatriated with their properties, and hunters who transgressed the boundary line would be punished.
3. Subjects of the two countries wtih passports could freely enter each other's territory for trade.
4. Deserters and fugitives would be extradited, and under no condition given refuge.
5. Citizens of either country now residing in the other should be allowed to remain.
6. With this peace settlement all past incidents should be disregarded.

This treaty—China's first such agreement with a "Western" power—was reached on the basis of equality between China and Russia, and on the whole both countries found it satisfactory. Russia gained control of Nerchinsk and some 93,000 square miles of undecided territory and was given a number of commercial privileges into the bargain, while China had the satisfaction of seeing the Russian problem at Albazin eliminated and the likelihood increased that Russia would remain neutral during China's struggle with Galdan. There was a noticeable gap in the treaty, however; the frontier between Mongolia and Siberia remained unsettled, for Golovin insisted that the issue was beyond his authority to negotiate. It was apparent that Russia was hedging from any settlement of this problem, because the Ch'ing dynasty was not in full control of Outer Mongolia.

With the Treaty of Nerchinsk signed and out of the way, K'ang-hsi turned to the problem of Ölöd. Several years of warfare followed, in which the imperial forces sustained considerable losses, both of men and ground—at one point Galdan penetrated as far as Ulan Butung, within 80 leagues (240 miles) of Peking. But the fighting remained for the most part inconclusive, neither side gaining much advantage. Finally, determined to crush Galdan, K'ang-hsi after several years of preparation sent out a grand expedition of 80,000 men in 1696. On June 12, Galdan was drawn into a confrontation at Jau Modo. His horsemen were completely overpowered by the Ch'ing artillery and musketeers, and Galdan, too proud to surrender, fled with a handful of followers. He died the next year, 1697, of a sudden illness—possibly he took his own life with poison. K'ang-hsi thus extended his rule to Outer Mongolia and Hami, and laid the foundation for his grandson, Emperor Ch'ien-lung, to complete the conquest of the entire Western Region (*Hsi-yü*)[25] in the 1750s.

The most significant fact about the Treaty of Nerchinsk was that it regularized Sino-Russian relations. A number of Russian trade and diplo-

25. Renamed Sinkiang, or New Territory, in 1768.

matic missions entered China after its enactment. The mission under
E. Izbrandt Ides in 1693 won permission for Russian caravans to be sent
to Peking once every three years. These caravans were restricted to 200
men and their stay was limited to 80 days; their goods—whether imports
or exports—were exempt from customs duties. Between 1698 and 1718
ten such caravans made the journey. In 1720 a diplomatic mission under
Ambassador Leon V. Izmailov arrived in Peking. The envoy performed
the full kowtow—three kneelings and nine knockings of the head on the
ground—on condition that future Chinese envoys to Russia would likewise conform to Russian court ceremonies.[26] He was favorably treated by
K'ang-hsi, but his request for the extension of trade and the establishment
of a consulate-general in Peking did not evoke a sympathetic response.
After three months in Peking he returned home, leaving his attaché Lo-
rentz Lange to continue the negotiations; seventeen months later (in
1722) Lang was expelled for arrogant behavior.

TULISEN'S MISSION TO THE TŪRGŪD TRIBE IN RUSSIA, 1714. If Russia sent
a number of missions to China, China also sent a couple to Russia. The
first of these was not to the court at St. Petersburg, however, but to the
Turgud tribe on the Volga. The Tūrgūd were an Ölöd tribe which originally lived in the Tarbagatai area but had migrated to Russia in 1630. By
1654 they had become Russian vassals, although their chieftains continued to send periodic tribute to China. In 1712 a Tūrgūd tributary mission came to Peking from the chief Ayüki, whose daughter was married
to the new Ölöd leader, Cewang Arabdan, Galdan's nephew. K'ang-hsi
decided to send a return mission, ostensibly to express his appreciation of
Ayüki's loyalty but in fact to strengthen China's ties with the Tūrgūd and
to forestall any alliance between Ayüki and Cewang Arabdan. Very likely
K'ang-hsi also wanted to persuade the Tūrgūd tribe to return to China (as
eventually it did, in 1770-71).

The mission was put under the charge of Tulisen, an assistant reader
in the Grand Secretariat. He left in 1712 and passed through Mongolia
and Siberia, where he was well received by the governor of Siberia. In
June 1714 the mission reached the Volga, where Tulisen and Ayüki met.
Except that their exchange of good will was friendly and that they talked
about returning Ayüki's nephew from China, little is known of their

26. Ripa, 105-7.

meeting. Tulisen returned with an account of his travels, entitled *I-yü lu* (Description of a foreign land).[27]

THE TREATY OF KIAKHTA, 1727. Tulisen's mission apparently strengthened China's ties with the Tūrgūd and possibly prevented an alliance between Ayüki and Cewang Arabdan. But the Ölöd threat to the Ch'ing dynasty was still present. The traffic that persisted between Cewang Arabdan and the Russians caused renewed fear among the Chinese that there might be secret plotting between them. In this light, the question of fixing the boundary between Outer Mongolia and Siberia—an issue which the Treaty of Nerchinsk had left unresolved—became doubly important. The new emperor in China, Yung-cheng, carrying on his father's policy to isolate Mongolia from Russia, decided it was necessary to settle all pending issues with Russia in a new treaty and thereby remove any excuse it might use for aiding the Ölöd or entering into an alliance with them.

The Russians were also anxious to settle a number of issues with China, such as frontier delimitation, extension of the overland trade, and the establishment of a religious mission in Peking. On pretext of congratulating Yung-cheng on his ascension to the throne in 1723, Catherine I, who succeeded Peter the Great in 1725, sent Sava Vladislavich Ruguzinskii as envoy extraordinary to China. The embassy, 100 men strong and escorted by 1,500 soldiers, arrived in Peking on October 21, 1726, after thirteen months of travel. Sava, a man of tact, patience, and vision, met with the Chinese negotiators—Tulisen and three others—thirty times in the six months between October 1726 and April 1727. The French Jesuit, Parrenin, served as a liaison between the two delegations and kept Sava informed of the current sentiments in the Chinese camp. Since there was no precedent for signing treaties in Peking—the first treaty with Russia was signed at the frontier town of Nerchinsk—the delegations moved to the border of the Boura River, a tributary of the Selinginsk, and there concluded a preliminary convention known as the Boura Agreement, which, when put into its final form on October 21, 1727, became the Treaty of Kiakhta. The important terms of this eleven-article treaty were as follows:

1. The Mongolian-Siberian frontier would be delimited by a joint Sino-Russian commission. The boundary was to run from the Sayan Mountains and Sapintabakha in the west to the Argun River in the east. The area from the Uda to the Stone Mountains in the east

27. Tulisen, *Narrative of the Chinese Embassy to the Khan of the Tourgouth, 1712-1715,* tr. by George L. Staunton (London, 1821).

was to remain undecided because of the lack of accurate information about it, but elsewhere the commission would demarcate the boundary on the spot.

2. In addition to the existing trade at Nerchinsk, the Russians were allowed to trade at Kiakhta on the frontier.
3. Deserters and fugitives from either country would be extradited.
4. Russian caravans of not more than 200 men would be allowed to come to Peking once every three years, free from import and export duties.
5. Russia would be permitted to maintain a religious mission with its own church in Peking, and Russian priests and students would be permitted to live in Peking.
6. Communications between China and Russia would bear the seals of both governments—in China that of the Li-fan yüan, and in Russia that of both the Senate and the governor of Tobolsk.

In the territorial settlement China lost some 40,000 square miles between the Upper Irtysh and the Sayan Mountains and in the area south and southwest of Lake Baikal, but it gained the security of seeing Russia partitioned off from the tribes of Mongolia. On the other hand Russia gained a number of trade concessions and the authorization to open a religious mission in China, but limited its border trade, hitherto conducted freely with the Mongols, to Nerchinsk and Kiakhta.

THE T'O-SHIH AND DESIN MISSIONS TO RUSSIA, 1729-32. Although the Treaty of Kiakhta settled many important issues, it raised a number of new problems because of the increased contacts which it permitted between the two countries. There were constant complaints from Russia of border raids by Mongolian bandits who were stealing horses, camels, oxen, and sheep; there were outcries over debts owed by Chinese merchants to Russian traders. Naturally this caused uneasiness over the question of whether Russia would remain neutral during the Ch'ing campaign against the new Ölöd leader, Galdan Cereng, a son of Cewang Arabdan, who had died in 1727. Seeking reassurance, Emperor Yung-cheng dispatched an embassy to Russia in 1729, a genuine diplomatic mission accredited to the court at St. Petersburg, and, as such, the first ever sent by China to a "Western" state.

The mission was led by T'o-shih, a Manchu titular vice-president of a Board, on the pretext of congratulating the coronation of the Russian tsar, Peter II (1727-30). Upon arrival in Moscow in January 1731, T'o-shih found that the tsar had died and that the new monarch was Anna

Ivanovna (1730-40), a niece of Peter the Great. The Russians greeted him with a thirty-one-gun salute and warmly received him in an imperial audience at the Kremlin. After presenting the Russian court with eighteen cases of luxurious gifts from the Chinese emperor and performing the celebrated kowtow to the tsarina, T'o-shih sent a formal message to the Senate requesting Russian neutrality in China's forthcoming campaign against the Ölöd. More specifically, he asked the Russians: (1) not to take a hostile attitude if the Chinese soldiers inadvertently crossed the Russian border during the attack; (2) to grant China the privilege of hot pursuit if the Ölöd troops fled to Russia; (3) to extradite the Ölöd leaders and nobles to China while keeping their tribesmen in Russia under strict control so that they would not make trouble for China in the future; and (4) to allow a Chinese delegation to visit the Tūrgūd tribe, which had settled in the Volga-Don Valley, and to expedite the repatriation of the tribe to its original homeland. In return, China offered Russia part of the territory that would be seized from the Ölöd.

The tsarina expressed her sincere desire to maintain peaceful relations with China and agreed to allow the Ch'ing forces the privilege of hot pursuit if the Ölöd soldiers escaped into Russian territory. Furthermore, she promised that the Ölöd tribesmen who took refuge in Russia would be kept under strict control, but she refused to extradite their leaders and noblemen to China. Permission was also granted to China to send a delegation to the Tūrgūd once, but no commitment was made on future missions because the Tūrgūd had officially become Russian vassals.

T'o-shih remained in the Russian capital for two months. En route home, he met with the Russian Senate secretary Bakunin at Tomsk and sought Russian help to arrange a Chinese mission to Turkey in an attempt to enlist Turkish support for China's management of her Moslem subjects. Bakunin, however, was noncommittal on this issue.

It is clear that the real objectives of the T'o-shih mission were to neutralize Russia in China's forthcoming war with the Ölöd, to win her cooperation in facilitating the return of the Tūrgūd, and, as a long shot, to secure her assistance in arranging a Chinese mission to Turkey. All these were aspects of a grand strategy of Ch'ing dynastic consolidation over the areas controlled by the Ölöd. Considering these objectives, the T'o-shih mission was only a partial success although it did improve mutual goodwill while reducing the possibility of Russian aid to the Ölöd. As it turned out, Russia was then preoccupied with the War of the Polish Succession and could not have helped the Ölöd anyway. The Ch'ing expeditionary forces, after encountering many difficulties, finally won a battle at Erdeni Tsu in

1732, which enabled China to negotiate a peace settlement but not to resolve the age-old Ölöd problem.

Even before T'o-shih returned home, Peking had decided to send a second mission with proper credentials to the tsarina. Led by Desin, a Manchu vice-president of the Board of Rites, the mission was courteously met at the Russian border but was denied permission to send a delegation to the Tūrgūd. Desin reached the new Russian capital, St. Peterburg, in 1732 and presented his credentials in an impressive audience, in which he performed the kowtow and offered nineteen boxes of expensive gifts from the Chinese emperor. He repeated the same message as T'o-shih and received similar answers from the Russians. For the reception of the two Chinese missions, the Russian court spent the considerable sums of 26,676 and 22,460 rubles, respectively.[28]

RUSSIA'S SPECIAL POSITION IN CHINA. Sino-Russian relations during the early Ch'ing period were markedly different from China's relations with Western European maritime states. Russia, in fact, occupied a very special position in China. It was the only foreign country with which China maintained treaty relations, the only "Western" state to which China sent diplomatic missions, and the only foreign power granted religious, commercial, and educational privileges in Peking. The early Ch'ing rulers recognized that Russian neutrality was essential to China's consolidation of its northern and northwestern frontiers, and that to gain this neutrality it was necessary to grant Russia certain considerations and privileges denied to other foreign states.[29]

Although the Ch'ing court insisted that Russian envoys kowtow to the Chinese emperor, and although Chinese records consistently described Russian emissaries as tribute bearers, Russia was not officially listed as a tributary state in any of the five editions of the *Collected Statutes of the Great Ch'ing Empire (Ta-Ch'ing hui-tien)*. In fact, K'ang-hsi explicitly noted that Russia should not be classified as such: "Although tribute from a foreign country [Russia] would be a magnificent thing, I am afraid that

---

28. Gaston Cahen, "Deux ambassades chinois en Russie au commencement du XVIIIe siècle, *Revue Historique*, 133:82-89 (1920); Li Ch'i-fang, "Ch'ing Yung-cheng Huang-ti liang-tz'u ch'ien-shih fu-O chih-mi—shih-pa shih-chi chung-yeh Chung-O kuan-hsi chih i-mu" (The mystery of two missions sent to Russia by Emperor Yung-cheng: An episode in the mid-18th-century Sino-Russian relations), *Bulletin of the Institute of Modern History*, Academia Sinica, Taipei, Taiwan, XIII:39-62 (June 1984).
29. Immanuel C. Y. Hsü, "Russia's Special Position in China during the Early Ch'ing Period," *Slavic Review*, 13:4:688 (Dec. 1964).

when it is carried on into later generations it may become a source of trouble." On many occasions K'ang-hsi extended Russia the consideration due an independent state. When he sent Tulisen to Russia in 1712, for example, he ordered him "to act in accordance with the ceremonies of that country." No such instructions had ever been given any Chinese emissary to the tributary states; on the contrary, all tributary kings were required to conform to Chinese etiquette when receiving a Chinese envoy. Again, this peculiar deference to Russia was manifested in K'ang-hsi's exchange with the Russian envoy Leon V. Izmailov in 1720. If he complied with Chinese court ceremonies, the emperor told him, and performed the kowtow, the Ch'ing government would see to it that future Chinese envoys to Russia would follow Russian ceremonies and perform whatever rituals were required of them. Upon Izmailov's compliance, K'ang-hsi favored him with a dozen audiences in three months, during which he alluded to Peter the Great as "his equal," "his good neighbor," and "a most great and honorable ruler in possession of a vast territory."[30] No such expressions—indeed, no such sentiments—had ever been applied to China's tributary kings or to any other foreign rulers. When T'o-shih and Desin went to Russia in 1731-32, they actually kowtowed to the tsarina, as no Chinese envoy had done before a tributary king.

One interesting ramification of China's special consideration for Russia was the treatment of Russian prisoners of war. These prisoners, about one hundred in all, taken in several battles before and during the Albazin siege, were pardoned and organized into a unit of the Ch'ing army—the Eleventh Company of the Fourth Regiment of the Manchu Bordered Yellow Banner. As bannermen, they were given the favor of ranks and the privilege of living in quarters by themselves. They received annual pensions and were allowed complete religious freedom. Emperor K'ang-hsi gave them a Buddhist temple, on the site of which they built an Orthodox church known as the Church of St. Nicolas, later renamed the Church of the Assumption. To the Chinese it was known as the *Lo-ch'a miao* (Temple of the Russians), more often called, incorrectly, the Northern Russian Hostel.

Russian traders also fared well. Beginning with the Ides mission in 1693, they were allowed to come to Peking every three years in groups of two hundred, and although they paid their own way, their goods were brought in duty-free. While in Peking they were lodged in the Southern Russian Hostel—the old *Hui-t'ung kuan* (Common Residence for En-

---

30. Gaston Cahen, *Histoire des relations de la Russia avec la Chine sous Pierre le Grand, 1869-1730* (Paris, 1911), 165.

voys) of the Ming dynasty. Officially they were supposed to conclude their business and leave Peking within eighty days, but this regulation was scarcely more than a token. The caravan under Liangusov and Sava-tiev in 1698, for instance, consisted of nearly 300 merchants and 200 secretaries, servants and employees. Between 1698 and 1718 ten such caravans came to Peking, averaging one every two, instead of every three, years as officially stipulated, and they were often permitted to remain in the capital longer than the legal eighty days. At times the Chinese court even advanced loans to distressed Russian merchants.

After the Treaty of Kiakhta in 1727, groups of Russian priests were allowed to come every ten years to minister to the Russians in Peking, and the Chinese government paid their traveling and living expenses. From 1729 and 1859 thirteen of these missions came to the capital. The priests lived in the Southern Russian Hostel, where they maintained a church called the Convent of Candlemas, later renamed the Church of the Purification of the Virgin. After 1729 the priests of the religious mission also conducted services at the Church of St. Nicolas.

The Treaty of Kiakhta permitted Russia to send students to Peking to learn Chinese and Manchu. In 1728 a language school for Russians was inaugurated as a separate institution within the Southern Russian Hostel. The students came for a ten-year period and the Chinese subsidized their traveling and living costs. They were required to wear Chinese clothes supplied them by the Li-fan yüan; the Board of Rites provided them with food, and the Imperial Academy (*Kuo-tzu chien*) assigned a Chinese and a Manchu instructor to teach them the languages. There were also private tutors attached to the school. By the same token the Chinese government felt the need for instruction in Russian. Twenty-four students were chosen by the Li-fan yüan from members of the Eight Banners to study Russian and Latin for five years. At the end of the period, examinations were held and the two best candidates were given official appointments of the eighth or ninth rank.

By virtue of these religious, educational, and commercial privileges, Russia, alone among nations, had an established foothold in the Chinese capital. These privileges, and their attendant special status, were not revoked even after Ch'ien-lung's successful consolidation of the empire in the 1750s made Russian neutrality no longer necessary. It was not until 1861, when Peking was opened to the diplomatic representatives of Britain, France, and the United States, that Russia's monopolistic position was broken.

The significance of Russia's special position in China cannot be overemphasized. Members of the Russian religious mission and the language school in Peking were able to see China from within and study her language, politics, and social and economic structure firsthand. They were able to detect the strength and weakness of the Ch'ing dynasty long before other Westerners. Possibly they were the only foreigners who understood the Chinese mentality. They witnessed the progressive decline of the Manchu power, and their reports to the home government helped guide Russia's policy toward China. When they returned home they started what was probably the first systematic Sinological study in Europe, preceding that of any other Western state by many decades.[31]

31. R. K. I. Quested, *The Expansion of Russia in East Asia, 1857-1860* (Kuala Lumpur, 1968), 24-29; Wu Hsiang-hsiang, *O-ti ch'in-lüeh Chung-kuo shih* (A history of the Russian imperialist aggression in China), (Taipei, 1957), 20-21. See also Eric Widmer, *The Russian Ecclesiastical Mission during the Eighteenth Century* (Cambridge, Mass., 1976).

Further Reading

Allan, Charles W., *Jesuits at the Court of Peking* (Shanghai, 1935).

Baddeley, John F., *Russia, Mongolia, China* (London, 1919), II.

Bernard, Henri, S.J., *Matteo Ricci's Scientific Contributions to China,* tr. by Edward C. Werner (Peiping, 1935).

Chang, T'ien-tse, *Sino-Portuguese Trade from 1514 to 1644: A Synthesis of Portuguese and Chinese Sources,* reprinted (Leiden, 1969).

Chang, Yin-lin 張蔭麟, "Ming-Ch'ing chih-chi Hsi-hsüeh shu-ju Chung-kuo k'ao-lüeh"明清之際西學輸入中國考略(A brief study of the introduction of Western learning into China during the Ming-Ch'ing transitional period), *Tsing-hua hsüeh-pao,* 1:1:38-69 (June 1923).

Ch'en, Agnes Fang-chih, "Chinese Frontier Diplomacy: the Coming of the Russians and the Treaty of Nerchinsk," *The Yenching Journal of Social Studies,* 4:2: 99-149 (Feb. 1949).

———, "Chinese Frontier Diplomacy: Kiakhta Boundary Treaties and Agreements," *The Yenching Journal of Social Studies,* 4:2:151-205 (Feb. 1949).

Ch'en, Fu-kuang 陳復光, *Yu-Ch'ing i-tai chih Chung-O kuan-hsi* 有清一代之中俄關係(Sino-Russian relations during the Ch'ing period exclusively), (Kunming, 1947), chs. 1-2.

Ch'en, Kenneth, "Matteo Ricci's Contribution to and Influence on Geographical Knowledge in China," *Journal of the American Oriental Society*, 59:325-59, 509 (1939).

Ch'en, Shou-yi 陳受頤, "Ming-mo Ch'ing-ch'u Yeh-su-hui-shih ti Ju-chiao-kuan chi ch'i fan-ying" 明末清初耶穌會士的儒教觀及其反應(The Jesuits' conception of Confucianism in the late Ming and early Ch'ing and its repercussions in China), *Kuo-hsüeh chi-k'an* 國學季刊, 5:2:1-64 (1935).

Ch'en, Vincent, *Sino-Russian Relations in the Seventeenth Century* (The Hague, 1966).

Cheng, Tien-fong, *A History of Sino-Russian Relations* (Washington, D.C., 1957), chs. 2-3.

Chu, Ch'ien-chih 朱謙之, *Chung-kuo ssu-hsiang tui-yü Ou-chou wen-hua chih ying-hsiang* 中國思想對於歐洲文化之影响(The influence of Chinese thought on European civilization), (Changsha, 1940).

Cranmer-Byng, J. L., "The Chinese Attitude Towards External Relations," *International Journal* (Canadian Institute of International Affairs), XXI:4:57-77 (Winter 1966).

Dunne, George H., S.J., *Generation of Giants: The Story of the Jesuits in China in the Last Decades of the Ming Dynasty* (Notre Dame, 1962).

Fang, Hao 方豪, *Li Chih-tsao yen-chiu* 李之藻研究(A study of Li Chih-tsao), (Taipei, 1966).

Fu, Lo-shu, *A Documentary Chronicle of Sino-Western Relations, 1644-1820* (Tucson, 1966), 2 vols.

Gallagher, Louis J., S.J., *China in the Sixteenth Century: The Journal of Matthew Ricci, 1583-1610* (New York, 1953).

Golder, F. A., *Russian Expansion on the Pacific, 1641-1850* (Cleveland, 1914).

Harris, George L., "The Mission of Matteo Ricci, S.J.: A Case Study of an Effort at Guided Culture Change in China in the Sixteenth Century," *Monumenta Serica*, XXV:1-168 (1966).

Hibbert, Eloise T., *Jesuit Adventure in China During the Reign of K'ang Hsi* (New York, 1941).

Hsiao, I-shan 蕭一山, *Ch'ing-tai t'ung-shih* 清代通史 (A general history of the Ch'ing period), rev. ed. (Taipei, 1962), I, chs. 22-25.

Hsü, Tsung-tse 徐宗澤, *Ming-Ch'ing chien Yeh-su-hui-shih i-chu t'i-yao* 明清間耶穌會士譯著提要(A synopsis of translations and writings of the Jesuits during the transitional period from the Ming to the Ch'ing), (Taipei, 1958).

Lach, Donald F., *Asia in the Making of Europe*, Vol. I: *The Century of Discovery* (Chicago, 1965).

———, *Asia in the Making of Europe*, Vol. II: *A Century of Wonder*, Book One: *The Visual Arts* (1970); Book Two: *The Literary Arts* (1978); Book Three: *The Scholarly Disciplines* (1978); (Chicago, 1970, 1978, and 1978 respectively).

Liu, Hsüan-min 劉選民, "Chung-O tsao-ch'i mao-i k'ao" 中俄早期貿易考 (A study of early Russo-Chinese commercial relations), *Yen-ching hsüeh-pao* (Yenching Journal of Chinese Studies), 25:151-212 (June 1939).

Mancall, Mark, *China at the Center: Three Hundred Years of Foreign Policy* (New York, 1984).

Mancall, Mark, *Russia and China: Their Diplomatic Relations to 1728* (Cambridge, Mass., 1971).

Meng, Ssu-ming, "The E-lo-ssu Kuan [Russian Hostel] in Peking," *Harvard Journal of Asiatic Studies*, 23:19-46 (1960-61).

Mungello, David E., *Leibniz and Confucianism: The Search for Accord* (Honolulu, 1977).

Quested, R. K. I., *The Expansion of Russia in East Asia, 1857-1860* (Kuala Lumpur, 1968), ch. 1.

————, *Sino-Russian Relations: A Short History* (London, 1984).

Ravenstein, E. G., *The Russians on the Amur, Its Discovery, Conquest, and Colonization* (London, 1861).

Ricci, Matteo, S.J., *The True Meaning of the Lord of Heaven (T'ien-chu shih-i)*, tr. by Douglas Lancashire and Peter Hu Kuo-chen, S.J., Chinese-English edition by Edward J. Malatesta, S.J. (St. Louis, 1985).

Rosso, A. S., O.F.M., *Apostolic Legations to China of the Eighteenth Century* (South Pasadena, 1948).

Rouleau, Francis A., S.J., "Maillard de Tournon, Papal Legate at the Court of Peking," *Archivum Historicum Societatis Iesu*, 31:264-323 (1962).

Rowbotham, Arnold H., *Missionary and Mandarin: The Jesuits at the Court of China* (Berkeley, 1942).

Sebes, Joseph, S.J., *The Jesuits and the Sino-Russian Treaty of Nerchinsk (1689)*, (Rome, 1961).

Souza, George Bryan, *The Survival of Empire: Portuguese Trade and Society in China and the South China Sea, 1630-1754* (Cambridge, Eng., 1986).

Spence, Jonathan D., *The Memory Palace of Matteo Ricci* (New York, 1984).

Wang, Chih-hsiang 王之相, and Liu Tse-jung 劉澤榮, *Ku-kung O-wen shih-liao* 故宮俄文史料 (Documents in Russian preserved in the National Palace Museum of Peiping), (Peiping, 1936).

Wang, P'ing 王萍, *Hsi-fang li-suan-hsüeh chih shu-ju* 西方曆算學之輸入 (The introduction of Western calendar and mathematics into China), (Taipei, 1966).

Widmer, Eric, *The Russian Ecclesiastical Mission in Peking during the Eighteenth Century* (Cambridge, Mass., 1976).

Wills, John E., Jr., *Embassies and Illusions: Dutch and Portuguese Envoys to K'ang-hsi, 1666-1687* (Cambridge, Mass., 1984).

Wills, John E., Jr., *Pepper, Guns, and Parleys: The Dutch East India Company and China, 1662-1681* (Cambridge, Mass., 1974).

Wu, Aitchen K., *China and the Soviet Union: A Study of Sino-Russian Relations* (New York, 1950).

Wu, Hsiang-hsiang 吳相湘, *O-ti ch'in-lüeh Chung-kuo shih* 俄帝侵略中國史 (A history of the Russian imperialist aggression in China), (Taipei, 1957), chs. 1-2.

Nurhaci (1559–1626),
founder of the Ch'ing dynasty.

Emperor K'ang-hsi (1662–72).

Emperor Yung-cheng (1723–35).

Emperor Ch'ien-lung (1736–95).

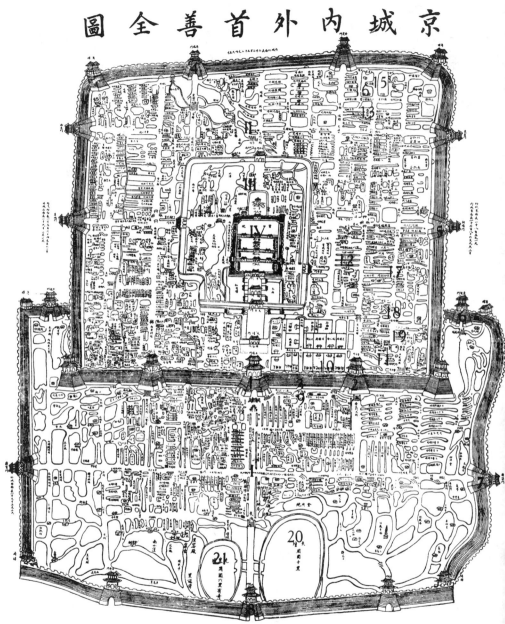

京城內外首善全圖

MAP OF PEKING

I. Native City.   III. Imperial City.
II. Tartar City.   IV. Forbidden City.

1. Ch'ienmen.
2. Hatamen.
3. Wall between the two gates fortified by the besieged.
4. Te-shengmen. (Court fled through this gate August 15, 1900.)
5. Ch'i Huamen. (Russian and Japanese relief entered at this gate in 1900.)
6. Tungpienmen. (The gate through which the Americans entered in 1900.)

7. Shakuomen. (Where the British entered in 1900.)
8. Yungtingmen. (Gate leading into the city from the station.)
9. Water Gate. (Here the relief troops entered the Legation.)
10. Legation quarters.
11. Methodist Mission.
12. American Board Mission.
13. Presbyterian Mission.

14. Peitang.
15. Lama temple.
16. Confucian temple.
17. Tsung Li Yamen.
18. Examination Halls.
19. Imperial Observatory.
20. Temple of Heaven, British headquarters in 1900.
21. Temple of Agriculture, American headquarters in 1900.

Map of Peking.

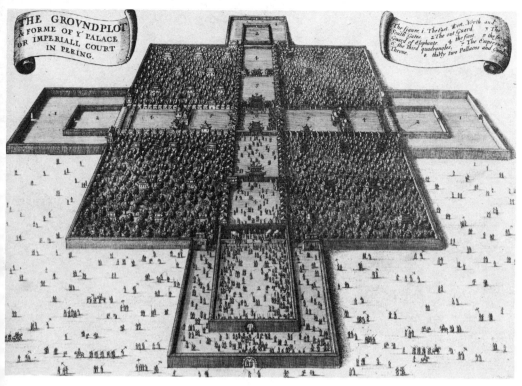

The ground plot of the Imperial Palace.

Imperial Palace, the Hall of Supreme Harmony (T'ai-ho tien).

A grand stone carving of dragons and clouds, 16.57 meters long and 3.07 meters wide, weighs about 250 tons, behind Pao Ho Palace. Restored in 1761.

A legendary Chinese animal, usual misnamed as a unicorn.

The Temple of Heaven, recently restored,
where the emperors used to worship.

Sacred Way to Mausoleum of Emperor Ch'ien-lung (Yu Ling).

A painting of bamboo and rocks, ca. 1740,
by Cheng Hsieh (Pan-ch'iao, 1693-1765),
one of the Eight Eccentrics of Yangchow.

An ancient king by Huang Shen (1687–1768), one of the Eight Eccentrics of Yangchow.

Gourd-shaped porcelain vase of the Ch'ien-lung period (1736–95).

Ku Yen-wu, great early Ch'ing scholar.

Matteo Ricci and Paul Hsü.

Adam Schall von Bell.

6

The Turn of Dynastic Fortune:
From Prosperity to Decline

We have noted that China before 1800 was a vast empire which stood resplendent and unrivaled in East Asia. Its territory stretched from the Central Asian massif to the coast of the China Sea, and from the Mongolian desert to the jungles and shores of the south. China in the middle 18th century was doubtless one of the most advanced countries on earth, and its secular political and social systems had won the admiration of not a few famous European philosophers.[1] But after 1775 the dynastic decline began to set in.

The Decline of the Manchu Power

When Emperor Ch'ien-lung abdicated the throne in 1795, the dynasty had already passed its apogee and the seeds of decay had long since been sown. His fifteenth son, who became Emperor Chia-ch'ing, inherited a country that was "externally strong but internally shriveled" (*wai-ch'iang chung-kan*). Indeed, Chia-ch'ing's twenty-five-year reign (1796-1820) was plagued with serious administrative, military, and moral problems which were unmistakable indices of the falling dynastic fortune.

ADMINISTRATIVE INEFFICIENCY. The suspicion which the Manchu court entertained toward Chinese officials and the resultant policy of mutual

1. Spinoza, Leibniz, Goethe, Voltaire, and Adam Smith.

check undermined administrative efficiency. A noted contemporary political scientist has remarked on this crippling effect: "Public functionaries were rarely given an opportunity to show initiative, independent judgment, or satisfactory performance of tasks through the exercising of adequate authority. On the contrary, all officials were subjected to a tight net of regulations, restrictions, and checks, and threatened with punishment for derelictions or offenses even in matters beyond their individual control. A situation eventually prevailed in which the most prudent thing for the average official to do was to assume as little responsibility as possible—to pay greater attention to formal compliance with written rules than to undertakings that were useful to the sovereign or beneficial to the people."[2] The accuracy of this observation can be seen in a piece of advice given by K'ang-hsi himself to a governor in 1711: "Now that the country is at peace, it is advisable that you avoid trouble. An act that is beneficial in one way may be harmful in another. The ancients said, 'More commitment is not as good as less commitment.' "[3] Thus the guiding principle in the officialdom was to avoid issues. A highly placed courtier once confided that the secret of success in government was to "kowtow more and talk less." It is therefore understandable that there developed a tendency toward compromise, superficiality, temporization—anything so as not to disturb the *status quo*. That these characters strangled the capacity of officials for energetic action and imaginative response to challenge did not trouble the court, for its primary concern was not for dynamic or even efficient administration, but for the dynasty's security. Large decisions were not the province of administrators, but the prerogative of the emperor. Under these conditions the state could prosper only in direct proportion to the emperor's capacity. Such a high concentration of power worked well enough with resourceful rulers such as K'ang-hsi, Yung-cheng, and Ch'ien-lung, but once leadership faltered, the ship of state drifted. After Ch'ien-lung, there was no great emperor.

WIDESPREAD CORRUPTION. The last twenty years of Ch'ien-lung's reign were very corrupt. Ho-shen, the imperial bodyguard whose meteoric rise to power was discussed in Chapter 2, bled the state for nearly a quarter of a century and amassed an incredible fortune of 800 million taels (about $1.5 billion), reputedly more than half the *actual* total state income for twenty years. The inventory of his estate revealed some interesting entries: 4,288 gold bowls and dishes, 600 silver pots, 119 gold wash basins,

2. Kung-ch'üan Hsiao, *Rural China*, 504.
3. Wang Hsien-ch'ien, *Tung-hua lu* (Tung-hua records), K'ang-hsi, 50th year (1711), 18:2b, edict to Governor Fan Tsung-lo.

5.8 million ounces of gold, 75 pawnshops with a capital of 30 million taels, 42 moneyshops with a capital of 40 million taels, and 800,000 *mou* of land at an estimated value of 8 million taels.[4] When Emperor Chia-ch'ing executed him in 1799, a popular saying circulated: "When Ho-shen fell, Chia-ch'ing feasted."[5]

It should be noted, however, that Ho-shen was an acute symptom rather than the cause of the widespread corruption, which was evident even before his rise. Nonetheless, he accentuated the trend and his evil influence continued to haunt the country. Graft, extortion, and irregular levies in both the civil government and the military services became commonplace, almost *de rigueur*. Metropolitan officials openly accepted "presents" from provincial officials, who in turn required them of their own subordinates. These officials led a way of life far beyond their salaries; many maintained luxurious residences with private staffs of servants, guards, and sedan-carriers, entertained permanent houseguests, and supported poor relatives. Their positions made demands which could not be met by their low salaries—which ranged from 180 taels annually for a first-rank official to 33 taels for a ninth-rank official—unless the salaries were supplemented by graft. Even the awarding of "anti-corruption fees" of 50 to 100 times the amount of a salary could not stop the practice of the "squeeze," which was, in fact, institutionalized. For instance, in the collection of the land-poll (*ti-ting*) tax, each locality was given a certain quota, any amount over which the magistrate could keep for himself. It was not uncommon for the tax collected to be many times the official quota. The major brunt of payment fell on the peasant, who under the pressure of the tax collector and local gentry often had to pay 50 percent to 80 percent more than the stated tax in cash and as high as 250 percent more in grain. Small wonder that an official who levied only 10 percent surtax was considered a good conscientious official. It was commonly estimated that a prefect's three-year term could, under normal conditions, yield a handsome 100,000 taels.

DEGRADATION OF THE MANCHUS AND THE BANNERMEN. The Manchus, as befitted the status of conquerors, were not permitted to engage in trade or farming, regardless of birth and social position. They hired the Chinese to till their land and received income from the rents. Their leisure and dependence on others bred laziness and irresponsibility. The bannermen, who constituted the backbone of the Manchu military power at the beginning of the dynasty, received stipends three times as high as those of

4. Hsiao I-shan, II, 264-67.
5. Actually only a small portion of Ho-shen's property was confiscated by the government and parceled out to the princes and nobles as gifts of the emperor. *Ibid.*, II, 268.

the Chinese soldiers. Their privileged position and its corollary soft life caused a marked flaccidity of their original martial spirit, and by the Yung-cheng period (1723-35) they had degenerated to a point where they could no longer fight. Instead of studying military arts, they led a debauched life of gambling, theater-going, watching the cock-fights, and ran usury and mortgaging businesses on the side. Not only could they not fulfill the duty of defending the dynasty, they had even become parasites on the society, and a great many parasites at that: the total number of Manchu, Mongol, and Chinese bannermen and their families was in the neighborhood of 1.5 million.

Corruption in the army was appalling. The Manchu general Fu-k'ang-an, reputedly a bastard son of Emperor Ch'ien-lung, deliberately prolonged his campaigns against the Chin-ch'uan rebels to increase his chances for embezzlement. Corruption was also rampant in the Chinese Green Standard army. Military funds for the suppression of the White Lotus Rebellion (1796-1804) went mostly into the private coffers of the officers in charge. The length of the campaigns and the size of the expenditures were testimonies to the degree of corruption and incompetence that prevailed in the army.

FINANCIAL STRINGENCY. The early Ch'ing rulers had laid very sound economic foundations for the empire. K'ang-hsi left behind 8 million taels, Yung-cheng 24 million, and Ch'ien-lung 70 million. But even while Ch'ien-lung was ruling, the trend toward luxury and massive spending had started. His Ten Perfect Accomplishments cost the state 120 million taels, and Chia-ch'ing's nine-year campaign against the White Lotus Sect and other secret societies cost 200 million. These inordinate military expenses plus the graft and corruption in the civil administration drained the treasury, resulting in a steady rise in the value of silver. Whereas in the early Ch'ien-lung period (1736-95) a tael of silver exchanged for 700 cash (copper coins), in the Chia-ch'ing period (1796-1820) it shot to 1,300 or 1,400 cash. By 1800 the economic foundation of the Ch'ing empire had been badly weakened.

POPULATION PRESSURE. Ch'ing population increased much faster than did the land acreage, causing a decline in the standard of living. In 1660 the population of China was probably somewhere around 100 to 150 million, and it rose to 300 million by 1800. Arable land, however, had not increased correspondingly. In 1661 there were 549 million *mou* of land and in 1812 only 791 million. The land increase, then, was less than 50 percent, whereas the population had increased by more than 100 percent.

The displaced, the poor, and the unemployed often turned to banditry or became recruits for rebel outfits.

INTELLECTUAL IRRESPONSIBILITY. Under the threat of frequent literary inquisitions, scholars shied away from politics and sought refuge in antiquarian studies, where learning could remain safely divorced from reality. They prided themselves on pursuing knowledge for its own sake and ceased to apply their knowledge to society. Those who did join the government through the civil service examinations had received their training in this atmosphere, and many of them were all too often spineless creatures without the potential to be statesmen. In 1799 Hung Liang-chi, a second-class compiler in the Hanlin Academy, sent off a memorial to the emperor in which he bluntly described the moral degradation of scholars and officials. He cited cases of high officials such as board presidents and vice-presidents kowtowing to grand councillors and grand secretaries in order to curry favor, of scholars befriending the servants of dignitaries for the same purpose, and of officials shamelessly bribing palace attendants and guards in hopes of gaining imperial attention. When intellectual irresponsibility and moral degradation fell to this level, it meant that scholars had become oblivious to their duties to the society and had lost track of the importance of the unity of knowledge and action. The society was deprived of real leadership. That the general decline of morality in the government was due at least in part to this very intellectual delinquency is an unavoidable conclusion.

All of these signs—administrative inefficiency, intellectual irresponsibility, widespread corruption, debasement of the military, pressures of a rising population, and a strained treasury—reflected the inner workings of the phenomenon known as "dynastic cycle." Indeed, by 1800 the ruling power had passed its peak and started to decline, making the country vulnerable to the twin evils of internal rebellion and external invasion (*neiluan wai-ho*), so characteristic of dynasties in their later years.

Rebellions by Secret Societies

In an autocracy such as the Ch'ing, where no "loyal opposition" was permitted, the only form of organized resistance apart from open rebellion was secret societies. After the suppression of the anti-Manchu movement on Taiwan in 1683, Ming loyalists went underground to form or join secret organizations by which to continue their fight. Foremost among them were (1) the Heaven and Earth Society (*T'ien-ti hui*), also known as the Triad Society (*San-ho hui* or *San-tien hui*); (2) the Ko-lao Brotherhood

Association (*Ko-lao hui*) in South China; and (3) the White Lotus Sect (*Pai-lien chiao*), and its branch the Heavenly Reason Sect (*T'ien-li chiao*) in North China.[6] Generally, the secret societies in South China called themselves *hui,* and those in the North called themselves *chiao.* The *hui* were secret political organizations with religious overtones, while the *chiao* were secret religious bodies with a nationalistic cast. They were all anti-Manchu.

The Heaven and Earth Society had its origin in the 1670s. Many of the Ming loyalists, convinced that their cause was lost, had retreated to the Shao-lin Monastery in Fukien as monks. In 1674 five of them—the "Five Former Progenitors"—secretly organized the Heaven and Earth Society to promote the overthrow of the Ch'ing and restoration of the Ming. The name of the society was derived from the saying, "Heaven is the father and earth the mother." In Western literature this organization was sometimes called the Triad Society because of its emphasis on the harmony of Heaven, Earth, and Man. The Triads maintained Five Grand Lodges and Five Minor Lodges, somewhat like those of Freemasonry, in the provinces.[7] Branches and affiliates of the Triads soon sprang up in the coastal areas—Taiwan, Kiangsu, Chekiang, Hunan, and Kwangtung— using secret names written with the three-dot "water radical 氵"; hence they were also known as the Three Dots Society (*San-tien hui*). Significantly, this three-dot radical also formed a component of the character *Hung,* which was part of the reign title of the first Ming emperor, Hung-wu.[8] Small wonder that members of the Heaven and Earth Society called their organization the Hung League (*Hung-men*).

These secret societies, bent on overthrowing Ch'ing rule and restoring the Ming, vowed to avenge the killing of Chinese by the Manchus. Whoever embraced these objectives was welcome to join, regardless of birth, education, and social position, but on the whole the societies attracted only the lower classes. New members were introduced by old members and had to learn the secret signs and esoteric language of the society. At

6. Since the XVIIth International Congress of Chinese Studies held at Leeds, England in July 1965, there has been organized an international project of studying Chinese secret societies, with Jean Chesneaux of the Paris Center as coordinator. Cf. *Ch'ing-shih wen-ti,* I:4:13-18 (Nov. 1966).

7. Jean Chesneaux, *Les sociétés secrètes en Chine (XIXe et XXe siècles),* (Paris, 1965), 50. (Avec la collaboration de Marianne Rochline.)

8. Wei Chü-hsien, *Chung-kuo ti pang-hui* (China's secret societies), (Chungking, 1945), Part II, 2-3. However, another interpretation suggests that the character *hung* is Han 漢 minus 圭 (central land), signifying that these Ming loyalists considered themselves Han (Chinese) men deprived of the heartland of China, which had been stolen by the Manchus.

the initiation the candidates vowed to keep the secrets of the fraternity, and read thirty-six oaths from a paper which was then burned and mixed into a bowl of chicken blood, tinted with wine, and sugar. They next pricked their left middle fingers, squeezed some blood into the bowl, and drank from it. With this, all became blood brothers, and after paying dues according to financial ability each received a membership card.[9]

The Ko-lao Brotherhood Association came into being during the Ch'ien-lung period (1736-95). Slightly more finicky than the Heaven and Earth Society, it denied membership to barbers, actors, sedan-carriers, and people born of "unclean blood," but otherwise accepted anyone whose interest was to destroy the Ch'ing and restore the Ming. The headman, or Dragon Head (*Lung-t'ou*), had complete authority over members, who were bound together as brothers. They pledged mutual support and whenever possible organized uprisings.

A much older cabal was the White Lotus Sect, first organized around A.D. 1250, or perhaps even earlier, as a quasi-religious secret body.[10] During the Yüan period (1280-1368), it dedicated itself to the objective of overturning the Mongol dynasty and restoring the Sung. Surviving into the Ch'ing period, it vowed to topple the Manchu dynasty and return the Ming to power. Members adopted Buddhistic as well as Taoistic ideas to win popular support. In 1781 one of its leaders, Liu Sung, was arrested and banished to the frontier; thereafter the government pursued a program of continuous harassment of the sect members which finally goaded them into revolt in 1793. This touched off a succession of mass arrests and persecution, against which White Lotus members in Central China rose up in protest in 1796, using as a pretext "official oppression forced people into revolt" (*kuan-pi min-fan*). The movement quickly spread to Szechwan, Hupeh, Shensi, Kansu, and Honan. Government troops were too corrupt to suppress the revolt, and out of self-defense the local gentry and officials organized militia and constructed fortresses of their own. The rebellion at length was suppressed, in 1804, after nine years of ineffectual and expensive campaigning.

There were a number of other uprisings on a smaller scale. Indeed, throughout the 25-year-reign of Chia-ch'ing, not a day passed without some trouble in the country. It was in this state of dynastic disrepair that

9. For details, see L. F. Comber, *Chinese Secret Societies in Malaya: A Survey of the Triad Society from 1800 to 1900* (Locust Valley, N.Y., 1959), ch. 1; Chesneaux, 29-43.
10. Comber, 19-20; Chesneaux, 57.

the Western powers, particularly Britain, with surplus energies generated by the Industrial Revolution, intensified their efforts to open China to international commerce and diplomacy.

The Western Advance and the Tributary System

The Ch'ing dynasty, though weakened by internal decay, still kept up the face of a great empire and cherished the glory of its former years. It clung to the fond, if fictitious, notion that China, as the Middle Kingdom on earth, was the center of the known civilized world and that all countries which desired relations with it must accept the tributary status.[11] The theory and practice of the tributary system reflected China's world view, and were highly significant in conditioning its relations with the advancing West.

By virtue of its cultural excellence, economic affluence, military power, and vast territorial expanse, China stood pre-eminent in East Asia for two millenia. Since early Ming times (1368-1643) there had been instituted a hierarchical system of "international relations" in East and Southeast Asia, with China occupying the position of leadership and Korea, Liuch'iu (Ryūkyū), Annam (Vietnam), Siam, Burma, and a host of other peripheral states in Southeast and Central Asia accepting the status of junior members.[12] The European term "family of nations" would appear to apply more aptly to this China-centered community of nations than to the Western international society, for in the former "international relations" were based on an extension of the Confucian idea of proper relations between individuals: just as every person in a domestic society had his specific status, so every state in an "international society" had its proper station. Two Korean terms illustrate the idea well: relations with China were described as *sadae*, serving the great, whereas relations with Japan were termed *kyorin*, neighborly intercourse. Thus the basic principle underlying this China-oriented family of nations was inequality of states rather than equality of states as in the modern West, and relations between the members were not governed by international law but by what is known as the tributary system.[13]

The tributary system is reminiscent of the ancient Chinese practice

---

11. For an excellent study of the Chinese view of the world, see John K. Fairbank (ed.), *The Chinese World Order: Traditional China's Foreign Relations* (Cambridge, Mass., 1968).
12. Japan also paid tribute to China for a short period—1404-1549.
13. For details, see Immanuel C. Y. Hsü, *China's Entrance into the Family of Nations*, chapter 1; John K. Fairbank and S. Y. Teng, "On the Ch'ing Tributary System," *Harvard Journal of Asiatic Studies*, 6:2:135-246 (June 1941).

whereby the emperor "invested" (*feng*) the feudal lords and vassals (*fan*) both inside and outside China, and received in return their offerings of local products (*fang-wu*) as "tribute" (*kung*), which was a sort of modified tax payments.[14] During Ming and Ch'ing times, tributary relations had been refined into a highly ritualistic performance, with clearly defined rights and duties on the part of each participant. To China fell the duty of keeping proper order in the East and Southeast Asian family of nations. It recognized the legitimacy of tributary kings by sending envoys to officiate at their investitures and by conferring on them the imperial patents of appointment. It went to their aid in times of foreign invasion, and sent relief missions and commiserative messages in times of disaster. On their part, the tributary states honored China as the superior state by sending periodic tribute, by requesting the investiture of their kings, and by adopting the Chinese calendar, i.e. recording events of their countries by the day, month, and year of the reign of the Chinese emperor.

The size, frequency, and route of the tributary mission were fixed by China—usually the closer the relationship the larger and more frequent the mission. For instance, Korea paid tribute four times a year, presenting it all at the end of the year, Liu-ch'iu twice every three years, Annam once every two years, Siam every three years, and Burma and Laos every ten years. Large numbers of traders were attached to the tributary missions, and their goods were brought into China duty-free. All travel expenses and maintenance of the missions in China were borne by the Chinese government, and when they arrived in Peking the members were lodged in the Common Residence for Tributary Envoys (*Hui-t'ung ssu-i kuan*). A felicitous day was chosen for the envoys to present the tribute and local products to the emperor, at which time they performed the full ceremony of the kowtow—three kneelings and nine knockings of the head on the ground. The envoys and the merchants were then allowed to open a market at their hostel for a few days—usually three to five—to sell their goods. The commercial transactions were highly profitable for the tributary missions. In addition, the emperor showed his benevolence by bestowing handsome gifts on the tributary kings and members of the missions. But on the whole, the value of his gifts was considerably less than the tribute and presents he received.

Tributary relations were costly to maintain. The dispatch of a mission to China was an arduous and expensive task. For instance, the Koreans had to make elaborate preparation in organizing a mission of 200 to 300 men and moving them 750 miles from Seoul to Peking, which took forty

14. Fairbank (ed.), *The Chinese World Order*, 7.

to sixty days. The tribute and gifts of "local products" in 1808 came to 100,000 copper taels,[15] roughly ten times as much as the Chinese emperor's bestowals to the Korean king and his family. Even more expensive was the reception of the Chinese investiture missions. As a rule, after his succession to the throne the tributary king dispatched a special envoy to Peking to request investiture. Imperial missions would then be sent, but only to the three important states of Korea, Liu-ch'iu, and Annam; lesser tributary kings received only the imperial patents of appointment carried back by their own envoys. The investiture mission usually consisted of 400 to 500 persons, for whose reception the Korean court spent an average of 230,000 copper taels, which was equal to one-sixth of its central government's annual expenses![16] The burden was proportionately more onerous for a smaller state, such as Liu-ch'iu, where the Chinese mission usually stayed for five months at the sumptuous Residence for the Celestial Envoy (*T'ien-shih kuan*). The Liu-ch'iuan government had to strain itself to meet the 320,000-silver tael expense involved in each investiture. During the ceremonies, the king had to perform no less than seven full kowtows—when he welcomed the imperial patent, when he greeted the imperial calligraphy housed in a moving pavilion, when he saluted the emperor, when he received the imperial gifts, when he offered thanks for the imperial grace, etc. But before the entire ceremony was over, he had yet to perform one more simple kowtow—one kneeling and three knockings of the head—to the Chinese envoy, who reciprocated the courtesy. The elaborate preparation and the vast expense involved in each investiture were so exacting that the Liu-ch'iuan king usually delayed the ceremonies until two years after his actual accession, and some waited as long as seventeen or eighteen years![17]

Tributary relations, in effect, entailed very considerable financial strain and physical exhaustion on the part of the smaller states, with no appreciable economic benefit for China. The expenses of maintaining the tributary missions—and there were many—while they were in China outweighed the excess value of the tribute and gifts the emperor received. Why then the system? There had to be reasons other than purely economic motivation. For the tributary king, the investiture legitimized his rule, raised his prestige before his people, offered him protection in times of foreign invasion and aid in times of natural disasters, brought him luxury articles

15. A copper tael was equal to one-third silver tael at the time; it was worth one-half silver tael during 1725-76.
16. Hae-jong Chun, "Sino-Korean Tributary Relations in the Ch'ing Period" in Fairbank (ed.), *The Chinese World Order*, 95-97, 104-06.
17. Ta-tuan Ch'en, "Investiture of Liu-ch'iu Kings in the Ch'ing Period" in Fairbank (ed.), *The Chinese World Order*, 136-37, 144, 148.

from the emperor, heightened the cultural link between his country and China, and allowed him to conduct profitable trade with the Middle Kingdom. For the Chinese emperor, it was an immense pleasure and satisfaction to see the myth of his universal overlordship acknowledged and to know that these peripheral states willingly served as an "outer fence" to shield China from barbarian attacks. All in all, tributary relations were maintained primarily to manifest the Confucian concept of propriety and to affirm the hierarchical world order in which China was assured of a superior status, security, and inviolability.[18]

It was this system of international relations that the West encountered when it intruded into East Asia. The Ch'ing court insisted that the tributary system applied not only to the peripheral states of Asia but also to all other states that wanted to establish relations with China. Indeed, during the splendid reigns of K'ang-hsi, Yung-cheng, and Ch'ien-lung, dozens of Asian states were enrolled in the system, and envoys from Portugal, Holland, and Russia kowtowed, albeit reluctantly, to the Chinese emperor. Although Russia and the Western European nations were not formally included in the system, the Chinese treated their missions as though they were tributary missions. To account for the sporadic nature of these missions (unbecoming a tributary state), it was explained in the *Collected Statutes of the Great Ch'ing Empire* that their great distance from China precluded the Western trading nations from maintaining a fixed schedule for bringing tribute. It is interesting to note that of the seventeen missions from the West between 1655 and 1795, all but one yielded to the Chinese demand and performed the kowtow to the emperor.[19] Ch'ing policy toward *official* missions from foreign countries was thus very strict, but its attitude toward *private* Western traders was more flexible. Private traders were allowed to reside in Macao and trade at Canton (after 1757) as a mark of imperial favor. These traders reaped quick and large profits from their transactions, but they were subject to a number of restrictive regulations regarding their movements and trade procedures (details will be given in the next chapter).

By the early 19th century both the governments and the private traders of the Western nations could no longer countenance the straitjacket of the Chinese system. The traders wanted greater freedom of action, and the Western governments, newly released from the Napoleonic Wars and greatly strengthened by the Industrial Revolution, would not suffer the

18. Wang Kungwu, "Early Ming Relations with Southeast Asia: A Background Essay" in Fairbank (ed.), *The Chinese World Order*, 61. See also 110-11, 160.
19. John K. Fairbank, *Trade and Diplomacy on the China Coast: The Opening of the Treaty Ports, 1842-1854* (Cambridge, Mass., 1953), I, 14.

tributary treatment. They insisted on international relations according to the law and diplomacy of Europe; but the Chinese would not sacrifice their cherished system. In effect they said, "We have not asked you to come; if you come you must accept our ways," to which the West's reply was, "You cannot stop us from coming and we will come on our terms." The story of Sino-Western relations thereafter is one of continuous conflict, leading to the ultimate humiliation of the Ch'ing empire.

Actually, by the time the West made a concerted effort to break down the Chinese institution of foreign relations, the tributary system had already worn itself out to a large extent. Since the middle of the 18th century it had been exposed to two disruptive influences: the rise of the Chinese junk trade with Southeast Asia (*Nan-yang*) and the growth of the European trade at Canton—both outside the range of the tributary system. Hundreds of Chinese junks, averaging 150 tons apiece, with the largest approaching 1,000 tons, sailed to Siam, Annam, the Malay Peninsula, Java, and the Moluccas to negotiate their own business. Many of the petty tributary states in these areas found that they no longer needed the tributary system to get along, and thereupon stopped sending tribute to China.[20] The independent European trade that had been permitted to go on at Canton was the other disruptive influence, and it was growing rapidly. Britain, the foremost industrial power and the leader of foreign trade, did the most to break down the existing Chinese system.

20. For details of the Chinese junk trade, see T'ien Ju-k'ang, "Shih-ch'i shih-chi chih shih-chiu shih-chi chung-yeh Chung-kuo fan-ch'uan tsai Tung-nan Ya-chou hang-yün ho shang-yeh shang te ti-wei" (The position of Chinese junks in shipping and trade with Southeast Asia from the 17th century to the middle of the 19th century), *Li-shih yen-chiu,* 8:1-21 (1956).

Further Reading

Abe, Takeo 安部健夫 , *Chūgokujin no tenka kannen* 中國人の天下觀念 (The Chinese world view), (Kyoto, 1956).

Chesneaux, Jean, *Les sociétés secrètes en Chine, XIXe et XXe siècles,* avec la collaboration de Marianne Rochline (Paris, 1965).

Chesneaux, Jean, Feiling Davis, and Nguyen Nguyet Ho, *Movements Populaires et Sociétés Secrètes en Chine aux XIXe et XX Siècles* (Paris, 1970).

Chu, Lin 朱琳 , *Hung-men chi* 洪門誌 (A record of the Hung Society), (Shanghai, 1947).

Comber, L. F., *Chinese Secret Societies in Malaya: A Survey of the Triad Society from 1800 to 1900* (Locust Valley, N.Y., 1959).

Fairbank, John K. (ed.), *The Chinese World Order: Traditional China's Foreign Relations* (Cambridge, Mass., 1968).

————, and S. Y. Teng, "On the Ch'ing Tributary System," *Harvard Journal of Asiatic Studies,* 6:2:135-246 (June 1941).

Hirayama, Amane 平山周, *Chung-kuo mi-mi she-hui shih* 中國秘密社會史 (A history of Chinese secret societies), (Shanghai, 1935).

Hsiao, I-shan 蕭一山, *Ch'ing-tai t'ung-shih* 清代通史 (A general history of the Ch'ing period), rev. ed. (Taipei, 1962), II, chs. 4-6.

————, *Chin-tai mi-mi she-hui shih-liao* 近代秘密社會史料 (Historical materials on the secret societies of modern times), (Peiping, 1935).

Inaba, Iwakichi 稻葉岩吉, *Shinchō zenshi* 清朝全史 (A complete history of the Ch'ing dynasty), (Tokyo, 1914), Chinese tr. by Tan T'ao under the title, *Ch'ing-ch'ao ch'üan-shih* 清朝全史 (Taipei, 1960), chs. 49-52.

Jones, Susan Mann, and Philip A. Kuhn, "Dynastic Decline and the Roots of Rebellion" in John K. Fairbank (ed.), *The Cambridge History of China* (Cambridge, England, 1978), Vol. 10, 107-162.

Morgan, W. P., *Triad Societies in Hong Kong* (Hong Kong, 1960).

Naquin, Susan, *Millenarian Rebellion in China: The Eight Trigrams Uprising of 1813* (New Haven, 1976).

Schlegel, Gustave, *Thian Ti Hwui: The Hung League or Heaven-Earth League* (Batavia, 1866).

T'ien, Ju-k'ang 田汝康, "Shih-ch'i shih-chi chih shih-chiu shih-chi chung-yeh Chung-kuo fan-ch'uan tsai Tung-nan Ya-chou hang-yün ho shang-yeh shang-te ti-wei" 十七世紀至十九世紀中葉中國帆船在東南亞洲航運和商業上的地位 (The position of Chinese junks in shipping and trade with southeast Asia from the 17th century to the middle of the 19th century), *Li-shih yen-chiu*, 8:1-21 (1956).

Viraphol, Sarasin, *Tribute and Profit: Sino Siamese Trade, 1652-1853* (Cambridge, Mass., 1977).

Ward, J. S. M., and W. G. Stirling, *The Hung Society or the Society of Heaven and Earth* (London, 1925-26), 3 vols.

Wei, Chü-hsien 衛聚賢, *Chung-kuo ti pang-hui* 中國的幫會 (China's secret societies), (Chungking, 1946).

II
Foreign Aggression
and Domestic Rebellions
1800-1864

7

The Canton System
of Trade

During the eighty-five years preceding China's opening to the West in 1842, Canton was the only port open to foreign trade, and Chinese foreign relations of this period essentially concerned Canton trade.

The Origin of the Single-Port Trade

Canton, located at the southern tip of the empire, had been an historic center of foreign trade since the T'ang period (A.D. 618-907). Subsequently, during the late Ming and early Ch'ing periods, its trade was virtually monopolized by the Portuguese, who had established themselves at Macao, as noted in Chapter 5. Ships and traders of other nationalities were denied admittance, only rarely succeeding in gaining entry. Barred from Canton, enterprising English traders sought opportunities elsewhere. They developed relations with the Ming loyalist Koxinga and his son on Taiwan, selling them munitions in exchange for trading rights there and at Amoy.

The Ch'ing court, troubled by Koxinga's raids along the coast, in 1662 ordered all ports closed to foreign trade and all coastal inhabitants evacuated to a distance of 30 to 50 *li* inland so as to cut off his sources of supply. Macao, however, was exempted from this rule as a favor to foreign traders, and Canton, though officially closed, was not strictly held to it. With the successful pacification of Taiwan in 1683, the court lifted the ban on foreign trade and in 1685 opened customs houses at Canton,

Chang-chou (in Fukien), Ningpo, and Yün-t'ai-shan (in Kiangsu). Among the ports, Canton was the most flourishing not only because it had the longest history of foreign trade, but also because it lay closest to Southeast Asia, which the Chinese called Nan-yang.

An old port, Canton was tradition-bound and corruption-ridden. The first ship of the East India Company that called in 1689 was assessed an exorbitant measurement fee of 2,484 taels. After much dickering with customs officials, it was whittled down to 1,500 taels, of which 1,200 represented the measurement fee and 300 a gratuity to the Hoppo, the superintendent of maritime customs. Such irregular exactions and the small demand for English woolens in semitropical Canton prompted the Company to seek trade at other ports to the north. It was thought that if trade were possible in the tea- and silk-producing areas of Kiangsu and Chekiang, their procurement costs might be lowered. So in 1698 the Company created a factory—a trading agency or business establishment—at Ting-hai, near Ningpo, with Allen Catchpoole as president. Ningpo, though, turned out no better than Canton, plagued as it also was by official interference, unreasonable levies, lack of demand for woolens, and, furthermore, the local traders' insufficient funds for conducting business. Shifting its interest back to Canton, where another factory had been established in 1699, the Company decided around 1715 to regularize its trade. A council of supercargoes was organized as the permanent staff of the factory until 1758, when it was replaced by a smaller and more efficient permanent Select Committee—composed of three senior supercargoes—which coordinated and directed the Company's business in China.

The arbitrary and whimsical exactions and the high costs of tea and silk at Canton once again renewed the Company's interest in Ningpo around 1753. Two ships were dispatched to Ting-hai in 1755 under Samuel Harrison and James Flint, the latter having learned Chinese. They were well treated by the local dignitaries, and the provincial authorities of Chekiang recommended to the court that since the ships of the "Red Hair"—the nickname for the English and Dutch—had not come for many years they should be received "with compassion."

The shift of business to Ningpo naturally reduced the calls of English ships at Canton: 27 in 1754, 22 in 1755, 15 in 1756, and 7 in 1757. The governor-general at Canton, fearful of deflection of trade to the north, petitioned the court in 1757 to increase the duties at Ningpo 100 percent. Though indifferent to occasional calls by foreign ships at Ningpo, Peking was concerned lest frequent visits turn it into another Macao. Moreover, the court feared, the northern ports at Ningpo, Shanghai, and Amoy, being more accessible from the ocean than Canton, would have

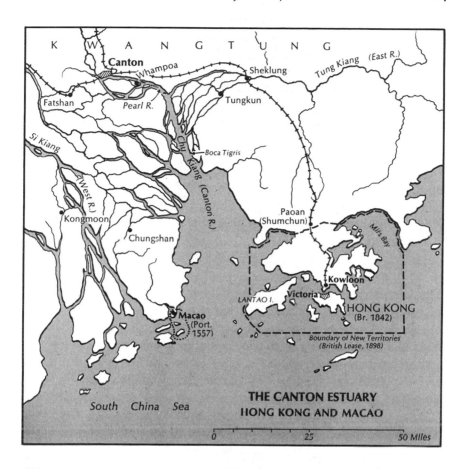

THE CANTON ESTUARY
HONG KONG AND MACAO

South China Sea

0   25   50 Miles

difficulty in controlling the movement of foreign ships, thus contributing to collusion between the aliens and traitorous natives; whereas at Canton, the established forts at Whampoa and the Bogue enabled the government to watch the foreigners and their ships. Furthermore, a substantial portion of the Canton populace traditionally lived on foreign trade; any shift of commerce to the north would seriously jeopardize their livelihood. Weighing these considerations, the court decided to increase the customs duties at Ningpo and the other northern ports, making them enough heavier than at Canton to discourage future trade. Foreigners were urged to desist from going north. Thus, although trade was not officially prohibited at Ningpo, Amoy, and Shanghai, the only port really left open to foreign trade after 1757 was Canton. It was a *de facto* if not *de jure* prohibition of trade in the north (*pu-chin chih-chin* 不禁之禁).

However, in 1759 James Flint defied custom and visited Ningpo on his own. When refused admission, he went to Tientsin to complain of the corruption and irregular exactions at Canton. For his temerity the court threw him into prison in Macao for three years, but also appointed an investigatory commission to Canton and dismissed the Hoppo. A more serious consequence of the Flint incident, however, was a new, explicit court decree (1759) that henceforth Canton was the only port open to foreign commerce. This order eliminated all possibility of extending the trade to other parts of China, thus perpetuating the Canton system until the end of the Opium War in 1842.

The Canton Trade

The Chinese attitude toward foreign trade was an outgrowth of their tributary mentality. It postulated that the bountiful Middle Kingdom had no need for things foreign, but that the benevolent emperor allowed trade as a mark of favor to foreigners and as a means of retaining their gratitude. Hence, trade was not a right to be insisted upon, but a privilege that could be withdrawn by China for any misbehavior. Moreover, since the Canton trade was conducted between private foreign and Chinese citizens, it required no formal diplomatic relations, only unofficial commercial transactions. Therefore, no direct contact was permitted between the foreign traders and Chinese government officials; the former could only *petition* the governor-general, governor, or Hoppo (customs superintendent) at Canton through the Chinese monopolistic merchants assigned to do business with them.

The chief characteristic of the Canton trade was its monopolistic structure. The court authorized "thirteen" commercial firms known as the *hongs* (a corruption of *yang-hang*) as sole agents of foreign trade. The proprietors, known as the hong (*hang*) merchants, secured their monopolistic privilege through handsome contributions to the court, reputedly in the neighborhood of 200,000 taels, or £55,000.

The origin of the hong merchants has sometimes been placed, erroneously, in 1720, which was actually the year they formed a guild in Canton; the hong merchants had existed long before. It is known that during the Wan-li period (1573-1619) of the Ming dynasty, some thirty-six hongs were trading with fourteen countries. The number of hongs dropped to thirteen toward the end of the Ming period (1368-1644), creating the designation of "The Thirteen Hongs," which persisted into the Ch'ing period. Actually the number of the hong merchants during the Ch'ing period fluctuated considerably, and only twice—in 1813 and 1837—actually totaled thirteen.[1]

Among the hongs there were three different groups: those specializing

1. Sixteen in 1720, 20 in 1757, 4 in 1781, and 5 in 1790.

in European and American trade, called the *Wai-yang hang;* those specializing in trade with Southeast Asia, called the *Peng-kang hang;* and those trading with Fukien and Ch'ao-chow, called the *Ch'ao-Fu hang.* Our discussion here is mainly concerned with the first group.

Juxtaposed with the Thirteen Hongs were the thirteen foreign "factories," or agencies, located outside the Canton city walls on the bank of the Pearl River. The factory grounds and buildings, spread over some twenty-one acres, were rented from the hong merchants at an average annual fee of 600 taels. The Chinese commonly dubbed these British, American, French, Dutch, Belgian, Swedish, Danish, Spanish, and other miscellaneous factories the "Barbarian Houses" (*I-kuan*).

The British trade, predominant over that of all other Western countries, was monopolized by the East India Company. Yet there was quite an active private English venture, too. The Company granted charters to private ships to sail from India to China under its license. This trade was known as the "country trade," and the ships "country ships," as opposed to the "Company ships." Six out of every ten of the country ships originated from Bombay, and two each from Bengal and Madras. The country traders were mostly Englishmen doing business in India, judging from their names; but they also included some Indians and Parsees. This country trade accounted for 30 percent of the total British trade at Canton between 1764 and 1800.

Another source of private trade originated from the Company's policy of allowing its ships' officers to carry a specified amount of gold and goods, supposedly to compensate for their small salaries—the captain's pay being only £10 a month, the first mate's £5. The captain of a 495-ton ship, for example, was allowed thirteen tons of private goods in 1730. In reality, the Company believed that when the ships' officers had a personal stake in the cargo, they would strive more diligently to make a speedy and successful voyage. Furthermore, they realized that no one could prevent the carrying of private goods anyway, so it seemed better to regulate than prohibit it. In addition to this type of private trade, the Company also allowed its junior supercargoes at Canton to engage in private transactions to compensate for their insubstantial wages. This private trade accounted for about 15 percent of the total British trade at Canton between 1764 and 1800, but it increased rapidly after the opening of the 19th century.[2]

THE HONG MERCHANTS. The hong merchants rose to prominence only after much arduous struggling. At one point they were nearly driven out

2. Earl H. Pritchard, *The Crucial Years of Early Anglo-Chinese Relations, 1750-1800* (Pullman, Washington, 1936), 170-74; "Private Trade between England and China in the 18th Century (1680-1833)," *Journal of Economic and Social History of the Orient,* I, 109 (Aug. 1957-Apr. 1958).

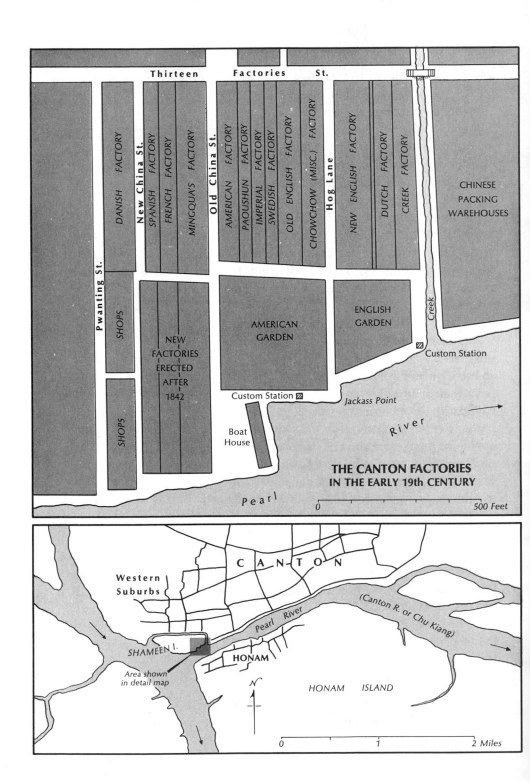

THE CANTON FACTORIES
IN THE EARLY 19th CENTURY

Thirteen   Factories   St.

DANISH FACTORY

New China St.
SPANISH FACTORY
FRENCH FACTORY
MINGQUA'S FACTORY

Old China St.
AMERICAN FACTORY
PAOUSHUN FACTORY
IMPERIAL FACTORY
SWEDISH FACTORY
OLD ENGLISH FACTORY
CHOWCHOW (MISC.) FACTORY

Hog Lane
NEW ENGLISH FACTORY
DUTCH FACTORY
CREEK FACTORY

CHINESE PACKING WAREHOUSES

Pwanting St.

SHOPS

SHOPS

NEW FACTORIES ERECTED AFTER 1842

AMERICAN GARDEN

ENGLISH GARDEN

Creek

Custom Station

Custom Station

Jackass Point

Boat House

River

Pearl

0   500 Feet

CANTON

Western Suburbs

(Canton R. or Chu Kiang)

Pearl River

SHAMEEN I.

HONAM

Area shown in detail map

N

HONAM   ISLAND

0   1   2 Miles

of business by the so-called "Emperor's Merchants" (*Huang-shang*), who, empowered to monopolize foreign trade, appeared in 1702 at Canton, Amoy, and Chusan. As it happened, the Emperor's Merchant at Canton—reputedly a former salt official who secured his new status from the crown prince through a donation of 42,000 taels—apparently possessed neither great capital nor a large stock of goods ready for delivery. Unable to fill orders without delay, he elicited complaints from foreign traders, which were echoed by hong merchants excluded from the profitable commerce. In 1704, five arriving English ships refused to deal with him, entering into clandestine arrangements with the hong merchants, who, having bribed the Hoppo, now openly competed with the Emperor's Merchant and eased him out of business. To strengthen their position, the hong merchants organized a guild in 1720 called the Co-hong (*Kung-hang*) and adopted a code of thirteen articles to regulate the prices and practices of transaction. The original membership consisted of sixteen hong merchants divided into three classes, stipulating that new members could join upon payment of 1,000 taels. Out of all business transactions, 3 percent was set aside as a reserve, called the Consoo fund, against cases of insolvency. The guild, though a private organization, received official patronage, for it served as a convenient buffer between the government and the foreign traders. Thus, the government officials, who did not understand foreign languages, and the foreign traders, who did not understand Chinese regulations, could avoid personal encounters. Upon the guild fell the double task of collecting customs duties for the government and of paying fees for the foreign merchants.

The guild monopolized the Canton trade to the exclusion of all non-members. The latter naturally protested, while some foreign traders also resented the strait-jacket arrangement. In the face of this opposition the guild disbanded after a year. In 1745 the Hoppo selected the five most substantial of the twenty or so hong merchants as "security merchants," to assume responsibility for all business transactions and to secure the proper conduct of all foreigners. By 1754 all hong merchants had become security merchants. On petition of the rich hong merchant Puankhequa, the Co-hong was revived in 1760, but it soon became ridden with serious problems of dissension and of debts to foreign traders. In 1771, by giving Puankhequa 100,000 taels to bribe the Chinese authorities, the East India Company succeeded in getting it disbanded, only to see it resurrected again in 1782, this time to survive until the end of the Opium War in 1842.[3]

3. Kuo T'ing-i, I, 343.

Among the hong merchants, the richest and most famous during the 18th and early 19th centuries were Puankhequa of the Tung-foo Company, Mowqua of the Kwonglei Company, and Howqua of the Ewo Company. Incidentally, the suffix to their names, *qua,* was a corruption of the character *kuan,* meaning official, which was an honorary designation acquired through notable contributions to the court in exchange for a brevet title. The staff of the hongs was comprised of (1) compradores, who served at once as brokers, accountants, and cashiers; (2) linguists, the indispensable go-betweens who actually, by foreign description, "knew no language but their own"; (3) schroffs, the silvermasters who assayed the quality of silver, ingots, or dollars, in their capacity of "teller"; and (4) scribes and clerks.

The rich hong merchants were subjected to merciless exploitation by officialdom. From 1786 on, they were required to pay a regular annual tribute of 55,000 taels and had, besides, to collect foreign clocks and watches to give to the governor and the Hoppo, who in turn presented them to the court. They were expected to proffer gifts on such festive occasions as imperial birthdays and marriages: on Emperor Chia-ch'ing's fiftieth birthday, for instance, they presented him with 120,000 taels. Frequently they were asked to contribute to military and river conservancy operations—such as Puankhequa's 200,000 tael donation in 1773 to the Chin-chu'an campaign and another 300,000 in 1787 toward the suppression of a Taiwan rebellion. For the campaign against the White Lotus Rebellion (1796-1804) the hong merchants collectively contributed 600,000 taels, and later an equal sum toward the suppression of the Moslem Rebellion under Jihangir in Sinkiang in the 1820s. Contributions for river conservancy included 550,000 taels in 1801, 200,000 in 1804, and 600,000 in 1811 and again in 1820. Between 1773 and 1832 they "contributed" nearly four million taels.[4] In addition, as rich residents of the province, these hong merchants were frequently called upon to donate to educational institutions, public charity, hospitals, and even clinics for smallpox vaccinations. And at times, they, as security merchants, were fined for crimes and uncivil acts of the foreign traders. The constant demands on the hong merchants and the high risks of their commercial ventures could well drive them into bankruptcy. Yet they could not easily quit business, because they were government-appointed agents for foreign trade. Many survived only through foreign loans.[5] On the whole, however, they fared quite well, and a number of them succeeded in amassing great wealth, as seen in the cases of Puankhequa, Mowqua, and Howqua.

4. Liang Chia-pin, *Kuang-tung shih-san-hang k'ao* (A study of the Thirteen Hongs of Canton), (Shanghai, 1937), 368 ff.
5. In 1782 their debts to foreign traders amounted to $3,808,075. Morse, I, 68.

THE TRADING PROCEDURE. The trading season began before the end of the southwest monsoon in the early fall and ended during the northeast monsoon in the winter, lasting roughly the three to four months from October through January. At the opening of the season the incoming ship had first to go to Macao to employ a pilot, a linguist, and a compradore who purveyed the ship and crew provisions. Then it proceeded to the Bogue for measurement and payment of fees, and thence, cleared of all obligations, it was permitted to anchor at Whampoa. Here the cargo was handed over to one of the hong merchants, who fixed the commodity prices without competition from others; similarly, foreign traders purchased goods only through the assigned hong merchant. All contracts for sales and purchases were made a year in advance.

The hong merchant taking the foreign consignment bore complete responsibility for the foreign ship. He undertook settling the traders in the proper factory and recommended to them the compradore, linguists, schroffs, and servants. He was not obligated to sell the complete consigned stock, but could take a part of it while farming out the remainder to other hong merchants. In fact, according to the original agreement of the Co-hong, a hong merchant could not acquire more than half of the total cargo of a foreign ship. For instance, Howqua, in dealing with the East India Company, usually took fourteen shares of the stock, leaving the rest to the others, some of whom took only one share or even half a share.[6] Conceivably, if the itinerary were followed strictly, a foreign ship could discharge its cargo and stow a new one within three weeks; but with the procedure described above, more often it would take a month or two. Once their business was concluded, the foreigners were required to leave Canton immediately, either returning home or going to Macao for the winter. However, for a consideration paid to "proper persons," some foreign traders were allowed to remain in Canton after the trading season.

EXACTIONS AND FEES. An incoming ship was subject to a variety of dues and levies, which fell into three major categories: the measurement fee, the presents and other gratuities (*kuei-li*), and the tariff on goods. The ship measurement was derived by multiplying the length between the mizzenmast and the foremast by the breadth of the ship at the gangway, and dividing the product by ten. According to this standard, ships were divided into three classes, with the large ones taxed at 7.777 taels per unit of measurement, medium ones at 7.142, and small ones at 5. The presents, on the other hand, were highly irregular levies of a very complex nature, including fees for opening the ship's hull, examining the

6. Li Shou-k'ung, 82.

hull, allowance for differences in scales and purity of silver, and a host of other impositions. Until 1726, when the government took over these fees, they went into the pockets of the Hoppo, the examining officers, the scribes, and the attendants—which accounted for their irregularity. In 1727, though, the gratuities were consolidated at 1,950 taels, where they remained for about a century. The gratuities and the measurement fee cost a first-class ship in 1810 something like 3,315 taels, and a second-class ship 2,666 taels. Ships going directly to Macao without anchoring at Canton paid about one half the measurement fees and gratuities, but had to pay an additional 2,520 taels to the Co-hong for trading privileges outside its territory. There were, in addition, all types of minor fees, such as $60 for the pilot for each of the inbound and outbound trips in the harbor, $400 for the compradore, $200 "expense money" for the linguist above his regular $75 pay and an allowance of $50 or $60 from the ship captain. Thus the total payment of a first-class ship during a three-month stay in Canton ran to about 4,500 taels.[7]

The regular customs dues were reasonably low, somewhere between 2 percent and 4 percent ad valorem, but frequently the customs officials charged twice as much, and not infrequently three or four times as much. The payment was usually made by the hong merchants for the foreign traders.

ARTICLES OF TRADE. By the late 18th century there was a flourishing triangular trade between Canton, India, and England. The most important exports to England were tea (accounting for 90 percent to 95 percent of the total), raw silk, chinaware, rhubarb, lacquered ware, and cassia; while imports from England included woolens, lead, tin, iron, copper, furs, linen, and various knicknacks. Exports to India consisted of nankeen cloth, alum, camphor, pepper, vermilion, sugar, sugar candy, drugs, and chinaware; while imports included raw cotton, ivory, sandlewood, silver, and opium.

The large volume of tea export may have resulted from several causes. The prohibition of rice export and the limitation of silk outflow to 140 piculs (175 bales) per ship made tea the logical staple item of export. There was a growing demand for tea in Europe, especially in England, since Europe produced no tea, having no idea of it until 1550. The first small quantity of Chinese black tea was brought to Europe by Dutch merchants in 1640 and soon appeared in England. Beginning in 1684 the East India Company annually purchased five to six chests of teas from Canton for presents in England, and in 1705 green tea made its initial

7. Morse, I, 77-78; Kuo T'ing-i, I, 457-72.

London appearance. Gradually, during the first quarter of the 18th century, the Company increased its tea purchase to 400,000 pounds a year, from which samples were presented to the crown and the nobles. Soon tea-drinking became a fetish among the polite society, later spreading to the populace—who drank it as a substitute for the heavily taxed liquor. So great was the national demand, that in 1800 the Company shipped 23.3 million pounds of tea, and after 1808 the annual British import averaged 26 million pounds, twice as much tea as shipped by other countries. By this time tea-drinking had become a national habit of England, and tea lovers went so far as to assert that its mild nature exercised a civilizing influence on character, whereas liquor often led to violence and misconduct. As its use increased, so did the English import duty—to an outrageous 100 percent, a rate sufficient to encourage smuggling from the continent (especially from Holland), reputedly in the neighborhood of 7 million pounds a year. Finally, in 1784, the Commutation Act reduced the tea import duty to 12.5 percent, putting an end to the lucrative smuggling. Even so, Chinese tea provided one-tenth of the English revenue.[8]

The tea-producing areas in China were Fukien (black tea), Anhwei (green tea), and Kiangsi (both). In February of each year, a thousand or more tea merchants came to Canton to make arrangements with the hong merchants for delivery. In 1755 a hundred catties[9] (*chin*) of tea cost 19 taels. It took a month or two to transport tea overland from its producing areas to Canton, roughly 2,400 *li*, or 800 miles, but much less time to ship along the coast. In 1813 some British steamers shipped a million pounds of teas from Foochow to Canton in thirteen days.

The flourishing state of the Canton trade is seen from the increasing number of ships that called, from 19 in 1751 to 81 in 1787 and then back to 57 in 1792, as shown in the following chart:

Year	British Company ships	Country ships	French	Dutch	Swedish	Danish	American	Others	Total
1751	7	3	2	4	2	1	–	–	19
1780	12	12	–	4	3	3	–	–	34
1787	29	33	3	5	2	2	2	5	81
1792	16	23	2	3	1	1	6	5	57

Evident in the last two decades of the 18th century was the increasing activity of the country trade, and the entry of the Americans into the China trade, signaled by the arrival of the *Empress of China* from New

8. Michael Greenberg, *British Trade and the Opening of China* (Cambridge, 1951), 3.
9. About 133⅓ lbs.

York in 1784. The Americans were free traders, as opposed to the monopolistic East India Company.

The balance of trade at Canton during the 18th century was very much in China's favor, because she needed few foreign products, while Western traders purchased large quantities of tea, silk, and rhubarb. Foreign ships had to bring silver bullion to purchase Chinese products; at times the cargo of the East India Company's ships from London consisted of 90 percent bullion. During 1775 and 1795 the Company's imports of goods and bullion into China amounted to 31.5 million taels, against an export of 56.6 million taels. The 25.1 million-tael deficit was partially relieved by the country trade and the private trade which enjoyed a favorable balance, the former showing a surplus of 13.6 million and the latter 1.7 million in the same period.[10] The proceeds from the country and private trade were transferred to the Company's treasury at Canton in return for bills of exchange payable in London. In the period mentioned above, the Company derived roughly a third of its funds for Canton purchases from the country trade.

Foreign Life at Canton

Since the Canton authorities governed aliens under the notion that trade was a privilege and not a right of foreigners, and since enjoyment of this imperial favor was contingent upon their good behavior, the foreigners were obliged to submit to certain rules of conduct periodically announced at the factories as a reminder. Violation of these rules could entail the stoppage of trade.

RULES OF BEHAVIOR. A set of Five Regulations was first promulgated by Governor-general Li Ssu-yao in 1759 in the wake of the Flint incident. It underwent many subsequent additions and revisions until it finally assumed the form of the following code of behavior in the early 19th century:

1. No foreign warships may sail inside the Bogue.
2. Neither foreign women nor firearms may be brought into the factories.
3. All pilots and compradores must register with the Chinese authorities in Macao; foreign ships must not enter into direct communication with Chinese people and merchants without the immediate supervision of the compradore.
4. Foreign factories shall employ no maids and no more than eight Chinese male servants.

10. Pritchard, *Crucial Years*, 180.

5. Foreigners may not communicate with Chinese officials except through the proper channel of the Co-hong.
6. Foreigners are not allowed to row boats freely in the river. They may, however, visit the Flower Gardens (*Hua-ti*) and the temple opposite the river in groups of ten or less three times a month—on the 8th, 18th, and 28th. They shall not visit other places.
7. Foreigners may not sit in sedan-chairs, or use the sanpan boats with flags flying; they may ride only in topless small boats.
8. Foreign trade must be conducted through the hong merchants. Foreigners living in the factories must not move in and out too frequently, although they may walk freely within a hundred yards of their factories. Clandestine transactions between them and traitorous Chinese merchants must be prevented.
9. Foreign traders must not remain in Canton after the trading season; even during the trading season when the ship is laden, they should return home or go to Macao.
10. Foreign ships may anchor at Whampoa but nowhere else.
11. Foreigners may neither buy Chinese books, nor learn Chinese.
12. The hong merchants shall not go into debt to foreigners.[11]

Except for item 4 regarding the employment of servants, all the rest were strictly enforced, particularly the regulation on women. In 1830 when three foreign women sneaked into the English factory, the Chinese authorities threatened to stop trade, and the women had to leave for Macao. In consequence of this strict rule against women in Canton, foreign traders usually left their families in Macao. Of the 4,480 foreigners in Macao in 1830, 2,149 were white females against 1,201 white males, the rest being slaves and servants. The foreign community in Canton, in contrast, consisted entirely of males; in 1836 there were 307 foreign men, of whom 213 were non-Asian.

The regulations governing foreign behavior doubtless caused discomfort to the traders, but the momentary pains were assuredly somewhat alleviated by the prospect of quick monetary gains. On the whole, life in the factories, with their spacious drawing rooms, was rather pleasant, and the relations between the foreign traders and the hong merchants were harmonious and friendly. William C. Hunter, an American who went to Canton in 1825 and stayed for many years, spoke of the hong merchants as a body of men "honourable and reliable in all their dealings, faithful to their contracts, and large-minded."[12] They shared a spirit of

11. Hsiao I-shan, II, 836-37.
12. William C. Hunter, *The "Fan Kwae" at Canton before Treaty Days, 1825-1844* (Shanghai, 1911), 40.

camaraderie with the foreign traders, each helping the other out in times of difficulty and insolvency. Howqua, who led a frugal life himself, was particularly known for his munificence and generosity. Once, when it came to his knowledge that an American trader who had suffered business reverses was stranded in Canton for three years and unable to return to his family, he called the American in and tore up his promissory note of $72,000, declaring the account settled. In his pidgin English—the business language—Howqua announced: "You and I are No. 1 'olo flen'; you belong honest man, only got no chance."[13] ("You and I are No. 1 old friends; you are an honest man; only you were unlucky.")

Pidgin—or pigeon—English was the *lingua franca* of the China coast trading communities. A mixture of English, Portuguese, and Indian words, it was spoken more or less in Chinese syntax without regard to English grammatical rules. Of Portuguese origin were such words as *mandarin*, from *mandar*, meaning to order; *compradore*, from *compra*, to buy; *maskee*, from *masque*, never mind. Of Indian origin were *bazaar*, a market; *schroff*, a money-dealer; *go-down*, a corruption of *ka-dang*, a warehouse; *lac*, one-hundred thousand; and *cooly* (coolie), a laborer. A typical sentence of pidgin English, as spoken by Howqua as he tore up the promissory note, was "Just now have settee counter, alla finishee; you go, you please."[14] ("Just now we have settled our account. All is finished. You may go as you please.")

THE PROBLEM OF JURISDICTION. The various restrictions upon foreign activities were one source of conflict; the problem of law enforcement was another. Chinese legal concepts and practices differed greatly from those of the West. There was no "due process of law" as understood by Westerners, nor advice of counsel in court. The judiciary was not an independent arm of government, the local judge being none other than the magistrate. Lawsuit and litigation were thought to be manifestations of one's lack of virtue, rather than assertions of one's legal rights. In a criminal case a defendant was considered guilty until proven innocent, and in cases of homicide the principle of "a life for a life" was followed. The Chinese sense of justice permitted a father to shield his son from justice, and vice versa, rather than surrendering him to judgment; and several families in a neighborhood might be held responsible for the crime of one. All this was "strange" and "barbarous" to foreigners at Canton and Macao.

The "doctrine of responsibility" was another source of friction. Just as

13. H. F. MacNair, *Modern Chinese History: Selected Readings* (Shanghai, 1913), I, 42.
14. MacNair, I, 42-43.

the emperor was theoretically responsible for all that happened under the sun, so the governor-general was culpable for all incidents within his jurisdiction, including the flooding of the river or disturbance of the peace by foreigners. To protect himself, he mercilessly governed foreigners with the strictest regulations. Extending the application of this doctrine, the hong merchants were responsible for "securing" the good behavior of the foreign traders, and the headmen of foreign communities were obliged to control their nationals and hand over criminals when demanded by the Chinese authorities, regardless of their personal judgments in the particular case.

The Chinese government insisted that foreigners committing crimes in China be tried according to Chinese law. On the other hand, foreigners demanded exemption from Chinese law. This was not so much because they denied the universal principle of territorial jurisdiction, but because of the "strange" way the Chinese court dispensed justice and because of the harshness of the sentences. Actually, civil cases involving foreigners were rare, as there was little contact between foreign traders and the Chinese public. Disputes between the hong merchants and the foreigners were mostly settled by negotiation and arbitration. Also, very few, if any, foreign traders were involved in criminal cases, as these seemed to be the specialty of the sailors. When criminal cases did occur, those involving foreigners in both parties were settled in one of three ways: (1) the Chinese court tried the case, but sent the convicted to his own country for punishment, as in the case of a Frenchman, who, having killed an English sailor in 1754, was sentenced by a Chinese court to death by strangulation, the execution being left to the French government after he had been returned home; (2) the Chinese court tried the case and executed the sentence in China, as it carried out the strangulation execution of an English sailor sentenced for the slaying of a Portuguese sailor; (3) where the guilty had fled, the Chinese court would pass sentence on him, forwarding it to the home country for execution, as in the case in 1830 of a group of Englishmen who had beaten a Dutch sailor to death and then fled to India. In this instance the governor-general at Canton sentenced the chief offender to death by strangulation and his accomplices to a hundred lashes each, but the sentence was transmitted to Britain for its execution.

In mixed cases, where the criminal was Chinese, justice was inexorably carried out with consummate speed and equity. A case in point was the killing of an English seaman in 1785 by a Chinese, who was summarily sentenced to death by strangulation. The same practice held if the culprit was a foreigner, as in the case of the British ship *Lady Hughes* in 1784.

On November 24 of that year, this country ship fired a salute and accidentally wounded three minor mandarins, two of whom subsequently died. The Canton authorities demanded the surrender of the gunner, and when told that he had absconded, seized George Smith, the supercargo of the ship, besieged the factory, and stopped trade. It was not until the gunner was found, on the *Lady Hughes,* and surrendered to the Chinese authorities that the supercargo was released and trade resumed. They then strangled the gunner.

The *Lady Hughes* incident, plus the Chinese explanation that the sentence was light because it only demanded one life for two, shocked the foreign community into a seizure of terror. Foreigners feared for their personal safety in future cases and deeply resented the Chinese practice of holding the supercargo or community chief responsible for crimes committed by others. Moreover, the harshness and apparent inhumaneness of Chinese sentences (*vide* the numerous "death by strangulation" penalties), the lack of a proper trial according to European justice, and the capricious stopping of trade or refusal of clearance to departing ships in order to force the surrender of the guilty—such irritants to the foreign sensibilities produced great anxiety and endless protest against the Canton authorities.

British Attempts to Change the Canton System

The *Lady Hughes* incident climaxed the foreigners' feeling of insecurity and heightened the general dissatisfaction with the Canton system of trade, i.e. the limitation to one port, the humiliating restrictions on personal freedom, and the numerous irregular exactions. The British felt much of the abuse at Canton was unknown to Peking. With a view to reducing the irritations, widening the trade, and placing British-Chinese relations on a regular diplomatic footing through direct contact with the central power, London decided to dispatch an official mission to China. Instrumental in this decision was Henry Dundas, president of the Pitt government's newly founded Board of Control in India. The East India Company, while not wholly pleased with this move lest it jeopardize the existing trade, agreed to bear the expense of the embassy and to furnish presents to the Chinese court. The ambassadorship went to Lieutenant Colonel Charles Cathcart, member of Parliament, quartermaster-general to the Bengal army, and a friend of Dundas. His instructions called for the improvement of British trade with China and removal of the present restrictions, dispersion of Chinese fears of British territorial designs and the assurance of the peaceful intention of trade, and acquisition of "a small tract of ground or detached island in some more convenient situation than Canton" as a depot for commerce under British jurisdiction. Failing in these objectives, he was to work toward relieving the immediate difficulties and embarrassments at Canton. But if his mission should terminate successfully, he was to request an exchange of permanent envoys between Britain and China.[15]

The embassy, which sailed December 21, 1787, struck an unpropitious note at its beginning. Cathcart was seriously ill with consumption, and the ship ran into a storm and unfavorable winds. In February and March of the following year dysentery plagued the sailors, and the ambassador in the last stage of consumption wrote of his inability to shake his cough. Still insisting that the constant change of air at sea would enhance his recovery, Cathcart succumbed on June 10 during the journey. The aborted mission returned to England.

Talks of a second mission failed to produce government action because of the East India Company's lukewarm attitude, the unsettled conditions in Europe caused by the French Revolution, the outbreak of war in India with the Tippoo Sultan in late 1789, and the difficulty of finding a man suitable to head such a mission. Not until June 1791, when Henry Dundas was promoted to Home Secretary in addition to his old position on the Board of Control, did the idea of another embassy revive. With the support of Pitt, who wished to satisfy the growing demands of industrialists for a wider market in China, Dundas chose his friend Lord Macartney, Baron of Lissanoure and a cousin of the Crown, to be ambassador to China.

THE MACARTNEY MISSION, 1793. Born near Belfast on May 14, 1737, Lord Macartney was a man of learning and dignity. An experienced colonial administrator and diplomat, he had been ambassador to Russia, member of the Irish and British Parliaments, chief secretary for Ireland, governor of the West Indian Island of Grenada, and governor of Madras. Having declined the post of governor-general of Bengal, he had been without a position since 1786. He was doubtless the best qualified and the most eligible man in England for the China mission. On May 3, 1792, he was officially designated "Ambassador Extraordinary and Plenipotentiary from the King of Great Britain to the Emperor of China." To add dignity to his mission, he was given the title of privy councillor and the rank of viscount. His lifelong friend, Sir George L. Staunton, was made "Secretary to the embassy and Minister Plenipotentiary in the absence of the Ambassador," with authorization to carry on the mission in the event of

15. Pritchard, *Crucial Years*, 255-58.

the death or incapacity of the leader. Macartney was stipulated a yearly compensation of £15,000 and Staunton £3,000.

On September 26, 1792, the mission set out from London with eighty-four members, including a machinist, a painter, a draftsman, an artificer, six musicians, and a number of military and naval officers. Impressive presents, costing £15,610, were prepared for the Chinese court, including a planetarium, globes, mathematical instruments, chronometers, a telescope, measuring instruments, chemical and electrical instruments, window and plate glass, carpets, Birmingham goods, Sheffield goods, copperware, and Wedgwood pottery.

In addition to collecting all available information about China—intellectual, political, military, social, economic, and philosophical—Macartney was instructed to achieve six specific objectives:

1. To acquire one or two places near the tea- and silk-producing and the woolen-consuming areas, where the British traders might reside and English jurisdiction be exercised.
2. To negotiate a commercial treaty with a view to extending trade throughout China if possible.
3. To relieve existing abuses at Canton.
4. To create a desire in China for British products.
5. To arrange diplomatic representation at Peking.
6. To open Japan, Cochin China, and the Eastern Islands to British commerce.

In short, the mission was entrusted with the task of opening the whole East to British trade and of placing relations with China on a regular treaty basis.

Macartney was instructed to conform to all the ceremonials of the Chinese court which did not compromise the honor of his king and the dignity of himself. To prepare for the arrival of this embassy, the East India Company dispatched a Secret and Superintending Committee to Canton in September 1792 to inform the governor-general of the event. Under the pretext of presenting felicitations from the English king to Emperor Ch'ien-lung on his eighty-third birthday, the mission arrived off Canton shores on June 19, 1793, in a man-of-war, the *Lion*, a brig, the *Jackal*, and a Company steamer, the *Hindustan*. It then proceeded north to Chusan and Taku.

Flattered by the first English "tributary" mission, which had come to admire his Celestial Empire and help celebrate his birthday, Emperor Ch'ien-lung was much pleased and ordered that the embassy be accorded an honorable welcome. An edict of July 24, 1793, stated that although

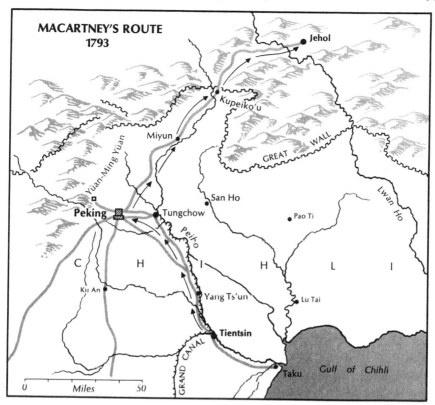

the reception of the English envoy need not be excessively elaborate, nevertheless—because it was his first visit and in view of his long ocean voyage—he should be treated differently from the tributary envoys of Burma and Annam. An order was issued[16] that Macartney be accorded proper courtesy and be given a reasonably good reception. Another edict, of August 1, reiterated the importance of receiving him in an appropriate manner—i.e. neither too servile nor too overbearing—so as to show China's benevolence toward men from afar. The court allocated a liberal daily allowance of 5,000 taels for the mission during its journey to Peking, and a daily maintenance subsidy of 1,500 taels while it stayed in the capital.[17]

At Tientsin the mission was warmly welcomed by the governor-general, and the 600 cases of presents were carried to Peking on an impressive train of wagons, barrows, horses, and coolies. Though a flag bearing the characters of "Tributary Envoy from England" was placed on his boat,

16. To Governor-general Liang K'en-t'ang at Tientsin, and to the Ch'ang-lu Salt Controller, Cheng-jui.
17. Kuo T'ing-i, I, 231-32; Hsiao I-shan, II, 811.

Macartney chose not to protest. At Peking he was lodged in the Summer Palace for five days, and on September 2 he left for Jehol, about a hundred miles north of Peking beyond the Great Wall, where the emperor was spending the summer.

An ostentatious and pompous old ruler, Emperor Ch'ien-lung was gratified at the coming of the English mission, but a little piqued at news that Macartney was reluctant to perform the kowtow. In an edict of August 14 he had announced that not only all tributary envoys but even their kings, when coming to China, must perform the kowtow; therefore Macartney, who had been sent by his king to congratulate the emperor on his birthday, should not resist the Chinese ceremony lest he fail to carry out his sovereign's intentions. The Chinese escorts of the mission were instructed to suggest that if the English envoys were perhaps physically unable to bend because of their knee buckles and garters, they should remove them temporarily in order to perform the kowtow. Macartney himself seemed to have no strong feelings about the kowtow, but he would do nothing that would compromise the dignity of his country or suggest a vassal status to China. He let it be known that he would pay the emperor the same obeisance which he did the English sovereign and that he would perform the kowtow only if an equally high ranking Chinese would perform the same ritual before the portrait of the English king. In the end Emperor Ch'ien-lung, who was in a good mood, yielded the point and allowed Macartney to fall on one knee during the audience as he would do to his own king, but exempted him from the English custom of kissing the hand of the monarch.

The celebrated audience took place on September 14, 1793, in a magnificent tent intended for extensive assemblies. Macartney and Staunton presented themselves in brilliant official costumes—the former wearing the mantle of the Order of the Bath over a rich embroidered velvet, with the collar and diamond badge and a diamond star, and the latter also wearing a rich embroidered velvet in addition to the scarlet silk habit of the Doctor of Laws of Oxford. They performed the modified ritual as agreed upon—kneeling on one knee—although Chinese records later bemusedly reported that Macartney, overcome by awe and excitement in the imposing presence of the emperor, "unconsciously fell on his two knees." Then came the presentation of the English king's letter, which Macartney personally delivered in a gold box to the emperor. Exchange of gifts followed, with the emperor bestowing on the English king, through Macartney, a whitish jade scepter (*ju-i*) about a foot and a half long, as a symbol of peace and prosperity, remarking that he hoped that the English king might live as long as he himself. The aged ruler then gave the two envoys a greenish carved agate each as a special mark of favor. In return Macartney presented him with a pair of beautiful enameled watches set with diamonds, while Staunton proffered a pair of elegant air guns. Imperial gifts were also bestowed on other members of the delegation. A most sumptuous banquet followed in honor of the visitors, at which time the emperor solicitously offered the envoys dishes from his own table and went so far as to personally pour them each a cup of warm wine. Macartney found Ch'ien-lung a bit condescending, but very affable, dignified, and vigorous, looking more like a man of sixty than eighty-three. The grandeur, luxury, and elegance of Ch'ien-lung's summer court called to Macartney's mind "King Solomon in all his glory."[18] Emperor Ch'ien-lung, on his part, celebrated the occasion with a poem of his own composition:[19]

> Formerly Portugal presented tribute;
> Now England is paying homage.
> They have out-traveled Shu-hai and Heng-chang;[20]
> My Ancestors' merit and virtue must have reached their distant shores.
> Though their tribute is commonplace, my heart approves sincerely.
> Curios and the boasted ingenuity of their devices I prize not.
> Though what they bring is meagre, yet,
> In my kindness to men from afar I make generous return,
> Wanting to preserve my good health and power.

On the following day Macartney was given a tour of the imperial Garden of Ten Thousand Trees (*Wan-shu yüan*) and another audience. In the next two days he was again taken sight-seeing, given presents, and invited to a puppet show and a comic drama. On September 17, the imperial birthday, Macartney was admitted to the full company of Chinese and Manchu courtiers to present his felicitations to the emperor. On September 26 the mission returned to Peking, four days before the emperor himself.

Macartney had attempted in vain to open discussion with Ho-shen, the powerful grand councillor and chief minister of state, about the extension of trade and an exchange of envoys. In both Jehol and Peking, Ho-shen was unresponsive, parrying all attempts at negotiation. Finally, when pressed by Macartney, who by this time was quite worn out and sick with the gout and rheumatism, Ho-shen vaguely indicated that the English

18. J. L. Cranmer-Byng, "Lord Macartney's Embassy to Peking in 1793," *Journal of Oriental Studies* (Hong Kong), IV:1-2:163 (1957-58).
19. Tr. by J. L. Cranmer-Byng, 164.
20. Mythological travelers.

envoy might present his requests in a note. This Macartney readily did on October 3 in the name of the Crown, requesting:

1. To extend trade to Chusan, Ningpo, and Tientsin.
2. To allow English traders a warehouse in Peking for the sale of their goods, as the Russians formerly had.
3. To assign a small unfortified island near Chusan for the residence of English traders, storage of goods, and outfitting of ships.
4. To assign a small place near Canton for the residence of English traders, and allow them freedom of movement between Canton and Macao.
5. To abolish the transit dues between Macao and Canton, or at least to reduce them to the tariff of 1782.
6. To prohibit the exaction of duties from English merchants over and above the rates set by the imperial tariff, a copy of which should be made available to them.

The Ch'ing court considered diplomatic negotiations completely out of order. As far as it was concerned, Macartney had come to congratulate the emperor on his birthday and his mission was accomplished when he had done so. As he had been well treated, he should be satisfied to return home with gratitude. Since no tributary mission ever tarried in Peking for more than forty days, the court was anxious that Macartney leave before October 9. Ho-shen intimated to him that with severe winter descending soon, the emperor was concerned about the ambassador's health. It was obvious that the host was hinting that the guest should leave. Macartney realized that any further attempt to remain would serve no useful purpose. Disappointed, he noted in his Journal: ". . . having been selected for this Commission to China, the first of its kind from Great Britain, of which considerable expectations of success had been formed by many, and by none more than by myself, I cannot help feeling the disappointment most severely. I cannot lose sight of my first prospects without infinite regret."[21]

The mission left Peking on October 7. Macartney remained in Canton from December 19, 1793, to January 10, 1794, then went to Macao, where he stayed until March 8, and finally returned to London on the following September 4. The whole experience of the embassy was wittily summed up in an epigram collected by Peter Auber, secretary to the Court of Directors of the East India Company: "It has just been observed that the Ambassador was received with the utmost politeness, treated with the

21. Cranmer-Byng, 176.

utmost hospitality, watched with the utmost viligance, and dismissed with the utmost civility."[22]

Although no reply had been made to Macartney directly, two edicts were issued to King George III. In the first celebrated message, dated October 3, 1793, Emperor Ch'ien-lung pompously stated that while China appreciated the English intention "to partake of the benefits" of Chinese civilization by dispatching a mission, the request for diplomatic residence in Peking could not be granted because it was contrary to the established practice of China: "Europe consists of many other nations besides your own: if each and all demanded to be represented at our court, how could we possibly consent? The thing is utterly impracticable. How can our dynasty alter its whole procedure and system of etiquette, established for more than a century, in order to meet your individual view?" Moreover, an envoy in Peking would be too far from Canton to control the traders. "If you assert that your reverence for Our Celestial dynasty fills you with a desire to acquire the rudiments of our civilization, you could not possibly transplant our manners and customs to your alien soil. Therefore, however adept the Envoy might become, nothing would be gained thereby." Apropos of the request for the extension of trade, Ch'ien-lung announced: "We possess all things. I set no value on objects strange or ingenious, and have no use for your country's manufactures." The message closed with the imperious statement: "It behooves you, O King, to respect my sentiments and to display even greater devotion and loyalty in future, so that, by perpetual submission to our Throne, you may secure peace and prosperity for your country thereafter."[23]

These were strong and provocative words to be addressed to the sovereign of a state which boasted of being the mistress of the seas, yet they unmistakably evinced the Chinese mentality on foreign relations at the close of the 18th century. "No one understands China until this document has ceased to seem absurd," remarked Bertrand Russell.[24]

In a separate mandate to George III, Ch'ien-lung rejected all six of Macartney's requests as impractical and unproductive of good results.

It may be, O King, that the above proposals have been wantonly made by your ambassador on his own responsibility; or peradventure you yourself are ignorant of our dynastic regulations and had no intention of transgressing them when you expressed these wild ideas and hopes . . . Above all, upon you, who live in a remote and inaccessible re-

22. Cranmer-Byng, 183.
23. MacNair, I, 2-4.
24. Cranmer-Byng, 182.

gion, far across the spaces of ocean, but who have shown your submissive loyalty by sending this tribute mission, I have heaped benefits far in excess of those I have accorded to other nations. But the demands presented by your embassy are not only a contravention of dynastic tradition, but would be unproductive of good results to yourself, besides being quite impracticable . . . It is your bounden duty reverently to appreciate my feelings and to obey these instructions henceforth for all times, so that you may enjoy the blessings of perpetual peace.[25]

The mission, which cost the British £78,522, was a complete diplomatic failure. It had achieved neither representation at Peking, nor the extension of trade, nor the opening of Japan, Cochin China, and the Eastern Islands. Nevertheless, it succeeded in collecting valuable firsthand information about the mysterious land called China. Macartney discovered the low state of her scientific and medical knowledge, the indifference of the literati class to material progress, the backwardness of an army which still used bows and arrows and lacked modern firearms, the poverty of the masses, and the widespread corruption and graft in government. Disbelieving, for example, that his mission had consumed the 1,500-tael daily allowance allocated by the court, Macartney surmised that part of it must have gone into the pockets of the officials in charge of his reception. The offspring of Confucius, he concluded, were no different from the descendants of Mammon. In regard to the future of the dynasty, he ventured a penetrating remark: "The empire of China is an old, crazy, first-rate Man of War, which a succession of able and vigilant officers have contrived to keep afloat for these hundred and fifty years past, and to overawe their neighbors merely by her bulk and appearance. But whenever an insufficient man happens to have the command on deck, adieu to the discipline and safety of the ship. She may, perhaps, not sink outright; she may drift some time as a wreck, and will then be dashed to pieces on the shores; but she can never be rebuilt on the bottom."[26] Whatever the diplomatic results, a leading officer of the East India Company remarked that "the information alone to be acquired from the embassy would far more than compensate for the expense."[27]

As for the British government, it was clearly disappointed at the lackluster outcome of the mission, though it placed neither censure nor honor on the ambassador. Macartney had done his best and failed; his only fault, perhaps, was the persistent view that the Chinese government was

25. MacNair, I, 4-9.
26. Cranmer-Byng, 181.
27. Pritchard, *Crucial Years*, 375.

not disinclined toward foreign intercourse. He now recommended that Staunton be sent on a follow-up legation to China as the king's minister and concurrent chief of the British supercargoes at Canton. Although the government was favorably disposed to the idea and steps were taken to put it into practice, Staunton's attack of paralysis and subsequent death in 1801 shelved the project. The lack of a suitable leader for the embassy and Britain's involvement in the Napoleonic Wars indefinitely postponed any move in that direction.

THE AMHERST MISSION, 1816. In the period that followed, the Canton trade continued much as before, but Sino-British relations were strained by several new incidents. The first arose out of British fear of a French seizure of Macao from the Portuguese, a move which would win for France a commanding position in Southeast Asian trade. To forestall this possibility, twice—in 1802 and 1808—British forces occupied Macao, despite Chinese protests that Macao was Chinese territory and not in danger of French seizure. The first British withdrawal was achieved with the news of the Peace of Amiens in 1802, but the second was much more complicated. When the British commander, Admiral Drury, refused to evacuate, the governor-general at Canton retaliated with a stoppage of trade, causing inconveniences and widespread complaint among all foreigners. Drury then proposed an interview with the governor-general, and when refused, defiantly forced his way past the Bogue with three warships, cast anchor at Whampoa, and demanded the interview. There followed an armed conflict with the Chinese, in which the British suffered some casualties. The situation remained tense until the Select Committee of the Company secured British withdrawal from Macao in December of that year by getting the Portuguese to pay a ransom of $600,000.

Other issues that strained Sino-British rapport included the British attack on the Chinese tributary of Nepal, and the seizure of the American steamer *Hunter* off Canton waters (by the British warship *Doris*) in April 1814, Britain then being at war with the United States. The Canton authorities protested the violation of Chinese jurisdiction and threatened to cut off the British trade unless the *Doris* left port. The British community at Canton refused to give in, and the Chinese bluff failed.

These incidents, along with the growing dissatisfaction over the Canton trade system, prompted the Company to request that London send another deputation to Peking. The restoration of peace in Europe after the Congress of Vienna in 1815 freed Britain from its European involvement. It decided to send Lord Amherst, the ex-governor of India, on a mission to the Ch'ing court, accompanied by two associate envoys, Henry

Ellis and Sir George Thomas Staunton, son of Macartney's secretary and president of the Select Committee in Canton. Lord Amherst's instructions called for the removal of grievances at Canton, the establishment of free trade between Chinese and British merchants, the abolition of the Co-hong system, freedom to reside at the factory without time limit and to employ Chinese servants, the establishment of direct communications between the factory and Chinese officials, the opening of more ports north of Canton, and the right to diplomatic representation in Peking. He was to remove Chinese misgivings about the British action in Nepal and explain the reasons for the *Doris* incident. Leaving Portsmouth on February 8, 1816, the delegation sailed straight to Tientsin without stopping at Canton, where it was feared the Chinese might block it from proceeding northward.

Unlike his expansive father Ch'ien-lung, Emperor Chia-ch'ing was reserved and hesitant to receive foreign envoys. Apprehensive of new British demands, he responded to the new embassy with the unenthusiastic comment: "All in all, I am not glad of this event." An order was issued that the reception of the mission need not be extravagant; that if Amherst was submissive, he might be allowed to come to Peking for an audience; but if he was headstrong and resistant to the ceremony of kowtow, he should be given a local reception at Tientsin only, and told that the emperor had gone on a hunting trip and would not return for several months.

On August 13, 1816, Amherst and his fifty-two pieces of "tribute" arrived in Tientsin, where he was received by the president of the Board of Public Works, who gave a dinner in his honor. When asked to express his thanks to the emperor by performing the kowtow, Amherst replied that he could not comply, but would remove his hat three times and bow his head nine times. Endless argument followed, but led to no solution. While the procession was en route to Peking, word came from the court that "if the English envoy refuses to comply with the ceremony no audience will be allowed." The procession stopped at Tungchow, about ten miles from Peking, and two high officials from the capital, the presidents of the Li-fan yüan and of the Board of Rites, came to reason with Amherst concerning the importance of kowtow. Amherst himself actually did not care one way or the other about the matter; he had been instructed in London to consider the kowtow as a matter of expediency and to perform it if it would expedite his mission. But the directors of the East India Company had advised him to resist the Chinese ritual lest it compromise the dignity and prestige of Britain. Division of opinion also existed between his associate envoys: Ellis favored accepting the Chinese demand, while Staunton firmly opposed it. Caught between contrary views, Amherst momentarily hesitated and wavered, but ultimately decided against the kowtow. He told the Chinese that he would bend one knee and bow his head three times, and repeat the performance three times so as to approximate the three kneelings and nine knockings of the head that were required. This the Chinese would not accept. The mission was stalled at Tungchow for ten days. Then an ameliorating order came from the court to the effect that since "outer barbarians" were unused to the kowtow, it should not matter too much if the envoy in question had not mastered the proper manner of kneeling and rising. Anxious to please the emperor, the president of the Li-fan yüan, who had been arguing with Amherst, reported on August 27 that "although his [Amherst's] kneeling and rising are somewhat unnatural, the ceremony may yet be performed."

On the evening of August 28 the mission was allowed to proceed to Peking. Emperor Chia-ch'ing, satisfied with the report of Amherst's "progress" in learning the kowtow, decided to accord him an audience the following day. The embassy was hurried on all night, and upon arrival in Peking early the next morning Amherst was told that the emperor was ready to receive him immediately at the Summer Palace. Exhausted from the bumpy journey and the summer heat, and unaccompanied by his credentials and costumes which had lagged behind, he pleaded for time to rest. After a heated argument with his Chinese escorts, Amherst left in anger. Presently the emperor sent for him. The president of the Li-fan yüan, unable to produce Amherst, falsely reported that the English envoy had fallen sick. The emperor then asked for the associate envoy, and was informed that he too was ill. Suspecting the envoys of falsehood, the exasperated ruler declared: "China is the universal overlord. How can she willingly submit to this kind of insult and insolence?" An edict was issued expelling the English embassy from the capital, rejecting its "tribute," and canceling the audience.

When the actual plight of the envoys became known to the emperor the next day, he was a little mollified, agreeing to recall part of the tribute and send some gifts to the king of England. He also sent an order to the governor-general at Nanking to avoid insulting Amherst, treating him with the courtesy due his rank. The embassy finally left for England from Canton on January 28, 1817.

Amherst's refusal to comply with Chinese ceremonies—the sole cause for his expulsion—received considerable attention in Europe. Napoleon Bonaparte, then in exile, chided him for applying the ceremonies of the court of St. James to the court at Peking. An envoy, in Napoleon's view, should respect the ceremonies and customs of the country to which he is accredited and should realize that he does not have the same privilege of discrimination as the sovereign who sent him; he should be satisfied with the treatment due a man of comparable rank at the local court. Hence,

British and Russian envoys, in his view, should accept the Chinese ceremonies if the Chinese government agreed to instruct its future deputies to
comply with the practices of London and St. Petersburg.

Since both the Macartney and Amherst attempts at peaceful negotiation had failed, the British throne faced three alternatives of action:
(1) abandon the China trade, (2) submit to the Chinese treatment, or
(3) change the situation by military means. For Britain, the most powerful state on earth and the ruler of the seas, the first two courses were un-
thinkable, leaving only the third alternative—force. On China's part, the
disrespect displayed by Amherst was utterly intolerable and wholly incompatible with its claim to universal overlordship. Emperor Chia-ch'ing
even considered severing relations with Britain and stopping the Canton
trade altogether, but was dissuaded from it by the governor-general at
Canton, who feared reprisals and possible war with Britain. The time was
fast approaching for a showdown between the two countries.

Meanwhile, the Canton trade had been undergoing a drastic metamorphosis in character as a result of the rapid growth of the private and
country trade and the phenomenal rise of opium-smuggling from India to
China. The private trade at Canton had risen from 688,880 taels in 1780-
81 to 992,444 taels in 1799-1800, and the country trade from 1,020,012
to 3,743,158 in the same period.[28] Their growth was even more rapid
after the turn of the century. By 1817-1834 they accounted for three-
quarters of the total British imports to China. Many of the private traders, to avoid the Company's intervention, secured consulships of other
European countries, and managed to stay in Canton and expand their
business. They served as agency houses for firms in London and India
and engaged in the lucrative illicit traffic of opium-smuggling at "outside"
anchorages, such as Lintin and Hong Kong, making transactions with
"outside" (i.e. non-hong) merchants for quick profit. So powerful had the
private traders become that they began to agitate for the abolition of the
Company's monopoly. By 1820 the complexion of the Canton trade had
changed: private trade had surpassed the company trade, and opium had
superseded regular articles as the chief item of import. These two developments contributed to the breakdown of the outworn Canton system
and precipitated the long-delayed clash between Britain and China. A
new page of history was about to be written.

28. Pritchard, *Crucial Years*, 401-2.

Further Reading

Auber, Peter, *China, An Outline of Its Government, Laws, and Policy: and of the British Embassies to, and Intercourse with, That Empire* (London, 1834).
Chang, Hsin-pao, *Commissioner Lin and the Opium War* (Cambridge, Mass., 1964), ch. 1.

Chang, Te-ch'ang 張德昌, "Ch'ing-tai Ya-p'ien chan-cheng ch'ien chih Chung-Hsi yen-hai t'ung-shang" 清代鴉片戰爭前之中西沿海通商 (Sino-Western coastal trade in the Ch'ing period before the Opium War). *Tsing-hua hsüeh-pao*, 10:1: 97-145 (Jan. 1935).

Cranmer-Byng, J. L., "Lord Macartney's Embassy to Peking in 1793," *Journal of Oriental Studies*, IV:1-2:117-183 (1957-58).

Danton, G. H., *The Cultural Contacts of the United States and China: The Earliest Sino-American Culture Contact, 1784-1844* (New York, 1931).

Dermingny, Louis, *La Chine et l'Occident: Le Commerce à Canton au XVIIIe siècle, 1719-1833* (Paris, 1964), 4 vols.

Goldstein, Jonathan, *Philadelphia and the China Trade, 1682-1846: Commercial, Cultural, and Attitudinal Effects* (University Park and London, 1978).

Greenberg, Michael, *British Trade and the Opening of China, 1800-42* (Cambridge, 1951).

Hou, Hou-p'ei 侯厚培, "Wu-k'ou t'ung-shang i-ch'ien o-kuo kuo-chi mao-i chih kai-k'uang" 五口通商以前我國國際貿易之概況 (The general condition of our country's international trade before the five-port trading period), *Tsing-hua hsüeh-pao*, 4:1:1,217-1,264 (June 1927).

Hunter, William O., The *"Fan Kwae" at Canton before Treaty Days, 1825-44* (Shanghai, 1911).

Liang, Chia-pin 梁嘉彬, *Kuang tung shih-san-hang k'ao* 廣東十三行考 (A study of the Thirteen Hongs of Canton), (Shanghai, 1937).

May, Ernest R., and John K. Fairbank (eds.), *America's China Trade in Historical Perspective: The Chinese and American Performance* (Cambridge, Mass., 1986).

Morse, H. B., *The International Relations of the Chinese Empire* (London, 1910), I, chs. 3-6.

———, *The Chronicles of the East India Company Trading to China, 1635 1834* (Oxford, 1926-29), 5 vols.

P'eng, Tse-i 彭澤一, "Ch'ing-tai Kwang-tung yang-hang chih-tu ti ch'i-yüan" 清代廣東洋行制度的起源 (The rise of the hongs in Kwangtung during the Ch'ing Dynasty), *Li-shih yen-chiu*, 1:1-24 (1957).

Pritchard, Earl H., *Anglo-Chinese Relations during the Seventeenth and Eighteenth Centuries*, University of Illinois Studies in the Social Sciences, 17:1-2:1-244 (March-June 1929).

———, *The Crucial Years of Early Anglo-Chinese Relations, 1750-1800*, Research Studies of the State College of Washington, 4:3-4:95-442 (Sept.-Dec. 1936).

———, "The Kowtow in the Macartney Embassy to China in 1793," *Far Eastern Quarterly*, II:2:163-203 (Feb. 1943).

———, "Private Trade Between England and China in the 18th Century (1680-1833)," *Journal of Economic and Social History of the Orient*, I, Parts 1-2 (Aug. 1957-April 1958).

Staunton, Sir George, *An Authentic Account of an Embassy from the King of Great Britain to the Emperor of China* (London, 1797), 2 vols.

Wakeman, Frederic, Jr., "The Canton Trade and the Opium War," in John K. Fairbank (ed.), *The Cambridge History of China* (Cambridge, Eng., 1978), Vol. 10, 163-212.

8

The Opium War

The Canton trade in the 18th century, as already noted, was heavily one-sided in China's favor. Foreign traders came to purchase tea, silk, rhubarb, and other articles, but they paid in gold and silver, the Chinese finding little need for the industrial products of the West—"We possess all things," as Emperor Ch'ien-lung told King George III. Frequently 90 percent—and sometimes as high as 98 percent—of the East India Company's shipment to China was gold, and only 10 percent commodities. Between 1781 and 1790, 16.4 million taels of silver flowed into China, and between 1800 and 1810, 26 million. This balance in China's favor continued until the mid-1820s when it settled into an equilibrium. After 1826 the balance began to slip the other way: between 1831 and 1833 nearly 10 million taels flowed out of China.[1] The reversal gathered further momentum as time went on. What could cause such a phenomenal inversion in a trade balance? One factor: opium.

The Opium Trade

The opium poppy was first introduced into China by the Arabs and the Turks in the late 7th or early 8th century. The Chinese called it *ying-su*, or *mi-nang*, or *a-fu-yung*, or simply *po-pi* (poppy), and used it chiefly as medicine to relieve pain and reduce tension; opium-smoking for pleasure was unknown until much later. In 1620 some Formosans were said to have mixed tobacco with opium for smoking, and the practice spread in

1. Hsin-pao Chang, *Commissioner Lin and the Opium War* (Cambridge, Mass., 1964), 41.

168

the 1660s to Fukien and Kwangtung, where the method of smoking was refined: the smoker burned the opium over a lamp and inhaled its fumes through a pipe. It rapidly became a fad with the leisure classes, and before long even the poor took it up. The demand for opium led to increased foreign importation and to native cultivation in Szechwan, Yunnan, Fukien, Chekiang, and Kwangtung. From a moral concern Emperor Yung-cheng (1723-35) prohibited the sale and smoking of opium in 1729, and Emperor Chia-ch'ing (1796-1820) outlawed its importation and cultivation in 1796. Later, in the 1820s and 1830s, economic considerations also entered the picture, for the trade was causing a rapid outflow of silver.

The British took over the lead in opium importation from the Portuguese in 1773, when the East India Company established a monopoly of the opium cultivation—from seedling to sale of the finished product by auction in Calcutta—under the Bengal government. But knowing the Chinese prohibition, the Company disengaged itself officially from the opium trade by leaving its distribution to the country ships which sailed under the Company's license. In the license a clause required such ships to carry the Company's opium, but in the public sailing order there was always a statement of prohibition against carrying opium "lest the Company be implicated."[2] Thus the East India Company perfected the technique of growing opium cheaply and abundantly in India, while piously disowning it in China. Legally and officially, it was not involved in the illicit trade.

In general there were three types of opium: the Patna (Bengal opium), the Malwa (West Indian opium), and the Turkish opium. Their prices varied with time and place. A chest[3] of Patna cost $560 to $590 in Macao in 1801, $2,075 in 1821, and $744 in 1835; a chest of Malwa cost about $400, $1,325, and $602 in the corresponding years. The annual importation of opium at the time of the first prohibition in 1729 was 200 chests, but by 1767 it rose to 1,000. The import growth was rapid and steady: between 1800 and 1820, the average annual importation was 4,500 chests, and between 1820 and 1830, over 10,000 chests. In the 1830s the volume rose enormously, reaching a peak of 40,000 chests in 1838-39. This sharp rise was caused by the abolition of the Company's monopoly of the China trade in 1834, the influx of private traders, and the extension of traffic beyond the Canton waters to the entire southeastern coast of China.

In contrast to the regular trade which was carried on by barter or on credit, the opium trade—forced into a subterranean existence because of

2. Greenberg, 110.
3. Opium was packed in chests, which weighed approximately 100 catties (*chin*), or 133⅓ lbs., for the Malwa, and 120 catties, or 160 lbs., for the Patna.

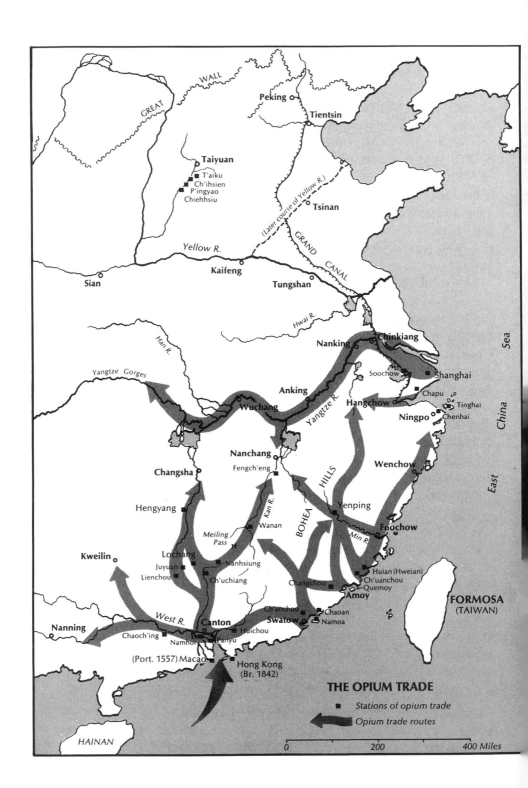

THE OPIUM TRADE

■ Stations of opium trade

◄ Opium trade routes

0   200   400 Miles

its illicit nature—was conducted on a cash basis. The lucrativeness of the trade drew nearly all foreign traders, except men like D. W. C. Olyphant, a "pious, devoted servant of Christ, and a friend of China." The leading British private firm, Jardine, Matheson and Company, handled 5,000 chests in 1829-30, roughly one-third of the opium total in China. But, said Jardine in 1839, "the father of all smuggling and smugglers is the East India Company."[4]

The American traders handled Turkish as well as Indian opium, but it was the latter which made up most of their total commodity—around 95 percent. Between 1800 and 1839 the Americans shipped 10,000 chests into China.

The opium-dealing organizations, known as the *yao-k'ao*, usually had capital anywhere from twenty thousand to one million dollars. They paid for the opium at the foreign factories and picked up the drug from the foreign "receiving ships" at Lintin in fast-moving small crafts called "smug boats," which were also known by such names as "fast crabs" (*k'uai-hsieh*) and "scrambling dragons" (*p'a-lung*). They were fully armed, rowed by sixty to seventy sailors, with twenty or more oars on each side, and they moved at an amazing speed. In 1831 there were something like a hundred or two hundred of them shuttling around the Canton waters. From there the opium was transported westward to Kwangsi and Kweichow, eastward to Fukien, and northward to Hunan, Kiangsi, Anhwei, and as far as Shensi. Opium dealers often maintained relations with the underworld—secret societies and brigands—as well as with the Shansi bankers for transmission of funds.

The rapid rise in opium imports naturally was connected with the growing demand for the drug in China. The addicts in the early 19th century were mostly young men of rich families, but gradually the habit spread to people of other walks of life: government officials, merchants, literati, women, servants, soldiers, and even monks, nuns, and priests. In 1838, in Kwangtung and Fukien provinces opium shops were as common as gin shops in England. The addict went to any length to acquire the drug, for its deprivation would cause restlessness, chills, hot flashes, nausea, muscle twitch, and bone aches. Though hungry, he could not eat, and though drowsy, he could not sleep. A common laborer made one-tenth to two-tenths of a tael a day; half of it would be spent on the drug if he was an addict. An average smoker consumed .05 of a Chinese ounce[5] of opium extract daily, while a good many smoked twice as much. The 40,000 chests imported in 1838-39 yielded 2.4 million catties of extract

4. Hsin-pao Chang, 31, 49; Greenberg, 137.
5. One Chinese ounce was about 1⅓ English ounces.

and supplied about 2.1 million consumers. It was said that between 10 percent to 20 percent of the central government officials and 20 percent to 30 percent of the local officials smoked opium. The total number of smokers was estimated at somewhere between two and ten million. The famous statesman Lin Tse-hsü stated that if one out of a hundred people in China smoked opium there would be four million addicts. Chinese estimates put the annual consumption of opium at 17 to 18 million taels between 1823 and 1831, 20 million taels between 1831 and 1834, and 30 million between 1834 and 1838.[6]

The economic repercussions of opium-smoking were most serious. Spending on opium caused a stagnation in the demand for other commodities, with a consequent general sluggishness in the market. Moreover, the constant inflow of opium caused a continuous outflow of silver. Between 1828 and 1836, the British exported $37.9 million from Canton, and in the year beginning July 1, 1837 they took $8.9 million. However, there was an offsetting factor: the Americans and other foreign traders brought silver and gold into China. Between 1818 and 1834 the Americans brought in $60 million of silver against $50 million shipped out by the British. But as the opium trade grew, less American cash flowed in, while more Chinese silver was taken out; between 1828 and 1833 the British shipped out $29.6 million of specie against the American inflow of $15.8 million. The drain was most acute in the middle and late 1830s, somewhere between 4 and 5 million dollars annually.[7] The silver depletion upset the domestic economy and rocked the exchange rate between silver and copper in the market. Whereas a tael of silver in 1740 exchanged for 800 copper coins, in 1828 it was worth 2,500 in Chihli and 2,600 in Shantung. To meet the economic crisis, the government debased the copper coins and increased their annual minting.

In spite of these serious economic repercussions, the opium traffic could not be stopped for lack of a well-organized customs service, an effective navy, and a sense of moral responsibility in public administration. Too often officials charged with suppressing the drug traffic connived with the smugglers and turned over "free samples" of opium to the government as intercepted contraband.

The ineffectiveness of the Chinese prohibition was matched by the enterprising promotion of the trade by the British. In 1832 the East India Company made 10 million rupees from its opium production, in 1837, 20 million, and in 1838, 30 million. Opium provided over 5 percent of the Company's revenue in India in 1826-27, 9 percent in 1828-29, and 12

6. Hsin-pao Chang, 35, 40; Kuo T'ing-i, II, 104-05.
7. Greenberg, 142; Hsin-pao Chang, 42.

percent in the 1850s, a sum close to £4 million sterling. The House of Commons' Select Committee reported in 1830 and 1832 that "it does not seem advisable to abandon so important a source of revenue as the East India Company's monopoly of opium in Bengal." In 1836 the British sold $18 million worth of opium in China as against the $17 million worth of Chinese tea and silk which they bought. Obviously, without the opium trade they would have suffered a severe deficit; opium had therefore become the economic panacea for the British trade doldrums. Small wonder that the astute and perceptive Duke of Wellington declared in May 1838 that Parliament had not only refused to frown upon the opium traffic but cherished it, extended it, and promoted it.[8]

The Napier Mission, 1834

A far-reaching event occurred in 1834 which greatly aggravated the Sino-British relations: the abolition of the East India Company's monopoly of the China trade. The doctrines of *laissez faire* and free trade had been gathering momentum in England since the middle of the 18th century, and the East India Company's monopolistic rights came under severe attack by the rising merchant class, which had been shut out of the lucrative Asiatic trade. When the Company's charter was up for renewal in 1813, Parliament, taking into consideration the clamor for liberation of trade, threw the Indian trade open to everyone, but continued the Company's monopoly of the China trade for another twenty years. This partial concession did not satisfy the increasingly influential manufacturers and entrepreneurs of Manchester, Glasgow, and London; the private traders in Canton also renewed their demand for free trade, citing the success of the American free traders as an example. There were debates in Parliament, and in 1830 a Select Committee was appointed to investigate the problem.

The news of the forthcoming expiration of the Company's monopoly reached Canton in 1830. The prospect of the dissolution of a Company which had operated in China for more than a century troubled the local authorities. They were concerned about the future control of foreign traders, who were supposedly greedy, violent, and unfathomable like "dogs and sheep." The governor-general at Canton[9] therefore asked in 1831 that England send a *taipan* (head merchant) to Canton when the Company's monopoly ended. The British Parliament, however, decided on August 28, 1833, that three superintendents of trade should be appointed. Reso-

8. Hsin-pao Chang, 48.
9. Li Hung-pin.

lutions were adopted to end the Company's monopoly of the China trade formally on April 22, 1834; to grant to all British subjects free trade between the Cape of Good Hope and the Straits of Magellan; and to create a court of justice for the trial of offenses committed by British subjects in China and within a hundred miles of the coast. On December 10, 1833, Lord William John Napier, a Scottish peer, was appointed chief superintendent of the British trade in China, with H. C. Plowden and John Francis Davis, the last president of the Company's Select Committee in Canton, as the second and third chiefs. Plowden did not take up the assignment, and Davis succeeded as the second chief, while Sir George B. Robinson, another Company man, became the third chief. Captain Charles Elliot was made Master Attendant to take charge of "all British ships and crews within the Boca Tigris (Bogue)."

These measures resulted in a fundamental change in Sino-British relations. The British government had replaced the East India Company in dealing with China, and official relations had been substituted for private relations. While commercial interests continued to dominate policy, considerations of national honor and prestige now assumed a greater importance than ever. This turn of events dealt a severe blow to the already faltering and distintegrating Canton system. The extension of the private opium trade along the coast had in fact spelled an end to single-port trade and to the monopolistic Canton trade system. The Co-hong and the East India Company no longer controlled the expanded commercial activity, and now the Select Committee had been replaced by newly appointed officials of the British Crown. Unfortunately, the full significance of these changes had eluded the Chinese, who made no preparations to meet the new situation.

It was under these conditions that Lord Napier set out on his mission. His instructions stressed a conciliatory and moderate approach to the Chinese problem. He was to "study . . . all practical methods to maintain a good and friendly understanding," to impress upon the British subjects their "duty of conforming to the laws and usages of the Chinese empire, so long as such laws shall be administered toward (British subjects) with justice and good faith and in the same manner" as toward the Chinese and the other foreigners. Specifically he was told (1) not to employ menacing language nor to offend Chinese sensibilities, (2) not to use military force unless absolutely necessary, and (3) to adjudicate cases involving British subjects in China. On January 25, 1834, Lord Palmerston, the foreign secretary, further instructed him to announce his arrival to the governor-general at Canton by letter, and to study the possibility of extending the trade beyond Canton. In short, Napier was given the contradictory orders of placing Britain on an equal footing with China, while adopting conciliatory and friendly methods.

However, the instructions were not alone self-defeating—Napier's haughty character and limited perception sufficed to foredoom the mission. As an officer of the Crown, he was overly anxious to defend his dignity and his country's honor. Arriving in China, he went straight to Canton, took up residence at the British factory, and dispatched a letter to the governor-general announcing his arrival. In doing so he violated the Chinese regulations on several counts: he did not wait in Macao for permission to come to Canton; he did not secure permission to move into the factory, and he did not address the governor-general by a "petition" (*ping*) through the hong merchants.

As expected, the governor-general[10] rejected his letter and ordered him to leave Canton at once. Taking this as an insult, Napier accused the governor-general of "ignorance and obstinacy," and announced that while Britain had no desire for war she was "perfectly prepared" for it; he added that it would be as difficult to stop his work as "to stop the current of the Canton river." The governor-general reacted by withdrawing all Chinese employees from the British factory, cutting off its food supply, and stopping trade. Napier called in two British frigates, threatening to move them right "under the wall of the town." To Earl Grey of India he wrote: "What can an army of bows and arrows and pikes and shields do against a handful of British veterans? I am sure they would never for a moment dare to show a front. The batteries at the Bogue are contemptible; and not a man to be seen within them." Napier fancied he would "hand his name down to posterity as the man who had thrown open the wide field of the Chinese Empire to the British Spirit and Industry."[11]

The governor-general sent troops to surround the factory, declaring that Napier alone was the culprit and that his departure would restore trade to normalcy. This divide-and-rule policy proved effective—a group of English traders from Whiteman, Dent, and Brightman privately requested the Hoppo to reopen the trade. Feeling deserted and betrayed by his countrymen, Napier on September 11 retreated to Macao, where he fell ill, and died on October 11, 1834. The "Napier fizzle"—as it was called at the time—having spent itself, the trade ban was lifted.

Napier's failure was caused as much by his personal pretensions as by his contradictory instructions. He behaved as if he were a royal emissary, whereas his title was only superintendent of trade. He did not comprehend that the Chinese had not asked for a British official to come to Can-

10. Lu K'un.
11. Hsin-pao Chang, 54-57.

ton, but only a *taipan,* a head merchant. The Chinese could not see why Napier should behave differently from the past presidents of the Select Committee. They could not see how the new *taipan* Napier dared to defy the established regulations and demand an equality of status with the governor-general. On Napier's part, his readiness to use force contradicted his instructions, which called for a conciliatory approach; and his ambitious desire to acquire fame in China drove him to precipitous actions which obviated any possibility of compromise. The Duke of Wellington aptly ascribed Napier's fiasco to "an attempt . . . to force upon the Chinese authorities at Canton, an unaccustomed mode of communication with an authority, with whose power and of whose nature they had no knowledge, which commenced its proceedings by an assumption of power hitherto unadmitted."[12]

The Lull Before the Storm

John Francis Davis assumed the superintendency of trade and pursued a quiescent policy. A long-time employee of the Company and the last president of its Select Committee in Canton, he had no sympathy for the free trade movement. Private traders lost no time in ridiculing and attacking him, asserting that "One brought up in the late School of monopoly can never . . . be a fit Representative and controller of the free traders." Before the end of 1834 some eighty-five merchants had petitioned London to send a diplomat to China, accompanied by warships and soldiers, to demand reparation for the insults to Lord Napier. Davis resigned under pressure, after having been in office but a hundred days.

Sir George B. Robinson became the new superintendent in January 1835. Also a Company man, he was never known for vigor or shrewdness, or even intelligence. To avoid running into trouble with the Chinese, Robinson moved his office aboard the *Louisa* at Lintin on November 25, 1835. His policy of "not rocking the boat" pleased the Canton authorities and the trade was regular, undisturbed, and prosperous. But Robinson's inactivity was not generally appreciated by the British traders, and consequently pressure was brought for his ouster. Elliot, who had been master attendant under Napier and third chief under Davis, became the logical choice as replacement.

Captain Charles Elliot, son of a Madras governor, was not satisfied with Napier's uncompromising, pretentious attitude; nor did he approve Robinson's meek, undynamic policy. He believed that a middle-of-the-road policy of confidence and strength combined with caution and concilia-

12. Hsin-pao Chang, 61.

tion, calculated to convince the Canton authorities that Britain meant no trouble for China and had no territorial designs, would win acceptance. He had been secretly communicating his views to the Foreign Office, which was impressed and so appointed him Chief Superintendent of Trade in June 1836. He was instructed to strive for direct and equal official communication with Chinese dignitaries, and to refrain from employing the humiliating superscription *ping,* or petition, in addressing them. However, Elliot deliberately used the petition form in his first message to the governor-general, Teng T'ing-chen, in order to create a good impression and to show British "magnanimity." The Chinese found his phraseology palatable and submissive, and allowed him to come to Canton.

Having won a foothold, Elliot proceeded to fight for direct and equal communication with the Canton authorities, and in this he partially succeeded. The governor-general allowed him to send and receive sealed documents via the hong merchants rather than through the Co-hong, and to come to Canton from Macao on business any time he wished, provided he notified the subprefect of Macao first. Elliot was elated to report to London that these arrangements put him in a different position from any foreigner in China before. However, his fight for the abolition of petition was unsuccessful; he rationalized his failure by noting that Chinese officers of his own rank also addressed the governor-general in the form of *ping.*

Early in his tenure of office, Elliot learned of a move on the part of some Chinese to legalize the opium trade. The idea originated with a group of scholars at the famous academy in Canton, the *Hsüeh-hai t'ang,* who were distressed with the ineffectiveness of the prohibitory law on the one hand and the drain of silver on the other. On May 17, 1836, Hsü Nai-chi, a subdirector of the Court of Sacrificial Worship and onetime associate of the academy, boldly proposed to the court that a legal tariff be imposed on opium imports as medicine, which should be purchased by barter in order to stop the silver outflow; and that domestic cultivation of opium be permitted to slacken the demand for foreign imports. While rather unconcerned with opium-smoking by common people, he urged that scholars, officials, and soldiers be strictly prohibited from smoking. Governor-general Teng, who had also been exposed to the views of the academy, supported the legalization of opium. Foreign traders on the whole were excited by this possibility, except for a few leading opium smugglers such as Jardine, who dryly admitted: "I do not think well of the plan as far as our interests are concerned." The prospect of legalization prompted foreign traders to intensify their opium imports.

Meanwhile, two powerful memorials against legalization reached the emperor. The first[13] argued that inability to suppress opium was no justification for lifting the ban. Laws were like dikes which should not be cast away simply because parts were broken. Indeed, prostitution, gambling, treason, and robbery existed in spite of prohibitory laws. The second memorial[14] contended that legalization would make it impossible to ban smoking among the populace. It recommended that severe punishment be meted to the hong merchants, opium dealers and brokers, operators of the "fast crabs," and military officers who accepted bribes from them. The memorialist identified nine foreign opium traders—Jardine, Innes, Dent, and others—and asked for their arrest. The emperor, though having no definite view of his own, was prompted by these two papers to reject the idea of legalizing opium. On September 19, 1836, he ordered Governor-general Teng to stamp out opium and to devise a long-range plan of control. The movement for legalization, which lasted from May to September 1836, came to an abrupt halt. Foreign traders, who had anticipated the legalization, suddenly found themselves stuck with an oversupply of opium for which they had sent from India during the interval.

Governor-general Teng, who assumed office in February 1836, was a hard-working and incorruptible official. He allowed the nine foreign traders, mentioned in the second memorial, four months in which to leave Canton. Prosecuting Chinese opium dealers and addicts inexorably, he succeeded in destroying all "fast crabs" and all native smuggling networks outside Canton by the end of 1837. As a result of his suppression, the price of opium in Canton fell off sharply: in February 1838, a chest of Patna cost only $450, and Benares and Malwa, $400. Opium export from Bombay dropped from 24.2 million rupees in 1836-37 to 11.2 million in 1837-38. By December 1838, two thousand Chinese opium dealers, brokers, and smokers had been imprisoned, and executions of addicts took place daily. Jardine reported that the governor-general had been "seizing, trying, and strangling the poor devils without mercy . . . We have never seen so serious a persecution, or one so general." The *Canton Press Price Current* of January 1839 reported that "There is absolutely nothing doing, and we therefore withdraw our quotations."[15] Foreign smuggling boats disappeared by the end of 1838, and as the new year began Canton was virtually cleared of all opium traffic. The stagnation of the opium traffic produced a disastrous effect on the British traders, but they would not easily concede so lucrative a trade.

13. By Chu Tsun, a subchancellor of the Grand Secretariat and vice-president of the Board of Rites.
14. By Hsü Ch'iu, a supervising censor of the Board of War.
15. Hsin-pao Chang, 111.

Commissioner Lin at Canton

As Governor-general Teng carried out his vigorous campaign in Canton, a grand debate erupted in Peking as to the best way to stamp out the illicit traffic, which had such a deleterious effect on the morality and health of the people and caused such a drain of silver from China. In a powerful memorial of June 2, 1838, Huang Chüeh-tze, director of the Court of State Ceremonial, demanded capital punishment for all addicts who did not reform within a year. The suggestion was judged too severe by most officials, but it won the support of a small minority including Lin Tse-hsü, governor-general of Hu-Kwang. In a hortatory memorial, which has been admired by patriots for over a century, Lin warned that in a few decades, if opium was not suppressed, China would have no soldiers to fight the enemy and no funds to support an army. "When I think of this, I cannot but tremble!" said Lin. He proposed a concrete six-point program for the destruction of the smoking equipment, the reform of the smokers within a set time limit, and the punishment of native opium dealers, traders, and consumers. Only about foreign smugglers was he rather reticent. Lin was not a talker, but a man of action. In his own jurisdiction of Hupeh and Hunan, he successfully enforced the program, confiscating 5,500 pipes and 12,000 ounces of the drug. Impressed with his arguments and achievements, the emperor appointed him imperial commissioner (*ch'in-ch'ai ta ch'en*) on December 31, 1838, charging him with suppression of the Canton opium traffic.

Commissioner Lin (1785-1850) of Hou-kuan, Fukien, was an exemplary product of Old China. A holder of the *chü-jen* degree in 1804 and the *chin-shih* in 1811, he served in various official capacities—among them that of Hanlin compiler, supervisor of the Yunnan provincial examination, circuit intendant and salt controller in Chekiang, judicial and financial commissioner in Kiangsu, governor of Kiangsu, and finally, in 1837, governor-general of Hu-Kwang. His uprightness and incorruptibility won him the honorable nickname of "Lin the Blue Sky" (*Lin Ch'ing-t'ien*). Appointed imperial commissioner at fifty-four, Lin was a man of wide experience and proven probity. Nineteen times the troubled emperor conferred with him on the opium problem. On January 8, 1839, Lin set out from Peking, reaching Canton on March 10.

Having established his headquarters at the Yüeh-hua Academy, Lin vowed that he would not quit until the opium problem had been solved. His policy was to deal severely and aggressively with Chinese opium dealers, brokers, and consumers, and to confront forbearingly, yet firmly, the foreign traders. He was aware of the prestige and power of Britain, and hoped to avoid a clash with her if possible; but opium had to be suppressed, even at the risk of war. His campaign against Chinese opium dealers was remarkably successful: by May 12, 1839, 1,600 violators of the prohibitory laws had been arrested and 42,741 pipes and 28,845 catties of opium confiscated. He tried and severely punished corrupt officers who connived with the smugglers.

Foreign smugglers posed a more difficult problem. Lin had sought to learn about the West by making translations of foreign newspapers in Macao and of foreign geographical works. He had also asked the American medical missionary, Dr. Peter Parker, to translate for him three paragraphs of Vattel's *Le Droit des gens* (International Law) dealing with the right of states to prohibit contraband and to declare war. Twice he wrote to Queen Victoria to seek her intercession. In his first letter, which was distributed to the Canton foreign community but which probably did not reach England, Lin urged the queen to stop poppy cultivation and manufacture. In his second and better-known letter, he stated in part:

> There appear among the crowd of barbarians both good persons and bad, unevenly. Consequently, there are those who smuggle opium to seduce the Chinese people and so cause the spread of the poison to all provinces. . . . The wealth of China is used to profit the barbarians . . . By what right do they in return use the poisonous drug to injure the Chinese people? . . . Let us ask, where is their conscience? I have heard that the smoking of opium is very strictly forbidden by your country . . . Why do you let it be passed on to the harm of other countries? Suppose there were people from another country who carried opium for sale to England and seduced your people into buying and smoking it; certainly your honorable ruler would deeply hate it and be bitterly aroused . . . Naturally you would not wish to give unto others what you yourself do not want . . . May you, O Queen, check your wicked and sift your vicious people before they come to China, in order to guarantee the peace of your nation, to show further the sincerity of your politeness and submissiveness.[16]

The letter was carried by Captain Warner of the *Thomas Coutts* in January 1840 to London, but the Foreign Office refused to recognize him (Warner).

Lin admonished foreign traders in Canton from the standpoints of natural law (*t'ien-li*), common sense, Chinese prohibitory regulations, and

16. S. Y. Teng and John K. Fairbank, *China's Response to the West: A Documentary Survey, 1839-1923* (Cambridge, Mass., 1954), 24-27, with minor changes. The excerpts throughout this book from *China's Response to the West* are reprinted by permission of Harvard University Press.

government policy. He announced that having come from the seacoast of Fukien himself, he was well aware of the barbarians' tricks and would not fall into their traps. On March 18, 1839, he ordered them to surrender all their opium in three days and sign a bond pledging not to engage in the illicit traffic in the future; violation of the bond would result in the death penalty and the confiscation of the drug. Lin offered a reward of five catties of tea for each chest of opium surrendered, but he never once mentioned monetary compensation; nor did he ever consider the British government's economic interest in the opium trade.

When the foreigners ignored his deadline of March 21, Lin threatened to decapitate two hong security merchants. The foreign traders surrendered 1,036 chests of opium as a token, which was, of course, unsatisfactory to the commissioner. Howqua and the elder Mowqua, the two leading hong merchants, were made to wear chains, and the former's son and the latter's brother were thrown into prison. Lin then turned to the British trader Dent, who was said to have been involved in more than half of the opium imports and silver exports. Dent was asked to surrender himself to the prefect of Canton, but he refused to do so unless the commissioner guaranteed him safe return. Howqua pleaded with the foreign merchants, reminding them that he would surely lose his head if Dent continued to resist. On March 23, Elliot came from Macao to join the traders at the factory; and on the 24th Lin ordered the stoppage of trade, the withdrawal of Chinese compradores and servants, and the siege of the British factory. Three hundred and fifty foreigners were confined to the factory compounds, inconvenienced by the loss of cooks, porters, and servants but never suffering from the lack of important provisions. Frequently the hong merchants, linguists, and former servants smuggled in bread, fowls, mutton, eggs, oil, and sugar. The greatest discomforts were the monotony, the muggy weather, and the uncertainty of the future. The detention lasted for six weeks. To Elliot, it was a piratical act against British lives, liberty, and property; but to Lin it was a rightful enforcement of Chinese laws and a just punishment for depraved smugglers.

Lin let it be known that when the first quarter of the opium was surrendered, the compradores, servants, and cooks would be returned; when the second quarter was surrendered, the passage boats between Whampoa and Macao would be allowed to resume activity; when the third quarter was surrendered, the siege of the factory would be lifted; and when the last quarter was given up, trade would be resumed.

It must be noted that there had been a stagnation of the opium trade for several months before the detention. On March 22, 1839, Matheson recorded that "not a chest of opium had been sold in Canton for the last five months." Some fifty thousand chests lay waiting for outlet, and more were on their way from Bombay. It occurred to Elliot that to surrender the opium to Lin would relieve the stagnant trade and would be a good way to hold the Chinese responsible for the cost. On March 27, 1839, he issued a notice in the name of his government ordering all British traders to surrender their opium to him for deliverance to Lin:

> Now I, the said Chief Superintendent . . . do hereby, in the name and on the behalf of Her Britannic Majesty's Government, enjoin and require all Her Majesty's subjects now present in Canton, forthwith to make a surrender to me, for the service of Her Said Majesty's Government, to be delivered over to the Government of China, of all the opium belonging to them or British opium under their control . . . and I . . . do now, in the most full and unreserved manner, hold myself responsible, for and on the behalf of Her Britannic Majesty's Government, to all and each of Her Majesty's subjects surrendering the said British-owned opium into my hands to be delivered over to the Chinese Government.[17]

With this proclamation the ownership of the opium changed hands: it was no longer the private property of the traders, but the public property of the British government. Elliot's decision was praised as "a large and statesmanlike measure" by Matheson, who also confessed that "the Chinese have fallen into the snare of rendering themselves directly liable to the British Crown. Had the Chinese declined receiving it . . . our position would have been far less favourable."[18] Elliot pledged to surrender 20,306 chests of opium to Lin, but he actually delivered 21,306 chests by May 18. Lin had originally planned to send the opium to Peking for inspection and destruction, but the complexity of transporting such a large amount caused the emperor to order him to destroy it locally. Three large trenches—150 feet long, 75 feet wide, and 7 feet deep—were dug for the purpose. Beginning June 3, in the presence of high officials and foreign spectators, opium balls were crushed to pieces and thrown into the trenches, where a profuse amount of salt and lime was scattered over two feet of water. The laborers stirred the opium in the mixture until it was completely dissolved and then flushed it to a nearby creek, which carried the last shred of debris to the ocean.[19] Lin, it seemed, had scored a complete moral and legal victory over opium, but the victory was chimerical, for Britain would never be content to rest her case there.

17. Hsin-pao Chang, 264-65.
18. *Ibid.,* 166.
19. The destruction lasted twenty-three days, until June 25.

After their liberation from detention, Elliot and the entire British community left for Macao on May 24, 1839, rather than accept Lin's demand for the bond. Elliot lost no time in urging London to start "prompt and vigorous proceedings" against China, and the traders also jointly petitioned Palmerston to protect British interests and to take steps to fulfill Elliot's promise of reimbursement for the surrendered opium. A special deputation under Jardine was sent to London to promote these views. Meanwhile, nearly three hundred firms in London, Manchester, and Liverpool connected with the China trade started a campaign for action. Numerous pamphlets and stories were circulated condemning the Chinese insult to the British subjects. One pamphleteer said: "You take my opium; I take your island in return, we are therefore quits; and henceforth, if you please, let us live in friendly communion and good fellowship."[20] On October 18, 1839, without prior consultation with Parliament, Palmerston informed Elliot that the government had decided to send an expeditionary force to blockade Canton and the Pei-ho.

The tense situation in the Canton-Macao area was further strained by the killing of a Chinese villager[21] by a group of English seamen in Kowloon on July 12, 1839. Commissioner Lin demanded the surrender of the culprits, stating: "He who kills a man must pay the penalty with his life; whether he be a native or a foreigner, the statute is in this respect quite the same." Elliot refused to submit British subjects to Chinese law; he tried the six suspects himself aboard the *Fort William*, sentencing two of them to three months' imprisonment at hard labor in England and a fifteen-pound fine, three more to six months' imprisonment and a twenty-five pound fine, and acquitting the last. But in fact, when the sailors returned to England they went unpunished, because the government ruled that Elliot had no authority to try them. Commissioner Lin, on his part, was irritated with Elliot's refusal to cooperate, and brought pressure to bear on the Portuguese authorities at Macao to expel the British. On August 26, 1839, all British subjects left for Hong Kong, a small barren island of some thirty square miles, about ninety miles from Canton. Commissioner Lin and Governor-general Teng then made a triumphant tour of Macao. Up to this point Lin had won at every stage of the conflict.

However, one issue remained unresolved: the signing of the bond. Elliot had persistently resisted it on the ground that the death penalty without a fair trial for the violators was uncivilized and contrary to the British concept of justice. In point of fact, the British had refused to submit to Chinese jurisdiction since 1784, and the Americans since 1821. While

20. Hsin-pao Chang, 192.
21. Lin Wei-hsi.

Elliot remained adamant, some British traders felt that he had no right to stop them from accepting the bond. Accordingly, the captains of the *Thomas Coutts* and the *Royal Saxon* signed it on their own, in defiance of Elliot's order. On November 3, 1839, when the *Royal Saxon* approached the Bogue in hopes of trading with the Chinese, Captain H. Smith of H.M.S. *Volage* fired a shot across its bow. In an attempt to protect the *Royal Saxon,* the Chinese navy under Admiral Kuan engaged the British ships at Ch'uan-pi. Of the twenty-nine Chinese war junks, one was blown to pieces immediately, three were sunk, and several more were seriously damaged. War had now broken out, although there was no formal declaration by the Chinese, but the Indian government did issue one on behalf of the British Crown on January 31, 1840.

The trade with the British was stopped "forever" on December 6, 1839, but certain venturesome British traders managed to continue business under the American flag. Many American firms had accepted the bond; Robert Forbes of Russell and Company declared that "I had not come to China for my health or pleasure, and . . . I shall remain at my post as long as I could sell a yard of goods or buy a pound of tea . . . We Yankees had no Queen to guarantee our losses."[22] It was not until June 1840, when the British reinforcement had arrived to renew the fighting, that the Americans left Canton for Macao.

The Opium War

The British expeditionary force arrived under Rear Admiral George Elliot. It consisted of sixteen warships mounting 540 guns, four armed steamers, twenty-seven transports, one troop ship, and 4,000 soldiers. For the British, the war was one of reprisal, a necessary action to defend their right to trade, to uphold their national honor, to correct the injustice inflicted upon the British officials and subjects in China, and to secure an open future. For the Chinese, the war was primarily a crusade against opium.

Admiral Elliot was appointed first commissioner, procurator, and plenipotentiary, while his cousin, Captain Elliot, assumed the second in command. Their instructions called for (1) satisfaction for the illegal detention of the British Superintendent of Trade and of British subjects generally; (2) the return of the surrendered opium or suitable compensation; (3) satisfaction for the affront and indignity heaped upon the British superintendent and subjects, and assurance of future security;

22. Hsin-pao Chang, 206.

(4) the cession of one or more islands; and (5) abolition of the monopolistic system of trade at Canton and repayment of the hong merchants' debts. Palmerston ordered the expedition to blockade all principal ports of China so as to impress the Chinese with British might; to demand compensation of military expenses; to occupy Chusan until the indemnity was fully paid; and to demand the reply of the Chinese government at the Pei-ho, although negotiations might be conducted elsewhere. Admiral Elliot was also instructed to deliver a letter from Palmerston to the Chinese officials at either Amoy, or Ningpo, or the Pei-ho, for transmission to the court.

THE FIRST STAGE. The war itself can be divided into three stages. The first lasted from the arrival of Admiral Elliot in June 1840 to the conclusion of the Ch'uan-pi Convention in January 1841. Commissioner Lin, in anticipation of an attack on Canton, had gathered a "water force" of some sixty warjunks, fortified the batteries at the Bogue with more than two hundred newly purchased foreign guns, and blockaded the river with huge iron chains. The British, however, did not attack Canton; they merely blockaded it and sailed north. The two Elliots attempted to deliver Palmerston's letter at Amoy on July 2, but were fired upon in spite of the white flag, which the Chinese apparently did not comprehend. They proceeded north and occupied Tinghai on the Chusan Islands on July 5. Unable to deliver the letter at Ningpo on July 10, they blockaded it, too, and sailed further north to the Pei-ho on August 29. There the letter was received by the governor-general, Ch'i-shan (Kishen).

Up to now, the emperor had had complete trust in Lin and endorsed his undertakings with the encouraging remark: "I do not worry about your aggressive prowess, but I admonish you against timidity." After the fall of Chusan and the blockade of the ports from Ningpo to the mouth of the Yangtze River, provincial officials began to criticize Lin for provoking the British into action, and the Manchu grand secretary and grand councillor, Mu-chang-a, also disapproved of Lin's hard, coercive policy. The emperor's confidence in Lin faltered, and when the British advanced to the Pei-ho, near Tientsin, threatening directly the security of Peking, his faith in Lin collapsed. Blaming him for creating complications without solving the opium problem, the emperor scolded Lin sternly: "Externally you wanted to eliminate the [opium] trade, but it has not been cut off; internally, you wanted to arrest the outlaws [smugglers], but they have not been cleared away. You have produced nothing more than empty excuses. Not only have you really accomplished nothing, you have, on the contrary, created many troubles. When I think of this, how angry I become! Let me see what explanation you have to make!" Lin sent a memorial saying that if China had used one-tenth of the customs revenue for making gunboats it would have no difficulty tackling the barbarian problem; to which the imperial reply was: "All nonsense." Since Palmerston's letter had complained, among other things, of Lin's injurious proceedings at Canton and demanded "from the emperor satisfaction and redress," the emperor took it to mean that he needed only to redress their grievances to reach a settlement. He authorized Ch'i-shan at Tientsin to receive the two Elliots and determine precisely what they wanted.

Ch'i-shan, a sly politician and a wily diplomat, knew well Peking's veiled anxieties over the British naval demonstration. As governor-general of the capital province of Chihli, he was responsible for safeguarding Peking; yet he was without means of defense. Chinese guns were obsolescent; those found at Shanhaikuan were left over from the Ming dynasty. In contrast, the British possessed powerful guns and speedy ships. With such inequality in weaponry and equipment, and with the disheartening news that the Yangtze and coast areas had all been blockaded, Ch'i-shan concluded not only that it was senseless to fight, but that it was essential to appease the barbarians. In view of the British complaint about Lin's mistreatment in Canton, Ch'i-shan, grasping at straws, came to believe that possibly the British had come north not to fight, but simply to plead a redress of grievances. In his mind the situation was not unlike a litigation between Captain Elliot and Commissioner Lin, awaiting adjudication by the emperor. On the basis of this diagnosis, Ch'i-shan treated Captain Elliot courteously, and, employing mollifying tactics and flattery, told him that the emperor, having learned of the British grievances, had dispatched a high official to Canton to investigate; and that it would be best for the British to return south, where the truth of the dispute could be ascertained and negotiations taken up. Encouraged by the prospects of negotiations and settlement, the two Elliots left the Pei-ho on September 15. Thus, without firing a gun or losing a soldier, Ch'i-shan rid North China of the enemy.[23] Impressed with his diplomacy, the emperor appointed him as imperial commissioner, while Lin was dismissed in disgrace and exiled to Ili, Sinkiang.

In the British hierarchy there was a change of command, too. Captain Elliot rose in power until he replaced Admiral Elliot as the first plenipotentiary on November 29, 1840, the latter said to have contracted a "sudden and severe illness." In his negotiation with Ch'i-shan at Canton

23. T. F. Tsiang, "New Light on Chinese Diplomacy, 1836-49," *The Journal of Modern History,* 3:4:578-91 (Dec. 1931); "Ch'i-shan yü Ya-p'ien chan-cheng" (Ch'i-shan and the Opium War), *Tsing-hua hsüeh-pao,* 6:3:1-26 (Oct. 1931).

during the latter part of December 1840, Captain Elliot demanded the cession of Hong Kong and an indemnity. Ch'i-shan realized that the situation was far more serious than a simple case of litigation between Elliot and Lin. Though conciliatory, he would not yield, for he knew the court would not approve the territorial cession. Captain Elliot then attacked the forts at Ch'uan-pi and threatened to take the Bogue. On January 20, 1841, he forced Ch'i-shan to agree to draft a "Ch'uan-pi Convention," which provided: (1) cession of Hong Kong, though the customs dues were still to be collected by the Chinese government; (2) an indemnity of $6 million; (3) direct, equal intercourse between the officials of the two countries; and (4) reopening of Canton to trade within ten days of the Chinese New Year, i.e., before February 1.

Ch'i-shan did not affix his seal to the convention but agreed to memorialize the throne for its approval. Meanwhile, he secured the British consent to evacuate Tinghai, return the forts near the Bogue, and limit trade to Canton. However, the British occupied Hong Kong even before the convention was ratified by the court. The emperor, so enraged by the terms of the convention, deposed Ch'i-shan and recalled him in chains to stand trial for his unauthorized cession of territory and agreement to pay an indemnity. According to the court, Ch'i-shan had been sent to Canton to investigate the situation caused by Lin's mismanagement and to correct the wrongs; he had no power to sign any agreement with foreigners. His punishment was confiscation of family property (estimated at £10 million) and death, which was later commuted to exile to the Amur in May 1842.

The British government was equally displeased with the terms of the convention. The indemnity was considered too small to cover the value of the surrendered opium; the evacuation of Tinghai was thought premature; and the cession of the sovereignty of Hong Kong was deemed incomplete. Palmerston informed the queen that Captain Elliot had not made full use of the military force at his disposal, and that he had accepted the "lowest" possible terms. On April 21, 1841, he administered a stern reprimand to Elliot: "You have disobeyed and neglected your instructions . . . Throughout the whole course of your proceedings, you seemed to have considered that my instructions were waste paper . . . and that you were at full liberty to deal with the interests of your country according to your own fancy. . . . You have agreed to evacuate the Island immediately. . . . You have obtained the cession of Hong Kong, a barren island with hardly a house upon it; and even this cession as it is called, seems to me, from the condition with which it is clogged, not to be a cession of the sovereignty of the island, which could only be made by the signature of the Emperor, but to be a permission to us to make a settlement there, upon the same footing on which the Portuguese have an establishment at Macao."[24]

That Elliot dared to ignore his instructions may be explained by the fact that for three years he had not had any and was forced to act on his own in situations of great difficulty and much delicacy. So used was he to freedom of action that when he at last was given specific instructions, he did not realize that he had to follow them explicitly. Elliot defended his position by saying that the evacuation of Tinghai was made necessary by the high rates of sickness and death among soldiers from dysentery, fever, and diarrhea; that the resumption of trade, after the conclusion of the draft convention, released 20,000 tons of shipping that had been held up, including the shipment of 30 million pounds of tea, which should net the British customs £3 million; that the restoration of commerce would promote an atmosphere of peace and demonstrate British magnanimity. However, before his defense reached London, the cabinet had decided on April 30, 1841, to dismiss him, disavow the convention, and appoint Colonel Sir Henry Pottinger as the new plenipotentiary to China.

THE SECOND STAGE. The repudiation of the Ch'uan-pi Convention by both governments ushered in a new phase of the war. The emperor appointed his nephew, I-shan, as imperial commissioner and barbarian-suppressing general in command of a large force against the British. Seizing the initiative, Captain Elliot, still in command before the arrival of Pottinger, took the Bogue forts in late February 1841, destroyed the Chinese defenses, occupied all the strategic points in the Pearl River, and besieged the city of Canton, where large Chinese forces were trapped. The hong merchants and the prefect of Canton offered a "ransom" of $6 million to save the city from destruction. Elliot accepted it to free his troops for the northern expedition, as he believed that pressure should be put on the court directly and not dissipated on the fringes of the empire. A second truce was reached on May 27, 1841, on the following terms: (1) payment of $6 million within one week to the British; (2) withdrawal of Chinese troops sixty miles outside of Canton within six days; (3) evacuation of the British troops from the Bogue; (4) exchange of prisoners of war; and (5) postponement of the question of the cession of Hong Kong. With the complete payment of the ransom, the British forces began to withdraw on May 31, 1841. At this point, a body of

24. George H. C. Wong, "The Ch'i-shan-Elliot Negotiations Concerning an Off-shore Entrepot and a Re-Evaluation of the Abortive Chuenpi Convention," *Monumenta Serica*, 14:539-73 (1949-55).

10,000 irate Cantonese, who had been organized by local gentry, launched a sudden attack at San-yüan-li, causing surprise but no great damage to the retreating British.[25] Marxist historians have hailed this incident as the first sign of Chinese nationalism.

THE THIRD STAGE. The arrival of Sir Henry Pottinger in Macao and the departure of Captain Elliot for England in August 1841, marked the beginning of the third stage of the war. Pottinger had been instructed to bypass Canton and go north to reoccupy Tinghai; to seize the important places on the Yangtze River; and if necessary to push north to the Pei-ho to open negotiations, at which time he was to demand monetary compensation, extension of trading ports, security of British subjects in China, and outright cession of Hong Kong. These terms were to be included in a formal treaty, which was to be approved by the Chinese emperor before being sent to the queen.

Pottinger carried out his instructions meticulously. After leaving a few ships to guard Hong Kong, he moved north on August 21, 1841, with 10 ships and 4 steamers carrying 336 guns and 2,519 men. Amoy was occupied on August 26, Tinghai on October 1, and Ningpo on October 13. As the alarmed court mobilized more troops and militia from the provinces, Pottinger had also received reinforcements from India in the spring of 1842: 25 warships carrying 668 guns, 14 steamers carrying 56 guns, 9 hospital and surveying ships, and troops for a total strength of 10,000 men, besides artillery. Moving swiftly, the British occupied Woosung on June 16, 1842, Shanghai on June 19, and Chinkiang on July 21—the last an important communication center at the crux of the Grand Canal and the Yangtze River, from which grains were shipped to North China. Its loss caused great anxiety among the provincial officials, who now requested the emperor to permit peace negotiations. The futility of war was obvious; furthermore it was imperative that the Manchu dynasty not lose any more face before the Chinese, lest they be encouraged to revolt. Ch'i-ying (Kiying), Tartar-General of Canton, was made imperial commissioner, and together with I-li-pu, the Deputy Lieutenant General of Chapu and former imperial commissioner, was ordered by the court to start peace negotiations. Pottinger, refusing to negotiate until Ch'i-ying produced his "full powers," poised his ships for an attack on Nanking on August 9. On the 17th the peace terms were accepted in principle by Ch'i-ying and I-li-pu, and after several more days of settling details and translating the text into Chinese, the formal Treaty of Nanking, consist-

25. One British private was killed, one officer and fourteen men wounded. For details of this incident, see Wakeman, *Strangers at the Gate*, 11-21.

ing of thirteen articles, was signed on the *Cornwallis,* on August 29, 1842,
the general tenor of which follows:

1. An indemnity of $21 million: $12 million for military expenses, $6
   million for the destroyed opium, and $3 million for the repayment
   of the hong merchants' debts to British traders.
2. Abolition of the Co-hong monopolistic system of trade.
3. Opening of five ports to trade and residence of British consuls and
   merchants and their families: Canton, Amoy, Foochow, Ningpo,
   and Shanghai.
4. Cession of Hong Kong. (The Chinese text of the treaty euphemistically states that the emperor graciously grants a place of rest and
   storage to the British after their long voyage to China.)
5. Equality in official correspondence.
6. A fixed tariff, to be established shortly afterwards.

This treaty was imposed by the victor upon the vanquished at gunpoint,
without the careful deliberation usually accompanying international agreements in Europe and America. A most ironic point was that opium, the
immediate cause of the war, was not even mentioned—the question of its
future status cautiously avoided by both sides. The emperor painfully approved the treaty on September 15, and Queen Victoria's ratification came
on December 28, 1842.

A supplementary Treaty of the Bogue was signed on October 18, 1843,
which fixed the import duty from 4 percent to 13 percent ad valorem,
averaging 5 percent, and the export duty from 1.5 percent to 10.75
percent.[26] It also allowed British consuls to try their own subjects (i.e.
extraterritoriality); allowed British warships to anchor at the five ports
to protect commerce and control sailors; and gave Britain the most-
favored-nation treatment, whereby China would grant Britain whatever
rights that might be conceded to other powers later.

Close on the heels of the British came the Americans and the French,
requesting similar treaties. Needless to say, after their defeat in the Opium
War the Chinese were anxious to avoid new conflicts. They reasoned
that denial of these requests would drive the Americans and French to
seek trade under British auspices, in which case the Chinese would have
difficulty distinguishing them, since they all looked alike and spoke
equally unintelligible languages. Added to this concern was the fear that
the French and Americans would be grateful to the British for the
privileges and not to the Chinese, who felt that American and French
good will might in the future protect China from collusion among the

26. Stanley F. Wright, *Hart and the Chinese Customs* (Belfast, 1950), 58.

three powers, and perhaps even obtain their aid against further foreign encroachments. Moreover, the struggle for profits among the foreigners might lead to conflict among themselves, which fitted well into the traditional Chinese policy of playing off the barbarians against one another (*i-i chih-i*). Since there was a limit to China's foreign trade potential, it mattered little whether the whole profit went to the British alone or was shared with the others. Granting the American and French demands would allow them to cut into British profit without injuring China. Because the British had confidently declared that they did "not desire to obtain for British subjects any exclusive privileges of trade which should not be equally extended to the subjects of any other Power," the Chinese saw no reason to deny France and America a share in the fruits of British labors. For all these reasons, China decided to comply with the American and French requests for treaties. On July 3, 1844, Caleb Cushing signed the Treaty of Wanghsia for the United States, and Théodore de Lagréné signed the Treaty of Whampoa for France on October 24, 1844. The American treaty specified the prohibition of the opium trade, extraterritoriality, the most-favored-nation treatment, the right to maintain churches and hospitals in the five ports, and treaty revision in twelve years. The French treaty stipulated in addition the free propagation of Catholicism.[27]

In these treaties three stipulations were particularly injurious to China—the fixed tariff, extraterritoriality, and the most-favored-nation clause. They were granted partly out of expediency and partly out of ignorance of international law and the concept of national sovereignty. The fixed tariff of 5 percent ad valorem, as suggested by the British, was readily accepted by the Chinese for the simple reason that it was higher than the existing imperial tariff, which averaged only 2 percent to 4 percent ad valorem, although the irregular fees had been high. Little did the Chinese realize that their assent to a fixed rate precluded a protective tariff in the future. Extraterritoriality was signed away under the expedient notion that the barbarians, who spoke different languages and had strange customs, should be allowed to govern themselves—to show Chinese magnanimity and to ease the task of governing them.[28] The most-favored-nation treatment was granted *pro forma* on the ground that the emperor looked upon men from

27. T. F. Tsiang, "The Extension of Equal Commercial Privileges to Other Nations than the British after the Treaty of Nanking," *The Chinese Social and Political Science Review* (CSPSR), 15:3:422-44 (Oct. 1931); Thomas Kearny, "The Tsiang Document, Elipoo, Keying, Pottinger, and Kearny and the Most Favored Nation and Open Door Policy in China in 1842-1844, An American View," CSPSR, 16:1:75-104 (April, 1932).
28. There was the precedent of Arab traders at Zayton (Ch'üan-chou) and at Canton during medieval times, when they were governed by their own chieftains.

afar with equal benevolence. The more practical considerations have been discussed in the preceding paragraph.

These British, American, and French treaties reinforced each other and formed the beginning of a treaty system, which was further enriched and enlarged by later agreements. Because they were not negotiated by nations treating each other as equals but were imposed on China after a war, and because they encroached upon China's sovereign rights, they have been dubbed "unequal treaties," which reduced China to semicolonial status. The Opium War introduced a century of humiliation for the Chinese people.

The outcome of the war was inevitable, considering the decay of the Ch'ing dynasty and the new power achieved by Britain after the Industrial Revolution. But in the conduct of the war, the emperor's vacillation between resistance and concession, war and peace; the erroneous assessment of London's commitment to overseas interests; and the lack of accurate information about the enemy—all these presaged defeat. Commissioner Lin was convinced that London would not support its traders over so vicious and infamous an issue as the opium trade. But he did not realize that without the illicit traffic the British could not conduct regular trade without incurring a tremendous deficit; nor did he know that the expansionist Victorian government was keen on defending its foreign interests. Some of the Chinese misconceptions of the enemy were appalling and ludicrous. Lin believed that the British could not live without tea and rhubarb, and that their soldiers' legs could not stretch because of the puttees. A censor suggested that any attack on their feet would be fatal, while Ch'i-ying reported that the barbarians could see but poorly at night!

In retrospect, it is apparent that opium was the immediate, but not the ultimate, cause of the war. Without it a conflict between China and the West would still have erupted as a result of their differing conceptions of international relations, trade, and jurisprudence. Far deeper than the opium question was the incompatibility of the Chinese claim to universal overlordship with the Western idea of national sovereignty; the conflict between the Chinese system of tributary relationships and the Western system of diplomatic intercourse; and the confrontation between self-sufficient, agrarian China and expansive, industrial Britain. Indeed, the Smithsian idea of free trade and the Chinese contempt for trade could not coexist. The power generated by the Industrial Revolution and the idea of progress through change propelled the West into overseas expansion. There was no way to stop it. It was unfortunate that the Manchu court and the Chinese scholars and officials had absolutely no recognition of these facts, and consequently China's confrontation with the West was rendered extremely painful.

The Opium War touched off explosive matters with far-reaching consequences. Politically, the cession of Hong Kong gave Britain a foothold in China for further advancement; the opening of the five ports extended foreign, particularly British, influence to the entire Eastern coast of China; and the loss of the three national rights mentioned above relegated China to a semicolonial state. Militarily, permission for foreign gunboats to anchor at the five ports, a concession later extended to the other ports opened along the Yangtze River, enabled foreign warships to navigate freely and legally in Chinese inland waterways, exposing the interior of the country mercilessly to alien powers. Economically, the fixed customs rates deprived China of a protective tariff and allowed an overabundant influx of foreign goods, which reduced Chinese handicraft industries to penury, causing social unrest and rebellion. Socially, the continuation of the illicit traffic deepened the opium problem, and the growth of foreign trade in the five ports introduced a new class of business entrepreneurs, sometimes derogatorily called the "compradore" class, who came to wield an increasing influence in society. Diplomatically, China entered into official contacts with the Western maritime powers and took the first step in its long journey to membership in the international society.

But the Opium War did not shock the Chinese people into realizing their backwardness. The fact that Commissioner Lin was dismissed before he had a chance to fight the enemy led many to believe that the defeat was an historical accident. They refused to acknowledge China's military inferiority and political retrogressiveness, and so allowed themselves to sleep another twenty years.

Only a few exceptionally alert men realized the need to learn about the West. Wei Yüan, an associate of Lin and an eminent scholar of the Modern Text School of classical learning, compiled the famous *Illustrated Gazetteer of the Maritime Countries* (*Hai-kuo t'u-chih*) in 1844, which was revised and enlarged in 1847 and 1852 into a hundred tomes (*chüan*). Another important work on world geography was compiled by the governor of Fukien, Hsü Chi-yü, in 1850, under the title, *A Brief Survey of the Maritime Circuit* (*Ying-huan chih-lüeh*). A humble beginning in Western studies was thus made, but greater efforts had to wait until more intense shocks stunned the Middle Kingdom.

Further Reading

Beeching, Jack, *The Opium Wars in China, 1834-1860* (London, 1975).

Chang, Hsin-pao, *Commissioner Lin and the Opium War* (Cambridge, Mass., 1964).

Ch'i, Ssu-ho 齊思和 et al. (eds.), *Ya-p'ien chan-cheng* 鴉片戰爭 (The Opium War), (Shanghai, 1954), 6 vols.

Ch'ing-shih wen-t'i (complete issue devoted to the Opium War), Vol. 3, No. 1 (Dec. 1977).

Fairbank, J. K., "Chinese Diplomacy and the Treaty of Nanking," *Journal of Modern History*, 12:1:1-30 (March 1940).

————, "The Manchu Appeasement Policy of 1843," *Journal of the American Oriental Society*, 59:4:469-84 (Dec. 1939).

————, *Trade and Diplomacy on the China Coast* (Cambridge, Mass., 1953), 2 vols.

Fay, Peter W., *The Opium War, 1840-1842* (Chapel Hill, 1975).

Fox, Grace, *British Admirals and Chinese Pirates, 1832-1869* (London, 1940).

Greenberg, Michael, *British Trade and the Opening of China, 1800-42* (Cambridge, 1951).

Grosse-Aschhoff, Angelus, *Negotiations between Ch'i-ying and Lagrené, 1844-1846* (New York, 1950).

Holt, Edgar, *The Opium Wars in China* (Chester Springs, Pa., 1964).

"Journal of Occurrences at Canton, 1839," Intro. by E. W. Ellsworth and notes by L. T. Ride and J. L. Cranmer-Byng, *Journal of the Hong Kong Branch of the Royal Asiatic Society*, 4:1-33 (1964).

Kearny, Thomas, "The Tsiang Document, Elipoo, Keying, Pottinger and Kearny and the Most Favored Nation and Open Door Policy in China in 1842-1844, an American View," *The Chinese Social and Political Science Review*, 16:1:75-104 (April 1932).

Kuo, P. C., *A Critical Study of the First Anglo-Chinese War, with Documents* (Shanghai, 1935).

Leonard, Jane Kate, *Wei Yuan and China's Rediscovery of the Maritime World* (Cambridge, Mass., 1984).

Morse, H. B., *The International Relations of the Chinese Empire* (London, 1910), I, chs. 6-12.

Owen, David E., *British Opium Policy in India and China* (New Haven, 1934).

Teng, Ssu-yü, *Chang Hsi and the Treaty of Nanking, 1842* (Chicago, 1944).

Tsiang, T. F., "New Light on Chinese Diplomacy, 1836-49," *The Journal of Modern History*, 3:4:578-91 (Dec. 1931).

————, "The Extension of Equal Commercial Privileges to Other Nations than the British after the Treaty of Nanking," *The Chinese Social and Political Science Review*, 15:3:422-44 (Oct. 1931).

————, "Difficulties of Reconstruction after the Treaty of Nanking," *The Chinese Social and Political Science Review*, 16:2:319-27 (July 1932).

————, *Chung-kuo chin-tai shih ta-kang* 中國近代史大綱 (An outline of Chinese modern history), (Taipei, 1959), ch. 1.

————, "Ch'i-shan yü Ya-p'ien chan-cheng" 琦善與鴉片戰爭 (Ch'i-shan and the Opium War), *Tsing-hua hsüeh-pao*, 6:3:1-26 (Oct. 1931).

————, *Chin-tai Chung-kuo wai-chiao shih tzu-liao chi-yao* 近代中國外交史資料輯要 (A collection of essential sources of modern Chinese diplomatic history), (Taipei, 1958), I, chs. 1-2.

Wakeman, Frederic, Jr., *Strangers at the Gate: Social Disorder in South China, 1839-1861* (Berkeley, 1966).

————, "The Canton Trade and the Opium War" in John K. Fairbank (ed.), *The Cambridge History of China* (Cambridge, Eng., 1978), Vol. 10, 163-212.

Waley, Arthur, *The Opium War Through Chinese Eyes* (London, 1958).

Wong, George, H. C., "The Ch'i-shan-Elliot Negotiations Concerning an Off-shore Entrepôt and a Re-Evaluation of the Abortive Chuenpi Convention," *Monumenta Serica,* 1:539-73 (1949-55).

Wright, Stanley F., *Hart and the Chinese Customs* (Belfast, 1950), ch. 2.

9

The Second Treaty Settlement

In the postwar period, Ch'i-ying (Kiying, Keying), the signer of China's first treaties with the West, emerged as the most colorful, spirited, and successful figure in Chinese foreign relations. Having pacified the British and saved the dynasty from a disastrous barbarian onslaught, he and his senior colleague, I-li-pu, enjoyed the dubious reputation of being the foremost experts on barbarian affairs. The court at Peking came to respect and rely upon them for management of the barbarians. On October 17, 1842, I-li-pu was made imperial commissioner and Tartar-general of Kwangtung, and Ch'i-ying was given the influential and lucrative post of governor-general of Liang-Kiang. That Ch'i-ying was kept at Nanking, rather than sent to Canton, indicated the need for a man of his experience to take charge of the opening of the ports, the development of trade regulations, and the general superintendence of Sino-Western relations in Kiangsu, Chekiang, and Fukien. His position was further enhanced upon the death of I-li-pu on March 4, 1843. The coveted title of imperial commissioner was conferred on him on April 6, and in this capacity he took over the general direction of Chinese foreign relations at Canton. Until his retirement in 1848, he was, in effect, China's "foreign minister."

Ch'i-ying's New Diplomacy

Desiring power and responsibility, Ch'i-ying actually had projected himself into the imperial commissionership. To the court he had maximized British confidence in him as a negotiator, intimating that there were matters only he could settle with them. Indeed, Ch'i-ying had evolved a new approach to foreign affairs which proved quite effective at times: the policy of friendship and personal diplomacy. Every effort was made to convince the foreign representatives of his sincerity, trustworthiness, and cooperation. In June 1843, shortly after his appointment as imperial commissioner, he was taken at his request in a British gunboat to Hong Kong, where he wined and dined and engaged in cordial social intercourse with the British. Meeting Pottinger again, he embraced him "with all the warmth and sincerity of an old friend and was even visibly affected by the strength of his emotion."[1] Ch'i-ying fulsomely praised the British gunboat and its captain, visited the admiral's flagship, participated in various banquets, sang operatic airs, played the "game of finger-guessing," drank to everybody's health, and showered affection on Pottinger.[2]

FRIENDSHIP WITH POTTINGER. Ch'i-ying's policy of friendship and personal diplomacy, while it was applied to foreign representatives in general, found its highest expression in his dealings with Pottinger. Having discovered that the British chieftain had an annual salary of $10,000, Ch'i-ying surmised that he must be a man of considerable importance in his own country, one who enjoyed great discretionary power in China and one who could probably exercise great influence in the high councils when he returned home. The cultivation of friendship and confidence with such a man was not only expedient, but essential. Ch'i-ying seized every opportunity to develop an intimate relationship with him. When shown a portrait of Pottinger's family, Ch'i-ying admired his son and declared that he should like to adopt him, as he had no son of his own. He then proposed to exchange his wife's picture for Lady Pottinger's—a most unusual gesture for a Manchu, prompted probably by his recent discovery that foreigners esteemed women. Not wishing to offend, Pottinger yielded on both counts, and his son's name became "Frederick Keying Pottinger." Following the establishment of this family relationship, Ch'i-ying exchanged gifts with Pottinger, giving him a gold bracelet and receiving in return an English sword and a belt. Ch'i-ying went so far as to say that when he himself returned to Peking in three or four years, he would recommend to the emperor that the famous Pottinger be invited back from England to receive the imperial favor of a double-eyed peacock feather! In their subsequent correspondence, the Manchu diplomat addressed the British chieftain as "my intimate friend."[3]

1. Fairbank, *Trade and Diplomacy*, I, 110.
2. *Ibid.*, I, 110, footnote *f*.
3. Fairbank, I, 111-12.

THE MEMORIAL OF 1844. Ch'i-ying's policy of appeasement, friendship, and personal diplomacy was designed to disarm foreigners' suspicion, to win their confidence and trust, and to subject them to a sort of psychological obligation to him. While it did reduce tension and friction here and there, it could not change the basic objectives of the foreigners. From Ch'i-ying's assumption of power in 1843 until his retirement in 1848, this approach was at its apogee, successfully maintaining relative peace and order in Chinese foreign relations. However, in the eyes of antiforeign, conservative officials, such a policy was obsequiousness to and ingratiation with the former enemy. Such opposition constrained Ch'i-ying to justify his action to the court in a memorial in November 1844, in which he stated that to make the outlandish barbarians conform to the requirements of Chinese civilization and ceremonies was to seek trouble unnecessarily, for they did not understand and could not appreciate such niceties; rather, it was necessary to humor them with material favors and outward sincerity, so as to win their trust and avoid quarrels:

> Throughout this period of three years the barbarian situation has undergone deceptive changes in many respects and has not produced a unified development. The methods by which to conciliate the barbarians and get them under control similarly could not but shift about and change their form. Certainly we have to curb them by skillful methods. There are times when it is possible to have them follow our directions but not let them understand the reasons. Sometimes we expose everything so that they will not be suspicious, whereupon we can dissipate their rebellious restlessness. Sometimes we have given them receptions and entertainment, after which they have had a feeling of appreciation. And at still other times we have shown trust in them in a broad-minded way and deemed it unnecessary to go deeply into minute discussions with them, whereupon we have been able to get their help in the business at hand.
>
> This is because the barbarians are born and grow up outside the frontiers of China, so that there are many things in the institutional system of the Celestial Dynasty with which they are not fully acquainted. Moreover, they are constantly making arbitrary interpretations of things, and it is difficult to enlighten them by means of reason . . .
>
> Moreover, the barbarians commonly lay great stress on their women. Whenever they have a distinguished guest, the wife is certain to come out to meet him . . . Your slave [minister] was confounded and ill at ease, while they on the other hand were deeply honored and delighted. Thus in actual fact the customs of the various Western countries cannot be regulated according to the ceremonies of the Middle

Kingdom. If we should abruptly rebuke them, it would be no way of shattering their stupidity and might give rise to their suspicion and dislike . . .

With this type of people from outside the bounds of civilization, who are blind and unawakened in styles of address and forms of ceremony, if we adhered to the proper forms in official documents and let them be weighed according to the status of superior and inferior, even though our tongues were dry and our throats parched [from urging them to follow our way], still they could not avoid closing their ears and acting as if deaf. Not only would there be no way to bring them to their senses, but also it would immediately cause friction. Truly it would be of no advantage in the essential business of subduing and conciliating them. To fight with them over empty names and get no substantial result would not be so good as to pass over these small matters and achieve our larger scheme.[4]

Ch'i-ying's methods worked well enough with Pottinger, who became rather proud of his own ability to make friends in China. Not being a merchant and therefore free from considerations of profit, Pottinger could take a more objective and sometimes even magnanimous attitude toward Sino-British relations. He saw the general tendency among foreigners to infringe upon Chinese rights in a way not tolerated elsewhere, and he declared to London that he would take "the most decided measures" to enforce the treaties on British subjects; he cautioned, moreover, that British officials in China should guard against the inclination to take advantage of the Chinese, lest the latter lose faith in British justice and moderation. The Foreign Office approved of this view and asked the Colonial Office, Admiralty, and India Board to instruct their servants in China accordingly. Thus, with Ch'i-ying's policy of friendship and Pottinger's sense of moderation, a period of relative harmony reigned in Sino-British relations.

The situation changed considerably when Pottinger was replaced in the middle of 1844 by John Davis, an old Company man and the former second superintendent of trade under Lord Napier. With characteristic Anglo-Indian arrogance toward the natives of the East, Davis spoke disparagingly of the Chinese "inability to comprehend the observance of good faith on the part of the strongest." He found Ch'i-ying's diplomacy "tiresome" and "childish," and was generally so unresponsive that Ch'i-ying finally gave it up in 1846.[5]

4. Teng and Fairbank, *China's Response*, 38-40.
5. Fairbank, *Trade and Diplomacy*, I, 269-70.

THE "CANTON CITY QUESTION." The most knotty issue in the postwar period was the question of the British right to enter the city of Canton. Of the five ports, all except Canton were opened on schedule to foreign trade, residence, and consulates: Shanghai in November and Ningpo in December, 1843, and Fuchow and Amoy in June 1844. The residents of Canton, however, steadfastly refused to admit the British into the city, allowing them to live only in the old factory area. They argued that although the treaty opened Canton, it did not specify that foreigners be allowed inside the city. Indeed, the treaty text did not spell out the point clearly, but none of the other four ports ever contested the British right to enter their walled cities. In fact, the foreigners at Shanghai, after having gained entry into the city, found the hygienic conditions and living quarters so undesirable that they voluntarily moved out of the city to found their own settlements. But in Canton the more they met resistance, the more the British insisted on their right of entry. The local populace would not yield; they considered the British entry an insult to their city. Hence the "Canton city question" became a point of disruptive contention.

Historically, Canton had a reputation for conflicts with foreigners: cases of massacre of the Arabs were known to exist in medieval times. More recently, during the Opium War, its people were subjected to British humiliation more than those of any of the other cities, and they were also the object of a "ransom" in 1841. In the postwar period, Canton suffered from losing part of the foreign trade to Shanghai, owing to the latter's proximity to the tea- and silk-producing areas. The volume of tea export from Canton declined from 69 million pounds in 1844 to 27 million in 1860; whereas that from Shanghai increased from 1.1 million pounds to 53 million in the corresponding period. Silk export from Canton dropped from 6,787 bales in 1845 to 1,200 in 1847; that from Shanghai increased from 6,433 bales to 21,176 in the same years.[6] The decline of the Canton trade adversely affected the livelihood of the local people, so they transferred their resentment to the British, the largest group of foreign traders. This popular discontent became a considerable force when organized by the gentry and armed with weapons originally supplied to the local militia by Commissioner Lin during the Opium War.

Ch'i-ying, as imperial commissioner and governor-general at Canton, was caught between the ever-increasing pressure of the British to enter the city and the stubborn resistance of the gentry and the people of Can-

6. Morse, I, 366.

ton. He knew, if the local populace did not, China's treaty obligations; and in January 1846, he boldly proclaimed the opening of the city. In doing so he exposed himself to public condemnation; numerous placards and notices were circulated to attack his appeasement policy and ridicule his ingratiation with the enemy. The mob staged an attack on the Canton prefect, who was supposedly pro-British, and burned his yamen (office building), plunging the city into disorder and confusion. Faced with this outburst of public wrath, Ch'i-ying had to modify his order. Fortunately, the British government did not want an immediate clash over the "Canton city question." In April 1846, Davis and Ch'i-ying reached an agreement: the British would postpone their entry into the city in exchange for a Chinese promise of nonalienation of the Chusan Islands to any other power (to block the rumored French design).

Elated by the British concession, the Canton populace became bolder than ever. Incidents of stoning and insulting British excursionists occurred repeatedly. In April 1847, Davis retaliated with his "famous" raid of Canton; with 900 soldiers in three armed steamers and a brig, he captured the Bogue forts, spiked 827 cannon, and occupied the Canton factories. Ch'i-ying hurriedly arranged an agreement with him on April 6, promising British entry to the city at the end of two years, and punishment of those Chinese who had offended the British. Other items gave British traders and missionaries the right to build warehouses and churches.

By his concessions to the British, Ch'i-ying's public image was irreparably damaged. Sensing that the "Canton city question" would sooner or later precipitate a clash which he was unable to prevent, and knowing his inability to cope with the growing problem of piracy along the coast, Ch'i-ying schemed to get out of the fix before it became uncontrollable. He asked the court that he be recalled on grounds of old age and infirmity. The request was granted, and in March 1848, Ch'i-ying left for Peking. The post of imperial commissioner and governor-general went to a xenophobic official, Hsü Kuang-chin (ca. 1786-1858), while the governorship of Kwangtung went to another, Yeh Ming-ch'en (1807-59). The appointment of these two, following Ch'i-ying's recall, marked the re-emergence in the government of the antiforeign element, which had been in eclipse since the defeat. Hsü and Yeh formed a team at Canton, adopting an unyielding and arrogant attitude toward foreigners, and secretly promoting antiforeign sentiment among the populace and encouraging them to block the British entry. There were frequent incidents of attacks, insults, stonings, and even killing of the British. Sino-British relations deteriorated rapidly.

The Hard Line at Canton, 1848-56

The change of Chinese personnel at Canton was paralleled on the English side. Davis was replaced by Sir S. George Bonham (1803-63) as governor of Hong Kong, envoy-extraordinary and minister-plenipotentiary, and superintendent of trade in China. Son of an East India Company ship captain, Bonham rose high early in his career. While still in his twenties, he was appointed resident councillor of Singapore, and in 1837 became governor of the Incorporated Settlement of Prince of Wales Island, Singapore, and Malacca. With some knowledge of the Chinese language, customs, and habits, and with a reputation for "practical common sense," he was appointed by Palmerston as governor of Hong Kong in 1848.

His first interview with the new Imperial Commissioner Hsü took place on April 29, 1848. While satisfied with the ceremonial, Bonham found Hsü "somewhat taciturn." On June 7 he wrote to Hsü suggesting preliminary arrangements be made to give effect to the Ch'i-ying-Davis agreement permitting British entry into Canton in 1849. Hsü replied that in view of the strong local opposition, the "temporary arrangement [of 1847] was by no means the way to insure perpetual protection or to secure lasting tranquillity to both sides." Palmerston, while not willing to give up the right of entry, doubted the practical value of going into a hostile city, and suggested that the right might be restricted to the British plenipotentiary or the consul, escorted by Chinese officials, when making business visits to the governor-general. On December 30, 1848, he authorized Bonham, in effect, to evade the issue.

On April 1, 1849, the Chinese commissioner communicated to Bonham an imperial rescript stating that the emperor could not ignore the spontaneous and unanimous opinion of the people of Canton. When all attempts by Bonham to see the commissioner failed, he informed the Chinese authorities on April 9, by letter, that "the question at issue rests where it was, and must remain in abeyance." The Cantonese believed that the awesome magnitude of their public demonstrations, involving a mob and militia of some 100,000, had intimidated the British into relinquishing the demand. When the jubilant Hsü and Yeh reported to the court that Bonham had agreed that "hereafter there will be no further discussion of entering the city," the delighted monarch rewarded Hsü with the title of viscount and Yeh with that of baron, and commended the people of Canton for their patriotism. Palmerston's wrath was irrepressible. He instructed Bonham to send Peking a message in which he reminded the high officials of the "mistake which was committed by their predecessors in 1839," and warned that "the forbearance which the British government has hitherto displayed arises, not from a sense of weakness, but from the consciousness of superior strength. The British government well knows that, if occasion required it, a British military force would be able to destroy the town of Canton, not leaving one single house standing, and could thus inflict the most signal chastisement upon the people of that city." The court contemptuously dismissed the warning with the remark that such a contumacious and insulting letter did not deserve a reply, lest the barbarians be encouraged to further insolence. Bonham then personally delivered a formal protest on August 24, 1849, in which he recited the whole series of events connected with the "Canton city question" and warned that "whatever may happen in future between the two countries that may be disagreeable to China, the fault thereof will lie upon the Chinese government."[7]

In 1850 the recalcitrant Emperor Tao-kuang died and was succeeded by his twenty-year-old son, Emperor Hsien-feng, who followed an even more uncompromising foreign policy. Advocates of appeasement, such as Mu-chang-a and Ch'i-ying, were dismissed, demoted, or replaced by stridently antiforeign officials. A xenophobic official suggested to the emperor that the hero of the Opium War, Lin Tse-hsü, be summoned to serve in the capital as a warning to the British: "The management of the barbarian affairs at Canton were begun by Lin and concluded by Hsü; both were most feared and respected by the British." However, Lin had been in bad health since the summer of 1849, and died on November 22 of the following year en route to Kwangsi to accept the new post as acting governor and imperial commissioner. When Hsü was transferred in 1852 to fight the Taipings (see next chapter), Yeh, who was even more antiforeign, stubborn, and arrogant, filled his office. Openly contemptuous of aliens, he refused to answer their communications or to meet with them, announcing that high officials of the Celestial Empire ought not debase themselves by receiving foreigners, but should preserve the dignity of their state by avoiding them. The French minister was unable to arrange an interview with him for fifteen months.

The stiffening of the Chinese attitude, however, was not reciprocated by the British. The new Liberal government was committed to a course of moderation, which was further confirmed when Bonham, having been granted a leave of absence, was replaced by John Bowring (1792-1872) in 1852. Bowring, a man of great learning and stature, had been a strong advocate of free trade, an editor of the *Westminster Review*, a private sec-

7. Morse, I, 395-98, 402.

cretary of Jeremy Bentham, and an intimate friend of George Villiers, later Lord Clarendon, the foreign secretary. Toward the end of his long career Bowring found himself financially straitened and applied for the post of consul in Canton. He won the appointment in 1849 and subsequently became fascinated with Chinese civilization as he saw it at Canton. When he replaced Bonham as superintendent and plenipotentiary in 1852, he was cautioned by Lord Granville not to begin irritating discussions with the Chinese authorities and not to use force without prior approval from home. When his request for an interview with Yeh ran into a stone wall—not unexpectedly—London instructed him "not to raise any question as to the admission of British subjects into the city of Canton, and not to attempt yourself to enter it."[8]

In addition to the "Canton city question," a number of other issues also strained Sino-Western relations: the ever-present foreign desire to extend trade beyond the five ports to all parts of China, the demand for resident ministers in Peking to bypass the stubborn Canton authorities, and the drive to reduce customs dues as a result of the general decline of commodity prices in the postwar period. These issues converged to generate a strong impetus among the foreigners for a treaty revision. According to the American and French treaties of 1844, a revision might take place in twelve years, i.e. 1856. Although the Treaty of Nanking of 1842 made no provision for treaty revision, the British claimed that the most-favored-nation treatment entitled them to similar revision in twelve years, i.e. 1854. Out of common interest the American and French ministers supported the British claim, and in 1854 the three ministers proposed discussion about it. Yeh replied point-blank that there was no need. After having failed to move Yeh at Canton or to open negotiations at Shanghai, the British and American representatives went north in October 1854, to demand satisfaction. At Taku they were met, not by the governor-general of Chihli, who was ordered by the court not to personally receive the barbarians, but by a lesser figure, Ch'ung-lun, the Ch'ang-lu Salt Controller. The two ministers demanded tariff revision, establishment of legations in Peking, opening of Tientsin, the right to purchase land in the interior, legalization of opium import, and abolition of the inland transit dues (*likin*). The court rejected these demands as unreasonable and urged the ministers to return to Canton.

In 1856, the ministers of the three powers once again raised the question of treaty revision. The court intimated that minor changes of a reasonable nature might be allowed, but no major items could be considered,

8. Fairbank, *Trade and Diplomacy*, I, 278; Morse, I, 403.

lest the Treaty of Eternal Peace (Treaty of Nanking) should lose its meaning. However, Yeh at Canton persistently refused to negotiate, even on minor issues, insisting that if he gave an inch the foreigners would want a foot. The American commissioner, Peter Parker, unwilling to concede, made a solo attempt to reach Peking. At Shanghai the Chinese frustrated his efforts to continue further. Under these conditions, foreign, and particularly British, patience approached exhaustion. Even the peace-minded Bowring was compelled to inform London that the extension and improvement of British relations with China would require ships of war.

The Arrow War

The occasion that provoked Britain into venting its wrath was the *Arrow* incident of 1856. The *Arrow* was a lorcha, a hybrid vessel with a European hull and Chinese sails; it was owned by a Chinese resident[9] of Hong Kong and registered with the British authorities of that Crown Colony for protection from coastal piracy, which the Chinese government was unable to suppress. On October 8, 1856, while lying off the city of Canton with British flags flying, between eight and eight-thirty in the morning, the *Arrow* was boarded by four Chinese officers and sixty soldiers for the alleged purpose of searching out one notorious pirate who was said to be aboard. They arrested twelve Chinese crew members, and in the turmoil the British flag was hauled down. The British consul at Canton, Harry Parkes, under instructions from Bowring, protested strongly on October 12 against the insult to the flag and the arrest of the crew without a warrant from the British consul. He demanded future respect for the flag, release of all twelve crewmen, and a written apology from the governor-general within forty-eight hours. Yeh caustically denied there was any flag flying at the time and questioned the right of the consul to intervene in a case which involved the arrest of Chinese nationals by Chinese police on a Chinese-owned vessel in a Chinese harbor. He might have added, but he did not know then, that the registration of the *Arrow* had lapsed at the time of the incident. Bowring was himself of the opinion that "after the expiry of the license, [British] protection could not be legally granted." However, an ordinance of Hong Kong provided that if the expiry occurred while the vessel was at sea, the registration remained valid until its return to Hong Kong. On the basis of this ordinance, Parkes insisted that the *Arrow*, while at Canton and before returning to Hong Kong, was still entitled to British protection, and that any British ship in Chinese waters

9. Fong Ah-ming.

was British soil, with full extraterritorial privileges. Yeh's reply was deemed unsatisfactory, and Parkes ordered the seizure of a Chinese war junk to enforce redress. After prolonged bickering, Yeh returned the twelve crewmen on October 22, but he emphatically refused to apologize. On October 23, British gunboats under Admiral Seymour moved in to bombard the city of Canton. Except for Sunday, October 26, which was declared a day of rest, the shelling continued with humiliating regularity: shots were fired at ten-minute intervals on Yeh's yamen. On the 28th, Yeh ordered an all-out attack on the barbarians; the British responded by marching through his yamen on the 29th. The aroused people of Canton, utterly powerless before the British armed forces, vented their wrath by burning the foreign factories on December 14 and 15.

In London, Her Majesty's Loyal Opposition severely criticized Parkes and Bowring for dragging Britain into a foreign war. Gladstone grandiloquently declared in Parliament on March 3, 1857: "You have turned a consul into a diplomatist, and that metamorphosed consul is forsooth to be at liberty to direct the whole might of England against the lives of a defenceless people."[10] In the House of Commons the Opposition succeeded in unseating the government by a vote of 263 to 247. Palmerston called an election, in which he stressed the importance of upholding British honor and overseas interests. He was returned with a majority of 85 in Parliament. His China policy having been vindicated, Palmerston sent Lord Elgin (1811-63), who had been governor-general of Canada in 1846, as plenipotentiary and the leader of an expedition to China.

The French government, capitalizing on the murder of a missionary, Abbé Auguste Chapdelaine, in February 1856, in Kwangsi province (which was not yet opened to the West), decided on a joint expedition with Britain by dispatching a task force under the command of Baron Gros, a veteran diplomat of thirty years' experience. The American and Russian governments abstained from joining this Anglo-French venture, but they sent representatives to participate in a "peaceful demonstration."

Lord Elgin's instructions called for (1) reparations for injuries to British subjects; (2) execution of treaty stipulations at Canton and other ports; (3) compensation to British subjects for losses sustained in the recent disturbances; (4) diplomatic representation at Peking, or at least the right to an occasional visit by a British minister, as well as the right of the British plenipotentiary to direct communication with high officials at Peking; and (5) revision of the treaties with a view to extending trade to the cities on the great rivers. Lord Clarendon, the foreign secretary, emphasized to El-

10. *Hansard Parliamentary Debate,* 144:1802 (1857).

gin that his chief mission was to liberate the trade from existing restrictions, and—since it was uncertain whether Yeh's conduct reflected his own xenophobia or orders from Peking—the matter of direct contact with Peking through diplomatic representation was also of utmost importance. Baron Gros' instructions required much the same things—including extension of trade, freedom for religious propagation, and diplomatic representation at Peking.

The plenipotentiary from the United States was William B. Reed, a politician from Philadelphia who had held state offices and taught American history at the University of Pennsylvania. He was told to cooperate with the French and British peacefully, but to make it clear to the Chinese that the United States had no territorial or political designs on China. His directives called for diplomatic residence in Peking, new ports, reduction of the domestic tariff, religious freedom, suppression of piracy, and extension of treaty benefits to all civilized nations. The other neutral state, Russia, sent Admiral Putiatin. Publicly he was to dissociate himself, before the Chinese, from the Anglo-French interventionists and emphasize the age-old friendship between Russia and China. But secretly he was to play the role of mediator between the Manchu empire and the European powers, so as to prevent the fall of the dynasty and the shifting of the political gravity from North China to South China—a shift which, if materialized, would benefit the British.

Lord Elgin arrived in Hong Kong on July 2, 1857, only to discover that the Sepoy Mutiny required diversion of his troops to India. Completing his Indian mission, he returned to Hong Kong in September and received authorization from Lord Clarendon to take Canton by force. Some delay in working out details of joint operations with the French ensued, but by early December 1857, the allied forces were ready for action. On December 12, Elgin and Gros demanded that Yeh peremptorily agree to direct negotiation and payment of an indemnity. Yeh was defiant. When the Allied ultimatum of the 24th went unanswered, the Anglo-French forces stormed the city on December 28, captured Yeh, and put him aboard H.M.S. *Inflexible*. Soon the British realized that this move rendered the warship useless for combat; they shipped him to Calcutta, where he died a year later. For governance of the city an Allied commission was established, with Harry Parkes as presiding officer, while the daily routines were left to the Mongol governor, Po-kuei. This puppet regime, which lasted for three years until the final treaty settlement in 1860, was probably the first of its kind in modern Chinese history.

The ease with which the Allies took Canton indicated that Yeh had made no special effort to strengthen its defenses. Story has it that he, a believer in oracular divination, consulted the planchette (*fu-chi*, a sort of ouija board) and was informed that the British would leave in fifteen days. Consequently he made no preparations for long-term resistance. After his fall a jingle appeared satirizing his superstitious approach to affairs of state: "He would not make war, would not make peace, would not make a defense, would not die, would not surrender, and would not flee." Actually Yeh was not the fool people believed. A calculating politician, Yeh's obduracy was the cover for his inward insecurity. He knew that militarily China could not withstand Britain. If he used force and brought on a disastrous war, he would end up in exile like Commissioner Lin; contrarily, if he followed a policy of appeasement, he might incur imperial displeasure, public condemnation, disgrace, and even banishment, such as befell Ch'i-shan and Ch'i-ying. Caught in a dilemma, he straddled the gulf of indecision, while erecting a façade of indifference, arrogance, and haughty contempt for the foreigners. Secretly he hoped the profit-minded barbarians would not prolong the disturbance at the expense of their trade. The price of his superstition and misjudgment was captivity, exile, and death in an alien land.[11]

THE TIENTSIN NEGOTIATIONS. Having settled the Canton question, Elgin and Gros proceeded north to demand satisfaction from the court. They arrived at the Gulf of Pechili off Tientsin in mid-April 1858, and after some preliminary encounters with the governor-general of Chihli, who was found unequipped with the "full powers" to negotiate, they took the Taku Forts and Tientsin. Shocked by the rapid enemy advance, the court hurriedly sent Kuei-liang, a 73-year-old grand secretary, and Hua-sha-na, a 52-year-old president of the Board of Civil Office, duly provided with the "full powers," to meet with Elgin and Gros. They arrived in Tientsin on June 3, and were shortly afterwards joined by a third negotiator, none other than the famous Ch'i-ying.

As may be recalled, Ch'i-ying returned to Peking in 1848 and was demoted to a fifth-grade mandarin in disgrace in 1850 upon the ascension to the throne of Emperor Hsien-feng. He lived in obscurity until 1858, when the resurgence of the insoluble barbarian trouble recalled to people's minds his clever diplomacy. The emperor summoned him from disgrace and sent him to Tientsin to manage the barbarians. Ch'i-ying, much deteriorated physically during his years of eclipse, and half blind, arrived in Tientsin on June 9. His presence caused concern among the Allies who rightly suspected a ruse, knowing, as Elgin did from the captured docu-

11. Yen-yü Huang, "Viceroy Yeh Ming-ch'en and the Canton Episode (1856-1861)," *Harvard Journal of Asiatic Studies*, 6:1:37-127 (March 1941).

ments in Yeh's yamen, how Ch'i-ying boasted to the court of his skills in managing the unfathomable barbarians. Intuitively Elgin discerned that Ch'i-ying had come with the knavish intention of reviving the old stratagem of "caressing" and "restraining" the barbarians; therefore he must not be allowed in Tientsin. Two young assistants, Horatio Lay and Thomas Wade, were sent by Elgin to see him. When Ch'i-ying began playing his old game of gentle restraint, personal charm, and endless praises for the two Englishmen, Lay dramatically produced a document—Ch'i-ying's famous memorial of 1844—and made Hua-sha-na read it aloud. The situation was embarrassing in the extreme, and Ch'i-ying could only respond in confusion with tears of shame, while the Englishmen left laughing elatedly. Finding himself unacceptable to the British, Ch'i-ying left the negotiations without imperial permission. For his unauthorized leave, he was taken to Peking in chains, given a trial, and sentenced to death by suicide. So ended the life of China's most colorful diplomat in the mid-19th century: one who strutted into the limelight in the 1840s because of his canny ability to manage the foreigners, then lost his life when this ability no longer charmed his wary adversaries.

THE TREATY OF TIENTSIN, 1858. The negotiations at Tientsin focused on four major issues: a resident minister in Peking, the opening of new ports along the Yangtze River, foreign travel in interior China, and the indemnity. Of the four, the resident minister issue was central in Elgin's thinking, for he had come to believe that pacific relations with China were impossible without abolishing the imperial commissioner at Canton as China's "foreign ministers" and forcing the court at Peking to take up foreign affairs itself so as to spare the local officials the dilemma of reporting unwelcome truths. His assistant, Horatio Lay, who carried on the major burden of negotiations, was even more emphatic about this point, insisting that the Canton system was the cause of foreigners' being "tossed to and fro like a shuttle between Imperial and Provincial authorities" and that without the right of diplomatic representation at Peking the new treaty "would not be worth the paper it was written upon."[12] The Chinese negotiators argued that diplomatic residence was incompatible with the established institutions of China (*t'i-chih*), whereupon Lay announced bluntly: "The provision will be for your good as well as ours, as you will surely see. The medicine may be unpleasant but the aftereffects will be grand. The more stern my attitude, the greater the service I render to you." He kept threatening, bullying, and insulting the Chinese negotiators. Totally

12. Horatio N. Lay, *Our Interests in China* (London, 1864), 49; and *Note on the Opium Question* (London, 1893), 12.

helpless before Lay's truculence, Kuei-liang pleaded for commiseration, saying that his acceptance of the term would cost him his head at the age of seventy-three. Lord Elgin could not help having some pity for the Manchu grandee, but in the end resolved not to relax his stand. "Being in the vicinity of Peking with an armed force," declared Elgin, "I might so demean myself as to make the Emperor think that he was under an obligation to his Plenipotentiaries for having made peace with me even on terms objected to."[13] On June 11, 1858, Lay warned that unless the term was accepted that day there would be a march on Peking. Kuei-liang had no alternative but to accede to British diplomatic representation at Peking. His last-minute tactics were to reach a settlement by whatever means, get the enemies out of North China, and then devise means to retrieve the lost rights.

Kuei-liang had made the concession without the prior approval of the court. The emperor, who viewed diplomatic representation as the end of the tributary system and a denial of China's universal overlordship, remained violently opposed to it. It taxed all the skills, courage, and manipulation of a deft and experienced politician such as Kuei-liang to convince him that there was no escape from the labyrinth. To the troubled ruler he divulged his secret tactics: "At present the peace treaties with the two nations of England and France should not be taken as true certificates and real contracts, but just a few pieces of paper by which the [enemy] warships could be made to withdraw temporarily from the harbor. In the future if the renouncement of the treaties and friendship is desired, your Highness needs only to charge your slaves [servants] with the crime of mismanagement: [the treaties] will immediately become waste paper." On another occasion he drolly told the throne that the barbarian envoys, once established in Peking, might not want to stay: "Barbarians dread most to spend money. Let them pay their own expenses. Furthermore, they fear wind and dust. With no advantages from residence [in Peking], they inevitably will leave on their own."[14]

On June 26, 1858, Frederick Bruce, Elgin's brother, warned that if the treaty were not signed by night, it would have to be signed in Peking itself. With a "knife" at their throats, Kuei-liang and Hua-sha-na concluded the Treaty of Tientsin with Britain that day, and with France a day later. The Treaties of Tientsin with Russia and the United States had been concluded earlier, on June 13 and 18 respectively. The French, Russian, and American treaties specified occasional visits of their diplomats to Peking, rather than the permanent-residence stipulation contained in the British treaty. But because of the most-favored-nation treatment, they

13. Hsü, *China's Entrance*, 52-54.
14. Hsü, *China's Entrance*, 67-68.

equally shared the fruits of the British labors. Other important items in the Treaties of Tientsin included (1) opening of ten new ports;[15] (2) foreign travel in all parts of China under passport issued by the consul and countersigned by Chinese authorities, but no passport required for travel within 100 *li* (33 miles) of the ports; (3) inland transit dues (*likin*) for foreign imports not to exceed 2.5 percent ad valorem; (4) indemnity of 4 million taels for Britain and 2 million taels for France; and (5) freedom of movement in all China for missionaries, Catholic and Protestant alike.

After the conclusion of the treaties, Frederick Bruce returned home as courier of the treaty for the queen's ratification. The exchange of the ratifications was to take place in China a year from the date of signing. The Allied troops withdrew from North China, and Lord Elgin sailed for Japan to negotiate a treaty. But he promised to return to Shanghai in a few months for the tariff conference, as stipulated in Article 26 of the Treaty of Tientsin.

THE SHANGHAI TARIFF CONFERENCE. During Elgin's absence, the emperor devised a secret plan by which he would offer to exempt the British from all customs dues in exchange for the abrogation of the Treaty of Tientsin, or at least its four most objectionable items: diplomatic representation, trade along the Yangtze, inland travel, and indemnity. One reason for the emperor's resistance to diplomatic residence was his ridiculous apprehension that the barbarian envoys would construct tall buildings from which they could spy on the activities of the palace through binoculars! He instructed Kuei-liang and Ho Kuei-ch'ing (1816-62), governor-general at Nanking, to propose the plan to Lord Elgin at the forthcoming Shanghai Tariff Conference. Shocked by this impractical plan, they vigorously opposed it, arguing that foreign traders and foreign officials were two distinctly separate entities; exemption of customs dues would win the gratitude of the former, but not necessarily of the latter, who would still insist on complete execution of the treaty. Artfully they pleaded that the barbarians, who paid no tribute, should pay taxes as a kind of offering, which could then be used to defray the indemnity costs. Furthermore, if tax exemption were allowed foreign traders but not Chinese merchants, the former would have an unfair advantage, driving the latter into bankruptcy. Because of their remonstration, the emperor finally relinquished his secret strategy.[16]

The Shanghai Tariff Conference of October 1858 was conducted in a

---

15. Nanking, Newchwang, Tengchow, Hankow, Kiukiang, Chinkiang, Taiwanfu, Tamsui, Swatow, and Kiungchow.
16. Hsü, *China's Entrance*, 71-75; T. F. Tsiang, "The Secret Plan of 1858," *CSPSR*, 15:2:291-99 (July 1931).

totally different atmosphere from the Tientsin negotiations four months earlier. Elgin was in a good mood, having returned from a triumphant mission to Japan. There was no Lay bullying nor Bruce threatening a march on Peking. Because Shanghai was more than 800 miles removed from the capital, there was no sense of the urgency from foreign threat, either. Under these relaxed conditions, Kuei-liang found a chance to display his diplomatic talents. By blending frank persuasion and earnest pleading, he succeeded in committing Elgin to a gentleman's agreement that, if the future British envoy bearing the ratification of the Treaty of Tientsin were properly received at Peking, Elgin would see that he resided in a place other than Peking and that his visits to the capital would be periodic business trips only.

As regards the new tariff, the principle of 5 percent ad valorem for both imports and exports was reaffirmed, except for opium, tea, and silk. The negotiations legalized opium importation at 30 taels per picul, which was about 7 percent to 8 percent of the average value. The existing tariff on tea, 2.5 taels per picul, which was about 15 percent to 20 percent ad valorem, was retained, as it represented only 1.5 pence per pound, in contrast to the much higher import duty of 1 shilling, 5 pence per pound in England. The agreement also maintained the old tariff of 10 taels per picul of silk, roughly below 5 percent of the value.

The Second Settlement

THE TAKU REPULSE. Frederick Bruce was appointed British Envoy Extraordinary and Minister Plenipotentiary to China on March 1, 1859. He was instructed to exchange the ratifications at Peking, but to make his residence at Shanghai. On arrival in China in May 1859, Bruce found the Chinese trying to oblige him to exchange the ratifications at Shanghai. Irritated by this stratagem, he declared that his visit to Peking was "a matter of right, not of favor." With ships and troops he charged to the north, reaching the Pei-ho on June 18. Having now resigned themselves to his journey to Peking, the Chinese asked him to take the back route via Pei-t'ang, north of Taku, but Bruce insisted that only the main route from Tientsin befitted his honor. The Chinese warned that the river leading to Tientsin had been blockaded with iron spikes, chains, and solid rafts, and that the forts on the two shores would permit no passage. The obstructions in the river were clearly visible, but Bruce could hardly take the Chinese warning too seriously. He ordered Admiral Hope to clear the blockade so as to open a way for him to proceed to Tientsin. On June 25, 1859, some 600 marines and an engineer company were sent to remove the ob-

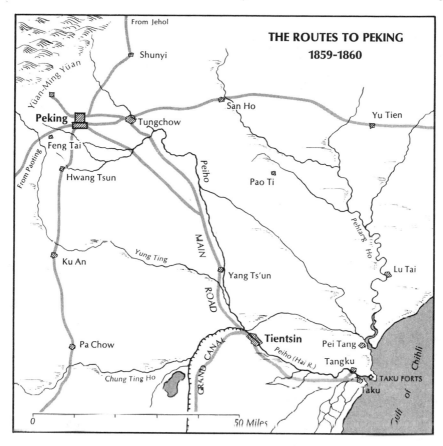

structions. Because of low tide and the soft deep mud, they were stuck and could not land. Suddenly the Chinese forts on the two shores opened fire with surprising accuracy, inflicting heavy losses on the British: 434 casualties, four ships sunk and two badly damaged, and Admiral Hope himself severely wounded. Commodore Tattnall of the United States fleet, a neutral observer of the scene, went to their aid in the belief that "blood is thicker than water." The Chinese suffered only nominal losses. The Taku repulse dealt a severe blow to British prestige and lent encouragement to antiforeign elements in China.

The British and French ministers retreated to Shanghai; but the American minister, John E. Ward, who had replaced William B. Reed, decided to accept the Chinese-assigned route to Peking. His party of twenty Americans and ten Chinese were conveyed in carts to Pei-t'ang, on July 20, 1859, and were then transferred to commodious boats which took them to Tungchow, whence they continued the journey by cart again, reaching

Peking on July 27. At Peking they were lodged in large, ostentatious houses, and their wants were provided for with "imperial munificence." They were, however, denied freedom of movement about the city, nor could they meet with the Russian ambassador, Nikolai Ignatiev, who had already established himself there. An audience with the emperor was arranged, but because of Ward's refusal to kowtow—he would kneel only to God and women—it did not materialize. President Buchanan's letter to the emperor was delivered to Kuei-liang for conveyance to the throne, and the exchange of the ratifications of the Treaty of Tientsin took place later at Pei-t'ang with the governor-general of Chihli.[17] On the whole the Chinese reception was "courteous rather than cordial or open." Ward himself said that he was treated with "high consideration and respect, with unceasing attention and courtesy" throughout the journey, and the American government also pronounced itself satisfied with the treatment. However, the British insisted that the American reception in Peking was not honorable.[18]

Bruce came under severe criticism in England for having exhibited "too much precipitancy" in his use of force. Foreign Secretary Lord Russell conceded that the Treaty of Tientsin did not specify the route to Peking for the exchange of the ratifications, and according to normal international usage, the inland rivers were not open to foreign warships in time of peace. On November 10, 1859, a reprimand was administered to Bruce: "Although the denial of a passage to the capital by the usual and most convenient route would have been evidence of an unfriendly disposition, yet it was a matter upon which you might have remonstrated and negotiated, without having recourse to force to clear the passage." Bruce acknowledged his poor judgment, admitting that he had no right to go to Peking under the old treaty, and, though the new treaty did give him the right, it was not yet operative.[19]

THE CONVENTIONS OF PEKING. In spite of British recognition of Bruce's tactical blunders, the Crown determined to enforce ratification at Peking. Its confidence in Bruce having been shaken, London decided to commit Lord Elgin to a second China mission, which he reluctantly accepted. His spirit was further dampened by a shipwreck at the Point de Galle en route. Rescue operations recovered some cases of champagne but not his credentials and decorations, which had to be sent for from London again. His expeditionary force consisted of 41 warships, 143 transports, and some

17. Heng-fu.
18. S. W. Williams, "Narrative of the American Embassy to Peking," *Journal of the North-China Branch of the Royal Asiatic Society*, 3:315-49 (Dec. 1859).
19. Hsü, *China's Entrance*, 95.

11,000 soldiers under the command of General Sir Hope Grant, Elgin's brother-in-law, in conjunction with 6,700 French troops under General de Montaubon.

The Allied forces skirted South China and pushed north to attack Pei-t'ang and the Pei-ho in August 1860, threatening once again the security of Peking. Kuei-liang was rushed to Tientsin, but he was unable to save the situation. Lord Elgin insisted on exchanging the ratifications in Peking in the company of 400 to 500 soldiers, and to prepare for his reception he sent a party under Harry Parkes to inspect the roads and living quarters. At Tungchow, some ten miles from Peking, Parkes ran into the new imperial commissioner, Prince I, whom he insulted in an argument. At this juncture, news arrived that the prefect of Tientsin had been kidnapped by the British soldiers. In retaliation, Prince I ordered the arrest of Parkes, who, in Chinese eyes, was the chief instigator of trouble in Canton, the specter of British imperialism, and probably the most hated foreigner in China. Elgin lost all patience and charged into Peking with his forces, driving the emperor to seek refuge in Jehol, Manchuria. Finding no court to negotiate with, Elgin toyed with the idea of replacing the Manchu dynasty with a Chinese one, and of burning the palaces as a punishment for the illegal detention of Parkes and the mistreatment of the prisoners of war. In the end he was persuaded by the Russian and French diplomats, General Ignatiev and Baron Gros, to abandon both ideas, burning the Summer Palace instead.[20]

In these critical days, the Russian ambassador Nikolai Ignatiev played an active role as mediator (*posrednik*) between the Anglo-French plenipotentiaries and Prince Kung, the emperor's younger brother who had been left in Peking to take charge of the peace settlement. When Kung was so terrified by the burning of the Summer Palace that he sought to flee the capital, it was Ignatiev who persuaded him to remain and accept the Allied terms, so as to avoid total destruction. Ignatiev's diplomacy will be discussed in the next section; suffice it to say here that by maneuvering at both ends he scored a great victory for Russia. On October 24, 1860, Lord Elgin dictated to Prince Kung the Convention of Peking, which established once and for all the British right to diplomatic representation in the Chinese capital. The indemnity was increased to 8 million taels each for Britain and France, and Tientsin was opened to foreign trade and residence. In addition, Britain acquired Kowloon Peninsula opposite Hong Kong, while France secured the right for Catholic missionaries to own properties in interior China. With this peace settlement, the Allied troops, urged by the Russian diplomat, left Peking around November 8, 1860.

20. China: *Dispatches*, Vol. 19, Doc. 26, Ward to Cass, Nov. 28, 1860 (National Archives, Washington, D.C.); Quested, *The Expansion of Russia*, 260-62.

THE RUSSIAN ADVANCE. On November 14, 1860, within a week of Allied evacuation of Peking, Ignatiev secured, as a reward for his mediation, a Supplementary Treaty of Peking, by which Russia won new territorial concessions east of the Ussuri River and legalized its previous acquisitions under the Treaty of Aigun of 1858. Ignatiev's work successfully crowned nearly two decades of Russian advance in the Amur Region under Nikolai Muraviev. Encouraged by the British success in the Opium War to intensify their own activities in China, the Russians under Nicholas I (1825-55) carried out a double-barreled penetration into Chinese Turkestan and the Amur Region. They secured a foothold in northern Sinkiang by the Treaty of Ili in 1851, which granted Russia the right to trade, to build warehouses, and to establish consulates in Ili (Kuldja) and Tarbagatai (Chuguchak). In the Amur Region, the advance was carried out by Muraviev, governor-general of Eastern Siberia since 1847. With headquarters at Irkutsk, Muraviev launched his forays along the Amur River, constructing fortified posts at strategic points and occupying its lower reaches. By 1858 he was in a position sufficiently powerful to intimidate the meek Manchu general I-shan into signing the Treaty of Aigun, which gave Russia the territory on the northern banks of the Amur and Sungari rivers, and joint possession with China of the land east of the Ussuri River to the sea. These three rivers were closed to ships of all nationalities except those of China and Russia. However, as this treaty completely ignored the boundaries established by the Treaty of Nerchinsk of 1689, the Ch'ing court adamantly refused to ratify it.

Ignatiev, once an aide-de-camp to the tsar, had succeeded Putiatin as ambassador to China in the summer of 1859. A shrewd schemer and a wily diplomat, he had been sent to perform the delicate task of seizing international leadership for Russia in Chinese affairs, getting the Treaty of Aigun approved, and preventing the downfall of the Manchu dynasty, with which Russia maintained favorable treaty arrangements. His ingenious means of achieving these objectives was to act as mediator between the Chinese and the Anglo-French interventionists. Arriving in Peking via the land route over Kiakhta, he first engaged in protracted and fruitless negotiations with Su-shun, president of the Li-fan yüan,[21] who steadfastly refused to accept the Treaty of Aigun, the extension of trade to interior China, and the re-demarcation of the Sinkiang border. In disgust, Ignatiev left Peking in May 1860 and went to Shanghai to denounce the Chinese obstructionist tactics to the Anglo-French plenipotentiaries, urging them to be positive and unyielding toward Peking. Ingratiating himself

21. From July to September 1859 and from December 1859 to April 1860.

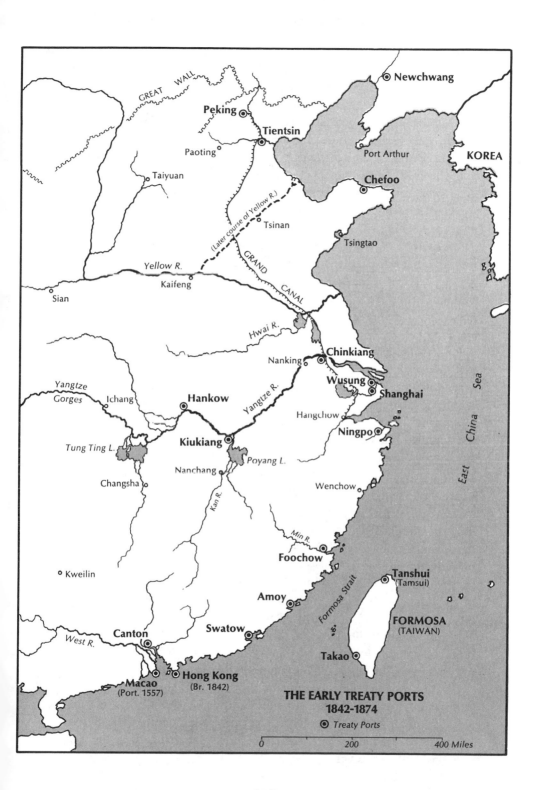

GREAT WALL

Newchwang

Peking

Tientsin

Paoting

Port Arthur

KOREA

Taiyuan

Chefoo

(Later course of Yellow R.)

Tsinan

Tsingtao

GRAND CANAL

Yellow R.

Kaifeng

Sian

Hwai R.

Chinkiang

Nanking

Wusung

Shanghai

Yangtze Gorges

Ichang

Hankow

Yangtze R.

Hangchow

Ningpo

Tung Ting L.

Kiukiang

Poyang L.

Sea

Nanchang

Changsha

Wenchow

East China

Kan R.

Min R.

Foochow

Kweilin

Tanshui
(Tamsui)

Amoy

Formosa Strait

FORMOSA
(TAIWAN)

Canton

Swatow

West R.

Takao

THE EARLY TREATY PORTS
1842–1874

Macao
(Port. 1557)

Hong Kong
(Br. 1842)

⊙ Treaty Ports

0   200   400 Miles

217

with the British, he informed them of the conditions in the Chinese capital, later guided them to land at Pei-t'ang, and provided General Grant with a map of Peking to facilitate the British attack.

Then returning to Peking on the heels of the Allies, Ignatiev cunningly presented himself to the Chinese as a needed friend. To Prince Kung he offered to mediate China's troubles with the Allies, to work toward a reduction of the indemnity, and to effect an early evacuation of the Allied troops from Peking, if Kung agreed to the following terms: (1) to approve the Treaty of Aigun; (2) to settle an eastern border along the Ussuri River to the limits of Korea, and in the north along the permanent Chinese picket line; and (3) to allow Russian consulates in Kashgar, Urga, and Tsitsikhar. Quite aware of the Russian's duplicity, Prince Kung was indisposed toward such expensive Russian mediation, yet he was apprehensive lest his refusal should drive the Russians into the Anglo-French camp and thus make China face three enemies simultaneously. In his desire to rid Peking of Allied troops, he succumbed to Ignatiev's offer of mediation.

After Lord Elgin had dictated the Convention of Peking on October 24, 1860, Ignatiev spared no efforts in promoting the idea that the severe winter of North China would soon descend to freeze the Pei-ho, in which case all foreigners would be trapped and exposed to possible Chinese mob attacks; he assured all that he would shortly leave for Tientsin to spend the winter. Under his influence, General Grant clamored for an early exodus from Peking, setting November 8 as the deadline.

Within a few days of the Allied evacuation, Ignatiev, now freed from all possible intervention, secured the Supplementary Treaty of Peking on November 14 as the reward for his service to China. Not only did it affirm the Russian gains in the Treaty of Aigun, including the land north of the Amur River, which had since become the Amur Province, but it went beyond to give Russia exclusive ownership of the land east of the Ussuri River to the sea, which had since become the Maritime Province. In addition, Urga and Kashgar were opened to Russian trade, consulates, and residences. Without a soldier or a shot, the Russians gained some 300,000 to 400,000 square miles of territory, plus impressive commercial concessions. Moreover, by the most-favored-nation treatment, they also shared the benefits of the British and French treaties.[22]

---

22. For Russian activities in China during this period, see Quested, chs. 2-4; A. Buksgevden, *Russkii Kitai: Ocherki diplomaticheskikh snoshenii Rossii s Kitaem—Pekinskii dogovor 1860 g.* (Russia's China: An account of the diplomatic relations between Russia and China—the Treaty of Peking, 1860), (Port Arthur, 1902). See also Hsü, *China's Entrance*, 103-05.

The second set of treaties reinforced the first signed after the Opium War, to form an iron-clad treaty system, from which China was not freed until 1943. Beyond a doubt, by 1860 the ancient civilization that was China had been thoroughly defeated and humiliated by the West. The maritime powers of Europe and America advanced, step by step, northward from Canton to Shanghai to Peking, while the Russian land power thrust southward from the Siberian-Manchurian border toward Peking. The Western states sought commercial interest and economic concessions through the creation of treaty ports and the extension of trade. The Russians stressed territorial acquisitions as well as commercial gains. The advance of the two from the south and from the north truly constituted a pincers movement, closing in ever more tightly on the declining Ch'ing dynasty. For the century that followed, the West and Russia constituted the two major sources of foreign impact, the effects of which are still discernible today in China.

Further Reading

Banno, Masataka 坂野正高, *Kindai Chūgoku gaikōshi kenkyū* 近代中國外交史研究 (A study of modern Chinese diplomatic history), (Tokyo, 1970).

Bonner-Smith, D., and W. R. Lumby, *The Second China War, 1856-1860* (London, 1954).

Buksgevden (Boxhowden), Baron A., *Russkii Kitai: Ocherki diplomaticheskikh snoshenii Rossii s Kitaem—Pekinskii dogovor 1860 g.* (Russia's China: an account of the diplomatic relations between Russia and China—the Treaty of Peking, 1860), (Port Arthur, 1902).

Chao, Chung-fu 趙中孚, *Ch'ing-chi Chung-O Tung-san-sheng chieh-wu chiao-she* 清季中俄東三省界務交涉 (Sino-Russian negotiations over the Manchurian border issue, 1858-1911), (Taipei, 1970).

Ch'en, Fu-kuang, *Yu-Ch'ing i-tai chih Chung-O kuan-hsi* (Sino-Russian relations during the Ch'ing period exclusively), (Kunming, 1947), ch. 3.

Chiang, Meng-yin 蔣孟引, *Ti-erh-tz'u Ya-p'ien chan-cheng* 第二次鴉片戰爭 (The second Opium War), (Peking, 1965).

Costin, W. C., *Great Britain and China, 1833-1860* (Oxford, 1937).

Dean, Britten, *China and Great Britain: The Diplomacy of Commercial Relations, 1860-1864* (Cambridge, Mass., 1974).

Dennett, Tyler, *Americans in Eastern Asia* (New York, 1941), chs. 17-18.

Drake, Fred W., *China Charts the World: Hsü Chi-yü and His Geography of 1848* (Cambridge, Mass., 1975).

Fairbank, John K., *Trade and Diplomacy on the China Coast: The Opening of the Treaty Ports, 1842-1854* (Cambridge, Mass., 1953), 2 vols.

———, "The Manchu Appeasement Policy of 1843," *Journal of the American Oriental Society*, 59:4:469-84 (Dec. 1939).

———, "The Manchu-Chinese Dyarchy in the 1840's and '50's," *The Far Eastern Quarterly*, XII:3:265-78 (May 1953).

————, "Synarchy under the Treaties" in John K. Fairbank (ed.), *Chinese Thought and Institutions* (Chicago, 1957), 204-31.

————, "The Creation of the Treaty System" in John K. Fairbank (ed.), *The Cambridge History of China* (Cambridge, Eng., 1978), Vol. 10, 213-63.

Gerson, Jack J., *Horatio Nelson Lay and Sino-British Relations 1854-1864* (Cambridge, Mass., 1972).

Graham, Gerald S., *The China Station: War and Diplomacy, 1830-1860* (New York, 1978).

Gulick, Edward V., *Peter Parker and the Opening of China* (Cambridge, Mass., 1973).

Hsü, Immanuel C. Y., *China's Entrance into the Family of Nations: The Diplomatic Phase, 1858-1880* (Cambridge, Mass., 1968), chs. 2-7.

Huang, Yen-yü, "Viceroy Yeh Ming-ch'en and the Canton Episode (1856-1861)" *Harvard Journal of Asiatic Studies*, 6:1:37-127 (March 1941).

Hurd, Douglas, *The Arrow War: An Anglo-Chinese Confusion, 1856-1860* (New York, 1968).

Lane-Poole, Stanley, and Frederick V. Dickins, *The Life of Sir Harry Parkes* (London, 1894), Vol. 1.

Lay, Horatio N., *Our Interests in China* (London, 1864).

Lin, T. C., "The Amur Frontier Questions between China and Russia, 1850-1860," *Pacific Historical Review*, 3:1-27 (1934).

Mancall, Mark, "Major-General Ignatiev's Mission to Peking, 1859-1860," *Papers on China*, 10:55-96, Center for East Asian Studies, Harvard University, 1956.

Quested, R. K. I., *The Expansion of Russia in East Asia, 1857-1860* (Kuala Lumpur, 1968).

————, "Further Light on the Expansion of Russia in East Asia: 1792-1860," *The Journal of Asian Studies*, XXIX:2:327-45 (Feb. 1970).

Shen, Wei-tai, *China's Foreign Policy, 1839-1960* (New York, 1932).

Tong, Te-kong, *United States Diplomacy in China, 1844-1860* (Seattle, 1964).

Tsiang, T. F., *Chung-kuo chin-tai shih ta-kang* (An outline of Chinese modern history), (Taipei, 1959), ch. 1.

————, "Tsui-chin san-pai-nien Tung-pei wai-chiao shih" 最近三百年東北外交史 (Diplomatic history of Manchuria in the last three hundred years), *Tsing-hua hsüeh-pao*, 8:1:1-70 (1932).

————, "The Secret Plan of 1858," *The Chinese Social and Political Science Review*, 15:2:291-99 (July 1931).

Wakeman, Frederic, Jr., *Strangers at the Gate: Social Disorder in South China, 1839-1861* (Berkeley, 1966).

Wong, George H. C., and Allan B. Cole, "Sino-Russian Border Relations, 1850-1860," *The Chung Chi Journal* (Hong Kong), 5:2:109-25 (May 1966).

Wong, J. Y., *Yeh Ming-ch'en: Viceroy of Liang Kuang, 1852-8* (Cambridge, Eng., 1976).

————, "Lin Tse-hsü and Yeh Ming-ch'en: A Comparison of Their Roles in the Two Opium Wars," *Ch'ing-shih wen-t'i*, III:1:63-85 (Dec. 1977).

Wu, Hsiang-hsiang 吳相湘, *O-ti ch'in-lüeh Chung-kuo shih* 俄帝侵略中國史 (A history of the Russian imperialist aggression in China), (Taipei, 1957), ch. 2.

The opium fleet at Lintin, with an opium boat, "fast dragon," in the foreground.

Howqua, a leading hong merchant (oil painting by George Chinnery R.H.A., 1852).

林文忠公燒燬鴉片

道光十九年。林文忠公督兩廣。比至。即查洋商所藏之鴉片。查得二萬二百八十三箱盡燒之于海口。時中輪入者。公乘月黑湖退時。後有泊舟外洋。出奇兵以援之。復燬其船二十三艘于長沙灣。遂以此釀成交涉之進口日多。今則英國政府已樂贊成。願除煙蠹由今思昔。有令人轉成為欷省。

Commissioner Lin superintending
the destruction of opium, 1839.

Sir Henry Pottinger.

Ch'i-ying, imperial commissioner and signer of China's first treaties with the West.

Commissioner Yeh (*Illustrated London News,* February 13, 1858, from a painting by a Chinese artist).

Lord Elgin.

Sir Frederick Bruce.

Kuei-liang.

Prince Kung.

10

The Taiping Revolution
and the Nien
and Moslem Rebellions

Mid-19th-century China was troubled not only by external wars but also by a series of debilitating internal convulsions. The Opium and the *Arrow* wars brought disasters and humiliation from the outside, while the domestic revolution and rebellions dealt serious blows to the ruling power from within. The largest of these upheavals, the Taiping Revolution, nearly toppled the dynasty. It lasted from 1850 to 1864, raging over sixteen provinces and destroying more than 600 cities. The Nien Rebellion lasted from 1851 to 1868, spreading over eight provinces. The Moslem Rebellion in Yunnan continued from 1855 until 1873, while that in the Northwest, known as the Tungan Rebellion, extended from 1862 to 1878. The fortune of the Ch'ing regime had reached its lowest point.

Causes for Social Upheaval

Traditional Chinese subscribed to the theory that domestic rebellion and foreign invasion occurred when the central power declined; they appeared together as symptoms of serious and upsetting internal weakness. If the ruling power had been strong, these troubles could have been met and stopped out of hand. Nowhere was this theory better manifested than in the case of the Manchu government in the middle of the 19th century. It had inherited many deep-rooted social and economic problems from earlier periods which made domestic convulsion well-nigh inevitable.

SOCIAL AND ECONOMIC FACTORS. For two thousand years preceding the mid-19th century the social structure and the mode of production in China had scarcely changed. It was for the most part an agrarian society, and social order and disorder depended to a great extent on the proper distribution of land. After each major disorder, enough people had been killed so that there was sufficient land for the survivors, but after a period of peace the population increase inevitably resulted in a decrease in per capita land cultivation. This caused difficulties in earning a livelihood, which led to banditry and uprisings, and these upsetting conditions were usually accompanied by administrative inefficiency, political corruption, and moral degeneration. A period of disorder followed, whereby the population was once again drastically reduced until, theoretically, a new balance was achieved between land and people. A period of peace and order then set in, signaling the beginning of a new cycle. In short, the alternation of order and disorder was nature's way of maintaining social equilibrium, and the Chinese had been at the mercy of this process since time immemorial. The philosopher Mencius (373-288 B.C.) perceptively observed that a period of order was perforce followed by a period of disorder, and the Chinese believed generally that a minor disturbance was to be expected every thirty years and a major one every hundred years. Western scholars sometimes described this phenomenon as "dynastic cycle," although it should more properly be called the theory of "natural evolution of history."

Applying this concept to the Ch'ing period, we find that the 150 years of peace and prosperity under K'ang-hsi, Yung-cheng, and Ch'ien-lung had nurtured a rapid growth in population, but that arable land had not increased correspondingly. The population rose from 143 million in 1741 to 430 million in 1850, a gain of 200 percent, whereas land rose from 549 million *mou* (*mou* = ⅙ acre) in 1661 to 737 (742?) million in 1833, an increase of only 35 percent. The discrepancy between population and land growth resulted in a sharp decrease in per capita cultivation. With 708 million *mou* in 1753, each individual could theoretically be allotted 3.86 *mou*, but with 791 million *mou* in 1812, only 2.19 *mou* per person could be figured. It became even worse: between 1812 and 1833 not only was there no increase but due to natural calamities there was actually a decrease in arable land, from 791 million *mou* to 737 (742?) million, whereas the population increased from 361 million to 398 million, lowering the per capita cultivation further to only 1.86 *mou*.[1]

1. Lo Erh-kang, "T'ai-p'ing t'ien-kuo ko-ming ch'ien ti jen-k'ou ya-p'o wen-t'i" (The population pressure in the pre-Taiping Rebellion years), *Chung-kuo she-hui ching-chi chi-k'an* (Collected writings on Chinese society and economics), Academia Sinica, 8:1:39 (Jan. 1939); also George Taylor, "The Taiping Rebellion: Its Economic Background and Social Theory," *The Chinese Social and Political Science Review*, 32:545-614 (1932-33); Ping-Ti Ho, *Studies on the Population of China*, 282.

Continuous shrinkage of individual landholdings could mean only increasing hardship for the peasant. When the yield of the small acreage could no longer sustain his life, he sold the land and became the tenant of a landlord. Once the land was sold, the peasant was not likely to buy it back, because the rich owner would not sell except at a very good price, which the peasant could not meet. The result of this spiral was the ever-increasing concentration of land among the rich. The Ho family of Chihli possessed a million *mou* in 1766, roughly $\frac{1}{700}$ of the total arable land of the country. Not only the landlords, but wealthy rice merchants, usurers, and pawnshop owners also manipulated land ownership, with the result that the land value increased several fold. A *mou* of land that used to cost one or two taels in the early Ch'ing period rose to seven or eight taels in the middle period.

The high concentration of arable land is illustrated by the fact that 50 percent to 60 percent of it was in the hands of the rich families. Another 10 percent was possessed by the bannermen and official villas, leaving only 30 percent for the rest of the 400 million. Sixty to 90 percent of the people had no land at all. The life of the landless peasant was wretched. He had to pay 50 percent of the yield for rent; and as the rent was not paid in kind but in commuted money, in the process of commutation usually another 30 percent was levied. For instance, a *mou* of land which produced 3 *shih* (*shih* = 133⅓ lbs.) should normally cost 1.5 *shih* for rent, but when commuted to money payment at 30 percent extra, the rent actually amounted to 1.95 *shih*, leaving the tiller only 1.05 *shih* for himself. Naturally he could not eke out a subsistence but had to borrow from usurers.[2] Many displaced and unemployed peasants drifted to the cities as porters, dockhands, and sailors, while others went abroad to seek a new life, and still others became idlers, rascals, and bandits. Had there been large-scale industries or big enterprises in China at the time, these surplus persons might have found their way into productive channels, but unfortunately there were no such industries, and the jobless became a source of unrest in the society.[3] They were ready material for uprising or revolution.

THE EFFECTS OF THE OPIUM WAR. Taking advantage of the fact that the Treaty of Nanking made no provisions against the import of opium, the foreign traders intensified their activities in this illicit but lucrative trade. The Chinese government, which had lost the war, dared not stop it. As a result, opium traffic practically became unrestrained and the volume of

2. P'eng Tse-i, *T'ai-p'ing t'ien-kuo ko-ming ssu-ch'ao* (The revolutionary thought-tide of the Heaveny Kingdom of Great Peace), (Shanghai, 1946), 14-15; Hsiao I-shan, III, 38-39.
3. Lo Erh-kang, 35.

import rose from 33,000 chests in 1842 to 46,000 chests in 1848, and to 52,929 chests in 1850. The year 1848 alone witnessed the outflow of more than 10 million taels of silver, accentuating the already grave economic dislocation and copper-silver exchange rate. A tael of silver, which had exchanged for 1,000 copper coins in the 18th century, had a market value of more than 2,000 in 1845. This 100 percent rise in the exchange rate virtually reduced a man's income by half, for although the silver tael and the copper coin were both common currencies of the state, it was the latter that was the basic medium of exchange in the market: rice was bought and wages were paid with the copper coin. A *shih* of rice formerly sold for 3,000 cash, which exchanged for 3 taels at the old rate of 1,000 to 1, but in 1851 it could only exchange for 1.5 taels at the inflated rate of 2,000 to 1. In effect, this meant that the farmer's land tax burden was doubled.[4]

The disruptive economic consequence of opium importation was further confounded by the general influx of foreign goods in the treaty port areas. Canton was particularly hard hit, because it had the longest history of foreign trade and the widest foreign contact. Local household industries were swept away and the self-sufficient agrarian economy suffered dislocation. Those who were adversely affected became a potential source of trouble.

POLITICAL CORRUPTION. As has been discussed in Chapter 6, government officials were characterized by superficiality, temporization, and irresponsibility. Little or no attention was paid to the people's welfare. Of the more "conscientious" officials who were relatively free from irregularities, some passed their time in literary activities, while others read Buddhist scriptures and dabbled in philanthropic works. They considered themselves lofty and refined, regarding those officials who busied themselves with administration as vulgar. Official irresponsibility was also reflected in the rampant selling of offices and extorted contributions. With 3,000 taels a man could purchase a magistracy; it would be rare that such a man did not try to recover this sum during his incumbency.

MILITARY DEGRADATION. The bannermen, who contributed much to the founding of the dynasty, had long since become enervated. As early as the K'ang-hsi period, they had degenerated to such a point as to be unable to suppress the Revolt of the Three Feudatories (1673-81), and the court had had to rely on the Chinese Green Standard army. By the time of the White Lotus Rebellion in 1796-1804, the Green Standard had lost its

4. Li Shou-k'ung, 143.

vigor, too, and the court was forced to use local militia. The bannermen and the Green Standard had forfeited the respect and fear of the people. Moreover, defeat in the Opium War exposed the military weakness of the dynasty. Secret societies and ambitious Chinese were encouraged to intensify their nationalistic and racial revolution against the Manchus.

NATURAL DISASTERS. The decades of the 1840s and '50s were full of natural calamities. Among the major ones were the severe drought in Honan in 1847, the flooding of the Yangtze River over the four provinces of Hupeh, Anhwei, Kiangsu, and Chekiang, the famine in Kwangsi in 1849, and the shifting of the course of the Yellow River from the southern to the northern route in Shantung in 1852, flooding a large area. Millions of people suffered from these natural disasters. Government relief at best was perfunctory, with much of the funds being embezzled at the same time. In disgust and desperation, the suffering masses were easily swayed to join a rebellion or uprising.

THE HAKKA AND CHRISTIANITY. The area in the south, the last to be conquered by the Ch'ing dynasty, was particularly vulnerable to uprising, because it was farthest from the seat of government (Peking) and because it had been exposed longest to foreign influence and contact. After the Opium War, many in the Canton area suffered by the shifting of foreign trade to Shanghai; former transportation workers connected with the shipment of tea and silk were thrown out of work.

The economic distress in the south was complicated and sharpened by the social conflict between the "natives" (original settlers) and the "guest settlers" known as the Hakka (*k'o-chia*) or *lai-jen* (men who had come). The Hakka were originally residents of Central China who had migrated to Kwangtung and Kwangsi during the Southern Sung (A.D. 1127-1278) period when the dynasty moved south under the barbarian threat. They were the social "out-group" and their different dialects, habits, and mode of life made it difficult for them to mix or assimilate with the natives. Collision between the two groups was bound to occur, and in areas where the "guest settlers" had gained ascendancy over the natives, conflict was exacerbated to the point of brutal fighting. By the middle of the 19th century, a new factor of friction was introduced: many Hakka took up Christianity, while the natives persisted in their worship of idols and spirits. The Hakka attacked the natives for their superstition, and the natives despised the Hakka for accepting a heterodox foreign faith. The tension between the two sharpened.

Men without deep social roots, the Hakka were on the whole more independent, daring, and prone to action than were the natives. Their major occupations were small farming, charcoal-making, and mining. It was here that potential revolutionary leaders recruited their followers.

From this description we get a picture of a country beset with social and economic problems, military degradation, political corruption, population pressure, natural calamities, and an explosive situation in Kwangtung. The country was ripe for an upheaval, and it is no coincidence that the largest and most significant convulsion, the Taiping Revolution, broke out in the south.

The Outbreak of the Taiping Revolution

Hung Hsiu-ch'üan (1814-64), leader of the Taiping Revolution, was the third son of a Hakka farming family in Hua-hsien, Kwangtung, located some thirty miles from Canton. As a boy he was proud, domineering, irritable, and short-tempered, but showed considerable promise in learning. His teachers and elders had hoped that he could bring honor to his family and home town through success in the civil service examinations. Four times in his life—in 1828, 1836, 1837, and 1843—he tried for the *hsiu-ts'ai* examinations at Canton, but each time he failed. During his second attempt, in 1836, two events took place that were to greatly influence his life later: (1) he became impressed with Confucian utopianism as envisaged in "The Evolution of Li" (*Li-yün*) and "The Grand Union" (*Ta-t'ung*), on which the famous scholar Chu Tz'u-ch'i was then lecturing in Canton; and (2) he met two Protestant missionaries in the street. One of them (Edwin Stevens) wore a robe and a long beard; the other handed Hung a set of nine tracts called "Good Words Exhorting the Age" (*Ch'üan-shih liang-yen*), prepared by the early convert Liang A-fa (1789-1855), an assistant of Dr. Robert Morrison of the London Missionary Society, who had settled at Canton to translate the Bible and spread the gospel. Preoccupied with his failure at the examinations, Hung gave no more than a casual glance at the tracts.

After his third failure in 1837, Hung became so distressed that he fell seriously ill. In his delirium he saw visions in which he was cleansed by an old woman, the Heavenly Mother, who told him: "Son of mine, you are filthy after your descent on earth. Let me wash you in the river before you are permitted to see your father."[5] Later, he was taken to heaven where a golden-bearded, venerable old man in a black robe gave him a sword to exterminate demons and a seal to overcome evil spirits.

5. Hsiang Ta et al. (eds.), *T'ai-p'ing t'ien-kuo* (The Taiping Heavenly Kingdom), (Shanghai, 1952), II, 632; see also Wakeman, ch. 12.

During similar visitations, he saw a middle-aged man, whom he called his Elder Brother, who instructed him in the annihilation of demons. Hung also saw Confucius confessing to the venerable old man his sin of not having explained the Truth clearly in his Classics. The delirium and visions went on intermittently for forty days; neither doctors nor sorcerers could cure him.

When he came out of the delirium, Hung's personality and appearance had drastically changed. He seemed larger in stature and his steps were more solemn, while his disposition had become much milder, more friendly and tolerant—he was, in fact, a different man. A modern psychologist who has studied Hung's visions suggests that the golden-bearded man in the dream must have been the missionary whom he had met earlier in the Canton street, and that the forty-day delirium was correlated to the period of Jesus' temptation in the wilderness, which Hung must have learned from the Christian tracts.[6]

For the next six years Hung continued his career as a village teacher. In 1843 he took the examinations for the fourth time, and again he failed. This occurred at a time when popular sentiments were running high over the "Canton city question." Hung's sympathy with this show of "nationalism" and his disgust with the existing system, which offered him no prospect for success, generated in him a strong incentive to foment a nationalistic-racial revolution against the Manchu dynasty. Yet, as with many revolutions in Chinese history, a religious aura would be helpful in sustaining such a movement.

One day, a cousin, Li Ching-fang, visited Hung and out of curiosity borrowed the Christian tracts on the shelves. Impressed with their unusual nature, Li urged Hung to read them. Hung did so, and came to believe that these tracts contained the key to his visions of six years before: the old man was God the Father, the middle-aged man was Jesus the Elder Brother, and Hung himself was the younger son of God and brother of Jesus. A new Trinity was born—at least in Hung's mind. He decided that the devils in his visions were the idols in the temples. Excited and overjoyed by this revelation, Hung and Li baptized themselves in the manner described in the tracts, promised God that they would abstain from idol-worship, and pledged to honor his Commandments. Among his first converts were a cousin, Hung Jen-kan (1822-64), and a neighbor and schoolmate, Feng Yün-shan (1822-52)—also a frustrated scholar. Shortly afterwards Hung's family was converted.

Hung Hsiu-ch'üan and Feng avidly studied the Christian tracts but

6. P. M. Yap, "The Mental Illness of Hung Hsiu-ch'uan, Leader of the Taiping Rebellion," *The Far Eastern Quarterly*, 13:3:287-304 (May 1954).

could not fully understand many of the concepts. They took "the Heavenly Kingdom" to mean China, and "God's selected people" to mean Hung himself and his countrymen. They went about to destroy idols in the temples and remove Confucian tablets from the schools; as a result they lost their positions as teachers in 1844. Influenced by the biblical statement that a prophet is not without honor save in his own country and in his own house, they went to the neighboring province, Kwangsi, to preach. They ignored the warning that their Christianity was based on a limited and private interpretation of a small part of the Bible and some tracts. A few months later, Hung returned to his home town, and during the next two years he continued to teach and to prepare religious tracts and odes, drawing ideas freely from the Bible and from the Confucian utopian writings, "The Evolution of Li" and "The Grand Union." He attacked opium-smoking, gambling, and drinking, and emphasized the egalitarian idea that all men are brothers and all women sisters. There is no doubt that Hung was using his new religion to build up a following for his revolutionary cause. Meanwhile, at Tzu-chin-shan (Thistlemont), some fifty *li* north of Kuei-p'ing, Kwangsi, Feng had organized an Association of the God Worshippers (*Pai Shang-ti hui*).

In 1847 Hung and his cousin Hung Jen-kan went to Canton to seek instruction in the Bible, as well as in Christian rituals and church organization from the Rev. Issachar J. Roberts (1802-71), an American Southern Baptist missionary. Hung's rapid progress aroused the jealousy of two of Roberts' Chinese assistants, who feared that they might be replaced by Hung. Taking advantage of Hung's naïveté, they persuaded him to ask Roberts for a subsidy for his baptism. Offended by Hung's venality, the missionary refused to baptize him, and Hung, conscious that he had been tricked, returned to Kwangsi without the baptism. By this time the Association of God Worshippers had already won more than 3,000 converts among the miners, charcoal-workers, and poor peasants, most of whom were Hakka. As the movement expanded, better educated and wealthier men also came into the fold. Among the first to join were Yang Hsiu-ch'ing, a charcoal-worker; Hsiao Ch'ao-kuei, a farmer and later brother-in-law to Hung; Wei Ch'ang-hui, a farmer of some education who had had previous experience in dealing with local officials; and Shih Ta-k'ai, a man of means, education, and fighting spirit. These four, together with Hung and Feng, formed the nucleus of a new religious and revolutionary movement. Hung, the second son of God, was acknowledged as the leader, while Feng was regarded as the third son of God; Yang, the fourth son; Hsiao, the fifth son. Ten Commandments, after Moses', were composed by Hung: (1) Thou shalt worship God; (2) Thou shalt not worship evil spirits; (3) Thou shalt not mention God's name superfluously; (4) Thou shalt worship God and praise him on the seventh day of the week; (5) Thou shalt have filial piety; (6) Thou shalt not kill or harm people; (7) Thou shalt not commit adultery and treachery; (8) Thou shalt not steal and rob; (9) Thou shalt not lie; and (10) Thou shalt not covet.[7] Hung's order of Christianity was primarily Protestant rather than Catholic, because Protestantism fitted better the nature of their movement, which was basically a "protest" against the existing order.

During the great famine of 1849-50, members of the Heaven and Earth Society (*T'ien-ti hui*) in Kwangsi rose to action under the pretext of "robbing the rich to aid the poor." The God Worshippers benefited from the disturbance. More Hakka joined them to seek refuge from the natives, and more poor people came to get protection from bandits and oppressive officials. Many naïvely believed that with their foreign religion the God Worshippers were immune to official intervention. By the spring of 1850 Hung had built up a following of 10,000. He picked the strategically located village of Chin-t'ien, Kwangsi, for his headquarters and sent for his family. In June 1850 all God Worshippers in the different areas were asked to sell their properties and bring the proceeds to a public treasury at Chin-t'ien, from which all drew their sustenance. The idea of mutual sharing had a great appeal to the poor.

Hung and his associates had by now completed their secret preparations for the revolution. In November 1850, when the government troops attempted to make irregular exaction from certain charcoal-workers who were God Worshippers, conflict broke out. On Hung's thirty-seventh birthday (38 *sui*), January 11, 1851, the God Worshippers celebrated the occasion at Chin-t'ien with a formal declaration of revolution.[8] Hung was proclaimed "Heavenly King" (*T'ien-wang*) of a new "Heavenly Kingdom of Great Peace" (*T'ai-p'ing t'ien-kuo*), and his five senior associates were given the rank of king, though without specific title.[9] The term *T'ai-p'ing* (Great Peace) appeared in the Chinese classics and had been the reign titles of several emperors before, while the term *"t'ien-kuo"* (heavenly kingdom) was taken from the Bible. Put together, *T'ai-p'ing t'ien-kuo* meant a heavenly kingdom of great peace on earth.

The Taipings maintained a delicate relationship with the secret societies. Hung found their idol-worship reprehensible and regarded their

7. Li Shou-k'ung, 161.
8. *Chin-t'ien ch'i-i* (Uprising at Chin-t'ien), (Nanning, 1975).
9. The older accounts, that these terms were not adopted until the Taipings had taken Yung-an in September 1851, have been refuted by more recent research; see S. Y. Teng, *New Light on the History of the Taiping Rebellion* (Cambridge, Mass., 1950).

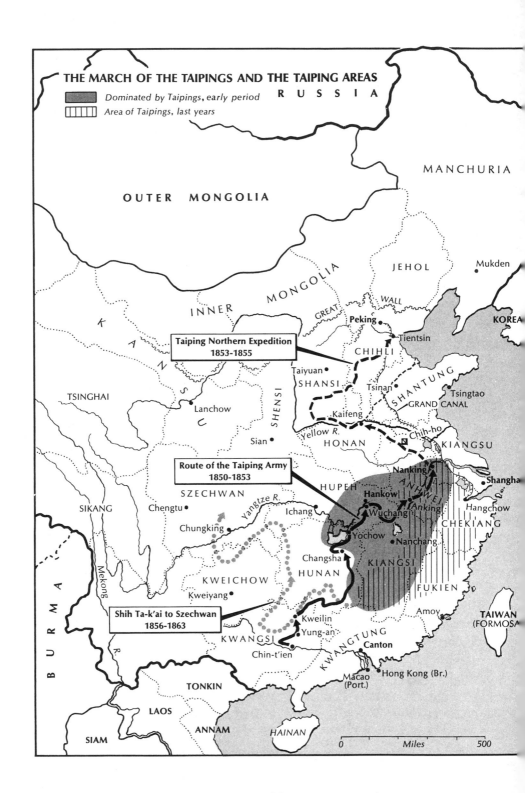

THE MARCH OF THE TAIPINGS AND THE TAIPING AREAS

Dominated by Taipings, early period

Area of Taipings, last years

RUSSIA

MANCHURIA

OUTER MONGOLIA

INNER MONGOLIA

MONGOLIA

JEHOL

Mukden

GREAT WALL

Peking

Tientsin

KOREA

**Taiping Northern Expedition
1853-1855**

Taiyuan

SHANSI

CHIHLI

Tsinan

SHANTUNG

Tsingtao

GRAND CANAL

TSINGHAI

KANSU

Lanchow

SHENSI

Sian

Kaifeng

Yellow R.

HONAN

Chih-ho

KIANGSU

**Route of the Taiping Army
1850-1853**

SZECHWAN

Chengtu

Yangtze R.

Ichang

HUPEH

Hankow

Wuchang

Nanking

ANHWEI

Shanghai

Anking

Hangchow

Chungking

Yochow

Nanchang

CHEKIANG

SIKANG

Mekong

Changsha

KIANGSI

KWEICHOW

HUNAN

FUKIEN

Kweiyang

**Shih Ta-k'ai to Szechwan
1856-1863**

Kweilin

Amoy

TAIWAN
(FORMOSA)

BURMA

Yung-an

KWANGSI

KWANGTUNG

Canton

Chin-t'ien

Macao
(Port.)

Hong Kong (Br.)

TONKIN

LAOS

ANNAM

HAINAN

SIAM

0   Miles   500

objective of restoring the Ming at cross purposes with his own scheme of creating a kingdom. However, their anti-Manchu stand was consonant with his revolutionary aim. Hung wanted to use them to advance his own cause without being used by them. He decided that members of the secret societies might join the Taipings provided they renounced idol-worship, took up God-worship, and accepted the Taiping Command-ments and discipline. Many secret society members who joined the Tai-pings found it hard to live up to these demands and withdrew, but a number of them remained, including the fierce fighters Lo Ta-kang and Lin Feng-hsiang.

The Taipings wore their hair long, in contrast to the prevailing style of shaving the front part of the head while keeping a long queue in the back; hence they were known as the "Long-haired banditti" or simply "Long Hair." The Taiping forces fought with all the vigor and spirit of religious zealots, and the imperial army, provincial forces, and local mili-tia melted away before them. From their base at Chin-t'ien, the Taipings charged north to take the important city of Yung-an on September 25, 1851, as a new base. There they halted for almost half a year to accumu-late enough provisions for three or four months and expand their army to 37,000 men. The various senior associates of Hung, who had been called kings, were now given specific titles: Yang became the East King, Feng the South King, Hsiao the West King, Wei the North King, and Shih the Assistant King. Various offices of the new kingdom were also set up, after the institutions of the Chou dynasty 3,000 years earlier. Most accounts state that it was at Yung-an that the establishment of the Heav-enly Kingdom of Great Peace was formally proclaimed, though this has been disputed by recent research.[10] The Taipings adopted a new calendar and issued a formal proclamation to the country, in which they denounced the Manchu oppression of the Chinese, attacked corruption within this alien rule, and called for the overthrow of the Ch'ing dynasty.

At Yung-an, the imperial forces laid a strong siege from which the Taipings could not break out for nearly half a year. At one point the morale sank so low that the movement might well have gone under had it not been for the clever tricks of the East King, who announced that God had revealed to him that the danger would pass after a hundred days. On April 3, 1852, the Taipings broke the siege and swept north. They then broke into Hunan with a view to taking the provincial capital of Changsha. In this campaign, they suffered two major losses in the deaths of the South King and the West King. Nevertheless the revolutionary

10. Teng, *New Light.*

forces rolled on, taking Yochow on December 13, 1852, where they discovered a huge arsenal of munitions and cannon sequestered more than a century and a half earlier by General Wu San-kuei. In addition they confiscated more than 5,000 vessels. Thus fortified, the Taipings moved on to the key triple cities of Wuhan[11] on the Yangtze River, where they seized 10,000 vessels, a million taels of silver from the provincial and granary treasuries, and a great supply of provisions and ammunitions. Now boasting half a million men, the Taipings were poised for an advance to Nanking, the capital of many dynasties and the base from which the first Ming emperor expelled the Mongol conquerors. Having looted Wuhan, the Taipings charged eastward along the Yangtze and swept into Nanking on March 19-21, 1853. Hung entered the city triumphantly in a sedan-chair carried by thirty-six men, as if he were emperor, while the East King's chair was carried by sixteen men. Nanking was renamed the Heavenly Capital.

The imperial forces could do no more than set up two large camps to threaten the security of the Taiping capital: one in the eastern suburb of Nanking called the Great Camp on the South of the River (*Chiang-nan ta-ying*), and the other in the outskirts of Yangchow called the Great Camp on the North of the River (*Chiang-pei ta-ying*).

Hung sent an army to Northern China under Lin Feng-hsiang and Li K'ai-fang, and another under Lo Ta-kang westward to Anhwei, Kiangsi, Hupeh, and Hunan. These generals, all former members of secret societies, were fierce fighters but were not close associates of Hung's. Possibly Hung intended to send them away to reduce the influence of secret societies in the Taiping movement.[12] The northern expedition got within twenty miles of Tientsin, but ultimately failed for lack of support; both leaders were captured and executed in Peking in 1855. The western expedition also ran into stiff resistance by the scholar-general Tseng Kuo-fan, as we will soon see.

Taiping Institutions

The Taiping kingdom was a theocracy, in which religion, civil and military administration, culture, and society in general were all interwoven. The capital was called the Heavenly Capital, the leader the Heavenly King, his palace the Heavenly Palace, his orders Heavenly Rescripts, and

11. Wuchang, Hankow, and Hanyang.
12. His decision to dissociate the Taipings from the secret societies offers an explanation for his refusal to aid the Small Sword Society, an offshoot of the Trial Society, which seized the walled city of Shanghai for a year and a half in 1853-54.

his treasury the Sacred Treasury. Opium-smoking, the use of tobacco and wine, prostitution, foot-binding, sale of slaves, gambling, and polygamy were all prohibited. There was a definite puritanical spirit in the early period of the Taipings, and the leaders conceived of many worthy institutions and innovations. The basic document of state was called "The Land System of the Heavenly Kingdom" (*T'ien-ch'ao t'ien-mou chih-tu*), which spelled out not only the land system, but also the military, civil, financial, judicial, and educational institutions. It was a sort of Taiping constitution.

THE LAND SYSTEM. Perhaps the most drastic of the Taiping innovations was the abolition of the private ownership of land and property. The spirit behind it was that all children of God must share his blessings, must be free from want, must have land to till, rice to eat, clothes to wear, and money to spend. To achieve this idealistic state, a basic change in the existing land system was imperative. The Taipings therefore divided land into nine classes according to yield.[13] All men and women over sixteen *sui* received a share of land, and everyone under sixteen received half a share. Thus, if a man received one *mou* of A+ land, his children up through the age of fifteen received half a *mou* of the same quality land. A family of six received an equal amount of good and poor land, that is, three members received good land and three received poor.[14] The distributed land did not become the property of the recipient; he was merely given the right to use it for production. All surplus products beyond one's needs had to be surrendered to the public storehouse. Savings and private property were prohibited.

The idea of communal utilization of land can be found in the ancient Chinese work *The Rites of Chou,* and the usurper of the Han power, Wang Mang, had put this into practice during his short-lived Hsin dynasty (A.D. 8-23). The Taipings revived the doctrine with great visionary idealism, but unfortunately, due to incessant warfare and the unsettled conditions in the countryside they could not put it into practice except in a few experimental pockets.

13. A *mou* of land with a yield of 1,200 chin (*chin* = 1⅓ lbs.) of product was classified as A+ (*shang-shang*), that with a yield of 1,100 *chin*, A (*shang-chung*); 1,000 *chin*, A— (*shang-hsia*); 900 *chin*, B+ (*chung-shang*); 800 *chin*, B (*chung-chung*); 700 *chin*, B— (*chung-hsia*); 600 *chin*, C+ (*hsia-shang*); 500 *chin*, C (*hsia-chung*); and 400 *chin*, C— (*hsia-hsia*).
14. One *mou* of the A+ land was equivalent to 1.1 *mou* of the A land, 1.2 *mou* of the A— land, 1.35 *mou* of the B+ land, 1.5 *mou* of the B land, 1.75 *mou* of the B— land, 2 *mou* of the C+ land, 2.4 *mou* of the C class, and 3 *mou* of the C— class.

THE UNITY OF THE MILITARY AND CIVIL ADMINISTRATION. The Taiping military institutions were derived from *The Rites of Chou* and the systems developed by the Ming general Ch'i Chi-kuang. The peculiar feature was the unity of the military with the civilian administration. Soldiers were farmers, and officers were assigned civil as well as military duties. For each 13,156 families there was an army commander, who was in charge of five divisions, each of which consisted of five brigades. Under each brigade commander were five captains, each of whom commanded four master sergeants, who in turn were in charge of five corporals each, who controlled four soldiers. An army thus consisted of 10,000 soldiers and 3,156 officers, making a total of 13,156 men. When more families were formed, new army units were organized.

The military officers were also civil administrators. Every twenty-five families formed a basic social unit, each with a public storehouse and a church under the charge of the master sergeant. He administered the civil, educational, religious, financial, and judicial matters of his twenty-five families and took charge of their litigations, marriages, and funerals. All the expenses of these affairs were paid out of the public storehouse, but there was a limit to what each event could cost. In time of peace the soldiers and corporals performed public works. The children of the twenty-five families went to church daily to receive instructions from the master sergeant on the Old and New Testaments and Christian tracts written by Hung. On Sunday the corporals led the people to church, where the master sergeant preached, and the men and women were seated separately. The hymnal was different from the Protestant hymnal, and although the ritual of the service in general followed the Protestant tradition, there were deviations, too, such as the use of drums and firecrackers and the serving of cakes and fruits, much as the Buddhists and the Taoists did.

The Taipings forbade ancestor-worship, and destroyed idols and temples whenever they were discovered. The officers usually preached to the inhabitants of any new area their army reached.

UNITY OF CULTURE AND RELIGION. In the Taiping kingdom, the inculcation of Christian ideas among the people was a primary undertaking. A new *Three-Character Classics* of 478 sentences and 1,434 words, based on Hung's interpretation of the Bible, was compiled for the children, with opening passages as: "God our Lord, creates Heaven [and] Earth, creates mountains [and] seas, provides all things; within six days, all was done . . ." There were also *Odes for Youths* and hymns glorifying God and Jesus as true saviors of mankind. In all these writings the vernacular was used, with punctuation, to facilitate easy reading and wide circulation. Some progressive writers today consider this plain style of writing a forerunner of the New Cultural Movement of the late 1910s and the early 1920s.

The Taipings held civil service examinations, too, in which the plain language was used in place of the classical style of writing required in the Ch'ing examinations. The theses in the Taiping examinations were not taken from the Confucian classics as in the Ch'ing examinations, but were selected from the Bible, Christian tracts, and Taiping proclamations, such as "God our Lord is the Only True Spirit" and "For Whom did God come to the Earth and why did Jesus sacrifice His Life?" The examinations, open to men and women alike, were at first given on the birthdays of Hung and his son, the crown prince; but later March 5 and 13 of the Taiping calendar were set aside for the annual examinations for the civil and military *hsiu-ts'ai* degree, respectively; May 5 and 15 for the civil and military *chü-jen* degree; and September 9 and 19 for the civil and military *chin-shih* degree.[15] Applicants for the examinations came from all walks of life, including fortunetellers and magicians. As might be expected, the standards of these examinations were not rigorous; it was said that in a certain Hupeh examination 800 out of 1,000 candidates passed. The examinations won the good will of many for this reason, but they compromised the original purpose of selecting talents for public service.

THE NEW CALENDAR. The calendar adopted by the Taipings was unique in that it was neither lunar nor solar but somewhere in between. The year was divided into 366 days, with 31 days to each odd month (January and every other month thereafter) and 30 days to each even month. The defect of this calendar was that it created three superfluous days every four years, or 30 extra days every 40 years. To remedy this situation, a "year of adjustment" was scheduled for every 40 years, at which time each month was divided into 28 days, making a total of 336 days for the year, just 30 days short of the regular year. January 1 of the first year of the Taiping calendar was February 4, 1852.

SOCIAL POLICIES. Men and women were equal in the Taiping kingdom. Women were allowed to serve in the civil and military administration,

15. To be more accurate, the names of these degrees were considered improper and the Taipings changed *hsiu-ts'ai* (beautiful talent) to *hsiu-shih* (beautiful scholar), *chü-jen* (employable man) to *po-shih* (erudite scholar), and *chin-shih* (advanced scholar) to *ta-shih* (all-round scholar). The title *kuo-shih* (national scholar) was substituted for the Han-lin.

and there were reportedly 100,000 female soldiers and officers under the command of Hung's sister. Women's Residential Halls (*Nü-kuan*) operated at Nanking in the early days of the Taipings for the young and unmarried, as well as for those women whose husbands had been killed in battle or were away. Governed by Hung's sister, they were relatively free of outside interference. Missionaries who visited Nanking were impressed with the freedom and ease with which members of the female sex walked or rode horses in the streets.

A number of social welfare measures were taken to support the disabled, the sick, the widowed and the orphaned. The egalitarian and ascetic spirit of the movement was also manifested, as noted earlier, in such policies as the prohibition of opium-smoking, foot-binding, slavery, and prostitution. In general, the Taiping society presented a marked contrast to the Ch'ing society.

Foreign Neutrality

The Taiping profession of Christianity evoked an early sympathy from foreigners, particularly the Protestant missionaries, although they were concerned about the new "Trinity," which smacked of blasphemy. Foreign traders were pleased with the prospect of extended commerce in the Taiping area, but were troubled by their strict prohibition of opium, which had become the single most lucrative article in the China trade. On the whole, foreign governments had mixed feelings about the Heavenly Kingdom of Great Peace, and the wise policy under such circumstances was to wait and see. The British plenipotentiary in China, Sir George Bonham, dispelled any impression of intended British aid to the Ch'ing army by announcing that Britain maintained neutrality in the Chinese civil strife and would not go beyond protection of the lives and properties of her subjects in Shanghai. For a firsthand knowledge of the Taipings, Bonham and his interpreter, Thomas T. Meadows, sailed in H.M.S. *Hermes* to Nanking in April 1853. The North King and the Assistant King granted an interview to Meadows and arranged for Bonham to see the East King, the prime minister. Bonham did not see him but sent a letter instead, in which he explained the British position of neutrality and asked the Taipings to recognize the British treaty rights and privileges. The Taiping reply was curiously condescending, stating that the Heavenly King, pleased with the coming of distant people from the "dependencies," graciously granted them permission to trade or to serve in Nanking. Accompanying the communication were several Taiping tracts which the British were asked to read to learn the Truth.

Bonham's visit, then, was not fruitful except in gathering information. He had come to ask for Taiping respect of British treaty rights, but the Heavenly King wanted to treat Britain as if she were an inferior state. Bonham went home, leaving a warning that if British lives and properties were violated, his government would take retaliatory measures such as they had a decade earlier in the Opium War. To London Bonham expressed doubt that the Taipings could replace the Ch'ing dynasty and recommended a policy of neutrality. Meadows' analysis of Taiping tracts and the movement as a whole sheds light on the mixed foreign feelings:

> The anthropomorphism displayed in the above pamphlet is very striking. The Deity is brought down from a state of distant superiority, and is represented as familiar with mortals, in a degree which to us appears somewhat revolting . . . There are some things good, very good, in the productions before us, leading us to infer that the authors were divinely taught, and to cherish the hope that not a few will, through the medium of these truths, find the road to heaven. There are, however, some things of which we most highly disapprove; not the least of which are the pretensions to new and immediate communications from the Deity; some of which afford representations of the Divine Being far different from what we have been accustomed to in the Christian Scriptures, and made to serve the end of personal aggrandizement and ambition . . .
>
> It would be sad to see Christian nations engaged in putting down the movement, as the insurgents possess an energy, and a tendency to improvement and general reform (as witness their calendar) which the Imperialists [Ch'ing] never have exhibited, and never can be expected to display. Questionable though it be, the form of Christianity which the insurgents profess is far better than the stupid idolatry hitherto practised by the Chinese; and it is possible that European nations, if engaged on the opposite side, would be going to war with some people in some respects better than themselves . . . The only policy that appears at present advisable, is to keep ourselves from being involved any further in the quarrel, and to avoid all Government connection with either party.[16]

The Americans also followed a policy of neutrality in the early years of the Taipings, although their sympathy was clearly not with the imperial government. France, as protector and propagator of the Catholic faith abroad, was not favorably disposed toward the Taiping Protestant ideology. However, when the French minister to China, M. de Bourboulon, visited Nanking in December 1853, he was impressed with the order and

16. McNair, I, 345-46.

discipline of the Taipings and he recommended a policy of neutrality to his government, too. Russia, whose trade with China was mostly in the border areas of Sinkiang, Mongolia, and Manchuria, was less impressed with the Taiping movement. Politically, a Taiping success would mean a shift of gravity from Peking, where Russia maintained a semidiplomatic, religious mission, to Nanking and the south, where the British influence was strong. It was therefore in Russia's interest to sustain the imperial cause. However, for the time being St. Petersburg also chose to stay neutral.

It has never been understood why the Taipings did not pursue an active policy of seeking foreign recognition; it is even more puzzling that on the one hand they professed equality of all men as God's children and on the other insisted on treating foreign representatives as delegates from inferior dependencies. Possibly the Taiping leaders distrusted foreigners and were afraid of their association. If the Taipings had had an active policy of foreign alignment against the Manchu government, with which the foreign powers were at odds, the future course of the revolution might have been different.

Tseng Kuo-fan and the Hunan Army

The early Taiping success was a result of many factors. First, there was general sympathy toward their nationalistic-racial revolution against the alien Manchu rule, and the secret societies extended them considerable help. Secondly, the Taiping forces were an ideological army with a "Messianic" mission. The soldiers believed that Hung had been commissioned by God to exterminate the demons from earth, and that if they died for the cause they would rise to heaven and dwell in the bosom of God for eternity. Unafraid of death and ready to become martyrs, they fought with the single-minded devotion and bravery of a crusading army. Thirdly, the Taiping military organization was based on the tested methods of the famous Ming general, Ch'i Chi-kuang. Their well-disciplined troops showed much greater consideration toward the people than did the corrupt and debauched government forces. The bannermen and the Green Standard army had long lost their fighting spirit; when they met the zealous revolutionaries they went to pieces.

Under these conditions, the court resorted to organizing militia for local defense, such as it had done during the White Lotus Rebellion (1796-1804). When the Taipings swept into Hunan and besieged the provincial capital of Changsha in mid-1852, the much-harassed court hurriedly ordered the scholar-official, Tseng Kuo-fan, out of mourning to organize a militia force for the defense of his home province.

Tseng Kuo-fan (1811-72) of Hsiang-hsiang, Hunan, was a persistent if not a brilliant scholar, known for his tenacity of purpose. After winning the *chin-shih* degree in 1838, he rose gradually in the civil service and in 1849 became a junior vice-president of the Board of Rites. Thereafter he served as a junior vice-president of the Board of Punishments in 1851 and as an acting senior vice-president of the Board of Civil Office a year later. While living in Peking he made friends with leading Neo-Confucian scholars (Sung School) and from them acquired an appreciation of "quietude" (*ching*), "persistence" (*nai*), and "discipline" (*yüeh*), which, when applied to practical affairs, meant cool-headedness in emergency, fearlessness in the midst of difficulties, and realistic self-control. These qualities were to benefit him greatly in his later career.

In the middle of 1852 Tseng was made chief commissioner of the provincial examinations in Kiangsi. En route to this assignment he learned of his mother's death, and in accordance with the social convention he retired to his home town for a period of mourning. It was during this time that the court summoned him to organize a militia for Hunan. A filial son and a correct Confucianist, he did not wish to cut short the mourning period but was ultimately persuaded by friends and the governor to place the interests of the country above his familial obligations; thus he went to Changsha.

Tseng knew well that the Green Standard army and the militia were no match for the Taiping revolutionary army. If he was to be effective, he would have to go beyond the imperial order and raise a new army. The Taipings were no White Lotus rebels, whom the militia had suppressed; they were an ideological army organized along the time-tested methods of the famed Ming general. To counter such a crusading force, it was necessary to turn the militia into a well-trained and well-indoctrinated army. For this purpose Tseng decided, interestingly enough, to adopt the military organization of the same Ming general, Ch'i Chi-kuang, and in addition to instill in his army a sense of mission: that of defending the Chinese cultural heritage in the tradition of Confucius and Mencius. Such an army would have an *esprit de corps* which could be assured by a careful recruitment policy based on common regional background.

With these ideas in mind, Tseng raised three battalions of troops, each comprising 360 men, making a total of 1,080. The officers were chosen from Confucian scholars, and the soldiers were selected from solid farm types rather than slick city dwellers. These officers and soldiers all came from the province of Hunan, as did Tseng himself—hence the name "Hunan army" (*Hsiang-chün*) or "Hunan braves" (*Hsiang-yung*)— "braves" being the name for irregular troops or temporary recruits as opposed to the regular, standing army. Gradually the Hunan army was expanded into thirteen battalions, each enlarged to 500 men, in addition to 180 service personnel who performed sundry chores. The soldier's pay was 4.5 taels a month, ten times the wages of an ordinary house servant. The battalion commander received a salary of 50 taels monthly plus 150 taels more for expenses. Along with the army organization, Tseng also raised a navy—or rather a "water force"—of ten battalions totaling 5,000 men, to contend with the Taipings on the Yangtze. Two hundred forty war junks were made by the two dockyards which he established. Tseng derived the support of his Hunan army and navy from (1) the inland transit dues (*likin*) which theoretically were 1 percent of the value of commodities but which actually amounted to 4 percent to 10 percent;[17] (2) customs dues; (3) Ch'ing treasury allocations; (4) contributions; (5) salt levies; (6) tribute grain; and (7) miscellaneous taxes.

The soldiers in the Hunan army were recruited by the officers, to whom they owed their allegiance; the officers in turn pledged their allegiance to Tseng, who had selected them. The Hunan army was therefore a private army. Up to this time no Ch'ing official, much less a private citizen, could maintain a personal army; the bannermen and the Green Standard army all belonged to the central government. But now, due to the exigency of circumstances, Tseng Kuo-fan, a scholar and an official in mourning—and a Chinese at that—broke the precedent.

The court had repeatedly urged Tseng to send his forces to rescue Hupeh province from the Taipings, but he refused to leave Hunan until he had cleared away the local bandits and completed the training of his army and navy. Impatient with the delay, the emperor reprimanded him for partiality toward his native province without regard for the general situation. In early 1854 the Taipings once again threatened Wuchang, and the anxious court "begged" Tseng to aid Hupeh, authorizing him freedom of action without control from Peking. By February of that year, Tseng set out for Hupeh with a force of 17,000 men. In leaving their native province, the Hunan army and navy shed what little traces were

17. In some provinces, it was as high as 20 percent. *Likin* originally meant a surtax of "a hundredth" of the value of commodities, but the definition was never literally interpreted—not even by its originator, Lei I-hsien, who wrote: "The amount which has been collected is in general only one part in a hundred [i.e. 1 percent of the value of the goods handled], and there are even cases in which it does not reach one part [in a hundred]." See Edwin George Beal, Jr., *The Origin of Likin* (Cambridge, Mass., 1958), 26.

left of a parochial militia and became a new fighting force of national stature. Tseng issued a proclamation in which he condemned the Taiping disruption of village life, the abolition of private ownership of land, the destruction of temples, and the violation of Confucian propriety and the Chinese way of life. The first two of these items, it is fairly evident, were meant for the peasants, and the last two were directed at the gentry. He skipped altogether the issue of nationalistic and racial revolution that the Taipings made much of, but stressed his own role as a defender of the cultural tradition. The literati flocked to his cause, not that they appreciated less the Taiping anti-Manchu stand, but that they loved more the preservation of Chinese heritage. Moreover, the Manchu dynasty had been established for more than 200 years, and the Chinese scholars had served it for all that time; were they to advocate racial revolution now, it would be unconvincing. In fact, the vested interest of the literati so intertwined with that of the dynasty that in supporting the imperial cause the literati were actually defending their own interests.[18] The peasants also responded to the call of Tseng because they were disheartened by the Taiping disruption of their peaceful country life and by the wanton destruction of the temples and shrines.

The initial encounter of the Hunan army and the Taipings was not encouraging for Tseng; he received several setbacks and was criticized by provincial authorities for having begun his campaign with a hollow bang. A change for the better came when his navy scored a major victory at Hsiang-t'an on May 1, 1854. Tseng's star began to rise and in October he recovered the important city of Wuchang, which had been occupied by the Taipings since June. The emperor was delighted with this turn of events and vested Tseng with the unqualified authority to conduct the campaign against the rebels. This marked the beginning of the shift of military power from the Manchus to the Chinese.

The victorious and confident Hunan army then entered Kiangsi to besiege the important city of Kiukiang. The Taipings not only resisted effectively but in fact succeeded in cutting the Hunan forces in two, nailing down Tseng's men in a most uncomfortable and immobile situation. Now the Taiping fortune rose again. They took Wuchang for the third time in April 1855, and overran Hupeh and Kiangsi, with Tseng quarantined in Kiangsi. Meanwhile, the Taipings at Nanking broke out to destroy the Imperial Great Camps on the South and North of the (Yangtze) River in mid-1856, driving the imperial commissioner Hsiang-jung to suicide. The entire Yangtze valley fell to the revolutionaries, and the future in-

18. Lo Erh-kang, *Hsiang chün hsin-chih* (A new study of the Hunan army), (Shanghai, 1939), 66.

deed looked good for them. But at the crest of their success, a blow fell from within which was to cripple the movement beyond recovery.

The Taiping Internal Dissension

In 1856 occurred a fratricidal conflict in Nanking which violently shook the Taiping kingdom. The basic cause of the trouble was the East King Yang's irrepressible ambitions, which had been evident from the start. Having seen through the falsity of Hung's divine commission and his new Trinity, Yang began going into trances himself and claimed that God had favored him with visitations. Hung dared not expose his tricks for fear of reprisals.

About six months before the revolution broke out at Chin-t'ien in January 1851, Yang suddenly fell sick, unable to speak or hear, and consequently unable to participate in the planning of the uprising. It was obvious that he was carrying out a passive resistance in order to force the other leaders to give him a higher place. He became commander-in-chief of the armed forces, next only to Hung in the movement. From then on, it was he who decided strategies and issued orders. As prime minister and commander-in-chief of the Heavenly Kingdom of Great Peace at Nanking, he made all the important decisions, issued orders, and controlled access to the Heavenly King, who was shrouded in mystery in the palace recesses. During audiences, the East King alone remained standing, while all others knelt, and he addressed himself to the Heavenly King by the proud and intimate cognomen of "your minister and younger brother." People in Nanking knew more about Yang than about Hung. It is not surprising, then, that his name should be as familiar and that the Taiping movement was frequently called the Hung-Yang Rebellion.

So proud was he of his part in the Taiping success—and it was unquestionably great—that the East King entertained hopes of replacing Hung himself. To facilitate the operation he sent the North King (Wei) and the Assistant King (Shih) away from Nanking, the former to Kiangsi and the latter to Hupeh. The East King began going into trances more often and angrily scolding Hung in the name of God. The climax of his scheming came after the 1856 destruction of the Imperial Great Camp on the South of the River. Elated with this success and confident of his leadership, the East King decided that the time had come to depose the Heavenly King. He instigated his followers to honor him, the East King, as the "Lord of Ten Thousand Years," a designation reserved for the Heavenly King and used otherwise only by emperors of dynasties. Hung knew that the hour of reckoning was coming on fast. He secretly sent for the North

King and Assistant King to rid the kingdom of the East King's menace. The North King hurried to Nanking by night, broke into the residence of the East King, and slaughtered him on September 2 along with some 20,000 followers. But the cure was worse than the disease, for the North King now behaved as overbearingly as the East King had done. When the Assistant King arrived and murmured about the unnecessary massacre, intimating that the guilt belonged to the East King alone and did not extend to his followers, the North King wanted to kill him, too. Although the Assistant King managed to escape under cover of darkness, his family and relatives were all slaughtered. Unable to bear the wanton killing, the Heavenly King had the North King executed, less than three months after the death of the East King. By now Hung had lost all faith in his associates; he turned over the reins of government to his two mediocre brothers, who could hardly keep the kingdom together. The Assistant King, the lone survivor of the original five kings under Hung, returned to Nanking to take charge of state affairs for a short while, but found himself distrusted by the Heavenly King. Uncertain of his safety and his future, the Assistant King left with a large following, roving through a number of provinces in the next seven years until he was finally killed in Szechwan in 1863.

The internecine strife of 1856 depleted the spirit and vigor of the Taiping movement so thoroughly that it was never able to recover. Hung gave himself over to indulgence in pleasure to forget his miseries, and leadership simply disappeared from his government. Many officers and men would have left but for the Ch'ing standing order that all Taipings who surrendered should be decapitated. A slight improvement in the morale came when Hung Jen-kan, a cousin and early convert of the Taiping leader, came to Nanking in 1859 after many years of sojourn at Hong Kong. He was made the Shield King (*Kan Wang*) and prime minister, but the decline of the regime was too far advanced to be arrested. Nonetheless, the ultimate downfall of the Heavenly Kingdom was delayed, largely because of the brilliant warfare waged by the young and talented general, Li Hsiu-ch'eng, better known as the Loyal King (*Chung Wang*). It was he who destroyed the resuscitated Imperial Great Camp on the South of the River in May 1860 for the second time, relieving Nanking of a thorn in its side. Riding the tide of victory, he swept all the way to the outskirts of Shanghai in August 1860, taking Soochow and Changchow on the way. Through him the Taipings recovered the entire province of Kiangsu, except for Shanghai and Chinkiang. It was a valiant effort of the Loyal King to keep the movement from crumbling, but no one man could arrest the disintegration of a doomed kingdom.

The Turning Point in the Campaign

Meanwhile, important changes were taking place in the Ch'ing command and in foreign attitudes toward the Chinese civil war. The second Taiping destruction of the Great Camp put an end to the imperial fighting forces, and the court had to rely even more on the Hunan army. In May 1860 Tseng Kuo-fan was given the coveted title of imperial commissioner and governor-general of Liang-Kiang, and was put in complete command of the operations against the Taipings. He had been fighting for years but without being given any specific titles such as those he had just received, and for this reason local authorities had not felt called upon to support or cooperate with him; indeed, many obstructed his plans. But now, with his new titles and authorization, Tseng was able to work out a unified strategy. It was also fortunate for him that the influential Manchu president of the Li-fan yüan, Su-shun, supported him and served as his spokesman before the emperor.

Tseng's rise to a high position of responsibility marked a turning point in the campaign. His Hunan army had grown to a powerful fighting force of 120,000 men, commanded by a number of able scholar-generals. In his headquarters, there were numerous planners, strategists, advisers, and administrative assistants, all destined for greater glory in later years. Tseng was doubtless the most powerful man in Southeast China, with jurisdiction over the four key provinces of Kiangsu, Anhwei, Kiangsi, and Chekiang. Yet the Taipings under the Loyal King were still active and effective. They enjoyed another brief spell of success in Chekiang and Anhwei in mid-1861, dealing an almost shattering blow to Tseng at Ch'imen, Anhwei. It was not until Tseng's brother, Tseng Kuo-ch'üan, took the important city of Anking in September that the tide was once again turned. Thereafter the Hunan army and navy gained ascendancy, recovering a number of cities on the Yangtze until they pressed near Nanking. In recognition of his success, the court made Tseng Kuo-fan a junior guardian of the crown prince in 1861 and an assistant grand secretary a year later. Tseng now placed Li Hung-chang, one of his chief assistants, in charge of military affairs in Kiangsu, and Tso Tsung-t'ang, another able assistant, in charge of military operations in Chekiang. The coveted assignment of attacking Nanking was awarded to his brother, who, with the help of the Hunan navy, penetrated to the vicinity of Nanking in June 1862. With 20,000 men he began a long siege of the Heavenly Capital.

At about this time foreign attitudes toward the Chinese civil war were undergoing a radical change. Foreigners had shown early sympathy toward the Taipings because of their supposed Christianity and the prospect of extended trade. But when they found the Taipings incapable of successful government, arrogant in their pretension of universal overlordship, adamantly opposed to opium importation, and continuously disturbing foreign trade and lives at Shanghai, they lost interest in the Heavenly Kingdom. Moreover, after the signing of the new treaties with the Ch'ing court in 1860, the Western powers realized that the enjoyment of the concessions was contingent on the continued existence of the dynasty.

Against this improved Ch'ing-Western relationship, the Taipings put up a poor show. Not only did they make no effort to lure the foreigners from the imperial cause, but their army continuously harassed Shanghai, the center of foreign trade and residence. At the same time, life in Nanking had degenerated to a new low. Alexander Michie, an Englishman who visited the Heavenly Capital in March 1861, wrote:

> I have no hope of any good ever coming of the rebel movement. No decent Chinaman will have anything to do with it. They do nothing but burn, murder, and destroy. They hardly profess anything beyond that. They are detested by all the country people, and even those in the city who are not of the "brethren" hate them. They have held Nanking eight years, and there is not a symptom of rebuilding it. Trade and industry are prohibited. Their land-taxes are three times heavier than those of the Imperialists; they adopt no measures to soothe and conciliate the people, nor do they act in any way as if they had a permanent interest in the soil. They don't care about the ordinary slow and sure sources of revenue; they look to plunder, and plunder alone, for subsistence, and I must say, I cannot see any elements of stability about them, nor anything which can claim our sympathy.[19]

Another description was furnished by the Rev. Issachar J. Roberts, who spent fifteen months in Nanking during 1861-62 at the invitation of the Heavenly King, who had formerly studied with him. His report of December 31, 1861 states, in part:

> As to the religious opinions of Tien Wang [Hung], which he propagates with great zeal, I believe them in the main abominable in the sight of God. In fact, I believe he is crazy, especially in religious matters, nor do I believe him soundly rational about anything . . . I do not believe they have any organized Government, nor do they know enough about Government to make one.
> He [Hung] wanted me to come, but it was not to preach the gospel

19. McNair, I, 349.

of Jesus Christ and convert men and women to God, but to take office, and preach his dogmas, and convert foreigners to himself. I would as lief convert them to Mormonism, or any other *ism* which I believe unscriptural, and, so far, from the devil. I believe that in their heart they feel a real opposition to Gospel, but for policy's sake they grant it toleration . . . And hence I am making up my mind to leave them . . .[20]

In short, early foreign sympathy toward the Taipings had given way to disappointment and a determination to assist the imperial regime, whose existence was recognized as essential to foreign interests in China.

The first foreign intervention in the Chinese civil war took place in 1860 when the Loyal King attacked Shanghai. Rich merchants and businessmen in the city financed the organization of a "foreign legion" and the American adventurer from Salem, Massachusetts, Frederick T. Ward, was engaged by the wealthy banker Yang Fang, better known by his firm's name "Taki," to recruit foreign deserters and discharged seamen to form a "Rifle Squadron." By taking the city of Sungkiang, this foreign mercenary army diverted the Taiping forces from Shanghai and scored its first major victory. In September 1861, Ward revamped his squadron by drafting 4,000 to 5,000 Chinese soldiers, who were drilled and clad in the European fashion and were commanded by 100 European officers. In addition, there were some 200 Filipinos in his army, which fought and pillaged in the Soochow-Sungkiang-T'ai-ts'ang area and won many battles. In March 1862, when they repulsed for the second time a Taiping attack on Shanghai, the emperor bestowed on them the flattering name the "Ever-Victorious army," and on Ward the rank of brigade-general. When Ward was fatally wounded on September 21, 1862 and died a day later, the leadership of the Ever-Victorious army went to another American adventurer, Henry A. Burgevine, a man of neither principle nor backbone. He argued with Taki for funds and forcibly seized 40,000 silver dollars. Consequently he was relieved of his duty and an English officer, the famous Charles G. (Chinese) Gordon, was installed as the new leader.

The Downfall of the Taiping Kingdom

On the recommendation of Tseng Kuo-fan, the court authorized Li Hung-chang to organize a new force to supplement the Hunan army. Tseng turned over 3,000 of his Hunan troops as the nucleus of the new army, while Li Hung-chang recruited several thousand more men, whom he

20. McNair, I, 349-51.

organized after the fashion of the Hunan army. Since most of these re-cruits came from the Huai River area in Anhwei province, they became known as the Huai (or Anhwei) army[21] During the Loyal King's second attack on Shanghai in 1862, Li led the Huai army to its rescue and scored a victory in the outskirts. Li was made governor of Kiangsu. In November 1863 the Huai army attacked the Taiping stronghold Soochow, backed up by the Ever-Victorious army which arrived from a different direction. The whole province of Kiangsu was pacified, except for Nanking and a few small pockets.

Li's success in Kiangsu was paralleled by Tso Tsung-t'ang's in Che-kiang; together they cut the sources of supply to the Heavenly Capital, which by now was under an ever-tightening siege—like "an iron barrel"—by Tseng Kuo-ch'üan. The Loyal King urged Hung to seek a new base in Kiangsi and Hupeh, but the latter replied that having been commis-sioned by God to be king on earth, he did not choose to leave. By early 1864 the food in Nanking had run out and the Heavenly King urged the people to subsist on "sweet dew," i.e. grass. Knowing that his cause was lost, Hung let state affairs lapse on the pretext of sickness, and muttered at times: "Since ancient times how can emperors be taken prisoners?" On June 1, 1864, he committed suicide at the age of fifty-two *sui*. His sixteen-year-old son Hung Fu was placed on the throne as the Young Heavenly King, with the Shield King as regent. On July 19, Tseng Kuo-ch'üan's army broke into Nanking and carried out a merciless massacre. Every Taiping officer and man resisted to the end and none surrendered. The Loyal King assisted the Young Heavenly King in a hasty flight from Nanking, and in the panic the young master's horse stumbled, throwing the rider to the ground. The Loyal King offered him his own horse, al-lowing himself to be captured. The Young Heavenly King managed to flee to Kiangsi, where he was ultimately discovered and executed. The Taiping Revolution came to an end in 1864.

In captivity the Loyal King was treated with courtesy by Tseng Kuo-fan, who had respect for his military talent. Tseng asked him to prepare an autobiographical deposition. From July 30 to August 7, 1864, he wrote several thousand words daily, recounting the history of the Taipings, pointing out the mistakes of the Heavenly King as well as those of the Ch'ing court, and praising the Tseng brothers and the Hunan army.[22]

21. For a study of the rise of the Huai Army, see Stanley Spector, *Li Hung-chang and the Huai Army: A Study in Nineteenth-Century Chinese Regionalism* (Seattle, 1964), chs. 2-3.
22. For an abridged English version of the deposition, see W. T. Lay (tr.), *The Auto-biography of the Chung-wang* (Shanghai, 1865).

Tseng sent to Peking an edited version of the deposition, omitting those portions that were critical of the Ch'ing court, while keeping the original in his family library.[23] At midnight of August 7, the Loyal King was executed at the age of 40 *sui*. So ended the life of a military genius, who singlehandedly upheld a tottering empire for eight years after 1856. But for him the Heavenly Kingdom would have crumbled long before.[24]

Tseng was rewarded with the title of marquis first class, while his brother and Li were made count first class. Tseng was probably the most honored and powerful man in the empire at this point. His Hunan army and navy boasted of 120,000 to 130,000 men, and his entourage consisted of more than eighty of the most brilliant and capable advisers, strategists, planners, generals, and secretaries. A single order of his was obeyed by thousands of officials. A correct Confucian and a loyal minister, he knew that any extraordinary power and fame enjoyed by a Chinese would be cause for suspicion by the Manchu overlords. Barely seventeen days after the recovery of Nanking, he was obliged to propose the disbandment of the Hunan forces, which had accomplished their original objectives and were beginning to show signs of fatigue. In statesmanship, in character, and in personal cultivation, Tseng had few equals. He was probably the most respected and the greatest scholar-official of 19th-century China. Yet he is denounced by Marxist scholars in China as a traitor and executioner, who betrayed and massacred his fellow countrymen in the interests of the alien Manchu rulers.[25]

23. This original was published nearly a hundred years later under the title, *Li Hsiu-ch'eng ch'in-kung shou-chi* (Li Hsiu-ch'eng's personal deposition in his own handwriting), (Taipei, 1962).
24. The Loyal King's behavior in captivity came under considerable criticism in mainland China during the centenary of the Taiping downfall in 1964. The historian Ch'i Pen-yu led the attack in anathematizing him as a shameless, treacherous traitor who begged for his life in captivity and ingratiated himself with his former enemies. On the other hand, Lo Erh-kang, a lifelong student of the Taiping movement, argued that the Loyal King made a "false surrender" in order to save the Taiping cause; that the deposition was a stratagem to fool Tseng Kuo-fan into believing that there was no urgent need to kill the Young Heavenly King and destroy the Taiping remnants. The party historians (Yüan Shu-i and Lu I-tzu) offered a qualified approval of the Loyal King by stating that although his plea for life was "counterrevolutionary and deplorable," it did no serious harm because the Taiping government had already fallen; that his earlier achievements outweighed his regrettable last act. What these critics did not know is the fact that the deposition on which they based their arguments was not the authentic version but the abridged one made by Tseng. Nowhere in the original version does the word "surrender" ever appear, and it is well known that the Loyal King was ready to die when he completed the deposition. For an interesting account, see Stephen Uhalley, Jr., "The Controversy over Li Hsiu-ch'eng: An Ill-timed Centenary," *The Journal of Asian Studies*, XXV:2:305-17 (Feb. 1966).
25. Fan Wen-lan, *Han-chien kuei-tzu-shou Tseng Kuo-fan ti i-sheng* (The life of the traitor and executioner Tseng Kuo-fan), (Shanghai, 1944).

Causes of the Taiping Failure

The Taiping Revolution affected sixteen of the eighteen provinces in China proper and lasted for fourteen years. Its rise was as vigorous and promising as its end was pathetic and pitiful. Historical hindsight reveals several major reasons for the ultimate failure of the movement.

STRATEGIC BLUNDER. After the capture of Nanking, the Taipings should have taken advantage of their impetus to sweep north all the way to Peking, and they might very possibly have dislodged the Manchu court. The northern expedition which they did send under Lin Feng-hsiang was not a main Taiping force, but a thin column which erred in overextending itself in enemy territory, courting annihilation. The Ch'ing court, thus saved from the onslaught, was able to continue as the legitimate center of political power and the focal point of resistance.

Failing to take Peking, the Taipings should have at least concentrated on destroying the imperial Great Camps on the two sides of the Yangtze so completely as to allow no chance of resuscitation, in order to make Nanking safe. They should also have taken the entire province of Kiangsu, including the important trading center of Shanghai, and established firm and friendly contacts with foreign representatives. It was a major mistake of Hung's to ignore the plea for assistance from the Small Sword Society, a branch of the Triad Society, which occupied the Chinese walled city of Shanghai for a year and a half in 1853-54, thereby losing a chance to deprive the Ch'ing of an important point of contact with foreigners and a base for operations.

IDEOLOGICAL CONFLICT. The anti-Manchu appeal of the Taiping cause was compromised by its Christian ideology. The destruction of the temples and idols and the disruption of village life alienated the sympathies of the literati and peasants. The Taiping concept of all men being brothers and all women sisters contradicted the Confucian ideas of propriety and social hierarchy, and their prohibition against the cohabitation of husbands and wives ran counter to basic human relationships. Moreover, the unorthodox nature of the Taiping Christianity excited foreign antipathy. Indeed, their religious ideology alienated both Chinese and Westerner alike.

Initially, Hung had used religion to support his nationalistic revolution against the Manchus. He claimed divine commission and created a new Trinity so as to construct an invincible, supernatural aura around himself. He convinced the soldiers that they would rise to heaven if they were killed, and achieved a fearless army. By these manipulations, he successfully used religion as a means to advance his revolution. But later, when he became engrossed in religion and refused to cooperate with the secret societies because their members were not Christians, and when he refused to aid the Small Sword Society when it occupied Shanghai in 1853, he lost sight of his primary objective and placed religious considerations above the nationalistic revolution. Indeed, when he subordinated revolution to religion, he blurred his image as a nationalistic revolutionary, allowing himself to degenerate to the order of "religious bandits" (*chiao-fei*), like the White Lotus rebels.

FAILURE IN LEADERSHIP. Of the original five leaders under Hung, the South and West Kings were lost in battles in 1852, and the East King and North King were killed in the fratricidal strife of 1856. Only the Assistant King survived, but he left Nanking to start a life of his own. Deprived of their support, Hung was lost. He had relied on the South King (Feng) for the organization of the God Worshippers and the initial uprising, as he had depended on the East King (Yang) for military operations and civil administration. After 1856, the only man of ability and courage was the Loyal King, but, as the Chinese described it, one pillar could not support a whole mansion. The Shield King, who came to Nanking in 1859, was more a man of ideas than of practical ability, and he had the shortcoming of being jealous of other talents.

The paucity of genuine leadership placed Hung in a quandary. He could not evolve long-range constructive policies or over-all military strategy; nor could he guide a civil administration adequately, so he simply withdrew from all responsibility. In contrast, the opponents of Hung, such as Tseng Kuo-fan, Li Hung-chang, and Tso Tsung-t'ang, were all men of learning, ability, and rationality.

INCONSISTENCIES IN TAIPING LIFE. The revolutionaries preached the abolition of private ownership, but the leaders themselves accumulated vast wealth. They advocated the separation of husbands and wives into different quarters, the equality of the sexes, and monogamy as the correct form of marriage; but Hung himself kept 88 concubines, the East King kept 36, the North King 14, and the Assistant King 7. When the Women's Residential Halls were disbanded, the members were matched to officials according to their ranks—the higher the rank, the greater the number of women.

While Hung forbade people to read works of Confucius and Mencius, which were labeled "bogey books," he himself read them freely, borrowed ideas from the *Rites of Chou,* and explained his Christianity in Confucian terms. In his last years, Hung had become neurotic and believed that God Almighty would solve all his problems and that he himself need do nothing. When Nanking was about to fall, Hung announced that his "heavenly soldiers," more numerous than "water," would guard his city as an "iron barrel."

POOR DIPLOMACY. The Taipings enjoyed at first the sympathy of foreign powers. But instead of using it as a starting point for winning foreign recognition and assistance, they insisted on treating foreign powers as dependencies, an attitude that quashed all chances for good relations. When the foreign powers discovered that the Taipings were no more accommodating than the Manchus and were in fact hurting the foreign commerce at Shanghai, they withdrew their sympathy, shifting it after 1860 to the imperial government. The subsequent foreign defense of Shanghai in 1860 and 1862 prevented the Loyal King from taking this rich city, depriving the Heavenly Capital of an important source of supply and contributing to its final collapse.

The Legacy of the Taiping Revolution

Though it ended in failure, the Taiping experience had profound consequences in China. Politically, it affected the transfer of government power from the Manchus to the Chinese. Officers of the Hunan and Huai armies were awarded important assignments in the post-Taiping era, and key governor-generalships and governorships formerly occupied by the Manchus now passed into Chinese hands.[26] A few examples will illustrate this point. Tseng Kuo-fan, as governor-general of Liang-Kiang (1860-65) who also superintended military affairs in Chekiang, was in charge of four rich and important provinces (Kiangsu, Kiangsi, Anhwei, and Chekiang), while Li Hung-chang became governor of Kiangsu, and Tso Tsung-t'ang governor of Chekiang. All three in time rose to be grand secretaries of state. Li, in particular, as governor-general of Chihli and high commissioner of the Northern Ocean,[27] was China's virtual "prime

26. During 1864-66 all fifteen governorships were occupied by Chinese and during 1867-69 by fourteen Chinese and one Manchu, as compared with seven Manchus and eight Chinese in 1840. Taking late Ch'ing (1851-1911) as a whole, 65.4 percent of the governors-general and 77.8 percent of the governors were Chinese, as compared with 34.6 percent and 22.2 percent Manchus (including Mongols and Chinese bannermen) respectively. See Lawrence D. Kessler, "Ethnic Composition of Provincial Leadership," 500, table 4; see also S. Y. Teng, "Some New Light on the Nien Movement," 65-66.
27. A post created in 1861 to take charge of the three northern ports of Tientsin, Newchwang (Yingkow), and Chefoo.

minister" from 1870 to 1895. Tso also became governor-general, first of Chekiang and Fukien (1863-66), and then of Shensi and Kansu (1867-80), crowning his long career with the recovery of Sinkiang from the Moslem rebels in the 1870s. Even in the innermost organ of the court, the Grand Council, more and more Chinese were appointed until finally they outnumbered the Manchus. In short, the locus of power in the government had shifted from the Manchus to the Chinese.

A corollary to this change was the growing influence of provincial officials in national affairs. Whereas in the early and middle Ch'ing periods the government was highly centralized, with the court deciding policies for the provinces, in the post-Taiping period the central power declined directly as the local power rose. The court often found it necessary to consult high provincial officials on national issues and defer to their opinions; not infrequently government offices in Peking solicited the views of local authorities in order to win support for their stands. At times powerful governors and governors-general would act independently of the central government. For instance, after the "Hundred-Day" Reform of 1898, Governor-general Liu K'un-i of Liang-Kiang vigorously opposed the empress dowager's plan of deposing the emperor; during the Boxer Rebellion of 1900 the provincial authorities in Southeast China refused to follow the court order of supporting the Boxers, and out of "self-preservation" independently entered into agreements with foreign powers. The most blatant instance of provincial independence occurred in 1911 when Dr. Sun Yat-sen's revolutionary army took Wuchang; the provincial authorities declared their support of the revolution and by so defying the court hastened the downfall of the Ch'ing dynasty.

Militarily, the Hunan and Huai armies were the forerunners of private armies that characterized the warlords of later periods. Tseng and Li recruited their officers on the four bases of (1) common provincial origin; (2) common scholastic background; (3) relatives and friends; and (4) teachers and students. The soldiers were raised and trained by the officers, to whom they owed undivided allegiance. "When the general dies, the army scatters. When the general lives, the army is complete"; so commented the author of the *Hsiang-chün chih* (A record of the Hunan army). This Tseng-Li tradition of the private army was inherited by Yüan Shih-k'ai, a onetime protégé of Li and later leader of the Peiyang warlords, who plagued China during the early years (1912-27) of the Republic.

Finally, the Taiping experience inspired later revolutionaries. The Taiping remnants who went underground to join the Heaven and Earth Society kept alive the idea of racial and nationalistic revolution against the Manchus. It became a source of inspiration to Dr. Sun Yet-sen (1866-1925), father of the Chinese Republic, who was born barely two years after the downfall of the Heavenly Kingdom. As a child he had heard stories about the Taipings and at the age of twelve decided to become a second Hung. His later revolution received support from the secret societies, and many of his early followers were members of the Ko-lao Brotherhood Association. Even his revolutionary philosophy—the Three People's Principles (*San Min Chu I*)—was influenced by the Taiping ideology. Sun believed Hung failed because he understood national independence but not popular sovereignty, and monarchy but not democracy. To repair these ideological defects, Sun advocated his first two principles of Nationalism (*Min-tsu*) and Democracy (*Min-ch'üan*). His third principle, Socialism (*Min-sheng*), which contained the ideas of "equalization of land" and "regulation of capital," was in part inspired by the Taiping land system and the communal ownership of property. The social revolution that the Taipings failed to realize was partly carried on by Dr. Sun and his followers.

Not only in China, but in Europe, the Taiping Revolution was a source of instruction. Disappointed by the failure of the 1848 revolution in Europe, Karl Marx found hope in the Taiping movement and gained a new perspective on the possibility of peasant revolution. Today, Chinese Marxist historians hail the Taiping experience as the first peasant revolution in the history of Modern China.[28]

The Nien and Moslem Rebellions

Although the Taipings had been suppressed by 1864, several other rebellions of a smaller order still raged over different parts of the country. The Nien Rebellion, which broke out in 1853 and lasted until 1868, focused its activities in the southern part of North China. The Moslem (Panthay) Rebellion in Yunnan lasted from 1855 to 1873, and the Tungan Rebellion in the northwest covered the period from 1862 to 1878. These long-lasting rebellions were extremely debilitating in their effects, but they established no rival governments to contest the court at Peking. They were not therefore as threatening as the Taipings.

"Nien" was the name for secret gangs in the Shantung, Honan, Kiangsu, and Anhwei area. The members were chiefly idlers, rascals, and bandits who lived off forced contributions and pillage. Several tens or sometimes

---

28. However, a recent study by Vincent Y. C. Shih, *The Taiping Ideology: Its Source, Interpretations and Influences* (Seattle, 1967), argues that the Taiping movement was not a revolution, much less a peasant revolution.

several hundreds of them formed a "nien," which literally meant "band." They were proscribed during the Chia-ch'ing period (1796-1820). When the Taipings established themselves in Nanking in 1853, uncoordinated bands of Nien rebels rose in sympathetic response. The most powerful leader of these bands was Chang Lo-hsing, whose forces were joined by the remnants of the Taiping northern expeditionary army after its failure. Chang was invested with the title of King Wo by the Heavenly King, and his men wore long hair and attempted to imitate the Taiping military systems. Frequently the Nien and the Taipings cooperated in their military operations. In 1855 the poorly coordinated, decentralized Nien units were organized into five bands, distinguished by the different colors of their banners: red, yellow, blue, white, and black—a structure reminiscent of the early Ch'ing banner system.

Basically a conglomerate of peripatetic bands, the Nien rebels relied for their strength on the swift movement of their cavalry. They adopted the guerrilla tactics of avoiding frontal and direct confrontation with the imperial forces, while launching unexpected attacks when the enemy was off guard. With their fast-moving horsemen, they forced the Ch'ing army into a kind of shadowboxing. After years of inconclusive warfare, the court sent the fierce Mongolian general, Prince Seng-ko-lin-ch'in, to fight them, and he succeeded in killing the rebel leader Chang in 1863. But the Nien movement continued and, in fact, became stronger after 1864 when the Taiping remnants joined its cause. When Seng-ko-lin-ch'in was killed in action in 1865, the court asked Tseng Kuo-fan to take charge of the operations.

Tseng fought over a year without achieving results, coming under sharp criticism from the censors. Feeling the effects of age and distressed by the loss of vigor of his Hunan army, Tseng recommended Li Hung-chang to take over his assignment in early 1867. By the end of that year, Li's Huai army had succeeded in suppressing the eastern band of the Nien, while the western band had come under the severe attack of Tso Tsung-t'ang, who had been appointed imperial commissioner in charge of military affairs in Shensi and Kansu. In August 1868, the Nien Rebellion was brought to an end.

The Moslem Rebellion in Yunnan, known in Western literature as the Panthay Rebellion—"Panthay" being a corruption of the Burmese term for Moslem—lasted from 1855 to 1873. It was generally believed that the Moslems in Yunnan had migrated from the Western Region (Sinkiang) during the Mongol period (1280-1368). They made up no more than 20 percent or 30 percent of the population of Yunnan, but being a closely knit group they represented a powerful minority. Despised by the Chinese and Manchus for their religion and different way of life, they were the object of social ostracism and political discrimination. Oppression by officials and infringement of rights by the Chinese people were frequent, and when the Moslems took their cases to court they were often denied justice.

In 1855 an open rebellion broke out over the issue of some mining properties which the Chinese and the Moslems both claimed. The rebel leader, Tu Wen-hsiu, occupied Tali and proclaimed himself generalissimo and Sultan Suleiman of a new Moslem kingdom. Provincial military forces could not suppress him and the Manchu governor-general committed suicide. By 1868 Tu was in command, reportedly, of 360,000 men and 53 cities. It was not until 1872 that the government troops, under the leadership of new provincial authorities, were able to check him. Tu's son went to England and Turkey to seek aid, but to no avail. The rebellion, which lasted for eighteen years, came to an end in January 1873, when Tu, in despair, killed his family, took poison himself, and then surrendered.

There was yet another rebellion that troubled the Ch'ing court. The Moslems in the northwest, known as the Tungan, numbered six million in Shensi and eight million in Kansu. They were Chinese-assimilated, having adopted Chinese customs, language, and dress, but they still suffered social and political discrimination. Among them the adherents of the New Sect[29] (*Hsin-chiao*) were particularly militant. They maintained their headquarters in Chin-chi-pao, Ninghsia, and in Chang-chia ch'üan, Kansu, as a rival counterpart of the Old Sect stronghold of Hochow. Resentment of social and political injustices, exacerbated by the conflict between the two sects themselves, led the New Sect to rebel in 1781 and 1783, but on both occasions the government suppressed them mercilessly.

In 1862 when the Taipings invaded Shensi, the Moslems—some of whose leaders had participated in the Yunnan Revolt—rose again in response. One fanatic leader was Ma Hua-lung, who came from a direct line of the founder of the New Sect. The Ch'ing court, preoccupied with the Taiping campaign, could spare no able generals or troops to fight the Tungan. By 1864 the whole northwest was ablaze, with Kansu, Shensi, Ninghsia, and Sinkiang fallen to the rebels. The threat was intensified in 1866 when the western band of the Nien rebels broke into Shensi and joined forces with the Moslems. The harassed court appointed Tso Tsung-t'ang governor-general of Shensi and Kansu, with the specific assignment of clearing the two provinces of the rebels. He was, as noted before, then fighting the Nien rebels and was unable to take up the new assignment

29. Founded by one Ma Ming-hsin in 1762. For details, see Immanuel C. Y. Hsü, *The Ili Crisis: A Study of Sino-Russian Diplomacy, 1871-1881* (Oxford, 1965), 22-24.

immediately. It was not until his suppression of the Nien rebels in August 1868, that he was able to turn his attention to the Moslem problem in Shensi and Kansu. Five years of hard campaigning ensued, and by 1873 he finally pacified the Moslem Rebellion in these two provinces.

With the suppression of these domestic rebellions, the Ch'ing government re-established its authority over most of the empire. The dynasty seemed to have reversed its downward course and experienced a restoration. The question is whether such an upturn marked the beginning of a second blossoming of the dynasty, or was simply a short respite during a general decline.

Further Reading

Beal, George E., Jr., *The Origin of Likin, 1853-1864* (Cambridge, Mass., 1961).

Boardman, Eugene P., *Christian Influence upon the Ideology of the Taiping Rebellion* (Madison, 1952).

Cheng, J. C., *Chinese Sources for the Taiping Rebellion, 1850-1864* (Hong Kong, 1963).

Chiang, Hsing-te 蔣星德, *Tseng Kuo-fan chih sheng-p'ing chi shih-yeh*曾國藩之生平及事業 (Tseng Kuo-fan's life and work), (Shanghai, 1935).

Chiang, Siang-tseh, *The Nien Rebellion* (Seattle, 1954).

Chiang, Ti 江地, *Ch'u-ch'i Nien-chün shih lun-ts'ung* 初期捻軍史論叢 (Essays on the early history of the Nien Army), (Peking, 1959).

———, *Nien-chün shih ch'u-t'an* 捻軍史初探 (A preliminary investigation into the history of the Nien Army), (Peking, 1956).

Chung, Wen-tien 鍾文典, *T'ai-p'ing chün tsai Yung-an* 太平軍在永安 (The Taiping army at Yung-an), (Peking, 1962).

Clarke, Prescott, and J. S. Gregory (eds.), *Western Reports on the Taiping: A Selection of Documents* (Honolulu, 1982).

Curwen, C. A., *Taiping Rebel: The Deposition of Li Hsiu-ch'eng* (Cambridge, Eng., 1977).

Fan, Wen-lan 范文瀾, *T'ai-p'ing t'ien-kuo ko-ming yün-tung*太平天國革命運動(The revolutionary movement of the Heavenly Kingdom of Great Peace), (Hong Kong, 1948).

———, *Han-chien kuei-tzu-shou Tseng Kuo-fan ti i-sheng* 漢奸劊子手曾國藩的一生 (The life of the traitor and executioner Tseng Kuo-fan), (Shanghai, 1949).

——— et al. (eds.), *Nien-chün* 捻軍 (The Nien Army), (Shanghai, 1953), 6 vols.

Feuerwerker, Albert, *Rebellion in Nineteenth-Century China* (Ann Arbor, 1975).

Giguel, Prosper, *A Journal of the Chinese Civil War, 1864,* Edited by Steven Leibo (Honolulu, 1985).

Gregory, J. S., *Great Britain and the Taipings* (New York, 1969).

Hail, William J., *Tseng Kuo-fan and the Taiping Rebellion* (New York, 1964).

Hsiang, Ta 向達 et al. (eds.) *T'ai-p'ing t'ien-kuo* 太平天國 (The Taiping Heavenly Kingdom), (Shanghai, 1952), 8 vols.

Hsia, Nai 夏鼐 , *T'ai-p'ing t'ien-kuo ch'ien-hou Ch'ang-chiang ko-sheng chih t'ien-fu wen-t'i*太平天國前後長江各省之田賦問題(Problems regarding the land tax in the provinces along the Yangtze River before and after the Heavenly Kingdom of Great Peace), *Tsing-hua hsüeh-pao*, 10:2:409-74 (April 1935).

Hsiao, I-shan, *Tseng Kuo-fan* (Chungking, 1944).

Jen, Yuwen, *The Taiping Revolutionary Movement* (New Haven, 1973).

Kuhn, Philip A., *Rebellion and Its Enemies in Late Imperial China: Militarization and Social Structure, 1796-1864* (Cambridge, Mass., 1970).

————, "The Taiping Rebellion" in John K. Fairbank (ed.), *The Cambridge History of China* (Cambridge, Eng., 1978), Vol. 10, Pt. I, 264-350.

Lay, W. T. (tr.), *The Autographic Deposition of Chung Wang, the Faithful King, at His Trial after the Capture of Nanking* (1865).

Liu, Robert H. T., *The Taiping Revolution: A Failure of Two Missions* (Lanham, Md., 1979).

Lo, Erh-kang 羅爾綱, *Hsiang-chün hsin-chih* 湘軍新誌 (A new study of the Hunan army), (Shanghai, 1939).

————, "T'ai-p'ing t'ien-kuo ko-ming ch'ien ti jen-k'ou ya-p'o wen-t'i 太平天國革命前的人口壓迫問題(The question of population pressure in the pre-Taiping Revolution years), *Chung-kuo she-hui ching-chi chi-k'an* (Collected writings on Chinese society and economics), Academia Sinica, 8:1:20-80 (Jan. 1939).

————, *T'ai-p'ing t'ien-kuo shih-kang* 太平天國史綱(An outline history of the Taiping Kingdom), (Shanghai, 1937).

Lo Yü-tung羅玉東, *Chung-kuo li-chin shih*中國厘金史 (A history of *likin* in China), (Shanghai, 1936).

Meadows, Thomas T., *The Chinese and Their Rebellions* (London, 1856).

Michael, Franz, "Military Organization and the Power Structure of China during the Taiping Rebellion," *Pacific Historical Review*, XVIII:469-83 (1949).

————, *The Taiping Rebellion: History and Documents*, Vols. 2 and 3 (Seattle, 1971).

————, and Chung-li Chang, *The Taiping Rebellion: History and Documents* (Seattle, 1966). Vol. 1.

Pai, Shou-i 白壽彝 (ed.), *Hui-min ch'i-i* 回民起義(The righteous uprising of the Moslems), (Shanghai, 1953), 4 vols.

P'eng, Tse-i彭澤益, *T'ai-p'ing t'ien-kuo ko-ming ssu-ch'ao* 太平天國革命思潮 (The revolutionary thought-tide of the Heavenly Kingdom of Great Peace), (Shanghai, 1946).

Perry, Elizabeth J. (ed.), *Chinese Perspectives on the Nien Rebellion* (Armonk, N.Y., 1981).

Porter, Jonathan, *Tseng Kuo-fan's Private Bureaucracy* (Berkeley, 1972).

Shen, Yüan 沈元, "Hung Hsiu-ch'üan ho T'ai-p'ing t'ien-kuo ko-ming," 洪秀全和太平天國革命 (Hung Hsiu-ch'üan and the Taiping Revolution), *Li-shih yen-chiu* 歷史研究 (Historical Research), 1:49-94 (1963).

Shih, Vincent Y. C., *The Taiping Ideology: Its Source, Interpretations and Influences* (Seattle, 1967).

Smith, Richard J., *Mercenaries and Mandarins: The Ever-Victorious Army in Nineteenth Century China* (New York, 1978).

So, Kwan-wai, Eugene Boardman, and Ch'iu Ping, "Hung Jen-kan: Taiping Prime Minister," *Harvard Journal of Asiatic Studies*, 20:1-2:262-94 (June 1957).

Spector, Stanley, *Li Hung-chang and the Huai Army: A Study in Nineteenth-Century Chinese Regionalism* (Seattle, 1964).

Taylor, George, "The Taiping Rebellion, Its Economic Background and Social Theory," *The Chinese Social and Political Science Review*, 32:545-614 (1932-33).

Teng, S. Y. 鄧嗣禹, "Hung Jen-kan, Prime Minister of the Taiping Kingdom and His Modernization Plans," *United College Journal*, 8:87-95 (Hong Kong, 1970-71).

——, *The Taiping Rebellion and the Western Powers: A Comprehensive Survey* (Oxford, 1971).

——, *New Light on the History of the Taiping Rebellion* (Cambridge, Mass., 1950).

——, *The Nien Army and Their Guerrilla Warfare, 1851-1868* (Paris, 1961).

——, *Historiography of the Taiping Rebellion* (Cambridge, Mass., 1962).

——, "T'ai-p'ing T'ien-kuo chih hsing-wang yü Mei-kuo chih kuan-hsi," 太平天國之興亡與美國之關係 (The rise and fall of the Taiping Rebellion and its relations with the U.S.), *Journal of the Institute of Chinese Studies*, Hong Kong, III:1:1-11 (1970).

Teng, Yüan-chung, "The Failure of Hung Jen-kan's Foreign Policy," *The Journal of Asian Studies*, XXVIII:1:125-38 (Nov. 1968).

Wagner, Rudolf G., *Reenacting the Heavenly Vision: The Role of Religion in the Taiping Rebellion* (Berkeley, 1984).

Wang, Shu-huai 王樹槐, *Hsien-T'ung Yün-nan Hui-min shih-pien* 咸同雲南回民事變 (The Yünnan Moslem Rebellion during the Hsien-feng and the T'ung-chih periods), (Taipei, 1968).

Wu, James T. K., "The Impact of the Taiping Rebellion upon the Manchu Fiscal System," *Pacific Historical Review*, 19:265-75 (Aug. 1950).

Yap, P. M., "The Mental Illness of Hung Hsiu-ch'üan, Leader of the Taiping Rebellion," *The Far Eastern Quarterly*, XIII:3:287-304 (May 1954).

III

Self-strengthening in an Age
of Accelerated Foreign Imperialism
1861-95

11

The Dynastic Revival
and the Self-strengthening Movement

The peace settlement with Britain and France in 1860 and the suppression of the Taiping Revolution in 1864 eliminated two deadly threats to the dynasty, one external and the other internal. The Ch'ing court had forged a momentary reprieve, and in the period that followed displayed a remarkable spirit of resurgence, as manifested in the suppression of the Nien and Moslem rebellions, the restoration of the traditional order and the Confucian government, the maintenance of peace with foreign powers, and the initiation of the Self-strengthening Movement through adoption of Western diplomatic practices and military and technological devices. Arrested, if only temporarily, was the dynastic decline characterized by the twin evils of "internal rebellion and external invasion."

Scholars and statesmen of the 1860s and 1870s were quick to refer to this dynastic "second blossoming" as a "T'ung-chih Restoration" (*T'ung-chih chung-hsing*). Here, "restoration" lacks the same connotation as the Meiji Restoration of Japan, which meant the return of state powers from the military dictator (*shogun*) and the feudal lords (*daimyo*) to the emperor; it refers rather to efforts at restoring the traditional order through reaffirmation of the old morality and application of knowledge to practical affairs (*ching-shih chih-yung*). Measures were taken to rehabilitate the devastated agricultural areas and to search out men of talent for public service. The government remitted or reduced agricultural taxes in the countryside, distributed seeds and tools to assist in the recovery, and stressed a program of personal austerity. Private academies and libraries were reopened, and government examinations resumed, particularly in those areas that had not held them during the years of civil disorder. These examinations, while continuing to be literary, stressed the practical problems of the day. The court increased the quotas for degrees in the various provinces to reward military and financial contributions, while limiting the sale of ranks and offices. Within the bureaucracy stricter discipline was enforced and cases of corruption punished. Meanwhile, Peking exercised great care in foreign affairs to maintain peace and good relations with Western powers and to afford the country the chance to engage in reconstruction and self-strengthening.

Previous successful restorations in Chinese history[1] usually involved strong, brilliant, and virtuous rulers, but Emperor T'ung-chih (1862-74) was a minor during eleven of his thirteen reigning years and a weakling in the remaining two. The power of state was grasped firmly in the hands of his mother, the Empress Dowager Tz'u-hsi, who controlled the court for 48 years until her death in 1908. Considering Emperor T'ung-chih's personality, his reign certainly does not deserve the designation of restoration. Yet the emperor was more an institution than a personality; unusual accomplishments by able supporters, which caused a tidal change, can be considered the constituents of a restoration.[2]

Nonetheless, the T'ung-chih restoration definitely stood at a lower level of revival in Chinese history. While it did stem the decline for a while, it failed to regenerate the dynasty to a degree sufficient to allow survival with honor in the modern world. Its imitation of Western armament, technology, and diplomacy was a superficial gesture toward modernization; the finer aspects of Western civilization—political institutions, social theories, philosophy, fine arts, and music—went totally untouched. Historical perspective indicates that it was hardly more than a flash of rejuvenation in an eventide of waning dynastic fortune—an Indian Summer. Nonetheless, the immediate effect of the T'ung-chih Restoration was to signal a brave and reasonably successful effort at reviving the old order and to initiate the beginning of a new.

New Leadership and the Coup of 1861

The emergence of a new political leadership at Peking played a vital role in forging a new era. As will be recalled, Prince Kung had been left in

1. Such as those of Emperor Hsüan-wang (827-782 B.C.) of the Western Chou period, Emperor Kuang-wu (A.D. 25-57) of the Eastern Han period, and Emperor Su-tsung (A.D. 756-62) of the T'ang period.
2. Mary C. Wright, *The Last Stand of Chinese Conservatism: The T'ung-chih Restoration, 1862-1874* (Stanford, 1957), 50.

Peking to deal with Lord Elgin and Baron Gros in September 1860, when Emperor Hsien-feng fled to Jehol in the face of the advancing enemy. That the prince achieved a settlement with the barbarians and effected their withdrawal from the capital—all this without the support of an army or a navy—was considered by many Chinese and Manchus as nothing short of a miracle. Prince Kung had emerged as the new leader at Peking, while the court still cowered at Jehol.

Following the evacuation of foreign troops from Peking, Prince Kung led a number of high officials in requesting the return of the emperor. The sovereign was reluctant, partially out of shame at his unheroic flight, and partially out of fear that the enemy forces might return to coerce him into receiving their envoys without the kowtow. It was not until Prince Kung, in December 1860, successfully committed the British and French plenipotentiaries to desist from demanding the audience that the ruler finally announced, on February 11, 1861, that he would return the following month. But the journey was not to materialize due to his failing health.

Emperor Hsien-feng, who ascended the throne in 1851, had never been physically robust. In his refuge at Jehol, he tried to escape his misery and shame by frequenting the pleasure quarters, so much so that he became physically depleted. His frequent companions and mentors in these escapades were Prince I, Prince Cheng, and Assistant Grand Secretary Su-shun, the last in particular having developed a strong hold on him. On August 21, 1861, the emperor lapsed into a fatal illness; on his deathbed he named his son Tsai-ch'ün, aged six *sui*, the heir apparent.

Su-shun and the two princes presently produced a valedictory edict appointing themselves and five other high officials advisers on state affairs and members of a Council of Regents for the boy emperor. Prince Kung was left out in the cold. He was further snubbed when the Council of Regents rejected his request to attend the imperial funeral in Jehol. Equally abused were the two empresses of the late ruler—the 27-*sui* Tz'u-an, who was childless, and the 25-*sui* Tz'u-hsi who bore him a son, the present boy emperor. They were made empresses dowager on August 23, but were denied their legal authority to approve the edicts of the regents. An intense power struggle followed between the dowagers and the regents at Jehol, and the former decided to isolate the latter by joining forces with Prince Kung, who maintained a third power center at Peking.

Tz'u-hsi was a woman of remarkable ability and sinister schemes. During the late emperor's illness she had assisted him in drafting comments on memorials while concealing her secret jealousy of Su-shun's influence on him. Imbued with great personal ambition, she managed to retain the imperial seal upon the death of the emperor, and, suspicious of the authenticity of the valedictory edict produced by Su-shun, succeeded in persuading the other empress dowager to join her in assuming the reins of government themselves "behind a silk screen" for "self-preservation." A special emissary, reputedly the eunuch An Te-hai, was dispatched to Peking to sound out Prince Kung, who readily agreed to cooperate in hopes of utilizing the dowagers to establish himself in place of the Council of Regents. He now obtained permission from the dowagers to come to Jehol to attend the imperial funeral.

At a secret meeting, the dowagers and the prince decided that they should attempt no action in Jehol, where the regents held sway, but should act in Peking where Prince Kung had the upper hand; and that the coffin of the late emperor was to be returned to the capital, at which time swift action would be taken to arrest the regents. Tz'u-hsi was concerned about foreign reaction, but Prince Kung, apparently assured of foreign support, confidently announced that he would deter the alien powers from intervening.

Indeed, the British played a secret role in supporting the prince. As signer of the Conventions of Peking, Prince Kung had impressed the foreign representations with his good will, quick intelligence, fine manners, and willingness to cooperate in executing the treaties. The British saw that it was in their interest to keep him in power. A dispatch from the British minister[3] to London on March 12, 1862, reveals the British role in this palace revolution: ". . . it is no small achievement within twelve months to have *created a party* inclined to and believing in the possibility of friendly intercourse, to have *effectually aided that party to power*. To have established satisfactory relations at Peking and become in some degree the advisers of a government with which eighteen months since we were at war." Prince Kung looked to the five or six thousand foreign troops in Tientsin as "a pillar of support . . . against his political opponents."[4]

On September 11, 1861, Prince Kung departed for Peking. Meanwhile, Grand Secretary Chou Tsu-p'ei and others in Peking, resentful of Su-shun's usurpation of power and conscious of the dowagers' desire to be regents themselves, asked the noted scholar Li Tz'u-ming to prepare a memorandum on the historical precedents for regency by the dowagers. Before Li had completed the study, a censor[5] had already seized the initiative by sending a memorial to Jehol, requesting the dowagers to expropri-

---

3. Frederick Bruce.
4. Masataka Banno, *China and the West: 1858-1861* (Cambridge, Mass., 1964), 241. Italics added.
5. Tung Yüan-chün.

ate the state administration and to appoint one or two princes of the blood to assist in the discharge of duties. Su-shun ridiculed the idea on the ground that there had never been the precedent of dowager administration in the Ch'ing dynasty.

Ignoring the strong opposition of the regents, the dowagers left for Peking with the boy emperor on October 26, accompanied by Princes I and Cheng, while the emperor's coffin set out shortly afterwards under the escort of Su-shun and Prince Chün. On November 1, the dowagers arrived in the capital and immediately received a joint memorial from Grand Secretary Chou Tsu-p'ei and the presidents of the Board of Revenue and Board of Punishment, requesting them to take charge of the state administration during the emperor's minority—apparently at the prompting of Prince Kung. They accused the eight regents of dominating rather than assisting the court. A day later, the dowagers summoned Prince Kung, the grand secretaries, and other high officials to the palace and announced the crimes of the eight regents and their immediate dismissal. Princes I and Cheng protested the illegality of the act, whereupon the dowagers issued a second edict stripping them and Su-shun of their nobility status and turning them over to the Imperial Clan Court for punishment. In a lightning movement, Prince Kung's guards arrested the two princes, while Su-shun, who was accompanying the coffin to Peking, was accosted en route and thrown into the prison of the Imperial Clan Court. The two princes were allowed to hang themselves; Su-shun was decapitated on November 8; and the other five regents were dismissed.

Emerging victorious from this *coup d'état,* the dowagers and Prince Kung became co-regents. Actually it was against the Ch'ing imperial family law to allow a mother-empress to administer state affairs. During the minority of the first emperor, Shun-chih (1644-61), his uncle Dorgan was made prince regent; and during the minority of Emperor K'ang-hsi (1662-1722) four regents under the leadership of Oboi assisted him. Never had there been any precedent of the imperial distaff serving as regent. But a lack of precedent could not deter the strong-willed Tz'u-hsi. She made Prince Kung her front man, with the titles of prince regent (*i-cheng wang*), grand councillor, chief minister of the Imperial Household, and head of the newly established Tsungli Yamen (Foreign Office).[6] The prince, on his part, needed the blessings of the dowagers to build his own power base. So they entered into an alliance founded in mutual expediency.

The reign title of the boy emperor was T'ung-chih (Co-eval rule)— probably connoting the contemporaneous rule of the two dowagers—

6. An abbreviation of Tsung-li ko-kuo shih-wu ya-men (Office for the general management of affairs concerning the various countries).

effective the following year, 1862. Tz'u-hsi did not like the designation of mother-empress (*mu-hou*), for it implied a status once removed from the legitimate source of power, preferring to be known as the Western Dowager, after the Western palace in which she lived. The other dowager, Tz'u-an, who lived in the Eastern palace, became known as the Eastern Dowager. Although both sat behind a silk screen while administering state affairs (*ch'ui-lien t'ing-cheng* 垂簾聽政) and receiving ministerial reports, it was usually the Western Dowager who read memorials, asked questions, and made decisions. The Eastern Dowager, known more for her virtue than ability, was rather reticent in disposition.

Foreign reactions to the coup were generally favorable. A. H. Layard, British Under-Secretary for Foreign Affairs, spoke before the House of Commons on March 18, 1862: "Within a very short time a great change had taken place; a coup d'état had been effected which led to a change of Ministers . . . Prince Kung and the two Empresses had called together a new Ministry and had inaugurated a new policy; for the first time a Chinese Government had admitted the rights of foreigners, and consented to treat them as equals."[7] Prince Kung and the Tsungli Yamen became symbols of progress and of good will toward foreigners.

The Cooperative Policy and Diplomatic Modernization

The bitter lesson of the 1860 peace settlement promoted a complete about-face in Prince Kung's attitude toward the foreigners. Whereas before this traumatic experience he was violently antiforeign, advocating unyielding resistance to foreign demands and the execution of Harry Parkes, after the settlement he arrived at a new conception of the barbarian problem. He came to respect and even admire the British power, convinced that China had no alternative but to learn to live with the West.

From his dealings with Lord Elgin and Baron Gros, Prince Kung had learned beyond doubt the superiority of Western weaponry. To his pleasant surprise, he discovered that the erstwhile enemies not only did not intend to deny China their military secrets, but openly offered to help China train its army and manufacture weapons after the Western fashion. The prompt evacuation of occupation troops from Peking after the peace further revealed that the foreign powers had no territorial designs on China and that they were not devoid of reason and good faith, as the Chinese were wont to depict them. Prince Kung came to the conclusion that if China kept its treaty obligations and treated foreigners with good will and open-mindedness, giving them no cause for complaints, peace would reign.

7. Banno, 240-41.

In this roseate new light, the treaties which were formerly considered a disgrace now became a useful instrument to specify the maximum concessions beyond which China would make no grant and the foreigners could not legally go. From this realization the twenty-eight-year old prince evolved a new policy for China: diplomatically it would accommodate the West to gain a lasting period of peace in which to build up, with Western aid, its military strength. Thus, peace through diplomacy became the immediate objective (*piao*) of the government, while self-strengthening loomed as the ultimate goal (*pen*). This double-pronged approach was wholeheartedly supported by the Manchu Grand Councillor Wen-hsiang in the capital, and by several powerful leaders in the provinces—such as Tseng Kuo-fan, Tso Tsung-t'ang, and Li Hung-chang.

Prince Kung was not alone in charting a new course for China, for so had the Westerners. They recognized the fact that enjoyment of treaty rights was predicated upon the continued existence of the government granting these rights. The Western powers therefore decided to sustain the Ch'ing court and helped it modernize, in the belief that a stable China was more conducive to ever-increasing foreign trade. With this shift in policy, British neutrality changed to active, if limited, support of the imperial cause against the Taipings, as already seen in the last chapter.

Foreign diplomats, who had taken up residence in Peking, also gained a better understanding of the Chinese viewpoint, and unconsciously acquired a certain degree of Sinicization. The American minister, Anson Burlingame, and the British minister, Frederick Bruce, now championed a "Cooperative Policy" toward China, which expounded (1) cooperation among Western powers; (2) cooperation with Chinese officials; (3) recognition of China's legitimate interests; and (4) enforcement of the treaty rights.[8]

The policy changes on the part of both China and the West resulted in a decade of relative peace, harmony, good will, and cooperation. It provided a good climate for the initiation of China's diplomatic and military modernization.

Diplomatic reform began with a recommendation by Prince Kung and Wen-hsiang on January 11, 1861, that a new office be established to direct foreign affairs; that a superintendent of trade be appointed at Tientsin to take charge of the three northern ports,[9] in addition to the one already stationed at Shanghai who took charge of the original five ports; that two linguists be sent from Canton and Shanghai for service in Peking; that intelligent Manchu boys below the age of thirteen or fourteen *sui* be se-

8. Wright, *The Last Stand*, 21-22.
9. Tientsin, Newchwang, and Chefoo.

lected from the bannermen's families to learn foreign languages; and that trade reports and foreign newspapers at the treaty ports be forwarded to the Tsungli Yamen. This state paper marked the beginning of the diplomatic phase of the Self-strengthening Movement.

THE TSUNGLI YAMEN. The Ch'ing court had not previously established a foreign office, because China never accorded recognition to another state on an equal, diplomatic level, but only on a tributary or trading basis. The court dealt with these tributary and trading affairs through various state organs, obviating the need for a foreign office. During the pre-Opium War period, tributary affairs, which basically reflected a ritualistic relationship, were directed by the Board of Rites. Russian and frontier affairs were governed by the Li-fan yüan (Court of Colonial Affairs), and trade with the Western maritime countries was deputed to the governor-general at Canton, who "managed" the foreigners through the Hoppo and the hong merchants. During the period between the Opium and the *Arrow* wars, 1842-56, the governors-general at Canton and Nanking were, for all practical purposes, China's unofficial foreign and deputy foreign ministers. After the Conventions of Peking in 1860 reaffirmed Western diplomatic residence in the Chinese capital, there was genuine need for a foreign office to centralize the direction of foreign affairs. Immediate attention was required of such matters as reception of foreign representatives, allocation of legation quarters, payment of indemnities, opening of the new ports, and a host of other questions relative to new treaty obligations. On recommendation of Prince Kung, the Tsungli Yamen (Office for General Management) was established in Peking on March 11, 1861. Though commonly known to the foreigners as the Foreign Office, in reality it functioned more like a subcommittee of the Grand Council than a regular board of the government. Its organization and characteristics may be summarized as follows:

1. It was intended to be a temporary office under the charge of a prince of the blood, supported by several ministers[10] who were concurrently high metropolitan officials—the grand councillors, grand secretraies, and presidents or vice-presidents of the boards. Under them were sixteen secretaries, divided equally between the Manchus and the Chinese. Prince Kung was the first and long-time presiding officer, while Wen-hsiang, grand councillor and vice-president of the Board of Revenue, was its principal minister until his death in 1876.

2. As a government agency, it had no statutory (i.e. "constitutional")

---

10. Usually 3 to 5 at first, but later increased to 9 to 11.

basis but was rather a makeshift creation necessitated by the exigency of circumstances. In theory it concerned itself only with the execution of foreign policy, not the making of it, since the ultimate power of decision rested with the emperor and his chief adviser, the Grand Council. In practice, however, the recommendations of the Tsungli Yamen were usually approved by the throne, Prince Kung and Wen-hsiang being both concurrently grand councillors.

3. It was organized into five bureaus: Russian, British, French, American, and Coastal Defense. In addition, two other offices were attached to it: the Inspectorate-general of Customs and the language school called the T'ung-wen kuan.

4. It engaged not only in foreign affairs but also in a number of modernization projects. Its promotion of modern schools, Western science, industry, and communication exposed it to frequent attacks by die-hard conservatives, while foreigners often criticized it for not progressing rapidly enough. Thus, the Tsungli Yamen occupied the uncomfortable position of a buffer between the two—accused by foreigners of procrastination and by xenophobes of selling out China's interest to the barbarians.

5. It played an active role in the 1860s, but its influence waned after 1869-70, when the dowager T'zu-hsi chastised Prince Kung for the second time; when the Alcock Convention failed to win British ratification (next chapter); and when Li Hung-chang shouldered both the governor-generalship of Chihli and the superintendency of trade for the three northern ports. In this double capacity, he eclipsed the Tsungli Yamen (next section).

6. Where it failed as an effective foreign office, the Tsungli Yamen succeeded reasonably well as a promoter of modernization. It was China's first major institutional innovation in response to the Western impact.[11]

THE SUPERINTENDENTS OF TRADE. Besides the Tsungli Yamen, a superintendent of trade for the three northern ports was established at Tientsin, with the Manchu nobleman Ch'ung-hou as the first incumbent until his replacement by Li Hung-chang in 1870. This post was created to parallel the existing commissionership at Shanghai, which took charge of the affairs of the five original ports and the new ports along the Yangtze River and along the seacoast to the south, opened by the recent treaties. In 1862 the commissioner at Shanghai received the title of Superintendent of Trade Affairs, a concurrent post for the governor of Kiangsu; later in 1866 it became a concurrent appointment for the governor-general at Nanking.

11. Ssu-ming Meng, *The Tsungli Yamen: Its Organization and Functions* (Cambridge, Mass., 1962).

These two superintendents of trade, one at Tientsin and one at Shanghai, became known respectively as the High Commissioner for the Northern Ocean (*Pei-yang ta-ch'en*) and High Commissioner for the Southern Ocean (*Nan-yang ta-ch'en*).[12]

One reason for the establishment of these trade superintendencies was to direct business away from Peking, in order to forestall much of diplomatic transactions in the capital. Knowing the court's fear and resentment of the imposed diplomatic residence in Peking, Prince Kung explained his secret motive in creating the new post: "If Tientsin can manage properly, then, even though the barbarian chieftains live in the capital, they must be depressed with having nothing to do and finally think of returning home."[13] So successful was the strategem that after Li Hung-chang undertook the Tientsin commissionership in 1870 he practically pre-empted the functions of the Tsungli Yamen. It was he who settled the case of the Tientsin Massacre in 1870, who recommended establishing an official relationship with Japan in 1871, and who settled the Margary murder incident in 1875-76 (next chapter). It was he, too, who conducted negotiations with the French over the Annamese question in 1884, and took charge of the opening of Korea in the early 1880s and negotiations with Japan after the war of 1894-95. Li's office at Tientsin became virtually China's foreign office for the quarter of a century following 1870. But foreign diplomats did not leave Peking.

THE T'UNG-WEN KUAN. Established in Peking in 1862 at the suggestion of Prince Kung, the T'ung-wen kuan, known to the foreigners alternately as the Interpreters College or the College of Foreign Languages, was originally intended as a school for the joint instruction of Western and Chinese languages, hence the name T'ung-wen (common languages). Initially it was created in response to the French and British Tientsin Treaty clauses specifying the English and French texts as the sole authentic versions of the treaties. China therefore needed to train able language experts to free herself from reliance on foreign interpreters and half-baked Canton linguists, who spoke only "pidgin" English.

Since no Chinese were qualified to instruct foreign languages, an English and a French missionary as well as a Russian interpreter at the legation initially were invited to teach their respective tongues at the T'ung-wen kuan; German was added later. Instruction in Chinese was also given.[14] In 1864 the American missionary-educator W. A. P. Martin joined the

12. The word "Ocean" in these titles really means "ports."
13. Hsü, *China's Entrance*, 107.
14. By Professor Hsü Shu-lin.

staff as professor of English. By 1866 astronomy and mathematics were introduced to the curriculum, over the opposition of the arch-conservative Grand Secretary Wo-jen; and in the following year the noted scholar-official Hsü Chi-yü was appointed director. The school gradually took on the appearance of a small liberal arts college.

In 1867 Martin returned to the United States for two years of advanced work in international law and political economy at the University of Indiana, where he earned a doctorate. Back in China in 1869, he was made president of the T'ung-wen kuan, with assurance of financial support from Robert Hart, the Inspector-general of Customs. Under Martin's direction, a variety of subjects were added to the eight-year curriculum, the first three being devoted to linguistic preparation and the next five to scientific and general studies. In 1879 the enrollment stood at 163, with 38 specializing in English, 25 in French, 15 in Russian, 10 in German, 33 in mathematics, 6 in astronomy, 7 in physics, 9 in international law, 12 in chemistry, and 8 in physiology.[15] The quality of students, however, was rather low, since few good Manchu or Chinese families would send their sons, with the result that a considerable portion of the student body consisted of middle-aged mediocrities enrolled for a pension.

Nonetheless, the T'ung-wen kuan marked the beginning of Western education in China. Since many of the foreign professors also engaged in translation with the help of their Chinese students, the school simultaneously functioned as a prototype research institute for dissemination of foreign knowledge. In 1873 a small printing office was set up as a primitive sort of "university press," from which seventeen major publications were issued in the fields of international law, political economy, chemistry, physics, and natural philosophy.

Similar schools of foreign languages and Western studies were established at Shanghai in 1863 (*Kuang fang-yen kuan*), at Canton in 1864, and at Foochow in 1866.[16] The T'ung-wen kuan at Peking lasted until 1902, when it was absorbed into the Imperial University. Among the more outstanding graduates of the Peking T'ung-wen kuan were two foreign ministers and a number of diplomats.

THE MARITIME CUSTOMS SERVICE. As head of the Tsungli Yamen, Prince Kung appointed Horatio N. Lay as inspector-general of customs on April 7, 1861, charging him with the duty of "exercising a general surveillance

15. W. A. P. Martin, *Calendar of the Tungwen College* (Peking, 1879), 10.
16. For details of T'ung-wen kuan and other schools at the Kiangnan Arsenal and the Foochow Dockyard, see an excellent study by Knight Biggerstaff, *The Earliest Modern Government Schools in China* (Ithaca, 1961).

over all things pertaining to the revenue, of aiding the Chinese superintendents to collect the revenue at the various ports, of preventing frauds upon the revenue, and of standing sponsor for the good conduct of the foreigners engaged in the Customs Service."[17] This appointment confirmed and institutionalized the foreign inspectorate of customs that had already been developed at Shanghai in 1854. It may be recalled that the Small Sword Society occupied the walled city of Shanghai in September 1853, and drove the Ch'ing customs superintendent[18] out of operation. Foreign traders gladly profited from this anarchy, paying no dues on their imports. However, the British consul in Shanghai, Rutherford Alcock, recalling Sir Henry Pottinger's instruction a decade earlier that British consuls see that their nationals pay the customs dues, cooperated with the American Commissioner in China, Humphrey Marshall, in devising a provisional system whereby the consuls of the two countries would collect customs dues from their respective countrymen for the Chinese government.

Fearful of the possibility of trade being deflected elsewhere, the British, French, and American consuls opened discussion with the governor-general of Liang-Kiang and obtained an agreement that a foreign Board of Inspectors be established at Shanghai to aid in the equitable collection of customs dues from all foreign traders; in return the Chinese abolished the inland customs dues.

On July 12, 1854, Thomas Wade of Britain, Lewis Carr of the United States, and Arthur Smith of France were sworn in as inspectors of customs at Shanghai, with full approval of the Chinese customs superintendent. The bulk of the work fell on Wade, who alone among the three understood Chinese and the customs procedure. Finding the work too burdensome to permit his Sinological studies, Wade resigned a year later. Horatio N. Lay, a twenty-three-year-old acting British vice consul at Shanghai, then secured the job for himself on June 1, 1855. His vigorous supervision resulted in more customs revenue for the Ch'ing government than under the former imperial customs superintendent. After a leave of absence in order to serve with Lord Elgin during the Treaty of Tientsin negotiations in 1858, Lay returned to the inspectorate and laid the groundwork for a new customs service. The Shanghai-style foreign inspectorate was to be extended to other treaty ports, each under the charge of a foreign inspector, or commissioner as he was later called, who took orders from the inspector-general at the head office in Shanghai, namely Lay himself. With this background, Prince Kung's appointment of Lay in

17. Wright, *Hart and the Chinese Customs*, 151.
18. Wu Chien-chang, the former Canton hong merchant Samqua.

April 1861 amounted to an official confirmation of a system already in operation.

Lay's response to the appointment was singularly curious and extremely insulting. He neither accepted nor rejected it, but took a trip to England on the grounds of poor health. During his absence, he designated G. H. FitzRoy, commissioner of customs at Shanghai, and Robert Hart, deputy commissioner of customs at Canton, as officiating inspectors-general until his return. Hart was sent to Peking to take Prince Kung's commands. In personality and manners, Hart was tactful and patient, the opposite of Lay. The Chinese responded to him much more warmly than to Lay, and when the latter proved to be too unruly and domineering over the Osborn flotilla issue (next section), Hart was named his replacement as inspector-general in 1863. Under Hart's leadership, an international customs service for China was developed, with 252 British and 156 other Western employees by 1875.[19]

In a circular dated June 21, 1864, Hart spelled out what amounted to a "code of behavior" for the foreign employees. He suggested that they should learn Chinese, act with patience "without affectation of superiority," labor "to convince rather than to dictate," and "introduce remedies without causing the irritation that attends the exposure of defects." In decidedly clear terms he told them:

> It is to be distinctly and constantly kept in mind that the Inspectorate of Customs is a Chinese and not a Foreign Service, and that, as such, it is the duty of each of its members to conduct himself towards the Chinese, people as well as officials, in such a way as to avoid all cause of offence and ill-feeling . . . It is to be expected from those who take the pay, and who are the servants of the Chinese Government, that they, at least, will so act as to neither offend susceptibilities, nor excite jealousies, suspicion, and dislike. In dealings, therefore, with native officials, and in intercourse with the people, it will be well for the Foreign employees of the Customs to remember, that they are *the brother officers of the one*, and that they, to some extent, accepted certain obligations and responsibilities by becoming, in a sense, *the countrymen of the others*: the man who cherishes such an idea, will be led to treat the one class with courtesy, and the other with friendliness.[20]

Hart's approach to proper behavior, his consideration, and his sense of proportion endeared him to the court, which took him into confidence as a

19. John K. Fairbank, K. F. Bruner, and E. M. Matheson (eds.), *The I. G. in Peking* (Cambridge, Mass., 1975), 2 vols.
20. MacNair, I, 384-85. Italics added.

trusted servant and an adviser on foreign affairs. As long as he lived, no other inspector-general was appointed; such was the regard of the Ch'ing court for its foreign employee, who on his part declined the position of British minister to China in the 1880s in order to remain with the customs service. Under his guidance many outstanding foreigners served in the Chinese international customs service; one of the more prominent was H. B. Morse, a graduate of Harvard in 1876, who after his retirement wrote several definitive and pioneer works on Chinese trade, administration, and foreign relations.

INTRODUCTION OF INTERNATIONAL LAW. Although before the Opium War Commissioner Lin Tse-hsü had asked the American missionary Peter Parker to translate three paragraphs of international law from Vattel, there was no complete text of the law of nations in Chinese. Ignorance of international law led early Chinese negotiators into many blunders: they conceded easily on such significant issues as tariff autonomy, extraterritoriality, and the most-favored-nation treatment, while struggling bitterly against such common and innocuous practices and issues as diplomatic residence and audience without the kowtow. Witnessing China's need for a guide to diplomacy, W. A. P. Martin took upon himself a translation of a text of international law, which he considered the best and most mature fruit of Christian civilization. With the help of Chinese scribes, he began to translate Henry Wheaton's *Elements of International Law* in 1862, hoping to show that Westerners had principles to govern their international relations and were not totally dependent on brutal force, and hoping that his translation might bring the atheistic Chinese government to recognize the spirit of Christianity.

Prince Kung, in his eagerness to learn about Western diplomacy, was secretly anxious to know international law. Through the good offices of the American minister, Anson Burlingame, Martin's translation was presented to the Tsungli Yamen in 1864. As its style and diction were rough and unidiomatic, the manuscript was put through a thorough editing by four secretaries of the Tsungli Yamen.

While the manuscript was being edited, the Tsungli Yamen had a chance to test its usefulness. The new Prussian minister, von Rehfues, arriving in the spring of 1864 in a man-of-war, found three Danish merchant ships off Taku, and immediately ordered their seizure as war prizes, Prussia and Denmark being at war in Europe. Prince Kung, armed with the new knowledge of international law, protested the extension of European quarrels to China and the seizure of ships in China's "inner waters," the Chinese expression for territorial waters. He refused to receive the

Prussian minister before the redress was made and chided him for beginning his duties in such an unbecoming manner. Embarrassed, von Rehfues released the three ships and paid a compensation of $1,500. Having proven the utility of the translation, Prince Kung distributed three hundred copies of it to provincial authorities.

Wielding this new knowledge in combination with other measures of diplomatic modernization, China managed to maintain peaceful relations with the foreign powers throughout the decade of the 1860s, thus furnishing the country with a much needed respite to begin the Self-strengthening programs.

Military Modernization
and Early Industrialization

Prince Kung's diplomatic modernization was paralleled by his efforts to create a modern navy and by those of the provincial leaders—Tseng Kuo-fan, Tso Tsung-t'ang, and Li Hung-chang—to introduce military modernization through the adoption of foreign ships and guns, organization of supporting industries, and opening of new training schools. Their endeavors marked the beginning of the Self-strengthening Movement which lasted through three and a half decades to 1895. However, the idea of learning from the West predated these leaders by more than two decades.

THE PIONEERS. It was Commissioner Lin Tse-hsü who first championed the idea of learning about the West. Orders were given to translate foreign newspapers from Macao, Singapore, and India, and gather information on Western geography, history, politics, and law. Under his sponsorship, passages of international law from Vattel were translated, and selections from Murray's *Cyclopaedia of Geography* were rendered into Chinese under the title of *Ssu-kuo chih* (A gazetteer of four countries) in 1841. With this rudimentary knowledge of the West, Lin developed a certain measure of respect for British power. The very fact that he purchased two hundred foreign guns to strengthen the Canton defense and ordered the translation of manuals of Western gun-making manifested his acute awareness of the superior barbarian weaponry and the need for China to unlock its mystery.

The defeat in the Opium War was dismissed by most Chinese scholars and officials as an historical accident. Nonetheless, a few farsighted private scholars associated with or influenced by Lin sensed the advent of a new era in China's relations with the outside world. Foremost among them was Wei Yüan (1794-1856), to whom Lin had turned over the materials on foreign countries. Wei compiled them into a large work of fifty tomes (*chüan*) in 1844[21] entitled *Hai-kuo t'u-chih* (An illustrated gazetteer of the maritime countries). The objective of the work was clearly explained in the preface: "Why did I compile this work? It is for the purpose of using barbarians to attack barbarians, using barbarians to negotiate with barbarians, and *learning the superior techniques of the barbarians* to control the barbarians."[22] The book falls into four parts: part one deals with the history, geography, and recent political conditions of Western countries; part two, the manufacturing and use of foreign guns; part three, shipbuilding, mining, and miscellaneous descriptions of the practical arts of the West; and part four, methods of dealing with the West, as suggested by the compiler and his contemporaries. It was the first significant Chinese work on the West.

Other pioneer works included a very complimentary account of the American political system, called *Ho-chung-kuo shuo* (On the United States), and a study of the recent foreign disturbance of China called *I-fen chi-wen* (A record of the barbarian miasma), both by Liang T'ing-nan. There followed the famous work of world geography by Hsü Chi-yü, entitled *Ying-huan chih-lüeh* (A brief survey of maritime circuit), and a vast compendium of eighty *chüan* on the geography, history, and politics of Russia and other northern countries called *Shuo-fang pei-sheng* (A manual for northern places) by Ho Ch'iu-t'ao. Except for the last-named, all the works stressed the importance of maritime defense against the Western sea powers. The essence of their message was that if China developed sufficient coastal defense, it could hold the enemy from the ocean at bay. There was little understanding or recognition of the fact that the overseas expansion of Europe and America, propelled by the forces of rising nationalism, capitalism, and rapid industrialization, could hardly be prevented by localized defense in China. This initial period of Western studies revolved around the limited subject of maritime defense, and the authors and compilers, being private scholars, had but a limited influence on their country.

The defeat in 1860 shocked the intelligentsia and officialdom into a greater awakening. Feng Kuei-fen (1809-74), also a onetime associate of Lin's, took the lead in promoting the idea of Self-strengthening (*Tzu-ch'iang*). In his famous work, *Protest from the Chiao-pin Studio* (*Chiao-pin-lu k'ang-i*), written about 1860-61, he realistically took notice of the vast difference between the old world that China had known in its past and the new world that had been thrust upon it, and urged that China

21. Enlarged to 100 *chüan* in 1852.
22. Italics added.

adopt Western ships and guns, and construct dockyards and arsenals in the trading ports. If the small country of Japan felt the need for fortifying itself along Western lines, he asserted, how much more precipitate should China be in strengthening herself! Bearing in mind the recently concluded peace with the Western powers, Feng warned that China must utilize this heaven-sent opportunity to strengthen herself, or it would live to regret missing the chance. As regards Wei Yüan's aspiration of using barbarians to control and negotiate with barbarians, Feng found it an impossibility; difficulty of foreign languages and unfamiliarity with their customs precluded China from sowing dissension among the barbarians. "Only one sentence of Wei Yüan is correct: *learn the superior techniques of the barbarians to control the barbarians*."[23] This dictum rang out as the motivating spirit of the Self-strengthening Movement (*Tzu-ch'iang yüntung* 自強運動) from 1861 until 1895.

PRINCE KUNG AND THE LAY-OSBORN FLOTILLA. The chief promoters of Self-strengthening in the capital were Prince Kung and Wen hsiang, who impressed upon the court that China lost the wars not because the soldiers did not fight hard but because they were not properly equipped. To guard against future humiliation, China must adopt Western firearms, ships, and army training. A first step in this direction was taken in 1862 when the prince directed Acting Inspector-general of Customs Robert Hart to commission Horatio Lay, then in England, to purchase and equip a steam fleet. Lay acquired eight ships and engaged the service of Captain Sherard Osborn of the Royal navy. Without the knowledge or prior approval of the court at Peking, Lay entered into an agreement with Osborn on January 16, 1863, whereby the latter was to be the sole commander-in-chief of this naval force as well as of native vessels manned by Europeans, and that he was to accept orders from no one but Lay as representative of the Chinese emperor. Behaving presumptuously in this matter, Lay justified his action on the ground that he was not bound by ordinary rules of conduct or normal procedure for transactions in China. "My position was that of a foreigner engaged by the Chinese government to perform certain work *for* them, not *under* them. I need scarcely observe that the notion of a gentleman acting *under* an Asiatic barbarian is preposterous."[24] Lay had the vision of erecting himself as the "First Lord of the Admiralty" in China and concurrently the Inspector-general of Customs; the one would give

---

23. Feng, Kuei-fen, *Chiao-pin-lu k'ang-i* (Protest from the Chiao-pin Studio), (Shanghai, 1897), 2:4b-6. Italics added. Here "Protest" really means "Straight Talk," in accordance with the meaning of *k'ang-i* in the *Hou Han-shu* (History of the Later Han).
24. Lay, *Our Interests in China*, 19.

him unrivaled military (naval) power, and the other, control of some seven million taels annual customs revenue. To befit his important status, he demanded a palace at Peking for residence. The Chinese found him unbearable.

Prince Kung informed Captain Osborn upon his arrival with the fleet in September 1863 that his official title was assistant commander-in-chief, with authority to control only the foreigners of the fleet, and that he must accept orders from the local governor-general and governor in whose jurisdiction he happened to be operating, i.e. Tseng Kuo-fan, the governor-general of Liang-Kiang, and Li Hung-chang, the governor of Kiangsu. Lay protested that he had come "to serve the emperor, not to be the servant of mere provincial authorities," especially under so "unprincipled an official" as Li. Prince Kung stood firm in his position. Under the circumstances Captain Osborn recommended that the fleet be disbanded, lest it fall into the hands of the Taipings, or the hostile daimyos of Japan, or even the American Confederacy.

Tseng Kuo-fan accepted the position that, rather than harboring an uncontrollable foreign navy which might cause unpredictable complications, China might best disband it and compensate the officers liberally. The American minister, Anson Burlingame, offered to mediate. Osborn was offered a special solatium of 10,000 taels and Lay was pensioned off with £14,000, as pay and allowances for the period he was quarreling with the Chinese government. He was then replaced by Hart as inspector-general of customs. The Chinese government had spent a total of £550,000 purchasing and disbanding a fleet, from which it got absolutely nothing but headaches. The first attempt at a modern navy was a complete fiasco.[25]

THE BEGINNING OF SELF-STRENGTHENING IN THE PROVINCES. The bulk of the Self-strengthening projects were promoted by provincial authorities such as Tseng Kuo-fan, Tso Tsung-t'ang, and Li Hung-chang. From their association with the Foreign Rifle Squadron and the Ever-Victorious army during the Taiping campaign, they had learned firsthand the superiority of Western guns and ships. But these contrivances were beyond the normal understanding of Confucian scholars and officials. Story has it that a contemporary of Tseng's, Hu Lin-i, was so astounded by the sight of two foreign steamers charging swiftly and effortlessly against contrary river currents that he sighed in resignation: "These are things we cannot com-

25. For details, see John L. Rawlinson, *China's Struggle for Naval Development, 1839-1895* (Cambridge, Mass., 1967), 34-37; Katherine F. Bruner, John K. Fairbank, and Richard J. Smith (eds.), *Entering China's Service: Robert Hart's Journals, 1854-1863* (Cambridge, Mass., 1986), 257, 316-17.

prehend!" The thought of facing such an unfathomable enemy in the future profoundly shocked him, and already weakened by overwork during the Taiping campaign, Hu died shortly afterwards.

If steamships made such a strong impact on Hu, he was not alone. To the farsighted, shipbuilding became a *sine qua non* for survival. Tseng Kuo-fan tried his hands at building one in Anking in 1862-63, but it could not move fast or freely. Baffled but not discouraged, he was the more determined to unlock the secrets of shipbuilding and gun-making in order to break the Western monopoly of power.[26]

Under Tseng's sponsorship, the Kiangnan Arsenal was established at Shanghai in 1865, with machines purchased from the United States by Yung Wing, the first Chinese graduate of Yale (1854), who had joined Tseng's staff in 1863. The arsenal not only manufactured guns and cannon, but also constructed ships and maintained a translation bureau. Its first ship, 185 feet long and 27.2 feet wide, was successfully completed in 1868. The arsenal turned out a total of five ships, the last in 1872 with 400 horsepower and carrying 26 guns. Its translation bureau completed 98 titles of Western works in less than ten years, of which 47 were in the field of natural science and 45 on military affairs and technology. Doubtless, the Kiangnan Arsenal shone as a major accomplishment in the early phase of the Self-strengthening Movement.

If Tseng opened the way to Westernization, Tso and Li bore the torch unflaggingly. Tso had been a firm believer in the unity of knowledge and action, and, in his train of thinking, was influenced by Lin Tse-hsü and Wei Yüan. His interest in shipbuilding blossomed into the establishment of the famous Foochow Dockyard in 1866, with two Frenchmen, Prosper Giquel (1835-86) and Paul d'Aiguebelle (1831-75) as chief engineer and supervisor. It turned out a total of forty ships. Its naval school graduated a number of able officers, including the extremely adept and perceptive Yen Fu (1853-1921), who studied in Britain and, in later years, translated a number of important Western works on thought, sociology, logic, and jurisprudence (see Chapter 17). The Foochow Dockyard was the second most important achievement of the Self-strengthening Movement.

The leading spirit of the Self-strengthening Movement was Li Hung-chang. His association with the Ever-Victorious army and with foreign officers such as Ward and Gordon made him cognizant of the awesome power of guns and ships. In superlatives he praised the Western cannon and explosive shells as "matchless weapons for offensive and defensive in the whole world." He believed, somewhat naïvely, that possession of steamships and guns with explosive shells alone would suffice to stop for-

26. Teng and Fairbank, *China's Response*, 62.

eign aggression. Li's adoration of the Western military system and arms is
seen in a letter to Tseng in February 1863:

> I have been aboard the warships of British and French admirals and I
> saw that their cannon are ingenious and uniform, their ammunition is
> fine and cleverly made, their weapons are bright, and their troops have
> a martial appearance and are orderly. These things are actually supe-
> rior to those of China. Their army is not their strong point, yet when-
> ever they attack a city or bombard a camp, the various firearms they
> use are all non-existent in China. Even their pontoon bridges, scaling
> ladders, and fortresses are particularly well prepared with excellent
> technique and marvelous usefulness. All these things I have never
> seen before . . . I feel deeply ashamed that Chinese weapons are far
> inferior to those of foreign countries. Every day I warn and instruct
> my officers to be humble-minded, to bear the humiliation, to learn one
> or two secret methods from the Westerners in the hope that we may
> increase our knowledge . . . If we encamp at Shanghai for a long
> time and cannot make use of nor take over the superior techniques of
> the foreigners, our regrets will be numerous![27]

If China did not catch up in shipbuilding and gun-making, Li warned,
Japan would soon imitate the West and take advantage of China. The urgency of the situation must compel China to institute Self-strengthening
programs at once. What China was facing, Li loudly pronounced in
1872, was *a totally unprecedented situation in its three thousand years of
history:* the West had advanced step by step from India to Southeast Asia
and to China. There was no way to stop the movement. China must meet
this challenge head-on, determined to strengthen itself through adoption
of Western guns and ships.

With Tseng Kuo-fan's death in 1872 and Tso Tsung-t'ang's involvement in the Moslem campaign in the Northwest and Sinkiang from 1868
to 1880, Li became the central spirit of Self-strengthening. His long tenure as governor-general of Chihli and as High Commissioner of the Northern Ocean for practically the quarter of a century after 1870 enabled him
to build up a substantial military and industrial empire in North China.
Although a provincial authority, he actually performed a number of functions for the central government and served as a sort of "coordinator" of
Self-strengthening programs throughout the country.[28] For thirty years he

27. Teng and Fairbank, *China's Response,* 69.
28. K. C. Liu, "Li Hung-chang in Chihli: The Emergence of a Policy, 1870-1875."
     Paper read before the 18th annual convention of the Association for Asian Studies,
     New York, April 4, 1966; later published in Feuerwerker, Murphey, and Wright
     (eds.), *Approaches to Modern Chinese History,* 68-104.

was the principal architect and instigator of "foreign matters" (*yang-wu*)[29] in China. Among his major achievements were the Nanking Arsenal in 1867, the China Merchants' Steam Navigation Company in 1872, a naval and a military academy at Tientsin in 1880 and 1885, respectively, and the Peiyang fleet in 1888.

However, Li's preoccupation with ships and guns and his negligence of Western political systems and culture very much limited the scope of the Self-strengthening Movement. His attitude stemmed partially from his belief that China surpassed the West in everything except weaponry,[30] and partially from the fact that he was in charge of military training and coastal defense. He saw China's immediate need for military strength, but not the larger and more distant requirement for political and social reform.

THE CONSERVATIVE OPPOSITION. Limited as was the scope of the Self-strengthening Movement and its leaders' grasp of the problems facing China, the advocates of *yang-wu* were already far ahead of their fellow officials and scholars, who were mostly ignorant of the modern world and blind to modernization. When Prince Kung memorialized the throne in 1867 for permission to add a department of astronomy and mathematics at the T'ung-wen kuan and to invite foreign professors to teach these subjects to students of sound background in Chinese studies, he was vehemently attacked by the arch-conservative Grand Secretary Wo-jen (d. 1871), a leading Neo-Confucian scholar. How could the Chinese forget the disgrace of the foreign occupation of Peking and the burning of the Summer Palace, Wo-jen asked. How could anyone suggest shifting good Chinese scholars to barbarian studies? Prince Kung's rebuttal was simple and pointed: if Wo-jen had a better way to save the country, let him show how:

> The grand secretary [i.e. Wo-jen] considers our action a hindrance. Certainly, he should have some better plans. If he really has some marvelous plan which can control foreign countries and not let us be controlled by them, your ministers should certainly follow the footsteps of the grand secretary . . . If he has no other plan than to use loyalty and sincerity as armor, and propriety and righteousness as a shield, and such similar phrases; and if he says that these words could accomplish diplomatic negotiations and be sufficient to control the life of our enemies, your ministers indeed do not presume to believe it.[31]

29. Such as ships, guns, railroad, telegraph, and other Western-style enterprises, as distinguished from "foreign affairs."
30. Li wrote: "Everything in China's civil and military system is far superior to the West. Only in firearms is it absolutely impossible to catch up with them." Teng and Fairbank, *China's Response*, 71.
31. Teng and Fairbank, *China's Response*, 76-9.

The Empress Dowager Tz'u-hsi, realizing the importance of Western learning but not wishing to offend the conservatives, approved the establishment of the science department on the one hand, and on the other authorized Wo-jen to set up a separate department of Chinese studies in the school. It was her game of using the conservatives to checkmate the progressives, lest the latter become too powerful for her to control. Wo-jen, reluctant to associate with the T'ung-wen kuan, intentionally fell while riding to the school, thus providing himself with an excuse not to attend. The conservative Confucian society and officialdom were so ill-disposed toward innovations that the Self-strengtheners had to fight every inch of the way to launch the movement.

Periods of Self-strengthening

THE FIRST PERIOD. According to the changing emphasis and the shifting philosophy, the Self-strengthening Movement can be divided into three periods. The first, roughly from 1861 to 1872, stressed the adoption of Western firearms, machines, scientific knowledge, and the training of technical and diplomatic personnel through the establishment of translation bureaus, new schools, and the dispatch of students abroad. Diplomatic innovations were introduced to insure good relations with Western powers so that China could discover their shipbuilding and armament secrets. As already indicated, the motivating force was the desire "to learn the superior *techniques* of the barbarians to control the barbarians"; there was no recognition of the need for anything else from the West. The embattled leaders of this period were Prince Kung and Wen-hsiang in the capital, and Tseng, Tso, and Li in the provinces, and their main accomplishments were as follows:

1861   Establishment of the Tsungli Yamen at Peking and the Superintendencies of Trade at Tientsin and Shanghai at the suggestion of Prince Kung.

1862   Establishment of the T'ung-wen kuan (Interpreters College) at Peking at the suggestion of Prince Kung.
Creation of three gun factories at Shanghai by Li Hung-chang, who also assigned his men to learn the use of cannon with explosive shells from British officers and the use of rifles from German officers.

1863   Establishment of a foreign-language school (Kuang-fang-yen kuan) at Shanghai by Li.
The arrival of the Lay-Osborn flotilla.

Dispatch of Yung-wing to the United States to purchase machines by Tseng.

1864 Creation of a small gun factory at Soochow by Li.

Establishment of a foreign-language school (T'ung-wen kuan) at Canton.

1865 Establishment of the Kiangnan Arsenal at Shanghai by Tseng and Li, with a translation bureau attached.

1866 Establishment of the Foochow Dockyard at Ma-wei, outside Foochow, by Tso Tsung-t'ang, with machines purchased from France. Attached was a naval school in two divisions: one specializing in French and shipbuilding, and the other in English and navigation. Dispatch of the Pin-ch'un exploratory mission to Europe.

1867 Establishment of the Nanking Arsenal by Li.

Creation of the Tientsin Machine Factory by Ch'ung-hou.

1868 Dispatch of Anson Burlingame as China's roving ambassador to the West, to assist the Manchu and Chinese co-envoys. (See next chapter.)

1870 Expansion of the Tientsin Machine Factory into four plants by Li.

1871 Planning for a Western-style fort at Taku.

1872 Dispatch of thirty teenage students to the United States to study at Hartford, Connecticut on recommendation of Tseng and Li. A total of 120 boys were sent in four installments, 1872-81.

Officers sent by Li to study in Germany.

Request by Li to open coal and iron mines.

The pre-eminent feature of this period of Self-strengthening was the emphasis on development of military industries, which had the following characteristics. First, they were all "government undertakings" (kuan-pan), which partook of all the usual bureaucratic inefficiency and nepotism of an official agency. They engaged in modern production but retained the old-style managerial and administrative procedures. Secondly, they relied on foreigners for operation and materials. There seemed to be a blind faith in the ability of foreigners, regardless of their training and experience. The Nanking Arsenal was put under the direction of an Englishman, Halliday Macartney, a medical doctor by profession. The Foochow Dockyard was supervised by two Frenchmen, Giquel and d'Aiguebelle, who had never before in their lives built a ship. The materials for construction were all imported. Because of poor leadership and bureaucratic corruption, the ships and guns produced were nowhere comparable in quality to their Western counterparts. Thirdly, these military industries formed the power bases of the provincial leaders sponsoring them, and consequently smacked of a strong regional and "feudal" flavor. As governor-general at Nanking, Li built the Nanking Arsenal; and as governor-general at Foochow, Tso constructed the Foochow Dockyard. There was little concerted effort or coordination between the various regional groups. Even after their transfer to other posts—Tso to the Northwest in 1868 and Li to North China (Tientsin), where they built up new power bases—they continued to maintain personal connections with their former projects.[32]

THE SECOND PERIOD. As the Self-strengthening Movement progressed, there was increasing recognition that wealth was the basis of power—one had to be rich in order to be strong. Modern defense cost far more than traditional defense; moreover, it had to be supported by better communication systems, industries, and enterprises. Li Hung-chang announced in September 1876: "China's chronic weakness stems from poverty." Therefore, in the second period, 1872 to 1885, while defense industries remained a chief occupation, greater attention was directed to the development of profit-oriented enterprises such as shipping, railways, mining, and the telegraph. These "foreign matters" (*yang-wu*) gradually came to be regarded as "current affairs" (*shih-wu*), as they had been thrust forward as the pressing problems of state.

In addition to the "government-sponsored" (*kuan-pan*) military industries, there now appeared another type of enterprise, modeled in organization after the traditional salt administration: "government-supervised merchant undertakings" (*kuan-tu shang-pan*).[33] Foremost among them were the China Merchants' Steam Navigation Company, the K'ai-p'ing Coal Mines, the Shanghai Cotton Cloth Mill, and the Imperial Telegraph Administration.[34] Capital for these undertakings came from private sources, although the government as patron might initially supply some funds or advance loans to be repaid later. But "profit and loss are entirely the responsibility of the merchants and do not involve the government," as Li Hung-chang decreed.[35] The merchant shareholders, who supplied the funds, were barred from the management, which was in the hands of government-appointed officials or private individuals (who might subscribe to shares later). For instance, the China Merchants' Steam Navigation

32. Mou An-shih, *Yang-wu yün-tung* (The "foreign matters" movement), (Shanghai, 1961), 79-86.
33. Often translated, rather incorrectly, as "government supervision-merchant management."
34. Albert Feuerwerker, *China's Early Industrialization: Sheng Hsüan-huai (1844-1916) and Mandarin Enterprise* (Cambridge, Mass., 1958), 9-10.
35. Kwang-ching Liu, "British-Chinese Steamship Rivalry in China, 1873-85" in C. D. Cowan (ed.), *The Economic Development of China and Japan* (London, 1964), 53.

Company's first promoter-manager was an official;[36] he was succeeded by a private individual who was the former compradore of the British firm of Jardine, Matheson and Company;[37] after 1884 it was again managed by an official.[38] These government-supervised merchant enterprises were a hybrid operation which smacked of strong official overtones and the usual bureaucratic inefficiency, corruption, and nepotism. Being profit-oriented, they discouraged private competition and tended to monopolize business through government favor or intervention. They also relied on foreign personnel for support: the "China Merchant" employed foreign marine superintendents, ship captains, and engineers.

During this second period, Li Hung-chang emerged as the leading proponent of modern industries and enterprises, Tseng having died in 1872 and Tso being involved in the Northwest with the suppression of the Moslem Rebellion. Prince Kung had lost much of his influence with the Dowager Tz'u-hsi after his two chastisements in 1865 and 1869 (next chapter), while Wen-hsiang had died in 1876. Li rose to be the unrivaled leader of the Self-strengthening Movement, and though a provincial official—governor-general of Chihli—he performed a number of central government functions due to his proximity to Peking and the dowager's trust in him. Over 90 percent of the modernization projects were launched under his aegis.

1872   Inauguration of the China Merchants' Steam Navigation Company as a "government-supervised merchant undertaking," supported by Li.

1875   Plans to construct iron-clad ships.
Dispatch of students from the Foochow Dockyard to study in France.

1876   Dispatch of seven officers to Germany by Li.
Sending of thirty students and apprentices from the Foochow Dockyard to Britain and France.
Diplomatic mission to Britain and France, followed by those to other countries in the next years.

1877   Creation of the Bureau for the K'ai-p'ing Coal Mines at Tientsin by Li.
Establishment of a machine factory in Szechwan by Ting Pao-chen.

1878   Establishment of a textile factory in Kansu by Tso.
Establishment of the Shanghai Cotton Cloth Mill by Li.

1879   Inauguration of a telegraph line between Taku and Tientsin.

36. Chu Ch'i-ang.
37. Tong King-sing.
38. Sheng Hsüan-huai.

1880   Establishment of a naval academy at Tientsin by Li.

Request for permission to build railways by Li.

Adoption of a plan for a modern navy, and beginning of purchasing foreign warships.

1881   Inauguration of the Imperial Telegraph Administration.

Opening of the first telegraph line from Shanghai to Tientsin.

Creation of a railway of twenty *li* (six miles) north of Tientsin.

Dispatch of ten naval students abroad to study.

1882   Beginning of construction of a harbor and a shipyard at Port Arthur, by Li.

1884   Sending of thirteen naval students and four apprentices by Li to study shipbuilding in Britain, France, and Germany, and nine students to Britain to learn navigation.

THE THIRD PERIOD. From 1885 to 1895, while emphasis on the military and naval build-up continued, as witnessed by the organization of the Board of Admiralty (*Hai-chün ya-men*) in 1885 and the formal establishment of the Peiyang fleet in 1888, the idea of enriching the nation through light industry gained increasing favor; as a result textile and cotton-weaving gathered momentum. Li continued to dominate the scene, but he now faced rising competition from Governor-general Chang Chih-tung at Wuhan, and Governor-general Liu K'un-i at Nanking. Meanwhile, Prince Chün, father of the emperor and head of the newly established Board of Admiralty, emerged as a powerful figure in the capital, with Prince Kung's descent into political eclipse after the French war, 1884-85.

Organizationally, two new types of industrial and mercantile enterprise—"joint government and merchant enterprises" (*kuan-shang ho-pan*) and incipient "private enterprises" (*shang-pan*)—bid for existence in revolt against the dominant bureaucratic "government-supervised merchant undertakings." But these two failed to prosper because of traditional official discrimination against, and jealousy of, merchants. Among the larger "joint government and merchant undertakings" were the Kweichow Ironworks established in 1891 and the Hupeh Textile Company in 1894. In both cases the officials welcomed private capital but resented private control. The struggle for domination of the Hupeh Textile Company was so acute that the merchant capital was ultimately forced out, leaving the company entirely a government operation. As regards private enterprises, they were very weak, representing but a small fraction of the total industrial effort and investment—a far cry from mobilizing private capital the way the Japanese did during the Meiji era. The major efforts of the decade from the mid-eighties to the mid-nineties included:

1885 Creation of a military academy at Tientsin by Li.

Inauguration of the Board of Admiralty in Peking, with Prince Chün as head, assisted by Li.

1886 Establishment of a textile mill by Chang Chih-tung at Canton.

1887 Establishment of mints by Chang and Li at Canton and Tientsin respectively.

Inauguration of the Mo-ho Gold Mines in Heilungkiang by Li.

1888 Establishment of the Peiyang fleet under Li's control.

1889 Creation of a cotton mill and an iron factory at Canton by Chang.

1890 Inauguration of the Ta-yeh Iron Mines, the Han-yang Ironworks, and the P'ing-hsiang Coal Mines by Chang.

1891 Establishment of the Lung-chang Paper Mill at Shanghai by Li.

Establishment of the Kweichow Ironworks as a "joint government and merchant undertaking."

1893 Inauguration of a general office for machine textile manufacturing by Li.

Establishment of four cotton and textile plants at Wuchang by Chang.

1894 Organization of two match companies in Hupeh province.

Creation of the Hupeh Textile Company, as a "joint government and merchant undertaking."

Limitations and Repercussions of the Self-strengthening Movement

The preceding lists may present quite an impressive picture of endeavors, but they really represent very superficial attempts at modernization. The scope of activity was limited to firearms, ships, machines, communications, mining, and light industries. No attempts were made to assimilate Western institutions, philosophy, arts, and culture. The Self-strengthening efforts barely scratched the surface of modernization, without achieving a breakthrough in industrialization. The basic weakness was exposed in the French war of 1884-85, when China, after twenty years of preparations, was unable to defend its tributary state, Annam. The failure of the movement was confirmed beyond doubt by the defeat in the Japanese war ten years later. Marxist historians stress the intrinsic contradiction in grafting modern capitalism and industry onto the agrarian Confucian social base. Perhaps the following points shed light on the lackluster performance of Self-strengthening.

LACK OF COORDINATION. The central power of the Ch'ing dynasty had declined after the Taiping Revolution, so much so that apart from a flash of vigor during the T'ung-chih period (1862-74), there was hardly any direction in the government. The major brunt of modernization was borne by provincial authorities without central direction, planning, and coordination. Although Li Hung-chang after 1870 performed some central government functions, he was basically a regional official who could not take the place of the central government. The provincial promoters of Self-strengthening rivaled rather than cooperated with each other and regarded their achievements as the foundation of personal power. Their sense of regionalism and their eagerness for self-preservation persisted so strongly that during the French war of 1884 the Peiyang and Nanyang fleets refused to go to the rescue of the Fukien fleet under enemy attack, and during the Japanese war of 1894-95 the Nanyang fleet maintained "neutrality" while the Peiyang fleet alone fought the Japanese navy. The results of both wars were, of course, disastrous.

LIMITED VISION. The advocates of Self-strengthening promoted modern projects primarily to enable their country to resist foreign aggression, to suppress domestic unrest, and to fortify their own positions of power. They never dreamed of remaking China into a modern state. In fact, they strove to strengthen the existing order rather than to replace it. They had absolutely no conception of economic development, industrial revolution, and modern transformation. Consequently, their endeavors resulted in no more than a handful of isolated modern enclaves scattered over an otherwise traditional country, in which the old institutions remained dominant.

Furthermore, the lack of popular participation restricted the scope of modernization. The leadership in Self-strengthening operated from the top down, with little grass-roots support as there was in Meiji Japan. Shackled by traditions, the Chinese officials were unable to shake off the age-old disdain for merchants, and continued to discourge private enterprise and competition. They also failed to instill private initiative in the government industries or the government-supervised merchant undertakings, which continuously suffered from the usual bureaucratic inefficiency, nepotism, and corruption.[39]

SHORTAGE OF CAPITAL. China was a poor country with a limited supply of capital. There was a shortage of bureaucratic as well as private capital, which restricted the initiation and growth of industries and enterprises. When the government raised taxes to support the new undertakings, it weakened the people's all too limited resources for investment. One need

39. Ch'uan Han-sheng, "Chia-wu chan-cheng i-ch'ien ti Chung-kuo kung-yeh-hua yün-tung" (China's industrialization movement before the Sino-Japanese War), *Li-shih yü-yen yen-chiu-so chi-k'an;* Academia Sinica, 25 (June 1954), 74.

only note the same persons associated with the various enterprises—the China Merchants' Steam Navigation Company, the Shanghai Cotton Cloth Mills, the Imperial Telegraph Administration, and the Han-Yeh-P'ing Mines—to know the small circle of entrepreneurs and the limited funds at their disposal.[40] Moreover, capital formation was difficult in these enterprises, since the profits, roughly 8 to 10 percent a year, were distributed to shareholders as dividends rather than reinvested for growth.

FOREIGN IMPERIALISM. The generation of Self-strengthening coincided with a period of intensified foreign imperialism, as evident in the Japanese invasion of Formosa in 1874 and annexation of the Liu-ch'iu Islands in 1879; the British attempt to open Yunnan in 1875; the Russian occupation of Ili in Sinkiang, 1871-81; the French seizure of Annam and the war of 1884-85; and the Japanese aggression in Korea and the war of 1894-95. These cataclysmic events not only divided the attention of the government and the modernizers, but also incurred vast military expenses and indemnities which siphoned away considerable sums that could otherwise have been applied to Self-strengthening.

TECHNICAL BACKWARDNESS AND MORAL DEGRADATION. Western machines and industrial management were alien to the traditional Chinese mentality. To overcome technical backwardness was a tremendous obstacle, the more so when the foreign advisers and teachers themselves were quite inexpert. The guns and ships turned out by the Self-strengthening projects were vastly inferior, necessitating continuous purchases from abroad. The nine large ships of the Peiyang fleet were all foreign-made, and the cannon at the Port Arthur and Weihaiwei naval bases were Krupp products.

Moreover, men of talent and integrity usually steered clear of foreign matters and enterprises; only the lesser characters were willing to associate with the modernization projects, resulting in frequent cases of corruption and irregularity. Even Li Hung-chang himself was not noted for high morals and character—he reportedly left behind an estate of 40 million taels! His followers squeezed and milked the factories and enterprises under their charge mercilessly. The most scandalous of all was the misuse of 30 million taels of naval funds to construct the Summer Palace (*I-ho-yüan*) for the amusement of the dowager in retirement.

SOCIAL AND PSYCHOLOGICAL INERTIA. The great majority of the scholar-official class regarded foreign affairs and Western-style enterprises as "dirty"

40. Feuerwerker, 249.

and "vulgar," beneath their dignity. So powerful was this conservatism that even the court could not ignore it. An excerpt from Li Hung-chang's letter to a friend[41] illustrates the difficulties of the modernizers:

> I had the honor of meeting Prince Kung and explained to him thoroughly the advantages of railways . . . Prince Kung agreed with my suggestion, but said that *nobody dared to promote such action.* I again begged him to watch for a chance to explain this point to the two dowager empresses, and he said that the two empresses could not initiate such a great plan either. Thereafter I spoke no more. . . .
>
> The gentry class forbids the local people to use Western methods and machines, so that eventually the people will not be able to do anything . . . Scholars and men of letters always criticize me for honoring strange knowledge and for being queer and unusual. *It is really difficult to understand the minds of some Chinese.*[42]

Cases of conservative opposition to modernization abound. In 1874 the British-built short railroad from Shanghai to Woosung was ripped off its bed by mobs because the locomotive ran over a spectator. Two years later the governor-general was pressured by the local gentry to buy this foreign railway and have it totally wrecked. In 1876 when Kuo Sung-tao went to Britain as a minister, the literati cruelly satirized him for leaving the land of the sages to serve the foreign devils. Kuo's diary, which praised the Western civilization as having a history of two thousand years, was condemned by the conservatives as heresy, and they forced the government to destroy its printing block. These few instances suffice to lay bare the unfavorable social and political atmosphere within which the advocates of Westernization had to operate. Considering the tremendous odds against them, it is really a wonder that they dared to espouse such an unpopular cause and that they achieved the record, however imperfect, they did!

For all its shortcomings, the Self-strengthening Movement marked the beginning of industrialization and sowed the seeds of modern capitalism in China, with many significant repercussions. First, most of the arsenals, dockyards, machine factories, schools, and modern enterprises were located in the treaty ports and cities along the coast or on the river, where foreign help was most readily available. They contributed to the development of great metropolises such as Shanghai, Nanking, Tientsin, Foochow, Canton, and Hankow. Secondly, the farming population in nearby

---

41. Kuo Sung-tao, minister to Britain and France, 1876-78.
42. Li Chien-nung, *The Political History of China, 1840-1928* (New York, 1956), ed. and tr. by S. Y. Teng and J. Ingalls, 108-09. Italics added.

agricultural areas was drawn to these centers to become industrial workers or laborers, swelling the size of these cities and gradually giving rise to
a new working class. Thirdly, the new industries and enterprises brought
into being new professional men such as engineers, managers, and entrepreneurs, while those who had studied abroad returned home to become
leaders in the army, navy, schools, and diplomatic service. They contributed to the rise of the new managerial and entrepreneurial class in China.

Further Reading

Banno, Masataka, *China and the West: 1858-1861: The Origins of the Tsungli
    Yamen* (Cambridge, Mass., 1964).

Bennett, Adrian A., *John Fryer: The Introduction of Western Science and Tech-
    nology into Nineteenth-Century China* (Cambridge, Mass., 1967).

Biggerstaff, Knight, *The Earliest Modern Government Schools in China* (Ithaca,
    1961).

Bruner, Katharine F., John K. Fairbank, and Richard J. Smith (eds.), *Entering
    China's Service: Robert Hart's Journals, 1854-1863* (Cambridge, Mass.,
    1986).

———— (eds.), *Robert Hart and China's Early Modernization: His Journals, 1863–
    1866* (Cambridge, Mass., 1991).

Carlson, C. Ellsworth, *The Kaiping Mines, 1877-1912* (Cambridge, Mass., 1957).

————, *The Foochow Missionaries, 1847-1880* (Cambridge, Mass., 1974).

Chan, Wellington K. K., *Merchants, Mandarins, and Modern Enterprise in Late
    Ch'ing China* (Cambridge, Mass., 1977).

Ch'en, Gideon, *Lin Tse-hsü, Pioneer Promoter of the Adoption of Western Means
    of Maritime Defense in China* (Peiping, 1934).

————, *Tseng Kuo-fan: Pioneer Promoter of the Steamship in China* (Peiping,
    1935).

————, *Tso Tsung-t'ang, Pioneer Promoter of the Modern Dockyard and the
    Woolen Mill in China* (Peiping, 1938).

Cheng, Ying-wan, *Postal Communication in China and Its Modernization, 1860-
    1896* (Cambridge, Mass., 1970).

Ch'i, Ssu-ho 齊思和 "Wei Yüan yü wan-Ch'ing hsüeh-feng" 魏源與晚清學風 (Wei
    Yüan and the late Ch'ing intellectual climate), *Yen-ching hsüeh-pao,* 39:
    177-266 (Dec. 1950).

Cohen, Paul A., *Between Tradition and Modernity: Wang T'ao and Reform in Late
    Ch'ing China* (Cambridge, Mass., 1974).

————, "Wang T'ao's Perspective on a Changing World" in Albert Feuerwerker,
    Rhoads Murphey, and Mary C. Wright (eds.), *Approaches to Modern
    Chinese History* (Berkeley, 1967), 133-62.

Cowan, C. D., *The Economic Development of China and Japan* (London, 1964).

Ch'üan, Han-sheng 全漢昇, "Ch'ing-chi ti Chiang-nan chih-tsao-chü" 清季的江南製
    造局 (The Kiangnan Arsenal of the late Ch'ing period), *Li-shih yü-yen yen-
    chiu-so chi-k'an* 歷史語言研究所集刊 (Bulletin of the Institute of History
    and Philology, Academia Sinica), 23:1:145-59 (Taipei, 1951).

————, "Chia-wu chan-cheng i-ch'ien ti Chung-kuo kung-yeh-hua yün-tung" 甲午戰爭以前的中國工業化運動(China's industrialization movement before the Sino-Japanese war), *Li-shih yü-yen yen-chiu-so chi-k'an* (Bulletin of the Institute of History and Philology, Academia Sinica), 25:59-79 (June 1954).

————, "Ch'ing-mo Han-yang t'ieh-ch'ang" 清末漢陽鐵廠(The Han-yang Iron and Steel Works at the end of the Ch'ing period), *She-hui k'o-hsüeh lun-ts'ung* (Journal of Social Sciences), 1:1-33 (April 1950).

Douglas, Robert K., *Li Hung-chang* (London, 1895).

Dow (Tou), Tsung-i 竇宗一, *Li Hung-chang nien (jih) p'u* 李鴻章年（日）譜 (A yearly (daily) biographical record of Li Hung-chang), (Hong Kong, 1968).

Elvin, Mark, and G. William Skinner (eds.), *The Chinese City between Two Worlds* (Stanford, 1971).

Eng., Robert Y., *Economic Imperialism in China: Silk Production and Exports, 1861-1932* (Berkeley, 1986).

Fairbank, John K., "The Creation of the Foreign Inspectorate of Customs at Shanghai," *The Chinese Social and Political Science Review*, 19:4:496-514 (Jan. 1936); 20:1:42-100 (April 1936).

Feuerwerker, Albert, *China's Early Industrialization: Sheng Hsüan-huai (1844-1916) and Mandarin Enterprise* (Cambridge, Mass., 1958).

Folsom, Kenneth E., *Friends, Guests, and Colleagues: The Mu-Fu System in the Late Ch'ing Period* (Berkeley, 1968).

Fu, Tsung-mou 傅宗懋 "Ch'ing-tai Tsung-li ko-kuo shih-wu ya-men yü Chün-chi-chu chih kuan-hsi," 清代總理各國事務衙門與軍機處之關係 (The relationship between the Tsungli Yamen and the Grand Council during the Ch'ing period), *Chung-shan hsüeh-shu wen-hua chi-k'an*, 12:285-323 (Nov. 12, 1973).

Gerson, Jack J., *Horatio Nelson Lay and Sino-British Relations, 1854-1864* (Cambridge, Mass., 1972).

Giquel, Prosper, *The Foochow Arsenal and Its Results, from the Commencement in 1867 to the End of the Foreign Directorate on the 16th February, 1874*, tr. by H. Lang (Shanghai, 1874).

Hao, Yen-p'ing, "A 'New Class' in China's Treaty Ports: The Rise of the Compradore-Merchants," *The Business History Review*, XLIV:4:446-59 (Winter 1970).

————, *The Compradore in Nineteenth Century China: Bridge Between East and West* (Cambridge, Mass., 1970).

————, *The Commercial Revolution in Nineteenth-Century China: The Rise of Sino-Western Mercantile Capitalism* (Berkeley, 1986).

Hou, Chi-ming, *Foreign Investment and Economic Development in China, 1840-1937* (Cambridge, Mass., 1965).

Hsü, Immanuel C. Y., *China's Entrance into the Family of Nations: The Diplomatic Phase, 1858-1880* (Cambridge, Mass., 1968), Pts. II and III.

Huang, I-feng 黃逸峯, and Chiang To 姜鐸, "Chung-kuo yang-wu yün-tung yü Jih-pen Ming-chih wei-hsin tsai ching-chi fa-chan shang ti pi-chiao" 中國洋務運動與日本明治維新在經濟發展上的比較(A comparison of the Chinese "foreign matters" movement and the Japanese Meiji modernization from the standpoint of economic development), *Li-shih yen-chiu* (1963) 1:27-47.

Huenemann, Ralph William, *The Dragon and the Iron Horse: The Economics of Railroads in China, 1876-1937* (Cambridge, Mass., 1984).

Kennedy, Thomas L., *The Arms of Kiangnan: Modernization in the Chinese Ordnance Industry, 1860-1895* (Boulder, 1978).

King, Frank H. H., *Money and Monetary Policy in China, 1845-1895* (Cambridge, Mass., 1965).

Kuo, Ting-yee, and Kwang-ching Liu, "Self-Strengthening: The Pursuit of Western Technology" in John K. Fairbank (ed.), *The Cambridge History of China* (Cambridge, Eng., 1978), Vol. 10, 491-542.

La Fargue, Thomas E., *China's First Hundred* (Pullman, Wash., 1942).

Le Fevour, Edward, *Western Enterprise in late Ch'ing China: A Selective Survey of Jardine, Matheson and Company's Operations, 1842-1895* (Cambridge, Mass., 1970).

Leung, Yuen-sang, *The Shanghai Taotai: Linkage Man in a Changing Society, 1843–90* (Singapore, 1990).

Li, Kuo-ch'i 李國祁, *Chung-kuo tsou-ch'i ti tieh-lu ching-ying* 中國早期的鐵路經營 (Early history of Chinese railway development), (Taipei, 1961).

Li, Lillian M., *China's Silk Trade: Traditional Industry in the Modern World, 1842-1937* (Cambridge, Mass., 1981).

Liu, Feng-han 劉鳳翰, *Hsin-chien lu-chün* 新建陸軍 (The newly established army), (Taipei, 1967).

Liu, Hsiung-hsiang 劉熊祥, *Ch'ing-chi ssu-shih-nien wai-chiao yü hai-fang* 清季四十年外交與海防 (Forty years of diplomacy and maritime defense in the late Ch'ing period), (Chungking, 1943).

Liu, Kwang-ching 劉廣京, "T'ang T'ing-shu chih mai-pan shih tai," 唐廷樞之買辦時代 (Tong King-sing: his compradore years), *The Tsing Hua Journal of Chinese Studies*, New Series, I:2:143-83 (June 1961).

———, *Anglo-American Steamship Rivalry in China, 1862-1874* (Cambridge, Mass., 1962).

———, "Li Hung-chang in Chihli: The Emergence of a Policy, 1870-1875" in Albert Feuerwerker, Rhoads Murphey, and Mary C. Wright (eds.), *Approaches to Modern Chinese History* (Berkeley, 1967), 68-104.

———, "The Confucian Patriot and Pragmatist: Li Hung-chang's Formative Years, 1823-1866," *Harvard Journal of Asiatic Studies*, 30:5-45 (1970).

———, "The Limits of Regional Power in the Late Ch'ing Period: A Reappraisal," *The Tsing Hua Journal of Chinese Studies*, New Series, X:2:176-223 (July 1974).

———, "The Ch'ing Restoration" in John K. Fairbank (ed.), *The Cambridge History of China* (Cambridge, Eng., 1978), Vol. 10, 409-490.

Lü, Shih-ch'iang 呂實強, *Chung-kuo tsou-ch'i ti lun-ch'uan ching-ying* 中國早期的輪船經營 (Early history of Chinese steamship development), (Taipei, 1962).

———, *Ting Jih-ch'ang yü tzu-ch'iang yün-tung* 丁日昌與自強運動 (Ting Jih-ch'ang and the Self-strengthening Movement), (Taipei, 1972).

Lutz, Jessie Gregory, *China and the Christian Colleges, 1850-1950* (Ithaca, 1971).

Meng, Ssu-ming, *The Tsungli Yamen: Its Organization and Functions* (Cambridge, Mass., 1962).

Mou, An-shih 牟安世, *Yang-wu yün-tung* 洋務運動 (The "foreign matters" movement), (Shanghai, 1961).

Murphey, Rhoads, *The Treaty Ports and China's Modernization: What Went Wrong?* (Ann Arbor, 1970).

Ocko, Jonathan K., *Bureaucratic Reform in Provincial China: T'ing Jih-ch'ang in Restoration Kiangsu, 1867-1870* (Cambridge, Mass., 1983).

Pong, David, "Keeping the Foochow Navy Yard Afloat: Government Finance and China's Early Modern Defense Industry, 1866-75," *Modern Chinese Studies,* 21:1:121-152 (1987).

Rowe, William T., *Hankow: Commerce and Society in a Chinese City, 1796-1889* (Stanford, 1984).

Saxton, Alexander P., *The Indispensable Enemy: Labor and the Anti-Chinese Movement in California* (Berkeley, 1971).

Spector, Stanley, *Li Hung-chang and the Huai Army* (Seattle, 1964).

Sun, Yü-t'ang 孫毓棠, "Chung-Jih Chia-wu chan-cheng ch'ien wai-kuo tzu-pen tsai Chung-kuo ching-ying ti chin-tai kung-yeh" 中日甲午戰爭前外國資本在中國經營的近代工業 (Modern industries operated by foreign capital in China before the Sino-Japanese War), *Li-shih yen-chiu,* 5:1-41 (1954).

Teng, Ssu-yü, and John K. Fairbank, *China's Response to the West* (Cambridge, Mass., 1954), I, chs. 5-14.

Teng, Yung-yüan T., "Prince Kung (I-hsin) and the Survival of Ch'ing Rule, 1858-1898" (Unpublished doctoral thesis, University of Wisconsin, Madison, Department of History, 1972).

Thomas, Stephen C., *Foreign Intervention and China's Industrial Development, 1870-1911* (Boulder, 1984).

Ts'ai, Shih-shan, "Chinese Immigration through Communist Chinese Eyes: An Introduction to the Historiography," *Pacific Historical Review,* XLIII:3:395-408 (Aug. 1974).

Tsiang, T. F., *Chung-kuo chin-tai-shih ta-kang* 中國近代史大綱 (An outline of Chinese modern history), (Taipei, 1959), ch. 3.

Wang, Erh-min 王爾敏, *Ch'ing-chi ping-kung-yeh ti hsing-ch'i* 清季兵工業的興起 (The rise of late Ch'ing military industry), (Taipei, 1963).

————, *Shang-hai Ko-chih Shu-yuan chih-lueh* (A Short History of the Shanghai Polytechnic Institute 1874-1911), (Hong Kong, 1980).

Wang, Hsin-chung 王信忠, "Fu-chou ch'uan-ch'ang chih yen-ko" 福州船廠之沿革 (The origin and development of the Foochow Dockyard), *Tsing-hua hsüeh-pao,* 8:1:1-57 (Dec. 1932).

Wright, Mary C., *The Last Stand of Chinese Conservatism: The T'ung-chih Restoration, 1862-1874* (Stanford, 1957).

Wright, Stanley F., *Hart and the Chinese Customs* (Belfast, 1950).

The interior of the Tsungli Yamen (Foreign Office)
with the leading ministers in discussion.

The Chinese Embassy to the United States
with its official guide, Anson Burlingame, in 1868.

Wei Yüan, an associate of Commissioner Lin
and a leading Modern Text scholar.

Wen-hsiang.

Tseng Kuo-fan.

Tso Tsung-t'ang.

Chang Chih-tung, long-time
governor-general at Wuhan.

Sir Robert Hart
(*Vanity Fair*, December 27, 1894).

Marquis Tseng.

Dr. William A. P. Martin, president of T'ung-wen Kuan.

Chinese students to the United States, 1872.

Chinese students' baseball club in front of the Chinese
Educational Mission in Hartford, Connecticut, 1878.

Li Hung-chang with British Prime Minister William Ewart Gladstone.

Empress Dowager Tz'u-hsi.

The Summer Palace (late Ch'ing dynasty, painting on paper), built with naval funds.

12

Foreign Relations and Court Politics, 1861-80

Foreign Affairs

During the period of the Self-strengthening Movement, China never lacked for advice and encouragement from foreign sources. Robert Hart, inspector-general of customs, and Thomas Wade, British minister at Peking, ceaselessly promoted "progress" in China. As a result of their constant urging, the Tsungli Yamen dispatched an exploratory mission to Europe in 1866.

THE PIN-CH'UN MISSION, 1866. In 1865 Hart submitted a memorandum to the Tsungli Yamen entitled "Observations by an Outsider" (*Chü-wai p'ang-kuan lun*), in which he stressed the advantages of railways, steamships, the telegraph, mining, and Western diplomatic practices. The last was particularly important, because the establishment of Chinese embassies abroad would enable Peking to bypass the headstrong foreign diplomats in China and make direct representations to foreign governments. "I regard representation abroad as of paramount importance," Hart wrote, "and as, in itself, progress, for while I thought that I saw in it one of China's least objectionable ways of preserving freedom and independence, I also supposed it would constitute a tie which should bind her to the West so firmly and commit her to a career of improvement so certainly as to make retrogression impossible."[1] A year later (1866) Thomas Wade

1. Robert Hart, "Notes on Chinese Matters," in Frederick W. Williams, *Anson Burlingame and the First Chinese Mission to Foreign Powers* (New York, 1912), 285.

also presented a communication to Prince Kung, entitled "A Brief Exposition of New Ideas" (*Hsin-i lüeh-lun*), which placed similar emphasis on the need for railways, the telegraph, mining, steamships, modern schools, Western-style army training, and diplomatic representation abroad. He admonished the Chinese to look not to the past for guidance but to the future for inspiration. In short, the message of Hart and Wade was progress through adoption of Western devices and products.

A direct consequence of their prompting was the decision by the Tsungli Yamen to send an informal exploratory mission to Europe under the guidance of Hart during his furlough in 1866. The mission was put under the leadership of Pin-ch'un, a 63-year-old ex-prefect who was Hart's secretary for Chinese correspondence at the time. He was given a temporary third civil service rank to add dignity to his mission, and was accompanied by several T'ung-wen kuan students. Prince Kung made it very clear that it was not a formal diplomatic mission, but only an informal information-gathering junket to the West. The mission visited London, Copenhagen, Stockholm, St. Petersburg, Berlin, Brussels, and Paris. Its novelty assured it a gracious welcome everywhere. Upon its return its members filled three diaries with detailed descriptions of what they had seen in Europe. Unfortunately their observations were mainly limited to Western social customs, tall buildings, the wonders of gaslight, elevators, and machines; only in passing did they touch upon the British Parliament and other political institutions.

THE BURLINGAME MISSION AND THE TREATY REVISION, 1868-70. While Western governments were committed to a "Cooperative Policy" in the 1860s, foreign traders and the Old China Hands in the treaty ports, especially Shanghai, never ceased to clamor for a more aggressive policy; they agitated to lay open all of China to Western commerce and to promote "progress" through adoption of railways, the telegraph, mining, and a host of other modern enterprises. Their pronouncements and the memoranda of Hart and Wade aroused fear in the Tsungli Yamen that the British might make many new demands on China during the forthcoming treaty revision—Article 27 of the Treaty of Tientsin with Britain specified that a revision might be made in ten years, i.e. 1868. To prepare for this ominous occasion, the Yamen anxiously polled the leading provincial authorities, who had become very powerful after the Taiping Revolution, for their views on such issues as were likely to arise: construction of telegraphs and railways, opening of mines, missionary activities, inland navigation, and Chinese diplomatic missions abroad.

Tseng Kuo-fan, the leading statesman of the period and governor-general at Nanking, suggested that China temperately but resolutely reject all foreign demands regarding railways, telegraphs, navigation of inland rivers, transportation of salt in Chinese waters, and opening of warehouses, because these activities would seriously hurt the livelihood of the Chinese people. On the other hand, mining was a potentially profitable venture, in which China might avail itself of foreign tools in the initial phase of operations. He definitely believed that China should open diplomatic missions abroad when suitable men and funds were available, but he showed no concern about missionary activities, believing that their alternate periods of success and decline—according to their funds, at the present on the ebb—made them generally ineffective and harmless. Tseng's views on these issues represented fairly well those of the more responsible and progressive officials.

Actually, the Tsungli Yamen's fear was unfounded. The British government looked without favor upon the Old China Hands' push for hasty and undue "progress" in China. On August 17, 1867 Lord Stanley, the foreign secretary, informed Minister Rutherford Alcock in Peking:

> We must not expect the Chinese, either the Government or the people, at once to see things in the same light as we see them; we must bear in mind that we have obtained our knowledge by experience extending over many years, and we must lead and not force the Chinese to the adoption of a better system. We must reconcile ourselves to waiting for the gradual development of that system, and content ourselves with reserving for revision at a future period, as in the case before us, any new arrangement which we may come to in 1868.[2]

The British government favored a "safe course" in China to consolidate the position already gained, and the use of moral influence, moderation, and patience to achieve future developments.

However, the Tsungli Yamen, without diplomatic agents in London, had no inkling of the British policy. But if it lacked intelligence reports, its common sense suggested application of the old principle of *i-i chih-i*—playing off the barbarians against one another. Prince Kung and Wen-hsiang, taking a hint from the retiring American minister Anson Burlingame that he would happily mediate as if he were China's envoy in cases of dispute with foreign powers, invited him to join a roving diplomatic mission[3] to the West to dissuade European and American governments from forcing the pace of Westernization in China. A born orator from

2. Hsü, *China's Entrance*, 167.
3. At a salary of £8,000 a year plus expenses.

Massachusetts, Burlingame declared: "When the oldest nation in the world, containing one-third of the human race, seeks, for the first time, to come into relations with the West, and requests the youngest nation through its representative, to act as the medium of such change, the mission is one not to be solicited or rejected."[4]

Burlingame, together with a Manchu and a Chinese co-envoy,[5] carried the mission to the United States in May 1868. The governor of California extended a warm welcome, describing Burlingame as "our guest, the son of the youngest, and representative of the oldest, government." In response Burlingame declared: "The hour has struck, the day has come" when China welcomed "the shining banners of Western civilization." In New York similar grandiloquence proclaimed the message that China was willing to invite missionaries to "plant the shining cross on every hill and every valley." Burlingame's eloquence and charm captivated the Americans, and perhaps himself; after a flattering audience with President Andrew Johnson, he signed a treaty with Secretary of State Seward on July 28, 1868—on his own authority, without the prior approval of the Chinese government. It committed the United States to a policy of noninterference in the development of China, and stipulated the sending of Chinese consuls and laborers to the United States, and reciprocal rights of residence, religion, travel, and access to schools in either country. Though not consulted in advance, Peking was too grateful to disown the treaty.

The mission moved on to London where it was received by Queen Victoria. Lord Clarendon, foreign secretary after Stanley, reaffirmed the policy of not forcing China "to advance more rapidly than was consistent with safety and with due and reasonable regard for the feelings of her subjects," and of deprecating any European pressure for adoption of new systems in China.[6] At Berlin, Burlingame committed Prince Bismarck to a statement that the North German Confederation would deal with China in whatever manner Peking considered in its best interest. At St. Petersburg, after an audience with the tsar, Burlingame contracted pneumonia and died on February 23, 1869. The mission was then carried on by the two co-envoys, visiting Brussels and Rome and then returning to China in October 1870.

Insofar as its immediate objectives were concerned, Burlingame's mission was a great success, for it did commit Western powers to a policy of restraint and moderation in the forthcoming treaty revision. Yet from a long-range standpoint, it encouraged the growth of conservatism in China.

4. *Foreign Relations of the United States,* 1868, I, 494.
5. Chih-kang and Sung Chia-ku.
6. Hsü, *China's Entrance,* 169.

The mandarins, who had spent 160,000 taels on the mission, came to believe that foreigners after all could be managed at a price. They became more complacent and less responsive to outside stimuli. The mission may have unwittingly produced a retarding effect on the modernization of China.

The actual negotiations on the treaty revision were carried out as between equals without the threat of guns and ships for the first time since the Opium War. The resultant Alcock Convention of 1869 allowed China to establish a consulate in Hong Kong, to increase the import duty on opium from 30 to 50 taels per picul (133⅓ lbs.) and export duty on silk from 10 to 20 taels, and to limit the most-favored-nation treatment so that the British must accept the conditions under which certain rights were granted to other powers, if they (the British) wished to claim the benefits of the same rights. These terms the British traders strongly opposed, especially the provision for the Chinese consul in Hong Kong, who was said to resemble a revenue officer and a spy. Great pressure subsequently led the British government to reject ratification of the Alcock Convention, an act which aroused deep Chinese disappointment and bitterness. The Tsungli Yamen felt betrayed in its trust in foreign good will, while conservatives and xenophobes were all too ready to point out that foreigners took but never gave, and that the minute a treaty slightly unfavorable to them was negotiated they disowned it. A new tide of anti-foreignism rolled in as the decade of the 1870s opened.

THE TIENTSIN MASSACRE, 1870. Even as Burlingame was inviting missionaries to plant the shining cross on every hill and every valley of China, a rash of anti-Christian activities broke out across the face of the country. Christianity, as a heterodox faith, was antithetical to Confucianism, and its practice of mixed congregations, which ran counter to the Chinese custom of avoiding open contact between men and women, instigated rumors of immoral and perverted behavior. Missionary protection of Chinese converts from local justice, and the construction of churches in disregard of time-honored concepts of geomancy (*feng-shui*) caused endless irritation to Chinese sensibilities.[7] Anti-Christian tracts appeared frequently; and a widely circulated one was entitled "A True Record to Ward Off Evil Doctrine" (*P'i-hsieh chi-shih*) written in the early 1860s by one who called himself "the most heartbroken man in the world."[8] Eruption of anti-

7. For a study of the missionary problem, see Paul A. Cohen, *China and Christianity: The Missionary Movement and the Growth of Chinese Anti-Foreignism, 1860-1870* (Cambridge, Mass., 1963), chs. 3-7.

8. John K. Fairbank, "Patterns Behind the Tientsin Massacre," *Harvard Journal of Asiatic Studies,* 20:3-4:501 (Dec. 1957).

missionary activities under gentry instigation was common, eliciting a ready reprisal from foreign representatives. The British Minister at Peking, Rutherford Alcock, said pompously: "Demands once made, there is no retreat possible without serious loss of prestige and influence, on which everything depends in the East."[9] Thus, in August 1868, when a mob in Yangchow plundered and set fire to the new missionary station established by Rev. J. Hudson Taylor of the China Inland Mission, Alcock sent Consul W. H. Medhurst and four ships to Nanking to pressure Governorgeneral Tseng Kuo-fan to cashier the Yangchow officials and pay a compensation. Such gunboat policy and humiliating chastisement produced quick results but invariably inflamed public feelings and excited xenophobic sentiments. Even London considered the Alcock-Medhurst action contrary to British policy.[10]

The incident that touched off a major anti-Christian riot was the Tientsin Massacre of 1870. It is no coincidence that Tientsin was the scene of the outburst, for it had been twice occupied by foreign troops during the negotiations of the Treaties of Tientsin in 1858 and the Conventions of Peking in 1860. Even after the peace settlement, the British and the French continued to station five to six thousand troops there as a guarantee of China's fulfillment of treaty obligations. Although the French troops evacuated in November 1861 and the British in May 1862, portions of the Anglo-French forces remained in Taku until 1865. The presence of foreign troops was always a cause of irritation, and additional fuel came from the French seizure in 1860 of the imperial villa in Tientsin[11] for a consulate. Then, in 1869, the church Notre Dame des Victoires which ran an orphanage was constructed on the site of a razed Buddhist temple. Because few Chinese would send orphans to a foreign establishment, the nuns offered a premium for each child, thereby giving incentive to rascals, known as the "child brokers," to kidnap children. Moreover, the nuns were particularly interested in baptizing the sick and dying children. The high mortality rate and the offer of premium inevitably aroused suspicion. Rumor spread that behind their high walls and closed gates, the foreigners bewitched the children, mutilated their bodies, and extracted their hearts and eyes to make medicine. An antiforeign riot loomed on the horizon. Ch'ung-hou, the superintendent of trade for the Three Northern Ports, inspected the orphanage and found no truth to the wild charges, but public feelings continued to run high. Then came the truculent French consul, Henri Fontanier, and his chancellor, M. Simon, armed with pistols,

9. *Ibid.*, 482-83.
10. *Ibid.*, 488.
11. Wang-hai lou, "Sea-viewing Pavilion."

to demand justice for the sisters. Angry at the sight of the seething mob, which the district magistrate tried to disperse, Fontanier fired a shot which missed the magistrate but killed his servant. The mob boiled out of control, killed Fontanier and his assistant, and burned the church and the orphanage. Ten sisters, two priests, and two French officials lost their lives. Three Russian traders were killed by mistake, and four British and American churches were destroyed. Foreign gunboats quickly anchored off Tientsin, and strong protests from seven foreign ministers were lodged with the Tsungli Yamen, demanding redress and punishment of the rioters.

The court appointed its most venerable servant, Tseng Kuo-fan, now governor-general of Chihli, to investigate the case. On sick leave at Paoting, the 60-year-old statesman doubted he had the stamina to survive the strenuous assignment. At Tientsin, he found the situation far more knotty than he had anticipated. The French chargé d'affaires, Count Julien de Rochechouart, demanded the lives of General Ch'en Kuo-jui and the Tientsin prefect and magistrate, while the conservative Chinese officials and literati clamored against any concession or appeasement. Tseng knew that to avoid a rupture with France he had to be impartial in his investigation, but that to be impartial in this case was to invite attacks from the unrelenting conservatives. In short, he had to choose between integrity and loss of reputation. Here again Tseng's character and courage were manifest. Rather than cater safely to public sentiment, he risked his political future with the candid recommendation that the absolute truth of the case be established. He advised the court that Britain, the United States, and Russia be indemnified first, to dissociate them from the French cause. Next he made a personal inspection of the orphanage and learned firsthand from the one hundred and fifty children that they were not kidnapped but had been sent by their families voluntarily. Tseng asked the court to restore the reputation of the nuns by issuing a proclamation denying any truth to the rumor about the mutilation of bodies and extraction of hearts and eyes.

To settle the case, Tseng recommended heavy penalties for those involved in the riot: dismissal of the circuit intendant, the Tientsin prefect, and the district magistrate; capital punishment for fifteen chief instigators and banishment for twenty-one. If these arrangements would not satisfy the French, Tseng stated, greater punishment might be imposed.

The conservatives immediately branded Tseng a traitor. Grand Secretary Wo-jen ridiculed the idea of bargaining with the French about the penalty, arguing pointedly that since the founding of the dynasty there had never been punishment without criminal evidence. The court, too, found Tseng's recommendations somewhat unpalatable. At this juncture, Li Hung-chang, governor-general at Wuhan, sent in a timely memorial adopting a median stance; he suggested that the civilized Christian state of France probably had no interest in imposing too heavy a penalty on Chinese officials, and that capital punishment of eight and exile of twenty would suffice. The court transferred Li to Tientsin to take over the investigation. Tseng was sent to Nanking as governor-general. Frustrated and embittered he lapsed into a state of great distress and agony. When he wrote to friends he often inscribed on the envelope: "I fear public criticism without, and am conscience-stricken within."

Li Hung-chang speedily settled the case with the French, agreeing to pay a compensation of 400,000 taels for the loss of lives and properties, to send a mission of apology, to banish the Tientsin prefect and magistrate, and to sentence eighteen to capital punishment and twenty-five to hard labor on the frontier. The apology mission, led by Ch'ung-hou, reached France, only to find the French government too involved in the Prussian war to receive it. On his way home via New York, however, Ch'ung-hou was summoned back to France, where the provisional president, M. Thiers, received him at Versailles on November 23, 1871. Thiers announced that France was not interested in decapitating the Chinese wrongdoers, but in lasting peace and order, and the case was officially closed with Thiers's acceptance of the Letter of Apology from the Chinese emperor.[12]

THE AUDIENCE QUESTION, 1873. Although foreign diplomats took up residence in Peking in 1861, they were continuously denied audience with the emperor. Prince Kung received them in his capacity as regent and explained that audience with the emperor was inadvisable during his minority, while that with the two dowagers would cause great inconvenience due to different social customs. The real reason for the Chinese postponement was that foreign diplomats would claim immunity from kowtow under the 1858 Treaties of Tientsin, which explicitly provided that envoys not be required to perform ceremonies derogatory to their honor and dignity. Foreign ministers themselves had averred repeatedly that they would not perform the kowtow in any future audience.

The Tsungli Yamen's tactics put off, but did not solve, the question of audience, for the boy emperor would grow up. During the discussion of treaty revision in 1867, the Yamen solicited views from leading provincial authorities on the question. Li Hung-chang, governor-general at Wuhan, declared that foreign diplomats should be allowed to perform such rituals

12. Knight Biggerstaff, "The Ch'ung Hou Mission to France, 1870-71," *Nankai Social and Economic Quarterly*, 8:3:633-47 (Oct. 1935).

as they would before their own rulers. Tseng Kuo-fan, governor-general at Nanking, asserted that just as Emperor K'ang-hsi (1662-1722) had treated Russia as an enemy state on equal footing rather than as an inferior dependent state, the court should also regard foreign ministers as envoys from enemy states of equal status exempt from Chinese customs. On the other hand, a number of conservative officials argued that China should not change its institutions and practices merely to suit the convenience of foreigners.

In 1872 the emperor reached his majority and was married, but no foreign diplomats were invited to take part in the celebration, thus avoiding protocol problems. In February of the following year he inaugurated his personal rule. Foreign representatives renewed their demand for an audience. Unable to defer the issue longer, the Tsungli Yamen conducted protracted discussions with the diplomats as to the proper rituals. They finally agreed that the foreign representatives should bow instead of kowtow during the audience. The Japanese foreign minister, Soejima Taneomi, who had arrived to exchange ratifications of the treaty of 1871 (next chapter), insisted that his ambassadorial rank entitled him to an audience before the Western diplomats, who all held the ministerial rank. It was an obvious attempt to impress others with Japanese mastery of European diplomatic practices and to assert Japanese equality with the Western powers.

On Sunday, June 29, 1873, the foreign diplomats were asked to convene at 5:30 a.m., but it was not until nine o'clock that they were received by Emperor T'ung-chih at the Pavilion of Violet Light. The Japanese foreign minister was received first, followed in order of seniority by the Russian minister (Vlangaly), the American minister (Low), the British minister (Wade), the French minister (de Geofroy), the Dutch minister (Ferguson), and the German interpreter (Bismarck). They laid their credentials on a table before the emperor, who expressed, through Prince Kung, his amicable feelings toward the foreign sovereigns represented there. The whole audience, for which the Western diplomats had waited twelve years, was over in half an hour.[13] It was an anticlimax, the more so when the foreign representatives later discovered that the pavilion in which they were received was also used for the reception of tributary envoys.[14]

13. For an interesting account of the audience, see *British Parliamentary Papers*, China, No. 1 (1874) *Correspondence respecting the Audience granted to Her Majesty's Minister and the other Foreign Representatives at Peking by the Emperor of China*.
14. Banquets were given there for the tributary envoys from Korea, Liu-ch'iu, Laos, Siam, and Annam in 1839-43, 1845-48, and 1864. See Fairbank (ed.), *The Chinese World Order*, 262.

THE MARGARY AFFAIR, 1875. The Great Depression in Europe in the early 1870s brought on by the tariff war adversely affected the China trade, which declined steadily after 1872. The Hong Kong and Shanghai Banking Corporation reported a loss and declared no dividend in 1874 and 1875 for the first time in its history. To brighten the trade prospects, the British concocted a scheme to open a back door to interior China by constructing a railway and trade route from Burma to Yunnan.

Lord Salisbury, head of the India Office in the Disraeli ministry of 1874, ordered the Indian government to undertake the survey of the proposed route and requested the Foreign Office to instruct the minister in Peking to seek Chinese permission for the entry of an exploratory mission from Burma.

Personally skeptical of the commercial possibilities of such a route, Wade found, to his surprise, that the Chinese government not only readily assented to his request but also agreed to let a British vice-consul, the 28-year-old Augustus Margary, travel up the Yangtze River to meet the mission. Though aware of the presence of guerrilla bands in the Chinese-Burmese border area and their hostility to foreigners, Margary ventured on to Bhamo on the frontier despite warnings by local Chinese officials, to await the mission from Burma under Colonel Horace A. Browne. There, on February 21, 1875, Margary was ambushed and killed.

International law provides that when a foreigner exposes himself to danger at his own risk, it is no responsibility of the host country to guarantee his safety. The British government, however, held the Chinese government responsible for the murder and instructed Wade to obtain redress. Capitalizing on this occasion, the ambitious Wade demanded an investigation of the murder, an indemnity for the bereaved family, another expedition, and trial of the acting governor-general of Yunnan and Kweichow in whose jurisdiction the incident took place; he also raised a number of extraneous issues such as the future audience procedure, transit dues, better etiquette in the treatment of foreign diplomats, and an apology mission to Britain. Peking readily assented to an investigation and an indemnity, but frowned on the other unrelated questions. Impetuously Wade withdrew his legation to Shanghai, threatening to break off relations. To avoid a rupture, the court on August 29, 1875, authorized the dispatch of an apology mission to Britain under the leadership of Kuo Sung-tao and sent its trusted foreign servant, Robert Hart, to Shanghai to persuade Wade to resume discussions. Hart tactfully intimated that if the negotiations were not reopened in China, Kuo might start diplomatic proceedings in London, where a settlement would exclude Wade from all claims to credit. Wade agreed to meet with Li Hung-chang at the summer resort of Chefoo. On September 13, 1876, the Chefoo Convention was concluded to settle the Margary case. Part I dealt with the dispatch of an apology mission to Britain and the payment of 200,000 taels to the bereaved family. Part II dealt with the preparation of an etiquette code between the Chinese government and foreign diplomats. Part III dealt with the opening of four new ports and the limitation of *likin*-free areas to treaty ports. However, this convention failed of British ratification until 1885, due to opposition from (1) the United States, Germany, France, and Russia, which criticized Britain's unilateral action; (2) the British mercantile communities, which clamored for a complete abolition of the *likin*; and (3) the Indian government, which protested the increase in opium tax.

The most significant outcome of the Margary incident was the dispatch of the mission of apology, which became the first resident Chinese legation abroad. The leader, Kuo Sung-tao, the progressive sixty-year-old friend of Li Hung-chang, was given the title of vice-president of the Board of War prior to his departure for Britain. After the presentation of the emperor's Letter of Apology to Queen Victoria on February 8, 1877, he set up the first Chinese embassy in London. In the next two years, other legations were also established in Paris, Berlin, Spain, Washington, Tokyo, and St. Petersburg. By 1880 China had belatedly taken her place in the family of nations.

China's slowness in reciprocating the Western practice of diplomatic representation may be attributed to several causes. Institutionally, it had never dispatched permanent, resident embassies abroad but only *ad hoc* missions, which were sent out either in times of strength and prosperity to spread the prestige of the Son of Heaven and to bring outlying states into the tributary system, or in times of weakness and disorder to beg for peace or alliance with barbarian tribes. Psychologically, the majority of mandarins eschewed foreign affairs as beneath their dignity and foreign assignment as a form of banishment; men of keen political acumen strove to steer clear of foreign associations. Burlingame's two associates fared badly after their return: one was sent to an obscure post in western China, and the other spent his life on a Mongolian frontier, as if they had been contaminated by their foreign trip. The censors, the Hanlin scholars, and the conservative gentry and officials ceaselessly harped on the theme that historically barbarians were always transformed by Chinese ways, not the Chinese by barbarian ways. They promoted conservatism against modernization and condemned foreign association as disgraceful. So powerful was the conservative atmosphere that the psychological inertia to innovation was immense. It took China more than fifteen years to overcome this barrier and reciprocate the Western practice of diplomatic representation.

The Imperial Woman and Her Politics

For nearly a half-century from 1861 to 1908, the Empress Dowager Tz'u-hsi ruled supreme over China. Endowed with large measures of determination, willpower, ability for quick decision, and not a little native intelligence, still she was without much education and mental breadth, and was totally uninformed on the nature of the modern world. Basically a narrow-minded, selfish woman, she placed her own interest above all else, irrespective of consequences to the dynasty and the state. To a large degree she must be held responsible for the failure to regenerate the dynasty and modernize the country. The question naturally arises as to how a single woman could wield such supreme power and remain at the pinnacle for so long, in contravention of dynastic laws and practices. The answer may partially be found in her consummate skill in political manipulation.

THE CHASTISEMENTS OF PRINCE KUNG. It has been stated in the previous chapter that the empress dowager and Prince Kung cooperated with each other out of expediency during and after the _coup d'état_ of 1861. Using him as a front man in dealing with foreign powers and winning domestic sympathies, the dowager thus bought time to learn statecraft. The prince, of course, needed her to secure a powerful and exalted position. An ambitious man, he had actually aspired to be sole regent to the boy emperor T'ung-chih—much as Dorgan had been at the beginning of the dynasty when Emperor Shun-chih was a minor—and let the two dowagers nominally administer the state behind a silk screen. But Tz'u-hsi was too shrewd to let him be the only regent. Cleverly she showered him with high honors and offices, while jealously holding the ultimate power of state in her own hands. His dream frustrated, the prince could not help harboring some disappointment behind his mask of apparent success.

Prince Kung,[15] sixth son of Emperor Tao-kuang (1820-50) and a younger brother of the late emperor Hsien-feng (1851-61), was a smart man with a quick but lightly cultivated intelligence. After the coup of 1861, he rose to leadership in the government, enjoying the "trust," however fleeting, of the two dowagers and the support of foreign diplomats. His power and status, though not as great and exalted as Dorgan's at the beginning of the dynasty, was unrivaled at court. As prince regent, grand

15. Personal name: I-hsin.

councillor, head of the Tsungli Yamen, and chief minister of the Imperial Household, he was the most sought-after man in Peking. Daily hundreds of officials and visitors queued up outside his office to await his decisions and favors, and not a few bribed their way into his circle. Elated with success and drunk in the enjoyment of power, his arrogance overrode his prudence, and even the dowagers found him overbearing during audiences. Attempts by friends to urge upon him caution, self-control, and restraint went unheeded, and an atmosphere of impending calamity encircled him. In 1865 a Hanlin compiler moved to impeach him, and the Dowager Tz'u-hsi, who felt capable enough in state administration by then, decided to chastise him. To the palace she summoned Grand Secretary Chou Tsu-p'ei and high officials of the boards of Civil Offices, Revenue, and Punishments to fix Prince Kung's guilt in accepting bribes, nepotism, usurpation of power, clique-formation, and overbearance. However, these officials dared not intervene in what they considered to be primarily a family quarrel between an imperial brother and sister-in-law, and begged the dowagers to decide the issue themselves. Tz'u-hsi, enraged at their timidity, drafted an edict herself—one noted for its countless wrong characters—dismissing the prince from all official posts. Suddenly deprived of forceful leadership, the government's normal operations were seriously handicapped. Princes Tun and Chün, as well as other dignitaries, interceded with the dowagers on Kung's behalf, pleading the importance of maintaining "family harmony" before the public. Tz'u-hsi, realizing that the objective of chastisement had been achieved and that she still needed Prince Kung to deal with the foreigners, restored him to the superintendency of the Tsungli Yamen. In this way, she demonstrated her leniency to Kung, preserved the "face" of those who interceded, and showed her absolute power. When Prince Kung penitently went to the palace to thank her for the partial restoration to favor, she in another gesture of generosity reinstated him as grand councillor. The title of prince regent (*i-cheng wang*), however, was withheld from him. Having been taught a lesson, the deflated prince lost some of his zeal in state affairs and became more restrained in conduct.

A second blow struck him in 1869, in connection with the eunuch An Te-hai, a confidant of Tz'u-hsi's since before the coup of 1861. It was no secret that he had a hand in the decision to deny Prince Kung the title of regent after the first chastisement. Kung became the more incensed by the increasing numbers of opportunistic officials flocking to An's door to curry favor. The chance for revenge came in 1869, when the eunuch in question left Peking[16] on a purchasing mission for Tz'u-hsi, contrary to the

16. His destination, according to some accounts, was Kwangsi, and according to others, Soochow.

dynastic rule that no eunuch was allowed to leave the capital on pain of decapitation. En route in Shantung, An was captured by Governor Ting Pao-chen, who sent for instructions from the court. Tz'u-hsi was caught off guard, while Tz'u-an, the other dowager, in consultation with the Grand Council under Prince Kung, ordered his immediate beheading. Tz'u-hsi blamed Kung for maneuvering behind the scenes, straining their relations more than ever. In frustration, Prince Kung resigned himself to a life of relative inactivity. When his able assistant, Wen-hsiang, passed away in 1876, the government was deprived of vital leadership.

With the death of the dowager Tz'u-an in April 1881—she was reputedly poisoned by Tz'u-hsi—Prince Kung lost a supporter and his position in government became even more tenuous. During the French war in 1884 the prince, accused by conservative officials of indecision, and four other grand councillors were summarily dismissed by her from the Grand Council. After this third chastisement, Prince Kung totally lost heart for state affairs and went into an eclipse. Prince Li became the nominal ranking grand councillor, while Prince Chün, brother-in-law of the dowager, became head of the new Board of Admiralty (*Hai-chün ya-men*) in 1885. Both were men of inconsequential ability; henceforth the government drifted without able leadership and effective direction.

MANIPULATION OF IMPERIAL SUCCESSION. In 1872 Emperor T'ung-chih reached his majority and selected as his empress a young woman recommended by the dowager Tz'u-an, rather than the one chosen by his mother Tz'u-hsi. The latter resorted to all manner of devices to block the emperor from visiting with his empress, encouraging him instead to frequent the quarters of the consort of her choice. Annoyed by her meddling, the emperor retaliated by boycotting both his empress and the consort, finding consolation instead in outside pleasure quarters. In February 1873 he began his personal rule and, tired of his mother's interference, struck upon the idea of reconstructing the Summer Palace (*Yüan-ming yüan*), which Lord Elgin had burned in 1860, as a place for her retirement. But the construction was forced to a halt in September 1874 by a notorious scandal involving an opportunistic Canton merchant and a French lumber dealer. Shortly afterwards, the young emperor became very ill and died on January 12, 1875, at the age of nineteen *sui*. During his illness, the dowager Tz'u-hsi did little to help him recover, but everything to hasten his end.

Emperor T'ung-chih died without issue, although it was known that his empress was pregnant. The question of imperial succession became a mat-

ter of great delicacy and intrigue. Tz'u-hsi quickly saw her opportunity to
regain the regency. Even before the late emperor's death she had begun
her machinations, instigating court officials to request that the two dowagers again administer state affairs behind the silk screen. She realized that
the choice of an adult prince for the throne would eliminate the need
for a regent, while selection of a minor prince one generation below the
deceased emperor would make her an "empress dowager-grandmother,"
i.e. twice removed from the legal source of power. Both courses of action
had to be avoided. To maintain her regency, the new ruler had to be a
minor, of the same generation as the late emperor, so that she would be
just once removed. With these considerations, she rejected the late emperor's deathbed choice[17] and dismissed Prince Kung's suggestion that
succession be delayed until after the empress had borne a child. At a
council of twenty-seven princes on January 12, 1875, she autocratically
announced her choice: her nephew, Tsai-t'ien, son of her sister and Prince
Chün (I-huan), a boy of four *sui* of the same generation as the late em
peror. The passing of the throne between members of the same generation
violated the dynastic laws of succession, yet none dared to challenge her.
Only a foolhardy Chinese secretary of the Board of Civil Offices, Wu
K'o-tu, committed suicide in protest—an act known as the "death remonstration."

Ironically, the reign (1875-1908) of the new emperor was designated
Kuang-hsü, or Glorious Succession. On January 15, 1875, the two dowagers graciously bowed to the "request" of the princes and high officials
to serve as co-regents again during his minority. An edict was issued to the
effect that the power of state would be returned to the emperor as soon as
he came of age and that his future son would be adopted as the son of
the late emperor. By this maneuvering, Tz'u-hsi assured herself of another
term of regency, exercising the power of state behind the silk screen once
more. But one thing she could not control: the growing up of the boy
emperor.

In 1886 Emperor Kuang-hsü reached the age of sixteen *sui* and announced his intention to begin personal rule the following year. Knowing
well the dowager's reluctance to renounce power, his father Prince Chün
tactfully advised postponement of the take-over. On February 7, 1887,
the emperor reached his long-awaited majority but was denied personal
rule for two years, during which period he was to learn from the dowager
the art of government (*hsün-cheng*). Finally, on March 4, 1889, the dow-

17. Tsai-chu.

ager officially "retired" to the Summer Palace, but no one doubted that she took with her the ultimate power of state. She forced the emperor to marry a cousin of her choice to insure her supervision of, and direct access to, policy matters. She maintained an iron control of the palaces through her confidant, the eunuch Li Lien-ying, and of the government through her trusted grand councillor, Sun Yü-wen. The emperor was but a figure-head.

To free the court from her manipulations, Emperor Kuang-hsü and Prince Chün revived the project of reconstructing the Summer Palace (*I-ho-yüan*), hoping that she might enjoy herself there and relinquish her hold on state affairs. The funds for the construction, some 30 million taels, came from the budget of the Board of Admiralty, which Prince Chün headed. Because of this misuse of funds, no new ships were bought after 1888; not surprisingly the Chinese navy was disgracefully defeated in the Japanese war of 1894-95.

One may ask how one woman could wield such great power and why the officials did not refuse to obey her. Her success can be attributed to three stratagems. First, while violating dynastic laws and precedents her-self she obliged all other Manchus to follow them strictly. She governed the royal members inexorably with the imperial family law, and mercilessly sent offenders to the Imperial Clan Court for punishment. She treated them with severity to strike terror into their hearts and reduce them to abject submission. Secondly, to the Chinese officials she stressed the Con-fucian concepts of proper relationship between the ruler and the subjects and the importance of filial piety. She said in effect: if the two boy em-perors obey and honor me so exactly, how much more should you officials! Thirdly, fully recognizing the Manchu degeneration, she relied on able Chinese such as Tseng Kuo-fan, Tso Tsung-t'ang, and Li Hung-chang, even though she was concerned about their rising power, their association with foreigners, and their control of the new army and navy and modern enterprises. To safeguard her position, she humored them with high posi-tions and honors but kept them in check by secretly encouraging conserva-tive forces to attack them. With these methods the dowager successfully dominated China for nearly half a century.[22] Yet her success was achieved at a high cost to the dynasty and the country. Under her despotic rule, the dynasty failed to achieve a regeneration, and China sank deeper and deeper into the slough of foreign imperialism. Barely three years after her death in 1908 the Ch'ing dynasty was overthrown.

22. Li Fang-ch'en, 381-84.

Further Reading

Biggerstaff, Knight, "The Official Chinese Attitude Toward the Burlingame Mission," *American Historical Review,* 41:4:682-702 (July 1936).

———, "The First Chinese Mission of Investigation Sent to Europe," *Pacific Historical Review,* 6:4:307-20 (Dec. 1937).

———, "The Ch'ung Hou Mission to France, 1870-71," *Nankai Social and Economic Quarterly,* 8:3:633-47 (Oct. 1935).

Bland, J. O. P., and E. Backhouse, *China under the Empress Dowager, Being the History of the Life and Times of Tzu Hsi* (London, 1910).

British Parliamentary Papers, China, No. 1 (1874), *Correspondence respecting the Audience granted to Her Majesty's Minister and the other Foreign Representatives at Peking by the Emperor of China.*

Buck, Pearl S., *Imperial Woman: Story of the Last Empress of China* (New York, 1955).

Chih-kang 志剛, *Ch'u-shih t'ai-Hsi chi* 初使泰西記 (The first embassy to the West), 4 chüan, 1877.

Cohen, Paul A., *China and Christianity: The Missionary Movement and the Growth of Chinese Antiforeignism, 1860-1870* (Cambridge, Mass., 1963).

Der Ling, Princess, *Old Buddha* (New York, 1932).

Fairbank, John King, "Patterns Behind the Tientsin Massacre," *Harvard Journal of Asiatic Studies,* 20:3-4:480-511 (Dec. 1957).

———, K. F. Bruner, and E. M. Matheson (eds.), *The I.G. in Peking: Letters of Robert Hart, Chinese Maritime Customs, 1868-1907* (Cambridge, Mass., 1975), 2 vols.

Frodsham, J. D., *The First Chinese Embassy to the West: The Journals of Kuo Sung-tao, Liu Hsi-hung, and Chang Te-yi* (Clarendon, 1974).

Haldane, Charlotte, *The Last Great Empress of China* (Indianapolis, 1965).

Hao, Yen-p'ing, and Erh-min Wang, "Changing Chinese Views of Western Relations, 1840-95" in John K. Fairbank and Kwang-ching Liu (eds.), *The Cambridge History of China* (Cambridge, Eng., 1980), Vol. 11, 142-201.

Hsü, Immanuel C. Y., *China's Entrance into the Family of Nations: The Diplomatic Phase, 1858-1880* (Cambridge, Mass., 1968), Pt. III.

Latourette, K. S., *A History of Christian Missions in China* (New York, 1929).

Li, Shih-yüeh 李時岳, "Chia-wu chan-cheng ch'ien san-shih-nien chien fan-yang-chiao yün-tung" 甲午戰爭前三十年間反洋教運動 (The antiforeign religion movement during the thirty years before the Sino-Japanese war), *Li-shih yen-chiu,* 6:1-15 (1958).

Liao, Kuang Sheng, *Antiforeignism and Modernization in China, 1860-1980* (New York, 1984).

Lü, Shih-ch'iang 呂實強, *Chung-kuo kuan-shen fan-chiao ti yüan-yin, 1860-1874* 中國官紳反教的原因 (The causes of the anti-Christian movement among Chinese officials and gentry, 1860-1874), (Taipei, 1966).

Michie, A., *The Englishman in China during the Victorian Era, As Illustrated in the Career of Sir Rutherford Alcock K.C.B., D.C.L., Many Years Consul and Minister in China and Japan* (London, 1900), 2 vols.

Miller, Stuart Creighton, *The Unwelcome Immigrant: The American Image of the Chinese, 1785-1882* (Berkeley, 1969).

Pelcovits, Nathan A., *Old China Hands and the Foreign Office* (New York, 1948).

Ross, John, *Chinese Foreign Policy* (Shanghai, 1877).

Tabohashi, Kiyoshi 田保橋潔, "Shin Dōchichō gaikoku kōshi no kinken" 清同治朝外國公使の觀見 (The audience of foreign ministers in the T'ung-chih period of the Ch'ing dynasty), *Seikyū gakuso*, 6:1-31 (Nov. 1931).

Tsiang, T. F., "Sino-Japanese Diplomatic Relations, 1870-1894," *The Chinese Social and Political Science Review*, 17:1:1-106 (April 1933).

Wang, S. T., *The Margary Affair and the Chefoo Convention* (New York, 1939).

Warner, Marina, *The Dragon Empress: The Life and Times of T'z'u-hsi, Empress Dowager of China, 1835-1908* (New York, 1972).

Williams, F. W., *Anson Burlingame and the First Chinese Mission to Foreign Powers* (New York, 1912).

Wright, Mary C., *The Last Stand of Chinese Conservatism: The T'ung-chih Restoration, 1862-1874* (Stanford, 1957), chapters 10-11.

Yüan, Ting-chung 袁定中, "Na-la-shih fan-tung ti i-sheng" 那拉氏反動的一生 (The reactionary life of the empress dowager), *Li-shih yen-chiu*, 10:31-41 (1958).

13

Foreign Encroachment in Formosa, Sinkiang, and Annam

The last three decades of the 19th century in China were a period of accelerated foreign imperialistic encroachments. Europe, experiencing "a generation of materialism," was propelled by the forces of nationalism, evangelism, capitalism, and Darwinism into heightened activity in Asia, Africa, and the Middle East. Economically, not only had Britain and France successfully industrialized, but also Germany, Italy, and the United States. This created a need for overseas markets for their surplus goods and a need for a source of raw materials. Culturally, Social Darwinism was the order of the day; it sanctioned overseas expansion with the philosophy that nations as well as species struggled to exist and that only the strongest was fit to survive. Religiously, the churches and denominations were fired with the zeal of a divine mission to evangelize the heathens. To all this was added the proud, self-righteous feeling of racial superiority, reflected in the term the "White Man's Burden."[1]

To be sure, most of these forces had previously existed, but several developments in the 1860s gave them effective direction and impetus: the end of the Civil War in the United States in 1865, the Meiji Restoration in Japan in 1868, the unifications of Italy and Germany in 1870, and the rise of the Third Republic in France in the same year. These epochal

---

1. For two excellent studies of imperialism, see William Langer, *The Diplomacy of Imperialism, 1890-1902* (New York, 1950); Carlton J. H. Hayes, *A Generation of Materialism, 1871-1900* (New York, 1941).

events liberated centrifugal energies for externally oriented action, while the completion of the Suez Canal in 1869 facilitated European expansion in Asia. Now not only the older aggressor nations—such as Britain, France, and Russia—but also the latecomers—notably Japan and Germany—executed their imperialistic designs. In contrast, China under the empress dowager was making little headway in self-improvement and regeneration; the dynastic strength steadily declined after a brief upsurge during the T'ung-chih (1862-74) period. Taking advantage of China's weakness, foreign powers nibbled away the frontier areas and tributary states, following with frontal thrusts to the heart of the "Sick Man of Asia." By the close of the 19th century, China faced the ominous prospect of partition.

Japanese Aggression on Formosa, 1871-74

Official Sino-Japanese relations had been held in abeyance for the three hundred years preceding 1871. Japan was a tributary state of China for a time during the Ming period (1368-1643). The Japanese shogun, Ashikaga Yoshimitsu, accepted the tributary status in order to enrich his coffers from trade—from 1433 to 1549 eleven tribute and trade missions sailed to China. Subsequently, however, nationalistic Japanese statesmen found such relations humiliating, and discontinued the practice after the middle of the 16th century, thus ending official contact with the mainland. But Japanese pirates, known as *Wakō* (*Wo-k'ou* in Chinese, meaning "dwarf pirates"), continued to disturb the China coast and made themselves a nuisance to the Ming dynasty. After the establishment of the Ch'ing dynasty in 1644, no resumption of official relations was made; the Manchu rulers, unlike the Ming emperors, never attempted to bring Japan into the tributary system.

With the opening of China and Japan to Western commerce and diplomacy in mid-19th century, Japanese traders began to arrive in Shanghai on British and Dutch ships. By 1870 the Meiji government had decided to establish official relations with China, and sent Yanagiwara Sakimitsu to Peking to seek a treaty. The Tsungli Yamen, though inclined to permit trade, was reluctant to sign a formal treaty.

Progressive officials such as Li Hung-chang and Tseng Kuo-fan favored treaty relations. Li opined that Japan, though a tributary state of the Ming dynasty, was never a Ch'ing tributary and its status was basically different from that of Korea and Annam. That Japan sought official relations without an introduction by, or the aid of, a Western power showed its independence and good will, and China should not begrudge it the request. If goaded into unfriendly relations, Li warned, Japan could cause worse trouble than the Western powers because of its proximity. Furthermore, one should not lose sight of the fact that China imported a considerable quantity of copper from Japan annually, and that there were large Chinese communities in Japan. On the basis of these considerations, Li recommended the establishment of equal treaty relations with Japan. Tseng Kuo-fan concurred in these views, stressing in addition the reciprocal nature of Sino-Japanese trade as opposed to the largely one-sided Sino-Western trade. He approved of treaty relations but recommended withholding the most-favored-nation treatment.

On the strength of their recommendations, the court authorized the conclusion of a commercial treaty with Japan on July 24, 1871, which contained the following important provisions: (1) nonaggression toward each other's territorial possessions; (2) mutual offer of good offices in case of conflict with a third power; (3) mutual consular jurisdiction; (4) trade and tariff in treaty ports only; and (5) no appointment of Japanese merchant consuls in China.

In 1873 the Japanese Foreign Minister Soejima came to Peking, ostensibly to exchange ratifications, but his real objective was to participate in the T'ung-chih audience and to sound out China's position on the Formosa incident. The latter involved the killing of 54 shipwrecked Ryūkyūan sailors by the aborigines of Formosa late in 1871. Japan seized upon this occasion to assert her *exclusive* right to speak for the Ryūkyūans, and in doing so she precipitated the question of Ryūkyū's status, which had been wrapped in a shroud of mystery and ambivalence for two and a half centuries.

Ryūkyū, or Liu-ch'iu in Chinese, had been a regular tributary state of China since 1372. During the Ch'ing period it paid tribute every other year and was one of the three most important tributary states—along with Korea and Annam. However, unknown to China, the Satsuma *han* (feudatory) of Japan subjugated Liu-ch'iu in 1609, putting the northern part under its direct administration while leaving the southern part to the Liu-ch'iuan king. Liu-ch'iu became a vassal of Satsuma, to which tribute was paid annually and also to the shogunal court at Edo (Tokyo) periodically. However, Satsuma directed that Liu-ch'iu continue its tributary relations with the Ch'ing dynasty so that Satsuma could reap the benefit of trade with China. Satsuma determined royal succession in Liu-ch'iu, but allowed the Chinese investiture mission to confirm the legitimacy of the king's rule. During the Ch'ing period a total of eight such missions came, the last in 1866, and throughout their stay in Liu-ch'iu, Satsuma took extreme care to remove its officials and things from sight, and to instruct the Liu-ch'iuans to answer Chinese queries in such a way as to hide the Japanese presence. Caught in double subordination, Liu-ch'iu regarded China as father and Japan as mother, using the Chinese calendar when dealing with China and the Japanese calendar when dealing with Japan. Although members of the Chinese investiture missions privately could not fail to detect some traces of Japanese influence on the islands, the Ch'ing court officially knew nothing of Liu-ch'iu's double status, and treated it as China's exclusive tributary state.[2]

Thus, when Soejima openly asserted the right to speak for the Liu-ch'iuans in 1873, the Tsungli Yamen pointedly told him that since Ryū-kyū was a Chinese tributary and Formosa part of China, the killing of the sailors of one by the aborigines of the other was no business of Japan. Moreover, China could not be held responsible for the behavior of the aborigines, because it always allowed them large measures of freedom and never interfered with their internal affairs. Soejima countered that sovereignty over a territory was evidenced by effective control; since China did not control the Formosan aborigines, they were clearly beyond its jurisdiction. Hence any action by Japan to chastise them would not constitute a violation of Chinese jurisdiction. With the support of the Home Minister Ōkubo Toshimichi, Soejima persuaded the Tokyo government to send an expedition to Formosa. This move manifested on the one hand the general Meiji foreign policy of expansion on the Asian mainland after the fashion of Western imperialism, and on the other a clever device to divert domestic demands for popular representative assemblies and to satisfy ex-samurais who had clamored for an expedition to Korea (next chapter). In April 1874 the Office of the Formosan Expedition was formed, with Ōkuma Shigenobu as director and Saigō Tsugumichi as commander-in-chief of the expeditionary force.

The Japanese army quickly landed in Formosa. Peking ordered Shen Pao-chen, director of the Foochow Dockyard, to defend the island. After careful examination of the situation with Li Hung-chang, Shen found effective defense impossible—the guns cast by Halliday Macartney in the Nanking Arsenals could fire nothing but salutes; real explosive shells would burst the gun, killing the gunner rather than the enemy. An agreement was arranged with the Japanese minister that exacted China's promise to control Formosa effectively, to secure bonds from the aborigines against future mistreatment of shipwrecked sailors, and to permit Saigō to punish the aborigines in two villages. However, Saigō refused to honor

2. For details of Liu-ch'iu's double subordination, see two excellent articles: Robert K. Sakai, "The Ryūkyū (Liu-ch'iu) Islands as Fief of Satsuma," in Fairbank (ed.), *The Chinese World Order*, 112-34; Ta-tuan Ch'en, 'Investiture of Liu-ch'iu Kings in the Ch'ing Period," *ibid.*, 135-64.

this agreement; the Home Minister Ōkubo then came to Peking himself on September 10, 1874.

Aided by the French jurist Gustave Boissonade, Ōkubo argued that absence of effective local Chinese administration on Formosa proved China lacked sovereignty. The Japanese landing, consequently, could not be construed as an invasion of Chinese territory. Prince Kung, however, insisted that Sino-Japanese relations be governed not by the general principles of international law, but by the specific treaty of 1871, which clearly stipulated nonaggression against each other's territorial possessions. To this Ōkubo retorted that the treaty concerned only Chinese-Japanese relations, not the Formosan aborigines who were beyond the pale of Chinese jurisdiction. With neither side willing to concede, a diplomatic impasse ensued. The British minister Thomas Wade offered mediation. Ōkubo's initial demand for an indemnity of $5 million was reduced to $2 million, a sum Wade considered not extravagant. After much dickering, the case was finally settled with Prince Kung agreeing to pay the aggressor half a million taels, of which 100,000 was for the Ryūkyūan victims and 400,000 for the purchase of Japanese barracks that had been constructed on Formosa. In addition, China agreed not to condemn the Japanese action—a concession which implied recognition to Japan's claim to sovereignty over Ryūkyū.[3] That China was willing to pay for being invaded— as the British minister in Japan Harry Parkes sarcastically described the case—was an invitation to further foreign encroachment.

Russian Occupation of Ili, 1871-81

Ili was a Chinese prefecture (*fu*) governing nine cities in northern Sinkiang (Chinese Turkestan) near the border of Russian Turkestan. One of the nine, Ning-yüan (I-ning), was known to the Russians and Westerners as Kuldja, which they often wrongly designated a province. The Ili valley was agriculturally and minerally rich and strategically important; its Muzart Pass, soaring 12,208 feet, controlled communication with southern Sinkiang. Control of Ili had always facilitated control of all Sinkiang, and many Western military experts described Ili as the fortress of Chinese Turkestan. A place of such commercial and military potential naturally attracted the attention of strong neighbors. In 1851 the Russians secured the Treaty of Ili from China, which allowed them to establish consulates and duty-free trade at Ili and Chuguchak (Tarbagatai) on the Mongolian border. The Ili trade grew rapidly thereafter, reaching a million pounds

3. In 1879 when China was involved with Russia over the Ili crisis in Sinkiang (next section), Japan annexed Ryūkū and renamed it Okinawa Prefecture.

sterling a year in the mid-1850s. The continuous expansion of the Russians in Central Asia brought them ever closer to Ili, and taking advantage of a Moslem rebellion in Sinkiang, General K. P. von Kaufman, the first governor-general of Russian Turkestan, schemed to make a new conquest.

THE CH'ING ADMINISTRATION IN SINKIANG AND THE MOSLEM REBELLIONS. The Moslem Rebellion in Sinkiang had its roots in the corrupt local Ch'ing administration. Ever since its conquest by Emperor Ch'ien-lung in 1759, Sinkiang had been governed as a military colony. The administration was headed by a military-governor at Ili, aided by a number of assistant military governors and imperial agents in various key points. Some 16,000 soldiers were deployed on the northern side of the Tienshan (Celestial) Mountains and 5,760 on the southern side. The high officials and officers were nearly entirely Manchus and bannermen, who ruled the local populace—mostly Turki-speaking, turban-wearing, Uighur Moslems— through 270 local chieftains known as the *begs*. The Manchu conquerors treated the subject Moslems with contempt, as if they were uncivilized aborigines, levying heavy taxes and exacting forced contributions to support their own unbridled extravagance. The discontent of the Moslems inspired a strong incentive to revolt, and their former rulers, the *khojas,* who had been banished by the Ch'ing to Khokand, were ever anxious to re-establish their personal rule. The *khojas* were the religious potentates who were descendants of the Prophet and ruled Kashgaria (southern Sinkiang) before the Ch'ing conquest in 1759. They perpetually encouraged their co-religionists in Sinkiang to revolt, while they themselves organized invasions. During the century after the Ch'ing conquest, no less than a dozen uprisings and invasions took place. In 1864, amid the dynastic decline and a Moslem rebellion in northwest China, the Moslems in Sinkiang struck again. The local Ch'ing administration was too weak to suppress them, while the central government in Peking was too preoccupied with the Taiping, the Nien, and other rebellions to undertake punitive measures.

During the disorder, Yakub Beg (1820-77), a Khokandian adventurer, invaded Sinkiang in 1865, and through a series of military and political manipulations established himself by 1870 as the ruler of Kashgaria and part of northern Sinkiang. The British in India, to block the southern extension of the Russian influence, encouraged his empire-building and sent missions to cultivate amicable relationships. Fearful of a Yakub Beg invasion of Ili under British sponsorship, disturbed by the interruption of trade, and anxious to expand Russian influence into Chinese Turkestan, General Kaufman ordered the occupation of Ili in July 1871. Disclaiming to the world any territorial designs, the Russians insisted that safeguarding their borders from Moslem raids necessitated the occupation, and that as soon as the Chinese imperial authority was re-established in Sinkiang, Ili would be returned. A magnanimous impression was created that the Russian stewardship was an act of kindness to China during a period of disorder. It was obvious that Russia never believed that the effete Ch'ing dynasty could recover Sinkiang. To perpetuate disorder so as to prolong their occupation of Ili, the Russians signed a commercial treaty with Yakub Beg in 1872; the British followed suit a year later, both countries granting him recognition in exchange for trade privileges.

The Chinese could not reach Yakub Beg before they had suppressed the Moslem Rebellion in Shensi and Kansu. In 1866 the court appointed Tso Tsung-t'ang, governor-general of Fukien and Chekiang, as governor general of Shensi and Kansu, with the specific assignment of suppressing the rebels there. However, before he assumed command, the court again transferred him to first fight the Nien rebels, as noted in Chapter 10. It was not until he had pacified the Nien Rebellion in 1868 that Tso was able to assume his earlier assignment. By efficient leadership, good strategy, and hard campaigning, he crushed the rebellion in these two provinces in 1873. The campaign had cost the government 40 million taels, and Tso's victorious army was poised to strike into Sinkiang. At this juncture, the Formosa crisis with Japan arose, and China's weakness as revealed in the settlement pointed up the urgent need for coastal defense. The nation now encountered the vexing question of whether it could support a bold naval program simultaneously with a costly Sinkiang campaign. A grand debate over the relative urgency and importance of the two ensued.

MARITIME DEFENSE VERSUS FRONTIER DEFENSE. Prince Kung and Wen-hsiang were the first to sound the note of alarm at the inadequacy of the coastal defense after a decade of Self-strengthening. High officials on the coast proposed the creation of a navy consisting of forty-eight ships, divided into three squadrons and stationed on the North, Central, and South China coasts. The threat of Japan, they felt, was more immediate than that of Russia. Li Hung-chang, leading spirit of this group, boldly asked the court to cancel the Sinkiang campaign and shift its funds to naval defense. He called for the purchase of foreign ships and guns, the training of officers and sailors, the recruitment of fresh talent by a new "foreign affairs" examination, the manufacture of munitions, and an increase in customs dues on opium imports to help pay for the naval expenses, estimated at 10 million taels annually.

The advocates of maritime defense advanced five arguments: (1) frontier defense was not as important and urgent as maritime defense, in view of Peking's proximity to the coast and Sinkiang's great distance from the capital; (2) financial exigency and the uncertainty of victory on the difficult terrain of Sinkiang compelled re-examination of the advisability of that campaign; (3) the barren land of Sinkiang, which was of little practical value to China, was not worth the cost of recovering it; (4) surrounded by strong neighbors, Sinkiang could not be effectively defended for long; and (5) to postpone the recovery of Sinkiang was not renunciation of territory conquered by former emperors, but simply a sensible way of preserving strength for the future.

On the other hand, many other officials, while not disputing the importance of naval defense, argued that it should not be undertaken at the expense of frontier defense. If China failed to suppress the rebels in Sinkiang, the Russians would continue their advance, and the Western powers might be encouraged to flare up along the coast in response. Russia, these officials argued, was a greater threat than Japan or any Western power because of its common frontier with China—Russia could reach China by land as well as by sea, whereas Japan and the Western countries could only reach it by sea. They compared the Russian trouble to a sickness of the heart, and the Western threat to that of the limbs. Tso Tsung-t'ang argued that Western powers fought for harbors, ports, and generally only commercial privileges, whereas Russia schemed to obtain both commercial and territorial concessions.

The advocates of frontier defense impressed upon the court five arguments, too: (1) Sinkiang was the first line of defense in the northwest; it protected Mongolia, which in turn shielded Peking. If Sinkiang were lost, Mongolia would be indefensible and Peking itself threatened; (2) the Western powers posed no danger of invasion at the moment, but the Russian advance in Sinkiang was an immediate threat; (3) the funds for frontier defense should not be shifted to coastal defense, since the latter had already been allocated its own standing fund; (4) the land conquered by the forefathers should not be given up; and (5) strategic spots such as Urumchi and Aksu should be recovered first. Tso Tsung-t'ang, the dominant figure of this group, warned that to halt the Sinkiang campaign now was to invite foreign domination of Sinkiang.[4]

The arguments of both groups were cogent and well reasoned. However, it was apparent that there was no immediate trouble along the coast, whereas there was a rebellion in Sinkiang, which needed to be suppressed, and an occupied Ili, which should be recovered. While not giving up the

4. Immanuel C. Y. Hsü, "The Great Policy Debate in China, 1874: Maritime Defense vs. Frontier Defense," *Harvard Journal of Asiatic Studies*, 25:212-28 (1965).

naval program, on April 23, 1875, the court appointed Tso imperial commissioner to conduct the Sinkiang campaign.

With headquarters at Lanchow, Kansu, Tso absorbed himself in an elaborate preparation for the campaign. His policy was "to proceed slowly but to fight quickly." By early 1876 he was ready to strike, and in March moved his headquarters to the advanced post of Suchow. General Liu Chin-t'ang struck hard and fast into Sinkiang, and by November had conquered its northern half. Yakub Beg, still established in southern Sinkiang, was apprehensive of his future; he sent an emissary to London in late spring 1877 to seek British mediation, indicating his willingness to accept the status of a tributary to China, like Burma. But Tso's army moved faster than discussions in London. Yakub Beg was soundly defeated and driven to suicide on May 29, 1877. His sons carried on the fight, but internecine strife among them precluded any effective resistance. By the end of 1877 all Sinkiang had been recovered except the small enclave of Ili, which was still under Russian occupation.

Having re-established the imperial authority in Sinkiang, China had fulfilled the Russian condition for the return of Ili. But the Russian minister in Peking[5] resorted to delaying tactics to postpone the issue. The Tsungli Yamen, then establishing Chinese legations abroad, charged the mission to Russia to negotiate for the return of Ili. The mission head, Ch'ung-hou, who had carried the apology to France in 1870, was given the title of imperial commissioner first class, i.e. ambassador, and authorized to act as he saw fit.

CH'UNG-HOU'S MISSION AND THE TREATY OF LIVADIA, 1879. Ch'ung-hou (1826-93), a pliable and pleasant Manchu noble of no great ability, embarked for Russia quite inadequately prepared for his task. Ignorant of Ili's geography and diplomatic intricacies, he arrived in St. Petersburg where Russian flattery apparently overwhelmed him and caused him to relax his vigilance. Moreover, he appeared anxious to conclude his business and return home. Speculation avers he feared the awesome Russians and earnestly desired to tend to urgent family affairs. His innocence and inattention allowed him to be duped into hastily signing the Treaty of Livadia, which returned Ili to China in name but ceded seven-tenths of the area to Russia, including the strategic Tekes valley and the Muzart Pass. In addition, it awarded to Russia an indemnity of five million rubles, the right to consulates in seven key places, and navigation on the Sungari River in Manchuria up to 600 *versts* (400 miles).[6] When these terms

5. Eugene K. Butzow.
6. To Potuna.

were telegraphed to Peking, the Tsungli Yamen was dumbfounded and cabled Ch'ung-hou not to sign the treaty. His curious reply was that the treaty had already been negotiated and the texts copied out; no changes or renegotiation was possible. On October 2, 1879, on his own authority he signed the treaty and returned home without imperial authorization.

Chinese officialdom responded with consternation. The Tsungli Yamen insisted that this type of restoral of Ili was worse than none. Tso Tsung-t'ang feared that the fruits of his arduous Sinkiang campaign were about to be snatched away by Ch'ung-hou's stupidity. He strongly urged the court to confront the Russians with a firm diplomacy supported by military readiness. "We shall first confront them [the Russians] with arguments . . . and then settle it on the battlefield," he confidently announced.[7] On the other hand, Li Hung-chang, never sympathetic to the Sinkiang campaign and the policy of pressing Russia for the return of Ili, was only superficially critical of the treaty and did not advocate its rejection: "The present mission of Ch'ung-hou had its origin in an imperial edict endowing him with full powers to act as he saw fit. We cannot say that he had no power to negotiate a treaty settlement. If we give assent first and then repudiate it later, we are at fault. Since time immemorial, the first essential of international relations is to decide whether a cause is just. If our cause is unjust, we only ask for insult."[8]

Li was in the unpopular minority. The prevailing sentiment among scholars and officials was for war to avenge the humiliation, even though their country was not ready for it. Barrages of memorials poured into the court demanding severe punishment of the signer and rejection of the treaty. The most eloquent of these came from a young librarian of the Supervisorate of Imperial Instruction, Chang Chih-tung (1837-1909). In beautiful prose he announced: "The Russians must be considered extremely covetous and truculent in making the demands, and Ch'ung-hou extremely stupid and absurd in accepting them . . . If we insist on changing the treaty, there may not be trouble; if we do not, we are unworthy to be called a state."[9] He demanded that Ch'ung-hou be decapitated to show China's determination to reject the treaty, even at the price of war. Because he spoke the mind of the literati and officials, Chang immediately garnered great public fame.

The court appointed Marquis Tseng Chi-tse, minister to Britain and France and son of the great statesman Tseng Kuo-fan, head of a second mission to Russia, with the specific assignment of renegotiating the treaty.

7. Hsü, *The Ili Crisis,* 62.
8. *Ibid.,* 64, with minor changes.
9. *Ibid.,* 71, with minor changes.

Meanwhile, Ch'ung-hou was sentenced to death by beheading after the Autumn Assizes. Strong protests were lodged by the representatives of Britain, France, Germany, and the United States over the inhumane treatment of a brother diplomat, and Queen Victoria even sent a personal plea to the empress dowager. On June 26, 1880, Ch'ung-hou was given a reprieve, but kept in prison to await the outcome of the second mission.

Irritated with China's denunciation of the treaty, its punishment of the signer, and its belligerent pronouncements, Russia sent twenty-three warships to China as a naval demonstration.[10] War clouds hung low over Peking; there was great fear of a Russian naval attack along the coast in concert with an army thrust from Siberia overland to Manchuria and Peking. The court did not intend to precipitate a clash, but was pushed by public sentiments into taking a stronger position than it really wanted. To prepare for the eventuality of war, it installed the Hunan army officers of Taiping fame—rather than those of Li's Huai army—in key defense positions, and through its trusted foreign servant, Robert Hart, invited Charles Gordon to China to help with defenses.

Gordon, the former leader of the Ever-Victorious army and a legendary figure of Victorian England, had been secretary to the viceroy of India since the spring of 1880; but finding the life of a desk officer "a living crucifixion," he resigned, and two days later received the telegraphic invitation from Hart. Gordon immediately seized this opportunity. After meeting with Li at Tientsin, Gordon agreed that China should not be led irresponsibly into a reckless war. He set out for Peking to warn that as long as it was the seat of government China could not afford to fight any first-rate power; the Taku forts could easily be taken from the rear, leaving Peking indefensible. If China must fight, he said, the court should move itself to the interior and be ready for a long war of attrition. Although such blunt counsels were unwelcome in the belligerent atmosphere of Peking, Gordon did make a powerful impression as to the inadvisability of war. He was used by Li both to discourage the war party from a disastrous venture and to show Russia that China did not lack friends in its hour of need.[11]

MARQUIS TSENG AND THE TREATY OF ST. PETERSBURG, 1881. As Gordon was counseling peace in China, Marquis Tseng was readying himself for the mission to St. Petersburg. To avoid his predecessor's mistakes, he had made extensive preparations for his diplomatic strategy and studied the

10. Under Admiral S. S. Lesovskii.
11. Immanuel C. Y. Hsü, "Gordon in China, 1880," *Pacific Historical Review,* 23:2: 147-66 (May 1964).

maps of Ili exhaustively. Determined to hold firm on the boundary issue, bargain on the question of trade, and be conciliatory on monetary compensations, Tseng set out for Russia with assurance from the British foreign office of unofficial assistance, and the British ambassador at St. Petersburg[12] became his secret adviser.

The Russians at first refused to open negotiations at St. Petersburg, insisting on moving the site to Peking as a punishment for China's bellicose attitude. Fearful of negotiating under the threat of the enemy fleet, Peking desperately urged Tseng to use all means to keep the negotiations in Russia. The Russians finally acquiesced and opened the discussions in their capital, but negotiations progressed slowly because they could not find a way to return Ili without losing face. They knew they were in no position to wage a distant war, due to their depressed economy caused by the Turkish war of 1876-77 and their international isolation after the Congress of Berlin in 1878. Yet they could not extricate themselves from China gracefully. After nearly half a year of fruitless argument, the tsar finally decided to end the quarrel by agreeing to return all of Ili, including the Tekes valley and the Muzart Pass, except for some territory in the western portion for the settlement of those Moslem refugees who refused to return to China. The number of Russian consulates was reduced to two,[13] while the indemnity, dignified under the name of "military compensation," was increased to nine million rubles, about five million taels. Since the Treaty of Livadia had been emptied of contents, these terms were incorporated into a new agreement, the Treaty of St. Petersburg, on February 24, 1881.

The peace settlement, generally considered a Chinese diplomatic victory, created two important repercussions. First, it encouraged an upsurge of conservatism in China. The thought of having won a round from a powerful Western state stimulated self-confidence and complacency, in spite of Marquis Tseng's warning against pride, optimism, and arrogance. The literati, who freely and irresponsibly made high-flown speeches to express their views (*ch'ing-i*), were encouraged to believe that the victory resulted from their firm stand, and overconfidently trusted in their ability to untangle China's problems in foreign relations.

The second significant outcome of the settlement was the new status accruing to Sinkiang. Traditionally known as the Western Region (Hsi-yü), Sinkiang had never been an integral part of China but remained a frontier area held by China when it was strong, lost when it was weak. After the Treaty of St. Petersburg, the Ch'ing court accepted the recommenda-

12. Lord Loftus Dufferin.
13. At Turfan and Suchow.

tion of Tso Tsung-t'ang and turned it into a regular province in 1884, with Liu Chin-t'ang, the brilliant young general who contributed much to its reconquest, as its first governor. This unprecedented institutional innovation constituted a significant milestone in Chinese frontier history.[14]

The Sino-French War Over Annam, 1884-85

No sooner had the Ili crisis been settled than the problem of French encroachment in the tributary state of Annam loomed. Known in ancient times as Vietnam, Annam first came under Chinese influence in the 3rd century B.C., and its northern part was conquered by the Han Wu-ti (140-87 B.C.) in 111 B.C. Its name was derived from the An-nan (South-pacifying) protectorate established during the T'ang dynasty (618-907) to govern the area. Though independent after the fall of the T'ang, Annam remained under strong Chinese cultural and political influence. It was an important tributary state during the Ming (1368-1643) and Ch'ing (1644-1911) periods. Between 1664 and 1881 some fifty tribute missions came to Peking.

Western influence in Annam arrived with the Jesuits in 1615, but church work progressed slowly in this predominantly Confucian state. The French East India Company had made an unsuccessful attempt at trade by the end of the 17th century. However, French influence began to wax by the end of the 18th century, when Nguyên Anh, the lone survival of the *ancien régime* that had been overthrown in 1788,[15] regained control of the country with the aid of some French officers. He was installed as Emperor Gia Long of the Nguyên dynasty, which lasted from 1802 until 1945.[16]

THE FRENCH AGGRESSION. Gia-Long and his successors were conservative Confucianists, who promoted Chinese studies and institutions and countenanced xenophobic riots against the missionaries and converts. Louis Napoleon, in his ambition to build a French Indo-Chinese empire and pose as a champion of Catholicism abroad, sent troops to Saigon in 1859 to punish missionary incidents—troops that he had withdrawn from China after the Treaty of Tientsin of 1858. A treaty was imposed on Annam in 1862, by which the French secured an indemnity of $4 million, the rights

14. Hsü, *The Ili Crisis*, 189-96.
15. By three Tây-son brothers: Nguyên Nhac, Nguyên Lu, Nguyên Huê.
16. For an outline of Vietnamese history of this and earlier periods, see D. G. E. Hall, *A History of Southeast Asia* (London, 1964), chs. 9, 22; also Truong Buu Lam, "Intervention versus Tribute in Sino-Vietnamese Relations, 1788-1790," in Fairbank (ed.), *The Chinese World Order*, 165-79.

to trade, to propagate religion, and to control Annamese foreign relations, as well as cession of three eastern provinces in south Annam, known to the French as Cochin China. Further discovery that the Red River in Tongking was a better route than the Mekong to China's Yunnan province aroused French ambitions to seize north Annam. In 1874 a new treaty was signed which confirmed the French possession of Cochin China, the right to direct Annamese foreign relations, and navigation on the Red River. With this document France reduced Annam to a protectorate, although recognizing its independence in name. Preoccupied with the Formosa crisis and the Margary murder case, China took no positive action to stop the French advance; it merely refused to honor the treaty of 1874 on the ground that Annam had always been a Chinese dependency.[17]

French empire-building in the East met with German encouragement. At the Congress of Berlin in 1878, Bismarck is reputed to have told the French delegates that Germany would fight any French attempt at recovering the lost territory in Europe but would gladly assist their overseas aggrandizement. It was therefore no surprise that the French intensified their activities in Annam, and by 1880 had stationed troops in Hanoi and Haiphong and established fortresses along the Red River. To counter the French advance, the Annamese government strengthened its ties with China, despite French protests, by continuing the tribute in 1877 and 1881 and by seeking the aid of the irregular Chinese Black Flag army[18] which had established itself at the Annamese border. By 1882 the Black Flag army had begun engaging the French troops; and in the following year the Ch'ing court, wishing to defend its suzerainty over Annam, yet unwilling to fight the French openly, quietly dispatched regular troops into Tongking.

Li Hung-chang, governor-general at Tientsin and animating spirit of the Self-strengthening Movement, admonished against challenging France before completion of the Chinese naval program and coastal defense. With neither the power to invalidate the French treaties with Annam nor the strength to expel the French from Annam, Li argued, China should not lightly talk of war lest it court disaster. It should fight only when attacked. Even then, he warned, the prospect was bleak, for any Chinese victory would only bring a renewed French effort to prolong the war, while a French victory would drive Chinese troops back to China. Li therefore favored a quick settlement through negotiations. Prince Kung,

---

17. For French activities in Annam, see John F. Cady, *The Roots of French Imperialism in Eastern Asia* (Ithaca, N.Y., 1967), ch. 16; *Southeast Asia: Its Historical Development* (New York, 1964), ch. 18; Hall, ch. 34.

18. Under Liu Yung-fu, a Taiping remnant associated with the Heaven and Earth Society.

head of the Tsungli Yamen and the leading member of the Grand Council, agreed that China should not prematurely challenge a first-rate Western power.

THE RISE OF THE CH'ING-I PARTY. The cautious attitude of Li and Kung was attacked and ridiculed as appeasement and defeatism by a coterie of young officials who were brilliant scholars and memorialists, former members of the Hanlin Academy, but who had had little practical experience or genuine knowledge of foreign and military affairs. They made ornate and fervid speeches to win public acclaim and imperial attention, and championed a belligerent course of action, as they did during the Ili crisis. They considered themselves the voice of the literati (*ch'ing-i*), calling themselves the party of the purists (*ch'ing-liu tang*). Two of the most vociferous members were Chang Chih-tung, who achieved instant fame during the Ili crisis, and Chang P'ei-lun, who emulated him in the present emergency.

The *ch'ing-liu* group disparagingly described France as a "spent arrow" and a country on the brink of bankruptcy. They advocated war to defend China's honor and its tributary state, and condemned appeasement as a sure way to encourage greater demands from the insatiable enemy. If China stood firm over Annam, they argued, the Japanese, the Russians, and the British would all be discouraged from adventures in Korea, Manchuria, and Burma. War was won, they pointed out, more by the human qualities of courage and virtue than by weapons: the spirit of men determined victory. They lashed at Li Hung-chang derisively: "The wily plans of the French are known even by lads and servants. Only Li Hung-chang does not know." "I fear that Li Hung-chang has been deluded by the French, and that the court has in turn been deluded by Li Hung-chang." Contemptuously they compared Li with the notorious historical traitor Ch'in Kuei (A.D. 1090-1155), and cowed other advocates of peace. Li complained to a friend: "I am plagued by the irresponsible talk of officials not in positions of authority . . . They discuss matters of policy, and after matters of policy, they discuss men. Most engage in bullying."[19]

The Ch'ing court vacillated between war and peace. It was caught on the alternate horns of honor, which demanded defense of its tributary, and fear of fighting a leading Western power. A report from Robert Hart's agent in London[20] led the court to believe that the French troops in Annam probably would not precipitate a full-scale war; and that if Hanoi

19. Lloyd E. Eastman, "Ch'ing-i and Chinese Policy Formation during the Nineteenth Century," *The Journal of Asian Studies*, XXIV:4:604-05 (Aug. 1965).
20. J. D. Campbell.

and the Red River were opened to trade and navigation, the basic cause of contention would be removed. The court therefore instructed Li Hung-chang to open negotiations with the French minister, A. Bourée. The resultant agreement, which turned Annam into a joint protectorate of China and France, was immediately rejected by Paris, which decided to dispatch an expedition to Annam. Defeat of the Black Flag army and the regular Chinese troops in Tongking and fear of French attack on China itself caused great anxiety to the empress dowager, who angrily dismissed Prince Kung and four other members of the Grand Council.[21] She again ordered Li to seek a settlement. The subsequent arrangement between Li and the French naval captain, F. E. Fournier, in 1884 called for Chinese recognition of all French treaties with Annam, withdrawal of Chinese troops from Tongking, and a French promise of no demand for indemnity, no invasion of China, and no undignified reference to China in any future treaties wtih Annam. The French parliament refused to ratify the agreement because the last condition resembled an implied recognition of Chinese suzerainty over Annam. On the other hand, the agreement so inflamed the *ch'ing-liu* party that forty-seven memorials poured into the court demanding Li's impeachment. Thus harassed, Li dared not report to the court the agreed-upon date of the withdrawal of Chinese troops from Annam.

THE OUTBREAK OF WAR. Not having received orders to withdraw, the Chinese troops in Tongking rejected a local French demand that they evacuate Langson. Hostilities were renewed, and the Chinese inflicted some casualties on the French troops. Paris accused China of bad faith and sent an ultimatum on July 12, 1884, demanding a large indemnity and immediate execution of the Li-Fournier agreement. Some more negotiating was done but led nowhere. Fearing a French attack on China itself, the court transferred the two leaders of the *ch'ing-liu* party to key defense positions: Chang Chih-tung as governor-general at Canton to guard the southern border, and Chang P'ei-lun as commander of the Fukien fleet. On August 23, twelve French ships under Admiral Courbet launched an all-out attack on Foochow. Within an hour they sank and damaged eleven Chinese warships and destroyed the Foochow Dockyard, built with French aid in 1866. Watching the fight from a hilltop, Chang P'ei-lun was among the first to flee. But his report to the court was so distorted with ambiguous and florid terms that Peking thought China had won a naval battle!

21. Lloyd E. Eastman, *Throne and Mandarins: China's Search for a Policy during the Sino-French Controversy, 1880-1885* (Cambridge, Mass., 1967), ch. 4.

When the truth was known a few days later, he was exiled to the frontier. The court at last stopped wavering and declared war on France.

THE PEACE SETTLEMENT. The empress dowager supported war resolutely for three months, from August to November 1884. In early December she began to vacillate again, as a result of her distress over the indecisive military outcome in Tongking, the French blockade of the Yangtze River and key ports, and the stoppage of tribute grain from South China. The expected aid from Britain and Germany did not materialize; and there was also threat of renewed Russian activities on the northern frontier and a Japanese advance in Korea. The dowager's inclination toward peace was reciprocated by a similar desire in France, where unstable political conditions and the difficulty of supporting a distant military operation began to weigh on the government. Through the good offices of Hart's agents in London, a preliminary peace was agreed upon in Paris, whereby China undertook to recognize the Li-Fournier agreement and France agreed to make no new demands. Fortuitously before this term was formalized into a treaty, the French army at Langson suffered a major defeat,[22] providing Peking with a good face-saving opportunity to pursue peace, and dampening the war spirit in France. In June 1885 Li Hung-chang and the French minister in China finally concluded a formal agreement: China recognized all the French treaties with Annam, and France evacuated its troops from Taiwan and the Pescadores. No indemnity was paid, but China suffered an economic loss in excess of 100 million taels and incurred some 20 million taels of debts.[23]

The indecision and vacillation of the court in the whole venture was pathetic. It had not wanted war but allowed itself to be harried into it by the *ch'ing-liu* party. If it had held firm from the very beginning and been determined for a long war, the French might not have dared to strike; and if it had followed a persistent policy of peace, the Fukien fleet and the Foochow Dockyard would have been spared. The price of ineffective leadership was the destruction of both and the loss of the tributary state of Annam. The *ch'ing-liu* pressure group must be held responsible to a large degree for espousing an unrealistic and emotional cause. Only one of them, Chang Chih-tung, survived the fiasco politically, while the rest faded into relative obscurity. Ironically, after a term of banishment Chang P'ei-lun returned to join Li Hung-chang's staff as his secretary, and later became his son-in-law.

22. By General Feng Tzu-ts'ai.
23. Shao Hsün-cheng, *Chung-Fa Yüeh-nan kuan-hsi shih-mo* (A complete account of Chinese-French relations concerning Vietnam), (Peiping, 1935).

The loss of Annam after a short and disastrous encounter with France signaled the failure of the twenty-year-old Self-strengthening Movement. The limited diplomatic, military, and technological modernization had not strengthened the country to a point where it could resist foreign imperialism. China's weakness prompted the British to emulate the French and detach Burma in 1885. A treaty was secured from China a year later, reducing Burma to a British protectorate but permitting it to continue paying tribute to Peking once every ten years. With the loss of these tributary states in the south, the fate of the leading tributary in the northeast, Korea, hung in a delicate balance, and this the Japanese were too astute not to notice.

Further Reading

Cady, John F., *Southeast Asia: Its Historical Development* (New York, 1964), chs. 12, 18.

——, *The Roots of French Imperialism in Eastern Asia* (Ithaca, N.Y., 1967).

Campbell, Robert R., *James Duncan Campbell: A Memoir by His Son* (Cambridge, Mass., 1970).

Ch'en, Ta-tuan, "Investiture of Liu-ch'iu Kings in the Ch'ing Period" in John K. Fairbank (ed.), *The Chinese World Order: Traditional China's Foreign Relations* (Cambridge, Mass., 1968), 135-64.

Chesneaux, Jean, and Marianne Bastid, *Histoire de la Chine*, Vol. I, *Des guerres de l'opium à la guerre franco-chinoise, 1840-1885* (Paris, 1969).

Ch'in, Han-ts'ai 秦翰才, *Tso-wen-hsiang-kung tsai Hsi-pei* 左文襄公在西北 (Tso Tsung-t'ang in the Northwest), (Chungking, 1945).

Chu, Wen-djang, *The Moslem Rebellion in Northwest China, 1862-1878: A Study of Government Minority Policy* (The Hague, 1966).

Eastman, Lloyd E., *Throne and Mandarins: China's Search for a Policy during the Sino-French Controversy, 1880-1885* (Cambridge, Mass., 1967).

——, "Ch'ing-i and Chinese Policy Formation during the Nineteenth Century," *The Journal of Asian Studies*, XXIV:4:595-611 (Aug. 1965).

——, "Political Reformism in China before the Sino-Japanese War," *The Journal of Asian Studies*, XXVII:4:695-710 (Aug. 1968).

Fletcher, Joseph, "The Heyday of the Ch'ing Order in Mongolia, Sinkiang, and Tibet" in John K. Fairbank (ed.), *The Cambridge History of China* (Cambridge, Eng., 1978), Vol. 10, 351-408.

Grousset, René, *The Empire of the Steppes: A History of Central Asia* (New Brunswick, 1970).

Hall, D. G. E., *A History of Southeast Asia* (London, 1964), chs. 9, 22, 34-35.

Hsü, Immanuel C. Y., *The Ili Crisis: A Study of Sino-Russian Diplomacy, 1871-1881* (Oxford, 1965).

——, "The Great Policy Debate in China, 1874: Maritime Defense vs. Frontier Defense," *Harvard Journal of Asiatic Studies*, 25:212-28 (1965).

——, "British Mediation of China's War with Yakub Beg, 1877," *Central Asiatic Journal* (Leiden), 9:2:142-49 (June 1964).

————, "Late Ch'ing Foreign Relations, 1866-1905" in John K. Fairbank and Kwang-ching Liu (eds.), *The Cambridge History of China* (Cambridge, Eng., 1980), Vol. 11, 70-141.

Jelavich, Charles and Barbara, *Russia in the East, 1876-1880* (Leiden, 1959).

Kiernan, E. V. G., *British Diplomacy in China, 1880 to 1885* (London, 1939).

Lamb, Alastair, *Britain and Chinese Central Asia* (London, 1960).

Langer, William, *The Diplomacy of Imperialism* (New York, 1950).

Lee, Robert, *France and the Exploitation of China: 1885–1901: A Study in Economic Imperialism* (Hong Kong, 1989).

Li, En-han 李恩涵, *Tseng Chi-tse ti wai-chiao* 曾紀澤的外交 (The diplomacy of Tseng Chi-tse), (Taipei, 1966).

Liu, Kwang-ching, "The Military Challenge: the Northwest and the Coast" in John K. Fairbank and Kwang-ching Liu (eds.), *The Cambridge History of China* (Cambridge, Eng., 1980), Vol. 11, 202-73.

McAleavy, Henry, *Black Flags in Vietnam* (New York, 1968).

Sakai, Robert K., "The Ryūkyū (Liu-ch'iu) Islands as Fief of Satsuma" in John K. Fairbank (ed.), *The Chinese World Order: Traditional China's Foreign Relations* (Cambridge, Mass., 1968), 112-34.

Shao, Hsün-cheng 邵循正, *Chung-Fa Yüeh-nan kuan-hsi shih-mo* 中法越南關係始末 (A complete account of Chinese-French relations concerning Vietnam), (Peiping, 1935).

———— et al., (eds.) *Chung-Fa chan-cheng* 中法戰爭 (The Sino-French war), (Shanghai, 1955), 7 vols.

Skrine, C. P. and Pamela Nightingale, *Macartney at Kashgar: New Light on British, Chinese and Russian Activities in Sinkiang, 1890-1918* (London, 1973).

Yüan, T'ung-li (tr.) 袁同禮譯, *I-li chiao-she ti O-fang wen-chien* 伊犁交涉的俄方文件 (The Russian documents on the Ili negotiations), (Taipei, 1966).

14

Acceleration of Imperialism:
The Japanese Aggression in Korea
and the "Partition of China"

By virtue of its proximity to North China, Korea was regarded by the Chinese as a valuable "outer fence" and a leading tributary state during Ming and Ch'ing times. The Yi dynasty (1392-1910) of Korea annually sent three regular tribute missions to the Ming court and four to the Ch'ing, apart from the numerous smaller embassies.[1] During the two and one-half centuries from 1637 to 1894, no less than 507 Korean missions came to Peking, while 169 Chinese missions went to Korea. So important was Korea to China that the Ming dynasty, despite its dwindling treasury and military power, sent 211,500 men and spent 10 million taels to defend it against a Japanese invasion in 1592, and a comparable sum for a second defense in 1597. These exertions so strained the Ming dynasty that they contributed to its ultimate downfall. The Koreans were of course grateful and respectful toward China. Living under its political and cultural shadow, they modeled their institutions and way of life after China's, and described relations with China as "serving the great" (sadae), as distinguished from the more equal "neighborly relations" (kyorin) with Japan. Since 1637 the Koreans had closed their country and maintained virtually no foreign intercourse other than sending tributary missions to China and

---

1. For details, see Hae-jong Chun, "Sino-Korean Tributary Relations in the Ch'ing Period" in Fairbank (ed.), *The Chinese World Order*, 90-111.

occasional delegations to Japan. To the Western world it came to be known as the Hermit Kingdom.

The Opening of Korea

Contact with the West first began in 1635 when a Dutch ship drifted to the Korean coast. Christianity began to spread during the second half of the 18th century, but was proscribed by the Korean court as a heterodox faith in 1786. In the following century, Christian missionaries and Korean converts suffered periodic persecution.

After the opening of China and Japan, the Hermit Kingdom came under increasing Western pressure for trade, religious propagation, and diplomatic relations. But, aside from caring for the shipwrecked, the Koreans refused to have any contact with the West. Jealously guarding their seclusion, they argued that their country was too small and too poor to engage in foreign trade and their people too "stupid" to understand Christianity. Korean intransigence was intensified after Taewongon (Great Lord of the Court), father of the minor king Kojong,[2] became regent in 1864. He promoted conservatism, resisted change, and in February 1866 renewed persecution of Christians, which resulted in a sweeping massacre of foreign priests. In October, the French minister in China, Bellonet, sent Admiral Roze on a punitive expedition of seven ships and 600 men. They captured Kangwha but suffered a defeat outside the city, sustaining 3 dead and 32 wounded. In August of the same year, the American merchant ship, the *General Sherman*, charged up the Taedong River to P'yongyong to demand trade, but was burned by the Koreans when it ran aground during low tide; all the crew was lost. The Department of State authorized Frederick F. Low, minister in China, to investigate the case in 1871, accompanied by five ships under Admiral Rodgers. Refused negotiations off the Kangwha Island, Low forced his way to the Han River that leads to Seoul. The Korean shore battery opened fire, and the Americans retaliated by bombarding the city of Kangwha in full force on June 10 and 11. Subsequently, they also withdrew for lack of authorization to fight. The Koreans congratulated themselves for having repulsed both the French and the Americans.

The Tsungli Yamen in China, itself just learning to adjust to the changing international order in East Asia, was aware of China's inability to defend Korea against the Western advance. Beginning from 1867 it tactfully advised Korea to reach an accommodation with the West, but not until

2. Li Hsi in Chinese.

1879-80 was positive action taken to urge Korea to enter into treaty relations with Western powers in order to counter the rising influence of Japan.[3]

Japanese relations with Korea during the Tokugawa period (1603-1867) had been under the charge of the feudal lords of the Tsushima Island,[4] but after the Meiji Restoration in 1868 the Tokyo government took upon itself the direction of policy. Three missions were dispatched to Korea to announce the political changes in Japan and to revise existing relations. Contemptuous of Japan's modernization and imitation of the West, Taewongon refused to alter relations and rejected as improper the Japanese state letter.

To this deliberate insult the Japanese leaders[5] reacted with a decision in 1873 to send a punitive expedition to Korea. Such a course of action served many other purposes, too: (1) to provide an outlet for the disgruntled samurai at home and shift critical attention from domestic problems to a foreign issue; (2) to win their country a leading position in Asia by successfully challenging China's supremacy in Korea; (3) to forestall Britain and Russia in securing a foothold near Japan; and (4) to avenge the failure of Hideyoshi's invasions of Korea in 1592 and 1597. However, a group of prudent statesmen returning from abroad[6] reversed this decision on the grounds that Japan's backward domestic conditions did not permit a foreign venture at this time, and that internal development and consolidation had to precede overseas expansion.

Though the expedition was not sent, a surveying team, accompanied by gunboats, was dispatched in 1875. When attacked at Kanghwa Bay, the Japanese returned fire and destroyed the Korean forts. Tokyo followed up the victory by sending six more ships to Korea[7] and an emissary to Peking[8] to sound out the Chinese response. Fearful of involvement, the Tsungli Yamen stated that though a tributary state Korea had always been allowed complete freedom in its domestic and foreign affairs. This timid disclaimer of responsibility encouraged the Japanese to force the opening of Korea, much as Commodore Perry did with Japan herself in 1854. Anxious to

3. For a succinct study of Korean intransigence and Chinese adaptability during this time, see Mary C. Wright, "The Adaptability of Ch'ing Diplomacy: The Case of Korea," *The Journal of Asian Studies*, XVII:3:363-81 (May 1958); for a study of China's involvement in Korea, leading to the war with Japan, see Wang Hsin-chung, *Chung-Jih Chia-wu chan-cheng chih wai-chiao pei-ching* (The diplomatic background of the Sino-Japanese war), (Peiping, 1937).

4. The Sō family.

5. Such as Saigō, Itagaki, and Soejima.

6. Such as Iwakura, Kido, and Itō.

7. Under Kuroda Kiyotaka and Inoue Kaoru.

8. Mori Arinori.

avoid a clash, the Ch'ing court, then engaged in the Margary affair, instructed Korea to enter into negotiations with Japan. On February 24, 1876, the Treaty of Kangwha was signed, stipulating: (1) recognition of Korea as an independent state on an equal footing with Japan; (2) exchange of envoys; (3) opening of three ports: Pusan (Fusan), Inchon, and Wonsan; and (4) Japanese consular jurisdiction in these ports. By not protesting the Korean independence, China in effect defaulted its exclusive claim to suzerainty.

Japan's forceful action in Korea was followed by its annexation of the Liu-ch'iu (Ryūkyū) Islands in 1879. Alarmed by these aggressive activities, the Chinese minister in Tokyo[9] and officials at home[10] urged the court that Korea be opened to Western powers to checkmate the rising Japanese influence. The Ch'ing government put Li Hung-chang in charge of Korean affairs, in place of the Board of Rites which had traditionally managed the tributary relations.

Li decided to throw Korea open to Western commerce and diplomacy. In 1882 he sent two of his subordinates, Ma Chien-chung, who had studied international law in France, and Admiral Ting Ju-ch'ang, to Korea in the company of three warships, to introduce Commodore R. W. Shufeldt of the United States for treaty negotiations. On May 22, 1882, the American-Korean treaty was signed, by which the two countries agreed to exchange diplomats, to establish consuls at trading ports, and to treat each other on the basis of equality. The United States recognized Korean independence, but the Koreans voluntarily issued a separate statement to the effect that Korea was a dependent state of China.

In the next few years, Ma Chien-chung introduced British, French, and German representatives to sign treaties with the Koreans, too. The Hermit Kingdom was finally opened to the West, and it tardily began some modernization after the Chinese fashion. Li's active diplomacy retrieved some of the ground lost by the Tsungli Yamen's *faux pas* in disclaiming responsibility for Korea.

Domestic Insurrections and International Politics

THE INSURRECTION OF 1882. After King Kojong began his personal rule in 1873, Queen Min gained increasing power at the expense of Taewongon. She supported reform and employed Japanese officers to train the army. Jealous and disgruntled, Taewongon was determined to curtail her influence and if possible eliminate her. The power struggle led to a head-on

9. Ho Ju-chang.
10. Such as Ting Jih-ch'ang, ex-governor of Fukien.

clash in 1882. Taewongon, capitalizing on the discontent of dismissed old soldiers who were victims of the queen's military reform, incited them to attack the palace and the Japanese legation. Queen Min narrowly escaped in disguise, while the Japanese legation was burned. Seven Japanese officers lost their lives, while the minister[11] fled home. The coup returned Taewongon to power.

The Chinese government once again sent Admiral Ting and Ma Chienchung to investigate the situation. A Korean courtier confidentially informed Ma that the root of all the trouble was Taewongon, who insulated the king from outside contact and executed officials connected with foreign affairs. If Taewongon were not properly disposed of, he warned, the Japanese would probably take punitive action. Ma quickly arrested Taewongon and brought him to China for detention.

Meanwhile, more Chinese and Japanese ships had arrived. On advice of Ma, the Korean king reached a settlement with Japan, agreeing to pay an indemnity of $50,000 for the slain officers and $500,000 for the Japanese government, to send a mission of apology to Tokyo, and to allow Japan to station troops and construct barracks at its legation. The settlement represented a significant victory for Japanese diplomacy, for it gave Japan the right to send troops to Korea. This condition, an oversight of Ma's in spite of his training in international law, was to cause great trouble later.

After the insurrection of 1882, Li Hung-chang began to take positive steps to strengthen China's position in Korea. A commercial treaty between the two countries was signed, giving China extraterritoriality, while loans and gifts of foreign-style guns were extended to the Korean government. Li arranged to appoint a Chinese commercial agent to supervise trade, and assigned a young officer, Yüan Shih-k'ai, to train the Korean army. Paul George von Mollendorf, the former German consul at Tientsin, was made customs commissioner in Korea and concurrently foreign affairs adviser. Six Chinese battalions were stationed in Korea to maintain order and to guard against future Japanese aggression. Chinese influence mounted to new heights in Korea under Li's positive policy.

THE INSURRECTION OF 1884. Yüan Shih-k'ai now allied himself with Queen Min to counter the rising Japanese influence. In the years that followed, struggle grew between the pro-Chinese and pro-Japanese Koreans. The head of the apology mission, having been warmly received in Tokyo, advised the Korean king to accept Japanese help in reform, and Kojong

---

11. Hanabusa Yoshitada.

engaged two Japanese advisers. In a good-will gesture, Tokyo offered to reduce its troops in Korea and to return part of the indemnity for the reform of the Korean administration. A new minister[12] was sent to Korea to promote friendship and guide the pro-Japanese group, now gathered under one Kim Ok-kyun.

The government of Korea at this point was dominated by Yüan Shih-k'ai and the pro-Chinese Koreans. In 1884, however, when China was at war with France and withdrew three battalions from Korea, the pro-Japanese group decided to stage a coup. At the inauguration dinner of the new post office at Seoul on December 4, 1884, all foreign representatives and high Chinese and Korean dignitaries were invited to attend, but the Japanese minister was conspicuously absent. Before the banquet was over, pro-Japanese Koreans set fire to the city, and aided by Japanese troops broke into the palace, captured the king, and wantonly killed pro-Chinese officials. Yüan Shih-k'ai's troops rushed to the palace on appeal from Queen Min. The Chinese soldiers overwhelmed the rebels and the Japanese troops and rescued the king. The plot a failure, the Japanese minister burned his legation and escaped to a seaport,[13] while the chief Korean instigator, Kim, fled to Japan.

Tokyo lost no time in sending an expedition and a high emissary[14] to Korea, forcing the Korean government to pay $110,000 for the loss of lives and properties, send a letter of apology, and pay $20,000 for the reconstruction of the legation. In a concerted move, Itō Hirobumi went to Tientsin to confer with Li Hung-chang. Preoccupied with the French war, Li compromised readily and concluded a Tientsin Convention with him on April 18, 1884, stipulating that (1) China and Japan should withdraw their troops from Korea within four months; (2) neither country would train Korean troops but would jointly urge the Korean government to engage instructors of a third nationality; and (3) before dispatching troops to Korea in the future, the signatories should notify each other in advance, and after the restoration of order, withdraw the troops at once. This agreement virtually reduced Korea to a co-protectorate of China and Japan, eliminated China's claim to exclusive suzerainty, and confirmed Japan's right to send troops.

International rivalry further confounded the situation. In 1885 Russia took ice-free Port Lazareff on the northeastern coast (lat. 39°N.), and the British retaliated by seizing Port Hamilton, an anchorage off the southern tip of Korea. Realizing the detrimental effect that Western in-

12. Takezoe Shinchirō.
13. Jinsen.
14. Inoue Kaoru.

fluence in Korea would have on its own interests, Japan adopted a new policy of encouraging China to strengthen its control, on the assumption that if China succeeded in curtailing foreign influences in Korea, Japan would have only China to deal with in the future. Oblivious to Japan's secret plans, Li Hung-chang proceeded to strengthen China's grip. He returned Taewongon to please the Koreans, and appointed Yüan Shih-k'ai Chinese Resident in Korea to direct all commercial and diplomatic affairs as well as to supervise the domestic administration. Young, energetic, and brash, Yüan quickly dominated the Korean court, the customs, the trade, and the telegraphic service. Expanding Chinese influence wherever and whenever possible, he rose to be the most powerful man in Korea from 1885 to 1893, all unaware that he was unwittingly serving Japan's interests. Neither he nor Li realized that this policy of exclusive control reversed their earlier policy of introducing Western influence to counter the Japanese. This period of Chinese supremacy in Korea coincided with the rapid economic and military growth in Japan, and by 1894 the Japanese had sufficiently modernized themselves to be ready to challenge China.

The incident that added fuel to the already tense situation was the assassination of Kim Ok-kyun, the pro-Japanese Korean leader of the 1884 coup who had fled to Japan. Repeated Korean requests to extradite him had been unsuccessful, but in March 1894 he was enticed—probably by one of Yüan's agents—to Shanghai, where he was assassinated by a Korean, son of a martyr of the 1884 coup. For want of a commercial steamship, the corpse was transported to Korea in a Chinese warship and there it was mutilated as a warning to traitors. The Japanese considered the incident a direct affront, and agitated for a war of chastisement. Foreign Minister Mutsu Munemitsu, however, explained in the Diet that the killing of one Korean by another in China did not legally concern Japan and could not constitute the _casus belli_. Nonetheless, Japanese feelings ran high, and secret societies such as the Genyōsha agitated for action. To create an excuse for sending troops, they encouraged the Tonghak rising in Korea.

THE TONGHAK INSURRECTION, 1894. The Tonghak movement was originally religious in nature, with some nationalistic cast but no political overtone; yet ultimately, because of official persecution, it took on political color. Its founder, Ch'oe Che-u (1824-64), was a frustrated scholar much like Hung Hsiu-ch'üan of the Taiping Revolution. Distressed by official oppression and the progress of Christianity at the expense of Buddhism and Confucianism, Ch'oe claimed, after years of meditation, to have received a "pill of immortality" and a commission to preach. The gospel he taught was supposedly a mixture of the quintessence of Buddhism, Confucianism, and Taoism, the three blending into a cult of "Eastern Learning" (*Tonghak*) as distinguished from Christianity, which was commonly dubbed the "Western Learning" (*Sohak*). Though Tonghak disciples emphasized "Eastern Learning," they also worshipped a divinity somewhat similar to that of the proscribed Catholics. The Korean government therefore banned the Tonghak as a heterodox sect intended to delude the minds of the people. In 1864 Ch'oe was arrested, given a trial, and decapitated. Although the cult went underground, it amassed some 100,000 secret adherents. Gradually, men with political ambitions also joined the organization. In 1892 the Tonghaks petitioned the government to lift the ban and clear the name of their founder, on the ground that the prohibition of Catholicism had been removed. The government not only did not comply, but ordered the dissolution of the sect.

Shortly afterwards, the Tonghaks staged a rebellion. The Korean court sent for help from Yüan. The Japanese minister, hoping to create an excuse for Japan to send troops, urged Yüan to take positive action against the rebels, giving the impression that Japan's sole interest was in safeguarding trade, without itself intending any military intervention. The Chinese minister in Tokyo also reported that the Japanese government was unlikely to start a war, as it had often been entangled in conflict with the Diet since the constitution came into effect in 1890. Li Hung-chang was therefore lulled into believing that Tokyo would not wage war, whereas in fact it was fully prepared to act the moment the Chinese moved their troops into Korea.

No sooner had the Chinese crushed the Tonghak uprising than 8,000 Japanese troops appeared in Korea. With the rebellion suppressed, the Japanese demanded reform of Korean internal administration. Li Hung-chang instructed the Korean government to stall the Japanese by declaring that reform might be carried out after the withdrawal of the Japanese troops.

The Outbreak of War

Determined to find a diplomatic solution, Li hoped to win Western sympathies and force Japan into a peace settlement. He had been assured by the Russian minister, Count Cassini, that St. Petersburg would intervene on China's behalf. However, the Russian government failed to act, having been warned by its minister in Tokyo that if Russia helped China, Britain might assist Japan. Li then turned to Britain for mediation. The latter suggested a simultaneous withdrawal of troops: the Chinese to the north and the Japanese to the south, leaving a neutral zone in the middle around the

Korean capital. Tokyo rejected the proposal but assured the British that in the event of war the neutrality of Shanghai and British commercial interests in China would be respected.

Li's diplomacy not only achieved nothing positive, but it delayed China's military preparations. Not until all hopes for a diplomatic settlement faded did he finally accede to Yüan's urgent request for reinforcement. Three British steamers were chartered to carry troops to Korea under the escort of three Chinese warships. On July 25, 1894, the Japanese navy sank the steamer *Kowshing* in the Korean Bay, drowning 950 Chinese soldiers. On August 1, China and Japan declared war on each other.

The war was, in effect, a significant contest between the two after a generation of modernization. On land the Japanese dealt a crushing defeat to Li's Huai army at P'yongyang, set up a puppet government under Taewongon, and declared Korea independent. At sea, the engagement was even more disastrous for China. Although the Chinese navy boasted 65 ships compared with Japan's 32 and held eighth place to Japan's eleventh in the world, not all of China's fleets were mobilized.[15] Only Li's Peiyang fleet fought the Japanese, whereas the Nanyang fleet and the other two provincial squadrons[16] remained "neutral" for self-preservation. The Japanese mobilized 21 ships, nine of which were constructed after 1889, capable of 23 knots. The Peiyang fleet possessed 25 ships in 1888, when it was formally established; two of them were large ironclads of 7,000 tons, as compared with the largest 4,000-ton Japanese ships. However, the speed of the Chinese ships was only fifteen or sixteen knots. In sum, the Chinese fleet was large, old, and slow, the Japanese, small, new, and fast.

The two fleets met on September 17, 1894, off the Yalu River in the Yellow Sea. After five hours of exchange, the Chinese had lost four ships and suffered casualties in excess of a thousand officers and men. The Japanese had lost but one ship. The surviving seven Chinese ships retreated to Port Arthur for repair, and then on October 18 moved to the naval base at Weihaiwei. In November the Japanese occupied Dairen (Ta-lien) and Port Arthur from the landward side, rendering ineffectual the numerous cannon in the forts. Li had invested millions of taels in building up these naval bases but got no use out of them. The fiasco was completed when the Japanese took Weihaiwei from the rear in February 1895, turning the guns in the forts on the Chinese ships in the harbor. Admiral Ting committed suicide, and his subordinates surrendered, turning over eleven ships to the Japanese.

15. T. F. Tsiang, *Chung-kuo chin-tai shih ta-kang* (An outline of Chinese modern history), (Taipei, 1959), 139.
16. At Canton and Foochow.

The humiliating defeat on land and at sea after more than thirty years of Self-strengthening exposed Li to severe criticism and impeachment. His defense was that victory was impossible when only his Peiyang fleet and Huai army fought against the whole might of the Japanese nation. Be that as it may, Li was dismissed, disgraced, and divested of the Yellow Jacket, a mark of imperial favor.

The Peace Settlement

Even before the total naval defeat, the court at Peking had initiated a peace move. The court sent Chang Yin-huan, a minister of the Tsungli Yamen and vice-president of the Board of Revenue, on a peace mission to Japan, with the American ex-Secretary of State J. W. Foster as adviser. When Itō and Mutsu met with Chang at Hiroshima on February 1, 1895, they deliberately snubbed him, insisting that his full powers were insufficient to negotiate a settlement. They indicated a preference for someone more exalted, such as Prince Kung or Li Hung-chang. The Peiyang navy by this time had surrendered, and the Ch'ing court was desperate for peace. On February 13 it appointed Li envoy first class to Japan.

The peace terms presented by the Japanese government represented a composite demand of diverse circles. The army insisted on cession of the Liaotung peninsula, which would facilitate domination of Korea and Peking. The navy wanted Taiwan (Formosa) as a base for future advance to South Asia, as well as a lease on the Liaotung peninsula. The treasury asked for a large indemnity in the neighborhood of 200 million taels. The Progressive Party suggested, in view of the impending partition of China, that Japan take over Shantung, Kiangsu, Fukien, and Kwangtung provinces, while the Liberal Party urged the cession of Manchuria and Taiwan. The Japanese government synthesized these views into a ten-point proposal, stressing the independence of Korea, an indemnity, the cession of territory, and future commercial and navigational privileges.

At the opening of negotiations at Shimonoseki, Li urged the Japanese negotiators, Itō and Mutsu, to bear in mind the larger interest of Asia in the age of Western imperialism, and pleaded that China and Japan, with their common cultural and racial backgrounds, should not exploit each other. Li compared his age—73 *sui*—with Itō's 55 and Mutsu's 52, in an attempt to gain a psychological advantage over his opponents.[17] In the ac-

---

17. "Ma-kuan i-ho Chung-Jih t'an-hua lu" (Minutes of Sino-Japanese peace negotiations at Shimonoseki), opening section, in Ch'eng Yen-sheng (ed.), *Chung-kuo nei-luan wai-huo li-shih ts'ung-shu* (A historical series on China's internal disorder and external trouble), (Shanghai, 1936), vol. 5.

tual negotiations, however, he had a hard time inducing them to be lenient, particularly on the issue of the indemnity, which was set at 300 million taels. In this difficult situation, a "blessing in disguise" suddenly achieved what his diplomacy could not: returning from the conference one day, Li was shot by a Japanese fanatic. The bullet lodged below his left eye but was not fatal. The incident greatly embarrassed the Japanese government, which voluntarily declared an armistice. The Japanese emperor sent his personal doctor to treat Li's wound, and the Japanese newspapers changed their tone from criticism to eulogy of Li's accomplishments. On the morrow of the accident, Foreign Minister Mutsu went to see Li's son, an attaché in the delegation, and declared that "the misfortune of Li is also the fortune of the Great Ch'ing Empire. From now on peace terms will be more easily arranged, and the Sino-Japanese war will be terminated."[18] On April 17, 1895, the Treaty of Shimonoseki was signed. It provided for (1) recognition of Korean independence and termination of tribute to China; (2) an indemnity of 200 million taels to Japan; (3) cession of Taiwan, the Pescadores, and the Liaotung peninsula; (4) the opening of Chungking, Soochow, Hangchow, and Sha-shih as ports; and (5) the right of Japanese nationals to open factories and engage in industry and manufacturing in China.

Back home the reaction was severely critical. Many Chinese scholars and officials accused Li and his son of selling out their country to preserve themselves. Chang Chih-tung, governor-general at Nanking, vigorously opposed the ratification of the treaty. On several occasions, hundreds of provincial graduates, gathering in Peking for the triennial metropolitan examinations, sent joint petitions to the court, urging rejection of the treaty, removal of the capital to the hinterland, and continuation of the fighting.[19] Despite all this display of anger, the court under Japanese pressure exchanged the ratifications on May 8, 1895.

The cession of Taiwan was met with strong resistance by the local populace. Taiwan, which had been made a province after the French war, had achieved considerable progress in modernization under its first governor[20] during 1885-91, and the Taiwanese now refused to cede their island to Japan. They declared independence on May 25, 1895, established a Republic of Taiwan, and offered the presidency to the incumbent governor.[21] On June 2, the Ch'ing court sent Li Ching-fang, son of Li Hung-chang, to

18. Li Shou-k'ung, 464-65.
19. They presented petitions seven times, the second on April 30 being the largest, involving 1,200 to 1,300 signatures.
20. Liu Ming-ch'uan.
21. T'ang Ching-sung.

Taiwan to turn over the island to Japan; large Japanese contingents also arrived to enforce the transfer. Finally, in October 1895 the local opposition was suppressed and the Republic of Taiwan went out of existence.

Causes of the Ch'ing Defeat

In reappraising the preceding events, China's defeat appears inevitable for many reasons. First, Japan had become a modern state in which a nationalistic consciousness bonded the government and people into a unified body. The war was fought with the wholly consolidated might of the Japanese nation. In China, the state polity was still basically medieval, with the government and people forming separate entities. The war hardly affected the people at all; it was fought mostly by Li Hung-chang's Peiyang fleet and Huai army. Western observers pithily described the war as one between Li and Japan.

Second, there was no clear demarcation of authority, no unity of command, and no nationwide mobilization in China. Li Hung-chang had the responsibility for directing diplomatic and military affairs regarding Korea, but not the authority to decide policy matters or to control the ships and troops outside his Peiyang command and his Huai army. True, the poor showing of his navy and army was unforgivable after so many years of training and preparation, but there was much truth in Li's defense that victory was impossible when only his regional forces were pitted against the entire might of the Japanese empire.

Third, corruption at court and in the Peiyang command doomed the Chinese effort from the start. The dowager's misuse of the naval funds for the construction of the Summer Palace (*I-ho-yüan*), her trust in the eunuchs, and the general degeneration of public morality predestined the defeat. For want of funds the British adviser's recommendation before the war that China purchase two fast ships went unheeded. Instead these two ships were purchased by Japan, and one of them, the *Yoshino*, established a splendid record in the naval battle.

Within the Peiyang command itself, corruption and irregularities were rampant. Li Hung-chang, himself not known for integrity and character, chose his subordinates on the basis of their personal loyalty and willingness to work rather than for their ability and uprightness. Many of his army and naval officers curried favor with the chief eunuch, Li Lien-ying, debasing themselves as his "disciples." They embezzled public funds to make presents to the eunuch, and he in turn protected their irregularities. The big ten-inch guns on the two ironclads, it was said, were allocated only three shells each, and the many smaller guns were assigned wrong size shot. The funds for the ammunition lined the pocket of the officer in charge of supply, none other than Li's nephew.[22] For all its outward brilliance—the newly painted ships and the neatly uniformed officers—the Peiyang was a comic-opera fleet good only for cruising the harbors, not fighting a modern war. Li knew its weakness only too well; hence he was reluctant to fight, relying instead on diplomacy to solve the Korean crisis.

Fourth, Li's diplomacy was limited by his obsession with the antiquated policy of playing off the barbarians against one another (*i-i chih-i*). He allowed himself to be misled by Cassini into believing that the Russian government would intervene on China's behalf to force peace on Japan. When the promised intervention failed to materialize, Li desperately turned to Britain and the United States for mediation, neither of whom could effectively influence Japanese policy. Li's diplomacy was a complete failure because he did not understand the essentials of modern international politics, and because he overestimated his personal powers of persuasion. When he finally recognized the inefficacy of his diplomacy, much time had already been lost in making military arrangements.

All in all, the defeat was an irrefutable testimonial to the failure of the Self-strengthening Movement, already evidenced in the French war ten years earlier. The limited diplomatic, military, and technological modernization, without corresponding change in institutions and spirit, was incapable of revitalizing the country and transforming it into a modern state. China's loss seemed all but inevitable.

The Repercussions of the War

The defeat signaled the impending demise of the Ch'ing dynasty, and ushered in a period of accelerated foreign imperialism and domestic political movements. Among its more important repercussions were the following.

INTENSIFICATION OF IMPERIALISM. The defeat exposed the decadence and helplessness of the Manchu dynasty and invited foreign powers to engage in a scramble for concessions (see next section). Foreign imperialists cut the China melon into leased territories and spheres of interest, within which they constructed railways, opened mines, established factories, operated banks, and ran all kinds of exploitive organizations. The intensification of imperialism plunged China ever deeper into a semi-colonial state, from which it was not freed until 1943.

22. Chang Shih-yen.

OPPRESSION OF NATIVE INDUSTRIES. The right to establish factories and industries in China, won by Japan in the peace treaty, was extended to all treaty powers through the most-favored-nation clause. It enabled imperialists to manufacture locally and thus avoid customs duties and reduce transportation costs. Foreign investors and developers, with their vast capital, technical know-how, and privileged position, had a distinct advantage over the nascent Chinese industrialists and businessmen. Foreign economic imperialism inhibited the spontaneous growth of native capitalism and relegated Chinese industries to a position of subordination and vassalage.

THE RISE OF JAPAN. Japan replaced China as the leading state in East Asia. With Taiwan in the south and Korea in the north, Japan had secured a solid base for future advance in Southeast Asia and a convenient springboard to Manchuria. The war paved the way for its challenge to Russia in 1904, its rise to the great-power status, its future aggression in China, and its domination of Southeast Asia during World War II.

NEW POLITICAL MOVEMENTS IN CHINA. The defeat demonstrated beyond doubt the inability of the Manchus to cope with the challenge of the times. Superficial modernization of the Self-strengthening type could not regenerate a rule deeply embedded in decadence. Furthermore, new crises of imperialism threatened the dismemberment of China. There was now a realization among thinking Chinese that China's salvation lay in a radical reform or even a revolution. The progressives advocated institutional reorganization after the fashion of Peter the Great and Emperor Meiji. The radicals demanded a revolution to replace the Manchu dynasty with a Chinese republic. These two currents constituted the main political movements in postwar China.

Postwar Foreign Relations

THE TRIPLE INTERVENTION. On April 23, 1895, barely six days after the signing of the Treaty of Shimonoseki, Russia, France, and Germany sent a joint note to Tokyo warning that the possession of the Liaotung peninsula by Japan would menace Peking, render illusory the independence of Korea, and threaten the general peace of the Far East. The instigator of this triple intervention was Russia, who felt threatened by Japan's acquisition of a foothold on the Asiatic mainland. In fact, the Russians themselves had cast covetous glances at the ice-free ports of Dairen and Port Arthur at the southern tip of Liaotung. Count Witte, Minister of Finance, openly stated that "it was imperative not to allow Japan to penetrate into the very heart of China and secure a footing in the Liaotung peninsula."[23] At the imperial conference of March 30, 1895, the Russians decided to seek preservation of the status quo *ante bellum* in Liaotung; to advise Japan to desist from seizing it; and, if Japan should ignore this warning, to take whatever action necessary from the standpoint of Russia's national interest, including the bombardment of Japanese ports. To the outside world Russia was to announce a disclaimer of territorial designs on China.[24] France, a partner in the Dual Alliance, was committed to support Russia, and Germany, anxious to keep Russia occupied in the East so as to lessen its pressure in Europe, joined in the intervention.

The Japanese government decided to return Liaotung to China at a price of 50 million taels, in addition to the original indemnity of 200 million taels. The three powers reduced the additional sum to 30 million taels, and on November 4, 1895, Li Hung-chang and Hayashi, the Japanese minister in China, signed a formal agreement retroceding Liaotung.

The Russians became heroes in the eyes of the Chinese, and won further gratitude by offering loans to pay the Japanese indemnity. The first payment of 50 million taels was due within six months, and the second installment of a like sum in the following six months. The Ch'ing court, with an annual revenue of 89 million taels, was in no position to meet these obligations except by contracting loans. For the first 50 million, plus the 30 million for the retrocession of Liaotung, Peking borrowed 400 million francs[25] from a Franco-Russian Banking Consortium[26] at 4 percent interest. Count Witte, who arranged the loan, pledged Russia's resources as security. Later in 1896 and 1898, China twice borrowed from a British-German Consortium 16 million pounds sterling at 5 percent and 4.5 percent respectively.

THE SINO-RUSSIAN SECRET ALLIANCE. Overwhelmed by the Russian tenderings of friendship, leading Chinese officials such as Chang Chih-tung and Liu K'un-i advocated an alliance with Russia as a safeguard against future Japanese and Western aggression. They won the support of Li Hung-chang, who, disappointed by the British failure to intervene on China's behalf during the war, more than ever considered an alliance with Russia

23. Abraham Yarmolinsky (tr. and ed.), *The Memoirs of Count Witte* (New York, 1921), 83.
24. *Ibid.,* 84.
25. About 100 million taels or 15,820,000 pounds sterling.
26. Including the Banque de Paris, Banque des Pays Bas, Crédit Lyonnais, and the Hotenger House.

as the cardinal principle of future Chinese diplomacy. Li had always been pro-Russian and anti-Japanese, as clearly indicated by his position during the great debate of 1874 and the Ili crisis in 1878-81. At length the empress dowager, too, fully sanctioned a Sino-Russian alliance.

On Russia's part, Count Witte welcomed closer relations with China in hopes of winning a concession to extend the Trans-Siberian Railway across Manchuria to Vladivostok, which, begun in 1891, had reached Transbaikalia. The question arose as to whether it should run to Vladivostok along the northern bank of the Amur River through some very difficult terrain, or whether it should cross over Manchuria at a saving of 514 *versts* (350 miles). Count Witte strongly favored the second alternative, to save time and money and to further his policy of *peaceful penetration* of China. Cassini, Russian minister to China, was instructed to explain to Li Hung-chang that such a railway would facilitate Russian troop movement for the defense of China. Some initial discussions apparently took place between the two, but no formal agreement was reached, despite the British *North China Daily News*'s report of a "Cassini Convention."

The Russian desire for a railway concession and the Chinese desire for an alliance materialized at the coronation of Nicholas II in 1896. The tsar himself was said to have sent a telegram to the empress dowager assuring her that the appointment of Li Hung-chang as emissary, would be most agreeable with him. So Li who had been in disgrace after the war, was designated imperial commissioner first class and head of the Chinese congratulatory mission to Russia. At the grand old age of 74 *sui*, Li left for the West for the first time in his life, to participate in the coronation of the tsar and to visit the rulers of Britain, France, Germany, and the United States.

At St. Petersburg Witte impressed upon Li that to uphold China's territorial integrity and render it armed assistance in case of emergency, Russia needed the shortest possible railway route from its European section to Vladivostok, across the northern section of Mongolia and Manchuria. Such a line, Witte assured his guest, would raise the productivity of the Chinese land it traversed and would not arouse Japanese opposition because it would serve to link Japan with Europe.[27] Witte and Li agreed on three principles:

    1. China would grant Russia permission to construct a railway along a straight line from Chita to Vladivostok; the operation of the railway would be managed by a private organization called the Chinese Eastern Railway Corporation.

27. Yarmolinsky, 85, 87.

2. China would cede a strip of land sufficient for the building and operation of the railway; within the limit of the land the corporation should have complete authority of control, including the right of maintaining police. The railway might be redeemed by China after 36 years at 700 million rubles, but would pass free to her after 80 years.

3. China and Russia agreed to defend each other against any Japanese attack on China, Korea, or Russian Far Eastern possessions.

It was apparent that the Russian emphasis was on the first two conditions, and the Chinese, on the third.

A rumor was circulated later that Li accepted a Russian bribe of $1.5 million, but Witte denied it. It appears that the bribery, even if a fact, was not decisive in Li's thinking, for he had come with the explicit secret mission of concluding a treaty of alliance. So proud was he of the policy of playing off the barbarians against one another—using Russia against Japan in this case—that he announced with satisfaction that the treaty would give China peace for twenty years.[28] But China was not to have peace for even two years.

THE SCRAMBLE FOR CONCESSIONS. After the triple intervention, Germany asked the Ch'ing court for a naval base as a reward, on the ground that all other major powers had a base in the Far East: Britain in Hong Kong, France in Tongking, and Russia's winter harbor in Kiaochow, Shantung province. The Chinese rejected the request. Then, in 1897 when the kaiser visited Russia he asked the tsar whether he would object to German occupation of Kiaochow, an excellent naval base which Admiral Tirpitz had selected for acquisition. Finding himself in an awkward position to object, the tsar gave a vague assent, knowing that Russia itself preferred a naval base more to the north. The Germans then capitalized on the murder of two missionaries in Shantung, in November 1897, to seize Kiaochow and compel the Chinese government to lease it for 99 years, along with a concession to build two railways in Shantung. Encouraged by the German success, Russian Foreign Minister Muraviev proposed to occupy Port Arthur or Dairen—a scheme that won the tsar's approval over the protests of Witte, who argued the importance of honoring the pledge to

28. T'ao-ch'i-yü-yin (ed.), "Li-fu-hsiang yu-li ko-kuo jih-chi" (A diary of State Minister Li's visit to the various countries), in Tso Shun-sheng (ed.), *Chung-kuo chin-pai-nien shih tzu-liao* (Materials relating to Chinese history of the last hundred years), supplementary volume, reprinted (Taipei, 1958), 387.

respect Chinese territorial integrity, and of the naval minister, who preferred a base in Korea. In December 1897 the Russians took these two ports on Liaotung peninsula, under pretext of protecting China from the Germans. On January 1, 1898, when General A. K. Kuropatkin became Minister of War, he insisted on extending the occupation zone to the adjacent areas of these two ports as well. Russia then imposed an agreement on China in March, acquiring the right to lease Port Arthur and Dairen for 25 years and to construct a Southern Manchurian Railway from the two ports to the Chinese Eastern Railway, from which a further branch was to be constructed to Yingkow and the Yalu River. In this negotiation Witte later admitted having bribed the two Chinese negotiators, Li Hungchang and Chang Yin-huan, with 500,000 and 250,000 rubles apiece.[29] The Russians now appropriated the Liaotung peninsula, which China had bought back from Japan three years earlier for 30 million taels.

With these precedents, the scramble for concessions raged like wildfire. Not to fall behind the Germans and the Russians, the British leased Weihaiwei for 25 years and Kowloon New Territories for 99 years, in addition to securing a promise that China would not alienate the Yangtze valley to any other power, thus turning the area into a British sphere of influence. A similar commitment to nonalienation of the Fukien province was granted to Japan. The French leased Kwangchow Bay for 99 years and established a sphere of influence in Kwangtung-Kwangsi-Yunnan. Only the Italian demand for San-meng-wan in Chekiang province was rejected with impunity, on advice of the customs inspector-general, Robert Hart. The United States, then involved in the Spanish War and the Cuban rebellion, did not participate in this mad scramble, although its navy at one point did covet the Samsah Bay.

The cutting of the China melon threatened the partition of the Ch'ing empire. Indeed, the deepening foreign encroachment precipitated a reform movement domestically and the declaration of the Open Door Policy by the United States.

THE OPEN DOOR POLICY. Britain kept a sphere of special interests in China but it also wanted an open door for trade where other powers had special influence. Being a party to the scramble for concessions, it could hardly promote the idea of preventing the closing of other doors to British trade, which accounted for some 35 million pounds sterling out of a total of 55 million China trade in 1899. The British therefore turned to the United

29. Yarmolinsky, 103.

States, the only major power with a "clean" record. Sir Julian Pauncefote, the British minister in Washington, twice approached the Department of State, in March 1898 and January 1899, to suggest joint sponsorship of a movement for equal commercial opportunity in China. But he met with no success. The Americans, however, became more responsive after the Spanish War and the annexation of the Philippines. Meanwhile, the British kept promoting the Open Door idea. Lord Charles Beresford wrote *The Breakup of China* and toured the United States to advocate the cause. Another British Old China Hand, A. E. Hippisley, of the Chinese customs service, successfully impressed similar ideas upon his American friend, W. W. Rockhill, formerly minister in China and presently adviser to Secretary of State John Hay on Far Eastern affairs. On the basis of a Hippisley memorandum, Rockhill prepared a note embracing the idea of equal commercial opportunity in China, which Secretary Hay delivered to Britain, Germany, Russia, France, Italy, and Japan in September 1899. It contained three main points:

1. Within its sphere of interest or leasehold in China, a power would agree not to interfere with any treaty ports or the vested interest of other powers.
2. Within its sphere of interest or influence, no power would discriminate against nationals of other countries in matters of harbor dues or railway charges.
3. Within each sphere of foreign influence, the Chinese treaty tariff should apply and the Chinese government be allowed to collect customs duties.

None of the powers committed itself to this pronouncement, stating equivocally that its acceptance was contingent on that of others. Hay nevertheless declared on March 20, 1900, that their assent was "final and definite." Only Japan challenged the American interpretation. Later, during the Boxer Rebellion, when the activities of the powers appeared to threaten the Open Door principle, the United States made a second declaration on July 3, 1900, extending its scope to include the preservation of Chinese territorial and administrative entity. Being merely a statement of intention, it did not solicit an answer from the other powers.

The Open Door was a declaration of principles rather than a formal policy of the United States, which had neither the will nor the power to enforce it militarily. Strangely, the partition of China tapered off after the declaration, not so much because the imperialists respected the American call, but because they feared rivalry and conflict among themselves. The resultant equilibrium saved the Ch'ing empire from an immediate collapse.

Further Reading

Anderson, David L., *Imperialism and Idealism: American Diplomats in China, 1861-1898* (Bloomington, 1985).

Chang, Yin-lin 張蔭麟, "Chia-wu Chung-kuo hai-chün chan-chi k'ao"甲午中國海軍戰績考(A study of the Chinese naval battle in 1894), *Tsing-hua hsüeh-pao*, 10:1 (Jan. 1935).

Ch'en, Edward I-te, "Japan's Decision to Annex Taiwan: A Study of Ito-Mutsu Diplomacy, 1894-95," *The Journal of Asian Studies*, XXXVII:1:61-72 (Nov. 1977).

Ch'en, Jerome, *Yüan Shih-k'ai, 1859-1916* (Stanford, 1961), chapters 1-2.

Chiang, Kai-shek, *Chiang Tsung-t'ung mi-lu* 蔣總統祕錄 (Secret memoirs of President Chiang), tr. from the Japanese by *Central Daily News*, Vol. 1 (Taipei, 1974).

Choe, Young, *The Rule of Taewon Gun, 1864-73* (Cambridge, Mass., 1972).

Chun, Hae-jong, "Sino-Korean Tributary Relations in the Ch'ing Period" in John K. Fairbank (ed.), *The Chinese World Order: Traditional China's Foreign Relations* (Cambridge, Mass., 1968), 90-111.

Cook, Harold F., *Korea's 1884 Incident: Its Background and Kim Ok-kyun's Elusive Dream* (Seattle, 1972).

Deuchler, Martina, *Confucian Gentlemen and Barbarian Envoys: The Opening of Korea, 1875-1885* (Seattle, 1978).

Hunt, Michael H., *Frontier and the Open Door: Manchuria in Chinese-American Relations, 1895-1911* (New Haven, 1973).

——, *The Making of a Special Relationship: The United States and China to 1914* (New York, 1985).

Iriye, Akira, *Pacific Estrangement: Japanese and American Expansion, 1897-1911* (Cambridge, Mass., 1972).

Joseph, Philip, *Foreign Diplomacy in China, 1894-1900* (London, 1928).

Kim, C. I. Eugene, and Han-Kyo Kim, *Korea and the Politics of Imperialism, 1876-1910* (Berkeley, 1967).

Kim, Key-hiuk, *The Last Phase of the East Asian World Order: Korea, Japan, and the Chinese Empire, 1860-1882* (Berkeley, 1980).

Langer, William, *The Diplomacy of Imperialism* (New York, 1950).

Lee, Yur-Bok, *West Goes East: Paul Georg von Mollendorff and Great Power Imperialism in Late Yi Korea* (Honolulu, 1988).

Lensen, George Alexander, *Balance of Intrigue: International Rivalry in Korea and Manchuria, 1884-1899*, 2 vols. (Honolulu, 1982).

Limley, Harry J., "The 1895 Taiwan Republic," *The Journal of Asian Studies*, XXVII:4:739-62 (Aug. 1968).

Lin, Ming-te 林明德, *Yüan Shih-k'ai yü Ch'ao-hsien* 袁世凱與朝鮮 (Yüan Shih-k'ai and Korea), (Taipei, 1970).

Lin, T. C., "Li Hung-chang, His Korea Policies 1870-1885," *Chinese Social and Political Science Review*, 19:2:200-33 (1935).

"Ma-kuan i-ho Chung-Jih t'an-hua lu" 馬關議和中日談話錄 (Minutes of Sino-Japanese peace negotiations at Shimonoseki), in Ch'eng Yen-sheng 程演生 (ed.), *Chung-kuo nei-luan wai-huo li-shih ts'ung-shu* 中國內亂外禍歷史叢書(A historical series on China's internal disorder and external trouble), (Shanghai, 1936), Vol. 5.

Mayo, Marlene J., "The Korean Crisis of 1873 and Early Meiji Foreign Policy," *The Journal of Asian Studies*, XXXI:4:793-820 (Aug. 1972).

McCordock, R. Stanley, *British Far Eastern Policy, 1894-1900* (New York, 1931).

Mutsu, Munemitsu 陸奥宗光, *Kenkenroku* 蹇蹇録 (Memoirs of Mutsu Munemitsu), (Tokyo, 1929).

Rockhill, William W., *China's Intercourse with Korea from XVth Century to 1895* (London, 1905).

Shao, Hsün-cheng 邵循正 et al. (eds.), *Chung-Jih chan-cheng* 中日戰爭 (The Sino-Japanese war), (Shanghai, 1956), 7 vols.

T'ao-ch'i-yü-yin 桃谿漁隱 et al. (eds.), "Li-fu-hsiang yu-li ko-kuo jih-chi" 李傅相游歷各國日記 (A diary of State Minister Li's visit to the various countries), in Tso Shun-sheng 左舜生 (ed.), *Chung-kuo chin-pai-nien shih tzu-liao, hsü-pien* 中國近百年史資料續編 (Materials relating to Chinese history of the last hundred years), II, 387-415.

Tsiang, T. F., "Sino-Japanese Diplomatic Relations, 1870-1894," *Chinese Social and Political Science Review*, 17:1-106 (1933).

Varg, Paul A., *The Making of a Myth: The U.S. and China, 1897-1912* (East Lansing, 1968).

———, *Open Door Diplomat: The Life of W. W. Rockhill*, Illinois Studies in the Social Sciences, 33:4 (Urbana, 1952).

Wang, Hsin-chung 王信忠, *Chung-Jih Chia-wu chan-cheng chih wai-chiao pei-ching* 中日甲午戰爭之外交背境 (The diplomatic background of the Sino-Japanese war), (Peiping, 1937).

Wang, Yün-sheng 王芸生 (ed.), *Liu-shih nien-lai Chung-kuo yü Jih-pen* 六十年來中國與日本 (China and Japan during the last sixty years (1856-1916), II (Tientsin, 1932-33).

Wright, Mary C., "The Adaptability of Ch'ing Diplomacy: The Case of Korea," *The Journal of Asian Studies*, XVII:3:363-81 (May 1958).

Yao, Hsi-kuang 姚錫光, *Tung-fang ping-shih chi-lüeh* 東方兵事紀略 (A brief record of the war in the East).

Yarmolinsky, Abraham (tr. and ed.), *The Memoirs of Count Witte* (New York, 1921).

IV
Reform and Revolution
1898-1912

15

The Reform Movement
of 1898

The frightful prospect of dismemberment precipitated a reform movement in China in 1898. The movement had actually been gathering momentum for ten years, for ever since China's defeat in the French war in 1885 the inadequacy of limited modernization had been obvious, and its defeat in the Japanese war in 1895 was irrefutable proof that the Self-strengthening Movement had failed. The need for a more extensive reform was recognized by scholars, officials, and even the emperor and the empress dowager, although they differed on the question of its nature, scope, and leadership. Li Hung-chang, the main figure in the Self-strengthening Movement, had gone into political eclipse; in his place rose Chang Chih-tung, the long-time (1889-94, 1896-1907) governor-general at Wuhan, and Weng T'ung-ho, the influential imperial tutor and president of the Board of Revenue (1886-98). Both advocated a conservative reform based on a limited administrative reorganization along with the adoption of Western methods to supplement the basic Chinese structure. A radical third force, advocating a drastic institutional change after the patterns of Peter the Great and Emperor Meiji, was led by the thinker-idealist K'ang Yu-wei. Emperor Kuang-hsü (1875-1908), first guided by Weng to his conservative reform, ultimately was won over by the dynamic K'ang. On the other hand, Empress Dowager Tz'u-hsi, who saw in the radical reform a threat to her own supremacy, threw her tremendous authority against it.

In this vortex of cross currents, one sees the power struggle between the

355

emperor and the dowager, the conflict between the conservatives and the progressives, the strife between the moderate reformers and the radicals, and the racial antipathy between the Manchus and the Chinese. The interplay of these forces was accentuated by the imminence of foreign partition of China. The Ch'ing empire in 1898 stood at a turning point in history: a successful reform might stave off the breakup, while failure could only foreshadow the extinction of the dynasty.

Early Advocates of Reform and Missionary Influence

The genesis of institutional reform may be traced to Feng Kuei-fen, whose *Protest from the Chiao-pin Studio (Chiao-pin-lu k'ang-i)* we have already briefly touched upon in Chapter 11. Feng argued that the new world which had been thrust upon China was drastically different from the world of ancient times, and that it behooved China to adopt the superior Western mathematics, physics, chemistry, and geography. China should initiate measures to reform its educational and examination systems, to abolish the eight-legged essay, to strengthen local political organizations, to encourage industrial manufacture, to reclaim virgin land, to open mines, and to improve agricultural tools. These farsighted views, first made after 1860, were too advanced for Feng's time. They were considered impracticable by Tseng Kuo-fan, and were only partially followed by Li Hung-chang.

Another advocate of progressive reform was Kuo Sung-tao, China's first minister to Britain and France (1876-78). His firsthand observation of the West resulted in an open, if unpopular, admission that Western countries had a distinctive history of their own of over 2,000 years and were in possession of illustrious political institutions and moral teachings. Deploring the limited scope of Self-strengthening, he praised the Japanese for sending students to Britain to study law and economics, and impressed upon Li Hung-chang the need to accept the Western educational system, political institutions, jurisprudence, and economics. These, rather than the military, Kuo argued, were the foundation of good government and a prosperous state. His plea went unheeded as Li insisted that having been entrusted with the country's defense he had no alternative but to stress the military aspect of modernization. If Kuo's progressive ideas were too liberal for Li, they were heresy in the eyes of Confucian moralists, who could not imagine any civilization unconnected with Confucius. To them Kuo was a betrayer of Chinese cultural heritage who rightfully deserved ostracism.

Among the nonofficials, Wang T'ao (1828-97) was remarkable for his progressive views. He had shown early sympathy toward the Taipings. Being suspect in the eyes of the Ch'ing court, he escaped to Hong Kong in 1862 and became an editor of the foreign-sponsored *Hong Kong News*. Invited by James Legge to Scotland in 1867 to assist him in the monumental task of translating *The Chinese Classics,* Wang's two-year sojourn in Europe acquainted him with Western culture and institutions. Back in Hong Kong in 1870, he became editor of a newspaper (*Hsün-huan jih-pao*) and later contributed to the influential paper in Shanghai, the *Shen-pao.* From this vantage point Wang launched his campaign for reform. Like Kuo Sung-tao, he praised the Japanese imitation of Western institutions and urged his countrymen to alter their methods of civil service examinations, military training, education, and legal practice. He attacked corruption in the Ch'ing administration, the sinecure posts in government, and the *likin* tax. He proposed the opening of mines, the building of textile mills, the construction of steamers, railways, telegraph lines, and the development of a navy. He warned against leaning too much on the superficial, external mechanics of the West, whose strength he said lay in its law, justice, political system, popular election, and constitutional government. However, Wang was not a radical; he did not urge an outright Westernization of the age-old Chinese institutions, but a gradual grafting of useful Western elements onto the Chinese foundation.

After the Sino-Japanese War (1894-95) the idea of institutional reform caught on with scholars, publicists, writers, and officials. The more famous among them were Cheng Kuan-ying, the onetime compradore at the British firm of Butterfield and Swire Company, who had become a writer-modernizer and author of "Warnings to the Seemingly Prosperous Age" (*Sheng-shih wei-yen*), and Ho Ch'i, author of several works on the need for reform. They went so far as to urge the adoption of such foreign institutions as parliament and constitutional monarchy.

The general awakening to the need for reform was partially a result of missionary influence. Ever since the 1870s, a number of the more enlightened British and American Protestant missionaries had adopted the view that they should "secularize" their work and extend it beyond religious propagation to include an introduction to Western knowledge and culture. Whereas formerly the emphasis had been on "saving the heathen from the sufferings of hell," now the new concern was "to save the heathen from the hell of suffering in this world."[1] They established schools, gave public lectures, opened libraries and museums, and published newspapers and magazines—the last being a foreign privilege denied to the Chinese.

1. Timothy Richard, *Forty-Five Years in China* (New York, 1916), 197.

The famous monthly *Wan-kuo kung-pao* (The Globe Magazine), published between 1875 and 1907 (except 1883-89) in Shanghai by Young J. Allen, devoted itself to "the extension of knowledge relating to geography, history, civilization, politics, religion, science, art, industry, and general progress of Western countries." By 1889 some 16,000 Chinese had studied at the mission schools.

With the establishment in 1887 in Shanghai of the *Kuang-hsüeh hui*, or Society for the Diffusion of Christian and General Knowledge among the Chinese (SDK), the missionaries reached out more effectively into the Chinese reading public and the upper class. Among the manifold activities in which the society engaged itself were an introduction of Western civilization through translations, promotion of the cause of reform, editorials on current topics, public addresses, and discussions conducted with scholars and officials. The leading members of the society, such as the Britishers Alexander Williamson and Timothy Richard, and the Americans Young J. Allen and W. A. P. Martin, all had a very good command of Chinese. In particular, Timothy Richard (1845-1919), secretary of the society since 1891 and its Peking representative in 1895, was dedicated to the cause of institutional reform. He published *New Views on Current Affairs* (*Shih-shih hsin-lun*), wrote on the works of Peter the Great and Emperor Meiji, and translated Robert MacKenzie's *The Nineteen Century: A History*. Through the missionary efforts, the mental horizons of Chinese intellectuals became broadened, and they developed a new respect for the foreigners. Timothy Richard was much sought after not only by prominent statesmen such as Prince Kung, Weng T'ung-ho, and Li Hung-chang, but also by such radical reformers as K'ang Yu-wei and Liang Ch'i-ch'ao. Many of K'ang's ideas on reform, in fact, came from the missionaries.[2]

The Conservative Reformers: Weng and Chang

A powerful figure in Peking was Weng T'ung-ho (1830-1904), the imperial tutor who had emerged as an advocate of conservative reform. After the Japanese war, he and Chang Chih-tung had replaced Li Hung-chang as leaders of modernization, one in Peking and the other in the provinces. There was much in common in their family and educational backgrounds. Weng, the son of a grand secretary and winner of the first honor in the

---

2. K'ang admitted to a foreign interviewer: "I owe my conversion [to reform] chiefly to the writings of two missionaries, the Reverend Timothy Richard and the Reverend Doctor Young J. Allen." Cyrus H. Peake, *Nationalism and Education in Modern China* (New York, 1932), 15.

chin-shih examination of 1856, became a tutor to Emperor T'ung-chih and an expositor of classical and historical works to the two empress dowagers. In 1876 he was appointed tutor to the boy emperor Kuang-hsü, a post which he retained for twenty years and which enabled him to develop an intimate relationship with the ruler. Thus favorably placed, Weng was in a position to influence the emperor and to maintain good connections with the empress dowager, Tz'u-hsi. Deeply concerned with the future of the dynasty, which he saw was declining rapidly, and with the ever-deepening foreign encroachment which threatened the partition of China, Weng, a Confucian traditionalist, came to the reluctant conclusion that China could not survive without a reform. He was too clever a court politician not to see in the reform a chance to wrest the leadership of modernization from Li Hung-chang and Chang Chih-tung. To succeed in this venture, he knew that the support of both the emperor and the empress dowager was absolutely necessary. His every move hence was calculated to win their approval and to secure for himself the leadership in the movement. With his steep Confucian background and his eagerness to avoid irritating the conservative-minded dowager, Weng cautiously promoted a moderate reform, which would entail a limited administrative reorganization along with adoption of some Western "implements" in the Self-strengthening tradition. He was too proud a Confucianist, and too shrewd a politician, to admit that reform should go beyond that. In 1889 when the emperor assumed personal rule and the dowager formally retired to the Summer Palace, Weng presented to them copies of Feng Kuei-fen's *Protest from the Chiao-pin Studio* to promote the idea of conservative reform, but he made it clear that Chinese moral principles and ethical teachings must remain as the foundation of state, which needed to be supplemented, but by no means replaced, by Western learning. Drawn to the idea of reform, the emperor began to read Western translations in 1889 and to learn English in 1891 from two graduates of T'ung-wen kuan who were W. A. P. Martin's students.

Weng's rival, Chang Chih-tung, was also a moderate reformer, a thoroughbred Confucianist, and a superb scholar. He passed first in the *chü-jen* examinations in Chihli in 1852, and ranked third (*t'an-hua*) in the palace examinations for the *chin-shih* degree in 1863. Immersed in the Chinese cultural and ethical tradition, he was described by a missionary as "a Chinese to the backbone," to whom "there is no country like China, no people like the Chinese, and no religion to be compared with the Confucian."[3] Although he had initiated many projects of modernization in his

---

3. Hsiao Kung-ch'üan, "Weng T'ung-ho and the Reform Movement of 1898," *Tsing-hua Journal of Chinese Studies*, New Series, 1:2:153 (April 1957).

jurisdiction, Chang never advocated the alteration of basic Chinese institutions and moral teachings. In fact, it was primarily to perpetuate the established institutions and way of life, rather than introduce progress, that he adopted foreign devices and implements, which he believed had made Western countries strong and wealthy.[4] Some administrative reorganization was necessary to raise efficiency, but the basic old order was not to be touched.

Chang wanted to save China by a renaissance of Confucianism, by education and industry, and by adoption of Western science and technical know-how. In his famous work, the *Exhortation to Learning* (*Ch'üan-hsüeh p'ien*), published in 1898, he impressed upon his countrymen the importance of "knowing" five things: (1) know the shame of having fallen behind Japan, Turkey, Siam, and Cuba; (2) know the fearful fate of Vietnam, Burma, Korea, Egypt, and Poland; (3) know the impossibility of improving the machines without first improving the "methods"; (4) know the essentials in Chinese and Western learning—the former being practical rather than antiquarian studies, and the latter being political systems rather than technology; and (5) know one's origin so as not to forget one's country when traveling abroad, not to ignore one's parents and relatives when becoming acquainted with foreign customs, and not to be so clever as to disregard the sages. The first two stressed the danger of foreign encroachment, the next two his approach to reform, and the last the importance of traditional morality. His message was essentially a reaffirmation of the superiority of China's moral tradition and the wisdom of supplementing, but not replacing, it with Western science and technology.

Chang's idea of reviving Confucianism as the moral basis of state and adopting Western devices for practical use crystallized into a crisp slogan, "Chinese learning for the foundation (*t'i*), Western learning for application (*yung*)."[5] Here, Chang was actually manipulating rather than correctly interpreting the concepts of *t'i* (substance, principle) and *yung* (usefulness, application). Chinese and Western learning each had its own *t'i* and *yung;* a hybrid mixture as Chang had suggested could not endure, for the latter was bound to affect the former. Chang's formula, more clever than correct, was an effective shield against conservative attack; not even the die-hard could accuse him of disloyalty to Confucianism and Chinese heritage. Having established this unassailable position, Chang went on to argue for the need for change.

---

4. Ch'en Ch'iu, "Wu-hsü cheng-pien shih fan-pien-fa jen-wu chih cheng-chih ssu-hsiang" (Political views of antireformers during the 1898 coup d'état), *Yen-ching hsüeh-pao*, 25:61 (June 1939).
5. *Chung-hsüeh wei-t'i, Hsi-hsüeh wei-yung.*

What was immutable, he said, was basic human relations and not laws and institutions; was the way of the sages and not machines and tools; was human mind and intention and not arts and crafts. There was no shame in learning from foreigners, for had not Confucius himself said that among three persons there must be one who could be his teacher? Chinese history itself was full of institutional changes: from feudal states to unified empire, from mercenary army to the *fu-ping* militia, from chariot warfare to cavalry and infantry warfare, from ancient script to modern script, and from barter trade to cash exchange. Reforms by Shang Yang (who died in 338 B.C.), Wang An-shih (A.D. 1021-86), and others in history were well-known, and even in the Ch'ing period innovations were not lacking. The early leaders relied on horsemen and archery when they were in Manchuria but shifted to cannon to suppress the Rebellion of the Three Feudatories. Emperor Ch'ien-lung (1736-95) partially modified the examination system, and Emperor Chia-ch'ing (1796-1820) created local militia outside the banner system and the Green Standard army. Other notable changes included the introduction of the *likin* (transit dues), the creation of the Yangtze naval force, the establishment of the Sinkiang province, the building of steamers, and the opening of telegraphic lines. All this showed that change could not be resisted. With this philosophy, reinforced by his powerful base at Wuhan, Chang challenged Weng T'ung-ho's leadership in a moderate reform.

With the exception of the ultraconservatives, reform seemed to be the consensus of officials and scholars after 1895. As reactionary a figure as the Grand Secretary Hsü T'ung (1819-1900), a leader of the so-called "Northern Party" (*Pei-p'ai*) at court, conceded its usefulness and attempted to bring Chang Chih-tung to Peking to lead the movement. Weng T'ung-ho, a leader of the "Southern Party" (*Nan-p'ai*), successfully blocked the move to save the leadership for himself. To buttress his position, Weng sought out promising young scholars and officials far his junior in status and age to assist him, men who would not threaten his leadership. Among them was one K'ang Yu-wei, whom he brought to the attention of the emperor with the intention of making him his chief assistant in the moderate reform. But K'ang was of a very different breed than Weng had thought—he was, in fact, a radical reformer with a program of his own.

The Radical Reformers: K'ang and Liang

K'ang Yu-wei (1858-1927) was a most unusual man whose intellectual development swung dramatically from one extreme to the other. Born into a well-to-do scholarly family in Nan-hai, Kwangtung, K'ang was a child prodigy, composing essays in the classical style at the early age of seven *sui*. His devotion to the sages and his frequent citation of their teachings won him the sobriquet "The Sage Wei." At eighteen *sui*, he became a student of the great Cantonese Neo-Confucian scholar Chu Tz'u-ch'i, who stressed the political history of China and the importance of uniting scholarship with public affairs. K'ang sat at his feet for a number of years and gained a solid foundation in Neo-Confucianism.

After leaving his teacher, K'ang retired to a mountain[6] to meditate, hoping to develop a school of thought of his own. His intellectual background up to this point had been entirely traditional and free from Western influence. Emerging from the self-imposed seclusion after two years,[7] he went to Peking. On his return trip, he visited Shanghai (1882), as he had Hong Kong earlier. The orderliness and efficiency of the municipal governments in these British-dominated cities made a deep impression on him. If Western colonial administration could produce such good results, he wondered, how much more progressive must be the mother countries themselves! His interest in the West thus aroused, K'ang eagerly purchased and read all available translations of works put out by the Kiangnan Arsenal and missionary organizations, including the *Wan-kuo kung-pao*. Abruptly a totally new vista was opened to him: he realized the backwardness of China and her dangerous position in the age of imperialism. He accepted the missionary view that progress as demonstrated by the Western nations was not only necessary but also desirable. Decisively he abandoned his plans to try for the civil service examinations in 1883, and turned his attention to new Western studies.[8]

In 1888, while yet a plebeian (i.e. without an official appointment), he attempted to present a memorial to the throne, in which he praised Japan's modernization along the lines of the Western powers, urged that China do likewise, and warned of the increasing threat of foreign encroachment. The Imperial College (*Kuo-tzu chien*), to which this memorial was submitted for transmittal to the throne, refused to forward it, suspecting its author to be insane. K'ang realized that to succeed in promoting reform he must (1) seize the intellectual leadership of the learned world, and (2) win over the emperor.

K'ang returned to Kwangtung to teach and write. His reputation as an unorthodox eccentric attracted young scholars, among whom was one Liang Ch'i-ch'ao (1873-1929), himself a child prodigy who attained the *chü-jen* degree at the age of seventeen *sui*. Liang was immediately im-

---

6. Hsi-chiao shan.
7. Some accounts say four years.
8. Lo Jung-pang (ed.), *K'ang Yu-wei: A Biography and a Symposium* (Tucson, 1967), 38.

pressed by K'ang and became his student. At the urging of his students, K'ang opened a school, the *Wan-mu ts'ao-t'ang* (The Grass Hut amid a Myriad of Trees) at Canton in 1891, in which he gave instruction in classical studies and promoted the idea of reform. Frequently he visited the library of the nearby Episcopal Mission to read books on representative government and constitutional monarchy.

THE MODERN TEXT MOVEMENT. K'ang's intellectual orientation by this time had undergone a radical change. Starting out as a Neo-Confucian scholar like many of his contemporaries, K'ang was now fired with the zeal for a Western-style political reform. The writings of Liao P'ing,[9] a proponent of the Modern Text school of classical learning, so impressed him with ideas useful to reform that he gave up all his past intellectual leanings. The Modern Text movement, he discovered, could be used as a vehicle to advance his cause.

"Modern Text" (*Chin-wen*) referred to the classics and their commentaries of the Ch'in (221-206 B.C.) and Han (202 B.C.-A.D. 220) periods, as opposed to the Ancient Text (*Ku-wen*) of earlier times. The book-burning of the First Emperor of Ch'in in 213 B.C. had supposedly destroyed all the ancient classics, and scholars of the succeeding dynasty, the Former Han, accepted classical texts written in the current script, the "seal characters," as authentic. These Modern Text scholars had dominated the intellectual world of the Former Han period, but toward the end of it a descendant of Confucius[10] claimed to have discovered, in the walls of his ancestral home, texts of the classics written in the ancient "tadpole characters." Although most scholars of the time doubted the authenticity of these Ancient Texts, one Liu Hsin (*ca.* 46 B.C.-A.D. 23) fought to establish them during the short-lived Hsin dynasty (A.D. 8-23) under the usurper Wang Mang. With the downfall of the Hsin dynasty and the restoration of the Han, now known as the Later Han (A.D. 25-220), the school of Ancient Texts declined. However, toward the end of the Later Han period, several great Ancient Text scholars appeared, including the master Cheng Hsüan (A.D. 127-200) who dominated the learned world. From then on, the Ancient Texts were ascendant at the expense of the Modern Texts.

The revival of antiquarian studies in the Ch'ing period and the concommitant interest in determining the authenticity of classical texts renewed the age-old issue of Ancient Texts versus Modern Texts. A favorite subject on which the Ch'ing Modern Text scholars focused their atten-

9. Liao P'ing, *Chin-ku hsüeh k'ao* (A study of Modern and Ancient Texts), (1886).
10. K'ung An-kuo.

tion was the *Kung-yang Commentary,* a lost subject for two thousand years.

K'ang Yu-wei determined to seize the leadership of the Modern Text movement, synthesize its key conceptions into works of his own, and invoke them to support his advocacy of institutional change. In 1891 he completed his first major book, *A Study of the Forged Classics of the Hsin Period (Hsin-hsüeh wei-ching k'ao),* in which he exposed as forgeries such classics as *The Rites of Chou,* the *Dispersed Rituals,* the *Tso Commentary,* and the *Mao Commentary on the Book of Odes.* Daringly K'ang argued: (1) the Ch'in book-burning did not impair the Six Classics; the Confucian texts had been transmitted intact to later generations; (2) consequently, there was no such thing as the Ancient Texts in the Former Han period; (3) the written character used at the time of Confucius was the same as that of the Ch'in and Han, namely the "seal character"; and (4) the so-called Ancient Texts were forged by Liu Hsin as part of a conspiracy to distort the Confucian "great principles hidden in esoteric language," with a view to helping Wang Mang usurp the Han throne.[11] Historical accuracy apart, K'ang's incisive argument, bold imagination, and piercing criticism struck the Ch'ing intellectual world like a hurricane. His attack on the Ancient Texts stimulated a spirit of doubt and pointed up the need for reappraising ancient books.

In 1897 his second book was completed, *A Study of Confucius on Institutional Reform (K'ung-tzu kai-chih k'ao).*[12] In it he boldly advanced the thesis that men in the past were mistaken to say that Confucius merely edited the Six Classics; he had, in fact, written them and had intended by them to promote institutional reform. Other philosophers of the Chou (1122-256 B.C.) and Ch'in (221-206 B.C.) had likewise advocated institutional reform—all of them, like the master, justifying their action on the pretext of imitating the past. They created an idealized golden past, irrespective of historical facts, to convince contemporary rulers of the wisdom of reform, as the ancient rulers Yao (2357-2256 B.C.) and Shun (2255-2206 B.C.) had done. By inference, K'ang in effect argued that since institutional reform had been championed by the sage Confucius and other great philosophers of the past, it could scarcely be morally wrong. With this skillful twist, K'ang sought the sanction of the most honored master as a shield against the antireformers.

Certain cryptic ideas of the Modern Text School were fully exploited by K'ang to advance his cause. The concept of *T'ung san-t'ung,* or "Going through the Three Periods of Unity," he interpreted to mean that the

11. Liang Ch'i-ch'ao, *Intellectual Trends,* 92.
12. Often translated as *Confucius as a Reformer.*

three ancient great dynasties—Hsia (2205-1766 B.C.), Shang (1766-1122 B.C.), and Chou (1122-256 B.C.)—were quite distinct from each other; hence changes and reform were inherent in the very nature of history. Another expression, *Chang san-shih,* or "Unfolding of the Three Epochs," was interpreted by him to mean that the world progressed from "The Epoch of Disorder" (*Chü-luan shih*) to that of the "Rising Peace" (*Sheng-p'ing shih*) and ultimately to that of "Universal Peace" (*T'ai-p'ing shih*). In short, more changes, more progress. Actually, K'ang did not originate these ideas but borrowed them from Liao P'ing.[13] Nevertheless, he synthesized and elucidated the existing Modern Text ideas to strike the learned world with his unusual interpretations and to prove that changes and reforms were inevitable in human development. If his first book was a hurricane, the second was an earthquake. The literati were shocked by his unorthodox exposition: die-hard conservatives accused him of "deluding the world and deceiving the people," while correct Confucianists branded his interpretations "wild and foxy."[14] Be that as it may, K'ang's reputation soared as a leading proponent of the Modern Text school, although his second book was banned.

These two books were basically reinterpretations of ancient works, but another of K'ang's work, the *Ta-t'ung shu* (The book of universal commonwealth), completed earlier in 1887, was a creative piece of his own, very progressive in its contents. Many ideas in this book, however, were influenced by the ancient work, *The Evolution of Li* (*Li-yün*), which said in part:

> When the Grand Course was pursued, a public and common spirit ruled all under the sky . . . men did not love their parents only, nor treat as children only their own sons. A competent provision was secured for the aged till their death, employment for the able-bodied, and the means of growing up to the young. They showed kindness and compassion to widows, orphans, childless men, and those who were disabled by disease, so that they were all sufficiently maintained. Males had their proper work, and females had their homes. [They accumulated] articles [of value], disliking that they should be thrown away upon the ground, but not wishing to keep them for their own gratification. [They labored] with their strength, disliking that it should not be exerted, but not exerting it [only] with a view to their

13. Ku Chieh-kang, *Tang-tai Chung-kuo shih-hsüeh* (The study of Chinese history today), rev. ed. (Hong Kong, 1964), 42.
14. For all his unconventional views, K'ang remained within the framework of Confucian school; he was a Confucian revisionist rather than a traditionalist. See Hsiao Kung-chüan, "K'ang Yu-wei and Confucianism," *Monumenta Serica,* vol. XVIII (1957), 100, 200.

own advantage . . . This was [the period of] what we call the Grand Union [Universal Commonwealth].[15]

Inspired by these utopian ideas, K'ang envisaged an ideal world in which:

1. there would be no nations: the whole world would be divided into different regions under a single government;
2. the central and regional governments would be popularly elected;
3. there would be no family or clans but rather cohabitation of men and women for the duration of one year, after which everyone would change mates;
4. institutions for prenatal education would be established for pregnant women, and nurseries for babies;
5. children would go to school from kindergarten up according to age;
6. adults would be assigned by government to work in agriculture, industries, and other productive enterprises;
7. there would be hospitals for the sick and Old Folks Homes for the aged;
8. there would be public dormitories and dining halls for the enjoyment of all classes according to their working income;
9. there would be special rewards for inventors, discoverers, and those who serve with distinction in the establishments for prenatal education, the nurseries, the kindergartens, the hospitals, and the Old Folks Homes;
10. the dead would be cremated and fertilizer factories erected in the neighborhood of the crematoria.[16]

This work of utopian socialism was shown to his students but kept from the public, as K'ang averred that the contemporary age was merely the disordered epoch, in which one could only speak of "Partial Security" (*Hsiao-k'ang*) and not "Universal Commonwealth" (*Ta-t'ung*). Students at the *Wan-mu ts'ao-t'ang* were greatly excited by these new ideas and engaged in daily discussions of them.

K'ANG'S DRIVE FOR RECOGNITION. K'ang had built up a resounding reputation for himself, but he still lacked the higher degrees that would qualify him for official appointment. His talented student Liang had already achieved the *chü-jen* degree in 1889, but K'ang did not win his until 1893. In 1895 the two of them went to Peking together for the triennial metropolitan examinations. It was a time of national humiliation, for

15. James Legge, *The Sacred Books of China*, Part III, *The Li Ki* (Oxford, 1885), 364-66, with minor changes.
16. Liang Ch'i-ch'ao, 96-97.

Japan had defeated China and was dictating the peace at Shimonoseki. K'ang and Liang, in righteous anger, prepared a ten thousand-word memorial and gathered the signatures of 603 provincial graduates[17] to protest the peace treaty. The occasion, known as the *Kung-che shang-shu,* or public vehicles presenting a memorial—"public vehicle" being the nickname for the provincial graduates who had come to Peking by public transportation for the metropolitan examinations—was considered by some as the first "mass political movement" in Modern China. They urged the Ch'ing court (1) to reject the peace treaty; (2) to move the capital and continue the war; and (3) to initiate institutional reform. Pointedly they stated: "If the institutional reform had been undertaken earlier, there would have been no disaster today; if the institutional reform is undertaken now, it can avert future disaster; if not, the future disaster will be worse than the present one."[18] The Censorate, to whom this memorial was addressed, refused to present it to the throne because of its blunt language and emotional overtone.

K'ang's daring mobilization of the provincial graduates, his devastating views on the classics, and his advocacy of reform were highly irritating to the conservatives. At the metropolitan examinations Hsü T'ung, the chief examiner, was determined to flunk him. Since all the papers were anonymous, Hsü could only look for the one with an eccentric style and unorthodox views, for which K'ang was noted. When the examination results were announced, the paper which Hsü had rejected turned out to be Liang's, whereas K'ang's was a model of virtue in strict conformity with Confucian morality and Chinese traditionalism. Although K'ang and Liang had succeeded in fooling Hsü T'ung, at the palace examinations that followed, the examiner[19] deliberately discriminated against K'ang. Consequently, although he won the *chin-shih* degree, K'ang was not appointed to the coveted Hanlin Academy but to the Board of Public Works, the least of the Six Boards, as a "secondary secretary" (*chu-shih*). Too proud to take up the assignment, K'ang decided instead to concentrate on capturing the imperial attention through a barrage of memorials. But the mere position of a sixth-rank secretary did not qualify him to address the throne directly; he still had to request his Board, or some other office, to forward the memorials.[20]

17. Often incorrectly given as 1,200 or 1,300. See Liu Feng-han, *Yüan Shih-k'ai yü Wu-hsü cheng-pien* (Yüan Shih-k'ai and the *coup d'état* of 1898), (Taipei, 1964), 197.
18. Li Shou-k'ung, 591.
19. Li Wen-t'ien.
20. His first two memorials, made in 1888 and 1895 as noted before, were not forwarded by the Imperial College and the Censorate to the throne.

K'ang's third memorial of May 29, 1895, suggesting methods of enriching the country, cultivating the people, educating students, and training the army, was forwarded by the Censorate to the throne on June 3. Impressed with the views expressed, the emperor ordered that copies be made for the empress dowager, the Grand Council, and the various provincial authorities. This marked the beginning of the imperial awareness of K'ang. However, his next memorial of June 30, 1895, suggesting the successive steps of reform and the opening of a parliament, was blocked by both the Censorate and the Board of Public Works.

K'ang and Liang now turned their attention to forming and participating in a number of "study societies" (*hsüeh-hui*) and newspapers. They joined the "Society for the Study of National Strengthening" (*Ch'iang-hsüeh hui*) in September 1895, other members including Sun Chia-nai, another imperial tutor and past president of several boards, Yüan Shih-k'ai, and several dozens of Britishers and Americans. The conservative reformers Weng T'ung-ho and Chang Chih-tung showed interest in the society, and Chang even made a contribution of 5,000 taels. The society sponsored lectures on reform every ten days, and engaged in a variety of other activities such as translation of Western and Japanese works, publication of newspapers, and establishment of libraries, museums, and an institute of politics. K'ang personally contributed funds toward the publication of a daily called the *Wan-kuo kung-pao*,[21] which had a circulation of 2,000 under the editorship of Liang. Many of the ideas in the paper relating to reform were borrowed from the publications of the missionary organization *Kuang-hsüeh hui* (SDK). K'ang himself met with Timothy Richard while Liang offered to serve as his secretary, and there developed a degree of mutual support between SDK and the reformers' organization.

At Shanghai, as many as thirty newspapers and magazines were in circulation to promote reform, and at Tientsin in November 1897 there appeared the prestigious *Kuo-wen pao* (National review) under the editorship of Yen Fu (1854-1921), a famous graduate of the naval school of the Foochow Dockyard and a translator of Western works (see Chapter 17). Here he published his translation of Thomas Huxley's *Evolution and Ethics* to introduce the Darwinian ideas of struggle and the survival of the fittest. In Hunan province, the progressive governor Ch'en Pao-chen invited Liang to be the chief instructor in a newly established School of Current Affairs (*Shih-wu hsüeh-t'ang*) at Changsha. Liang's views on misgovernment, the need for reform, and popular sovereignty found full expression there. The progressives then established a China Reform Asso-

---

21. Named after the missionary magazine as a tribute.

ciation to stimulate group discussion, and published The Hunan Daily (*Hsiang-pao*) and The Hunan Journal (*Hsiang hsüeh-pao*). The inland province of Hunan, long known for its conservatism, was transformed overnight into a progressive center.

As for K'ang himself, he traveled, lectured, and promoted the cause of reform in several provinces. Within three years, he had influenced the creation of many study societies, schools, and newspapers, most of which were in Hunan, Kiangsu, Kwangtung, and Peking.

THE RISE OF K'ANG YU-WEI. The German lease of Kiaochow in 1897 and the subsequent scramble for concessions by other powers precipitated a new national crisis. K'ang Yu-wei hurried to Peking to present his fifth memorial, warning of the danger of partition and the pressing need for reform. He recommended that the emperor pursue three courses of action: (1) proclaim a national policy on reform after the fashion of Peter the Great and Emperor Meiji—"May your Highness adopt the heart of Peter the Great of Russia and the administration of Meiji of Japan!" (2) gather all the talents of the country to prepare for an institutional reorganization; and (3) allow provincial authorities to initiate institutional reform within their own jurisdictions. The memorial closed with a warning that any delay would invite further foreign encroachment and ultimate extinction of the dynasty. The president of the Board of Public Works refused to forward the memorial because of its bluntness; its contents, however, soon became common knowledge in Peking and Shanghai. Nonetheless, the memorial could not reach the throne. In disgust, K'ang thought of returning to the south but was persuaded to remain by Weng T'ung-ho, who, baffled by his own limited knowledge of foreign affairs and threatened by Chang Chih-tung's bid for leadership in the conservative reform, had secretly hoped to enlist K'ang as an aide in his own movement. Weng supported the recommendation of the Supervising Censor Kao Hsieh-tseng on January 11, 1898, that K'ang be granted an imperial audience, remarking to the emperor that K'ang's ability was a hundred times superior to his own and that it behooved the emperor to hear him (K'ang) on matters of reform. Emperor Kuang-hsü was then ready to grant K'ang an audience, but Prince Kung reminded him that court rules permitted no interviews for officials below the fourth civil rank. Reluctantly the emperor gave in, but ordered that K'ang be received by high officials at the Tsungli Yamen.

The celebrated interview, the first official airing of K'ang's views, took place on January 24, 1898. The highlights of the meetings, as given by K'ang himself, included the following exchanges:

Jung-lu:[22] "The institutions of the ancestors cannot be changed."

K'ang: "The institutions of the ancestors are used to govern the realm
that had been theirs. Now we cannot preserve the realm of the
ancestors; what is the use for their institutions? . . ."

Liao Shou-heng, president of the Board of Punishment: "How should
the institutions be reformed?"

K'ang: "We shall change the laws and regulations; the governmental
system [*kuan-chih*] should be the first [to be reformed]."

Li Hung-chang: "Shall we, then, abolish all the Six Boards and throw
away all the existing institutions and rules?"

K'ang: "The present is a time in which countries exist side by side;
the world is no longer a unified one. The laws and governmental
system [as they now exist in China] are institutions of a unified
empire. It is these that have made China weak and will ruin her.
Undoubtedly, they should be done away with. Even if we could
not abolish them all at once, we should modify them as circumstances require. Only so can we carry out reform."[23]

The interview lasted until dusk. Jung-lu was the first to leave, apparently
disgusted with what he had heard. Weng T'ung-ho, also present at the
meeting and somewhat disturbed by K'ang's radical views, described him
as "high-flown" and "very crazy."

When the report on the interview reached the emperor, he was eager
to meet K'ang but was again blocked by Prince Kung. However, Kuang-
hsü ordered on January 29 that K'ang be allowed to present memorials
any time without obstruction or delay by court officials. K'ang's access to
the ruler was thus assured. One of his earlier unforwarded memorials now
reached the emperor, who was deeply moved by a blunt statement in it to
the effect that without reform the sovereign might not even have the
chance of becoming a commoner in the future, and might very well end
up in the same pathetic way as the last emperor of the Ming dynasty who
hanged himself. The emperor remarked that only a man of complete devotion could have made such a straightforward statement at the risk of his
life! Kuang-hsü's trust in K'ang rose steadily.

On January 29 K'ang presented his sixth memorial, in which he asked
that the emperor decide on a national policy, select talents for public service, and create a "Bureau of Government Institutions" (*Chih-tu chü*) to
assist in reform and to draft a constitution. In addition, twelve administrative bureaus, each resembling a European ministry, should be established:

22. Manchu general and a confidant of the dowager; president of the Board of War in
    1895 and formerly commandant of the Peking Gendarmerie.
23. Hsiao Kung-ch'üan, "Weng T'ung-ho," 175-76.

Jurisprudence, Finance, Education, Agriculture, Industry, Commerce, Railway, Postal Service, Mining, Cultural and International Exchange, Army, and Navy. In the provinces, a "Bureau of People's Affairs" (*Min-cheng chü*) should be created in each circuit, with branches in the districts. The director of the circuit bureau should have the equivalent status of the governor-general and governor, while the district branch officer should take charge of all administrative matters such as education, public health, agriculture, and police, leaving only lawsuits and revenue to the regular magistrate. Emperor Kuang-hsü, much impressed with these novel ideas, asked the princes and the ministers of the Tsungli Yamen to deliberate on them.

K'ang followed up with a seventh memorial in February 1898, repeating the suggestion that the emperor follow the examples of Peter the Great and Emperor Meiji. To acquaint the ruler with the reforms of foreign countries, K'ang presented his own works, "A Study of Meiji Reform" and "A Study of the Reform of Peter of Russia," as well as Timothy Richard's translation of "An Outline of New Western History" and other works on the reforms of the various countries. Reading these manuals daily, the emperor was more than ever determined to effect an institutional change.

On the death of Prince Kung on May 30, 1898, K'ang urged Weng T'ung ho to forge ahead with reform at once. The latter, greatly annoyed by K'ang's fast-rising reputation and his growing influence with the emperor, now considered K'ang a threat to his own position. Weng urged him to leave Peking to escape the conservative attack and impeachment, but K'ang was unconcerned. On June 8 he presented his eighth memorial, followed by another shortly afterwards, requesting again that the emperor take the decisive step of proclaiming a national policy. On June 11, 1898, Emperor Kuang-hsü acceded to the request and issued the first reform decree, urging the princes, the officials, and the commoners alike to strive to learn the useful foreign knowledge without sacrificing the basic Chinese moral teachings. The Hanlin reader Hsü Chih-ching then recommended that the emperor receive K'ang in person. The audience took place on June 16. Some highlights of the five-hour interview, as given by Liang Ch-i-ch'ao, were as follows:

> After the emperor had asked about his [K'ang's] age and his qualifications, K'ang stated: "The four barbarians are all invading us and their attempted partition is gradually being carried out: China will soon perish."
> The emperor: "Today it is really imperative that we reform."
> K'ang: "It is not because in recent years we have not talked about re-

form, but because it was only a slight reform, not a complete one; we change the first thing and do not change the second, and then we have everything so confused as to incur failure, and eventually there will be no success."

"The prerequisites of reform are that all the laws and the political and social systems be changed and decided anew, before it can be called a reform. Now those who talk about reform only change some specific affairs, and do not reform the institutions."

The emperor consented to K'ang's suggestions that a bureau be established to study the various systems, and stated: "Your reform program is very detailed."

K'ang: "Your Majesty's sagacity has already noted it. Why not vigorously carry it through?"

The emperor glanced outside the screen and then said, with a sigh, "What can I do with so much hindrance?"

K'ang: "According to the authority which Your Majesty is now exercising to carry out the reforms, if he works on only the most important things, it will be sufficient to save China, even though he cannot make a complete reform. Nevertheless, today most of the high ministers are very old and conservative, and they do not understand matters concerning foreign countries. If Your Majesty wishes to rely on them for reform it will be like climbing a tree to seek for fish."

After a long pause the emperor nodded and said, "You should withdraw and take a rest. . . . If you have something more to say you may prepare memorials to state your suggestions in detail and send them here."

As K'ang rose to leave, the emperor's eyes escorted him to the door. The palace attendants opined that there had never been an audience as long.[24]

On the same day, June 16, K'ang was appointed a secretary of the Tsungli Yamen. Three days later, he again presented a memorial, through the Yamen, requesting the adoption of a national policy on reform and the establishment of a bureau for governmental institutions. Completely won over by K'ang, the emperor ordered that hereafter he need not present memorials through any agency but should send them directly to the throne. Furthermore, he asked for several of K'ang's works, "The Partition of Poland," "A Study of Reform in France," "A Study of Reform in Germany," and "A Study of Reform in Britain." Kuang-hsü was now fully convinced of the urgency of institutional change. At the age of forty, K'ang had captivated the emperor and become the leader of a radical reform.

24. Teng and Fairbank, *China's Response*, 177-79, with minor changes.

The "Hundred-Day" Reform

K'ang's views on reform, which had been evolved over a ten-year period prior to 1898, may be summarized as follows: The existing political institutions and administrative procedures of China, he believed, were designed at a time when China was a world in itself, free from involvement with Western powers. The primary consideration of the ruling dynasty then was to prevent domestic rebellion and uprising; hence the cumbersome system of checks and surveillance in central and local administrations and the impractical nature of the civil service examinations. Now that times had changed and internal security was no longer the sole concern of state, the old imperial system had become totally outdated. The government must consider new problems of foreign relations and industrialization and modernize its structure accordingly. But to effect this basic change, the emperor must wrest power from the empress dowager, who in K'ang's view was the major obstacle to progress. As early as 1888, in his first memorial, K'ang had said, "The affairs of the empire remained in a sorry state as a result of the evil influences of eunuchs and palace maids." In his fourth memorial of June 30, 1895, he again urged the emperor to tidy his administration and "make decisions according to his own sage wishes alone." The reforms of Peter the Great and Emperor Meiji were repeatedly cited as examples for Kuang-hsü; in particular, the Japanese experience was stressed as worthy of following in view of the geographical proximity and cultural and social affinity between China and Japan. In more concrete terms, K'ang proposed (1) revision of the examination system and the legal code; (2) establishment of a governmental institution bureau and creation of twelve new bureaus to render useless the Grand Council, the Six Boards, and other existing offices; (3) establishment of bureaus of people's affairs in the circuits and branches in the districts as an embryonic form of local self-government; (4) creation of a parliament in Peking; (5) establishment of a national assembly (*kuo-hui*); and (6) adoption of a constitution and the principle of division of power between the executive, the legislative, and the judiciary. In short, K'ang envisaged a constitutional monarchy to replace the age-old "imperial Confucian" system.[25]

Such grandiose plans were far beyond the dream of the conservative reformers. Weng T'ung-ho was shocked beyond himself. Having lost to K'ang the leadership of reform and imperial patronage, Weng was even more distressed by the conservatives' attack on him for having introduced K'ang to the emperor. Weng therefore turned to block K'ang's work. On

---

25. Hsiao Küng-ch'üan, "Weng T'ung-ho," 162-64.

May 26, 1898, when the emperor asked him to gather together a set of K'ang's writings, Weng spoke derogatorily: "I do not associate with K'ang . . . this man's intentions are unpredictable." When the emperor asked why he had not mentioned this before, Weng replied: "Your servant discovered this recently upon reading his *Confucius as a Reformer*." The emperor could not understand Weng's sudden change of attitude and was deeply hurt by his contemptuous remarks about K'ang, for whom he, the emperor, had now developed a deep respect and fondness. The long, trusting, and affectionate relationship between Weng and his imperial pupil was discernibly strained. The emperor became receptive to the motion to dismiss him, engineered by K'ang's supporters to clear the way for their leader. They impeached Weng on charges of crimes ranging from accepting bribes to unconscionable behavior in the Board of Revenue, of which he was the long-time president. With the approval of the empress dowager, who hated Weng for having led the emperor astray and for having introduced K'ang to him, Weng was relieved of all official duties on June 15.

The emperor and K'ang now forged ahead in their bold program of reform. Kuang-hsü did not dare appoint K'ang to the all-important Grand Council for fear of the empress dowager, but he placed K'ang's assistants in several key positions: Liang Ch-i-ch'ao was given the sixth civil rank on July 3 to take charge of a translation bureau, and on September 5 four progressives—Yang Jui, Liu Kuang-ti, Lin Hsü, and T'an Ssu-t'ung—were given the fourth civil rank and made secretaries in the Grand Council. Two of them, Lin and T'an, were K'ang's students. These four secretaries became the link between the emperor and K'ang; they were the *de facto* executives of reform, drafting all important decrees and reading all important memorials, relating to institutional changes, while the unsympathetic grand councillors were bypassed.

The spirit of the "New Deal" (*Hsin-cheng*) was manifested in the September 12, 1898, edict, which took the unprecedented and liberal view that the basic principles of government were the same in China as in the West and that reform was merely putting into effect those methods and principles that had proved sound, valid, and useful in the West:

> In revitalizing the various administrative departments our government adopts Western methods and principles. For, in a true sense, there is no difference between China and the West in setting up government for the sake of the people. Since, however, Westerners have studied [the science of government] more diligently (than we), their findings can be used to supplement our deficiencies. Scholars and officials of today whose purview does not go beyond China, [regard Westerners] as practically devoid of precepts or principles. They do not know that the

science of government as it exists in Western countries has very rich and varied contents, and that its chief aim is to develop the people's knowledge and intelligence and to make their living commodious. The best part of that science is capable of bringing about improvements in human nature and the prolongation of human life.[26]

For 103 days, from June 11 to September 20, some forty to fifty reform decrees were issued in rapid succession in the areas of education, government administration, industry, and international cultural exchange:

I. *Education.*
   A. Replacement of the eight-legged essay in the civil service examinations by essays on current affairs (June 23, 1898).
   B. Establishment of an Imperial University at Peking (June 11, August 9).
   C. Establishment of modern schools in the provinces devoted to the pursuit of both Chinese and Western studies. Transformation of large private academies (*shu-yüan*) in the provincial capitals into colleges, of those in the prefectural capitals into high schools, and of those in the districts into elementary schools (July 10).
   D. Establishment of a school for the overseas subjects (August 6).
   E. Creation of a medical school under the Imperial University (September 8).
   F. Publication of an official newspaper (July 26).
   G. Opening of a special examination in political economy (July 13).
II. *Political Administration.*
   A. Abolition of sinecure and unnecessary offices, including:
      1. The Supervisorate of Imperial Instruction, the Office of Transmission, the Banqueting Court, the Court of State Ceremonial, the Imperial Stud, the Court of Sacrificial Worship, and the Court of Judicature and Revision.
      2. The governorships of Hupeh, Kwangtung, and Yunnan.
      3. The director-generalship of the Yellow River, the circuit intendants for grain transport, and the salt intendants (August 30).
   B. Appointment of the progressives in government (September 5).
   C. Improvement in administrative efficiency by eliminating delays and by developing a new, simplified administrative procedure (June 26).
   D. Encouragement of suggestions from private citizens, to be for-

26. Hsiao Kung-ch'üan, "Weng T'ung-ho," 165.

warded by government offices on the day they are received
(September 11).

    E. Permission for the Manchus to engage in trade (September
14).

III. *Industry.*

    A. Promotion of railway construction (June 25).

    B. Promotion of agricultural, industrial, and commercial developments (June 20).

    C. Encouragement of invention (July 5).

    D. Beautification of the Capital (September 5).

IV. *Others.*

    A. Tour of foreign countries by high officials (June 12).

    B. Protection of missionaries (June 12).

    C. Improvement and simplification of legal codes (July 29).

    D. Preparation of a budget (September 16).

Although Emperor Kuang-hsü and K'ang Yu-wei vigorously pushed the reform program, it was boycotted by most of the high officials in the central and provincial administrations. The abolition of the eight-legged essay met with strong opposition from the Board of Rites, which was in charge of the civil service examinations. The Tsungli Yamen, though more liberal, frowned upon the proposal of twelve new bureaus. As to the provincial authorities, all but the governor of Hunan, Ch'en Pao-chen, ignored or delayed the orders for reform. These central and local officials dared to challenge or disregard the emperor's orders in full knowledge of the fact that the real power of state was not in his hands but in those of the empress dowager, who was ill-disposed toward the reform.

The Empress Dowager and the Coup D'État

Though in retirement at the Summer Palace since 1889, the Empress Dowager Tz'u-hsi still held the reins of government tightly. So conscious was she of her ultimate power that she could not tolerate any move—be it conservative or radical—to undermine her supreme status.

Any radical change in the political and social systems that affected the precepts of the Confucian ethical code, particularly the concept of filial piety, on which her position and authority rested, was a threat. She therefore warned against burning the "ancestral tablets," rash action, and imitating Japan—the last was simply too humiliating.[27] A conservative reform

27. Upon learning that the emperor was contemplating inviting the visiting Japanese statesman, Itō Hirobumi, to be his chief adviser in reform, the dowager became fearful lest the experienced and able Japanese turn the "New Deal" into a success and

such as advocated by Weng T'ung-ho or Chang Chih-tung would be more to her taste, and the latter's attractive slogan, "Chinese learning for foundation, Western learning for practical application," suited her mentality. At the outset of the "Hundred-Day" Reform, she reportedly told the emperor: "So long as you keep the ancestral tablets and do not burn them, and so long as you do not cut off your queue, I shall not interfere."[28] In short, she would accept a moderate reorganization that did not upset the basic institutions or threaten her authority.

However, as the reform progressed, the dowager became alarmed by the abolition of the eight-legged essays, by the elimination of the sinecure offices and the three governorships, and by a host of other radical changes that swept away the ancestral institutions and traditional procedure in administration. Intuitively she came to regard the reform as a concealed scheme of wresting power from her, which indeed was just what K'ang and the progressives intended it to be. The issue of reform now became a power struggle between the emperor and the empress dowager, and the conflict was sharpened after the death in June 1896 of the emperor's mother, the dowager's sister, who had served as a cushion between the two. With the dismissal in June 1898 of Weng T'ung-ho, who had attempted to reconcile the two, any hope of compromise was shattered. The dowager was determined to teach the emperor a lesson, and in this she was influenced by the chief eunuch, Li Lien-ying, whose corrupt practices at the inner palace would not be tolerated by the progressives.

With the support of Jung-lu, she instigated a censor to request that she and the emperor review troops at Tientsin in October, at which time Jung-lu and his armed forces were to stage a *coup d'état* to depose Kuang-hsü.[29] The emperor vowed that he would not go to the review. Rumor was rife in Peking that the emperor would soon be deposed. K'ang suggested that the emperor establish a new capital at Shanghai, cut his queue, change his attire, and adopt a different reign title to mark a new start. Plans were also made to approach Yüan Shih-k'ai, Jung-lu's subordinate who was training a new army of 7,000 men near Tientsin and who had previously shown symapthy toward reform and the *Ch'iang-hsüeh hui*. On September 14 Yüan arrived in Peking, and two days later was received

---

liberate the emperor from her control. She therefore insisted on sitting behind a hidden screen during the emperor's audience with Itō. Thus inhibited, the emperor could not discuss matters of substance but merely exchanged formalities with the visitor. See Hsiao I-shan, IV, 2, 122-23.

28. Hsiao Kung-ch'üan, "Weng T'ung-ho," 142-43, 145.

29. A recent study suggests that the story of deposing the emperor was fabricated by K'ang's party to discredit the conservatives. See Liu Feng-han, 169.

by the emperor, who praised him for his accomplishments in army-training and sponsorship of new schools, and conferred on him the title of expectant vice-president, with a hint that hereafter he could act independently of Jung-lu.

On September 18, Jung-lu transferred troops to Tientsin and Peking, and ordered Yüan to return. Yüan temporized with a reply that he was still awaiting an audience with the emperor. Sensing the urgency of the situation, the reformers sent T'an Ssu-t'ung to see Yüan that night (September 18), urging him to protect the emperor at the forthcoming review. T'an prevailed upon Yüan (1) to besiege the Summer Palace, and (2) to kill Jung-lu, while he himself (T'an) would undertake to send assassins to dispose of the "Old Fogey" (i.e. dowager). Yüan tactfully parried any commitment, cautioning against hasty action, and stalled T'an with a vague suggestion that during the forthcoming review at Tientsin the emperor should hurry to his camp and give the order to kill Jung-lu.[30] When T'an informed K'ang of what transpired, it was clear that Yüan would not cooperate. K'ang decided to flee the capital.

On September 20, during an audience with Yüan, the emperor appears to have given him a very secret decree appointing him governor-general of Chihli after he had completed his mission. At three o'clock in the afternoon Yüan returned to Tientsin and unfolded the whole plot to Jung-lu, who quickly took the five-o'clock train for Peking.[31] Fearing that any delay might cause complications, the dowager's party advanced the date of the coup. On September 21, she raided the emperor's palace and intercepted all reform documents. On that very day, she announced publicly that a serious illness had incapacitated the emperor, making it imperative that she take over the administration. For the third time in her life the dowager returned to administer state affairs behind a silk curtain, while the emperor was put under detention on a small island in the Imperial Garden west of the palace. The reform came to an abrupt end after 103 days.

Orders were quickly issued to arrest K'ang and the reformers. K'ang had already left Peking a day earlier, taking an English steamer from Taku to Shanghai. Meanwhile, the British government had instructed its consul at Shanghai to rescue K'ang. Under the protection of a British war-

---

30. Yüan later published his diary to whitewash himself, saying that he was too stunned to agree to T'an's suggestion of sending troops to besiege the Summer Palace and kill Jung-lu; that he temporized with T'an because the latter was a new dignitary and had come with a "protruding object" in his pocket, i.e. a pistol. Yüan implied that whatever promise he gave was exacted under duress. See Yüan Shih-k'ai, *Wu-hsü jih-chi* (My diary of 1898), (1909, 1922).

31. Liu Feng-han, 152, 172, 174-75.

ship, he reached Hong Kong safely on September 29. From there he sailed for Japan, after having been assured of protection by the Tokyo government.[32] Liang Ch'i-ch'ao fled to the Japanese legation in Peking and with its help also managed to escape to Japan. T'an Ssu-t'ung, who could have fled, offered to be a martyr, declaring that since time immemorial no revolution had succeeded without bloodshed. The four reformers at the Grand Council, appointed only sixteen days earlier, as well as the censor Yang Shen-hsiu, and K'ang's younger brother K'ang Kuang-jen were all summarily executed without a trial. They were collectively known as the "Six Gentlemen" or "Six Martyrs" (*Liu chün-tzu*). The progressive governor of Hunan, Ch'en Pao-chen, who recommended the appointment of the four reformers to the Grand Council, and Weng T'ung-ho, who had already been dismissed for having introduced K'ang to the emperor, were permanently disqualified for official appointments. A total of twenty-two reformers were arrested, imprisoned, dismissed, banished, and stripped of their properties. K'ang's writings were banned.

Most of the reform measures were reversed. The seven sinecure offices and three governorships abolished during the "Hundred-Day" Reform were reinstated; so was the eight-legged essay. The government press was closed; formation of societies was prohibited; and newspaper publishers and editors in Shanghai, Hankow, and Tientsin were ordered arrested. Private citizens were forbidden to submit memorials on state affairs.

However, there was to some degree a continuation of moderate reform. The Imperial University at Peking and the colleges at provincial capitals were allowed to continue, while the high schools and elementary schools at the prefectural and district levels could also operate if they suited local conditions. Provincial authorities were instructed to abolish or amalgamate superfluous offices and dismiss sinecure appointees. Some associates of Weng T'ung-ho were retained in important positions: Sun Chia-nai, president of the Board of Civil Office, was made associate grand secretary and put in charge of the Imperial University, and Wang Wen-shao, formerly governor-general of Chihli, was made a grand councillor, president of the Board of Revenue, and concurrently a minister of the Tsungli

32. On October 1, 1898, K'ang asked the Japanese consul in Hong Kong, Ueno, to inquire of his government whether he would be welcomed and protected in Japan. Premier Ōkuma cabled Ueno on October 9: "Inform K'ang that he will receive proper protection in Japan." Ueno in addition presented K'ang with the passage fare of 350 *yen*. See Teshirogi Kōsuke, "Bojutsu yori kōshi ni itaru kakumeiha to hempōha no kōshō—tōji no Nisshin kankei no ichi dammen" (The negotiations between the revolutionary party and the reform party from 1898 to 1900—an aspect of the then Japanese-Ch'ing relations), *Kindai Chūgoku kenkyū* (Studies on Modern China), Toyo Bunko, 7:175-76 (1966).

Yamen. Feng Kuei-fen's *Protest from the Chiao-pin Studio* was reprinted to spread the idea of a conservative, limited reform. The empress dowager made it clear that reform itself was not bad but that K'ang Yu-wei had carried it out badly.

Since the emperor had promoted a radical reform in total disregard of ancestral institutions, had relied on suspicious characters, and had attempted to wrest power from her, he must pay for his folly. Tz'u-hsi announced to the country that he was very sick. There was widespread speculation that the emperor would soon be done away with. Foreign diplomats in Peking warned that any untoward and underhanded incident involving the emperor would bring about an intervention. Under foreign pressure, the court admitted a French doctor to the emperor, and it was on his testimony that foreigners believed that the emperor was still alive. Still bent on revenge, the dowager solicited views from provincial authorities as to the advisability of deposing the emperor. Governor-general Liu K'un-i at Nanking vigorously opposed the idea with an incisive statement: "The relationship between the emperor and his ministers has already been fixed, while the mouths in and out of China cannot be easily silenced." Though frustrated, the dowager was undaunted. Another attempt at deposing the emperor was made in the following winter (1899) when, against all dynastic practices, she chose an heir apparent[33] for him. There was renewed fear for Kuang-hsü's life. Foreign diplomats boycotted her invitation to celebrate the choice, while 1,200 social leaders at Shanghai cabled the Tsungli Yamen to demand the protection of the emperor. All this, of course, greatly aggravated the Imperial Woman.

Causes and Effects of the Failure of the Reform

Among the principal causes for the failure of the reform were the inexperience of the reformers and their ill-considered strategy, the reluctance of the empress dowager to give up power, and the powerful conservative opposition.

THE REFORMERS' INEXPERIENCE. In 1898 K'ang Yu-wei was only forty and his chief supporter Liang Ch'i-ch'ao twenty-five, both without previous experience in government service. Neither had been abroad before the reform, and neither had more than a superficial understanding of Western culture and institutions. Their knowledge of the West was limited to what they read in the missionary publications and what they observed in the colonial administrations in Hong Kong and Shanghai. Small wonder

33. P'u-ch'ün.

that Chang Chih-tung ridiculed them for not having a real grasp of Western learning and institutions.

K'ang, in particular, was an idealist and philosopher rather than a practical statesman. He had very little knowledge of the reality of power politics and no power base from which to operate. He was able to win over the emperor as the legal source of power, but he ignored the obvious fact that the real power of state rested with the dowager. Impatient to achieve quick results, he hardly considered the effects of the reform decrees on others. Naïvely he believed that with the support of the emperor he could overcome all difficulties. Little did he realize that the radical reform was in effect a war on the whole Confucian state and society, one which could not but arouse strong opposition from many quarters. The abolition of the eight-legged essay hurt the future of all students who, having spent their lives preparing for the civil service examinations, suddenly discovered that what they learned was not what the government wanted. They vowed to "eat" K'ang. The elimination of sinecure offices and the three governorships, and the suggestion that twelve new bureaus be created caused fear of dismissal among all office holders. The decree which called for the appointment of men of practical knowledge instead of the promotion of incumbents on a seniority basis created insecurity in officialdom. The military reform jeopardized the privileges of the Manchu bannermen and the Chinese Green Standard army, and the attack on corruption doomed the practice of the squeeze, most avidly pursued by the chief eunuch, Li Lien-ying. The order to turn temples and shrines into schools irritated the monks and priests. The fact that all the reformers except the emperor were Chinese aroused fear among the Manchus. All these elements—scholars, officials, army officers, eunuchs, monks, and the Manchus in general—sought to undo the reform.

The progressives were not without inkling of the danger that lay in wait for them. K'ang's brother had urged him to quit before it was too late, but Weng T'ung-ho, determined to destroy K'ang, persuaded him to stay under the pretext that the emperor could not bear to see him leave. K'ang himself was too overwhelmed by the imperial favor to want to go, declaring that life and death were decided by Providence, beyond human control. K'ang's brother then planned with Liang to have him sent to Japan as envoy, but the emperor sent instead another progressive.[34] He was simply too dependent on K'ang to allow him to leave, and K'ang was too proud to quit halfway. The reform movement was in full swing for only 103 days when the dowager struck.

34. Huang Tsun-hsien.

THE POWER OF TZ'U-HSI. For thirty-seven years since 1861, the empress dowager had been the ultimate power of state. She was too experienced, and too well entrenched, to be uprooted by a handful of inexperienced reformers. Though in retirement since 1889, she was in firm control of political and military affairs. Her confidants in the Grand Council reported to her all policy decisions, the eunuchs in the palaces watched every move of the emperor, and Jung-lu, her henchman at Tientsin, was in charge of the Peiyang army. There was not a thing that escaped her notice. Jung-lu's troops, stationed at Taku, Tientsin, Tungchow, and the vicinity of Peking, were ready to defend her interest. These soldiers might be useless before foreign aggressors, but sufficient to frustrate any domestic venture by the reformers. Indeed, Jung-lu had been the bodyguard and protector of the dowager ever since the coup of 1861. The emperor and the idealistic reformers, without any army at their direct command, could turn only to Yüan Shih-k'ai, but the latter was too shrewd and opportunistic not to know the inevitable outcome of a contest between the emperor and the dowager. He chose the winning side and hastened the collapse of the reform.

THE CONSERVATIVE OPPOSITION. In their self-appointed role as defenders of Confucian morality and traditionalism, conservative scholars[35] attacked the reformers of having respect for "neither the sovereign nor the fathers" (*wu-chün wu-fu*) and of confusing the basic Three Bonds of human relations when they advocated popular sovereignty and individual equality. K'ang's interpretation of Confucius as a reformer and his casting doubt on the authenticity of the classics were nothing less than blasphemy and heresy in the eyes of these guardians of Confucian virtues. One of them, Yeh Te-hui, accused K'ang of using the sage to advance his own interests, and sneered: "K'ang Yu-wei's face is Confucian . . . but his heart is barbarian . . . The Kung-yang [Modern Text] school of today is not the same as that of the Han period; the latter honored China whereas the former, the barbarians." Contemptuously he announced: "Even if [K'ang's] words might be accepted, he as a person should never be used."[36]

Even the moderate reformers and those who sympathized with the institutional change found it difficult to accept K'ang's interpretations. Weng T'ung-ho, who brought K'ang to the attention of the emperor, remarked of him after reading *A Study of the Forged Classics of the Hsin Period*: "truly a 'wild-fox' meditator among the commentators of the classics! No end to my astonishment." Ch'en Pao-chen, the progressive governor of

35. Such as Yeh Te-hui and Wang Hsien-ch'ien.
36. Li Shou-k'ung, 546.

Hunan, commented that K'ang's *A Study of Confucius on Institutional Reform* went beyond the usual academic interpretation of Confucian teachings, with dangerous and undesirable political implications. Sun Chia-nai, a sympathizer of reform and president of the Imperial University, was also critical of this work, as he informed the throne:[37]

> In the eighth *chüan* . . . there is a section entitled, "Confucius formed institutions and assumed the title of king." K'ang tries to establish, on questionable grounds, that Confucius assumed the kingly title when he projected his reforms . . . It is feared that if this view is taught [to scholars], every one [of them] would entertain the idea of altering the institutions, every one would believe that he could be a "su-wang" [uncrowned king]. As a consequence, schools which are established to educate talented men would instead confuse and poison the minds of the people. That would lead the empire into disorder.

Thus, while K'ang's intellectual exertion won him great fame as an exponent of the Modern Text school, it also alienated a number of moderate and prudent scholars, who simply could not accept the dangerous implications of his works. It is a poignant irony that K'ang's self-styled title *ch'ang-su*, which he took after *su-wang* (the uncrowned king, i.e. Confucius), had the double meaning of "a constant follower of the uncrowned king," or simply "the permanently uncrowned."

All in all, K'ang was too much of a philosopher to be a practical statesman. His radical reform was a gallant attempt to save the dynasty, but it also represented a sharp departure from the general trend of gradual change that had begun with the Self-strengthening Movement of the 1860s. It was clearly too advanced for his time. Given the general decadence and senility of the dynasty, one wonders whether his program, even if it were carried out in full, could have saved it.

The repercussions of the failure of 1898 were many and far-reaching. First, it proved that progressive reform from the top down was impossible. Secondly, under the empress dowager and die-hard conservatives who had returned to power, the court was totally incapable of leadership. It encouraged antiforeignism and fostered the Boxer movement, which incurred the eight-power occupation of Peking in 1900. It followed an anti-Chinese policy to punish the reformers, and thereby widened the cleavage between the Manchus and the Chinese. The reactionary grand secretary, Kang-i, said: "Reform benefits the Chinese but hurts the Manchus. If I have properties, I would rather give them to my friends than let the slaves share the benefit." Thirdly, an increasing number of the Chinese came to

37. Hsiao Kung-ch'üan, "Weng T'ung-ho," 158, 174.

feel that their future lay in the complete overthrow of the Manchu dynasty, and that such an occurrence could not be realized by peaceful change; only a bloody revolution from below could effect it. Dr. Sun Yat-sen took the lead in promoting this approach.

Further Reading

Ayers, William, *Chang Chih-tung and Educational Reform in China* (Cambridge, Mass., 1971).

Barnett, Suzanne Wilson, and John K. Fairbank (eds.), *Christianity in China: Early Protestant Missionary Writings* (Cambridge, Mass., 1985).

Bohr, Paul Richard, *Famine in China and the Missionary: Timothy Richard as Relief Administrator and Advocate of National Reform, 1876-1884* (Cambridge, Mass., 1972).

Burt, E. W., "Timothy Richard: His Contribution to Modern China," *International Review of Missions,* 293-300 (July 1945).

Cameron, Meribeth E., *The Reform Movement in China, 1898-1912* (Stanford, 1931).

Candler, W. A., *Young J. Allen* (Nashville, 1931).

Chan, Sin-wai (tr.), *An Exposition of Benevolence: The Jen-hsüeh of T'an Ssu-t'ung* (Hong Kong, 1984).

Chang, Chih-tung 張之洞, *Ch'üan-hsüeh p'ien* 勸學篇 (An exhortation to learning), (1898), 2 *chüan.*

Chang, Hao, *Liang Ch'i-ch'ao and Intellectual Transition in China, 1890-1907* (Cambridge, Mass., 1971).

——, "Intellectual Change and the Reform Movement, 1890-8" in John K. Fairbank and Kwang-ching Liu (eds.), *The Cambridge History of China* (Cambridge, Eng., 1980), Vol. 11, 274-338.

Ch'en, Ch'iu 陳黌, "Wu-hsü cheng-pien shih fan-pien-fa jen-wu chih cheng-chih ssu-hsiang" 戊戌政變時反變法人物之政治思想 (Political views of antireformers during the 1898 coup d'état), *Yen-ching hsüeh-pao,* 25:59-106 (June 1939).

Ch'i, Ssu-ho 齊思和, "Wei Yüan yü wan-Ch'ing hsüeh-feng" 魏源與晚清學風 (Wei Yüan and the late Ch'ing intellectual climate), *Yen-ching hsüeh-pao,* 39: 177-226 (Dec. 1950).

Chien, Po-tsan 翦伯贊 et al. (eds.), *Wu-hsü pien-fa* 戊戌變法 (The reform of 1898), (Shanghai, 1953), 4 vols.

Ch'ien, Mu 錢穆, "K'ang Yu-wei hsüeh-shu shu-p'ing" 康有為學術述評 (A critical study of K'ang Yu-wei's scholarship), *Tsing-hua hsüeh-pao,* 11:3:583-656 (July 1936).

Chong, Key Ray, *Americans and Chinese Reform and Revolution: 1898-1922: The Role of Private Citizens in Diplomacy* (Lanham, Md., 1984).

Ch'üan, Han-sheng 全漢昇, "Ch'ing-mo ti Hsi-hsüeh yüan-ch'u Ching-kuo shuo 清末的西學源出中國說 (The late Ch'ing hypothesis of the Chinese origin of Western learning), *Ling-nan hsüeh-pao,* 4:2:57-102 (June 1935).

——, "Ch'ing-mo fan-tui Hsi-hua ti yen-lun" 清末反對西化的言論 (Anti-Westernization views at the end of the Ch'ing dynasty), *Ling-nan hsüeh-pao*, 5:3-4: 122-66 (Dec. 1936).

Cohen, Paul A., "Christian Missions and Their Impact to 1900," in John K. Fairbank (ed.), *The Cambridge History of China* (Cambridge, Eng., 1978), Vol. 10, 543-90.

——, and John Schrecker (eds.), *Reform in Nineteenth Century China* (Cambridge, Mass., 1976).

Fairbank, John King (ed.), *Missionary Enterprise in China and America* (Cambridge, Mass., 1974).

Forsythe, Sidney A., *American Missionary Community in China, 1895-1905* (Cambridge, Mass., 1971).

Ho, Ping-ti, "Weng T'ung-ho and the 'One Hundred Days of Reform'," *Far Eastern Quarterly*, X:2:125-35 (Feb. 1951).

Hsiao, Kung-ch'üan, *A Modern China and a New World: K'ang Yu-wei, Reformer and Utopian, 1858-1927* (Seattle, 1975).

Hu, Pin 胡濱, *Wu-hsu pien-fa* 戊戌變法 (The reform of 1898), (Shanghai, 1956).

Hummel, William F., "K'ang Yu-wei, Historical Critic and Social Philosopher, 1857-1927," *Pacific Historical Review*, 4:4:343-55 (Dec. 1935).

Hunter, Jane, *Gospel of Gentility: American Women Missionaries in Turn-of-the-Century China* (New Haven, 1984).

Hyatt, Irwin, *Our Ordered Lives Confess: Three Nineteenth Century American Missionaries in East Shantung* (Cambridge, Mass., 1976).

Kamachi, Noriko, "American Influences on Chinese Reform Thought: Huang Tsun-hsien in California, 1882-1885," *Pacific Historical Review*, XLVII:2: 239-60 (May 1978).

——, *Reform in China: Huang Tsun-hsien and the Japanese Model* (Cambridge, Mass., 1981).

Kwang, Luke S. K., *A Mosaic of the Hundred Days* (Cambridge, Mass., 1984).

Levenson, Joseph R., *Liang Ch'i-ch'ao and the Mind of Modern China* (Cambridge, Mass., 1953).

Liang, Ch'i-ch'ao 梁啓超, *Wu-hsü cheng-pien chi* 戊戌政變紀 (An account of the 1898 coup).

——, *Intellectual Trends in the Ch'ing Period* (*Ch'ing-tai hsüeh-shu kai-lun*), tr. by Immanuel C. Y. Hsü (Cambridge, Mass., 1959), Pt. III.

Liu, Feng-han 劉鳳翰, *Yüan Shih-k'ai yü Wu-hsü cheng-pien* 袁世凱與戊戌政變 (Yüan Shih-k'ai and the coup d'état of 1898), (Taipei, 1964).

Liu, Jen-ta 劉仁達, "Wu-hsü pien-fa yün-tung chung K'ang Yu-wei so t'i-ch'u ti cheng-chih kang-ling" 戊戌變法運動中康有為所提出的政治綱領 (K'ang Yu-wei's political platform for the reform movement of 1898), *Li-shih yen-chiu*, 4:1-10, (1958).

Lo, Jung-pang (ed.), *K'ang Yu-wei: A Biography and a Symposium* (Tucson, 1967).

Onogawa, Hidemi 小野川秀美, "Kō Yū-i no hempōron" 康有為の變法論 (K'ang Yu-wei's ideas of reform), in *Kindai Chūgoku kenkyū* 近代中國研究 (Studies on modern China), ed. by the seminar on modern China, Toyo Bunko, Tokyo, 2:101-88 (1958).

Pusey, James Reeve, *China and Charles Darwin* (Cambridge, Mass., 1983).

Rabe, Valentin, *The Home Base of American China Missions, 1880-1920* (Cambridge, Mass., 1978).

Soothill, W. E., *Timothy Richard of China* (London, 1924).

T'ang, Chih-chün 湯志均, *Wu-hsü pien-fa chien-shih* 戊戌變法簡史 (A short history of the reform of 1898), (Peking, 1960).

Thompson, L. G. (tr.), *Ta T'ung-shu: The One-World Philosophy of K'ang Yu-wei* (London, 1958).

Wang, Shu-huai 王樹槐, *Wai-jen yü Wu-hsü pien-fa* 外人與戊戌變法 (Foreigners and the Reform of 1898), (Taipei, 1965).

Wong, Young-tsu, "The Significance of the Kuang Hsü Emperor to the Reform Movement of 1898," in *Transition and Permanence: Chinese History and Culture* (Dec. 1972). A Festschrift in honor of Dr. Hsiao Kung-ch'üan.

Woodbridge, Samuel I., *China's Only Hope: An Appeal by Her Greatest Viceroy, Chang Chih-tung* (New York, 1900).

Yüan, Shih-k'ai 袁世凱, *Wu-hsü jih-chi* 戊戌日記 (My diary of 1898), (1909, 1922).

16

The Boxer Uprising, 1900

The *coup d'état* of 1898 reversed the entire power structure, restoring reactionary Manchus to office at the expense of both radical and moderate Chinese. Jung-lu, Yü-lu, and Ch'i-hsiu entered the Grand Council, while the die-hard conservative grand secretary, Kang-i, gained increasing favor with the empress dowager, on whom he had developed an even greater hold than Jung-lu. Blind to the realities of international politics, these men rejected diplomacy and mutual accommodation, advocating instead a policy of hard resistance. Under their influence, the dowager determined to make no more concessions to foreign powers. The test came in February 1899 when the Italians demanded the cession of the Sanmen Bay in Chekiang. Ordering the governor of Chekiang to fight enemy landings without hesitation, she saw the new policy of intransigency vindicated when the Italians backed down in October. Proudly, the Imperial Woman instructed the provincial authorities on November 21, 1899, to entertain no more illusions of peace.[1]

The Background of the Boxer Movement

Strong antiforeign sentiment permeated not only the court under the empress dowager, but the scholars, the officials, the gentry, and the people at large. Half a century of foreign humiliation, in war as well as in peace,

1. Chester C. T'an, *The Boxer Catastrophe* (New York, 1955), 32.

387

had deeply wounded their national pride and self-respect. The presence of haughty foreign ministers, fire-eating consuls, aggressive missionaries, and self-seeking traders constantly reminded them of China's misfortune. This gnawing sense of injustice generated a burning desire for revenge until it burst out in a vast antiforeign movement. There were, of course, larger social, economic, political, and religious factors which contributed to such an outbreak.

ANTIPATHY TOWARD CHRISTIANITY. Imbued with the teachings of Confucianism, Taoism, and Buddhism, the Chinese resented the invasion of Christianity under the protection of gunboats. The Treaties of Tientsin in 1858 had allowed its free propagation in the interior, and the Conventions of Peking in 1860 granted missionaries the right to rent and buy land for the construction of churches. Protected by the flag and the treaties, the missionaries moved about freely in China, although they had great difficulty winning converts. They resorted to the practice of offering converts monetary subsidies and protection against official or unofficial interference and insult.[2] The Chinese disparagingly called these native Christians men who "eat by religion" (*ch'ih-chiao*)—i.e. they lived on income from the church. Indeed, those who accepted pecuniary compensation in exchange for their belief were seldom men of high purpose; most were poor material drawn from low social strata who sometimes took advantage of their association with the missionaries to bully their fellow countrymen and to evade the law. When these converts became involved in trouble and lawsuits, the missionaries often came to their aid, interceding with the magistrates on their behalf. The public demonstration by the missionaries of their protective power, influence, and wealth attracted the weak and the opportunistic to the church but repelled the strong and the proud.

The gentry regarded Christianity as a socially disruptive, delusive, heterodox sect. The converts' failure to kowtow to the idols, to worship Confucius and ancestors, and to participate in local festivals honoring the spirits greatly irritated the gentry. As self-appointed guardians of Confucian propriety, they resented the effrontery of inroads by any foreign religion or philosophy, and not infrequently they were the secret instigators of religious incidents. Christianity, as a "heterodox" faith in China, became a basic cause and focus for antiforeignism.

2. *Hai-kuo t'u-chih* (An illustrated gazetteer of the maritime countries) mentioned 130 taels as the amount of business allowance for each convert, while the *Chung-Hsi chi-shih* (A record of Sino-Western affairs) reported subsistence subsidies of 4 taels apiece.

PUBLIC ANGER OVER IMPERIALISM. As the pace of foreign encroachment accelerated in 1897-98, a sense of imminent extinction grew. K'ang Yu-wei, speaking before the National Protection Society in Peking on April 17, 1898, warned of the danger of becoming a second Burma, Annam, India, or Poland. The progressives proposed national salvation through a radical institutional reform, but the reactionaries and the ignorant yearned to vent their wrath by killing foreigners.

HARDSHIP OF LIFE AS A RESULT OF FOREIGN ECONOMIC DOMINATION. The influx of foreign imports after the Opium War created a depressant effect on the native economy, and the fixed 5 percent ad valorem customs duty ruined China's protective tariff. Foreign cotton cloth sold for only one-third of the price of the Chinese cloth, driving native weavers and textile manufacturers into bankruptcy. Handicraft household industries fared especially badly in the face of foreign competition, casting many workers into unemployment. The hardships of life accelerated during the Taiping period; with widespread famine and starvation, destitute people became bandits, vagrants, or troublemakers. While many of those in extremity at first blamed their misfortune on the Taipings, they ultimately transferred their hatred to foreigners for having inspired the rebels with the alien Christian ideology.

In the post-Taiping era, further expansion of foreign trade resulted in an ever-increasing foreign domination of the Chinese markets, and during the Self-strengthening period (1861-95) large numbers of foreign-style enterprises and industries, as well as considerable foreign capital, were introduced. In 1899 China suffered a trade deficit of 69 million taels and a government budgetary imbalance of some 12 million taels (101 million expenditure versus 89 million revenue). To meet the deficit, the court increased taxes and solicited provincial contributions, the burden of which ultimately fell on the people. When life became unbearable for the all too hard-pressed people, they sought alleviation in banditry and secret societies.

Moreover, the foreign device of the railway worked havoc on the traditional communication systems. The two old north-south trunk lines—the Grand Canal and the land route from Hankow to Peking—lost out in competition with railways, and thousands of bargemen, carters, innkeepers, and businessmen were thrown out of work. With the commutation of tribute rice from the south to cash payment in 1900, the Grand Canal became all but obsolete, effecting the decline of the cities and the livelihood of the people along its banks.

By the end of the 19th century, the country was beset by bankruptcy of village industries, decline of domestic commerce, rising unemployment, and a general hardship of livelihood. Many attributed this sorry state of affairs to evil foreign influence and domination of the Chinese economy. It was not surprising that hostility developed toward foreigners and things foreign.

NATURAL CALAMITIES. Added to the economic hardship, a series of natural disasters intensified further the difficulty of life. The Yellow River, which shifted its course from Honan to Shantung in 1852, and flooded frequently after 1882, broke loose again in 1898. It inundated hundreds of villages in Shantung, affecting more than a million people. Similar floods occurred in Szechwan, Kiangsi, Kiangsu, and Anhwei. As if the torrential suffering was not enough, a severe draught followed in 1900 in most of North China, including Peking. Victims of natural calamities as well as superstitious scholars and officials blamed the misfortune on the foreigners, who, they insisted, had offended the spirits by propagating a heterodox religion and prohibiting the worship of Confucius, idols, and ancestors. Foreigners were accused of damaging the "dragon's vein" (*lung-mai*) in the land when they constructed railways, and of letting out the "precious breath" (*pao-ch'i*) of the mountains when they opened mines. The gentry held foreigners responsible for destroying the tranquillity of the land and interfering with the natural functioning of the "wind and water" (*feng-shui,* geomancy), thus adversely affecting the harmony between men and nature. Such an evil influence, they argued, must be eliminated if China was to have a peaceful, good life. The question was, how could they rid the country of the foreigners who possessed big ships and powerful guns? As a poor and weak country, China could not possibly expel them by military means; but some naïvely clung to the notion that she could invoke the supernatural powers to neutralize the effect of guns!

It was in this atmosphere of superstition, economic depression, extreme privation, public anger over foreign imperialism, and resentment of the missionaries that a major antiforeign riot broke out in 1900.

The Origin of the Boxers

"Boxers" was the name given by foreigners to a Chinese secret society called the I-ho ch'üan, or the "Righteous and Harmonious Fists," since members of this organization practiced old-style calisthenics. The I-ho ch'üan was an offshoot of the Eight-trigram Sect (*Pa-kua chiao*), which was associated with the White Lotus Sect, an anti-Ch'ing secret society, which fomented the rebellion of 1796-1804. The first official mention of I-ho ch'üan appeared in an edict of 1808, which described the appearance of sword-carrying rascals in Shantung, Honan, and Kiangnan [Kiangsu and Anhwei] provinces, who gathered under the name of I-ho ch'üan and Pa-kua chiao, and set up gambling tents in markets and fairs to take advantage of the local people. Despite official prohibition, the I-ho ch'üan spread to Chihli province by 1818 and continued its activities. In the 1890s, this antidynastic secret body took on an antiforeign cast, vowing to kill foreigners and their Chinese collaborators. The conservative governor of Shantung, Li Ping-heng, encouraged their activities, and his second successor, Yü-hsien, equally reactionary, changed their name in 1899 to I-ho t'uan, the "Righteous and Harmonious Militia."[3]

The Boxers called foreigners "Primary Hairy Men" (*Ta mao-tzu*), Chinese Christians and those engaged in foreign matters "Secondary Hairy Men" (*Erh mao-tzu*), and those who used foreign articles "Tertiary Hairy Men" (*San mao-tzu*). All "Hairy Men" were subject to extermination.

The Boxer's pantheon included both legendary and historical figures. Numbered among their gods were the Jade Emperor (Taoist diety), Kuan Kung (the god of war), Chu-ko Liang (the wise strategist) and Hsiang Yü (the Hegemon King of the Western Ch'u State).

Elemental in the Boxers' program, and of primary appeal to the superstitious populace, was the practice of magic arts, by which they claimed immunity to bullets after a hundred days of training, and the power to fly after four hundred days of work. They used charms, incantations, and rituals to invoke the supernatural powers. In the battlefields they burned a small yellow paper with the image of a footless man, while murmuring some magic formulae, which purportedly could bring down divine generals and soldiers. Being antiforeign, the Boxers shunned the use of guns, preferring old-style swords and lances.

Originally anti-Ch'ing, the Boxers in the 1890s became prodynastic and antiforeign.[4] They vowed to get "one dragon, two tigers, and three hundred lambs"—the dragon signified Emperor Kuang-hsü who sponsored the reform of 1898, two tigers meant Prince Ch'ing and Li Hung-chang who engaged in foreign affairs, and the three hundred lambs denoted metropolitan officials who had anything to do with the foreigners. Only eighteen court officials, the Boxers claimed, deserved to live; these were, of course, the die-hard reactionaries who supported the Boxers.

3. This version of the Boxers' origin is based on the authoritative source, Lao Nai-hsüan, "I-ho ch'üan chiao-men yüan-liu k'ao" (A study of the origin of the Boxers), (1899).
4. For a detailed study of this point, see Victor Purcell, *The Boxer Uprising* (Cambridge, 1963), chs. 9, 10.

The Court Patronage of the Boxers

In the 1890s the Boxers were particularly active under the name of the Big Sword Society (*Ta-tao hui*) in Shantung, where they received secret encouragement from the reactionary governor, Li Ping-heng. He cleverly shielded the incidents they created and recommended a policy of pacification rather than suppression.[5] However, when two German missionaries were killed in 1897, the court was pressured by the German minister to dismiss Li.[6] It was this missionary case that gave Germany an excuse to demand the occupation of Kiaochow, thereby touching off a scramble for concessions by the other powers.

In March 1899 Yü-hsien was appointed governor of the now thoroughly disturbed Shantung province. As antiforeign as Li Ping-heng, he continued to patronize the Boxers and the Big Sword Society, ordering prefects and district magistrates to ignore the petitions and complaints of the missionaries and converts as just so much wastepaper. Under his aegis, the Boxers raised the banner of support for the Ch'ing and extermination of the foreigners (*fu-Ch'ing mieh-yang*). The governor subsidized them with silver and invited them to set up training centers to teach his soldiers boxing. More than eight hundred such centers sprang into being, concentrating in the area west of the Grand Canal where people suffered most from the floods. As noted earlier, Yü-hsien dignified the Boxers with a new name, the I-ho t'uan, or the "Righteous and Harmonious Militia." Emboldened by official support, the Boxers stepped up their attacks on the missionaries and the converts.

However, in December 1899 foreign pressure again forced the court to remove Yü-hsien. He came to Peking praising the dependability of the Boxers and condemning any act of suppression as hurting China's own interests. Impressed with his presentation, the reactionary Prince Tuan, Prince Chuang, and Grand Secretary Kang-i recommended the use of the Boxers to the empress dowager, who, in her frustration with the foreigners, readily embraced the idea. Yü-hsien was rewarded with the governorship of Shansi; his successor in Shantung, the acting governor Yüan Shih-k'ai, who stood for a vigorous policy of suppression, was repeatedly admonished by Peking to refrain from punishing the Boxers. On January 3, 1900, it instructed him to use persuasion and pacification rather than suppression, but Yüan refused to acquiesce, and the Boxers were suppressed in Shantung.

However, the court continued to favor the Boxers. On January 12, 1900,

5. T'an, 46.
6. Shortly afterwards promoted to be governor-general of Szechwan.

it decreed that people drilling themselves for self-defense and for protection of their villages should not be considered bandits. On April 17, the court announced that organization of militia (*t'uan*) by peaceful and law-abiding villagers to preserve themselves and their families was in line with the ancient principle of "keeping mutual watch and giving mutual aid"; hence such activity should not be prohibited. The Boxers became more daring, burning and destroying railways and telegraph lines as symbols of foreign enslavement.

In early May 1900 the court contemplated organizing the Boxers into a militia, only to be blocked by Yü-lu and Yüan Shih-k'ai. The reactionaries in power would not give up; Kang-i repeatedly impressed upon the dowager that the Boxers were favored by the gods and immune to bullets—exactly the type of men China should rely upon to expel the foreigners. The dowager secretly asked him to summon the Boxers to Peking and when their invulnerability to firearms was "confirmed" in a palace demonstration, she commended their leaders[7] and ordered court attendants, including the women, to learn boxing. The princes and nobles now invited the Boxers to their residences as guards and set up tables to burn incense to the Boxer gods. Half of the regular government troops joined the Boxers, and the distinction between the two was lost. Boxing had become a craze.

On May 28 the rising tide of antiforeignism alerted the foreign diplomats in Peking to the precautionary measure of calling in the legation guards from the ships off Tientsin harbor. First disapproving, then reluctantly assenting to the move, the Tsungli Yamen tried to limit the number of such guards to thirty for each legation. However, the first detachment that arrived in Peking on June 1 and 3 consisted of 75 Russians, 75 British, 75 French, 50 Americans, 40 Italians and 25 Japanese.

The Boxers found encouragement in yet another court decree of May 29, which cautioned provincial officials not to attack them indiscriminately, for there were both good and bad elements among those who practised boxing. Their pride enkindled by such official approbation, the Boxers cut the railway between Peking and Tientsin on June 3, and the situation rapidly got out of control.

By now, the court was completely dominated by the reactionaries. Prince Tuan had replaced Prince Ch'ing as head of the Tsungli Yamen, to which Hsü T'ung and Ch'i-hsiu were also appointed as ministers. Foreign diplomats came to the conclusion that the court intended to kill all foreigners in the capital. The British minister sent for urgent help from

---

7. Li Lai-chung and Ts'ao Fu-t'ien.

Admiral Seymour at Tientsin. An international force of 2,100 men left Tientsin by train on the morning of June 10, and encountered the Boxers at Lang-fang, halfway between Peking and Tientsin. Heavy fighting took place, blocking the foreign expedition. The telegraphic lines between Peking and Tientsin were cut, leaving the fate of the foreigners in the capital a mystery and a matter of grave concern. On June 10, the Boxers burned the British summer legation in the West Hills; a day later, the chancellor of the Japanese legation, Sugiyama, was killed by the troops of the reactionary Moslem general Tung Fu-hsiang, who earlier boasted to the dowager that he had no ability other than killing foreigners. The war dogs had been unleashed and would now, uncontrolled, run ravening and slavering about northern China.

On June 13 the court announced that since the embassies had been adequately protected by the legation guards, there was really no need for more foreign troops to come to Peking. On the same day, large bodies of rampaging Boxers swarmed into Peking. They burned churches and foreign residences, and killed Chinese converts on sight or buried them alive. They exhumed the graves of missionaries, including those of the early Jesuits such as Matteo Ricci, Schall von Bell, and Ferdinand Verbiest. On June 14 they made several attacks on the legation guards, and on June 20 killed the German minister, Clemens von Ketteler.

At Tientsin, the Boxers were equally uncontrollable. They burned churches and shops that sold foreign merchandise and books, and killed Chinese Christians. They broke into prison, released the inmates, and coerced the governor-general into allowing them a free pick of weapons from the government arsenals. Facing such fanatic disorder, foreign officers on the ships outside the harbor decided to take the Taku Forts, which they overpowered on June 16 and occupied a day later. Meanwhile the Seymour expedition, blocked from reaching Peking, decided to fight its way back to Tientsin.

Prince Tuan and Kang-i now advocated an all-out attack on the legations as the only way to expunge the national humiliations of half a century, and in this the dowager concurred. On June 16 the first of four imperial councils was called to deliberate on war or peace. Yüan Ch'ang, a director of the Court of Sacrificial Worship, pointing out the fake immunity of the Boxers to guns, cautiously opposed opening hostility against the legations. The dowager cut him short with the remark: "If we cannot rely upon the supernatural formulae, can we not rely upon the hearts of the people? China has been extremely weak; the only thing we can rely upon is the hearts of the people. If we lose them, how can we maintain our country?" The meeting was inconclusive, but a decree was issued to recruit the "young and strong" Boxers into the army.

At the second imperial council on June 17, the dowager ordered that the foreign ministers be informed that if their countries meant to fight they should go home. On June 18, the third imperial council was called but again reached no decision. On the following day, a belated report came from Yü-lu that foreigners had demanded the surrender of the Taku Forts. Assuming that fighting had formally broken out, the dowager called the fourth imperial council on the same day to announce the break-off of diplomatic relations. She had made up her mind to fight the powers with the help of the Boxers. Hsü Ching-ch'eng (1845-1900), ex-minister to Russia and a vice-president of the Board of Civil Office, was given the assignment of telling the foreign diplomats to leave Peking within twenty-four hours under Chinese military escort. Emperor Kuang-hsü, never sympathetic to the Boxers, held Hsü's hand, murmuring: "We should deliberate this matter more carefully." Immediately the dowager shouted: "Emperor, release his hand. Do not spoil the situation!" On June 21, another memorial arrived from Yü-lu, giving an ambiguous but rather favorable picture of the first three days' fighting at Taku and Tientsin. Feeling confident, the court declared war on the foreign powers that day.[8]

The court now formally ordered provincial authorities to organize the Boxers to fight the foreign invasion. At Peking, the Boxers were officially designated as "righteous people" (*i-min*) and rewarded with a rice subsidy and silver. Prince Chuang and Kang-i assumed official command of some 30,000 Boxers, while Prince Tuan directed a total of 1,400 bands, each consisting of 100 to 300 men. Marshaling them together with the government troops under General Tung Fu-hsiang, they launched vehement attacks on the legations and the Northern Roman Catholic Cathedral. For each foreign male captured alive Prince Chuang offered a reward of 50 taels, for each female, 40, and for each child, 30. Kang-i announced: "When the legations are taken, the barbarians will have no more roots. The country will then have peace." The attack on the legations was made with the full knowledge and support of the dowager; needless to say, it afforded the reactionaries in and out of the government the greatest emotional satisfaction. They saw in the destruction of the legations a way to vent their wrath on the barbarians, to rid the capital of the foreign menace, to kill evidence of the court's sponsorship of the Boxers, and to stimulate general patriotism among the people.

8. Hsiao I-shan, IV, 2,196-98.

In the legation grounds there were about 450 guards, 475 civilians including 12 foreign ministers, 2,300 Chinese Christians, and some 50
servants, who put up a stiff resistance. The Boxers, adopting the mannerisms and dress appropriate to their magical-supernatural associations,
wore wild, loose hanging hair and moved with the formulated steps ascribed
to witches.

Independence of Southeast China

When the court issued the declaration of war on June 21, the southeastern provincial authorities—Li Hung-chang at Canton, Liu K'un-i at Nanking, Chang Chih-tung at Wuhan, and Yüan Shih-k'ai in Shantung—collectively refused to recognize its validity, insisting that it was a *luan-ming*,
an illegitimate order issued without proper authorization of the throne.
They suppressed the declaration from the public, as they did the order of
the same day that they should organize the Boxers to fight foreign invasion. Chang Chih-tung cleverly twisted an edict of June 20 which ordered
that the governors-general "should be united together to protect their territories" to mean that they should cooperate to suppress the Boxers and protect the foreigners. On the suggestion of Sheng Hsüan-huai, director of
Railways and Telegraphs, Chang and Liu—the Yangtze valley governors-
general—entered into an informal pact with foreign consuls at Shanghai to
the effect that they, as the highest authorities in their provinces, would
protect foreign lives and properties and suppress the Boxers within their
jurisdictions, while the foreign powers would refrain from sending troops
into their regions. Li Hung-chang, Yüan Shih-k'ai, and the governor-general of Fukien and Chekiang subscribed to this agreement. Hence the
whole of southeast China was exempt from the Boxer disturbance and
foreign invasion.

The Allies held the Ch'ing government responsible for the foreign lives
in the legations, while organizing an international force to relieve the siege.
On July 14, foreign troops took Tientsin and threatened to march on Peking. On the same day, thirteen southeastern provincial authorities collectively urged the court to suppress the Boxers, protect the foreigners and
compensate them for the losses sustained in the recent disturbance, and
send a letter of apology to Germany for the death of von Ketteler. Under
their pressure, the court turned somewhat conciliatory for a moment. The
Tsungli Yamen was allowed to invite foreign diplomats and their families
to move to the Yamen for safety, pending further arrangements for a safe
return home. The suspicious foreign ministers replied that they could not
understand "why they should be safer in the Yamen than in the lega-

tions." On July 18, Li Hung-chang was ordered by the court to ask the Chinese diplomats abroad to inform the respective governments that their representatives in Peking were safe. A day later, the apprehensive Tsungli Yamen renewed the offer to send foreign ministers to Tientsin under Chinese military escort. Still suspicious, the foreigners asked the Yamen to explain "why, if the Chinese government cannot insure the protection of the foreign envoys in Peking, they feel confident of their power to do so outside the city, on the way to Tientsin."[9] They preferred to remain in the legation quarter to await the relief. On July 20 and 26, the Yamen twice sent cartloads of vegetables, watermelons, rice, and flour to the legations. During this brief period of conciliation (July 14-26), attacks on the legations were suspended for twelve days.

The war-storm broke again, however, with the arrival in Peking of the reactionary official, Li Ping-heng, on July 26. Encouraged by Kang-i and Hsü T'ung, he forcefully, and successfully, impressed upon the dowager that one could negotiate a settlement only when one could fight. The policy of war and extermination of foreigners was reaffirmed. High officials who dared to counsel peace met evil days, and five of them were executed.[10] The terrifying state of affairs is reflected in a telegram from Yüan Shih-k'ai to Sheng Hsüan-huai on August 2: "It is hopeless; better say less."[11]

Allied reinforcements arrived at Taku in late July and on August 4 set out from Tientsin for Peking. This international force consisted of 18,000 men, of whom the Japanese numbered 8,000; the Russians, 4,800; the British, 3,000; the Americans, 2,100; the French, 800; the Austrians, 58; and the Italians, 53. The Germans arrived too late to join this international force. The powerful Allied forces stormed across the Tientsin-Peking route, driving and dispersing before it the erratic Boxers and government troops. So quickly and emphatically did the Western powers defeat the Chinese that Yü-lu and Li Ping-heng committed suicide in humiliation on August 6 and 11, respectively. The Allied forces charged into Peking on August 14 and relieved the beleaguered legations.[12] The fact that some 450 guards, 475 civilians, and 2,300 Chinese Christians were able to with-

---

9. T'an, 102.
10. Hsü Ching-ch'eng, a vice-president of the Board of Civil Office and ex-envoy to Russia; Yüan Ch'ang, director of the Court of Sacrificial Worship; Hsü Yung-i, president of the Board of War; Lien-yüan, sub-chancellor of the Grand Secretariat, and Li Shan, president of the Board of Revenue.
11. T'an, 106.
12. Following their occupation of Peking, Allied troops, particularly the Russians, engaged in free looting and pillaging of the palaces and private residences. One Russian lieutenant general returned home with ten trunkfuls of valuables.

hold the assault of an infinitely larger number of government troops and Boxers for nearly two months was a miracle. However, this miracle was made possible by Jung-lu, the commander-in-chief of the Peiyang forces, who had no sympathy for the Boxers but lacked the courage to oppose the dowager. He carried out the attack halfheartedly, firing noisy but empty guns and withholding the new and large-caliber cannon from use. As a result, the legation defense was not broken.

On the morrow of the Allied advance into Peking, the dowager, the emperor, and a small entourage fled in disguise. The emperor had actually wanted to remain in Peking to negotiate a peace with the powers and to take over the reins of government himself, but the dowager, shrewd as ever in her extremity, would not let him re-establish himself at her expense. She ordered, at the last minute of her departure, that the emperor's favorite consort,[13] who counseled him to stay, be thrown into a well, and forced the emperor to flee with her. Clad in coarse commoner's clothes to avoid identification, they escaped westward under pitiful conditions. After a long and hard journey, the court was re-established in Sian on October 23.

The Boxer catastrophe which had swept over North China, Inner Mongolia, and Manchuria, had at last been stilled, leaving in its wake 231 foreigners dead and many more Chinese Christians slain. Shansi, in particular, where Yü-hsien had become governor, had suffered greatly from the disturbance.

The Peace Settlement

In the aftermath of the Boxer Uprising, the venerable elder statesman Li Hung-chang, governor-general at Canton, was given the assignment of mending the situation. Twice, on July 3 and 6, the court urged him to come north without delay, following, on July 8, with appointments as governor-general of Chihli and superintendent of trade for the northern ports, posts which previously he had held from 1870 to 1895. Only then, and tardily, did he sail for Shanghai, arriving on July 21. There he accepted the British government's advice that he wait in Shanghai until the foreign ministers had been safely conducted to Tientsin. On August 7, the court appointed him plenipotentiary to negotiate with the powers. Still he would not go north.

On August 20, the court in flight displayed signs of penitence by admitting responsibility for having brought on the misfortune. Repeatedly it "begged" Li to go to Peking to search for a settlement with the powers.

13. Consort Chen.

Li's delaying tactics stemmed from the belief that the court would not follow his recommendation to suppress the Boxers and that, unless the siege
of the legations was lifted and the foreign ministers given safe conduct to
Tientsin, there was no prospect for peace. Of some consolation was the
knowledge that the powers did not consider themselves at war with China;
they sent expeditionary forces merely to suppress the rebels. When Russia
offered to withdraw its troops, diplomats, and citizens to Tientsin in preparation for the opening of negotiations, and indicated confidentially that it
would set a tone of moderation at the conference to forestall excessive
demands by the other powers, Li decided that it was time to go north, and
he requested the court to appoint Prince Ch'ing and Jung-lu to join him
in the peace endeavors. When the court complied, Li went north under
Russian protection, arriving in Tientsin on September 18.

The court in exile still reigned under the influence of such reactionaries as Prince Tuan and Kang-i, who advocated a long drawn-out war of
attrition. To checkmate them, Li petitioned that Jung-lu, who had been
found unacceptable as a negotiator by the Allies because of his association with the attack on the legations, be allowed to join the court. On
November 11, Jung-lu reached Sian and resumed his role as a member of
the Grand Council.

Meanwhile, the Allied representatives in Peking declined to open negotiations before "the return of the court," by which they meant "the return
of the emperor to power." They raised the issue as leverage to gain satisfaction of their other demands. The dowager, clinging to her power, refused to return on the grounds that she feared untoward treatment and
unacceptable terms being imposed on her, indicating clearly that the court
would return after, not before, the peace settlement. The southeastern
provincial leaders now adopted the tactics of shifting the Allied attention
to punishing the guilty ministers. Particularly anxious for this approach
was Yüan Shih-k'ai, who knew that the return to power of the emperor,
whom he had betrayed during the 1898 reform, would be most detrimental to his own interests. These southeastern leaders, as well as Prince
Ch'ing and Li Hung-chang, put great pressure on the court to accept the
Allied demand for the punishment of nine pro-Boxer ministers plus Yü-
hsien and General Tung Fu-hsiang, who led the attack on the legations.
On December 3, 1900, the court reluctantly stripped Tung of his ranks
and sent him to Kansu. During all the discussion of guilt, no mention
was ever made of the two chief culprits; the dowager, who was most guilty,
and Jung-lu, who probably could have prevented the rise of the Boxers,
went unpunished.

During the Peking negotiations, the Allied representatives, working at cross purposes, had a hard time agreeing on the terms. In a vengeful spirit Germany demanded stern punishment. The kaiser spoke of a severe punitive action and even destruction of Peking; when dispatching a 7,000-man expedition, he declared: "May the name of Germany become known in such a manner in China that no Chinese will ever again even dare to look askance at a German."[14] Because of von Ketteler's murder, the kaiser secured the appointment of Field Marshal Count von Waldersee, onetime assistant to Moltke on the Grand General Staff, as commander-in-chief of the Allied forces in China. Arriving in Peking on October 17, some two months after it had been occupied by the Allies, Waldersee took the dowager's palace, the I-luan t'ien, as his quarters. The British supported the Germans in an attempt to check the Russian advance in China, while the Russians ingratiated themselves with the Chinese in hopes of gaining concessions in Manchuria, which they (the Russians) had already occupied during the turmoil. The Japanese, disturbed by Russian ambitions, adopted the policy of winning Chinese good will by offering to withdraw part of their troops to Tientsin. The French announced that they did not desire a break-up of China and entertained no secret designs on it. The United States announced the second Open Door Policy on July 3, 1900, supporting "Chinese territorial and administrative entity" and "permanent safety and peace."

After much niggling debate and argument among themselves, the Allies finally agreed on December 24, 1900, on a joint note of twelve articles. On the basis of this note, discussions were conducted toward a final settlement, which consisted of the following main features:

1. Punishment of the Guilty. The Allies had originally demanded the death penalty for twelve officials, including Princes Chuang and Tuan, Kang-i, Yü-hsien, Li Ping-heng, Hsü T'ung, and General Tung Fu-hsiang.[15] In the final settlement, Prince Chuang was ordered to commit suicide, Prince Tuan to be banished to Sinkiang for life imprisonment, and Yü-hsien to be executed. General Tung was deprived of office. Kang-i, Hsü T'ung, and Li Ping-heng, who had already died, received posthumous degradation.[16] In the provinces, a total of 119 officials received penalties ranging from capital punishment to mere reprimand.

2. Indemnity. A penal compensation of 40 million pounds sterling was proposed on March 21, 1901, by the United States commissioner-plenipo-

---

14. Morse, III, 309.
15. Others were Duke Lan, Yin-nien, Chao Shu-ch'iao, Hsü Ch'eng-yü, and Ch'i-hsiu.
16. Duke Lan was sentenced to banishment to Sinkiang for life imprisonment; Ch'i-hsiu and Hsü Ch'eng-yu were executed, while Ying-nien and Chao Shu-ch'iao were asked to commit suicide.

tentiary in Peking, W. W. Rockhill, but the German representative asked
for 63 million pounds sterling instead. On April 25, the Allies fixed the
indemnity at 67 million pounds to include the occupation cost up to July
1, 1901. On May 7 the figure was further revised to 67.5 million pounds,
or 450 million taels. The payment was to be completed in 39 years (i.e.
1940) at 4 percent annual interest, with the maritime customs, *likin,* native customs, and salt gabelle as security. To help meet the payment, it
was agreed to increase the existing tariff from an *actual* 3.18 percent to 5
percent, and to tax hitherto duty-free merchandise. The detailed breakdown of the indemnity was as follows:

Russia	130,371,120 taels	29% of total
Germany	90,070,515	20%
France	70,878,240	15.75%
Britain	50,620,545	11.25%
Japan	34,793,100	7.7%
United States	32,939,055	7.3%
Italy	26,617,005	5.9%
Belgium	8,484,345	1.9%
Austria	4,003,920	.9%
Others	1,222,155	.3%

3. *Other Important Stipulations.* In addition to the above two items, a
number of other terms were agreed upon, including:
   a. Apology missions to Germany and Japan.
   b. Establishment of a permanent legation guard.
   c. Destruction of the Taku and other forts from Peking to the sea.
   d. Prohibition of the importation of arms for two years.
   e. Stationing of foreign troops in key points from Peking to the sea.
   f. Suspension of official examinations for five years in some 45 cit-
      ies, where the Boxers had been active.

These items were formalized into The Boxer Protocol of twelve articles
and nineteen annexes, and signed by Li Hung-chang, Prince Ch'ing, and
the representatives of eleven powers on September 7, 1901, a year and
twenty-four days after the relief of the siege of the legations. The Allied
troops evacuated Peking on September 17, though the court did not return until January 7, 1902.

Russian Occupation of Manchuria

Peace had finally been restored between the Allies and China, but the
question of Russian occupation of Manchuria had yet to be resolved. Un-

der the pretext of restoring order and suppressing the "rioters" in Manchuria, the Russians had sent 200,000 troops in July 1900, with the ambitious design of reducing it to a second Bukhara. They gained control over all Manchuria through the course of three months of military operations. On November 30, Admiral Alexeiev, Russian governor-general of Liaotung Peninsula, coerced Tseng-ch'i, the Manchu military-governor of Mukden, into signing a nine-article "provisional agreement," which virtually preempted Chinese rule in Manchuria: Tseng-ch'i was to disarm and disband all his troops in Manchuria, surrender all munitions in the arsenals, dismantle forts and defenses, and agree to the appointment of a Russian Resident in Mukden. The Ch'ing court, angered, fearful, and humiliated, refused to recognize the validity of this agreement, which it insisted Tseng-ch'i had no authority to sign.

Negotiations then opened in St. Petersburg. General Kuropatkin and Count Witte, ministers of War and Finance, advocated a separate Manchurian pact, independent of the general agreement then being negotiated at Peking, with the intention of excluding other foreign influence and investment from Manchuria and the areas beyond the Great Wall. On February 16, 1901, the Russians proposed a twelve-article treaty (to replace the Alexeiev-Tseng agreement), which returned Manchuria to China in name but which, in effect, legalized the occupation of Manchuria by Russian troops disguised as "railway guards." It prohibited China from sending arms to Manchuria, or granting railway and mining privileges to other powers without Russian consent. The culminating insult, however, was the stipulation that China pay for the Russian occupation costs and damages to railways and properties of the Chinese Eastern Railway Company, as well as granting Russia the right to construct a line from the said railway to the Great Wall in the direction of Peking.

The Russian aggression in Manchuria aroused grave apprehension among the powers, especially Japan, whose interests conflicted with those of Russia. The Japanese minister in Peking[17] warned Prince Ch'ing that any concession on the Russian occupation of Manchuria could lead to the partition of China: Britain was certain to follow with the occupation of the Yangtze valley, Germany with the Shantung province, and Japan would have no choice but to reserve to itself freedom of action. Admonitory messages also came from Britain and Germany against any separate territorial or financial treaty with Russia before the signing of the general agreement with the Allies in Peking. The United States, Austria, and Italy, too, urged China to resist the Russian demand. On the other hand,

17. Komura Jutarō.

Witte threatened that a rejection of the proposed treaty would lead to Russian incorporation of Manchuria. The hapless Ch'ing court, still in exile at Sian, could come to no definite stand. It dared not offend either the powers or Russia; all it could do was to order Prince Ch'ing and Li Hung-chang to devise a way that would neither arouse the anger of the Russian court nor aggravate the indignation of the various powers. Li Hung-chang allowed his pro-Russian leanings to get the better of him, and advised the court to sign the treaty to avoid a perilous break. However, other powerful provincial figures, such as Chang Chih-tung and Liu K'un-i—the Yangtze governors-general—vigorously opposed the treaty. Liu argued that Russia would not return Manchuria whether China accepted the treaty or not, while Chang warned against a possible partition of China if it succumbed to the Russian threat. Caught between these opposing views, and pressured in diverse directions by Russia, Britain, and Japan, the court was totally incapable of making up its mind. It abjectly passed the decision to the Chinese minister in Russia, Yang Ju, who was authorized to act as he saw fit. Now Li asked him to accept the treaty, while Chang and Liu urged him to reject it, lest he become the target of public condemnation. Embroiled in this dilemma and in great anxiety, Yang seriously wounded his leg in an accident on March 22, 1901. On the following day he telegraphed the court that he would not sign the treaty without its express instructions. By this time, Chinese ministers in Tokyo, London, and Berlin had sent Peking a barrage of admonitions against signing. Most emphatic was the diplomat in Japan who argued that Russia most assuredly dared not face the combined forces of Britain and Japan, and that any Chinese concession at this point could only earn British and Japanese enmity and complicate the pending general settlement at Peking. Under such pressure, the court finally decided on March 23 to reject the Russian treaty. Facing powerful international opposition, the Russians did nothing more than issue a disgruntled statement on April 6 that, much as they would like to evacuate Manchuria, the realities of international politics did not permit them to do so at the moment. The tense negotiations in St. Petersburg, having hung so perilously for months, suddenly ended in an anticlimax, without the much anticipated dire consequences to China. Dunned by the Russians from without and ridiculed by his countrymen from within, Li Hung-chang, old, weak, and ashamed, passed away suddenly on November 7, 1901, at the age of 78.

Li's unfinished work was carried on by Prince Ch'ing and Grand Councillor Wang Wen-shao. The international situation was very much in Russia's disfavor, especially after the signing of the Anglo-Japanese alliance on January 30, 1902. Ultimately, the Russians signed an agreement with

China on April 4, promising to evacuate Manchuria in three stages at six-month intervals. On its part, China agreed to protect the Russian-dominated Chinese Eastern Railway, its employees and properties, as well as all its allied enterprises. The first stage of evacuation was carried out on schedule, but when the second stage came due in April 1903, the Russians did not leave but resorted to the subterfuge of changing the uniforms of the troops to those of "railway guards." In addition, they demanded new monopolistic rights and reoccupied some of the evacuated cities, such as Mukden and Newchwang. This Russian incursion into Manchuria foreshadowed the war with Japan in 1904.

Repercussions of the Boxer Uprising

In retrospect, it becomes apparent that the Boxer movement was propelled by the combined forces of the reactionary Manchu court, the die-hard conservative officials and gentry, and the ignorant and superstitious people. It was a foolish and unreasoned outburst of emotion and anger against foreign imperialism, yet one cannot overlook the patriotic element inherent in it. Marxist historians today consider the Boxer movement a primitive form of a patriotic peasant uprising, with the right motive but the wrong methods.

The Boxer Uprising and its final settlement left behind many significant consequences:

1. The Allied occupation of Peking and the Russian advance into Manchuria threatened the partition of China and sharpened international jealousy and rivalry. There developed a growing fear among the powers of conflict between themselves, and a deep concern over the future of equal commercial opportunity in China, resulting in a general international desire to reduce tension and maintain the *status quo* in China. The United States declared the second Open Door Policy on July 3, 1900, with a view to preserving "Chinese territorial and administrative entity" and to safeguarding "for the world the principle of equal and impartial trade with all parts of the Chinese Empire." The declaration was followed by the Anglo-German agreement of October 16, 1900 (to which other powers were invited to adhere) which stipulated that the signatories would refrain from seizing territory in China. The subsequent stalemate in imperialistic activities prevented an immediate break-up of China. Nonetheless, its international position in the society of nations plummeted ignominiously to rock bottom.

2. The Boxer Protocol infringed upon Chinese sovereignty severely. Article 5 which stipulated the prohibition of the importation of arms, Ar-

ticle 8 which stipulated the destruction of the Taku and other forts, Article 7 which provided for stationing of foreign troops in the legation quarter, and Article 9 which gave foreign powers the right to deploy troops from Peking to the sea—all these compromised China's power of self-defense and restricted the free exercise of its sovereign rights. Article 10, which suspended government examinations in many parts of the country for five years as a punishment to the gentry class, was a blatant interference with the internal administration of China.

3. The indemnity of 450 million taels ($330 million) and its accrued interest over 39 years at 4 percent annually amounted to a grand total of 982,238,150 taels, more than twice the original amount. The payments, which had to be made in foreign currencies rather than in Chinese taels, incurred an additional loss of several million taels annually in the exchange, especially during the years when the value of the silver suffered a sharp decline. For instance, in 1903 China had to pay 53.5 million taels instead of 42.5 million as originally agreed upon.[18] The outflow of such large capital inhibited, if not incapacitated, China's economic growth.

4. Foreign ministers in Peking now organized themselves into a powerful diplomatic corps, functioning above the Manchu court as a sort of super-government. The prestige of the Ch'ing dynasty sank to a nadir.

5. The barbarous conduct of the Boxers exposed China in an uncivilized light in the community of nations. On the other hand, the brutal demonstration of power by the foreign expeditionary forces created such an image of invincibility and superiority that Chinese pride and self-respect were shattered. The Chinese attitude toward foreigners swung from one of disdain and hostility to one of fear and toadying.

6. In a struggle for survival, the Manchu court instituted some half-hearted, superficial reform toward a constitutional government; while many Chinese, witnessing the hopelessness of the Manchu leadership, turned to revolution as the only hope for their country. Dr. Sun Yat-sen's advocacy of a forceful overthrow of the Ch'ing dynasty, hitherto regarded by respectable Chinese as an unlawful movement to eschew, now received

---

18. It should be noted, however, that in an act of justice and good will, the United States, on recommendation of W. W. Rockhill, later returned the excessive portion of the indemnity. The total American private claims, amounting to only $2 million, had been paid by 1905, and in 1908 the United States government returned to China $10,785,286, while retaining $2 million for possible future adjustments. In 1924 the rest of the indemnity was waived. This Boxer refund was specified to be used for educating Chinese students in the United States. Remissions by other countries followed suit: Britain, 1922; Russia, 1924; France, 1925; Italy, 1925 and 1933; Belgium, 1928; Netherlands, 1933. See Chi-ming Hou, *Foreign Investment and Economic Development in China, 1840-1937* (Cambridge, Mass., 1965) 26; also Paul A. Varg, *Open Door Diplomat: The Life of W. W. Rockhill* (Urbana, 1952), 48, 81-82.

increasing sympathy and support. His image reversed from that of a disloyal rebel to that of a high-minded, patriotic revolutionary. As a result, the pulse of revolution quickened, precipitating the ultimate downfall of the Manchu dynasty in 1911.

Further Reading

Campbell, Charles S., *Special Business Interests and the Open Door Policy* (New Haven, 1951).

Chao, Chung-fu 趙中孚, *Ch'ing-chi Chung-O Tung-san-sheng chieh-wu chiao-she* 清季中俄東三省界務交涉 (Sino-Russian negotiations over the Manchurian border issue, 1858-1911), (Taipei, 1970).

Chien, Po-tsan 翦伯贊 et al. (eds.), *I-ho t'uan* 義和團 (The Boxer Movement), (Shanghai, 1951), 4 vols.

Davis, Fei-ling, *Primitive Revolutionaries of China: A Study of Secret Societies in the Late Nineteen Century* (Honolulu, 1977).

Esherick, Joseph W., *The Origins of the Boxer Uprising* (Berkeley, 1987).

Fairbank, John K., "'American China Policy' to 1898: A misconception," *Pacific Historical Review*, XXXIX:4:409-20 (Nov. 1970).

Fleming, Peter, *The Siege at Peking* (New York, 1959).

Hart, Robert, *These from the Land of Sinim, Essays on the Chinese Question* (London, 1903).

Ho, Ping-ti 何炳棣, "Ying-kuo yü men-hu k'ai-fang cheng-ts'e chih ch'i-yüan" 英國與門戶開放政策之起源 (Britain and the origin of the Open Door Policy), *Shih-hsüeh nien-pao*, 2:321-40 (1938).

Hunt, Michael H., *Frontier Defense and the Open Door: Manchuria in Chinese-American Relations, 1895-1911* (New Haven, 1973).

Joseph, Philip, *Foreign Diplomacy in China, 1894-1900* (London, 1928).

Kuo, Pin-chia 郭斌佳, "Keng-tzu ch'üan-luan" 庚子拳亂 (The Boxer Rebellion of 1900), *Kuo-li Wu-han ta-hsüeh wen-che chi-k'an* (Quarterly Journal of Literature and Philosophy), National Wuhan University, 6:1:135-82 (1936).

Lensen, George A., *The Russo-Chinese War* (Tallahassee, 1967).

Li, Kuo-ch'i 李國祈, *Chang Chih-tung ti wai-chiao cheng-ts'e* 張之洞的外交政策 (Chang Chih-tung's foreign policy), (Taipei, 1970).

Lo, Tun-yung 羅惇曧, "Keng-tzu kuo-pien chi" 庚子國變記 (The national crisis of 1900), in Tso Shun-sheng 左舜生 (ed.), *Chung-kuo chin-pai-nien shih tzu-liao, ch'u-pien* 中國近百年史資料初編 (Materials relating to Chinese history of the last hundred years), (Shanghai, 1926), I, 517-35.

Malozemoff, Andrew, *Russian Far Eastern Policy, 1881-1904* (Berkeley, 1958).

McKee, Delber, *Chinese Exclusion versus the Open Door Policy, 1900-1906* (Detroit, 1977).

Purcell, Victor C., *The Boxer Uprising* (Cambridge, 1963).

Quested, R. K. I., *"Matey" Imperialists? The Tsarist Russians in Manchuria, 1895-1917* (Hong Kong, 1982).

Ronning, Chester, *A Memoir of China in Revolution from the Boxer Rebellion to the People's Republic* (New York, 1974).

Schrecker, John E., *Imperialism and Chinese Nationalism: Germany in Shantung* (Cambridge, Mass., 1971).

Tai, Hsüan-chih, *I-ho t'uan yen-chiu* (A study of the I-ho t'uan), (Taipei, 1963).

————, "A Monograph on the Yi Ho Boxers," *Synopsis of Monographical Studies on Chinese History and Social Sciences,* China Committee for Publication Aid and Prize Awards, I:31-58 (Taipei, 1964).

T'an, Chester C., *The Boxer Catastrophe* (New York, 1955).

Varg, Paul A., *Open Door Diplomat: The Life of W. W. Rockhill* (Urbana, 1952), chs. 4-6.

————, *Missionaries, Chinese, and Diplomats: The American Protestant Missionary in China, 1890-1952* (Princeton, 1952).

————, "William W. Rockhill's Influence on the Boxer Negotiations," *Pacific Historical Review,* 18:3:369-80 (Aug. 1949).

————, *The Making of a Myth: The United States and China, 1897-1912* (East Lansing, 1968).

Vladimir (Zenone Volpicelli), *Russia on the Pacific and the Siberian Railway* (London, 1899).

Wang, Wen-shao 王文韶, "Keng-tzu liang-kung meng-ch'en chi-shih" 庚子兩宮蒙塵紀實 (The true story of the flight of the empress dowager and the emperor in 1900), in Tso Shun-sheng 左舜生 (ed.), *Chung-kuo chin-pai-nien shih tzu-liao, hsü-pien* 中國近百年史資料續編 (Materials relating to Chinese history of the last hundred years), II, 501-04.

Wang, Yen-wei 王彥威, *Hsi-hsün ta-shih chi* 西巡大事記 (Journal of the imperial western tour), (Peiping, 1933).

Wehrle, Edmund S., *Britain, China, and the Antimissionary riots, 1891-1900* (Minneapolis, 1966).

Wu, Yung, *The Flight of An Empress* (New Haven, 1936).

Young, L. K., *British Policy in China, 1895-1902* (Oxford, 1970).

Young, Marilyn B., *The Rhetoric of Empire: American China Policy, 1895-1901* (Cambridge, Mass., 1968).

Yün, Yü-ting 惲毓鼎, "Ch'ung-ling ch'uan-hsin lu" 崇陵傳信錄 (A true record of Emperor Kuang-hsü), in Tso Shun-sheng (ed.), *Chung-kuo chin-pai-nien shih tzu-liao, ch'u-pien* (Materials relating to Chinese history of the last hundred years), (Shanghai, 1926), I, 454-88.

17

Reform and Constitutionalism
at the End
of the Ch'ing Period

To the empress dowager, the Boxer catastrophe proved a traumatic experience. She often wept, and ruefully declared: "I had not expected to become the object of ridicule by the emperor!" Her astute political acumen and shrewd instinct dictated that it would be difficult for her to regain foreign esteem and domestic respect unless she showed some semblance of repentance and instituted measures of political reform. On August 20, 1900, while still in flight, she overcame pride, and issued a decree blaming herself for China's misfortune. After the court had been re-established at Sian, she proclaimed the desire to institute a reform of her own.

The Ch'ing Reform, 1901-05

In a statement of January 29, 1901, the dowager solicited advice on reform from ministers of state, provincial authorities, and envoys abroad. She allowed them two months in which to make detailed recommendations. They were to base their suggestions on Chinese and Western political systems, in order to indicate how best to renovate existing governmental institutions, administrative procedure, people's livelihood, methods of education, the military organization, and the financial system.

On February 14, 1901, the court reaffirmed its determination to institute reform, and accepted responsibility for the Boxer calamity. On April 21, a

Superintendency of Political Affairs (*Tu-pan cheng-wu ch'u*) was instituted to formulate a legitimate program. Prince Ch'ing, Jung-lu, Li Hung-chang, and three others were appointed directors; while Chang Chih-tung and Liu K'un-i received appointment as associates.

The Yangtze governors-general, Chang and Liu, jointly presented three memorials in July 1901 in response to the court's call. In the first memorial, they stressed loyalty to the existing system, but indicated the need for educational reform to cultivate native talents, recommending:

1. the institution of modern schools at all levels, with a mixed curriculum of Chinese classics and Western history, geography, politics, science, and technology;
2. a change in the contents of the civil service examinations to include questions on both Chinese and Western subjects;
3. the termination of military examinations;
4. encouragement of foreign study and travel.

The memorial decisively concluded, in Chang's eloquent, pithy prose: "Unless we cultivate talents, we cannot expect to exist. Unless we promote education we cannot cultivate talents. Unless we reform civil and military examinations, we cannot promote education. Unless we study abroad, we cannot make up deficiencies of education [at home]."[1]

The second paper continued with a discussion of the essentials of good government, investigating methods of acquiring wealth and power. The memorialists recommended frugality, the recruitment of unusual talents, and an increase in the anticorruption subsidy to end irregularities in government. It also suggested termination of sales of office, and the reduction of the obsolete Green Standard army, as well as the dismissal of useless scribes and clerks in government offices.

The memorialists concluded their prospectus with a third paper which suggested the adoption of "Western methods," among which they recommended the expansion of military appropriations, the introduction of a foreign-style drill, promotion of agriculture, encouragement of industry and technology, and an organized compilation of regulations with regard to mining, railroad, and commerce. They also suggested adoption of the silver dollar, the use of an official revenue stamp, improvement of the postal service, and the active translation of foreign books. They tendered their proposals with a view to "redressing the Chinese system in order to implement the Western."

Predominantly on the basis of their recommendations, the dowager ini-

1. Li Shou-k'ung, 707.

tiated an institutional reform, which differed little in content from the reform of 1898. It lasted over a more extended period, commencing in 1901 and terminating in 1905. The dowager reluctantly admitted that China could not be saved by patchy, piecemeal reform; and that complete reorganization and self-strengthening provided the only hope for the future. Salient features of her program were:

I. *Abolition of old offices*
   A. Dismissal of useless clerks and attendants in government offices. (May 1901)
   B. Termination of the sale of office. (August 1901)
   C. Incorporation of the Supervisorate of Imperial Instruction (*Chan-shih fu*) into the Hanlin Academy. (August 1901)
   D. Abolition of the governorships of Yunnan and Hupeh (December 1904) and Kwangtung (July 1905), as well as the director-generalship of the Conservancy of the Yellow River and the Grand Canal. (February 1902)

II. *Creation of new offices*
   A. The Superintendency of Political Affairs. (April 1901)
   B. The Ministry of Foreign Affairs to replace the Tsungli Yamen. (July 1901)
   C. The Ministry of Commerce, which absorbed the old Bureaus of Railways and Mining. (August 1903)
   D. The Bureau of Military Training. (December 1903)
   E. The Ministry of Police. (October 1905)
   F. The Ministry of Education. (December 1905)

III. *Military reform*
   A. Termination of military examinations. (August 1901)
   B. Reduction of the Green Standard Army and Braves by 20 to 30 percent within a year. (August 1901)
   C. Creation of provincial military academies. (August 1901)
   D. Training of the bannermen in Peking by T'ieh-liang and Yüan Shih-k'ai.
   E. Establishment of the Bureau of Military Training. (December 1903)

IV. *Educational reform*
   A. Opening of the state examinations in political economy for the Hanlin members above the compilers. (May 1901)
   B. Recruitment of Chinese students abroad for service at home by the envoys. (June 1901)
   C. Replacement of the "eight-legged essay" by current topics in provincial and metropolitan examinations, to begin in 1902. (August 1901)
   D. An order to transform provincial academies into colleges, prefectural schools into middle schools, and district schools into

elementary schools, with a mixed curriculum, including the Confucian Four Books, Five Classics, Chinese history, as well as the study of foreign governments. (September 1901)

    E. Orders to provincial authorities to select students to study abroad. (September 1901, October 1902)

    F. An order to the Imperial Clan Court to select bannermen's children to study abroad. (January 1902)

    G. An order to Hanlin compilers and other holders of the *chin-shih* degree to study in the various departments of the Imperial University. (December 1902)

    H. Annual examinations for returned students from abroad. (July 1905)

    I. Abolition of the government examinations. (August 1905)

  V. *Social reform*

    A. Permission for marriages between the Manchus and the Chinese. (February 1902)

    B. Liberation of women from foot-binding. (February 1902)

    C. Prohibition of opium. (September 1906)

  VI. *Other reforms*

    A. Revision of regulations on the tribute rice, and promotion of railway construction. (June 1901)

    B. Provincial taxes on tobacco and liquor. (December 1903)

    C. An order for drafting a commercial law. (December 1901)

    D. Establishment of refugee camps to absorb vagrants and the unemployed. (June 1905)

    E. Reduction of expenses in the palaces. (June 1904)

The program was a shrewd effort on the part of the dowager to disguise her shame over her role in the Boxer catastrophe. Her insincerity was revealed in the fact that while she openly asked for suggestions from officials in the central and provincial governments, she secretly intimated her profound distaste for things foreign. The Grand Council therefore tactfully advised officials not to speak freely of adopting Western ways. Distressed, Chang Chih-tung commented on imperial duplicity in a cable to a grand councillor, dated March 24, 1901: "I have heard that the inner circle [i.e. the dowager] does not like to speak of Western ways. Your telegram also advises us not to imitate the superficialities [lit. "skin and hair"] of Western methods so as to avoid criticism. I cannot but respond with a long sigh of resignation. If the situation is really so, then the two words 'institutional reform' have not yet hit the proper target. It is still useless, and ultimately, China will perish."[2]

2. Li Shou-k'ung, 713. In the light of this evidence, it is hard to accept the thesis that the dowager was a sincere convert to reform, as advocated in Meribeth E. Cameron, *The Reform Movement in China, 1898-1912* (Stanford, 1931), chapter 3, "The Empress Dowager's Conversion," also pages 199, 201.

The dowager's reform program was essentially a noisy demonstration without much substance or promise of accomplishment. Only three concrete improvements were actually made, namely (1) the abolition of the civil service examinations; (2) the establishment of modern schools; and (3) the sending of students abroad.

In addition to the dowager's insincerity, anti-Chinese discrimination and inept Manchu leadership also contributed to the ineffectiveness of the program. Important appointments were given to Manchus to an increasing extent. The Superintendency of Political Affairs,[3] for instance, was controlled by Jung-lu, a Manchu, and the newly formed Ministry of Foreign Affairs was placed under Prince Ch'ing, who controlled the Bureau of Military Training as well. This one-sided distribution of offices became even more evident after the deaths of the elder Chinese statesmen Li Hung-chang in 1901 and Liu K'un-i in 1902. The prospect for successful reform became even more remote.

The Constitutional Movement, 1905-11

In 1905 a dramatic change occurred in the Ch'ing reform program, following Japan's spectacular victory over Russia. To many Chinese the defeat of the large autocratic Western power by a tiny Oriental constitutional monarchy was proof of the effectiveness of constitutionalism. They were further impressed by the discovery that nearly all the leading Western powers operated on the basic principles of constitutional government, and that the Russians themselves were moving in the direction of constitutionalism, with renewed popular demands for the convocation of the Duma (assembly). The floundering Chinese believed that at long last they had found a formula for survival. The famous scholar-turned-industrialist, Chang Chien, announced triumphantly that "the victory of Japan and the defeat of Russia are the victory of constitutionalism and the defeat of monarchism." The idea of constitutionalism suddenly caught fire and spread rapidly among intellectuals, social leaders, and the forward-looking governors-general and governors in the country.

The persuasive voice of the reformer Liang Ch'i-ch'ao greatly contributed to the national clamor for constitutionalism. In exile in Japan since the failure of the "Hundred-Day" Reform, Liang came into contact with Japanese modernizers and widely read translations of Western philosophy and political thought. He fervently embraced nationalism and such concepts as liberty and equality as the inalienable rights of the people. Per-

3. Although it had three Chinese and three Manchu directors.

sistently he expounded these ideas in his journals, *The Public Opinion* (*Ch'ing-i pao*), 1898-1902, and *The New People's Miscellany* (*Hsin-min ts'ung-pao*), 1902-07, in an attempt to instill his countrymen with these ideas. His diagnosis of China's weakness showed that the Chinese people owed personal allegiance to the ruler, but not to the state; that Confucianists talked about universal rule without first providing effective emphasis on the importance of the Chinese nation; that despotism and autocracy lay at the roots of corruption and weakness of China. He ardently insisted that the Chinese had to accept nationalism as a prerequisite to the exercise of such rights as equality, liberty, and sovereignty. However, he did not believe that the China of his time was ready for a truly democratic and representative government, but considered constitutional monarchy more effective as an immediate goal. He advocated gradual political change and deprecated violent revolution. Liang employed a mixture of classical and colloquial diction in a new style of writing which won an immediate following among the reading public. His journals were eagerly pursued by young students who rushed to bookstores for recent issues, in order to imbibe such new concepts as people's sovereignty, nationalism, and constitutionalism. Liang rose to become a glittering star of Chinese journalism and political philosophy during the early years of the twentieth century.[4]

Radicals under Dr. Sun Yat-sen, however, launched a powerful counterattack to Liang's concepts of constitutional monarchy. They forcefully contended that it was essential for China to overthrow the Manchu dynasty and establish a republic in order to inaugurate a new era. They founded *The People's Tribune* (*Min-pao*) in 1905 to debate with Liang. The empress dowager, whose hatred for revolution exceeded her distaste for constitutionalism, determined to lend support to the constitutional movement, which she considered a lesser evil. She assented to send Manchu princes and nobles abroad to investigate foreign political systems as a prelude to introducing a constitution. She knew that this undertaking would prove time-consuming, and therefore work in her favor.

An investigatory mission of five members was created under the leadership of Tsai-tse, a Manchu noble. Three of the members were to visit Japan, Britain, France, and Belgium, while two others were to proceed to the United States, Germany, Austria, and Italy. The delegation set out on December 11, 1905, returning home the following July.

The mission reported favorable impressions of the British and German systems of government, but concluded that the Japanese constitution was more suitable to China because of greater similarity between the two

4. Liang Ch'i-ch'ao, 102.

countries. The Manchu leader of the mission personally proposed adoption of a constitution within five years. He indicated that a well-designed constitution could become an instrument of executive power, providing concentrated leadership in the central government. The recommendation was approved by a royal commission, and endorsed by the dowager on September 1, 1906. But she shrewdly omitted to specify the date of promulgation.

Divergent views with regard to the constitution were held by different factions in government. The dowager considered it a convenient device by which to conciliate the public without actually compromising her own power. The Manchus saw in it a chance to centralize government control and exclude the Chinese from inner circles, thereby wresting power from the provincial governors-general, who were predominantly Chinese. Thus constitutionalism became an anti-Chinese device of the Manchus. On the other hand, to many Chinese constitutionalism provided hope of liberation from unfair, oppressive Manchu discrimination and domination.

Having assented to the principle of constitutionalism, the court appointed a group of officials to deliberate on the reform of governmental institutions on September 2, 1906, as a first step toward establishment of a constitutional monarchy. Conflict of interests and fear of criticism led to the decision to exclude five offices from discussion: the Grand Council, the Department of Imperial Household, the Eight Banners, the Hanlin Academy, and the eunuchs. A report on administrative reorganization, which was finally submitted, stressed concentration of responsibility, elimination of inveterate governmental weakness, and the increase of efficiency. On the grounds of this study, the court issued a decree of reform on November 7, 1906. It achieved little more than expanding Six Boards into eleven modern-sounding ministries. It created the image of modern constitutionalism but retained the essence of the old governmental procedure. It provided an institutional reshuffle, which was retrogressive in that it increased Manchu power in proportion to that of the Chinese. The latter accounted for less than one-third of the top echelons in government after the reorganization. The widened schism between Manchus and Chinese disappointed many proponents of constitutionalism.

Manchu power consolidation took place in local government, too. The court curbed the powers of the governors-general and governors in 1907 by directly appointing provincial judicial, police, and agricultural-industrial-commercial commissioners. Carefully worked out measures then followed to withdraw the two most coveted powers of local authorities when the court appointed provincial financial commissioners and transferred provincial forces to the new Ministry of Army. Yüan Shih-k'ai lost four of his six Peiyang divisions. The court delivered a coup de grace in August 1907, when it transferred the two most powerful Chinese governors-general, Chang Chih-tung and Yüan Shih-k'ai, to Peking as grand councillors, with the latter serving concurrently as minister of foreign affairs. Under the guise of constitutionalism, the Manchus successfully carried out their anti-Chinese policy and achieved unprecedented concentration of power.

A few promising aspects of the constitutional movement existed, however. These provided for the establishment of a Bureau of Constitutional Compilation, in August 1907; the dispatch of three officials, in September 1907, to Japan, Britain, and Germany to study constitutionalism; the appointment of two individuals, one Chinese and one Manchu, to inaugurate a National Assembly; and the order to establish provincial, prefectural, and district assemblies.

The reformers of 1898, still in exile in Japan, were apparently heartened by developments in China. Hoping to be invited to join the constitutional movement, Liang Ch'i-ch'ao suspended publication of his *New People's Miscellany* and proceeded to organize a Political Information Society (*Cheng-wen she*) in Japan, to promote (1) responsible parliamentary government; (2) legal reform to insure judiciary independence; (3) local self-government and clear demarcation of authority vis-à-vis the central government; and (4) cautious diplomacy to strive for equal rights in the international community. Members of the society showed interest in co-operating with the Ch'ing court, but Yüan Shih-k'ai who betrayed the reformers in 1898 refused to have anything to do with them; so did the dowager, who hated K'ang and Liang.

Revolutionaries under Dr. Sun Yat-sen, on the other hand, ridiculed Liang and associates for their flirtation with the reactionary court. Rejected by both the court and the revolutionaries, the Political Information Society was left in midstream. Some members of the society, however, secretly returned to China in an attempt to goad social leaders, students, and overseas groups into demanding the early establishment of parliament and the immediate promulgation of a constitution. Dozens of so-called "Constitution-Protection Clubs" sprang up in the provinces, and waves of delegates came to Peking to petition for the early promulgation of the constitution. The tide became so powerful that even Manchu bannermen joined the cause. Under such pressure, the court, on August 27, 1908, issued an "Outline of Constitution" (*Hsien-fa ta-kang*), a parliamentary law, and prescribed a nine-year tutelage period before the constitution became effective.

The empress dowager never genuinely contemplated introducing constitutional monarchy in China. The Ch'ing "Outline" actually gave the throne even greater power than the Japanese model. It specified that executive, legislative, and judiciary power resided in the emperor, who, sacred and inviolable, would continue to rule the empire in the unbroken line of ten thousand generations. Parliament could consider, but not decide, questions of government; the laws and regulations passed by it would not become effective without the approval of the sovereign. Furthermore, provisions with regard to the rights and duties of the citizens were little more than meaningless formalities. The "Outline of Constitution" was an instrument of imperial procrastination, in the attempt to consolidate dynastic power and prolong the Manchu rule. In spite of these safeguards, the dowager was reluctant to put the "Outline" into practice, and sought to delay the introduction of a constitution in China during her lifetime by requiring a nine-year gestation period, following the Japanese pattern.[5]

The dowager, already 73, apparently had great confidence in her longevity and delaying tactics; but a serious illness hit her in less than three months and ended her life on November 15, 1908. An announcement regarding the strangely coincidental death on the preceding day of the 37-year-old emperor, Kuang Hsü, followed the dowager's demise. Despite accounts of his having Bright's disease, court sources close to the emperor concurred that he had enjoyed excellent health and had seldom been sick in his life. Legend reveals that he secretly, if imprudently, rejoiced over the dowager's impending death. The Imperial Woman then vengefully vowed: "I cannot die before him!" Indications point to the possibility that she poisoned him the day before she died. Widespread rumor circulated that Yüan Shih-k'ai participated in the plot because he had betrayed the emperor in 1898 and dreaded his return to power, but there was no evidence to substantiate this story.

The dowager's three-year-old grandnephew, P'u-i,[6] succeeded to the throne, with his father, the second Prince Chün,[7] acting as regent. The prince appeared bent on eliminating Yüan for his betrayal of the late emperor, but was restrained by fear of mutiny among the Peiyang forces. The Chinese statesman Chang Chih-tung also reportedly cautioned him against killing high officials during the period of imperial mourning. Prince Chün then insisted that Yüan was suffering from a leg ailment, from which he needed to recuperate in quiet retirement. On January 2, 1909, Yüan was forced out of the government.

Having successfully avenged the late emperor's betrayal and enforced

5. In 1881 the Japanese emperor promised a Diet (parliament) in 1890.
6. Later known as Henry Pu-yi.
7. Tsai-feng, half-brother of the late Emperor Kuang-hsü.

an inherently anti-Chinese policy, Prince Chün now posed, purportedly, as the instrument of constitutional monarchism. On February 17, 1909, he ordered the establishment of provincial assemblies, which were inaugurated October 14. With the creation of these popular bodies, the demand for the convocation of parliament gained rapid momentum. Three times in the following year—on January 26, June 22, and October 3, 1910—representatives of sixteen provinces went to Peking to petition the early convening of parliament. The court reprimanded them for interfering with state affairs and ordered them to go home. Thus insulted, these representatives, mostly chairmen and vice-chairmen of their respective provincial assemblies, met in a secret conclave and reputedly decided to shift their sympathies quietly to the revolutionaries.[8] In spite of the tremendous pressure from the provincial assemblies and private constitutionalists, all Prince Chün did was to announce on November 4, 1910, that he would shorten the period of constitutional preparation from nine to six years. At the same time, he furthered his anti-Chinese policy by organizing a "Royal Cabinet" on May 8, 1911, with five imperial relatives among the thirteen appointees. There were eight Manchus and one Mongol bannerman, but only four Chinese, in this cabinet. When provincial assemblies protested against royal domination of the cabinet, they were pointedly reminded of the throne's absolute control of appointments, designated in the "Outline of Constitution." The Chinese became increasingly convinced that genuine constitutionalism was impossible under Manchu leadership.

Disillusion and disappointment generated mounting anti-Manchu sentiment and swung public feeling toward the revolutionary cause. Within a few months, Dr. Sun Yat-sen's party swept the Ch'ing dynasty into the oblivion of history.

8. P'eng-yüan Chang, "The Constitutionalists," Mary C. Wright (ed.), *China in Revolution*, 160-70.

Further Reading

Adshead, S. A. M., *Province and Politics in Late Imperial China: Viceregal Government in Szechuan, 1898-1911* (London, 1984).

Ayers, William, *Chang Chih-tung and Educational Reform in China* (Cambridge, Mass., 1971).

Bland, J. O. P., *Recent Events and Present Policies in China* (Philadelphia, 1912).

———, and E. Backfouse, *China under the Empress Dowager* (Philadelphia, 1910).

Cameron, Meribeth E., *The Reform Movement in China, 1898-1912* (Stanford, 1931).

Chang, P'eng-yüan 張朋園, *Liang Ch'i-ch'ao yü Ch'ing-chi ko-ming* 梁啟超與清季革命 (Liang Ch'i-ch'ao and the late Ch'ing revolution), (Taipei, 1964).

Chang, Yü-fa 張玉法, *Ch'ing-chi ti li-hsien t'uan-t'i* 清季的立憲團體 (The constitutional bodies established in the late Ch'ing), (Taipei, 1971).

Ch'en, Jerome, *Yüan Shih-k'ai, 1859-1916* (Stanford, 1961), chs. 3-7.

Ch'i, Ping-feng 亓冰峯, *Ch'ing-mo ko-ming yü chün-hsien ti lun-cheng* 清末革命與君憲的論爭 (The debate between the revolutionists and constitutional monarchists during the late Ch'ing), (Taipei, 1966).

Chu, Samuel C., *Reform in Modern China: Chang Chien, 1853-1926* (New York, 1965).

Chuang, Chi-fa 莊吉發, *Ching-shih ta-hsüeh-t'ang* 京師大學堂 (Former Peking University: An historical study), (Taipei, 1970).

Der Ling, Princess, *Old Buddha* (*Empress Tzu Hsi*) (London, 1929).

Franke, Wolfgang, *The Reform and Abolition of the Traditional Chinese Examination System* (Cambridge, Mass., 1960).

Haldane, Charlotte, *The Last Great Empress of China* (Indianapolis, 1965).

Ichiko, Chuzo, "Political and Institutional Reform, 1901-11" in John K. Fairbank and Kwang-ching Liu (eds.), *The Cambridge History of China* (Cambridge, Eng., 1980), Vol. 11, 375-415.

Israel, Jerry, *Progressivism and the Open Door: America and China, 1905-1921* (Pittsburgh, 1971).

Kent, Percy Horace, *The Passing of the Manchus* (London, 1912).

Levenson, Joseph R., *Liang Ch'i-ch'ao and the Mind of Modern China* (Cambridge, Mass., 1953).

Rankin, Mary Backus, *Elite Activism and Political Transformation in China: Zhejiang Province, 1865-1911* (Stanford, 1986).

Reid, John G., *The Manchu Abdication and the Powers, 1908-1912* (Berkeley, 1935).

Sun, E-tu Zen, "The Chinese Constitutional Missions of 1905-1906," *Journal of Modern History*, 24:3:251-68 (Sept. 1952).

Tai, Hung-tz'u 戴鴻慈, *Ch'u-shih chiu-kuo jih-chi* 出使九國日記 (Diary of my diplomatic mission to nine countries), (Peking, 1906).

Tsai-tse 戴澤, *K'ao-ch'a cheng-chih jih-chi* 考察政治日記 (My diary of political studies abroad), (Peking, 1908).

A Boxer poster.

Portrait of the young Emperor Kuang-hsü.

Yüan Shih-k'ai.

K'ang Yu-wei.

Liang Ch'i-ch'ao.

Timothy Richard.

Yün Shih-k'ai, center, as provisional president of the Republic of China, 1912.

Wu P'ei-fu, left, with two of his generals.

Sun Yat-sen with wife Soong Ch'ing-ling, a graduate of Wesleyan College
for Women in Macon, Georgia.

Huang Hsing, powerful revolutionary associate of Dr. Sun Yat-sen.

Pu-yi, the "Last Emperor."

18

Late Ch'ing Intellectual, Social, and Economic Changes, with Special Reference to 1895-1911

The late Ch'ing was a period of drastic transformation, with the pace of change accelerating after 1895. There occurred not only political reforms as described in the previous chapters, but also great changes in intellectual, social, and economic life. Intellectually, in addition to the Modern Text Movement (Chapter 15), there was a fundamental reorientation of outlook and activity as a result of shifting trends in traditional learning and the influx of Western ideas. Socially, the individual emerged as the basic unit of society, replacing the family and the clan, while two new classes, the compradores and militarists, gained prominence, and the cities grew enormously. Economically, there was an increasing stricture upon government finances, a growing trade imbalance, and a deepening foreign control of the modernized section of the Chinese economy. Seldom had China seen such drastic socioeconomic and intellectual changes in so short a time.[1]

Intellectual Reorientation

THE METAMORPHOSIS OF TRADITIONAL LEARNING. Late Ch'ing intellectual trends differed markedly from the middle period. The double challenge of domestic rebellion and foreign invasion forced the scholars to re-examine

1. The idea of a rapidly changing China after 1900 has been convincingly presented in Mary C. Wright (ed.), *China in Revolution, The First Phase, 1900-1913* (New Haven, 1968), 1-63, "Introduction: The Rising Tide of Change."

their role in society. The Han School of Empirical Research (*K'ao-cheng hsüeh*), with its preoccupation with antiquarian studies and its pride in pursuing knowledge for the sake of knowledge, struck a discordant note in the rapidly changing times. Two new currents became apparent: the revival of the idea of "practical statesmanship" (i.e. unity of knowledge and practice) and the trend toward intellectual tolerance and integration. Pressed with the vital problems of foreign invasion and domestic upheaval, the scholars felt a moral obligation to contribute their share to social and political stabilization. Even the Han scholars renounced their traditional apathy toward public affairs. All late Ch'ing scholars shared the conviction that they had an integral role to play in public affairs.

Scholars also took interest in a broad spectrum of subjects and approaches. For instance, the statesman Tseng Kuo-fan had attempted to integrate (Sung) philosophy, (Han) textual criticism, literature, and practical statesmanship into one comprehensive, basic learning, called the *li-hsüeh*, to reflect the Confucian concept of *li*, or propriety. K'ang Yu-wei moved from Neo-Confucian studies to Modern Texts and to Western works of political reform. Broad interest in and a syncretic approach to learning characterized the age. Hence, the late Ch'ing intellectual world, having moved from the dominance of one school (Han) to the juxtaposition of many, moved again from division toward integration. In this process, the mental horizon of Ch'ing scholars was rendered much broader than before, reaching out of the traditional boundary into Western studies.[2]

THE "NEW LEARNING." The influx of Western ideas began with the translation of the Bible and religious tracts in the pre-Opium War period. Of the 795 titles translated by Protestant missionaries between 1810 and 1867, 86 percent were in religion, and only 6 percent in the humanities and sciences. During the 1861-95 Self-strengthening Movement, translations extended into diplomacy, military arts, science, and technology. Of 567 works translated between 1850 and 1899, 40 percent were in applied sciences, 30 percent in natural sciences, 10 percent in history and geography, 8 percent in social sciences, and about 3.5 percent in religion, philosophy, literature, and the fine arts.[3] During this stage, emphasis was on science and technology, the chief sources of information being Anglo-American works, which accounted for 85 percent of all translations as opposed to 15 percent from Japanese texts.

After the Japanese war of 1894-95 the trend shifted. The narrowness

2. Hsiao I-shan, IV, 1,746, 1,748, 1,951-60.
3. Tsuen-hsiun Tsien, "Western Impact on China through Translations," *Far Eastern Quarterly*, XIII:3:311, 315 (May 1954).

of China's modernization program became fully apparent: men of foresight realized clearly that China must broaden its understanding of the West beyond merely military and industrial techniques to include studies of political institutions, economic systems, social structures, and scientific as well as philosophical thought. Translations of Western works in these fields became a paramount prerequisite to reform and renovation. After the Boxer Rebellion, the Peking Imperial University absorbed the old T'ung-wen kuan translation bureau, completing a number of translations and compilations of textbooks in mathematics, physics, trigonometry, education, and philosophy. In 1907 the Ch'ing court formally created a Bureau of Translation and Compilation, to which many proud products of the old literary examinations were appointed. Among them was Wang Kuo-wei, a solid scholar who took great interest in Kant, Schopenhauer, and Nietzsche. On the whole, however, the official translation bureaus had less impact on China's culture than did the private translators. Among the latter, two were particularly remarkable: Yen Fu and Lin Shu.

Yen Fu (1854-1921) of Hou-kuan, Fukien, pioneered a new direction in China's endeavor to understand the modern West. Having received in childhood a thorough grounding in the Chinese classics, at fourteen Yen entered the naval academy at the Foochow Dockyard; there he acquired a new education in English, arithmetic, algebra, geometry, trigonometry, physics, chemistry, mechanics, geology, astronomy, and navigation. He graduated with high honors in 1871. Selected in 1876 to study in an English naval academy, he arrived in the British Isles the following year at a time when the great masters and thinkers—Darwin, Huxley, and Spencer—were shaking the world with their ideas of evolution and the social application of the struggle for survival. Yen's immediate attraction to Darwinism came not so much from its biological import as its stress on the assertive energy of men and the "actualization of potentialities within a competitive situation."[4] Not surprisingly, Yen began to examine China's problems and her position in the world in the light of social Darwinism.

Eager to discover the sources of Western (particularly British) wealth and power, Yen Fu diligently studied the British political systems, economic institutions, social philosophy, and legal concepts. He came to believe that the basis of British power was the legal concept of impartial justice.[5]

Yen returned home in 1879 and was made dean of instruction at Li Hung-chang's Peiyang Naval Academy at Tientsin, where he remained

4. Benjamin I. Schwartz, *In Search of Wealth and Power: Yen Fu and the West* (Cambridge, Mass., 1964), 46.
5. *Ibid.,* 29.

for nearly twenty years. Though he rose to its superintendency in 1890, he was never taken into confidence by Li, and his naval career never truly blossomed; while his Japanese contemporaries in Britain, such as Itō and Tōgō, all became leaders of a modernization which turned their country into a powerful state.[6] Frustrated by his inability to help his country, especially after the Peiyang fleet's fiasco in the Japanese war, in which many of his former colleagues and students were casualties, Yen began to lash out at China's weakness through writing and translation. In this he finally found his true vocation as a publicist, free and able to air his pent-up ideas.

A key to Western development, he loudly proclaimed to his countrymen, was the "different vision of reality" which involved ideas and values; it was thought, not military power, which made a country strong and wealthy. In order to acquaint his people with firsthand knowledge of Western ideas, he spent the next fifteen years translating a number of important works, including T. H. Huxley's *Evolution and Ethics* (1900), J. S. Mill's *On Liberty* (1903) and *Logic* (1905), Herbert Spencer's *A Study of Sociology*, Montesquieu's *De l'esprit des lois* (1909), Edward Jenk's *A History of Politics,* and William S. Jevon's *Logic.* For the first time the Chinese met the ideas of evolution, free trade, the principles of sociology, and the division of power in government.

The crux of Yen's message in all his writings was that the basic difference between China and the modern West lay in their dissimilar attitudes toward human energy. The West exalted action, assertiveness, struggle, and dynamism in order to actualize the unlimited human potentialities. Government and society provided favorable conditions—liberty, rising equality of opportunity, self-government, public spirit, and impartial justice—to facilitate the liberation of the individual's inner energies, and channeled them toward collective goals. The government promoted rather than suppressed the individual's constructive self-interest, so that the public and private interests reinforced each other. Thus, when Britain fostered ideas, values, and proper environment for the fulfillment of human potential, and when she elevated her people's capability, intelligence, and morality, she became rich and powerful.[7]

In China, he contended, the opposite held true. The ways of the Sages discouraged the development of the people's capacities and inhibited the free flow of their vital energies. The traditional rulers since the Ch'in dynasty (221 B.C.-206 B.C.) had all been "robbers," skimming off the cream

6. Itō Hirobumi became premier, while Tōgō Heihachirō rose to be an admiral and distinguished himself in the Sino-Japanese and Russo-Japanese wars. See Edwin Albert Falk, *Togo and the Rise of Japanese Sea Power* (New York, 1936).
7. Schwartz, 70-75, 238-43.

of the populace and failing to elevate their intelligence. This was the basic trouble of China, Yen pointed out in a ten-thousand-word memorial which he prepared for, but did not have time to submit to, Emperor Kuang-hsü during the short-lived 1898 reform. He bluntly announced that 70 percent of China's troubles were internal, only 30 percent external. What China needed was not a piecemeal improvement, but a radical change in its view on domestic peace and order. The traditional rulers, he insisted, had always tried to keep the people weak and ignorant so as to facilitate their control of the country. They deprecated competition and innovation and admonished men to follow ancestral paths in order to achieve stability. They encouraged thrift and discouraged the development of resources. They honored antiquity and despised modernity. They deplored aggressiveness and promoted contentment, and they inculcated in the people a habit of meekness to forestall rebellion. All this, Yen proclaimed, reversed the Western promotion of progress and improvement through competition, release of energies, and elevation of human capacity and intelligence.[8]

Yen's exaltation of Western assertiveness and dynamism, and his deprecation of Chinese passivity and enervation led later to the description of Western civilization as activity-oriented (*tung-te wen-hua*) and Chinese civilization as stability-oriented (*ching-te wen-hua*).

If the traditional Chinese methods of maintaining domestic order led to poverty, ignorance, and weakness, Yen argued, discard them, even though they were the work of the Sages. On the other hand, he insisted, if Western methods could alter the deplorable situation, adopt them; for knowledge knows no national boundary. China must change its old ways and compete for survival in the modern world. It must develop patriotism and nationalism, foster universal technical and scientific education, promote popular economic self-interest, and create "the organs of a rationalized national state."[9] This in essence was Yen's message to his countrymen.

Yen was noteworthy not only for his ideas but also for his excellent writing style. In his translation he followed the threefold criteria of faithfulness (*hsin*), comprehensiveness (*ta*), and elegance (*ya*). Because of the syntactical differences between Chinese and Western languages, Yen's translations were basically not literal renditions but summations or paraphrasings of the original. His method was to immerse himself in the original, capture its spirit and quintessence, and then communicate its meaning in idiomatic, classical Chinese. For instance, he conveyed the essential meaning of "the struggle for existence" and "the survival of the fittest" by

8. Hsiao I-shan, IV, 2,021-24.
9. Schwartz, 185.

rendering them as: "Things struggle; nature selects. The superior is victorious; the inferior vanquished."

His lofty, abstruse, terse, and elegant style was very much admired but it militated against popular acceptance, appealing to only a small group of educated elite, such as Liang Ch'i-ch'ao. Consequently his influence had too little circumference. Nonetheless, Yen Fu lived to become a monument in the annals of Sino-Western cultural exchange. It was he who first made a penetrating, comparative study of the two diverse civilizations and came up with bold answers to the perennial questions: "What does the West have that China does not?" and "What are the sources of Western power and wealth?"

Contemporary with Yen Fu was Lin Shu (1852-1924), another great translator, who excelled in rendering Western novels. Having attained the first and second literary degrees in 1872 and 1882, but having repeatedly failed the examinations for the coveted third degree, he resigned himself, a disgruntled literatus, to teaching.

A tubercular, Lin was sensitive, tense, sentimental, and impulsive. A series of family deaths—his mother in 1895, his wife in 1897, and two of his children in the next two years—threw him into despondency and loneliness. To raise him from despair, a friend,[10] who had been a cadet at the Foochow Naval Academy and a law student at the University of Paris, suggested that they jointly translate Alexander Dumas's *La Dame aux camélias*. Aware of Lin's unfamiliarity with foreign languages, this friend translated orally as Lin composed it into acceptable Chinese. So successful was this "oral translation" that it set the pattern for Lin's later projects. Lin moved his pen extremely fast, often completing his written version simultaneously with the oral rendition.[11] The scope of his translation was vast, ranging from romantic fiction to social novels, fables, biographies, plays, and detective stories. His most famous translations included the above-mentioned *La Dame aux camélias;* Charles Dickens's *Oliver Twist, David Copperfield, The Old Curiosity Shop, Dombey and Son,* and *Nicholas Nickleby* (all published between 1907 and 1908); H. Rider Haggard's *King Solomon's Mines, Montezuma's Daughter,* and *Beatrice;* Sir Walter Scott's *Ivanhoe, The Talisman,* and *The Betrothed.* In his lifetime Lin completed no less than 159 titles, in 12 million words.

Though readily admitting the inaccuracies of oral translation, Lin, because of his sensitivity and excellent literary endowment, was able to grasp the spirit, mood, and humor of the original by instinct. Hence he came remarkably close to the essence of the novels he translated. He achieved

10. Wang Tzu-jen.
11. Two of his most constant collaborators were Wei I and Tseng Tsung-kung.

the result by spontaneously merging himself with the characters, as he explained: "People in a book become at once my nearest and dearest relatives. When they are in difficulties, I fall into despair; when they are successful, I am triumphant. I am no longer a human being but a puppet whom the author dangles on his strings."[12] He attained such success in communicating the original feeling in a restrained, classical style that at times his translations were considered superior to the originals. Arthur Waley, a leading contemporary English translator of Eastern works, remarked after comparing Lin's versions with Dickens's own works: "The humor is there, but is transmuted by a precise, economic style; every point that Dickens spoils by uncontrolled exuberance, Lin Shu makes quietly and effectively."[13] On the other hand, errors and distortions were also present in Lin's editions—for instance, when he rendered some of Shakespeare's plays into prose narratives. On the whole, however, his translations convey more of the original spirit of the Western literature than a beginning Chinese student of foreign languages could possibly attain from a direct reading.

Through Lin, Western literature was introduced into China, and through his translations the Chinese gained invaluable insights into Western customs, social problems, literary currents, ethical concepts, familial relations, and the glittering world of literature itself. In addition to the translations, Lin often promoted patriotism, nationalism, social progress, and better human relationships in the prologues and introductions of his works. His influence on the younger generation cannot be overemphasized. Though his tenacious adherence to the ancient style caused him to lag behind the times, his contributions rank him with Yen Fu as one of the twin luminaries in the firmament of Chinese translators at the turn of the century.

JAPANESE TRANSLATIONS. In addition to Western works, numerous Japanese translations of Western subjects were also rendered into Chinese. K'ang Yu-wei and Liang Ch'i-ch'ao, during the "Hundred-Day" Reform, vigorously promoted the use of the Japanese media as a convenient short cut to the essentials of Western learning, since the Japanese had already translated many of the more important Western masterpieces, and because learning Japanese was easier than learning a Western language.

Although the failure of the reform program obviated en masse transla-

---

12. Leo Ou-fan Lee, "Lin Shu and His Translations: Western Fiction in Chinese Perspective," *Papers on China*, East Asian Research Center, Harvard University, 19:186 (December 1965).
13. *Ibid.*, 187.

tion of Japanese works, K'ang and Liang continued to promote the idea during their exile in Japan, and they influenced a vast number of Chinese students in that country. The Ch'ing court, as well as provincial authorities and private parties, had been sending increasing numbers of students to Japan, until the count had reached some 13,000 in 1906. The students not only delivered into Chinese a great variety of Japanese books and translations of Western works, but also borrowed many Japanese terms for key subjects, such as *che-hsüeh* (philosophy), *ching-chi-hsüeh* (economics), and *she-hui-hsüeh* (sociology). The late Ch'ing new educational system was modeled upon that of Japan, as were most of the textbooks. Between 1902 and 1904 translations from Japanese sources accounted for 62.2 percent of the total 573 works, while British sources dwindled to 10.7 percent, and American to 6.1 percent. Of the total, social sciences occupied 25.5 percent, history and geography 24 percent, natural science 21 percent, applied sciences 10.5 percent, philosophy 6.5 percent, and literature 4.8 percent.[14] Clearly Japan had replaced Britain and the United States as the chief supplier of information, and the emphasis had shifted from science and technology to social sciences, philosophy, and literature. China's interest in the West had definitely moved from military science to social studies and the humanities.

The translation of Western and Japanese works resulted in widespread dissemination of foreign ideas among the educated Chinese. Democracy, parliamentary government, constitutionalism, division of power, liberty, equality of the sexes, Darwinism, and other imported concepts invaded the discussions and idiom of the intelligentsia. Such ideas could not but exert a considerable impact upon the society.

Social Changes

The kinship society of China, with its age-old customs, values, and emphasis upon the family and clan as basic units, was shaken to its foundations during the last decade of the dynasty. The Confucian concepts of family loyalty, filial piety, chastity, Three Bonds, and Five Relationships gave way to Western ideas of individualism, freedom, and equality of the sexes. Realization grew that the individual was more a member of the society and state than of the family, vested with inalienable rights upon which nobody, not even the family elders, could infringe. Young Chinese began to assert their independence from their families and to condemn Confucian teachings on proper relationships as obsolescent and feudal. The omnipotence of the family head was challenged.

14. Tsuen-hsuin Tsien, 319.

THE DISINTEGRATION OF THE FAMILY-CENTERED SOCIETY. Until the late Ch'ing, the traditional Chinese family resembled a miniature kingdom in which the head occupied the place of the sovereign, with authority to enact family law and make life and death decisions for the members. The government, recognizing this familial omnipotence, never intervened in the domestic relations between father and son, husband and wife, and brother and sister. However, with the influx of foreign doctrines and political philosophy, New Scholars of Western Learning began to promote the radical notion that the power of the family head logically belonged to the state, that the inalienable rights of the individual were beyond the control of the family head, and that as basic components of the state, male and female should be equal. These ideas, striking at the very roots of family relations, won currency among the young. Furthermore, the opening of modern schools at the turn of the century in effect meant that the government had taken over from the family the responsibility for educating the youth. Thus, when the state intervened in family relationships, it struck away the political prop for a kinship society.

Concomitantly, the legal support of the family-centered society also crumbled. The old juridical system, devised to support the kinship social structure, recognized the special position of the family head, the inequality of the sexes, the inability of women to inherit property, the exclusion of sons of concubines and illegitimate marriages from family succession, the collective punishment of family members for crimes committed by one, and the so-called Ten Unpardonable Crimes.[15] It inflicted heavier punishment upon wives than husbands involved in fighting, and sanctioned the Confucian teaching that father and son should shield each other from justice. Reflecting as they did feudal social relationships, they were clearly out of tune with the rapidly changing times. When the new legal codes of the late Ch'ing,[16] as well as those of the early republican period, took into consideration the inalienable rights of the individual, the equality of the sexes, the right of women to inherit property, etc., the old juridical basis of the kinship society was shattered.

Similarly, the economic foundations of the old society tottered. The influx of foreign commodities under preferential customs and the right of foreigners after 1895 to engage in local manufacturing disastrously affected the native handicraft industries and agrarian economy. Foreigners dominated Chinese public utilities, communication, mining, banking, and other modern enterprises. Their factories, by virtue of vast capital and

15. Parricide, unfilial behavior, incest, lack of harmony, insubordination, rebellion, conspiracy against the ruler, treason, inhuman offenses, and sacrilege.
16. Enacted under leadership of Shen Chia-pen.

mass production, outsold Chinese rivals even in distant villages. As common an agricultural product as cotton was marketed by foreigners more cheaply than the Chinese could produce it. Farm women who traditionally weaved as a secondary vocation were thrown out of work, and farmers had increasing difficulty eking out a living.

Such economic distress created an adverse impact on familial relationships. The clan and family could no longer provide help and comfort to those members who became unemployed, sick, and destitute. The displaced handicraft worker or the peasant left for the city, where he slipped from family and clan control; if lucky enough to find a new life, his meager income hardly sufficed to support his own dependents, let alone the clansmen. The ties between such a man and his clan became attenuated. Very often the wife and children of such a man were forced to work in a different city just to scrabble out a living, thereby scattering further not only clansmen but even immediate family members. Little wonder that old familial relationships broke down under the impact of the foreign economic invasion.

Shorn of its political, legal, and economic props, the kinship society simply could not survive. At the same time, it became socially fashionable and economically expedient to adopt the Western style small family system. The big-family system and kinship society gradually passed out of existence as China moved from an agrarian, premodern state toward a proto-industrial, modern society.[17]

THE NEW CLASSES. The second major social change was the rise of two new social forces, the compradores and the militarists. The former were the new rich, the latter the new power. Both threatened to eclipse the influence of the scholar-official class. Hence the traditional social stratification of the four classes—scholar-official (*shih*), farmer (*nung*), artisan (*kung*), and merchant (*shang*)—could no longer adequately reflect the principal functional orders of society.

The compradores assumed the function of business agents or managers for foreign establishments. Their familiarity with local conditions, as well as their language facility, made them indispensable to foreign banks, trading companies, industrial concerns, and factories. They assisted their foreign employers in finding business sites, recruiting factory workers, selling finished goods, buying raw materials, making investments, and arranging loans to the Chinese government and private parties. They were compensated with handsome salaries and lucrative commissions on a contractual

17. Hsiao I-shan, IV, 1,455-59.

basis. As middlemen they were able to manipulate the terms of transactions between the foreigners and Chinese, reaping a quick and handsome profit. Maintaining wide connections in official and private circles, they led the high life of the former hong merchants. Their association with influential foreign firms, their power to manipulate, their connections and new wealth made them a new social force compelling recognition.

In their business transactions the compradores were contractually bound to work for the benefit of their foreign employers; not infrequently they had to work against the interests of their own country and people. They helped foreign banks to make loans to the Chinese at usurious rates; they opposed any patriotic movement to boycott foreign imports; and they helped the foreigners to squeeze the utmost profit from the Chinese market. Consequently, the compradore class has been dealt much scathing criticism by modern Chinese historians and Marxist scholars alike as unpatriotic, traitorous, and parasitic. Yet many of them, once having acquired sufficient capital and managerial skill, turned to promoting modern industries and enterprises themselves.[18] Obviously one cannot lightly dismiss the compradores as a class of sinners and parasites, for many did contribute to their country's economic development.

The second new social force was that of the militarists. In traditional Chinese society soldiers were held in contempt, as reflected in the popular saying: "Just as good iron is not made into nails, so good men do not become soldiers."[19] However, a new military type arose by the close of the Ch'ing period—men who had received some modern military training and were not illiterate and foolhardy in the traditional image. They were connected with the new army the court had trained after the collapse of Li Hung-chang's Huai army in the Japanese war. Among the leading trainers was Yüan Shih-k'ai, organizer of a new army of 7,000 men fashioned more or less on the German model at Hsiao-chan, some 70 *li* (23 miles) from Tientsin. His officers, largely chosen from Tientsin Military Academy graduates, owed him personal allegiance. His reputation as an able officer soared when, as governor of Shantung in 1900, he kept that province free from Boxer disturbances. Upon the death of Li Hung-chang in 1901, Yüan succeeded to the important governor-generalship of Chihli. And even though he was later forced by Manchu jealousy to shed command of his new army, his loyal subordinates retained control of these forces. Numbered among these officers were Tuan Ch'i-jui, Feng Kuo-chang, Chang Hüan, and Ts'ao K'un, all destined to be leading power-holders in the late 1910s and 1920s. Yüan and his cohorts, known as the

18. As in the case of Tong King-sing.
19. *Hao-t'ieh pu ta ting, hao-nan pu tang ping.*

Peiyang clique, exercised tremendous political influence through their military power. That Yüan's original brigade produced five future presidents or acting chief executives, one premier, and most of the warlords in North China testifies to the rise of the militarists as a new social and political class. Appropriately, Yüan was dubbed "father of the warlords."[20]

THE GROWTH OF THE CITIES. A third new social phenomenon was the rise of the great metropolises. The government sponsored Self-strengthening projects centered mostly along the coast and in the treaty ports where foreign assistance was most readily available. Foreigners and their establishments—such as banks, trading companies, and factories—also located mainly in these ports and leased territories. The greater safety of these places and the concentration of foreign capital therein induced Chinese businessmen to migrate there, while displaced peasants also came to the cities seeking employment, usually ending up in factories operated by foreigners or Chinese entrepreneurs. Increasingly the treaty ports became the financial, industrial, and population centers of China. Such sites as Shanghai, Nanking, Canton, Hankow, and Tientsin expanded into urban centers of considerable size and wealth. The growth of cities and city-centered industries delineates the emerging capitalism of modern China.

Economic Plight

BUDGETARY DEFICIT. Government finances in the late Ch'ing presented a totally different picture from early and middle Ch'ing, when revenue usually exceeded expenditures. In the K'ang-hsi period (1662-1722), in spite of repeated tax remissions totaling in excess of 120 million taels, there was a treasury surplus of 8 million. The Ch'ien-lung period (1736-95) saw the reserve increase to 70 million, despite vast spending and costly military campaigns.

Beginning with the 19th century, however, the situation deteriorated. Domestic rebellions, foreign wars, drought, floods, opium importation, and silver outflow reduced the treasury reserve to a mere 8 million taels by 1850, which was further cut two years later to 3 million taels as a result of military expenses against the Taipings. Normal channels of revenue could no longer sustain the costly campaign, so a new commercial transit tax, the *likin,* was instituted in 1853, with an annual yield of 10 to 20 million taels. In the ensuing two decades, a total of 70 million taels was

20. Ralph C. Powell, *The Rise of Chinese Military Power, 1895-1912* (Princeton, 1955), 76-80.

expended suppressing the Taiping, the Nien, and the Moslem rebellions. These extraordinary outlays set government finances so far back that budgetary imbalance became the order of the day. By the T'ung-chih period (1862-74), the average deficit had grown to 10 million annually—60 million income against 70 million expenditure.

During the Kuang-hsü period (1875-1908), even though the government income increased rapidly, its expenditures grew even more rapidly, causing an ever-widening gap. This sharp increase in expenditures resulted from foreign wars, indemnities, costs of foreign loans, and new Self-strengthening projects. A few major outlays of funds illustrate the burden of the government: military expenses in Sinkiang from 1875-1881, 52 million taels; the Ili indemnity, 5-6 million; the French war of 1884-85, 30 million; the Japanese war of 1894-95, 60 million; the Japanese indemnity, 230 million; the Boxer indemnity, 450 million; river conservancy, 10 million; natural disaster relief, 30 million; and sundry other reparations for "church incidents" and damages to foreign properties. In addition, there was the naval expense on the order of 5 million a year. In 1899, the government expenditure rose to 101 million, against an income of 88.4 million.[21] The deficit was caused largely by the cost and payment of foreign loans, 24 million taels, which amounted to 30 percent of the revenue. From 1874 to 1911, 171.4 million pounds sterling[22] in loans were contracted, of which only 32.3 million pounds sterling had been paid off by the end of the dynasty in 1911, with 139 million still outstanding. The practice of borrowing money to pay former debts plunged the government into a hopeless mire, leaving the new republication government a tremendous financial burden at its birth in 1912.

TRADE IMBALANCE. Foreign trade provides an equally discouraging picture. Imports continuously exceeded exports, causing a steady outflow of capital. The following chart gives a bird's-eye view of foreign trade at ten-year intervals:

Year	Import	Export	Balance
1865	55,715,458 taels	54,103,274	−1,612,184
1875	67,803,247	68,912,929	+1,109,682
1885	88,200,018	65,005,711	−23,194,307
1895	171,696,715	143,293,211	−28,402,504
1905	447,100,082	227,888,197	−219,212,549
1911	471,503,943	377,338,166	−94,165,777

21. Hsiao I-shan, IV, 1,534-36; Chi-ming Hou, *Foreign Investment*, 239-40.
22. One pound was equal to U.S. $4.86 at that time.

Within the short period of half a century, imports rose nearly nine times, from 55 million taels to 471 million, and exports seven times, from 54 million to 377 million. Except for the short span from 1872 to 1876 when there was a slight favorable balance between 2.5 and 10 million, the entire late Ch'ing period suffered from trade imbalance, with 1905 at the worst when the deficit reached 219 million taels.[23]

The imbalance of payment was somewhat relieved by remittances from Chinese overseas subjects, amounting to some 50 million annually, and by the spending of foreign legations and missionaries and other organizations, totaling 10 million in 1893, 26 million in 1894, and 30 million in 1895, with the average thereafter at 10 to 15 million annually.[24] Thus, the ill effects of the late Ch'ing trade deficit was partially cushioned by these remittances.

FOREIGN INVESTMENT AND DOMINATION. An unusual, if not abnormal, development in late Ch'ing economy was the dominating role exercised by foreigners in modern Chinese industries and enterprises. Their degree of control and scope of activities are seldom seen in independent states; hence late Ch'ing economy has been appropriately dubbed "semicolonial." The following brief survey delineates the foreign participation in several key sections of modern Chinese economy.

1. Banking. The old-style Chinese banks, or money shops, never financed foreign trade, so foreign banks and their branches in the treaty ports monopolized the financing of imports and exports in China for nearly half a century—from its opening in 1842 to the establishment of the first modern Chinese bank in 1898. The first foreign bank in China was the British-chartered Oriental Banking Corporation, which established a branch in Hong Kong in 1845 and another in Shanghai in 1848. The most powerful foreign banks were the Chartered Bank of India, Australia, and China, and the Hong Kong and Shanghai Banking Corporation— which began their operations in 1853 and 1864-65 respectively. These two British banks exercised a virtual monopoly over China's foreign trade financing until 1889, when the Deutsch-Asiatische Bank entered upon the scene. Not to be excluded from the rewards, other foreign banks soon followed suit: the Japanese-owned Yokohama Specie Bank at Shanghai in 1892; the Russo-Chinese Bank (founded in 1895 to finance the construction of the Chinese Eastern Railway in Manchuria); the American-owned

23. Chi-ming Hou, 231-32; Yu-kwei Cheng, *Foreign Trade and Industrial Development of China* (Washington, D.C., 1956), 258-59.
24. Hsiao I-shan, IV, 1,591.

Cathay Company (dominated by the Guaranty Trust of New York) and the International Banking Corporation;[25] and, of course, French, Belgian, and Italian banks, too.

These enterprising foreign institutions performed not only the normal banking functions, but assumed more unusual roles as well, such as serving as treasury agents for their respective governments, receiving deposits of the Chinese maritime customs and salt gabelle that had been pledged as security against foreign loans, and even issuing their own bank notes. The latter were issued without any explicit permission of China, but the foreign banks insisted that their extraterritoriality entitled them the right, and the feeble Ch'ing court was helpless to stay them. In reality, the notes were nothing more than promissory notes, "an interest free loan from the Chinese public to the foreign banks."[26] Playing both ends against the middle, with these notes the foreign banks bought Chinese commodities, and with the Manchu and Chinese private and official deposits the banks made highly profitable investments in China. Naturally, when bankruptcy occurred, as during World War I, the bank notes became useless and Chinese deposits were lost. It has been estimated that by 1910 the total amount of foreign bank notes in circulation was between 35 and 100 million Chinese silver dollars.[27]

To compete with foreign banks, the Ch'ing court approved, in 1898, the creation of a private Chinese bank, the Commercial Bank of China (*Chung-kuo t'ung-shang yin-hang*),[28] with an initial capital of 5 million taels. In 1905 the Hu-pu Bank (*Hu-pu yin-hang*) began operations with a capital of 10 million taels; three years later it changed its name to Ta-Ch'ing Bank (*Ta-Ch'ing yin-hang*), and again, after the establishment of the Republic in 1912, changed its name to Bank of China. In 1907 the Bank of Communications was organized by the government, and by 1914 there were 59 Chinese banks.[29]

2. Shipping. In conjunction with trade, foreign firms often established shipping companies, which led to a rapidly expanding and highly competitive international industry in Chinese coastal and inland waters—a right usually denied to foreigners in an independent state, but forced upon China by the unequal treaties. The first such outfit, the Shanghai Steam Navigation Company, was founded in 1862 by the American firm

25. Later, after 1927, known as the National City Bank of New York.
26. Chi-ming Hou, 57.
27. *Ibid.*, 58.
28. Originally called the Imperial Bank of China.
29. L. S. Yang, *Money and Credit in China* (Cambridge, Mass., 1952), 90; Frank M. Tamagna, *Banking and Finance in China* (New York, 1942), 35-37.

of Russell and Company, with an initial capital of one million taels (U.S. $1,356,000), and for fifteen years it was the largest shipping company in China.

However, the largest share of foreign shipping was British. Butterfield and Swire Company invested 970,000 taels in the China Navigation Company in 1872, followed by Jardine, Matheson and Company's investment of 325,000 taels a year later in the China Coast Steam Navigation Company. The fast, large, efficient foreign ships quickly pre-empted much of the business of the slower, older Chinese junks. To protect national interests, in 1872 Li Hung-chang sponsored the creation of the China Merchants' Steam Navigation Company, which had an initial investment of 476,000 taels, and in 1877 it acquired the entire American fleet of Shanghai Steam Navigation Company. Facing heightened competition, Jardine, Matheson and Company established two more shipping firms, the Yangtze Steam Navigation Company in 1879, and the Indochina Steam Navigation Company in 1881—the first with a capital of 300,000 taels, the second with 1,370,000 taels.[30] The Japanese entered as a late but vigorous competitor by consolidating, in 1907, four shipping companies into the giant Nisshin Kisen Kaisha under heavy government subsidy. The Japanese and German shares of shipping grew rapidly after 1900, while the British maintained a consistent lead throughout the late Ch'ing period, and the Americans commanded a respectable position only during 1868-76:[31]

Foreign Shipping in China

Year	Tonnage (in millions)	Britain	U.S.	Japan	Germany	Others
1868	6.4	52.2%	35%	0.1%	7.3%	5.4%
1872	8.5	46.8	41.1	0.1	7.2	4.9
1877	8.0	81.1	6.9	1.4	6.2	4.3
1892	22.9	84.4	0.3	2.8	6.4	6.1
1902	44.6	60.4	1.1	16.5	16.2	5.9
1907	63.4	52.5	1.6	24.6	10.5	10.8

As a consequence of increasing foreign domination, the Chinese share of shipping declined drastically from 30.4 percent in 1880 to 19.3 percent in 1900.[32]

3. Railways. The frenzied scramble for railway concessions after the

30. K. C. Liu, *Anglo-American Steamship Rivalry in China, 1862-1874* (Cambridge, Mass., 1962), 11.
31. Chi-ming Hou, 61.
32. *Ibid.*, 138.

Japanese war was probably the most blatant form of economic imperialism. Unable to resist, in 1895 Peking granted France the right to construct a 289-mile line from Indochina to Yunnan. In the following year Russia obtained rights for construction of the Chinese Eastern Railway across Manchuria as an extension of the Trans-Siberian Railway to Vladivostok, totaling 1,073 miles. Two years later it extorted another concession to build a 709-mile branch to Port Arthur and Darien under the name of Southern Manchurian Railway, which was transferred to Japan after Russia's defeat in 1905. Germany, too, was no laggard, acquiring the rights in 1897 to construct 285 miles of rail-line between Kiaochow and Tsinan in Shantung province. These four major foreign railways alone totaled 2,356 miles of track, or 41 percent of the entire railway mileage in China in 1911.[33] In addition, many Chinese lines were built with foreign loans, and were therefore not free from foreign control or influence.[34]

Added to the insult of imperialism was the injury of economic loss: the foreign powers obtained their rights as concessions, hence paying nothing and permitting no Chinese government agency to collect taxes on the railway properties and income. Not only were foreign-owned railways instruments of economic imperialism, but also they were political and military bludgeons to further foreign influence and facilitate troop movements in times of conflict.

4. Mining and Manufacturing. Foreigners in China did not limit themselves to banking, shipping, and railway operations, but also engaged in mining and manufacturing. Among the largest and best known of foreign-operated mines were the Japanese-dominated (since 1902) Han-Yeh-P'ing Mines and Ironworks, and the British-dominated (since 1900) Kaiping Coal Mine in Chihli, which in 1912 amalgamated with the Chinese-operated Lanchow Mining Company to form the Kailan Mining Administration. As regards foreign manufacturing and allied activities, the best illustration is the multifarious Jardine, Matheson and Company, which, besides foreign trade, was engaged in tea-processing, silk-reeling, ship-repairing, engineering, breweries, cotton textile, insurance, packing, cold storage, and loans—indeed a ubiquitous industrial and economic complex!

33. Chi-ming Hou, 65.
34. The Peking-Hankow railway: a Belgian loan of £4.5 million in 1899 at 5 percent interest, and two Anglo-French loans of £5 million at 5 percent in 1908 and £450,000 at 7 percent in 1910; the Shanghai-Nanking Railway: a British loan of £2.9 million at 5 percent in 1904-7; the Canton-Hankow Railway: a British loan of £1.1 million at 4½ percent in 1905; the Shanghai-Ningpo Railway: a British loan of £1.5 million at 5 percent in 1908; and the Tientsin-Pukow Railway: two Anglo-German loans of £5 million in 1908-9 and £3 million in 1910, both at 5 percent. Morse, III, 449.

Other foreign activities included shipbuilding and repairing, textile manufacturing, sugar-refining, spinning, and weaving, tobacco, and public utilities. Not a single phase of China's modern economy was immune to the encroachment of foreign capital, influence, and control. And out of a total of 636 foreign firms in business in China by 1897, more than half, 374, were British.[35]

Foreign control extended even to the postal service in China for more than a quarter of a century. The inefficiency of the traditional Chinese postal stations led the foreigners in 1860 to establish their own service in the treaty ports, even though the Ch'ing government never granted them the right. As China yawned wider before the increasing alien inroads, the foreign postal service extended along the coast and into the interior. It was not until 1896 that the Manchu court established the Imperial Postal Service under the charge of the inspector-general of Maritime Customs, Sir Robert Hart. Finally in 1911 the new Ministry of Posts and Communications took over general direction of postal functions; only then did China divest the postal service of foreign control.

TWO SIDES OF IMPERIALISM. The total foreign investment in China reached U.S. $788 million in 1902 and U.S. $1,610 million in 1914.[36] The degree of foreign domination is seen in the fact that 84 percent of shipping, 34 percent of cotton-yarn spinning, and 100 percent of iron production were under foreign control in 1907, while 93 percent of railways were foreign-dominated in 1911. The scope of foreign influence was as wide as the modern sector of Chinese economy, which had been reduced to "semicolonial." Nationalistic Chinese historians and economists, as well as Marxist scholars, point to such a high degree of foreign dominance as proof of naked imperialism. They charge that foreigners stifled the native industry and exercised a depressant and oppressive effect on the Chinese economy. They expostulate on the unfair advantages that foreigners had over their Chinese competitors due to their vast capital, technical knowledge, special privileges under unequal treaties, and immunity from Chinese laws, taxes, and official interference. All this, of course, is true. Foreigners invested in China to make money; few thought in terms of aiding China's economic development. The fact that they reaped profit to the extent of better than 10 percent annually and controlled most of the modern sector of China's economy, certainly made it difficult for the Chinese to make money and win their rightful place in business.

Yet imperialism was not without its beneficial side effects. Foreign in-

35. Chi-ming Hou, 103.
36. Other estimates: U.S. $1,509.3 million in 1902 and U.S. $2,255.7 million in 1914. See Chi-ming Hou, 211, 235.

vestors introduced modern technology and the entrepreneurial spirit, and financed many modern industries. Their success created an environment in which profit from industrial undertakings was demonstrably possible, thereby prompting the Chinese to follow their example. Additionally, the employment and training of Chinese in foreign factories and business establishments produced a native pool of technical knowledge of production and managerial skills which later were to be profitably tapped by and for the Chinese. It was not unusual for the compradores, after having learned foreign business methods and having accumulated considerable capital, themselves to invest in industry or serve in government-sponsored enterprises. Such a one was Tong King-sing, the director of the China Merchants' Steam Navigation Company, and formerly compradore with Jardine, Matheson and Company. Nor should one lose sight of the fact that foreign-leased areas and treaty ports provided a certain degree of peace and order necessary for industrial growth; and that foreign establishments had already borne most of the cost of "social overhead," such as public utilities, roads, and communication facilities, which eased the development of Chinese industry. Clearly, foreign investment produced an "imitation" effect on the Chinese and provided the preconditions essential for the economic modernization of China.[37]

In sum, imperialism was both baneful and beneficial. On the one hand it inhibited the growth of native industry, on the other it stimulated patriotism—by inciting a desire for national economic protection and competitive equality—and provided the incentive for economic modernization. Corroboration for the latter aspect is evident in the fact that during the height of imperialism many Chinese factories and enterprises were born. In 1904-8, 227 modern Chinese companies were registered with the government, and by 1912 there were 20,749 native factories in operation, though the majority were of small or medium size, and only 750 employed workers in excess of a hundred.[38] While it is true that they had to struggle for survival in the shadow of the giant foreign firms, the fact remains that they emerged under foreign stimulation.

The decade and a half following the Japanese war was a very turbulent era, in which the old intellectual, social, and economic order passed away and the new struggled toward birth. The rapid transition presaged a major political upheaval in the offing. The Manchu dynasty, already two and a half centuries old, stood at a critical point in history. If it could not keep abreast of the times and offer an alternative to violent change, it would be doomed to extinction.

37. Chi-ming Hou, 217, 221.
38. Feuerwerker, 3-5.

Further Reading

Bastid-Bruguiere, Marianne, "Currents of Social Change" in John K. Fairbank and Kwang-ching Liu (eds.), *The Cambridge History of China* (Cambridge, Eng., 1980), Vol. 11, 535-602.

Bays, Daniel H., *China Enters the Twentieth Century: Chang Chih-tung and the Issues of a New Age, 1895-1909* (Ann Arbor, 1978).

Bonner, Joey, *Wang Kuo-wei: An Intellectual Biography* (Cambridge, Mass., 1986).

Brière, O., S.J., *Fifty Years of Chinese Philosophy, 1898-1950* (London, 1956).

Chan, Wellington K. K., "Government, Merchants, and Industry to 1911" in John K. Fairbank and Kwang-ching Liu (eds.), *The Cambridge History of China* (Cambridge, Eng., 1980), Vol. 11, 416-62.

Chang, Hao, *Liang Ch'i-ch'ao and Intellectual Transition in China, 1890-1907* (Cambridge, Mass., 1971).

————, *Chinese Intellectuals in Crisis: The Search for Order and Meaning, 1890-1911* (Berkeley, 1987).

Chou, Ku-ch'eng 周谷城, *Chung-kuo she-hui chih pien-hua* 中國社會之變化 (Changes in Chinese society), (Shanghai, 1931).

Feuerwerker, Albert, *The Foreign Establishment in China in the Early Twentieth Century* (Ann Arbor, 1976).

————, "Economic Trends in the Late Ch'ing Empire, 1870-1911" in John K. Fairbank and Kwang-ching Liu (eds.), *The Cambridge History of China* (Cambridge, Eng., 1980), Vol. 11, 1-69.

Hao, Yen-p'ing, "A 'New Class' in China's Treaty Ports: The Rise of the Compradore-Merchants," *The Business History Review*, XLIV:4:446-59 (Winter 1970).

Hou, Chi-ming, *Foreign Investment and Economic Development in China, 1840-1937* (Cambridge, Mass., 1965).

Hsiao, Liang-lin, *China's Foreign Trade Statistics, 1864-1949* (Cambridge, Mass., 1974).

Huang, Philip C., *Liang Ch'i-ch'ao and Modern Chinese Liberalism* (Seattle, 1972).

Johnson, David, Andrew Nathan, and Evelyn Rawski (eds.), *Popular Culture in Late Imperial China* (Berkeley, 1985).

Lee, En-han, *China's Quest for Railway Autonomy, 1904-1911: A Study of the Chinese Railway-Rights Recovery Movement* (Singapore, 1977).

————, "The Chekiang Gentry-Merchants vs. The Peking Court Officials: China's Struggle for Recovery of the British Soochow-Hangchow-Ningpo Railway Concession, 1905-1911," *Bulletin of the Institute of Modern History*, Academia Sinica, III:1:223-68 (July 1972).

————, 李恩涵, "Chung-Mei shou-hui Yüeh-Han lü-ch'üan chiao-she: Wan-Ch'ing shou-hui tieh-lü li-ch'üan yün-tung yen-chiu chih-i," 中美收回粵漢路權交涉···晚清收回鐵路利權運動研究之一 (Sino-American negotiations on the recovery of the Canton-Hankow Railway: A case study of the movement for recovery of railway rights in the late Ch'ing period), *Bulletin of the Institute of Modern History*, Academia Sinica, I:149-215

————, *Wan-Ch'ing ti shou-hui k'uang-ch'üan yün-tung* 晚清的收回鑛權運動 (Late Ch'ing movement for recovery of the mining rights), (Taipei, 1963). (April 1969).

Lee, Leo Ou-fan, "Lin Shu and His Translations: Western Fiction in Chinese Perspective," *Paper on China*, Harvard East Asian Research Center, 19: 159-93 (Dec. 1965).

Levy, M. J., and Shih Kuo-heng, *The Rise of the Modern Chinese Business Class* (New York, 1949).

Liang, Ch'i-ch'ao, *Intellectual Trends in the Ch'ing Period* (*Ch'ing-tai hsüeh-shu kai-lun*), tr. by Immanuel C. Y. Hsü (Cambridge, Mass., 1959), Part III.

Liu, Kwang-ching, *Anglo-American Steamship Rivalry in China, 1862-1874* (Cambridge, Mass., 1962).

Lo, Yü-tung 羅玉東, "Kuang-hsü ch'ao pu-chiu ts'ai-cheng chih fang-ts'e" 光緒朝補救財政之方策 (The government policies of meeting the financial crisis during the Kuang-hsü period (1875-1908)), *Chung-kuo chin-tai ching-chi-shih yen-chiu chi-k'an* 中國近代經濟史研究集刊 (Studies in modern economic history of China), 1:2:189-270 (May 1933).

McElderry, Andrea Lee, *Shanghai Old-Style Banks* (*Ch'ien-chuang*), *1800-1935* (Ann Arbor, 1976).

Powell, Ralph L., *The Rise of Chinese Military Power, 1895-1912* (Princeton, 1955).

Schwartz, Benjamin I., *In Search of Wealth and Power: Yen Fu and the West* (Cambridge, Mass., 1964).

Skinner, G. William (ed.), *The City in Late Imperial China* (Stanford, 1977).

Skinner, William, and Mark Elvin (eds.), *The Chinese City Between Two Worlds* (Stanford, 1974).

Sun, E-tu Zen, *Chinese Railways and British Interests, 1898-1911* (New York, 1954).

Sun, Yü-t'ang 孫毓棠, "Chung-Jih Chia-wu chan-cheng ch'ien wai-kuo tzu-pen tsai Chung-kuo ching-ying ti chin-tai kung-yeh" 中日甲午戰爭前外國資本在中國經營的近代工業 (Modern industries operated by foreign capital in China before the Sino-Japanese War), *Li-shih yen-chiu*, 5:1-41 (1954).

Tsien, Tsun-hsuin, "Western Impact on China Through Translations," *Far Eastern Quarterly*, XIII:3:305-27 (May 1954).

Van der Valk, M. H., *Conservatism in Modern Chinese Family Law* (Leiden, 1956).

Vevier, Charles, *The United States and China, 1906-1913: A Study of Finance and Diplomacy* (New Brunswick, 1955).

Wang, Erh-min 王爾敏, *Wan-Ch'ing cheng-chi ssu-hsiang shih lün* 晚清政治思想史論 (On late Ch'ing political thought), (Taipei, 1969).

Wang, Sha 王璽, *Chung-Ying K'ai-p'ing kuang-ch'üan chiao-she* 中英開平礦權交涉 (The Anglo-Chinese negotiations on the Kaiping mining rights), (Taipei, 1962).

Wang, Tsao-shih 王造時, "Chung-Hsi chieh-ch'u-hou she-hui-shang ti pien-hua" 中西接觸後社會上的變化 (Social changes after the Sino-Western contact), *Tung-fang tsa-chih*, 31:2:31-40 (1934).

Wang, Y. C., *Chinese Intellectuals and the West, 1872-1949* (Chapel Hill, 1966).

Wang, Yeh-chien, *An Estimate of Land Tax Collection in China, 1753 and 1908* (Cambridge, Mass., 1973).

————, *Land Taxation in Imperial China, 1750-1911* (Cambridge, Mass., 1974).

Wright, Tim, *Coal Mining in China's Economy and Society: 1895-1937* (New York, 1985).

19

The Ch'ing Period
in Historical Perspective

The foregoing survey of Ch'ing history inevitably raises the question as to the significance, accomplishments, and failings of the period. Indeed, historical perspective affords the Ch'ing, the last in a succession of twenty-five imperial dynasties, a distinctive and crucial position in Chinese history. It proves to have been the most durable period of foreign rule in China, spanning 268 years as opposed to the Yüan (Mongol) dynasty's 89. It saw the rise of the second largest empire in Chinese history, next only to the Yüan, and provided the country with a prolonged period of peace and prosperity. This *Pax Sinica* precipitated, among other things, an unprecedented population growth from 150 million in 1650 to 430 million in 1850. These territorial and demographic legacies underlie the bases of China's strength today.

Moreover, the Ch'ing period witnessed the epochal transition from traditional China to its modern counterpart. The Confucian state and society and the old ways of life which persisted during the early and middle stages of the dynasty went through a radical transformation under the Western impact after the mid-19th century. The labored emergence of a new order can only be understood through a knowledge of the Ch'ing history, thereby facilitating comprehension of China's extreme difficulty in adjusting to the modern world.[1]

The Ch'ing dynasty provided China with both brilliant accomplish-

1. Ping-ti Ho, "The Significance of the Ch'ing Period in Chinese History," *The Journal of Asian Studies,* XXVI:2:189-95 (Feb. 1967).

ment and ignominious disaster. But on the whole their rulers seem to have performed better than their Ming counterparts.[2] The conclusive lesson which emerges from historical analysis of the Ch'ing period indicates that survival hinges on the ability to respond constructively and creatively to the challenge of the times. Manchu success during the 17th century appears to have been largely a result of such adaptation; while its failure, two and a half centuries later, appears to have resulted from a lack of parallel adjustment.

The Manchus seized power through the enterprise of outstanding leaders like Nurhaci and Abahai, who appeared at the critical moment when the reigning Ming dynasty was beset by political corruption, eunuch dominance, heavy taxation, and debilitating rebellions. Their leaders were flexible enough to overcome Manchu tribal mentality and organization, enlisted Chinese collaboration, and incorporated existing Ming institutions. A succession of farsighted and able rulers followed the formal establishment of the dynasty in 1644, and K'ang-hsi, Yung-cheng, and Ch'ien-lung inaugurated wise and far-reaching policies.

A prolonged period of peace, affluence, and military glories was ushered in. The Confucian order was retained, perceptively, as it was conceived to be essential to the success of alien rule. In addition, a system of governmental dyarchy was instituted, providing for the appointment of both Manchus and Chinese to positions of administrative responsibility in an attempt to absorb Chinese talent and reduce racial antipathy, and simultaneously introducing a system of mutual checks and balances. As it emerged, the new Ch'ing empire rested upon the foundations of an administrative system inherited from the Ming with supplementary Manchu innovations such as the Li-fan yüan and the Grand Council. Neo-Confucianism was promoted, with proper emphasis on such concepts as loyalty and maintenance of the *status quo,* for the primary purpose of stabilizing existing society. The Manchus strove to preserve their own identity within the larger circumference of Chinese society, specifically creating institutions such as the Imperial Clan Court to keep a close watch over Manchu nobles. They also prevented Chinese immigration to their Manchurian homeland. Intermarriages were banned, whether Chinese-Manchu or Chinese-Mongol. Power was centralized in the hands of the sovereign to an unprecedented degree, and divisive tendencies were forestalled by thwarting movements toward feudalism among Manchu nobles and by forbidding them and bannermen to develop provincial connections. The rise of eunuchs was banned, the influence of imperial

2. Hsiao I-shan, *Ch'ing-tai shih* (A history of the Ch'ing dynasty), 312.

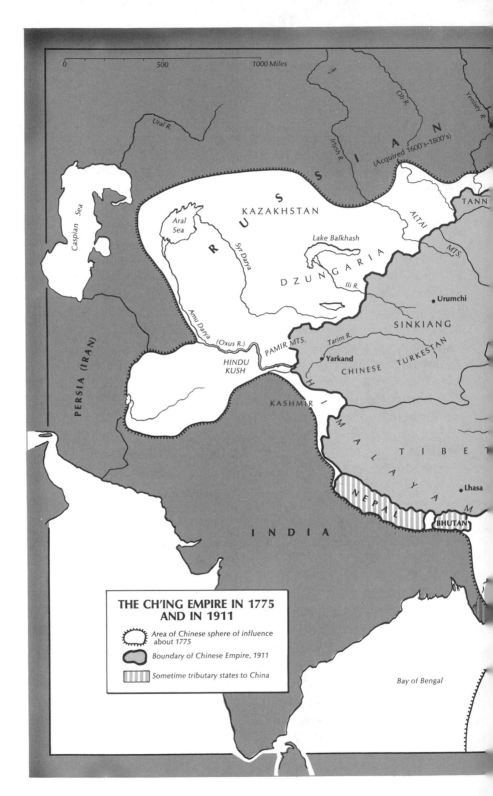

THE CH'ING EMPIRE IN 1775
AND IN 1911

Area of Chinese sphere of influence
about 1775

Boundary of Chinese Empire, 1911

Sometime tributary states to China

EMPIRE

STANOVOI MTS.

SAKHALIN
(1853–1875)

Lake
Baikal

(Acquired by Russia, 1858–1860)

Amur R.

MANCHURIA

Sungari R.

OUTER
MONGOLIA

Vladivostok

JAPAN

Mukden

Toyko

INNER MONGOLIA

KOREA

Kyoto

Dairen
Peking
Tientsin
Port Arthur

CHIHLI

SHANSI

GREAT WALL

Chefoo

Shimonoseki

Yalu R.

SHENSI

KANSU

Lanchow

SHANTUNG

Yellow R.

HONAN

KIANGSU

Nanking

Shanghai

C H I N A

HUPEH

Hankow

ANHWEI

Ningpo

Chengtu

SZECHWAN

Yangtze R.

Changsha

HUNAN

KIANGSI

CHEKIANG

Foochow

RYUKYU IS.

Pacific

Ocean

KWEICHOW

FUKIEN

Taipei

TAIWAN
(FORMOSA)

Amoy

PESCADORES

YUNNAN

KWANGSI

Canton

KWANGTUNG

Hong Kong
(Br. 1842)

Macao (Port.)

TONKIN

HAINAN

PHILIPPINE IS.

LAOS

VIETNAM

ANNAM

Mekong R.

SIAM

South China
Sea

CAMBODIA

COCHIN CHINA

443

distaff curbed, and official cliques and factions prohibited. Furthermore, through literary inquisitions the Ch'ing struck terror into recalcitrant Chinese scholars on the one hand and enticed them on the other to enter government service whenever possible through literary examinations and attractive appointments.

Such exertions contributed to a successful consolidation of the Manchu position in China. Yet the height of fortune is also the beginning of decline—"When the sun reaches midday it begins to set, and when the moon is fullest it begins to wane," the Chinese saying goes. The scintillation and the splendor of Ch'ien-lung's reign already contained the seeds of decay. China had indulged excessively in luxury, neglecting essential problems. Population growth outpaced land increase, resulting in a decrease of per capita acreage. Military training was lax. Corruption and irregularity became widespread within the imperial bureaucracy, contributing to the recurring historical phenomena known as the "dynastic cycle." By 1775, the Ch'ing dynasty had begun its downward course.

Experience dictates that internal decay invites both domestic rebellion and foreign invasion; and in Chinese history they often occurred simultaneously when imperial power declined. True to dictum, the diminution of Ch'ing authority witnessed the outbreak of the White Lotus Rebellion (1796-1804), as well as an intensified Western drive to open China to trade and diplomacy. The double threat, both internal and external, plagued the imperial court throughout the 19th century. The dynasty was able to quell domestic rebellion, a known quantity, but completely failed to stem Western intrusion, an unknown quantity. Dynastic disintegration intensified after the Taiping Revolution (1850-64), witnessing a shift in political power from the capital to the provinces, and from Manchu to Chinese dominance. The future of the Ch'ing dynasty was doomed, even though the T'ung-chih Restoration (1862-74) temporarily arrested the downward drift.

The ultimate failing of Ch'ing rule lay in its incapacity to adequately cope with the Western impact by initiating far-reaching reform in order to swiftly transform China into a modern state. Whereas in the 17th century Manchu leaders showed flexibility, adopting Chinese institutions and the Confucian order, their late 19th- and early 20th-century descendants were totally inept and unable to transcend tradition. They could not successfully innovate an alternative to revolution in creative response to contemporary challenge. Indeed, political mismanagement, domestic rebellion, and foreign humiliation had so degraded the Ch'ing dynasty that its impending fall appeared evident in the late 19th century, just as the imminent fall of the Ming had seemed inevitable in the early 17th century.

After 268 years, the Manchu dynasty lost its "Mandate of Heaven" and reached what the Chinese call its "preordained finale."

The Ch'ing experience presents a sharp and notorious contrast to the experience of Meiji Japan, affording opportunity for much discussion and a variety of interpretations from conflicting points of view. The more obvious reasons for China's hampered advance appear to have been factors which restricted the dissemination of new ideas, such as vast territorial jurisdiction, poor communications, self-sufficiency, the almost total lack of a tradition of borrowing from abroad, and a conservative intellectual posture. Among other reasons for Ch'ing decay, the following are worth consideration.

FEEBLE LEADERSHIP AND FAULTY INSTITUTIONS. The Ch'ing autocracy concentrated state power in the person of the emperor, making dynamic leadership mandatory. Resourceful rulers like K'ang-hsi, Yung-cheng, and Ch'ien-lung inaugurated eras of brilliance, providing China with much splendor and accomplishment. Mediocre emperors followed in trepidation, exerting themselves to preserve, rather than magnify, past glory. Unfortunately for China and the Ch'ing dynasty, powerful and creative leadership failed to appear when it was most needed. The last of the great emperors, Ch'ien-lung, was followed by Chia-ch'ing and Tao-kuang, both methodical and average men. Hsien-feng's eleven-year reign was disrupted by the debilitating Taiping Revolution and the humiliation of the *Arrow* War with Britain and France. T'ung-chih and Kuang-hsü were two boy emperors during whose reigns state power rested with the empress dowager, Tz'u-hsi, who ruled supreme for nearly half a century. While not without native intelligence and decisiveness, she was basically ignorant, conservative, venal, and selfish, placing her own interests above those of the dynasty and the country at large. She supported the Self-strengthening Movement not with a view to transforming China into a modern state, but to preserving the old order and her own position. All she sought from the Movement was to suppress domestic revolt and resist foreign imperialism. She permitted piecemeal improvement instead of full-scale regeneration, partly because she feared subversion by those Chinese who were in charge of modernization. Tz'u-hsi was therefore responsible, to a large extent, for failing to provide constructive leadership.

Ineffective imperial leadership might have been offset by vigorous, farsighted statesmanship on the part of the central bureaucracy. In a Confucian state, though ministers were supposed to serve and not lead the emperor, they could offer counsel and influence official policy. During the late Ch'ing period, however, the majority of scholars and officials were a "saturated class," too content with their privileges and vested interests to want to change the existing order. Advisers and advocates on "foreign matters" (*yang-wu*) were lone exceptions in the conservative officialdom. These progressives did not combine to form a "creative minority" (to quote Toynbee), but appeared as isolated individuals who sought to adopt Western devices in answer to the challenge of the times. They did not function as a closely knit group, as in the case of Meiji Japan, but responded separately without reference to a centrally coordinated master plan. Of the early Self-strengthening leaders, only Prince Kung and Wen-hsiang were in the central government, while Tseng Kuo-fan, Tso Tsung-t'ang, and Li Hung-chang were all provincial personalities. The slender thread of central direction disappeared after the dowager's repeated chastisements of Kung and Wen-hsiang's death in 1876. Li Hung-chang became, to some extent, a "coordinator" of modernization projects, performing some functions of the central government after his appointment in 1870 as governor-general of Chihli and imperial commissioner in charge of the northern ports. Nevertheless, he faced jurisdictional restrictions, receiving no authority to command provinces other than his own. Later modernizers, Chang Chih-tung and Liu K'un-i, also conducted their projects in piecemeal fashion within their own provinces. In short, there was regional but not national planning, contrary to governmental policy in Meiji Japan. Moreover, these regional efforts were largely directed from the top down, with little popular participation, thereby precluding wide proliferation of modern industries and ideas.

Without the benefit of either effective imperial leadership or a creative liberal minority, it became increasingly difficult, if not impossible, to channel national energy toward the collective goal of rejuvenating the country.

MANCHU SUSPICION OF THE CHINESE. Despite the announced court policy of nondiscrimination and appointment of both Chinese and Manchus to governmental office, the Manchus in reality remained conquerors and the Chinese were merely tolerated as "outsiders." Until the time of the Taiping Revolution, key appointments, both civil and military, went to the Manchus, and even those positions allotted to the Chinese could be filled by Manchus, but not vice versa. Although the practice relaxed after the Taiping era, Manchu suspicion of the Chinese never abated. The life of Li Hung-chang provides an excellent illustration.

Li, who became the leading spirit of the Self-strengthening Movement, met with opposition and frustration from all sides. He was constantly ridiculed by conservatives for betraying his country's interest to foreigners.

The empress dowager realized the need for his services, but feared his growth in stature at the expense of the court. She therefore halfheartedly supported his modernization projects, allowing reactionaries to attack him with impunity. She also resorted to the strategy of divide and rule by permitting or even encouraging conservative literati opinion (*ch'ing-i*) to stall the progressives. In 1874, when Li asked Prince Kung to speak to the two dowagers about the desirability of railways, Kung replied that even they could not reach a decision on the matter because of powerful popular opposition.[3] When a Board of Admiralty was organized in 1885, the directorship went to an ignorant Manchu, Prince Chün, instead of to a more capable Chinese like Li Hung-chang or Tseng Chi-tse. Manchu fear of Chinese subversion and Chinese fear of Manchu jealousy precluded effective cooperation between the two, and far-reaching reform became impossible. K'ang Yu-wei's program for institutional reform in 1898 was condemned by the Manchus as a plot to benefit the Chinese at their expense. The last decade of the Ch'ing rule saw an intensified effort on the part of the Manchus to curb Chinese influence. Grand Secretary Kang-i announced: "When the Chinese become powerful, the Manchus extinguish; when the Chinese are tired, the Manchus grow fat."[4] Ethnic antipathy impeded genuine modernization, because the Manchus contended that reform and constitutionalism would undermine their power. Doubtless, the Manchu-Chinese schism precluded engendering an effective joint enterprise to regenerate the nation.

IGNORANCE OF THE NATURE OF THE WESTERN CHALLENGE. The Western advance was characterized by factors such as ships and guns, trade and evangelism, and imperialism and nationalism. It was sustained by a modern, dynamic civilization, superior to the Chinese in many respects. It introduced novel phenomena into China, understood by few. The unprecedented challenge caught China ill-prepared, and unable to achieve a solution. Until the time of the reform in 1898, the majority of scholars and officials drew instruction from Chinese history, where barbarian invasion had always been transitory. Hence the 19th-century Western advance, which figured in this category, was merely regarded as unfortunate and passing. Even China's repeated defeat by the West was interpreted as accidental. The actual nature, extent, and scope of the Western impact thus were misunderstood, even by progressive advocates of Self-strengthening measures. Li Hung-chang, for instance, recognized that the situa-

3. Hsü, *China's Entrance*, 205.
4. Sun Chen-t'ao, *Ch'ing-shih shu-lün* (A critique of Ch'ing history), (Hong Kong, 1957), 128.

tion was without parallel in 3,000 years of Chinese experience, and yet he continued to exhibit an exceedingly limited view of the Western potential. His modernization programs concentrated largely on military and diplomatic improvement. As regards the imperial court, it engaged in the Self-strengthening program in a defensive manner. When foreign pressure subsided, action slowed down, and a long-range platform, expressing national and foreign policy, failed to emerge. Ch'ing efforts were haphazard, like patching old clothes with new cloth and filling old bottles with new wine. They produced piecemeal endeavors which fell short of accomplishing a major breakthrough in economic development. Evidently modern capitalism and political reform could not be successfully grafted onto an outdated Confucian foundation.

Ch'ing ignorance of the contemporary world is not hard to understand when one considers the general mentality of scholars and officials prior to the 1898 reform. The majority of them lived in the past. They existed in a dream world of Chinese "culturalism," looking to antiquity for guidance instead of to the future for inspiration. Ancient ways were glorified and contemporary example despised. Machines, ships, guns, and telegraphic and railroad communications were considered artful contrivances beneath their dignity. They revealed ethnocentric pride, but little nationalistic spirit, citing historical sayings to justify the attitude that it was well and proper to Sinicize barbarians, but outrageous to imitate their ways. It was inconceivable to them that China, the Celestial Empire, should be transformed in the image of the West.

It was precisely against such narrow, retrogressive views that Yen Fu agitated for the adoption of new values in life, suggesting a "different vision of reality" through the study of Western thought. Similarly, Liang Ch'i-ch'ao advocated a "renovation" of China after the turn of the century. Their exertions provided the seeds of future intellectual fermentation.

NATIONAL AND INTERNATIONAL TURMOIL, AND INSUFFICIENT CAPITAL. Modernization and economic development require extended peace and adequate capital, but both factors were lacking in late Ch'ing China. The country was constantly plagued by internal rebellion, foreign conflict, missionary incidents, and natural calamity, causing a widespread breakdown of law and order, and sharp increases in government expenditure. After 1830, China went through the Opium War; the *Arrow* War; the Taiping Revolution; the Nien Rebellion; the Panthay (Moslem) Rebellion; the Tungan (Moslem) Revolt; the Tientsin Massacre; the Formosa crisis; the Margary Murder; the Ili crisis; the French war; the Japanese war; and the Boxer Rebellion. Peace and stability scarcely existed.

In addition to continuous disturbance which inhibited economic development, war expenses and indemnities caused the continual outflow of capital. Fiscal welfare became dependent upon foreign loans and funds extracted from the provinces. From 1842 to 1895, China paid out a total of 300 million taels for indemnities and interest. For the Boxer indemnity of 450 million taels, the Chinese government paid 225 million taels between 1902 and 1910, of which 164 million (72 percent) came from provincial assessments, 33 million (16 percent) from maritime customs, and 27 million (12 percent) from the national treasury. Such drainage of resources naturally impeded economic development, and the prospect of successful modernization grew increasingly dimmer with the growing fiscal stringency of the central government, as discussed in the last chapter.[5]

THE FOREIGN ROLE. Inasmuch as foreign influence was a major shaping force in late Ch'ing China, it requires at least a general analysis. Foreign governments and their representatives, while they wished to see China move in the direction of "progress," and while they continually impressed on imperial agencies the urgency of accepting Western institutions and products, they nevertheless implicitly shared the view that China should be kept dependent on the West. A moderately progressive and prosperous but weak China, dependent on foreign advice, good will, trade, and aid would be much more to the interest of the West than a completely independent and assertive China. China must not be allowed to modernize too far and acquire the strength to repel the West. Sir Thomas Wade, British minister to China, for example, made a policy statement with regard to the role of the Chinese Maritime Customs under Sir Robert Hart, in support of this attitude, stating:

> . . . in the well-being of which [Foreign Inspectorate of Customs] we, the English, are specially concerned, not only for its regulation of trade, but as the one instrument by which progress is being introduced into China, as it were, unknown to herself, and therefore without provoking her suspicion; and lastly, *unless I am greatly mistaken, for every measure of precaution possible against her* [China's] *acquisition of a fleet or her organization of an army.*[6]

5. Feuerwerker, 45; Chi-ming Hou, 164.
6. Foreign Office, 418/1/242, Wade to Granville, *very confidential*, July 25, 1880, Public Record Office, London. Italics added.

If this view sheds light on British policy toward China, one is not surprised at the perennial weakness of Ch'ing forces. It lends credence to the Marxist view that, by relying on foreign collaboration instead of struggling against it, the Ch'ing dynasty failed to achieve genuine strength.[7]

These domestic and foreign factors subverted successful modernization. Both Manchu and Chinese leaders were to be blamed for not circumventing obstacles, and the price of failure was the extinction of the dynasty. Lord Macartney's prophetic words, spoken in 1794, after his visit to the court of Ch'ien-lung, now became doubly meaningful. He stated:

> The Empire of China is an old, crazy, first-rate Man of War, which a fortunate succession of able and vigilant officers have contrived to keep afloat for these hundred and fifty years past, and to over-awe their neighbours merely by her bulk and appearance. But, whenever an insufficient man happens to have command on deck, adieu to the discipline and safety of the ship. She may, perhaps, not sink outright; she may drift some time as a wreck, and will then be dashed to pieces on the shore; but she can never be rebuilt on the old bottom.

Indeed, China could not be rebuilt on the old foundation; only a revolution could hope to regenerate it.

7. Huang I-feng and Chiang To, "Chung-kuo yang-wu yün-tung yü Jih-pen Ming-chih wei-hsin tsai ching-chi fa-chan shang ti pi-chiao" (A comparative study of China's "foreign matters" movement and Japan's Meiji modernization from the standpoint of economic development), *Li-shih yen-chiu* (Historical Research), 1:40 (1963).

Further Reading

Feuerwerker, Albert, *China's Early Industrialization: Sheng Hsüan-huai (1844-1916) and Mandarin Enterprise* (Cambridge, Mass., 1958).

Ho, Ping-ti, "Salient Aspects of China's Heritage," in Ping-ti Ho and Tang Tsou (eds.), *China in Crisis*, Vol. I, *China's Heritage and the Communist Political System*, Book I, 1-37.

———, "The Significance of the Ch'ing Period in Chinese History," *The Journal of Asian Studies*, XXVI:2:189-95 (Feb. 1967).

Hou, Chi-ming, *Foreign Investment and Economic Development in China, 1840-1937* (Cambridge, Mass., 1965).

Hsü, Immanuel C. Y., *China's Entrance into the Family of Nations: The Diplomatic Phase, 1858-1880* (Cambridge, Mass., 1960), ch. 13, "The Imperial Chinese Tradition in the Modern World."

Hu, Shih, *The Chinese Renaissance* (Chicago, 1934).

Huang, I-feng 黃逸峯 and Chiang To 姜鐸, "Chung-kuo yang-wu yün-tung yü Jih-pen Ming-chih wei-hsin tsai ching-chi fa-chan shang ti pi-chiao" 中國洋務運動與日本明治維新在經濟發展上的比較 (A comparative study of China's "foreign matters" movement and Japan's Meiji modernization from the stand-point of economic development), *Li-shih yen-chiu* (1963), 1:27-47.

Liu, Kwang-ching, "Nineteenth-Century China: The Disintegration of the Old Order and the Impact of the West," in Ping-ti Ho and Tang Tsou (eds.), *China in Crisis,* Vol. I, *China's Heritage and the Communist Political System* (Chicago, 1968), Book I, 93-178.

Meng, Shen 孟森, *Ch'ing-tai shih* 清代史(The Ch'ing history), (Taipei, 1960).

Sun, Chen-t'ao 孫甄陶, *Ch'ing-shih shu-lun* 清史述論(A critique of Ch'ing history), (Hong Kong, 1957).

Wakeman, Frederic, Jr., *The Fall of Imperial China* (New York, 1975).

———, (ed.), *Conflict and Control in Late Imperial China* (Berkeley, 1975).

20

Revolution, Republic, and Warlordism

The pressing question facing China after the Japanese war was what it could do to achieve national salvation in the face of accelerating foreign imperialism and dynastic decline. Two major political movements developed, each representing a different approach to the problem. One was the progressive reform of 1898, led by K'ang Yu-wei, from which evolved the Ch'ing reform and constitutional movements of the 1900s, as seen in the previous chapters. The other was a revolutionary movement led by Western-trained Dr. Sun Yat-sen, who advocated the complete overthrow of the Manchu dynasty. At first, the progressive reformers played the more prominent role. However, as Ch'ing endeavors proved insincere and discriminatory against the Chinese, the revolutionaries gained increasing support from the younger intellectuals, the secret societies, and the overseas Chinese communities. The momentum of their movement grew steadily until it finally swept the age-old imperial institution out of existence and replaced it with a republic—an epochal change in the long Chinese history.

Background and Characteristics
of the Revolution

CH'ING DECADENCE. Ever since the mid-19th century, Chinese history was largely one continuous record of national humiliation. The long list of unequal treaties from the Treaty of Nanking in 1842 to the Boxer Protocol of 1901, the loss of the tributary states in the 1880s and the 1890s, and the lack of vigor in domestic administration testify to the utter Ch'ing inability to defend China's honor in the modern world. What formerly had been the proud Middle Kingdom was now reduced to a semicolony. The Manchus, who had entered China in 1644 as conquerors, had completely lost face before the Chinese public. The death knell was sounded when the court, in a desperate struggle for survival, instituted an anti-Chinese policy under the pretext of reform and constitutionalism. Such a flagrant show of discrimination amidst rapid dynastic decline exacerbated opposition from the ruled.

THE TRADITION OF NATIONALISTIC REVOLUTION. Anti-Manchu sentiment never disappeared throughout the 268-year dynasty. Chinese thinkers in the early Ch'ing period such as Ku Yen-wu and Wang Fu-chih ceaselessly promoted the "anti-Ch'ing, revive Ming" idea. Although their activities did not result in an outright overthrow of the alien rule, the germ of revolution was kept alive in underground organizations and secret societies. The various Ming loyalist movements, the Revolt of the Three Feudatories, the activities of the Heaven and Earth Society, the White Lotus Rebellion, and the Taiping Revolution demonstrate the unending thread of nationalistic-racial protest. Dr. Sun's revolution was very much in this tradition.

FOREIGN INFLUENCE. The great revolutions of the modern West—the Glorious Revolution in England, the American Revolution, and the French Revolution—all exerted a profound influence upon the Chinese. The ideas of democracy, independence, human rights, equality, and freedom swept through the minds of young Chinese. Moreover, the success of the national unification movements of Italy and Germany in 1870 served as shining examples for forward-looking Chinese, prompting them to take similar action. Nationalism, democracy, and republicanism now became the motivating forces for revolutionary change in China.

NEED FOR POLITICAL INNOVATION. The monarchical institution of China, in the view of Dr. Sun, was responsible for the succession of imperial dynasties in the past 2,000 years without ever changing the substance of government. Chinese history revolved around the cycle of division, disorder, unification, and despotism, noted Sun, and each period of disorder was followed by a lengthy and merciless struggle for the throne by many contenders until one ultimately won out. In the process the country and the people suffered needlessly and the historical pattern repeated itself periodically. To break this cycle and to give a proper outlet to the ambitions of men, it was necessary to replace the monarchy with a republic—a federal republic—in which all could fulfill their dreams, exercise their rights, and become leaders of the provinces and the nation. To achieve this goal, Sun urged all freedom-loving people of China to participate in a National Revolution to bring about the downfall of the imperial system and the Manchu dynasty, and to introduce a modern republic, free from foreign intervention and interference.

THREE REVOLUTIONS IN ONE. For all their prosperity, independence, and democracy, the Western powers, observed Dr. Sun, were beset with the problems of industrialism. Labor-management disputes, strikes, demands for higher wages, and the unequal distribution of wealth between the capitalist minority and the worker majority foreboded a social revolution. Though China had not been industrialized enough to witness these same problems, the seeds of capitalism had been sown ever since the Self-strengthening Movement in the 1860s. To forestall the evils of capitalism, Sun proposed the regulation of capital to prevent the concentration of wealth among the few.

Moreover, in view of the perennial land problem caused by the faster pace of population increase over land accretion, Sun advocated a program of land equalization to the end that the ancient utopian dream, "land to the tiller," might be fulfilled.

In short, Sun envisioned a three-in-one revolution propelled by all the people of China: a nationalistic revolution to overthrow the Manchu dynasty and the imperial institution, a democratic revolution to establish a republic and popular sovereignty, and a social revolution to equalize the land rights and to prevent the ills of capitalism. Seldom had any revolution in the world been conceived in such a grandiose manner.

Dr. Sun and the Revolution

Sun Yat-sen (1866-1925),[1] father of the Chinese Revolution, was born on November 12, 1866, in Hsiang-shan, near Canton, of peasant parentage. He was one of six children, of whom two boys and two girls survived. Because of the poor soil, the people of Hsiang-shan had a tradition of seeking their livelihood away from home. At the age of fifteen, Sun's elder brother (Sun Mei) left for Honolulu where he was to build up a prosper-

1. His personal name was Wen, while Yat-sen was his style (secondary name). In China he is better known by his other name, Chung-shan, the Chinese pronunciation of the Japanese name, Nakayama, which he adopted while a political refugee there at the age of 31.

ous business. Sun himself entered school at six and had studied the traditional primers and the Confucian Four Books and the Five Classics by the
age of twelve. But because of the family's poverty, he did not receive a
thorough grounding in Chinese classical studies. Born only two years after
the downfall of the Taiping kingdom, Sun in his childhood often heard
the stories of the revolution and secretly aspired to be a second Hung
Hsiu-ch'üan.

THE INFLUENCE OF HONOLULU AND HONG KONG. In 1879 Sun went with
his mother to Honolulu to join his brother. For the first time the boy saw
the wonders of ships, the prosperous good life, and the fair taxes of the
islands. He entered the Anglican missionary Iolani School and later graduated from Oahu College in 1883 at the age of seventeen. His ambition
to continue his studies in the United States was thwarted by his brother,
who feared for his conversion to Christianity. Sun returned to Hong Kong
and after spending a year at the Diocesan Home to improve his English,
he transferred to the Queen's College, where he was Christianized after
all. In 1885 he married, and after a short trip to Honolulu he returned in
time to witness China's defeat by France. Thoroughly disgusted with the
Ch'ing decadence, he began to develop ideas about overthrowing the
dynasty.

At twenty Sun enrolled in the Po-chi Medical School at Canton, while
improving his Chinese studies by reading the twenty-four dynastic his-
tories. Among his schoolmates was one Cheng Shih-liang, who had wide
connections with the secret societies. The two of them frequently engaged
in lengthy discussion about the need for a revolution, and Cheng volun-
teered to enlist help from his underground friends. In 1887 Sun transferred to the College of Medicine for Chinese in Hong Kong because of
its better curriculum and the freedom that the British colony afforded for
revolutionary activity. While on the one hand Sun received a sound training in science and medicine from the strict English dean, Dr. James
Cantlie, on the other he used the school as the headquarters for his revolutionary activities, traveling back and forth between Hong Kong and
Macao to promote the cause. After five years of study, he graduated first
in his class and began practice in Macao in 1892. A year later, he moved
to Canton, freely donating his services and medical supplies to the needy
in order to win friends and to make new contacts. Here he ran into an old
Taoist priest, who advised him that for the revolution to succeed he must
seek the help of the secret societies. From him Sun learned of the organization and the locations of these secret bodies, and Cheng Shih-liang was
instructed to develop contacts with them.

It is apparent, then, that Hawaii and Hong Kong had a strong influence on Sun during the formative period of his life. What he saw in these places, and the contrasts they presented to his native district of Hsiang-shan, could not help but make a deep impression upon his young mind. Hawaii during his stay (1879-83), still an independent island kingdom, was rapidly being invaded by American influence, bringing with it the ideas of democracy, a modern legal system, modern schools, and the need for industrial development. The progressives of the islands were advocating the overthrow of the monarchical system in favor of an American-style democracy, while the conservatives rejected foreign intrusion and republicanism. Hawaii was experiencing problems not unlike those confronting China. Although Hawaii did become a republic in 1893, it was constantly under the threat of American annexation.[2] From this historical lesson Sun became convinced that it was insufficient to merely overthrow the Manchu dynasty and establish a republic; it was imperative to instill in the people a strong sense of nationalism with which to reconstruct the country and preserve their independence.

Hong Kong was no less a source of inspiration and instruction. The efficiency of the British colonial administration, the modern hygienic developments, and the orderliness provided a sharp contrast to Sun's native place. Why should they be so different when they were only fifty miles apart, he asked. Later he was to discover that the provincial capital and the metropolis of Peking were even more corrupt than his home district. For all its 4,000 years of civilization, China had no cities as well governed as Hong Kong which had been under the British rule for only a few decades. The contrast kindled in Sun's heart a burning desire to overthrow the inefficient Ch'ing government.

However, Dr. Sun was a realist endowed with a large measure of "tactical flexibility" and a "proficiency at focusing simultaneously upon conflicting goals."[3] Before 1894, while planning for the overthrow of the Manchu dynasty, he also considered reform as a possible means of saving China. Influenced by two respected reformists, the famous journalist Wang T'ao and the founder of his medical college, Ho Ch'i (Ho Kai), Sun entertained the idea of joining the reformist group. As an ex-peasant and a Christian convert who had a Western education but no traditional degree, Sun realized only too well that he was an "outsider" barred from the inner circles of the traditional society. But joining the camp of the gentry-reformists would enable him to drive a wedge into the elitist estab-

2. It did take place in 1898.
3. Harold Z. Schiffrin, *Sun Yat-sen and the Origins of the Chinese Revolution* (Berkeley, 1968), 27.

lishment. He decided to reach Li Hung-chang, the epitome of gentry-reformists.

In the summer of 1894 Sun and a companion, Lu Hao-tung, went north to see the state of affairs in the capital and to seek an interview with Li. In a letter Sun advised Li that the wealth and power of European states were not achieved by battleships and cannon but by the full development of human talents, the full exploitation of the earth's resources, the full utilization of material devices, and the free exchange of goods. China must develop its talents through universal free education, vocational guidance, and promotion of science and agriculture. Describing himself as one who had traveled abroad and studied foreign languages, literature, politics, mathematics, and medicine, Sun went on to say: "I paid particular attention to their [the West's] methods of achieving a prosperous country and a powerful army and to their laws for reforming the people and perfecting their customs."[4] Li Hung-chang, however, was too preoccupied with the Japanese war either to see him or accept his service. The deep disappointment that ensued, coupled with the firsthand observation of the Manchu decadence in Peking, more than ever strengthened Sun's determination to overthrow the dynasty.

THE REVIVE CHINA SOCIETY, 1895. Sun resolved to return to his original goal of revolution and seek aid from those whom he knew best—the Chinese overseas, the secret societies, the Christian converts, and the missionaries—men existing on the fringes of Chinese society.[5] He went to Honolulu in the fall of 1894. With the help of his brother he organized the Revive China Society (Hsing-Chung hui) on November 24, 1894, with an initial membership of 112. Planning to expand his activities to the United States, Sun was urgently called back to China to take advantage of the war situation. He returned to Hong Kong and established a headquarters for the Revive China Society there on February 21, 1895, with branches in the provinces. Members of the Society took an oath to "expel the Manchus, restore the Chinese rule, and establish a federal republic."[6] And with that, the first revolutionary body was born.

On March 16 the Revive China Society set out to raise 3,000 men to capture Canton in order to establish it as their revolutionary base. Lu Hao-tung designed a "Blue Sky-White Sun" flag for the revolutionaries, which has since become the national flag of the Republic of China. Canton at

4. Schiffrin, 37.
5. Schiffrin, 40.
6. A recent survey disputes this oath, which is said to have been added retroactively. See Chün-tu Hsüeh, *Huang Hsing and the Chinese Revolution* (Stanford, 1961), 29.

the time was seething with unrest, caused by the sudden disbandment of troops that had been raised to reinforce the Japanese war. Sun developed connections with the San-yüan-li militia and scheduled an uprising for October 26. However, the plot was discovered, resulting in the loss of munitions and forty-eight lives, including that of Lu Hao-tung—the first martyr of the revolution.

Sun fled to Hong Kong, only to find that the British authorities had complied with a Ch'ing request to ban him for five years. Following the advice of Dr. Cantlie, Sun and a follower, Ch'en Shao-po, escaped to Japan. Upon their arrival at Kobe, they were pleasantly surprised by the local journalistic description of the Canton uprising as a "revolution" rather than an unlawful revolt. Flattered, Sun ordered that henceforth all uprisings should be called "revolutions." At Yokohama a Revive China Society branch was established, and the revolutionaries began developing connections with Japanese sympathizers, among whom were Sone Toshitora and the Miyazaki brothers (Yazō and Torazō). Sun was a changed man; he cut off his queue, took up Western attire, and left for Honolulu to promote the revolution.

THE LONDON KIDNAP. The trip to Honolulu proved unfruitful. Many of Sun's supporters had grown indifferent after the abortive Canton uprising. To the United States Sun went to seek support from the Chinese communities, only to find them even less politically conscious. Their Hung League organizations had forgotten their original "anti-Ch'ing, revive Ming" objectives and had become little more than fraternal and social clubs. It was not until after repeated lectures by Sun that these organizations revived their old dedications to revolution.

Arriving in London on October 1, 1896, Sun was lodged at the Gray's Inn as arranged by Dr. Cantlie. On October 11, on his way to church, Sun was lured to the Chinese legation, where he was kidnapped and detained on the third floor. The Ch'ing minister had secured the approval of Tsungli Yamen to send him home secretly on a chartered ship at a cost of £7,000. Sun, however, was able to slip a message out to Dr. Cantlie via the English porter of the legation. Having unsuccessfully sought intercession from Scotland Yard, Cantlie brought the case to the Foreign Office, and *The London Globe* in striking headlines exposed the illegal kidnap on October 22. Shocked by the incident, the Foreign Office obliged the Ch'ing legation to release Sun the next day. The kidnap had the unexpected result of making him famous overnight. It was, in a way, a blessing in disguise.

Sun remained in England for nine months to study firsthand the recent political and social developments. Witnessing the growing trend toward social reform and revolution in the various industrialized countries, Sun wanted to save China from similar problems of strikes and labor-capital disputes in the future. In 1897 he developed the idea of a social revolution to complement his earlier nationalistic and democratic revolution. Here was the basis of his famous Three People's Principles (*San-min chu-i*)—People's National Consciousness, or Nationalism (*Min-tsu*); People's Rights, or Democracy (*Min-ch'üan*); and People's Livelihood, or Socialism (*Min-sheng*)—which Sun proudly compared with the famous Lincolnian expression "of the people, by the people, and for the people."

The Three People's Principles became the revolutionary philosophy for Sun and his followers. The first principle, nationalism, called for not only the overthrow of the alien Manchu rule, but also the removal of the foreign imperialistic yoke. The second principle, democracy, aimed at achieving the Four Rights for the people—initiative, referendum, election, and recall—and the Five Powers for the government: executive, legislative, judicial, control (supervisory), and examination—the last two reflecting the traditional functions of the censorate and the civil service examinations. The third principle, socialism, stressed the need for regulating capital and equalizing land. Here we see traces of the ancient Chinese utopian idea of "land to the tiller" as well as the influence of the Taiping land revolution. But more immediately and positively, it was from the famous Single Taxer Henry George and John Stuart Mill that Sun gained the idea that all increment in land value after it had been fixed (after the revolution) should go to the government. Thus, Sun's idea of a social revolution, first conceived in 1897 in an embryonic form, became a full-fledged third principle of his revolution by 1905-06.[7] These Three People's Principles are still the abiding creed of the Nationalist government on Taiwan today.

A DIFFICULT PERIOD, 1896-1900. Sun had built up fame and had evolved a revolutionary philosophy, but still lacked concrete success. Realizing the small numbers of Chinese students and merchants in Europe, he returned in mid-1897 to Japan, where the overseas Chinese communities were much larger and where it was more convenient to direct the revolutionary work on the mainland. Befriended by Inukai Ki, leader of the Japanese Liberal Party, Sun was introduced to Premier Okuma Shigenobu and Soejima Taneomi, vice-president of the Privy Council. Other private figures, notably the Miyazaki brothers and Hirayama Shū, became devoted supporters of Sun.

These Japanese and Sun shared the common feeling of Asia's grievance

7. Martin Bernal, "The Triumph of Anarchism over Marxism, 1906-1907" in Mary C. Wright (ed.), *China in Revolution*, 103-04.

against Western imperialism. Many of them believed that China, once a great civilization, was only in temporary doldrums from which it could lift itself if proper outside help and a new leadership were available. Japan, having achieved modernization first, must repay its ancient cultural debt to China by assisting it to reform, modernize, and gain freedom from foreign imperialism. These ideas, persuasively set forth in the so-called Ōkuma Doctrine of 1898, won wide currency among the Japanese *shishi*—"men of high purpose"—many of whom considered Sun the man of destiny to regenerate China in the cause of Pan-Asianism.[8]

In contrast to these warmhearted Japanese sympathizers, the Chinese communities in Japan were largely apolitical and conservative. Out of 10,000, only a hundred or so supported Sun. Revolutionary work in China progressed even more slowly because of the general fear of involvement in anti-Manchu activity. Although the secret societies were an exception, they lacked the necessary education, cohesion, and sense of direction to offer any leadership.

To add to the frustration, there was the hostile influence of the Emperor-Protection Society (*Pao-huang tang*) under the leadership of K'ang Yu-wei and Liang Ch'i-ch'ao, who had fled to Japan after the ill-fated "Hundred-Day" Reform. They and their followers vehemently attacked the ideas of revolution and republicanism. Sun took a rather conciliatory attitude toward K'ang, since they were both political refugees in a foreign country. However, his proposal for cooperation was contemptuously rejected by K'ang, who still regarded himself as an imperial tutor too dignified to be associated with a rebel. Inukai's good-will mediation succeeded no further than arranging a meeting between the two, but at the appointed time K'ang did not show up. Liang, on the other hand, was far less arrogant and showed a receptiveness to revolutionary ideas, but was kept from cooperating by his teacher. The monarchist reformers and the revolutionaries clashed like "water and fire." It was not until K'ang was ordered by the Japanese government to leave the country that a change took place. Liang and Sun began discussing cooperation and even the possible amalgamation of the two groups, with Sun as director and Liang vice-director. K'ang, then traveling in Britain and Canada, quickly transferred Liang to Hawaii to take charge of the local branch of the Emperor-Protection Society. Still inclined toward reconciliation, Liang proposed that Emperor Kuang-hsü be made president of the future republic—an interesting idea but totally unacceptable to Sun.

Taking advantage of the Boxer Rebellion in 1900, Sun dispatched

8. Marius B. Jansen, *The Japanese and Sun Yat-sen* (Cambridge, Mass., 1954), 53.

Cheng Shih-liang to organize an uprising at Waichow (Hui-chou), north of Hong Kong, while Shih Chien-ju went to Canton to plot a sympathetic movement. Sun himself planned to go to Hong Kong in the company of a dozen or so Japanese supporters and officers, in the hopes of leading a revolutionary army northward. Unfortunately, the plot was again discovered and with Hong Kong authorities still refusing him admission, Sun fled to Formosa. Once there, he was befriended by the Japanese governor, Kodama, who promised help. Meanwhile, the revolutionaries had initiated activities along the coastal areas of Kwangtung, aided by the secret societies. They met with initial successes, but were stalled while anxiously waiting for reinforcements and supplies from Sun and the Japanese. Quite unexpectedly, there was a sudden change of government in Japan, and the new premier, Prince Itō, banned officers from serving in Sun's revolutionary army and ordered Governor Kodama to halt all assistance. Sun was not even allowed to leave Formosa. Without the reinforcements and supplies, the revolutionary army could not hold out for long; it ultimately disbanded and its leader, Cheng, fled to Hong Kong. The lone Japanese participant, Yamada, was captured and killed by the Ch'ing forces—the first foreigner lost in the Chinese Revolution. Meanwhile, Shih Chien-ju's attempt to blow up the governor-general's office at Canton led to his capture and loss of life at the tender age of twenty-one. Thus, the Waichow uprising ended in a fiasco.

Yet Sun's image improved dramatically at this point. The mismanagement of the Ch'ing court during the Boxer catastrophe led many to look upon him with favor. He was no longer regarded as a rebel or an outlaw, but rather as a patriotic, devoted revolutionary working for the betterment of his country and people. Students at home and in Japan enthusiastically supported him. Those in Japan published the *Citizen's Tribune* (*Kuo-min pao*) and the *Twentieth Century China* to promote the revolutionary cause and to advocate the assassination of Ch'ing officials. Back home several well-established scholars published the *Kiangsu Tribune* (*Su-pao*), and a young revolutionary, Tsou Jung, contributed a 20,000-word treatise, "The Revolutionary Army" (*Ko-ming chün*) in 1903, attacking the Ch'ing court and favoring revolution. The editor of *Su-pao*, Chang Ping-lin, was imprisoned for two years and the author died in jail at only 20 years of age.

In addition to these publications, a number of societies sprang up to support the revolution. At Shanghai, a prominent scholar, Ts'ai Yüan-p'ei, organized the Recovery Society (*Kuang-fu hui*), and at Changsha, Huang Hsing, who had secretly studied military arts in Japan, formed the China Revival Society (*Hua-hsing hui*) in 1903 with an initial membership of 500, including Sung Chiao-jen who later distinguished himself as a leading revolutionary figure. The constituents of this latter organization were mostly intellectuals and members of secret societies, particularly the Ko-lao Brotherhood Association which enrolled more than 100,000 men. After an abortive attempt to seize Changsha in 1904, Huang escaped to Japan, where he built up a strong following.

THE T'UNG-MENG HUI, 1905. The fortunes of the revolution turned considerably for the better during the period 1902-5, providing a sharp contrast to the dark days of the immediate past. Sun traveled widely in Vietnam, Japan, Honolulu, and the United States rallying support for his cause. Encouraged by the enthusiastic response of Chinese students in Japan, he developed the idea of forming a revolutionary party.[9] Many of these students had aspired to military studies but were prevented by the Ch'ing embassy. However, through Sun and Inukai, two Japanese officers were engaged to give instructions secretly to a group of fourteen students in the methods of weapon-making, military tactics, and guerrilla warfare. The students took an oath before Sun to "expel the Manchus, restore the Chinese rule, establish a republic, and equalize the land."

In Honolulu, since the Emperor-Protection Society had appropriated much of his former power base, Sun adopted the advice of his maternal uncle[10] to enter the Hung League (*Hung-men*), and was elected the "Hung Rod," i.e. generalissimo. With this new title and stature, he was warmly welcomed into the Hung League in the United States in 1904 as "Elder Brother Sun." He succeeded in revising the charter of the League by stressing its original anti-Ch'ing objective while interpolating the new purpose of "expelling the Manchus, restoring the Chinese rule, establishing a republic, and equalizing the land." In doing so, he swayed the Chinese communities in the United States over to his side from the Emperor-Protection Society.

In the spring of 1905 Sun was invited by Chinese students in Europe for a visit. Mutual discussion led to the decision of not only seeking support from students and the secret societies but also from the Ch'ing New Army. At Brussels Sun initiated thirty students into a revolutionary society, and at Berlin and Paris he initiated twenty and ten more, respectively, all pledging the four objectives mentioned above. However, the largest revolutionary organization was in Tokyo, where several hundred students were recruited, representing seventeen of the eighteen provinces of China—Kansu having sent no students to Japan at that time. The seeds

9. Leonard S. Hsü, *Sun Yat-sen: His Political and Social Ideals* (Los Angeles, 1933), 61.
10. Yang Wen-na.

of a new revolutionary party were sown, and Sun was encouraged to believe that the revolution would succeed in his lifetime.[11]

Through the intercession of Miyazaki, who praised Sun as a most remarkable man whose equal could not be found in the Western or Eastern hemispheres, Huang Hsing and Sung Chiao-jen met with Sun on July 28, 1905, at the office of their magazine, the *Twentieth Century China*. Sun stressed the need for unifying all revolutionary groups into one organization to avoid duplication of efforts and a struggle for power among themselves. After several meetings, they decided to join hands in a unified organization, The Chinese United League (*Chung-kuo T'ung-meng hui*), or T'ung-meng hui in abbreviation, on August 20, 1905. Sun, 37, was elected chairman; Huang Hsing, 31, became chief of the executive department with the authorization to act for the chairman during his absence; and Sung Chiao-jen, 23, was made a member of the judicial department. About seventy persons joined the T'ung-meng hui at its inauguration. After taking an oath to pledge their support of the four principles already mentioned, they were instructed by Sun in the secret handshake and three sets of passwords: "Chinese, Chinese things, and world affairs." Sun then shook hands with each, pronouncing happily: "From now on you are no longer subjects of the Ch'ing dynasty!" Just as he was speaking, the wooden partition of the room fell with a bang. Sun quipped: "This symbolizes the downfall of the Manchus!"

Sun worked out a detailed procedure for his revolution. Initially, there would be a military rule of three years in the areas liberated by the revolutionary forces. During this period the military government would control all military as well as civil affairs at the district (*hsien*) level. Meanwhile it would cooperate with the local people toward eliminating the old political and social evils, such as slavery, foot-binding, opium-smoking, and bureaucratic corruption. The second stage would be a period of political tutelage, lasting not more than six years, during which time local self-government would be instituted and popular elections for local assemblies and administrators would be held. However, the military government would still retain control of the central government. During this period, there would be a provisional constitution to specify the rights and duties of the military government and the people. When the period of tutelage ended, the military government would be dissolved and the country would be governed by a new constitution. In short, Sun envisioned a three-stage revolution to lead the country into constitutionalism.

11. Leonard S. Hsü, 62-63.

Sun's Three People's Principles were accepted as the revolutionary philosophy of the T'ung-meng hui, although the majority of the members focused only on the first two principles of nationalism and democracy. This was because the China Revival Society and the Restoration Society, both of which emphasized the overthrow of the Manchus and the establishment of a republic, contributed most of the members, while Sun's direct followers constituted but a small percentage. Huang Hsing now emerged as the strong man of the party, his name often appearing alongside that of Sun's as co-leaders. The membership of T'ung-meng hui grew rapidly to 963 by 1906, of whom 863 joined in Japan, the rest coming from Europe, Hawaii, Hong Kong, and Malaya.[12] Branches were established in China as well as in key overseas communities.

Huang Hsing turned over his *Twentieth Century China* as the official publication of the T'ung-meng hui. It engaged in heated debates with Liang Ch'i-ch'ao, who deprecated the ideas of revolution and republicanism in favor of constitutional monarchy. Not long afterwards, the *Twentieth Century China* was banned because it offended the sensitivity of the Japanese government by publishing an article entitled "The Japanese Politicians' Exploitation of China." The revolutionaries changed the name of the publication to the *People's Tribune* (*Min-pao*), with the first issue appearing on November 26, 1905. It boasted many gifted contributors, such as Chang Ping-lin, Hu Han-min and Wang Ching-wei. Their aggressive enthusiasm and combined talents overwhelmed Liang, who, for all his persuasive pen and flowing style, simply could not hold the ground alone for the Emperor-Protection Society. Moreover, Liang was secretly sympathetic with the revolutionary cause. His stress on the need for a constitution exposed the ineptness of the Manchu government, thereby indirectly advancing the revolutionary cause.[13] More and more, the younger generation turned to the side of the revolution.

The founding of the T'ung-meng hui constituted a milestone in the Chinese revolution, for it materially changed the character and style of the revolution. No longer did Sun operate on the periphery of society; he had moved into the "main stream of Chinese nationalism," receiving new support from the returned students, disaffected literati, and progressive army officers—groups that traditionally provided the leadership in China. The social base and potential areas of operations had been substantially enlarged. The T'ung-meng hui was multiprovincial and multiclass, as compared with the predominantly Cantonese makeup of the earlier Hsing-

12. Chün-tu Hsüeh, 44.
13. Chang P'eng-yüan, *Liang Ch'i-ch'ao yu Ch'ing-chi ko-ming* (Liang Ch'i-ch'ao and the late Ch'ing revolution), (Taipei, 1964), 325-26, 330-33.

Chung hui, capable of instigating uprisings along the coast as well as in the *interior* of China. Above all, it provided a unified central organization that resembled a modern political party, which served as a rallying point for all revolutionary and progressive forces in the country.[14] As such, it fittingly received the tribute "the mother of the Chinese revolution."

The pulse of the revolution now quickened. One uprising followed another between 1906 and 1911—six times in Kwangtung and one each in Kwangsi and Yunnan, making a total of ten, counting the first two attempts at Canton and Waichow in 1895 and 1900. The last revolutionary attempt in April 1911, aiming at capturing the important provincial capital of Canton, created such a sensation and shock to the Ch'ing court that it presaged the success of the next attempt at Wuchang half a year later. This tenth failure produced the famous seventy-two martyrs, many of whom were students who had returned from Japan. They were later buried at the Yellow Flower Mound (*Huang-hua kang*) in the northern suburb of Canton.[15]

The Rise of the Republic

The ten unsuccessful attempts at revolution all took place in the south and the southwest, where proximity to Hong Kong and Hanoi offered greater freedom of plotting and organization. However, powerful elements within the T'ung-meng hui now advocated to skip these peripheral areas and hit where it hurt the dynasty most—either in Peking or in the heartland of Central China along the Yangtze. If the tri-cities of Wuhan could be captured, they reasoned, the revolutionaries would be in a good position to respond to action in the south or to advance north to the capital. Thus, on July 13, 1911, a Central China Bureau of the T'ung-meng hui was established in Shanghai with Sung Chiao-jen as leader. The central provinces of Hupeh and Hunan emerged as the prime target.

In Hupeh there existed already two organizations affiliated with, but not a part of, the T'ung-meng hui. One was the Common Advancement Society (*Kung-chin hui*), founded in August 1907, which consisted largely of returned students from Japan and secret society members. The other was the misnamed Literary Society (*Wen-hsüeh hui*), an offspring of the Military Study Society (*Chen-wu hsüeh-she*), which came into being on January 30, 1911, and consisted generally of members of the Ch'ing New

14. Schiffrin, 8-9.
15. The common expression, "Huang-hua kang 72 martyrs" who died on "March 29" is inexact. Actually, more than 82 revolutionaries lost their lives on April 27, 1911, which was the 29th day of the third lunar month. See Chün-tu Hsüeh, 93.

Army in Hupeh who had been won over to the revolutionary cause. Of the two groups, the former was more prestigious, while the latter was more powerful due to its infiltration of the New Army. On June 1, 1911, the two societies agreed to cooperate in a joint action at Wuchang, and invitations were extended to Huang Hsing and Sung Chiao-jen in Shanghai—Sun being abroad—to come to direct the revolution. So quick and so successful was the subversion of the New Army that an immediate outbreak was irrepressible; the occasion that touched it off was the turmoil created by the railway controversy.

NATIONALIZATION OF THE RAILWAYS. Railway construction, which ran into such opposition in the 1870s and 1880s, became a craze after the Sino-Japanese War, and Sheng Hsüan-huai was appointed by the court as director-general of a new railway company in 1896. He had hoped to raise funds from government and private sources as well as from foreign loans, but since the first two were unable to contribute much, the major supply of capital came from the foreigners. In the decade that followed, a number of lines were constructed with foreign loans, the most famous being the Peking-Hankow and the Shanghai-Nanking railways. In 1898 a loan with the American-owned China Development Company was negotiated to construct the Canton-Hankow line. However, strong gentry and merchant opposition led Chang Chih-tung, governor-general at Wuhan, in 1905 to redeem the right from the American company with a payment of U.S. $6.75 million, financed by a new £1.2 million loan from the Hong Kong government. The people of Kwangtung, Hunan, and Hupeh, through whose provinces the proposed line would cross, were allowed to build it themselves; in addition the people of Szechwan were given the right to construct the line from Hankow into their province.

Provincial ability and resources, however, proved inadequate. In spite of new taxes on land, rice, property, and salaries, Hunan was able to raise only 5 million taels against a 60-million-tael construction cost. Kwangtung gathered only half of the needed amount. The Szechwan gentry and merchants found few subscribers to the shares of their railway company, and the situation was further confused by a 2-million-tael embezzlement among the directors of the company. Under such conditions, the court in 1908 put Chang Chih-tung in charge of the Canton-Hankow Railway as well as the Hupeh portion of the Szechwan-Hankow line. In June 1909 he began negotiations for a £6 million loan from the British-French-German-American banking consortium, but progress was delayed by his death four months later on October 5.

In line with its policy of centralizing power (as noted in Chapter 17), the Ch'ing government in the spring of 1911 approved the proposal of a junior metropolitan censor[16] that the main railways be nationalized while the minor or branch lines be left to private operations. On May 9, the court formally ordered the nationalization of the Canton-Hankow and Szechwan-Hankow lines, and on May 20 Sheng Hsüan-huai signed a contract with the four-power banking consortium for a 40-year loan at 5 percent interest.

The gentry and the people of the four provinces vigorously protested the nationalization policy and the invasion of foreign capital. Having invested considerable, if still insufficient, sums in these lines, they organized "railway-protection clubs" to defend their vested interests and mobilized the provincial assemblies to fight for their rights. Delegations were dispatched to Peking to appeal to the court, and demands were made to dismiss Sheng for selling out China's interests to foreigners. So powerful was the sense of injustice that popular uprisings in Szechwan and Hunan were all but inevitable.

On June 17 the court offered to indemnify those people who had invested in the railways: for Hunan and Hupeh, a 100 percent compensation; for Kwangtung, 60 percent, with the remaining 40 percent to be paid in government bonds redeemable within ten years after the railway had become profitable; for Szechwan, due to the proven embezzlement, the government would give only redeemable bonds at 6 percent interest to cover the railway capital of 7 million taels and the actual construction cost of 4 million taels. The treatment of the four provinces was thus unequal; Hunan and Hupeh received the best terms, Kwangtung next, and Szechwan the worst. Small wonder that the people of Szechwan were incensed, while those of the other three provinces were less agitated!

The Provincial Assembly of Szechwan, representing the interests of the gentry, the rich landlords, and the wealthy merchants, took the lead in protesting this unfair treatment. It attacked Peking for betraying the interests of Szechwan to the foreigners, and reacted strongly against the high-handed, despotic manner in which the government negotiated the loan and proclaimed the nationalization policy without first consulting the provincial assemblies. Aroused by Yüan Shih-k'ai's special emissary, T'ang Shao-i, and encouraged by the secret sympathy of the ex-governor-general, Wang Jen-wen, the leaders of the Provincial Assembly organized a mass movement of students and people to demand postponement of the na-

16. Shih Ch'ang-hsin. He was supported by Sheng Hsüan-huai, now Minister of Posts and Communications.

tionalization project and the impeachment of Sheng Hsüan-huai.[17] On August 24, 1911, more than 10,000 Szechwanese staged a rally in the provincial capital of Chengtu. Overcome by emotion, they wailed and screamed; they resolved to stop tax payments, to sponsor strikes at the schools and markets, and to mourn before the placard of the late Emperor Kuang-hsü who had granted them the right to construct the railway. The new governor-general, Chao Erh-feng, anxious to please the court and keep his position, ordered the arrest of gentry representatives, and an open conflict broke out between the troops and the demonstrators, resulting in thirty-two deaths among the latter. Thereafter, fighting between the people of Szechwan and the government troops intensified.

It must be noted that at this point although the Szechwan gentry agitated against the court, they did so to protect their own interests without any idea of overthrowing it—most of the members of the Provincial Assembly believed in constitutional monarchy.[18] But when their demands were ignored by the government, they turned their sympathy to the revolutionaries. A Szechwanese leader[19] declared: "Domestic politics is hopeless, and the government apparently does not care for the people. To save the country there is no other way but revolution. We Szechwanese have already made proper preparation, and would coordinate with other provinces for joint action."[20] The railway controversy and the revolution now fused into one pressing issue.

THE WUCHANG REVOLUTION. To control the unrest in Szechwan, the court transferred part of the Hupeh New Army there—an act which placed the strategic city of Wuchang in a vulnerable position, which the revolutionaries were quick to take advantage of. Huang Hsing, still in Shanghai, had hoped to take action by the end of October, but on October 9 a bomb accidentally exploded in the revolutionary headquarters located in the Russian Concession of Hankow. Subsequent police raids resulted in the arrest of thirty-two revolutionaries and the seizure of weapons, explosives, and important documents including the lists of names of the members of the New Army who had been won over. To protect themselves, the engineering and artillery battalions of the New Army decided to strike the following day.

On the morning of October 10, the engineering unit took the lead in

17. Chūzō Ichiko, "The Railway Protection Movement in Szechuan in 1911," *Memoirs of the Research Department of the Toyo Bunko*, Tokyo, 14:50-57 (1955).
18. Chūzō Ichiko, 68-69.
19. Liu Sheng-yüan or perhaps P'u Tien-chün.
20. Li Shou-k'ung, 736-37.

seizing the government munition depot in Wuchang, and the artillery joined in a combined attack on the office of the governor-general, who fled along with the military commander.[21] The New Army rebels met little resistance and had complete control of the city by noon. Having no genuine revolutionary leaders present—Sun being abroad and Huang still in Shanghai—they drafted the reluctant Ch'ing brigade commander, Li Yüan-hung, to be the military governor of Hupeh. Meanwhile, T'ang Hua-lung, the former chairman of the Hupeh Provincial Assembly who had long shown sympathy with the revolution, was appointed the civilian executive chief of the revolutionary government, charged with the duty of setting up an initial administrative body. It was he who on the one hand sent out telegrams to the other provinces urging them to declare independence of the Ch'ing court, and on the other successfully convinced the foreign consuls in Hankow that they should stay neutral during the turmoil. Thus, when the escaped Ch'ing governor-general requested the foreign consuls to call in gunboats to bombard the revolutionaries, the French and Russian consuls simply replied that the situation was totally different from the Boxer Rebellion, while other consuls proclaimed strict neutrality.[22] On October 12, Hanyang and Hankow fell to the revolutionaries.

The quick success was indeed "providential," as Dr. Sun later recalled, for if the Manchu governor-general had not been scared away, the military commander would have stayed and probably would have crushed the thin revolutionary forces, estimated at a little more than 2,000 men. Foreign neutrality, of course, helped the revolutionary cause. What was most encouraging was the rapid succession of declarations of independence by the provinces and important municipalities: Changsha, October 22; Yunnan, October 31; Shanghai, November 3; Chekiang, November 5; Fukien and Kwangtung, November 9 and Szechwan, November 27. Within a month and a half, fifteen provinces, or two-thirds of all China, seceded from the Ch'ing dynasty.

In order to appease the public anger, the court dismissed Sheng Hsüan-huai on October 26 and released the imprisoned Szechwanese gentry. Meanwhile, counterattacks by the government's Peiyang forces had succeeded in recovering Hankow on November 2 and Hanyang on November 27. However, these temporary Ch'ing victories were more than offset by the loss to the revolutionaries of Shanghai in early November and Nanking on December 4, 1911. At the latter place, a provisional revolutionary government was established, electing Huang Hsing generalissimo

21. Jui-cheng and Chang Piao, respectively.
22. P'eng-yüan Chang, "The Constitutionalists" in Mary C. Wright (ed.), *China in Revolution*, 175-76.

and Li Yüan-hung vice-generalissimo. However, both declined the appointments, awaiting Sun's return from abroad.

Traveling in Denver, Colorado, Sun learned of the success of the Wuchang revolution from a local newspaper account. His first thought was to rush home as fast as he could to have the personal satisfaction of directing the revolution, but his better judgment dictated that he engage in diplomacy instead. Knowing that British support was essential to the future of the revolution, he traveled east to New York, from whence he sailed for London. Successfully he committed the British government to stop all loan negotiations with the Ch'ing government, to prevent Japan from aiding the Peking regime, and to lift the ban on his entering British territories and possessions so that he could return home freely. A promise was also secured from the president of the four-power banking consortium that as soon as the revolutionary government was recognized by the powers, the consortium would negotiate with it. With these diplomatic accomplishments, Sun went to France, where he was warmly greeted by Premier Clemenceau and the French people. Back in Shanghai on December 25, Sun was elected four days later by a nearly unanimous vote of the provincial delegates to be the provisional president of the Republic of China.[23] Li Yüan-hung became the provisional vice-president, and Huang Hsing the minister of war. The new government adopted the solar calendar in place of the lunar one, and designated January 1, 1912, as the first day of the republic. After some twenty-seven years of struggle,[24] Sun's lifelong dream came to a glorious fulfillment. The question now facing the Nanking government was how to terminate the Ch'ing dynasty and achieve national unification.

THE MANCHU ABDICATION. In a dying struggle for survival, the court sent Minister of War Ying-ch'ang and Admiral Sa Chen-ping to chastise the revolutionaries at Wuchang, and appointed Yüan Shih-k'ai governor-general of Hunan and Hupeh. Still smarting from his summary dismissal in 1908 and dissatisfied with the limited appointment, Yüan refused to end his retirement on the grounds that his "leg ailment"—the Ch'ing pretext for his forced retirement—had not yet recovered. Ying-ch'ang's army, commanded by officers who were mostly former subordinates of Yüan, fought halfheartedly and suffered repeated defeats, while Admiral Sa was persuaded by Li Yüan-hung to defect on November 11. Under these conditions, the court had no choice but to turn to Yüan again. Yüan demanded (1) inauguration of a national assembly in a year; (2) organiza-

23. Sun received sixteen out of seventeen possible votes, and Huang Hsing received one.
24. Since 1885.

tion of a responsible cabinet; (3) pardon for the revolutionaries; (4) lifting the ban on parties; (5) full power to control the army and the navy; and (6) guarantee of sufficient military funds. The first four of these terms were aimed at mollifying the public and the revolutionaries, while the next two were designed to make Yüan the most powerful man in the country. By item two—perhaps the most important of all—Yüan did not really mean a genuine "responsible" cabinet; it was his subterfuge to preempt the power of the regent, Prince Chün, who had retired him earlier, and to eliminate the "Royal Cabinet."

Under the pressure of military defeat and the rapid secession of the provinces, the regent gave in to Yüan's demands. On October 27, 1911, Yüan was appointed imperial commissioner in full charge of the army and the navy, and his two chief lieutenants, Feng Kuo-chang and Tuan Ch'i-jui, were given command of the First and Second armies respectively. Still unsatisfied, Yüan continued to bargain and refused to come out of retirement. However, to show his power and ability to control the situation, he ordered Feng to deal a severe blow to the revolutionaries, which he did by taking Hankow on November 2.

About this time, a dramatic event occurred in North China. On October 29, two leaders[25] of the Ch'ing 20th Division stationed at Luan-chou (halfway between Mukden and Peking) demanded of the court the inauguration of a constitutional monarchy within a year. Fully expecting a rejection which would give them an excuse to march on Peking and effect a "central revolution," they found to their surprise that the court, stunned by the secession of Shansi on the same day, meekly bowed to their demand. Prince Chün declared himself unfit to govern as regent and Prince Ch'ing resigned as premier. On November 1, Yüan was made premier; it was only then that he came out of retirement and went south to take charge of the campaign against the revolutionaries. Two days later the court hurriedly promulgated a nineteen-article "principle of constitution" in an attempt to appease the public.

Yüan assumed the premiership, formed his cabinet, and placed his henchmen in full control of the capital area and the imperial guards. On December 4 the regent was retired at an annual pension of 50,000 taels. What was left of the Ch'ing court was merely the boy emperor and the widowed dowager.[26] With them as his puppets, Yüan started to flirt with the revolutionary forces for his personal future.

Three times before November 10 he sent emissaries to Li Yüan-hung

25. Chang Shao-tseng and Lan T'ien-wei, both graduates of a Japanese military academy and secret members of the T'ung-meng hui.
26. Lung-yü, wife of the late emperor Kuang-hsü.

to propose peace talks, while his son, Yüan K'o-ting, went to see Huang Hsing, the commander-in-chief of the revolutionary forces at Hanyang, suggesting collaboration and joint action. However, both attempts failed because the revolutionaries knew well Yüan's favorite trick of playing both ends against the middle. Thus snubbed, Yüan ordered his troops to shatter the revolutionaries' defense at Hanyang, which fell on November 27. Having shown his power, he halted any further attack to show his leniency, and persuaded the British minister, John Jordan, to instruct the British consul at Hankow to mediate a truce, which was arranged on December 1. His peace emissary, T'ang Shao-i, then went to Shanghai to negotiate with the revolutionary representative, Wu Ting-fang. Huang Hsing now cabled Yüan that if he would support the republic and force the abdication of the Ch'ing emperor, the future presidency of the republic would be his. So eager was he for this new position that when Sun was elected provisional president on December 29, Yüan was incensed and broke off peace negotiations.

Strangely enough, most of the revolutionaries at this point considered Yüan indispensable: he was the only man who could save the country from a civil war and who could force the Ch'ing court out of existence. Sun did not favor compromise but as an idealist he cared little whether he or Yüan were president, as long as the Manchu dynasty was overthrown and the principle of a republic was firmly established. Furthermore, he was a bit irked with his own followers, who ignored his three-stage revolutionary procedure and his principles of democratic reconstruction and people's livelihood. They emphasized only nationalism to overthrow the Manchus. With this state of mind and the knowledge of Yüan's superior military power, Sun was willing to step down. He humored Yüan with the explanation that he had accepted the *provisional* presidency in order to keep the *regular* presidency for him. Still unpacified, Yüan ordered more than forty of his generals to oppose the republic in favor of a constitutional monarchy, and under the pretext of raising military funds to fight the revolutionaries, he exacted 80,000 ounces of gold from the helpless dowager. Sun had to reassure Yüan that if he could avert the civil war, a "just reward" would be awaiting him. When a group of Ch'ing diplomats abroad[27] urged the abdication of the throne on January 3, 1912, Yüan knew that the days of the dynasty were numbered. He informed the Nanking government that he would induce the voluntary abdication of the Ch'ing throne if the presidency of the republic were offered to him. To guard against duplicity on the part of Yüan, Sun specified, through the

27. Under the leadership of Lu Cheng-hsiang, minister to Russia.

news media, the procedure for the transfer of power: (1) Yüan must notify foreign ministers or consuls of the Ch'ing abdication; (2) Yüan must publicly declare his support of the republic; (3) Sun would resign upon receiving notification from the diplomatic or consular corps of the Ch'ing abdication; (4) the parliament would elect Yüan provisional president, and (5) Yüan must pledge to honor the constitution to be prepared by the parliament, and until he did so he would not be given military power.

Yüan mobilized his friend Prince Ch'ing to impress upon the court that rather than lose everything, it would be wise to abdicate gracefully under the favorable conditions that the revolutionaries were willing to offer. Between January 17 and 19, three imperial conferences were held to deliberate the question; most of the Manchu and Mongol princes were opposed to abdication. Yüan then instigated some fifty of his generals to announce their support of the republic. Tuan Ch'i-jui went so far as to inform the court that if the Manchu nobles had doubts about the republic he would bring troops to Peking to argue with them. Feng Kuo-chang also spoke openly to his troops in favor of the republic. In collaboration with these moves, Yüan's emissaries visited the palace repeatedly to urge an early abdication. They tactfully advised the dowager that since Emperor Kuang-hsü started, but did not live to see, the constitutional movement, it behooved her to carry on his work and to accept republicanism. The dowager is reported to have answered: "I know that the country is public property and not the private possession of the Manchus, but since the Manchus have a heritage of more than 200 years, I only ask that the tomb of Emperor Kuang-hsü be maintained and repaired, and that the status of the imperial family be not degraded." On January 30 Prince Chün and Prince Ch'ing, the former regent and premier, advised that "since the government troops have lost the will to fight, it would be better to abdicate under favorable conditions." On February 1, 1912, the dowager summoned Yüan to the palace and sobbingly announced: "I leave the various matters to your judgment and have no request other than the preservation of the dignity and honor of the emperor."[28]

The Nanking government offered to treat the deposed Ch'ing emperor with the same courtesy as a foreign sovereign, subsidize him with 4 million taels annually,[29] and allow him to live in the Summer Palace with the usual number of guards and attendants. On February 12, Sun warned that these favorable terms would be withdrawn if the abdication did not take place within two days. On that very day, Yüan made public the previously prepared imperial rescript, countersigned by himself as premier

28. Hsiao I-shan, IV, 2,725, 2,727.
29. To be changed to 4 million silver dollars after the new currency was issued.

and all the cabinet ministers, announcing the formal abdication of the Ch'ing throne. And with that the 268-year Ch'ing rule, the last of China's twenty-five dynasties, came to an end.

The imperial rescript continued a statement which authorized Yüan to organize a provisional republican government and to negotiate for a national unification with the revolutionaries. Such a statement did not appear in the original version, prepared by the famous scholar Chang Chien for the Nanking government and accepted by Yüan, but was later secretly inserted by Yüan to show that he derived the provisional presidency from the deposed Ch'ing emperor and not from the Nanking regime. Sun was exasperated but could do nothing about it since it was already published.

On the same day Yüan pledged his support of the republic, which was a prerequisite to his assumption of the presidency: "The republic is universally recognized as the best form of state . . . Now that the Ch'ing emperor has clearly announced his abdication in a rescript which has been countersigned by me, the date of such announcement is the end of the imperial administration and the beginning of the republic. Let us henceforth forge ahead and endeavor to reach a state of perfection. *Never shall we allow the monarchical system to reappear in China.*"[30] On February 13, Sun resigned as the provisional president of the republic and recommended that Yüan be named his successor, contingent upon the latter's acceptance of three conditions: (1) that Nanking remain the capital; (2) that he come to Nanking to assume the provisional presidency; and (3) that he observe the provisional constitution to be drafted by the provisional parliament. On the following day, the provisional parliament formally elected Yüan the provisional president and Li Yüan-hung the provisional vice-president. A delegation of eminent leaders was then dispatched to Peking on February 18 to escort Yüan to Nanking.

Yüan was in no mood to leave his power base in the North for the South, where the revolutionaries were strong. He instigated riots by his soldiers to justify the need for his continued presence in Peking. The revolutionary leaders had no choice but to allow him to inaugurate in Peking, which took place on March 10. A day later, Sun promulgated the Provisional Constitution of fifty-six articles—the first such document in China. On April 1, 1912, Sun formally relinquished his duties as provisional president, and by a vote of the parliament on April 5 Peking was made the national capital. The United States was the first to recognize the new republic of China, followed by Brazil, Peru, Austria, Portugal, and others.

30. Italics added.

THE SIGNIFICANCE. The rise of the republic was an epochal event in Chinese history, for it brought an end to more than two thousand years of imperial dynasties. China no longer belonged to any "Son of Heaven" or any imperial family but to all the people. The success of the revolution fulfilled not only the dreams of the two-and-a-half-century nationalistic revolutionary tradition, but went beyond narrow racial considerations to liberate political power from one ethnic group, the Manchus, and extend it to all the people of China: the Chinese, the Manchus, the Mongols, the Moslems, and the Tibetans. The rapidity of the success—from the Wuchang Revolution of October 10, 1911, to the establishment of the republic on January 1, 1912, a total of eighty-three days—was seldom equaled by any other great revolutions of the world.

Yet the revolution was an incomplete one with many unfortunate repercussions, much to the chagrin of Sun. Most of his followers devoted themselves to the overthrow of the Manchus and the establishment of the republic; few paid attention to the more important task of democratic reconstruction and the problem of people's livelihood. When the dynasty was overthrown and the republic established, they felt that their prime objectives had been achieved. So anxious were they for peace that they were willing to compromise with so unprincipled a man as Yüan, over the opposition of Sun, who, outvoted, was regarded as an impractical idealist. Of the Three People's Principles, they discarded the second and third totally and accepted only part of the first—nationalism against the alien Manchu rule—without realizing that after the establishment of the republic they must continue to struggle against foreign imperialism. They ignored Sun's three-stage revolutionary program altogether. Their readiness to cooperate with the old elements, and their favorable treatment of the deposed emperor, paved the way for future warlordism and attempts to revive the imperial system—by Yüan in 1915 and by Chang Hsün in 1917. Sun's disillusionment with his own party was a major cause for his resignation as provisional president. He asked, "Without revolutionary reconstruction, what's the use of a revolutionary president?"

Yüan's Betrayal of the Republic

Once elected the provisional president, Yüan started to make a travesty of the republic. In this first cabinet, the four substantive ministries—Foreign Affairs, Internal Affairs, War, and Navy—all went to his henchmen, while the four lesser ministries—Education, Justice, Agriculture, and Forestry—were allocated to the T'ung-meng hui members. Huang Hsing, the choice of the revolutionaries for the Ministry of War, was merely made the resident-general of Nanking, and since Yüan refused to pay his 50,000 troops, Huang was soon obliged to disband them. The premier, T'ang Shao-i, one of the boys who went to the United States to study in 1872, genuinely desired to lead the nation toward the rule of law, in apparent opposition to Yüan's secret wishes. To humiliate him, Yüan ordered that the military governor-designate of Chihli[31] be sent to Nanking to help disband the troops, without the premier's countersignature as required by the provisional constitution. T'ang resigned in protest on June 16, 1912, as did the four T'ung-meng hui cabinet ministers.

The next premier was an ineffective diplomat, Lu Cheng-hsiang, a former minister to Russia. His lack of policy and sense of direction led to his impeachment by the parliament, and after July 27 Lu stopped going to the office on the pretext of illness. Yüan's confidant Chao Ping-chün, the minister of internal affairs, served as acting premier and later became premier on September 24. Under him, the cabinet was nothing but a puppet of the president. Within five months, Yüan had succeeded in reducing the "responsible cabinet" to a shambles.

However, to the southern revolutionary leaders Yüan displayed a great outward deference, cordially inviting Sun and Huang to visit him. They accepted the invitation but did not go together, for fear of being trapped simultaneously. Sun went first and during his 26-day stay in Peking was warmly welcomed by Yüan, who thirteen times listened attentively to his views on land reform, the single tax theory, the importance of transferring the capital from Peking to the interior, and the need for constructing 200,000 miles of railway. On September 9, Yüan appointed Sun director of railways with the full power to draw up a plan for a national railway system. Sun went away with the belief that Yüan was a man of ability and sincerity, "indispensable to the presidency in the next ten years."[32] Then came Huang Hsing, "the Napoleon of the Chinese Revolution," who was given the same cordial treatment and the same airing of his views on a variety of subjects, including his exposition of the need for industrial development and the usefulness of an efficient parliamentary system. Huang was appointed director-general of the Canton-Hankow and Szechwan railways. Having pacified the two revolutionary leaders, Yüan became bolder than ever in his search for dictatorial powers.

THE SECOND REVOLUTION. According to the provisional constitution, a parliament was to be elected within six months of the formation of the gov-

31. General Wang Chih-hsiang.
32. Chün-tu Hsüeh, 141.

ernment. Election laws and regulations on the organization of the parliament were promulgated by the provisional government in August 1912, including the adoption of a bicameral system. By the time of the election in December, the T'ung-meng hui had absorbed four splinter parties to form the Nationalist Party (*Kuomintang*) under the effective guidance of Sung Chiao-jen. Sung had studied parliamentary theories in Japan, had won the support of Huang Hsing, and was respected by prominent constitutionalists outside the party. Though not opposed to Yüan's election as president, he strongly advocated party government and a responsible cabinet to guide the country into constitutionalism and to check the abuse of the president.

Against the Nationalist Party were a number of smaller parties, such as the Unification Party (*T'ung-i tang*), the Republican Party (*Kung-ho tang*), and the Democratic Party (*Min-chu tang*), the last under the chairmanship of Liang Ch'i-ch'ao. The elections gave the Nationalists a landslide victory, taking 269 seats out of a total 596 in the Lower House, and 123 out of 274 in the Upper House. The Nationalist Party commanded more votes than the other three parties combined, which now merged into one Progressive Party (*Chin-pu tang*), in general support of the Yüan government.

The Nationalist victory was largely the work of Sung Chiao-jen. His organizing ability and frequent public advocacy of using the responsible cabinet and the loyal opposition systems to check the excesses of the president irritated Yüan greatly. Failing to win him over by bribery Yüan decided to eliminate him through assassination, and Chao Ping-chün, fearful of losing the premiership to Sung, joined in the plotting. On March 20, 1913, just as he was leaving the Shanghai railway station to take up his new assignment as the Nationalist representative in Peking, Sung was shot; he died two days later, at the age of thirty one. The captured documents and the subsequent investigations implicated Premier Chao and possibly President Yüan. However, after a hearing before the Shanghai Mixed Court, the assassin died suddenly in prison. Premier Chao refused to be subpoenaed to the court on the pretext of illness. He was later transferred to the military governorship of Chihli, where he mysteriously died by poisoning on February 17, 1914. Others involved in the case were either killed or poisoned, and the trial dragged on inconclusively, without ever reaching a clear verdict. It was nevertheless generally assumed that Yüan was behind the Sung assassination.

To bolster his position against the Nationalists, in April 1913 Yüan negotiated a so-called "reorganization loan" of £25 million from the Five-

Power Banking Consortium.[33] Sun and Huang Hsing urged the parliament to reject this "illegal" loan, whereupon Yüan's acting premier, Tuan Ch'i-jui, surrounded the parliament building with troops and declared presumptuously: "It being a *fait accompli,* there is no need for further discussion!" When the Nationalist members of the parliament impeached the government, an irreparable schism developed between Yüan and the revolutionaries. In a lightning manner, Yüan dismissed the Nationalist military governors in Kiangsi, Kwangtung, and Anhwei, and his army readied for an attack on the south.

On July 12, 1913, the military governor of Kiangsi[34] declared independence and in less than a month six other provinces followed suit,[35] starting what is known as the "Second Revolution." Yüan had little trouble crushing these poorly equipped southern armies. Within a couple of months the fighting was over; Yüan's generals took over control of the Yangtze area as provincial warlords.

YÜAN'S MONARCHICAL DREAM. The easy suppression of the Second Revolution elated Yüan, whose personal ambitions now knew no limits. No longer satisfied with the title of provisional president, he yearned for it to be changed to president with a lifelong tenure, preparatory to his ultimate goal of emperorship. In his dream for glory, Yüan had completely swept aside his 1912 inaugural pledge that he would uphold the republic against any reappearance of the monarchy.

The first step in Yüan's scheme was to prompt the parliament to issue the presidential election law on October 5, 1913, before the completion of the constitution. A day later, the two houses of the parliament held the presidential election, amidst the hue and cry of the so-called "citizen corps"—Yüan's disguised soldiers, police, and plainclothesmen—who besieged the building, shouting: "If you do not elect the president we want, do not expect to leave." In spite of the threat, Yüan failed to win the necessary votes on the first two ballots,[36] and it was only on the third that he was elected by a plurality vote. On October 10, 1913, Yüan was formally inaugurated as president, and the provisional government became the regular government.

Within three weeks, on October 31, the parliament promulgated the

33. Britain, France, Germany, Russia, and Japan.
34. Li Lieh-chün.
35. Kiangsu, Anhwei, Kwangtung, Fukien, Hunan, and Szechwan.
36. Receiving 471 votes on the first ballot and 497 on the second, out of a total 759 parliamentarians present.

T'ien-t'an Constitution, which adopted the cabinet rather than the presidential system, to check Yüan's powers. Incensed, Yüan asked his generals to attack it as a document incompatible with the national conditions and as a Nationalist device to dominate the parliament. When the parliament stood firm, Yüan simply dissolved the Nationalist Party on November 4 and revoked the credentials of 358 of its parliamentarians (eighty more later) on the pretext of their involvement in the Second Revolution. As 1914 opened, the parliament could not meet for lack of a legal quorum. Having brushed aside the constitution, the parliament, and the opposition party, Yüan achieved a virtual dictatorship.

Mindful of the importance of legality, Yüan called a national conference on March 18, 1914, to revise the 1912 provisional constitution. Each of the twenty-two provinces contributed two delegates, while four each came from the capital and the national chamber of commerce, and eight from Mongolia, Tibet, and Chinghai, making a total of sixty. The upshot of the conference was the shift from the cabinet to the presidential system and the authorization of the president and the parliament to prepare a new constitution. The new document, known as the Constitutional Compact, was promulgated on May 1, 1914, and extended the presidential term to ten years, renewable by re-election without limit. Moreover, the president had the right to nominate his own successor. With this constitution Yüan was assured of the lifelong tenure as well as the right to pass it on to his offspring. For all intents and purposes he had become an emperor, without the title. Yet he was still unsatisfied. He wanted to be a *de facto* as well as a *de jure* monarch. His eldest son, Yüan K'o-ting, anxious to become the crown prince and future sovereign, did his best to fan his father's vanity and desire for glory. By 1915 Yüan was fully prepared to betray the republic, much as Napoleon III did France.

To forestall foreign opposition Yüan agreed to accept the infamous Twenty-one Demands from Japan,[37] and signed agreements with Russia and Britain recognizing their special interests and positions in Outer Mongolia and Tibet respectively. He was further heartened by an intriguing, if noncommittal, statement of the Japanese premier, Ōkuma, to the effect that should China become a monarchy her political system would be identical with Japan's; that since Yüan was already in full control of China's

---

37. In five groups: (1) recognition of Japan's position in Shantung; (2) special position for Japan in Manchuria and Inner Mongolia; (3) joint operation of China's iron and steel industries; (4) nonalienation of coastal areas to any third power; and (5) control by Japan of China's several important domestic administrations. For details, see next chapter.

political power a change to the monarchy would bring the situation more in accord with reality. Yüan took it to mean a Japanese endorsement of his monarchical dreams.

Yüan's American adviser on constitutional matters, Dr. Frank J. Goodnow, who was the president of the Johns Hopkins University, published an article in which he stated that Americans had long doubted the fitness of a republic for China, where the tradition of autocracy would make constitutional monarchy a far more suitable institution, if it met no opposition. Yüan's Japanese adviser also stressed constitutional monarchy as the source of national strength, as in Japan and Britain. With these expert endorsements, the hush-hush monarchical movement broke out into the open. Yang Tu, chief organizer of the movement, publicly advocated national salvation through constitutional monarchy, and on August 21, 1915, the Peace-Planning Society (*Ch'ou-an hui*) was organized to draft Yüan for emperor. The famous translator of Western thought, Yen Fu, who had doubts about China's readiness for democracy, was listed, against his will, as one of the six directors. The movement quickly swept into full swing, although Yüan himself remained conspicuously aloof, denying continuously any imperial aspirations.

Nevertheless, the monarchical movement grew more pronounced every day. Numerous "petitions" reached the government favoring change in the national polity. The National People's Representative Assembly, which was called to deliberate the issue, approved monarchy by an overwhelming majority on November 20, 1915. On December 11, representatives of the provinces petitioned in the name of the people that Yüan consent to become the emperor of China. After a polite declination on the grounds that he lacked virtue and merit, Yüan "reluctantly" acceded to the second petition on December 12. A day later he decreed that the next year, 1916, would mark the start of his new reign, to be called ironically the Glorious Constitution (*Hung-hsien*).

Like many dictators before and after him, Yüan was overtaken by megalomania, too confident to know when to stop. He did not seem to see that in spite of all the uncertainties in the early republican period, one thing was definite: the imperial system could never return. His betrayal of the republic and his shameless drive for the emperorship went beyond the point of tolerance of his countrymen—not only his critics but even his own followers.

Sun had, in the meantime, fled to Japan after the failure of the Second Revolution. Convinced that the internal disunity was a major cause for his defeat, he reorganized the Nationalist Party into a tighter structure under the name of Chinese Revolutionary Party (*Chung-kuo ko-ming tang)* on July 8, 1914. Members were required to owe him personal allegiance and to fingerprint their written pledge. Sun retained strict control of the central and branch organizations as well as the power of appointments at all levels—the embryo of a principle later known as "democratic centralism." Now appointed generalissimo of a Chinese Revolutionary army, Sun set out to fight Yüan's illegal destruction of the parliament, the provisional constitution, and his abject betrayal of the republic.

In Yunnan, a National Protection army came into being to fight the monarchist movement, under a group of revolutionaries including the former military governor Ts'ai Ao. He and his former teacher, Liang Ch'ich'ao, vowed to fight Yüan, one with guns and the other with the pen, in order to save the republic and to preserve the honor and character of China's 400 million people. The Yunnan revolutionaries dedicated the National Protection army *(Hu-kuo chun)*[38] to the "elimination of the country's thief, defense of the republic, upholding democracy, and developing the spirit of popular sovereignty." On December 23, an ultimatum was delivered to Yüan, allowing him two days to cancel his monarchist movement. When Yüan refused, Yunnan declared its independence on December 25, and the National Protection army, some 10,000 strong, set out in a three-direction campaign. On December 27 Kweichow declared independence. Pressed by these developments, Yüan delayed his enthronement scheduled for January 1, 1916. Two of his leading generals, Tuan Ch'i-jui and Feng Kuo-chang, each declined an appointment as commander of the expedition against the National Protection army on the pretext of illness. On March 15, Kwangsi declared independence, while a separate antimonarchist army rose in Shantung. The Japanese government served notice that in view of Peking's inability to keep domestic peace and to win support of the powers, it had forfeited its right to represent China, and that henceforth Japan would treat the north and the south as equal belligerent parties.

Facing these discouraging domestic and foreign developments, Yüan had no choice but to forsake his monarchical dream and the reign of "Glorious Constitution" on March 22, 1916. Yet he still hoped to hang on to his presidency by reviving the cabinet system to appease the revolutionaries. However, events moved too fast for him: Kwangtung declared its independence on April 6 and Chekiang on April 12. By May 5, the various revolutionary groups had unified into one Military Affairs Council, which refused to recognize Yüan as president, as did prominent citizens of nineteen provinces. Even K'ang Yu-wei twice urged him to retire

38. Coincidentally, the name of the monastery where they met happened to be *Hu-kuo ssu.*

and take a trip abroad.[39] By then Yüan's cause was all but lost; his followers one after another began to disown him. When he asked Feng Kuo-
chang to mobilize generals and military governors to support him for the
presidency, Feng simply asked him to resign. On May 9 Shensi declared
independence, followed by Szechwan on May 22 and Hunan on May
27.[40] Deserted by his henchmen and overcome with shame, anxiety, and
grief, Yüan suddenly died of uremia on June 6, 1916, at the age of fifty-
six. The tragicomic drama of monarchism came to an abrupt end.

Commenting on the life of Yüan, Liang Ch'i-ch'ao remarked that he
(Yüan) made no distinction between men and animals, assuming that all
could be bought with gold and intimidated by the sword. His mockery of
the constitution, his illegal manipulation of the parliament, his methods
of bribery, coercion, murder, and enslavement were an irreparable affront
to public character and morale, and laid the groundwork of lawlessness
and disorder in the decade that followed.

Period of Warlordism, 1916-27

The disappearance of a strong power-holder generated centrifugal forces,
plunging the country into a period of chaos and disorder. The warlords
fought against each other for power and self-aggrandizement without any
sense, logic, or reason, rendering this period the darkest in republican
history.

On June 7, 1916, Vice-President Li Yüan-hung took over the presidency. A question of "legality" immediately arose as to whether he had
succeeded to the office according to the 1912 constitution or *acted* for the
deceased president in accordance with Yüan's 1914 constitution. In short,
which of the two constitutions was valid? The revolutionaries in the south
insisted on the former, arguing that the very purpose of the antimonarchist
movement and the civil war was to protect the legality of the 1912 constitution, whereas Premier Tuan Ch'i-jui in Peking favored continuation
of the 1914 constitution which had been in effect for two years. The conflict was resolved when the naval commander at Shanghai[41] declared inde-

---

39. Pai Chiao, *Yüan Shih-k'ai yü Chung-hua min-kuo* (Yüan Shih-k'ai and the Chinese
    Republic), (Shanghai, 1936), 341-42, 350-71.
40. It was said that Yüan fainted upon reading the telegram from his confidant governor
    of Szechwan, Ch'en Huan, which said: "From today, Szechwan severs all relations
    with Yüan Shih-k'ai." Yüan later sighed: "Now, even Ch'en Huan is like this. What
    is there for me to say! Please reply to him and tell him that I will retire." See Jerome
    Ch'en, *Yüan shih-kai, 1859-1916* (Stanford, 1961), 232.
41. Li Ting-hsin.

pendence of the Peking regime on June 25, throwing his support to the south. Feng Kuo-chang, who had built up a power base at Shanghai and was fearful of losing it, put pressure on Peking to restore the 1912 constitution. On August 1, President Li complied with the request, re-established the old parliament that had been dissolved illegally by Yüan on January 10, 1914, and renamed Tuan premier according to the 1912 constitution. The revolutionaries agreed to abolish their Military Affairs Council in the interest of national unification.

RESTORATION OF THE MANCHU EMPEROR, 1917. The question of whether China should enter the war against Germany now loomed large. Premier Tuan, under American prodding, declared war on Germany on May 14, 1917, without the approval of the president and the parliament. To disarm parliamentary opposition, he employed Yüan's tactics of instigating some 3,000 "citizens" from business, political, and military circles to surround the parliament and demand passage of the war declaration. Tuan's generals and military governors bluntly demanded that President Li dissolve the parliament, while the latter in retaliation urged Li to relieve Tuan of the premiership. On May 23, Li took the bold step of dismissing Tuan, only to find a rash of declarations of independence by his henchmen in the provinces—Shensi, Shansi, Chekiang, Shantung, Chihli, Fukien, etc. They organized a general staff at Tientsin and decided to march on Peking. In desperation President Li sought the good offices of Chang Hsün, the military governor of Anhwei. Chang came to the capital with 5,000 soldiers on June 7, 1917, but he demanded the dissolution of the parliament as a prerequisite to mediating. Li had no choice but to comply on June 12, in full knowledge of its illegality according to the 1912 constitution.

Once established in Peking, Chang, with the support of K'ang Yu-wei and secret concurrence of the Peiyang leaders Tuan and Feng, restored the last Manchu emperor, P'u-i, to the throne on July 1.[42] Ch'ing institutions were revived and ranks and appointments were awarded. Chang Hsüan was made the chief minister of the cabinet and concurrently governor-general of Chihli, a post taken from the warlord Ts'ao K'un, while Tuan was left out of the distribution of offices. Feeling deceived, Tuan and Ts'ao gathered their Peiyang forces against the 20,000 long-queued soldiers of Chang, driving them out of Peking on July 12 and quickly ending the restoration movement.

42. For an intimate and interesting account of the life of P'u-i after his abdication in 1912, see his autobiography which appears in English under the title, *The Last Manchu*, tr. by K. Y. P. Tsai and ed. by Paul Kramer (New York, 1967), chs. 1-8.

THE CIVIL WAR AMONG THE WARLORDS. Once again Tuan was the premier, supported by the so-called Research Clique (*Yen-chiu hsi*) under Liang Ch'i-ch'ao, who was now the finance minister. The Research Clique argued that since the restoration movement had officially put an end to the republic, it behooved the country to construct a new republic under the leadership of Tuan, and the first step in that direction was to call a new provisional parliament. When Tuan did so on November 10, rather than reconvening the old parliament dissolved by President Li on June 12, the revolutionaries in the south accused him of violating the 1912 constitution. Sun Yat-sen once again established a military government at Canton to launch a Constitution Protection Movement (*Hu-fa yün-tung*).

To crush domestic opposition, Tuan negotiated foreign loans under the pretext of entering the world war. Using methods reminiscent of Yüan's, he manipulated the provisional parliament to revise the election and organization laws of the 1912 constitution, and set about organizing an An-Fu Club[43] to rally the support of his military and civilian followers. In the re-elected parliament which convened on August 12, 1918, the An-Fu Clique controlled more than 330 votes, and the Research Clique about 20. This "An-Fu Clique Parliament" easily passed the resolution to declare war on Germany on August 14 as Tuan wanted, enabling him to negotiate the so-called "Nishihara loans"[44] of some 145 million yen under the pretext of sustaining China's war effort.

Thus replenished, Tuan set out to destroy the southern military government. Troops were sent to Hunan to exert pressure on the revolutionaries in Canton, and to Szechwan to check any possible revolt by Yunnan. In doing so Tuan precipitated another civil war. However, President Feng Kuo-chang, successor to Li Yüan-hung, favored a peaceful solution to the domestic squabble. His clash with Tuan, a former colleague under Yüan, split the Peiyang Clique in two: the group under Tuan of Anhwei became known as the Anhwei Clique, and the one under Feng of Chihli became known as the Chihli Clique. Feng's followers sabotaged Tuan's campaign against the Constitution Protection army, causing a failure of Tuan's military policy and his resignation on November 22. What followed was a period of mad fighting between the two cliques. The Chihli group ultimately won out with the support of yet another clique from Manchuria—the Fengtien army under the leadership of a former bandit, Chang Tso-lin. In April 1922, fighting broke out between the two groups themselves, resulting again in the victory of the Chihli Clique. However,

43. Named after the An-Fu Street in Peking.
44. Named after the Japanese negotiator, Nishihara. The Japanese yen was worth about one-half of the American dollar at the time.

Chang Tso-lin was able to retain control of Manchuria, independent of the Peking regime.

The victorious Chihli Clique offered the presidency to Li Yüan-hung in the hopes of achieving national unification through a peaceful settlement with the Canton government. It encountered opposition from a powerful wing within the clique, and by the middle of 1922 a split took place: (1) the Lo-yang faction under Wu P'ei-fu favored a military conquest of China and support of President Li; and (2) the Tientsin-Paoting faction which opposed Wu favored Ts'ao K'un for president. In the end, President Li was driven out of office in a most demeaning manner, and Ts'ao K'un had himself elected president in October 1923 by bribing some 500 members of the parliament with payoffs of an alleged 5,000 Chinese dollars apiece. Public morale hit rock bottom, and the people were disgusted with politics in the north. The only hope lay with the revolutionary government at Canton.

Yet Sun had enough troubles of his own in the south. His Constitution Protection Movement had made little progress, for ever since the establishment of the military government at Canton on August 25, 1917, he had been handicapped by not having direct control of the armed forces, despite his title of generalissimo. The real power of command lay with the southwestern provincial leaders such as Lu Jung-t'ing of Kwangtung and Kwangsi. With ambitions of his own, Lu had forced Sun out of the military government in May 1918. Fleeing to Shanghai in deep disappointment and frustration, Sun led a life of resignation, engaging mostly in writing his "Outline of National Reconstruction" (*Chien-kuo fang lüeh*) and planning the reorganization of the party. On October 10, 1919, he tightened the Chinese Revolutionary Party and renamed it the Chinese Nationalist Party (*Chung-kuo kuo-min-tang*). To chastise the rebels at Canton rather than to fight Tuan in the north, Sun directed his forces southward. Through a series of maneuverings, he was able to recover Canton and to revive the military regime; the formal establishment of a republican government followed on April 2, 1921, with Sun as president, in rival existence with the warlord government in Peking.

On February 3, 1922, Sun set out northward to continue his Constitution Protection campaign, only to be turned back by an unexpected mutiny in Canton, led by a former supporter, Ch'en Chiung-ming. Caught in his presidential headquarters under heavy bombardment, Sun narrowly escaped to a loyal warship, and later with British and Russian help, he reached Shanghai. Thus, his Constitution Protection campaign really never got off the ground.

Following Ts'ao K'un's disgraceful election to the presidency in October 1923, the Fengtien forces advanced from Manchuria toward Peking, precipitating a second Chihli-Fengtien war. Most unexpectedly, when the commander-in-chief[45] of the 170,000-man Chihli army went to the front, his Third Army commander Feng Yü-hsiang mutinied and occupied Peking on October 23, 1924, bringing about a total collapse of the Chihli forces. Supported by his own "National People's army" (*Kuo-min chün*), Feng reorganized the cabinet and forced President Ts'ao K'un out of office on November 2, 1924.

Now, in the interest of national unification, the "National People's army," the Fengtien Clique, and the Anhwei Clique jointly asked Tuan Ch'i-jui to be the executive of a provisional government, and invited Sun Yat-sen to Peking to discuss the problem of peace and unification. Though his health was failing, Sun made the trip and arrived in Peking on December 31, 1924. He was heartened by the warm welcome of more than 100,000 people, though annoyed with Tuan's apparent insincerity. His condition turned worse after January 20 and death overtook him on March 12, 1925. At the last minute, he was still murmuring "peace, struggle . . . save China." A will, signed by him a day earlier, urged his comrades to carry on his unfinished work. So ended the career of the father of the Chinese Revolution, who had devoted forty years of his life to the betterment of his country and his people.

Sun died a disappointed man. The revolution and the republic had not brought the anticipated peace and order: if anything, the republican period saw more misery and lawlessness than before. It resembled the traditional disorder and chaos that always followed the fall of a dynasty. Yet Sun had laid the foundation for progress, from which his disciples could carry on. In 1926, a young general, Chiang Kai-shek, resumed the unfinished "Northern Expedition" against the warlords and succeeded to a large extent in his mission. In 1928 a Nationalist government was established in Nanking, and the long eluded objective of unification was finally achieved, even if only superficially.

45. Wu P'ei-fu.

Further Reading

Anschel, Eugene, *Homer Lea, Sun Yat-sen and the Chinese Revolution* (New York, 1984).

Barlow, Jeffrey G., *Sun Yat-sen and the French, 1900-1908* (Berkeley, 1979).

Belov, E. A., *Uchanskoe vosstanie v Kitae (1911 g.)* (The Wuchang Revolt in China, 1911), (Moscow, 1971).

Bergère, Marie-Claire, *La bourgeoisie Chinoise et la révolution de 1911* (The Hague, 1969).

Cantlie, Sir James, and C. Sheridan Jones, *Sun Yat-sen and the Awakening of China* (New York, 1912).

Ch'ai, Te-keng 柴德賡 et al. (eds.), *Hsin-hai ko-ming* 辛亥革命(The revolution of 1911), (Shanghai, 1957), 8 vols.

Chang, Ch'i-yün 張其昀, *Chung-hua min-kuo ch'uang-li shih* 中華民國創立史(A history of the founding of the Chinese Republic), (Taipei, 1953).

Chang, P'eng-yüan 張朋園, *Liang Ch'i-ch'ao yü Ch'ing-chi ko-ming* 梁啓超與清季革命 (Liang Ch'i-ch'ao and the late Ch'ing revolution), (Taipei, 1964).

————, 張朋園, *Liang Ch'i-ch'ao yü min-kuo cheng-chih* 梁啓超與民國政治(Liang Ch'i-ch'ao and the politics of the Republic of China), (Taipei, 1978).

Chang, Yü-fa 張玉法. *Ch'ing-chi ti li-hsien t'uan-t'i* 清季的立憲團體(Constitutionalist groups of the late Ch'ing period), (Taipei, 1971).

————, *Ch'ing-chi ti ko-ming t'uan-t'i* 清季的革命團體 (Revolutionary groups of the late Ch'ing period), (Taipei, 1975).

Chen, Stephen, and Robert Payne, *Sun Yat-sen* (New York, 1946).

Ch'en, Jerome, *Yüan Shih-k'ai, 1859-1916* (Stanford, 1961).

Ch'i, Hsi-sheng, *Warlord Politics in China, 1916-1928* (Stanford, 1976).

Chiang, Kai-shek, *Chiang Tsung-t'ung mi-lu* 蔣總統祕錄 (Secret memoirs of President Chiang), tr. from the Japanese by *Central Daily News*, Vol. II (Taipei, 1975).

Des Forges, Roger V., *Hsi-liang and the Chinese National Revolution* (New Haven, 1973).

Feng, Tzu-yu 馮自由, *Chung-hua min-kuo k'ai-kuo ch'ien ko-ming shih* 中華民國開國前革命史(A history of the revolution before the establishment of the Chinese Republic), (Chungking, 1944), 3 vols.

Esherick, Joseph, *Reform and Revolution in China: The 1911 Revolution in Hunan and Hubei* (Berkeley, 1976).

Eto, Shinkichi, and Harold Z. Shiffrin (eds.), *The 1911 Revolution in China* (Tokyo, 1984).

Fewsmith, Joseph, *Party, State, and Local Elites in Republican China: Merchant Organizations and Politics in Shanghai, 1890-1930* (Honolulu, 1984).

Friedman, Edward, *Backward toward Revolution: The Chinese Revolutionary Party, 1914-1916* (Berkeley, 1974).

Fung, Edmund S. K., *The Military Dimension of the Chinese Revolution: The New Army and Its Role in the Revolution of 1911* (Vancouver, 1980).

Gasster, Michael, *Chinese Intellectuals and the Revolution of 1911* (Seattle, 1969).

————, "Reform and Revolution in China's Political Modernization" in Mary C. Wright (ed.), *China in Revolution: The First Phase, 1900-1913* (New Haven, 1968), 67-96.

————, "The Republican Revolutionary Movement" in John K. Fairbank and Kwang-ching Liu (eds.), *The Cambridge History of China* (Cambridge, Eng., 1980), Vol. 11, 463-534.

Gillin, Donald G., *Warlord Yen Hsi-shan in Shansi Province, 1911-1949* (Princeton, 1967).

Ho, Hon-wai, *Ching-Han tieh-lu ch'u-ch'i shih-lueh* (The early history of Peking-Hankow Railway), (Hong Kong, 1979).

Hsieh, Winston, *Chinese Historiography on the Revolution of 1911* (Stanford, 1975).

Hsü, Leonard S., *Sun Yat-sen: His Political and Social Ideals* (Los Angeles, 1933).

Hwang, Yen Ching, *The Overseas Chinese and the 1911 Revolution: With Special Reference to Singapore and Malaya* (New York, 1977).

Ikei, Masaru, "Japan's Response to the Chinese Revolution of 1911," *The Journal of Asian Studies*, XXV:2:213-27 (Feb. 1966).

Jansen, Marius, "Japan and the Chinese Revolution of 1911" in John K. Fairbank and Kwang-ching Liu (eds.), *The Cambridge History of China* (Cambridge, Eng., 1980), Vol. 11, 339-374.

———, *The Japanese and Sun Yat-sen* (Cambridge, Mass., 1954).

Kuo, Pin-chia 郭斌佳, "Min-kuo erh-tz'u ko-ming shih"民國二次革命史(The second republican revolution), *Kuo-li Wu-han ta-hsüeh wen-che chi-k'an* 國立武漢大學文哲季刊(Quarterly Journal of Literature and Philosophy), National Wuhan University, 4:3 (1935).

Leng, Shao-chuan, and Norman D. Palmer, *Sun Yat-sen and Communism* (New York, 1960).

Li, Nai-han 黎乃涵, *Hsin-hai ko-ming yu Yüan Shih-k'ai* 辛亥革命與袁世凱(The revolution of 1911 and Yüan Shih-k'ai), (Shanghai, 1949).

Li, Tien-yi, *Woodrow Wilson's China Policy, 1913-1917* (Lawrence, Kansas, 1952).

Li, Yü-shu 李毓澍, *Chung-Jih Erh-shih-i-t'iao chiao-she*, Vol. I, 中日廿一條交涉 (上) (The Sino-Japanese negotiations on the Twenty-one Demands), (Taipei, 1966).

Liang, Chin-tung, *The Chinese Revolution of 1911* (New York, 1962).

Liew, K. S., *Struggle for Democracy: Sung Chiao-jen and the 1911 Chinese Revolution* (Berkeley, 1971).

Linebarger, Paul, *Sun Yat-sen and the Chinese Republic* (New York, 1925).

———, *The Gospel of Chung Shan* (Paris, 1932).

Ma, L. Eve Armentrout, *Revolutionaries, Monarchists, and Chinatowns: Chinese Politics in the Americas and the 1911 Revolution* (Honolulu, 1990).

MacKinnon, Stephen R., *Power and Politics in Late Imperial China: Yuan Shikai in Beijing and Tianjian, 1901-1908* (Berkeley, 1980).

Nathan, Andrew J., *Peking Politics, 1918-1923: Factionalism and the Failure of Constitutionalism* (Berkeley, 1976).

Pai, Chiao 白蕉, *Yüan Shih-k'ai yü Chung-hua min-kuo* 袁世凱與中華民國(Yüan Shih-k'ai and the Chinese Republic), (Shanghai, 1936).

Powell, Ralph L., *The Rise of Chinese Military Power, 1895-1912* (Princeton, 1955).

Power, Brian, *The Puppet Emperor: The Life of Pu Yi, The Last Emperor of China* (New York, 1986).

Price, Don C., *Russia and the Roots of the Chinese Revolution, 1896-1911* (Cambridge, Mass., 1974).

Price, Frank W. (tr.), *San Min Chu I* (Three People's Principles), (Shanghai, 1927).

Pugach, Noel, "Embarrassed Monarchist: Frank J. Goodnow and Constitutional Development in China, 1913-1915," *Pacific Historical Review*, XLII:4:499-517 (Nov. 1973).

P'u-i, Henry, *The Last Manchu: The Autobiography of Henry Pu Yi, Last Emperor of China*, tr. by Kuo Ying Paul Tsai, and ed. with intro. by Paul Kramer (New York, 1967).

Rankin, Mary B., *Early Chinese Revolutionaries: Radical Intellectuals in Shanghai and Chekiang, 1902-1911* (Cambridge, Mass., 1971).

Reed, James. *The Missionary Mind and American East Asia Policy, 1911-1915* (Cambridge, Mass., 1983).

Rhoads, Edward, *China's Republican Revolution: The Case of Kwangtung, 1895-1913* (Cambridge, Mass., 1975).

Scalapino, Robert A., "Prelude to Marxism: The Chinese Student Movement in Japan, 1900-1910" in Albert Feuerwerker, Rhoads Murphey, and Mary C. Wright (eds.), *Approaches to Modern Chinese History* (Berkeley, 1967), 190-215.

Schiffrin, Harold, *Sun Yat-sen: Reluctant Revolutionary* (Boston, 1980).

———, *Sun Yat-sen and the Origins of the Chinese Revolution* (Berkeley, 1968).

Schoppa, R. Keith, *Chinese Elites and Political Change: Zhejiang Province in the Early Twentieth Century* (Cambridge, Mass., 1981).

Sharman, Lyon, *Sun Yat-sen, His Life and Its Meaning* (New York, 1934).

Shen Tsu-hsien, et al. (eds.), *Jung-an ti-tzu chi* 容庵弟子記 (An account of Yüan Shih-k'ai by his disciples), reprinted (Taipei, 1962).

Sheridan, James E., *Chinese Warlord, the Career of Feng Yü-hsiang* (Stanford, 1966).

Sutton, Donald S., *Provincial Militarism and the Chinese Republic: The Yunnan Army, 1905-25* (Ann Arbor, 1980).

Wilbur, C. Martin, *Sun Yat-sen: Frustrated Patriot* (New York, 1976).

Wong, J. Y., *The Creation of an Historic Image: Sun Yatsen in London, 1896-1897* (Hong Kong, 1986).

——— (ed.), *Sun Yatsen: His International Ideas and International Connections* (Sydney, 1986).

Wou, Odoric Y. K., *Militarism in Modern China: The Career of Wu P'ei-fu* (Canberra, 1978).

Wright, Mary (ed.), *China in Revolution: The First Phase, 1900-1913* (New Haven, 1968).

Young, Ernest P., *The Presidency of Yuan Shih-k'ai: Liberalism and Dictatorship in Early Republican China* (Ann Arbor, 1977).

Yu, George T., *Party Politics in Republican China: The Kuomintang, 1912-1924* (Berkeley, 1966).

V

Ideological Awakening
and the War of Resistance
1917-45

General Chiang Kai-shek and Yen Hsi-shan.

Feng Yü-hsiang.

Hu Shih as head of the Academia
Sinica, ca. 1960.

Lu Hsün, the famous writer and
social critic at the age of 50
at Shanghai.

Meeting place of the First National Congress of the Communist
party of China in Shanghai, July 1921.

Ch'en Tu-hsiu, co-founder of the
Chinese Communist Party.

Li Ta-chao, co-founder of the
Chinese Communist Party.

Mao Tse-tung as a young revolutionary.

Yenan: Chinese Communist capital, 1936–47.

Mao's cave residence in Yenan.

21

The Intellectual Revolution, 1917-23

The founding of the republic had not brought peace, order, and unity. Instead, the early republican years were characterized by moral degradation, monarchist movements, warlordism, and intensified foreign imperialism. Obviously, political face lifting through the adoption of republican institutions was insufficient to regenerate the nation; something far more fundamental was needed to awaken the country and the people.

The new intellectuals, Western-trained or Western-influenced, advocated a radical change in the philosophical foundations of national life. They called for a critical re-evaluation of China's cultural heritage in the light of modern Western standards, a willingness to part with those elements that had made China weak, and a determination to accept Western science, democracy, and culture as the foundation of a new order. At the same time, they launched a campaign to introduce a new literature based on the vernacular language instead of the classical. This intellectual outburst dealt a shattering blow to Confucianism—including traditional ethics, customs, human relations, and social conventions—and ushered in a new iconoclastic attitude toward China's past. In terms of depth and scope, the intellectual transformation that resulted surpassed that of the 1895-1911 period (Chapter 18). Indeed, in the opinion of some, nowhere in Chinese history since the Spring and Autumn and the Warring States periods (722-221 B.C.) had social and intellectual changes been so drastic and fundamental.[1]

1. Kuo Chan-po, *Chin-wu-shih-nien Chung-kuo ssu-hsiang shih* (A history of Chinese thought during the last fifty years), reprinted (Hong Kong, 1965), 1.

493

This intellectual revolution, taking place somewhere between 1917 and 1923, hailed a New Cultural Movement which has sometimes been described, perhaps exaggeratedly, as a "Chinese Renaissance." A high point in this turbulent period was the gigantic student demonstration in Peking on May 4, 1919, which quickly evoked nationwide response. Hence this period is also commonly known as that of the May Fourth Movement.

The Background

This stirring age of intellectual ferment could not have come to pass without certain significant developments abroad and at home. Externally, sentiments of nationalism and democracy were particularly strong during World War I, and the Wilsonian ideals of national self-determination and abolition of secret diplomacy appealed to Chinese intellectuals. Moreover, rolling events of epochal significance were occurring in different parts of the world: the Bolshevik Revolution in Russia in 1917; the socialist revolts in Finland, Germany, Austria, and Hungary; and the rice riots in Japan in 1918. In contrast, China was plagued by chaos and warlordism. Chinese intellectuals felt deeply committed to revive their strife-ridden and civil war-torn country.

These intellectuals approached the task with fiercely nationalistic and patriotic sentiments, stimulated partly by Japan's humiliating Twenty-one Demands of 1915.[2] Divided into five groups, the first four called for Japanese control of Shantung, Manchuria, Inner Mongolia, the southeast coast of China, and the Yangtze valley. The fifth, the most sinister of all, required employment of Japanese advisers in Chinese political, financial, military, and police administrations, as well as the purchase of at least 50 percent of China's munitions from Japan.

These terms inflamed the Chinese public. Yet under the pressure of a Japanese ultimatum on May 7, 1915, Yüan accepted the first four groups while putting a reservation on the fifth. Then, without the consent of the legislature, he concluded a treaty with Japan on May 25.

In protest, groups of Chinese students in Japan returned home, while merchants in China organized a widespread boycott of Japanese goods. The Twenty-one Demands had the unexpected effect of precipitating a fear of imminent extinction and a consequent outburst of nationalism.

Contributing to the rise of the new nationalism was the rapid emergence of a politically conscious merchant-enterpreneur class and a labor force which numbered between two and three million by 1919. Indeed,

2. Delivered by the Japanese minister, Hioki Eki, to President Yüan Shih-k'ai on January 18, 1915. See p. 479, footnote 37.

the World War I period had witnessed an unprecedented expansion of Chinese industry and commerce—especially in the fields of textiles, flour mills, silk, matches, cement, cigarettes, and modern banks and joint-stock corporations—as a result of favorable internal and external conditions. Domestically, the replacement of the imperial dynasty by a new republic in 1912 marked the inauguration of a new era. No longer did the government regard industrialists and merchants as suspect; and no longer did it prohibit the formation of private "cliques" and associations as under the Ch'ing. The scholar-turned-industrialist Chang Chien, as minister of agriculture and industry, promulgated a series of regulations to encourage and protect industrial and commercial development.

Externally, the World War I period witnessed a rapid decline of Western imperialism in China. The war had so adversely affected European industries and trade with Asia that it created a golden chance for China's native industries to develop unhindered. Between 1913 and 1918, British imports fell from 96 million taels to 49 million; French imports, from 5.2 million to 1.5; and German imports, from 28 million to 0. In reverse proportion, the Chinese foreign trade deficit was cut from 166 million customs taels in 1913 to 16 million in 1919, while silk export rose from 87,517 *tan* in 1914 to 131,506 *tan* in 1919.[3] Similarly, native industries and commerce grew by leaps and bounds: textile companies rose from 22 in 1911 to 54 in 1919, and 109 in 1921; flour mills from 67 in 1916, to 86 in 1918; modern banks from 7 in 1911, to 131 by 1923; steamships from 893 (total tonnage 141,024) in 1913, to 2,027 (236,622 tons) in 1918; coal production from 12.8 million tons in 1913, to 20.1 in 1919; and iron from 1 million tons in 1914, to 1.8 million in 1919.[4]

These new industries and enterprises gave rise to new merchant and labor classes, which, unlike the old-style apolitical tradesmen and inert peasants, were sensitive to China's predicament under imperialism. They were determined to defend their country's interests. Most of them lived in the cities, where they contributed to the expansion of the urban centers and their economy. Peking, Shanghai, Wuhan, Nanking, Tientsin, and Canton all became large metropolises which nourished the growth of a new intelligentsia. From 1907 to 1917 at least 10 million members of these classes had received some sort of modern education, and were imbued with a strong nationalist determination "to save their country" (*chiu-kuo*) from the twin scourge of foreign imperialism and domestic disorder.

3. Tan = one picul = 133⅓ lbs.
4. Chou Hsiu-luan, *Ti-i-tz'u shih-chieh ta-chan shih-ch'i Chung-kuo min-tsu kung-yeh ti fa-chan* (The development of Chinese national industries during World War I), (Shanghai, 1958), chs. 1-2.

The returned students—those who had studied abroad—were particularly eager to introduce reforms. From 1903 to 1919, 41.51 percent of these students studied in Japan, 33.85 percent in the United States, and 24.64 percent in Europe.[5] France, as the cradle of modern Western civilization, attracted a considerable number of Chinese work-study students (*ch'in-kung chien-hsüeh*) during World War I and a large labor force of some 200,000 by 1918-19. The latter group worked on roads, docks, factories, and munition dumps, and at least 28,000 of them were educated. The United States, which had a tradition of educating Chinese youths since 1872, drew some 1,200 by 1915. But Japan, because of geographic proximity and lower costs of living, attracted the largest number of Chinese students—13,000 by 1906.[6]

Among the most prominent returned students were Ch'en Tu-hsiu and Ts'ai Yüan-p'ei from France, Kuo Mo-jo and Lu Hsün (Chou Shu-jen) from Japan, and Hu Shih and Chiang Monlin from the United States. Ch'en, Ts'ai and Hu rapidly became the guiding spirit of the intellectual revolution.

Ch'en Tu-hsiu (1879-1942) of Anhwei had received a thorough training in Chinese classical studies in youth and had passed the first Ch'ing civil service examinations in 1896. In 1902 and 1906 he traveled to Japan, staying only for a short time. He went to France in 1907 and came strongly under its political and literary influence. Returning home in 1910 he participated in the republican revolution, though not a T'ung-meng hui member. Later, as a result of his involvement in the Second Revolution, he fled to Japan. In 1915 he returned home in protest to the Twenty-one Demands.

Ts'ai Yüan-p'ei (1876-1940) of Chekiang, after winning the second and third degrees (in 1889 and 1892, respectively) and a coveted membership in the Hanlin Academy, went to Germany to study at the University of Leipzig in 1907. Four years later he returned home in time to take part in the republican revolution and was appointed minister of education in Dr. Sun's government, a post from which he resigned after Yüan Shih-k'ai took over the presidency. In the summer of 1912 he again went to Germany where he stayed for about a year. His next three years were spent in France, where he took charge of the work-study program for Chinese students and laborers. In 1916, after declining the governorship of Chekiang,

5. Tse-tsung Chow, *The May Fourth Movement: Intellectual Revolution in Modern China* (Cambridge, Mass., 1960), 26, 31.
6. Estimates range from 8,000 to 13,000. See Robert A. Scalapino, "Prelude to Marxism: The Chinese Student Movement in Japan 1900-1910" in Feuerwerker, Murphey, and Wright (eds.), *Approaches to Modern Chinese History*, 192.

he returned home to become chancellor of the National University of Peking.

Hu Shih (1891-1962), a scion of the famous early Ch'ing scholar Hu Wei (1633-1714), also received a classical education in his youth. After graduating from the Chinese Public Institute in 1909, he won a government scholarship to study in the United States, earning the B.A. and Ph.D. in philosophy from Cornell and Columbia universities, respectively. Influenced by John Dewey and Thomas Huxley, he firmly believed in pragmatism, scientific methods of thought, and the evolutionary improvement of society. His seven-year sojourn in the United States thoroughly exposed him to American literary and social movements, for it was a time of liberation, characterized by a craze for new things: new humanism, new nationalism, new history, new art, new poetry, and new women. Influenced by Harriet Monroe's *Poetry: A Magazine of Verse*, which promoted verse-writing in plain language, Hu's own idea of substituting the vernacular language for the classical in literary writing[7] assumed greater importance in his mind. While still a student at Cornell in 1915, he and Y. R. Chao boldly started a movement to introduce the *pai-hua* (plain language) style of writing.

These new intellectuals were products of a transitional period—all thoroughly grounded in Chinese classical studies and yet well acquainted with Western civilization. Liberalism, socialism, pragmatism, science, and democracy had left their indelible mark. When they returned home—Ch'en in 1915, Ts'ai in 1916, and Hu in 1917—they functioned as leaven in transforming the literary and intellectual personality of China. Their call for a critical re-evaluation of the national heritage and the introduction of Western thought and ideologies sparked an intellectual revolution which dealt a shattering blow to traditionalism and ushered in the period of the New Cultural Movement.

The Unfolding of the New Cultural Movement

CH'EN TU-HSIU AND THE NEW YOUTH. Back from Japan in 1915, Ch'en Tu-hsiu founded a monthly periodical in Shanghai, the *Youth Magazine* (*Ch'ing-nien tsa-chih*), later renamed the *New Youth* (*Hsin ch'ing-nien*) or *La Jeunesse*. It was dedicated to arousing the youth of the country to destroy the stagnant old traditions and forge a new culture. In the first issue Ch'en called on the young generation to struggle against the old and rotten elements of society and to reform their thought and behavior in

7. First conceived during his high school days at the Chinese Public Institute from 1906 to 1909.

order to achieve a national awakening.[8] The youth were asked to choose the fresh, vital elements from all the civilizations of the world in order to create a new culture for China. In this monumental task, Ch'en suggested six guiding principles: (1) to be independent and not servile; (2) to be progressive and not conservative; (3) to be aggressive and not retrogressive; (4) to be cosmopolitan and not isolationist; (5) to be utilitarian and not impractical; and (6) to be scientific and not visionary.

Ch'en vehemently attacked conservatism and traditionalism as the roots of China's evils. Confucianism, in particular, fared badly in his writings. It was, he said, the product of an agrarian and feudal social order, totally incompatible with modern life in an industrial and capitalistic society. Confucianism must be rooted out because (1) it advocated "superfluous ceremonies and preached the morality of meek compliance," making the Chinese people weak and passive, unfit to struggle and compete in the modern world; (2) it recognized the family and not the individual as the basic unit of society; (3) it upheld the inequality of the status of individuals; (4) it stressed filial piety which made men subservient and dependent; and (5) it preached orthodoxy of thought in total disregard of freedom of thinking and expression.[9] Loudly Ch'en called for the destruction of conservatism in order to make room for constructing a new culture.

> We indeed do not know which of our traditional institutions may be fit for survival in the modern world. I would rather see the ruin of our traditional "national quintessence" than have our race of the present and future extinguished because of its unfitness for survival . . . The world continually progresses and will not stop. All those who cannot change themselves and keep pace with it are unfit for survival and will be eliminated by the processes of natural selection. Therefore, what is the good of conservatism?[10]

Ch'en's bold attack on traditionalism opened a new vista in the musty intellectual world, and quickly won him an enthusiastic following among the educated youth.

TS'AI YÜAN-P'EI AND THE PEITA. The New Cultural Movement received a great impetus when Ts'ai Yüan-p'ei took over the chancellorship of the National University of Peking, or Peita in abbreviation, in December 1916. This government-supported institution of higher learning had a conservative tradition, with its professors drawn mostly from the official-

8. Tse-tsung Chow, 46.
9. Tse-tsung Chow, 302; Kuo Chan-po, 103.
10. Tse-tsung Chow, 46.

dom. Students did not take their education seriously, but regarded it as a stepping stone to official appointments. The frivolous atmosphere of the university and the loose morals of students and faculty were notorious.

Upon assuming the chancellorship, Ts'ai admonished them that the university was a place of learning and not a short cut to wealth and position. His administration rested upon three principles: (1) the university should be an institution of research—dedicated not merely to the introduction of Western civilization but to the creation of a new Chinese culture; not to the preservation of national quintessence but to its re-evaluation by scientific methods; (2) a university education was not a substitute for the old civil service examinations; and (3) absolute academic freedom was to be allowed, and free expression of divergent thories and viewpoints guaranteed, as long as they could be sustained on rational grounds.

Under Ts'ai's guidance, the Peita became an exciting institution of higher learning, with professors of different political persuasions—liberals, radicals, socialists, anarchists, conservatives, and reactionaries—composing the faculty. The university boasted of an incredibly productive and intellectual life, as the most vital and promising scholars of the country flocked to join the staff. In 1917 Ch'en Tu-hsiu was made dean of the School of Letters, and Hu Shih, returning from the United States, became a professor of literature. The next year, Li Ta-chao was appointed librarian, and in his employ was a young assistant named Mao Tse-tung.

HU SHIH AND HIS CONTRIBUTIONS. Hu Shih was an energetic proponent of scientific thinking, pragmatism, and the vernacular style of writing. Because of the Huxley and Dewey influences, the main sources of Hu's inspiration were agnosticism and pragmatism, which became his principal approaches in evaluating traditional ethics and ideas. Truth, according to the pragmatist, is changeable in proportion to its utility based on experimentation. Such an attitude, distinctly a product of an industrial capitalistic society, was diametrically opposed to the Confucian concept that truth is eternal and unchangeable. Confucianism was therefore in Hu's eyes totally out of touch with the realities of the modern world.[11] He invented the pejorative phrase "Confucius and Sons Incorporated," and his followers shouted "Down with Confucianism."

If Hu was against Confucianism, he was for liberalism, individualism, science, and democracy. Drawing from pragmatism, he preached a gradual, bit-by-bit improvement of society through study of its problems,

11. Kuo Chan-po, 124-25.

experimentation, and solution. Under his aegis, "Mr. Science" and "Mr. Democracy" became the catchwords of the age. Since both originated in the West, Hu in effect advocated a complete Westernization. "Go West!" was his message.

Hu's philosophy is best explained in his own words:

> The spirit of the new thought tide is a critical attitude. The methods of the new thought tide are the study of problems and the introduction of academic theories. . . . The attitude of the new thought tide toward the old civilization is, on the negative side, to oppose blind obedience and to oppose compromise, and on the positive side, to reorganize our national heritage with scientific methods. What is the sole aim of the new thought tide? It is to recreate civilization.[12]

Hu Shih's most important single contribution was perhaps the introduction of plain language (*pai-hua*) in writing. Condemning the traditional emphasis on style rather than on substance, he maintained that the classical style of writing was dead and that a dead language could not produce a living literature. He proposed to write in the vernacular language, and succeeded in creating a very lucid, vivid style, which won immediate acceptance among liberal and forward-looking men. He advised students to shun classical allusions, time-worn literary phrases, and parallel sentences; to avoid imitating the ancients; and to write with meaning, substance, and genuine feelings.

Conservative opposition was not lacking. Upholders of traditionalism published *The National Heritage* (*Kuo-ku*) to defend the classical style of writing, but the magazine had little appeal and ceased to exist after only four issues. Nevertheless, Yen Fu and Lin Shu, the two famous translators around the turn of the century, continued to boycott the movement. In a letter to Chancellor Ts'ai, Lin ridiculed the vernacular style of writing as the work of "roadside peddlers." Yen chided the substitution of "vulgar" vernacular for the elegant classical style as retrogression, which could not survive the law of evolution and competition. Ts'ai's reply was remarkable for its simplicity: the plain language differed from the classical only in form and not in content; the works of Huxley, Montesquieu, and Adam Smith, as well as the fiction of Dickens, Dumas fils, and Hardy—which Yen and Lin translated—all appeared in the plain language. Could they say, in all fairness, that the translations, which appeared in the classical style, surpassed the originals? The case for the *pai-hua* was officially vindicated when the government in 1920 adopted it for use in the schools.

12. Tse-tsung Chow, 219.

From the historical standpoint, the success of the plain-language movement stemmed, at least partially, from the fact that after the abolition of the "eight-legged essay" in 1902, students in China lacked definite models to follow. In their search for the new and unusual, they were first briefly attracted to Liang Ch'i-ch'ao's semiclassical and semicolloquial journalistic style. But with the advent of *pai-hua* they readily joined the new trend.

In 1918, students at Peita organized a magazine called the *New Tide* (*Hsin-ch'ao*), which was governed by three criteria: a critical spirit, scientific thinking, and a reformed rhetoric. The *New Youth* and the *New Tide*, along with a host of others including the *Weekly Critic* (*Mei-hou p'ing-lun*),[13] launched an all-out attack on the bastions of traditionalism— old literature, old ethics, old human relations, and Confucianism. These magazines ridiculed old patterns of thought, old customs, personal loyalty of officials, filial piety, superstition, the double standard of chastity for men and women, the big family system, and above all, monarchism and warlordism. They attacked the unquestioned acceptance of the national heritage and demanded a critical reappraisal of all the classics and ancient works, and the creation of a new culture. Science, democracy, technology, agnosticism, pragmatism, liberalism, parliamentarianism, and individualism found new favor with them.

These magazines were an intellectual bombshell. For the first time in China important national and social problems were being publicly discussed and debated. The youth of the country could not wait to read each new issue. John Dewey, upon visiting China in 1919, commented: "There seems to be no country in the world where students are so unanimously and eagerly interested as in China in what is modern and new in thought, especially about social and economic matters, nor where the arguments which can be brought in favor of the established order and the status quo have so little weight—indeed, are so unuttered."[14] The explosive nature of this social and intellectual ferment sparked a massive national outburst.

The May Fourth Movement, 1919

On May 4, 1919, about 5,000 students in Peking held a huge demonstration against the verdict of the Versailles Peace Conference on Shantung. It was at once an explosion of public anger, an outburst of nationalism, a

13. Under the editorship of Hu Shih.
14. Tse-tsung Chow, 183.

deep disappointment in the West, and a violent indictment of the "traitorous" warlord government in Peking. So powerful and so far-reaching
was this incident that it evoked an immediate national response and pressured the Chinese delegation at Versailles to reject the peace treaty. Nationalism, public opinion, and mass demonstration had emerged as new
forces in Chinese politics. Some historians hailed the May Fourth incident
as the first genuine mass movement in modern Chinese history.

It will be recalled that in 1898 Germany leased from the Ch'ing government the naval base of Kiaochow in Shantung province for 99 years.
When World War I broke out, China was a neutral, while Japan joined
the Allies and ousted the Germans from Kiaochow; subsequently Japan
took over most of Shantung. To legalize its occupation, Japan included
in the Twenty-one Demands provisions which recognized its position in
Shantung, and to further bolster its claim it entered into a series of treaties
with the great powers. By the Russo-Japanese agreement of February 20,
1917, Russia recognized the Twenty-one Demands, while Japan agreed
to recognize the Russian gains in Outer Mongolia during 1912-15. The
Anglo-Japanese agreement of a day later obligated Britain to support the
Japanese position in Shantung at the forthcoming peace conference and to
second its claims to German possessions in the Pacific north of the equator;
in return Japan agreed to support the British claims to German islands in
the Pacific south of the equator. Similar secret agreements were made with
France and Italy. Then, in November 1917, the Lansing-Ishii Agreement
was concluded by which the United States recognized that "geographical
propinquity creates special relations between nations"—i.e. Japan had a
special position in China—while Japan paid lip service to the Open Door
Policy.

The *coup de grâce* was the secret pacts of September 1918 between
Peking and Tokyo. By granting the Chinese warlord government a loan
of 20 million yen, Japan won the right to build two railways in Shantung,
to station troops at various key points, and to train and direct Chinese
railway guards. Under instructions from Peking, the Chinese minister in
Tokyo, Chang Tsung-hsiang, "gladly agreed" (*hsin-jan t'ung-i*) to these
terms.

Armed with these secret treaties, the Japanese came to Versailles confident of winning their case on Shantung. Needless to say, retention of
Shantung would indirectly acknowledge the validity of the Twenty-one
Demands and the viability of the secret treaties with the Peking regime,
agreements which gave Japan far greater concessions in Manchuria and
other parts of China than in Shantung. The Japanese repertoire of treaties
evinced a pragmatic approach to international relations which appeared in
stark contrast to the naïve Chinese faith in Western ideals.

The Chinese delegation[15] had come to what they believed a just tribunal dedicated to the principles of democracy, self-determination, and protection of the weak. Indeed, Wilsonian idealism and the Fourteen Points had caught the Chinese fancy; many believed that the long-awaited age of world democracy had finally arrived, and that Wilson was about to forge a new world out of the fragments of the old. On November 17, 1918, 6,000 Chinese paraded in Peking to celebrate the victory of Western democracy over German despotism and militarism. It was in this state of high expectation that the Chinese delegation had come to Versailles, pledged to seek the recovery of Shantung and the complete abolition of the unequal treaties. But their exuberant optimism rapidly turned to dismay. They were coldly told that the peace conference had not been called to adjust all the international grievances of the past, but to settle problems arising from the conclusion of the war. Consequently, only Shantung belonged on the agenda.

The Chinese delegation pleaded that Shantung, the birthplace of Confucius and Mencius, was the Holy Land of China—and that the German rights which the Japanese had claimed to inherit had ceased to exist when China entered the war in 1917 and abrogated all treaties with Germany. Moreover, Article 5 of the 1898 agreement on Kiaochow stipulated that "Germany engages at no time to sublet the territory leased from China to another power." Similarly, the Twenty-one Demands were invalid because the Chinese parliament had never ratified them. Moreover, China's entry into the war in 1917 had so changed its status—from a neutral to a belligerent—that it was qualified to invoke the international law principle of *rebus sic stantibus*[16] to nullify the Twenty-one Demands. In rebuttal, the Japanese delegation calmly divulged the 1918 secret agreements with Peking, pointing out that they had been "gladly agreed" to by China *after* its entry into the war. No amount of Chinese argument could alter this fact, and the fate of Shantung was sealed.

The Allies were bound by secret treaties to support the Japanese position, which left Wilson as the lone champion of the Chinese cause. Japan threatened to raise the issue of racial equality for discussion and to withdraw from the conference if its demands were not met. It was clear that Japan could not be denied on both the Shantung and the racial issues. Ultimately, Wilson was persuaded by the Allied representatives as well as his own advisers[17] that it was important to first establish the League of

15. Consisting of members from both the Peking and Dr. Sun's Canton governments in order to give an appearance of national unity.
16. This principle suggests that when the objects of a treaty, or the conditions under which it is concluded, no longer exist, the treaty becomes null and void.
17. Such as Colonel House.

Nations with Japan in it, and to secure justice for China later. On April 28, 1919, the peace conference adjudicated the Shantung question in favor of Japan.

When news of the Paris decision reached Peking, Chinese faith in Wilson and the tenets of his idealism was shattered. Enraged by what they saw as Western betrayal, students vowed to defend Shantung by blood. The influential newspaper *Shen-pao* editorialized: "At the outset of the Paris Conference, we heard a lot of what was called 'the triumph of right and justice,' 'the upholding of the rights and privileges of small and weak nations,' but what do we get? Whoever expects help from others is doomed to be disappointed. Let our countrymen understand today once and for all that their only course is to act by themselves. Had our countrymen not abandoned their own interests, who could have infringed upon them?"[18]

On May 4, several hundred returned students met to discuss what they could do in this period of national crisis and humiliation. They decided to send telegrams to the Versailles Conference to protest the unjust verdict and to the Chinese delegation to urge the rejection of the treaty if the terms on Shantung were not revised. It was also resolved to stage a mass demonstration and to present petitions to the foreign legations for transmittal to Paris.

The demonstration was joined by large groups of students from the thirteen universities and colleges in Peking, swelling the number to 5,000. Huge banners floated above the crowd with such inscriptions as "Reclaim Kiaochow unto death," and "Punish the traitor Ts'ao Ju-lin."[19] The orderliness of the parade evaporated when it passed the house of Ts'ao, at which time the students went out of control and broke into it. Since Ts'ao had escaped, they beat up his houseguest—who was none other than the Chinese minister to Japan who had "gladly agreed" to the 1918 pacts—and set fire to the house. With the belated arrival of the police, most of the demonstrators had gone; only ten of them were arrested.

The immediate response to the arrest was a general strike by all students in Peking and the resignation of Ts'ao as chancellor of Peita. The strike quickly spread to students in other major cities, and was joined by shopkeepers, industrial workers, and employees in commercial establishments all over the country. A concerted boycott of Japanese goods followed; people stopped buying Japanese products and taking Japanese steamers, and dockhands refused to unload Japanese goods. Under increasing pressure from the public, the Peking regime released the students on May 7.

18. *North China Herald*, May 17, 1919, p. 415, with minor changes.
19. The foreign minister.

Meanwhile, thousands of telegrams were sent to the Chinese delegation at Paris, asking them to reject the treaty and threatening them with punishment if they did not. Perhaps most representative was the one sent by the Society for China's Salvation: "The whole nation is indignant over the failure of the Shantung question. Never sign the treaty. We demand your immediate withdrawal from the Conference. Better to have forced occupation than voluntary submission. Otherwise sole responsibility rests on you."[20] The Peking warlord regime, confused and unable to take a definite stand, left the decision of signing to the delegation itself. Lest the delegates yield under foreign pressure or secret government order, Chinese students in Paris organized an around-the-clock vigil to see that none of them left their quarters. At the signing ceremony on June 28, there were no Chinese representatives. Visibly distressed, President Wilson was heard muttering: "That is most serious. It will cause grave complications . . . this is most unfortunate, but I don't know what we can do."[21]

Wilson sacrificed China in order to lure Japan into the League of Nations; yet he was unable to get his own country into the international organization. Ironically, Japan was among the first to withdraw from the League, in 1933. As to China, although it rejected the peace treaty with Germany, it signed the treaty with Austria, and by virtue of that act automatically became a member of the League of Nations.

Expansion of the New Cultural Movement

The May Fourth incident served as a catalyst for the intellectual revolution in China. While interest in the West continued in the days that followed, a split appeared among Chinese intellectuals. Those who were bitterly disappointed by the Versailles Conference began to turn to Marxist socialism under the influence of the Bolshevik Revolution; others who were tradition-bound blamed Western materialism as the cause of World War I and suggested Chinese spiritualism as an antidote. These different strands of thought—together with the grand debates on the relative value of Eastern and Western civilizations, of science and metaphysics, and attempts to re-evaluate the Chinese national heritage by modern methods and standards—propelled the New Cultural Movement to greater heights.

FOREIGN VISITORS. John Dewey and his wife visited China from May 1, 1919, to July 11, 1921. With Hu Shih as interpreter, Dewey gave a number of public lectures on his social and political philosophy of pragmatism; on his own ideas about education, methods of thought, and ethics;

20. *North China Herald*, May 17, 1919, p. 413.
21. *Foreign Relations of the United States*, 1919, XI, 602.

and on his views of the three leading contemporary philosophers: Bergson, Russell, and James. His lecture halls were always packed with large crowds, including high school and college students. Dewey told his audiences: "China could not be changed without a social transformation based upon a transformation of ideas. The political revolution was a failure, because it was external, formal, touching the mechanism of social action but not affecting conceptions of life, which really control society."[22] Impressed with the eagerness of Chinese youth to listen to his exposition of philosophy and social ideas which seemingly only bored American students, Dewey enthusiastically reported: "There is an eager thirst for ideas—beyond anything existing, I am convinced, in the youth of any other country on earth."[23]

Bertrand Russell stayed for the better part of a year, from October 1920 to July 1921. With Y. R. Chao interpreting, he gave a series of public lectures, but the tenor of his message was quite different from Dewey's. Rather than telling the Chinese what they should do to get along in the modern world, Russell, an avid pacifist, extolled the value of the tranquil, humane, tolerant, and pacific Chinese outlook on life. The Confucian concept of filial piety, he said, in spite of its many shortcomings, was "less harmful than its Western counterpart, patriotism," which led more easily to imperialism and militarism.[24] He was attracted to the Taoist ideas of "production without possession, action without self-assertion, [and] development without domination," which approximated his own ideas of promoting creative impulses while eliminating the possessive tendency. Apologetically he commented: "In so far as there is a difference of morals between us and the Chinese, we differ for the worse, because we are more energetic, and can therefore commit more crimes *per diem*." The essence of Russell's message was that the West should learn from China "the just conception of the ends of life," while China should "acquire Western knowledge without acquiring the mechanistic outlook"—by which was meant taking men as raw material to be molded by scientific manipulation.[25]

Russell's advice did not strike a very responsive note with Chinese intellectuals, who, in their eagerness to be modern, wanted to be patriotic, nationalistic, and active rather than pacific, filial, and passive. They were more anxious to destroy Confucianism and to promote Westernization than to teach the West how to acquire the Chinese humane conception

22. John Dewey, "New Culture in China," *Asia*, XXI:7:581 (July 1921).
23. Dewey, 586.
24. Betrand Russell, *The Problem of China* (London, 1922), 41.
25. *Ibid.*, 81-82, 192-94.

of life. The latter was a yoke upon their efforts to be like the moving, dynamic West, and had to be thrown off in the name of progress. There was no place in the velocity and rhythm of Western-style change for the tranquillity of the Confucian past.

Other visitors included the American educator Paul Monroe in 1921, the German philosopher Hans Driesch in 1923, and the Indian Nobel prize laureate R. Tagore in 1924. Plans to invite Bergson and Eucken did not materialize.

In addition to the contributions of foreign visitors, Western thought and ideologies were also eagerly pursued by Chinese intellectuals themselves, whose tastes reflected a gradual shift from Anglo-American to German-Russian sources. The works of the French philosopher Bergson, were transmitted by Carsun Chang, and those of the German philosophers Schopenhauer and Nietzsche by Wang Kuo-wei. Ch'en Tu-hsiu and Li Ta-chao introduced Marx and Engels, while Li Ta wrote on the dialectical methods and the thought of Lenin, Bukharin, and Plekhanov. Li Shih-tseng introduced the Russian anarchist Kropotkin, and popularized his ideas on "mutual assistance" and "unity" as the basic forces of progress—in direct refutation of Darwin's idea of "struggle." Many Chinese intellectuals and scholar-politicians adopted anarchist views, and after the May Fourth incident Marxism and Bolshevism gained increasing favor among the radicals. A grand debate soon erupted over the relative merits of gradual social reforms versus rapid fundamental changes.

PROBLEMS AND "ISMS." Hu Shih, the high priest of pragmatism in China, vigorously advocated an evolutionary "drop-by-drop" improvement of society through the study and solution of specific, practical problems. Li Ta-chao, and shortly afterwards Ch'en Tu-hsiu, argued for an immediate and thoroughgoing sociopolitical transformation, after the Soviet fashion. In an article entitled "More Study of Problems and Less Talk of "Isms',"[26] Hu Shih urged his countrymen to shun the high-sounding, all-embracing "isms" as nothing but "the dreams of self-deceived and deceptive persons, iron-clad proof of the bankruptcy of Chinese thought, and the death-knell of Chinese social reform!" Forcefully he argued:

> Civilization was not created *in toto*, but by inches and drops. Evolution was not accomplished overnight but in inches and drops. People nowadays indulge in talk about liberation and reform, but they should know that there is no liberation *in toto*, or reform *in toto*. Liberation means the liberation of this or that system, or this or that idea, or of

26. *Weekly Critic,* July 20, 1919.

this or that individual; it is reform by inches and drops. The first step in the re-creation of civilization is the study of this or that problem. Progress in the re-creation of civilization lies in the solution of this or that problem.[27]

Hu cautioned against blind activism and rudderless revolutions, proposing instead spontaneous and gradual reform to eliminate the five enemies of social progress—poverty, sickness, illiteracy, corruption, and disorder.

Li Ta-chao, the leading convert to Marxism, replied that "isms" were necessary to provide a "common direction" in solving social problems. Speaking in equally forceful terms, he argued: "It is first necessary to have a fundamental solution, and then there will be hope of solving concrete problems one by one. Take Russia as an example. If the Romanoffs had not been overthrown and the economic organization not reformed, no problems could have been solved. Now they are all being solved."[28]

Hu's rebuttal was that there was no panacea for all the ills of China; each must be attacked and solved individually, and "isms" were only romantic hypotheses for solving social problems. Though conceding to this last point, Li nevertheless championed political action: "The solution of the economic problem is the fundamental solution. As soon as the economic problem is solved, then all political and legal problems and the problems of the family system, women's liberation, and the worker's liberation can be solved."[29] Ch'en Tu-hsiu, less committed to Marxism than Li in mid-1919, conceded that "it is better to promote the practical movement of education and emancipation of workers than vaguely to talk anarchism and socialism." But by the end of 1920 he too became a firm convert to Bolshevism and to the efficacy of political action, arguing that "isms" in social reforms performed the same necessary function as the destination in a voyage. Still, he conceded that revolution and social reforms could not be accomplished *in toto* overnight.[30]

On the surface, the debate ended in Hu's favor. Yet it was a hollow victory, for it was the vogue among youth to discuss "isms," and even Hu himself constantly spoke of liberalism, pragmatism, experimentalism, etc. A witty critic described Hu and the pragmatists as saying: "You should give up all 'isms' and accept our 'isms,' because, according to our 'ism,' no 'ism' should be accepted as a creed."[31]

27. Maurice Meisner, *Li Ta-chao and the Origins of Chinese Marxism* (Cambridge, Mass., 1967), 107.
28. Meisner, 107.
29. *Ibid.,* 111.
30. Tse-tsung Chow, 220.
31. Tse-tsung Chow, 222.

Paradoxically, after preaching "more study of problems," Hu and his followers delved into the less practical pursuits of textual criticism, ancient history, and archeological investigations in the 1920s, at a time when social and political problems pressed for urgent and immediate solution. On the other hand, many advocates of "isms" and fundamental change went to the peasants and workers and studied their problems firsthand. It is apparent that Hu failed to see that pragmatism was the product of a stable American society that permitted free examination of problems and the implementation of reforms, whereas China of the warlord period totally lacked the sociopolitical conditions prerequisite to experimentation and gradual reform.[32]

GO EAST! GO WEST! The vast destruction wrought by World War I had greatly disillusioned many Chinese. Liang Ch'i-ch'ao blamed Western imperialism and blind worship of science as the roots of conflict and suggested that Chinese spiritualism might redress the imbalance. Liang Souming, author of *East and West: Their Civilizations and Philosophies*, also argued against science and democracy in an effort to defend the integrity of the Chinese civilization. The life of a people depends on their basic spirit, he declared; to sacrifice China's own spirit in favor of a foreign ethos and institutions was to undermine its destiny. Rather, it should develop its forte solely from its own standpoint.[33] The two Liangs deprecated Western materialistic civilization in direct proportion to their eulogy of Chinese spiritual civilization. They urged their countrymen: "Go East!"

In opposition, Hu Shih and other advocates of Westernization said: "Go West!" Wu Chih-hui berated Liang Sou-ming as a "useless creature of the 17th century." Hu Shih announced that China lagged behind the West not only in science and technology, but also in everything else—politics, literature, music, arts, morality, and even physical stature.[34] Nevertheless, the advocates of Westernization still demonstrated interest in a scientific and critical re-evaluation of the Chinese cultural heritage. Utilizing Western approaches and methods of research, Hu completed *An Outline History of Chinese Philosophy*, in which he advanced the bold and unprecedented thesis that the School of Logicians in ancient China was not strictly a school and that each of the Hundred Schools had its own methods of logical thinking. In a similar example of modern scholarship, Liang Ch'i-ch'ao re-examined the work of the ancient philosopher, Mo-tzu, and authored, among other works, *The Political Thought of the*

32. Meisner, 108-9.
33. Kuo Chan-po, 317.
34. *Ibid.*, 318.

Pre-Ch'in Period, and *Intellectual Trends in the Ch'ing Period.*[35] Also noteworthy were the "Antiquity-doubters" (*I-ku p'ai*) at Peita,[36] who after exhaustive studies of ancient classics and history raised doubt as to their authenticity and rejected the traditional view that Confucius was their editor. Needless to say, the re-evaluation of the national heritage constituted another major achievement of the New Cultural Movement and greatly expanded its scope and dimensions.

Concluding Remarks

The intellectual revolution of 1917-23 represents China's third stage of response to the Western impact. The first stage—the Self-strengthening Movement from 1861 to 1895—saw superficial attempts at diplomatic and military modernization, and the second—the era of reform and revolution from 1898 to 1912—witnessed the acceptance of Western political institutions. The intellectual awakening of 1917-23 marked a further shift away from the traditional Chinese base toward complete Westernization. By 1920 China was very much a part of the modern world.

Appraisals of the significance of the New Cultural Movement varied according to different viewpoints. Liberals proclaimed it as a movement of emancipation from old thought, old ethics, old values, and affirmation of human rights. The birth of a new literature with a new style of writing, and the official adoption of the Plain Language encouraged some to regard the May Fourth Movement as a Chinese Renaissance. Conservatives, however, attacked the movement for its corrupting influence on the youth and lack of respect for traditionalism, although they conceded its usefulness in stimulating nationalism. Radicals eulogized the movement. Li Ta-chao praised it as not only a patriotic movement but "a part of human liberation," and Mao Tse-tung described it as essentially an "anti-imperialist and anti-feudal bourgeois-democratic revolution of China," propelled by a united front of workers, students, and national bourgeoisie under the leadership of the intelligentsia.[37] Similarly, the Chinese Communist Party and its historians regard May 4, 1919, as the watershed which separated the eighty-year period of the "Old Democracy" from the period of "New Democracy." During this latter period, the proletariat had become a conscious, independent political force, and communism had developed into an increasingly powerful ideological tool in the social, political, and cultural revolution of China.

35. For an English rendition of the latter work, see translation by Immanuel C. Y. Hsü (Harvard University Press, 1959).
36. Such as Ch'ien Hsüan-t'ung (classics) and Ku Chieh-kang (history).
37. Tse-tsung Chow, 347, 349.

Regardless of these different viewpoints, the fact remains that the May Fourth Movement was essentially a socio-politico-intellectual revolution aimed at achieving national independence, individual emancipation, and creation of a new culture through a critical and scientific re-evaluation of the national heritage and selected acceptance of foreign civilization. Leaders of the movement regarded a radical change in the "thought base" as a prerequisite to successful modernization and national regeneration. Old ethics, customs, literature, social relations, and economic and political institutions came under disparaging attack to make way for the new. Yet a new culture was slow to emerge. The May Fourth Movement had been far more effective at destroying the past than at constructing the future.

Nonetheless, three main achievements are indisputable. First, the literary revolution led to the establishment of the Plain Language in 1920 and the rise of a new literature in vernacular style—based on humanitarianism, romanticism, realism, and nationalism. Literature now assumed a didactic role of instilling social consciousness in the public—"from literary revolution to revolutionary literature."

Second, the influx of diverse foreign ideas and ideologies caused the emergence of two opposing views on social reconstruction and national re generation: the pragmatic, evolutionary method expounded by Hu Shih and later partially accepted by the Nationalist Party; and the Marxist revolutionary approach adopted by the Chinese Communist Party. The contemporary history of China from 1921 onward is primarily a story of the struggle between these two parties and their different approaches.

Third, the intensification of nationalism stimulated the rise of a Young China, extremely sensitive to its perilous position in the modern world and jealous of guiding its own destiny. Such an attitude generated psy chological reconstruction and national confidence which partially compensated for the sense of inadequacy and inferiority that had built up over the decades. The result was a violent reaction against foreign imperialism and an intense drive to end the unequal treaties.

Yet, in historical perspective, for all its bombastic characteristics, the intellectual revolution succeeded primarily in introducing Western thought and destroying Chinese traditionalism, rather than creating new systems of thought and new schools of philosophy. The avowed purpose of forging a new culture through a critical re-evaluation of Chinese and Western civilizations stirred up a series of debates and polemics without really creating a new culture as such. Nonetheless, a foundation had been laid to adapt foreign ideas and institutions creatively to the Chinese situation. Whether by the evolutionary or revolutionary route, the ultimate goal remained the same: national salvation through the creation of a New China—thoroughly modernized, yet distinctly Chinese.

Further Reading

Alitto, Guy S., *The Last Confucian: Liang Shu-ming and the Chinese Dilemma of Modernity* (Berkeley, 1978).

Brière, D., S.J., *Fifty Years of Chinese Philosophy, 1898-1950* (London, 1956).

Ch'en, Jerome, *China and the West: Society and Culture, 1815-1937* (Bloomington, 1980).

Chen, Joseph T., *The May Fourth Movement in Shanghai: The Making of a Social Movement in Modern China* (Leiden, 1971).

Ch'en, Tuan-chih 陳端志, *Wu-ssu yün-tung chich shih ti p'ing-chia* 五四運動之史的價評 (A historical review of the May 4th Movement), (Shanghai, 1936).

Chiang, Monlin, *Tides from the West: A Chinese Autobiography* (New Haven, 1947).

Chou, Hsiu-luan 周秀鸞, *Ti-i-tz'u shih-chieh ta-chan shih-ch'i Chung-kuo min-tsu kung-yeh ti fa-chan* 第一次世界大戰時期中國民族工業的發展 (The development of Chinese national industries during World War I), (Shanghai, 1958).

Chou, Min-chih, *Hu Shih and Intellectual Choice in Modern China* (Ann Arbor, 1984).

Chow, Tse-tsung, *The May Fourth Movement: Intellectual Revolution in Modern China* (Cambridge, Mass., 1960).

De Francis, John, *Nationalism and Language Reform in China* (Princeton, 1950).

Dewey, John, *Lectures in China, 1919-1920*, tr. from the Chinese and ed. by Robert W. Clopton and Tsuin-chen Ou (Honolulu, 1973).

———, *Letters from China and Japan* (New York, 1921).

———, "Old China and New," *Asia*, XXI:5:445-56 (May 1921).

———, "New Culture in China," *Asia*, XXI:7:581-86 (July 1921).

Duiker, William J., *Ts'ai Yüan-p'ei: Educator of Modern China* (University Park, Penn., 1977).

Feng, En-jung 馮恩榮, *Ch'üan-p'an Hsi-hua yen-lun hsü-chi* 全盤西化言論續集 (A supplementary collection of articles on complete Westernization), (Canton, 1935).

Fifield, Russell H., *Woodrow Wilson and the Far East, The Diplomacy of the Shantung Question* (New York, 1952).

Furth, Charlotte, *Ting Wen-chiang: Science and China's New Culture* (Cambridge, Mass., 1970).

Goldman, Merle (ed.), *Modern Chinese Literature in the May Fourth Era* (Cambridge, Mass., 1977).

Grieder, Jerome B., *Hu Shih and the Chinese Renaissance: Liberalism in the Chinese Revolution, 1917-1937* (Cambridge, Mass., 1970).

Hay, Stephen N., *Asian Ideas of East and West: Tagore and His Critics in Japan, China, and India* (Cambridge, Mass., 1970).

Hu, Shih, *The Chinese Renaissance* (Chicago, 1934).

Hua Kang, 華崗, *Wu-ssu yün-tung shih* 五四運動史 (A history of the May 4th Movement), (Shanghai, 1951).

Huang, Sung-k'ang, *Lu Hsün and the New Cultural Movement of Modern China* (Amsterdam, 1957).

Keenan, Barry C., *The Dewey Experiment in China: Educational Reform and Political Power in the Early Republic* (Cambridge, Mass., 1977).

King, Wunsz, *China at the Paris Peace Conference in 1919* (New York, 1961).

Kuo, Chan-po 郭湛波, *Chin-wu-shih-nien Chung-kuo ssu-hsiang shih* 近五十年中國思想史 (A history of Chinese thought during the last fifty years), reprinted (Hong Kong, 1965).

Kwok, D. W. Y., *Scientism in Chinese Thought* (New Haven, 1965).

Lau, Joseph S. M., C. T. Hsia, and Leo Ou-fan Lee (eds.), *Modern Chinese Stories and Novels, 1919-1949* (New York, 1981).

Lee, Leo Ou-fan, *The Romantic Generation of Modern Chinese Writers* (Cambridge, Mass., 1973).

Levenson, Joseph R., *Liang Ch'i-ch'ao and the Mind of Modern China* (Cambridge, Mass., 1953).

————, *Confucian China and Its Modern Fate*, Vol. I: *The Problem of Intellectual Continuity* (Berkeley, 1958), chs. 8-9.

Lin, Yu-sheng, *The Crisis of Chinese Consciousness: Radical Antitraditionalism in the May Fourth Era* (Madison, 1978).

Lü, Hsüeh-hai 呂學海, *Ch'üan-p'an Hsi-hua yen-lun chi* 全盤西化言論集 (A collection of articles on complete Westernization), (Canton, 1934).

McDougall, Bonnie S., *The Introduction of Western Literary Theory into Modern China, 1919-1925* (Tokyo, 1971).

Meisner, Maurice, *Li Ta-chao and the Origins of Chinese Marxism* (Cambridge, Mass., 1967).

Roy, David T., *Kuo Mo-jo: The Early Years* (Cambridge, Mass., 1971).

Russell, Bertrand, *The Problems of China* (London, 1922).

Schneider, Laurence A., *Ku Chieh-kang and China's New History: Nationalism and the Quest for Alternative Traditions* (Berkeley, 1971).

Schwarcz, Vera, *The Chinese Enlightenment: Intellectuals and the Legacy of the May Fourth Movement of 1919* (Berkeley, 1986).

Schwartz, Benjamin I. (ed.), *Reflections on the May Fourth Movement: A Symposium* (Cambridge, Mass., 1972).

Wang, Y. C., *Chinese Intellectuals and the West, 1872-1949* (Chapel Hill, 1966).

Wu-ssu yün-tung lun-ts'ung 五四運動論叢 (A collection of articles on the May 4th Movement), (Taipei, 1961).

Yü, Ying-shih, *Chung-kuo chin-tai ssu-hsiang shang ti Hu Shih* (Hu Shih in Modern Chinese Intellectual History), (Taipei, 1984).

22

National Unification
Amidst Ideological Ferment
and Anti-imperialistic Agitation

In the wake of the intellectual revolution, two major political events developed as repercussions of the Bolshevik Revolution in Russia. One was the rise of the Chinese Communist Party, and the other the reorganization of the Nationalist Party. Both developments played a major role in shaping the course of the contemporary history of China.

The Birth of the Communist Party, 1921

Chinese awareness of Marxism probably began around 1905 when the *Min-pao* published a biography of Karl Marx in its second issue. In early 1908 the anarchist journal, *T'ien-i pao* (Journal of natural justice), published translations from the Japanese of Friedrich Engels's 1888 "Introduction to the Communist Manifesto," the first chapter of the *Manifesto* itself, and excerpts from Engels's *The Origin of the Family*. Although there was an incipient recognition of Marx and Engels as the founding fathers of "scientific socialism," the influence of Marxism remained small until the May Fourth period when the success of the Bolshevik Revolution dramatized the power of such an ideology. Many Chinese intellectuals had lost faith in the West after the Versailles pronouncement on Shantung, and found it difficult to accept the West as teacher and oppressor simultaneously. Therefore, ideas and ideologies critical of the West found new favor, and powerful elements among the intellectuals were drawn to the utopian socialism of Saint-Simon, the anarchism of Kropotkin and Bakunin, and the revolutionary philosophy of Marx. Socialism was appealing because it provided a practical philosophy with which to reject "both the traditions of the Chinese past and the Western domination of the present."[1] Moreover, it represented a goal as yet unrealized in western Europe and America, and its acceptance in China would put it ideologically ahead of the capitalist states. This subtle psychological satisfaction, buttressed by a general disappointment with the West and a secret desire to surpass it, made Marxism especially attractive.

The intellectual and psychological appeal of Marxism was further strengthened by the Soviet offer of friendship and the enticing Leninist theory of imperialism. Anxious to win friends and create a new image, Moscow twice announced—in 1918 and 1919[2]—its readiness to renounce the old tsarist special rights and privileges in China. Although it altered its position somewhat in 1920 and proposed to *negotiate* the abolition of the unequal treaties—as a means to win Chinese recognition—the Soviet overture nevertheless created a favorable impact, for it represented an unsolicited, unilateral expression of friendship and a radical departure from the haughty and rapacious behavior of imperialist powers.

Added to this olive branch was the encouraging Leninist theory of imperialism. Lenin had proclaimed that imperialism was the inevitable product of the last stage of capitalism; when capitalism grew to a high point, as in the late 19th and early 20th centuries, it had to seek overseas markets to sell its surplus goods and to buy raw materials. At this point mutual jealousy and rivalry among the capitalist states would inevitably lead them into conflict and eventual extinction. The downtrodden people of Asia and other underdeveloped areas should thus rise against foreign imperialism and hasten the passing away of the foreign yoke. This Leninist theory offered comfort to the Chinese intellectuals, for not only did it blame the West for China's ills and predict the imminent demise of capitalism, but it also gave Asia a place in the world revolution—in refutation of the previous stand of most European Marxists who insisted that the problems of the world could be solved only in and by the West.

In effect, the intellectual appeal of Marxism-Leninism, the voluntary offer of friendship by the Soviet regime, and the practical success of the Bolshevik Revolution combined to create a powerful ideological impact in China. Marxist and Leninist study groups began to spring up, and the National University of Peking, where intellectual curiosity and freedom

1. Bernal, 111, 137; Meisner, 100.
2. Through Foreign Commissar G. V. Chicherin on July 4, 1918, and Assistant Foreign Commissar Leo Karakhan on July 15, 1919.

of expression were most pronounced, became a hotbed of radicalism. As early as the middle of 1918, the librarian Li Ta-chao professed his conversion to Marxism and hailed the Bolshevik Revolution as a "great, universal, and elemental force" comparable in importance to the French Revolution. But he envisaged an even greater experience of rebirth in China, and founded the New Tide Society (*Hsin-ch'ao she*) in the autumn of 1918, followed shortly by the Marxist Research Society. Li celebrated "The Victory of Bolshevism" in the November 1918 issue of *New Youth*, and edited a whole issue of it on Marxism in 1919. His library office humorously became known as the "Red Chamber" (*Hung-lou*), frequented by young and eager followers including his students Ch'ü Ch'iu-pai, Chang Kuo-t'ao, and his library assistant Mao Tse-tung—all destined to be the future leaders of the Chinese Communist movement.

No less potent than the Marxist impact was the stunning effect of the May Fourth incident on the Chinese intellectuals. Whereas formerly many of them embraced Western democracy, liberalism, and internationalism without being overly concerned with the question of imperialism, they now decisively cut off their dependence on the West and vowed to take China's fate into their own hands. Political activism was the new catchword. Among those rudely awakened and fervently militant intellectuals, the foremost was Ch'en Tu-hsiu. Deeply struck by the student role in the May Fourth incident, he committed himself to subsequent demonstrations and was landed in prison on June 11, 1919. After his release in September, Ch'en resigned from the university under conservative pressure. Making Shanghai his new home, he became increasingly absorbed in Marxism. By mid-1920 his faith in the West was completely demolished, and democracy to him was no more than a tool used by the bourgeoisie "to swindle mankind in order to maintain political power."[3] Ch'en became the second most important convert to Marxism, and organized a Marxist Study Society in May 1920 and a Socialist Youth Corps in August, which were the forerunners of the Chinese Communist Party.

Meanwhile, in Peking, another group was gathering around Li Ta-chao, whose Marxist Research Society was replaced by the Society for the Study of Socialism in December 1919. By March 1920 the various Marxist groups in Peking had united to form the Peking Society for the Study of Marxist Theory. Two Russians, A. A. Muller and N. Bortman, had offered Li help in 1919 but concrete steps toward forming a party did not begin until the arrival in early 1920 of Grigorii Voitinsky, an agent of the Third (Communist) International, or Comintern in abbreviation. In

3. Meisner, 113.

March he conferred with Li about organizing a party, and shortly afterwards left for Shanghai to confer with Ch'en. The upshot of these critical conferences was the decision to establish a branch party in Shanghai under Ch'en and another in Peking under Li. Only the consolidation of these two branches remained for the unification of communism in China.

In July 1921 the founding meeting of the Chinese Communist Party—since called the First Congress of the Party—was held secretly at a girls' boarding school[4] in the French Concession of Shanghai. It was attended by twelve delegates[5] representing fifty-seven members, but neither Ch'en nor Li was present—Ch'en was at Canton and his group was represented by Chou Fu-hai, while Li's group was represented by Chang Kuo-t'ao. In spite of their absence, Ch'en and Li, 41 and 32 respectively, were honored as the co-founders of the party. Although the central party headquarters was set up in Shanghai, Li's Peking group maintained virtual independence. The expression, "Ch'en in the south, Li in the north" (*Nan-Ch'en pei-Li*), underscored the absence of a tight unified party organization at birth.

Not only did they form two regional foci, but Ch'en and Li also differed considerably on the revolutionary role of the workers and the peasants. Ch'en subscribed to the general European Marxist emphasis on the workers and the implicit disdain toward the inert peasant mass. He believed that the progressive urban elements should spearhead the movement while the backward peasantry followed meekly: "The peasants are scattered and their forces are not easy to concentrate, their culture is low, their desires in life are simple, and they easily tend toward conservatism . . . These environmental factors make it difficult for the peasants to participate in the revolutionary movement."[6] Ch'ü Ch'iu-pai likewise rejected the idea that the agrarian sector could take the lead in reforming the Chinese society. On the other hand, Li Ta-chao, imbued with a more romantic attitude toward social change, took the opposite view to stress the importance of the peasantry: "In economically backward and semicolonial China, the peasantry constitutes more than ninety per cent of the population; among the whole population they occupy the principal position, and agriculture is still the basis of the national economy. Therefore, when we estimate the forces of the revolution, we must emphasize that the peasantry is the important part."[7] Impassioned by an innate love for

4. Po-wen Middle School for Girls.
5. Including Mao Tse-tung, but not Chou En-lai who was in France, or Chu Teh who was in Germany.
6. Meisner, 242.
7. Meisner, 239.

the purity of the countryside and a deep aversion to the corruption of city life, Li urged young intellectuals to go to the villages to liberate the peasants and stimulate their revolutionary energies, in the spirit of the Russian Populist (*narodnik*) movement. Indeed, he saw in the liberation of the peasantry the liberation of China.[8]

Although the party supported Ch'en's position, Li's views offered a powerful alternative and strongly influenced the thinking of his young assistant, Mao, whom he introduced to Marxism in 1918 and whom he successfully inspired with the Populist, nationalistic views on the peasant role in the revolution. After Li's execution by the warlord Chang Tso-lin on April 28, 1927, it was Mao who carried on the peasant struggle and put his mentor's ideas into practice.

The Nationalist Party Reorganization, 1923-24

The Bolshevik Revolution not only influenced the establishment of the Chinese Communist Party (CCP), but also prompted the reorganization of the Nationalist Party, or Kuomintang (KMT). Dr. Sun, father of the Chinese Revolution, had long been disappointed by the lack of unity and discipline within his party, and by the Western reluctance to assist him in developing China. Ever since the founding of the republic in 1912, he had encountered obstruction and disobedience within his party and precious little cooperation. Nor was the latter quality any more prevalent after two major reorganizations: from the T'ung-meng hui to the Chinese Revolutionary Party (*Chung-kuo ko-ming-tang*) in 1914, and to the Chinese Nationalist Party (*Chung-kuo kuo-min-tang*) in 1919. Sun was continuously frustrated by flagrant acts of insubordination such as Ch'en Chiung-ming's mutiny in 1922 and the open obstruction of the southwestern military governors who had earlier pledged their allegiance.

Equally annoying was the Western support of the warlords and their lack of interest in Sun's international development plan for China. As early as 1913 the Western imperialists had patronized Yüan Shih-k'ai with a £25 million loan from the Five-Power Banking Consortium with which he crushed the Second Revolution. The British minister in particular, John Jordan, supplied Yüan with munitions and barred Sun and Huang Hsing from landing in Hong Kong. After Yüan's death, the imperialists supported the various warlords, fomented civil strife, and turned a deaf ear to Sun's pleas for assistance. The Paris Peace Conference, which ignored China's rightful claims in Shantung, and the Washington Conference of 1922 (p. 532), which smoothed Anglo-American relations with

8. *Ibid.*, 81.

Japan more than it solved China's problems, were further proof of Western insincerity.

Throughout the republican period, Sun was plagued by the threefold problem of foreign imperialism, party disunity, and civil strife, from which he could find no escape and solution. In his frustration, he found the sparkling success of the Bolshevik Revolution doubly inspiring, and the Soviet offer of friendship and abolition of the unequal treaties gratifying and refreshing. Just as he attributed the Russian success to good party organization and strict discipline, Sun blamed his failure on poor discipline, slack organization, and inadequate indoctrination. He was anxious to reorganize the KMT after the successful Soviet model and to seek Soviet aid for his National Revolution.

Two other factors also influenced him. One was the founding of the Chinese Communist Party, which had developed close ties with labor and agrarian organizations, and the other was the fervent nationalism and buoyant public spirit of the younger generation after the May Fourth incident. Since both forces shared his objectives of "anti-imperialism and anti-warlordism," he was ready to introduce new blood into his somewhat aged organization.

However, Sun had to wait to learn the secrets of Soviet success and to reorganize his party. As head of the revolutionary regime at Canton rather than that of the legal government at Peking, he was not the first choice of Moscow. The Russians had sent M. I. Yurin and A. K. Paikes to Peking in 1920 to negotiate a treaty, but the warlord government was advised by the British and the Japanese to decline the overture. The Soviets next turned to the powerful warlord Wu P'ei-fu, whom they conveniently transformed into a "bourgeois nationalist," but under British pressure Wu also proved unresponsive. It was only then that the Soviets "rediscovered" Sun, who had reputedly sent congratulations to Lenin in 1918 which heartened that Bolshevik leader.[9]

In the spring of 1921 the Comintern's Dutch agent, H. Maring, met with Sun in Kwangsi and was most impressed with his nationalist spirit and ideas on revolution. Sun, on his part, was gratified to learn of the Soviet New Economic Policy which he likened naïvely to his own Industrial Plan (*Shih-yeh chi-hua*). Maring soon became convinced that the KMT was the mainstream of Chinese nationalism and that the nascent CCP should expand its influence through utilizing the established base of the KMT. He urged the CCP members to join the KMT on the grounds that

9. C. Martin Wilbur and Julie Lien-ying How, *Documents on Communism, Nationalism, and Soviet Advisers in China, 1918-1927: Papers Seized in the 1927 Peking Raid* (New York, 1956), 138.

it was not a bourgeois party per se but a coalition of all classes. Ch'en and Li reluctantly gave in to his pressure, and in August 1922 the CCP Central Committee resolved to permit individual Communists to enter the KMT. Li Ta-chao took the lead in joining the Nationalist Party through the introduction of its senior member, Chang Chi.

Sun was willing to accept the Communists for a number of reasons. Idealistically, he felt that *all* Chinese, including the Communists, had a right to participate in his National Revolution (*Kuo-min ko-ming*). Practically, he wanted to utilize the CCP's ties with the labor and agrarian movements and Soviet aid in reorganizing the KMT. Furthermore, he very realistically believed that any rapid, independent growth of the CCP under the Soviet aegis, with its commitment to class struggle, would ultimately undermine his own cause of National Revolution; it would therefore be wise to absorb them into his party and assimilate them in time. Lastly, Sun had considerable concern about possible Soviet aid to some warlords not friendly to him; Li Ta-chao and Ch'en Tu-hsiu had been instructed by the Soviets to develop connections with Wu P'ei-fu and Ch'en Chiung-ming, both avowed enemies of the National Revolution. A policy of friendship and alliance with the Soviets and the CCP would undercut these warlords.[10]

The Comintern dispatched Adolf Joffe to China to work out the basis of Soviet-KMT-CCP cooperation. Arriving in Peking on August 12, 1922, Joffe received a most cordial welcome from the New Tide Society and thirteen other organizations, much to the jealousy of the Western consular corps and the displeasure of the warlord government. Subsequently, he engaged in lengthy correspondence and negotiations with Sun, who by this time had decided on the policy of "alliance with the Soviets; admission of the Communists" (*Lien-O yung-Kung*). This policy, approved by fifty-three Nationalist leaders at a Shanghai conference on September 4, 1922, became the cardinal principle in the reorganization of the KMT. A nine-man committee including Ch'en Tu-hsiu was appointed to take charge of the reorganization, and a manifesto drafted by Hu Han-min was announced on January 1, 1923.

On January 12, the Comintern instructed the Chinese Communists to enter the Nationalist Party and take part in Sun's *bourgeois democratic* revolution. Ch'en accepted the order most reluctantly, for he feared for the KMT's corrupting influence on the worker and peasant members of the CCP. "It was only due to the pressure of the Third Internationale that the Chinese Communist Party grudgingly recognized the necessity

10. Chiang Yung-ching, *Pao-lo-t'ing yü Wu-han cheng-ch'üan* (Borodin and the Wu-han regime), (Taipei, 1963), 2-3.

of carrying on its activities within the Kuomintang," remarked Ch'en.[11] However, the CCP itself was not dissolved; the Communists entered the KMT as individuals rather than as a bloc and they agreed to accept the order and discipline of the Nationalist leaders. Publicly, the CCP acknowledged the KMT as the leader and central force of the National Revolution.

The Sun-Joffe negotiations led to a joint manifesto on January 26, 1923, which included four main points: (1) it is not possible to carry out Communism or the Soviet system in China at present; (2) the Soviet government reaffirms its earlier announcement of September 27, 1920, regarding the renouncement of special rights and privileges in China; (3) a mutual understanding is reached with regard to the future administration and reorganization of the Chinese Eastern Railway; and (4) the Soviets disavow any imperialistic intentions or policies in Outer Mongolia.[12]

In his negotiations with Joffe, Sun demonstrated hardheaded practical statesmanship. For all his eagerness to seek Soviet aid, he refused to substitute Communism for his Three People's Principles; nor would he surrender the power of leadership to Marxist discipline and order. He left no doubt that the KMT occupied the leadership position in the National Revolution, and that it was the Communists who entered the Nationalist Party, not vice versa. Moreover, they had entered as individuals, not as a group, so as to avoid the embarrassing situation of "bloc within" or a "party within a party." From all appearances Sun had achieved his objectives very much on his own terms.

Following the Sun-Joffe agreement, the Soviets sent Mikhail Borodin (Grusenburg), an experienced diplomat, to help Sun reorganize the KMT, and General Galen (Blücher) to help train a party army. In addition, some forty Soviet advisers came with them. In August 1923, Sun dispatched a young general, Chiang Kai-shek, to study firsthand the Soviet military system, the political indoctrination of the Red army, and the methods of discipline in the Bolshevik Party. After a three-month visit, Chiang returned home where he was soon commissioned by Sun to found the Whampoa Military Academy outside Canton.

At the first National Congress of the KMT held from January 20 to 30, 1924, and attended by 165 delegates, Sun stressed the importance of party unity and the development of a strong organization for national unification and reconstruction. He called on the members to sacrifice their per-

11. Benjamin I. Schwartz, *Chinese Communism and the Rise of Mao* (Cambridge, Mass., 1958), 53, 60.
12. Complete text in Conrad Brandt, Benjamin I. Schwartz, and John K. Fairbank, *A Document History of Chinese Communism* (London, 1952), 70-71.

sonal freedom and contribute their talents unselfishly to the revolutionary objectives. It was during the meeting that the news of Lenin's death on January 25 arrived, and the Congress, to show its grief and respect, recessed for three days—a public affirmation of the new policy of friendship and alliance with the Soviet Union.

There were, of course, KMT members who, though not averse to the Soviet alliance, were reluctant to accept the CCP. To smooth the path of cooperation, Sun patiently explained that since both the KMT and the CCP were committed to anti-imperialism and anti-warlordism, it behooved them to join hands in the common struggle. On January 28, 1924, Li Ta-chao tactfully declared that members of his party had entered the KMT in order to devote themselves to the revolution, without any ulterior motives to advance the Communist cause. Moreover, they entered as individuals, not as a bloc; therefore one could not accuse them of forming "a bloc within" the KMT party, even though they held double memberships. Li reiterated that as long as the Communists remained in the KMT, they would obey the latter's orders and accept its disciplinary action. He emphatically disclaimed any intention of infiltrating or subverting the KMT from within.[13] In spite of his explanation, the fact remained that the Communist Party did not dissolve itself, nor did its members who entered the KMT lose their Communist membership. There was, in fact, a Communist bloc within the Nationalist Party.

Sun had admitted Communists in the interest of the revolution without seeming to realize all the implications of his action. Still idealistic, he assumed that since the Comintern had favored such a collaboration it would help him control the Communist members, and perhaps even instruct them to obey him. He also entertained some hopes that in due course the small number of Communists might be effectively submerged within the substantially larger Nationalist ranks. What he did not realize was that the real intention of Moscow was to graft the young CCP onto the established body of the KMT so that it could subvert it from within, seize the proletarian hegemony, and squeeze out the rightists like "lemons."[14]

Meanwhile, the Congress had created a Presidium of five members, including Li Ta-chao.[15] It closed with a manifesto emphasizing its anti-imperialist and anti-warlord stand, its dedication to the Three People's Principles and the Five-Power Constitution, and its determination to

13. Wilbur and How, 149.
14. Schwartz, 80.
15. Others included Sun himself, Hu Han-min, Wang Ching-wei, and Lin Shen.

abolish the unequal treaties externally and to establish local self-government internally.

The KMT-CCP collaboration was a marriage of convenience, each needing but distrusting the other. The KMT desired Soviet aid in revitalizing the party, in developing a party army, and in carrying out the National Revolution; it also aspired to the utilization of Communist ties with the workers, peasants, and the masses. On the other hand, the Comintern and the CCP wanted to use the KMT base to expand their influence and eventually to subvert it from within. In this tenuous relationship, cooperation lasted as long as it was in the interest of both; each hoped to emerge the victor when the other had outlived his usefulness. Sun's stature and prestige were decisive factors in holding together the various elements, but once he passed away, divisive forces were unleashed and loomed increasingly large on the horizon.

The Northern Expedition and the KMT-CCP Split

Having reorganized the party, Sun was eager to resume the much-delayed Northern Expedition to wipe out the warlords and frustrate their imperialist supporters. But his death on March 12, 1925, aborted the move.

Sun's political mantle fell on Wang Ching-wei and Hu Han-min, the left- and right-wing leaders of the KMT, respectively. But the military power rested with Chiang Kai-shek, the superintendent of the Whampoa Military Academy who was in charge of developing an officer corps to staff the new party army. His cadets were given political indoctrination as well as military training so that they could correctly instruct the soldiers in the political mission of the revolution. Within the academy, Sun's confidant Liao Chung-k'ai served as the chief party representative, while Ho Ying-ch'in was the head military instructor. All orders and regulations in the academy and in the party army, to be valid and enforceable by the superintendent, had to be countersigned by the party representative. The deputy head of the Political Education Department at the academy was none other than the young Communist Chou En-lai, and among the students of the fourth graduating class was one Lin Piao.

The cadets rapidly became a powerful military factor. They suppressed the Hong Kong–Canton Merchants' Volunteers Uprising in October 1924,[16] drove away the rebel governor Ch'en Chiung-ming, and frustrated the various southwestern warlords. With Canton relatively safe from hos-

16. Organized by Ch'en Lien-po, a compradore of the Hong Kong and Shanghai Banking Corporation.

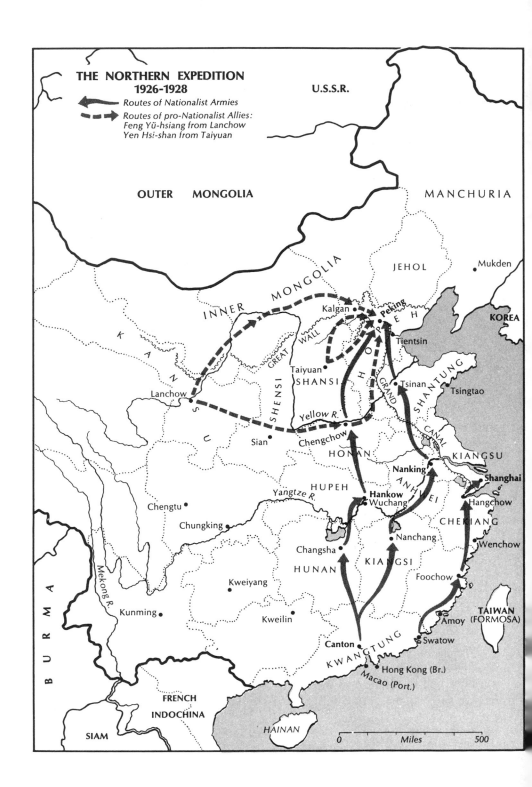

THE NORTHERN EXPEDITION
1926-1928

→ Routes of Nationalist Armies

⇢ Routes of pro-Nationalist Allies:
Feng Yü-hsiang from Lanchow
Yen Hsi-shan from Taiyuan

U.S.S.R.

OUTER MONGOLIA

MANCHURIA

Mukden

INNER MONGOLIA

JEHOL

KOREA

Kalgan

Peking

H O P E H

Tientsin

GREAT WALL

K A N S U

Taiyuan

SHANSI

Tsinan

SHANTUNG

Tsingtao

Lanchow

SHENSI

Yellow R.

GRAND CANAL

Sian

Chengchow

H O N A N

KIANGSU

Nanking

ANHWEI

Shanghai

HUPEH

Yangtze R.

Hankow
Wuchang

Hangchow

Chengtu

Chungking

CHEKIANG

Nanchang

Wenchow

Changsha

KIANGSI

HUNAN

Foochow

Kweiyang

Kunming

Kweilin

Amoy

TAIWAN
(FORMOSA)

Canton

Swatow

K W A N G T U N G

Hong Kong (Br.)

Macao (Port.)

BURMA

Mekong R.

FRENCH
INDOCHINA

SIAM

HAINAN

0   Miles   500

tile forces, a Nationalist Government was established on July 1, 1925, in rival existence with the Peking warlord government, and Wang Ching-wei was made president. A series of pacification campaigns followed in Kwangtung and Kwangsi, and by February 1926 all opposition in the two provinces had been suppressed. Determined to resume the Northern Expedition, the Nationalist Government on June 25 appointed Chiang Kai-shek commander-in-chief of the National Revolutionary army, consisting of 6,000 Whampoa cadets and 85,000 troops. On July 27 Chiang set out on his celebrated campaign against the northern warlords, who were deployed as follows:

1. The Chihli warlord Wu P'ei-fu had control of Honan, Hupeh, part of Chihli and Hunan, and the Peking-Hankow Railway.
2. The Fengtien (Manchurian) warlord Chang Tso-lin had established himself as generalissimo in Peking, controlling Manchuria, Chihli, Shantung, and the Fengtien-Peking and Tientsin-Pukow railways.
3. Sun Ch'uan-fang, who had seceded from the Chihli Clique, was established at Nanking, dominating the five southeastern provinces of Kiangsu, Chekiang, Fukien, Kiangsi, and Anhwei.

In addition, two independent forces were established in the Northwest which belonged neither to these warlord groups nor to the revolutionary army, although they were somewhat sympathetic to the latter:

1. Feng Yü-hsiang's "National People's Army" had retreated to the Northwest under the pressure of the Chihli and Fengtien forces.
2. Yen Hsi-shan had established a firm base in Shansi, without participating in the civil strife.

Chiang's strategy was to attack Wu P'ei-fu first, and then Sun Ch'uan-fang and Chang Tso-lin. Strengthened by Soviet supplies[17] and aided by CCP advance agents who mobilized peasant and worker organizations and fomented strikes and sabotage in the cities, the Northern Expeditionary forces struck a blitzkrieg from Canton to Central China, taking Wuhan in September 1926, Nanchang in November, Foochow in December, and Shanghai and Nanking in March 1927. Within nine months the southern half of China was conquered. The campaign had been a spectacular success; the future indeed looked bright. But at this point an ominous KMT-CCP split developed, which threatened to wreck the party and interrupt the Northern Expedition.

17. Soviet supplies to the KMT reached two million rubles from October 1924 to December 1925. See Wilbur and How, 169.

The bone of contention from the outset was the question of dual membership and its corollary of the "bloc within." The Nationalists had admitted the Communists as individuals and expected them to accept the KMT leadership and obey its orders, but the Communist Party demanded that its members take orders from itself and form a secret bloc within the KMT. Holders of dual membership, in short, were expected to be nominal KMT members but real CCP members. Conflicting orders naturally led to friction which involved the sensitive question of discipline. Yet during Sun's lifetime, no open break took place, although tension continued to rise.

After the assassination of Liao Chung-k'ai in August 1925, some fifteen rightist KMT Executive and Supervisory Committee members[18] left Canton for Western Hill (Hsi-shan), outside Peking, to hold a Fourth Central Executive Committee meeting on November 23 in front of Sun's coffin. Here, they issued a proclamation calling for the expulsion of the Communists from the KMT and the dismissal of Borodin as adviser. The KMT left wing at Canton accused the Western Hill group of lacking the legal quorum to pass valid resolutions. It called its own Fourth Central Executive Committee meeting, and adopted resolutions to censure the Western Hill faction and to call a Second National Congress on January 1, 1926. At this congress, Borodin's domination was formidable. The Communists won new memberships on the KMT Supervisory Committee and increased memberships on the Central Executive Committee. Of the latter's nine-man Standing Committee, three were Communist and three were fellow travelers. At least five or six of the nine ministries of the KMT central party headquarters came under Communist control: Organization, Propaganda, Workers, Farmers, Overseas, and Youth.[19] Facing these developments, the Western Hill group set up its own party headquarters in Shanghai, to signify a split with Canton.

Another ingredient added to the tension was the "Warship *Chung-shan* incident" of March 20, 1926. On that day the captain of *Chung-shan*, under Communist influence, unsuccessfully attempted to kidnap Chiang Kai-shek, who in return dismissed the captain and all Soviet advisers and party representatives in the First Army and its affiliated military establishments. It was in a sense the first step in Chiang's break with the Communists, yet in the interest of the impending Northern Expedition no open schism was announced. Nonetheless, some restrictive measures were quickly taken against the Communists. On May 15, 1926, the

18. Including Tai Chi-t'ao, Lin Shen, Chü Cheng, Chang Chi, Tsou Lu.
19. Chiang Yung-ching, 10-11.

KMT Central Executive Committee passed nine resolutions to limit Communists to no more than a third of all committee memberships, to exclude them from department directorships in the central party headquarters, and to prohibit KMT members from accepting Communist memberships. Although the CCP Central Executive Committee rejected these decisions and resolved to organize its own military forces, Stalin, not wishing to precipitate a split at this point, ordered it to tolerate the resolutions in order to remain within the KMT.[20]

It was only after the Nationalists had successfully imposed these restrictions on the Communists that Chiang set out on his Northern Expedition in July 1926. His troop movement was fast, as noted before, and having pacified Central China, the KMT decided to move the government from Canton to Wuhan by January 1, 1927. Meanwhile, the CCP received an order from Stalin dated November 30, 1926, instructing it to intensify its political work in the revolutionary army, and to improve its military knowledge so as to be ready for important positions in the army.

The Wuhan government was dominated by Borodin and the KMT left wing, and the two important ministries of workers and farmers were put under the charge of the Communists.[21] The latter actively carried out Stalin's new order of March 3, 1927, calling for the intensification of mass movements, arming the workers and peasants, and mobilizing the masses to embarrass and attack the KMT rightists. These stepped-up activities were most evident in the areas under the control of Wuhan, which included Hupeh, Hunan, and Kiangsi.

At the same time Chiang Kai-shek, conducting a successful military campaign, was rapidly building a power base in Eastern and Southeastern China. He had deliberately disregarded Borodin's advice to skip Shanghai in favor of the North, driving instead straight to the gates of that financial center. There the Communist-dominated General Labor Union had already staged a debilitating strike, mobilized its armed pickets, fought the local garrison, and won control of the city from within. Not knowing whether they should cooperate with Chiang, they waited for orders from Moscow. Still hoping to prevent a split, Stalin asked the Shanghai workers to "bury their weapons" and to "avoid any clashes" with Chiang, whose troops thus entered the city unopposed on March 22.[22] Riding the tide of victory, the Northern Expeditionary forces went on to conquer Nanking on March 24 and won domination of Fukien, Chekiang, and greater parts

20. Conrad Brandt, *Stalin's Failure in China, 1924-1927* (Cambridge, Mass., 1958), 76.
21. Su Chao-cheng and T'an P'ing-shan respectively.
22. Brandt, 112-113.

of Kiangsu and Anhwei. It was apparent that Wuhan and Nanking formed two power centers within the KMT hierarchy, and that a split was imminent.

With the support of the Shanghai-Nanking financial circles, Chiang became more determined to persecute the Communists. A "purge committee" (*ch'ing-tang*) was organized on April 10, 1927, and orders were issued to dissolve the political department of the National Revolutionary army. From April 12 on, wholesale liquidation of the Communists began—first in Shanghai, and then in Nanking, Hangchow, Foochow, Canton, and other places. Nationalist troops, police, and secret agents raided Communist cells, shot down suspects on sight, disarmed the workers' pickets, and eliminated the labor unions. When they had finished, a devastating blow had been dealt to China's proletarian vanguard. Surprisingly, during all this time, Chiang continued to profess friendship with Moscow—his quarrels were only with the local Communists.

Buffeted by Communist protest, the Wuhan government on April 17 dismissed Chiang as commander-in-chief of the National Revolutionary army. Chiang could not have cared less; with the help of Hu Han-min he organized his own Nationalist government at Nanking a day later. The split between the two power centers had widened into an unbridgeable gulf.

To counter Chiang's success at Shanghai and Nanking, Borodin proposed that Wuhan itself launch a "Second Northern Expedition"[23] to Peking and sought the collaboration of Feng Yü-hsiang and Yen Hsi-shan. The plan was opposed by M. N. Roy, the new Comintern representative, on the ground that it was too risky to depend on the support of Feng and Yen. Sharp and derogatory exchanges took place between Roy and Borodin. On April 18, the Wuhan government under Wang Ching-wei decided on a double-barrel approach of launching the Northern Expedition first, to be followed by an Eastern Expedition. The military plans projected the occupation of Peking three months after the Wuhan army joined forces with Feng on the Peking-Hankow Railway.[24]

As planned, the Wuhan forces struck successfully into Honan, inflicted heavy casualties on the Fengtien army, and met Feng at the important railway center of Chengchow. Once established in Honan, however, Feng proved to be far more independent than Borodin had expected. Yen was even more of a problem; he refused to cooperate altogether on the ground that Wuhan represented the Communist regime while Nanking

23. The campaign from Canton to Wuhan was considered the "First Northern Expedition."
24. Chiang Yung-ching, 196-99; 202.

was the true Nationalist government. Meanwhile, Chiang Kai-shek carried on his own Northern Expedition successfully along the Tientsin-Pukow Railway, taking Hsuchow on June 2, 1927.

Feng now proposed a joint Northern Expedition by Chiang, Wuhan, and himself. To Borodin and Ch'en Tu-hsiu, joint expedition meant "joint extermination of the CCP," and they spurned the proposal outright. Feng then visited Chiang at Hsuchow on June 20-21, ostensibly to attempt a reconciliation between Nanking and Wuhan but actually to concert action against communism.[25] The conference ended with Feng's public demand that Wuhan expel Borodin and the Communists.

Feng's apostasy and Yen's refusal to collaborate not only shattered Borodin's Northern Expedition but also placed Wuhan in a pocket of hostile forces. Compounding the plight was the effect of Stalin's power struggle with Trotsky. In the wake of Chiang's success, Trotsky had accused Stalin of flawed leadership in China and violation of a cardinal Leninist principle that temporary agreement or even alliance with bourgeois elements was permissible, only if the Communists retained their organizational independence and freedom of action. In the KMT-CCP collaboration, Trotsky asked, where was the Communist freedom of action? To vindicate his China policy, Stalin badly needed a victory. On June 1, 1927, he sent a telegram to Borodin and the CCP asking them (1) to organize a new armed force of 20,000 Communist members and 50,000 workers and peasants; (2) to reorganize the KMT at Wuhan; (3) to increase the worker and peasant members in the KMT Central Committee; (4) to confiscate land at the local level without waiting for orders from the Wuhan government; and (5) to set up a KMT special court to try the counterrevolutionaries without involving the Communists. It was, in effect, a call to raise a separate army and to transform Wuhan into a Communist regime under the intended puppet Wang Ching-wei. Realizing the impossibility of the order, Borodin and Ch'en Tu-hsiu asked Roy to execute it. On June 5, Roy revealed the telegram to Wang in an attempt to show his good will and complete trust. Not until then did Wang realize Stalin's real intention to destroy the KMT left wing and turn the Wuhan regime into a Communist puppetry. Yet he took no immediate action to stop the plot. Instead, he went to see Feng at Chengchow on June 6; Feng offered to mediate between Wang and Chiang.

On July 13, Borodin announced that the Communists would leave the regime, though not the KMT party. The CCP moved its headquarters to Kiu-kiang, Kiangsi, and stepped up its attack on Wuhan. Wang retaliated

25. Chiang Yung-ching, 381.

by announcing on July 14 that Communist members of the KMT guilty of violating Nationalist policies and ideology by word or by deed would be punitively sanctioned. Two days later he further announced that if the Communists left the Wuhan government, they might as well leave the KMT party, army, and all levels of government.

Although Wang seemed to have split with the Communists, he was still tolerant of them. There was no immediate liquidation, nor forcible dismissal of Communists from the KMT party and army. It was not until July 26 that the Wuhan presidium, under increasing Communist excoriation, ordered the ousting of Communists from KMT party and government posts unless they resigned from their CCP membership. They were also ordered to stop obstructing the National Revolution. Meanwhile, no KMT members were allowed to take membership in other parties. Under these adverse conditions, Borodin had no choice but to leave Wuhan on July 27, 1927, returning to Russia via Mongolia.

The final *coup de grâce* fell after the Nanchang Uprising of August 1, which was carried out by the Communists under the name of the Nationalist left wing. Wang finally ordered an all-out liquidation of the Communists and reorganization of the front organizations such as the General Labor Union, the Farmers' Association, the Women's Association, and the Merchants' Association.

Now that Nanking and Wuhan had both liquidated the Communists, the Western Hill faction at Shanghai proposed reconciliation. A special Central Committee was established at Nanking to exercise the power of the party headquarters, and on December 10 all differences between Wuhan and Nanking were formally resolved: Chiang was reappointed commander-in-chief of the National Revolutionary army, while Wang announced plans to go abroad.

The Wuhan government was dissolved in February 1928, although a branch Political Council continued to exist. With the domestic conflict finally resolved, Chiang resumed his Northern Expedition. Though blocked by Japanese troops at Tsinan in Shantung province, Chiang was able to surmount the obstacle. With the help of Feng-Yü-hsiang and Yen Hsi-shan, he marched on Peking, then occupied by the Fengtien warlord Chang Tso-lin. Fleeing to Manchuria, Chang was killed in a train explosion at Huangkutun, near Mukden, engineered by the Japanese, on June 4, 1928.[26] His son, the Young Marshal Chang Hsüeh-liang, pledged

26. The mastermind of the plot was the Kwantung Army's Lt. Col. Kōmoto Daisaku, who wanted to forge a new political order in Manchuria out of the confusion. Tokyo was not informed of the plot in advance. When notified of the incident, Premier Tanaka sighed: "What fools! They [the Kwantung army] behave like children. They have no idea what the parent has to go through." Takehiko Yoshihashi, *Conspiracy at Mukden: The Rise of the Japanese Military* (New Haven, 1963), 50-51.

allegiance to the Nationalist Government in July, and later, on December 31, he endorsed the Three People's Principles, "renounced" his regional control of Manchuria, and signified his support of the Nationalist Government by accepting its flag. As the year 1929 opened, China, or the greater part of it, was united by Chiang Kai-shek, after thirteen dismal years of civil anarchy. With Nanking as the new seat of government, the old capital Peking was renamed Peiping, Northern Peace.

In retrospect, one cannot but conclude that the KMT-CCP split testifies to the complete failure of Stalin's policy in China. He had wanted to seize the proletarian hegemony within the KMT and squeeze out the rightists like "lemons," but little did he realize that the reorganized KMT was no longer the loose, inefficient collectivity he once knew. The party structure had been revitalized by Borodin, and the party army had been trained with the assistance of Galen. Above all, Stalin failed to see that the CCP did not control the army. Moreover, Chiang's political acumen appeared to be at its height and he acted with speed and decisiveness, squeezing out the Communists before Stalin, thousands of miles away, had a chance to strike.[27] Commenting on the event a decade later, Mao Tse-tung told an American journalist that Borodin was indecisive, Roy was a dunce who talked but did not act, and Ch'en Tu-hsiu was guilty of rightist opportunism.[28]

The Diplomacy of Nationalism

The era of ideological and political ferment was also the age of surging nationalism in China. In diplomacy and on the domestic front, Chinese behavior was dominated by a vigorous outburst of nationalistic sentiment. At the Washington Conference of 1921-22, the Chinese fought hard for independence and international respect, and in the post-conference period they struggled ceaselessly with the imperialist powers for tariff autonomy, revocation of extraterritoriality, and relinquishment of foreign concessions. Their intense drive to eliminate these national stigmas led to numerous clashes with foreign police and mercenaries, who reacted all too frequently with highhanded and needlessly harsh measures of repression, with the result that the decade of the 1920s was filled with what the Chinese outrageously called "cases of atrocious murder" (*ts'an-an*). Nationalism, the moving spirit of 19th-century Europe, had finally caught fire with the Chinese, and propelled them forward in a new mission of saving their country from the double scourge of imperialism and warlordism.

27. Schwartz, 80.
28. Edgar Snow, *Red Star Over China* (New York, 1938), 165.

THE WASHINGTON CONFERENCE. The failure of the Paris Peace Conference to settle the Shantung question equitably and to tackle many problems of the Pacific weighed heavily on the Americans. To rectify the wrongs and resolve the unfinished business of the Paris Conference, the United States began making plans in 1920 for another international meeting, which materialized in the Washington Conference of November 12, 1921-February 6, 1922. It was attended by nine powers which had interests in the Far East and the Pacific: Britain, the United States, France, Italy, Japan, China, Belgium, the Netherlands, and Portugal.

The Chinese delegation came with high expectations. It presented a nine-point proposal which asked the participants to honor China's territorial integrity and political independence, to desist from concluding treaties among themselves that would affect China, to respect its rights of neutrality in future wars, to remove all limitations on its political, jurisdictional and administrative freedom, to review all foreign special rights, immunities, and concessions in China, and to set time limits to its commitments. The proposal evoked a warm and sympathetic response from the American and European delegations.

Under the sponsorship of the United States, the Chinese proposal was consolidated into four general principles which ultimately found expression in a Nine-Power Treaty of February 6, 1922. The signatories agreed to respect China's territorial integrity and political independence, to renounce further attempts to seek spheres of influence, to respect its neutrality in time of war, and to honor equal commercial opportunity for all. Separately, the powers agreed to close on January 1, 1923, all foreign post offices in China except those in the leased territories, and to let China increase the import tariff from the actual 3.5 percent to 5 percent ad valorem.

As regards Shantung, direct negotiations between the Chinese and Japanese took place under the good offices of the United States and Britain. World public opinion, particularly official and unofficial American pressure, obliged Japan to relinquish Shantung while retaining some economic rights. It was allowed to keep certain properties in Shantung needed by the Japanese community there, such as consular buildings, public schools, cemeteries and shrines. Japanese nationals were to be appointed advisers in various utilities, stockyards, and vital enterprises; they were also to serve for five years as the chief engineer, the traffic manager, and the chief accountant of the Tsingtao-Tsinan Railway, which the Chinese were to purchase with a Japanese loan. All in all, China accomplished most of its objectives although it did not score a clean sweep.

Elsewhere in the Conference, two important international agreements were concluded. By the Four-Power Treaty of December 13, 1921, intended to replace the Anglo-Japanese Alliance, Britain, the United States, Japan, and France agreed to settle disputes in the Pacific by peaceful consultation. By the Five-Power Naval Treaty of February 5, 1922, these four powers and Italy agreed to maintain a military *status quo* in the Far East and refrain from building new fortifications and naval installations east of the 110th meridian east longitude. The naval ratio of capital ships of the five powers was fixed at 5 each for Britain and the United States, 3 for Japan, and 1.75 each for France and Italy. This ratio entitled Britain and the United States to 15 capital ships of 525,000 tons each, Japan 9 ships of 315,000 tons, and France and Italy 175,000 tons apiece.

On the surface this naval ratio seemed to favor Britain and the United States at the expense of Japan, but it actually benefited the latter, too, in several ways. For one thing, both Britain and the United States maintained two fleets in the Atlantic and in the Pacific, whereas Japan maintained only one fleet in the Pacific. Moreover, at the time of the treaty the Japanese naval strength was 50 percent of that of the United States; the ratio of 5 : 3 actually allowed Japan a 10 percent increase in sea power.[29] All in all, the naval treaty assured Japan of a dominant position in the western Pacific and relative security from British or American attack.

UPSURGE OF CHINESE NATIONALISM. The Nine-Power Treaty was basically an expression of good will on the part of the signatories toward China's *future* development, but it lacked enforcement power. It neither invalidated the existing privileges of the powers in China, nor bound them to defend the Open Door or Chinese independence by force, and as such it did little to alleviate China's sense of injury to self-respect. Foreigners continued to bestride the Chinese scene with haughty arrogance, and filled high posts in the Chinese Maritime Customs, the Salt Revenue, and the Postal Service. Foreign settlements and municipal concessions existed as usual, and Chinese in the Shanghai International Settlement paid taxes without representation on the Municipal Council. The Japanese still ran the Southern Manchurian Railway and used it as an instrument of encroachment, and the British continued to dominate South China trade through Hong Kong. To Chinese patriots, these humiliating signs of imperialism were a constant irritant and a reminder of China's semicolonial status, which should no longer be endured. Fired with nationalism, they set out to "save their country" (*chiu-kuo*) from imperialism, capitalistic exploitation, and warlordism. In this endeavor, the young students and

29. Tang Tsou, *America's Failure in China, 1941-50* (Chicago, 1963), 17.

the growing labor class in the large cities played a major role; they vowed to eliminate these foreign and domestic evils by force if necessary.

Foremost among the nationalistic outbreaks was the "May 30th Incident" of 1925, which had its origin in a Chinese workers' strike in February of that year in protest of low wages at a Japanese cotton weaving mill in Shanghai. Mediation by the Chamber of Commerce and other civic bodies resulted in a preliminary settlement, which the Japanese owner subsequently rejected. The workers went on a second strike, and on May 15 sent eight representatives to negotiate with the management. The confrontation ended in a violent clash, resulting in the killing of one worker and wounding of the other seven. The British-dominated Shanghai Municipal Council not only did not persecute the Japanese who opened fire, but arrested a number of Chinese workers on charges of disturbing peace and order. On May 22, a large body of college students and workers held a public memorial service for the slain and engaged in roadside speeches attacking the Japanese owner. The arrest of many of them by the police incited a 3,000-student demonstration on Nanking Road on May 30, to protest British and Japanese atrocities. At this critical point, a British police lieutenant ordered his men to open fire, killing eleven Chinese and wounding several dozens. In addition, some fifty students were arrested.

This "May 30th Atrocious Incident" (*Wu-san ts'an-an*) provoked nationwide protests, strikes, and boycotts by students, workers, and merchants alike. Not until December, when the British police inspector-general and his lieutenant were fired and the Municipal Council paid an indemnity of Ch\$75,000 to the deceased and the wounded, did the public anger subside. Following the May 30th incident, a rash of nationalistic agitation broke out in different parts of the country, and clashes with the imperialists occurred in many places.

Amid these outbursts, the Chinese in the Shanghai International Settlement strenuously protested "taxation without representation." In 1926 the foreign voters conceded with a resolution permitting three Chinese to be elected to the Municipal Council, which had hitherto been dominated by nine foreigners. The offer was not accepted until the Chinese membership was increased to five in 1930, while the foreign membership remained the same.

In other areas of anti-imperialistic activity, the Chinese succeeded in recovering a number of foreign municipal concessions and to a large extent the tariff autonomy (see next chapter). By the end of the 1920s, the nationalistic revolution had made considerable progress: foreign imperialism had been dealt a severe blow and domestic warlordism considerably reduced by the Northern Expedition.

The Nanking Government

With the success of the Northern Expedition in 1928, the military phase of Dr. Sun's three-stage revolution was completed, and the second stage of political tutelage was due to be introduced. On October 3, 1928, the KMT Central Executive Committee adopted a provisional constitution called "An Outline of Political Tutelage" (*Hsün-cheng kang-ling*), which legalized the party's guidance of the government. The KMT was entrusted with the double duty of tutoring the people in the exercise of their four rights—election, recall, initiative, and referendum—and of supervising the government's exercise of its five powers—executive, legislative, judicial, control, and examination. The highest organ of the party was the National Party Congress, which when not in session delegated its powers to the Central Executive Committee, which in turn maintained a Standing Committee—the real seat of power. In juxtaposition with the Central Executive Committee was the Central Supervisory Committee, which was concerned with matters of discipline and supervision of the finances.

The dominant feature of the government was its five-yüan structure[30] under the president of the republic. Of the five the most important was the Executive Yüan (*Hsing-cheng yüan*), the highest organ of administration commonly dubbed the "cabinet." It consisted of ten ministries,[31] each headed by a minister and two vice-ministers, and a number of special commissions in charge of national reconstruction, overseas affairs, Tibetan-Mongolian affairs, etc. Contrary to Western practice, the Executive Yüan was not responsible to the legislative branch of government, but to the party and the president of the republic.

The Legislative Yüan (*Li-fa yüan*) was composed of 49 to 99 members chosen for two years on a more or less geographical basis. It was not a counterpart of the Western parliament or congress, but was essentially a law-drafting organization which translated the legislative principles adopted by the KMT Central Executive Committee into law. Its duties included deliberations on laws, budget, amnesty, declaration of war, and treaties of peace.

The Judiciary Yüan (*Ssu-fa yüan*) was the highest judicial organ of state and was in charge of interpreting laws and orders, initiating pardons, reprieves, and restitution of civil rights, and coordinating the court systems. However, it could not interfere with the court decisions.

---

30. Created on recommendation of Hu Han-min in accordance with Dr. Sun's "Outline of National Reconstruction."
31. Internal affairs, foreign affairs, war, finance, agriculture and mining, industry and commerce, education, communication, railway, and hygiene.

The Examination Yüan (*K'ao-shih yüan*) was made a separate branch of the government largely because of the tradition of the civil service examinations. Included under its administration were two structures: an Examination Commission which conducted different types of government examinations, and a Ministry of Personnel which took charge of civil service ratings.

The Control Yüan (*Chien-ch'a yüan*) approximated in function the old Censorate. It was composed of 19 to 29 members who supervised government operations, audited the budget, and impeached derelict officials.

Each of these five yüan was headed by a president and a vice-president who were usually senior members of the KMT. The initial slate of key appointments in the National Government was as follows:

> President of the republic: Chiang Kai-shek (Chiang Chung-cheng).
> Executive Yüan: T'an Yen-k'ai, president; Feng Yü-hsiang, vice-president.
> Legislative Yüan: Hu Han-min, president; Lin Shen, vice-president.
> Judiciary Yüan: Wang Ch'ung-hui, president; Chang Chi, vice-president.
> Examination Yüan: Tai Chi-t'ao, president; Sun Fo (Sun K'o), vice-president.
> Control Yüan: Ts'ai Yüan-p'ei, president; Ch'en Kuo-fu, vice-president.

The new government was dedicated to the fulfillment of Dr. Sun's legacy—The Three People's Principles, the Five-Power Constitution, the Fundamentals of National Reconstruction, and Sun's deathbed admonition: "The revolution is not yet complete; comrades must strive on!" Pledging to carry on the unfinished work of revolution, it vowed externally to fight for the complete abolition of the unequal treaties and win for China a position of equality with the leading powers, and internally to initiate democratic reconstruction and social reforms. Hopefully, by the end of the Period of Political Tutelage, which was fixed to last six years from 1929, the country would be ready for the introduction of a constitution.

The challenge of the threefold revolution—nationalistic, democratic, and social—was indeed great, and the responsibility of the government heavy. It remained to be seen whether its ability was equal to its task.

Further Reading

Adshead, S. A. M., *The Modernization of the Chinese Salt Administration, 1900-1920* (Cambridge, Mass., 1970).

Bernal, Martin, *Chinese Socialism to 1907* (Utica, 1976).

Bianco, Lucien, *Origins of the Chinese Revolution, 1915-1949* (Stanford, 1971).

Borg, Dorothy, *American Policy and the Chinese Revolution, 1925-1928* (New York, 1947).

Brandt, Conrad, *Stalin's Failure in China, 1924-1927* (Cambridge, Mass., 1958).

————, Benjamin Schwartz, and John K. Fairbank, *A Documentary History of Chinese Communism* (London, 1952).

Chan, F. Gilbert, and Thomas H. Etzold, *China in the 1920's: Nationalism and Revolution* (New York, 1976).

Ch'en, Jerome, "The Left Wing Kuomintang—a Definition," *Bulletin of the School of Oriental and African Studies,* University of London, XXV:Part 3:557-76 (1962).

Chesneaux, Jean, *The Chinese Labor Movement, 1919-1927,* tr. from the French by H. M. Wright (Stanford, 1968).

Chiang, Kai-shek, *China's Destiny* (New York, 1947).

————, *Soviet Russia in China; A Summing Up at Seventy* (New York, 1957).

Chiang, Yung-ching 蔣永敬, *Pao-lo-t'ing yü Wu-han cheng-ch'üan* 鮑羅廷與武漢政權 (Borodin and the Wuhan regime), (Taipei, 1963).

Dirlik, Arif, *Revolution and History: Origins of Marxist Historiography in China, 1919-1937* (Berkeley, 1978).

Fewsmith, Joseph, *Party, State, and Local Elites in Republican China: Merchant Organizations and Politics in Shanghai, 1890-1930* (Honolulu, 1984).

Hofheinz, Roy, Jr., "The Autumn Harvest Insurrection," *The China Quarterly,* 32:37-87 (Oct.-Dec. 1967).

————, *The Broken Wave: The Chinese Communist Peasant Movement, 1922-1928* (Cambridge, Eng., 1977).

Holubnychy, Lydia, *Michael Borodin and the Chinese Revolution, 1923-1925* (University Microfilms International for the East Asian Institute, Columbia University, 1979).

Honig, Emily, *Sisters and Strangers: Women in the Shanghai Cotton Mills, 1919-1949* (Stanford, 1986).

Hsiao, Tso-liang, *Chinese Communism in 1927: City vs. Countryside* (Hong Kong, 1970).

Iriye, Akira, *After Imperialism: The Search for a New Order in the Far East, 1921-1931* (Cambridge, Mass., 1965).

Isaacs, Harold R., *The Tragedy of Chinese Revolution* (Stanford, 1951).

Jacobs, Dan N., *Borodin: Stalin's Man in China* (Cambridge, Mass., 1981).

Jordan, Donald A., *The Northern Expedition: China's National Revolution of 1926-1928* (Honolulu, 1976).

Kagan, Richard C., "Ch'en Tu-hsiu's Unfinished Autobiography," *The China Quarterly,* 50:295-314 (April-June 1972).

Kasanin, Marc, *China in the Twenties* (Moscow, 1973).

King, Wunsz, *China at the Washington Conference, 1921-1922* (New York, 1963).

Kovalev, E. F., "New Materials on the First Congress of the Communist Party of China," *Chinese Studies in History,* VII:3:19-36 (Spring 1974).

Kuo, Hua-lun 郭華倫, *Chung-kung shih-lun* 中共史論 (On the history of Chinese Communism), 4 vols. (Taipei, 1969-71).

Kuo, Thomas C., *Ch'en Tu-hsiu (1879-1942) and the Chinese Communist Movement* (South Orange, N.J., 1975).

Kuo, Warren, *Analytical History of the Chinese Communist Party* (Taipei, 1968), 2 vols.

Kwei, Chung-gi, *The Kuomintang-Communist Struggle in China, 1922-1949* (The Hague, 1970).

Landis, Richard B., "The Origins of Whampoa Graduates Who Served in the Northern Expedition," *Studies on Asia*, 149-63 (1964).

Lee, Chong-sik, *Revolutionary Struggle in Manchuria: Chinese Communism and Soviet Interest, 1922-1945* (Berkeley, 1983).

Lee, Feigon, Chen Duxiu: *Chen Duxiu: The Founder of the Chinese Communist Party* (Princeton, 1983).

Leong, Sow-theng, *Sino-Soviet Diplomatic Relations, 1917-1926* (Honolulu, 1976).

Li, Yu-ning, *The Introduction of Socialism into China* (New York, 1971).

Loh, Pichon P. Y., *The Early Chiang Kai-shek: A Study of His Personality and Politics, 1887-1924* (New York, 1971).

MacFarquhar, Roderick L., "The Whampoa Military Academy," *Papers on China*, Harvard University, Vol. 9 (1955).

McDonald, Angus W., Jr., *The Urban Origins of Rural Revolution: Elites and the Masses in Hunan Province, China, 1911-1927* (Berkeley, 1978).

Meisner, Maurice, *Li Ta-chao and the Origins of Chinese Marxism* (Cambridge, Mass., 1967).

Nathan, Andrew, *Peking Politics, 1918-1923* (Berkeley, 1976).

North, Robert C., *Kuomintang and Chinese Communist Elites* (Stanford, 1952).

——, *Moscow and Chinese Communists* (Stanford, 1953).

——, and Xenia J. Eudin, *M. N. Roy's Mission to China* (Berkeley, 1963).

Rea, Kenneth W. (ed.), *Canton in Revolution: The Collected Papers of Earl Swisher, 1925-1928* (Boulder, 1977).

Roy, M. N., *My Experience in China* (Calcutta, 1945).

——, *Revolution and Counter-Revolution in China* (Calcutta, 1946).

Schwartz, Benjamin I., *Chinese Communism and the Rise of Mao* (Cambridge, Mass., 1958).

Snow, Edgar, *Red Star over China* (New York, 1938).

So, Wai-chor, *The Kuomintang Left in the National Revolution, 1924–1931* (Hong Kong, 1991).

T'ang, Leang-li, *The Inner History of the Chinese Revolution* (London, 1920).

——, *The Suppression of Communist Banditry in China* (Shanghai, 1934).

Thornton, Richard C., "The Emergence of a New Comintern Strategy for China: 1928" in M. M. Drackhovitch and B. Lazitch (eds.), *The Comintern: Historical Highlights* (New York, 1966), 66-110.

——, *China, the Struggle for Power, 1917-1972* (Bloomington, 1973).

Trotsky, Leon, *Problems of the Chinese Revolution*, 3rd ed. (New York, 1966).

Wang, Lü-chün 王聿均, *Chung-Su wai-chiao ti hsü mu—ts'ung Yu-lin tao Yüeh-fei* 中蘇外交的序幕···從優林到越飛 (Prelude to Sino-Soviet diplomacy: from Yurin to Joffe), (Taipei, 1963).

Whiting, Allen S., *Soviet Policies in China, 1917-1924* (New York, 1954).

Wilbur, C. Martin, "Military Separatism and the Process of Reunification under the Nationalist Regime, 1922-1937" in Ping-ti Ho and Tang Tsou (eds.), *China in Crisis,* Vol. I, *China's Heritage and the Communist Political System* (Chicago, 1968), Book I, 203-63.

————, *The Nationalist Revolution in China, 1923-1928* (New York, 1985).

————, and Julie Lien-ying How (eds.), *Documents on Communism, Nationalism, and Soviet Advisers in China, 1918–1927: Papers Seized in the 1927 Peking Raid* (New York, 1956).

————. *Missionaries of Revolution: Soviet Advisers and Nationalist China, 1920–1927* (Cambridge. Mass., 1989).

Willoughby, W. W., *China at the Conference* (Baltimore, 1922).

Wu, Hsiang-hsiang 吳相湘, *O-ti ch'in-lüeh Chung-kuo shih* 俄帝侵略中國史 (A history of the Russian imperialist aggression in China), (Taipei, 1957), Vol. 2, chs. 1-2.

Wu, Tien-wei, "Chiang Kai-shek's March Twentieth Coup d'état of 1926," *The Journal of Asian Studies,* XXVII:3:585-602 (May 1968).

Yoshihashi, Takehiko, *Conspiracy at Mukden: The Rise of the Japanese Military* (New Haven, 1963).

23

The Nationalist Government: A Decade of Challenges, 1928-37

From its inception in 1928 to the outbreak of the Sino-Japanese War in 1937, the Nationalist government at Nanking hardly enjoyed a day of peace from domestic squabbles and foreign aggression. No sooner had it been established as the legal government of China than it found itself challenged by dissident politicians within the KMT and by the rebellious "new warlords." Compounding the disorder were the two larger threats of rising Communist opposition in the southeast and Japanese aggression in Manchuria, Shanghai, and North China. The decade in question was indeed fraught with "internal troubles and external invasion" (*nei-yu wai-huan*), as the traditional phrase has it. Partly because of the overwhelming circumstances, the Nationalists failed to carry out the much-needed social and economic reforms to alleviate the plight of the peasant—a negligence which was to have far-reaching consequences a decade later. Yet, in spite of all odds the government was able to score some progress in modernization—particularly in the fields of finance, communication, education, defense, and light industry. While a definitive account of this Nationalist decade has yet to be written, pending the opening of new archival materials, the key developments of the period can be traced with some perspective and accuracy.

The "New Warlords" and the Dissident Politicians

The unification achieved by the Northern Expedition was more apparent than real, for although many of the northern warlords had been wiped out, a number of others had maintained themselves by nominally supporting the Expedition. In his eagerness to achieve a national unification, Chiang Kai-shek negotiated with them for a mutual accommodation, granting them appointments which confirmed their semi-independent regional status while receiving in return their recognition of Nanking as the central government of China.

Some of these warlords were in fact quite "progressive" in outlook, promoting modernization in their own jurisdictions. But they lacked the sense of national commitment that would make them surrender their semi-independence. Cooperation with Nanking was possible as long as their interests did not collide; if they did the warlords would act as they saw fit or perhaps challenge Nanking to a contest. Dubbed the "New Warlords," they maintained their regional power bases as follows:

1. Li Tsung-jen and Li Chi-ch'en headed the Kwangsi Clique which dominated the provinces of Kwangsi, Kwangtung, Hunan, and Hupeh.
2. Feng Yü-hsiang and his "National People's Army" occupied a preponderant position in the northern and northwestern provinces of Shantung, Honan, Shensi, Kansu, Chinghai, and Ninghsia.
3. The Young Marshal Chang Hsüeh-liang controlled the Northeast (Manchuria) and Jehol.
4. Yen Hsi-shan had established a strong base in Shansi, reaching out into Hopeh, Suiyuan, and Chahar.

Each of them maintained a large army for territorial aggrandizement as well as for self-protection; collectively, they drained a good portion of the country's meager resources desperately needed for national reconstruction. In March 1929 the KMT Third National Congress called for the amalgamation of splinter military groups into one national command, the reorganization of all troops into one national army, and the centralization of local financial administrations to prevent the provinces from siphoning off receipts that legally belonged to the central government. The new warlords saw in these resolutions artful devices to reduce their power and demanded that Nanking take the first step in military cut. But Chiang considered his Whampoa-trained staff and troops the backbone of a new national army and insisted that any cuts should start with the provincial forces. The obstinace of the one side merely reinforced the intransigence

CHINA IN 1930

Main railways (in China)

0   Miles   1000

S.

R.

Lake Baikal

Amur R.

HEILUNGKIANG

Ulan
Bator

Harbin

MANCHURIA

KIRIN

ONGOLIA

Vladivostok

M O N G O L I A

CHAHAR

JEHOL

Mukden

L I A O N I N G

SUIYUAN

Kalgan

Peking

Tientsin

Dairen (Jap.)

K O R E A

JAPAN

NGSIA

HOPEH

Taiyuan

SHANSI

S H A N T U N G

Tsingtao

chow

Yellow

R.

SHENSI

Sian

Kaifeng

HONAN

KIANGSU

Nanking

ANHWEI

Shanghai

nglu

Yangtze R.

HUPEH

Hankow

Ichang

HWAN

CHEKIANG

king

Changsha

Nanchang

Wenchow

HUNAN

KIANGSI

RYUKYU IS.

Pacific

WEICHOW

iyang

FUKIEN

Foochow

Kweilin

Amoy

TAIWAN
(FORMOSA)

KWANGSI

K W A N G T U N G

Canton

Swatow

Ocean

Macao
(Port.)

Hong Kong (Br.)

CH

CHINA

HAINAN

PHILIPPINE IS.

of the other and the resolutions of the Congress were rendered inoperative from the beginning.[1]

Quite apart from the problem of the new warlords, the KMT was plagued with factional strife. The right wing, led by party elder Hu Han-min and the West Hill group,[2] was in constant conflict with the left wing headed by Wang Ching-wei.[3] Meanwhile Chiang Kai-shek, the new strong man holding the military power, represented a third force. Junior to both Hu and Wang in party status, Chiang charted his course alternately favoring each with his support according to the dictates of political necessity and expediency. In his new government at Nanking, Chiang collaborated with Hu, who was made president of the Legislative Yüan, while Wang and his left-wing followers were out of office. The latter group retaliated by accusing Chiang of betraying the principles and ideas of Sun, and demanded a reorganization of the KMT in the spirit of the 1924 manifesto—hence their nickname "The Reorganizationists."

When Nanking called a KMT national congress to deliberate on the provisional constitution for the Tutelage Period (*Hsün-cheng yüeh-fa*), Wang, along with Eugene Ch'en and others, protested with the creation of a separatist government at Canton in May 1931. Confronted by a rival government and hostile public opinion, Chiang resigned in December as president of the Nanking government. He was replaced by the mild elder statesman, Lin Shen, while Sun Fo (Dr. Sun's son) became the head of the Executive Yüan. With this reorganization, the Canton regime agreed to dissolve itself. The new leaders now pleaded with Chiang and Wang to compose their differences in the interest of the country, and the two of them met at Hangchow and went to Nanking jointly to symbolize a *rapprochement*. On January 25, 1932, Sun Fo resigned the presidency of the Executive Yüan in favor of Wang, while Chiang accepted the chairmanship of the Military Commission. It is noteworthy that the Wang-Chiang reconciliation was possible only because the latter had split with Hu Han-min in March 1931.[4]

The political realignment restored some peace within the KMT, but not the country. The year 1933 witnessed another insurrection in Fukien, engineered by the commanders[5] of the Cantonese Nineteenth Route army, which had heroically fought the Japanese at Shanghai the year before

1. Ch'ien Tuan-sheng, *The Government and Politics of China* (Cambridge, Mass., 1950), 101.
2. Including Wu Chih-hui, Chang Chi, Sun Fo, Lin Shen, and Tai Chi-t'ao.
3. Including Madame Sun Yat-sen and Eugene Ch'en.
4. Over the question of the provisional tutelage constitution, which Chiang wanted but which Hu insisted was unnecessary.
5. Generals Ts'ai T'ing-k'ai and Chiang Kuang-nai.

(see next section). This army had been transferred to Fukien to fight the Communists after the Shanghai truce in May 1932, but once there, its commanders were won over by Communist propaganda and by ambitious southern politicians.[6] The army leaders created a "People's Revolutionary Government" at Foochow, renamed their forces the "People's Revolutionary Army," and in November 1933 rebelled against the central government. They called for war with Japan and collaboration with the Communists and the Soviet Union. Despite its leftist orientation, the movement failed to receive aid from the Communists who themselves were hard pressed by the Nationalists (section below). Thus deprived of critical support, the Fukien insurrection was suppressed in January 1934, and the Nineteenth Route army was reorganized as the Seventh National army.

From the above survey it is clear that throughout its first decade of existence, the Nanking government was seriously beset with internal strife and civil wars. Though it survived these crises, its energy and resources, which could otherwise have been devoted to the urgent task of national reconstruction, had been much spent. The fate of the government would have been very different had it not been for the help it twice received from the Young Marshal. Yet in transferring troops to North China, Chang Hsüeh-liang left Manchuria in a vulnerable position. Of this, the Japanese were quick to take notice.

Japanese Aggression in Manchuria

Manchuria, the rich Northeast of China, is noted for its abundant agricultural products and mineral resources. Japan had coveted the area ever since its defeat of China in 1895, and its ambitions were further aroused by the acquisition of the former tsarist rights in Manchuria after the Russo-Japanese War. With the annexation of Korea in 1910, many Japanese came to regard Manchuria as the next "logical" target for conquest. Three times—in 1912, 1916, and 1928—they schemed to foment a "Manchuria-Mongolia Autonomous Movement," and although these attempts failed, the idea that "to conquer the world it is necessary to conquer China first, and to conquer China it is necessary to conquer Manchuria and Mongolia first,"[7] gained increasing momentum.

6. Li Chi-ch'en and Ch'en Ming-shu.
7. Often attributed to the "Tanaka Memorial" of 1927, which actually did not exist; however, the ideas contained in the alleged memorial were quite common among the Japanese. In fact, the Dairen Conference of 1927 adopted resolutions that contained these ideas. Cf. Liang Ching-tun, *Chiu-i-pa shih-pien shih-shu* (A historical account of the September 18, 1931 Incident), (Hong Kong, 1964), 2-3, 197, 199, 218.

Japanese activists generally assumed that chaos and disorder in China would facilitate their scheme; hence any attempt at unification had to be prevented. They received the sympathy and encouragement of the Kwantung army in Manchuria. The Kwantung army was something of an anomaly in the Japanese military establishment. Its origin can be traced to the period immediately after Japan's defeat of Russia in 1905. As part of the peace settlement, Japan took over the Russian leasehold in the Liaotung Peninsula and tsarist railway and economic rights in Manchuria. In 1906, the Japanese renamed the southern base of Liaotung, including Port Arthur and Dairen, the Kwantung Leased Territory, to be administered by a governor-general whose jurisdiction included the railway zone in Manchuria. For thirteen years, the position of governor-general was held by a general who simultaneously served as commander of the local army, but in 1919 the office of the governor-general became a civilian administration, with a separate Kwantung Army Command to guard the leased territory and the Manchurian railway zone. Following the Russian trick of posing as "railway guards," the Kwantung army firmly established itself in Manchuria—so firmly, in fact, that it moved its headquarters from Port Arthur to Mukden in 1928. Virtually free from home control, the Kwantung army enjoyed a semiautonomous status and took upon itself the task of wresting Manchuria from China.[8]

This self-appointed mission of the Kwantung army received a new impetus with the arrival of Lt. Colonel Ishiwara Kanji in October 1928 and Colonel Itagaki Seishirō in July 1929. These two strategists quickly became the ideological and political mentors of the Kwantung army, relegating the commanding general and the chief of staff to nominal leadership. Itagaki and Ishiwara openly advocated the occupation of Manchuria, which they proposed to use as a bulwark against a Soviet southern advance and as a supply base in the event of war with the United States. Moreover, Manchuria's vast territory and natural wealth could alleviate the crowded conditions and limited resources of Japan, provide business opportunities, and relieve the unemployment problem at home. To justify the seizure, they argued that the thirty million suffering people of Manchuria were eagerly awaiting Japanese liberation from the misrule of warlords and greedy bureaucrats.[9] They considered 1931 propitious for action. China was deeply mired in domestic turmoil and natural disaster. The Communist threat loomed large, and the central government was involved

8. Sadako N. Ogata, *Defiance in Manchuria: The Making of Japanese Foreign Policy, 1931-1932* (Berkeley, 1964), 3-4; Takehiko Yoshihashi, *Conspiracy at Mukden,* 37, 130-31.
9. Ogata, 42-45.

in a succession of costly campaigns (next section). Aggravating the plight were the devastating floods of the Grand Canal and the Yangtze and Huai rivers, which drowned 140,000 and left 250,000 homeless in the ten central provinces.

Internationally, the situation was no less favorable. The Western powers, hard hit by the Depression, were too involved with domestic problems to block Japanese aggression, and the League of Nations was too powerless to intervene. The Nine-Power Treaty of 1922 which guaranteed China's political and territorial integrity, and the Kellogg-Briand (Paris) Pact of 1928 which outlawed war as an instrument of national policy, had all but become mere paper shorn of the power to enforce their idealistic goals.

Within Japan itself, ominous signs of economic and social distress were appearing even before depression hit the country. Industry, which had undergone vast expansion in the 1920s, was suffering the effects of overproduction, which caused business failures and rising unemployment. In 1927, thirty-five banks including the large Bank of Taiwan went out of business, and between July 1929 and June 1930 some 660,000 men were thrown out of work.[10] Furthermore, the world-wide Depression had sharply reduced Japan's trade with the United States, Britain, and China. Capitalizing on the economic and social unrest thus generated, many expansionists advocated that the conquest of Manchuria would lift the country out of its predicament, and they received the endorsement of the army and the *zaibatsu* (financial-industrial overlords).

Until the 1920s the militarists had traditionally remained aloof from politics. But with the election of General Tanaka Giichi as president of the Seiyūkai (Political Friends Association) in 1925 and his assumption of the premiership in 1927, the military leaders emerged as a powerful force in national policy at the expense of party government. They accused the civilian leaders of incompetency as shown in their acceptance of the "humiliating" naval ratio of 5 : 5 : 3 at the Washington Conference (1922) and in their ratification of the London Naval Agreement (1930) which reaffirmed that ratio. They criticized, ridiculed, and bullied the civilian government, and loudly clamored for a positive policy that would lead to Japan's domination of China and ultimately the world. The first step, they insisted, was the conquest of Manchuria. Inasmuch as 75 percent of foreign investment there was already Japanese, the *zaibatsu* also favored intensified activity, although they preferred peaceful penetration to outright military conquest.

Tokyo's military authorities had set the spring of 1932 as the target date

10. Yoshihashi, 12, 116.

for the occupation of Manchuria, but the Kwantung army would not wait. Their sense of urgency was in some measure precipitated by the possibility of the routine reassignment of Itagaki and Ishiwara in the summer of 1931. In June, Major Hanaya Tadashi, dispatched to Tokyo to plead for an immediate invasion of Manchuria, succeeded in winning the approval of the most influential members of the military establishment.

Worried about the military's unruly behavior, the Japanese emperor repeatedly urged prudence and restraint on September 10 and 11. On the 15th, Minister of War Minami dispatched General Tatekawa Yoshisugu of the General Staff to Mukden to "caution the Kwantung army against rash action and to warn that support could not be expected from the government." When word of this mission was secretly transmitted to the Kwantung army by the Second Division of the General Staff, Itagaki and Ishiwara abruptly decided to schedule the incident before the emissary could deliver his restraining message. Arriving in Mukden on September 18, Tatekawa was immediately whisked off to a lavish dinner party by the wily Itagaki, and then wined and dined into a state of complete intoxication. Story has it that in secret sympathy with the Kwantung army's plot, Tatekawa allowed himself to be tricked into delaying the delivery of the message.[11]

At 10:00 p.m. a bomb exploded on the Southern Manchurian Railway track outside Mukden. The damage was actually so minimal that it did not disrupt normal railway service, but the Japanese patrol claimed that Chinese soldiers opened fire from the fields after the explosion and that it had no choice but to fight back in "self-defense." By 3:40 the following morning the walls of Mukden had been breeched and the city occupied. Changchun was taken on September 19, Antung and Yingkow the 20th, and Kirin the 21st.

Both Tokyo and Commander Honjō Shigeru knew of the contrived invasion plot but took no action to stop it, thus allowing field-grade officers of the Kwantung army to take the fate of Japan into their own hands and lead her onto the road of militarism, conquest, and ultimately, destruction. In the view of many, the Mukden Incident of September 18, 1931, sowed the seeds of World War II.

News of the incident reached Tokyo at 2:00 a.m. September 19. The government was split on a future course of action. The Minister of War and the General Staff urged the support of the Kwantung army on the ground that the patriotism of its officers should not be blunted. The civilian cabinet under Wakatsuki in principle opposed the military conquest

11. Ogata, 58-59.

of Manchuria and was distressed at the headstrong conduct of the Kwantung army, but was unable to counter it. An irrevocably divided situation emerged at midday on September 19 when, just as the cabinet declared a policy of nonaggression, the Minister of War announced that the army need not consult the cabinet about future measures but would rely on the discretion of the Kwantung army. Although the cabinet denied Honjō's request for three reinforcement divisions and prohibited the Korean Command to send troops to Manchuria, the Kwantung army continued to advance on its own, while the Korean Command defiantly dispatched reinforcements to Manchuria on September 21. Wakatsuki struggled briefly to withhold funds from the Korean expedition but gave in to army pressure on September 23, thus in effect approving the Manchurian incident. At a time when the country was in desperate need of political leadership, Wakatsuki failed to provide it.[12] What followed was a period of "credibility gap," with the civilian government repeatedly pronouncing its policy of nonexpansion of hostilities, while the military continued to advance in Manchuria. Step by step Wakatsuki was made to accept the *fait accompli* of the army, and the embarrassment that followed led to the cabinet's downfall in December 1931.

International sanction was slow to come. The new British government, in office but a month, was beset with domestic problems. British public opinion was surprisingly lenient toward Japan, whose action in Manchuria it considered not "entirely unjustified." The London *Times* stated that "Japan had a strong case, but had put herself regrettably and unnecessarily in the wrong." The United States took the easy position that Tokyo could not be held responsible for the violation of the Paris Pact since the Kwantung army had acted without its authorization. The Soviet Union also took no action as long as its Siberian border remained unviolated.[13] Thus, China was left to face the enemy alone.

Actually, the Japanese attack was not entirely unexpected in China. On September 11, 1931, Chiang Kai-shek warned the Young Marshal not to engage the Japanese, and on September 15 the bulk of the Northeastern forces at Mukden were transferred. When hostilities broke out on September 18, the Young Marshal again asked for instructions from Peking, where he lay sick, and was told once more not to resist. Deeply embroiled in civil strife, Chiang Kai-shek could not afford a foreign war. He decided to appeal to the League of Nations, with full knowledge that it was powerless to intervene and that the Western powers were disinclined to help, yet he could find no other source of support. By appealing to the interna-

12. Ogata, 65-69; Yoshihashi, 9, 235.
13. Ogata, 71-73.

tional organization for justice, he hoped to gain time to organize his defense and await a favorable turnabout in Japanese domestic politics. For some unknown reason, he did not pursue direct negotiations with Tokyo. The policy of the Nanking government, so often oversimplified as one of nonresistance, was actually a combination of "nonresistance, noncompromise, and nondirect negotiation." In retrospect, one cannot help feeling that such a negative approach could hardly achieve positive results. If the government had authorized the Northeastern army to resist the invader, the glamour of aggression might have been dimmed, thus providing a chance for the more moderate civilian government in Tokyo to have had a greater voice in the China affair. Moreover, if Nanking had pursued an active policy of negotiations with Tokyo, it might have reaped more positive results.[14] Unfortunately, it followed neither course. Instead, it placed its reliance on protests to Tokyo and on appeals to the League of Nations. On December 10 the League decided to dispatch an investigatory mission to Manchuria.

The United States, on its part, announced on January 7, 1932, the "Non-Recognition Doctrine" through Secretary of State Henry Stimson. By the terms of this statement, the United States declared that it would not recognize any situation, treaty, or agreement created by means contrary to the covenants and obligations of the Kellogg-Briand (Paris) Pact of 1928, which outlawed war as an instrument of national policy.

Without effective international sanction and concerted Chinese resistance, the Japanese army overran Manchuria in five months. The only flicker of Chinese heroism came from a local general—Ma Chan-shan, acting governor of Heilungkiang—who stubbornly resisted the enemy in spite of all the odds against him. His ability to frustrate the invader inspired the rise of local militia and "righteous volunteers" who fought the Japanese as best they could. Yet in the end these sporadic, uncoordinated resistance movements failed to stop the enemy.

On January 28, 1932, the Japanese opened a second front at Shanghai to divert international attention from Manchuria. There they ran into stiff resistance by the Cantonese Nineteenth Route army and Nanking's modernized Fifth army. But after holding the enemy at bay for more than a month, the Chinese defenses crumbled, and the Nanking government retreated to Loyang in Central China. Later, through international mediation a truce was arranged on May 5, 1932, by which the Japanese agreed to evacuate the occupied areas of Shanghai and Woosung.

14. Liang Ching-tun, prefaces, iii, vi.

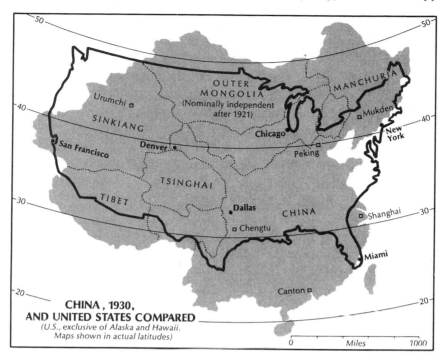

CHINA , 1930,
AND UNITED STATES COMPARED
*(U.S., exclusive of Alaska and Hawaii.
Maps shown in actual latitudes)*

To legitimize their naked aggression, the Japanese on March 9, 1932, created the puppet state of Manchukuo—the Manchu state—to impress the forthcoming International Commission of Inquiry with the "local character" of the Manchurian incident. The last Ch'ing emperor, P'u-i, deposed in 1912, was made Chief Executive, with a group of leftover old literati as ministers.[15] The Commission, led by the Acting Viceroy of India, Lord Lytton, spent six weeks (April 21-June 4) in Manchuria, and in September submitted its report. Undeceived by the façade, it condemned Japan as an aggressor and rejected its claim that Manchukuo was a spontaneous development of the Manchus. The report refuted the Japanese argument that military operations in Manchuria were necessitated by self-defense, and branded Manchukuo a puppet state under the domination of Japanese military and civilian officials. As a result, the League refused to recognize the legality of Manchukuo; but apart from this moral sanction, it could do nothing else.

Japan's reaction was singularly arrogant and insulting: it withdrew from the League.

The Japanese advance was finally brought to a halt by the T'ang-ku

15. Including Cheng Hsiao-hsü, the poet and calligrapher, as premier.

Truce on May 31, 1933, which turned Eastern Hopeh into a demilitarized zone, from which both Chinese and Japanese troops were to be evacuated. In effect, the defense of Peiping and Tientsin was lost by default.

Having completed the conquest of the four Northeastern provinces of China, the Japanese enthroned P'u-i on March 1, 1934, with a reign title of "K'ang-te," Prosperity and Virtue.

The Communist Challenge

Even as the Nationalist government was threatened by Japanese invasion from without and dissident politicians and new warlords from within, it was confronted by a much greater and more fundamental challenge from the Communists. The latter, after their split with the KMT in 1927, had bifurcated into two distinct entities: the party's Central Politburo under the leadership of Moscow-trained Chinese Communists went underground in Shanghai, while Mao Tse-tung pursued an independent course in the countryside of Hunan and Kiangsi. The Politburo followed the Comintern tactics of strikes, sabotage, and uprisings in the cities, but Mao organized peasant support and developed Soviet areas far away from KMT control. Mao's unorthodox approach placed him in the position of "opposition" vis-à-vis both Moscow and the CCP Politburo; yet of all the tactics used by the Communists, his eventually reaped the greatest success.

THE COMINTERN POLICY. The CCP revolutionary strategy was largely determined by Stalin several thousand miles away, and his orders were at times a product of fantasy and at times a result of his feud with Trotsky. After the KMT-CCP split, Trotsky announced that the Chinese revolutionary tide had receded to a low ebb which required a cautious policy of peaceful penetration. Stalin, however, insisted that China was experiencing a high revolutionary tide which justified armed uprisings, seizure of power, and establishment of the Soviets. Stalin's triumph over Trotsky in the Soviet power struggle assured the dominance of his line, and orders were issued to the CCP to carry out armed insurrections.

With the support of a newly organized 15,000-man peasant-worker army, a group of Communists[16] staged a coup at Nanchang, Kiangsi on August 1, 1927. They seized control of the city for three days, and then the Nationalists closed in on them. On August 5 the rebels broke the siege and fled to the border areas of Kwangtung, Kiangsi, and Fukien. The Nanchang uprising, a brainchild of Stalin's, had been a total failure.

16. Including Ho Lung, Chu Teh, and Chou En-lai.

At this time, a radical reshuffling of the CCP leadership was taking place. At an emergency meeting in Hankow on August 7, 1927, Ch'en Tu-hsiu was dismissed from the party leadership because of his "surrenderism," and Ch'ü Ch'iu-pai, a Stalin protégé, took over as secretary-general of the Central Politburo, a new organ which had superseded the Central Committee. Li Li-san, another choice of Stalin's, was put in charge of propaganda. Under the guidance of the new Comintern representative, B. Lominadze, these leaders accepted Moscow's diagnosis that China was ripe for armed insurrections and creation of the Soviets. Ch'en Tu-hsiu, in disgrace, organized his own Third Party with the support of T'an P'ing-shan.

Mao Tse-tung on his part had entered Hunan to foment popular resentments which culminated in the Autumn Harvest Uprising of September 7, 1927. Under his inspiration and direction, the rebellious peasants destroyed sections of the Canton-Hankow Railway, seized control of a number of places in the province, and carried out "liquidation struggles" and land revolution. This first uprising of Mao's, however, fell far short of success. Under the attack of government troops, Mao was obliged to flee to Ching-kang-shan in the border area of Kiangsi and Hunan to regroup his forces. Unsympathetic to the peasant movement, Ch'ü Ch'iu-pai sponsored a resolution at the November meeting of the Central Politburo stating that "a purely peasant uprising without the leadership and help of the working class cannot achieve conclusive victories."[17] The failure of the Autumn Harvest Uprising cost Mao his membership in the Politburo.

To step up armed uprising, Moscow sent Heinz Neumann[18] to China, and under his guidance an insurrection broke out in Canton on December 11, 1927.[19] The Communists won control of the city for three days, and established a "Canton Commune" and a Soviet regime. But the success was short-lived. Under the joint attack by government troops and workers from the city labor union, the uprising was abruptly suppressed.

As these urban uprisings failed one after another, Mao's activity in the countryside began to assume importance. At Ching-kang-shan he was joined by Chu Teh and Ch'en Yi on January 23, 1928, and their combined forces formed the Fourth Red army, with Chu as the commander and Mao as the "party" representative. Here was the birth of the celebrated Chu-Mao leadership. In July they moved their headquarters to

17. Schwartz, *Chinese Communism*, 104.
18. A German agent, alias A. Neuberg.
19. Led by Chang T'ai-lui and Yeh T'ing.

Juichin, Kiangsi, where a Soviet regime was established.[20] In Shensi, another Communist base was being created by Liu Chih-tan and Kao Kang. These two centers in the border areas operated outside the jurisdiction of the CCP Central Politburo.

At the CCP Sixth Party Congress in July 1928—held in Moscow partly to avoid the KMT raid and partly to coincide with the Comintern's International Congress which had been called to eliminate the Trotskyite influence—Ch'en Tu-hsiu was condemned for his "rightist opportunism," while Ch'ü Ch'iu-pai was attacked for his "leftist deviationism." The Congress called for (1) the overthrow of the Nationalist government and the destruction of its military power; (2) the establishment of Soviets in China; (3) land revolution and confiscation of the holdings of landlords; and (4) the unification of China through the expulsion of the imperialists. Chu-Mao activities in the Kiangsi-Hunan hinterland were recognized as legitimate though not models for the Chinese Communist movement. Instead, the Congress elected Hsiang Chung-fa and Li Li-san new leaders, with the former serving as secretary-general and the latter director of propaganda. The party headquarters remained in underground Shanghai.

Of the two, Li was more lively and eloquent. He emerged as the strong man of the party between June 1929 and September 1930. In October 1929 he was told by the Comintern that he should ready himself for the advent of a new revolutionary tide. Not long afterward, a large-scale civil war broke out in Central China in July 1930, and taking advantage of the situation Li fomented strikes and sabotage, sending the newly organized Red army under P'eng Te-huai to attack Changsha, the capital of Hunan. The city was taken, but the success again was fleeting. Within three days the government troops recovered Changsha and inflicted heavy losses on the rebels. The failure of the precipitous "Li Li-san line" led the Comintern representative, Pavel Mif, who resented Li's iron control of the party, to ask Moscow to dismiss him. The Comintern dispatched Ch'ü Ch'iu-pai to investigate the case, but he could not attack Li without criticizing the Kremlin's policy. Caught on the horns of a dilemma, Ch'ü's attack was halfhearted and ineffectual. However, the Comintern and the "Returned Student Clique" (see below) launched a devastating campaign against Li, accusing him of (1) "opportunist passivity" because of his reliance on the prospect of world revolution; (2) "petty bourgeois chauvinism" and "great Chinaism" because of his exaggeration of the importance

20. This was, however, not the first Soviet government in China. The earliest was established in November 1927, in the Hailufeng area near Canton. See Etō Shinkichi, "Hai-lu feng: The First Chinese Soviet Government," *The China Quarterly*, 8:163-83 (Oct.-Dec. 1961); 9:149-81 (Jan.-March 1962).

of the Chinese revolution; (3) "adventurism" because of his misunderstanding of the meaning of "upsurge" and "direct revolution"; and (4) Trotskyite proclivity because of his reference to the imminent transformation of the Chinese Revolution into a socialist revolution.[21] A victim of Stalin's failure in China, Li was sent to Moscow for recantation. There, he was severely condemned by the Presidium of the Executive Committee of the Comintern and sent to Lenin University to study and correct his mistakes. Ch'ü Ch'iu-pai fared even worse; he was attacked for his double-dealing, factionalism, "wily oriental diplomacy," and wrong views on agrarian and peasant questions under the influence of Borodin, Ch'en Tu-hsiu, and other undesirable characters.[22] Ch'ü was dropped from the CCP Politburo a month later.

The party leadership now fell to Wang Ming (Ch'en Shao-yü) and Po Ku (Ch'in Pang hsien), who headed the "international wing" (*kuo-chi pai*) of the CCP consisting of twenty-eight returned students who had studied at the Sun Yat-sen University at Moscow from 1926 to 1930. Returning home in early 1930, they became known as the "Twenty-eight Bolsheviks" and "China's Stalin Section." They took over the Politburo in January 1931, with the support of Mif, the Comintern representative.

MAO'S INDEPENDENCE. Operating outside the jurisdiction of the CCP central organization, Mao and Chu were relatively unaffected by the party squabble. They had developed an independent and comparatively unorthodox activity by organizing the peasants and creating Soviets in the hinterland of Kiangsi and Hunan. They had practiced guerrillaism to perfection and initiated a rather "egalitarian" land revolution by parceling out the redistributed land to the rich and poor peasants alike. They had evolved a self-sufficient territorial base without relying on the help or guidance of the Comintern or the party leaders at Shanghai, with whom they maintained an attitude of "outward obedience and inward disobedience." The CCP Central Politburo never really approved of Mao's activity, while Moscow merely tolerated it because all other CCP-led uprisings had failed. Mao's growing power and independence contrasted sharply with the state of the party central organization, which was plagued by unstable leadership, lack of financial support from the Soviet Union which itself was absorbed in economic reconstruction, and Nationalist persecution. The KMT raids had become so devastating that the Communist secret service chief in Hankow[23] was arrested and forecd to divulge names.

21. Schwartz, 151-63.
22. John E. Rue, *Mao Tse-tung in Opposition, 1927-1935* (Stanford, 1966), 241.
23. Ku Shun-chang.

This led to the capture and subsequent execution of the secretary-general, Hsiang Chung-fa, on June 24, 1931. The party fortunes had fallen to a low ebb.

Sensing the desperation of the Politburo, Mao boldly invited its members to attend the First All-China Congress of the Soviets to be held in Juichin on November 7, 1931. The Twenty-eight Bolsheviks condescendingly arrived in Mao's capital, not with the intention of supporting his movement but of chastising his unorthodox conduct. Prior to the Congress, they called a party conference which adopted resolutions to condemn Mao for his failure to adopt a strong "class and mass line," for his guerrilla tactics, and for his "rich peasant" mentality in the land revolution. They childed his narrow empiricism, his "opportunistic pragmatism," and his "general ideological poverty." The conference closed with a call for the proletarian leadership in agrarian reforms, the expansion of the Red army, and the adoption of the regular warfare in place of guerrillaism. Starkly, the Twenty-eight Bolsheviks rejected the Maoist approach and intended to replace his machine.[24]

However, at the Congress the Maoists completely dominated the scene. Having defeated the first two KMT campaigns (see below), they were flushed with success and confidence. With their control of the majority votes, they easily outmaneuvered the Twenty-eight Bolsheviks. Mao was elected chairman of the Central Executive Committee of the All-China Soviet Government and retained his position as Chief Political Commissar of the First Front Red army. A number of the former party leaders were absorbed into his government, while the Twenty-eight Bolsheviks were deliberately left in the cold during the distribution of offices. Only three of them, including Wang Ming who was absent, were given places on the Central Executive Committee, while Po Ku was not appointed to anything.

In this power struggle, Mao was able to score an impressive victory and won increased recognition of his activity, not because of Soviet patronage but because of a very realistic strategy he had evolved himself on the basis of five important elements: (1) the peasant mass support; (2) a party and government apparatus of his own; (3) an independent military force; (4) a secure territorial base far away from KMT control; and (5) self-sufficiency.[25] But Mao's success was not complete; the Twenty-eight Bolsheviks retained their iron control of the Politburo to which he could gain no admission. Mao remained an "outsider," in opposition to the CCP cen-

24. Rue, 247-48.
25. Schwartz, 189-90.

tral organization. After the Congress, Po Ku and most of the other Politburo members returned to Shanghai, in preparation for the next round of
dueling with Mao.

The Bolsheviks and the Maoists split over a number of issues too fundamental to be reconciled. On the question of land reform, Mao favored
equal distribution of all grades of land to small landlords, rich peasants,
and poor peasants alike, whereas the Politburo members insisted on complete deprivation of the landlords, and relocation of land to favor the poor
at the expense of the rich. They condemned the Maoists of backward
peasant mentality, with unsteady and weak class consciousness. On military strategy, Mao opted for the mobile guerrilla tactics of luring the enemy deep into the base area where the Red army could "amass superior
forces to attack the enemy's weak spots" and assure elimination of "a part,
small or large, of the enemy's forces by picking them off one at a time."
By "circling around in a whirling motion," the Red army could confuse
the enemy and win the battle.[26] The Politburo, on the other hand, insisted
on positional warfare, holding the Communist base, and invading the enemy territory rather than waiting for the enemy to invade the Red territories. On the sensitive issue of Japanese aggression, Mao announced his
readiness to form a United Front and a coalition army of all military forces
willing to fight the enemy, whereas the Politburo rejected collaboration
with reformist groups, and argued for the rapid expansion of the Red army
so that it could perform the sublime duty of safeguarding the Soviet Union
against imperialist attack, should the need arise. So wide was the gulf between the Maoists and the Bolsheviks that a *rapprochement* seemed impossible.

Mao's problem was not the Politburo alone; he had to fight the KMT
invasion, too. Indeed, Chiang Kai-shek was relentlessly organizing expeditions against the Communists.

THE KMT CAMPAIGNS. After his dismissal of the Russian military advisers
in 1927, Chiang increasingly sought to obtain German aid for the development of his army. Beginning with the appointment of Colonel Max Bauer,
an associate of General Ludendorff during World War I, as his adviser in
1928, a German military mission in China gradually took shape. In 1933
the famous strategist General Hans von Seeckt arrived to coordinate the
campaign against the Communists and in the following year he assumed
the direction of the mission. When poor health obliged him to resign in

26. Rue, 272.

March 1935, the mission leadership then went to General Alexander von Falkenhausen. Through the efforts of these advisers, Chiang developed a German-style Central army of more than half a million men.

From 1930 to 1934 Chiang launched a total of Five Campaigns of Encirclement and Extermination against the Communists.[27] The first four, lasting from December 19, 1930, to April 29, 1933, all ended in failure. At this point, a critical power struggle developed within the Communist camp. In late 1932 or early 1933, Po Ku and other CCP Politburo members arrived in Juichin in the company of a Comintern military adviser, Li Te[28] with the intention of discrediting Mao and replacing his men in the army and the party. Excoriating Mao's "egalitarian" approach, they pressed for a radical land investigation drive to eliminate the landlords, attack the rich peasants, neutralize the middle peasants, and ally the poor peasants and the landless laborers with the party. The funds realized from this campaign were to be used for the expansion of the Red army. Mao reluctantly acquiesced as he could not dispute the need for funds to support and expand the army.

Meanwhile, the Nationalists were organizing a fifth campaign, which was opened in October with 700,000 men. Counseled by his German advisers, Chiang adopted a "strategically offensive but tactically defensive" posture, moving his troops gingerly and relying on encirclement and progressive economic strangulation. His troops constructed fortresses and pillboxes as they advanced, tightening the blockade ever more until all outside supplies to the Red areas were cut off. Proclaiming the Communist problem as "70 percent political, 30 percent military" in nature, Chiang stressed rural reconstruction and neighborhood organization (*pao-chia*) in the reconquered areas. The campaign progressed slowly but steadily.

Mao faced an extremely critical situation at this point of his career—not only were the Nationalists hitting hard, but the Politburo members were doing their best to destroy him from within. At the Second All-China Soviet Congress, he nearly lost his grip on the Chinese Communist movement. Though re-elected chairman of the Soviet government in January 1934, he lost control of the Central Executive Committee which was placed under a seventeen-man presidium dominated by the Twenty-eight Bolsheviks, one of whom, Chang Wen-t'ien, took over from

27. The dates of these campaigns and the strength of the Nationalist troops employed vary with different accounts. My information has been largely drawn from the official Nationalist source; see Wang Chien-min, *Chung-kuo Kung-ch'an tang shih-kao* (A draft history of the Chinese Communist Party), (Taipei, 1965), II, ch. 20.

28. His real name was Otto Braun (1900-74); see his memoirs, *A Comintern Agent in China 1932-1939* (Stanford, 1982), 278. Tr. from German by Jeanne Moore.

Mao the chairmanship of the Council of People's Commissars—a sort of premiership in the Soviet government. Although Mao retained the chairmanship of the Central Executive Committee, the post was reduced to something of an honorary title by the Bolsheviks, who intended to make Mao a figurehead. The final *coup de grâce* came in July 1934 when Po Ku at Juichin and Wang Ming in Moscow conspired to obtain an order from the Comintern that Mao be put on probation and barred from party meetings. For three months from July he was placed under house arrest at Yü-tu, some sixty miles west of Juichin. It was not until the Long March began in October that he was released.[29]

THE LONG MARCH AND THE TSUNYI CONFERENCE. The KMT Fifth Campaign routed the enemy from their seven-year-old base in Kiangsi. The Communist defeat was largely the result, from a military standpoint, of Li Te's wrong strategy of positional warfare instead of Mao's test-proven guerrilla warfare. Throughout the first half of 1934 the Red army suffered incalculable losses and by the midyear it was nearly crushed. Mao wanted the Red Army to break through the siege and split into small groups to fight guerrilla war, but the Revolutionary Military Council under the dominance of Li Te ordered the Red Army to break through the siege as a united force and not split into smaller guerrilla groups. The able-bodied were allowed to join the exodus, while the wounded and dependent were ordered to stay behind. On October 15, 1934, the Long March officially began with 85,000 soldiers, 15,000 government and party officials, and 35 women who were wives of high leaders. A number of Maoists and ex-party leaders unacceptable to the Twenty-eight Bolsheviks were left behind to defend the base, including Su Yü, Ch'en Yi, and Ch'u Ch'iu-pai. Mao's two children were also left behind. On November 10, 1934, Juichin fell to the Nationalists.

Initially, a three-man Military Group composed of Li Te, Po Ku, and Chou En-lai directed the Long March. Troop morale was very low, and the heavy shelling of KMT forces drove some political and military leaders to become disillusioned with Li and Po's inept leadership. These officials wondered why the Communists had successfully fought against the KMT during the first three campaigns but failed miserably in the fifth. Furthermore, they resented Li's arrogance and his high-handed manner of operation. In their view, Li behaved as if he were the commander-in-chief when in fact his title was only that of a Military Adviser from the Comintern. To make matters worse, Po Ku, the party general secretary charged with

29. Rue, 263-64.

overall responsibility, colluded with him to gang up on the others. The feeling was strong that they had to be removed from power.

Wang Chia-hsiang, a key member of the Politburo and director of the Political Department of the Red Army, first expressed this sentiment. Vowing that Li and Po had to be "thrown out," Wang confided his concerns to Mao. Mao concurred but urged caution and careful preparation for a showdown. Wang then lobbied and won the support of several key officials, including Chang Wen-t'ien, chairman of the People's Commissars; Chu Teh, commander of the Red Army; and Chou En-lai, vice-chairman of the Central Revolutionary Military Council—all of whom harbored similar doubts about Li's leadership. At a Politburo meeting at Liping on December 18, 1934, party leaders decided to convene an enlarged Politburo Conference soon in order to review the military situation arising from the KMT Fifth Campaign and the Western (Long) March. Two trends became apparent at this time: first, the majority of Politburo members desired a change in leadership, and second, Mao's star was rising because he symbolized the correct line of opposition to Li and Po.

On January 7, 1935, the Red Army broke into Tsunyi, the second largest city in Kweichow Province, and two days later the party center moved in. After several days of intensive preparation, an enlarged Politburo Conference convened in a former warlord's (Po Hui-cheng) residence from January 15-18, with some 18 regular participants and 2 observers. Those who participated were all powerful leaders in the Party and the Red Army, i.e. full and alternate members of the Politburo and leading army corps commanders and political commissars.[30] Po Ku, the presiding officer, delivered the political report first, and Chou En-lai followed with a supplementary military report. Then Mao delivered a blistering attack on the erroneous military leadership of Li and Po, accusing them of "leftist adventurism," which Mao insisted was responsible for the defeat in the Fifth Campaign, the loss of the base area, and the near destruction of the Red Army.

No sooner had Mao finished than Wang Chia-hsiang spoke out vigorously in support of Mao. Others, including Chang Wen-t'ien, Chou En-lai, and Chu Teh, echoed similar views. In the face of this barrage of criticism, Po Ku could not argue against the stark fact of defeat; he made a feeble defense on the basis of objective difficulties (i.e. imperialist sup-

---

30. Benjamin Yang, "The Zunyi Conference as One Step in Mao's Rise to Power: A Survey of Historical Studies of the Chinese Communist Party," *The China Quarterly*, 106:241 (June 1986). Li Te attended as an observer with an interpreter, Wu Hsiu-ch'uan. Current official Chinese list of participants includes Teng Hsiao-p'ing, editor of *The Red Star*.

THE LONG MARCH
1934-1935

Communist areas, 1934-1936

Route of main Communist forces from Juichin area

Route of Communist forces from other areas

U.S.S.R.

MANCHUKUO (MANCHURIA)

OUTER MONGOLIA

INNER MONGOLIA

JEHOL

Mukden

KANSU

Kalgan

Peking

KOREA

Tientsin

TSINGHAI

Lanchow

SHANSI

Taiyuan

Tsinan

Tsingtao

SHENSI

Yenan

Sian

Yellow R.

Nanking

Shanghai

SZECHWAN

Chengtu

Yangtze R.

Hankow

Ichang

SIKANG

Chungking

Nanchang

Mekong R.

Tsunyi

Changsha

KIANGSI

FUKIEN

KWEICHOW

HUNAN

Juichin

BURMA

Kweiyang

TAIWAN (FORMOSA)

Kunming

Kweilin

Amoy

YUNNAN

KWANGSI

Canton

Swatow

KWANGTUNG

FRENCH INDOCHINA

Hong Kong (Br.)

Macao (Port.)

SIAM

HAINAN

0   Miles   500

561

port of the KMT, superior enemy forces, etc.) Li Te refused to admit mistakes; he sat by the door with an interpreter, chain-smoking cigarettes in desperation.[31] Only one Politburo alternate member, Kai Feng, came to their defense, but to no avail. The die was cast; Li and Po had to go. No vote was recorded at the conclusion of the conference although Mao later said that Wang cast the "pivotal vote," perhaps referring to his role in arranging the Tsunyi Conference. Also, no resolution was passed at that time although one was later drafted by Chang Wen-t'ien to summarize the conference proceedings.[32]

Mao became a member of the Politburo Standing Committee and an assistant to Chou En-lai in military affairs.[33] On February 5, 1935, Chang Wen-t'ien replaced Po Ku as "the person with overall responsibility," and in March a new three-man Military Group was formed, with Mao, Chou, and Wang as members. Of the three, Wang was quite ill and Chou deferred to Mao, who now became first among equals; thus, in essence, he was the real power. Soon, with Chang's help, Mao won absolute control of the military, which became the foundation of his power, one that he never relinquished.

The Tsunyi Conference did not give Mao a complete victory, but it was a giant step in his quest for supreme power. Yet even in his moment of triumph, Mao was not spared from challenge. The party elder, Chang Kuo-t'ao, did not attend the conference and refused to accept its decisions. He questioned the choice of northern Shensi as the terminus for the Long March and argued for moving south or west in the direction of Sikang or Tibet. A split developed, with Chang leading his troops to Sikang while Mao and the majority of the Politburo and their troops moved toward northern Shensi, where Kao Kang and Liu Chih-tan had built up a Soviet base.

In October 1935, Mao's group reached Wuch'ichen in Paoan County after an extremely difficult period of mountain-climbing and river-crossing. At the end of this epic 25,000 li (6,000 miles) Long March, Mao's forces could count only 8,000 survivors. Later, the arrival of other units under

---

31. Wu Hsiu-ch'uan, "Sheng-ssu yu-kuan ti li-shih chuan-che (A historical turning point of life-and-death significance), *Hsiang-ho liao-yuan ts'ung-k'an*, I: 19, 26 (1982), published by Chinese People's University, Peking, China.

32. Po Ku accepted defeat rather "gracefully" and later made a self-criticism at the Seventh Party Congress in 1945. He was killed in an airplane crash in 1946. Li Te, however, was unrepentant and without work for a while; later, he was assigned to train cavalry in Yenan and to teach at the Red Army University. In 1939 he flew in the same plane that carried Chou En-lai to Moscow for medical treatment. He died in East Germany in 1974.

33. Other members were Chang Wen-t'ien, Ch'en Yun, Chou En-lai, and Po Ku.

Ho Lung, Chang Kuo-t'ao, and Chu Teh, along with local Communist units, swelled the ranks to 30,000. In December 1936, the CCP head-quarters was moved to Yenan; there Mao rebuilt the party and the army around himself and engaged in theoretical writings. He was now the *de facto* leader of Chinese Communism. Mao received two additional ac-knowledgments of his authority in 1938: a Soviet encyclopedia recognized him as the leader of the CCP, and he was selected as the chairman of the preparatory committee for the convocation of the Seventh Party Congress, which was held in 1945. Then and only then was Mao's victory complete: he became chairman of the CCP Central Committee, the Politburo, the Secretariat, and the Military Commission; furthermore, his Thought was accepted as the guiding principle of the Party.

In historical perspective, the Tsunyi Conference must be considered as a political meeting rather than a military putsch.[34] Yet the presence of so many military leaders who were full or alternate Politburo members and army commanders lent great support to Mao's attacks on Li's erroneous military strategy. In the strictest sense, this extended Politburo meeting was unprecedented because it was not held in conformity with the party constitution or its regulations and bylaws; nevertheless, it was a landmark in his history of the party and in Mao's rise to the pinnacle of power.

Sian Incident and the United Front

The radical change in the position of the Chinese Communists coincided with a basic shift in the Comintern's world revolutionary strategy. Facing the rise of Nazi Germany and Fascist Italy in Europe and militarist Japan in Asia, the Comintern at its Seventh Congress in August 1935 adopted a resolution urging the various national Communist parties to form alli-ances with leftist and anti-Fascist groups against the threat of these avowed enemies of Bolshevism and Marxism. In the case of China, a policy of the United Front would have the added benefit of relieving the Communists of Nationalist attacks.

Beginning in 1936, the CCP started to promote collaboration with all parties, groups, and armies in a grand alliance against Japan. Under its sponsorship, popular organizations such as "The National Liberation Anti-Japanese Association," "The People's Anti-Japanese League," and "The National Salvation Society" sprang into existence. New persuasive slogans such as "Chinese must not fight Chinese" and "Immediate war with Ja-

34. Yang, 250.

pan; stop fighting the Communists" were widely circulated to evoke a sentimental response among patriotic Chinese, especially the youth in Peiping, Nanking, and Shanghai. Popular pressure mounted feverishly demanding an end to the civil war and the turning of guns against the Japanese.

The Nanking government, as noted before, had decided on a policy of domestic consolidation before an external war. With the Communists pushed into a pocket in the Northwest, Chiang Kai-shek was anxious to finish with them once and for all. Confidently, he ordered the Northeastern army[35] under Chang Hsüeh-liang and the Northwestern army under Yang Hu-ch'eng to mount an offensive against the Communists. But the fighting was ineffective. Homesick and weary of a civil war, the Northeastern officers and men became susceptible to the United Front propaganda. Communist agents began to infiltrate the Northeastern officers' training corps, and by the summer of 1936 the two commanders were won over to the United Front.

On December 3 Chiang flew to Sian, the headquarters of Chang and Yang, with a view to stabilizing the restless situation and increasing the effectiveness of the campaign. There, at daybreak on December 12, a mutiny broke out, engineered by the Northeastern 105th division and the second battalion of Chang's personal guard. With Chiang as his captive, the Young Marshal named eight demands:

1. Reorganization of the Nanking government to include all parties and groups responsible for national salvation.
2. Termination of all civic strife.
3. Immediate release of patriotic leaders who had been arrested in Shanghai.
4. Release of all political prisoners.
5. Protection of the people's right to assembly.
6. Freedom to organize the people's patriotic movement.
7. Faithful fulfillment of Dr. Sun's will.
8. Immediate convocation of a National Salvation Conference.

On December 14, the Northeastern, Northwestern, and Communist forces formed a United Anti-Japanese command headed by a Military Commission. That Chang served as the chairman of the Commission implied that he might have had some secret ambitions to head the United Front.

The Sian mutiny and the kidnap of Chiang stunned the country and the world. Right-wing Nationalist leaders at Nanking quickly decided

---

35. It had moved from the Peiping-Tientsin area to Shensi after the T'ang-ku Truce, 1933.

upon a punitive expedition and sent its airplanes to Sian in a demonstration of power. The country was once again on the verge of a civil war. At this point the Communists discovered that the mutineers were more anti-Chiang than anti-Japanese, and came to the conclusion that any large-scale Nationalist attack would inevitably involve them (the Communists) and hurt their cause. Moscow also realized that disorder in China could only benefit Japan and that Chiang should be spared to lead the fight against Japan. A Sino-Japanese war would surely relieve Japanese pressure on the Soviet Union and the Nationalist pressure on the Communists. Prompted by these considerations, Chou En-lai emerged from behind the mountains to offer mediation. The Communist positions had shifted overnight from "anti-Chiang against Japan" to "ally with Chiang against Japan."

Bewildered by this change, and pressured by public opinion, the Young Marshal finally agreed to release his prized prisoner. On Christmas day, 1936, Chiang flew back to Nanking in the company of his erstwhile captor, who had offered himself up for punishment. A special military court sentenced him to ten years' imprisonment and five years' loss of civil rights, but through Chiang's intercession on the basis of Chang's quick repentence, the sentence on imprisonment was remitted. Nevertheless, he was put under house arrest.[36]

Although Chiang insisted that he did not sign any agreement as to the conditions of his release, he did promise that the Communists could participate in the future war against Japan if they pledged their support of the Three People's Principles. The anti-Communist campaign was terminated, though the government blockade of the Red area in the Northwest continued.

The Sian incident might be considered a blessing in disguise. It helped unite the country and put an end to the civil strife. No longer was Chiang considered the obstacle to fighting the Japanese, but a national hero with a new mandate of leading the country in a United Front against the aggressor.

Success or Failure: A Decade in Review

Saddled though it was with endless domestic and foreign problems, the Nationalist government struggled to carry out Dr. Sun's legacy of national reconstruction. The record at the end of the first decade revealed some progress in the fields of finance, communication, industrial development, and education. On the other hand, the government neglected the much-

36. Now living on Taiwan. Interviewed in 1991 at age 90 by Nicholas D. Kristof, *The New York Times*, Feb. 20, 1991.

needed basic social and economic reforms, and carried on an irresponsible fiscal policy of deficit spending—both of which left far-reaching consequences of a very fundamental nature which ultimately proved disastrous. The following is a quick review of the achievements and failures of the decade.[37]

FINANCIAL REFORM. The outstanding accomplishments were the substitution of the silver dollar (*yüan*) for the tael (*liang*) and the introduction of paper currency, *fa-pi*, as the legal tender. Although the silver dollar had been introduced in 1914 as a basic unit of currency, the tael was continuously used in commercial transactions because of tradition and supposed convenience. The juxtaposition of the two media of exchange caused confusion and complications, since their exchange rates varied with places and seasons. On April 4, 1933, the government decisively abolished the tael and substituted for it the silver dollar at an exchange rate of 0.715 (tael) for 1 (dollar).

No sooner had this reform been introduced than a new problem arose: the sharp rise in silver value in the world market was causing a rapid outflow of the metal from China, undermining the very basis of the new currency. The continuous drain caused inflation, high interest rates, tight money, decline in the stock market, stagnancy in real estate, and bankruptcies of enterprises. On November 3, 1935, the government finally took the bold step of nationalizing silver and introduced a new paper money, *fa-pi*, to be issued by four national banks on 25 percent silver reserve. Later, in February 1936, a decimal system of nickel coins in denominations of 5, 10, and 20 cents, as well as copper coins of ½ and 1 cent, were circulated to supplement the paper currency.

The four national banks were assigned different duties. The largest of them, The Central Bank, with a capital of 100 million Chinese dollars in 1934, became the central bank of the nation charged with maintaining currency stability. The Bank of China, with a capital of Ch$40 million, directed foreign exchanges, and the Bank of Communication, with a capital of Ch$20 million, was entrusted with assisting domestic industries and enterprises. The Farmers' Bank of China handled farm credit and land mortgages up to Ch$50 million. The first three banks were authorized to buy and sell foreign currencies in unlimited amounts in order to stabilize the rates of exchange. Thus, for the first time in Chinese history, foreign exchanges were controlled by government banks.

37. Information in this section is largely extracted from *K'ang-chan ch'ien shih-nien chih Chung-kuo* (China during the ten years before the Sino-Japanese War, 1937), compiled by Chung-kuo wen-hua chien-she hsieh-hui (China cultural reconstruction association), reprinted (Hong Kong, 1965).

TARIFF AUTONOMY. The fixed tariff of 5 percent ad valorem, imposed after the Opium War, had been a constant reminder of China's semicolonial status and a major irritant to the rising national consciousness of its people. Abolition of the tariff restriction had been a cardinal goal of the Nationalist government since its inception. Supported by the rising nationalism, it announced on July 7, 1928, two guiding principles in which treaties and agreements that had expired would be replaced by new ones, while those not yet expired would be abolished and renegotiated according to legal procedures. The United States was the first to enter into an equal and friendly tariff agreement with China on July 24, followed swiftly by Germany (August 17), Belgium (November 22), Italy (November 27), Britain (December 20), France (December 22), and Japan (May 6, 1929). With these agreements, the great powers recognized China's tariff autonomy and, furthermore, agreed *in principle* to give up their consular jurisdiction.

RECOVERY OF FOREIGN CONCESSIONS. In concert with the struggle for tariff autonomy, the Nationalists succeeded in revoking a number of foreign municipal concessions. The British, who bore the major brunt of Chinese pressure, agreed to relinquish their concessions at Hankow and Kiukiang in February 1927, at Chinkiang in February 1929, at Weihaiwei in April 1930, and at Amoy in September of the same year. The Belgian concession at Tientsin was also recovered in January 1931. However, restitution of China's lost rights was not complete until 1943 when the United States and Britain took the lead in voluntarily abolishing all unequal treaties with China, thereby ending the century-old national humiliation.

COMMUNICATION. Improvement of the communication systems was another positive accomplishment of the government. In 1928 a Ministry of Railways was established to direct the improvement of existing lines and the construction of new ones. Among the most prominent projects were the extension of the Lung-Hai Railway—the East-West trunk line—to Sian in 1934 and to Pao-chi in 1935, and the completion in 1936 of the Canton-Hankow Railway—the major south-central trunk line. Other noteworthy achievements included the development of the ferry system at Nanking, which linked the Tientsin-Pukow and the Shanghai-Nanking railways, and the construction of the iron bridge over the Ch'ien-t'ang River in 1937, which connected the Chekiang-Kiangsi and the Shanghai-Hangchow-Ningpo lines. These national accomplishments were matched in the provinces by completion of a number of smaller projects. From 1928 to 1937, the railway network grew from 8,000 kilometers to 13,000.

Even more impressive was highway construction, due to its lower costs—

about one-twentieth that of the railway. In 1936 the highway network accounted for 115,703 kilometers, as compared with a mere 1,000 in 1921.

Modern airlines were also initiated. The China National Aviation Corporation was organized in 1930 with Chinese and American capital, and it operated four lines between Shanghai and Chengtu, Shanghai and Peiping, Shanghai and Ch'üan-chou, and Szechwan and Kunming. The second largest was the Eurasia Aviation Corporation, a Sino-German enterprise which opened for business in 1931. It also operated four lines between Shanghai and Sinkiang, Peiping and Canton, Peiping and Lanchow, and Sian and Chengtu. A third company, the Southwestern Aviation Corporation, was established by the Southwestern provincial authorities in 1933, with flights in Kwangtung and Kwangsi, and from there to Kunming and Foochow.

Postal service and telecommunication were much improved and expanded during the decade. In 1921 post offices numbered less than 10,000 over 400,000 *li* of postroads. By 1935-36 they had increased to 14,000 over 584,800 *li*. Telegraph lines, which had suffered severe damage during the warlord period, underwent rapid restoration and construction; by 1936 they totaled 95,300 kilometers. Correspondingly, long-distance telephone lines grew from 4,000 kilometers in 1925 to 52,200 in 1937.

INDUSTRIAL DEVELOPMENT. There was a general recognition that economic development was essential to the creation of a modern state. Despite the loss of Manchuria and the Japanese attack on Shanghai, which wrought havoc with foreign trade at that key port, the importation of heavy machinery never abated. Over a ten-year period between 1927 and 1937, the total importation of industrial equipment reached Ch$500 million, which, though small by Western standards, represented considerable effort in a war-torn, poverty-stricken country. Although no spectacular breakthrough was achieved in industrialization, good progress was scored in a number of light industries such as cotton weaving, flour production, matches, cement, and chemical manufacturing.

EDUCATION. Notable progress was also achieved in the field of education. The Ministry of Education reorganized and amalgamated a number of public universities, colleges, and professional schools into thirteen national universities,[38] five technical colleges, and nine provincial universities. It extended subsidies to private institutions of higher learning[39] for the dual

38. The most famous were Peking, Tsing-hua, and Central universities.
39. The most famous were Yenching, Soochow, Shanghai, Lingnan, and St. John's universities, all Christian institutions.

purpose of establishing new professorships and purchasing equipment. Of the 20 private universities and 33 private colleges, 32 received this help in 1934 and 1935, and 40 in 1936. Not to be outdone, secondary education underwent a four-to-fivefold growth during the decade. By 1937 there were 2,042 middle schools, 1,211 normal schools, and 370 professional schools, with a total enrollment of 545,207.

THE NEW LIFE MOVEMENT. To revitalize the moral fibre of the people and achieve a spiritual awakening, the government in 1934 promoted a New Life Movement, which stressed hygienic practices, promptness, truthfulness, courtesy, and the four traditional virtues of politeness (*li*), righteousness (*i*), integrity (*lien*), and self-respect (*ch'ih*). Scholars and officials were urged to read the writings of the 19th-century statesman Tseng Kuo-fan for their spirit of loyalty and devotion to public service. Although the young generation did not take to these old virtues too seriously,[40] the New Life Movement and the related activties such as military training for the able-bodied and military instructions at schools did provide something of a psychological uplift and the feeling of doing something in the face of Japanese aggression.

THE WORLD OF LITERATURE. Literary activities were extremely lively during the decade under review, with most of the creative writings reflecting social realities of the times. A powerful organization was the League of Chinese Left-wing Writers, founded in 1930 under the auspices of the CCP with a view to capturing the literary scene of China. Members of the League attacked the Nationalist government, scorned the rightist writers and lovers of the traditional art and literature, criticized the Anglo-American school of writers, and eulogized Soviet literature and leftist programs. The guiding spirit of the League was none other than Ch'ü Ch'iu-pai, the deposed party leader, although its spokesman was the famous writer Lu Hsün. Through its numerous publications,[41] it succeeded to a considerable extent in dominating the literary circles.

However, two groups stood out adamantly in opposition and won acclaim through their own merits. One was led by Lin Yü-t'ang, whose hu-

---

40. A famous joke about the New Life Movement was its admonition that one should always walk on the left side of the street. The warlord Han Fu-ch'ü, governor of Shantung, reputedly remarked: "If everybody walks on the left side, who walks on the right?"

41. *World Culture, Sprout, The Pathfinder, The Big Dipper, Modern Fiction, Mass Literature, Literature Monthly, Literature News,* and *Literature* (Wen-hsüeh). Cf. C. T. Hsia, *A History of Modern Chinese Fiction, 1917-1957* (New Haven, 1961), 125.

morous, satirical, and somewhat playful publications, *The Analects, This Human World,* and *The Cosmic Wind* continuously enjoyed public favor. The second group which centered mostly around the faculty members of the universities and colleges in Peiping, published the *Literature Quarterly* and the *Literary Supplement* of the *Ta Kung Pao.* They enjoyed wide circulation because of their avant-garde, critical attitude and their use of the advanced techniques and strategies of Western writers. The crosscurrents of these three major groups contributed to an unusually lively literary atmosphere, making the decade "the richest literary period in modern China."[42]

The foremost writer of the decade was Lu Hsün (1881-1936),[43] although he had perhaps passed his peak of creativity. Known for his sharp, satirical indictment of the decadence and injustice of the old as well as the existing order, he attacked the hypocrisy and cruelty of the traditional life.

Lu Hsün's most famous work was perhaps "The True Story of Ah Q," in which the hero symbolized a national disease. Ah Q, a crude country lad of a very low social status who lived at the end of the Ch'ing dynasty, was continuously bullied by his fellow villagers. Unable to fight them, he developed a dream world for himself. Whenever humiliated, he would put on an air of superiority and pretend to have won a "spiritual victory." In an attempt to boost his prestige, he went to town to engage in thievery, boasting to the villagers upon his return of his new association with the revolution. When the real revolutionaries came to the village, they collaborated with the gentry and put Ah Q on trial for robbery. The moral of the story was that Ah Q epitomized China's national disease and that the revolution had compromised with the old elements at the expense of its professed goals of social improvement.

Another major leftist writer was Mao Tun (1896-1981),[44] editor of *The Short Story* and author of a number of novels, including the trilogy of *Disillusion, Vacillation,* and *Pursuit.* His major work, *The Twilight,* described the futile effort of a nationalistic industrialist in Shanghai, who, facing economic recession and business failures in the wake of Communist uprisings, plunged into the stock market to recoup his losses. Outmaneuvered by a foreign-supported compradore, he ended in a dismal failure that sent him into bankruptcy. By his inability to grasp the Marxist oracle, the author inferred, the industrialist was doomed to failure.[45]

42. Hsia, 138-39.
43. Real name: Chou Shu-jen.
44. Real name: Shen Yen-ping.
45. Hsia, 156-57.

If Mao Tun looked to Marxism for solution to China's problems, Lao She (1898-?),[46] believed in patriotism and individual duty. Having spent five years (1925-30) in London, he was influenced by English writers. The best known of his works was *Camel Hsiang-tzu* (1937), or *Ricksha Boy* in English, which portrayed the endless struggle of a ricksha-puller who dreamed of improving his station through sheer personal effort—hard work, marrying the boss's daughter, etc.—only to find the social obstacle too great to overcome. Dejected and resigned to his fate, he took up smoking and drinking, and finally made peace with the old order by living the degenerate life of a funeral mourner. The moral of the story was clear: in a sick society individual endeavor was futile; only collective action could alleviate the life of the poor.

Another famous non-leftist writer was Pa Chin (1904-   ),[47] a prolific writer with a dozen novels and four collections of short stories to his credit by 1937. Born into a well-to-do Szechwanese family, Pa Chin wrote with sentimentalism, frequently employing such themes as love versus revolution, good versus evil, heroes versus weaklings, and bravery versus cowardice. His *Love Trilogy—Fog, Rain,* and *Lightning*—made an immediate hit with the younger generation, while his autobiographical trilogy—*Family* (1937), *Spring* (1938), and *Autumn* (1940)—won wide acclaim for its moving descriptions of the tribulations of the younger members of a large family who struggled to break away from their elders, only to court stubborn opposition and tragic outcomes. In this work the author drove home the message that the bad systems in China must be held responsible for the evils of the society.

Regardless of their political persuasions, the writers of the 1930s had two things in common: they were imbued with a strong sense of their didactic functions, and their works reflected social realism. With satire, sarcasm, and pity, they portrayed the decadence and backwardness of the old society. Perhaps, in a transitional period when revolutionary changes were taking place and when tradition constantly fought with modernity, it was inevitable that the writers involve themselves in social issues. From this standpoint, their works represented a legitimate indictment and a social protest of the existing order.

NEGLECT OF SOCIAL AND ECONOMIC REFORMS. Against the record of accomplishments in finance, communication, tariff autonomy, industrial de-

---

46. Real name: Shu Ch'ing-ch'un. Committed suicide during the Cultural Revolution.
47. Real name: Li Fei-kan. The pen name Pa Chin was made up of the Chinese renditions of the first and last syllables of the Russian anarchists, *Ba*kunin and Kropot*kin*. Cf. Hsia, 238.

velopment, and education, the Nationalist government was seriously remiss in ignoring the age-old problem of landlordism and the misery of the
peasant, who constituted more than 80 percent of the total population.
This failure was in part a result of compromise with the "new warlords"
after the Northern Expedition. Chiang Kai-shek, in his eagerness to win
a quick victory and unify the country, negotiated with the more "progressive" warlords and absorbed them into his system. These warlords had
little concern for the welfare of the masses and the suffering of the peasants. Their inclusion in the KMT hierarchy diluted its social consciousness. Moreover, a considerable percentage of the KMT generals and
officials were themselves connected with the landed interests; hence they
were not anxious for any radical reform that would jeopardize their own
position. The middle class—mostly merchants, traders, businessmen, and
usurers—was no better motivated in this regard. Living in the treaty ports
or operating in the villages as loan sharks, they were the beneficiaries of
the existing order and hardly desired any change that would rock the
boat. It was these people—the warlords, the generals, the officials, the
merchants, the traders, and the money-lenders—upon whom the Nationalist government relied for support. Small wonder that it could not implement its professed social and economic programs. In fact, a general feeling
prevailed over many complacent KMT personnel that since the peasant
had suffered for ages, it mattered little if they were asked to wait a little
longer—until the government had solved the more pressing problems of
domestic insurrections and foreign aggression.

But the plight of the peasant had reached the point of desperation. A
League of Nations study revealed that tenant and semitenant farming
comprised 60-90 percent in South China, and in addition to paying 40-
60 percent of the annual crops as rental, they had to pay for their landlords' regular land tax and surtax as well—the latter varying from 35 percent to 350 percent of the former.[48] The peasant had been exploited to
the limit; only a revolution could give him relief. Yet all the KMT did
was to pass a resolution in 1930 to reduce the land rent to 37.5 percent
of the main crops, and even this modest step was never really put into
practice. Dr. Sun's ideal of "land to the tiller" was never fulfilled.

FISCAL IRRESPONSIBILITY. That the Nationalist government was indifferent to the land problem is again seen in the fact that it relegated the land
tax, the most basic of revenues in the old dynastic days, to provincial ad-

48. Swarup, 52.

ministrations, while relying on customs revenues and commercial taxes for its own sustenance. Established in the coastal areas and using Western-trained financiers such as T. V. Soong and H. H. Kung to chart the economic course, the Nanking government was never close to the peasant and the soil, and probably did not care about or understand the severity of the land problem. From 1928 to 1935, the government derived 42.23 percent of its income from customs revenues, 17.13 percent from the salt tax, and 9.16 percent from commodity taxes. Yet the total receipts covered only 80 percent of the expenditures, which largely consisted of military expenses (40.3 percent) and debts service (25-37 percent).[49] Throughout these years, the government never achieved a fiscal balance but subsisted on deficit spending. The chronic ill of budgetary imbalance led to abusive issuance of notes, which later was to cause severe inflation during the Japanese and civil wars and precipitated the economic collapse of the government in 1949.

On balance, at the end of its first decade the Nationalist government appeared stronger than it really was. On the surface, it looked as though it were forging a new order out of chaos—having pacified or reached working arrangements with the new warlords and the dissident politicians, quarantined the Communists in the Northwest, trained a German-style Central army, carried out some modernization programs in the several fields mentioned above, and formed a United Front with the various parties and groups against Japanese aggression. A superficial observer might readily say that a new China was emerging on the horizon. Yet beneath the veneer of progress lay the serious fundamental problems of social and economic injustices and the chronic ill of deficit spending. Of the three goals it set out to achieve in 1928—nationalistic revolution, democratic reconstruction, and social reform—the government by 1937 had made considerable progress toward the first, modest advance toward the second, but failed miserably in the third. Moreover, its extension of the Political Tutelage Period beyond the original six years from 1929, under the pretext of foreign invasion and domestic insurrections, disenchanted the liberals, who came to regard the delay as an artful device of the Nationalists to prolong their monopoly of power at the expense of constitutionalism.

The decade under review may be summed up in a neat Chinese expression, *wai-ch'iang chung-kan*, which suggests that the government was "strong on the outside but weak inside."

49. Shun-hsin Chou, *The Chinese Inflation, 1937-1949* (New York, 1963), 40-42.

Further Reading

Alitto, Guy S., *The Last Confucian: Liang Shu-ming and the Chinese Dilemma of Modernity* (Berkeley, 1984).

Atwell, Pamela, *British Mandarins and Chinese Reformers: The British Administration of Weihaiwei (1898-1930) and the Territory's Return to Chinese Rule* (New York, 1986).

Bedeski, Robert E., "The Tutelary State and National Revolution in Kuomintang Ideology, 1928-31," *The China Quarterly*, 46:308-30 (April-June 1971).

Bertram, James M., *China in Crisis: the Story of the Sian Mutiny* (London, 1937).

Bisson, Thomas A., *Japan in China* (New York, 1938).

Borg, Dorothy, *The United States and the Far Eastern Crisis of 1933-1938* (Cambridge, Mass., 1964).

Braun, Otto, *A Comintern Agent in China, 1932-1939*, tr. from the German by Jeanne Moore (Stanford, 1982).

Chan, Anthony B., *Arming the Chinese: The Western Armaments Trade in Warlord China, 1920-1928* (Vancouver, 1982).

Chang, Maria Hsia, *The Chinese Blue Shirt Society: Fascism and Developmental Nationalism* (Berkeley, 1985).

Ch'i, Hsi-sheng, *Warlord Politics in China, 1916-1928* (Stanford, 1976).

Christopher, J. W., *Conflict in the Far East: American Diplomacy in China from 1928-33* (Leiden, 1950).

Chu, Pao-chin, *V. K. Wellington Koo: A Case Study of China's Diplomat and Diplomacy of Nationalism, 1912-1966* (Hong Kong, 1981).

Coble, Parks M., Jr., "Chiang Kai-shek and the Anti-Japanese Movement in China: Zou Tao-fen and the National Salvation Association, 1931-1937," *The Journal of Asian Studies*, XLIV:2:293-310 (Feb. 1985).

———, *Facing Japan: Chinese Politics and Japanese Imperialism, 1931–1937* (Cambridge, Mass., 1992).

———, *The Shanghai Capitalists and the Nationalist Government, 1927-1937* (Cambridge, Mass., 1980).

Crow, Carl (ed.), *Japan's Dream of World Empire, The Tanaka Memorial* (London, 1943).

Eastman, Lloyd E., *The Abortive Revolution: China under Nationalist Rule, 1927-1937* (Cambridge, Mass., 1974).

———, *Seeds of Destruction: Nationalist China in War and Revolution, 1937-1949* (Stanford, 1984).

Eastman, Lloyd (ed.), *The Nationalist Era in China, 1927–1949* (Cambridge, Eng., 1991).

Forbes, Andrew D. W., *Warlords and Muslims in Chinese Central Asia: A Political History of Republican Sinkiang, 1911-1949* (Cambridge, Eng., 1986).

Friedman, I. S., *The Relations of Great Britain with China, 1933-1939* (New York, 1939).

Furth, Charlotte (ed.), *The Limits of Change: Essays on Conservative Alternatives in Republican China* (Cambridge, Mass., 1976).

Furuya, Keiji, *Chiang Kai-shek: His Life and Times* (New York, 1981). English abridged version by Chun-ming Chang.

Galbiati, Fernando, *P'eng P'ai and the Hai-lu-feng Soviet* (Stanford, 1985).

Gamble, Sidney D., *Ting Hsien: A North China Rural Community* (New York, 1954).

Gillin, Donald G., *Warlord Yen Hsi-shan in Shansi Province, 1911-1949* (Princeton, 1967).

Hall, J. C. S., *The Yunnan Provincial Faction, 1928-1937* (Canberra, 1976).

Heinzig, Dieter, "The Otto Braun Memoirs and Mao's Rise to Power," *The China Quarterly,* 46:274-88 (April-June 1971).

Hsia, C. T., *A History of Modern Chinese Fiction, 1917-1957* (New Haven, 1961).

———, *Twentieth Century Chinese Stories* (New York, 1971).

Hsia, T. A., "Ch'ü Ch'iu-pai's Autobiographical Writings: The Making and Destruction of a 'Tender-Hearted' Communist," *The China Quarterly,* 25: 176-212 (Jan.-March 1966).

Hsiao, Tso-liang, *The Land Revolution in China, 1930-1934* (Seattle, 1969).

———, *Power Relations within the Chinese Communist Movement, 1930-1934* (Seattle, 1961).

Huang, Philip C. C., et al., *Chinese Communists and Rural Society, 1927-1934* (Berkeley, 1978).

Huang, Sung-k'ang, *Lu Hsün and the New Cultural Movement of Modern China* (Amsterdam, 1957).

Hung, Chang-tai, *Going to the People: Chinese Intellectuals and Folk Literature, 1918-1937* (Cambridge, Mass., 1985).

Isaacs, Harold R., *The Tragedy of the Chinese Revolution,* revised (Stanford, 1951).

Israel, John, *Student Nationalism in China, 1927-37* (Stanford, 1966).

Jones, F. C., *Manchuria Since 1931* (London, 1949).

Kahn, Winston, "Doihara Kenji and the 'North China Autonomy Movement,' 1935-1936," *Occasional paper,* No. 4 (Center for Asian Studies, Arizona State University, Tempe, Nov. 1973).

K'ang-chan ch'ien shih-nien chih Chung-kuo 抗戰前十年之中國 (China during the ten years before the Sino-Japanese War, 1937), compiled by Chung-kuo wen-hua chien-she hsieh-hui 中國文化建設協會 (Chinese cultural reconstruction association), reprinted (Hong Kong, 1965).

Kapp, Robert A., *Szechwan and the Chinese Republic: Provincial Militarism and Central Power, 1911-1938* (New Haven, 1973).

Kataoka, Tetsuya, *Resistance and Revolution in China: The Communists and the Second United Front* (Berkeley, 1974).

Kiang, Wen-han, *The Chinese Student Movement* (New York, 1948).

Kim, Ilpyong J., *The Politics of Chinese Communism: Kiangsi under the Soviets* (Berkeley, 1973).

Kirby, William C., *Germany and Republican China* (Stanford, 1984).

Kuo, Hua-lun 郭華倫, *Chung-kung shih-lun* 中共史論 (On the history of Chinese Communism), 4 vols. (Taipei, 1969-71).

Kuo, Warren, *Analytical History of the Chinese Communist Party* (Taipei, 1968), 2 vols.

Lang, Olga, *Pa Chin and His Writings: Chinese Youth between the Two Revolutions* (Cambridge, Mass., 1967).

Lao, She, *Rickshaw: The Novel of Lo-t'o Hsiang Tzu*, tr. by Jean M. James (Honolulu, 1979).

Lary, Diana, *Region and Nation: The Kwangsi Clique in Chinese Politics, 1925-1937* (Cambridge, Mass., 1974).

Lattimore, Owen, *Manchuria, Cradle of Conflict* (New York, 1935).

Lau, Joseph S. M., *Ts'ao Yu, The Reluctant Disciple of Chekhov and O'Neill* (Hong Kong, 1970).

League of Nations, *Report of the (Lytton) Commission of Enquiry* (1932).

Lee, Feigon, *Chen Duxiu: The Founder of the Chinese Communist Party* (Princeton, 1983).

Lee, Leo Ou-fan, *The Romantic Generation of Modern Chinese Writers* (Cambridge, Mass., 1973).

———, *Voices from the Iron House: A Study of Lu Xun* (Bloomington, 1987).

———, (ed.), *Lu Xun and His Legacy* (Berkeley, 1985).

Li, Jui, *The Early Revolutionary Activities of Comrade Mao Tse-tung*, trans. from Chinese and intro. by Stuart Schram in 34 pp. (White Plains, N.Y., 1977).

Liang, Ching-tun 梁敬錞, *Chiu-i-pa shih-pien shih-shu* 九一八事變史述 (An historical account of the September 18, 1931, incident), (Hong Kong, 1934).

Linebarger, Paul, *Government in Republican China* (New York, 1938).

Liu, F. F., *A Military History of Modern China, 1924-1949* (Princeton, 1956).

Lötveit, Trygve, *Chinese Communism, 1931-1934: Experience in Civil Government* (Lund, Sweden, 1973).

Louis, William Roger, *British Strategy in the Far East, 1919-1939* (Oxford at Clarendon, 1971), chs. 4, 5.

Lyell, William A., Jr., *Lu Hsün's Vision of Reality* (Berkeley, 1976).

McCormack, Gavan, *Chang Tso-lin in Northeast China, 1911-1928: China, Japan, and the Manchurian Idea* (Stanford, 1978).

Nieh, Hua-ling, *Shen Ts'ung-wen* (New York, 1972).

North, Robert C., *Moscow and Chinese Communists* (Stanford, 1963).

Ogata, Sadako N., *Defiance in Manchuria: the Making of Japanese Foreign Policy* (Berkeley, 1964).

Peattie, Mark, *Ishiwara Kanji and Japan's Confrontation with the West* (Princeton, 1975).

Pickowicz, Paul G., "Ch'ü Ch'iu-pai and the Chinese Marxist Conception of Revolutionary Popular Literature and Art," *The China Quarterly*, 70:296-314 (June 1977).

———, *Marxist Literary Thought: The Influence of Ch'ü Ch'iu-pai* (Berkeley, 1981).

Rappaport, Armin, *Stimson and Japan, 1931-33* (Chicago, 1963).

Ristaino, Marcia R., *China's Art of Revolution: The Mobilization of Discontent, 1927 and 1928* (Durham, 1987).

Rue, John E., *Mao Tse-tung in Opposition, 1927-1935* (Stanford, 1966).

Salisbury, Harrison, *The Long March: The Untold Story* (New York, 1986).

Schram, Stuart, *Mao Tse-tung* (New York, 1966).

Schwartz, Benjamin I., *Chinese Communism and the Rise of Mao* (Cambridge, Mass., 1958), chs. 4-13.

Scott, A. C., *Literature and the Arts in Twentieth Century China* (New York, 1963).

Semanov, V. I., *Lu Hsün and His Predecessors* (White Plains, N.Y., 1980).

Sheng, Yüeh, *Sun Yat-sen University in Moscow and the Chinese Revolution: A Personal Account* (Lawrence, Kansas, 1971).

Sih, Paul K. T., *The Strenuous Decade: China's Nation-Building Efforts, 1927-1937* (Jamaica, N.Y., 1970).

Smith, Sara M., *The Manchurian Crisis, 1931-1932* (New York, 1948).

Stimson, Henry J., *The Far Eastern Crisis* (New York, 1936).

Strong, Anna Louise, *China's Millions: the Revolutionary Struggles from 1927 to 1935* (New York, 1935).

Thomson, James C., Jr., *While China Faced West: American Reformers in Nationalist China, 1928-1937* (Cambridge, Mass., 1969).

Thornton, Richard C., *The Comintern and the Chinese Communists, 1928-1931* (Seattle, 1969).

Tien, Hung-mao, *Government and Politics in Kuomintang China, 1927-1937* (Stanford, 1972).

Ting, Lee-hsia Hsu, *Government Control of the Press in Modern China, 1900-1948* (Cambridge, Mass., 1975).

Tong, Hollington K., *Chiang Kai-shek, Soldier and Statesman* (Shanghai, 1937), 2 vols.

Tung, William L., *The Political Institutions of Modern China* (The Hague, 1964).

Waller, Derek J., *The Kiangsi Soviet Republic: Mao and the National Congresses of 1931 and 1934* (Berkeley, 1973).

Wang, Chi-Chen (tr.), *Ah Q and Others: Selected Stories of Lusin* (New York, 1941).

Wang, Chien-min 王健民, *Chung-Kuo Kung-ch'an-tang shih-kao* 中國共產黨史稿 (A draft history of the Chinese Communist Party), (Taipei, 1965), 3 vols.

Wei, William, *Counterrevolution in China: The Nationalists in Jiangxi During the Soviet Period* (Ann Arbor, 1985).

Wilbur, C. Martin, *The Nationalist Revolution in China, 1923-1928* (New York, 1985).

Willoughby, W. W., *The Sino-Japanese Controversy and the League of Nations* (Baltimore, 1935).

Wong, Wang-chi, *Politics and Literature in Shanghai: The Chinese League of Left-wing Writers, 1930–36* (Manchester, Eng., 1991).

Wright, S. F., *China's Struggle for Tariff Autonomy: 1843-1938* (Shanghai, 1938).

Wu, Tien-wei, *The Sian Incident: A Pivotal Point in Modern Chinese History* (Ann Arbor, 1976).

Yakhontoff, Victor A., *The Chinese Soviets* (New York, 1934).

Yang, Benjamin, "The Zunyi Conference as One Step in Mao's Rise to Power: A Survey of Historical Studies of the Chinese Communist Party," *The China Quarterly*, 106:235-271 (June 1986).

Young, Arthur N., *China's Nation Building Efforts, 1927-1937: The Financial and Economic Record* (Stanford, 1971).

24
The Sino-Japanese War, 1937-45

The prospect of a China united against foreign aggression worried the Japanese militarists and extremists, who feared for the future of their expansion policy on the continent. As in 1931, those anxious to strike before China became too strong were again the young officers of the Kwantung army; encouraged by the easy conquest of Manchuria, by the lack of international sanctions, and by the rise of Nazism and Fascism in Europe, they were eager to turn North China into a second Manchukuo and there build up a continental base for Japan. Barely half a year after the Sian incident and the adoption of the United Front policy in China, these officers manufactured an incident at the Marco Polo Bridge (Lukou-ch'iao) about ten miles west of Peiping, on July 7, 1937, precipitating a clash with the Chinese garrison. Once hostilities began and all hope for a peaceful settlement had faded, the Chinese government became fiercely determined to fight for its survival to the bitter end. What the Japanese had intended to be a short war to conquer North China turned out to be a long war of attrition which lasted until 1945. This second Sino-Japanese war in less than half a century[1] produced serious and far-reaching repercussions in both countries: it led to the defeat of Japan for the first time in its modern history and thoroughly exhausted the Nationalist government in China, while giving the Communists a chance to expand their army and party in preparation for the ultimate seizure of power.

1. The first was fought in 1894-95.

The Rise of the Japanese Militarists

Though it appeared on the surface as a locally inspired affair master-minded by the Kwantung army, the "China Incident" of 1937 was actually a well-planned, premeditated plot climaxing a series of clashes between the Japanese military and the civilian government, as well as between the different groups of the military itself. Ever since their successful venture in Manchuria, the militarists had catapulted themselves into national politics at the expense of the civilian government. Ignoring the traditional admonition that the military should refrain from meddling in politics, the young officers openly attacked party politicians for their mishandling of the affairs of state which was said to have lowered Japan's international status. Spurred by chauvinistic zeal, they ridiculed the bureaucrats for their inefficiency and corruption, and censured the *zaibatsu* for their alleged role in bringing about the Depression. As the self-appointed saviors of the country, the young officers vowed to eliminate these evil elements and to effect a "Showa Restoration" whereby the emperor, through the army, would re-establish a direct relationship with the farmers and the people at large. So "sacred" was this mission that the young officers succeeded in creating an image that nothing could be allowed to undermine the prestige and position of the army. Their ruthless drive for power, their open advocacy of expansion, and their ready recourse to conspiracies, plots, intimidation, and assassinations bespoke a certain type of "abnormal behavior."[2] Indeed, their unruly, unrestrained behavior was hard to swallow even for the older, well-disciplined officers. Yet the young officers could not be stopped, for they claimed to represent the voice of the people and the future of Japan, and they enjoyed the support of extremist politicians and secret societies. Throughout the period of 1932-36 the militarists rose steadily in national politics until they totally eclipsed the party government. It was a tragedy for modern Japan.

THE COUP OF MAY 15, 1932. The meteoric rise of the militarists was achieved partly through the brutal methods of coups and assassinations. Following two abortive coups in the spring of 1931, the young officers employed political assassination as a tool and effectively eliminated the former finance minister and governor of the Bank of Japan as well as the leader of the Mitsui *zaibatsu*[3] in the spring of 1932. Then, on May 15,

2. For an illuminating study of Japanese behavior of this period, see Maruyama Masao, *Gendai seiji no shisō to kōdō* (Contemporary political thoughts and actions), I (Tokyo, 1961), 7-148.
3. Inoue and Baron Dan respectively.

1932, a group of army and navy officers attacked Tokyo's police station, banks, and party headquarters and succeeded in murdering Premier Inukai Tsuyoshi, who had disapproved of the military action in China and had favored a negotiated settlement. In brazen defiance of the consequences, the assassins then voluntarily surrendered themselves to the police. The trial that followed turned out to be a public airing of the conspirators' philosophy—to save the country through the elimination of the weak politicians, corrupt bureaucrats, and selfish *zaibatsu*. The argument evoked widespread sympathy to the point where even the prosecutor and the newspapers treated the conspirators as heroes rather than assassins. The party government, in existence since 1918, was dealt a mortal blow from which it did not recover until after World War II.

THE COUP OF FEBRUARY 26, 1936. The army, though united in its struggle against civilian rule, had its own inner conflicts. A number of older, more responsible officers subscribed to the traditional injunction to shun politics, but many others had become political-minded and meddlesome; among these latter were two groups. One was the Imperial Way Faction (*Kōdō ha*), consisting of young activist, field-grade officers under the leadership of Minister of War Araki, vice-chief of the General Staff Mazaki, and commander-in-chief of the Gendarmerie Hata. They demanded a military dictatorship, control of the national budget, expansion of the army and the navy, nationalization of the essential industries, territorial aggrandizement in Asia, and direct action in China. The other group was the Control Faction (*Tōsei ha*), consisting of older and better disciplined high-ranking officers, such as Generals Nagata Tetsuzan, Abe Nobuyuki, and strangely enough Tōjō Hideki. They too wanted a firm foreign policy and expansion of Japan's hegemony in Asia, but they disapproved of direct action and terroristic methods, preferring to gain influence through legal means and proper channels.

The inner conflicts of the army generated instability and turmoil. In July 1933 a coterie of young officers, disappointed with Premier Saitō's inability to carry out the reforms demanded by the Imperial Way Faction, plotted to kill all cabinet ministers and party leaders. Although this fantastic conspiracy was uncovered in time to prevent a vast bloodshed, the forty-four defendants, amid the rising tide of militarism and chauvinism, escaped trial until 1937 and were released in 1941. Nonetheless, the failure of this coup enabled the Control Faction to score a point in the power struggle. General Araki was replaced by General Hayashi Senjūrō as minister of war. In an attempt to reduce tension between the two groups, Hayashi replaced in July 1935 General Mazaki as inspector-general of military education with a milder, more amenable officer, Lieutenant General Watanabe Jutarō. The mastermind of this reshuffle was General Nagata of the Control Faction, whose moments of glory lasted but a month—on August 12, 1935, a young military radical, Lieutenant Colonel Aizawa, assassinated him. However, the sentence of death for the assassin hardly blunted the razor-sharp fanaticism of the young officers. When the minister of finance slashed the army budget in January 1936, they decided to strike again.

On February 26, a group of young officers and 1,400 soldiers, led by Captain Ando Teruzo, seized control of central Tokyo, occupying the Diet building, the police headquarters, and the War Department. They invaded the premier's residence, killing his brother-in-law by mistake. Others that were killed included a former premier,[4] the minister of finance,[5] and the inspector-general of military education.[6] It was not until the loyal troops had surrounded the mutineers and proclaimed the emperor's order that they return to their camps that the coup was finally brought under control. The subsequent trial sent thirteen young officers to execution and Generals Araki and Mazaki to the reserves.

With the resignation of Admiral Okada as premier in March 1936, Hirota Kōki took over the helm of government. He had been foreign minister since autumn 1933 and was known for his extremist connections and aggressive China policy. Supported by the militarists, he promised to "reform" the government and appointed many army-approved personnel to his cabinet.[7]

HIROTA'S CHINA POLICY. Hirota's policy aimed at isolating China from the rest of the world so as to coerce it into submission. The Japanese in North China sponsored an autonomous movement of the five provinces of Hopeh, Chahar, Suiyuan, Shansi, and Shantung, and in December 1935 created an Eastern Hopeh Autonomous Council. The Chinese government countered with the establishment of a Hopeh-Chahar Political Council with its headquarters in Peiping. But, friction mounted as Japanese *ronin* (rascals) as well as Koreans and Formosans blatantly engaged in large-scale silver- and narcotics-smuggling. To all appearances, North China was beginning to look like a second Manchuria.

4. Admiral Saitō.
5. Takahashi.
6. General Watanabe.
7. For instance, his minister of finance, Baba Eiichi, was an army stooge. The new president of the Privy Council, Baron Hiranuma, was an avowed Fascist and extremist. His army and navy ministers, General Terauchi and Admiral Nagano, were both expansionists.

Chinese popular demands for resistance against Japan mounted daily, and a nationwide boycott of Japanese imports succeeded in cutting trade by two-thirds. Hirota, acting on the wishes of the Diet which had a Minseitō (Democratic Party) majority, opened negotiations with China in the summer of 1936, proposing: (1) an end to anti-Japanese activities in China; (2) recognition of Japan's special position in North China; (3) Sino-Japanese collaboration against communism, especially in Outer Mongolia; (4) Sino-Japanese economic cooperation; and (5) Japanese advisers in all branches of the Chinese government. These terms, especially the last which smacked of the Twenty-one Demands of 1915, were rejected by the Chinese government, which counterproposed: (1) termination of Japanese, Korean, and Formosan smuggling; (2) withdrawal of Japanese troops from Hopeh and Chahar; and (3) suppression of the Japanese-sponsored autonomous movements. Negotiations broke down in December 1936.

Internationally, Hirota had adopted a policy oriented toward isolating the Soviet Union, preparing for war against the United States and Britain, and collaborating with Germany and Italy. Japan signed the Anti-Comintern Pact with Germany in 1936, and with Italy a year later, moving ever closer to the brink of war.

Hirota's government fell in December 1936, and in February of the following year General Hayashi became premier. The new foreign minister, Sato Naotake, proposed a *rapprochement* with China through the restoration of economic relations and an initial settlement of minor issues; North China, however, was to be maintained as a special area. China by this time had already been committed to a United Front and was in no mood for further concessions. The Hayashi cabinet fell after four months; the next administration under Prince Konoe Fumimaro was entirely dominated by the army. General Tōjō, chief of staff of the Kwantung Army, advocated the application of force against China. Although his ideas were rejected by Tokyo, the Japanese army in North China decided to go ahead on its own and provoked a clash on July 7, 1937.

The Undeclared War, 1937

Invoking the Boxer Protocol of 1901 which permitted foreign signatories to station troops between Peking (Peiping) and the sea, the Japanese garrison in North China in early July 1937 held a field exercise outside Peiping, near the Marco Polo Bridge. On the pretext that a soldier was missing, the Japanese demanded to enter the nearby city of Wanping before midnight of July 7 to conduct a search. When refused by the local

Chinese garrison—the 29th Army under General Sung Che-yüan—the Japanese army bombarded the city and occupied it at 4:30 a.m. on the morning of July 8, thus precipitating an undeclared war between the two countries.

Once hostilities began, Japanese reinforcements from Manchuria and the home islands poured into North China, occupying all the strategic points outside Peiping. Obviously, the Marco Polo Bridge incident was but the beginning of a much larger design. The Nanking government, having committed itself to the United Front against Japanese aggression, was determined to fight. On July 17, 1937, at a summer conference at Kuling, Chiang Kai-shek resolutely announced that when pushed to the ultimate extreme, China had no choice but "to throw the last ounce of energy" into "a struggle for national survival." The war of resistance, delayed since 1931, finally assumed coherence and meaning. The Chinese people and the various parties—the KMT, the CCP, the Youth, etc.—all enthusiastically pledged their support of the war.

If China was prepared for a long fight, Japan had no intention of becoming bogged down on the Asiatic mainland; Russia was still the principal enemy in the minds of the General Staff. Japan wanted a quick victory to seize North China and force Nanking into economic cooperation. Disparaging China's ability and will to wage a full-fledged war, the Japanese military allowed three months to conclude the China affair. From a strictly military point of view, it looked as if their prediction was accurate.

The modernized Japanese army proved more than a match for the Chinese. Having inflicted heavy losses on the 29th Army, the Japanese were poised to attack Peiping in late July. Any effective defense of this city of ancient treasure and culture would undoubtedly have caused untold damage to the priceless historic relics and art objects. The Nationalists decided to spare Peiping this terrible fate and ordered its evacuation on July 28. Two days later Tientsin also fell.

On August 13 the Japanese opened a second front in Shanghai, the financial center of the nation, to destroy China's economic capacity for war. There, unexpectedly, Chiang threw in some of his best German-trained troops—the 87th and 88th divisions—which succeeded brilliantly in stalling the enemy advance for three months. But the Japanese tactic of outflanking the defender ultimately worked, causing an unexpectedly rapid disintegration of the Chinese defenses. The road to Nanking was left wide open, and the enemy swiftly advanced to the gate of the Chinese capital.

Chiang moved his capital to Chungking, Szechwan, where the rugged terrain, the precipitous gorges, and the rapid currents in the narrowing

Yangtze would make it all but impossible for the enemy to penetrate. Chiang himself, as commander-in-chief of the Chinese forces, remained in strategic Wuhan to direct military operations, while schools, factories, and other establishments in the occupied areas were encouraged to migrate inland. The southwest became a new base of resistance which dashed the Japanese dream of a quick settlement.

The fall of Nanking was followed by the indiscriminate massacre of approximately 100,000 civilians, accompanied by innumerable cases of molestation of women. So notorious was "The Rape of Nanking," as it came to be known, that even the Japanese militarists concealed it from the public at home. When the facts were finally revealed in the postwar International War Crimes Tribunal in Tokyo, the Japanese people were deeply ashamed of the atrocities.

After their conquest of Nanking, the main Japanese forces moved northward toward the important communication junction of Hsuchow. But at T'ai-erh-chuang, near Hsuchow, they encountered heroic resistance by Chinese forces in late March and early April of 1938, sustaining 30,000 casualties. This was a major Chinese victory since the fall of Nanking, but ultimately Hsuchow had to be evacuated on May 19. Later, in June, the Chinese broke the Yellow River dikes to slow the enemy advance.

The next major battle was fought at Wuhan, where Chiang had established his headquarters and where twelve Japanese divisions had closed in from two directions along the Yangtze and the Huai rivers. After several hundred large and small encounters over a period of four and one-half months, Wuhan was finally relinquished on December 25, 1938. Its loss, coupled with the fall of Canton on October 21, drove some of the unfirm Nationalist leaders to despair and despondency, but Chiang carried on the fight as he had pledged.

The fall of Wuhan marked the end of the first phase of the war, which lasted sixteen months. During this period the Chinese traded space for time and enticed the enemy deep into the hinterland. The Japanese army was mired deep in the abdomen of China from which it could not extricate itself.

International sanctions were slow in coming, for Europe itself was threatened by Nazism and Fascism, and the United States still clung to its neutrality. Yet, in spite of everything, the Japanese could not quickly win the war. Tokyo finally resigned itself to a stalemate; it adopted the policy of living off the conquered land with the help of puppet governments. On October 29, 1937, a Mongolian Autonomous Government was created in Chahar and Suiyuan, with the Inner Mongolian Prince Teh as the figurehead ruler. On December 14, another puppet "provisional government" was established in Peiping, with Wang K'e-ming as the front man; it governed the five northern provinces of Hopeh, Chahar, Suiyuan, Honan, and Shantung. On March 28, 1938, a third puppet government was set up at Nanking under the formal leadership of Liang Hung-chih, with jurisdiction over the three eastern provinces of Kiangsu, Chekiang, and Anhwei. But none of the three leaders had the national stature necessary to achieve unification; so the Japanese intensified their search for a man of greater prominence.

Wang Ching-Wei's Peace Movement

On November 3, 1938 in commemoration of the birthday of Emperor Meiji, Prince Konoe, premier of Japan, proclaimed a "New Order in East Asia," based upon six principles: (1) permament stability for East Asia; (2) neighborly amity and international justice; (3) joint defense against communism; (4) economic cooperation; (5) creation of a new culture; and (6) world peace. It was a kind of Japanese "Monroe Doctrine," manifesting the historical Japanese aspiration to dominate Asia as reflected in Hideyoshi's attempts to conquer Korea and China in the late 16th century, in the Pan-Asiatic ideas among Japanese leaders in the early 20th century, and in the alleged "Tanaka Memorial" of 1927.

To those Chinese politicians who had been frustrated by the absence of prospects for victory or peace, by the progressive inflation, and by the mounting difficulties of life, the Konoe statement appeared to offer a ray of hope for a quick settlement. Wang Ching-wei, among others, found the Japanese idea intriguing. He flew out of Chungking to Hanoi on December 18, 1938, to start a peace movement. Four days later, Konoe announced Japan's decision to destroy the Nationalist government and to adjust Sino-Japanese relations with a "new" Chinese regime on the basis of (1) friendship and amity: not only would Japan make no demands of territory or indemnity but it would return to China all concessions and leased territories and abolish extraterritoriality; (2) mutual defense against communism in the same spirit as the anti-Comintern Pacts between the Axis powers; and (3) economic cooperation without any intention on the part of Japan to monopolize China's economy.

Wang urged the Chungking government to accept them as the basis for a settlement. Chiang dismissed the notion summarily and persuaded the KMT to expel Wang. The latter then signed eight agreements with Japan, including recognition of Manchukuo and permission for Japan to station troops in China in order to facilitate a joint defense against communism. Other aspects of the agreements acknowledged Japanese control over China's natural resources in an act of "economic cooperation," and Japanese authority to appoint advisers in Chinese education and cultural affairs.

In March 1940, under Japanese aegis, Wang set up a five-yüan government in Nanking, which absorbed the older puppet regimes at Peiping and Nanking. The regime was recognized by Manchukuo, and the three Axis powers and their satellites,[8] but not by any of the major Western powers.

The question naturally arises as to why Wang, the close follower of Sun and the second-ranking member in the Nationalist Party and government, should want to defect to the enemy and risk his reputation? It appears that first of all he was driven by defeatism to conclude that China could not win the war; it would be more realistic to negotiate a peace before total defeat. Second, he had been engaged in a power struggle with Chiang whom he thought had usurped his place as successor to Dr. Sun. Third, he was concerned about the welfare of the people in the Japanese-occupied areas. A fourth and intriguing interpretation was offered by his chief aide, Chou Fu-hai, who artfully explained in a news conference that Wang's movement could not hurt China; if Chiang won the war, Wang's agreements with Japan would naturally be nullified; if Chiang could not win the war, his future settlement with Japan could not possibly surpass the terms achieved by Wang. Then, in a move calculated to evoke sympathy, Chou asked the searching question: If Wang had not offered his services, who would have taken care of the teeming millions in the occupied areas? One may surmise that Wang, driven by defeatism and jealousy of Chiang, decided to offer himself as a buffer between the harsh Japanese conquerors and the helpless people of China. Perhaps he truly believed that his peace mission could not hurt China in the long run and could actually reduce a good deal of misery in the short run. The truth of his motivation may never be known, for he died a few months before Japan's surrender in 1945. Though he was spared a public trial, his senior followers were shot as traitors soon after the war.

The Nationalist Program of Resistance and Reconstruction

During the early years of the war, a number of important developments took place within the KMT and the government. Of great significance was the convocation at Wuhan of the KMT Provisional National Congress in April 1938 and its adoption of four major resolutions: (1) crea-

---

8. Such as Rumania, Bulgaria, and Denmark.

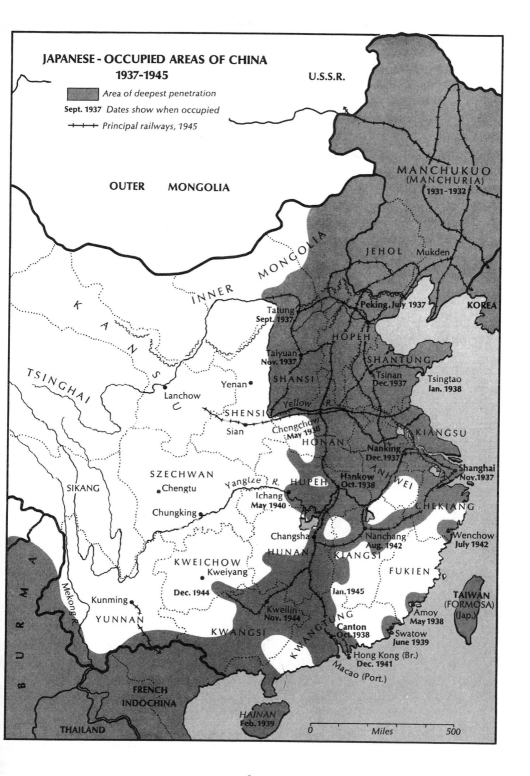

JAPANESE - OCCUPIED AREAS OF CHINA
1937-1945

Area of deepest penetration
Sept. 1937 Dates show when occupied
+++++ Principal railways, 1945

U.S.S.R.

OUTER MONGOLIA

MANCHUKUO
(MANCHURIA)
1931-1932

INNER MONGOLIA

JEHOL   Mukden

KOREA

KANSU

TSINGHAI

Lanchow

Yenan

SHENSI

Sian

Tatung
Sept. 1937

Taiyuan
Nov. 1937

SHANSI

Yellow R.

Chengchow
May 1938

HONAN

Peking, July 1937

HOPEH

SHANTUNG

Tsinan
Dec. 1937

Tsingtao
Jan. 1938

KIANGSU

SIKANG

SZECHWAN

Chengtu

Chungking

Yangtze R.

Ichang
May 1940

HUPEH

Hankow
Oct. 1938

Nanking
Dec. 1937

ANHWEI

Shanghai
Nov. 1937

CHEKIANG

Changsha

HUNAN

Nanchang
Aug. 1942

KIANGSI

Wenchow
July 1942

KWEICHOW

Kweiyang

Dec. 1944

FUKIEN

Jan. 1945

TAIWAN
(FORMOSA)
(Jap.)

BURMA

Mekong R.

Kunming

YUNNAN

KWANGSI

Kweilin
Nov. 1944

KWANGTUNG

Canton
Oct. 1938

Amoy
May 1938

Swatow
June 1939

Hong Kong (Br.)
Dec. 1941

Macao (Port.)

FRENCH
INDOCHINA

THAILAND

HAINAN
Feb. 1939

0   Miles   500

tion of the new post of Director-general (*Tsung-ts'ai*) as the party Leader, with Chiang as the first appointee; (2) creation of a Three People's Principles Youth Corps to train young men as a basic force in the war of resistance and national reconstruction; (3) creation of a People's Political Council to replace the National Defense Advisory Council, as the highest wartime popular organ of state; and (4) adoption of an "Outline of Resistance and Reconstruction."

Following the KMT Congress, a People's Political Council was organized, comprised of members of all political persuasions including Mao Tse-tung and a number of other Communist leaders. Its inaugural meeting, convened at Wuhan between July 7 and 15, 1938, was attended by 162 delegates who staunchly pledged that all Chinese, regardless of parties, religions, creeds, and professions, were committed to supporting the war until the final victory was won.

The country was united in the struggle against Japan, yet behind the brave façade of national solidarity were critical cleavages and the seeds of discord, especially with regard to the question of the Communist Party.

The United Front and Its Decline

Shortly after the outbreak of the Sino-Japanese War, the CCP issued on September 22, 1937, an appealing manifesto entitled, "Together We Confront the National Crisis" (*Kung-fu kuo-nan*), to explain its position during the war:

1. The CCP will struggle to fulfill completely Dr. Sun's Three People's Principles which best answer China's needs today.
2. The CCP will abolish the policy of sabotage and Sovietization which aims at the overthrow of the KMT government, and will stop the forcible confiscation of the holdings of landlords.
3. The CCP will abolish all existing Soviets in favor of democratic government, so as to achieve unified political administration throughout the country.
4. The CCP will abolish the name and insignia of the Red Army, which will be reorganized as the National Revolutionary Army and is to be subject to control by the government's Military Commission; it is ready to march forward and fight the Japanese at the front.[9]

Welcoming the Communist pledges, Chiang expressed the hope that the CCP would prove its sincerity through actual contribution to the

9. The basis of this manifesto was contained in a telegram of February 10, 1937, from the CCP Central Committee to the KMT Central Executive Committee. See Schram, 183.

war and to the cause of National Revolution. The 30,000-man Red army was reorganized as the 8th Route army—later renamed the 18th Route army—under the command of Chu Teh and P'eng Te-huai. The army was sent to fight the Japanese in northern Shansi. Later, in December 1937, Communist units south of the Yangtze were organized as the New Fourth army under the command of Yeh T'ing and Hsiang Ying, with a strength of 10,000 men. A further symbol of the KMT-CCP *rapprochement*, as already noted, was the election of Mao and other Communist leaders to the newly created People's Political Council in 1938. Thus, the two parties once again collaborated in the face of Japanese aggression.

Yet this alliance, like the earlier one of 1923-27, was ill-fated. From the beginning the Communists regarded it as nothing more than a means to carry out the orders of the Comintern, to be freed from Nationalist attacks, and to build up strength during the war. Lest any member of his party misunderstand these secret objectives, Mao admonished the cadres that they should fully utilize the opportunity of the Japanese war for self-expansion. "Our fixed policy," he said, "should be 70 percent expansion, 20 percent dealing with the Kuomintang, and 10 percent resisting Japan."[10] The cadres were asked to act accordingly if they lost contact with the party headquarters. Mao made it very clear that temporary cooperation with the KMT was not a betrayal of one's principles, nor a surrender to the enemy, but a realistic way to heal the battle fatigue and preserve the revolutionary strength for the future. He reassured his followers that reorganizing the Red army into a National Revolutionary army and substituting border governments for the Soviet regimes were merely changes in form and not in substance. Collaboration with the KMT, in short, afforded the CCP a chance to recoup, to expand, and to initiate a new approach in place of the old one that had not proved very successful.

So that the United Front might be exploited to full advantage, Mao mapped out a three-stage strategy: first, to achieve a compromise with the KMT in order to safeguard the existence of the CCP; second, to struggle for parity with the KMT; and third, to infiltrate into central China and build up a new base from which to launch a counterattack and seize the supreme power of state.

In light of these explanations, it is not hard to see why the Communists

10. This statement has been quoted frequently. See F. F. Liu, *A Military History of Modern China, 1924-1929* (Princeton, 1956), 206; Chiang Kai-shek, *Soviet Russia in China: A Summing-Up at Seventy* (New York, 1957), 85; and Arthur N. Young, *China and the Helping Hand, 1937-1945* (Cambridge, Mass., 1963), 58.

concentrated on self-development and expansion, striving toward the goal of "a million Red soldiers and a million party members."[11] In March 1939 the Communists created their own Shensi-Kansu-Ninghsia border government, and later established the Shansi-Hopeh-Chahar-Suiyuan border government. Clashes between the Communist and Nationalist forces began to occur with increasing frequency.

The growing tension between the KMT and the CCP was catalyzed by rapidly shifting international alignments. The signing of the German-Soviet Non-Aggression Pact in August 1939—followed by the Japanese-Soviet Neutrality Pact in April 1941—had removed the doctrinal basis and expediency of the United Front. Conflict between the KMT and CCP became more serious. The situation was particularly critical in western Shantung and in Kiangsu where the New Fourth army was stationed. On January 5, 1941, a major clash took place between the New Fourth army and the Nationalist 40th division, resulting in the government decision, on January 17, to disband the New Fourth Army and to arrest its commander for court martial. The CCP simply appointed another commander[12] and retaliated by increasing the strength of the New Fourth army to seven divisions. This "New Fourth army incident" all but destroyed the United Front, and Communist members of the People's Political Council refused to attend. Resumption of talks between the two parties took place in March 1943, only to break down again over Communist demands for legal status and military expansion to four armies of twelve divisions. By the end of 1943 the CCP negotiator, Chou En-lai, had left Chungking.

Throughout the remainder of the war, the KMT-CCP conflict was never resolved. Although Chiang repeatedly announced that the Communist problem, being political in nature, should be resolved by political means,[13] he sent a large body of his best troops to blockade the Communist areas in the Northwest, with the intention of using them in the event of a civil war following the Japanese war. On their part, the Communists ceaselessly expanded their military forces and popular organizations, and carried out a number of far-reaching programs in their base areas in preparation for the future confrontation.

11. On April 24, 1945, Mao Tse-tung claimed that during 1943 and 1944 his troops engaged 64 percent and 56 percent of the Japanese forces, respectively, and 95 percent of the "puppet army." Cf. Mao Tse-tung, III, 1,043-44.
12. Ch'en Yi.
13. *United States Relations with China, With Special Reference to the period 1944-1949* (Washington D.C., 1949), 135.

The Yenan Experience and Foreign Observation

The Yenan period of wartime resistance (1937-45) provided Mao and the CCP with the much needed time to restructure the party and the army, organize the masses, and develop new social, political, and economic institutions. Mao was at the peak of his creativity, ingeniously reconciling the universalist Marxist-Leninist principles with the particularist demand of Chinese conditions and the Chinese revolutionary experience. Hence, the Yenan experience was of seminal importance to the development of Chinese Communism; in it was planted the seed of Mao's ultimate success.

The heart of the Yenan Way was the perfection of the mass line and the sharpening of revolutionary nationalism in the countryside, which became the twin pillars of Maoism.[14] To be sure, these ideas were first developed during the Kiangsi period,[15] but they were precluded from full expression by repeated Nationalist attacks from without and by incessant party squabbles from within. The Moscow-trained Chinese Communists such as Li Li-san, Wang Ming, and Po Ku opposed Mao's policies and advocated the urban-oriented Soviet model of proletarian revolution. Now, at Yenan, freed from external attack and internal dissension, Mao was able to carry out his own strategy and develop his own work style that was to become the benchmark of Chinese Communism.

In accordance with his mass line approach, Mao vigorously addressed himself to the needs of the peasants, carried out land reform and rent reduction programs, and brought the peasants into full participation in the political, economic, and military organizations in the base areas. Indeed, the poverty of Shensi and the border areas stimulated rather than impeded the birth of "peasant radicalism,"[16] and the Japanese war gave new impetus to revolutionary nationalism. The Yenan period was therefore one of growth and preparation for the ultimate seizure of power.

To activate the inexperienced peasantry, Mao created the poor-peasant corps and the farm labor-union under the *hsiang* level and encouraged them to participate actively in land confiscation and redistribution move-

---

14. James P. Harrison, *The Long March to Power* (New York, 1972), 514.
15. Ilpyong J. Kim, *The Politics of Chinese Communism: Kiangsi under the Soviets* (Berkeley, 1973); "Mass Mobilization Policies and Techniques Developed in the Period of the Chinese Soviet Republic," in A. Doak Barnett (ed.), *Chinese Communist Politics in Action* (New York, 1969), 78-98.
16. Mark Selden, *The Yenan Way in Revolutionary China* (Cambridge, Mass., 1971), 28, 90, 100.

ments. The direct involvement of the peasantry in the process of mass sociopolitical mobilization against the endemic problems of rural poverty and oppression not only sharpened their class consciousness but also forced them to shed their traditional timidity. Moreover, during 1937-41 all peasants who were sixteen or older were drawn into the mainstream of political activities in the border areas through the institution of universal, direct, and equal suffrage by secret ballot. Mao believed that "all men could transcend the limitations of class, experience, and ideology to act creatively in building a new China."[17] He also formulated the "Three-thirds system" (*San-san chih*), which limited party-member participation in government and councils of the base areas to one-third, leaving the other two-thirds to progressive leftists and independents. On the surface at least, these United Front policies gave the border areas a democratic overtone.

CAMPAIGNS TO ACHIEVE SELF-SUFFICIENCY. On the basis of cooperative and participatory principles, six major campaigns were launched during the Yenan period; in them we find the central features of many of Mao's later policies:

1. Adoption of the principle of "crack troops and simple administration" (*ching-ping chien-cheng*) to reduce army and government bureaucracy.
2. Introduction of the "To the Village" (*Hsia-hsiang*) campaign to mingle intellectuals and party cadres with workers and peasants.
3. Reduction of rent and interest in areas where there was no land reform by 25 to 40 percent so that they would not exceed one-third of the total yield of the land.
4. Introduction of the mutual-aid cooperative movement to reorganize the village economy.
5. Introduction of an "organizational economy" to make every organization and cadre participate in managerial as well as manual work.
6. A new education movement for social, economic, and cultural transformation of the rural society.[18]

Thus, the mass line approach to politics, economics, war, and revolution forged a close link between the leaders and the people, and formed the core of the Yenan experience.

During the Yenan period, Mao devoted much of his time to thinking,

17. *Ibid.*, 210.
18. *Ibid.*, 210-11, 212-74; Jerome Ch'en, *Mao and the Chinese Revolution* (London, 1965), 204.

theorizing, and writing on the problems confronting the party and the country and on the strategy of laying the foundation for ultimate victory. Extremely resourceful in his forties, he worked 13 or 14 hours a day, frequently into the small hours of the morning. It was a most productive period in his life. Once in 1938 he worked almost without interruption for nine days and nights to complete an essay, *On the Protracted War,* and at the end of his work was physically exhausted. Many other important works came from his pen during this period: *Problems of Strategy in China's Revolutionary War* (December 1936), *Urgent Tasks of the Chinese Revolution since the Formation of the KMT-CCP United Front* (September 1937), *Interview with James Bertram* (October 1937), *Problems of Strategy in Guerrilla War against Japan* (May 1938), *On Protracted War* (May 1938), and *Problems of War and Strategy* (November 1938). In about two years he had written 200 pages on strategy, 165 pages on politics, and 55 pages on philosophy.[19] He also authored many other famous works including: *On the New Democracy* (January 1940), *Rectifying the Party's Style of Work* (February 1942), *Opposed Stereotyped Party Writing* (February 1942), *On Coalition Government* (April 1945), and *On the Chungking Negotiations* (October 1945). In July 1949 he wrote another major work: *On the People's Democratic Dictatorship.*

The existence of "another China" with its own territory, government, disciplined party and army, and prominent leadership attracted the curiosity of foreigners who wanted to see for themselves how different this separate political entity was from the KMT region. As a result a number of foreign visitors entered the Communist areas and wrote reports about their discoveries.

FOREIGN OBSERVERS. In July 1936 Edgar Snow broke the KMT news blockade and entered the Communist area. He had previously described Chinese Communism as a form of "Agrarian Communism," but after visiting Yenan and talking with Mao he promoted the Chinese Communists to the status of dynamic Marxist revolutionaries. Snow contested the idea that the CCP was simply a subservient puppet of Moscow and

---

19. Jerome Ch'en, 209, 216-17. Ch'en's calculation of the volume of Mao's writings during this period is underestimated, for he did not include Mao's many other works such as: *The Tasks of the Chinese Communist Party in the Period of Resistance to Japan* (May 1937); *On Practice: On the Relations between Knowledge and Practice, between Knowing and Doing* (July 1937); *On Contradiction* (August 1937); *The Role of the Chinese Communist Party in the National War* (October 1938); *The Chinese Revolution and the Chinese Communist Party* (December 1939), and a host of miscellaneous writings. See Mao Tse-tung, *Selected Works* (Peking, 1967), Vols. I and II.

asserted that the Chinese had developed a unique and indigenous brand of communism. In his *Red Star Over China*, a journalistic classic which had a resounding impact on the popular American conception of Red China, Snow portrayed the Communists as austere and patriotic and the Nationalists as corrupt and unreliable. His wife was still more outspoken in her contrast between the two parties. Once in Yenan in 1937 she called her trip "a journey of discovery . . . of a new mind and a new people, creating a new world in the heart of the oldest and most change-less civilization on earth." To her the Communists were a "new species" of Chinese who were "extremely humanitarian"—a character trait that appealed to her "very much."[20] She went so far as to say that the Chinese Communists seemed "more like us [Americans]." They were "reaching out for a bridge to the Western world through their Marxian concept and were trying to become men of their own century." She praised the CCP for its efforts to "destroy feudalism and establish a modern society," and concluded that the Chinese Communists "belonged to the same human race as myself."[21]

The feeling of revulsion to the KMT and attraction to the CCP was shared by most foreign reporters with the notable exception of one Catholic priest.[22] Although some of the visitors were predisposed to a favorable view of Red China, on the whole they represented a wide spectrum of political convictions; they were impressed by the Communists' activity, hope, honesty, and concern for the masses. T. A. Bisson distinguished the Nationalist *"feudal* China" from the Communist *"democratic* China." Gunther Stein of the Associated Press and *The Christian Science Monitor* called Chungking a "pathetic city" and "a nightmare." Flying from Yenan to Chungking was like traveling "from one Chinese world to another."[23] Theodore White of *Time-Life* described the Yenan people as "ruddier and healthier" than elsewhere in China, while Harrison Forman of the United Press and the *New York Herald Tribune* considered the Red soldiers in 1943 "about the best nourished troops I had yet seen."[24]

20. Nym Wales (Helen Foster Snow), *Inside Red China* (Garden City, N.Y., 1939), xi, 38.
21. Kenneth E. Shewmaker, *Americans and Chinese Communists, 1927-1945: A Persuading Encounter* (Ithaca, N.Y., 1971), 338-39.
22. These visitors included Brooks Atkinson, James M. Bertram, T. A. Bisson, Evans F. Carlson, Israel Epstein, Harrison Forman, Phillip J. Jaffe, Ralph Lapwood, Michael Lindsay, Agnes Smedley, Edgar Snow, Helen Snow, Gunther Stein, Anna Louise Strong, and Theodore H. White, and others. The Catholic priest was Cormac Shanahan of the *China Correspondent* and various other Catholic publications.
23. Gunther Stein, *The Challenge of Red China* (New York, 1945), 5, 88, 460; Shewmaker, 340.
24. Jerome Ch'en, 248.

The yardstick of comparison was not Communist China and the United States but Chungking and Yenan. The former represented "Old China"—inert, decadent, selfish, suffering, indifferent to the common people, poor, inhumanitarian, and nepotistic—and the latter represented "New China"—hopeful, young, efficient, vigorous, Spartan, and enthusiastic. Edgar Snow spoke of a Red Star rising over China, and Theodore White thought that the KMT regime was losing the Mandate of Heaven by default, rotting away through moral degeneration and political misrule. Although White "distrusted Communist intentions and had no desire to see China engulfed in a Red tide," he considered the KMT "decadent" and the CCP "dynamic"—the latter had "shone by comparison."[25]

American reporters felt that the Communists were "better people" or "the lesser evil" than the Nationalists because of their pro-poor and anti-rich attitude. Coming from a society accustomed to thinking of ideals of social betterment and democratic progress, they looked into the future with expectations of a rising level of improvement in the well-being and intelligence of the masses. Hence, the traditional values of the KMT were less compatible to them than were those of the CCP, as the Marxist historic optimism and utopian materialism struck a somewhat responsive cord with them.[26] In their reports the Chinese Communists appeared as "superior human beings" and as "Prince Valiants in Straw Sandals"; their encounter with the CCP has been described as "persuading" and that with the KMT as "repelling."[27]

On the basis of their observations of the border regions, foreign visitors were impressed enough by the "Three-thirds" system, the electoral procedures, the mass participation in political processes, and the extension of civil rights to conclude that Red China qualified as "a species of democracy."[28] They described the Chinese Communists as "social reformers and patriots" or "peasant reformers" who practiced "representative democracy," "agrarian or peasant democracy," or a "functioning popular democracy." The myth of the Chinese Communists as "agrarian reformers" and different from Russian Communists became widespread in American minds.

In retrospect, one is inclined to think that the foreign reporters were journalists who wrote about what they saw rather than trained political scientists with the ability to perceive accurately the theoretical nature and ultimate goals of the CCP. These reporters were unable to distinguish

25. Shewmaker, 340-46.
26. *Ibid.,* 345-46.
27. *Ibid.,* 183, 191-99, 339.
28. *Ibid.,* 211, 215.

between democratic means and Communist ends, and failed to see that the United Front policy was but a deferral, not the abandonment, of the fundamental principle of seizure of power and world revolution.[29]

Chiang Kai-shek, as expected, denounced these reports as "unfair . . . biased."[30] Mao also rejected the notion that the Chinese Communists were not genuine Marxists. He made it clear to foreign interviewers that the CCP "was, is, and will ever be, faithful to Marxism-Leninism," and that the Chinese Communists were internationalists favoring the world Communist movement. Snow therefore reported that the CCP's reformist orientation was "only a very provisional affair" while its ultimate goal remained "a true and complete Socialist State of the Marxist-Leninist conception."[31] Despite Mao's protestations and Snow's reporting, the image of Chinese Communists as "agrarian reformers" could not be removed from American minds.

The question of Russian connections with the Chinese Communists naturally interested foreign observers, but they found only one *Tass* correspondent in Shansi during 1937-38 and three Russians in Yenan in the summer of 1944: two *Tass* correspondents and one surgeon, all there with the permission of the Nationalist government. Western visitors were satisfied that there was little evidence of Soviet aid to the Chinese Communists. Indeed, Mao told the CCP Seventh Congress in April 1945 that the Chinese system must be created in the light of Chinese history, just as the Russian system was created by Russian history.[32]

THE DIXIE MISSION. The first American officer to enter the Communist area with the permission of Chiang Kai-shek was U.S. Marine Corps Captain Evans F. Carlson, a former intelligence officer in China during 1927-28 and 1933-35. He was sent shortly after the outbreak of the Sino-Japanese war in July 1937 to observe Communist military operations. Sympathetically comparing Communist forces to the minutemen of the American Revolutionary War, he praised their "ethical indoctrination" which to him meant high political consciousness, moral conduct, and democratic camaraderie between officers and soldiers. Equally enthusiastic was his report of the close bond and "organic connection" between the army and the people: "the Eighth Route Army is like the fish and the people like the water."[33]

29. *Ibid.*, 215-16, 227.
30. Jerome Ch'en, 242.
31. Edgar Snow, *Red Star Over China* (New York, 1961), 188; Shewmaker, 249, 251, 255-56.
32. *Ibid.*, 231-32, 238.
33. *Ibid.*, 105, 194-95, 197.

In early 1943 John P. Davies, a political adviser to General Joseph Stilwell (next section), recommended the dispatch of an American Military Observers Mission to the Communist area, but this suggestion was not acted upon by Stilwell. Davies resubmitted his proposal on June 24, 1943, and on January 15, 1944, to Stilwell as well as to the Department of State; he pointed out that the CCP army was the most cohesive, disciplined, and aggressive anti-Japanese force in North China, and that it was there that the future Soviet entry into the war would most likely take place. An American mission to Yenan could collect military intelligence, determine Soviet intentions, and perhaps neutralize Russian influence on the Chinese Communists.[34] Though attracted to this proposal, President Franklin D. Roosevelt, due to Chiang's opposition, let the matter drop. Another initiative was made by Foreign Service Officer John S. Service who reported that the CCP was the most dynamic force in China under the impact of World War II and suggested that Washington use it as a lever against Soviet influence in China and East Asia. Finally, on June 23, 1944, Roosevelt secured the consent of Chiang to send a military observer mission to Yenan.[35]

The first American contingent arrived in the Communist capital on July 22 and the second on August 7. This group was known as the Dixie Mission, and it consisted of 18 members under the command of Colonel David D. Barrett, who had been a Chinese language officer and a military attaché in Peking. The main objectives of the mission were to become "informed about the people," "to determine their future war potential" if equipped by American arms, and to evaluate the "potential contribution of [the] Communists to the war effort."[36]

Barrett found that the "training methods [of the Chinese Communists] were largely formalized and by our standards of little value." He concluded that "they were excellent guerrilla fighters, but as far as large-scale operations were concerned . . . they were never able to stand toe-to-toe and slug it out with a strong Japanese force." However, with American training and equipment they could engage in regular operations against the Japanese.[37]

The Communist leaders received the Dixie Mission cordially, and by the end of August 1944 Service had his famous and long-suppressed interview with Chairman Mao. Mao stated that he wished to avoid civil war,

---

34. David D. Barrett, *Dixie Mission: The United States Army Observer Group in Yenan, 1944* (Berkeley, 1970), 22-23.
35. *Ibid.*, 26-27.
36. *Ibid.*, 13, 27-28.
37. *Ibid.*, 34, 41, 91.

but that only the Americans could intervene to compel Chiang to accept a compromise. Such intervention was critical because without American aid the KMT could not suppress the CCP forcibly. Civil war was "inevitable but not quite certain," but ultimately for the Americans to decide. In any event, Mao added, the Americans alone would have to liberate China from Japan, and at that point the aid of his armies would be crucial. The impact of the war would limit Soviet help, both militarily and in the postwar period.[38] Washington read and ignored Service's accounts. It continued to sustain Chiang as the head of the legal government of China while hoping that he would reform his regime in order to outflank the Communists. However, if possible, the United States wished to incorporate Communist forces into the war against Japan.

Davies and Service were convinced that Chiang was not going to wipe out the Communists and that quite possibly the reverse might happen unless the KMT undertook drastic reforms. At the beginning of November 1944 Davies, who had replaced Service in Yenan, came to the conclusion that "the Communists are in China to stay. And China's destiny is not Chiang's but theirs."[39] Davies and Service now believed that the Communists' aid was much more significant than that of the Nationalists. They considered every option, including separating the Communists from the Russians as much as possible on the basis of their Chinese nationalism. However, in the end, they knew that Chiang would not reform his regime, and they were afraid that the Americans might cut him off. Davies advised Washington that "we should not now abandon Chiang Kai-shek. To do so at this juncture would be to lose more than we could gain."[40]

In this ambiguous context the Americans in China considered covert relations with the Communists, despite Chiang's well-known opposition. The Army Mission in Yenan called the Communists' area "a different country," and Yenan "the most modern place in China." The Americans repeatedly noted Communist nationalism and pragmatism, and during the fall of 1944 predicted the very real possibility of their ultimate triumph. Even Roosevelt's special emissary to China, Patrick Hurley, known for his pro-Nationalist stand, remarked after a visit to Yenan in November 1944 that the Communists were "the only real democrats in China" and that they were "not in fact Communists; they are striving for

---

38. John S. Service, *The Amerasia Papers: Some Problems in the History of U.S.-China Relations* (Berkeley, 1971), 172-73.
39. *Ibid.*, 162.
40. *United States Relations with China: With Special Reference to the Period 1944-1949* (Washington, D.C., 1949), 574.

democratic principles." Ambassador Clarence Gauss also believed that it was likely that they would eventually win, and favored "pulling the plug and allowing the show [the Nationalist Government] to go down the drain."[41]

It was in this setting that some of General Albert C. Wedemeyer's staff, in conjunction with the O.S.S., decided to propose to the Communists a plan to arm 25,000 guerrillas and many more militia. Special American units would train and lead attacks on special points to be selected by Wedemeyer, and the entire Communist army would cooperate with him. However, the United States Naval Intelligence, the most conservative of American intelligence groups in China and affiliated with the Nationalist secret police, broke the news to Chiang. Both Wedemeyer and Hurley claimed ignorance of the exploration, placing the blame on Barrett, who was subsequently denied promotion to brigadier general and forced to suffer many other humiliations.[42]

The Yenan experience was vitally important in the history of Chinese Communism. Internally, it instituted a new social and political system based on the mass line, while Mao creatively laid the theoretical foundation of his revolutionary movement. Externally, it attracted the presence of an American military mission, Foreign Service officers, the visit of the American presidential emissary, and a flow of foreign reporters. For all intents and purposes, it had achieved a status of quasi-international recognition. By 1945 Yenan was in control of 18 base areas, with a total of one million square kilometers of land and nearly 100 million people. It had almost a million party members, and an equal number of armed forces. Mao had in fact created another China in competition with the Nationalist government for the supreme power of the Chinese state. In the view of a noted historian, no policy of Mao's was more responsible for his ultimate success than "that of the second United Front in the context of the Resistance War."[43]

Wartime Diplomacy and U.S. Involvement in China

From the outbreak of war in July 1937 to the Japanese attack on Pearl Harbor in December 1941, China fought alone. While it received sym-

41. Herbert Feis, *The China Tangle* (Princeton, 1953), 222; U.S. Senate, Committee on Judiciary, Subcommittee to Investigate the Administration of the Internal Security Act and Other Internal Security Laws, *Morgenthau Diary: China* (89th Congress First Session) (Washington, D.C., 1965), 1380, 1381, 1247-48, 1304-08, 1318-21.
42. Charles Romanus and Riley Sunderland, *Time Runs Out in CBI* (Washington, D.C., 1959), 73-76, 250-54; Barrett, 91-92.
43. Jerome Ch'en, 213, 255.

pathy, moral support, and some small loans from the Western powers, the Soviet Union was the only country that extended China substantial material aid. Relieved of direct Japanese pressure because of the China war, the U.S.S.R. offered China a nonaggression pact in August 1937, sent "volunteer" pilots, and granted three loans totaling U.S. $250 million —U.S. $50 million each in 1937 and 1938 and U.S. $150 million in 1939 —at the very low interest rate of 3 percent. By the end of 1939, the Soviet Union had supplied 1,000 planes and dispatched some 2,000 pilots and 500 military advisers. In fact, some of Russia's best military talent was connected with the China aid program.[44]

Western aid during the same period was pitifully small—a result of isolationism in the United States and troublesome developments in Europe. Of the total Western contribution of U.S. $263.5 million— which barely surpassed Russia's U.S. $250 million—the United States provided U.S. $120 million for nonmilitary purchases and U.S. $50 million for currency stability, while Britain and France provided a meager U.S. $78.5 million and $15 million respectively. However, American purchase of Chinese silver before and after the outbreak of war in 1937, amounting to 350 million ounces at U.S. $252 million, indirectly helped to alleviate the crushing burden of war expenses. Yet, paradoxically, until the termination of the Japanese-American commercial treaty in July 1939, the United States was a heavy purchaser of Japanese silk and a major supplier of oil, scrap iron, and automobile parts; it also met nearly 40 percent of Japan's total needs for metals, cotton, and wood pulp.[45] There can be no doubt that Japan's access to the American market directly and indirectly supported its war efforts in China.

The outbreak of the war in Europe in September 1939, however, substantially altered the foreign aid picture. Russian assistance to China slackened and finally ceased, while France and Britain leaned backward to avoid irritating Japan. Under Japanese pressure, France discontinued the rail service from Vietnam to Yunnan in June 1940, and Britain closed the Burma Road a month later, thus totally isolating China from the outside world. The situation was somewhat improved as United States aid to China gathered momentum in the wake of worsening Japanese-American relations. In March 1941, President Roosevelt made the Lend-Lease available to China, and although the amount was only U.S. $26 million in 1941, or 1.7 percent of the total to all countries, it represented a significant start. In addition, a number of other American and British

---

44. Such as Marshal Klimenti Voroshilov, General Georgi Zhukov, and General Vassily Chuikov, all destined to win fame during World War II.
45. Young, 144, 206–7, 350, 440–41.

credits were extended to help stabilize Chinese currency and foreign exchange.

The attack on Pearl Harbor changed the character of the China war and transformed the foreign aid picture. Anglo-American declarations of war against Japan and similar Chinese action against the Axis powers turned the war in Asia into part of a world-wide struggle against aggression and totalitarianism. The Allied powers established a China-Burma-India theater of war, with Chiang Kai-shek as the supreme commander of the China theater, effective January 5, 1942. General Joseph Stilwell, who had been a language officer in Peiping earlier, was sent to Chungking as Chiang's chief of staff. Moreover, the group of American "volunteer" pilots —the Flying Tigers—who had been operating in Kunming since August 1941 were incorporated into the United States Fourteenth Air Force on July 4, 1942, with General Claire L. Chennault as commander.[46] American aid from then on expanded substantially. From 1942 to the end of the war in 1945, United States credits to China reached the unprecedented mark of U.S. $500 million. Correspondingly Lend-Lease aid rose to U.S. $1.3 billion, which when combined with the U.S. $26 million in 1941 and U.S. $210 million in 1946, made a grand total of U.S. $1.54 billion, or 3 percent of the total Lend-Lease to all countries.

During the early phase of the Pacific war, the Japanese achieved spectacular successes, taking Hong Kong, Singapore, Burma, and the Philippines one after another. The lackluster Allied performance contrasted sharply with the long Chinese resistance which now gained new Western respect. Secretary of War Henry Stimson told Roosevelt that "the brilliant resistance to aggression which the Chinese have made and are making, and their contribution to the common cause, deserve the fullest support we can give." Not only did Washington grant China a $300 million loan for currency stabilization, but it persuaded London to issue a joint renunciation of all unequal treaties of the past century on January 11, 1943. Furthermore, Roosevelt and Secretary of State Cordell Hull were determined to make China one of the Big Four despite British and Soviet opposition. British Foreign Minister Anthony Eden "did not like the idea of the Chinese running up and down the Pacific," while the Soviet Foreign Minister, V. Molotov, argued that China had no admissible interest in Europe. In the end they both bowed to American persuasion and accepted China as one of the cosigners of the Moscow Declaration of November 1, 1943. This critical document was a pledge by the Four Powers to prosecute the war unceasingly until the final vic-

---

46. On May 3, 1991, the U.S. government finally acknowledged that the "Flying Tigers" were a covert American operation in China under the direction of the White House and the War Department (*Los Angeles Times*, July 6, 1991).

tory was won; it expressly disavowed any intention of signing separate peace treaties with the enemies.[47]

CAIRO CONFERENCE, 1943. Roosevelt was fond of personally meeting with world leaders and making major decisions on war aims and future peace plans. Likewise, the various national leaders were eager to meet wit' him in order to secure greater American aid. In this context, Roosevelt was anxious to talk with Chiang and Stalin, but the Chinese leader was reluctant to face his Russian counterpart, embittered as he was by the Japanese-Soviet Neutrality Pact of 1941 and the alleged Soviet support of the Chinese Communists. Chiang asked that he be given the first chance to see the President separately, and if this could not be arranged he would rather postpone the meeting. Roosevelt and Churchill then arranged two meetings, with Chiang at Cairo and with Stalin at Teheran.

In the Allied grand strategy, Europe came first, the Pacific second, and China third. Apprehensive lest the President's fondness for China might sway him into making extensive promises to Chiang at the expense of the European war, Churchill asked to hold prefatory discussions with Roosevelt first. Fearful that such a move would arouse Chinese and Russian suspicion, Roosevelt went straight to Cairo. With Madame Chiang, a Wellesley graduate who has a remarkable command of English, as interpreter, Chiang and Roosevelt held lengthy and cordial talks, to the chagrin of Churchill, who remarked: "The talks of the British and American staffs were sadly distracted by the Chinese story. . . . The president . . . was soon closeted in long conferences with the Generalissimo [Chiang]. All hopes of persuading Chiang and his wife to go see the Pyramids and enjoy themselves until we return from Teheran fell to the ground, with the result that the Chinese business occupied first instead of last place at Cairo."[48] Chiang's request for the prompt return of all lost territories won Roosevelt's endorsement and was later subscribed to by Churchill and Stalin. The President further agreed to increase supply shipments to China over the Hump (Himalayas), carry out long-range bombing of Japan, and give China a high place in the future United Nations organization. By giving China a handsome reward, the President felt, it would fight harder in the war.

The Cairo Declaration of December 1, 1943, demanded for the first time the "unconditional surrender" of Japan, the complete restoration of Chinese territories lost to Japan, and the return of Japanese possessions outside Japan proper, i.e. Sakhalin and Kurile islands to Russia and some

47. Feis, 20-21, 96, footnote.
48. Feis, 103.

of the Japanese mandatories in the Pacific to the United States. In his Christmas message to the American people, the President warmly announced: "Today we and the Republic of China are closer together than ever before in deep friendship and in unity of purpose."[49]

THE STILWELL CRISIS. A thorn in the side of Sino-American relations was the personality of General Stilwell, Chiang's American chief of staff. Nicknamed "Vinegar Joe," Stilwell was blunt and stubborn, lacking the qualities of a military diplomat that his office demanded. On September 6, 1943, he made the suggestion—politically sensitive, if militarily sound—that Chiang lift the military blockade of the Communist areas in the Northwest and permit the 18th Route army to fight the Japanese alongside the Nationalist troops. The American embassy in Chungking estimated that at least twenty divisions, or perhaps as many as 400,000 of Chiang's best troops, were relegated to blockading the Communist areas when they could have been fighting the Japanese. Incensed by Stilwell's meddling in Chinese politics, Chiang was ready to ask for his recall but was dissuaded by Madame Chiang on grounds that such a move would be unpopular in the United States. The strained relations between the two were exacerbated by the disagreements over strategy in Northern Burma, where Stilwell had been training Chinese soldiers to open a new supply line to Free China.[50] The bickering came to a head in the face of a Japanese general offensive in 1944 ("Operation Ichigo"), which aimed at clearing the "continental corridor" from North to South China and Indochina. In this drive, the enemy penetrated to the important city of Kweilin in Kwangsi province. Not only were American airbases lost—earlier used by the B29's for raids on Japan—but Chungking itself was threatened. Stilwell renewed his proposition to use the Communist troops, while Chiang stubbornly refused. The relations between the two had deteriorated beyond repair.

Deeply concerned with the Communist question in China and the overall Sino-Soviet relationship, Roosevelt sent Vice-President Henry Wallace to China and instructed Ambassador Averell Harriman in Moscow to impress upon Stalin the need for friendly relations with China. Stalin and Molotov rejoined that the Chinese Communists were not real Communists, but "margarine Communists," "cabbage Communists," and "radish Communists"—meaning red on the outside but white inside.[51]

---

49. *United States Relations with China*, 37.
50. Largely due to Stilwell's efforts, the Lido and Burma roads were finally completed in January 1945; they were renamed the "Stilwell Road."
51. Feis, 140-41, 180.

When told of these descriptions, Chiang pointedly remarked that the Chinese Communists posed as innocent agrarian democrats when in fact they were more communistic than the Russians. He assured Wallace that he would use political means to solve the Communist question, while hoping that the CCP would give up its independent army and territory and merge into the Nationalist government. Regarding Sino-Soviet relations, he pledged that he would go more than halfway to meet Stalin if President Roosevelt agreed to serve as an "arbiter" or "middleman."[52]

Chiang complained bitterly of Stilwell's lack of cooperation and judgment. In an attempt to establish a direct contact with the White House, he requested that the President send a special personal representative to Chungking.[53] By this maneuver he hoped to bypass the State and War departments, which had sustained Stilwell. Wallace noted: "I am deeply moved by the cry of a man in distress."

As the Japanese offensive rolled closer to Chungking, the United States Joint Chiefs were persuaded by Stilwell to get the President to ask that Chiang hand over the command of all Chinese troops, including the Communist forces, to Stilwell. His pride deeply wounded, Chiang told Roosevelt that he would accept this "exacting but sincere" suggestion if three conditions were met: (1) a clear definition of Stilwell's authority; (2) noninclusion of the Communist troops in his command; and (3) complete control and distribution of the lend-lease by Chiang himself. The President noted to the Joint Chiefs: "There is a good deal in what the Generalissimo says."[54]

General Patrick Hurley, secretary of war in the Hoover Administration, was sent to Chungking as the presidential emissary to harmonize Stilwell's relations with Chiang and facilitate the former's installment as commander of the Chinese troops. Suave and persuasive, Hurley secured Chiang's consent to turn over the command to Stilwell, although the Generalissimo insisted on retaining the final say on major strategic decisions so that Stilwell would have no greater power than he. At this point—September 19—"Vinegar Joe" arrived at the Chinese leader's quarters and delivered, against Hurley's advice, a strongly worded, reproving message from Roosevelt which asked Chiang to give Stilwell "unrestricted command" of Chinese forces at once or assume "personal responsibility" for the rapidly deteriorating military situation in China. The message struck Chiang as if he "had been hit in the solar plexus," observed Hurley.

---

52. *United States Relations with China*, 558.
53. Churchill kept a personal representative in Chungking, Gen. Carton de Wiart.
54. Feis, 153, 172.

Chiang told the President that while he would accept an American commander and reorganize his military chain of command, he could not assign such a heavy responsibility to Stilwell, who wanted to command rather than cooperate with him. Chiang bluntly asked for Stilwell's recall. Awed by the prospect of losing his command, Stilwell softened his stand and agreed to drop the use of Communist forces. But the die was cast; Chiang would not alter his position.

Although the Joint Chiefs continued to sustain Stilwell in this controversy, whatever doubts the President may have had were dissolved by a masterly report from Hurley, who wrote that he believed Chiang was open to persuasion and leadership but that Stilwell was convinced that the Generalissimo would not act except under force; hence every move of Stilwell's was calculated to subjugate rather than to collaborate with Chiang. "There is no issue between you and Chiang except Stilwell," Hurley shrewdly told the President; "my opinion is that if you sustain Stilwell in this controversy, you will lose Chiang Kai-shek and possibly you will lose China with him." The report closed with the belief that if another American general were appointed to replace Stilwell, Chiang would cooperate with him and devise a way to stop the Japanese advance. Hurley's advice proved to be decisive and Stilwell was recalled on October 19, 1944.[55]

Stilwell's replacement, Lieutenant General Albert C. Wedemeyer, was made commanding general of American forces in China and chief of staff to Chiang, but not commander of the Chinese forces. Wedemeyer's mild and conciliatory manner, which provided quite a contrast to Stilwell's headstrong behavior, won him immediate acceptance by Chiang, and Sino-American relations changed for the better overnight. Strangely, the Japanese offensive also tapered off by itself during this time, as a result of the transfer of troops to fight the American campaign in the Pacific. Thereafter, Japanese forces in China mounted no further offensives of great magnitude.

HURLEY'S MEDIATION, 1944-45. Since the KMT-CCP friction compromised China's war efforts and threatened future national unity and reconstruction, Hurley sought ways to bring about a reconciliation between the two parties. On November 7, 1944, with the approval of Chiang and the American general staff, he flew to Yenan for a two-day conference with Mao. Impressed with his initiative in making the trip, the Communists gave him a warm welcome. The ensuing discussions led to a Five-Point

55. Feis, 191, 198.

Draft Agreement on November 10. It called for the formation of a coalition government, representation of the CCP on a United National Military Council, legal status for the CCP, civil and political freedoms, and unification of all armed forces under a coalition national government. Mao signed it as chairman of the CCP Central Committee, and Hurley did so as the "personal representative of the President of the United States," although the Department of State later insisted that he had signed merely as a "witness."[56] Showing his appreciation of the American effort, Mao wrote to Roosevelt on November 10: "It has always been our desire to reach an agreement with President Chiang Kai-shek which will promote the welfare of the Chinese people. Through the good offices of General Hurley we have suddenly seen hope of realization."[57] Hurley's visit to Yenan convinced him that "there is very little difference, if any, between the avowed principles of the National Government the Kuomintang and the avowed principles of the Chinese Communist Party."[58]

However, Chiang Kai-shek looked at the Communist issue from an entirely different angle. To him, a coalition government implied a failure of KMT tutelage and opened the door for Communist infiltration of the government. His rejection of the Five-Point Draft Agreement was unmistakable when he set forward his own Three-Point Plan, which asked the Communists to accept Dr. Sun's Three People's Principles and turn over their troops to the Nationalist government, which in return would grant the CCP a legal status, a place on the National Military Council, and some political and civil liberties. In short, he asked Mao to turn over his guns and confide in Nationalist sincerity during the future redistribution of political power. Mao commented in his *Coalition Government*:

> These people [i.e. Chiang and his followers] said to the Communists: "If you give up your army, we shall give you freedom." If these words were sincere, then the parties which had no army should have enjoyed freedom long ago . . . yet neither of them enjoyed any freedom . . . Just because they [the workers, peasants, students, intellectuals and bourgeoisie] had no army, they lost their freedom.[59]

Pressured by public opinion and American advice, Chiang agreed to call a National Affairs Conference, which was to be composed of repre-

56. Tang Tsou, 290; Jerome Ch'en, *Mao*, 266.
57. *Foreign Relations of the United States, China* (Washington, D.C., 1967), vol. 6 (1944), 689.
58. *Foreign Relations of the United States, China,* 748. Hurley to Secretary of State Stettinius, Dec. 24, 1944.
59. Mao Tse-tung, III, 1073; Carsun Chang, *The Third Force in China* (New York, 1952), 136; Tang Tsou, 292.

sentatives of all parties and independents. Ostensibly, the conference would study problems relating to the termination of the KMT tutelage, the introduction of a constitution, the preparation of a common political program, and the participation of all parties in government *before* the inauguration of the constitution. In actuality, however, coalition government was anathema to Chiang and the KMT; they secretly contrived ways to block it. On March 3, 1945, without prior consultation with the Communists, the Nationalist government announced the convocation of the National Assembly on November 12 to adopt a new constitution. Since the delegates to this assembly had been elected in 1936 under Nationalist sponsorship, Chiang could count on the adoption of a constitution favorable to the KMT. Chou En-lai denounced the Nationalist move as "deceitful," and Mao refused to recognize the legality of the 1936 National Assembly. Deadlocked, negotiations broke down again.[60]

Hurley, who was appointed ambassador on November 17, 1944, following the resignation of Ambassador Clarence E. Gauss, did not enjoy complete support of the embassy staff, many of whom had grown openly critical of Chiang and his regime. They urged Washington to bypass the Nationalist government and work directly with the CCP and the other parties in fighting the Japanese. The idea was overruled by the President, who supported Hurley's policy of exclusive and unconditional support of Chiang. By May 1945, however, the idea of bringing pressure to bear on Chiang to reach a settlement with the Communists and thereby to broaden his government was meeting with greater favor in Washington. Meanwhile, important decisions were being made at the international level regarding the climax of the war against Japan.

THE YALTA CONFERENCE, 1945. By the end of 1944, Germany's defeat was in sight and the Allied leaders shifted their strategic focus to Japan. Washington had decided to attack Japan directly from the Pacific rather than from China as previously planned. The Joint Chiefs estimated that the defeat of Japan could be achieved within eighteen months after the defeat of Germany, which was predicted to occur somewhere between July 1 and December 1, 1945. This overestimation of Japanese strength led to the decision to invite the U.S.S.R. into the war so as to shorten it and save Allied lives. General MacArthur, having reconquered the Philippines, estimated that as many as sixty Soviet divisions would be needed to destroy the Japanese army in Manchuria. To fix the terms of the Soviet entry into the Pacific war, a meeting between the Big Three was called at Yalta in February 1945.

60. Jerome Ch'en, *Mao*, 269.

There, Stalin agreed to enter the war against Japan within two or three months after Germany's defeat, on the condition that all former Russian rights violated by the Japanese attack in 1904, as well as Russian privileges in Manchuria, be restored to the Soviet Union. Specifically, he asked for the Kurile Islands, Southern Sakhalin, warm-water ports such as Dairen and Port Arthur, the Chinese Eastern Railway and the Southern Manchurian Railway, and support of the *status quo* in Outer Mongolia. "It is clear," he told Roosevelt, "that if these conditions were not met, it would be hard for him and Molotov to explain to the Soviet people why Russia was entering the war against Japan."[61] Since many of the conditions touched upon the sovereignty of China, which was not represented at the conference, it devolved upon Roosevelt to secure Chiang's approval of these terms. On his part, Stalin agreed to respect Chinese sovereignty in Manchuria and to sign a treaty with Chiang·recognizing him as the sole leader of China.

Already ill and very tired, Roosevelt did not drive a hard bargain at Yalta. He felt that he had accomplished the main objectives of the conference, namely Stalin's agreement (1) to enter the war three months after Germany's defeat; (2) to support Chiang as the Chinese leader; and (3) to recognize Chinese sovereignty in Manchuria. Nonetheless, he did "sign away" Chinese sovereign rights in Manchuria without authorization. British Foreign Secretary Anthony Eden maintained that there was no need to pay Russia such a high price for intervention since it would probably enter the war on its own anyway. However, his advice against signing the Yalta Agreement was overruled by Churchill, who desired to manifest his faith in the President's judgment and to safeguard British interests in the Far East.

The exact terms of the Yalta Agreement were kept from Chiang and Hurley, although both had learned something about them indirectly. Feeling bypassed and insulted, Hurley decided to confront the President when he (Hurley) returned to Washington in March 1945. To his great astonishment, Hurley found Roosevelt's hand, when stretched out for greeting, to be nothing but "a very loose bag of bones," and the skin on his face "seemed to be pasted down on his cheekbones." "As you know," Hurley testified later, "all the fight that I had in me went out."[62] On April 12, Roosevelt died and Harry Truman assumed office in complete ignorance of the Yalta Agreement, which was kept in Admiral Leahy's special file.

The Soviet Union had notified Japan on April 5 that the 1941 neutrality pact between them had lost its meaning; hence the "impossibility" of

61. Feis, 243.
62. Feis, 279.

its continuation. Actually, by the terms of the pact, it was to remain in force for one year after such notice had been served, but it was obvious that the U.S.S.R. was not going to wait. Events now moved rapidly. Hitler committed suicide on May 1, and Germany surrendered a week later. Soviet troops began to move from Europe to Asia.

Chiang sent his brother-in-law, T. V. Soong, to Moscow to work out an agreement with Stalin before the Soviet troops poured into Manchuria. Stalin offered China a thirty-year treaty of friendship and alliance against future Japanese aggression. He promised to support Chiang as the leader of China, to abstain from aiding his enemies, to begin evacuating Soviet troops in Manchuria three weeks after Japan's surrender, and to complete this withdrawal in two or three months. In return, China was to grant the U.S.S.R. many vital concessions in Manchuria: the right to station naval and air forces in a military zone including Port Arthur, Dairen, and adjacent areas, ownership of the Manchurian railways and connected enterprises, and Chinese recognition of the independence of Outer Mongolia. Before the terms were written into a formal treaty, Stalin left for Potsdam to meet with Churchill and Truman. Significantly, he left with the knowledge that Japan, on July 6, had requested him to mediate for a settlement with the Allies.

On the evening of the first day of the Potsdam Conference, July 16, 1945, the news reached Truman of the successful detonation of the first atomic bomb in New Mexico.[63] The President, who had felt insecure before Churchill and Stalin, was "tremendously pepped up by it [the news] . . . and said that it gave him an entirely new feeling of confidence." Learning of the bomb, Churchill spoke in poetic grandeur: "What was gunpowder? Trivial. What was electricity? Meaningless. This atomic bomb is the second coming in wrath."[64] The British leader was convinced that the war would be over in one or two violent shocks and that there was no more need to ask the Soviets to enter the war. The American military chiefs concurred but maintained that the Soviet entry would end the war sooner with a corresponding saving of lives. At any rate, a feeling persisted that Soviet control of Manchuria could not be prevented unless the United States was willing to go to war to defend it; barring this, the Americans had best allow the Russians to earn their reward.

The Potsdam Declaration of July 26, 1945, demanded Japan's "unconditional surrender or prompt and utter destruction." When Tokyo ignored the warning, the first atomic bomb was dropped on Hiroshima on August

63. A plutonium bomb of the implosion type, called the "Fat Boy."
64. Herbert Feis, *Japan Surrendered: The Atomic Bomb and the End of the War in the Pacific* (Princeton, 1961), 72-73, 75.

6. Two days later the Soviet Union entered the war. On August 9, the second atomic bomb fell on Nagasaki, and a day later the Japanese government made a conditional acceptance of the Potsdam Declaration. On that day Stalin warned T. V. Soong that if the treaty of alliance were not signed soon, Manchuria would be in danger of falling to the Chinese Communists; the treaty was therefore signed on August 14.[65] The Soviet Union agreed to give the Chinese central government under Chiang moral, military, and material aid; to respect Chinese sovereignty in Manchuria; to evacute troops from Manchuria within three weeks of Japan's defeat and to complete the move in three months; to refrain from intervention in Sinkiang; and to acknowledge the political independence and territorial integrity of Outer Mongolia. On its part, China agreed to allow self-determination for Outer Mongolia by means of a plebiscite, and to acknowledge joint control with the U.S.S.R. of the Chinese Eastern Railway and the Southern Manchurian Railway for thirty years, after which time they would automatically revert to China without compensation. Dairen was to be a free port for all treaty nations for thirty years, while Port Arthur was to become a joint naval base for China and the Soviet Union.

The treaty, though costly, was accepted by Chiang with satisfaction, for it secured peace for China on its northern border, and committed the Soviet Union to recognition of Chinese sovereignty in Manchuria and Sinkiang and to nonsupport of the Communists against the Nationalist government. Indeed, peace with Russia was essential to China's postwar reconstruction, and Chiang felt that if the Soviets failed to observe their obligations, the treaty could be used as a yardstick to judge their behavior.[66]

On August 14, 1945, the Japanese emperor issued an imperial rescript to end the war, and on September 2 the Instrument of Surrender was signed on board the U.S.S. *Missouri* in Tokyo Bay. After eight years of fighting, China had finally emerged victorious. Chiang's prestige was never higher, for he had led the country through the darkest days of war to ultimate victory. China's international position was also never more honorable—it had fought the longest fight against aggression and totalitarianism.

The country rejoiced over the end of war and eagerly looked forward to a period of peace and reconstruction. Yet under the veneer of jubilation and excitement there was a deep concern over the still unresolved Communist problem and its ominous implications. Indeed, Mao Tse-tung had waited patiently to make his bid for power.

---

65. Signed not by Soong, who feared for his political future, but by Foreign Minister Wang Shih-chieh for China.
66. Chiang Kai-shek, 228.

The War Consequences

The war had generated far-reaching repercussions in China, Japan, and East Asia. Among the most important were the following.

A NEW INTERNATIONAL ORDER IN ASIA. The end of the Pacific war ushered in a new day in East Asia. Through its long years of struggle against aggression, China replaced Japan as the leading power. It emerged from its prewar semicolonial status to become one of the Big Five and a chartered member of the United Nations, with a permanent seat and veto power in the Security Council. Never before in its modern history was China's international prestige higher than at this point. In contrast, Japan ceased to be a major force in international politics and turned inward toward economic reconstruction under American occupation and guidance. The old European colonial powers—Britain, France, and the Netherlands— though victors in war, were shorn of much of their former prestige, as they had been expelled from their Asian possessions by the Japanese during the war. Their former colonies, India, Burma, Indochina, and Indonesia, all clamored for independence. The age of European colonialism in Asia, which began in the 16th century, finally came to an end. On the other hand, by virtue of its dominant role in defeating Japan, the United States had emerged as the most powerful state on the Pacific. This turn of events heralded a totally new chapter in the evolution of international relations in Asia.

NATIONALIST EXHAUSTION. Although the victory over Japan was primarily won by the Americans, China's contributions could not be overlooked. Throughout the war it had pinned down a substantial percentage of the Japanese armed forces, which might otherwise have been employed elsewhere. From 1937 to 1941 when it fought alone, China engaged between 500,000 and 750,000 enemy troops in China proper—roughly half of the total Japanese strength—in addition to the 200,000- to 700,000-man Kwantung army in Manchuria. At the end of the war in 1945, 1.2 million out of a total 2.3 million overseas Japanese armed forces were tied down in China. The China campaigns consumed 35 percent of total Japanese war expenditures—U.S. $12 billion out of a total U.S. $34 billion—and resulted in 396,040 Japanese killed and a much larger number wounded.[67] On its part, China mobilized 14 million men, sustained total casualties of 3,211,419—including 1,319,958 killed, 1,761,355 wounded, and 130,126 missing—and incurred an awesome war debt of Ch$1,464 billion.[68] Civil-

67. Young, 417-18.
68. Chiang Kai-shek, 131.

ian casualties and property losses were incalculable. The Nationalist government, which bore the major brunt of the fighting, was so depleted physically and spiritually that it was manifestly incapable of coping with the new challenges of the postwar era.

ECONOMIC DISTRESS. The chronic ill of deficit spending, which had plagued the Nationalist government since its inception in 1928, was exacerbated during the war as a result of mounting military expenditures and the loss of customs revenues from the coastal provinces that had fallen to the enemy. The vast discrepancy between income and expenditures is alarmingly evident in the following statistics from three typical years:[69]

	War Costs	Revenues
	(million Ch$)	
1937	1,167	870
1941	10,933	2,024
1945	1,268,031	216,519

There was no way for the government to bridge the gap except through the admittedly unwise course of increasing note issues, with the full knowledge that such a measure would inevitably bring on inflation. The note issues rocketed from Ch$1.9 billion at the outset of war in 1937 to Ch$15.81 billion by the end of 1941, and to Ch$1,031.9 billion in 1945. The consequence of the abusive issue of paper money was rampant inflation and the rapid rise in average retail prices:[70]

	Retail Price Rise (percent)
1937 (first 9 months after the war)	29
1938	49
1939	83
1940	124
1941	173
1942	235
1943	245
1944	231
1945 (to August)	251
1945 (from August to end of year)	230

69. Young, 435.
70. Young, 436.

Ultimately, inflation damaged army morale, destroyed administrative efficiency, ruined civilian lives, and reduced the middle class to destitution. The economic distress caused by inflation alienated large segments of the Chinese people, especially the intellectuals, who blamed the government for mismanagement and irresponsibility. If inflation was a necessary evil to sustain the war, it had become a curse in the postwar period and undermined the very economic foundations of the government.

PSYCHOLOGICAL WEARINESS. Having patiently endured all hardships during eight years of war, the Chinese people were too weary to undertake any kind of struggle once the victory had been won—least of all a civil war between the Nationalists and the Communists. They longed for peace and recuperation, and when these eluded them they blamed the government and the party in power. Mao Tse-tung, who had correctly predicted this turn of events in the early phase of the war, was quick to exploit this mass discontent. No sooner had peace returned than he began to challenge Nationalist supremacy. Civil war clouds once again hovered ominously on the horizon, portending a future fraught with turmoil for the exhausted nation.

Further Reading

Barrett, David D., *Dixie Mission: The United States Army Observer Group in Yenan, 1944* (Berkeley, 1970).

Bisson, Thomas A., *Japan in China* (New York, 1938).

Borg, Dorothy, *The United States and the Far Eastern Crisis of 1933-1938* (Cambridge, Mass., 1964).

Boyle, John Hunter, *China and Japan at War, 1937-1945: The Politics of Collaboration* (Stanford, 1972).

Bunker, Gerald E., *The Peace Conspiracy: Wang Ching-wei and the China War, 1937-1941* (Cambridge, Mass., 1972).

Chang, Kia-ngau, *The Inflationary Spiral: The Experience in China, 1939-1950* (New York, 1958).

Ch'en, Yung-fa, *Making Revolution: The Communist Movement in Eastern and Central China, 1937-1945* (Berkeley, 1986).

Chiang, Kai-shek, *Chiang Tsung-t'ung mi-lu* 蔣總統祕錄 (Secret memoirs of President Chiang), tr. from the Japanese by *Central Daily News*, Vol. 1 (Taipei, 1974).

———, *China's Destiny* (New York, 1947).

———, *Soviet Russia in China: A Summing Up at Seventy* (New York, 1957).

Chou, Shun-hsin, *The Chinese Inflation, 1937-1949* (New York, 1963).

Clemens, Diane Shaver, *Yalta* (New York, 1970).

Clifford, Nicholas R., *Retreat from China: British Policy in the Far East, 1937-1941* (Seattle, 1967).

Colegrove, Kenneth, "The New Order in East Asia," *Far Eastern Quarterly*, I:1: 5-24 (Nov. 1941).

Denning, Margaret B., *The Sino-American Alliance in World War II: Cooperation and Dispute among Nationalists, Communists, and Americans* (Berne, 1986).

Doenecke, Justus D., *The Diplomacy of Frustration: The Manchurian Crisis of 1931-1933 as Revealed in the Papers of Stanley K. Hornbeck* (Stanford, 1981).

Esherick, Joseph W. (ed.), *Lost Chance in China: The World War II Dispatches of John S. Service* (New York, 1974).

Feis, Herbert, *The China Tangle* (Princeton, 1953).

———, *Japan Surrendered: The Atomic Bomb and the End of the War in the Pacific* (Princeton, 1961).

Fishel, W. R., *The End of Extraterritoriality in China* (Berkeley, 1952).

Foreign Relations of the United States, 1945, The Conference at Malta and Yalta (Washington, D.C., 1955).

———, *1942, China* (Washington, D.C., 1956).

———, *1943*, Vol. VI, *China* (Washington, D.C., 1957).

———, *The Conferences at Cairo and Teheran* (Washington, D.C., 1961).

———, *1944*, Vol. VI, *China* (Washington, D.C., 1967).

———, *1945*, Vol. VII, *The Far East—China* (Washington, D.C., 1969).

Garver, John W., *Chinese-Soviet Relations, 1937–1945* (New York, 1988).

———, "The Origins of the Second United Front: The Comintern and the Chinese Communist Party," *The China Quarterly*, 113:29-59 (March 1988).

Gunn, Edward M., *Unwelcome Muse: Chinese Literature in Shanghai and Peking, 1937-1945* (New York, 1980).

Head, William P., *America's China Sojourn: America's Foreign Policy and Its Effects on Sino-American Relations, 1942-1948* (Lanham, Md., 1983).

———, *Yenan: Colonel Wilbur J. Peterkin and the Dixie Mission, 1944-1945* (Chapel Hill, 1987).

Israel, Jerry, "Mao's Mr. America: Edgar Snow's Images of China," *Pacific Historical Review*, XLVII:1:107-122 (Feb. 1978).

Jansen, Marius B., *Japan and China: From War to Peace, 1894-1972* (Chicago, 1975).

Jones, F. C., *Manchuria Since 1931* (London, 1949).

Jordan, Donald A., "The Place of Chinese Disunity in Japanese Army Strategy during 1931," *The China Quarterly*, 109:42-63 (March 1987).

Kataoka, Tetsuya, *Resistance and Revolution in China: The Communists and the Second United Front* (Berkeley, 1974).

Lee, Bradford A., *Britain and the Sino-Japanese War, 1937-1939: A Study in the Dilemmas of British Decline* (Stanford, 1973).

Lensen, George A., *The Strange Neutrality: Soviet-Japanese Relations during the Second World War, 1941-1945* (Tallahassee, 1972).

Liang, Chin-tung 梁敬錞, *Shih-ti-wei shih chien* 史迪威事件 (The Stilwell incident), (Taipei, 1971).

———, *General Stilwell in China, 1942-1944: The Full Story* (Jamaica, New York, 1972).

Linebarger, Paul M. A., *The China of Chiang K'ai-shek; A Political Study* (Boston, 1941).

Liu, James T. C., "Sino-Japanese Diplomacy during the Appeasement Period, 1933-1937" (Ph.D. thesis, University of Pittsburgh, 1950).

Lohbeck, Donald, *Patrick J. Hurley* (Chicago, 1956).

Lowe, Peter, *Great Britain and the Origins of the Pacific War: A Study of British Policy in East Asia, 1937-1941* (New York, 1977).

McLane, Charles, *Soviet Policy and the Chinese Communists, 1931-1946* (New York, 1958).

Miles, Milton E., U.S.N. *A Different Kind of War: The Little Known Story of the Combined Guerrilla Forces Created in China by the U.S. Navy and the Chinese during World War II* (Garden City, N.Y., 1967).

Morton, William Fitch, *Tanaka Giichi and Japan's China Policy* (New York, 1980).

North, Robert, *Moscow and the Chinese Communists* (Stanford, 1953).

Reardon-Anderson, James, *Yenan and the Great Powers: The Origins of Chinese Communist Foreign Policy, 1944-1946* (New York, 1980).

Romanus, Charles F. and Riley Sunderland, *Stilwell's Command Problems* (Washington, D.C., 1956).

———, *Stilwell's Mission to China* (Washington, D.C., 1953).

Rosinger, Lawrence K., *China's Wartime Politics, 1937-1944* (Princeton, 1945).

Schaller, Michael, *The U.S. Crusade in China, 1938-1945* (New York, 1979).

Schran, Peter, *Guerrilla Economy: The Development of the Shensi-Kansu-Ninghsia Border Region, 1937-1945* (New York, 1976).

Selden, Mark, *The Yenan Way in Revolutionary China* (Cambridge, Mass., 1971).

Service, John S., *The Amerasia Papers: Some Problems in the History of U.S.-China Relations* (Berkeley, 1971).

Sheng, Michael M., "Mao, Stalin, and the Formation of the Anti-Japanese United Front: 1935–37," *The China Quarterly*, March 1992, 149–83.

Shewmaker, Kenneth E., *Americans and Chinese Communists, 1927-1945: A Persuading Encounter* (Ithaca, 1971).

Sih, Paul K. T. (ed.), *Nationalist China during the Sino-Japanese War, 1937-1945* (Hicksville, N.Y., 1977).

Snow, Edgar, *Journey to the Beginning* (New York, 1972).

———, *Red Star over China* (New York, 1938).

———, *Random Notes on Red China, 1936-1945* (Cambridge, Mass., 1957).

Stewart, Roderick, *Bethune* (Don Mills, Ontario, 1973).

Thorne, Christopher, *Allies of a Kind: The United States, Britain and the War against Japan, 1941-1945* (Oxford, 1978).

Truman, Harry S., *Memoirs*, Vol. I: *Year of Decisions* (Garden City, 1955).

———, *Memoirs*, Vol. II: *Years of Trial and Hope* (Garden City, 1956).

Tuchman, Barbara W., *Stilwell and the American Experience in China, 1911-1945* (New York, 1970).

Van Slyke, Lyman P., *Enemies and Friends: The United Front in Chinese Communist History* (Stanford, 1967).

Varg, Paul A., *The Closing of the Door: Sino-American Relations, 1936-46* (East Lansing, 1973).

Vartabedian, Ralph, "One Last Combat Victory: The Flying Tigers," *Los Angeles Times*, July 6, 1991.

Vincent, John Carter, *The Extraterritorial System in China: Final Phase* (Cambridge, Mass., 1970).

White, Theodore H. (ed.), *The Stilwell Papers* (New York, 1948).

————, and Annalee Jacoby, *Thunder Out of China* (New York, 1946).

Wylie, Raymond F., "Mao Tse-tung, Ch'en Po-ta and the 'Sinification of Marxism,' 1936-38," *The China Quarterly*, 79:447-480 (Sept. 1979).

————, *The Emergence of Maoism* (Stanford, 1980).

Young, Arthur N., *China and the Helping Hand, 1937-1945* (Cambridge, Mass., 1963).

VI

The Rise
of the Chinese People's
Republic

25

The Civil War,
1945-49

The collapse of Japan, after two atomic shocks, had come much sooner than expected. It left the Nationalist government totally unprepared for the consequences of the sudden termination of war. A number of pressing problems now clamored for Chiang Kai-shek's immediate attention, and foremost among them was the Communist threat to move into the Japanese occupied territories and take over the enemy arms. No less ominous was the situation in Manchuria where Soviet forces had plunged deep into the hinterland and refused to stop with the Japanese surrender. Despite Stalin's promise of evacuation in three months, their intentions remained shrouded in secrecy. Thus, the end of the war had created an extremely critical military situation for the Nationalists.

Following the Japanese surrender, a mad race took place between the KMT and CCP forces, each trying to reach the occupied territories first to receive the Japanese surrender and thereby harvest the vast quantity of enemy arms and military supplies. In the contest the Communists seemed to enjoy a distinct geographical advantage. They were in control of eighteen "liberated areas" in North, South, and Central China, with a population of 100 million, and boasted of one million regular troops and two million militiamen,[1] who were deployed in the countryside of the Yellow, Yangtze, and Pearl river valleys. The big metropolises of Peiping, Tientsin, Shanghai, Nanking, Hankow, and Canton, which were located

1. Mao Tse-tung, *Mao Tse-tung hsüan-chi* (Selected works of Mao Tse-tung), (Peking, 1963), IV, 1157.

in these valleys, became urban islands in a Communist-dominated rural ocean. To make full use of this favorable situation, Mao Tse-tung declared on August 9, 1945—a day after the Soviet entry into the war—that the collapse of Japan was in sight and that the hour had arrived for the CCP to mount a general offensive. On August 10, Chu Teh, commander-in-chief of the People's Liberation army (PLA), ordered his troops to seize all towns, cities, and communication centers under Japanese occupation, and to receive the enemy's surrender and military supplies. On August 11, Lin Piao led a 100,000-man army along the Peiping-Mukden Railway, striking into Manchuria. Within two weeks of the Japanese surrender, the Communists expanded their territory from 116 to 175 counties.[2]

The Nationalist forces, scattered along the several battlefronts and in Western China, were less favorably situated in the race, but Chiang was determined not to let the fruits of victory slip from his fingers. On August 10 he appealed to the Communist leaders to refrain from independent actions, and ordered the Japanese and puppet forces to hold out against non-Nationalist troops. Denouncing Chiang's action as "beneficial to the Japanese invader and traitors," Chu Teh directly asked the Japanese commander-in-chief in China, Okamura Yasuji, to surrender to the Communist representatives. To overcome the Communist geographical advantage, Chiang requested American help to airlift and sealift his troops to the occupied areas.

The United States quickly came to the aid of the Nationalists. It authorized the transportation of their troops to the occupied areas and the landing of 50,000 American marines in key ports and communication centers to await the arrival of Nationalist forces. Three government armies were airlifted to Peiping, Tientsin, Shanghai, and Nanking, and subsequently a total of half a million troops were transported to the various parts of the country. In addition, Washington's General Order No. 1 to Tokyo explicitly required that the surrender of Japanese forces in China (exclusive of Manchuria),[3] Taiwan, and French Indochina north of the 16° parallel be made to Chiang and his representatives. On August 15 Chiang himself ordered Okamura to maintain order and keep all military supplies inside occupied territory. On August 22 Okamura was further told to allow passage only of Nationalist troops to the occupied territory. The Japanese commander complied fully.

With American assistance and Japanese cooperation, the Nationalists won the first round of competition. The government regained control of nearly all the important cities and communication centers in Central, East,

2. Mao Tse-tung, III, 1,119; Jerome Ch'en, *Mao*, 261.
3. In Manchuria, the Soviet forces were authorized to receive the Japanese surrender.

and South China, while the Communist forces temporarily retreated to the countryside. Yet in spite of this setback, the CCP managed to score some gains during the first two weeks of contest, winning control of fifty-nine cities and vast countryside, especially in North China.[4]

Manchuria presented a particularly explosive picture. The Soviet forces under Marshal Rodion Malinovsky had swept in with amazing speed on August 8 and were joined by additional striking forces from Outer Mongolia two days later. The Soviet advance did not stop with the Japanese surrender on August 14; nor did it halt at the geographical limit of Manchuria. It penetrated deep into Jehol and Chahar, and facilitated the entry of the CCP forces into Manchuria, where the Soviets turned over to them considerable quantities of surrendered Japanese arms.[5] However, the Russians did not set the Chinese Communists up for the take-over of Manchuria.

In an effort to resolve these knotty problems and effect a *rapprochement* with the CCP, Chiang three times invited Mao to a conference in Chungking. Though hesitant and fearful of a Nationalist ruse, Mao ultimately decided to come, after the American envoy, Patrick Hurley, had gone to Yenan to vouch for his safety. On August 28, 1945, Mao flew to Chungking. The people of China, both eager and weary, held their breath for this historic meeting, praying for an amicable outcome so that a civil war might be averted.

Mao in Chungking

Mao's strategy in the negotiations had been carefully worked out before he left Yenan. Despite the initial Nationalist success in regaining control of the big cities, Mao was confident that the CCP would eventually dominate the areas north of the Lower Yangtze and Huai rivers, most of Shantung, Hopeh, Shansi, Suiyuan, all of Jehol and Chahar, and part of Liaotung. However, for the immediate future he foresaw many difficulties; therefore he decided to adopt a flexible and conciliatory course of action in Chungking but hold firm on matters of basic importance.[6] Consequently, during his stay in the Nationalist wartime capital, Mao made every effort to appear reasonable and willing to make concessions—a posture calculated to win world public opinion and the sympathy of the middle-of-the-roaders. He appeared every bit an amiable, warm human being rather than a

4. Mao Tse-tung, IV, 1,151. Report of Mao on August 26, 1945.
5. Mao Tse-tung, IV, 1,134; Tang Tsou, 315-16. 300,000 rifles, 138,000 machine guns, 2,700 pieces of artillery, etc.
6. Mao Tse-tung, IV, 1,151-54.

tough, fire-eating revolutionary. Outwardly, there was an expression of friendliness and civility on the part of both the host and the guest, which buoyed popular hopes for reconciliation and peace.

In the formal negotiations, Mao cultivated the image of being reasonable and ready to accommodate. No longer did he insist on the coalition government, but asked instead for the calling of a National Affairs Conference[7] which would study the problems relating to the formation of such a coalition, the convocation of a National Assembly, and the introduction of a constitution.

On the issue of the relative strength of the KMT and CCP forces and their integration into a national army, Mao offered to keep only 20 to 24 divisions if the KMT agreed to cut its forces to 120 divisions.[8]

On the question of political control of the liberated areas, Mao essentially wanted a free hand in North China, Inner Mongolia, and some important cities. When the Nationalists balked at the idea, he proposed a temporary *status quo* of the liberated areas pending the adoption of a constitution, which would stipulate popular election of the local government. It was clear that Mao strove to maintain control of local affairs, but on this point the Nationalists stubbornly refused to yield.

As regards the question of receiving the Japanese surrender, the Nationalists insisted on the exclusive right to disarm the enemy, while the Communists claimed similar privileges in areas where they had been active or where they had already encircled the enemy. No agreement was reached on this point either.

Six weeks of negotiations left no doubt that little progress had been made. Despite Mao's conciliatory appearance, he would not yield on those basic points that touched the fundamental position of the CCP. On the other hand the Nationalists, negotiating from a position of strength, stubbornly refused to compromise their privileged status. Chiang's prestige was at its zenith, having led the country to victory against seemingly insuperable odds. Furthermore, he enjoyed American aid and support and had signed a treaty of friendship and alliance with Stalin, in addition to maintaining a vastly superior military strength over his enemy.[9] He was hardly interested in the kind of "transient arrangements" for peace that Mao was willing to make at this juncture.

Chiang had endured the hardships of retreat for eight years and now demanded most, if not all, of the fruits of victory. He saw no point in sharing his glory with the Communists. Perhaps if Hurley had played a

7. Later to be called the Political Consultative Conference.
8. Mao Tse-tung, IV, 1,155-64.
9. Allegedly 11 to 1, according to the Nationalist minister of war.

more active role in persuading Chiang to accept a *modus vivendi* along Mao's lines, the Communists could have been confined to North China. But Hurley chose to be a passive peacemaker and scrupulously maintained his neutrality; the most he would do was to urge both leaders to strive for agreement on the "basic over-all principles" first and work out the "details" later. But it was precisely on these matters of details that the two parties were unable to come together.

The final communiqué, issued by Chiang and Mao on October 10, stressed their agreement on the convocation of a Political Consultative Conference and on the importance of peaceful reconstruction. The depth of their disagreement was not communicated to the public, but it was obvious that the talks had not produced concrete results. Upon his return to Yenan, Mao called upon his followers to redouble their efforts for "peace" by mobilizing the masses and expanding the people's army to build a new China. His emboldened posture was in part an outcome of new developments in Manchuria.

Soviet Operations in Manchuria

Soviet activities in Manchuria completely belied Stalin's promise at Yalta and at the Sino-Russian treaty negotiations that he would evacuate troops from Manchuria within three weeks of the occupation and complete the withdrawal within three months. It appears that when he made the promise in February and July-August 1945, he had not expected the imminent rise of the Chinese Communists to power. He did not seem to mind American mediation in China, and had in fact advised Mao to work out some agreement with Chiang. Mao superficially heeded Stalin's admonition, but secretly decided to proceed with the military contest with the Nationalists.[10] On his own, Stalin later admitted his mistaken diagnosis of the China situation.[11]

Apparently, Stalin's position underwent a radical change shortly after the war, when the vigor and resourcefulness of the CCP strongly impressed him. Once the special rights and privileges in Manchuria—granted by the Yalta Agreement and the Sino-Soviet treaty—had been confirmed by actual Soviet occupation, Stalin saw no need to honor his promise. The Soviet forces looted the Manchurian industrial plants and moved their valuable equipment to Russia as "war booty" at a replacement cost of $2

10. Lin Piao recalled in 1960: "Some well-intentioned friends at home and abroad [i.e. Stalin] . . . were worried about us," but Chairman Mao correctly assessed the situation and branded all reactionaries "paper tigers." See Tang Tsou, 326.
11. To Eduard Kardelj, an aide to Tito.

billion.[12] They employed all kinds of pretexts designed to prevent the entry of Nationalist troops into Manchuria.

Chiang was determined to recover Manchuria, which he said was the *raison d'être* of China's eight-year war with Japan. General Wedemeyer, doubting the Nationalist capacity to take Manchuria, had advised him to first consolidate the areas south of the Great Wall and north of the Yangtze and safeguard the communication lines in North China. Rejecting the counsel, Chiang committed nearly half a million of his best-equipped troops to Manchuria—a decision he was to regret later.[13] Finally, the Soviet commanders allowed Nationalist units to be airlifted into major Manchurian cities, and government forces entered Changchun on January 5, 1946, and Mukden three weeks later. By then the CCP forces had almost completely dominated the vast countryside outside of these pockets, thus confronting the Nationalists with an untenable position. The Soviet forces finally left Manchuria in May 1946.

Marshall in China

By November 1945 Washington had adopted a new policy which called for continuing American support of the Nationalist government, on the condition that it not employ American arms to conduct a civil war and that it strive to reach a settlement with the Communists. In effect, this shift represented a repudiation of the former policy which had espoused unconditional support of Chiang's government. Disillusioned, Hurley resigned in protest on November 27, charging career officials of the State Department with plotting behind his back and siding with the Chinese Communists. President Truman then appointed General George C. Marshall, the most distinguished American soldier of World War II, as a special presidential ambassador to China.[14] He was instructed to assist the Nationalist government in re-establishing its authority as far as possible, including Manchuria, but not to involve the United States in any direct military intervention. He was also to urge Chiang to call a national conference of all major parties to deliberate on the cessation of the civil war and the unification of the country, to the end that a "strong, united, and democratic China" might emerge. Finally, he was told to make clear to Chiang that large-scale American aid was contingent on the achievement of a truce and national unity.[15]

12. According to the estimates made by Edwin Pauley, American member of the Inter-Allied Reparation Commission.
13. Chiang Kai-shek, 232-33.
14. On suggestion of the secretary of agriculture, Clinton Anderson, in the November 27, 1945 cabinet meeting.
15. *United States Relations with China*, 133, 605-7.

Arriving in China in mid-December 1945, Marshall found both parties receptive to his mediation and prepared to endorse his three immediate objects: (1) a cease-fire in the civil war; (2) the convocation of a Political Consultative Conference to deliberate the formation of a coalition government; and (3) the integration of the KMT and CCP forces into a national army. The polite welcome and the pledge of support by the contending parties were heartening gestures, but it was evident that they could not have done otherwise. Marshall's high prestige, his apparent sincerity, his professed goal of helping China achieve peace, unity, and democracy, and above all the enormous power of his country were sufficient inducements for cordial behavior by both the KMT and the CCP. Nonetheless, beneath the veneer of warmth and appreciation, the extremists in both parties harbored feelings of antipathy toward what they considered an example of American meddling.

Deeply distrustful of each other, the Nationalists, with at least a five-to-one military superiority over the Communists in early 1946, were confident of their ability to crush the enemy in a quick bout. On the other hand, the Communists sneered at the Nationalist "paper tiger" which they felt sure could be torn apart in a prolonged contest. Each of the two insisted on a different set of conditions for collaboration. The Nationalists demanded that the Communists surrender their troops[16] *before* the establishment of the constitutional government, whereas the Communists insisted that such integration should come *after* its establishment. The KMT advocated the presidential system in the coalition government; the CCP argued for a cabinet system. Since the KMT would most likely dominate the central government, especially its executive branch, the CCP adamantly demanded a large degree of provincial autonomy and a strong legislature to checkmate the executive. If the rising spirit of belligerency was to be contained, these key issues had to be thrashed out to the satisfaction of both sides.

Marshall's active mediation achieved rapid and impressive results. On January 10, 1946, he committed the KMT and the CCP to calling a Political Consultative Conference, an immediate cease-fire, and a restoration of communications. A tripartite Executive Headquarters was created, consisting of one Nationalist, one Communist, and one American member, with the last-mentioned as chairman; its decisions required unanimous agreement. To supervise the cease-fire, teams reflecting a similar three-party composition were sent into the field.

The Political Consultative Conference was convened between January 10 and 31 during the truce and was composed of thirty-eight members:

16. Under the euphemism of "unity of military command," i.e. the integration of the CCP forces *into* the KMT forces.

eight from the KMT, seven from the CCP, nine from the Democratic League, five from the Youth Party, and nine independents. Its lengthy deliberations resulted in the resolution that the supreme organ of state should be a multiparty State Council vested with both legislative and executive powers. The Council was to consist of forty members, of whom one half would be nominated by the KMT and the other half by the other parties and the independents. Decisions of the State Council involving a change in the resolutions of the Political Consultative Conference (PCC) would require ⅔ votes; hence any party or group able to muster ⅓ votes— 14 to be exact—enjoyed a veto power. The Communists and their sympathizers, notably the Democratic League, were confident that they could line up the necessary votes to block any KMT attempts at revision.

The PCC adopted the cabinet system of government in which the Executive Yüan was to be responsible to the Legislative Yüan. It further resolved that the future constitution should recognize the province as the highest organ of local government, with a popularly elected governor and a constitution of its own to insure the proper division of power between the central and provincial governments.

The work of the PCC, more favorable to the Communists than to the Nationalists, reflected the general desire for peace and democratic rule. Though he did not take part in its deliberations, Marshall approvingly described its work as "a liberal and forward-looking charter." For the first time since the end of the war, there appeared a ray of hope for peace and reconstruction.

A further monumental achievement by Marshall was the agreement on February 25, 1946, regarding the relative strength of the KMT and the CCP forces and the integration of the two into a national army. It was resolved that within a year the KMT forces were to be reduced to 90 divisions and the CCP forces to 18, followed in the next six months by a further reduction of forces to 50 and 10 divisions respectively. These military cuts were to be distributed as follows: Manchuria: 14 KMT divisions to 1 CCP division; North China: 11 KMT divisions to 7 CCP divisions; Central China: 10 KMT divisions to 2 CCP divisions; South (including Formosa) and Northwest China: 6 and 9 KMT divisions respectively, and no Communist forces.[17] Obviously, the Nationalists came out very well in this military arrangement, as Communist influence was drastically reduced in Manchuria and North China. Similarly, the Nationalists were enabled to

17. *United States Relations with China* 141. However, Chiang had serious doubts as to the Communist willingness to put the military reorganization plans into practice. To Marshall he described the integration of KMT-CCP forces as a task "as difficult to achieve as 'to negotiate with the tiger for its skin.'" Chiang Kai-shek, 162.

take over the Communist base in the Northwest, thus blocking direct contacts between the Chinese Communists and their Soviet comrades.

Marshall's quick success led President Truman to announce on February 25, 1946, the establishment of a United States Military Mission in China, staffed by 1,000 officers and men under General Wedemeyer. It was understood, from Marshall's earlier promise, that the Communist forces would be included in American training programs and would receive American equipment before their integration into the Nationalist forces. On March 11, 1946, a relieved and satisfied Marshall returned to the United States to arrange a loan of $500 million from the Export-Import Bank. It was during his brief absence from China that the sincerity of the Nationalists and the Communists was put to a severe test.

Neither the KMT nor the CCP trusted each other since both were revolutionary parties committed to different causes. Cooperation was nearly impossible, except on a temporary and expedient basis. While Marshall's early success was largely a result of his active persuasion and respected stature, it was also true that the two contending parties found it bad politics not to accommodate him. Secretly, the extremist elements in both parties found him standing in their way to victory. The CC Clique[18] in the Nationalist Party—dubbed by Marshall as the "selfish irreconcilables"—strongly felt that the agreements with the Communists were imposed upon the KMT by Marshall, and that without his intervention the Nationalists could have scored a victory over the enemy long before. The Communists, on their part, temporized with Marshall as long as the agreements were basically in their favor or not insufferably detrimental, but they never diminished the secret expansion of their army and areas of operation. Only Marshall's presence kept the two parties from ripping apart the façade of cooperation. But once he was out of sight, they ignored the truce and scrambled to improve their positions on the battlefield, so that they would be better situated should the final settlement fall short of realization. What began as local clashes soon grew into large-scale fighting in April 1946. The resolutions of the Political Consultative Conference remained unfulfilled dreams.

Fighting in Manchuria was particularly bitter. The CCP forces dealt a crushing blow to the Nationalist army and occupied the strategic city of Changchun on April 18, 1946. With this major victory, the CCP demanded an upward revision of the military deployment ratio in Manchuria from one to five CCP divisions vis-à-vis KMT's fourteen. Angrily rejecting the demand, Chiang ordered an all-out attack which resulted in

18. Named after its leaders, Ch'en Li-fu and Ch'en Kuo-fu.

the recovery of Changchun in May. Fighting would have rapidly gotten out of hand but for the remonstrance of Marshall, now back in China. A fifteen-day truce was arranged on June 6. But war fever had gripped both parties and the feeling was gaining ground that Marshall obstructed their ultimate victory. In early July 1946, Chiang told him that "it was first necessary to deal harshly with the Communists, and later, after two or three months, to adopt a generous attitude." On another occasion he said: "If General Marshall were patient, the Communists would appeal for a settlement and would be willing to make the compromises necessary for a settlement."[19] The Communists, on their part, were equally confident of an ultimate victory. Accusing the United States of playing the double role of aiding the Nationalists while posing as an impartial mediator, they demanded the withdrawal of all American troops. It appears that sometime in mid-1946, both the KMT and the CCP had decided to embark upon a new course of action irrespective of Marshall, whose influence had fallen to a nadir.

Riding the tide of victories, the Nationalist government announced unilaterally on July 4, 1946, that they would convoke the National Assembly on November 12, in open disregard of the PCC resolution that no such Assembly should be called before the formation of the coalition government. The CCP and the Democratic League proclaimed their boycott of this "illegal" Assembly; in addition Mao called for a war of self-defense. The split had widened into an unbreachable gulf. Marshall appealed to the Chinese people to exert their pressure on both parties for a compromise, but it was a lone cry in the wilderness which won much sympathy but no results. His warning to Chiang of possible economic collapse and a Communist victory had little effect. The Generalissimo still believed that the inflation, though fierce and threatening, would not bring on an economic disaster, because the agrarian economy of China was governed by forces different from those of the industrial Western states.[20]

From July to September 1946, Chiang's forces won practically every battle, a fact which seemed to strengthen the view that Marshall had delayed the Nationalist victory. The Communists, in temporary retreat, openly accused the United States of using mediation as a smoke screen while underwriting Chiang's civil war. With his integrity thus questioned, Marshall warned Chiang on October 1 that unless the fighting stopped he would terminate the mediation and return home. Still winning in the battlefield, Chiang refused to stop. Then, in a grand gesture of magnanimity

19. Tang Tsou, 425.
20. *United States Relations with China*, 212.

he called off the offensive on November 8—a few days before the convocation of the National Assembly—to allow the Communist Party and the Democratic League to reconsider their position. So confident was he of the justness of his cause and of his ability to win that he told Marshall on December 1, 1946, that the enemy forces could be wiped out in eight to ten months.[21]

Marshall realized that he had failed miserably in his mission. On January 6, 1947, President Truman announced his recall. In his farewell message to the Chinese people, a bitterly disappointed Marshall blamed the KMT "irreconcilable groups" for their "feudal control of China" and lack of interest in implementing the PCC resolutions; he also criticized the Communists for their "unwillingness to make a fair compromise." China's hope lay with the liberals, he said, but they lacked the power to exercise a "controlling influence."[22] The prospect for peace and unity was indeed bleak, and on this note the American dream of mediation in China came to an end.

Marshall returned home to become the secretary of state, but embittered by his recent experience he was incapable of evolving a positive China policy.[23] He adopted an attitude of "wait and see," hoping that things might work themselves out in China. The only flicker of positive action was the dispatch of Wedemeyer on a fact-finding mission in July 1947, at the suggestion of Republican Congressman Walter Judd, a former medical missionary in China and a staunch supporter of Chiang Kai-shek. Wedemeyer stayed in China for a month, tried in vain to impress Chiang with the need for reform, and came back with a report that recommended "sufficient and prompt military assistance" to the Nationalist government under the supervision of 10,000 American officers and men. Further, the report called for economic aid for five years, and a five-power guardianship of Manchuria under the United States, the Soviet Union, France, Britain, and China, or failing that, a United Nations "trusteeship" for Manchuria. The report was politely received and quietly shelved by Marshall, who was unable to see how the United States could spare 10,000 men when rapid demobilization had left only one and one-third divisions in the country.[24] It was sadly apparent that the Wedemeyer mission had made little impression on either Chiang or Marshall.

21. *United States Relations with China,* 212.
22. "Personal Statement by the Special Representative of the President (Marshall), Jan. 7, 1947," in *United States Relations with China,* 686-89.
23. Tang Tsou, 445.
24. Out of a total strength of 925,163 men in June 1947. Cf. Tang Tsou, 459.

The Civil War

Chiang opted to resolve the Communist problem by military means after mid-1946, in order to prove that he could easily wipe out the enemy if unimpeded by American mediation. A victory would vindicate the correctness of his judgment and prove the impracticality of the romantic American dream of a coalition government in China. In spite of repeated warnings that the United States would not underwrite his civil war, Chiang could not persuade himself to believe that Washington would prefer the Communists to himself. There was a feeling among the Nationalists that the United States could not afford to see China slip to the Communists; therefore, its warning could not be taken at face value. If the situation grew desperate enough the Americans would have no choice but to come to the aid of the Nationalists.

In the early phase of the civil war, the government troops reaped victories at every encounter. On the other hand, the Communists foresaw many difficult days ahead before a final victory. Mao predicted in 1946 that it would take five years to settle accounts with Chiang, and the Communists were prepared for a long and hard campaign.[25]

From July to December 1946, the Nationalists captured 165 towns and 174,000 square kilometers of territory from the Communists. The crowning success came in March 1947 when they seized the Communist capital of Yenan. Chiang confidently told the American ambassador, Leighton Stuart, that the enemy could be totally defeated or driven to the hinterland by August or September. Indeed, Mao and the CCP central organization found themselves in temporary retreat. They evacuated Yenan on March 18 and fled into hiding, hotly pursued by some 400,000 Nationalist troops.[26] At the end of the first year of civil war in June 1947, the Communist "liberated areas" had shrunk by 191,000 square kilometers and 18 million population.[27]

Buoyed by the chain of military victories, Chiang confidently launched his political offensive. The National Assembly was convened on November 15, 1946, despite the boycott of the Communist Party and the Democratic League. Its 1,744 delegates adopted a new constitution on Christmas Day, consisting of fourteen chapters and 175 articles. This document, promulgated on New Year's Day, 1947, reaffirmed the Three People's Principles as the basic philosophy of the state, the five-yüan government, and the people's four rights—initiation, referendum, election, and recall.

25. Mao Tse-tung, IV, 1,364; Jerome Ch'en, *Mao*, 291-92.
26. *United States Relations with China*, 238; Jerome Ch'en, *Mao*, 283 84.
27. Jerome Ch'en, *Mao*, 299-300.

The president of the republic was to be elected by the National Assembly for a six-year term. Also stipulated in the document was the right of the chief executive to appoint the president of the Executive Yüan with the consent of the Legislative Yüan, as well as the ministers in the Executive Yüan on the recommendation of the Yüan president. Members of the Legislative Yüan were to be elected on geographical and professional bases for a three-year term. The Judicial Yüan enjoyed the right to interpret the constitution, thus establishing the viability of judicial review in the Chinese legal system. Essentially, this government structure followed neither the presidential nor the cabinet system exclusively but was a mixture of both. The Executive Yüan, for instance, with the consent of the president of the republic, could veto the resolutions of the Legislative Yüan, but if the latter overruled the veto by a ⅔ vote, the Executive Yüan had to accept it or resign. With regard to local government, provisions were made for the popular election of provincial governors and district (*hsien*) magistrates.

As expected, the Communists loudly attacked the constitution as illegal. Unruffled by these charges, the Nationalists proceeded with the election of a new National Assembly and the selection of members of the Legislative Yüan in November 1947. The Assembly, convened on March 29, 1948, elected Chiang Kai-shek president of the republic on April 19, with Li Tsung-jen as vice-president. With this election, the twenty-year political tutelage of the KMT—which was originally scheduled to last only six years—formally came to an end. But even as Chiang accepted the mantle of office, the civil war had entered a critical stage for the Nationalists.

Mid-1947 seemed to mark a turning point in the fighting. The victory-laden Nationalist military machine began to sputter, partly because of increased assignment of soldiers to garrison duties in reconquered areas, with a corresponding reduction in the actual fighting force. In contrast, the Communist army had been expanding steadily, reaching 1.95 million in June 1947 as compared with KMT's 3.73 million.[28] The Communists went on a general offensive in the second half of 1947, scoring victories in Honan and northern Hopeh.

By far the severest blow to the Nationalists occurred in Manchuria. Within three months of Christmas 1947, Lin Piao's army had inflicted losses of 150,000 on the crack Nationalist army. The remainder were pressed into a small triangular area between Mukden, Changchun, and

---

28. By June 1948, the CCP forces reached 2.8 million vs. KMT's 3.62. In November, the CCP forces actually surpassed the KMT: 3 million vs. 2.9 million. In June 1949, the CCP had achieved an overwhelming superiority over the Nationalists: 4 million vs. 1.5 million. Cf. Jerome Ch'en, *Mao*, 374.

Chinchow, which represented less than 1 percent of Manchuria. It was hopeless to hold such an untenable position, yet Chiang decided to fight to the bitter end. By mid-1948 Lin had so tightened the encirclement that he practically smothered the Nationalist defenders. Having destroyed 100,000 government troops, he conquered Chinchow on October 14, Changchun on October 18, and Mukden on November 2. The Manchurian campaign cost Chiang 470,000 of his best troops[29] and dealt a mortal blow to the morale of the entire government army. In the opinion of General David Barr, it "spelled the beginning of the end" for the Nationalist cause.[30]

Operating simultaneously with the Manchurian battles, another Communist field army under Ch'en Yi conquered Shantung after fierce fighting at Tsinan on September 26, 1948. This accomplished, the CCP forces, 550,000 strong, moved on to attack the historic battle site of Hsuchow, at the junction of the Tientsin-Pukow and Lung-Hai railways. Chiang had deployed 400,000 of his mechanized troops, equipped with tanks, heavy artillery, and armored cars to defend this gateway to Nanking. But many of his officers had become demoralized under the relentless hammer blows of the enemy. They were further frustrated by rain, snow, and sleet which immobilized their mechanized units. No sooner had the Battle of Huai-Hai[31] begun in October 1948 than two entire Nationalist divisions defected. From November 11 to 22, 100,000 government troops were destroyed. Hsuchow fell on December 15. By the time the Battle of Huai-Hai was over in January 1949, the Nationalists had lost no less than 200,000 men and two well-known commanders,[32] who were captured by the enemy. Flushed with success, Mao confidently predicted victory in one year.[33] His forces now pressed toward Nanking, the seat of the Nationalist government.

Meanwhiles, Lin Piao's 800,000-man army, freed from Manchurian engagements, together with the Communist North China Army Group,[34] formed a pincer movement against Peiping-Tientsin in December 1948. The Nationalist defender, General Fu Tso-yi, who had earlier defeated Communist forces in Suiyuan, had 500,000 men under his command. But all expectations of resistance evaporated when his defense plans were stolen by a Communist agent operating in his headquarters.[35] Deprived

29. The Nationalists admitted to a loss of 300,000 men.
30. *United States Relations with China*, 335. General Barr was head of the American Army Advisory Group in China.
31. The combined names of *Huai* River and *Lung-Hai* Railway.
32. Generals Tu Yü-ming and Huang Wei.
33. Mao Tse-tung, IV, 1,164.
34. Under General Nieh Jung-chen.
35. Teng Pao-shan.

of their strategy and hopelessly outnumbered, the garrisons of Tientsin and Peiping capitulated on January 15 and 23, 1949, respectively. General Fu himself surrendered, along with 200,000 troops. From September 1948 to January 1949 the government had lost one and one-half million men.[36] Under such staggering losses, the Nationalist forces simply collapsed.

What of the future of the government? Chiang was forced by the peace faction within his party to resign on January 21, 1949, and Vice-President Li Tsung-jen took over the reins of government as acting president. Still hoping to hold the southern half of China below the Yangtze, Li tried to initiate negotiations with the Communists, but to no avail. With victory so close at hand, Mao saw no reason to compromise. On April 21, his forces crossed the Yangtze, and three days later occupied Nanking, driving the refugee government to seek asylum in Canton. The Communist advance now accelerated in all directions and simply could not be stopped. Even before all China had been conquered, Mao proclaimed the establishment of the People's Republic on October 1, 1949. When the Nationalist government fled from Canton to Chungking on October 13 and to Taiwan on December 8, the Communist conquest of mainland China was complete. After twenty-eight years (1921-49) of struggle, Mao rose to the pinnacle of power.

The Role of the United States

What did the United States do during the Chinese civil war and what were its "sins" of commission or omission? It must be stated at the outset that when he dispatched Marshall to China in December 1945, President Truman made it very clear that large-scale aid to China was to be contingent on the achievement of national unity. Marshall himself had repeatedly warned Chiang in mid-1946 that the United States was not prepared to underwrite a Chinese civil war and that the spiraling inflation might precipitate an economic collapse. When Chiang ignored these warnings and went ahead with the fighting, the die was cast.

Washington's chief mistake was its inability to evolve a positive policy toward China. It neither disowned the Nationalist regime nor extricated itself entirely from China, but followed a course of partial withdrawal and limited assistance to the Nationalist government—such as the granting of $27.7 million for economic aid in October 1947 and the establishment of a small Army Advisory Group to offer Chiang counsel. This policy of drift prompted Chiang's friends in Washington and the "China Lobby" to engineer a move to block the European Recovery Program unless a meaningful China aid program was initiated. General MacArthur

36. Jerome Ch'en, *Mao*, 307.

pressed for greater China aid and sneered at the American pressure for KMT reforms while fighting a civil war: "The two issues are as impossible of synchronization as it would be to alter the structural design of a house while the same was being consumed by flame."[37]

In response to an urgent Nationalist request by the end of 1947 for a four-year aid program of $1.5 billion—of which $500 million of economic aid and $100 million of military aid were to be administered for the first year—Truman recommended on February 18, 1948, a grant-in-aid of $570 million for fifteen months to retard the Chinese economic collapse. The China aid bill limped through Congress with a 13 percent cut at $400 million, but was not implemented until the second half of 1948 when the Nationalist cause was all but lost. The aid was too little and too late. On July 30, 1948, Mao declared the demise of the KMT regime "not too far away." On August 13 Marshall was reported to have said: "I wash my hands of the problem which has passed altogether beyond my comprehension and my power to make judgments."[38]

The misfortune of the Nationalists was compounded by their entanglement in American politics during an election year. Disappointed with the Democratic Administration, Nationalist diplomats cultivated the Republicans on the assumption that the 1948 election would result in a change of administration. Governor Thomas Dewey of New York, the Republican candidate for the Presidency, declared on June 25, 1948, that if elected he would extend massive financial and military aid to China. But Truman confounded the world—and Chiang—with a resounding victory in the election. As president, he twice turned down Nationalist pleas for aid in November and December 1948.[39] The feeling was strong in Washington, after Chiang's resignation in January 1949, that the United States ought to get out of China as fast as possible.

In retrospect, the United States, though "guilty" of many acts of omission and default, cannot be held responsible for the "loss" of China which it never "owned." Chinese Communism was an internal development of great vitality spanning thirty years, and it is unlikely that foreign intervention could have altered its course. Active American armed intervention before the spring of 1948 might have delayed the Communist ascent temporarily but most certainly could not have stopped it permanently. Such an intervention would have required, in the view of a China expert,[40] 150,000 American troops, although a million or two seems more

37. Tang Tsou, 466, 468.
38. Tang Tsou, 446, 473, 478.
39. First request for an American military mission; second request for an aid program of $3 billion for three years, tendered by Madame Chiang herself.
40. Professor Nathaniel Peffer of Columbia University.

realistic given the later Korean and Vietnam experiences. The question then arises, how long could American soldiers have been kept in China, when demobilization and return to normalcy were the order of the day at home?

In point of fact, the United States government never intended to be involved in the Chinese civil war. Washington explicitly stated that massive intervention was not "practicable or desirable" because it would "require our [American] participation in the civil war and our taking over the direction of military operation and administration."[41] China's strategic value could not justify a large-scale American intervention, and even the prospect of a Russian-dominated China could not alter Washington's position.[42] Washington found the rise of Communist China not desirable but tolerable, since the Chinese were not likely to present a threat to United States security for years or even decades. The failure of mediation left the Americans no choice but to accept the realities of the Chinese situation.

A Reappraisal of the American Policy

America's China policy during the 1940s must be viewed in the larger context of United States global strategy, and more particularly in the light of its approach to East Asia and the Pacific. Despite the official rhetoric of friendship for China, the long-range goals of American policy definitely were not based on altruism or sentimental attachment to the Chinese culture or people, but rather on the pragmatic consideration of the strategic and economic interests of the United States in postwar East Asia.[43]

Washington wanted to create a new balance of power in the Pacific and East Asia, one in which the United States would occupy a dominant position. However, with Europe as the focal point of postwar global consideration, the Americans wanted to achieve their East Asian goals with a minimum of resource commitment. Such a policy required a strong alliance with a major state in the region: either China or Japan. During 1944 Secretary of State Cordell Hull assumed that China would emerge "at the center of any arrangement that was made," but he thought that China had only a fifty-fifty chance of becoming a great power.[44] With his retirement from the State Department, Acting Secretary of State Joseph Grew, a former U.S. ambassador in Tokyo with pronounced pro-Japanese

41. *Foreign Relations of the United States, 1947,* Vol. VII, *The Far East: China* (Washington, D.C., 1972), 855.
42. *Ibid.,* 854.
43. *Ibid.,* 790.
44. Cordell Hull, *The Memoirs of Cordell Hull* (New York, 1948), 1586-87.

sympathies, was inclined to view a revitalized Japan as a desirable option open to the Americans should China prove unable to fulfill its assigned role in the United States strategy.[45] By the spring of 1945 Secretary of the Navy James Forrestal pointedly asked Secretary of War Henry Stimson: "What is our policy on Russian influence in the Far East? Do we desire a counterweight to that influence? And should it be China or should it be Japan?"[46] No decision was reached at that time, but the essential problem remained for American East Asian policy-makers: If China could fit into their scheme, it deserved American support; if not, a revitalized Japan could also serve as an anchor for United States interests in East Asia. It is primarily for this reason that President Truman insisted at the Potsdam Conference in July 1945 that the occupation of Japan should be a sole American enterprise and not a zonal undertaking with the Soviet Union and other powers.[47]

The long-term objective of the United States was to encourage the development of a relatively strong and friendly China capable of serving as a counterweight to the Soviet Union and open to the penetration of American capital.[48] Washington recognized that neither a Communist-ruled China nor a weak China that would invite Russian penetration could fulfill these goals. Hence President Roosevelt promoted the idea of China as one of the great powers and State Department advocated the policy of helping develop "a united, democratically progressive, and cooperative China."[49] Such an objective was partly based on the understanding, as the Department put it in 1945, that China would provide "a large-scale market for American goods and capital."[50] Indeed, as early as 1939 Treasury Department officials unabashedly stated that the Sino-Japanese war provided a wonderful opportunity for the Americans to "get a firm foothold on future Chinese business and we will get the bulk of reconstruction work in China. . . . China under peace time conditions and a revitalized Central Government will make a wonderful future market for American goods and enterprise."[51]

45. U.S. Congress, Senate, Committee on the Judiciary, Internal Security Subcommittee, Hearings, *Morgenthau Diary* (China), 89th Congress, First Session (Washington, D.C., 1965), 1394.
46. Walter Millis (ed.), *The Forrestal Diaries* (New York, 1951), 52.
47. Harry Truman, *Memoirs,* Vol. I, *Year of Decisions* (Garden City, 1955), 551-52.
48. *Foreign Relations of the United States, 1945, The Conference at Malta and Yalta* (Washington, D.C., 1955), 353.
49. Charles Romanus and Riley Sunderland, *Time Runs Out in CBI* (Washington, D.C., 1959), 337.
50. John W. Dower, "Occupied Japan and the American Lake," in Edward Freeman and Mark Seldon (eds.), *America's Asia* (New York, 1969), 167.
51. *Morgenthau Diary,* 7.

It was for these reasons that the United States extended to the Nationalist government Lend-Lease supplies, "currency stabilization" loans, and other forms of military and economic credits. Washington saw these loans as a way to stop the Chinese "defeatists" from defecting to the Japanese, and to encourage Chinese war effort, thereby tying down a considerable number of Japanese troops on the mainland which might otherwise be transferred elsewhere to fight the Americans.

Washington never had a high regard for the Chinese military effort, except for the early phase of the war when China alone withstood the Japanese assault while the Allies faced defeat after defeat by the Imperial Japanese army. At the end of 1943 the Joint Chiefs of Staff, estimating Chinese capabilities, concluded: "We feel that, at most, not more than one-fifth of the Chinese Army is currently capable of sustained defensive operations and then only with effective air support (by the Americans)."[52] Furthermore, Secretary of the Treasury Henry Morgenthau spoke of Chinese officials in charge of finances as "just a bunch of crooks," and Harry Truman later also referred to Kuomintang leaders as "grafters and crooks."[53]

In the early phase of the war, the Pentagon seriously considered using China as a staging-area from which to attack the home-islands of Japan. However, this strategy came under subsequent question, and was dropped at the Cairo Conference in November 1943. George Marshall, head of the Joint Chiefs of Staff, explicitly instructed Stilwell in 1944 that the function of the China-Burma-India theater of war was to distract and draw off Japanese troops from American operations in the central and southern Pacific: "Japan should be defeated without undertaking a major campaign against her on the mainland of Asia if her defeat could be accomplished in this manner."[54] Clearly, the United States wanted to avoid a land war against Japan in China where the price of victory would be measured in manpower rather than in matériel. Indeed, it was this desire that was the central limiting factor of American political and military strategy in East Asia during the war; it was also the reason behind American efforts at developing contacts with the Chinese Communists and at drawing the Russians into the war against Japan. Furthermore, the Americans were apprehensive that Chiang was attempting to draw United States forces into his civil war with the Communists. Stimson stated that

52. *Foreign Relations of the United States, The Conference at Cairo and Teheran, 1943* (Washington, D.C., 1961), 242.
53. *Morgenthau Diary,* 133; Joyce Kolko and Gabriel Kolko, *The Limits of Power* (New York, 1972), 554-55.
54. Charles Romanus and Riley Sunderland, *Stilwell's Command Problems* (Washington, D.C., 1956), 363-64.

this was "the very thing that I am resolved that we should not do unless it is over my dead body."[55]

The dread of being embroiled in China led to the American tactic of mediating a political settlement in the dispute between the KMT and the CCP. At the same time Washington vigorously urged Chiang to renovate his government in order to outflank the Communists via reform. Chiang, however, not only refused to heed the advice but also rejected a political settlement with the Communists on any terms but his own. American observers in China realistically warned Washington that Chiang's regime could not withstand the impact of a military solution to China's profound internal problems; Foreign Service Officers John P. Davies and John S. Service boldly stated their belief that China's destiny was not Chiang's but the Communists'. In April 1945 the State Department advised Truman that the United States should continue to support Chiang, for he still "offers the best hope for unification and for avoidance of chaos in China's war effort," but if "the possible disintegration of the authority of the existing government" occurred, the long-term interests of the United States in China warranted "flexibility to permit cooperation with any other leadership in China which may give greater promise."[56] In June the State Department in a policy paper advocated a united China without mentioning the Kuomintang or Chiang; all it hoped for was "an effective and stable government"—one that would "safeguard the principle of equal opportunity for the commerce and industry of all nations in China."[57] After Marshall's recall in early 1947, Washington had all but given up hope for Chiang and the Kuomintang. With Europe as America's primary consideration and as the area where the United States would invest the bulk of its manpower and financial resources, China occupied at best a tertiary position in the American priorities of global commitments; indeed, by May of 1947 the Joint Chiefs of Staff had placed "China very low on the list of countries which should be given such assistance."[58] Washington had decided by the second half of 1947 to rebuild Japan as an alternative base for American power in East Asia. From the American perspective, the rebuilding of Japan would entail far less commitment of resources than would be the case for China, and offer more assurance of success. Thus, Japan replaced China as the cornerstone of American interests in East Asia and the Pacific.

55. Quoted in Gabriel Kolko, *The Politics of War: The World and United States Foreign Policy, 1943-1945* (New York, 1968), 535.
56. Harry S. Truman, *Memoirs*, Vol. I, 102-3.
57. *Foreign Relations of the United States, The Conference of Berlin (The Potsdam Conference), 1945* (Washington, D.C., 1960), I, 858.
58. *Foreign Relations of the United States, 1947, op. cit.* Vol. VII, 853-54.

American mediation in China won the goodwill of neither the Nationalists nor the Communists. The former blamed Washington for spoiling their optimum chances for destroying the opponent, while the latter attacked the United States for ostensibly posing as a neutral mediator while actually aiding the Nationalists. In pursuing his military solution to China's political problems, Chiang undermined his chance of serving as the anchor for American interests in East Asia. Instead of consolidating his existing position and outflanking the Communists via reform, as the Americans advised him to do, he followed a strategy that ripped apart the fragile fabric of postwar Chinese society and opened the door to a Communist victory. Chiang assumed that the Americans would not tolerate the Communists as his successors, but there was nothing in American policy that should have realistically led him to believe that the Americans would not abandon his regime if it became too expensive to uphold.

America's venture in China not only failed in its objective of using the Nationalist government to secure a hold for the United States in East Asia, but it also alienated the Communists. Inability to assess correctly the potential of the Communists in the postwar world and to work out a détente with them in order to undercut their dependence on the Soviet Union represented a misjudgment and "a lost chance in China."[59] In the final analysis, America's failure stemmed from its incapacity to bridge the gap between its goals and Chinese realities.

Causes of the Nationalist Defeat

For an event as important as the fall of the Nationalist government on mainland China, historians have the unshirkable responsibility of assessing its causes. While a definitive study may be premature, pending a more complete opening of the archives, some tentative interpretations may be offered at the risk of oversimplification. The most important near cause for the downfall of the Nationalists was the eight-year Japanese war, which completely exhausted the government militarily, financially, and spiritually. Had there been no Japanese war, the situation in China would have been very different. Hence, many of the disastrous repercussions of the war discussed in the last chapter continued to plague the Nationalists during their struggle with the Communists. The price the Nationalists paid to win the Japanese war was also the first installment toward its eventual downfall.

59. Joseph W. Esherick (ed.), *Lost Chance in China: The World War II Dispatches of John S. Service* (New York, 1974).

DECEPTIVE MILITARY STRENGTH. Although the Nationalist army emerged from the Japanese war better equipped and trained than ever before, it was a tired and weary force. Already exhibiting signs of fatigue during the last stage of the Japanese war, it was held together by nationalism, patriotism, and the prospect of an imminent Allied victory. Japan's surrender gave the troops a sense of relief and a feeling of having accomplished the mission, and they longed for a rest. The thought of fighting another civil war was abhorrent to them. Though they fought when ordered, their spirit was unwilling and their flesh weak. Their credible performance before mid-1947 represented a last desperate thrust before the final collapse.

The Communists, on the other hand, during the Japanese war had vastly expanded their military forces. The end of war was also the hour of recognition for them, now fresh, vigorous, and confident of the future. Ideologies apart, the difference in stamina contributed to the outcome of the bout.

Leaving aside the question of war weariness, the Nationalist strategy also left much to be desired. Against American advice, Chiang sent large bodies of his troops to Manchuria, only to have 470,000 of them slaughtered or captured, when he should have concentrated his men to defend areas south of the Great Wall. The ill-fated decision to take Yenan and pursue the fleeing Communist leaders to the strategically unimportant mountainous Northwest drained another 400,000 troops. The battles of Huai-Hai and Tientsin-Peiping were poorly directed, causing yet another irreparable loss in manpower. In the short period from September 1948 to January 1949, the Nationalists lost well over one million men; the heart of their army was destroyed and what was left could no longer fight.

INFLATION AND ECONOMIC COLLAPSE. Even more disastrous than war weariness and mistakes in strategy was the galloping inflation which was already rampant during the Japanese war, and became completely uncontrollable after the war. The single most important cause of this inflation was the flagrant increase in note issues which grew from 1.3 billion yüan in January 1937 to a fantastic 24,558,999 billion by the end of 1948, with the result that prices increased by 30 percent per month during 1945-48. During the brief span from August 1948 to April 1949, note issues increased by 4,524 times, and the Shanghai price index rose an astronomical 135,742 times. Inflation and financial mismanagement destroyed the livelihood of hundreds of millions of Chinese and totally discredited the government. Small wonder that the majority of the people were not averse to, and even looked forward to, a change in administration.

LOSS OF PUBLIC CONFIDENCE AND RESPECT. In addition to fiscal irresponsibility which brought on rampant inflation, the obnoxious conduct of Nationalist officials who returned to the Japanese-occupied areas after the war did permanent damage to Nationalist prestige. They returned as conquerors and treated the people with contempt, as if they had been disloyal citizens or traitors. The officials were more interested in taking over enemy properties for selfish purposes than for the welfare of the people who had suffered so grievously during the Japanese occupation. They monopolized profitable commodities and enterprises in open competition with the people, and publicly auctioned relief materials for personal gains. Worst of all, they forced the conversion of the Japanese-supported puppet currency in South and Central China into *fapi* at the exorbitant rate of 200 to 1, when a more equitable rate would have been half that much.[60] When the savings and cash reserves of the people were so suddenly and drastically reduced, their immediate reaction was deep resentment, the more so when a few years earlier the puppet government had forced them to convert their *fapi* to the puppet currency at the rate of two to one. The two conversions slashed the cash reserves of the people by 400 times! These citizens within occupied territories, who had waited for eight years for the return of Nationalist rule, were so mercilessly milked and so contemptuously treated that they wondered whether life would not have been better under the Japanese. The net result of the misbehavior of the Nationalist officials was the alienation of millions of suffering people.

FAILURE OF AMERICAN MEDIATION AND AID. The course of events in postwar China could have been different if the United States had followed a different course during the Japanese war. First, had its China aid been more substantial during the first four years of the war, 1937-41, it might have beefed up Nationalist finances to the point where inflation could have been checked in its early stages. By nipping the trouble in the bud, the later runaway situation might never have occurred, thus avoiding the ultimate economic collapse. Second, if the United States had retained the original strategy of attacking Japan from the Chinese mainland, American soldiers would have landed in the coastal provinces of China, seizing territory from the Japanese and turning it over to the Nationalist government. The plan was discarded, however, by a change in the Allied strategy in 1943-44, which called for the invasion of Japan from the Pacific, by-passing China altogether. This decision placed Nationalist

60. Shun-hsin Chou, 24.

China in a strategically insignificant position, and when the war suddenly ended it was ill-prepared for the consequences of the unexpected peace.

Quite apart from these economic and military considerations of what might have happened, the United States lost at least three chances to exert a decisive influence on China diplomatically. First, if Hurley had played a more active role of mediation during Mao's visit to Chungking in August-October 1945, he might have prevailed upon Chiang to accept the "transient arrangements" that Mao was offering, thereby averting the immediate outbreak of the civil war. This was a golden chance carelessly thrown away. Second, if Marshall had been more forceful in "pressuring" the KMT and Chiang into honoring the Political Consultative Conference resolutions, hostilities might have been checked. Third, when the Nationalists were in a critical retreat during the spring of 1948, the United States had one last chance to intervene militarily but it did not choose to do so. In retrospect, it seems that the United States lost all these chances by default.

RETARDATION OF SOCIAL AND ECONOMIC REFORMS. Aside from these immediate causes of the Nationalist downfall, a far more fundamental failing was the continuous retardation of badly needed social and economic reforms. This neglect might have been partially caused by overwhelming circumstances beyond Nationalist control. From the outset, it was challenged by the "new warlords" and dissident politicians, and hardly had it resolved these problems when it was deluged by the mounting threats of Japanese aggression and Communist insurrection. It took all the energy, resources, and skills the Nationalists could muster to fend off a war with Japan while launching five campaigns against the Communists. There was little time or inclination left to tackle the seemingly less imminent, if more basic, problems of economic justice and social reforms. Not only was the Principle of People's Livelihood—regulation of capital and equalization of land—never implemented, but even the far more moderate resolution of farm rent reduction to 37.5 percent of the yearly crop never materialized. Dr. Sun's ideal of "land to the tiller" remained only a fond dream. Once the Japanese war broke out, military affairs took precedent over all others, relegating the long-overdue social reforms further to the background.

Despite these overwhelming circumstances, it was nevertheless correct to say that the Nationalists lacked the necessary motivation to initiate social and economic reforms. Established in the coastal regions far away from the hinterland, the Nanking government relied on customs dues and city commercial taxes for its sustenance, paying little attention to agrarian problems. It did not understand the peasants, saw no urgency in solving their problems, and was unsympathetic to their plight. Ironically, the Nationalist officials continued to live under the shadow of the Confucian distinction between the rulers and the ruled, and looked down upon the peasants as an inert nonentity. They failed to see the revolutionary potential of the peasant masses and consequently never attempted to organize them. It was precisely in this area of neglect that the talent of Mao found its highest and most successful expression. The stone that one builder had rejected became the cornerstone of the other's house.

Further Reading

Beal, John R., *Marshall in China* (Garden City, N.Y., 1970).

Beloff, Max, *Soviet Policy in the Far East, 1944-1951* (London, 1953).

Borg, Dorothy, and Waldo Heinrichs (eds.), *Uncertain Years: Chinese-American Relations, 1947-1950* (New York, 1980).

Chan, Lau Kit-ching, *The Chinese Youth Party, 1923-1945* (Hong Kong, 1972).

Chang, Carsun, *The Third Force in China* (New York, 1952).

Chang, Kia-ngau, *The Inflationary Spiral: The Experience in China, 1939-1950* (New York, 1958).

Chassin, Lionel M., *The Communist Conquest of China: A History of the Civil War, 1945-49* (Cambridge, Mass., 1965).

Ch'en, Jerome, *Mao and the Chinese Revolution* (London, 1965).

Chiang, Kai-shek, *Soviet Russia in China: A Summing Up at Seventy* (New York, 1957).

Ch'ien, Tuan-sheng, "The Role of the Military in the Chinese Government," *Pacific Affairs*, 21:239-51 (1948).

Chou, Shun-hsin, *The Chinese Inflation, 1937-1949* (New York, 1963).

Fairbank, John K., *China Bound: A Fifty-Year Memoir* (New York, 1982).

Fairbank, Wilma, *America's Cultural Experiment in China, 1942-1949* (Washington, D.C., 1976).

Foreign Relations of the United States, 1947, Vol. VII, *The Far East: China* (Washington, D.C., 1972).

————, *1946*, Vol. IX, *The Far East: China* (Washington, D.C., 1972).

————, *1946*, Vol. X, *The Far East: China* (Washington, D.C., 1972).

————, *1948*, Vol. VII, *The Far East: China* (Washington, D.C., 1973).

————, *1948*, Vol. VIII, *The Far East: China* (Washington, D.C., 1973).

Grasso, June M., *Truman's Two China Policy* (Armonk, N.Y., 1987).

Griffith, Samuel B., II, *The Chinese People's Liberation Army* (New York, 1967).

Harding, Harry, and Yuan Ming (eds.), *Sino-American Relations, 1945-1955: A Joint Reassessment of a Critical Decade* (Wilmington, Del., 1989).

Ho, Kan-chih, *A History of the Modern Chinese Revolution* (Peking, 1959).

Hu, Ch'iao-mu, *Thirty Years of the Communist Party of China* (Peking, 1951).

Johnson, Chalmer A., *Peasant Nationalism and Communist Power, The Emergence of Revolutionary China* (Stanford, 1962).

Koen, Ross Y., *The China Lobby* (New York, 1960).

Kwei, Chung-gi, The Kuomintang-Communist Struggle in China, 1922-1949 (The Hague, 1971).

Lee, Chong-sik, Revolutionary Struggle in Manchuria: Chinese Communism and Soviet Interest, 1922-1945 (Berkeley, 1983).

Levine, Steven I., Anvil of Victory: The Communist Victory in Manchuria, 1945-1948 (New York, 1987).

Liao, Kai-lung, From Yenan to Peking (Peking, 1954).

Lippit, Victor D., The Economic Development of China (Armonk, N.Y., 1987).

Liu, F. F., A Military History of Modern China, 1924-1949 (Princeton, 1956).

Loh, Pichon P. Y., The Kuomintang Debacle of 1949: Conquest or Collapse? (Boston, 1965).

Mao, Tse-tung, Mao Tse-tung hsüan-chi (Selected works of Mao Tse-tung), (Peking, 1963), III and IV.

May, Ernest R., The Truman Administration and China, 1945-49 (Philadelphia, 1975).

Melby, John F., The Mandate of Heaven: Record of a Civil War, China 1945-49 (Garden City, N.Y., 1971).

Pepper, Suzanne, Civil War in China: The Political Struggle, 1945-1949 (Berkeley, 1978).

Porter, Brian E., Britain and the Rise of Communist China: A Study of British Attitudes, 1945-1954 (London, 1967).

Purifoy, Lewis McCarroll, Harry Truman's China Policy: McCarthyism and the Diplomacy of Hysteria, 1947-1951 (New York, 1976).

Rea, Kenneth W., and John C. Brewer (eds.), The Forgotten Ambassador: The Reports of John Leighton Stuart, 1946-1949 (Boulder, 1981).

Schaller, Michael, The U.S. Crusade in China, 1938-1945 (New York, 1979).

Sheridan, James E., China in Disintegration: The Republican Era in Chinese History, 1912-1949 (New York, 1975).

Tsou, Tang, America's Failure in China, 1941-50 (Chicago, 1963).

United States Relations with China, With Special Reference to the Period 1944-1949 (Washington, D.C., 1949).

Van Slyke, Lyman P. (ed.), The Chinese Communist Movement: A Report of the U.S. War Department, July 1945 (Stanford, 1968).

———— (intro.), Marshall's Mission to China, December 1945-January 1947: The Report and Appended Documents (Arlington, Va., 1976).

Wedemeyer, Albert (General), Wedemeyer Reports! (New York, 1959).

Young, Arthur N., China and the Helping Hand, 1937-1945 (Cambridge, Mass., 1963).

————, China's Wartime Finance and Inflation, 1937-1945 (Cambridge, Mass., 1965).

A Chinese painting of Mao Tse-tung proclaiming the establishment of
the People's Republic of China on October 1, 1949.

Great Hall of the People, Peking, Tien-an-men Square.

China's Big Four, from left, Chou En-lai, Liu Shao-ch'i, Chu Teh,
and Mao Tse-tung.

Premier Chou En-lai with his
Chinese signature.

Celebrating the Spring Festival in 1966, prior to the onset of
the Cultural Revolution. From left, Teng Hsiao-p'ing, Chou En-lai,
Soong Ch'ing-ling, Liu Shao-ch'i

Lin Piao on May Day, 1967.

Ta-ch'ing General Petrochemical plant.

A tractor factory at Loyang.

26

The People's Republic:
Its First Decade

With the conquest of the country nearly complete, Mao Tse-tung summoned a People's Political Consultative Conference on September 12, 1949, to prepare for the formation of a new government. It met for twelve days and adopted an Organic Law of the Central People's Government, a Common Program which was basically a statement on national purposes, and a national flag consisting of a red background with a large yellow star in the upper left corner surrounded by four smaller stars. The large star symbolized CCP leadership and the smaller ones represented the four-class coalition of workers, peasants, petty bourgeois, and national bourgeois. On October 1, the Chinese People's Republic was formally established, with a capital in Peiping, now renamed Peking. It was recognized by the Soviet Union a day later and by the other Communist states in a rapid succession.[1] Among the non-Communist countries that granted recognition were India, Burma, Pakistan, Ceylon, Britain, and France.[2] The United States, however, did not recognize the government until January 1, 1979.

1. Bulgaria and Rumania, October 3, 1949; Czechoslovakia, Poland, Hungary, and Yugoslavia, October 4; East Germany, October 27; and Albania, November 23.
2. In 1968, 51 countries recognized the People's Republic of China, while 65 recognized Nationalist China. However, by October 1974 more than 80 countries had recognized the People's Republic of China, compared with 32 which maintained diplomatic relations with the Republic of China on Taiwan. In 1988, 150 countries recognized the People's Republic of China, and about 23 recognized the Nationalist government on Taiwan.

The theory and practice of the new government reflected to a large extent ideas expressed in Mao's *New Democracy* and the Rectification (or Intra-Party Struggle) Movement (*Cheng-feng*) developed in Yenan. The *New Democracy,* a major theoretical work of 1940, was a creative adaptation of Marxism-Leninism to the Chinese situation during the critical transition from semicolonialism and semifeudalism to socialism. It stipulated that the economic structure should consist of three sectors: the state economy, in which the government should control big industries, mines, enterprises, and public utilities; the agricultural economy, in which individual farms should develop into collective farms; and the private economy, in which the middle and small capitalists should be allowed to operate. Of the three, the state sector was to assume the position of leadership and strive to increase production faster than the private sector so as to eliminate possible competition. It was also to be responsible for guiding the other sectors toward socialism. As regards the political structure, the work set forth the principle of "Democratic Centralism" and the coexistence of the four classes under the leadership of the proletariat and its party, the CCP. Culturally, Mao favored the development of a scientific, antifeudal, mass culture. Selective acceptance of useful elements of foreign cultures was desirable but the culture of the New Democracy should be national and anti-imperialistic, able to advocate the dignity and independence of the Chinese nation. "It belongs to our nation and bears the characteristics of our nation."[3]

Mao's second theoretical contribution during the Yenan period was the Rectification Movement of 1942, to combat (1) subjectivism and unorthodox tendencies; (2) sectarianism within the party ranks; and (3) formalism in literature. It was a drive aimed at inculcating in the party members a correct understanding of Marxism-Leninism, the thought of Mao, and the general party line in order that they might avoid "leftism" and "rightism." The struggle was one of principle and not of personality, with a view to winning back deviants to orthodoxy through indoctrination, thought reform, and realization of mistakes.[4] With this process of ideological rectification, the Chinese Communists hoped to avoid the permanent purge that had characterized the Soviet experience.

Political Organization

The Organic Law of 1949 made it very clear that the Chinese People's Republic was not a "Dictatorship of the Proletariat" as in the Soviet

3. Mao Tse-tung, II, 655-704.
4. Mao Tse-tung, III, 813-30.

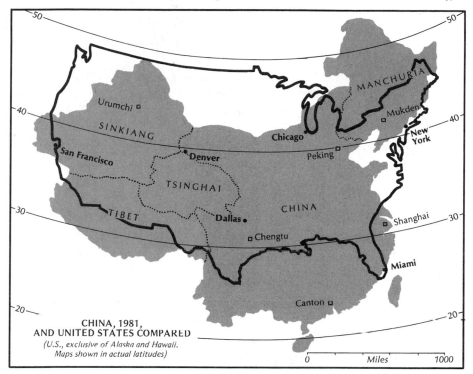

CHINA, 1981,
AND UNITED STATES COMPARED
*(U.S., exclusive of Alaska and Hawaii.
Maps shown in actual latitudes)*

Union, but a "Democratic Dictatorship" led by the CCP on the basis of a four-class alliance. The coexistence of the four classes endowed the government with a "democratic" character, while its uncompromising and unyielding attitude toward the counterrevolutionaries gave it the attribute of a "dictatorship." A cardinal principle followed by the new government was "Democratic Centralism," which provided for popularly elected bodies at different levels of government. These assemblies would elect their own representative officials, pending the approval of the higher authorities. The "election" part of this process was "democratic," while obedience to higher authorities suggested "centralism." By extension, the term also came to mean free discussion in the formation of policy and tight, unswerving compliance to a decision once it had been made, regardless of one's original stand.

The period of New Democracy lasted until 1953, when a program of Socialist Transformation was launched. By 1956 a new period of Socialist Construction set in. These three stages give some indication of the progression of Chinese Communism.[5]

---

5. Chalmers Johnson, "The Two Chinese Revolutions," *The China Quarterly* (July-Sept. 1969), 39:17.

THE GOVERNMENT STRUCTURE. Under the Organic Law the supreme organ of state was the Central People's Government Council which exercised executive, legislative, and judicial powers. It met twice a month to deliberate on high policies of state. Its members included the chairman (Mao), the six vice-chairmen, and fifty-six others elected by the People's Political Consultative Council. When not in session its powers were delegated to a State Administrative Council, whose twenty-odd members constituted a sort of cabinet responsible to the Government Council, or, when the latter was not in session, to the chairman of the state, Mao. The State Administrative Council was headed by a premier (*tsung-li*), Chou En-lai, and a number of vice-premiers; under them were four committees: Political and Legal Affairs, Finance and Economics, Culture and Education, and People's Supervision. Each of these committees directed a determined number of the thirty various ministries, commissions, and boards. On a par with the State Administrative Council were the People's Supreme Court, and the Procurator-general's Office.

Under the central government but above the provincial administrations were six unique Great Administrative Areas,[6] each with jurisdiction over several provinces. Presumably, these intermediate setups were created to help consolidate the central government's hold on the provinces, but, as it later turned out, they developed centrifugal tendencies at the expense of the central power. Subsequently, they were abolished in 1953, and the political structure of the nation reverted to the three traditional tiers of national, provincial, and district (or county) administrations.

The Organic Law remained in force for five years while steps were undertaken to introduce a constitution. In 1953 a census was taken and an election law promulgated, allowing voting privileges to all citizens eighteen and above, except landlords and counterrevolutionaries. In early 1954 elections were held, with the village and township congresses electing their representatives to the district congresses, which then elected their delegates to the provincial congresses, which in turn elected their delegates to the National People's Congress. The last-named body was convened between September 15 and 28, and adopted a new constitution of four chapters and 106 articles. Chapter 1 reiterated the principle of "Democratic Centralism" and the alliance of the four classes, as well as the four types of ownership: state, cooperative, individual, and capitalist. Chapter 2 described the government structure, purposely omitting the Great Administrative Areas. Chapter 3 was the usual bill of rights, with the exception

6. The Northeast, North China, East China, Central-South China, Northwest China, and Southwest China.

that the government reserved the right to "reform traitors and counter-revolutionaries"—thus voiding the legal guarantees for those who had the misfortune of disagreeing with the government. Chapter 4 designated Peking as the capital and described the national flag (as mentioned before).

Under the constitution, the highest organ of state was the National People's Congress, which was supposed to meet briefly each year to ponder major policy decisions and elect top government officials. The position of the chairman of the republic (Mao) emerged stronger than before, since the six vice-chairmanships had been reduced to one (Chu Teh). Other major central offices included a State Council (*Kuo-wu yüan*), a National Defense Council, a Supreme People's Court, and a Supreme People's Procuratorate. Below the central government were provincial and district (or county) administrations.

POLITICAL PARTIES. The leading party was of course the CCP, which claimed a membership of 4.5 million in 1949, 17 million in 1961, and about 46 million in 1988. The party was organized in four channels: representative, executive, administrative, and control. According to the 1956 party constitution, the highest organ in the representative sector was the National Party Congress, which convened once a year and whose members were elected for five-year terms. In the executive sector, there was the powerful Central Committee,[7] also elected for five years, which was headed by a chairman and four vice-chairmen; in 1958 a fifth vice-chairman was added. Convened twice a year, the Central Committee when not in session delegated its powers to the Politburo,[8] which in turn maintained a Standing Committee of China's seven most powerful men. Until the Cultural Revolution, these seven included the Central Committee's chairman (Mao), five vice-chairmen (Liu Shao-ch'i, Chou En-lai, Chu Teh, Ch'en Yün, and Lin Piao), and the party general secretary (Teng Hsiao-p'ing).[9] The Central Committee maintained six regional bureaus and a number of departments such as Organization, Training, Propaganda, and Social Affairs.

In the administrative sector of the party, there was the Central Secretariat, and in the control sector, the Central Control Commission. An over-all view of the party structure appears in the following chart.

7. 97 regular and 73 alternate members in 1956; grew to 170 and 109 respectively, in 1969, and 195 and 124 in 1973.
8. 17 full and 7 alternate members in 1956; 21 and 4 respectively in 1969 and 1973.
9. Franz Schurmann, *Ideology and Organization in Communist China* (Berkeley, 1966), 143, 146.

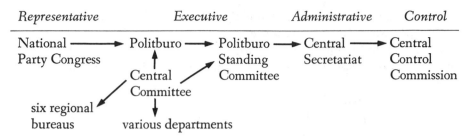

Significantly, party and government were closely interlocked, and all important party members filled key positions in government as well as in semiofficial organizations such as trade unions, farm groups, and mass organizations. Mao in 1949 held the chairmanships of the CCP Central Committee, the Politburo, the Central Secretariat, the People's Republic, the Revolutionary Military Council, and the National People's Congress. After the 1956 party reorganization, however, he relinquished the chairmanship of the Central Secretariat; in 1959 he resigned from the chairmanship of the People's Republic (State Chairmanship) in favor of Liu Shao-ch'i, who concurrently held the position of first vice-chairman of the CCP Central Committee. After Liu's disgrace in 1966, the State Chairmanship was not filled until 1983, when Li Hsien-nien was elected to it.

As a "People's Democratic Dictatorship" rather than a "Dictatorship of the Proletariat," the Chinese People's Republic permitted a number of non-Communist parties to exist. The most important ones included (1) the KMT Revolutionary Committee, which consisted of a group of leftist KMT members, who chose to stay in mainland China rather than join Chiang on Taiwan; (2) the Democratic League, which had always sympathized with the CCP during its struggles with the KMT; (3) the Third Party, which had unsuccessfully attempted to effect a reconciliation between the KMT and the CCP; (4) the Chih-kung tang—constituted largely of Chinese overseas—which had antecedents in the secret Hungmen Society of old; (5) the Democratic Reconstruction Association; and (6) the People's Salvation Association, of anti-Japanese fame in the 1930s.

Since a Communist dictatorship recognized neither the concept of a "loyal opposition" nor the freedom associated with political parties as understood in the West, these non-Communist parties were essentially window dressing. They possessed only the right to agree and cooperate with the CCP and the government.

UNIQUENESS OF CHINESE COMMUNISM. The differences between the Chinese and Russian revolutionary experiences accounted for many unique features of Chinese Communism. First of all, the Chinese Revolution was led to success by a group of professional revolutionaries who received wide support from the peasantry; in Russia, it was the urban workers who gave decisive support to the professional revolutionary leaders. The Chinese leaders were mostly of intellectual and middle-class backgrounds; those who were of true proletarian origin accounted for a very small fraction. The success of the Chinese experience suggested that possession of correct doctrine was more important than formal organic ties with the proletariat. Nonetheless, the CCP leaders insisted that they were spiritual, if not social, proletarians.[10]

Second, the classical Marxist pattern of societal development from feudalism to capitalism to socialism did not apply in China, for the stage of capitalism did not properly exist there. In its place the Chinese Communists substituted a semifeudal, semicolonial period, from which China moved into an intermediary stage called the New Democracy before advancing to the stage of socialism.

Third, during New Democracy four classes were to coexist and non-Communist parties were permitted to operate, albeit in a limited fashion. In contrast, the dictatorship of the proletariat in the Soviet Union permitted only one class and one party.

Fourth, the Chinese Communist seizure of power had its base in the countryside and achieved success after a prolonged period of struggle. The Russian Revolution was characterized by strikes, sabotage, and uprisings in urban centers, and won victory in a much shorter time than the Chinese.

Fifth, in place of the Soviet permanent purge which accounted for the killing of 70 percent of the Central Committee members elected in 1934,[11] the Chinese stressed ideological remolding and thought reform. Until the Cultural Revolution in the mid-1960s, the Chinese leadership had been remarkably stable and closely knit. The only purge of note was connected with the antiparty activity in 1953-54 of Kao Kang, party boss of Manchuria, and Jao Shu-shih, party first secretary in East China.

The success of the Chinese Communist Revolution led Mao to assert that the Chinese experience should provide a prototype of revolution in Asia. By implication he denied the universal applicability of the Soviet model and reduced it to a European variety. The Chinese claim had far-reaching consequences in Sino-Soviet relations, as we will see in a later chapter; it might even have inadvertently inspired the Eastern European satellite countries' assertion of "different paths to socialism."

10. Benjamin I. Schwartz, "On the 'Originality' of Mao Tse-tung," *Foreign Affairs*, (Oct. 1955), 74.
11. Chün-tu Hsüeh, "The Cultural Revolution and Leadership Crisis in Communist China," *Political Science Quarterly*, LXXXII:2:184 (June 1967).

Economic Development

The government in 1949 inherited a badly disrupted economy. Inflation had rocketed beyond control; floods had affected 30-40 percent of the arable land; and industrial and food output had plummeted to 56 and 70-75 percent of the prewar peak, respectively. Thus, the first order of business was to rehabilitate the economic life of the nation and restore industrial and agricultural production to prewar levels.

To promote financial stability, the government issued a People's Currency (*Jen-min p'iao*) in May 1949 and banned the circulation of foreign currencies as a medium of exchange. Strenuous efforts were made to achieve price and wage stabilization through a drastic reduction of the paper money in circulation and the introduction of a "wage-point" system for payment of workers, based on the prices of five basic items—rice, oil, coal, flour, and cotton cloth. As the prices of these articles fluctuated from week to week, the "wage-point" rose and fell accordingly, so that the average salary of workers varied in money value but not in actual purchasing power. Similar methods were used to safeguard savings and bank deposits. Furthermore, concerted efforts were made by the Liberation Army to restore communication lines in order to facilitate the exchange of commodities. Also put into practice was a new taxation system involving agricultural, industrial, commercial, sales and incomes taxes. With these measures, by 1950 inflation was controlled and the government budget balanced.

LAND REVOLUTION AND AGRICULTURAL COLLECTIVIZATION. In addition to efforts to eliminate inflation and restore fiscal stability, the government launched a vigorous agrarian revolution in an attempt to cure the age-old problem of landlordism.[12] The government promulgated in June 1950 the Agrarian Reform Law, which called for the abolition of the "land ownership system of feudal exploitation" and the confiscation of landowners' holdings and farm implements for redistribution to landless peasants. The agrarian population was classified into five categories: (1) landlords: those who possessed large land properties and who did no manual work themselves but lived on usury and the exploitation of others; (2) rich peasants: those who owned land but worked it themselves while also hiring farm hands, lending money, and renting part of the land to poor peasants; (3) middle peasants: those who owned land but worked it themselves without

12. Actually the average landlord's holding was only forty acres. See T. J. Hughes and D. E. T. Luard, *The Economic Development of Communist China, 1949-1960* (London, 1962), 143.

exploiting others; (4) poor peasants: those who owned little land or farm implements and who had to sell part of their land to make ends meet, or who had to rent land from others; and (5) hired hands: those who owned no land and had to live on labor or loans.

Theoretically, the government allowed the landlords to keep their portions of the redistributed land, and exempted from confiscation the rich peasants' land that they themselves cultivated. But in practice many injustices and acts of violence were committed in local "accusation meetings," where virulent denunciations of landlords and rich peasants took place under the guidance of overzealous party cadres and vengeful peasants. Both landlords and rich peasants suffered grievous losses at these meetings, and many were summarily shot after a brief public trial. The gentry, formerly the dominant elite and the backbone of the traditional society, was destroyed.

By December 1952 the agrarian revolution had been completed, and some 700 million *mou* (⅙ acre) of land had been redistributed to 300 million peasants. On the average, in East and South China—where the population density was the highest—each head received one *mou*; in Central China, 2 to 3 *mou*; in North China, 3 *mou*; and in Manchuria, 7 *mou*. On the whole, the land revolution favored the poor peasants and the hired hands at the expense of the landlords and the rich peasants, while the middle peasants were affected least of all.

No sooner had the land revolution been completed in December 1952 than the government started a second phase of agrarian reform—a drive toward collectivization in 1953, with a view to raising production, preventing the re-emergence of rich peasants, achieving greater agricultural specialization, and proceeding faster toward the goal of socialist transformation. Collectivization involved several stages, the lowest being the "mutual aid" teams where the peasants pooled or loaned their implements and worked jointly and seasonally, as during the spring planting and the autumn harvest. The second stage was the semisocialist agricultural producers' cooperatives, in which the members pooled not only their implements and labor but land as well, although theoretically retaining individual ownership. The third stage was the fully socialized cooperative, similar to the Soviet collective farm, in which all members collectively owned the land. By the end of 1956, some 96 percent of all peasant households had officially become members of the semisocialist producers' cooperatives. When the collectivization campaign was completed in 1957, there were a total of 760,000 to 800,000 cooperative farms, each averaging 160 families, or 600 to 700 persons. A further move toward socialist transformation was the introduction of the people's communes in 1958.

INDUSTRIAL EXPANSION. Lenin stated, "There is only one real foundation for a socialist society, and it is large industry." Recognizing the critical role of industrialization in building a socialist state, the CCP spared no effort to achieve this goal. By 1952 not only had the prewar industrial and agricultural peaks been matched, but those of 1949 surpassed by 77.5 percent. Preparatory work for a First Five-Year Plan began in 1951, and in the autumn of the following year a State Planning Committee was established under the direction of Kao Kang, chairman of the Northeast Administrative Area. This Five-Year Plan was supposed to start in 1953, but inexperience and lack of statistical knowledge, planning technique, and machinery plagued it with delay and constant revision. When the Plan was finally put into practice in February 1955—some two years after the official beginning—it was in effect only a Two-and-a-half-Year Plan. It called for the construction of 694 industrial projects, of which 156 plants were to be built with Soviet aid. By the end of the planned period, the industrial output was supposed to double, the introduction of the cooperative farms was to be effected, and the incorporation of private industry and commerce into state organizations was to be completed, so that a "socialist transformation" might became a reality. Of the total outlay of capital, 58.2 percent was earmarked for industrial construction, 19.2 percent for transport, posts, and telecommunication, 7.6 percent for agriculture, forestry, and water conservancy, and 7.2 percent for culture, education, and public health.

The year 1956 marked a spectacular advance in industrial output that topped the previous year by 25 percent, matched by an increase of 60 percent in capital investment. Although the pace slowed down somewhat in 1957, the First Five-Year Plan still overfulfilled the original targets by 17 percent according to the "fixed prices of 1952." Steel production reached 5.3 million tons, iron 5.8 million tons, electric power 19,030 million kwh—each representing a 25 percent increase over the original quotas. The coal production of 122 million tons was an 8 percent overfulfillment, while grain output was 11.6 percent above quota.

The success of the First Five-Year Plan prompted the government to launch a more ambitious Second Five-Year Plan for 1958-62. It called for an overall increase of 75 percent in both industrial and agricultural production by 1962 and a 50 percent increase in national income. A sample of the target figures for key industries and agricultural products indicates the confidence of the planners: coal, 190-210 million tons; steel, 10.5-12 million tons, electricity, 40,000-43,000 million kwh; crude oil, 5-6 million tons; grain, 275 million tons; and cotton, 2.4 million tons.[13]

13. Hughes and Luard, 31, 64-65.

In harmony with the rapid economic growth, the government radically revamped the system of higher education with a view to producing larger numbers of engineers and technicians in a shorter time. Liberal arts education was discouraged in favor of technical education, and many technical institutes were created at the expense of general universities. The curriculum was revised, and departments within the universities and institutes were reorganized to allow the student greater concentration on a specialty. Thus, specialized knowledge in a narrow field was preferred to general education. According to the study of a noted American scientist, 90 percent of China's quarter of a million scientists and engineers in 1960 had been trained since the Communist take-over in 1949, and in 1960, China graduated about 75 percent as many engineers as the United States.[14]

THE "GREAT LEAP" AND THE COMMUNE. No sooner had the Second Five-Year Plan begun than the government plunged ahead in a new feverish drive to accelerate the expansion of the already overheated economy. In early February 1958, the National People's Congress announced a "Great Leap Forward" Movement for the next three years, calling for a 19 percent increase in steel production, 18 percent in electricity, and 17 percent in coal output for 1958. Mao talked about catching up with or even surpassing the British industrial capacity in fifteen years, i.e. 1972. Buoyed by optimism, the exuberant planners repeatedly revised the production targets upward in the ensuing months in hopes of achieving an unprecedented rate of growth. The steel quota was raised from 6.2 million tons in February 1958 to 8-8.5 million in May and to 10.7 million in August. A general increase of 33 percent in industrial output was confidently predicted for the year.[15] To achieve this phenomenal development record, everyone was urged to participate in industrial production, and in so doing everybody, regardless of his background—government official, peasant, student, professor, worker, etc.—became a proletarian. By the fall of 1958, some 600,000 backyard furnaces had sprung up all over the country.

Along with this frenzied drive for industrialization, the government took a further step toward socialist transformation by the creation of People's Communes. In the spring of 1958, piecemeal amalgamation of agricultural producers' cooperatives had already begun in Hopeh, Honan, and parts of Manchuria; by July the movement reached a "high tide," and the term "People's Commune" formally appeared. Mao and other high offi-

14. John A. Berberet, *Science and Technology in Communist China* (Santa Barbara, 1960), 3.
15. Hughes and Luard, 66-69.

cials inspected some of the early models in Honan and Hopeh, and on August 29 the CCP Central Committee officially announced the birth of People's Communes. By November 1958 there were 26,000 communes embracing 98 percent of the farm population. On the average each rural commune consisted of some thirty cooperatives of about 5,000 households, or 25,000 people. It assumed the administrative functions of the villages; controlled the area's agricultural as well as industrial resources; collected taxes; and operated schools, banks, nurseries, public kitchens, old folks homes, public cemeteries, etc. It appropriated all private properties such as land, houses, and livestock. However, the family institution was not destroyed and members continued to live under the same roof. Only those who were single, widowed, or childless lived in the communal facilities. The size of the commune was later reduced and by the early 1960s there were 74,000 of them, each maintaining numerous production brigades and teams to increase agricultural and industrial output.

In parallel existence with rural communes were urban communes, of which the Red Flag Commune of Chengchow in North China stood out as a model. It was created in August 1958 with 4,134 households—comprising 18,729 people—and centered around the Chengchow Spinning and Weaving Machinery Factory. Collective living began with the moving of the workers to the factory area, around which the commune developed clothing stores, public dining halls, child care centers, nurseries, hospitals, schools, parks, banks, and movie theatres. There were also old folks homes, savings banks, and farms where vegetables were grown and pigs and poultry were raised for the public mess halls. Since 80 percent of the women were employed, "livelihood service stations" and neighborhood service units were indispensable. They were run mostly by elderly persons, who performed chores for a small fee, such as paying bills, mending clothes, cleaning the houses, baby-sitting, and caring for the sick. Organizationally, the head of the Chengchow Spinning and Weaving Machinery Factory served also as the head of the commune, while the CCP branch at the factory was concurrently the party committee of the commune, over which it maintained close control. The various departments of the commune included industry, agriculture, finance, planning, civil security, welfare, sanitation, and culture. There were, of course, production teams—organized along military lines as regiments, battalions, and platoons—to raise industrial, agricultural, and every other aspect of output.[16]

Historically, only two commune experiments had ever been attempted

16. Janet Salaff, "The Urban Communes and Anti-city Experiment in Communist China," *The China Quarterly* (January-March 1967), 82-110.

and both soon expired as ignominious failures. They were the Paris Commune of 1871, which lasted seventy-three days from March 17 to May 28, and the peasant communes in the Soviet Union during the early revolutionary period. In 1930 Stalin pronounced communes unfit for the socialist present, though ideal for the distant future. Mao was of course not unaware of these experiences, but he seems to have been more influenced by the late Ch'ing reformer K'ang Yu-wei's work, *Ta-t'ung shu* (The Book of Universal Commonwealth), which was inspired by the ancient work "The Evolution of Li" (*Li-yün*) of the *Book of Rites* (*Li-chi*). In the *Ta-t'ung shu*, K'ang argued for the creation of a utopia in which there would be no private property, no private ownership, no sale of land, no private industry, no private commerce; in which there would be public hospitals, public maternity wards, public welfare, public education, public homes for the aged, and public cemeteries. A basic feature of this utopia was the destruction of the family and the emancipation of women from servitude in the kitchen.[17] It was no coincidence that Mao described the characteristic of the commune as *ta* (grand) and *kung* (public), which were the key concepts in the opening passage of the "Evolution of Li": "When the *Grand* Course was pursued, a *public* and common spirit ruled all under the sky."[18] Hailing the introduction of the commune as "the morning sun above the broad horizon of East Asia," the Chinese Communists confidently boasted that "the attainment of communism in China is no longer a remote event."[19]

As a result of the Great Leap and the introduction of the communes, the government proudly announced at the end of 1958 that industrial production for the year had surpassed that of 1957 by 65 percent. Machine tools had trebled; coal and steel had doubled; oil had increased by 50 percent and electricity by 40 percent. Even given the unavoidable exaggeration of the figures, the progress made was considerable. Yet, much of the quality was sacrificed for quantity as the government itself later admitted. In August 1959, 3 million of the 11 million tons of steel produced in 1958 was pronounced unfit for industrial use—backyard furnaces simply did not perform the same function as the giant steel mill. Amid the utopian dreams of instant development, a new feeling of pragmatism be-

---

17. For contents of the *Ta-t'ung shu*, see Liang Ch'i-ch'ao, *Intellectual Trends*, 95-98. For K'ang's influence on Mao, see Wen-shun Chi, "The Ideological Source of the People's Communes in Communist China," *Pacific Coast Philology*, II (April 1967), 62-78.
18. Translation by James Legge, *The Sacred Books of China*, Part III, *The Li Ki* (Oxford, 1885), 364. Italics added.
19. Benjamin I. Schwartz, "China and the Communist Bloc: A Speculative Reconstruction," *Current History*, 35:208:326 (Dec. 1958).

gan to emerge which stressed realism in planning and expertise in technological operations. It was becoming apparent that authentic economic progress needed far more than ideological power to become a reality.

Western sources generally agreed that Chinese economic growth in the 1950s was quite impressive but less so in the 1960s. A noted American economist estimated that China's Gross National Product (GNP) rose from 73.8 billions of *yüan* in 1952 to 123.4 billions *yüan* in 1959, an increase of 70 percent, compared with 30 percent between 1959 and 1970 (171.4 billions). Taking the period 1952-70 as a whole, the annual growth rate was about 4 to 4.5 percent, a respectable though not spectacular performance.[20]

Social and Psychological Control

A basic ingredient of Maoism is the continuous organization of mass movements for the attainment of specific objectives predetermined by the party. Indeed, surges of mass campaigns punctuate the daily rhythm of life in Communist China, and the Chinese people, once described as a pile of loose sand, are now more tightly organized than any other national population in the world. Practically everybody belongs to some mass organization through which the party and the government exercise their control and carry out national policy. In addition, their monopoly of the communications media and the omnipresence of their security police and party cadre have combined to make the society a watertight compartment unprecedented in the history of China. Under such strict control freedom is anathema except when it serves the interests of the state.

The mass organizations are actually semigovernmental bodies of gigantic size. Foremost among them in 1953 were the All-China Federation of Democratic Youth, with a membership of 18 million; the All-China Federation of Trade Union, 10.2 million; the All-China Democratic Women's Federation, 76 million; and the All-China Students' Federation, 3.29 million. In addition, the Young Pioneers, which included children between the ages of 9 and 14, claimed a membership of 8 million, and the Democratic Youth League, which spanned the age group 14 to 25, boasted 12 million. Through these gigantic organizations, the government indoctrinated the people and organized them for demonstrations, parades, and drives, such as the Resist-America Aid-Korea Campaign in 1951, the Three-Anti (*San-fan*) Movement in 1951 to combat corruption, waste,

20. Value indicators in 1952 prices; exchange rate at 2.46 *yüan* to the dollar in 1952. Alexander Eckstein, "Economic Growth and Change in China: A Twenty-Year Perspective." *The China Quarterly* (April-June 1973), 54:232, 234-35.

and bureaucratism, and the Five-Anti (*Wu-fan*) Movement in 1952 to fight bribery, tax evasion, fraud, theft of government property, and leakage of state economic secrets. The greatest of the mass campaigns was perhaps the organization in 1966 of hundreds of thousands of students into the Red Guards to combat the Anti-Maoists.

THE NEW SOCIALIST MAN. The new society under the Communists encouraged the forging of a new life style (*tso-feng*) and the creation of a new Socialist Man. People were urged to mind not only their own business but to check on one another's thoughts and actions, to attend numerous political gatherings, and to participate in "learning" and "struggle" sessions. The Socialist Man was to have no regard for face, be prepared to make public confessions, and put the state before his family. His whole being was irrevocably dedicated to advancing the cause of the proletarian revolution rather than to seeking individual advancement or bringing honor to his family, as in the feudal past. Article 42 of the Common Program demanded that the new man have Five Loves—love of the fatherland, love of people, love of labor, love of science, and love of public property. Although there was fear that such a Socialist Man might be too austere and unnatural—one worthy of awe more than of love—this misgiving was somewhat relieved by the discovery that he still retained vestiges of humanness.

PSYCHOLOGICAL CONTROL. In connection with the molding of a new style of life, there developed a subtle if devastating indoctrination program known in journalistic jargon as "brainwashing" (*hsi-nao*).[21] It is a process of psychological coercion based on Pavlov's theory that environmental conditioning can alter human will and remold the character of the individual. Thus, brainwashing is used not only to convert enemies and extract confessions, but also to indoctrinate party cadres and transform intellectuals so that they can serve the state rather than be liquidated as in postrevolutionary Russia.

The indoctrination process usually lasts from several months to a year, depending on the intensity of the objective, but in all cases it takes place in a remote controlled camp where the individual, completely isolated from the outside world, is deprived of all sense of security. Upon arriving at the camp, the trainees are grimly impressed with the impossibility of

21. Edward Hunter, *Brainwashing in Red China* (New York, 1951); Richard L. Walker, *China under Communism: The First Five Years* (New Haven, 1955), ch. 3; Robert J. Lifton, *Thought Reform and the Psychology of Totalism: A Study of "Brainwashing" in China* (New York, 1961).

retreat. They are divided into small groups, each under the guidance of an activist. They are assigned heavy physical labor to insure fatigue, thereby weakening their will to resist. In this condition they study and criticize each other's background and life history. During this initial period of about two months, food and living quarters are very poor.

In the next stage of three or four months, food and living conditions improve, while physical exertion is somewhat reduced but remains sufficient to guarantee fatigue at the end of the day. There are more study sessions and group meetings, where the insignificance of the individual and the omnipotence of the party are stressed. The works of Marx, Lenin, Stalin, and Mao become the new Bible. The past is depicted as dark, corrupt, and decadent, while the new life under the Communist regime represents liberation and progress and provides chances for a new and meaningful existence. Class struggle and the inevitability of ultimate party victory are continuously impressed upon the trainees.

At the end of the second period, the trainee in all likelihood experiences an emotional crisis, whereupon he comes to the conclusion that there is no point in hiding or resisting; the party will win anyway. So he seeks liberation by releasing his feelings and accepting wholeheartedly the party and what it represents. With this, a heavy burden is lifted and he is reborn. He finds new meaning in the Communist jargon and propaganda, and is anxious to help others have the same experience, partly to justify his own conversion. Consolidation of this state of mind requires about four months. When the indoctrination is completed, a quarter of the graduates are sent for further schooling, while the remainder go into society to organize and lead the public. The whole society, in fact, is a laboratory of mass control.

Through mass organizations, secret police, mass communications media, and indoctrination, the government has succeeded in controlling and remolding the society and the people to an extent unknown in Chinese history. The once individualistic Chinese have become the most regimented people in the world.

Foreign Relations

Though deeply committed to international communism, Mao and his followers were national communists at heart, keenly aware of China's misfortunes of the past century. Like Chinese of all persuasions, they were fired with a burning desire to restore China's rightful position under the sun, to achieve the big-power status denied it since the Opium War, and to revive the national confidence and self-respect that had been lost during a century of foreign humiliation. The new regime from its inception adopted a bold stance toward the Western powers, to show that it was not afraid of them as were the Manchu and the Nationalist governments, which meekly bowed before the imperialist gunboats.

A major source of strength during the early years of the People's Republic was its close ties with the Soviet Union. In 1949 Mao unequivocally announced his "lean to one side" policy: "The Chinese people must lean either to the side of imperialism or to the side of socialism. There can be no exception. There can be no sitting on the fence; there is no third path." This policy was prompted not only by ideological affinity but also by practical considerations: the infant People's Republic needed Soviet aid and protection to forestall Western intervention, such as that which occurred in Siberia after the Bolshevik Revolution. Thus, Mao went to the Soviet Union shortly after his rise to power in 1949—his first trip abroad—to seek a treaty of friendship and alliance. The Moscow-Peking Axis, formally established on February 14, 1950, became the cornerstone of the People's Republic's foreign policy for the greater part of its first decade. Stalin granted Mao a military alliance, $300 million in credit, and a promise to provide experts to help with Chinese industrialization and military modernization. Mao in 1952 eulogized the Axis as "lasting, unbreakable, and invincible," while Liu Shao-ch'i paid tribute to it in the following glowing terms: "There is no deceit, competition, mutual exclusion or extortion between each other, or oppression and plunder of the one by the other—as is inherent among the capitalist countries." Despite their own urgent needs, the Soviets dispatched to China large numbers of scientists, technicians, and military advisers: 1,000 to 25,000 each in the Chinese air force and navy, 5,000 to 10,000 in the Chinese army in 1953, and some 400,000 in various industrial enterprises and plants by February 1954. When Khrushchev and Bulganin visited Peking in 1954 they agreed to help build 156 production enterprises. By 1955 Moscow accorded China the position of associate leader in the international Communist movement, as Foreign Minister Molotov noted before the Council of Ministers: "The most important result of the Second World War was the organization, side by side with the world capitalist camp, of the world camp of socialism and democracy headed by the Soviet Union—to speak more truly, headed by the Soviet Union and the Chinese People's Republic."[22]

The mutual profession of friendship and cordiality was reflected in many ways. Acts of Soviet friendship included the return to China of the Manchurian "loot" in 1952 and the Soviet share of the joint ownership

22. Peter S. H. Tang, *Communist China Today* (New York, 1957), 378-81, 383.

of the Chinese Eastern Railway in 1953, the renunciation of the Soviet right to use Port Arthur as a naval base, and the abandonment of Soviet "joint stocks" in Sinkiang enterprises in 1955. A further expression of generosity was made in 1957 when the Soviets agreed to assist China in its nuclear development, and a year later sent a heavy-water-type reactor. Peking, on its part, helped maintain Soviet primacy in the socialist bloc by mediating the differences between Moscow and its East European satellites in 1956, as we will see in a later section.

Peking's relations with other Asian states reflected its intense drive for leadership in Asia. In this respect, China's historical relations with the smaller states on its periphery, especially Korea and Vietnam (Annam)— the leading tributary states during the Ming and Ch'ing periods—clearly influenced Mao and his associates. They unhesitatingly sent a million "volunteers" to aid North Korea against the American "invasion" in 1950 when they themselves had barely set up their government. They acted as a "big brother" to North Vietnam at the 1954 Geneva Conference, and extended to it substantial aid in its war against South Vietnam and the Americans.

As regards the non-Communist states of Asia, such as India, Pakistan, Indonesia, Burma, Laos, and Cambodia, Peking attempted to neutralize them by stressing the theme of coexistence: (1) mutual respect for each other's territory and sovereignty; (2) mutual nonaggression; (3) mutual noninterference in each other's domestic affairs; (4) equality of relationships and mutual benefit; and (5) peaceful coexistence. With these five principles, Peking succeeded to a large extent in keeping these states from aligning with the West. The theme of peaceful coexistence was repeated at the Bandung Conference of twenty-nine Afro-Asian states in 1955, where Chou En-lai won recognition as the upholder of Asian-African nationalism against Western imperialism. The Geneva and Bandung conferences, in effect, accorded China leadership position in the Afro-Asian bloc. As the underdog who has found a viable formula to reverse its destiny and to achieve quick elevation of national status, Peking has become an inspiration to some underdeveloped nations.

From the above survey one gathers that there were at least five factors underlying the Chinese foreign policy: (1) nationalism and the drive for big-power status; (2) the politics of the International Communist movement; (3) domestic politics considerations; (4) Marxist-Leninist-Maoist ideology; and (5) "a strategic-political imagery based on a traditional spatial-ideological world order."[23]

23. Albert Feuerwerker, "Chinese History and the Foreign Relations of Contemporary China," *The Annals of the American Academy of Political and Social Science,* July 1972; *China in the World Today,* 5.

First Signs of Strain

While the first decade of the People's Republic abounded in accomplishments, it was not without stress and strain. The first notes of discord were struck in 1953-54, when two powerful regional authorities challenged the second- and third-ranking leaders in the central government. Kao Kang, party boss of Manchuria,[24] and Jao Shu-shih, chairman of the East China Military and Administrative Committee,[25] jointly espoused the adoption of the Soviet system of economic development, the separation of party officials from industries, and the adoption of a "single director" system at all levels of industrial management. The proposal was in effect an indictment of the current policy of Liu Shao-ch'i and Chou En-lai. The latter countered with the proposition of "collective leadership" in industrial management. The clash was climaxed by the challengers' demand that they and the party general secretary, Teng Hsiao-p'ing, resign. The three defendants, with the help of the mayor of Peking, P'eng Chen, managed to have the challengers expelled from the Central Committee in 1954 on charges of antiparty activities and attempts to create "independent kingdoms." This first power struggle ended with Kao's suicide and Jao's imprisonment.

A second challenge to the state came in 1956 from the public. The seven strict years of Communist rule had generated considerable resentment and repressed emotion, which erupted after the Hungarian Revolt. "Tens of thousands of persons went out to the street to oppose the People's Government," as Mao described it.[26] Partly to afford the people a chance to let off steam—lest there be a Hungarian-type revolt in China—and partly to ferret out the real critics, Mao cleverly declared: "Let hundred flowers blossom; let hundred schools contend!" Many intellectuals naïvely mistook the statement to mean a liberation of expression and spoke their minds. The severe criticism that ensued surpassed the government's expectation. Finding it unbearable and detrimental, Mao clamped down with lightning speed. The critics were caught and although they regretted their imprudence, it was too late to recant. Many were sent to corrective camps or were forced to sign a "socialist self-reform pact" to renew their pledge of allegiance. In the wake of this antirightist campaign, the government in 1957 initiated a "socialist education movement" among the industrial and agrarian population, followed by the dispatch of military

24. Concurrently a vice-chairman of the People's Republic and the chairman of the important State Planning Committee.
25. Also director of the Organization Department of the CCP's Central Committee.
26. Gene T. Hsiao, "The Background and Development of 'The Proletarian Cultural Revolution'," *Asian Survey*, VII:6 (June 1967), 393.

and civil leaders to physical labor as an example to the people. The importance of "redness," i.e. ideology over "expertise," was very much emphasized.

On balance, the first decade of the People's Republic closed with considerable success. Domestically, it had consolidated the control of the country and scored a respectable economic growth. Externally, it had maintained close ties with the Soviet Union and the East European satellite states, fought the United Nations' forces in Korea to a standstill (which was a sort of victory), played the big-power role at Geneva and Bandung, and mediated between the Soviets and the satellites. Nuclear development had also begun. It was in this rosy state of euphoria that Mao introduced the communes in hopes of putting China on a higher ideological plane than the Soviet Union and, eventually, of overtaking it in the race toward the portals of true communism.

Yet beneath this veneer of pride and success, internal dissension over economic policy and divisive tendencies in the Moscow-Peking Axis were already appearing. As the second decade opened in 1959, the government faced the double challenge of worsening relations with the Soviets and an incipient power struggle among its top leaders.

Further Reading

Agunsanwo, Alaba, *China's Policy in Africa, 1958-71* (Cambridge, Eng., 1974).

Arkush, R. David, *Fei Xiaotong and Sociology in Revolutionary China* (Cambridge, Mass., 1981).

Bao, Ruo-wang (Jean Pasqualini), and Rudolph Chelminski, *Prisoner of Mao* (New York, 1973).

Bennett, Gordon, *Yundong: Mass Campaigns in Chinese Communist Leadership* (Berkeley, 1976).

————, *Huadong: The Story of a Chinese People's Commune* (Boulder, 1978).

Bianco, Lucien, *Origins of the Chinese Revolution, 1915-1949*, tr. from the French by Muriel Bell (Stanford reprint, 1972).

Boardman, Robert, *Britain and the People's Republic of China, 1949-1974* (New York, 1976).

Buck, David D., *Urban Change in China: Politics and Development in Tsinan, Shantung* (Madison, 1978).

Burki, Shabid J., *Study of Chinese Communes, 1965* (Cambridge, Mass., 1970).

Chai, Winberg, *The Foreign Relations of the People's Republic of China* (New York, 1972).

Chan, Anita, Richard Madsen, and Jonathan Unger, *Chen Village: The Recent History of a Peasant Community in Mao's China* (Berkeley, 1984).

Chan, Leslie W., *The Taching Oilfield: A Maoist Model for Economic Development* (Canberra, 1974).

Chao, Kang, *Agricultural Production in Communist China, 1949-1965* (Madison, 1971).

————, *Capital Formation in Mainland China, 1952-1965* (Berkeley, 1974).

Ch'en, Jerome (ed.), *Mao Papers: Anthology and Bibliography* (London, 1970).

Ch'en, Theodore H. E., *Thought Reform of the Chinese Intellectual* (Hong Kong, 1960).

————, *The Maoist Educational Revolution* (New York, 1974).

Chesneaux, Jean et al., *China: The People's Republic, 1949-1976* (New York, 1979).

Chiu, Hungdah, *The People's Republic of China and the Law of Treaties* (Cambridge, Mass., 1972).

Chow, Gregory C., *The Chinese Economy* (Hong Kong, 1985).

Clark, M. Gardner, *The Development of China's Steel Industry and Soviet Technical Aid* (Ithaca, 1973).

Clough, Ralph N. et al., *The United States, China, and Arms Control* (Washington, D.C., 1975).

Cohen, Arthur A., *The Communism of Mao Tse-tung* (Chicago, 1964).

Cohen, Jerome A., *The Criminal Process in the People's Republic of China, 1949-1963: An Introduction* (Cambridge, Mass., 1968).

———— (ed.), *The Dynamics of China's Foreign Relations* (Cambridge, Mass., 1970).

———— et al., *China Trade Prospects and U.S. Policy*, ed. by Alexander Eckstein (New York, 1971).

Davin, Delia, *Woman-Work: Women and the Party in Revolutionary China* (Oxford, 1976).

Dittmer, Lowell, *China's Continuous Revolution: The Post-Liberation Epoch, 1949-1981* (Berkeley, 1987).

Domes, Jürgen, *The Internal Politics of China, 1949-1972* (New York, 1973).

————, *The Government and Politics of the PRC* (Boulder, 1985).

Dreyer, June Teufel, *China's Forty Millions: Minority Nationalities and National Integration in the People's Republic of China* (Cambridge, Mass., 1976).

Dulles, Foster Rhea, *American Policy Toward Communist China, The Historical Record: 1949-1969* (New York, 1972).

Eckstein, Alexander, *Communist China's National Income* (New York, 1961).

————, "Sino-Soviet Economic Relations: A Re-appraisal," in C. D. Cowan (ed.), *The Economic Development of China and Japan* (London, 1964), 128-59.

————, *Communist China's Economic Growth and Foreign Trade* (New York, 1966).

————, "Economic Growth and Change in China: A Twenty-Year Perspective," *The China Quarterly*, 54:211-41 (April-June 1973).

Feuerwerker, Yi-tsi Mei, *Ding Ling's Fiction: Ideology and Narrative in Modern Chinese Literature* (Cambridge, Mass., 1982).

Fitzgerald, Stephen, *China and the Overseas Chinese: A Study of Peking's Changing Policy, 1949-1970* (Cambridge, Eng., 1973).

Fokkema, D. W., *Literary Doctrine in China and Soviet Influence, 1956-1960* (The Hague, 1965).

Frolic, B. Michael, *Mao's People: Sixteen Portraits of Life in Revolutionary China* (Cambridge, Mass., 1980).

Galbiati, Fernando, P'eng P'ai and the Hai-lu-feng Soviet (Stanford, 1985).

Gamberg, Ruth, Red and Expert: Education in the People's Republic of China (New York, 1977).

Gittings, John, The Role of the Chinese Army (London, 1967).

Griffin, P., Chinese Communist Treatment of Counterrevolutionaries (Princeton, 1976).

Gurley, John R., China's Economy and the Maoist Strategy (New York, 1976).

Hao, Yufan, and Zhai Zhihai, "China's Decision to Enter the Korean War," The China Quarterly, March 1990, 94-115.

Harding, Henry, Organizing China: The Problem of Bureaucracy, 1949-1976 (Stanford, 1981).

Harrison, James P., The Communists and Chinese Peasant Rebellions: A Study in the Rewriting of Chinese History (New York, 1971).

————, The Long March to Power: A History of the Chinese Communist Party, 1921-72 (New York, 1972).

Hinton, William, Fanshen: A Documentary of Revolution in a Chinese Village (New York, 1966).

Ho, Kan-chih, A History of the Modern Chinese Revolution (Peking, 1959).

Howe, Christopher, Wage Patterns and Wage Policy in Modern China 1919-1972 (London, 1973).

Hsiao, Gene T., The Foreign Trade of China: Policy, Law, and Practice (Berkeley, 1977).

Hsiao, Katharine Huang, Money and Monetary Policy in Communist China (New York, 1971).

Hsiung, James Chieh, Ideology and Practice: The Evolution of Chinese Communism (New York, 1970).

————, Law and Policy in China's Foreign Relations: A Study of Attitudes and Practices (New York, 1972).

———— (ed.), The Logic of "'Maoism'": Critiques and Explication (New York, 1974).

Hsü, Kai-yü, Chou En-lai: China's Gray Eminence (New York, 1968).

Huang, Joe C., Heroes and Villains in Communist China: The Contemporary Chinese Novel as a Reflection of Life (New York, 1973).

Huck, Arthur, The Security of China: Chinese Approaches to Problems of War and Strategy (New York, 1970).

Israel, John and Donald K. Klein, Rebels and Bureaucrats: China's December Wars (Berkeley, 1976).

Johnson, Cecil, Communist China and Latin America, 1959-1967 (New York, 1970).

Johnson, Chalmers, "Building a Communist Nation in China" in Robert A. Scalapino (ed.), The Communist Revolution in Asia (Englewood Cliffs, 1965), 47-81.

———— (ed.), Ideology and Politics in Contemporary China (Seattle, 1973).

Kahn, Harold, and Albert Feuerwerker, "The Ideology of Scholarship: China's New Historiography," The China Quarterly, 22:1-13 (April-June 1965).

Karnow, Stanley, Mao and China: from Revolution to Revolution (New York, 1973).

Kau, Ying-mao, The People's Liberation Army and China's Nation Building (White Plains, N.Y., 1973).

Kierman, Frank A., Jr., and John K. Fairbank (eds.), *Chinese Ways in Warfare* (Cambridge, Mass., 1974).

Kirby, R. J. R., *Urbanization in China: Town and Country in a Developing Economy, 1949-2000 A.D.* (New York, 1985).

Kuo, Leslie T. C., *Agriculture in the People's Republic of China: Structural Changes and Technical Transformation* (New York, 1976).

Lall, Arthur, *How Communist China Negotiates* (New York, 1968).

Lardy, Nicholas R., *Agriculture in China's Modern Economic Development* (Cambridge, Eng., 1983).

———, and Kenneth Lieberthal (eds.), *Chen Yun's Strategy for China's Development* (Armonk, N.Y., 1983).

Larkin, Bruce D., *China and Africa, 1949-1970: The Foreign Policy of the People's Republic of China* (Berkeley, 1971).

Lee, Rance P. L., and Lau Siu-kai (eds.), *The People's Commune and Rural Development* (Hong Kong, 1981).

Lewis, John Wilson, *Leadership in Communist China* (Ithaca, 1963).

——— (ed.), *Party Leadership and Revolutionary Power in China* (Cambridge, Eng., 1970).

Leys, Simon, *Chinese Shadows* (tr. from French ed., 1974), (New York, 1977).

Li, Choh-ming, *Economic Development in Communist China: An Appraisal of the First Five Years of Industrialization* (Berkeley, 1959).

———, *The Statistical System of Communist China* (Berkeley, 1962).

Li, Tien-min, *Chou En-lai* (Taipei, 1970).

Li, Victor H. (ed.), *Law and Politics in China's Foreign Trade* (Seattle, 1977).

———, *Law without Lawyers: A Comparative View of Law in the United States and China* (Boulder, 1978).

Li Jui (Rui), *Lu-shan hui-i shih-lu* (*Lushan huiyi shilu*) [A true account of the Lu-shan Conference] (Changsha, Hunan, 1989).

Lifton, Robert J., *Thought Reform and the Psychology of Totalism: A Study of "Brainwashing" in China* (New York, 1961).

Liu, Alan P. L., *Communications and National Integration in Communist China* (Berkeley, 1971).

———, *Political Culture and Group Conflict in Communist China* (Santa Barbara, 1976).

Liu, Ta-chung, *The Economy of the Chinese Mainland: National Income and Economic Development, 1933-1959* (Princeton, 1965).

Lowe, Donald M., *The Function of "China" in Marx, Lenin, and Mao* (Berkeley, 1966).

Lyons, Thomas P., *Economic Integration and Planning in Maoist China* (New York, 1987).

Ma, Laurence J. C., and Edward W. Hanten (eds.), *Urban Development in Modern China* (Boulder, 1981).

MacFarquhar, Roderick, *The Hundred Flowers Campaign and the Chinese Intellectuals* (New York, 1960).

———, *The Origins of the Cultural Revolution*, vol. 2, *The Great Leap Forward, 1958-1960* (New York, 1983).

Mao, Tse-tung, *New Democracy* (New York, 1945).

———, *Selected Works of Mao Tse-tung*, 4 vols. (London, 1954).

"Mao's 2 Telegrams on Korea," *The New York Times*, Feb. 26, 1992, A4.

Madsen, Richard, *Morality and Power in a Chinese Village* (Berkeley, 1984).

Marshall, Marsh, *Organization, and Growth in Rural China* (New York, 1985).

McDougall, Bonnie S. (ed.), *Popular Chinese Literature and Performing Arts in the People's Republic of China, 1949-1979* (Berkeley, 1984).

Meisner, Maurice, *Mao's China: A History of the People's Republic* (New York, 1977).

Moody, Peter R., *Opposition and Dissent in Contemporary China* (Stanford, 1977).

Mozingo, David, *Chinese Policy toward Indonesia, 1949-1967* (Ithaca, 1976).

Mu, Fu-sheng, *The Wilting of the Hundred Flowers: The Chinese Intelligentsia under Mao* (New York, 1962).

Mueller, Peter G., and Douglas A. Ross, *China and Japan: Emerging Global Powers* (New York, 1975).

Munro, Donald J., "Chinese Communist Treatment of the Thinkers of the Hundred Schools Period," *The China Quarterly*, 24:119-40 (Oct.-Dec. 1965).

Nee, Victor, and David Mozingo (eds.), *State and Society in Contemporary China* (Ithaca, 1983).

Nelsen, Harvey W., *The Chinese Military System: An Organizational Study of the Chinese People's Liberation Army* (Boulder, 1977).

Oksenberg, Michel (ed.), *China's Developmental Experience* (New York, 1973).

Orleans, Leo A., *Every Fifth Child: The Population of China* (Stanford, 1972).

———, *China's Experience in Population Control: The Elusive Model* (Washington, D.C., 1974).

——— (ed. with intro.), *Chinese Approaches to Family Planning* (White Plains, N.Y., 1980).

Parish, William L., and Martin King Whyte, *Village and Family in Contemporary China* (Chicago, 1978).

Perkins, Dwight H., *Market Control and Planning in Communist China* (Cambridge, Mass., 1966).

——— (ed.), *China's Modern Economy in Historical Perspective* (Stanford, 1975).

——— (ed.), *Rural Small-scale Industry in the People's Republic of China* (Berkeley, 1977).

Perry, Elizabeth J., *Rebels and Revolutionaries in North China: 1845-1945* (Stanford, 1980).

Pincus, Fred L., *Education in the People's Republic of China* (Baltimore, 1975).

Price, Jane L., *Cadres, Commanders and Commissars: The Training of the Chinese Communist Leadership, 1920-1945* (Boulder, 1976).

Printz, Peggy, and Paul Steinle, *Commune: Life in Rural China* (New York, 1973).

Rádvanyi, János, "The Hungarian Revolution and the Hundred Flowers Campaign," *The China Quarterly*, 43:121-29 (July-Sept. 1970).

Rice, Edward E., *Mao's Way* (Berkeley, 1972).

Schram, Stuart R., *The Political Thought of Mao Tse-tung* (New York, 1963).

———, *Mao Tse-tung* (New York, 1966).

———, "Mao Tse-tung and the Theory of Permanent Revolution, 1958-69," *The China Quarterly*, 46:221-44 (April-June 1971).

——— (ed. and intro.), *Chairman Mao Talks to the People: Talks and Letters, 1956-1971* (New York, 1974; English edition: *Chairman Mao Unrehearsed*).

Schurmann, Franz, *Ideology and Organization in Communist China* (Berkeley, 1966).

Schwartz, Benjamin, I., "On the 'Originality' of Mao Tse-tung," *Foreign Affairs* (Oct. 1955), 67-76.

———, "The Maoist Image of World Order" in John C. Farrell and Asa P. Smith (eds.), *Image and Reality in World Politics* (New York, 1968), 92-102.

———, *Communism and China: Ideology in Flux* (Cambridge, Mass., 1968).

Selden, Mark (ed.), *The People's Republic of China: A Documentary History of Revolutionary Change* (New York, 1979).

Shabad, Theodore, *China's Changing Map: A Political and Economic Geography of the Chinese People's Republic* (New York, 1956).

Sit, Victor F. S. (ed.), *Chinese Cities: The Growth of the Metropolis since 1949* (Hong Kong, 1985).

Solinger, Dorothy J., *Regional Government and Political Integration in Southwest China, 1949-1954: A Case Study* (Berkeley, 1977).

———, *Chinese Business Under Socialism: The Politics of Domestic Commerce, 1949-1980* (Berkeley, 1984).

Stacey, Judith, *Patriarchy and Socialist Revolution in China* (Berkeley, 1984).

Starr, John Bryan, "Revolution in Retrospect: The Paris Commune Through Chinese Eyes," *The China Quarterly*, 49:106-25 (Jan.-March, 1972).

———, *Continuing the Revolution: The Political Thought of Mao* (Princeton, 1979).

Teiwes, Frederick C., *Politics and Purges in China: Rectification and the Decline of Party Norms, 1950-1965* (White Plains, N.Y., 1980).

Townsend, James R., *Political Participation in Communist China* (Berkeley, 1967).

———, *Politics in China* (Boston, 1974).

Tsou, Tang, *Embroilment over Quemoy: Mao, Chiang, and Dulles* (Salt Lake City, 1959).

Tucker, Nancy Bernkopf, *Patterns In The Dust. Chinese-American Relations and The Recognition Controversy, 1949-1950* (New York, 1983).

Vogel, Ezra F., *Canton under Communism: Programs and Politics in a Provincial Capital, 1949-1968* (Cambridge, Mass., 1969).

Vohra, Ranbir, *Lao She and the Chinese Revolution* (Cambridge, Mass., 1974).

Wakeman, Frederic, Jr., *History and Will: Philosophical Perspectives of Mao Tse-tung's Thought* (Berkeley, 1973).

Waller, Derek J., *The Government and Politics of Communist China* (Garden City, N.Y., 1970).

Watson, James L. (ed.), *Class and Social Stratification in Post-Revolution China* (New York, 1984).

White, Lynn T., III, *Careers in Shanghai: The Social Guidance of Personal Energies in a Developing Chinese City, 1949-1966* (Berkeley, 1978).

Whiting, Allen S., *China Crosses the Yalu: The Decision to Enter the Korean War* (Stanford, 1968).

Whitson, William W., *Chinese Military and Political Leaders and the Distribution of Power in China, 1956-1971* (New York, 1972).

Wittfogel, Karl A., "Some Remarks on Mao's Handling of Concepts and Problems of Dialectics," *Studies in Soviet Thought*, III:4:251-77 (Dec. 1963).

Wolf, Margery, and Roxane Witke (eds.), *Women in Chinese Society* (Stanford, 1975).

Wong, John, *Land Reform in the People's Republic of China: Institutional Transformation in Agriculture* (New York, 1973).

Yang, C. K., *The Chinese Family in the Communist Revolution* (Cambridge, Mass., 1959).

Yin, John, *Government of Socialist China* (Lanham, Md., 1984).

Young, Marilyn B., *Women in China* (Ann Arbor, 1973).

Young, Kenneth T., *Negotiating with the Chinese Communists: The United States Experience, 1953-1967* (New York, 1968).

Yu, George T., *China's African Policy: A Study of Tanzania* (New York, 1975).

Zhai, Qiang, "China and the Geneva Conference of 1954," *The China Quarterly,* March 1992, 103-122.

27

The Sino-Soviet Split

The Moscow-Peking Axis, characterized as a formidable front of international proletarian solidarity, deteriorated rapidly toward the end of the 1950s. It is an enigma that a relationship once described as "lasting, unbreakable, and invincible" degenerated so quickly into a bitter ideological dispute and fierce border clashes, which shattered the myth of monolithic world Communism and threatened the disintegration of the Socialist bloc. Historians are intrigued by the problem of explaining this complex development.

The Historical Roots of the Conflict

From the infancy of Chinese Communism, Mao's contact with Moscow was neither pleasant nor gratifying. His unorthodox method of revolution, based on peasant mobilization in the countryside, was tolerated by Moscow as legitimate only because all other types of Communist insurrection in China had failed. Mao's approach was never endorsed by Stalin as proper for revolutionizing China. Stalin continued to favor Chinese who had studied in the Soviet Union, such as Ch'ü Ch'iu-pai, Li Li-san, and the Twenty-eight Bolsheviks, putting them in charge of the Chinese Communist Party while keeping Mao out of the Central Committee. Even after Mao became the de facto leader of Chinese Communism in 1936 after the Long March, Stalin remained reluctant to accept his leadership until 1938. Even then Stalin continued to regard Wang Ming as a potential alternative to Mao until 1945.

Although respecting Stalin as a builder of socialism, Mao had witnessed the ineptness of the Comintern's China policy in the 1920s and early 1930s. Therefore, he had no confidence in Moscow's judgment and rejected Stalin's "authority as a political and military strategist for the Chinese revolution."[1] In 1962, recalling events three decades earlier, Mao stated: "It is we Chinese who have achieved understanding of the objective world of China, not the comrades concerned with Chinese questions in the Communist International. These comrades in the Communist International simply did not understand, or we could say they utterly failed to understand, Chinese society, the Chinese nation, or the Chinese revolution."[2] Only a Chinese who understood the Sinification of Marxism could lead the Chinese revolution to success. Thus, Mao fought the battle in his own way and achieved success not because of, but in spite of, Stalin's support. There was no rapport between them; in fact, there was hidden mutual distrust and antipathy. Stalin considered Mao defective in his understanding of Marxism, lacking an international perspective, and limited in his revolutionary experience.[3] Nikita Khrushchev recalled that "Stalin was always fairly critical of Mao," calling him a "margarine Marxist" (*peschany marksist*).[4] Other Soviet references to Chinese Communists as "cabbage Communists" and "radish Communists"—red on the outside but white on the inside—belittled Chinese Communism and hurt Mao's pride. As early as 1936 Mao told Edgar Snow: "We are certainly not fighting for an emancipated China in order to turn the country over to Moscow!"[5]

Toward the end of World War II, Mao and Chou En-lai looked to the future with the goal of avoiding complete dependence on the Soviet Union. In January 1945, they secretly proposed a visit to Washington to confer with Roosevelt as "leaders of a primary Chinese party." They wanted to convince the President that they, not the KMT, represented the future of China. They desired a coalition government, access to American aid following the model of Tito, recognition by the United States as a major party, not an outlaw, and the belligerent status allowing them to participate in the postwar China arrangements and the organization of the United Nations. Most importantly, if Chiang refused coalition, they

1. Benjamin Schwartz, "China's Developmental Experience, 1949-72" in Michel Oksenberg (ed.), *China's Developmental Experience* (New York, 1973), 19.
2. Stuart R. Schram, "Introduction: The Cultural Revolution in Historical Perspective" in Stuart R. Schram (ed.), *Authority Participation and Cultural Change in China* (Cambridge, Eng., 1973), 15-16.
3. *Ibid.*, 16-17.
4. Nikita Khrushchev, *Khrushchev Remembers* (Boston, 1970), 462.
5. Schram, 18.

wanted to know the possibilities for American support of the Chinese Communist Party.[6]

Mao's request was relayed to Roosevelt after some delay by Ambassador Patrick J. Hurley with the comment that military cooperation with Yenan would constitute "recognition of the Communist Party as an armed belligerent" and lead to the "destruction of the National Government. . . . chaos and civil war, and a defeat of America's policy in China."[7] The President, receiving the message on January 14, 1945, was preoccupied with preparation for the Yalta Conference and problems arising from the approaching Allied victory, including the postwar treatment of Germany, war crimes, the Soviet demand for 16 seats in the United Nations, and the Polish border question. Moreover, only five months after Stilwell's recall he wanted no new complications in his relations with Chiang Kaishek. The President therefore was not receptive to Mao's request.[8]

Although the visit to Washington did not materialize, Mao continued to talk with John S. Service, political officer of the Dixie Mission in Yenan, about the need for American aid. On March 13, 1945, Mao stated:

> China's greatest post-war need is economic development. She lacks the capitalistic foundation necessary to carry this out alone. . . . America and China complement each other economically: they will not compete.
>
> China needs to build up light industries to supply her own market and raise the living standards of her own people.
>
> America is not only the most suitable country to assist this economic development of China: she is also the only country fully able to participate. For all these reasons, there must not and cannot be any conflict, estrangement or misunderstanding between the Chinese people and America.[9]

The designation of the United States as the "most suitable" and "the only country" to assist in China's postwar economic development underscored Mao's desire not to rely solely on Soviet assistance. The Americans were unresponsive to his suggestion.

Near the end of World War II Soviet troops poured into Manchuria, stripping industrial facilities as "war booty" to a replacement cost of $2

6. Barbara W. Tuchman, "If Mao had come to Washington: An Essay in Alternatives," *Foreign Affairs* (Oct. 1972), 44, 50-51, 58.
7. *Ibid.*, 55.
8. *Ibid.*, 50-51, 56.
9. *Foreign Relations of the United States, Diplomatic Papers, 1945, Vol. VII, The Far East: China* (Washington, 1969), 273 ff, report by John S. Service on a conversation with Mao Tse-tung, March 13, 1945.

billion and confiscating $3 billion in bullion and $850 million in Manchurian *yüan*.[10] Although Soviet troops turned over a large quantity of captured Japanese weapons to the Chinese Communists, they did not set the latter up for control of the Northeast and, in fact, sometimes conflicted with them.[11] Mao was also displeased by the Soviet treaty of friendship with Chiang Kai-shek, which undercut his position, and by Stalin's advice that the Communists cooperate with the Nationalists. Bitterly Mao recalled the event at the 8th CCP Central Committee meeting on September 28, 1962:

> The roots [of the conflict between the Soviet Union and Communist China] were laid long before. They [the Soviet Communists] did not allow China to make [Communist] revolution. This was in 1945, when Stalin refused to permit the Chinese [Communist] revolution by saying that we should not engage in any civil war and that we must collaborate with Chiang Kai-shek. Otherwise, the Republic of China will collapse. At that time, we did not adhere to that, and the revolution was victorious.[12]

Even as Mao was completing the conquest of the mainland in 1949, Stalin schemed to have Sinkiang declare independence, much as Outer Mongolia did in 1921, with his guarantee of diplomatic recognition and subsequent incorporation into the Soviet Union as an autonomous republic. The plot failed for lack of cooperation from the Nationalist Commander in Sinkiang.[13] Sinkiang, therefore, became part of the People's Republic of China, narrowly escaping the fate of Outer Mongolia.

In December 1949, shortly after the Communist victory, Mao travelled to Moscow to celebrate Stalin's 70th birthday and to seek aid and alliance. Stalin ignored him for days until Mao threatened to leave.[14] Only then did the Soviet leader agree to talk, but the ensuing negotiations involved a long, hard struggle. The Treaty of Friendship, Alliance, and Mutual Assistance signed on February 14, 1950 as well as related agreements, presented the façade of the solidification of a strong Moscow-Peking Axis based on a monolithic structure of world communism; but, in fact, the alliance was more an act of necessity than a demonstration of proletarian internationalism.[15] Mao received far less and gave much more than he had

10. Tai Sung An, *The Sino-Soviet Territorial Dispute* (Philadelphia, 1973), 62.
11. James B. Harrison, *The Long March to Power* (New York, 1972), 379.
12. Quoted in Tai Sung An, 63.
13. General Tao Shih-yüeh.
14. "Khrushchev's Last Testament: Power and Peace," *Time*, May 6, 1974, 44.
15. John Gittings, "The Great-Power Triangle and Chinese Foreign Policy," *The China Quarterly* (July-Sept. 1969), 39:44-45.

anticipated. In return for the alliance and a modest five-year loan of $300 million, Mao had to accept the independence of Outer Mongolia, joint Sino-Soviet exploitation of mineral resources in Sinkiang, joint administration of the Changchun Railway (i.e. the combined Chinese Eastern and Southern Manchurian Railways), and joint use of Port Arthur and Dairen. Stalin behaved like a "new Tzar," and even Khrushchev considered his action "unwise" and "an insult to the Chinese people."[16] The Soviets did not return the Manchurian "loot" until 1952, the Manchurian Railway until 1953, and Port Arthur, Dairen, and the "joint stocks" in the Sinkiang enterprises until 1955. Mao's bitterness over Stalin's demand for special interests in China is made manifest in his report to the 8th Central Committee meeting on September 28, 1962:

> Even after the victory of the Chinese Communists Stalin feared that China would become a Yugoslavia and I would become a Tito. Later on, I went to Moscow in December 1949 to conclude the Chinese-Soviet Treaty of Alliance and Mutual Assistance [of February 14, 1950], which also involved a struggle. Stalin did not want to sign it, but finally agreed after two months of negotiations. When did Stalin begin to have confidence in us? It began in the winter of 1950, when our country became involved in the Resist-America Aid-Korea Campaign [the Korean War]. Stalin then believed that we were not Yugoslavia and not Titoist.[17]

The cost of Stalin's "trust" was high: China sent a million "volunteers" to intervene in the Korean War and had to pay the entire $1.35 billion for the Soviet equipment and supplies necessary for the venture, and Mao lost a son in the war.

During this early stage of the Sino-Soviet alliance, the seed of ideological dispute was already sown. The Soviets unequivocally rejected the Chinese portrayal of Mao as an original contributor to the treasure-house of Marxism-Leninism and that "the road of Mao Tse-tung" was the prototype of Asian revolutionary movements. On November 23, 1949, at the Trade Union Conference of Asian and Australasian Countries in Peking, Liu Shao-ch'i delivered a speech which included the following statement:

> The road taken by the Chinese people in defeating imperialism and in founding the Chinese People's Republic is the road that should be taken by the peoples of many colonial and semi-colonial countries in

16. Khrushchev, *Khrushchev Remembers*, 463.
17. Quoted in Tai Sung An, 66.

their fight for national independence and people's democracy. . . . *This road is the road of Mao Tse-tung.* It can be the basic road for liberation of peoples of other colonial and semi-colonial countries, where similar conditions exist. . . . This is the inevitable road of many colonial and semi-colonial peoples in the struggle for their independence and liberation.[18]

The Chinese were especially pertinacious about two elements of "Mao's road": (1) a broad nationwide anti-imperialist united front led by the working class and the Communist Party; and (2) a national army led by the Communist Party, engaging in a protracted armed struggle and engulfing the cities from the rural areas.

On July 1, 1951, the 30th anniversary of the founding of the CCP, Lu Ting-i, director of the Department of Propaganda of the CCP Central Committee, stated:

> Mao Tse-tung's theory of the Chinese revolution is a new development of Marxism-Leninism in the revolutions of the colonial and semi-colonial countries and especially in the Chinese revolution. Mao Tse-tung's theory of the Chinese revolution has a significance not only for China and Asia—*it is of universal significance for the world Communist movement.* It is indeed a new contribution to the treasury of Marxism-Leninism. . . .
>
> The classic type of revolution in imperialist countries is the October Revolution. *The classic type of revolution in colonial and semi-colonial countries is the Chinese revolution.*[19]

Subsequently, the deputy director of the Department of Propaganda, Ch'en Po-ta, praised Mao for furthering "the development of Marxism-Leninism in the East," and lauded his successful approach to revolution as "the new Marxist conclusion arrived at in colonial and semi-colonial countries." A Chinese journal proudly predicted that "China's today then is the tomorrow of Vietnam, Burma, Ceylon, India, and the other various Asian colonial and semi-colonial nations."[20]

The Soviet propagandists, disputing Mao's claim for doctrinal originality and the applicability of his "road" to the rest of Asia, expostulated that the Chinese victory was merely the logical result of the application of the universal truths of Marxism-Leninism in conjunction with Stalin's pre-

18. Philip Bridgham, Arthur Cohen, and Leonard Jaffe, "Mao's Road and Sino-Soviet Relations: A View from Washington, 1953," *The China Quarterly* (Oct.-Dec. 1972), 52:678, italics added.

19. *Ibid.,* 681, italics added.

20. *Ibid.,* 681-82.

cepts on the national-colonial question in general and on China in particular.

At the Moscow Scientific Conference on November 12, 1951, the principal speaker, Ye. Zhukov, warned that "it would be risky to regard the Chinese revolution as some kind of 'stereotype' for people's democratic revolutions in other countries in Asia." As far as the Soviet writers were concerned, there was no "Mao's road" or "Mao's ideology"; the only road to be followed by Asian peoples was that pointed to by Marx-Lenin-Stalin. There was a deliberate Soviet effort to limit Mao's importance and contribution, and also to undercut his authority on the subject of Asian and world revolution. However, the debate on "Mao's road" became muted after the Moscow Scientific Conference, apparently by mutual agreement.

If Mao's encounter with Stalin was bitter, his experience with Khrushchev was vitriolic. At least Stalin commanded some respect for his seniority and accomplishment of building socialism, but Khrushchev possessed neither of these attributes. Twice, in 1954 and in 1958, Mao and Chou attempted to discuss with him the status of Outer Mongolia but were unable to elicit any response. On the question of the Sino-Soviet border, Khrushchev considered the Chinese map "so outrageous that we threw it away in disgust."[21] On another occasion, when Mao remarked that the combined might of China and Russia was greater than that of the capitalist West, Khrushchev lectured him:

> Comrade Mao Tse-tung, nowadays that sort of thinking is out of date. You can no longer calculate the alignment of forces on the basis of who has the most men. Back in the days when a dispute was settled with fists or bayonets, it made a difference who had the most men and the most bayonets on each side. . . . Now with the atomic bomb, the number of troops on each side makes practically no difference to the alignment of real power and the outcome of a war. The more troops on a side, the more bomb fodder.[22]

Mao considered Khrushchev a coward.

A further irritation arose in 1959 when Khrushchev went to China and requested the right to operate a radio station on Chinese soil to maintain contact with Soviet submarines; he also asked for the privilege of refueling and repairing his ships in China as well as shore leaves for the sailors. Mao furiously rejected the request, stating: "For the last time, *no*, and I don't want to hear anything more about it." When Khrushchev persisted, Mao declared: "No! . . . we don't want you here. We've had the British

21. Khrushchev, *Khrushchev Remembers*, 474.
22. *Ibid.*, 470.

and other foreigners on our territory for years now, and we are not ever going to let anyone use our land for their own purposes again." Khrushchev indicated his loss of patience and recalled that as early as 1954 he predicted that "conflict with China is inevitable."[23]

The preceding sketch of events indicates that for thirty years Mao had unpleasant experiences with Soviet leaders. Initially, he was excluded from the leadership of the Chinese Communist movement by Stalin and then treated disrespectfully by both Stalin and Khrushchev. The sense of personal injury and grievance could only have an adverse effect on state relations.

The Ideological Dispute

Outwardly, the immediate cause of the Sino-Soviet split was precipitated by Khrushchev's denunciation of Stalin and his attack on the "personality cult" at the 20th Congress of the Soviet Communist Party in 1956. Mao, who had little respect for Khrushchev and himself practiced the "personality cult," was not favorably disposed to the Soviet development. Yet, on the surface he cooperated with Moscow by reorganizing the CCP Central Committee into a "collective leadership" patterned after the Soviet example. The façade of Axis collaboration was maintained, but Mao was convinced that he, not Khrushchev, represented the torch-bearer of Marxism-Leninism and that the Chinese experience provided the example of revolution for Asian and other colonial and semicolonial countries. By implication he reduced the Soviet path to socialism to a European variety and refuted its universality.

Mao was determined to be the ideological leader of international communism and make Peking the new center in the Socialist bloc. The failure of the Russian Communists to produce a dominant, charismatic leader after Stalin's death facilitated his scheme, while the unrest in the East European satellite states in the wake of de-Stalinization provided an excellent opportunity for manipulation. The Hungarian Revolt of 1956 was followed by the Polish demand for the right of "people's democracies" to pursue "different paths to socialism." Seizing the opportunity to serve as bloc mediator, Mao sent his ebullient premier and foreign minister, Chou En-lai, on a moderating mission to eastern Europe. Chou stressed the need both for Socialist solidarity under Soviet leadership in the face of the capitalist threat and the importance of recognizing the different conditions in the satellite countries. The Maoist approach, though a stopgap device,

23. *Ibid.*, 466, 472-73.

arrested the disintegrative tendency in the Communist world and maintained Soviet primacy in the bloc. For the first time Peking extended its influence beyond Asia and provided an alternative voice to Moscow. Encouraged by this success, in 1957 Mao launched a vehement attack on Tito's revisionism and made himself the defender of Marxist-Leninist doctrinal purity. By 1958 Mao had made Peking an alternative center to Moscow. The monolithic structure of the Socialist camp crumbled and the indisputable leadership of the Soviet Union was broken.

The rise of Mao and the challenge of Communist China to Soviet hegemony raised the question of whether the Socialist world could tolerate two voices and two centers. Regardless of the answer, the question generated tension between the two, especially with respect to global strategy and approaches to problems of world revolution. When the Soviets launched their first intercontinental ballistic missile and orbited a satellite in August and October 1957, an elated Mao proclaimed, "The east wind prevails over the west wind." He considered the time opportune to dynamically advance the cause of international socialism, but Khrushchev hesitated. In 1958, during the Quemoy-Matsu crisis, the Soviets failed to support Peking's efforts to capture these offshore islands. Most damaging to Sino-Soviet relations was Khrushchev's interference in China's domestic politics; in 1959, he encouraged the visiting Chinese Defense Minister P'eng Te-huai to oppose Mao, and attempted to weaken Mao's position by unilaterally cancelling the October 1957 pact for New Technology and National Defense a week after P'eng's return to China in June.[24] The 1957 agreement provided for Moscow to furnish China with a sample atomic bomb, scientific data, and technical personnel to help develop the bomb. After sending a heavy-water reactor in 1958, Khrushchev regretted his promise. Not only was the sample bomb never delivered but in 1959 the flow of scientific information was restricted and Soviet technicians were withdrawn from China, complete with blueprints of half-completed projects. Deeply embittered, Mao accused Khrushchev of submitting to American propaganda, following "revisionism," and of holding heretical views on the nature of war and the grand strategy of world revolution. By implication, he questioned Khrushchev's fitness to lead the international Communist movement. The polemic of ideological dispute festered.

The Chinese Communists subscribed to the classical Marxist-Leninist concept that war between the socialist and capitalist worlds was inevitable. In Peking's view, war enhanced rather than hindered the cause of

24. For details, see next chapter, section on the dismissal of P'eng Te-huai.

communism. World War I made possible the emergence of Soviet Russia; World War II gave rise to Communist China; and World War III would bring communism to power in the United States and the end to the capitalist world.

Peking claimed that it was not afraid of war; although a nuclear holocaust might kill 300 million Chinese, the more advanced industrial nations of the West would fare worse. Reportedly, Chou En-lai stated that after the next war there would be "twenty million Americans, five million Englishmen, fifty million Russians, and three hundred million Chinese left."[25] This bold stance was probably designed to forestall fear of war and to prevent the dampening of the revolutionary fires of the world communist movement.

Khrushchev disputed the Chinese view. He argued that war had become too devastating to be inevitable and that World War III would decimate the earth, rendering meaningless the victory of communism. On the other hand, because of its superior system, the Socialist camp could outproduce the capitalist world and emerge victorious through peaceful competition. To Peking these views were heresy. In April 1960 the Chinese vehemently accused Khrushchev of "emasculating, betraying, and revising" Marxism-Leninism—acts which would only lead the international Communist movement to disaster. Khrushchev accused Mao of "being like Stalin, of being oblivious of any interests but his own, of spinning theories detached from the realities of the modern world." He called Mao "an ultra-leftist, an ultra-dogmatist, and a left revisionist."[26] The intensity of argument between Moscow and Peking precipitated a grand conference of eighty-one Communist parties in Moscow in November 1960 to adjudicate the dispute.

The Chinese delegation ridiculed the idea of peaceful coexistence as a mirage, argued for increased support for wars of liberation and national independence movements, and belittled the chances of a peaceful takeover. The Soviets retorted that the dangers of global war were too great and that, since the Communist bloc had not achieved a decisive strategic superiority, it should act with caution. Moreover, the economic race would be the decisive factor and time favored the Socialist camp.[27] The final communiqué of the conference, issued on December 6, 1960, en-

25. Zbigniew K. Brzezinski, *The Soviet Bloc: Unity and Conflict* (Cambridge, Mass., 1960), 403.
26. Donald S. Zagoria, "The Future of Sino-Soviet Relations," *Asian Survey*, 1:2:3-14 (April 1961); Edward Crankshaw, "Khrushchev and China," *The Atlantic Monthly* (May 1961), 43-47.
27. For a penetrating analysis of the Moscow Conference, see Donald S. Zagoria, *The Sino-Soviet Conflict, 1956-1961* (Princeton, 1962), ch. 15.

dorsed the Soviet view on peaceful coexistence with minor changes to appease the Chinese. Peking reluctantly subscribed to the communiqué to preserve bloc solidarity but, beneath the surface, bitterness remained. Although the Soviets scored a victory at the conference, it was clear that Peking could challenge Moscow without the risk of expulsion from the bloc. The days of Soviet hegemony were over.

Sino-Soviet relations were further strained by Moscow's unilateral action during the Cuban missile crisis in 1962, by its acceptance of a nuclear test ban treaty without consulting the Chinese, and by its refusal to support China in the 1962 border war with India.[28] The animus had become so bitter that in 1964, when completion of a Chinese nuclear device seemed imminent, Khrushchev considered destroying Peking's atomic installations. The scheme dismayed other Soviet leaders who quickly notified the Chinese leadership. Mao threatened to invade Outer Mongolia to prove that the U.S.S.R. could not protect its satellites in Asia. A clash was averted at the last minute when, on October 15, 1964, the more pragmatic Soviet leaders, Leonid Brezhnev and Alexei Kosygin, succeeded in mustering enough support in the Central Committee to relieve Khrushchev as Chairman of the Council of Ministers.[29] A day later the first Chinese atomic device was exploded in Sinkiang—a feat achieved by China's own scientists after the withdrawal of Soviet personnel.

Although Sino-Soviet relations temporarily improved, it soon became evident that the new Soviet leadership had not deviated basically from the course charted by Khrushchev. Peking renewed its attack on Soviet revisionism, and the schism widened. A number of Communist parties around the world evinced sympathy for Peking's dynamic stance, while others split into pro-Chinese and pro-Russian factions. Repeated Soviet attempts to expel the Chinese from the international Communist movement have been unsuccessful.

The Territorial Controversy

Compounding the problems of ideological dispute and struggle for leadership over the international Communist movement was the territorial controversy over the 4,150-mile Sino-Soviet border and the deep sense of injury inflicted by Tsarist Russia and Stalin. Mao was intent upon recovering all lost territories and rights.

Chinese grievances over lost territories are historically deep-rooted.

28. Statement by Foreign Minister Ch'en Yi to a group of Scandinavian newsmen, as reported in *The Christian Science Monitor*, May 27, 1966.
29. Harold C. Hinton, *Communist China in World Politics* (New York, 1966), 478-82.

Even though the first two treaties with Russia, that of Nerchinsk of 1689 and of Kiakhta of 1727, were generally considered equal, China lost 93,000 and 40,000 square miles of land respectively. The remaining treaties with Russia were definitely unequal: the Treaty of Aigun of 1858 ceded to Russia 185,000 square miles of territory in the Amur River Valley, which became Amursky Province; the Treaty of Peking of 1860, which confirmed the Aigun treaty, ceded in addition some 133,000 square miles of land east of the Ussuri River, which became the Maritime Province of Russia; and the Treaty of St. Petersburg of 1881 detached 15,000 square miles of Chinese territory. During the Republican Revolution of 1912, Russia promoted the Mongolian independence movement and recognized the "autonomy" of Outer Mongolia, turning it into a de facto Russian protectorate. Although after the Bolshevik Revolution China regained control of Outer Mongolia in 1919, Soviet forces invaded Outer Mongolia in July 1921, restoring its "independence" while annexing Tannu Tuva.[30] In 1924, the Mongolian People's Republic was established, becoming the first Soviet satellite, with a capital at Urga, now renamed Ulan Bator (Red Hero). Then, by the Yalta Agreement of February 1945, later confirmed by the Sino-Soviet Treaty of August 14, 1945, the Soviets regained all the Tsarist special rights and interests in Manchuria lost to Japan after Russia's defeat in 1905. Even after the rise of the People's Republic of China, Stalin exacted from Mao many humiliating concessions.

Like all patriotic Chinese, Mao had always wanted to rectify the injuries China had suffered in the past. As early as 1936 he told the American journalist Edgar Snow that "it is the immediate task of China to regain all our lost territories," including Outer Mongolia, which should become a part of the "Chinese federation."[31]

On March 8, 1963, Peking published a list of lost territories, including part of Southern Siberia, the Maritime Province, and at least 500,000 square miles of land in Russian Central Asia. It demanded that the Soviets acknowledge, for the record, that the current Sino-Soviet frontier was a product of "unequal" and therefore "illegal" treaties. Moscow denied having territorial problems with any neighboring state and refused to admit the illegality of the old treaties with China. On July 10, 1964, Mao told a visiting Japanese Socialist Party delegation: "About a hundred years ago, the area to the east of [Lake] Baikal became Russian territory and since then Vladivostok, Khabarovsk, Kamchatka, and other areas have become Soviet territory. We have not yet presented our bill for this list."[32] Mos-

30. Tai Sung An, 50-51.
31. Edgar Snow, *Red Star Over China* (New York, 1961), 96.
32. Tai Sung An, 76, 82.

cow denounced Mao's statement as reminiscent of Hitler's *Lebensraum* and Khrushchev retorted publicly that if Tsarist Russia was expansionist, so was Imperial China; both countries had taken land from other people and their actions should cancel out each other. Hence, the border negotiations that began in Peking on February 25, 1964, made no progress and broke down on October 15 of the same year. What followed were more frontier incidents of increasing frequency and intensity. By November 1968 the Brezhnev-Kosygin leadership decided to take a hard line toward the territorial dispute and major clashes seemed inevitable.

According to the Chinese claim, from the breakdown of the negotiations in 1964 to March 1969 the Soviets violated China's border 4,189 times. Frontier tension escalated to such a degree that two large clashes erupted on March 2 and 14-15, 1969 on Chen-pao (Treasure) or Damansky Island in the Ussuri River. Claimed by both China and the Soviet Union, Chen-pao Island is located at 133° 51' E. longitude and 46° 51' N. latitude and is about a mile long and a third of a mile wide. The Chinese argued that historically this island was part of the Chinese bank, that it could be reached by foot during low-water in late summer, and that it was used by Chinese fishermen for net-drying. The Soviets argued that they had maintained a frontier outpost on it since 1922 but abolished it in 1950 when the People's Republic of China was established; furthermore, the fact that the Chinese periodically applied for use of Damansky Island was proof of Soviet ownership.

On the night of March 1-2, some 300 camouflaged Chinese soldiers reached Chen-pao Island and dug foxholes in a wooded area in preparation for an ambush. At approximately 11 a.m. the following morning the Chinese, seemingly unarmed, marched toward the Russians; when they had reached within twenty feet of the enemy the first row quickly scattered to the side and the second row opened fire, killing seven Russians, including their leader. Other Chinese soldiers rushed out of hiding and overwhelmed the entire Russian unit, taking nineteen prisoners and considerable amounts of Soviet equipment. Russian reinforcements finally arrived and drove the Chinese out. Both sides claimed victory while accusing the other of provocation.

The second clash of March 14-15 was initiated by the Russians as an act of retaliation and involved the use of tanks and many soldiers. Beginning at 10 a.m. a battle ensued for nine hours, resulting in 60 Russian and 800 Chinese casualties.[33] Both sides made feverish propaganda out of the clashes, held exhibitions to illustrate the other side's atrocities, and

33. Thomas W. Robinson, "The Sino-Soviet Border Dispute: Background Development, and the March 1969 Clashes," *The American Political Science Review*, LXVI:4: 1199 (Dec. 1972).

organized vast demonstrations to stir up national sentiments and war scares. The imminence of war was further heightened by subsequent clashes on Pa-cha (Goldinsky) Island in the Amur River and along the Sinkiang border.

The Danger of War

Coming in the wake of the Soviet invasion of Czechoslovakia in 1968 and Brezhnev's announcement that the Soviet Union had the right to intervene in the internal affairs of other Communist states which were deemed deviants from the Socialist cause, these border clashes brought on the ominous prospect of a Soviet attack on China. In fact, the more venturesome military and political leaders in Moscow seriously proposed a preemptive nuclear strike on China's atomic installations and explored the possibility of enlisting American cooperation or tacit approval. Overtures were made by the Soviet military attachés in the Tokyo and Canberra Embassies and later at a higher level to the United States for a joint venture against China. But President Nixon ordered "a sharply, angrily negative response."[34] On September 11, 1969, while flying home from Hanoi after attending the funeral of Ho Chih-minh, Soviet Premier Alexei Kosygin was ordered in mid-air near Irkutsk to turn back to Peking, where he met with Chou En-lai at the airport for three hours. They reached an understanding that an agreement should be signed on non-use of force, maintenance of the status quo of the frontiers, the prevention of military conflicts and clashes, the separation of forces in disputed regions, the solutions of all frontier questions through talks, and non-aggression. After this meeting, border disputes subsided and high-level negotiations resumed on October 20, 1969 in Peking. However, in July 1970 the Soviet Union once again proposed to the United States a joint agreement against any "provocative actions" by China or any other nuclear power. The proposal, which involved not only China but also the nuclear powers among the NATO alliance, was rejected by Washington.[35]

At the reconstituted border negotiations in Peking, the Chinese did not demand an outright return of all the territories lost to Russia; however, they did insist on using the old treaties as the basis for negotiating an overall settlement of boundary questions and on gaining Soviet acknowledgment that the old Tsarist treaties were unequal and illegal. The Chinese demanded that all Soviet gains in excess of the old Tsarist treaties

---

34. Joseph Alsop, "Thoughts out of China—(I) Go versus No go," *The New York Times Magazine*, March 11, 1973, 31.
35. John Newhouse, *Cold Dawn: The Story of SALT* (New York, 1973), 188-89.

must be restored to China unconditionally, including 600 of the 700 islands in the Ussuri and Amur Rivers, totalling some 400 square miles of land, and 12,000 square miles of land in the Pamir mountain area in Sinkiang. To justify its argument Peking utilized the international law principle of the *thalweg* which takes the deepest point of the main channels of boundary rivers—the Amur and the Ussuri in this case—as the dividing line and which would award China the 600 islands, including Chen-pao and Pa-cha. As regards the Pamir sector, the Chinese claimed that it was illegally occupied by Russian troops in violation of the 1884 boundary protocol.[36]

The Soviet delegates argued that the treaties signed by Manchu China and Tsarist Russia were not "unequal" and that the Sino-Russian boundary had been fixed legally and historically. Although they saw no reason to alter the boundary now, the Soviets were willing to demonstrate their goodwill by making some minor adjustments in particular frontier areas, reportedly in the strategic Sinkiang and Vladivostok regions. As regards the Pamir sector and the Amur and Ussuri Rivers, the Soviets denied the illegal occupation of the first and rejected the doctrine of *thalweg* for the second, insisting that the 600 islands belonged to Russia. There was no agreement or acceptable compromise at the negotiations.

Apparently, both sides used the negotiations to gain time for other purposes. By mid 1973 the Soviets increased their strength on the Chinese border to more than a million men, equipped with missiles and nuclear weapons, and stationed at least 150 ships in the Pacific. They also deployed some 100 ABM's, known as the Galosh system, around Moscow and Leningrad, for a first strike advantage against China.[37] Meanwhile, they worked diligently toward reaching a détente with Western Europe, especially West Germany, in order to avoid confrontation on two fronts. The Chinese reinforced their frontiers with at least one million men and made a major foreign policy shift in order to achieve a détente with the United States and Japan, breaking out of imperialist encirclement and forestalling possible Soviet-American collusion. Avoidance of a two-front war was Mao's major concern. In the meantime, to prepare for the worst, the Chinese built extensive networks of tunnels as bomb shelters in major cities, complete with water, food, and medical supplies. Nuclear installations at Lop Nor in Sinkiang were quietly moved inland to some unknown

36. Tai Sung An, 109, 114-115; Robinson, 1180-81.
37. Peter S. H. Tang, "Russian Threat to China," *The Christian Science Monitor,* Sept. 12, 1972; Shinkichi Etō, "Motivations and Tactics of Peking's New Foreign Policy," 33-34. A paper read before the German Association for East Asian Studies, International Conference, Schloss Reisenburg, June 24-30, 1973.

spots in Tibet, while the old sites were camouflaged as genuine atomic facilities. Chinese short-range nuclear missiles were deployed for strikes against Vladivostok, Irkutsk, and other Soviet cities near the border, and middle-range missiles capable of 1,200 to 2,300 miles were set up for targets in Siberia and Soviet Central Asia. By 1973 the intermediate range 3,500-mile missiles capable of reaching Moscow and Leningrad were in production and in 1980 the 6,000-mile ICBM's became available.[38] The Chinese were rapidly developing a second strike capability. Militarily, the time for a Soviet nuclear blitz with impunity was long past and diplomatically the international community would not tolerate such a strike. Reportedly, when Nixon visited Moscow in May 1972 he warned Brezhnev "in the frankest possible terms" that a Soviet attack on China would threaten world peace and would be considered against the national interest of the United States. Brezhnev bluntly questioned the qualification of the United States to be the arbiter of affairs among Communist states. Nonetheless, the Soviet desire for American wheat, technology, and a détente generally precluded Moscow from totally ignoring the American warning. Powerful European Communist parties, such as those of Italy, Rumania, and Yugoslavia, also strongly disapproved of a Soviet attack on China. Brezhnev then conceived of an Asian Collective Security System to outflank China by strengthening Soviet ties with India, Japan, North Korea, and countries in Indochina, but the plan had been ineffective, except with India and Vietnam.

In the final analysis, the Chinese were too preoccupied with socialist transformation, industrial development, and problems of leadership and "succession" to want a war, and the Soviets were too obsessed with the China question to know exactly what to do about it. The Kremlin seemed to waver between threat and conciliation; in mid-June 1973 it offered a nonaggression pact, which was rejected by Peking.[39] Then in early November 1974 the Chinese suprisingly proposed a nonaggression pact and separation of forces in disputed border areas on the basis of the September 1969 understanding between Kosygin and Chou En-lai. But Brezhnev spurned it. The state of Sino-Soviet relations was reflected in China's refusal in 1980 to renew the treaty of friendship and alliance signed thirty years earlier. Nevertheless, the danger of war had receded but tension between the two countries persisted, and it only benefited the other pro-

38. By mid-1973, the Chinese reportedly had 15 missiles of 1,200-mile range and 15 to 20 missiles of 2,300-mile range. In May 1980, the Chinese successfully tested two CSS-X-4 missiles capable of 6000-7000 miles.

39. Robert A. Scalapino, "China and the Balance of Power," *Foreign Affairs* (Jan. 1974), 361.

tagonist in the triangular relationship: the United States. When the bear and the dragon contended, the eagle reaped the benefit.

For most of the decade of the 1980s, Sino-Soviet relations were clouded by three obstacles: (1) Soviet support for a Vietnamese invasion and occupation of Cambodia, (2) Soviet invasion of Afghanistan, and (3) Soviet troops stationed near the Chinese border. China insisted that normalization of relations would be predicated upon the removal of these three conditions. China was the only Communist state that did not send an official party delegation to the 70th anniversary celebrations of the Bolshevik Revolution in October 1987. However, China kept a close watch over the progress of Soviet leader Mikhail S. Gorbachev's policies of openness (*glasnost*) and political restructuring (*perestroika*), as both countries were attempting the same to overhaul the domestic economy. Gorbachev considered improvement of relations with China a top priority and in November 1987 asked for a meeting with the Chinese leader Teng Hsiao-p'ing, but Teng turned down the overture because the Soviets had not brought about a withdrawal of Vietnamese troops from Cambodia. However, some warming trends were visible after the Soviets began to withdraw their troops from Afghanistan. In May 1988 China and the Soviet Union signed a cultural exchange agreement, under which the Bolshoi Ballet would perform in Peking and the Hermitage Museum in Leningrad would exchange exhibitions with Peking's National Palace Museum. In late 1988 all signs pointed to a summit meeting between Gorbachev and Teng in May 1989 and restoration of official relations between the two countries shortly afterwards.

Further Reading

Alsop, Joseph, "Thoughts out of China—(I) Go versus No go," *The New York Times Magazine,* March 11, 1973, 31, 100-108.

An, Tai-sung, *The Sino-Soviet Territorial Dispute* (Philadelphia, 1973).

Boorman, Howard L. et al. (eds.), *Moscow-Peking Axis: Strengths and Strains* (New York, 1957).

Bridgham, Philip, Arthur Cohen, and Leonard Jaffe, "Mao's Road and Sino-Soviet Relations: A View from Washington, 1953," *The China Quarterly,* 52: 670-98 (Oct.-Dec. 1972).

Brzezinski, Zbigniew K., *The Soviet Bloc: Unity and Conflict* (Cambridge, Mass., 1960).

Clubb, Edmund O., *China and Russia: The "Great Game"* (New York, 1971).

Dittmer, Lowell, *Sino-Soviet Normalization and Its International Implications, 1945-1990* (Seattle, 1992).

Djilas, Milovan, *Conversations with Stalin,* tr. by Michael B. Petrovich (New York, 1962).

Doolin, Dennis J., *Territorial Claims in the Sino-Soviet Conflict: Documents and Analysis* (Stanford, 1965).

Fitzgerald, C. P., "Tension on the Sino-Soviet Border," *Foreign Affairs*, 45:4: 683-93 (July 1967).

Garthoff, Raymond, *Sino-Soviet Military Relations* (New York, 1966).

Garver, John, "New Light on Sino-Soviet Relations: The Memoirs of China's Ambassador to Moscow, 1955-62," *The China Quarterly*, June 1990, 303-07.

Ginsburgs, George, *Sino-Soviet Territorial Dispute* (New York, 1975).

Gittings, John, *Survey of the Sino-Soviet Dispute* (Oxford, 1968).

———, "The Great-Power Triangle and Chinese Foreign Policy," *The China Quarterly*, 39:41-54 (July-Sept. 1969).

Goncharov, Sergei N., John W. Lewis, and Xue Litai, *Uncertain Partners: Stalin, Mao, and the Korean War* (Stanford, 1994).

Jackson, W. A., *The Russo-Chinese Borderlands* (Princeton, 1962).

Jukes, Geoffrey, *The Soviet Union in Asia* (Berkeley, 1973).

Kao, Ting Tsz, *The Chinese Frontier* (Chicago, 1980).

Khrushchev, Nikita, *Khrushchev Remembers* (Boston, 1970).

———, *Khrushchev Remembers: The Last Testament*, tr. by Strobe Talbott (Boston, 1974).

Liu, Xiao, *Ch'u-shih Su-lien pa-nien* (Eight years of ambassadorship in the Soviet Union) (Beijing, 1986).

Low, Alfred D., *The Sino-Soviet Confrontation Since Mao Zedong: Dispute, Détente, or Conflict?* (Boulder, 1987).

Maxwell, Neville, "The Chinese Account of the 1969 Fighting at Chenpao," *The China Quarterly*, 56:730-39 (Oct.-Dec. 1973).

Mayers, David Allan, *Cracking the Monolith: US Policy Against the Sino-Soviet Alliance, 1949-1955* (Baton Rouge, 1986).

Murphy, George G. S., *Soviet Mongolia: A Study of the Oldest Political Satellite* (Berkeley, 1966).

Patterson, George N., *The Unquiet Frontier: Border Tensions in the Sino-Soviet Conflict* (Hong Kong, 1966).

Robinson, Thomas W., "The Sino-Soviet Border Dispute: Background Development, and the March 1969 Clashes," *The American Political Science Review*, LXVI:4:1175-1202 (Dec. 1972).

Salisbury, Harrison E., *War Between Russia and China* (New York, 1970).

Simon, Sheldon W., "The Japan-China-USSR Triangle," *Pacific Affairs*, 47:2: 125-38 (Summer 1974).

Tuchman, Barbara W., "If Mao had come to Washington: An Essay in Alternatives," *Foreign Affairs*, 51:1:44-64 (Oct. 1972).

Vucinich, Wayne S., *Russia and Asia: Essays on the Influence of Russia on the Asian Peoples* (Stanford, 1972).

Zagoria, Donald S., "Mao's Role in the Sino-Soviet Conflict," *Pacific Affairs*, 47:2: 139-53 (Summer 1974).

———, *The Sino-Soviet Conflict, 1956-1961* (Princeton, 1962).

———, "The Strategic Debate in Peking," Tang Tsou (ed.), *China in Crisis: China's Policies in Asia and America's Alternatives* (Chicago, 1968), II: 237-68.

28

The Great Proletarian Cultural Revolution

The Great Leap, the Communes, and the Sino-Soviet split increased tension and divisiveness within the Chinese leadership. A number of senior party members, more pragmatic and less idealistic than Mao, began to question what they believed to be the precipitous and careening course chartered by their Leader. They were disappointed with the initial results of the Great Leap and the Communes and critical of Mao's handling of the dispute with the Soviet Union, whose continued aid they considered important to China's economic, military, and scientific development. Their disillusionment was intensified by doubts about the wisdom of the breakneck pace of agricultural collectivization and industrial expansion, the emphasis on ideology (i.e. "redness") over expertise in technology and weaponry, and the incessant mass campaigns which sapped national energies. Because of their long association with Mao and their actual control of the daily operations of the party and the government, these leaders believed they were in a position to assess the needs of the nation more accurately than Mao, who secluded himself in majestic isolation. Deeply concerned with what they considered to be Mao's romantic, hasty approach to Socialist transformation, they decided to subtly block, divert, or modify the implementation of his policies. In the mid- and late 1960s the collision of the two lines resulted in a gigantic social, political, and cultural upheaval—the Great Proletarian Cultural Revolution.

These leaders' pertinaciousness in proposing an alternative to Mao's direction arose from the fact that they had acquired great influence over the years, particularly after the party reorganization in 1956. The reorganization created four vice-chairmanships of the Central Committee, a Standing Committee of the Politburo, and a general secretaryship of the party. The chairman of the Central Committee was no longer the concurrent head of the Secretariat, which was put under the supervision of the party general secretary. The Politburo's Standing Committee, the seat of power, was comprised of the chairman, the four vice-chairmen of the Central Committee, and the party general secretary—representing a collective leadership of six. Mao retained the chairmanship of the Central Committee, but Liu Shao-ch'i became first vice-chairman and exercised certain functions for the chairman, and Teng Hsiao-p'ing was appointed party general secretary. With this reshuffle, Mao retreated to the "second line" while advancing Liu and Teng to the "first line," enhancing their prestige and status so that, as Mao recalled ten years later (1966), "When I have to see God, the country can avoid great chaos."[1]

With their newly elevated positions, Liu and Teng felt confident enough to speak out vigorously on important policy matters. They favored a less precipitous course of action in order to cultivate a more stable social order and priority for industrial development over agricultural collectivization, while Mao continued to push for mass movements, increased speed in collectivization, the Great Leap, and the Communes. By 1958 the Central Committee had come under the tight control of Liu and Teng, and it soon found the idea of Communes "unsound." At its Sixth Plenum on December 10, 1958, two resolutions were adopted, one to amend the Commune system, and the other to "accept" Mao's request not to be the State Chairman (i.e. chairman of the People's Republic) for another term, on the grounds that he needed more time to pursue the theoretical works of Marx and Lenin and to devote himself to party and state affairs. Every effort was made to convey the impression that Mao had made the move voluntarily, when in fact Liu had quietly "eased him out" of the state chairmanship. Mao himself commented on the event in October 1966: "I was extremely discontented with that decision, but I could do nothing about it."[2] He further complained that he was treated like a "dead man at his own funeral."[3]

The year 1959 was particularly critical for the politics of economic development. The results of the Great Leap and the Communes were deeply discouraging. Indigenous production methods entailed an enormous waste

1. Gene T. Hsiao, "The Background and Development of 'The Proletarian Cultural Revolution,'" *Asian Survey*, VII:6:392 (June 1967).
2. *Ibid.*, 395.
3. James Pinckney Harrison, *The Long March to Power* (New York, 1972), 477.

of materials and a high cost of operation, while yielding low-quality products. The steel turned out by the backyard blast furnaces could not meet industrial requirements, and many of these makeshift furnaces dissolved in rainstorms. Agricultural production was adversely affected by deficient water conservation projects and faulty realkalinization of the soil. Intensive use of machinery without proper maintenance—in order to keep up with the high production quotas—resulted in rapid deterioration and damage to the tools, while eager recruitment of inexperienced workers caused managerial problems.[4] The Communes were responsible for lower, not higher, agricultural output. The grain output declined from 200 million metric tons in 1958 to 165 million in 1959 and to 160 million in 1960. The general agricultural production index, using 1957 at 100, dropped from 108 in 1958 to 86 in 1959 and to 83 in 1960. The industrial production index fared slightly better: using 1957 as 100, it was 131 in 1958, 166 in 1959, 161-163 in 1960, and 107-110 in 1961. The total picture of the Gross National Product (GNP) was disappointing, dropping from $95 billion in 1958 to $92 billion in 1959, to $89 billion in 1960 and to $72 billion in 1961.[5] All in all, the results of the first years of the Great Leap and the Communes were highly unsatisfactory, and there was great discontent among the people.

The pragmatists grew increasingly critical of the Maoist policies. They felt that Mao ran the country with a guerrilla mentality out of tune with the changing times, and that he was more effective in his earlier role of a revolutionary than as the chief administrator of a large state. Tactful allusions to his failing faculties began to appear. Indeed, the economic disruptions and the social turmoils stirred up by the "Three Red Banners Campaign"—the General Line of the Party (i.e. Mao's will), the Great Leap Forward Movement, and the Commune—cast increasing doubts in the minds of the pragmatists about the Maoist approach. Boldly, they proposed an alternative course, even at the risk of offending the Leader.

The Dismissal of P'eng Te-huai

At the Lushan Conference of the Central Committee in August 1959, Ch'en Yün, the fifth-ranking party member and a leading economic planner, argued for several major changes, including a return from Com-

4. Kang Chao, "Economic Aftermath of the Great Leap in Communist China," *Asian Survey*, IV:5:851-58 (May 1964).
5. *People's Republic of China: An Economic Assessment*, A compendium of papers submitted to the Joint Economic Committee, Congress of the United States (Washington, D.C., 1972), 5. (Hereafter to be referred to as *Joint Economic Committee Report*.)

munes to cooperatives, a more realistic schedule of economic development, limited cooperation with the Soviets, and greater emphasis on technical expertise than on political and ideological purity ("redness"). Ch'en's statement probably had the prior approval of Liu Shao-ch'i, while it undoubtedly offended the Maoists.[6]

A far blunter critic was Defense Minister P'eng Te-huai, an old comrade-in-arms of Mao's and a prominent soldier. He had recently returned from a three-week goodwill trip to Eastern Europe, during which time he met Nikita Khrushchev in Tirana, Albania. The two of them discussed the deterioration of Sino-Soviet relations, and P'eng, in an attempt to dissuade Khrushchev from withdrawing his pledge of nuclear aid to China, indiscreetly divulged the confusing conditions in China and the shortcomings of the Great Leap. Khrushchev encouraged him to oppose Mao, and a week after his return home the Soviets announced the termination of the nuclear agreement with China signed two years earlier, perhaps hoping to strengthen P'eng's position against Mao. At Lushan, the foolhardy minister circulated a "Letter of Opinion" criticizing the Maoist approach to Socialist transformation as "hasty and excessive," pointing to the confusion and chaos resulting from the introduction of the Communes and the waste of two billion dollars (yüan) from the "backyard furnace." Sarcastically he remarked:

> Bewitched by the achievements of the Great Leap Forward and the passion of the mass movement, some leftist tendencies emerged, since we had always wanted to enter communism in a single step. . . . In the view of some comrades, putting politics in command could be a substitute for everything. . . . But putting politics in command is no substitute for economic principles, much less for economic measures.[7]

He equated the Great Leap with "petty bourgeois fanaticism" and called it "a rush of blood to the brain . . . a high fever" of unrealism—an unmistakable allusion to the condition of Mao's health. P'eng oppugned the idea of surpassing the British industrial output in fifteen years and deprecated China's economic "blunder after blunder." He attacked party control of the army and asked for the creation of a modern professional army with up-to-date equipment. By inference he favored a closer relationship with the Soviet Union to effect the supply of modern weapons, questioned Mao's views on the "people's war," and decried the constant diver-

6. Henry G. Schwartz, "The Great Proletarian Cultural Revolution," *Orbis* (Fall 1966), 813-14.
7. Stanley Karnow, *Mao and China: From Revolution to Revolution* (New York, 1972), 117-19.

sion of soldiers to political duties. Vitriolic as his criticisms were, P'eng did not urge the removal of Mao but only a revision of his policy.[8]

Ch'en Yün considered such an attack on Mao to be counterproductive; a more effective way to influence him would be to present facts and figures showing the poor results of his policies and allow him to make his own decision. In this way Mao, unchallenged, would probably terminate or temper the Great Leap on his own.[9]

As expected, the vehemence of P'eng's attack shocked Mao, who described it as one that "nearly levelled half of Lushan." Had P'eng not contacted the detested Khrushchev, his harsh and unprecedented criticism might have been tolerated in view of his well-known propensity for bluntness. But collusion with Soviet social imperialism was unforgivable and automatically made him a revisionist. P'eng's call for a professional army with modern weapons would entail a greater reliance on Soviet aid and diversion of China's limited resources to the military, necessitating a reordering of national priorities. Mao's later accusation that Khrushchev participated in a "behind-the-scenes factional activity against a fraternal Party" and his attack on P'eng of having "sown discord at the bidding of a foreign country" in "betrayal of the fatherland," indicated the bitter exasperation felt by Mao over the "seditious" relationship between P'eng and Khrushchev.[10] P'eng apparently had counted on the support of Liu Shao-ch'i and Teng Hsiao-p'ing but when neither sustained him, his cause was lost.

Mao was furious with P'eng but conciliatory toward the others. Apologetically he accepted responsibility for the initial poor showing of the Communes and the Great Leap: "Everybody has shortcomings. Even Confucius made mistakes. So did Marx. He thought that the revolution would take place in Europe during his lifetime. I have seen Lenin's manuscripts, which are filled with changes. He, too, made mistakes." A master tactician, Mao blamed himself for, yet absolved himself from, the economic disasters: "I devoted myself mainly to revolution. I am absolutely no good at construction, and I don't understand industrial planning. So don't write about my wise leadership, since I had not even taken charge of these matters. However, Comrades, I should take primary responsibility for 1958 and 1959. It is I who am to blame."[11] By implication, his

---

8. For an account of the Lushan confrontation, see David A. Charles, "The Dismissal of Marshal P'eng Teh-huai," *The China Quarterly* (Oct.-Dec. 1961), 8:63-76; Robert S. Elegant, "Mt. Lu Meeting Paved Way for Turbulence in China," *Los Angeles Times*, Dec. 17, 1967; Karnow, ch. 6, "The Marshal vs. Mao."
9. *Joint Economic Committee Report*, 55.
10. Karnow, 116.
11. *Ibid.*, 121.

failure was one of supervision; others were responsible for compounding his problems. Then, in a clever twist, Mao declared that since the party had a collective leadership, the responsibility had to be shared.[12]

The setbacks, Mao insisted, were temporary—"Come back in ten years to see whether we were correct."[13] In an overpowering move, he warned that if outvoted he would go to the countryside to raise a new peasant army to overthrow the government. "I think, however, that you will follow me." Thereupon, the committee members pledged their support, and Mao triumphantly closed the Lushan Conference. Marshal P'eng was dismissed as minister of defense and replaced by Marshal Lin Piao on September 17, 1959.[14] Mao proclaimed that the dizzy pace of the Great Leap was over and that there would be "no more pompous exaggeration." He acquiesced—at least temporarily—to a policy of moderation.

Retrenchment and Liberalization

The Lushan Conference was followed by three years of retrenchment and pragmatism necessitated by poor harvest, bad weather, drop in agricultural and industrial production, and the withdrawal of Soviet technicians. The country was facing serious economic dislocation. During an inspection trip to Hunan in the spring of 1961, Liu Shao-ch'i, although hardened by a lifetime of struggle, was deeply touched by the destitution of the people. There was general weariness, bitterness, and apathy. Liu privately commented that "the problems were not caused by natural calamities. They were man-made," meaning Mao-made.

The pragmatic leaders then decided on a crash program of aid to agriculture and a policy of liberalization to induce greater incentive for work. Under Liu's direction the Central Committee in January 1962 adopted the domestic line of "Three Privates and One Guarantee" (*San-tzu i-pao*) to counter the effects of Mao's "Three Red Banners Campaign." The "Three Privates" included permission for peasants to cultivate their own modest plots (about 5 percent of arable land), operate small private handicraft enterprises, and sell their products at rural free markets. The "One Guarantee" called for the fulfillment of agricultural production quotas set by the government. Externally, the Central Committee adopted the policy of "Three Reconciliations and One Reduction" (*San-ho i-shao*):

12. Richard H. Solomon, *Mao's Revolution and the Chinese Political Culture* (Berkeley, 1971), 395, 400.
13. Karnow, 119.
14. P'eng's associate, Huang K'o-ch'eng, the army chief-of-staff, was replaced by Lo Jui-ch'ing, formerly a secret service man.

the former connoting moderation of conflicts with "imperialists, reactionaries, and revisionists," and the latter reduction of aid to foreign national liberation movements.[15]

Mao was alarmed by the prospect of the return of capitalism, the rise of a new madarin class, and a compromise with Soviet revisionism. If these trends persisted, his lifelong work of politicizing the masses would disappear, the party would be cut off from the grass roots, and the country would once again fall into reliance upon the Soviet Union.

The reluctance of the party ranks to perpetuate the revolution worried Mao. He saw in them a tendency toward enjoyment of the easy life, the seeking of material incentives and wage privileges which would undoubtedly lead to an increasing bureaucratization of the party and the rise of individualism—with a corresponding loss of revolutionary fervor and the emergence of a new mandarin class similar to the Soviet party *apparatchiks*. These bureaucrats would strive to maintain the status quo and betray the true spirit of proletarian revolution, undermining the collective economy and nurturing an elitism apart from the masses. Mao found such tendencies appalling. China must not follow the Soviet pattern, and the pernicious influence of Khrushchev's "liberalization" of the government and peaceful coexistence with the capitalist West had to be eradicated.[16]

Mao was equally concerned with the lack of revolutionary challenge among the young and the unhealthiness of the educational system. Danger signals also existed in the literary and artistic circles: party intellectuals were not creating works that reflected the Socialist transformation and many seemed to be regressing back to the old cultural tradition. In November 1962, Chou Yang, deputy director of the CCP Propaganda Department, sponsored a "Forum on Confucius" in Shantung to commemorate the 2,440th anniversary of the death of the philosopher. To the Maoists, it was a deliberate attempt to glorify the old virtue of "human heartedness" (*jen*) and benevolent government, at the expense of "Mao's efforts to sharpen political confrontation."[17]

To obviate these developments Mao ordered a Socialist Education Movement in September 1962, stressing class struggle. Officials and intellectuals were sent downward (*hsia-fang*) to the countryside to learn from the masses. They were urged to produce works reflecting the realities of the Socialist transformation. But many found the *hsia-fang* movement a hardship to be endured rather than an experience to be cherished. Al-

15. According to Chou-En-lai's report to the Third National People's Congress in late 1964, quoted in Solomon, 419.
16. Mao told André Malreaux in 1967, quoted in Daubier, 11.
17. Solomon, 449.

though under the aegis of Mao's wife, Chiang Ch'ing, new forms of revolutionary art did appear in 1964—for example, the play *The Raid on the White Tiger Regiment* and the ballet, *The Red Detachment of Women*—there was resistance to the new movement. P'eng Chen, mayor of Peking, openly professed his fondness for Peking operas and sponsored theatrical works antithetical to Chiang Ch'ing's revolutionary models.

The Socialist Education Movement was ineffective because it was directed by a party unsympathetic with Mao's views and unresponsive to his guidance. Mao became convinced that it was not his policies that were wrong; rather, it was those in high party positions who were subtly foot dragging, blocking, distorting, and diluting their implementation. Mao insisted that the main causes of the difficulties following the Great Leap were the faulty party line, compounded by the bad harvests of 1959-61, and the Soviet withdrawal of technical aid in summer 1960.[18]

The question of authority also became a sensitive issue. The behavior of the Liuists challenged a fundamental belief of Mao's that "within the national society there should be a central supreme authority—a 'Center'—with overwhelming decision-making power." In the 1960s Mao was undoubtedly the living embodiment of that central authority.[19] But, if the party dared to ignore his will and Liu pressed for an alternative course of action, the erosion of Mao's absolute power was unmistakable.

All in all, Mao felt that his authority, his approach to social progress and his style of leadership developed over a lifetime were being covertly sabotaged by his chosen successor, Liu. The party had become increasingly aloof from him and the masses, and resistant to the struggle against "bourgeois rightism" among the intellectuals, the peasant's "spontaneous tendencies toward capitalism," and the drift toward Soviet revisionism. Mao could not persuade himself that the party and Liu could be trusted to carry on the revolutionary work after his passing. He was determined to give them a thorough shakeup, to change the power structure, and to effect an "irreversible transformation" of the people's thought and behavioral patterns.[20]

18. *Joint Economic Committee Report*, 21, 54.
19. Benjamin Schwartz, "Thoughts of Mao Tse-tung," *The New York Review of Books*, Feb. 8, 1973, 29.
20. Stuart R. Schram, "Introduction: The Cultural Revolution in Historical Perspective" in Stuart R. Schram (ed.), *Authority Participation and Cultural Change in China* (Cambridge, Eng., 1973), 85.

The Beginnings of the Cultural Revolution

According to Lin Piao, then very close to Mao, as early as 1962 Mao was "first to perceive the danger of the counter-revolutionary plots of Liu Shao-ch'i and his gang," and that Mao attempted unsuccessfully to regain the "first-line" control of the party.[21] By mid-1964 he had developed serious doubts about Liu's fitness to be his successor—believing that Liu might disown him as Khrushchev had repudiated Stalin.[22] When Liu proposed resurrecting the Sino-Soviet alliance because of the intensification of the Vietnam war, Mao decided at a party meeting on January 25, 1965, that Liu must be disinherited and the party shattered in order to be subsequently reconstructed.[23]

Mao wanted to re-establish the supremacy of his authority, his line of revolution, his work-style, to revitalize the youth, politicize the masses, and combat old customs, old habits, old culture, and old thinking—and to do so with a single, immediate expurgation. The sense of urgency might have been prompted by considerations of health. In 1964 he was 71, suffering from Parkinson's disease, and possibly the victim of a stroke in the fall of that year. He told Edgar Snow that "he was soon going to see God," and in August 1965 André Malraux noticed a nurse at his side. It is highly possible that Mao felt that he had only a limited time in which to rectify the problems he saw.

The first salvo of the Cultural Revolution was fired by the editor-in-chief of the Shanghai branch of the *Liberation Army Daily*, Yao Wen-yüan. The November 10, 1965 issue of *Wen-hui Pao* printed Yao's "Comment on the Newly Composed Historical Play 'Hai Jui Dismissed From Office'"—an article that attacked Wu Han, the deputy mayor of Peking and a former university professor.[24] Wu had earlier written a story, "Hai Jui Scolds the Emperor," under a pseudonym in the *People's Daily* on June 16, 1959, not long before the dismissal of Defense Minister P'eng. Hai Jui, a mid-16th century Ming official, supposedly reprimanded the emperor in these words: "For a long time the nation has not been satisfied with you. All officials, in and out of the capital, know that your mind is not right, that you are too arbitrary, that you are perverse. You think that you alone are right; you refuse to accept criticism; and your mistakes are

21. Harrison, 492; Solomon, 453.
22. Solomon, 459.
23. Edgar Snow, "Mao Tse-tung and the Cost of Living: Aftermath of the Cultural Revolution," *The New Republic*, April 10, 1971, 19.
24. Not a Communist member, but one of the Democratic League.

many."[25] The story was later rendered into a historical play entitled "Hai Jui Dismissed from Office," which appeared in the January 1961 issue of *Peking Literature and Art.* In this new version, Hai Jui was portrayed as an honest official who lost his governorship because the emperor disliked his proposal of returning to the peasants land that had been seized by rich landlords. It required little imagination to see that the author implied the emperor to be Mao and the dismissed official to be P'eng. Naturally, Mao did not miss the point, as he later stated on December 21, 1967: "P'eng Te-huai is 'Hai Jui' too."[26]

Two other members of the Peking Municipal Government were even more scathing in their criticism of the leadership. They were Teng T'o, secretary of the Peking Municipal Committee, editor of its theoretical journal, the *Frontline,* and former editor-in-chief of the *People's Daily* (1954-59); and Liao Mo-sha, director of the United Front Department in the Peking Municipal Committee. Under the pseudonym Wu Nan-hsing,[27] the three of them jointly published sixty-seven articles in the *Frontline* between October 10, 1961 and July 1964, criticizing the Leader by implications or historical analogies. Particularly outspoken, Teng wrote 153 articles in the *Peking Evening News* between March 1961 and September 1962 under the general title, "Evening Chats at Yenshan." One of these "Evening Chats," dated June 15, 1961, sarcastically compared, by allusion, the Great Leap with building a castle in the air. Another article, "Special Cure for Amnesia," which appeared in the *Frontline* of July 25, 1962, read in part:

> There are all sorts of illnesses in this world . . . one of which is called "amnesia." Anyone who has this disease has a lot of trouble, because it cannot easily be cured. . . . The man who has this illness . . . often forgets what he has said or done. Gradually he will become temperamental . . . easily angered, and finally mad. . . . Another symptom is that he often faints. If not cured in time, he will become an idiot. Once one of these symptoms is discovered, he must take a full rest and stop talking or doing things, otherwise the result will be disastrous.[28]

The frequent appearance of sarcastic writings using historical analogies or allusions to criticize Mao and his policies suggested a coordinated effort

25. Chün-tu Hsüeh, "The Cultural Revolution and Leadership Crisis in Communist China," *Political Science Quarterly,* LXXXII:2:173 (June 1967).
26. Solomon, 479.
27. Wu for Wu Han; Nan for Nan-tsun, Teng's pen name; and Hsing for Fan Hsing, Liao's pen name.
28. Quoted in Chün-tu Hsüeh, 175.

directed by individuals very high in authority. Since all three contributors were members of the Peking Municipal Government, it was clear that the mayor, P'eng Chen, or someone higher in the party hierarchy, was behind them. Mao was ready to launch a counterattack but Peking was so tightly controlled by its mayor that Mao could not find room "to put in a needle."[29] In the summer of 1965, he disappeared to Shanghai, where at a branch meeting of the Central Committee in September, he called for attacks on "reactionary bourgeois ideology." The groundwork was laid for a gigantic counteroffensive. Under his and his wife's instructions, the editor-in-chief of the Shanghai branch of the *Liberation Army Daily*, Yao Wen-yüan, fired the first shot of the Cultural Revolution on November 10, 1965, when he attacked the play "Hai Jui Dismissed from Office." Wu Han was denounced for his "humanism" and lack of "class view," since he held that Hai Jui of the official ruling class could understand and help another class—namely, the peasants. Other attacks quickly descended on Wu Han, Teng T'o, and Liao Mo-sha, now dubbed the "Black Gang." They were accused of falsifying historical figures to satirize the present Leader, of deceiving people into learning the virtues of the feudal past, of blurring the class struggle, and of urging the restoration of land and private economy.

Mao's selection of Wu Han as the first target was well thought out. As deputy mayor of Peking and a leading intellectual associated with Teng and Liao, Wu's chastisement would inevitably involve the other two and perhaps Mayor P'eng Chen. If it were proven that P'eng was part of this "counterrevolutionary and revisionist" gang, then his patron, Liu Shaoch'i, would be implicated. Since their criticism of the Maoist mismanagement of the economy and practice of the "personality cult" resembled Khrushchev's denunciation of Stalin, they could be branded as revisionists. Thus, a literary and cultural revolution exploded into a bitter power struggle among the top leaders.

The Full Swing of the Purge

Under mounting pressure, Wu Han recanted on December 30, 1965, admitting that he had failed to use Mao's theory of class struggle in his play. Unsatisfied, the Maoists insisted on getting at the "truth." Apparently unaware of the depth of the attack and confident of his hold on the party machinery, Liu left Peking on March 26, 1966, on a prescheduled state visit to Pakistan and Afghanistan. That very day, Mayor P'eng Chen, a

29. Mao's own expression, quoted in Gene T. Hsiao, 397.

vice-premier and the eighth-ranking member of the Politburo, disappeared.

On April 18, the *Liberation Army Daily* editorialized: "Hold High the Great Red Banner of Mao Tse-tung's Thought and Actively Participate in the Great Socialist Cultural Revolution." It was a clear declaration of the army's support for Mao and his policy. During Liu's absence, the Maiosts closed in on the Peking Municipal Party Committee and extracted confessions from the Black Gang. In early May, Mao's spokesman in Shanghai, Yao Wen-yüan, charged that the Liu faction intended to replace Premier Chou En-lai with P'eng Chen and to rehabilitate former Defense Minister P'eng Te-huai and his revisionist line at the expense of Lin Piao. On May 16, an article in the *Red Flag*, edited by Mao's private secretary Ch'en Po-ta, asked: "Who has been sheltering Teng T'o and his group?" On June 1, Maoist forces seized control of the *People's Daily*, which then belatedly joined the Cultural Revolution.

Meanwhile, Mao's health apparently had improved: in July 1966 he demonstrated his vigor by swimming in the Yangtze River and urged the youth to relive the revolutionary experience with him by "advancing in the teeth of great storms and waves."[30] Mao proved himself physically capable of conducting the Cultural Revolution.

Mao's reliance on the army to crush the party catapulted Lin Piao to new heights of power and prestige. He became the strong man and the main pillar of support for Mao when Mao returned on July 18, 1966, to Peking—which had been made secure by Lin and his army. On August 1, Mao named him the first vice-chairman of the Central Committee, i.e. the second-ranking member in the hierarchy, while Liu fell to eighth. Side by side, Mao and Lin reviewed the Red Guards (then being formed), while Mao fondly addressed Lin as his "closest comrade-in-arms." On August 5 Mao wrote the first wall poster: "Bomb the (Liu-Teng) Headquarters!"

The Eleventh Plenum of the Central Committee adopted several key resolutions. It designated as the target of attack "those within the party who are in authority" and who were "taking the capitalist road." It announced the creation of the Red Guards (*Hung-wei ping*) as a "shock force" to carry the movement from the capital to the provinces—bypassing the party machinery and its Youth League, which were under Liu's control. It called for the establishment of permanent "cultural revolutionary groups, committees and congresses" at all levels, and the application of Mao's ideas on the mass line, the class struggle, and the theory of contradictions. On November 22, 1966 a seventeen-member Central Cultural

30. Solomon, 464, 476.

Revolutionary Committee was formed, with Mao's secretary Ch'en Po-ta (editor of the *Red Flag*) as chairman, and Mao's wife, Chiang Ch'ing, as first vice-chairwoman. This Revolutionary Committee, together with the military under Lin Piao, and the State Council under Chou En-lai, became the ruling triumvirate, under Mao's guidance. In particular, Lin seemed to have taken over the purge during the second half of the Cultural Revolution.

The youthful Red Guards envisioned themselves as "revolutionary successors" and "revolutionary rebels," dedicated to the elimination of old thought, old culture, old customs, and old habits. Vowing to uphold the Thought of Mao, they were determined to expunge bourgeois influences and revisionist tendencies. They wrote big-character wall posters (*Ta-tzu pao*), ransacked private property, rampaged cities, renamed streets, attacked those with modern attire and haircuts, and humiliated foreign diplomats. By the spring of 1967 the Red Guard disturbances reached alarming proportions.[31]

The Red Guards attacked Liu Shao-ch'i as a revisionist and a Chinese Khrushchev, and pressured him and his wife into public self-criticism. Thousands of them marched by his house demanding his dismissal. In November 1968 the CCP Central Committee announced that Liu had been ousted from all party and government posts.

Other prominent officials who had been attacked, humiliated, dismissed, and purged included Teng Hsiao-p'ing, the party general secretary; Chu Teh, a founder of the Red Army; Po I-po, a vice premier and chairman of the State Economic Commission; and several hundred government and party leaders.

Throughout the Cultural Revolution, Premier Chou En-lai played the role of a "conflict manager" superbly. He supported Mao, maintained a working relationship with the military under Lin Piao and the Cultural Revolutionary Group under Chiang Ch'ing—while exerting his moderating influence in an attempt to keep the turmoil within bounds. Many times he spoke out against the excesses of the Red Guards.

It was due to Chou's mediation that a three-way alliance between the PLA representatives, "revolutionary" party cadres, and representatives of the "revolutionary" masses was created in the Revolutionary Committees at the various levels—thus achieving some coherence among the different groups which might otherwise work at cross purposes. Chou also protected officials and generals who came under Red Guard attacks. He defended Nieh Jung-chen, head of the National Defense Science and Technology

31. William W. Whitson, *The Chinese High Command: A History of Communist Military Politics, 1927-71* (New York, 1973), 392.

Commission which directed advanced weapon projects, against the militant's threat to disturb the atomic research and weapon programs in winter and spring of 1968. All in all, Chou worked diligently to keep the Cultural Revolution under control.

Yet, the help of the army was still needed. In January 1967, Mao instructed the military to intervene and restore order. In retrospect, it was a momentous decision, for it gave Lin Piao and the PLA a golden opportunity to fill the vacuum created by the decimation of the party organs at different levels and to penetrate deep into industry, plants, and other important institutions. Unprecedentedly, the military became a powerful political force.

The violence and chaos wrought by the Red Guards may have surprised their original organizers. In July 1968 Mao summoned the five student leaders representing major Red Gaurd groups in Peking to reprimand them for their "ultra-leftism, their sectarianism, and the mad fratricidal combats." With tears in his eyes he said that they had let him down.

Meanwhile, intensive preparations were made for the convocation of a party congress, the first since 1958. The Ninth Party Congress, which finally opened in April 1969, unanimously elected Mao chairman of the party and the Central Committee, with Lin Piao as vice-chairman. The new party constitutions reaffirmed the Thought of Mao as the guiding policy of the party and the state, and designated Lin as Mao's successor. The Central Committee was enlarged to include 170 regular members and 109 alternates, a substantial percentage of whom were army commanders and leaders of the Cultural Revolution.

The Cultural Revolution in Retrospect

The Maoists proclaimed the Cultural Revolution a great victory because it re-established the supremacy of Mao's authority and of his thought and ideology, which were deemed essential to China's progress. It was a planned upheaval conducted by Mao from the start and seen as part of a *continual* rectification movement that must erupt periodically to insure the purity of the party and the correctness of its line.[32] In 1967 Mao warned that in the future there would be "one, two, three, or four Cultural Revolutions."

Obviously the Cultural Revolution benefitted Lin Piao and the military, but it was significant that Mao's wife also rose to national prominence

32. Solomon, 476.

during the prolonged period of upheaval. In the early sixties, with Mao's blessing Chiang Ch'ing developed an active interest in reforming the arts and produced eight "model operas." She then took over the media as a prelude to winning control of the national culture and the people's minds. "She felt she needed such command over popular consciousness, including recognition by the masses, as the basis of her personal power and authority," remarked her American biographer.[33] That she was a power behind Mao in precipitating the Cultural Revolution was beyond dispute. She accompanied him to Shanghai in 1965, jointly directing Yao Wen-yüan to fire the first shot of the Cultural Revolution, and a year later she was catapulted to the post of first vice-chairperson of the Central Cultural Revolutionary Committee. Her meteoric rise could only be the result of Mao's sponsorship. In purging his senior associates he unwittingly, or wittingly, made way for her and her followers to build up a power base, and she exploited the opportunity to the hilt. Asked whether the Cultural Revolution involved struggle over succession and civil war, she readily acknowledged: "There's some truth in that."[34] One might say that whatever other motives Mao may have had in launching the Cultural Revolution he certainly opened a way to thrust his wife into the forefront of national politics, positioning her for the ultimate bid for succession.

In retrospect, the Cultural Revolution ushered in a decade of turmoil and civil strife that drove the country to utter chaos and the brink of bankruptcy. The party had been decimated and many of its leaders purged or dismissed. Industrial and agricultural productions suffered severe setbacks, and the disruption in education caused the loss of a generation of trained manpower. In fact, not only was the younger generation deprived of education but a great many middle-aged and senior scholars and scientists were sent to the countryside to do menial chores, denying them for years the opportunity for research and teaching. The damage in effect invovled three generations. Poignantly, the Cultural Revolution turned out to be anticultural, anti-intellectual, and antiscientific, for knowledge was considered the source of reactionary and bourgeois thought and action. Countless officials and individuals were wrongfully accused of anti-revolutionary activities and driven to suicide or imprisonment. Yet for the revolutionary purists, no price was too high and no sacrifice too great for the perpetuation of the Maoist vision and approach to socialist transformation. In 1981, an authoritative assessment of the Cultural Revolution was made (see chapter 36).

33. Roxane Witke, *Comrade Chiang Ch'ing* (Boston, 1977), 380.
34. *Ibid.*, 297.

Further Reading

Ahn, Byung-joon, *Chinese Politics and the Cultural Revolution: Dynamics of Policy Processes* (Seattle, 1976).

Barcata, Louis, *China in the Throes of the Cultural Revolution* (New York, 1968).

Baum, Richard, *Prelude to Revolution: Mao, the Party, and the Peasant Question, 1962-1966* (New York, 1975).

Bernstein, Thomas P., *Up to the Mountains and to the Villages: The Transfer of Youth from Urban to Rural China* (New Haven, 1977).

Bettelheim, Charles, *Cultural Revolution and Industrial Organization in China: Changes in Management and the Division of Labor* (New York, 1974).

Bridgham, Philip, "Mao's 'Cultural Revolution': Origin and Development," *The China Quarterly*, 29:1-25 (Jan.-March 1967).

———, "Mao's Cultural Revolution: The Struggle to Seize Power," *The China Quarterly*, 41:1-25 (Jan.-March 1979).

Brugger, Bii (ed.), *China: The Impact of the Cultural Revolution* (New York, 1978).

Chan, Anita, *Children of Mao: Personality Development and Political Activism in the Red Guard Generation* (Seattle, 1985).

Chao, Kang, "Economic Aftermath of the Great Leap in Communist China," *Asian Survey*, IV:5:851-58 (May 1964).

Charles, David A., "The Dismissal of Marshal P'eng Teh-huai," *The China Quarterly*, 8:63-76 (Oct.-Dec. 1961).

Chen, Jack, *Inside the Cultural Revolution* (New York, 1975).

Chen, Jo-hsi, *The Execution of Mayor Yin and Other Stories from the Great Proletarian Cultural Revolution* (Bloomington, 1978).

Cheng, Chu-yüan, "The Root of China's Cultural Revolution: The Feud between Mao Tse-tung and Liu Shao-ch'i," *Orbis*, XI:4:1160-78 (Winter 1968).

Daubier, Jean, *A History of the Chinese Cultural Revolution*, tr. by Richard Seaver (New York, 1974).

Dittmer, Lowell, *Liu Shao-ch'i and the Chinese Cultural Revolution: The Politics of Mass Criticism* (Berkeley, 1975).

Domes, Jürgen, *Peng Te-huai: The Man and the Image* (London, 1985).

Fokkema, D. W., *Report From Peking: Observations of a Western Diplomat on the Cultural Revolution* (Montreal, 1972).

Gao, Yuan, *Born Red: A Chronicle of the Cultural Revolution* (Stanford, 1987).

Goldman, Merle, "The Fall of Chou Yang," *The China Quarterly*, 27:132-48 (July-Sept. 1966).

Gray, Jack, and Patrick Cavendish, *Chinese Communism in Crisis: Maoism and the Cultural Revolution* (New York, 1968).

Hsüeh, Chün-tu, "The Cultural Revolution and Leadership Crisis in Communist China," *Political Science Quarterly*, LXXXII:2:169-90 (June 1967).

Joffe, Ellis, "The Chinese Army after the Cultural Revolution: The Effects of Intervention," *The China Quarterly*, 55:450-77 (July-Sept. 1973).

Joseph, William A., *The Critique of Ultra-Leftism in China, 1958-1981* (Stanford, 1984).

Karnow, Stanley, *Mao and China: From Revolution to Revolution* (New York, 1972).

Klein, Donald W., "Victims of the Great Proletarian Cultural Revolution," *The China Quarterly*, 27:162-65 (July-Sept. 1966).

Laing, Ellen Jornston, *The Winking Owl: Art in the People's Republic of China* (Berkeley, 1988).

Lee, Hong Yung, *The Politics of the Chinese Cultural Revolution: A Case Study* (Berkeley, 1978).

———, "Mao's Strategy for Revolutionary Change: A Case Study of the Cultural Revolution," *The China Quarterly*, 77:50-73 (March 1979).

Leys, Simon (Pierre Ryckmans), *The Chairman's New Clothes: Mao and the Cultural Revolution* (tr. from French ed., 1972), (New York, 1977).

Lifton, Robert J., *Revolutionary Immortality: Mao Tse-tung and the Chinese Cultural Revolution* (New York, 1968).

Lo, Ruth Earnshaw, and Katharine S. Kinderman, *In the Eye of the Typhoon, An American Woman during the Cultural Revolution* (New York, 1987).

MacFarquhar, Roderick, *The Origins of the Cultural Revolution, Vol. 1: Contradictions among the People, 1956-1957* (London, 1974).

———, *The Origins of the Cultural Revolution*, Vol. 2, *The Great Leap Forward, 1958-1960* (New York, 1984).

Michael, Franz, *Mao and the Perpetual Revolution* (Woodbury, N.Y., 1977).

Nelsen, Harvey, "Military Forces in the Cultural Revolution," *The China Quarterly*, 51:444-74 (July-Sept. 1972).

Oksenberg, Michel et al., *The Cultural Revolution, 1967, in Review* (Ann Arbor, 1968).

Raddock, David M., *Political Behavior of Adolescents in China: The Cultural Revolution in Kwangchow* (Tucson, 1977).

Robinson, Thomas W. (ed.), *The Cultural Revolution in China* (Berkeley, 1971).

Rosen, Stanley, *Red Guards Factionalism in China's Cultural Revolution: A Social Analysis* (Boulder, 1981).

Seybolt, Peter J., *The Rustication of Urban Youth in China: A Social Experiment*, ed. with an intro. by Thomas Bernstein (White Plains, N.Y., 1977).

Schram, Stuart R., "Introduction: The Cultural Revolution in Historical Perspective," in Stuart R. Schram (ed.), *Authority Participation and Cultural Change in China* (Cambridge, Eng., 1973), 1-108.

Solomon, Richard H., *Mao's Revolution and the Chinese Political Culture* (Berkeley, 1971).

Starr, John Bryan, "Conceptual Foundations of Mao Tse-tung's Theory of Continuous Revolution," *Asian Survey*, XI:6:610-28 (June 1971).

Thurston, Ann F., *Enemies of the People: The Ordeal of the Intellectuals in China's Great Cultural Revolution* (Cambridge, Mass., 1988).

Tsou, Tang, "The Cultural Revolution and the Chinese Political System," *The China Quarterly*, 38:63-91 (Apr.-June 1969).

20 *Years On: Four Views on the Cultural Revolution, The China Quarterly*, 108: 597-651 (Dec. 1986).

Whitson, William W., *The Chinese High Command: A History of Communist Military Politics, 1927-71* (New York, 1973).

Wang, James C. F., *The Cultural Revolution in China: An Annotated Bibliography* (New York, 1976).

Wu, Tien-wei, *Lin Biao and the Gang of Four* (Carbondale, Ill., 1983).

Yue, Daiyun, and Carolyn Wakeman, *To the Storm: The Odyssey of a Revolutionary Chinese Woman* (Berkeley, 1985).

President Nixon toasting Premier Chou En-lai at a banquet in the Great
Hall of the People, Peking. February 21, 1972.

Chairman Mao receives Japanese Prime Minister Kakuei Tanaka in his
study in Peking. September 1972.

Chairman Mao receives President Nixon and Henry Kissinger in his study
in Peking. Also present: Premier Chou En-lai at left. February 1972.

Chiang Ch'ing and Chang Ch'un-ch'iao in 1973.

Wang Hung-wen in 1973.

Yao Wen-yüan in 1973.

Archeological finds in China: terra-cotta figures in battle formation guarding the Tomb of Ch'in Shih-huang-ti ca. 210 B.C., outside Sian.

Archeologist measures tomb figures.

Close-up of figure from the Tomb of Ch'in Shih-huang-ti.

Bronze warriors, horses, and chariot found in a tomb of Eastern Han dynasty (25–220 A.D.) at Wūwei, Kansu Province.

Bronze chimes of the Warring States Period unearthed in Fuling, Szechuan Province.

Chiang Kai-shek and his son, Chiang Ching-kuo, in May 1974. Picture first appeared in Japanese newspaper *Sankei Shimbun* on August 15, 1974.

29

The Fall of Lin Piao
and Its Aftermath

Lin Piao emerged from the Ninth Party Congress victorious and seemingly invincible. As vice-chairman of the party and successor to Mao Tsetung, he occupied a unique position in the history of Chinese Communism. Never before had the party constitution prescribed succession and the change reflected Lin's great influence and Mao's patronage of him. Fondly describing him as his "closest comrade-in-arms," Mao appeared to have chosen Lin to be groomed for future leadership.

Lin's ostensible strength was buttressed by the powerful People's Liberation Army (PLA) which came to occupy a pivotal position in the Chinese political structure because of its role during the Cultural Revolution. Of the 170 full and 109 alternate members of the Central Committee, 44.1 percent were military men. In the 21-man Politburo there were four marshals, six generals, and Lin's wife, Yeh Ch'ün—enough for a majority if they voted in consensus. In the provinces the military representation was no less formidable. Nineteen of the chairmen and twenty of the vice-chairmen of the provincial revolutionary committees were members of the PLA, thus accounting for more than 50 percent of the top positions. Further, military commanders or commissars at both provincial and local levels frequently doubled as party secretaries and administrative chairmen in the government.[1]

1. Ralph L. Powell, "Party Still Striving to Retain Control of 'the gun' in China," *The Christian Science Monitor*, Sept. 21, 1973, 11. Other accounts give slightly different figures.

The Rise of Lin

Born in 1907, Lin became a Communist in his teens and before the age of twenty was a member of the fourth graduating class of Whampoa Military Academy. During the Long March he led the vanguard and during the Civil War (1945-49) he commanded the famous Fourth Field Army that swept into Manchuria and down to Southeast China, winning every battle along the way. In 1954 he was named a vice-premier, a year later one of the ten marshals of the PLA and a member of the Politburo, and in 1958 a member of its Standing Committee. The following year he replaced P'eng Te-huai as minister of defense and was concurrently made the de facto head of the Party's Military Affairs Committee, a position connoting superintendency of all the armed services. He used this supervisory position to improve the combat readiness of the armed forces and to strengthen their political reliability. Indoctrination of army personnel made it clear that politics was in command and that the party had "absolute control" of the army. Lin insured that the army's political commissars enjoyed great power and distinctive status. He won praise from the party, commendation for his "creative employment" of the Thought of Mao, and received the unique title of *Tsung* (chief) in 1961.[2] The successful campaign against India in 1962 further enhanced his and the PLA's prestige and stature.

With each success Lin increased his glorification of Mao, praising the Chairman's Thought as the ever-inspiring guide for all time. Lin's influence penetrated nearly all of Chinese life, and army personnel supervised party activities, especially those concerning production and economic management. Thus, the army, reversing the party's practice of infiltrating the military, extended its influence into the party and the government. The development of a nuclear capability further elevated the status of Lin and the prestige of the army. Most conspicuous was the statement following the fourth test in October 1966 emphasizing that all of the personnel involved in the project had "enthusiastically" responded to "the call of Comrade Lin Piao."[3] With the publication in September 1965 of his 20,000-word treatise, "Long Live the Victory of the People's War," Lin's reputation soared beyond the military ranks. Cleverly exploiting the Maoist idea of organizing the peasants against the bourgeoisie, he advanced the thesis that the underdeveloped nations in the "countryside" of the world could encircle and defeat the capitalist industrial states in the

2. Ralph L. Powell, "The Increasing Power of Lin Piao and the Party-Soldiers, 1959-1966," *The China Quarterly* (April-June 1968), 34:44.
3. *Peking Review*, 44, Oct. 28, 1966, special supplement, iii, 41, Oct. 7, 1966, 31.

"city" areas.[4] With this essay Lin became a leading theorist and the interpreter and propagator of the Thought of Mao.[5]

During the Cultural Revolution, the apotheosis of Mao intensified. Placing him first in the Communist pantheon—above Marx, Engels, Lenin, and Stalin—conscious efforts were made to characterize Mao as having *creatively developed* Marxism to new peaks. In 1967 Lin Piao made a glowing eulogy of Mao:

> It is Comrade Mao Tse-tung, the great teacher of the world proletariat of our time, who in the new historical conditions, has systematically summed up the historical experience of the dictatorship of the proletariat in the world, scientifically analyzed the contradictions in the socialist society, profoundly shown the laws of class struggle in socialist society and put forward a whole set of theory, line, principles, methods and policies for the continuation of the revolution under the dictatorship of the proletariat. With supreme courage and wisdom, Chairman Mao has successfully led the great Proletarian Cultural Revolution in history. This is an extremely important landmark, demonstrating that *Marxism-Leninism has developed to the state of Mao Tse-tung's thoughts.*[6]

Mao's January 1967 decision to order the PLA to intervene in the Cultural Revolution and to restore order was momentous for Lin and the military. This involvement allowed them to further infiltrate industry, agriculture, and education while increasing their power throughout the country. The vacuum created by the decimation of the party was filled quickly by the military. By April 1967 Lin was generally portrayed as the "most loyal supporter" and "best successor" to Mao. At the Ninth Party Congress in April 1969 Lin's victory was virtually "complete," when he was designated vice-chairman of the party and successor to Mao.

A clever schemer, Lin was well aware that his rise contradicted the Maoist dictum that "the party commands the gun, and the gun will never be allowed to command the party." He was conscious of his tenuous position and realized that his success would not be total or secure until he actually became the Leader. Before that moment came, he must do everything possible to safeguard his gains and insure his right to succession. Thus, in public he made every effort to show his absolute deference to Mao, walking ever so meekly behind him as a humble pupil. But, beneath the surface of this sycophantic behavior, he steadily expanded his control

4. Text in *Peking Review*, 36, Oct. 3, 1965, 9-30.
5. *People's Daily*, Dec. 29, 1964; Chu-yüan Cheng, "Power Struggle in Red China," *Asian Survey*, VI:9:472-74 (Sept. 1966).
6. *Peking Review*, Nov. 10, 1967; italics added.

of the armed forces, the party, the government, and the provinces. His position became so strong that any challenge to his right to succession would likely precipitate vast conflict and great upheaval. To all appearances, Lin's position looked impregnable in 1970; all he needed to do was wait patiently for the mantle to pass into his hands. But, the history of Chinese Communism is full of the unexpected, and only time could tell what was in Mao's mind.

The Fall of Lin

Traditionally, the Chinese Communist armed forces refrained from political participation, strictly observing the aforementioned Maoist dictum. From the very beginning when he joined Mao in 1928, Chu Teh willingly submitted to Mao as the political leader, although militarily, Chu was the stronger of the two. In the years and decades that followed, there was never any question about politics being in command. Even during the Cultural Revolution, the army did not intervene voluntarily but was ordered by Mao to play a political role, albeit the new role fitted in well with Lin's plans.[7] Mao did so reluctantly as he had no choice but to rely on the army to shatter the party, which was subtly bypassing or resisting the implementation of his policies.

But, Mao was already suspicious of Lin's intention. In a letter to his wife dated July 8, 1966, Mao questioned some of Lin's views:

> My friend [Lin] . . . specialized in discussing the problems of coup d'état. This kind of thinking has never been done before and it made me feel *very uneasy*. I have never believed that a few small books of mine could have that kind of supernatural power, but under his sponsorship the whole country came to praise them. This is like Mrs. Wang selling melons for self-advertisement. I am forced to go along [with him] as it seems I have no choice but to agree. On questions of great importance, *this is the first time that I agree with others against my own will.*[8]

As Lin continued to exploit Mao's name and vastly expand his own power, Mao secretly became alarmed but took no immediate action. Nonetheless, after the Ninth Party Congress Mao began to shift his position and relied increasingly on Chou En-lai for rebuilding a civilian party. There is also evidence that in autumn 1968 differences of opinion between Mao and Lin already developed over the magnitude of the purge

7. Ellis Joffe, "The Chinese Army after the Cultural Revolution: The Effects of Intervention," *The China Quarterly* (July-Sept. 1973), 55:451-52.
8. Mao's letter to Chiang Ch'ing, dated July 8, 1966, and made public in 1972.

of party members.[9] Describing himself as a 'center-Leftist," Mao believed that the army had been too zealous and violent in the purge and treatment of party cadres.

Because of his growing distrust of Lin, in March 1970 Mao decided to abolish the state chairmanship. The post was not included in the new draft state constitution, thus precluding Lin's capture of that position. Sensing the possibility of disinheritance, Lin persistently demanded the retention of that position, only to be told by Mao on six occasions that he saw no need for it. Mao also deleted Lin's proposal for a provision to extol Mao as "genius," remarking: "I am no genius. I studied Confucian books for six years and works on capitalism for seven years before I read Marxism-Leninism in 1918. How can I be a genius? . . . A genius does not rely on one or more men but on the party as the vanguard of the proletarian class. A genius relies on the mass line for collective wisdom."[10]

These questions—the practicality of the state chairmanship and what constitutes genius, along with a host of others—were openly debated for the first time at the Second Plenum of the Ninth Central Committee held at Lushan in August 1970. The Mao-Chou group, having drafted the new constitution, were challenged on such issues as the correct ideology, the power structure, and the political line. The antagonists, Lin Piao and Ch'en Po-ta,[11] were supported by an array of prominent military figures, including Chief of the General Staff Huang Yung-sheng, Air Force Commander Wu Fa-hsien, the First Political Commissar of the Navy Li Tso-p'eng, and Deputy Chief of Staff Ch'iu Hui-tso. As Mao recalled in a report dated March 17, 1972: "At the 1970 Lushan Conference . . . they first conspired covertly and then launched a surprise attack. . . . A certain person, anxious to become chief of state, tried to split the party and was eager to seize power."[12] The "surprise attack" continued from August 23 to mid-day August 25 and then failed. The nature of this attack was

9. Claude Julien, "The Lin Piao 'mystery,'" *Le Monde*, Dec. 30, 1971, quoted in Philip Bridgham, "The Fall of Lin Piao," *The China Quarterly* (July-Sept. 1973), 55:429.
10. CCP Central Committee document, *Chung-fa*, No. 12, March 17, 1972, entitled "Summary of Chairman Mao's Talks with Leading Comrades during His Inspection Tour to Outlying Areas (mid-August to Sept. 12, 1971)," reproduced in full in *Chung-yang jih-pao* (*Central Daily News*), Taipei, Aug. 10, 1972. Translation is mine. An English version may be found in *Studies on Chinese Communism*, 6:9: 18-24 (Sept. 10, 1972).
11. Ch'en, Mao's long-time private secretary and speech writer, rose to be the head of the Central Revolutionary Committee during the Cultural Revolution. Reportedly, he was eying the premiership occupied by Chou En-lai, and turned to the rising star, Lin Piao, in the power struggle against the Mao-Chou group.
12. CCP Central Committee document, *Chung-fa*, No. 12, March 17, 1972. Translation is mine.

not revealed until the Tenth Party Congress in August 1973 when Chou En-lai characterized it as the first of two abortive attempts to kill Mao.

The Lushan Conference signalled an open split between Mao and Lin, with the latter desperately attempting to hold on to his right of succession and the former vigorously trying to withdraw it. Mao knew that he could not destroy the opposition hastily without risking a rebellion; he had to proceed cautiously. Instead of confronting Lin directly, he attacked Ch'en Po-ta as an "ultra-leftist" and "political swindler," and asked the military leaders below Lin to recant and to critize Ch'en's views. When the leaders complied, Ch'en Po-ta disappeared and Lin lost an important supporter. Chou En-lai subsequently conducted a short campaign criticizing the "Five Great Generals"—the four mentioned above and Lin's wife, Yeh Ch'ün—and then declared the problems resolved.

To weaken Lin's position, Mao used three tactics colorfully described as: (1) "throwing a stone," i.e. the purge of Ch'en Po-ta as an admonition to Lin; (2) "mixing gravel" with mud, i.e. infiltrating the Military Affairs Committee with his own men to spy on Lin; and (3) "digging the cornerstone," i.e. reorganizing the Peking Military Region in January 1971 to undermine Lin's power base.[13]

With Ch'en Po-ta deposed and the leading generals deflated, Lin knew that he would be the next target. Between the winter of 1970 and the spring of 1971, he gradually concluded that the only viable alternative was a military coup.[14] He put his son, Lin Li-kuo, an air force major general and deputy commander of the Air Operations Command, in charge of drawing up plans for a coup that was to be supported by a small group of determined recalcitrants. Between March 22-24, 1971, the conspirators secretly gathered in Shanghai and prepared the "5-7-1 Engineering Outline"[15]—the Chinese pronunciation of "5-7-1," *wu-ch'i-i*, being homonymous with the characters for "armed uprising." Mao was given the code name "B-52," Lin "the chief," and his followers "the combined fleet."

The plan stated that "at the present both our enemy and ourselves are riding a tiger and finding it hard to dismount. It is a situation of life-or-death struggle; either we destroy them or they destroy us."[16] Portraying

13. *Ibid.* Translation is mine.
14. Bridgham, 436.
15. This "5-7-1" project, revealed in a CCP Central Committee document, *Chung-fa*, No. 4, Jan. 13, 1972, was later disclosed in summary form by Chou En-lai to foreign friends. It was reprinted in *Chung-yang Jih-pao* (*Central Daily News*), Taipei, April 13, 1972. Excerpts in English appear in *Studies on Chinese Communism*, 6:7:7-12 (July 1972). Western observers generally accepted its authenticity as Peking's secret file. See Robert E. Elegant, "China Politics still haunted by Lin Piao," *Los Angeles Times*, April 1, 1973.
16. "5-7-1" document, translation is mine.

Mao as a "sadist and a man afflicted with suspicion-mania," the plan continued,

> B-52 always plays off one faction against another. . . . He may talk sweetly today to those whose support he solicits, but tomorrow they may be condemned to death on trumped-up charges. . . . Once he thinks someone is his enemy, he won't stop until the victim is put to death; once you offend him, he'll persist to the end—passing all blame to the victim, held responsible for crimes committed by himself.[17]

In terms of the operational methods, "5-7-1" continued,

> If B-52 falls into our hands, the enemy battleships [senior leaders] will also be in our hands. They will fall into our trap. We could use a high level meeting to seize them or we could cut off B-52's right-hand men and force him to submit to a palace revolution. We could also use extraordinary methods like chemical or bacteriological weapons, bombing, 543 [a presumed secret weapon, nature unspecified], and arranged car accident, assassination, kidnapping or urban guerrilla units.[18]

The plan was said to have had the support of Moscow, and this seems highly possible since Lin had argued for a rapprochement with the Soviet Union, in opposition to the Mao-Chou policy of détente with the United States, on the grounds that any compromise with capitalistic Washington would constitute a betrayal of proletarian internationalism.[19]

Lin had counted on the support of many of the provincial military leaders who owed him their appointments. To forestall a military confrontation, Mao, the consummate master of struggle, made an inspection tour to Nanking and Canton from mid-August to September 12, 1971, impressing upon the regional generals the need to relinquish their political role. Overpowered by Mao's imposing presence and sensing the futility of opposition, these regional military heads were helplessly neutralized. During the tour, an unsuccessful attempt was made on Mao's life. The storm was about to break.

Basically, Lin Piao's strength depended on Mao's patronage and the

17. Translation in *Free China Weekly*, Taipei, April 16, 1972; Bridgham, 439.
18. Robert Elegant, *Los Angeles Times*, April 1, 1973.
19. According to one source the White House amazingly learned of Lin's plot and his secret relations with Moscow through an Israeli intelligence source which had good contacts inside the Kremlin. President Nixon then decided to inform Mao and the message was sent by Kissinger to Chou En-lai during their first meeting in July 1971. This fantastic information came from the Washington correspondent of the *Evening Standard* of London, Jeremy Campbell, reproduced in *Free China Weekly*, Jan. 30, 1972.

support of the army. When he lost the first and much of the second, there was little left except for a handful of sworn followers around him. Desperately scheming to put the "5-7-1" plot into action, Lin was suddenly betrayed by one of the conspirators, Li Wei-hsin, deputy secretary of the Political Department of the Fourth Air Force. Lin hurriedly ordered a plane to his summer residence at Pehtaiho, some 175 miles east of Peking. Mao grounded all planes. Lin's son managed to locate a Trident Jet No. 256 and although it lacked adequate fuel, a navigator, or even a radio operator, the three Lins and six others boarded and took off for Mongolia in the direction of the Soviet Union. The plane attempted a forced landing in order to refuel at an airstrip near Under Khan in Outer Mongolia, but its wings hit the ground and all perished in the crash on September 13.[20]

Thus, the anticipated showdown between Lin and Mao ended anticlimactically. Judging from the confession of the betrayer, it seems that Lin's indecision was a main cause of his downfall. After Ch'en Po-ta was purged and the "Five Generals" criticized, Huang Yung-sheng and Lin's wife sensed the imminent danger and agitated for action, but Lin hesitated and lost the initiative. Instead, he and his son considered mobilizing the Fourth and Fifth Air Force to seize Shanghai and turning it into a stronghold against possible military intervention by the Nanking military commander, Hsü Shih-yu. Should the attempt fail, Lin's forces would retreat to the mountains in Chekiang to engage in guerrilla warfare.[21] But, all these plans had no chance to materialize, for at the last minute there was no confrontation. Lin died at 64 years of age, a "renegade and traitor,"[22] while his chief supporters all disappeared from public life.

Mao described the Lin affair as the tenth and the most serious struggle of the party in its fifty-year history and warned that many more confrontations would follow. Struggle was a way of life with the CCP, and "great disorder" was inevitable every seven or eight years (a seven-year itch?). Viewed in this light, Lin's case was an unavoidable by-product of the law of struggle beyond the control of human will. By implication, it absolved all who had tolerated so bad a man as Lin.

20. Chou En-lai's statement to a twenty-two-man delegation of the American Society of Newspaper Editors, *The Christian Science Monitor,* Oct. 12, 1972. However, an unofficial version of Lin's death insisted that he was murdered by Mao's agents right in Peking. See Ming-le Yao, *The Conspiracy and Murder of Mao's Heir* (London, 1983).
21. The confession of Li Wei-hsin was attached to the "5-7-1" document, printed in *Chung-yang Jih-pao* (Central Daily News), April 13, 1972.
22. Chou En-lai's statement, *The New York Times,* Oct. 12, 1972.

The Tenth Party Congress, 1973

The fall of Lin Piao and his cohorts left many vacant posts in the party and government. Of the twenty-one-man Politburo only ten were left and of its five-man Standing Committee, only three, namely, Mao, Chou, and the sickly K'ang Sheng. Therefore, it was necessary to call another party congress to elect new members to the unoccupied positions and to condemn Lin Piao formally as a traitor, conspirator, political swindler, and a right opportunist who "waved the red flag to defeat the red flag" (i.e. posing as a leftist to seize power and effect his rightist designs).

THE NEW POWER STRUCTURE. At the Tenth Party Congress, convened between August 24-28, 1973, the Cultural Revolutionary Group made an intense bid for power with the apparent support of Mao. Chiang Ch'ing and Yao Wen-yüan were elected to the Politburo and Chang Ch'un-ch'iao to its Standing Committee. Most surprising was the selection of a 37-year-old former Shanghai cotton mill worker, Wang Hung-wen, a protégé of Chiang Ch'ing, as the second vice-chairman of the party, next to Mao and Chou.

ASSESSMENTS OF THE DOMESTIC AND INTERNATIONAL SITUATION. Two major reports were presented at the Tenth Party Congress. The first was delivered by the 75-year-old Vice-Chairman Chou En-lai, presumably representing the moderate civilian officials and party cadres, and the second by the 37-year-old second Vice-Chairman Wang Hung-wen, speaking for the radical Cultural Revolutionary Group. Chou scathingly condemned Lin Piao for antirevolutionary activities—including the conspiracy of engaging in "a wild attempt to assassinate our great leader Chairman Mao and set up a rival central committee," betraying the line and policies of the Ninth Party Congress, and playing the treasonous role of a "super-spy" for the Soviet Union which had tried to reduce "China to a colony of Soviet revisionist social imperialism."

Wang Hung-wen, making his debut as a keynote speaker at a party congress, announced that the smashing of the "two bourgeois headquarters, the one headed by Liu Shao-ch'i and the other by Lin Piao," was the main accomplishment of the Great Proletarian Cultural Revolution. Thus, he turned the Cultural Revolution into an ongoing process, extending from the mid-1960s, to the Ninth and Tenth Party Congresses, and into the future, for in the words of Mao, every seven or eight years a great disorder was inevitable because monsters and demons were bound to appear as a result of their class nature. Wang stressed the revolutionary spirit of *daring to go against the tide* and the importance of training young leaders. With

Basic Data of Ten National Party Congresses of the CCP, 1921-1973*

Congress	First	Second	Third	Fourth	Fifth	Sixth	Seventh	Eighth	Ninth	Tenth
Date	July 1921	July 1922	June 1923	January 1925	April 1927	June 1928	April 1945	September 1956	April 1969	August 1973
Duration (days)	5	1-0					49	13	24	5
Place	Shanghai	Shanghai	Canton	Shanghai	Wuhan	Moscow	Yenan	Peking	Peking	Peking
Delegates	12	12	27	20	80	84	547	1,026	1,512	1,249
Members of Presidium								63	176	148
Vice-chairmen or Standing Committee of Presidium								13	1	5
Major reports							3	3	1	2
Full members of Central Committee	3	5	9	10	29	31	44	97	170	195
Alternate members of Central Committee	3		5		11		33	73	109	124
Full members of Politburo						7	9	17	21	21
Alternate members of Politburo								7	4	4
Standing Committee of Politburo								6	5	9
Vice-chairmen of Central Committee								5	1	5
Plenums of Central Committee						6	7	12	2	3
Interval between Congresses (year-month)		1-0	0-11	1-8	2-3	1-2	16-10	11-5	12-7	4-4
Party membership	57	123	432	950	57,967	40,000	1,211,128	10,734,384	—	28,000,000

* Data adapted with minor changes from "Chung-kung shih-t'zu ch'ü-kuo tai-piao ta-hui tzu-liao chien-piao." Chung-Hua yüeh-pao [The China Monthly], No. 697 (October 1973), 39.

Source: Chinese Law and Government: The Tenth CCP Congress, Analysis and Documents (Spring-Summer 1974, pp. 106-7). Reproduced by special permission of the International Arts and Science Press, Inc.

the future of the country in the hands of the young, struggle and *continued revolution* would punctuate Chinese political life.[23]

23. Delivered Aug. 24, 1973 and adopted Aug. 28. Full text in *Peking Review*. Nos. 35-36, Sept. 7, 1973, 29-33. Italics added.

Further Reading

Bao, Ruo-wang, and Rudolph Chelminski, "The Case Against Confucius," *Saturday Review/World*, May 18, 1974, 12.

Barnett, A. Doak, *China After Mao* (Princeton, 1967).

Bridgham, Philip, "The Fall of Lin Piao," *The China Quarterly*, 55:427-49 (July-Sept. 1973).

CCP Central Committee document, *Chung-fa*, No. 4, Jan. 13, 1972, revealing the "5-7-1" Project, was reprinted in *Chung-yang Jih-pao* (*Central Daily News*), Taipei, April 13, 1972. Excerpts in English appear in *Studies on Chinese Communism*, 6:7:7-12 (July 1972).

————, No. 12, March 17, 1972, entitled "Summary of Chairman Mao's Talks with Leading Comrades during His Inspection Tour to Outlying Areas (mid-August to Sept. 12, 1971)," reproduced in *Chung-yang Jih-pao* (*Central Daily News*), Taipei, August 10, 1972. An English version may be found in *Studies on Chinese Communism*, 6:9:18-24 (Sept. 10, 1972).

————, No. 34, Sept. 8, 1973, reproduced in *Chung-yang Jih-pao* (*Central Daily News*), Feb. 6, 1974.

Chang, Parris, "The Anti-Lin Piao and Confucius Campaign: Its Meaning and Purposes," *Asian Survey*, XIV:10:871-86 (Oct. 1974).

Domes, Jürgen, "The Chinese Leadership Crisis: Doom of an Heir," *Orbis*, XVII:3:863-79 (Fall 1973).

————, *China After the Cultural Revolution: Politics Between Two Party Congresses* (Berkeley, 1977).

Feng, Yu-lan, "Criticism of Lin Piao and Confucius and the Party's Policy Towards Intellectuals—My Understanding," *Peking Review*, 12:14-16 (March 22, 1974).

Joffe, Ellis, "The Chinese Army after the Cultural Revolution: The Effects of Intervention," *The China Quarterly*, 55:450-77 (July-Sept. 1973).

Kau, Ying-mao, *The Case of Lin Piao: Power Politics and Military Coup* (New York, 1974).

Oksenberg, Michel, "The Political Scramble in Mao's China," *Saturday Review/World*, May 18, 1974, 10-15.

P'i-Lin p'i-K'ung wên-chang hui-pien, Vol. II, 批林批孔文章匯編 (A collection of essays on the criticism of Lin Piao and Confucius) (Peking, 1974).

Powell, Ralph L., "The Increasing Power of Lin Piao and the Party-Soldiers, 1959-1966," *The China Quarterly*, 34:38-65 (Apr.-June 1968).

————, "Party Still Striving to Retain Control of 'the Gun' in China," *The Christian Science Monitor*, Sept. 21, 1973.

Wich, Richard, "The Tenth Party Congress: The Power Structure and the Succession Question," *The China Quarterly*, 58:231-48 (Apr.-June 1974).

Yao, Ming-le, *The Conspiracy and Murder of Mao's Heir* (London, 1983).

30

China Rejoins
the International Community

The Sino-American Détente

On July 15, 1971, President Richard Nixon announced a dramatic change in Sino-American relations by revealing that Foreign Affairs Advisor Dr. Henry Kissinger had secretly travelled to Peking between July 9-11 and that he himself had accepted an invitation to visit the People's Republic. The announcement was a great surprise to the international community, especially Japan, which had been precluded by the United States from developing closer relations with China. The "Nixon shock" was a major diplomatic breakthrough, mollifying twenty-two years of hostility toward China and viewed in historical perspective it constituted a watershed in American policy toward the People's Republic.

THE VICISSITUDES OF AMERICAN POLICY. When the Communists emerged victorious in the Chinese Civil War in 1949, President Truman appeared resigned to their conquest of Taiwan and considered recognition of the People's Republic of China.[1] From the American perspective, the rise of Communist China, though not desirable, was acceptable because it was not deemed a military threat to the security and hegemony of the United States. But, the outbreak of the Korean War in 1950 and the ensuing Chinese participation altered the perceptions of the architects of American foreign policy. China and Russia were portrayed as part of an international monolithic Communist conspiracy determined to destroy the democratic

1. Warren I. Cohen, *America's Response to China: An Interpretive History of Sino-American Relations* (New York, 1971), 201.

718

system of the West. Therefore, a Maoist conquest of Taiwan was no longer regarded by Washington as the logical conclusion of the Chinese Civil War, but rather as part of a larger design involving Communist expansion in Asia. Truman dispatched the Seventh Fleet to interpose between Taiwan and the mainland, and in effect prevented the Communist takeover of the island—thus re-involving the United States in the Chinese Civil War. As American soldiers died in Korea, Washington became increasingly hostile toward China. Truman extended the policy of containment to Asia, intensified the reconstruction of Japan as a counterpoise to Russia and China, and rendered reconciliation with Peking virtually impossible. China, for its part, proclaimed the policy of "lean-to-one-side" and showed indifference to United States recognition or United Nations membership.

The Eisenhower administration displayed even greater enmity toward Peking. Secretary of State John Foster Dulles believed in the immorality and the potential danger of communism, especially the Chinese variety. The United States not only refused recognition to the People's Republic but pertinaciously opposed its participation in the United Nations. A policy of military encirclement of China was initiated, involving U.S. bases in Korea, Japan, Okinawa, Taiwan, South Vietnam, Burma, and Thailand. These were reinforced by a series of alliances with South Korea, Thailand, Australia, and New Zealand in the Southeast Asia Collective Defense Treaty and by the 1954 defense pact with the Nationalist government on Taiwan. In the view of a leading China expert, the "Dullesian cold war against Peking in the 1950s was fundamentally mistaken and unnecessary, based on an utter misconception of Chinese history and the Chinese revolution."[2]

Domestically, the Eisenhower administration allowed and secretly encouraged McCarthyism, which left a deep imprint on the China policy. Those in the Department of State attacked for the "loss of China"[3] were dismissed, disgraced, or transferred to insignificant posts that were tantamount to banishment. A sense of fear inimical to the free expression of views gripped the Washington officialdom and university-based China experts. Under such conditions, any proposal for reconciliation with Peking was to court political disaster and risk the epithet of appeasement and "soft on communism." Although Eisenhower later attempted to extricate himself from McCarthyism, its legacy was powerful enough to stifle any

2. John K. Fairbank, "The New China and the American Connection," *Foreign Affairs* (Oct. 1972), 37.
3. Such as John S. Service, John Paton Davis, John Carter Vincent, O. Edmund Clubb and others.

positive approach toward China. Dulles, noted for employing such male-dictions as "agonizing reappraisal" and "massive retaliation," was so brazen that he refused to shake hands with Premier Chou En-lai at the Geneva Conference in 1954. During the offshore crises over Quemoy and Matsu in 1954 and 1958, Washington supported the Nationalists and threatened to use nuclear weapons against the Communists.[4] Thus, the Eisenhower administration obstinately refused to improve relations with Peking.

The Kennedy administration, despite its generally liberal outlook, continued the legacy of "containment and isolation of China." Eisenhower warned the new president that any change in the China policy would bring him out of retirement fighting.[5] Kennedy also considered the time inauspicious for innovation or revision and in 1961 he assured the Nationalists that the United States would continue to veto the admission of the People's Republic to the United Nations. His secretary of state, Dean Rusk, considered China a "Slovanic Manchukuo," more venturesome and more dangerous to world peace than the Soviet Union. The Sino-Indian border clash in 1962 and the quick Chinese victory were depicted as confirming Washington's fears. Therefore, not only did the Kennedy Administration not attempt to reduce Sino-American antipathy, but it actually increased the animosity by intensifying American presence on China's southern border through increased United States activity in Vietnam.

When the Sino-Soviet split became overt in the early 1960s, it exploded the myth of a monolithic international Communist conspiracy, disproved the idea that China was a sycophant of Russia, and provided a new opportunity for rapprochement with Peking. Yet, Washington ignored the new possibility as policy makers believed that improvement of relations with the one would inevitably antagonize the other. In view of Mao's attack on Khrushchev's policy of peaceful coexistence with the West and the greater Soviet military threat to the United States, it was deemed more expedient and feasible to reach a détente with Moscow than with Peking.[6] Kennedy reportedly took "an exceedingly dark view" of China's nuclear development, and even considered a joint nuclear attack with the Soviet Union in order to destroy China's atomic capability.[7] Although nothing materialized from such consideration, Kennedy's unfavorable attitude toward China was unmistakable.

4. Allen S. Whiting, "Statement on U.S.-China Relations," delivered before the Senate Committee on Foreign Relations, June 28, 1971, 17.
5. James C. Thomson, Jr., "On the Making of U.S. China Policy, 1961-9: A Study in Bureaucratic Politics," *The China Quarterly* (April-June 1972), 50:220-21.
6. Cohen, 220.
7. Joseph Alsop, "Thoughts out of China—(I) Go versus No-go," *The New York Times Magazine,* March 11, 1973, 31.

Nonetheless, in the State Department some administrative reorganization began, making manifest a degree of recognition of the growing importance of China. In 1962 a new desk for "Mainland China Affairs" was established and staffed by younger experts of the post-McCarthy era, in contrast to the desk for "Republic of China Affairs" staffed by older China hands. Initially, the new desk occupied a less important position, but one year later it was elevated to a fully independent "Office of Asian Communist Affairs." On November 14, 1963 Kennedy announced at a press conference that "We are not wedded to a policy of hostility to Red China." Reportedly he had reserved the China matter for his second term.[8]

The Johnson administration, bound by the exigencies of Vietnam, assigned a low priority to the China question—a situation exacerbated by China's turn inward during the Cultural Revolution. Secretary of State Dean Rusk expressed his "apocalyptic vision" of "a billion Chinese on the mainland, armed with nuclear weapons." Yet, Johnson was surprisingly flexible on some ideological matters and seemed to desire a trans-Pacific statecraft that would overshadow his Vietnam difficulties. On July 12, 1966, he called for a policy of "cooperation and not hostility" with China. He hinted at the possibility of seating Peking in the United Nations and was the first president since "the loss of China" to speak conciliatorily about Peking, favoring a policy of "containment but not necessarily isolation of China." However, Johnson would not act unless prompted by his secretary of state, and Dean Rusk would not advise improved relations with China. Such an act would expose the administration to a Republican attack of appeasement, and Johnson allowed his own feelings to be subjugated by domestic political considerations and Rusk's attitude. However, he was said to have dreamed of crowning his presidential years with summit meetings in Moscow and perhaps in Peking, but the Soviet invasion of Czechoslovakia in the summer of 1968 and the United States response shattered such predilections.[9]

From this brief survey it is evident that for twenty years United States recalcitrance and Chinese aloofness precluded renewed diplomatic relations. The only contact that existed was the more than 130 ambassadorial talks held at Geneva and Warsaw between 1955 and 1967, but these only served the purpose of informing the adversaries of their respective positions on critical issues.[10] But, when Nixon was inaugurated in 1969 the time for change appeared more propitious. Internationally, there was grow-

8. Thomson, 226, 229.
9. *Ibid.,* 240-42.
10. Kenneth T. Young, *Negotiating with the Chinese Communists: The United States Experience, 1953-1967* (New York, 1968).

ing recognition of China as a nuclear power, the rise of Japan as an economic rival of the United States, and the intensification of the Sino-Soviet split which offered opportunities for exploitation to the advantage of the United States. Domestically, an environment conducive to improved relations with Peking was created by the anti-Vietnam war movement, the general call for a reassessment of the China policy by liberal politicians, academicians, powerful business interests desiring trade with China, as well as by the administration's decision to withdraw from Asia in recognition of the fact that the United States could no longer be the "policeman of the world." Politically, President Nixon, with a strong anti-Communist background, could afford to adopt a conciliatory policy toward China without being accused of appeasement. Under the tutelage of Henry Kissinger, former Harvard professor and expert on the diplomacy of the nineteenth-century Austrian statesman Clemens von Metternich, who advanced the principle of equilibrium among states and a system of limited security rather than the total security of any one state, Nixon terminated the policy of containment and instituted a new one whose main goal was the establishment of equilibrium between China, the Soviet Union, and the United States together with the maintenance of good relations with Japan and Western Europe. Thus, by reinterpreting the concept of balance of power, he envisioned a five-power centered world with China as one locus of power, drawn out of the isolation that was partly self-imposed and partly the result of American and Soviet encirclement. Because of a concomitant policy change in Peking, rapprochement indeed became possible.[11]

THE NEW BALANCE OF POWER. The Nixon-Kissinger world perspective was a product of Realpolitik. It postulated that the bi-polarization of the post-World War II era, characterized by American and Soviet domination, had drawn to a close. In the next decade, and possibly the rest of this century, there would exist five power centers in the world: the United States, the Soviet Union, China, Japan, and Western Europe. Of the five only China was isolated from the world community and therefore it was "imperative" that it be reintegrated. Since the Soviet Union was entangled in hostilities with China, the initiative was left to Washington. The United States' position was enhanced in that the Strategic Arms Limitation Talks (SALT) with the Soviet Union were progressing and the prospects of nuclear confrontation lessening. Further, the United States had compatible relationships with Japan and Western Europe. Therefore, if only it could "open

11. Robert A. Scalapino, "China and the Balance of Power," *Foreign Affairs* (Jan. 1974), 356.

the door" to China, it would be the maker of a new age of diplomacy.[12] Naturally, a China active in international politics could more easily checkmate Russia, and an economically powerful Japan could counterbalance China. In this "novel" pattern of diplomacy, the United States would call the tune and display world leadership. Therefore, China was pivotal to President Nixon's diplomatic offensive.

As early as his January 1969 inaugural address, the President articulated the idea of shifting from "confrontation to negotiation," and two weeks later instructed Dr. Kissinger to investigate the means for reconciliation. Overtures to the Chinese ambassador at the Warsaw Talks proved futile, although Peking did not reject them outright. The Chinese were wary; they were waiting for more demonstrative manifestations of a change in America's policy. The President reiterated his desire by references to the People's Republic of China (not Red China) during a toast to the visiting Romanian President Nicholae Ceausescu in October 1970, and again in his "State of the World Report" in February 1971. Ceausescu relayed to Peking secret messages expressing America's desire to open a dialogue. Chou En-lai then publicly noted President Nixon's use of the "proper name" of his country. Meanwhile, in October 1970 the American journalist Edgar Snow, who visited China for six months, was seen strategi-

---

12. Briefing by President Nixon for midwestern newspaper and broadcasting executives in Kansas City, Mo., on July 6, 1971. Full text in *U.S. News & World Report*, Aug. 2, 1971, 46-47.

It should be noted that the concept of a five-power centered world, though brilliantly conceived, is not without deficiencies. Primarily, the idea is based on international power factors without proper consideration of economic forces which often intertwine with political issues. The exclusion of the Arab countries, whose ability to cripple the economy of Japan and Western Europe and seriously injure that of the United States—three of the Five Centers—during the "oil crisis" of 1973 is a case in point. The omission of the Third World upon which the major powers increasingly depend for raw material also contributes to the problematic nature of the policy. Even in political terms, the five centers are not balanced in the classical fashion, and the forces governing today's diplomacy are vastly different from those of the 19th century. Public opinion and popularly elected representative bodies, rather than secret decisions by autocratic monarchs, play a major role in the formulation of modern foreign policy. Hence, former Under Secretary of State George Ball remarked of the balance of power concept with skepticism: 'That was only viable for the autocratic government of the nineteenth century, which made decisions without reference to public opinion or Congress." Hypothetically, a Sino-Soviet rapprochement could shatter the balance and render the system inefficacious. Thus, the durability of the Nixon-Kissinger idea of international relations may be questionable. For views of Edwin O. Reischauer of Harvard, Zbigniew Brzezinski of Columbia, and George Ball, see "The Kissinger Revisionists," *Newsweek*, July 30, 1973, 12; Douglas D. Adler, "Kissinger: A. Historian's View," *The Christian Science Monitor*, Jan. 30, 1973; and Max Lerner, "Kissinger's World May Be Coming Unhinged," *Los Angeles Times*, March 28, 1974.

cally placed between Mao and Chou on a public reviewing stand. He was told by Mao that the Sino-American difficulties could only be obviated through direct negotiations and that he would welcome a visit by Nixon either as president or tourist.[13]

Soon after came the "Ping-Pong diplomacy" and Chou En-lai's warm reception of the American players. Emphasizing people's diplomacy, he declared that the visit of the American team "opened a new page in the relations between the Chinese and American peoples" and that he was "confident that this beginning again of our friendship will certainly meet with the support of our two peoples." President Nixon quickly responded with a five-point easing of the trade embargo with China. This was followed by secret moves to dispatch a high emissary to China, for he had been assured by Peking that it would welcome such a mission. The finale was the legendary Kissinger visit on July 9-11 and the dramatic announcement by the President on July 15 that he had accepted Premier Chou En-lai's invitation to visit China. It was a major diplomatic victory for the President. He showed that it was a Republican, not a Democrat, who was attempting to open a door to China and perhaps discover a corridor through Peking and Moscow leading to peace in Vietnam. Moreover, such a visit, to be viewed by millions on television, would be politically beneficial in an election year. But the President knew that the door to China stood ajar only because Peking permitted it.

THE CHINESE MOTIVES. Peking welcomed Nixon's visit for both practical and psychological reasons. Although the Chinese previously had denounced him as an imperialist warmonger, they now found him to represent a potentially useful historical force. He was the man with whom the Chinese wanted to discuss United Nations membership, United States recognition, resolution of the Taiwan problem, potential Japanese rearmament, purchase of American airplanes and scientific instruments, and above all an improved international position to deter possible Soviet attack. Unquestionably, the visit would also provide immense psychological satisfaction for the Chinese.

The Soviet threat. Sino-Soviet relations had been nothing less than maverick. After a decade of cooperation during 1949-58, tension began to surface over ideological differences, revolutionary strategy, sharing of nuclear data, Soviet economic and technical aid, leadership in the international Communist movement, and territorial dispute. So bitter and acrimonious had the situation become that certain factions in the Soviet hierar-

13. Edgar Snow, "A Conversation with Mao Tse-tung," *Life*, April 30, 1971, 47.

chy favored a "pre-emptive" strike against China. Russian military buildup on Chinese borders, which began in the middle 1960s, became more ominous after the Soviet invasion of Czechoslovakia in 1968. The situation became precarious with the announcement of the "Brezhnev Doctrine" which declared the Soviet Union's right to intervene in the internal affairs of other Communist states deemed deviants from the Socialist cause. To the Chinese this doctrine connoted the possibility of their being the next target of Soviet invasion—a prospect hardly to be relished, particularly when one remembers that similar to China, Czechoslovakia had been isolated and without United Nations membership.

Border conflicts on the Ussuri River and in Sinkiang in 1969 provided further substance to the Chinese suspicion of Soviet intentions and the danger of war appeared imminent. The Soviets made various attempts in 1969 to seek an American agreement to a preventive attack on Chinese nuclear capabilities, but the President responded angrily and negatively. The North Vietnamese government was also concerned about a potential war between the two big brothers, and when Premier Alexei Kosygin visited Hanoi in mid-September he was urged to avoid conflict with China. It is quite possible that the Soviet's decision to retreat from war was due primarily to Nixon's forceful disapproval and secondarily to Hanoi's pleas.

It seems that in the summer of 1969 the Soviet ground forces stationed on the Chinese border were insufficient to launch a successful attack. Therefore, tactical nuclear weapons would have to be used with the subsequent danger of fallout on Japan, Korea, and perhaps the United States. Since the retreat from war in 1969, the Soviets have greatly increased their ground forces along the Manchurian border and complemented their potential with missiles and rockets equipped with nuclear warheads. Soviet troop and logistic movement on the Trans-Siberian Railway had been so intense that civilian traffic was interrupted several times between 1970-72.[14] By 1973-74 the Soviet strength on the Chinese border reached 45 to 49 divisions (one million men), reinforced by 150 warships in the Pacific. Undoubtedly, the *Ostpolitik* of West German Chancellor Willy Brandt and his détente with Moscow enabled the Soviets to transfer troops from Eastern Europe to China's borders.[15]

The Chinese deployed at least an equal number of troops along the border and dug extensive tunnels in major cities in preparation for possible

14. Alsop, 100.
15. *Strategic Survey 1973* (The International Institute for Strategic Studies, London, 1974), 67; Shinkichi Etō, "Motivations and Tactics of Peking's New Foreign Policy," 33. Paper presented at the German Association for East Asian Research, International China Conference at Schloss Reisensburg, June 24-30, 1973.

enemy attack. They considered Brezhnev and Kosygin "much worse than Khrushchev" because of the troop distribution on the Chinese border and the preoccupation with organizing support among Eastern European Communist states for an invasion of China. Additionally, Peking recalled Khrushchev's lecture to German Chancellor Konrad Adenauer in Moscow in 1955 concerning "the Yellow Peril"—the expression of the last German emperor.[16]

Therefore, it appears that the imminence of a Soviet attack induced China to seek American connection, United Nations membership, and extensive diplomatic recognition. To end isolation would create an international climate antithetical to any rash Soviet action. Moreover, since China was determined not to fight on two fronts simultaneously, the United States became crucial to China's security, while Russia metamorphosed into China's principal enemy.

Yet, collaboration with the capitalistic United States could be labeled ideological perfidy. Peking, therefore, effusively explained the policy by reference to the negotiations with the Nationalists in 1945. Mao's article "On the Chungking Negotiations" (Oct. 1945) was widely redistributed to justify a change in tactics but not in the ultimate strategy of world revolution. There was no betrayal of principle.

Taiwan. According to Edgar Snow, the reintegration of Taiwan and the mainland was Chairman Mao's "last national goal of unification." Mao insisted that Taiwan was a province of China that had to be liberated, but that he would be reasonably lenient in dealing with the issue—"perhaps even granting a degree of autonomy to Chiang Kai-shek if he should wish to remain governor there for his lifetime."[17] But, first of all, Mao wanted the withdrawal of United States military forces from Taiwan and the Straits and the recognition in principle of Taiwan as part of the People's Republic of China. In this way, Peking would increasingly isolate the Nationalists, coercing them into negotiation and eventual submission.

Japanese rearmament. Peking viewed Japan's vast economic growth as the basis for renewed military power and was deeply concerned with the burgeoning Japanese influence on Taiwan and Korea. Having suffered from Japanese invasion since 1894, China was extremely sensitive to potential resurgence of Japanese militarism. Since Japan was protected by the American nuclear umbrella, it was necessary to discuss the problem with Nixon.

These three main issues, coupled with the desire for United Nations

16. Alsop, 102-3, reporting conversations with Premier Chou En-lai.
17. Edgar Snow, "China Will Talk from a Position of Strength," *Life*, July 30, 1971, 24.

membership, American diplomatic recognition, and increased trade and technical exchanges, prompted the invitation for the Nixon visit. Undoubtedly, such an unprecedented visit by an incumbent president[18] of the most powerful Western nation would afford the Chinese great psychological satisfaction. Some Asians viewed Nixon's visit as a pilgrimage, for historically no Chinese emperor ever left the country; only the tributary kings and foreign envoys sojourned to pay him homage. Nixon's visit would give China a new sense of dignity as well as great-power status.

Both Mao and Chou were consummate revolutionaries and astute students of history. They knew the importance of seizing the propitious moment and directing it to their ends. America's probing for an opening to China was harmonious with their own Grand Design, and Nixon became a welcome guest in Peking.

NIXON IN PEKING. Nixon left for China on February 18, 1972 in a freshly painted blue-and-white presidential jetliner, the Spirit of '76. After a two-day rest in Hawaii and an overnight on Guam, he arrived in Peking on February 21. His party, including Mrs. Nixon, Secretary of State William Rogers, Henry Kissinger and others, were greeted at the airport by Premier Chou En-lai and a group of Chinese dignitaries, but there were no Mao, no crowds, and no diplomatic corps. The welcome was correct and formal but low-keyed and mildly austere, as the Chinese treated the occasion as a semiofficial affair. As he stepped down from the plane, smiling and refreshed after the 16,000-mile trip, Nixon extended to Chou the historic handshake, which Dulles had shunned at Geneva in 1954. Nixon's was a firm, friendly gesture, as if to amend Dulles' inimical conduct. The President then shook hands with the other Chinese officials and to the orchestration of both national anthems, inspected an impressive honor guard comprised of members of the Liberation Army, Navy, and Air Force. The fifteen-minute airport ceremony was described by the American news media as "not warm, not cold"[19]—just right from the Chinese perspective since there were no formal diplomatic relations between Washington and Peking.

The Nixon party was lodged at a large state guest house five miles west of the center of Peking. There an American flag flew for the first time in twenty-two years. A few hours later, Nixon and Kissinger were received by Mao in his study for an hour-long, unscheduled meeting. The Presi-

18. President Ulysses S. Grant visited Peking in 1879 during his world tour after retirement from office.
19. *San Francisco Examiner*, Feb. 21, 1972, A.

dent and his foreign policy advisor entered the room as if "they were calling on something more than just a man." Mao spoke "superbly," with authority and a confident, earthy wit, while Chou remained silent. Nixon "felt the force of Mao," and was "humbled and awed in a rare way" never captured by the news media.[20] The occasion signified Mao's approval of détente and the subdued and somewhat rigid Chinese attitude immediately became warm and enthusiastic.

In the Great Hall of the People a grand state banquet followed, hosted by Chou in Nixon's honor. The President was ebullient and Chou a cordial and perfect host. An air of rapport and joviality filled the hall. Chou mounted the rostrum to speak first, stressing the unprecedented nature of the occasion and the traditional friendship between the American and Chinese people. Now that "the gates to friendly contact have finally been opened" after more than twenty years, Chou called for the normalization of relations. Obviously enjoying the moment, Nixon responded by quoting Mao's famous expression "seize the day, seize the hour" and declared:

> Let us, in these next five days, start a *long march* together, not in lock step, but on different roads leading to the same goal—the goal of building a world structure of peace and justice in which all may stand together with *equal dignity* and in which each nation, large or small, has a right to determine its own form of government free of outside interference or domination. . . . There is no reason for us to be enemies. Neither of us seeks domination of the other. Neither of us wants to dominate the other.

Between the negotiations, Nixon viewed the Great Wall, the Ming Tombs, the scenic West Lake, and the Shanghai industrial exhibitions, while his wife visited schools, hospitals, and shops. During the entire time, the President made certain that his Chinese hosts were cognizant of his admiration for what he saw and of his appreciation for the warm reception. Nixon's attitude, his use of such familiar phrases as the "long march" and "seize the hour" and the burst of geniality during the banquet, were calculated to create a new image of him—a considerate, appreciative guest, totally different from the pompous foreign overlords and imperialists of the past. His behavior was perceived as a deliberate efforts designed to signal the end of a century of Western exploitation and domination of China. In this respect, Nixon's performance was highly creditable.

Although politely conducted, the tenacity of the negotiations was made manifest in that no agreement was reached at the end of five days of talks

20. Hugh Sidney, "The Visit to Mao's House," *Life*, May 17, 1972.

in Peking. Not until the sixth day, at the West Lake in Hangchow, was the impasse resolved and a communiqué was issued in Shanghai on the last day of the visit. The Chinese appeared to have profited the most, benefiting from a stronger bargaining position since it was Nixon who came to renew relations. Openly, they claimed to have made no concessions, but because of their anxiousness for rapprochement, the final result represented a compromise. Although Nixon seemed to have conceded more, he established direct relations with China, reduced international tension, and improved the prospect for world peace.

THE SHANGHAI COMMUNIQUÉ. This document of February 28, 1972 expressed hope for the future and was unique in that it articulated areas of disagreement as well as agreement. The essential features of this 1,750-word statement were as follows:

1. On the question of Taiwan, the United States declared that it

> acknowledges that all Chinese on either side of the Taiwan Strait maintain there is but one China and that Taiwan is a part of China. The United States government does not challenge that position. It reaffirms its interest in a peaceful settlement of the Taiwan question by the Chinese themselves. With this prospect in mind it affirms the ultimate objective of the withdrawal of all U.S. forces and military installations from Taiwan. In the meanwhile, it will progressively reduce its forces and military intallations on Taiwan as the tension in the area diminishes.

Obviously, Nixon made a concession but perhaps it was the minimum required for a détente, which in the long run benefitted the United States, improved the prospects for a more peaceful world, and reduced the risk of a Soviet attack on China. But, it is imperative to note that while Nixon agreed that "there is but one China and that Taiwan is a part of China," he avoided referring to Taiwan as part of the People's Republic of China; nor was he committal as to which government speaks for China. But, since the mainland is so much larger than Taiwan and occupies the territory traditionally known as China, the implication is clear that mainland China is the real China. For their part the Chinese conceded by not insisting that the United States declare Peking as the legal government of all China and abrogate the 1954 mutual defense treaty with the Nationalists. In an earlier agreement Canada "took note" of Peking's claim to sovereignty over Taiwan and agreed to sever diplomatic relations with the Nationalist government. Even more definitively in September 1972 the Japanese openly recognized the People's Republic as the legal government of China. The Nixon concession, in effect, recognized "One China—but not now."[21]

2. The United States endorsed the Five Principles of Peaceful Coexistence first espoused by Peking at Bandung in 1955: (1) respect for the sovereignty and territorial integrity of all states; (2) nonaggression against other states; (3) noninterference in the internal affairs of other states; (4) equality and mutual benefits; and (5) peaceful coexistence. Alone, these principles were innocuous, but the United States had refused to accept them because they were formulated by the "nonaligned" nations hostile to imperialism, colonialism, and the United States policy of the containment of communism. Nixon's endorsement was largely a psychological concession. But it also reduced the chances of a Soviet assault on China.

3. Both parties agreed not to seek "hegemony" in the Asia-Pacific region. By implication they also opposed Soviet domination of the area.

4. Both sides pledged to reduce the danger of international military conflict and agreed that "it would be against the interests of the peoples of the world for any major country to collude with another against other countries, or for major countries to divide up the world into spheres of interest." This statement forestalled any possible U.S.-U.S.S.R. collusion against China.

5. Each country agreed to facilitate exchanges in science, technology, culture, journalism, and sports.

6. Through various methods, including the occasional sending of a senior United States diplomat to Peking, both nations agreed to continue seeking further normalization of relations. In early April 1973 an advance group from the Department of State travelled to Peking to establish a liaison office. On May 14 David K. E. Bruce, the 75-year-old former ambassador to London, Paris, and Bonn, entered China as ambassador to the office, and two weeks later Huang Chen, the 66-year-old member of the Party's Central Committee and China's most prestigious diplomat and former ambassador to France, came to head the Chinese liaison office in Washington, D.C.

The communiqué reflected a cooperative atmosphere, goodwill, and mutual friendship. As he left Shanghai, a former bastille of foreign imperialism, Nixon exuberantly proclaimed that "never again shall foreign domination, foreign occupation, be visited upon this city or any part of

---

21. Robert A. Scalapino, "First Results of the Sino-American Détente," 14. Paper presented at the German Association for East Asian Research, International China Conference at Schloss Reisenburg, June 24-30, 1973.

China or any independent country of his world." Boldly, he hailed his seven-day visit as "a week that changed the world."[22] Nixon's performance in China was possibly his finest while in office.

THE ACCOMPLISHMENTS OF DÉTENTE. The Chinese were clearly satisfied with their accomplishments, as evidenced by the rousing welcome accorded Chou En-lai upon his return to Peking. Nixon's "pilgrimage" was especially heartwarming to the Chinese who had suffered abuse for a century, and the visit significantly promoted China's international position. Even prior to Nixon's trip, the repercussions of the quest for détente were manifest in Peking's admission to the United Nations on October 15, 1971. The United States' endorsement of the Five Principles of Peaceful Coexistence and its opposition to any power seeking hegemony in the Asia-Pacific area circumscribed, if not eliminated, the possibility of a Soviet attack on China. Thus, China achieved the measure of security that was the primary objective of its diplomatic offensive.

As regards Taiwan, China scored a substantial if not complete victory. It won the American acceptance of "One China" and the promise to withdraw its troops and military installations from Taiwan as tension in that area diminished. But Peking did not receive outright American recognition as the sole, legal government of China, and the United States was not committed to abrogating its mutual security pact with Taiwan. Nonetheless, the new American posture appeared to circuitously promise the future recognition of Peking.

The détente enabled Peking to purchase American airliners, scientific instruments, and chemical, industrial, and agricultural products needed for China's modernization. The exchange of scholars, journalists, athletes, scientists, and officials facilitated the mutual flow of ideas and knowledge, reversing the trend of twenty-two years of noncommunication.

On the other hand, reconciliation with the capitalistic United States appeared ideologically unsound and suggested a compromise of the principle of world revolution. It raised the question of China's credibility before other Communist states, especially those in Asia. Also, the détente could dampen the revolutionary fires of the world's Socialist nations.

For the United States, the reconciliation initiated direct relations with China and reduced the possibility of war between Russia and China, thereby enhancing the prospects for world peace. A prosperous and friendly China had always been considered in the interest of the United

22. Full text of the Shanghai Communiqué in The Department of State, *Selected Documents, No. 9, U.S. Policy Toward China, July 15, 1971-January 15, 1979* (Washington, D.C., 1979), 6-8.

States. The concept of the five-power centered world appeared to be materializing and China's promise to peacefully settle international disputes suggested that Peking would not intervene militarily in Vietnam or forcibly liberate Taiwan. Finally, the possibility of release for Americans detained in China increased; and indeed one CIA agent,[23] imprisoned for more than twenty years, and two pilots, shot down over China during the Vietnam war, were freed in March 1973.

Materially, the most conspicuous gain for the United States was the growing Sino-American trade which helped reduce the American balance-of-payment problem. China desired American scientific, technological, and agricultural products much more than the United States needed Chinese goods; hence, the trade balance heavily favored the United States. Shortly after Nixon's visit, China purchased two earth satellite stations from RCA Global Communications, Inc. Later, they purchased ten Boeing 707's, forty replacement jet engines from Pratt & Whitney, twenty tractors for towing airplanes on the ground, and large quantities of wheat, corn, and cotton. China exported to the United States only limited quantities of tin and tin alloys, hog bristles, silk, vegetable oil, and art objects. The following table shows the trend of Sino-American trade:

Year	U.S. Exports to China (millions)	Chinese Exports to U.S. (millions)
1972	$ 60	$ 32
1975	304	156
1980	3,755	1,059
1985	3,855	3,840

On balance, Nixon achieved much from the détente. It weakened the Soviet international position, making the U.S.S.R. more anxious to reach agreement with the United States on issues already under negotiation. China tied down a million Russian troops on the Manchurian-Siberian border, thereby correspondingly reducing Soviet military pressure elsewhere. Thus, in the new triangular relationship, the United States distinctly held the balance. Just as the British statesman George Canning in 1825 "called the New World into existence to redress the balance of the old," Nixon, in a sense, called China into play to redress the balance of the world. In his desire to fashion a new era of international relations,

23. John Downey.

Nixon emulated President Franklin Roosevelt who laid much of the groundwork of the pattern of diplomacy for the post-World War II period.

The Sino-Japanese Rapprochement

The Japanese were startled by Nixon's July 15, 1971, announcement to visit China and by the contents of the Shanghai Communiqué of February 28, 1972. They believed that the lack of prior consultation concerning such important decisions was insulting and perhaps even necessitated reciprocal action. To demonstrate their independence and to protect their interests, the Japanese decided to normalize relations with Peking in a more lucent fashion than had the United States. Undoubtedly, Tokyo had intended to improve relations with Peking gradually even before the President's July announcement, but the "Nixon shocks" greatly expedited the process.

For at least two or three years the Japanese news media had been promoting closer ties with China regardless of the position of the United States. The attitude was prompted by a pragmatic assessment of the changing international situation, prospects of closer economic ties with China, old cultural bonds, and leftist propaganda favoring rapprochement. A number of Japanese businesses accepted the Four Principles of Trade announced by Peking on April 19, 1970: (1) firms trading with China must not trade with Taiwan or South Korea; (2) they must not invest in these two places; (3) they must not export weapons for American use in Indochina, and (4) they must not affiliate as joint ventures or subsidiaries of American firms in Japan. In spite of the ideological and social differences, closer relations between the two countries seemed beneficial and inevitable, and the "Nixon shocks" provided the impetus that hastened the realization of that prospect.

Domestically, the Japanese Communists and Socialists adamantly advocated the necessity of resuming diplomatic relations with China. A prominent member[24] of the Japanese Socialist Party demanded that Japan apologize to China for its past malefactions in order to facilitate winning Chinese trust and normalizing relations. In early March 1972 Foreign Minister Takeo Fukuda concurred that "we must convey our self-criticism and apology to China" for all the mistakes perpetrated during the Manchurian Incident of 1931 and the Sino-Japanese War of 1937-1945.[25] Indeed, the time appeared propitious for an initiative toward Peking. Yet, Prime Minister Sato, a politician of the old school sentimentally attached

24. Susumu Kobayashi.
25. *Los Angeles Times*, March 1, 1972, Part I, 7.

to the Nationalist government because of Chiang K'ai-shek's postwar benevolence toward Japan, found it difficult to respond objectively. Powerless to alter the international and domestic tides, he offered to resign on
June 17, 1972, after seven years and eight months in office. With Sato's
resignation the way was cleared to elect a new prime minister predisposed
to normalizing relations with the People's Republic of China.

TANAKA'S NEW CHINA POLICY. Immediately after assuming the premiership
on July 6, 1972, Kakuei Tanaka was besieged by demands to re-establish
relations with Peking. Pressure was exerted by a variety of interests: pro-
Chinese left-wing politicians, the news media, businessmen anxious to
trade with China, and finally Peking.

The Chinese attitude toward Japan had changed dramatically. They no
longer attacked Japanese economic imperialism and rising militarism, but
courted Tanaka with smiles and words of praise, and welcomed his visit.
On August 11, Tanaka made a formal request to visit China, and a day
later Chou announced his welcome. The date of the visit was set for the
week of September 25. To further pave the way for rapprochement, Chou
En-lai made it clear to the Japanese government that (1) China would
waive any claim for war indemnity (in unofficial circles in Japan there had
been talk of several billion dollars of compensation); (2) China would not
consider the United States-Japan security pact and the Sato-Nixon communiqué of November 21, 1969, an obstacle to establishing diplomatic
relations between Peking and Tokyo; and (3) China would conclude a
new treaty of peace and friendship with Japan which should invalidate
Japan's peace treaty with the Nationalists on Taiwan.[26]

The Japanese government was willing to recognize People's Republic
of China as the legal government of China. But, it was not prepared to
make a definite statement concerning Taiwan, because, having forfeited
its claim to Taiwan in the San Farncisco Peace Treaty, Japan believed
such a move to be superfluous. The Japanese were also reluctant to formally abrogate the peace treaty with Taiwan because of a sentimental
feeling of indebtedness to Chiang K'ai-shek and a desire to retain or enhance trade with and investments in Taiwan. Tanaka needed time to resolve the dichotomies and the Foreign Office cautioned against hasty reconciliation.

In August 1972 Dr. Kissinger made a surprise one-day visit to Japan.
Washington wanted to be sure that Japan would make no commitments
to Peking compromising the United States-Japan Security Treaty and its

26. Gene T. Hsiao, "The Sino-Japanese Rapprochement: A Relationship of Ambivalence," *The China Quarterly* (Jan.-March 1974), 57:109-110.

application to Taiwan. This treaty, signed in 1951 and renewed in 1960, allowed the United States to mobilize American troops stationed in Japan for use elsewhere in East Asia in the name of security. In 1969 Nixon and Sato issued a joint communiqué containing a "Taiwan Clause," which declared Taiwan's security to be important to Japan's security and thereby extended the applicability of the 1960 pact to Taiwan. When Kissinger arrived, the Japanese government reassured him that the Sino-Japanese rapprochement would not impair the United States-Japan Security Treaty, but it refused commitment on the Taiwan question, insisting that Nixon's visit to China so drastically altered the Far Eastern situation that the Taiwan issue had to be resolved "realistically."

Tanaka had to surmount three obstacles before his Peking trip: (1) to allay American fear of Japan's rapid rapprochement with China; (2) to secure a Nationalist Chinese understanding of the Japanese position without prejudicing Japan-Taiwan cultural and economic relations; and (3) to overcome conservative opposition within his party in order to create unity favoring his China venture. To solve the first problem, Tanaka was to meet Nixon in Honolulu; for the second, he dispatched the deputy director of his party, Etsusaburo Shiina, who had numerous friends on Taiwan, as a special envoy to the Nationalist government to explain the Japanese position; and for the third, he relied on the persuasive talent of a former foreign minister[27] to reconcile the right-wing party members.

When Tanaka was in Honolulu from August 31 to September 1, 1972, he hoped to "buy" American support for his China policy. The fact that Japan sent 30 percent of its exports to the United States and enjoyed an embarrassingly high surplus of $3.8 billion in the bilateral trade enabled Tanaka to be generous.

After two days of discussion, a joint communiqué, which reaffirmed the significance of the Japanese-American security pact, did not refer to Taiwan and therefore left each party free to form its own interpretations. To the Americans the treaty applied to Taiwan, but to the Japanese it did not because the changed situation in East Asia had invalidated the "Taiwan Clause." Indeed, they believed that the Sino-American détente virtually eliminated the prospect of Taiwan's forced liberation by Peking, and thus the question of using American troops to defend Taiwan became academic.

Tanaka skirted the Taiwan issue and won Nixon's endorsement of his China trip as a further step in the direction of relaxing tension in Asia. In return, Tanaka granted large trade concessions by agreeing to increase Jap-

27. Zentaro Kosaka.

anese purchases of American products by $1 billion to help balance the United States' trade deficit.

If Tanaka succeeded by largess, his special envoy to Taiwan was not nearly as fortunate. Etsusaburo Shiina and his aids arrived on Taipei on September 17. They were immediately confronted by several hundred protesters who halted their motorcade and kicked and beat their cars with sticks. Three days of patient and self-effacing talks with the Nationalist leaders resulted in no progress at all. The Japanese offer of aid to develop Taiwan's economy, including a $40 million nuclear power plant and a $17 million motorway, was rejected. Shiina asked Premier Chiang Ching-kuo what Taiwan wanted from Japan, and the reply was that Japan should stop betraying friends and save itself from being communized. Taiwan was determined to sever relations with Japan if it recognized Peking. Shiina returned to Japan frustrated, but many Japanese adopted the perspective that after the emotional outburst subsided, Taiwan would have to reach accommodation because of its dependence on Japanese trade and economic cooperation.[28]

Tanaka's third task—pacifying the conservatives—was more easily accomplished. The party accepted a right-wing recommendation that "consideration should be given to continue the past relations between our country and the Republic of China." But the recommendation was not binding on the party; it was merely a device to enable the right wing to claim credit for supporting Taiwan without blocking the rapprochement with Peking.

Having removed the three obstacles, Tanaka was ready to embark upon his trip to China.

TANAKA IN CHINA. The fifty-member Japanese delegation, led by Premier Tanaka, Foreign Minister Ohira, and Chief Cabinet Secretary Susumu Nikaido, arrived in Peking in the late morning on September 25, 1972, for a five-day visit. The flag of Japan fluttered over the airport for the first time since it was hauled down after Japan's surrender twenty-seven years ago. After a brief welcoming ceremony by Chou En-lai and other Chinese dignitaries and the review of the honor guard—all reminiscent of Nixon's visit—the Japanese, tense and ill at ease, were driven to the State Guest House. Serious and frank negotiations with Chou followed in the afternoon. By the evening Tanaka seemed more relaxed while enjoying the banquet given in his honor, and the Chinese military band's rendition of "Sakura" (Cherry Blossoms), a popular Japanese song. Chou took the ros-

28. Gene T. Hsiao, 113.

trum first, speaking briefly of Japanese aggression, stressing that "the Chinese people make a strict distinction between the very few militarists and the broad masses of the Japanese people." He announced that "now is the time for us to accomplish this historic task of restoration of diplomatic relations."

Tanaka, in his toast, accentuated the importance of future relations and apologized for past mistakes:

> It is regrettable that for several decades in the past the relations between Japan and China had unfortunate experiences. During that time our country gave great troubles to Chinese people, for which I once again make profound self-examination. After World War II, the relations between Japan and China remained in an abnormal and unnatural state. We cannot but frankly admit this historical fact.

Tanaka's "profound self-examination" drew loud applause from the Chinese, although his expression of regret and repentance was guarded in order to appease the right-wing of his party which remained critical of his China policy. However, in the final communiqué, the apology was more explicit: "The Japanese side is keenly aware of Japan's responsibility for causing enormous damage in the past to the Chinese people through war, and deeply reproaches itself."

Subsequent to the banquet, Tanaka was received by Mao in his book-lined study for approximately an hour. The visitor, a former cavalry officer in Manchuria, appeared nervous. Sensing his uneasiness, Mao spoke half-jokingly: "Well, have you two ('Tanaka and Chou') finished your quarrel?" Politely, Tanaka answered that the talks had been amicable, whereupon Mao responded: "But you should quarrel. Good friends are made only after a quarrel." Mao gave Tanaka a collection of ancient Chinese poetry—the songs of Ch'u, two giant pandas, and an ornate fan. Tanaka presented a modern painting to Mao, a silk tapestry to Chou, and 2,000 cherry and larch saplings to the people of Peking.

The cordial mood was reflected in the negotiations. The Chinese dropped all claims to war reparations and did not insist on the inclusion in the final communiqué a statement of Japan's abrogation of the 1952 peace treaty with the Nationalist government. The Japanese accepted the following terms:[29]

1. The abnormal state of affairs which has hitherto existed between the People's Republic of China and Japan is declared terminated on the date of publication of this statement.

29. Complete text of the Sino-Japanese communiqué in *The New York Times*, Sept. 30, 1972; *Peking Review*, 40:12-13, Oct. 6, 1972.

2. Japan recognizes the People's Republic of China as the sole legal Government of China.
3. China reaffirms that Taiwan is an inalienable part of the territory of the People's Republic of China. Japan fully understands and respects this stand of the Government of China and adheres to its stand of complying with Article 8 of the Potsdam Proclamation.[30]
4. The governments of China and Japan have decided upon the establishment of diplomatic relations as of September 29, 1972.
5. The Government of China declares that in the interest of the friendship between the peoples of China and Japan, it renounces its demand for war indemnities from Japan.
6. The two governments agree to establish durable relations of peace and friendship on the basis of the Five Principles of Peaceful Coexistence.
7. The normalization of relations between China and Japan is not directed against third countries. Neither of the two countries should seek hegemony in the Asia-Pacific region and each country is opposed to efforts by any countries to establish such hegemony.

When Chou proposed a toast at the farewell banquet, Tanaka mistakenly raised the Japanese saké instead of the Chinese *mao-tai* (a gin-like liquor). Chou discreetly motioned to him and Tanaka, slapping his forehead in self-reproach, quickly switched drinks. The encounter between the 73-year-old Chou and the 54-year-old Tanaka sharply contrasted the negotiations which followed China's defeat in 1895. Li Hung-chang, 73 years old, went to Shimonoseki, Japan, to beg for peace before Prime Minister Itō, 55 years old, and Foreign Minister Mutsu, 52 years old. The arrogance, pomposity, and unyielding attitude of Itō and Mutsu was the opposite of Tanaka's submissive and contrite atttitude toward Chou and his reverence toward Mao.

Although it is said that the communiqué manifested an 85 percent victory for China,[31] the Japanese were pleased. On the issue of Taiwan, Japan was able to state that it "fully understands and respects" China's claim to the island, while not explicitly acknowledging that Taiwan is a part of the People's Republic of China. The Japanese also gained the uncontested right to continue economic relations with Taiwan and South Korea. Moreover, in the communiqué Japan did not officially abrogate its peace treaty with Taiwan, although Foreign Minister Ohira announced at a Peking press conference that the treaty had become obsolete and therefore was rescinded. The legality of such a statement is debatable, but am-

30. This article reaffirms the Cairo Declaration of 1943 that Taiwan be returned to China after the end of war.
31. Scalapino, 13.

biguity was resolved when the Nationalist government, upon learning of the communiqué, broke off relations with Japan.

From an international perspective, Japan secured continuation of its security pact with the United States and an enhancement of its position in the forthcoming territorial and peace treaty negotiations with the Soviet Union.[32] Japan would receive Chinese support and both were committed to opposing hegemony in the Asia-Pacific region by either the Soviet Union or the United States.

Domestically, the powerful leftist "China Lobby" in Japan could no longer constrain the government into being obligatory to the Chinese government. The official apology to China relieved years of guilty conscience, and the period of penitence terminated. Japan could begin a new and promising relation with China, independent from American dominance.

China achieved most of what it desired, making only minor concessions. The recognition by Japan had an immense psychological impact upon other countries, especially the Asian states. The new relationship intensified Taiwan's isolation and ended any attempt by Japanese conservatives to support a Taiwan independence movement. Reconciliation also signalled a greater role for Japanese investment and technical assistance in China's economic development[33] and enhanced its position in its confrontation with the Soviet Union. In the historical context, the new agreement signified the end of a century of Japanese exploitation of China and the beginning of an era in which China, free of foreign shackles, could conduct its affairs on its own terms.

The party most adversely affected by the Sino-Japanese exchange was Taiwan. The Nationalists were first shocked by Nixon's détente with Peking, and now they were stunned and angered by Tanaka's action, which they assailed as perfidious, unfaithful to treaty commitments, and ungrateful to Chiang Kai-shek's postwar generosity. The Japanese embassy in Taipei, threatened by demonstrators, had to be protected by 300 police and secret service agents. The Nationalist government immediately severed relations with Japan and reaffirmed its unwavering anti-Communist stand.[34] Taipei called for survival in an unfavorable international environment

32. Japan wants the USSR to return four northern islands: Habomai, Kunashiri, Etorofu, and Shikotan. See Elizabeth Pond, "Japan and Russia: The View from Tokyo," *Foreign Affairs* (Oct. 1973), 145.

33. In 1972 Japan exported to China $609 million and imported from China $490 million, netting a surplus of $119 million. On Aug. 30, 1973 Tokyo and Peking granted each other the most-favored-nation treatment. In 1973 the two-way trade grew to $1.8 billion as opposed to $1.1 billion in 1972.

34. In April 1974 when Tokyo and Peking signed an aviation agreement, the Nationalist government banned all Japan Air Lines flights to or over Taiwan and stopped its own China Airlines flights to Japan.

through greater unity and self-reliance. However, after the initial outburst of anger and the sense of betrayal eased, realism returned. Taiwan, a small island, needed the large trade with Japan to survive, and there were many Chinese living in Japan and several thousand Japanese living in Taiwan, all needing protection. Moreover, Japanese friends of Taiwan wished to maintain some form of cultural, economic, technical, and scientific exchange. For these reasons, pro-Taiwan Japanese organized a "Japan Inter-exchange Association" (JIA) in Tokyo on December 1, 1972, with a main branch in Taipei and a second one in Kaohsiung. A day later a similar organization—the "Association of East Asian Relations" (AEAR)— was established in Taipei, with a main branch in Tokyo and two other branches at Osaka and Fukuoka. Both organizations were designed to care for citizens living in the other's jurisdiction and to promote cultural, economic, and scientific exchanges. Though "unofficial," these organizations were fully supported by the governments and staffed by disguised officials enjoying quasi-legation status and performing regular consular functions.[35] Thus, while official relations between Japan and Taiwan were severed, semiofficial contacts remained unbroken.

With Japan as the 78th country to recognize the People's Republic of China, a deluge of others followed: West Germany, maintaining no diplomatic ties with Taiwan, in October 1972, New Zealand and Australia in December 1972 and January 1973 respectively, Spain in March 1973, and Malaysia in May 1974. Concomitantly, Taiwan became increasingly isolated; by October 1988 it had diplomatic ties with only 23 countries.

The Mao-Chou Diplomatic Grand Design was highly efficacious in that it greatly elevated China's international status, won a permanent seat in the United Nations as a great power and a spokesman for the Third World, and attracted a succession of visiting heads of state and leaders from various circles. Peking became an international crossroads. No longer isolated, China had brilliantly broken out of Soviet and American encirclement. Once again China had entered the family of nations, this time not as a weak, semifeudal, semicolonial state, but as a respected power with an honored status.

---

35. Gene T. Hsiao, 118-20. The board chairman of JIA, Osamu Itagaki, was ambassador to Taipei (1969-71) and the director of its Taipei branch, Hironori Itō, was a minister of the Japanese embassy in Taipei until September 1972, while the director of the Kaohsiung branch was a former Japanese consul general. As regards the AEAR, its director of the Tokyo branch, Ma Shu-li, was a high-ranking Nationalist official in charge of overseas Chinese affairs.

Further Reading

Alsop, Joseph, "Thoughts Out of China—(I) Go versus No-go," *The New York Times Magazine*, March 11, 1973, 31, 100-108.

Bachrack, Stanley D., *The Committee of One Million: "China Lobby" Politics, 1953-1971* (New York, 1976).

Barnds, William J. (ed.), *China and America: The Search for a New Relationship* (New York, 1977).

Barnett, A. Doak, *Communist China and Asia: Challenge to American Policy* (New York, 1960).

Brandon, Henry, "The Balance of Mutual Weakness: Nixon's Voyage into the World of the 1970s," *The Atlantic*, Jan. 1973, 35-42.

Brodine, Victoria, and Mark Selden (eds.), *Open Secret: The Kissinger-Nixon Doctrine in Asia* (New York, 1972).

Buss, Claude A., *China: The People's Republic of China and Richard Nixon* (San Francisco, 1974).

Caute, David, *The Great Fear: The Anti-Communist Purge under Truman and Eisenhower* (New York, 1978).

Clark, Ian, "Sino-American Relations in Soviet Perspective," *Orbis*, XVII:2:480-92 (Summer 1973).

Cohen, Warren I., *The Chinese Connection: Roger S. Greene, Thomas W. Lamont, George E. Sokolsky and American-East Asian Relations* (New York, 1978).

———, *America's Response to China: An Interpretative History of Sino-American Relations* (New York, 1971).

Davis, Forrest, and Robert A. Hunter, *The Red China Lobby* (New York, 1963).

Dulles, Foster Rhea, *American Policy toward Communist China, 1949-69* (New York, 1972).

Fleming, D. F., *America's Role in China* (New York, 1969).

Friedman, Edward, and Mark Selden (eds.), *America's Asia: Dissenting Essays on Asian-American Relations* (New York, 1971).

Garver, John, *China's Decision for Rapprochement with the United States, 1968-1971* (Boulder, 1982).

Gladue, E. Ted, Jr., *China's Perception of Global Politics* (Lanham, Md., 1983).

Griffith, William E., *Peking, Moscow, and Beyond: The Sino-American-Soviet Triangle* (Washington, D.C., 1973).

Hinton, Harold C., *China's Turbulent Quest: An Analysis of China's Foreign Relations Since 1949* (New York, 1972).

Hsiao, Gene T. (ed.), *Sino-American Détente and Its Policy Implications* (New York, 1974).

———, "The Sino-Japanese Rapprochement: A Relationship of Ambivalence," *The China Quarterly*, 57:101-23 (Jan.-March 1974).

Hsiung, James Chieh, *Law and Politics in China's Foreign Relations: A Study of Attitude and Practice* (New York, 1972).

Hudson, Geoffrey, "Japanese Attitudes and Policies Towards China in 1973," *The China Quarterly*, 56:700-707 (Oct.-Dec. 1973).

Iriye, Akira, *The Cold War in Asia: A Historical Introduction* (Englewood Cliffs, N.J., 1974).

———— (ed.), *The Chinese and the Japanese: Essays in Political and Cultural Interactions* (Princeton, 1979).

Kalicki, J. H., *The Pattern of Sino-American Crisis: Political-Military Interactions in the 1950's* (New York, 1975).

Kintner, William R., *The Impact of President Nixon's Visit to Peking on International Politics* (Philadelphia, 1972).

Kissinger, Henry A., *American Foreign Policy: Three Essays* (New York, 1969).

Lampton, David M., "The U.S. Image of Peking in Three International Crises," *The Western Political Quarterly*, XXVI:1:28:50 (March 1973).

Levine, Laurence W., *U.S.-China Relations* (New York, 1972).

MacFarquhar, Roderick, *Sino-American Relations, 1949-71* (New York, 1972).

May, Ernest R., and James C. Thomson, Jr. (eds.), *American-East Asian Relations: A Survey* (Cambridge, Mass., 1972).

McCutcheon, James M., *China and America: A Bibliography of Interactions, Foreign and Domestic* (Honolulu, 1972).

Nixon, Richard, *U.S. Foreign Policy for the 1970's: Building for Peace*, A Report to the Congress (Washington, D.C., 1971).

————, *U.S. Foreign Policy for the 1970's: The Emerging Structure of Peace*, A Report to the Congress (Washington, D.C., 1972).

————, *A New Road for America* (New York, 1972).

Overholt, William H., "President Nixon's Trip to China and Its Consequences," *Asian Survey*, XIII:7:707-21 (July 1973).

Parker, Maynard, "Vietnam: The War that Won't End," *Foreign Affairs*, LIII:2: 352-74 (Jan. 1975).

Pfaltzgraff, Robert L., Jr., "Multipolarity Alliances and U.S.-Soviet-Chinese Relations," *Orbis*, XVII:3:720-36 (Fall 1973).

Pollack, Jonathan D., "Chinese Attitude Towards Nuclear Weapons, 1964-69," *The China Quarterly*, 50:244-71 (April-June 1972).

Pond, Elizabeth, "Japan and Russia: The View from Tokyo," *Foreign Affairs*, 52:1: 141-52 (Oct. 1972).

Pye, Lucian W., "China and the United States: A New Phase," in *The Annals of the American Academy of Political and Social Science*, 402:97-106 (July 1972).

Quester, George H., "Some Alternative Explanations of Sino-American Détente," *International Journal* (Canada), XXVIII:2:236-50 (Spring 1973).

Ravenal, Earl C., "Approaching China, Defending Taiwan," *Foreign Affairs*, 50:1: 44-58 (Oct. 1971).

Rhee, T. C., "Peking and Washington in a New Balance of Power," *Orbis*, XVIII: 1:151-78 (Spring 1974).

Rice, Edward E., "The Sino-U.S. Détente: How Durable?" *Asian Survey*, XIII: 9:805-11 (Sept. 1973).

Robinson, Thomas, "The View From Peking: China's Policies Toward the U.S., and Soviet Union and Japan," *Pacific Affairs*, XLV:3:333-53 (Fall 1972).

Scalapino, Robert A., "The Question of 'Two Chinas'" in Ping-ti Ho and Tang Tsou (eds.), *China in Crisis: China's Heritage and Communist Political System* (Chicago, 1968), 109-20.

————, *Asia and the Major Powers: Implications for the International Order* (Stanford, 1972).

————, "China and the Balance of Power," *Foreign Affairs*, 52:2:349-85 (Jan. 1974).

Schaller, Michael, *The United States and China in the Twentieth Century* (New York, 1980).

Schwartz, Benjamin I., "The Maoist Image of World Order" in John C. Farrell and Asa P. Smith (eds.), *Image and Reality in World Politics* (New York, 1968), 92-102.

Sidney, Hugh, "The Visit to Mao's House," *Life*, May 17, 1972.

Snow, Edgar, "Talks with Chou En-lai: The Open Door," *The New Republic*, March 29, 1971.

————, "A Conversation with Mao Tse-tung," *Life*, April 30, 1971.

————, "China Will Talk from a Position of Strength," *Life*, July 30, 1971.

Solomon, Richard H., "America's Revolutionary Alliance with Communist China: Parochialism and Paradox in Sino-American Relations," *Asian Survey*, VII: 12:832-50 (Dec. 1967).

Starr, John B., "China and the New Open Door" in Alan M. Jones, Jr., (ed.), *U.S. Foreign Policy in a Changing World: The Nixon Administration, 1969-73* (New York, 1973), 67-82.

Steele, A. T., *The American People and China* (New York, 1966).

Syed, Anwar Hussain, *China and Pakistan: Diplomacy of an Entente Cordiale* (Amherst, 1974).

Thomas, John N., *The Institute of Pacific Relations: Asian Scholars and American Politics* (Seattle, 1974).

Thomson, James C., Jr., "On the Making of U.S. China Policy, 1961-9: A Study in Bureaucratic Politics," *The China Quarterly*, 50:220-43 (April-June 1972).

Van der Linden, Frank, *Nixon's Quest for Peace* (Washington, D.C., 1972).

Van Ness, Peter, *Revolution and Chinese Foreign Policy* (Berkeley, 1973).

Weng, Byron S. J., *Peking's UN Policy: Continuity and Change* (New York, 1972).

Wilson, Francis O. (ed.), *China and the Great Powers: Relations with the United States, the Soviet Union, and Japan* (New York, 1974).

Wu, Fu-mei Chiu, *Richard M. Nixon and China* (Washington, D.C., 1978).

Yahuda, Michael, "Kremlinology and the Chinese Strategic Debate, 1965-66," *The China Quarterly*, 49:32-75 (Jan.-March 1972).

Zagoria, Donald, "The Strategic Debate in Peking" in Tang Tsou (ed.), *China in Crisis: China's Policies in Asia and America's Alternatives* (Chicago, 1968), II:237-68.

31

The Nationalist Rule
on Taiwan

In rivalry with the People's Republic on the mainland is the Nationalist government on Taiwan, each disputing the other's claim of constituting the legal government of China. Taiwan, once the base of Ming loyalist opposition to the Ch'ing,[1] has become the citadel of a new resistance movement.

Taiwan, or Formosa ("beautiful" in Portuguese), lies about 100 miles east of the Asian continent and 695 miles south of Japan. The island measures 240 miles long and 98 miles wide at its broadest points, covering an area of 13,844 square miles—larger than the Netherlands, a trifle smaller than Switzerland, and about the size of Massachusetts, Rhode Island, and Connecticut combined.[2] Ceded to Japan by the Ch'ing dynasty after the war of 1894-95, Taiwan was restored to China in 1945 in accordance with the terms of the 1943 Cairo Declaration and the 1945 Potsdam Proclamation in which the Allies pledged to return Manchuria, Formosa, and the Pescadores to China after the defeat of Japan.

With the rapid deterioration of the Nationalist's military situation in China toward the end of 1948, Chiang Kai-shek looked to Taiwan as a refuge. In preparation for this eventuality, he appointed General Ch'en Ch'eng, a confidant, governor of Taiwan on December 29, 1948. After resigning as president on January 21, 1949, Chiang retired to his home

---

1. Under the leadership of Cheng Ch'eng-kung, better known as Koxinga, and his son, from 1661 to 1683. See Chapter 2.
2. Chiao-min Hsieh, *Taiwan: Ilha Formosa* (London, 1964), 3-6.

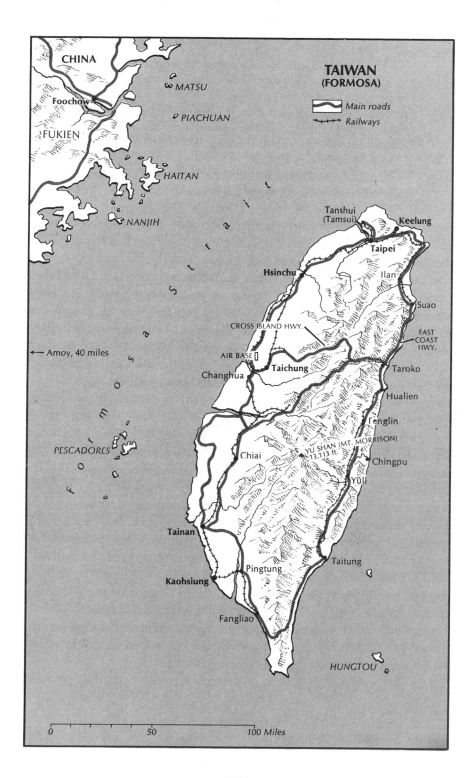

town,[3] near Ningpo, to make contingency plans for withdrawal to the island. He could do so because, though officially out of the government, he retained the KMT director-generalship and control over government troops and funds. With the fall of Nanking in April 1949, all indications pointed to an imminent Nationalist collapse. In order to establish a new base of resistance, Chiang directed the evacuation of government troops and military equipment as well as the removal of $300 million in gold reserves and foreign currencies to Taiwan. He assumed de facto leadership on Taiwan even though the acting president, Li Tsung-jen, remained the official head of state on the mainland. When the Nationalist government finally moved to Taiwan in December 1949, Li went to the United States for "medical treatment." Chiang resumed the post of president of the republic on March 1, 1950.

To bolster the defense of Taiwan, Chiang made "strategic" withdrawals from several outposts. In April and May 1950 government troops were evacuated from the Hainan and Chusan Islands, and in 1953-54 Nationalist guerrilla units stranded in North Vietnam and the Yunnan-Burmese border region were repatriated. Added strength came when 14,209 Chinese Communist prisoners of war, captured during the Korean War, joined the Nationalists in January 1954. A year later, in January 1955, a considerable number of civilians and troops on the Ta-ch-'en Islands off the Chekiang coast were repatriated. With these withdrawals and new programs of military training and recruitment, Taiwan developed a respectable force of 600,000 troops, a fifth of whom are stationed a few miles from the mainland on the islands of Quemoy and Matsu.

United States Policy Toward Taiwan

United States policy toward the Nationalists on Taiwan has gone through a cycle: from indifference to active support, back to benign neglect and then to de-recognition in January 1979. In the latter half of 1949, Washington, observing the Nationalist collapse on the continent, was resigned to the probable fall of Taiwan. President Truman announced a policy of noninvolvement on January 5, 1950:

> The United States has no predatory designs on Formosa or on any other Chinese territory. The United States has no desire to obtain special rights or privileges or to establish military bases on Formosa at this time. Nor does it have any intention of utilizing its armed forces to interfere in the present situation. The United States Government

3. Ch'i-k'ou.

will not pursue a course which will lead to involvement in the civil conflict in China.

Similarly, the United States Government will not provide military aid or advice to Chinese forces on Formosa.[4]

However, this policy was reversed after the North Korean attack on South Korea on June 25, 1950. Truman declared on June 27 that "communism has passed beyond the use of subversion to conquer independent nations and will now use armed invasion and war. . . . In these circumstances the occupation of Formosa by Communist forces would be a direct threat to the security of the Pacific area and to the United States forces . . . in that area."[5] He ordered the Seventh Fleet to defend Taiwan against any Communist invasion and urged the Nationalists to desist from attacking the mainland so as to prevent the extension of hostilities. With this change of policy, Washington dispatched a chargé d'affaires to Taipei on July 28, 1950, and three days later General MacArthur visited Chiang Kia-shek to discuss joint defense plans. On August 4 MacArthur's deputy chief of staff[6] arrived in Taipei to set up a permanent liaison with the Nationalist government.

Thus, the Korean War marked a turning point in United States–Nationalist relations. Washington's "hands-off" policy was jettisoned in favor of "neutralizing" Taiwan while preventing it from becoming a Communist trophy. A further stiffening of the United States position followed the entry of Chinese Communist "volunteers" into the Korean War in October 1950. Washington boycotted diplomatic recognition of the People's Republic of China and objected to its admission into the United Nations. Meanwhile, American military shipments to Taiwan were resumed and economic aid was initiated, amounting to $98 million from June 1, 1950 to June 30, 1951.[7] In addition, a 116-man Military Assistance Advisory Group was established on Taiwan in April 1951 under Major General William C. Chase, and by May 1952 this mission had grown to 400 men. Taiwan had acquired a new strategic importance in United States defense plans, and MacArthur described the island as "an unsinkable aircraft carrier and submarine tender ideally located to accomplish offensive strategy and at the same time checkmate defensive or counteroffensive operations by friendly forces based on Okinawa and the Philip-

4. Quoted in Joseph W. Ballantine, *Formosa: A Problem for United States Policy* (Washington, D.C., 1952), 120.
5. Ballantine, 127.
6. Major General Fox.
7. Economic aid in 1951-52: $81 million; 1952-53: 105 million; 1953-54: 116 million; 1954-55: 138 million; 1955-56: 79 million; 1956-57: 90 million; 1957-58: 61 million; 1958-59: 74 million; 1959-60: 70 million; 1961-62: 134 million.

pines."[8] A further boost to the Nationalist cause occurred when Japan decided on April 28, 1952, to sign a peace treaty with Nationalist China rather than with the People's Republic on the mainland. Taking into consideration the Nationalist claim of legal jurisdiction over all of China, the Japanese specified in the treaty that its terms applied to all territories now, as well as those "which may hereafter be," under Nationalist control.

The improved status of Taiwan was reflected in the elevation of the United States chargé d'affaires to ambassadorial rank[9] in January 1953 and in the signing of a mutual defense pact with the United States in December 1954. Attempts by the Communists to take Quemoy and Matsu in the autumn of 1954 and in 1958 were repeatedly frustrated by Nationalist resistance under American encouragement. President Eisenhower accepted the Nationalist view that giving up these offshore islands, whatever their military value, was tantamount to abject surrender. President Kennedy announced that should an attack on these islands constitute, in the opinion of the United States, a prelude to an attack on Taiwan, Washington would take appropriate actions for its defense. Thus, Tiawan's security was protected by the United States. Until 1971 the Nationalist government represented China at the United Nations.

However, President Nixon's announcement on July 15, 1971, of his proposed visit to Peking inaugurated a series of diplomatic maneuvers that adversely affected the international status of Taiwan. The first major blow was the admission of the People's Republic of China into the United Nations in October 1971 and the subsequent withdrawal of the Nationalist delegation in protest. Next came the Sino-American détente which was followed by Japanese recognition of the People's Republic of China as the sole, legal government of China and by its cancellation of the peace treaty with Taiwan signed in 1952. These events produced a bandwagon psychology among other nations, and one country after another deserted Taiwan in favor of Peking's claim to represent the legal government of China. By October 1974 Taiwan maintained diplomatic ties with only 32 countries, as compared with 65 in 1969, and the number of nations that recognized Nationalist China continued to decline. The United States had been moving gradually in this direction too, as evidenced by the establishment of liaison offices in Peking and Washington in May 1973. Finally, in January 1979 the United States recognized Peking as the legal government of China. By 1988 only twenty-two countries recognized the Nationalist government on Taiwan, although 151 countries maintained trading relations with it.

8. Quoted in Ballantine, 153.
9. First ambasador, Karl L. Rankin.

Political Structure

With the surrender of Japan in August 1945 Taiwan was restored to and made a province of China, with Ch'en Yi[10] as governor. The choice of Ch'en was unfortunate, for he was not a conscientious administrator. Corrupt and discriminatory, his term of office was marred by numerous scandals, including lucrative public auctions of confiscated Japanese properties, and by outrageous discrimination against the Taiwanese, who were treated as colonial subjects unfit for executive and managerial posts in either the government or large enterprises. Those Taiwanese who had originally welcomed the Nationalist take-over quickly lost faith in Ch'en's administration which they came to regard as worse than Japanese colonial rule. Finally, public indignation could not be contained and on February 28, 1947, a violent uprising erupted. Ch'en temporized to gain time while calling for reinforcements from the mainland, and when they arrived he carried out a ruthless massacre of the Taiwanese. Although his subsequent dismissal[11] mollified the situation somewhat, his misrule greatly damaged the Nationalist cause and embittered the Taiwanese against the mainlanders.[12] It was not until Ch'en Ch'eng became governor in January 1949 that the acrimonious relationship began to improve, largely through increased Taiwanese participation in provincial and local governments, although high appointments in the central government continued to elude them. By 1974 most of the provincial posts, a third of the cabinet positions, and several important mayoralities were filled by Taiwanese.

The Nationalist government on Taiwan is headed by a president and assisted by a secretary-general and a chief of staff who help him with civil and military affairs. The five-yüan structure is maintained, although the functions and legal positions of some of the yüans have changed. The powers of the Legislative Lüan have been expanded at the expense of the Executive Yüan. The appointment of the president (premier) of the Executive Yüan is subject to approval by the Legislative Yüan, which also has the right of interpellation. Members of the Legislative Yüan are elected, and they select a president and vice-president from among themselves. The Legislative Yüan can initiate legislation, interpellate ministers, review the budget, and conduct independent investigations—a far cry

---

10. Not to be confused with the Communist foreign minister, Ch'en Yi, whose personal name is composed of a different character but has a similar pronunciation.
11. Later shot in January 1949 for conniving and colluding with Communist agents.
12. The ratios between the Taiwanese and the mainlanders are 6.8 million vs. 524,940 in 1950, and approximately 10 million vs. 3 million in 1968. In February 1992 the Executive Yuan issued an investigatory report on the February 28, 1947, incident (*Erh-erh-pa shih-chien yen-chiu pao-kao*), and President Lee Teng-hui formally apologized to the families of the victims.

from the Legislative Yüan of the Political Tutelage period, which was hardly more than a law-drafting bureau of the KMT Central Executive Committee.

The composition of the Judiciary Yüan has also changed considerably. It consists of (1) a Council of Grand Judges who interpret the constitution, laws, and decrees, consisting of seventeen members, all appointed by the president of the republic with the concurrence of the Control Yüan; (2) the Supreme Court; (3) the Administrative Court; and (4) the Disciplinary Commission.

The president and vice-president of the Control Yüan are elected from among the members of the Yüan itself, who are elected for six years by provincial and municipal assemblies. The Examination Yüan has a president and a vice-president and nineteen commissioners, also elected for six years.

Except for the ministries of foreign affairs and national defense, the five-yüan government duplicates many of the functions of the provincial government, which has its own departments of civil affairs, finance, education, agriculture and forestry, communication, public health, public safety, etc. From the standpoint of administrative efficiency, the presence of two parallel governments on Taiwan is a luxury which the small island can ill afford. Yet, this juxtaposition is a political necessity, for the existence of the central government not only substantiates its claim to jurisdiction over all of China but also offers hope for an eventual return to the mainland.

Although the Republic of China claimed to be a constitutional democracy, it did not grant full freedom of speech and assembly as in the United States on the grounds that some restrictions were necessary during a period of general mobilization against "Communist insurgency." The government strictly controlled the press and the media, permitting no one to disseminate Marxist literature and allowing no one except specially authorized persons to read Communist publications. Critics of Chiang, his family, and KMT rule, as well as the promoters of the Taiwanese Independence Movement, were liable to arrest by secret police. Military preparation for counterattacking the mainland and extreme vigilance against Communist infiltration put the island under martial law which creates a feeling of uneasiness and caution among the populace. The Nationalist government had been subjected to foreign criticism for its curtailment of civil liberties, but it justified its restrictive measures on the grounds of national security and invites its critics to compare the degree of freedom on Taiwan with that on the mainland. By and large, the people of Taiwan seemed to accept the restrictions as the price they had to pay for living a relatively free and prosperous life.

A sensitive issue in the political life of Taiwan was the presidency of Chiang Kai-shek. The constitution specified that presidents were elected for six-year terms and could be re-elected for a second term. Initially elected president in 1948 in Nanking, Chiang was re-elected in 1954 on Taiwan. As 1960 approached the delicate and unprecedented question arose with respect to the possibility of the incumbent president serving for a third term. Since Chiang disfavored amendment of the constitution, the question was resolved by the National Assembly in February 1960, which voted to suspend the constitutional stipulation that limits the number of presidential terms during periods of "general mobilization and rebellion-suppression." On March 21, 1960, Chiang was elected to a third term, with Ch'en Ch'eng as vice-president, and in 1966 to a fourth term, with Yen Chia-kan as vice-president and premier.

In 1972 Chiang was elected to a fifth term, with Yen as vice-president and Chiang Ching-kuo, his eldest son, as premier. It was a trying time for the Republic of China, as its government was faced with a series of international adversities: the expulsion of the Nationalist U.N. delegation, the Washington-Peking détente, and the Japanese recognition of the People's Republic of China. Domestically, President Chiang, 86 years old, was failing in health and was not seen in public until May 1974. The major work of government was conducted by the vice-president and the premier; they realized that the future of Taiwan lay in reaching an accommodation with the Taiwanese, who had been barred from high councils for 25 years. Premier Chiang, 62 years old and destined to be his father's successor, rectified the situation by appointing an unprecedented number of Taiwanese to key government posts: six cabinet members, the governor of Taiwan,[13] and the mayor of Taipei.[14] The December 1972 elections to fill 53 new seats in the Legislative Yüan further increased Taiwanese representation in government. It appeared that Yen and Chiang held three major keys to the future stability of Taiwan: (1) the question of succession to Chiang Kai-shek; (2) the admission of Taiwanese to high positions; and (3) the liberalization of the existing political process with a corresponding easing of the grip by the old guard. Under them the government braced itself in quiet dignity against international setbacks and Peking's diplomatic and psychological offensive. The Nationalist government exhorted its people to strive for "survival through self-reliance" and to "over-

13. Shieh Tung-min.
14. Chang Feng-hsü.

come" international appeasement "with unwavering conviction of ultimate victory."

The Washington-Peking détente had the unexpected effect of bringing the 4 million mainlanders and the 12 million Taiwanese closer together; as Peking considered both the Nationalist government and the Taiwan Independence Movement illegal, international isolation meant that they faced the same fragile future. The mainlanders, who hitherto had slighted the Taiwanese, and the Taiwanese, who had resented Nationalist control, suddenly found their fate intertwined and realized that it was to their advantage to band together. United they stood a better chance for survival; disunited they were vulnerable to submission to Peking. This realization narrowed the political, social, and psychological distance between the two groups; intermarriage was also on the rise. Actually, except for a small minority of aborigines in the mountains, the Taiwanese and the mainlanders descended from the same ethnic stock and are "essentially Chinese in their social and political outlooks as in their ancestry."[15] Among today's Taiwanese 75 percent were descendants of immigrants from Fukien province and about 13 percent are Hakkas from southern China.[16] The distinction between the mainlanders and the Taiwanese was an artificial one, and the government and the people were making a conscious effort to submerge it.

According to Edgar Snow, Mao Tse-tung considered the return of Taiwan to mainland sovereignty as his "last national goal of unification," and insisted that Taiwan, as a province of China, must be liberated; but he would be reasonable in dealing with the issue, "perhaps even granting a degree of autonomy to Chiang Kai-shek if he should wish to remain governor there for his lifetime."[17] To Chiang and his followers, the idea was preposterous and insulting; they steadfastly refused to negotiate with the Communists. Peking, on the other hand, facing a Soviet threat on its northern border and having reached a détente with the United States, seemed to be in no hurry to liberate Taiwan by force. Convinced that time was on its side, Peking sought to demoralize the Nationalist government through diplomatic maneuvers that were aimed at the international isolation of Taiwan and by bombarding the island with the propaganda of the inevitability of the unification of Taiwan with the mainland. On the other hand, the Nationalist government seemed to follow a policy of striving to

15. Sheldon Appleton, "Taiwanese and Mainlanders on Taiwan: A Survey of Student Attitudes," *The China Quarterly* (Oct.-Dec. 1970), 44:56.
16. Mark A. Plummer, "Taiwan's Chinese Nationalist Government," *Current History* (Sept. 1971), 171.
17. Edgar Snow, "China Will Talk from a Position of Strength," *Life*, July 30, 1971, 24.

maintain the status quo as long as possible, while concentrating on economic development and keeping international ties intact.

Economic and Social Development

The bitter lesson of defeat on the mainland taught the Nationalists that they could not ignore the pressing problems of social and economic reform. Once established on Taiwan, they endeavored to succeed where earlier they had failed. With determination, American advice (through the Joint Commission on Rural Reconstruction [JCRR]), and considerable concentration of brainpower and technical skill, the Nationalist government successfully implemented a three-stage program of land reform and thereby fulfilled Sun Yat-sen's "land to the tiller" ideal.

The moving spirit behind this rural reconstruction program was Governor Ch'en Ch'eng, who began the first stage of reform in 1949 by making compulsory a reduction of the annual land rent from the prevailing 50-70 percent to 37.5 percent of the main crop.[18] In places where the existing rent was below 37.5 percent it was to remain as it was. Moreover, the old practice of giving an "oral lease," which offered no legal protection to tenant farmers, was replaced by a written lease that was valid for a period of at least six years. By these measures, the livelihood of 300,000 farm families was substantially improved, and the increased income enabled them to purchase cattle and houses which were dubbed "37.5 cattle" and "37.5 houses."[19]

The second stage of the land reform program was inaugurated in June 1951 with the sale of 430,000 acres of public land. This property, representing 20 percent of the arable land on Taiwan, had originally been set aside by the Japanese colonial administration for settlement by Japanese immigrants. The Nationalist government allowed each farmer to purchase a sufficient amount of this acreage to support a family of six: seven acres of paddy land or fourteen acres of dry land. The purchase price was set at two and one-half times the annual yield of the main crop and was repayable in twenty semiannual installments at 4 percent interest. No single installment was to exceed the prevailing rent. This second stage of reform enabled 139,688 farmers to become landowners.[20]

---

18. This rate of 37.5 percent was first adopted by the KMT in 1930 but was never implemented.
19. W. G. Goddard, *Formosa: A Study in Chinese History* (East Lansing, Mich., 1966), 191.
20. Goddard, 192; Chiao-min Hsieh, 285-86.

The third stage of land reform was launched in January 1953 with the compulsory sale of private and tenanted land to the government, which resold it to farmers at the same price while charging a 4 percent annual interest. No less than 193,823 families benefited from these measures, bringing the total number of landowning families to 400,000—or two and one-half to three million individuals. The completion of the agrarian reform program reduced tenancy from 39 to 15 percent on all farmland.[21] By early 1968 tenants operated only 10 percent of the land, while 90 percent was tilled by owners.[22]

As a result of these agrarian reform policies main crop productivity increased substantially. The general well-being of the farmers was reflected in the large number of houses they constructed or repaired and the number of bicycles and sewing machines they owned.

Most impressive has been the rate of industrial development. After an initial period (1945-52) of rehabilitation of war-torn industrial machinery, the government launched a Four-Year Economic Development Plan for 1953-56 and took the lead in promoting medium- and small-size basic industries which required no great outlay of capital and which utilized local raw materials. In order to improve employment in rural communities the government also encouraged the revitalization of cottage and handicraft industries. Backed by technical and managerial personnel from the mainland, American economic assistance, and a fierce determination to make Taiwan a showcase in the Pacific, the First Four-Year Plan was a remarkable success. Progress was made in nearly every line of industrial activity: aluminum, alkali, textiles, electricity, leather, chemicals, paper, jute, sugar, pineapple, mushrooms, handicrafts, etc. By the end of 1956 some 2,000 factories were in operation, a third of which were built after 1952. The number of industrial workers increased from 274,000 to 340,000 during this period, while the production index in 1956 was more than double that of 1951. In 1956 per capita income rose by 42 percent over 1953.[23]

The dynamic achievements of the First Four-Year Plan prompted the government to subsequently launch many more four-year plans, all of which were successful. For 1963-73, Taiwan averaged an annual economic growth rate of 9.7 percent; in 1964 the highest annual rate of growth was achieved at 14.2 percent and in 1966 the lowest annual growth rate occurred at 8.07 percent, but all per annum growth rates surpassed the original target of 7 percent. The Gross National Product (GNP) for 1973 reached $9.39 billion, as compared with $1.2 billion in 1952. Government

21. Goddard, 193; Chiao-min Hsieh, 286.
22. *Free China Weekly*, Taipei, Feb. 4, 1968.
23. Chiao-min Hsieh, 309-10.

officials proudly announced in 1973 that Taiwan had achieved a yearly per capita income of $467 and an individual daily intake of 2,697 calories, and that the standard of living of its people was second only to Japan among Asian nations.[24] With a record GNP of $14.1 billion and a per capita income of $702 in 1974, Taiwan enjoyed relative economic prosperity despite the phasing out of American aid in mid-1965. Taiwan benefited greatly from American procurement for the Vietnam War between 1965 and 1972; but, even after the settlement of the war its economy continued to boom. In 1973 foreign trade rose to $8.26 billion—a 50.2 percent increase over 1972—with exports totalling $4.47 billion and imports totalling $3.79 billion. In 1973 Taiwan enjoyed trade surpluses with all other nations except Japan, where a deficit of $603 million was registered. Overall, Taiwan's foreign trade enjoyed a surplus of $682.6 million.[25] These figures show that Taiwan, despite its uncertain political future, was economically robust as an independent entity. But, this also made it more attractive to the mainland government as a prize to win.

Cultural Life

By 1967 Taiwan attained the remarkably high literacy rate of 97.15 percent. This impressive record was made possible by a constitutional requirement that 15 percent of the national, 25 percent of the provincial, and 35 percent of the district budgets be devoted to education. Beginning with the fall term of 1968 free education was extended from six to nine years. In 1973 more than a quarter of the total population were students that were distributed in 2,307 elementary schools, 948 middle schools, and 99 universities and colleges. The increased number of educational institutions underscored the vast improvement in Taiwan's educational system since the end of Japanese rule; in 1946 there were only 1,130 elementary schools, 215 middle schools, and one university and three colleges on the island.[26] By mid-1974, there were 278 students out of every 1,000 population, a triple increase over 1950.[27]

Accompanying the growth of Taiwan's educational system is the increasing scope and quality of its research organizations. National Taiwan University, the leading institution of higher learning, maintains a number of graduate programs and since 1960 has awarded the doctorate in conjunction with the Ministry of Education. Among pure research organiza-

24. *Free China Weekly,* Dec. 23, 1973, May 5, 1974.
25. *Free China Weekly,* Jan. 20, March 24, 1974.
26. *Chung-yang Jih-pao (Central Daily News),* Taipei, Oct. 24, 1974.
27. *Free China Weekly,* July 21, 1974.

tions, the most prestigious is the Academia Sinica located at Nankang, outside Taipei. Its beautiful site and idyllic surroundings provide a haven for serious scholars. It maintains a number of institutes, such as Mathematics, History and Philology, Chemistry, Zoology, Ethnology, and Modern History. The last-named institute, one of the youngest, was formed in 1955 and has turned out a number of sound monographs by able young scholars. Other research organizations of note are the Tsing-hua University Atomic Research Institute and the Chiao-t'ung University Electronics Research Institute.

Scholars on Taiwan consider the island the repository of the Chinese cultural heritage. Many art collections formerly housed in the museums of Peking and Nanking are now on Taiwan. From the Peking Palace Museum came 231,910 pieces of exquisite art work and rare books, and from the Central Museum at Nanking came 11,729 priceless artifacts. Exhibitions of this treasury of national art are shown regularly at the magnificent Palace Museum outside Taipei.

This rapid survey of major developments on Taiwan points up the fact that the Nationalist government, which failed miserably on the mainland, has succeeded in turning the island into a "model" province and a showcase in Asia. Materially, the people enjoy a general well-being and high standard of living unequalled in Chinese history. Yet, for all its outward prosperity, Taiwan is a small island with limited possibilities. It is not the spiritual home of China. The older mainlanders felt socially rootless, intellectually isolated, and spiritually empty. Young people were discouraged by the lack of outlets for their talents. The greatest desire of many was to emigrate elsewhere to seek a new life. Although the mainlanders still longed to return to the continent, the practical-minded knew that it was a dream hardly realizable during their lifetime. The slogan "going back to the mainland" has been muted while portrayal of Taiwan as a "Treasure Island" (*Pao-tao*) of enduring value is stressed. No longer a way station before the eventual return to the continent, Taiwan offers an alternative to the Communist rule. The coexistence of material prosperity and spiritual anxiety testifies to the biblical truth that man does not live by bread alone. He needs hope to live a meaningful life.

The Nationalist cause was dealt a severe psychological blow by the death of Chiang Kai-shek on April 5, 1975. Long a symbol of anticommunism in Asia, Chiang was the last surviving major allied leader of World War II. His passing was mourned by the people on Taiwan with a deep sense of loss, but it had little effect on the power structure or the policy of the government, which had prepared for the occasion since 1972. In accordance with constitutional procedures, Vice-President Yen Chia-kan was sworn in as the new president on April 6, serving as the titular head while Premier Chiang Ching-kuo exercised the real power of government.

In his will, dated March 29, 1975, Chiang asked his people not to be discouraged by his passing but to devote themselves to the realization of Dr. Sun's Three People's Principles, the recovery of the mainland, the rehabilitation of the cultural heritage, and adherence to democracy. The style of his testament appears remarkably similar to that of Dr. Sun's in 1925. President Ford eulogized Chiang as "a man of firm integrity, high courage and deep political conviction." For many Westerners, Chiang's death marked the end of an era in Chinese history, but, for his followers on Taiwan, the task of fulfilling his will had just begun.

Further Reading

Ahern, Emily M., and Hill Gates (eds.), *The Anthropology of Taiwanese Society* (Stanford, 1981).

Appleton, Sheldon, "Taiwanese and Mainlanders on Taiwan: A Survey of Student Attitudes," *The China Quarterly* 44:38-60 (Oct.-Dec. 1970).

Ballantine, Joseph W., *Formosa: A Problem for United States Foreign Policy* (Washington, D.C., 1952).

Barclay, George W., *Colonial Development and Population in Taiwan* (Princeton, 1954).

Bueler, William M., *U.S. China Policy and the Problem of Taiwan* (Boulder, 1971).

Chiu, Hungdah (ed.), *China and the Question of Taiwan: Documents and Analysis* (New York, 1973).

Chiu, Hungdah, and Shao-chuan Leng (eds.), *China: Seventy Years after the 1911 Hsin-hai Revolution* (Charlottesville, 1984).

Clough, Ralph N., *Island China* (Cambridge, Mass., 1978).

Cohen, Jerome Alan et al., *Taiwan and American Policy: The Dilemma in U.S.-China Relations* (New York, 1971).

Davidson, James W., *The Island of Formosa, Past and Present: History, People, Resources, and Commercial Prospects* (New York, 1989).

Dickson, Bruce J., "The Lessons of Defeat: The Reorganization of the Kuomintang on Taiwan, 1950-52," *The China Quarterly*, March 1993, 56-84.

Erh-erh-pa shih-chien yen-chiu pao-kao (A research report on the February 28 [1947] Incident) (Taipei, 1992). Issued by the Executive Yuan of the Republic of China on Taiwan.

Freedman, Ronald, and John Y. Takeshita, *Family Planning in Taiwan: An Experiment in Social Change* (Princeton, 1969).

Galenson, Walter (ed.), *Economic Growth and Structural Change in Taiwan: The Postwar Experience of the Republic of China* (Ithaca, 1979).

Goddard, W. G., *The Makers of Taiwan* (Taipei, 1963).

———, *Formosa: A Study in Chinese History* (East Lansing, 1966).

Gordon, Leonard H. D. (ed.), *Taiwan: Studies in Chinese Local History* (New York, 1970).

Grasso, June M., *Truman's Two-China Policy, 1948-1950* (Armonk, N.Y., 1987).

Gregor, A. James, with Maria Hsia Chang and Andrew B. Zimmerman, *Ideology and Development: Sun Yat-sen and the Economic History of Taiwan* (Berkeley, 1982).

Ho, Yhi-min, *Agricultural Development of Taiwan, 1903-1960* (Nashville, 1966).

Hsieh, Chiao-min, *Taiwan: Ilha Formosa* (London, 1964).

Huang, Chia-mo 黃嘉謨, *Mei-kuo yü Tai-wan—1784-1895* 美國與台灣 (United States and Taiwan—1784-1895), (Taipei, 1966).

Huang, Shu-min, *Agricultural Degradation: Changing Community Systems in Rural Taiwan* (Lanham, Md., 1982).

Huebner, Jon W., "The Abortive Liberation of Taiwan," *The China Quarterly,* 110:256-75 (June 1987).

Jacobs, J. Bruce, "Recent Leadership and Political Trends in Taiwan," *The China Quarterly*, 45:120-54 (Jan.-March 1971).

Joint Commission on Rural Reconstruction, *A Decade of Rural Progress, 1948-1958* (Taipei, 1958).

Kerr, George, *Formosa Betrayed* (Boston, 1965).

———, *Formosa: Licensed Revolution and the Home Rule Movement, 1895-1945* (Honolulu, 1974).

Kirby, E. S. (ed.), *Rural Progress in Taiwan* (Taipei, 1960).

Kung, Lydia, *Factory Women in Taiwan* (Ann Arbor, 1983).

Lai, Tse-han, *A Tragic Beginning: The Taiwan Uprising of February 28, 1947* (Stanford, 1991).

Lasater, Martin L., *The Taiwan Issue in Sino-American Strategic Relations* (Boulder, 1984).

Lee, Teng-hui, *Intersectoral Capital Flows in the Economic Development of Taiwan, 1895-1960* (Ithaca, 1971).

Lien, Heng 連橫, *Tai-wan t'ung-shih* 台灣通史 (A general history of Taiwan), (Shanghai, 1947), 2 vols.

Mendel, Douglas, *The Politics of Formosan Nationalism* (Berkeley, 1970).

Peng, Ming-min, "Political Offenses in Taiwan: Laws and Problems," *The China Quarterly*, 47:471-93 (July-Sept. 1971).

Rankin, Karl L., *China Assignment* (Seattle, 1964).

Ravenal, Earl C., "Approaching China, Defending Taiwan," *Foreign Affairs*, 50:1: 44-58 (Oct. 1971).

Riggs, Fred W., *Formosa under Chinese Nationalist Rule* (New York, 1952).

Shen, T. H., *Agricultural Development on Taiwan Since World War II* (Ithaca, 1964).

———, *The Sino-American Joint Commission on Rural Reconstruction: Twenty Years of Cooperation for Agricultural Development* (Ithaca, 1970).

Shieh, Milton J. T., *Taiwan and the Democratic World* (Taipei, 1951).

Sih, Paul K. T. (ed.), *Taiwan in Modern Times* (New York, 1973).

Simon, Denis Fred, *Taiwan, Technology Transfer, and Transnationalism* (Boulder, 1983).

Stolper, Thomas E., *China, Taiwan, and the Offshore Islands* (Armonk, N.Y., 1985).

Tsurumi, E. Patricia, *Japanese Colonial Education in Taiwan, 1895-1945* (Cambridge, Mass., 1977).

Vander Meer, Canute and Paul Vander Meer, "Land Property Data on Taiwan," *The Journal of Asian Studies*, XXVIII:1:144-50 (Nov. 1968).

Wilson, Richard, *Learning to Be Chinese: Political Socialization of Children in Taiwan* (Cambrdge, Mass., 1970).

Wolf, Margery, *Women and the Family in Rural Taiwan* (Stanford, 1972).

Wu, Rong-I, *The Strategy of Economic Development: A Case Study of Taiwan* (Louvain, 1971).

Yang, Martin M. C., *Socioeconomic Results of Land Reform in Taiwan* (Honolulu, 1970).

VII

China After Mao:
The Search for a New Order

32

The Smashing
of the Gang of Four

Nineteen seventy-six was a year of agony for China. Deep bereavement was felt in every corner of the land over the loss of three of its great leaders: Premier Chou En-lai in January, Marshal Chu Teh in July, and Chairman Mao Tse-tung in September. Added to human grief was a series of natural disasters: in July a major earthquake demolished the industrial city of T'ang shan, 105 miles southeast of Peking; and during the next two months the Yellow River flooded seven times. Compounding the human misery and political instability was the succession crisis precipitated by Mao's wife Chiang Ch'ing and her associates, later dubbed the Gang of Four. Indeed, the tumultuous year was marked by what the Chinese call "natural disaster and human misfortune" (*t'ien-tsai jen-huo*). Ia was a time of sorrow, yet like darkness before dawn, also a time of hope. Out of disorder a new order was struggling to be born, and with it the promise of greater stability, progress, and a better life for the people.

The Deaths of National Leaders

CHOU EN-LAI (1898-1976). The death of Chou En-lai on January 8, 1976, was an irreparable loss. A pillar of strength in both party and government, he was the moderating influence through numerous political storms. Chou had saved the country from the utter chaos during the upheaval of the Cultural Revolution and had helped thwart the Gang of Four's grasp for supreme power.

Born to a gentry family of Shaohsing toward the end of the Ch'ing dy-

763

nasty, Chou attended the Nankai University Middle School before going to Japan for further studies. In 1920 he went to France as a work-study student and spent the next four years in Europe, where he, along with fellow-student Teng Hsiao-p'ing, joined the Chinese Communist Youth Corps and later the Chinese Communist Party. He became a loyal supporter of Mao after the Tsunyi Conference of January 1935. As premier from 1949 until his death, Chou submitted to Mao's oracular leadership, running the machinery of government unobtrusively while quietly moderating certain of Mao's excesses.

Chou exuded a disarming charm and an urbane sophistication. His savvy manner and vast knowledge of world affairs struck all foreign visitors. President Nixon said: "Only a handful of men in the 20th century will ever match Premier Chou's impact on world history. . . . none surpass him in keen intellect, philosophical breadth and the experienced wisdom which made him a great leader."[1]

Afflicted with cancer as early as 1972, Chou appears to have engineered the rehabilitation of Teng Hsiao-p'ing in 1973 as vice-premier and groomed him for succession. In response, the radicals launched the Anti-Lin Piao, Anti-Confucius Campaign to harass Chou by allusion. Chou continued to work even following his hospitalization in the summer of 1974, administering state affairs from his sickbed, receiving visitors, and making occasional public appearances. He attended the celebration of the 25th anniversary of the founding of the People's Republic and delivered the keynote speech at the Fourth National People's Congress in January 1975. This speech laid the groundwork for what has since become known as the Four Modernizations: a comprehensive modernization of agriculture, industry, national defense, and science and technology that would put China in the front ranks of the world by the end of the century.

Dead at 78, Chou was mourned by his 900 million countrymen as a beloved elder-protector, hero both in the struggle of revolution and in the management of the affairs of state. He must be credited for long years of invaluable service to the state, especially during the "decade of catastrophe" otherwise known as the Cultural Revolution.

THE EMERGENCE OF HUA KUO-FENG. Chou had carefully groomed Teng Hsiao-p'ing to be his successor. In the last year of Chou's hospitalization, Teng had amassed considerable power and was the de facto premier directing the day-to-day work of the State Council and receiving foreign leaders. Yet, shortly after eulogizing Chou at the memorial service on

1. *The New York Times*, Jan. 9, 1976, 11-12.

January 15, 1976, Teng dropped out of sight without explanation; and on February 6 the *People's Daily* carried a front page article attacking the "unrepentent" powerholders who took the "capitalist road." The Cultural Revolutionary Group under Chiang Ch'ing were promoting Chang Ch'un-ch'iao, second vice-premier, for the premiership, but Chang was unacceptable to many party seniors and military leaders. On February 7 the Chinese government announced the startling appointment of Hua Kuo-feng, the sixth-ranking vice-premier and minister of public security, as acting premier. This choice pleased neither the radicals nor the moderates but was not challenged due to Mao's enormous prestige.

In the waning months of his life, Mao, under pressure from his wife, had wanted to favor the Cultural Revolutionary Group but hesitated to offend the senior cadres and military leaders. In this dilemma, Mao appears to have selected Hua, a faithful follower who was ideologically safe and who would adequately serve the state until Chiang Ch'ing's group could assume power. The choice of Hua, a former party first secretary in Mao's home province of Hunan and an agricultural expert, was clearly a compromise as he had close ties with neither the pragmatists nor the radicals. At the same time, in naming Hua, Mao masterfully blocked Chou's scheme to place Teng in line for succession.[2]

Shortly after Hua's appointment, a momentous event took place during the Ch'ing-ming Festival, when the Chinese traditionally visit the ancestral tombs. Between March 29 and April 4, an increasing number of people went daily to Tien-an-men Square to pay tribute to the late Premier Chou and to lay wreaths at the Monument to the Martyrs of the Revolution, which had become the symbolic tomb of Chou. When the wreaths were removed by police and security guards, people were enraged. On April 5, 100,000 gathered at Tien-an-men Square in a protest demonstration chanting "the era of Ch'in Shih Huang is gone," hoisting signs in support of Teng, and singing the praises of Chou and Teng while criticizing Mao by allusion.[3] Emotion soon reached a feverish pitch, and the demonstration got out of control. Frenzied demonstrators set fire to four motorcars and smashed the windows of a military barrack before they were dispersed by police, security guards, and militiamen.

Mayor Wu Te of Peking linked the violent outburst with the opponents of the "Antirightist deviationist struggle." Teng was openly accused of being a ringleader, a capitalist roader, and "the general behind-the-scene

2. Ch'en Yung-sheng, "The 'October 6th Coup' and Hua Kuo-feng's Rise to Power," *Issues and Studies,* XV: 10:81-82 (Oct. 1979).

3. Mao had likened himself to this first emperor of the Ch'in dynasty who unified China in 221 B.C.

promoter of the Rightist Deviationist attempt to reverse correct verdicts."
Two days later (April 7) on Mao's recommendation, the Central Committee ordered Teng, "whose problem has turned into one of antagonistic contradictions," dismissed from all party and government posts, but allowed him to "keep his party membership so as to see how he will behave himself in the future."

Meanwhile, the appointment of Hua as premier and first vice-chairman of the party was announced.[4] On April 8, 100,000 people paraded in Tien-an-men Square in a counterdemonstration to support the new leadership. Hua emerged as the dark horse winner in the succession struggle for the premiership.

On April 30 Mao gave Hua three crucial handwritten instructions: (1) "Carry out the work slowly, not in haste"; (2) "Act according to past principles"; and (3) "With you in charge, I am at ease."[5] Viewing the instructions as implying Mao's intention to designate him his heir, Hua forwarded the first two to the Politburo in the presence of an infuriated Gang of Four, who now considered him no longer a possible ally but a new enemy. Hua remained unperturbed, keeping secret the third instruction and reserving it for future use.

THE T'ANG-SHAN EARTHQUAKE. As if the deaths of national leaders and the political confusion of the succession struggle were not enough punishment for the country, a gigantic earthquake measuring 8.2 on the Richter scale occurred on July 28, 1976, in T'ang-shan, a mining center of 1.6 million inhabitants. It leveled the entire city and caused considerable damage to the nearby metropolis of Tientsin, China's third largest city with a population of 4.3 million. T'ang-shan itself was reduced to a desert of rubble. A confidential government report listed 655,237 dead and 779,000 injured, although later figures given by the Chinese Seismology Society were considerably lower.[6] Following tradition the Chinese people viewed such massive natural disasters as portents of social and political upheaval.

MAO TSE-TUNG (1893-1976). For years Mao had been afflicted with Parkinson's disease, a slowly degenerative sickness causing muscular rigidity and tremors. His health failed rapidly in the last two or three years of his life due to a stroke, which affected the left side of his body impairing his

---

4. Text of Central Committee announcement in English carried by *The New York Times*, Apr. 8, 1976, 16.

5. *Peking Review*, Dec. 24, 1976, 8. See also Richard C. Thornton, "The Political Succession to Mao Tse-tung," *Issues and Studies*, XIV:6:35 (June 1978).

6. There were 240,000 dead and 164,000 injured. *Los Angeles Times*, June 11, 1977.

speech. Each day moments of clarity and well-being alternated with lapses into a less lucid state, hence the strange meeting times and abrupt notices given to foreign dignitaries who awaited audiences with him. Death came on September 9, 1976.

A full assessment of Mao must wait until history has had time to digest his impact on China and the world; for now, only preliminary remarks are in order. For China, Mao was Lenin and Stalin combined. He was a great revolutionary, the most successful of the mid-20th century. His greatest achievement was the seizure of power through the creative adaptation of Marxist-Leninist theory to the realities of the situation in China. Influenced by Li Ta-chao, he came to believe in the liberation of the peasant as the prelude to the liberation of China. He evolved the strategy of organizing the peasantry to encircle the cities and created a successful model of revolution for the Third World. He envisioned the ultimate application of this strategy to the international scene, urging the Third World to unite, engulf, and effect the eventual downfall of the Western bourgeois societies.

Throughout his life, Mao was motivated by a perpetual restlessness. He rebelled against his father, against landlord and capitalist, against Nationalist rule, against Soviet domination and revisionism, and finally against his own party establishment and senior associates. Impatient for change, he wanted to transform the state, the society, and human nature in one stroke. "Ten thousand years are too long; seize the day, seize the hour!" A purist at heart, he kept up the momentum of revolution by creating incessant upheaval, exhausting both country and people. Much national energy was spent on mass movement and internecine strife, which impeded national progress. His twenty-seven-year rule brought little improvement in people's living standard.

It thus appears that after the success of revolution in 1949, the genius that was in Mao was largely spent. The ingredients that led him to the seizure of power could not lead him to successfully administer the sprawling state. After the first years of liberation, Mao's leadership faltered. The Antirightist Campaign (1957) did irreparable damage to the intellectuals whose knowledge and skills China sorely needed. The rush to commune was too hasty; the Great Leap Forward went backward; the fight with P'eng Te-huai was ill-conceived; and the decimation of the party during the Cultural Revolution was an unmitigated disaster. The fostering of Chiang Ch'ing as a national leader and possible successor worked against the wishes of the people and Mao's senior associates. In his last years Mao spun himself farther and farther into a cocoon of his own making, insensitive to the feelings of the masses he had always claimed to represent. He died a lonely and unhappy man, his dream of transforming human nature and turning China into a powerful modern state unfulfilled. Historical perspective will in due time allow a full assessment of Mao's achievements and mistakes. For now, my own view of his life might be summed up in the following words:

> As a revolutionary,
>   Mao had few peers.
> As a nation-builder,
>   He was unequal to the task.

革命有餘
建國不足

The Gang of Four

THE PLOT OF THE GANG. The absence of a constitutional mechanism for the peaceful transfer of power led to a succession crisis when the incumbent leader died. The intense power struggle that erupted following Mao's death was led by his wife, Chiang Ch'ing, who aspired to succeed him as chairman, to make Wang Hung-wen chairman of the Standing Committee of the National People's Congress, and to install Chang Ch'un-ch'iao premier of the State Council. Yao Wen-yüan, already in charge of the party's propaganda department, was probably designated to be a "cultural tsar" with added titles. These four, the hard core of the Cultural Revolutionary Group, conspired to seize power, but their major obstacle was Hua Kuo-feng. As the first vice-chairman of the party, premier of the State Council, and the object of Mao's instruction ("With you in charge, I am at ease"), Hua had a firm claim to succession. Hua also had the support of Wang Tung-hsing, Mao's chief bodyguard and head of the 20,000-man 8341 special unit.

Chiang Ch'ing's trump card, Mao, was gone. Still in her deck were control of the media and of the urban militia in key places such as Shanghai, Peking, Tientsin, Shenyang, and Canton.[7] Before Mao's death the Four had schemed to distribute weapons and ammunition to the Shanghai militia, establishing a sort of National General Militia Headquarters to rival the Military Commission in Peking. The day after Mao's death, six million rounds of ammunition were issued to the Shanghai militia.[8]

The senior party cadres and military leaders, who loathed Chiang Ch'ing and her cohorts but had been powerless against them as long as Mao lived, decided secretly after the Tien-an-men Square Incident that only a coun-

7. Chien T'ieh, "The Chiang Ch'ing Faction and People's Military Forces," *Issues and Studies*, XII:1:25 (Jan. 1976).
8. *Peking Review*, Feb. 4, 1977, 5-10; Andres D. Onate, "Hua Kuo-feng and the Arrest of the 'Gang of Four,'" *The China Quarterly*, 75:555-56 (Sept. 1978).

tercoup could stop the Four from seizing power. They entrusted to Yeh Chien-ying, minister of defense, the delicate task of cultivating friendship with Hua and of promising him their support as Mao's successor.[9] Hua knew only too well the Gang's record and ambitions.

Another anti-Ch'iang Ch'ing force was also in secret operation. Teng Hsiao-p'ing, dismissed in April and hunted by the Gang, had fled to Canton under the protection of Yeh Chien-ying and Hsü Shih-yu. These three, in a secret meeting also attended by several others including Chao Tzu-yang (later General Secretary) decided to fight the Four by forming an alliance with the Foochow and Nanking Military Regions, with headquarters in Canton. Should Chiang Ch'ing gain power, they would establish a rival provisional Central Committee to contest her. After Mao's death Teng secretly returned to Peking to await developments.[10]

The Gang meanwhile were plotting to assassinate Politburo members, with Hua, Yeh, and several others as the main targets. Facing a common threat, the two became close allies and made the necessary preparations for a coup, which included the winning over of Wang Tung-hsing. A three-way coalition thus formed, with Yeh as the mastermind, Hua laying out the plan of action, and Wang implementing it. Shanghai would be secured first, and for this purpose the help of General Hsü Shih-yu of the Canton Military Region, who had extensive connections in the Shanghai-Nanking area, was enlisted to obtain the cooperation of the Shanghai garrison commander and win control of the key city *before* the urban militia could act.[11] In Peking, Hua had the firm support of Commander Ch'en Hsi-lien, Mayor Wu Te and the cooperation of the garrison forces, the army, and special unit 8341.

At the memorial service on September 18, Hua quoted Mao's famous dictum, "Political power grows out of the barrel of a gun," implying that he had the support of the military. Hua also quoted Mao's command, the "Three do's and don't's": (1) "Practice Marxism and not revisionism"; (2) "Unite, don't split"; and (3) "Be open and aboveboard, and don't intrigue and conspire."[12] The significance of this quotation could not be lost, for on the occasion of the original command, Mao had warned his wife and her cohorts "not to function as a gang of four."

THE OCTOBER 6 COUP. Several stormy Politburo meetings took place toward the end of September. Chiang Ch'ing indicated that Hua was in-

9. Ch'en Yung-sheng, 78.
10. Testimony by Chang Ping-hua, former director of the Propaganda Department of the CCP Central Committee, quoted in Ch'en Yung-sheng, 85-86.
11. Onate, 556.
12. Complete text of the speech in *Peking Review*, Sept. 24, 1976, 12-16.

competent to lead the party and demanded that she be made chairman of the Central Committee. Hua retorted that he was not only competent but knew how to "solve problems"—in retrospect an ominous reference to his intention to remove the Four.

In the early morning of October 5, a secret meeting was held in the headquarters of the commander of the People's Liberation Army, with five participants: Hua, Yeh, Wang, Peking Garrison Commander Ch'en Hsi-lien, and Vice-Premier Li Hsien-nien (an ally of Teng Hsiao-p'ing). They decided to act decisively and quickly before the Gang could stage their coup by arresting all four leaders in a single swoop. Hua and Yeh assumed overall direction with Ch'en Hsi-lien assigned the duty of safeguarding Peking and Wang Tung-hsing, the job of arresting the Four. Meanwhile, the Canton Military Region was alerted to ready two divisions for airlift to Peking on instant notice.

Hua next invited the Four to attend an emergency Politburo meeting at midnight, October 5, at the party headquarters in Chung-nan-hai. Wang Hung-wen arrived first. He resisted arrest and killed two guards but was himself wounded and subdued. Then came Chang Ch'un-ch'iao and Yao Wen-yüan; both fell into the trap. Chiang Ch'ing was in bed when her captors arrived. She shouted: "How dare you to rebel when the Chairman's body is not yet cold!"[13] In the small hours of October 6, the Gang was felled in one clean sweep. The Four were placed in solitary confinement in separate locations in Peking.

On October 7 Hua Kuo-feng and Wang Tung-hsing each delivered two reports and Yeh Chien-ying, one, to the Politburo. These reports, containing detailed charges against the Four, must have been prepared in extreme secrecy some time prior to the arrest. The Gang had been so smoothly and resolutely smashed that no question of civil strife arose; the success must be credited to the three protagonists who had long years of experience in security and military matters. The grateful Politburo named Hua chairman of the party Central Committee and concurrently chairman of the Military Commission, and put him in charge of editing the fifth volume of the *Selected Works of Mao Tse-tung*.[14]

---

13. Reported in *Central Daily News,* Taipei, Sept. 23, 1980.
14. The selection of Hua, although supposedly made on October 7, was possibly made later. In the initial Politburo announcement of October 7, there was no mention of Hua's appointment. The October 29, 1976, issue of *Peking Review* carried an article, "Great Historic Victory" (p. 14) which belatedly stated: "In accordance with the arrangements Chairman Mao had made before he passed away, the October 7, 1976 resolution of the Central Committee of the Communist Party appointed Comrade Hua Kuo-feng chairman of the Central Committee of the Communist Party of China and Chairman of the Military Commission of the C.P.C. Central Committee."

On October 24 a million soldiers and civilians held a victory rally at Tien-an-men Square to celebrate the smashing of the Gang of Four. A smiling Hua appeared, accompanied by top military leaders indicating the key role the military had played in the "palace revolution" and its continued support of Hua. Hua was hailed as a "worthy leader" of the party, a "worthy helmsman" to succeed Mao, and a brilliant leader who most nearly possessed the merits of Mao and Chou.[15] The following day, the *People's Daily*, the *Red Flag*, and the *Liberation Army Daily*, under new management, jointly editorialized a "Great Historic Victory." Later, the Third Plenary Session of the Tenth Central Committee (July 1977) described the smashing of the Gang of Four as the 11th major struggle in the history of the party, of almost equal importance with the Tsunyi Conference (1935), and credited Hua with saving the revolution and the party.[16]

Several factors accounted for Hua's success. As first vice-chairman of the party and premier of the State Council, he had all the advantages of an incumbent leader. He enjoyed the support of the military and party leaders, and had won the cooperation of Wang Tung-hsing and the 8341 unit. He had taken the pulse of the country and knew the people's hatred of Chiang Ch'ing and the Gang; in smashing them he was expressing the "common aspiration" of the people. And finally, having been a member of the committee investigating the Lin Piao affair, he knew that indecision was the chief cause of Lin's downfall and therefore acted decisively and swiftly to surprise and overwhelm the conspirators.

On the other hand, the Gang's failure must be traced first and foremost to the death of Mao. Under Mao's patronage the Four issued orders in his name and rode roughshod over the uncooperative. They mistreated thousands of respected elders and leaders, and used terrorists and secret agents to browbeat unsympathetic intellectuals and the people. In addition, the Gang led decadent, privileged, bourgeois private lives. Chiang Ch'ing, for example, kept a "silver" jet for her own use, enjoyed the most expensive of German photographic equipment, wore silk blouses, and received guests in lavish settings.[17] While these excesses alienated the masses and mocked the ideals of the Chinese revolution, Mao's patronage of the Gang effectively stilled criticism.

The second major source of the Gang's weakness was the imbalance between military strength and media control. The Gang did not control the

---

15. "Comrade Hua Kuo-feng Is Our Party's Worthy Leader," and "Great Historic Victory," both in *Peking Review*, No. 44:14-16, Oct. 29, 1976; No. 45:5-6, Nov. 5, 1976.
16. *Peking Review*, No. 47, Nov. 19, 1976.
17. Roxane Witke, *Comrade Chiang Ch'ing* (Boston, 1977), 37-38.

army but only the militia, which lacked organization and firepower; so they relied heavily on their control of the media and cultural scene to mold public opinion and to give an exaggerated image of power. Perhaps the loud support and broad coverage they received lured them into believing that they were stronger than they actually were. Moreover, Chiang Ch'ing was overly confident that as Mao's wife, nobody would dare oppose her. But the fact was, the minute she became his widow, her fate was sealed.

The Four were expelled from the party, removed from all official posts, and branded as conspirators, ultrarightists, counterrevolutionaries, and representatives of Kuomintang.

Mao and the Gang

There is no way to dissociate Mao from the Gang. Without him there could have been no Gang, for without his wife they would have had no safely protected leader. Chiang Ch'ing, a former left-wing movie actress, came to Yenan after the Long March and became Mao's secretary. They fell in love and Mao asked for her hand, much to the consternation of Mao's third wife who protested violently and refused to divorce him. Senior cadres also disapproved of the marriage, but nonetheless the two were married in 1939. Reportedly, before the marriage senior officials exacted a promise from Mao that his wife not be active in politics for life or at least 20 years.[18]

Mao realized his wife's "wide ambitions" to become chairman, and he also knew of the countless number of people she had wronged, harmed, arrested, or killed during the decade of the Cultural Revolution. On July 17, 1974, Mao had warned the Gang: "You'd better be careful; don't let yourselves become a small faction of four." In May 1975 he admonished them with the "Three do's and don't's," ending with, "Don't function as a gang of four; don't do it anymore."[19] Mao was thus aware of the Gang's excesses and could have restrained their leader by a simple order. That this was not done reflected his failings as party chairman and the Great Helmsman.

When the American playwright Arthur Miller visited China in 1978, he met with Chinese writers, artists, movie directors, and stage managers. He learned that many of the country's leading artists and intellectuals had been killed or imprisoned and tortured. To Miller it was inconceivable that Chiang Ch'ing could have committed such injustices without the sup-

18. Witke, 148-57, 335.
19. Joint editorials of *People's Daily*, *Red Flag*, and *Liberation Army Daily*, Oct. 25, 1976.

port of Mao. Quoting Mao, "People are not chives; their heads do not grow back when they are cut off," Miller concluded: "It has become impossible to believe that a 'faction' could have swung the People's Republic around its head without the consent of the Great Helmsman."[20] For Miller, Mao's lack of leadership could not be blamed on his physical infirmity, for people were jailed and killed in the 1960s when he was still strong enough to swim six and one-half kilometers in the Yangtze River. Miller's final judgment: the Gang of Four was "merely a screen for the still-sacrosanct name of Mao."[21]

Although deified before 900 million people, Mao in private life was an aged and doddering husband. As he increasingly submitted to Chiang Ch'ing's pressure, he lost all sense of proportion in state affairs. A communism tainted with familial favoritism smacked of "socialist feudalism." Yet from the dramatic events of 1976 and the defeat of extremism, there came a promise of greater stability, prosperity, and a new drive for modernization.

20. Inge Morath and Arthur Miller, *Chinese Encounters* (New York, 1979), 21, 40.
21. *Ibid.,* 7.

Further Reading

Bonavia, David, *Verdict in Peking: The Trial of the Gang of Four* (New York, 1984).

Chang, David W., *Zhou Enlai and Deng Xiaoping in the Chinese Leadership Succession Crisis* (Lanham, Md., 1983).

Ch'en, Yung-sheng, "The 'October 6th Coup' and Hua Kuo-weng's Rise to Power," *Issues & Studies,* XV:10:75-86 (Oct. 1979).

Cheng, J. Chester, *Documents of Dissent: Chinese Political Thought Since Mao* (Stanford, 1981).

Chi, Hsin, *The Rise and Fall of the "Gang of Four"* (tr. from *The Seventies Magazine*), (New York, 1977).

Domes, Jürgen, "The Gang of Four and Hua Kuo-feng: An Analysis of Political Events in 1975-76," *The China Quarterly,* 71:473-497 (Sept. 1977).

"Great Historic Victory," *Peking Review,* 44:14-16 (Oct. 29, 1976).

"How the 'Gang of Four' Used Shanghai as a Base to Usurp Party and State Power," *Peking Review,* 6:5-10 (Feb. 4, 1977).

Leng, Shao-chaun, with Hungdah Chiu, *Criminal Justice in Post-Mao China: Analysis and Documents* (Albany, N.Y., 1985).

Liu, Alan P. L., "The Gang of Four and the Chinese People's Liberation Army," *Asian Survey,* XIX:9:817-837.

Morath, Inge, and Arthur Miller, *Chinese Encounters* (New York, 1979).

Nee, Victor, and James Peck (eds.), *China's Uninterrupted Revolution* (New York, 1975).

Oksenberg, Michel, and Sai-choung Yeung, "Hua Kuo-feng's Pre-Cultural Revolution Hunan Years, 1946-66: The Making of a Political Generalist," *The China Quarterly*, 69:3-53 (March 1977).

Oksenberg, Michel, "Evaluating the Chinese Political System," *Contemporary China*, III:2:102-111 (Summer 1979).

Onate, Andres D., "Hua Kuo-feng and the Arrest of the 'Gang of Four,' *The China Quarterly*, 75:540-565 (Sept. 1978).

Roots, John McCook, *An Informal Biography of China's Legendary Chou En-lai* (New York, 1978).

Schram, Stuart R., *Mao Zedong, A Preliminary Reassessment* (Hong Kong, 1983).

Teiwes, Frederick C., *Leadership, Legitimacy, and Conflict in China: From a Charismatic Mao to the Politics of Succession* (Armonk, N.Y., 1984).

Terrill, Ross, *White-Boned Demon: A Biography of Madame Mao Zedong* (New York, 1984).

Tsou, Tang, "Mao Tse-tung Thought, the Last Struggle for Succession, and the Post-Mao Era," *The China Quarterly*, 71:498-527 (Sept. 1977).

Wang, Hsueh-wen, "The 'Gang of Four' Incident: Official Exposé by a CCPCC Document," *Issues & Studies*, XIII:9:46-58 (Sept. 1977).

Wilson, Dick (ed.), *Mao Tse-tung in the Scales of History: A Preliminary Assessment Organized by the China Quarterly* (Cambridge, Eng., 1977).

———, *Chou: The Story of Zhou Enlai, 1898-1976* (London, 1984).

Witke, Roxane, *Comrade Chiang Ch'ing* (Boston, 1977).

Wong, Paul, *China's Higher Leadership in the Socialist Transition* (New York, 1976).

33

Teng Hsiao-p'ing and China's New Order

Following the downfall of the Gang of Four, Chairman Hua Kuo-feng faced three pressing issues: (1) his legitimacy as Mao's successor; (2) the rehabilitation of Teng Hsiao-p'ing; and (3) the reordering of economic priorities to promote modernization. Regarding the succession, Mao's instruction to Hua ("With you in charge, I am at ease") was regarded by Yeh and Teng supporters[1] as reflecting Mao's personal view rather than the will of the party, whose constitution has specific provisions governing the election of the party chairman. By implications, Hua's assumption of the chairmanship of the Central Committee and of its Military Commission was deemed unconstitutional; but, if he would agree to the reinstatement of Teng, this question of legitimacy could be negotiated or even withdrawn. Thus, the two issues came into balance. As a result of mediation by Marshal Yeh and Vice-Premier Li Hsien-nien, who desperately desired a smooth transition to the post-Mao era, Hua agreed in principle to rehabilitate Teng, and to revise the five-year economic plan to accelerate the Four Modernizations. In late November 1976 Hua announced that Teng's reinstatement would be discussed at the next Central Committee meetings in July 1977. In return, he received support from Yeh, Li, and others for chairmanship of the Central Committee and its Military Commission.

At the Third Plenum of the Tenth Central Committee meeting, three

1. Such as General Hsü Shih-yu and Wei Kuo-ch'ing, both Politburo members; Hsü was also commander of the Canton Military Region, and Wei, party first secretary in Kwangtung.

resolutions were passed to confirm earlier Politburo decisions. First was the approval of Hua as chairman of the party and of the Military Commission; next was the acceptance of Hua's recommendation that Teng be restored to his former posts—Politburo Standing Committee member, vice-chairman of the Central Committee, first deputy premier of the State Council, vice-chairman of the Military Commission, and chief of the General Staff of the Liberation Army—all top positions in the military, the party, and the government. The third resolution condemned the antiparty activities of the Gang of Four and accused them of "conspiring to overthrow Comrade Chou En-lai," of "violently attacking and falsely accusing Comrade Teng Hsiao-p'ing," of being "extremely hostile and thoroughly opposed to" Mao's choice of Hua, and of "plotting to overthrow the party Central Committee headed by Comrade Hua Kuo-feng and bring about a counterrevolutionary restoration." The Four were officially expelled from the party.[2]

Teng's Drive for Political Dominance

ENLARGEMENT OF THE POWER BASE. Teng was intent upon enlarging his power base by rehabilitating men who had suffered under Mao and the Gang in the name of "righting the wrong" (p'ing-fan). He took a strong stand against leaders associated with the Cultural Revolution and the Gang of Four, especially those who had criticized him and blocked his succession of Chou En-lai. These included Hua (who rose under Mao's patronage), Wang Tung-hsing (head of Mao's bodyguard), Wu Te (mayor of Peking), and Chi Teng-kuei (a doctrinaire Politburo member). Teng attacked not Hua but his associates, chiselling away at his political periphery so that the center would be rendered hollow. Meanwhile, Teng also cultivated able, younger followers, placing them in key positions so that they could perpetuate his economic policies.

Teng, however, did not limit himself to attacks on individuals or the appointments of "young blood"; he simultaneously eroded the ideological power base of his former adversaries by combatting the embedded supremacy of "Mao Thought." To this end he announced in May and June of 1978 two clever guiding principles: "Practice is the sole criterion of truth" and "Seek truth from facts." By implication Mao Tse-tung's thought was no longer the standard by which a policy or an action must be judged; in fact, the thought itself must be subject to the scrutiny of facts, practice, and truth. The problem of Mao's leadership and his responsibility for the

---

2. Richard C. Thornton, "The Political Succession to Mao Tse-tung," *Issues and Studies,* XIV:6:47-49 (June 1978); *Red Flag,* 8:7-8 (1977).

Cultural Revolution and China's ills became a major concern of the post-Gang government.

"ECONOMICS IN COMMAND": REMOVAL OF OPPONENTS AND INTRODUCTION OF "NEW BLOOD." The Fifth Plenum of the Eleventh Central Committee (February 23-29, 1980) marked the end of the transitional period from Mao's death to the meteoric rise of Teng as the most powerful figure in Chinese politics. The party rejected Mao's "politics in command" for Teng's "economics in command" hoping to turn China into an advanced nation by the year 2000. Any activity or person deemed unsympathetic to this course would be curtailed or removed. Thus, though divergent views were tolerated to a degree, acts of dissidence such as posters on "Democracy Wall" attacking the government were not tolerated.[3] Both the government and the party were fearful that the delicate stability might be disturbed by too large a dosage of unaccustomed freedom; yet they were determined not to stifle the creativity, initiative, and enthusiasm that the new national goals generated. Their compromise resulted in a "restrained democracy" with moderate controls.

The trend toward eliminating dissension was not limited to Democracy Wall. Four Politburo members who were lukewarm or unsympathetic toward Teng and his policy were relieved of their high party and government posts. On the other hand, two of Teng's dynamic protégés were appointed to the Politburo Standing Committee—Chao Tzu-yang, an effective party first secretary in Szechwan, and Hu Yao-pang, Teng's righthand man in party affairs. Hu also became head of the newly reorganized party Secretariat in charge of the party's daily affairs.

Meanwhile, the rehabilitations that so benefited Teng continued. To clear the name of the former Chief of State Liu Shao-ch'i, who was disgraced and discredited along with Teng during the Cultural Revolution, the party resolved that he be posthumously restored to honor. On May 17, 1980, a national memorial service was held, and Liu was praised as a great proletarian fighter. The occasion was viewed as a negation of the values of the Cultural Revolution and a denial of Mao's infallibility.

HUA'S RESIGNATION FROM THE PREMIERSHIP. At the Third Plenum of the Fifth National People's Congress (August 29-September 10, 1980) the Tengists rose to the pinnacle of power. Teng had long urged separating party and government functions as well as ending lifelong appointments

3. The leader of the democratic movement, Wei Ching-sheng (Wei Jingsheng), was jailed for fifteen years and released in September 1993 one-half year before completion of the full term as part of an all-out effort by the Chinese government to win the Olympic site for Peking in A.D. 2000.

to cadres. The Congress approved his reorganization plan, insuring an orderly transfer of power to a collective leadership of relatively young pragmatists committed to modernization regardless of the fates of Teng and other aging leaders. Hua resigned as premier and nominated Chao Tzu-yang as his successor. Teng and six other vice-premiers resigned for reasons of old age, other important appointments, or "voluntary" withdrawal.

It should be noted that the National Congress was only concerned with government appointments. Those who retired or resigned did not lose their party positions. Hua remained chairman of the Central Committee and of its Military Commission; and Teng still held his party vice-chairmanship, and the four former vice-premiers, their seats on the Politburo. With Chao as premier and Hu as party general secretary, the pragmatists were in firm control of both government and party. For the first time an orderly transfer of power seemed to have been arranged while the previous incumbents were still healthy, creating a precedent for future leaders which might avoid the wrenching political turmoil and uncertainty of the past.

The Demystification of Mao

During the last fifteen years of his life, Mao, the Chinese "Lenin and Stalin combined," was sanctified as an all-knowing, all-wise demigod who could do no wrong.

Once Mao was dead and the Gang of Four smashed, Mao's image quickly became tarnished. His responsibility for the rise of the Gang was common knowledge; yet no one dared to debunk him as Khrushchev had Stalin. Leaders gingerly invoked Mao's sayings of the 1950s to refute his later policies, but de-Maoification had to be handled with care because Hua, until the 1977 Party Congress had confirmed his status, derived the legitimacy of his position largely from Mao's patronage. Hua honored Mao's legacy in order to consolidate his own position while reinterpreting Mao to suit his need in the changing times and circumstances.

The foremost question facing the nation was how to deal with the question of Mao's responsibility for China's recent ills. Before any answers could be offered, the party had elevated Chou En-lai to a position of near-parity with Mao, ending the solitary eminence of the Great Helmsman. Chou's wife, Teng Ying-ch'ao, was appointed to a vice-chairmanship of the National People's Congress. That Mao's wife was in jail and Chou's in high honor symbolized a national consensus reflecting the demystification of Mao.

The first year after Mao's death witnessed a growing sense of relief and a movement toward a new beginning. The structural references introduced by Mao or the Gang apparently no longer fit the realities of life where stability, unity, discipline, and economic progress were the new order. The revolutionary rhetoric and cultural intolerance which had rendered China an intellectual desert of artistic insipidity gave way to some degree of relaxation and freedom of expression. The cultural straightjacket dictated by the Gang (e.g. that China needed only eight model operas, or "more knowledge means more reactionism") was now condemned as absurd and counterproductive. Beethoven, Mozart, and Shakespeare, once symbols of "bourgeois decadence and running dogs of imperialism," reappeared in mid-1977; so did the works of the great T'ang poets Li Po and Tu Fu, "products of the feudal past."

With the rehabilitation of Teng in July 1977, Mao's desanctification was accelerated. First by indirect and later by open criticism, Mao's pedestal was chipped away. At the Eleventh Party Congress in August, Hua declared an end to the Cultural Revolution in contradiction of Mao's assertion that cultural revolution was a continuing process to be renewed every seven or eight years. Teng's "economics in command" triumphed as the new line.

On July 1, 1978, the 57th anniversary of the founding of the Chinese Communist Party, a speech made by Mao in 1962 was reprinted to show that he confessed to mistakes and an ignorance of economic planning, industry, and commerce: "In socialist construction, we are still acting blindly to a very large degree. . . . I myself do not understand many problems in the work of economic construction . . . [or] much about industry and commerce. I understand something about agriculture but only relatively and in a limited way. . . . When it comes to productive forces, I know very little."[4] The underlying message could not have been more clear—Mao was not an omniscient deity, but a fallible human being.

The second anniversary of Mao's death, September 9, 1978, passed without observance. Shortly after, the Red Guard, a symbol of Mao's support of the Cultural Revolution, was dissolved; both the "Little Red Book" and Mao's quotations on newspaper mastheads disappeared. Throughout the second half of 1978 wall posters and articles continued to criticize Mao's mistakes, implying a concerted effort to demystify him and to erode his image as a god-hero. Increasingly the editorials of the *People's Daily* referred to Mao as comrade rather than chairman, and criticisms of his role in the Cultural Revolution—now dubbed "Ten Years of Great Catastrophe"—became pronounced.

Teng's two principles, "Practice is the sole criterion of truth" and "Seek

4. *People's Daily*, July 1, 1978, p. 3. Tr. mine.

truth from facts," struck at the very heart of the Thought of Mao. Actually, verification of truth through practice is Marxist theory; and Mao's thought, until successfully practiced, could only be theory, not truth.[5] Mao himself had said: "We must believe in science and nothing else, that is to say, we must not be superstitious. . . . What is right is right and what is wrong is wrong—otherwise it is superstition."[6]

A poster displayed on November 22 in front of Tien-an-men Square applied Teng's slogans to Mao's achievements:

> We do not question the great achievements of Chairman Mao, but that does not mean he did not make mistakes. . . . Mao was a human, not a god. We must ascribe to him the status he deserved. Only so can we defend Marxism-Leninism and the Thought of Mao. Without an accurate understanding of Mao, freedom of speech is empty talk. It is the time for all Chinese to shake off the shackles on their thought and behavior.[7]

In applying Teng's precepts to Mao's actions and in invoking Marx and early Mao to refute later Mao, a clever way of demystifying Mao was discovered, one which also undermined the position of those whose political lives depended upon his status.

In September 1979 the third anniversary of Mao's death passed unnoticed. By spring of the following year, most of Mao's portraits in public places had been removed, as had the billboards bearing his quotations at street intersections. In March 1980 the party posthumously attacked Mao's secret service head, K'ang Sheng. By mid-year Mao's treasured models of production, the Ta-chai agricultural commune and the Ta-ch'ing oil field, lost their "paragon model" status—Ta-chai was declared a failure and Ta-ch'ing, inefficient and unscientific. Even Yenan, Mao's revolutionary cradle (which the author visited in May 1980), was left in a state of benign neglect. It was preserved as a revolutionary shrine of the past while current attention was being focused on the Four Modernizations and their success in the future.

These acts of de-Maoification were outer manifestations of an intense, continuous debate within the party over the quality of Mao's leadership

---

5. Marx said in his "Theses on Feuerbach": "The question whether objective truth can be attained by human thinking is not a question of theory but is a practical question. It is in practice that man must prove the truth, that is, the reality and power, the temporal nature of his thinking. The dispute over the reality or unreality of thinking which is isolated from practice is a purely scholastic question." See the article "Practice is the Sole Criterion of Truth," *People's Daily*, May 12, 1978.

6. "Science and Superstition," *People's Daily*, Oct. 2, 1978.

7. Reprinted in *Central Daily News*, Jan. 3, 1979. Tr. mine.

and over the assessment of his responsibility. The party had scrutinized Mao's thought in light of "truth according to facts" and, due to his failure to modernize China during his 27-year rule, gave him an "abstract affirmation but a concrete negation." On the other hand, the "Whateverists"—those who obeyed whatever Mao ordered—still carried Mao's banner and wanted to place revolution in command of modernization. To them, "truth according to facts" was just another of Teng's clever slogans intended to cut down Mao's banner.

While disagreement over Mao's waning reputations continued, the speech delivered by Marshal Yeh on the thirtieth anniversary of the People's Republic on October 1, 1979, was a measured indictment of Mao's leadership and misgovernment:

> Of course, the Mao Tse-tung Thought is not the product of Mao's personal wisdom alone; it is also the product of the wisdom of his comrades-in-arms, the party, and the revolutionary people. Mao himself has said: "It is the product of the collective struggle of the party and the people."

Surveying the history of the past thirty years, Yeh made clear the mistakes committed by the party under Mao's guidance:

> Amidst the immense victories we became imprudent. In 1957, while it was necessary to counterattack a small group of bourgeois rightists, we made the mistake of enlarging the scope [of attack]. In 1958, we violated the principle of carrying out an in-depth investigation, study, and examination of all innovations before giving arbitrary direction, being boastful, and stirring up a "communist storm." In 1959, we improperly carried out the struggle against the so-called right opportunism within the party.

Yeh charged that the Cultural Revolution was "the most severe reversal of our socialist cause since the establishment [of the People's Republic] in 1949." Then, pointedly, he announced:

> Leaders are not gods. It is impossible for them to be free from mistakes or shortcomings. They should definitely not be deified. We should not play down the role of the collectives and the masses; nor should we indiscriminately exaggerate the role of individual leaders.[8]

In this way the party renounced the personality cult of Mao and moved him from the lofty status of demigod to the humble one of human. Still, an important issue remained unresolved: how far the criticism of Mao

8. Yeh Chien-ying, "Speech Celebrating the 30th Anniversary of the Founding of the People's Republic of China," *People's Daily*, Sept. 30, 1979. Tr. mine.

should go. In February 1980 Yeh made an impassionate plea against a complete repudiation of Mao:

> We can pass resolutions to admit our party's mistakes. We can clear the name of Liu Shao-ch'i and give him a very high and positive assessment. But we should not reject Mao and dig too deeply into our own cornerstones. . . . The Soviets removed Stalin's tomb, and we whip the corpse of Mao. Wouldn't that prompt people to ask, what is right with socialism and what is good about communism? We can occasionally slap our own faces, but we cannot, nor do we have time to, start from scratch. Those who opposed Mao were not necessarily all wrong, just as those who supported him were not necessarily all right. His opponents and his supporters were all his followers. Was it right or wrong to follow him? Who elevated Mao to such heights and who gave him so much power? Was it the people of the entire country? It was given by the party, the party center, and the army under the leadership of the party. . . . If we want to trace the responsibility to the end, we will find that it lies not with Mao alone. It lies with all of us.[9]

Yeh's sentiments of moderation were shared by a large segment of party members, especially those in rural areas and those who had joined the party during or after the Cultural Revolution who accounted for half of the 38 million members. They were opposed to harsh criticism of Mao; he was, after all, human and not a god. Complete repudiation of him would risk negating the party itself.

Certainly, the party would neither deny Mao's contributions nor hide his mistakes, especially his part in the Cultural Revolution, the "decade of great catastrophe." An official assessment of Mao was to be made at the party meetings in mid-1981. Meanwhile, volume five of Mao's *Selected Works* edited by Hua was to be revised, implying dissatisfaction with the editor and with his selections. Thus not only the position of Mao but also that of his anointed successor Hua hung in suspense.

9. Reprinted in *Central Daily News,* Apr. 30, 1980. Tr. mine.

Further Reading

Burns, John P., and Stanley Rosen (eds.), *Policy Conflicts in Post-Mao China* (Armonk, N.Y., 1986).

Bush, Richard C., "Deng Xiaoping: China's Old Man in a Hurry" in Robert B. Oxnam and Richard C. Bush (eds.), *China Briefing*, 1980 (Boulder, 1980), 9-24.

Chang, Parris H., "The Rise of Wang Tung-hsing: Head of China's Security Apparatus," *The China Quarterly*, 73:122-137 (Mar. 1978).

Chi, Hsin, *Teng Hsiao-ping, A Political Biography* (Hong Kong, 1978).

Ching, Frank, "The Current Political Scene in China," *The China Quarterly*, 80: 691-715 (Dec. 1979).

Ch'iu, Hungdah, "China's New Legal System," *Current History*, 79:458:29-32, 44-45 (Sept. 1980).

Cohen, Jerome Alan, "China's Changing Constitution," *The China Quarterly*, 76: 794-841 (Dec. 1978).

Dittmer, Lowell, "Death and Transfiguration: Liu Shaoqi's Rehabilitation and Contemporary Chinese Politics," *The Journal of Asian Studies*, XI:3:455-79 (May 1981).

Fang, Hsüeh-ch'un, "Teng Hsiao-p'ing: Supporters and Possible Successors," *Issues & Studies*, XV:4:47-60 (Apr. 1979).

Hua, Guofeng, "Report on the Work of the Government," *Beijing Review*, 27:5-31 (July 6, 1979).

Jain, Pagdish Prasad, *After Mao What? Army Party Group Rivalries in China* (Boulder, 1976).

Joffe, Ellis, *The Chinese Army After Mao* (Cambridge, Mass., 1987).

Lampton, David M., "China's Succession in Comparative Perspective," *Contemporary China*, III:1:72-79 (Spring, 1979).

Lee, Leo Ou-fan, "Recent Chinese Literature: A Second Hundred Flowers" in Robert B. Oxnam and Richard C. Bush (eds.), *China Briefing, 1980* (Boulder, 1980), 65-73.

Lieberthal, Kenneth, "Modernization and Succession in China," *Contemporary China*, III, 1:53-71 (Spring 1979).

"Man of the Year: Visionary of a New China, Teng Hsiao-p'ing Opens the Middle Kingdom to the World," *Time* Magazine, Jan. 1, 1979, 13-29.

McDougall, Bonnie S., "Dissent Literature: Official and Nonofficial Literature in and about China in the Seventies," *Contemporary China*, III:4:49-79 (Winter 1979).

McGough, James P. (tr. and ed.), *Fei Hsiao-t'ung: The Dilemma of a Chinese Intellectual* (White Plains, N.Y., 1980).

Montaperto, Ronald N., and Henderson, Jay (eds.), *China's Schools in Flux: Report by the State Education Leaders Delegation, National Committee on United States-China Relations* (White Plains, N.Y., 1980).

Munro, Robin, "Settling Accounts with the Cultural Revolution at Beijing University, 1977-78," *The China Quarterly*, 82:304-333 (June 1980).

"On Policy towards Intellectuals," *Beijing Review*, 5:10-15 (Feb. 2, 1979).

Pepper, Suzanne, "An Interview on Changes in Chinese Education after the Gang of Four," *The China Quarterly*, 72:815-824 (Dec. 1977).

————, "Chinese Education After Mao: Two Steps Forward, Two Steps Back and Begin Again," *The China Quarterly*, 81:1-65 (Mar. 1980).

Shambaugh, David L., *The Making of a Premier: Zhao Ziyang's Provincial Career* (Boulder, 1984).

Teng Hsiao-ping and the "General Program" (San Francisco, 1977).

"The Communiqué of the Third Plenum of the Tenth Central Committee of the

Chinese Communist Party," full text in Chinese in *Hongqi* (Red Flag), No. 8:5-9 (1977).

Thornton, Richard C., "The Political Succession to Mao Tse-tung," *Issues & Studies*, XIV:6:32-52 (June 1978).

Wakeman, Frederick, Jr., "Historiography in China after Smashing the Gang of Four," *The China Quarterly*, 76:891-911 (Dec. 1978).

Official portrait of Mao Tse-tung.

Chou En-lai, one of his last
official portraits.

The Democracy Wall in Peking, 1979.

Tomb of the Unknown Soldier in Tien-an-men Square, Peking.

Teng Hsiao-p'ing, vice-chairman of the party and main
architect of China's new order.

Military statesman Marshal Yeh Chien-ying.

Hua Kuo-feng, immediate successor to Mao as Party chairman
(and premier).

Hu Yao-pang, new party chairman.

Premier Chao Tzu-yang.

Chiang Ch'ing at the Trial of the Gang of Four.

Wang Hung-wen in the dock during the Trial.

President Chiang Ching-kuo visits with the people.

34

The Normalization of Relations
Between China
and the United States

Following the Nixon visit to Peking in 1972, there was little progress in Sino-American relations due to unfavorable conditions in both countries. In China the radical Gang of Four, experiencing the heights of their influence, were scheming to seize power in hopes of succeeding Mao; their line was firmly antiforeign and suspicious of any *rapprochement* with the capitalist Americans. In the United States, recognition of China faltered on the Taiwan issue as Peking insisted on the fulfillment of three conditions: (1) terminate diplomatic relations with the Republic of China on Taiwan; (2) abrogate the United States–Taiwan defense treaty of 1954; and (3) withdraw all American forces from Taiwan. In a global perspective, to accede to these conditions might be perceived as abandoning Taiwan and cast doubt on the credibility of American commitments to other allies. The Taiwan issue had become a mirror of America's international self-image, and thus its resolution took on added significance beyond the problems of normalization.

President Nixon was reportedly prepared to recognize Peking but was kept from doing so by the Watergate scandal. His political survival came to depend increasingly on the support of conservatives in Congress who opposed normalization, and the preservation of his presidency seemed far more urgent than the diplomatic recognition of China. He was too deeply mired in the fight for his political life to take action on China.[1]

1. Warren I. Cohen, *America's Response to China: An Interpretative History of Sino-American Relations,* 2nd ed. (New York, 1980), 244.

After Nixon's resignation, interim President Gerald Ford was first immobilized by the debacle of Vietnam's collapse and then by his growing aspirations to seek election in 1976. Though in favor of normalization in principle, Ford, too, realized his need for conservative support and made no moves toward recognizing China. Jimmy Carter, Ford's successor, also favored normalization in principle, but his first year in office was occupied with the Panama Canal Treaties, Strategic Arms Limitation Talks with the Soviets (SALT II), Russian-Cuban activities in Africa, and the Middle East problems. These pressing issues required the support of the conservatives in Congress who often considered themselves "friends of Taiwan."

Indeed, Taiwan became a sensitive issue in American domestic politics standing tenaciously in the way of normalization. American public opinion opposed breaking relations with Taiwan, although it favored the recognition of China.[2] The problem before the United States became one of safeguarding Taiwan if normalization were to occur. Politicians agreed that if arms sales to Taiwan could continue, the United States would not appear to be abandoning a faithful ally, and this would minimize the questions of America's credibility and its commitments to other countries.

An easing of the domestic conditions in both countries was necessary before either would feel ready to move toward normalization. Ultimately, the breakthrough came largely as a result of changes in Chinese policy. These changes came in the form of subtle concessions for which three American presidents had waited nearly seven years.

The Normalization of Diplomatic Relations

During the three years following Watergate and the resignation of President Nixon, the Chinese had frequently expressed impatience with the lack of progress toward normalization. President Carter saw no urgent reason to accommodate Peking, especially when he could not seem to find an expedient solution to the Taiwan issue. However, he experienced increasing pressure from his national security advisor, foreign policy staff, and liberal Democrats to jettison formal ties with Taiwan in favor of recognizing China. Biding his time, Carter dispatched Secretary of State Cyrus Vance on an "exploratory mission" to Peking, in reality a mission of "contact" without substance.

VANCE'S VISIT. Vance was in China from August 21 to 25, 1977. Although he was the first high official of the Carter administration to visit China, the

2. *The New York Times*, Dec. 15, 1978, 8.

Chinese gave him a lukewarm reception.[3] Vance suggested the establishment of an American embassy in Peking and a liaison office in Taipei, but the Chinese rejected the idea, insisting on the fulfillment of the three conditions previously mentioned. The issue of continued American arms sales to Taiwan after normalization was not even discussed.[4] The Chinese would not commit themselves to a nonviolent method of liberating Taiwan.

From the American perspective, the mission ventured little and gained little. With the Panama Canal Treaties before Congress for ratification, Washington saw no need for "hasty" action on China. The Chinese, however, considered Vance's visit a step backward in Chinese-American relations.

BRZEZINSKI'S VISIT. If Vance's reception in China was less than effusive, the visit made by National Security Advisor Zbigniew Brzezinski (May 20-22, 1978) was a study in contrast. As Washington perceived in the Soviet paranoia of a Sino-American axis a powerful weapon for SALT negotiations, concessions to China were suddenly rendered "practical." Washington let it be known that it was ready to accede to the three Chinese demands, while expecting China not to take Taiwan by force and not to object to continued American arms sales to Taiwan after normalization.

In Peking, Brzezinski announced: "The president of the United States desires friendly relations with a strong China. He is determined to join you in overcoming the remaining obstacles in the way of full normalization of our relations." He remarked that the United States shared China's resolve to "resist the efforts of any nation which seeks to establish global or regional hegemony," adding that "neither of us dispatches international marauders who masquerade as non-aligned to advance big-power ambitions in Africa. Neither of us seeks to enforce the political obedience of our neighbors through military force."[5]

To demonstrate American sincerity, Brzezinski divulged to the Chinese the contents of two secret documents: Presidential Review Memorandum 10 (the U.S. assessment of the world situation), and Presidential Directive 18 (the president's security policy implementation plan). Other American experts consulted with their Chinese counterparts on defense, technology, and bilateral relations.

Although no public announcement was made on normalization, Brzezinski privately informed Hua and Teng that Ambassador Leonard Wood-

3. *The Wall Street Journal,* Aug. 23, 1977.
4. According to Deputy Premier Teng Hsiao-p'ing, *The Christian Science Monitor,* Sept. 8, 1977, 3.
5. *The New York Times,* May 21, 24, 1978.

cock would be ready to begin serious negotiations to that end.[6] Satisfied with Brzezinski's visit, the Chinese called it "two steps forward."

TOWARD NORMALIZATION. In October Carter made his most important decision concerning normalization. Feeling politically secure after his success as a mediator in the Camp David peace talks between Egyptian President Anwar el-Sadat and Israeli Prime Minister Menachem Begin, the president decided he could finally afford to break America's commitments to Taiwan and set January 1, 1979, as the deadline for diplomatic recognition of China. It was calculated that by that time the Egyptian-Israeli treaty would have been signed, and in its euphoric wake any criticism of the handling of Taiwan would be defused.[7] On the other hand, if the Middle East agreement failed, a successful normalization with China would serve to assure the American electorate of Carter's statesmanship as a world leader. The president wanted to appear decisive, to use China to speed up the SALT negotiations with the Russians, and to outplay liberal advocates of China's recognition such as Senator Edward Kennedy.[8]

Woodcock, the former president of the United Automobile Workers, was an experienced and skillful negotiator. In November he presented to the Peking government the draft of a joint communiqué to which the Chinese responded by asking for certain clarifications. Then, unexpectedly, Deputy Premier Teng announced that he would like to visit the United States—a signal of his willingness to deal. On December 4 the Chinese presented their version of a joint communiqué, and on December 11 Teng was officially invited to visit the United States.

On December 15, 1978, a somber President Carter made a hastily arranged television appearance to announce that the United States and the People's Republic of China had agreed to establish full diplomatic relations on January 1, 1979, including the exchange of ambassadors and the establishment of embassies on the following March 1. The United States would break official relations with Taiwan and abrogate the 1954 Mutual Defense Treaty on January 1, 1980, in accordance with the treaty's termination provision that one year's advance notice was required. The president pledged that Taiwan "won't be sacrificed": the United States would continue to maintain commercial, cultural, and other relations with Taiwan through informal representatives, and the relationship would include

6. Martin Tolchin, "How China and the U.S. Toppled Barriers to Normalization," *The New York Times*, Dec. 18, 1978, A12.
7. Fox Butterfield, "After Camp David, Carter Set a Date for China Ties," *The New York Times*, Dec. 18, 1978, A12.
8. *Ibid.*

arms sales. Then, with obvious exhilaration, he announced that Deputy Premier Teng would visit the United States in January 1979.

The majority of Americans, while regretting "dumping" the Nationalist government on Taiwan, found it hard to oppose the simple mathematics of the possibility of relations with 900 million people on mainland China compared with the 17 million on Taiwan.[9] In bringing normalization to fruition, Carter projected the image of a determined president and politically he gained more than he lost in popular support.[10]

Simultaneously in Peking Chairman Hua called an unprecedented news conference for foreign and Chinese journalists to announce the normalization. He specifically pointed out that China did not like the continued American sales of arms to, and maintenance of cultural and commercial links with, Taiwan, but it would not let these issues stand in the way of normalization.

The salient features of the joint communiqué are as follows:[11]

1. The United States of America and the People's Republic of China have agreed to recognize each other and to establish diplomatic relations as of January 1, 1979.

2. The United States recognizes the government of the People's Republic of China as the sole legal government of China. Within this context, the people of the United States will maintain cultural, commercial, and other unofficial relations with the people of Taiwan.

3. The United States and China reaffirm the principles agreed to by the two sides in the Shanghai Communiqué and emphasize again that:

   a. Both wish to reduce the danger of international military conflict.

   b. Neither should seek hegemony in the Asia-Pacific region or in any other region of the world and each is opposed to efforts by any other country or group of countries to establish such hegemony.

   c. Neither is prepared to negotiate on behalf of any third party or to enter into agreements or understandings with the other directed at other states.

   d. The United States acknowledges the Chinese position that there is but one China and Taiwan is part of China.

   e. Both believe that normalization of Sino-American relations is

9. The Gallop poll showed 58 percent approved the President's action and 24 percent disapproved, with 18 percent no opinion. *The New York Times,* Jan. 14, 1979.

10. Stanley D. Bachrack, "The Death Rattle of the China Lobby," *Los Angeles Times,* Dec. 20, 1978.

11. The Department of State, Selected Documents No. 9: *U.S. Policy Toward China, July 15, 1971-January 15, 1979,* Office of Public Communication, Jan. 1979, 45-46.

not only in the interest of the Chinese and American peoples but also contributes to the cause of peace in Asia and the world.
4. The United States and China will exchange Ambassadors and establish Embassies on March 1, 1979.

Separately, the United States issued a statement on Taiwan:[12]

1. On that same date, January 1, 1979, the United States will notify Taiwan that it is terminating diplomatic relations and that the Mutual Defense Treaty between the United States and the Republic of China is being terminated in accordance with the provisions of the Treaty. The United States also states that it will be withdrawing its remaining military personnel from Taiwan within four months.
2. In the future, the American people and the people of Taiwan will maintain commercial, cultural, and other relations without official government representation and without diplomatic relations.
3. The United States is confident that the people of Taiwan face a peaceful and prosperous future. The United States continues to have an interest in the peaceful resolution of the Taiwan issue and expects that the Taiwan issue will be settled peacefully by the Chinese themselves.

Obviously, the initiative for breaking the Taiwan impasse came from China with Teng as the chief mover. Normalization would give him the success that had eluded Mao and Chou, facilitate his visit to the United States, increase trade, and make available to the Chinese American science, technology, capital, and credit. In this light, Taiwan paled into relative insignificance. In any case, China was well aware it lacked the naval capacity to launch an attack on the island and was clearly too absorbed in the Four Modernizations to want a costly, nasty, and prolonged war over Taiwan. Accepting the status quo was expedient because it gave China an American recognition of its title to Taiwan, though not immediate possession of it.[13]

Peking accepted the new view that China's relations with the United States were more important than Taiwan in the present world setting. The Soviet-Vietnamese treaty of November 1978 with its overtones of military alliance might have prodded the Chinese to seek a closer tie with the United States. Ironically, China's growing preoccupation with its two erstwhile allies may have prompted its *rapprochement* with its former enemies in the West. It is even possible that China was already contemplating

12. *Ibid.*, 48.
13. Linda Mathews, "Is the U.S. About to Take a Dragon by the Tail?" *Los Angeles Times*, Feb. 11, 1979.

a military confrontation with Vietnam over the worsening situation in Cambodia and that it was counting on a friendly United States to deter Soviet involvement. At any rate, the Soviet press blasted away at China's motives in seeking American and Western connections.

RELATIONS WITH TAIWAN. To soften the blow to Taiwan, Carter sent a high-level delegation to Taipei on December 27, 1978, led by Deputy Secretary of State Warren Christopher.[14] He carried the message that despite the termination of formal relations, the United States hoped that trade and cultural ties would continue to expand. The Americans were greeted by 10,000 angry demonstrators.

President Chiang told the delegation that his government and people were enraged by Washington's failure to consult Taipei in advance of the agreement with Peking. He insisted that future relations between Taiwan and the United States be conducted on a government-to-government basis, that his government be recognized as the one in actual control of Taiwan, and that it would continue to present itself as the legal government of China. The Americans, however, were only prepared to negotiate a framework for unofficial, nongovernmental contacts. Two days of attempts at talks ended inconclusively, and the delegation returned home with nothing resolved.

American ties with Taiwan were extremely complex. Apart from the defense treaty, the United States maintained 59 lesser treaties and agreements with the government of the Republic of China. These protected the special relationship of the two countries in agricultural commodities, atomic energy, aviation, claims, controlled drugs, economic and technical cooperation, education, investment guarantees, maritime matters, postal matters, taxation, and trade and commerce.[15]

American investment in Taiwan was considerable. Leading American corporations doing business in Taiwan included such giants as Bank of America, Chase Manhattan Bank, Citicorp, American Express, Ford, RCA, Union Carbide, Zenith, and Corning Glass. In 1978, 220 American corporations had over 500 million dollars invested in Taiwan.[16] Taiwan enjoyed a brisk foreign trade of $23.7 billion in 1978 and a third of it ($7.3 billion) was American. Obviously, American economic ties with Taiwan could not be easily reduced; if anything, they were expected to continue to expand, regardless of withdrawal of diplomatic recognition.

---

14. Accompanied by legal advisor Herbert Hansell and Commander-in-Chief of U.S. forces in the Pacific, Ad. Maurice Weisner.
15. *The New York Times,* Dec. 18, 1978, A10.
16. Ross Terrill, *The Future of China* (New York, 1978), 201.

After an initial spasm of angry outrage, Taiwan's leaders calmed down and weathered the political storm with dignity, dedication, and self-reliance. They realized they could not afford to irritate the United States too much, for the American tie, albeit unofficial, was a vital one.[17]

Since the American embassy in Taipei, and the Republic of China embassy in Washington were scheduled to close on March 1, 1979, it was imperative that substitute offices be designated to handle continuing relations. The Nationalist government struggled for some sort of official status, while the American negotiators insisted on unofficial relations. On February 15, 1979, it was finally agreed that there should be an American Institute in Taipei to replace the embassy and a Coordinating Council for North American Affairs in Washington, D.C., to take care of Taiwan's interests, with consulate-like branches in nine major cities. The American Institute would be staffed by "retired" State Department and other government personnel who would work without official titles.

Meanwhile, Congress produced a number of resolutions expressing concerns for the future of Taiwan. On March 10, 1979, the Senate and the House overwhelmingly supported two slightly different versions of the legislation (the American-Chinese Relations Act, or the Taiwan Relations Act) ratifying the normalization of relations with China and approving the machinery for unofficial relations with Taiwan. The legislation spelled out American determination to maintain extensive relations with the people of Taiwan and "to consider any effort to resolve the Taiwan issue by other than peaceful means a threat to the peace and security of the Western Pacific area and of grave concern to the United States." Passing two versions of the bill necessitated a compromise by a joint conference committee, and the final bill was passed in the Senate (85-4) and the House (339-50) on March 28.

THE VIETNAMESE INVASION OF CAMBODIA. An unacknowledged but possible repercussion of normalization was the Vietnamese invasion of Cambodia under Soviet patronage. Relations between Vietnam and Cambodia had been deteriorating, and in November 1978 Vietnam and Russia had signed what was, in effect, a military alliance. The American announcement of recognition of China was followed ten days later by the Vietnamese invasion of Cambodia. On January 7, 1979, after a chillingly effective blitzkrieg of fifteen days, the Vietnamese forces took the Cambodian capital of Phnom Penh, destroying the Chinese-supported Pol Pot regime. Cambodia appealed to the United Nations Security Council for

17. *Los Angeles Times*, Dec. 25, 1978.

intervention while Peking took the invasion as proof of Soviet hegemony in Asia and moved troops toward its border with Vietnam.

While a cause-and-effect relationship between American recognition of China and the Vietnamese invasion of Cambodia was impossible to prove, many secretly opined that normalization had goaded the Soviets and the Vietnamese into action against Cambodia.[18] China faced the difficult decision of how to deal with the aggressors. Teng's visit to the United States would surely help the Chinese assess the American position.

TENG'S VISIT. China's dynamic, diminutive Deputy Premier Teng Hsiao-p'ing flew into Washington, D.C., on January 28, 1979, for a nine-day visit. This being the first visit by a senior official from the People's Republic of China in thirty years, it warranted a more lavish and regal reception than Washington usually provided. Though ranked third on China's official protocol list, Teng was beyond doubt China's most powerful leader. Washington was anxious for him to see the country, to get a sense of its creativity and diversity, and to understand the important role Congress plays in the formulation of national policy. The administration secretly hoped that Teng would speak softly on Taiwan and not make statements irritating to the Soviets.

The first day after his arrival, Teng was officially welcomed on the White House lawn with a 19-gun salute and review of the honor guard. Hailing the visit as a "time of reunion and new beginnings" for the two countries, President Carter said: "It is a day of reconciliation when windows too long closed have been reopened." Teng was gracious in his response but would not let the occasion pass without a veiled attack on the Soviet Union: "The world today is far from tranquil. There are not only threats to peace, but the factors causing war are visibly growing." Following the formal welcome, the two leaders and their aides conferred privately for four hours.

At the White House dinner reception attended by hundreds of corporate executives, members of Congress, and other prominent Americans, Teng delivered another oblique but unmistakable swipe at the Soviet threat to world peace: "In the joint communiqué on the establishment of diplomatic relations, our two sides solemnly committed ourselves that neither should seek hegemony and each was opposed to efforts by any other country or group of countries to establish such hegemony."

After the glittering dinner party, the group moved to the Kennedy Center Opera House for an evening of American music and dance and the

18. CBS radio broadcast by Marvin Kalb, Jan. 9, 1979.

basketball wizardry of the Harlem Globetrotters. It was Teng, however, who was the evening's biggest hit. Charming performers and audience alike, he went on stage to shake hands and kiss the foreheads of the children in a choir very much in the style of an American politician running for office. Vice-President Mondale quipped, "It's a good thing you're not an American citizen, because you'd be elected to any office you sought."[19]

In meetings with senators and representatives, Teng took Capitol Hill by storm. On Taiwan he indicated that China no longer used the expression "liberation" but only "unification" with the motherland:

> Until Taiwan is returned and there is only one China, we will fully respect the realities on Taiwan. We will permit the present system on Taiwan and its way of life to remain unchanged. We will allow the local government of Taiwan to maintain people-to-people relations with other people like Japan and the United States. With this policy we believe we can achieve peaceful means of unification. We Chinese have patience. However, China cannot commit herself not to resort to other means.

This was reassuring to legislators like Senator Henry Jackson who would have preferred stronger assurances of nonviolence, but respected Teng's caution in exercising China's options.

As for the Soviet Union, Teng's criticism was stinging. Though not opposed to any strategic arms agreement the United States might reach with the Soviet Union, he stressed, "You can't trust the Russians," bringing nods of agreement from the lawmakers. When alone with newsmen Teng was more forceful in denouncing the Soviets, urging the formation of a common front between the United States, Japan, Western Europe, and China to block Russian expansion the world over. He condemned Soviet support of Vietnam's invasion of Cambodia and suggested that the United States denounce both of them, or at least the Vietnamese, those "Cubans of the Orient," who, Teng insisted, "must be taught some necessary lessons."

Teng played the role of a goodwill ambassador superbly. He struck up a warm friendship with Carter and managed to charm Congress with his quick wit, humor, and controlled self-confidence. His adroit showmanship—shaking hands, hugging, kissing, beaming, laughing, and teasing—endeared him to the American public, persuading them that in a cowboy hat even a Communist was hard to hate. The Chinese people, via television satellite, followed Teng's every movement with pride and delight. The blunt, feisty, irascible man many reporters had portrayed was no-

19. *U.S. News & World Report*, Feb. 12, 1979, 26.

where to be seen. Teng projected himself as a warm human being rather than as a fiery revolutionary. He made it clear that while China might be poor and backward, it was no international beggar. It needed foreign technology and capital but could also offer a rewarding market for American products. Partly because of Teng's captivating personality and mastery of mass psychology, and partly because of America's taste for the novel and tendency to glamorize new celebrities, Teng's striking success opened the mind and heart of America to the People's Republic of China.

Perhaps an equally constructive, if less tangible, part of the visit was Teng's personal observation of the workings of American democracy and of the operations of a modern economy. The executive branch, though powerful, had its limitations, and Teng witnessed Congress's distinct role in forming national policy. In visits to a Ford assembly plant, the Hughes Tool Company, and the Johnson Space Center, Teng saw the efficiency of American business operations which with space-age technology and hard-working employees could provide the clearly comfortable American standard of living. It is possible that much of what Teng learned could prove useful in shaping China's future.

Teng himself was satisfied with the results of his trip. He and Foreign Minister Huang Hua signed three agreements with Carter and Vance on science and technology, cultural exchanges, and consular relations. The last permitted China to establish consulates in San Francisco and Houston, and the United States to do the same in Canton and Shanghai. In a farewell message to Carter, Teng said that the visit was a "complete success" and expressed the belief that Chinese-American relations "will witness major progress under the new historical conditions." Teng's optimism was surpassed only by the excitement and hopes of the American and Chinese peoples, who could now view one another more freely and with open interest for the first time in thirty years.

THE CHINESE INVASION OF VIETNAM. On February 17, 1979, barely a week after Teng's return, a large Chinese invasion force struck into Vietnam. In name the invasion was in retaliation for numerous Vietnamese incursions into China, but in fact it was China's punishment for Vietnam's invasion of Cambodia and for its blatant ingratitude after it had accepted more than 25 years of Chinese assistance.

As early as 1950 Mao had offered Ho Chi-minh military, political, and economic aid—the famous battle of Dienbienfu (1954) was fought largely with Chinese weapons and under Chinese direction. During the height of American involvement in Vietnam (1964-71), China dispatched 300,000 technical personnel and troops to Vietnam to help in air defense, engi-

neering work, railway construction, road repairs, and logistics supplies;
some 10,000 of them lost their lives. Chinese economic aid to Vietnam
between 1950 and 1978 totaled somewhere between $15-20 billion, which
represented considerable Chinese sacrifice.[20]

For all these acts of friendship, China had expected Vietnam's gratitude
and goodwill but received precious little once Hanoi gained control over
all Vietnam. Perhaps fearful of China's vast influence, Vietnam rejected
its dependence on China and turned instead to the Russians for help.
Gradually, increasing mistreatment of Chinese residents in Vietnam and
of Vietnamese of Chinese descent was followed by a wave of persecution,
and 160,000 of them were forced to flee. The crowning insult was Vietnam's conclusion of a 25-year Friendship and Mutual Defense Treaty
with Russia, which served the Soviet purpose of encircling China and
represented a stinging Vietnamese rejection of Peking. China's patience
was strained beyond tolerance by Vietnam's invasion of Laos and Cambodia, and the consequent collapse of the Chinese-supported Pol Pot
regime.

During his stay in Washington, Teng openly spoke of "teaching the
Vietnamese some necessary lessons," but he never specified the type of action China might take. The Chinese wanted a quick war—a repeat of their
invasion into India in 1962—lightning success and rapid retreat before the
Russians could decide on a proper response. It was a calculated risk which
Teng thought worth taking. With China's new international connections
he anticipated no Soviet military intervention. To calm world public opinion, China declared at the outset of the invasion that it would be a limited
operation of short duration, with no design on Vietnamese territory.

The magnitude of the Chinese invasion, involving 250,000 troops and
hundreds of tanks, fighter planes, and artillery striking in ten directions
along a 450-mile front, suggested a well-prepared military operation. The
Chinese forces advanced swiftly and successfully at first, taking four Vietnamese provincial capitals near the border by the end of a week.[21] However, their movement was soon slowed considerably, due largely to the lack
of modern weapons. The Chinese had hoped to draw the enemy forces
into a major battle and destroy them decisively, but the Vietnamese deliberately avoided a direct confrontation. Of Vietnam's 600,000 troops,
about two-thirds were stationed in Cambodia and South Vietnam performing "occupation duties." Rather than risk the security of these areas
by removing troops, Vietnam's plan was to employ only regional forces

20. *Central Daily News*, Taipei, July 31; Dec. 7, 1979; Oct. 15, 1981.
21. Lao Cai, Lai Chau, Cao Bang, and Mong Cai.

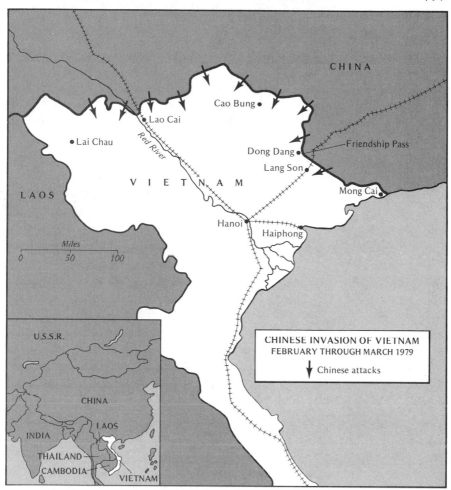

CHINESE INVASION OF VIETNAM
FEBRUARY THROUGH MARCH 1979

↓ Chinese attacks

and militia to fight the Chinese, thereby preserving its best troops from annihilation.

As the war dragged on, the prospect for a quick Chinese success faded and the danger of Soviet retaliation increased correspondingly. On March 1 the Chinese proposed peace talks while stepping up their costly assault on Lang Son. There was a general belief, or even hope, that once the Chinese overran Lang Son, they could claim victory and return home. By March 2 the city was in utter ruins, and the Vietnamese abandoned it to take positions in the surrounding hills. The Chinese finally took Lang Son, but the victory was labored and unspectacular—a far cry from the lightning blitz that the Chinese had dreamed of. Having taken nearly all the important towns and provincial capitals in northern Vietnam, China declared its objectives fulfilled and called for a ceasefire. On March 5 Peking announced the withdrawal of its troops after 17 days of fighting, and on the following day Hanoi agreed to hold peace talks. By March 16 the Chinese had completed their withdrawal.

Western military experts wondered why China risked so much, including military confrontation with the Soviets, for so little. They failed to see that China felt both humiliated and betrayed in its rejection by a former supplicant of favors, and that the feeling of betrayal was exacerbated by Vietnam's growing arrogance toward China and other smaller neighbors, as well as by its alignment with Russia. Such open hostility, the Chinese felt, had to be dealt with, or China's credibility would be at stake. Teng also wanted to show the world that China did not fear war or the threat of Soviet intervention.

If China taught Vietnam a lesson, it also learned one: that China could not fight a modern war effectively without streamlining the military and that the economic consequences of war could be disastrous. In the 17-day operation, China sustained 46,000 casualties, lost 400 tanks and armored vehicles, and spent $1.36 billion. Draining the country's scarce resources had an immediate and adverse effect on the Four Modernizations, necessitating a cutback in the 1980 military budget by $1.9 billion. At the same time, the goal of scientific and technological modernization of the military became increasingly imperative.

The Normalization of Trade and Other Relations

The diplomatic recognition of China was followed by a series of negotiations for the normalization of commercial, cultural, scientific, and to some extent even military relations. China lacked the most-favored-nation status, making it extremely difficult for Chinese exports to the United States to be competitive. For Washington to grant China this status, it was necessary first to settle the question of blocked American assets in China and of frozen Chinese assets in the United States.

CLAIMS SETTLEMENT AND THE TRADE PACT. The day after the American liaison office in Peking became an official embassy on March 1, 1979, Treasury Secretary Michael Blumenthal, who had lived in Shanghai as a youth, initialed an agreement with Chinese Minister of Finance Chang Ching-fu settling the disputes of "frozen assets and blocked claims." The disputes dated back to the early years of the Korean War when in 1950, responding to China's entry into the war, President Truman froze $80.5 million in Chinese assets in the United States. China retaliated by seizing property in China owned by American churches, corporations, schools, and individuals valued at $196.9 million. In reality, the American property in question had been in Chinese hands since the establishment of the People's Republic in October 1949. There were 384 American claims, the largest coming from the Boise Cascade Corporation.

The agreement initialed in Peking allowed China to retain the American property but pay $80.5 million to settle the total American claims—roughly 41 cents on the dollar. China was to pay $30 million on October 1, 1979, and the rest in installments of $10.1 million each October until 1984. For its part, the United States would "unfreeze" the $80.5 million in Chinese assets, but it was not known how much of it belonged to the Chinese government and how much to banks, corporations, schools, and individuals, both inside and outside of China. A noteworthy point, rarely seen in international settlements, was that the Chinese payment, though equal to the amount of the frozen assets, was not tied to them—American claimants would receive their reimbursements quickly. It appeared to be a favorable agreement when compared with other such international settlements, and its compensations were more favorable than the Americans might have hoped for.[22]

The initialed accord had only to be officially signed by representatives of the two governments and was not subject to Congressional ratification. To sign the accord and to negotiate a new trade agreement, Washington sent Secretary of Commerce Juanita Kreps to China. On May 11, 1979, she signed the agreement on "frozen assets-blocked claims" with Chinese Finance Minister Chang, finally settling the long-standing dispute.

The way was then clear for negotiating a trade pact that would grant each country most-favored-nation status, permitting businessmen to establish offices in each country, providing reciprocal protection for patents, trade marks, and copyrights, as well as enabling regulated banking transactions. Each of these issues required detailed discussion, and it was only on the last day of Kreps' visit that an agreement was reached and initialed.

Two months later the trade agreement was signed, but Congressional approval was delayed due partly to a fear of an influx of Chinese textiles.

It was impossible to sum up the benefits of normalization so soon—they continued ad infinitum, depending on one's interests and political persua-

22. Anthony M. Soloman, "When 41¢ on the dollar is a good deal," *The Christian Science Monitor,* Mar. 28, 1979. Mr. Soloman was Undersecretary for Monetary Affairs, U.S. Treasury. A similar settlement with the USSR was 12 cents on the dollar; with Hungary, 30 cents; with Poland and Rumania, 40 cents.

sion. There were, however, benefits so basic that they were equally appreciated by all. First and foremost, a stable China was in the best interests of the United States and of world peace. In spite of great concern for Taiwan's security, the island after normalization appeared to be in no greater danger of a forceful takeover than before the mainland was recognized. China's repeated calls for the unification of Taiwan with the mainland had reduced the frenzied tension many felt. The expansion of cultural ties enabled China to carefully choose the models for and means of modernization while opening wide the vast store of Chinese wisdom, skills, and arts, both traditional and modern, to an interested and appreciative America. Improved accessibility to advancements in Western technology and science would ease China's "New Long March" toward the Four Modernizations, while the expanding, if limited, market for American goods came at a time when the United States needed to increase its exports. And finally, a well-equipped Chinese military might well safeguard what remained of Asian peace, while the Sino-American connection spelled relief for the world from the vacillations of the Soviet-American polarity, with a new, more balanced triangle of power.

Further Reading

Alexiev, Alex, "Prospects for Accommodation," *Contemporary China*. III:2:36-46 (Summer 1979).

"An Interview with Teng Hsiao-p'ing: Calling for Stronger U.S.-China Ties and a United Front against Moscow," *Time* Magazine, Feb. 5, 1979, 32-33.

Barnett, A. Doak, "Military-Security Relations between China and the United States," *Foreign Affairs*, LV:3:584-597 (Apr. 1977).

Bellows, Thomas J., "Normalization: Process and Prognosis," *Sino-American Relations*, V:3:11-21 (Autumn 1979).

Butterfield, Fox, and William Safire, "China: Unraveling the New Mysteries," *The New York Times Magazine*, June 19, 1977, 32-34, 48-59.

———, "After Camp David, Carter Set a Date for China Ties," *The New York Times*, Dec. 18, 1978, A12.

Chang, Jaw-ling Joanne, *United States-China Normalization: An Evaluation of Foreign Policy Decision* (Denver, 1986).

Chay, John (ed.), *The Problems and Prospects of American-East Asian Relations* (Boulder, 1977).

Chiu, Hungdah (ed.), *Normalizing Relations with the People's Republic of China: Problems, Analysis, and Documents* (University of Maryland, School of Law, Occasional Papers, 1978).

Cohen, Jerome Alan, "A China Policy for the Next Administration," *Foreign Affairs*, LV:1:20-37 (Oct. 1976).

Cohen, Warren I., *America's Response to China: An Interpretative History of Sino-American Relations*, 2nd ed. (New York, 1980).

Copper, John Franklin, *China's Global Role* (Stanford, 1981).

Department of State, "Diplomatic Relations with the People's Republic of China and Future Relations with Taiwan" (Washington, D.C., Dec. 1978).

———, *U.S. Policy Toward China, July 15, 1971-January 15, 1979,* Selected Documents No. 9, Office of Public Communication (Washington, D.C., 1979).

Garrett, Banning, "The China Card: To Play or Not to Play," *Contemporary China,* III:1:3-18 (Spring 1979).

———, "Explosion in U.S.-China Trade?" *Contemporary China,* III:1:32-42 (Spring 1979).

Gurtov, Melvin, "China Invades Vietnam: An Assessment of Motives and Objectives," *Contemporary China,* III:4:3-9 (Winter 1979).

Harrison, Selig S., *The Widening Gulf: Asian Nationalism and American Policy* (New York, 1978).

Hsiao, Gene T., and Michael Witunski, *Sino-American Normalization and Its Policy Implications* (New York, 1983).

Johnson, Chalmers, "The New Thrust in China's Foreign Policy," *Foreign Affairs,* LVII:1:125-37 (Fall, 1978).

Kim, Samuel S., *China, the United Nations, and World Order* (Princeton, 1978).

Kim, Se-jin, "American Moral Psyche in Political Perspective," *Sino-American Relations,* VI:1:8-18 (Spring 1980).

Larkin, Bruce D., "China and Asia: The Year of the China-Vietnam War," *Current History,* 77:449:53-56, 83 (Sept. 1979).

Lawson, Eugene K., *Sino-Vietnamese Conflict* (New York, 1984).

Lieberthal, Kenneth, "The Foreign Policy Debate in Peking, as seen through Allegorical Articles, 1973-76," *The China Quarterly,* 71:528-54 (Sept. 1977).

Luttwak, Edward N., "Against the China Card," *Contemporary China,* III:1:19-31 (Spring 1979).

Martin, Edwin W., *Southeast Asia and China: The End of Containment* (Boulder, 1977).

Mendl, Wolf, *Issues in Japan's China Policy* (London, 1978).

Middleton, Drew, *The Duel of the Giants: China and Russia in Asia* (New York, 1978).

Nathan, Andrew J., "Prospects for Sino-American Relations and the Effects on Korea," *Contemporary China,* II:4:14-22 (Winter 1978).

Okita, Saburo, "Japan, China and the United States: Economic Relations & Prospects," *Foreign Affairs,* LVII:5:1090-1110 (Summer 1979).

Oksenberg, Michel, "China Policy for the 1980s," *Foreign Affairs,* LIX:2:304-22 (Winter 1980/81).

———, and Robert B. Oxnam, *Dragon and Eagle: United States-China Relations: Past and Future* (New York, 1978).

Ray, Heman, *China's Vietnam War* (New Delhi, 1983).

Rothenberg, Morris, "The Kremlin Looks at China," *Contemporary China,* III:2:25-35 (Summer 1979).

Segal, Gerald, "China and the Great Power Triangle," *The China Quarterly,* 83:490-509 (Sept. 1980).

Solomon, Richard H., "Thinking Through the China Problem," *Foreign Affairs,* LVI:2:324-56 (Jan. 1978).

Sutter, Robert G., *Chinese Foreign Policy after the Cultural Revolution, 1966-1977* (Boulder, 1978).

————, *China-Watch: Towards Sino-American Reconciliation* (New York, 1978).

————, *Chinese Foreign Relations: Development After Mao* (New York, 1986).

Tretiak, Daniel, "China's Vietnam War and Its Consequences," *The China Quarterly*, 80:740-767 (Dec. 1979).

Yu, George T. (ed.), *Intra-Asian International Relations* (Boulder, 1978).

35

The Four Modernizations

If there was such a thing as a national consensus in China, it focused on the commitment to the Four Modernizations—of agriculture, industry, science and technology, and national defense. The avowed goal was to turn China into a leading modern state by the year 2000. The Four Modernizations had been written into the party constitution (Eleventh Congress, August 18, 1977) and the state constitution (Fifth National People's Congress, March 5, 1978); hence the program should not be affected by changes in leadership.

The Ten-Year Plan

At the first session of the Fifth National People's Congress in February 1978, Chairman Hua unveiled a grandiose ten-year modernization program for 1976-85; as two years had already passed, it was actually an eight-year plan. It detailed the major goals to be achieved in the four sectors.

THE INDUSTRIAL SECTOR. Investment for capital construction in industry was to equal or surpass that of the entire previous 28 years, which was estimated at US$400 billion, and the annual rate of industrial growth was set at 10 percent. Hua called for the completion of 120 major projects, including: 10 iron and steel complexes, 6 oil and gas fields, 30 power stations, 8 coal mines, 9 nonferrous metal complexes, 7 major trunk railways, and 5 key harbors. It was hoped that by the end of the century Chinese industrial output in the major sectors would "approach, equal, or outstrip that of the most developed capitalist countries."[1]

Steel. In 1952 steel production (1.55 million tons) had already surpassed the pre-Liberation peak, and it rose to 18.67 million tons by 1960. The Great Leap cut the output back to 8 million tons in 1961, and the Cultural Revolution provided further inhibition. It was not until 1970 that steel production recovered, reaching 25.5 million tons by 1973. Yet production decreased again under the Gang of Four—in 1976 only 21 million tons were produced. In short, from 1960 to 1976, only small gains were achieved.

The Ten-Year Plan called for increased production to 60 million tons by 1985 and 180 million tons by 1999. To achieve such major increases, a giant steel complex at Chi-tung (eastern Hopei), capable of producing 10 million tons a year, was planned under contract with German firms at a cost of $14 billion; a 6-million-ton complex was to be constructed at Pao-shan (a suburb of Shanghai) under contracts with Japanese firms with an estimated initial cost of $2 billion. A number of other sizable plants were to be built elsewhere, and existing plants were to be renovated.

Oil. Before 1957 China's petroleum production was insignificant (1.46 million tons of crude oil per year). Vast advances were made in the 1960s, with new discoveries and the establishment of the Ta-ch'ing Oil Field in Manchuria, the Sheng-li Oil Field in Shantung province, and the Ta-kang Oil Field in the Tientsin harbor area. Crude output doubled between 1960 and 1965 and again by 1969. By 1978 it had reached 104 million tons. The Ten-Year Plan called for the construction of 10 new oil and gas fields costing $60 billion.[2]

Coal. Coal provided 70 percent of China's primary energy supply, but most of the mines were small and antiquated. The Ten-Year Plan called for 8 new mines along with the renovation of existing ones in hopes of doubling production to 900 million tons a year. This meant an annual growth rate of about 7.2 percent compared with 6.3 percent in 1970-77.

Electric Power. Surprisingly, the production of electricity was the weakest link in the modernization plan. In 1978 production totaled 256.6 billion kilowatt hours, ranking China ninth in the world in electricity production, but per capita consumption remained extremely low below both India and Pakistan. The Ten-Year Plan called for the construction of 30 power stations, 20 of them to be hydropower. The largest projects included the 2.7 mililon kilowatt Ko-hou-pa hydropower station on the Yangtze

1. *People's Daily,* Mar. 9, 1978, 1-5.
2. Chu-yüan Cheng, "The Modernization of Chinese Industry" in Richard Baum (ed.), *China's Four Modernizations: The New Technological Revolution* (Boulder, 1980), 26.

River near I-chang (Hupeh) and the 1.6 million kilowatt Lung-yang Gorge Station on the upper reaches of the Yellow River near Sining (Tsinghai). The 30 new plants would increase production by 6 to 8 million kilowatts per year—far short of the 13-14 percent growth rate needed to sustain the goal of a 10 percent annual increase in industry, and leaving nothing with which to increase personal consumption.

THE AGRICULTURAL SECTOR. Agriculture was the foundation of the Chinese economy. Yet, since 1949, agriculture had consistently received less investment than industry and defense. Collectivization and the commune did not materially raise agricultural production. The 1963 movement "In agriculture, learn from Ta-chai" was nothing more than a propaganda gimmick, and the Cultural Revolution drove agriculture to the brink of bankruptcy. On August 8, 1977, the *People's Daily* frankly stated that "whenever farms are hit by disastrous natural calamities, drastic reduction in output resulted; in the event of smaller disaster, smaller reduction; even with perfect weather conditions there was not much increase."[3] A Chinese leader admitted that "in 1977, the average amount of grain per capita in the nation was the same as the 1955 level; in other words, the growth of grain production was only about equal to the population growth plus the increase in grain requirements for industrial and other uses."[4] Agricultural modernization was vital to the success of the Four Modernizations.

When he announced the Ten-Year Plan, Hua called for maximizing farm production through mechanization, electrification, irrigation, and higher utilization of chemical fertilizers. Specifically, the targets included:

1. Increase of gross agricultural product by 4-5 percent annually.
2. Increase of food output to 400 million tons by 1985 (from 285 million tons in 1977, a 4.4 percent annual growth).
3. Mechanization of 85 percent of major farming tasks.
4. Expansion of water works to assure one good *mou* (⅙ acre) or dependably irrigated land per farming capita totalling 800 million *mou* (121 million acres).
5. Establishment of twelve commodity and food base areas throughout the nation.

To bolster the slow agricultural growth rate of 2 percent annually since 1957, the government laid down several new guidelines. The "production

3. *People's Daily*, Aug. 8, 1977, "Report on National Conference on Fundamental Farm Construction."
4. Hu Ch'iao-mu, "Observe Economic Laws, Speed Up the Four Modernizations," *Peking Review*, 47:8 (Nov. 24, 1978).

team," hitherto the basic accounting unit responsible for any surplus or deficit, was replaced by the larger "production brigade." Next, the principle "to each according to his work" was adopted to stimulate farm initiative and enthusiasm; hence, "more pay for more work and less pay for less work" had become a basic rural economic policy. In addition, encouragement of household "sideline production" should work to supplement the larger economy. Rural families did not own the communally distributed "private" plots but had the right to farm them. They could not rent, sell, or transfer the land, but they owned its products. "Sideline" production made up some 25 percent of total agricultural and subsidiary production. Finally, it was hoped that through intensive development, commune- and brigade-operated enterprises would be able to support large industries and the export trade.

SCIENTIFIC MODERNIZATION. Science and technology were considered basic to successful modernization of the other three sectors. At the National Science Conference in March 1978, a Draft Outline National Plan for the Development of Science and Technology was presented by Vice-Premier Fang Yi, calling for: (1) achieving or approaching the 1970 scientific levels of advanced nations in various scientific and technological fields; (2) increasing professional scientific researchers to 800,000; (3) developing up-to-date centers for scientific experiments; and (4) completing a nationwide system of scientific and technological research. The Outline identified 108 items in 27 fields as key projects for research.[5] It was hoped that by 1985 China would be only ten years behind the most advanced nations, with a solid foundation for catching up to the advanced nations by the end of the century.

MILITARY MODERNIZATION. China had the largest regular armed force in the world, numbering some 4,325,000. The army alone included 3,250,000 troops, and China's naval and air forces ranked third internationally in terms of numbers.[6] But, except for pockets of intensive development in the

5. These 27 fields include natural resources, agriculture, industry, defense, transportation, oceanography, environmental protection, medicine, finance, trade, culture, and education, in addition to a number of basic and technical sciences.
6. According to a study of the Royal Institute of International Affairs, London, by Lawrence Freedman entitled *The West and the Modernization of China* (London, 1979), 5, Chinese armed forces consisted of the following main units:

> *Strategic Forces:* Medium-range ballistic missiles: 30-40 CSS-1, 600-700 miles. Intermediate-range ballistic missiles: 30-40 CSS-2, 1750 miles. [Long-range ballistic missiles: some CSS-3, 3500 miles, first tested in 1976, and a

strategic sector (e.g. nuclear bombs and ballistic missiles), Chinese military technology remained some twenty to thirty years behind the West. Troops were well trained, highly motivated, and politically indoctrinated but equipped with woefully inadequate weapons. The situation, brought about by a lack of funds and by an underdeveloped technology, worsened with Mao's emphasis on spirit over weapons. His idea of "people's war," employing large numbers of politically motivated, well-trained guerrillas to harass and drive out the invader was primarily a defensive notion, lacking offensive punch. The unspectacular Chinese invasion of Vietnam in 1979 clearly illustrated this. Su Yu, a brilliant strategist and former chief-of-staff, stated that Mao's concepts had "seriously shackled the people's minds and obstructed the development of military ideas."[7]

With extensive Soviet aid in the 1950s, the Chinese built up a nearly self-sufficient defense-manufacturing industry, and some of their products (e.g. the AK-47 rifle) rank among the world's best.[8] Yet, by and large, Chinese military technology was two or three decades out of date. Truly swift modernization would require massive purchases of foreign weapons and instruments, but that would be prohibitively expensive and also place China at the mercy of foreign suppliers. As the paramount consideration in China's long-term military planning was still the indigenous control of production capabilities, only selective purchases of high-technology systems and weaponry with special contracts for production in China were planned.

Chinese leaders recognized the urgent need to update obsolete equipment on a massive scale but also saw its astronomical cost. Although China's defense budget was a state secret, Western estimates put it at $32.8 billion in 1976, the third largest globally.[9] A British source put China's 1978 defense spending at 7-10 percent of the GNP, or about $35 billion.[10] The production of new equipment, spare parts, and mainte-

---

small number of CSS-X-4, 6000-7000 miles, first tested in May 1980—author.]

Army: 3,250,000 men, 10 armored divisions, 121 infantry divisions, and 150 independent regiments.

Navy: 30,000 men, 30,000 Naval Air Force with 700 shore-based aircraft, 38,000 Marines, 23 major surface combat ships, and a rather large number of submarines and destroyers with missile-launching capability.

Air Force: 400,000 men, 5,000 combat aircraft including some 4,000 MIG 17/19, and a small number of MIG 21 and F-9 fighters.

7. Cited in Lawrence Freedman, 6.
8. Jonathan Pollack, "The Modernization of National Defense" in Baum (ed.), 247.
9. Jonathan Pollack, 243.
10. Lawrence Freedman, 19-20.

nance accounted for 58 percent of that figure. To modernize fully even a portion of China's military would cost an impossible $300 billion by 1985.[11] Since such an expenditure would require massive infusions of foreign capital and equipment, military modernization occupied a low priority.

The ultimate irony might be that after straining to acquire current state-of-the-art technology and weaponry, it would take the Chinese five to ten years to integrate such modern equipment into existing structures. By that time new strides would have been made in the more advanced countries, and China would yet remain behind by ten to fifteen years. While this would represent an improvement over present capabilities, it would have to be seen as falling short of true modernization goals.

Retrenchment and Revised Priorities

The original TenYear Plan was more of a political wish than an economic blueprint, and it lacked careful study as to its feasibility. During the first year of the program, some 100,000 construction projects were launched by the government costing $40 billion; with military and scientific procurements the total reached 24 percent of the 1978 national income of $198 billion. Large foreign contracts were also negotiated including the Pao-shan steel complex ($2 billion), the Chi-tung steel complex ($14 billion), and a hotel construction project with the U.S. Intercontinental Hotel Corporation ($500 million). In addition, regional organizations contracted a large number of sizable agreements with foreign suppliers, which, together with local construction projects, raised the total investment for 1978 to 36 percent of the national income, quite close to the 40 percent rate of the disastrous Great Leap years. Such zealous overspending was clearly insupportable.[12]

Economic realities soon set in to force a critical reassessment. A debate at the highest level took place regarding the scope and priorities of investment. In July 1978, Hu Ch'iao-mu, president of the Chinese Academy of Social Sciences, called for greater emphasis on agricultural production which reflected the results of the top echelon's reassessments.[13] Similar sentiments were expressed in December 1978 at the meetings of the Eleventh Central Committee (Third Plenum).

China's limited financial and scientific resources forced leaders to reas-

11. *Ibid.*, 19.
12. Chu-yüan Cheng, "Industrial Modernization in China," *Current History*, Sept. 1980, 24.
13. *Peking Review*, No. 47:17-21 (Nov. 24, 1978).

sess the Ten-Year Plan critically. It was decided that the top priority should be agriculture, the foundation of the economy, followed by light industry, which could meet domestic demands and earn foreign exchange, and then heavy industry. Capital investment in agriculture was increased from $26 billion (Ch$40 billion) to $50 billion (Ch$90 billion), and light and export industries also received new allocations. Within heavy industry, steel production targets were slashed from 60 million to 45 million tons; but coal, electric power, petroleum, and building industries retained priorities for investments.[14] Projects that could be completed quickly and earn foreign exchange were encouraged, and bank loans rather than government appropriations were planned for future investment projects. On the other hand, projects requiring huge amounts of capital and facing problems in resources, raw materials, location, transportation, technical capabilities, or energy supply were delayed or suspended.

At the Fifth National People's Congress (second session, June 1979), Hua Kuo-feng announced a three-year period (1979-81) for the "adjustment, reconstruction, consolidation, and improvement" of the national economy. The immediate effect of the retrenchment was the halting of 348 important heavy industrial projects (including 38 steel and metallurgical plants) and 4,500 smaller ones. Capital investment for 1979 was reduced to 34.8 percent of state expenditures. Specifically, investments in the steel, machinery, and chemical industries were most deeply cut, losing from 30-45 prcent of their investment allotment in 1979-80.[15] Construction also suffered, with a 33 percent cut in Shanghai and a 40 percent cut for Inner Mongolia. Simultaneously, investment in agriculture increased from 10.7 percent of the state budget in 1978 to 14 percent in 1979 and 16 percent in 1980, while in textile and light industries investment rose from 5.4 percent in 1978 to 5.8 percent in 1979 and perhaps 8 percent in 1980.

The retrenchment was necessitated not only by China's limited foreign credit, financial resources, and absorptive power but also by the unexpectedly high cost of invading Vietnam in 1979. In addition, original estimates of oil production and its export potential were far too optimistic, and disappointing performance in the energy sector dampened China's hope of using oil exports to finance modernization. The 1978 budgetary deficit was $6.5 billion and climbed to $11.3 billion in 1979.[16] Clearly, more sophisticated and thorough economic planning was required. Saburo Okita, chairman of the Japan Economic Research Center and an architect of the Japanese economic miracle, was invited to China as a consultant.

14. Chu-yüan Cheng, in Baum (ed.), 41; *Los Angeles Times*, May 10, 1979.
15. Chu-yüan Cheng, "Industrial Modernization," *Current History*, 25.
16. John Bryan Starr, "China's Economic Outreach," *Current History*, Sept. 1979, 50-51.

As a result of retrenchment, the new scaled-down targets for 1985 and the actual output of the five major industries in 1985 appeared as follows:[17]

	1978	1979	1985 (Ten-Year Plan)	1985 (revised)	1985 (actual)
Steel (million tons)	31.8	34.5	60	45	46.66
Coal (million tons)	618	635	900	800	850
Crude Oil (million tons)	104	106.2	500	300	125
Electricity (billion kwh)	256.6	282	n.a.	n.a.	407
Cement (million tons)	65.2	73.9	100	100	142.46

Of note was the very small growth in coal and oil production, the two main energy sources. Coal output grew 12.5 percent in 1978 but only 2.75 percent in 1979. Crude oil registered a 1.9 percent increase in 1979 compared to 11 percent in 1978 and an annual 22.5 percent between 1957 and 1977. This vast decrease may suggest that oil output at current producing sites had already peaked, and thus indicate the necessity for new exploration. Electricity output also fell from an annual growth rate of 13 percent between 1957 and 1978 to 9.9 percent in 1979 and 2.9 percent in 1980.[18]

These figures clearly indicate that transportation and energy remained major obstacles in the modernization plan. Oil, coal, and electricity production fell far short of meeting new demands. While freight volume increased 9.7 times from 1950 to 1978, railway mileage increased only 1.4 times—transport lines were strained to the limit. Unless the energy and transportation bottlenecks were eased, China's modernization would be constrained. The small increase in oil output had drastically reduced China's ability to earn foreign currency to finance the purchase of foreign high technology.

It is possible that some additional income might be acquired from the textile and light industries which could more easily meet consumer de-

17. Chu-yüan Cheng, "Industrial Modernization," *Current History*, Sept. 1980, 26; James T. H. Tsao, *China's Development Strategies and Foreign Trade* (Lexington, Mass., 1987), 151.
18. Cheng, 27.

mands and earn foreign exchange. As a result of increased state investment, bank loans, and better material, textile output rose 30 percent in the first quarter of 1980 over the same period in 1979, and light industry rose by 21 percent.[19] But it was questionable whether these sources would generate enough funds to hasten materially the rate of modernization.

The Consequences of Rapid Modernization

Just as there were problems in achieving modernization, there were also problems created by its accelerated achievement. First and foremost was inflation, which was almost nonexistent in earlier periods when the government deliberately adopted a policy of low wages and low commodity prices. When the people had little purchasing power, the demand for goods was kept low and prices were stable. With the increase in wages and government procurement prices for farm products (up 20-50 percent by 1977-79), the state correspondingly raised sale prices on various commodities creating inflation officially computed at 5.8 percent in 1979 but more likely reaching 15 percent. The upward spiral of price increases had continued unabated reaching an annual rate of 15-30 percent in 1980, while the light industry growth was only 9.7 percent. When prices increased faster than productivity, an inflationary psychology set in, resulting in a black market and speculation. Government budgets also revealed growing deficits: $11.3 billion in 1979, $10-12 million in 1980, and perhaps $6 billion in 1981. To offset the deficits, the Ministry of Finance planned in the spring of 1981 to issue $3.3 billion in ten-year maturity bonds at 4 percent interest per annum. Government enterprises, administrative organizations, communes, and the army had been urged to buy according to their ability, but individuals seemed free to purchase the bonds as they chose.[20] The floating of bonds indicated the financial difficulties China was facing.

Another immediate consequence of inflation and overzealous spending was the decision to cut back major construction projects by 40 percent in 1981. Many large projects involving foreign companies had been abruptly terminated. The Japanese, who entered the China market early, had fared the worst: total Japanese losses were estimated at $1.5 billion including the Pao-shan steel complex and three petrochemical projects. The Germans fared less badly while the Americans, who entered the China market late, suffered the least. The Chinese simply explained that they could not afford to go ahead with these costly projects at this time and agreed to

19. *Ibid.*, 27.
20. *Far East Times*, San Francisco, Mar. 10, 1981.

compensate for losses incurred without setting definite figures.[21] Foreigners understood China's financial dilemma but had to question the nation's international credibility when agreements entered into in good faith were unilaterally cancelled. There was no doubt that China's reputation as a reliable trader had been affected.

Cancellations of construction projects had led to the layoff of numerous workers, intensifying the already serious unemployment problem. China once boasted that its socialist system guaranteed employment for every able body, declaring proudly that there was no unemployment in China—there were only those awaiting job assignment!

Another anomalous phenomenon of modernization was the emergence of new classes in a so-called classless society. Modernization had given new prestige to the scientists, engineers, technicians, plant managers, writers, artists, and other intellectuals who would lead China's "Great Leap Outward." There was a new feeling that "among all activities, only science and technology are lofty." Scientists and intellectuals along with high party members now constituted a privileged upper class; urban workers of productive enterprises and lower-echelon cadres formed the second class, with farmers and those living in the countryside at the bottom of the totem pole. The selective sending of scholars and students abroad for advanced education, many of whom were blood relatives of high party members, further strengthened the elitist trend and widened the class cleavages.

Another deepening problem arose from the increasing disparity between the city and the countryside and among various industrial enterprises themselves. Since the government had opted for an "enclave" strategy of locating key industries in selected urban areas, these areas were more likely to enjoy the fruits of modernization—higher wages, greater upward mobility, and a higher standard of living. An average industrial worker in a city earned about $40 a month with an additional bonus, while the average monthly cash income of a peasant was only $5-7. It was not unusual for a city worker to earn six to eight times more than a peasant, and scientific or technical personnel over ten times more. Within the industrial sector, the profit was vastly uneven: in 1978 the oil industry enjoyed a 40 percent profit margin; electricity, 31 percent; metallurgy, 13 percent; and coal mining, only 1 percent. Since profit decided not only levels of investment but also the size of bonuses and fringe benefits, it deeply affected the worker's life-style. Differences in rewards led to different degrees of enthusiasm for work.

Beneath all the adverse consequences of rapid modernization lay what

21. *The Christian Science Monitor*, Feb. 20, 1981. Reportedly, the Japanese firm Mitsubishi Heavy Industries asked for an $84 million reparation.

might be the most serious of China's problems—a crisis of confidence. After thirty years of socialist construction, the country remained poor and backward. Past reports of achievements had often been exposed as pure propaganda, and many, especially the young, had lost faith in the superiority of socialism. There was a conspicuous lack of confidence in achieving a true modernization. Young people were especially critical of the party cadre's privileged status and of bureaucratism, and on the basis of past performance doubted both their capability and sincerity in implementing modernization programs. Indeed, many middle- and lower-echelon party members in responsible positions but lacking scientific expertise were threatened by the new demands of modernization. They secretly resisted, sabotaged, and slowed new undertakings which ran counter to their interests.[22] There was now a popular saying: "The two ends are hot, but the middle is cold"—meaning the leaders and the people wanted modernization but the middle-level bureaucrats resisted change. Chinese newspapers and journals openly discussed China's triple crises: a lack of faith, confidence, and trust in the party and the government.[23]

Foreign Value and Chinese Essence

Modernization had been a goal in China for more than a century. But the radical Communists considered learning from the West and Japan dishonorable. In September 1975 a spokesman for the Gang of Four proclaimed:

> Politically, "wholesale Westernization" meant loss of sovereignty and national humiliation, a total sell-out of China's independence and self-determination. . . . Ideologically, "wholesale Westernization" was meant to praise what was foreign and belittle what was Chinese. . . . Economically, "wholesale Westernization" was aimed at spreading a blind faith in the Western capitalist material civilization so as to turn the Chinese economy into a complete appendage of imperialism.[24]

Thus the Maoists advocated self-reliance. Mao recognized China's backwardness as much as he was aware of its limited resources. He feared the effects of modernization on such sensitive issues as income distribution, worker status, and the revival of elitism and bureaucratism at the expense of egalitarianism. Yet self-reliance and the rejection of foreign technology

22. *Hongqi* (Red Flag), 14:25-27 (1980).
23. *People's Daily*, July 1, 1980; Nov. 11, 1980; Feb. 24, 1981; *Kuang-ming Daily*, Mar. 28, 1981.
24. Liang Hsiao, "The Yang Wu Movement and the Slavish Comprador Philosophy," *Historical Research*, No. 5 (Oct. 20, 1975).

for nearly two decades (1958-76) left China in an undeveloped abyss of poverty, while other countries through technological innovations charged ahead by leaps and bounds.

With Mao's death and the demise of the Gang, the way was cleared for a new start to make up for lost time. A crash program for modernization had been launched, with Teng as the main spirit of the New Leap Outward. The Chinese leaders assumed that science, technology, and the dynamics of technological change were basically politically neutral and classless, and that they could be transplanted without injury to Chinese social and cultural institutions.[25]

Chinese leaders proclaimed that they did not intend to ape the West but would forge a "Chinese-styled modernization." Yet the knowledge and skills associated with foreign technology would inevitably influence the thinking and behavior of those who acquired them. The late Ch'ing debate on the "fundamentals vs. application" (*t'i-yung*) dichotomy would reappear in a different form. Western scientists in China and "returned" students trained in advanced countries would undoubtedly exert new influences on Chinese life and thinking.

The cultural consequences of contact with foreign ideology, institutions, and ways of life cannot be totally contained, despite party admonitions and exhortations to the contrary. It is hoped that the Chinese could achieve a happy medium whereby they would indeed become modern in thought and in their specialties without sacrificing the distinctiveness of their Chinese origin. While the meeting of a particular timetable for China's modernization could not be assured, the leadership's increasingly pragmatic goals seemed to point toward the eventuality of the successful modernization of the Chinese nation—perhaps thirty years into the next century.

25. Genevieve C. Dean, "A Note on Recent Policy Changes" in Baum (ed.), 105.

Further Reading

Andors, Stephen, *China's Industrial Revolution: Politics, Planning, and Management, 1949 to the Present* (New York, 1977).

Baum, Richard (ed.), *China's Four Modernizations* (Boulder, 1980).

Chang, Arnold, *Painting in the People's Republic of China: The Politics of Style* (Boulder, 1980).

Chen, Kuan-I, "Agricultural Modernization in China," *Current History*, 77:449:66-70, 85-86 (Sept. 1979).

Cheng, Chu-yüan, *China's Petroleum Industry: Output Growth and Export Potential* (New York, 1976).

Cheng, Chu-yüan 鄭竹園, "Chung-Kung hsien-tai-hua ti tun-ts'o chi chan-wang," 中共近代化的頓挫及展望 (The frustration of and prospect for Communist China's modernization), *Hai-wai hsüeh-jen* 海外學人 (Overseas Scholars) Sept. 1980, 25-32.

Chou, S. H., "Industrial Modernization in China," *Current History*, 77:449:49-52, 87-88 (Sept. 1979).

Dean, Genevieve C., *Science and Technology in the Development of Modern China: An Annotated Bibliography* (London, 1974).

———, and Fred Chernow, *The Choice of Technology in the Electronics Industry of the People's Republic of China: The Fabrication of Semiconductors* (Palo Alto, 1978).

Field, Robert Michael, "A Slowdown in Chinese Industry," *The China Quarterly*, 80:734-739 (Dec. 1979).

Freedman, Lawrence, *The West and the Modernization of China*, Chatham House Papers (The Royal Institute of International Affairs, 1979).

Fureng, Dong, "Some Problems Concerning the Chinese Economy," *The China Quarterly*, 84:726-736 (Dec. 1980).

Gelber, Harry G., *Technology, Defense, and External Relations in China, 1975-1978* (Boulder, 1979).

Godwin, Paul H. B., "China and the Second World: The Search for Defense Technology," *Contemporary China*, II:3:3-9 (Fall 1978).

———, "China's Defense Dilemma: The Modernization Crisis of 1976 and 1977," *Contemporary China*, II:3:63-85 (Fall 1978).

———, *PLA-Military Forces of the PRC* (Boulder, 1981).

Goldman, Merle, "Teng Hsiao-p'ing and the Debate over Science and Technology," *Contemporary China*, II:4:46-69 (Winter 1978).

Hardy, Randall W., *China's Oil Future: A Case of Modest Expectations* (Boulder, 1978).

Harrison, Selig S., *China, Oil, and Asia: Conflict Ahead?* (New York, 1977).

Hu, Ch'iao-mu, "Observe Economic Laws, Speed Up the Four Modernizations," Pt. I, *Peking Review*, 45:7-12 (Nov. 10, 1978); Pt. II, *Peking Review*, 46:15-23 (Nov. 17, 1978); Pt. III, *Peking Review*, 47:13-21 (Nov. 24, 1978).

Huang, Chih-chien 黃志堅, "Chiu-ching ying-tang ju-ho jen-shih che-i-tai ch'ing-nien" 究竟應當如何認識這一代青年 (How shall we recognize this generation of youths?), *Renmin Ribao* (*People's Daily*), Peking, Feb. 24, 1981.

Huang, Philip C. C. (ed.), *The Development of Underdevelopment in China* (White Plains, N.Y., 1980).

Klatt, W., "China's New Economic Policy: A Statistical Appraisal," *The China Quarterly*, 80:716-733 (Dec. 1979).

———, "China Statistics Up-dated," *The China Quarterly*, 84:737-743 (Dec. 1980).

Kokubun, Ryosei, "The Politics of Foreign Economic Policy-making in China: The Case of Plant Cancellation with Japan," *The China Quarterly*, 105:19-44 (March 1986).

Lardy, Nicholas R., "China's Economic Readjustment: Recovery or Paralysis?" in Robert B. Oxnam and Richard C. Bush (eds.), *China Briefing, 1980* (Boulder, 1980), 39-51.

———, *Economic Growth and Distribution in China* (Cambridge, Eng., 1978).

——— (ed.), *Chinese Economic Planning: Translations from Chi-hua ching-chi* (White Plains, N.Y., 1979).

Li, Hung-lin 李洪林, "Hsin-yang hui-chi shuo-ming liao shih-mo?" 信仰危機説明

了什麼 (What does the crisis of confidence mean?), *Renmin Ribao* (*People's Daily*), Peking, Nov. 11, 1980.

Myers, Ramon H., *The Chinese Economy: Past and Present* (Belmont, Cal., 1980).

National Foreign Assessment Center, *China: Economic Indicators* (Washington, D.C., Dec. 1978).

————, *China's Economy* (Washington, D.C., Nov. 1977).

————, *China: Gross Value of Industrial Output, 1965-77* (Washington, D.C., June 1978.

————, *China: In Pursuit of Economic Modernization* (Washington, D.C., Dec. 1978).

————, *China: Post-Mao Search for Civilian Industrial Technology* (Washington, D.C., 1979).

————, *China: The Continuing Search for a Modernization Strategy* (Washington, D.C., April 1980).

————, *China: A Statistical Compendium* (Washington, D.C., July 1979).

Nelsen, Harvey, *The Chinese Military System* (Boulder, 1981).

Pollack, Jonathan. "The Modernization of National Defense" in Richard Baum (ed.), *China's Four Modernizations* (Boulder, 1980), 241-261.

Prybyla, Jan. S., *The Chinese Economy: Problems and Policies* (Columbia, S.C., 1978).

————, "Feeding One Billion People: Agricultural Modernization in China," *Current History*, 79:458:19-23, 40-42 (Sept. 1980).

Reardon-Anderson, James, "Science and Technology in Post-Mao China," *Contemporary China*, II:37-45 (Winter 1978).

Segal, Gerald, and William Tow (eds.), *Chinese Defense Policy* (Champaign, 1984).

Sigurdson, Jon, *Rural Industrialization in China* (Cambridge, Mass., 1977).

Smil, Vaclav, *China's Energy Archievements, Problems, Prospects* (New York, 1976).

————, "The Energy Cost of China's Modernization," *Contemporary China*, II:3: 109-114 (Fall 1978).

Stavis, Benedict, *Making Green Revolution—The Politics of Agricultural Development in China* (Ithaca, 1974).

Stover, Leon E., *The Cultural Ecology of Chinese Civilization: Peasants Elites in the Last of the Agrarian States* (Stanford, 1979).

Suttmeier, Richard P., et al., *Science and Technology in the People's Republic of China* (Paris, 1977).

————, *Science, Technology and China's Drive for Modernization* (Stanford, 1981).

Teng, Li-ch'un 鄧力群, "Kung-ch'an chu-i shih ch'ien-ch'iu wan-tai ti ch'ung-kao shih yeh" 共產主義是千秋萬代的崇高事業 (Communism is a lofty calling for a myriad of generations), *Kuang-ming Daily*, March 28, 1981.

Ullerich, Curtis, *Rural Employment and Manpower Problems in China* (White Plains, N.Y., 1979).

Volti, Rudi, *Science and Technology in China* (Boulder, 1981).

Wiens, Thomas B., "China's Agricultural Targets: Can They Be Met?" *Contemporary China*, II:3:115-127 (Fall 1978).

Young, Graham (ed.), *China: Dilemmas of Modernization* (Dover, 1985).

36

The End
of the Maoist Age

*The Trial of the Gang of Four
and the Lin Piao Group*

An unprecedented legal and political event took place in China from November 1980 to January 1981: the trial of the Gang of Four and the associates of Lin Piao. In the past, political dissidents or opponents defeated in power struggles were summarily purged, imprisoned, liquidated, or made "non-persons." But the new leadership, anxious to project an image of proper regard for the rule of law, established a special court to try the Chiang Ch'ing and Lin Piao groups for crimes allegedly committed against the state and the people.

The four-year interval between the arrest of the Gang and the opening of the trial indicated the sensitivity and complexity of the issues at stake and the intensity of intraparty debate over the wisdom of and procedure for such an undertaking. The crux of the matter was Mao's intimate involvement in the rise of the Gang and its activities. Should he be implicated in the trial it would be necessary to have an official party assessment of his role during and after the Cultural Revolution in order to get at the truth of the matter. Yet no quick consensus could be reached due to vast differences in the opinions held by various leaders; to await such an assessment would further delay the trial. In addition, the position of Chairman Hua Kuo-feng was extremely sensitive due to his role as minister of public security (and later premier) while the Gang held sway over national policies. It was entirely possible, even inevitable, that he would be named as a witness at the trial. Thus, the question of separating Mao and

Hua from the trial became a key issue in the high councils of state, one which could not be resolved without prolonged debate, intense negotiating, and many compromises.

After their arrest in October 1976, the Four were repeatedly interrogated by government investigators in hopes of collecting evidence, confessions, and any relevant information as a basis for formal charges. But all four were crafty politicians and skillfully dodged questions, passing all responsibility for their actions to Mao. In May 1980 the party held a secret pretrial hearing of the Four. Chiang Ch'ing vigorously protested her innocence by insisting that her every act was carried out under express orders from Mao with the approval of the party Central Committee. Mao's only mistake, she said, was his choice of Hua as premier, for it whetted Hua's appetite for higher positions; and in the end he betrayed Mao's teaching and surrendered to capitalist countries in the manner of Li Hung-chang a century earlier. She insisted that Hua had been not only fully aware and supportive of her activities but was actually deeply involved (as minister of public security) in the suppression of the April 5, 1976, Tien-an-men Square Incident; hence he must be called as a witness at the trial.[1]

Wang Hung-wen and Chang Ch'un-ch'iao likewise attributed all responsibility for their actions to Mao, indicating further that Hua, as an insider, knew the full story. Yao Wen-yüan's defense differed only slightly, holding the party center responsible and criticizing the current leadership for deviation from Mao's line in pursuing the Four Modernizations with the cooperation of foreign capitalists. In short, all four involved Mao and Hua in their defense.

Within the party leadership two lines of thought quickly emerged. General Secretary Hu Yao-pang and party Vice-Chairman Ch'en Yün argued that only by first assessing Mao's contributions and mistakes could the crimes of the Gang be properly fixed. If Mao had not ignored "party democracy" and sponsored the Gang's rise to power, how could the Four have done such atrocious harm to the country? On the other hand, Hua and his supporters argued that any assessment of Mao's responsibility before the trial would lighten the responsibility of the Gang for its crimes; the inevitable inference would be that Mao and the party would bear the ultimate burden. They asked for an assessment of the Gang's crimes before an assessment of Mao.

To avoid further delay, the leadership finally decided that the trial should go ahead as announced without first making an assessment of Mao's role. Several guiding principles were adopted, the most basic of

1. *Central Daily News*, Nov. 16, 1980. Based on Nationalist intelligence reports.

which was, according to Vice-Chairman Teng, to distinguish "political mistakes or misjudgments" from the actual crimes of murder, illegal detention, and torture. Mao's role in the Cultural Revolution was seen as a "mistake," not a "crime"; hence he could not be indicted. The other principles adopted were:

1. The trial was to be held in secret in order to prevent the disclosure of "state secrets" that might be revealed during the proceedings.
2. Great effort would be made to separate Mao from the Gang; the less he was mentioned the better. The crimes of the Four would be determined first so as to retain flexibility in the assessment of Mao.
3. Although the Four behaved as a group and committed similar crimes, the degree of responsibility varied and hence their penalties must be graded. Chiang Ch'ing was the chief culprit, followed by Chang Ch'un-ch'iao and Yao Wen-yüan, while Wang Hung-wen, a young upstart who joined the Gang seeking rapid promotions (and who showed repentance during interrogations) was assessed last in the order of crimes and punishment.[2]

A Special Court of thirty-five judges was created in which there were two sections: a civil tribunal for the Gang of Four and a military tribunal for the six associates of Lin Piao. Since the Lin group was accused of plotting against Mao, while the Gang was charged with usurping state power and party leadership under the aegis of Mao, the logic of combining the two groups in one trial seemed questionable. But the government rationalized that the two groups had conspired together during the Cultural Revolution in an attempt to overthrow the proletarian dictatorship, and in doing so they severely hurt the country, the people, and the present leaders. Ironically, both groups happened to have been favored by Mao at one time or another and both had failed in attempts to seize supreme power.

THE TRIAL. On November 20, 1980, the long-awaited trial was formally opened by Chiang Hua, chief justice of the Supreme Court. The other thirty-four judges included military men, politicians, and well-known intellectuals;[3] seven of them had no legal training but were "special assessors" brought in to express the people's condemnation of the Cultural Revolution and the Gang. No foreign correspondents were admitted, and only 880 selected Chinese representatives chosen from provincial and government organizations, the party, and the army were admitted in a system of rotation.

The spotlight, of course, was on the star defendant, Chiang Ch'ing, 67,

2. *Ibid.,* Aug. 12, 1980. Intelligence reports.
3. Including the famous sociologist Fei Hsiao-t'ung.

who strode into the courtroom haughtily. Wearing a shiny black wig, she was immaculately clothed in black—a color said to symbolize the injustice that had been inflicted upon her, and perhaps also her grief over the demise of the leftist ideology she had once personified. Of the remaining Gang members, Chang Ch'un-ch'iao, 63, looked weary, defiant, and older than his years; Yao Wen-yüan, 49, had grown fatter; while Wang Hung-wen, 45, appeared hesitant, perhaps a result of his cooperation with the prosecution in what might be called "plea bargaining." Other co-defendants included Ch'en Po-ta, 76, Mao's former political secretary who defected to Lin Piao and was purged in 1970, and five generals associated with Lin.[4] Sitting behind the iron bars of the prisoners' dock with downcast faces and uneasy demeanors suggesting long years in prison, they looked unkempt, haggard, and old. Only Chiang Ch'ing appeared proud and strong, staring at the judges and prosecutors with utter contempt.

First, Special Prosecutor Huang Huo-ch'ing, head of the State Procuratorate, named six deceased persons who, if alive, would have been prosecuted as co-defendants: Lin Piao, his wife and son, a follower killed in the 1971 air crash, K'ang Sheng (Mao's former security head), and his successor, Hsieh Fu-chih. Pointedly unnamed was Mao, who nevertheless was viewed by many as an "unindicted defendant" of a "Gang of Five."

Then Huang read a 20,000-word indictment charging the defendants with usurpation of state power and party leadership. Their chief crimes fell into four major categories:

1. Framing and persecuting party and state leaders and plotting to overthrow the proletarian dictatorship.
2. Persecuting, killing, and torturing a large number of cadres and masses in excess of 34,375 people.
3. Plotting an armed uprising in Shanghai after Mao's death, with Wang Hung-wen in charge of distributing 300 cannon, 74,000 rifles, and 10 million rounds of bullets to the militia in August 1976.
4. Plotting to assassinate Mao and to stage an armed counterrevolutionary coup.

The first two categories applied to all ten defendants; the third to the Gang, and the fourth to the Lin Piao group. Forty-eight specific charges

4. The five generals were: Huang Yung-sheng, 70, former chief-of-staff of the Liberation Army; Wu Fa-hsien, 65, former air force commander; Li Tso-p'eng, 66, former political commissar for the navy; Ch'iu Hui-tso, 66, former head of the army logistics department; and Chiang T'eng-chiao, 61, former air force commander in Nanking.

of crimes as legally defined, which did not include ideological or political mistakes, were made.[5]

During the trial, Chiang Ch'ing chose to speak for herself, and this former actress came close to the best performance of her life. She assumed a studied posture of innocence and composure mingled with a pride and arrogance that suggested a "regal" disdain for the entire proceedings. She tried to project the image of a revolutionary martyr—a Chinese Joan of Arc—whose only crime was defeat in a political struggle. All her actions, she insisted, were carried out on the express orders of Mao with the approval of the Central Committee. How else could she have done as she did? Many Chinese who called her a "witch" and "the most hated in the world" secretly agreed with her, and even conceded in private that she had an "indomitable spirit." Mao was the real culprit, they reasoned, for without him she could never have become what she was.

Chiang was accused of being the chief instigator of a plot to send Wang Hung-wen to see Mao in October 1974 to falsely accuse Premier Chou En-lai of suspicious meetings with other leaders and to block his appointment of Teng as first vice-premier. She responded to the charge contemptuously: "No, I don't know (anything about that). How would I know that?" The prosecutor then called Wang to testify. He admitted that the Four did meet at Chiang Ch'ing's residence in Peking (the Angler's Guest House) in October 1974 to plot the defamation of Chou and Teng, adding, "It was Chiang Ch'ing who called (us) together and the purpose was to prevent Teng from becoming first vice-premier." Wang further implicated Yao Wen-yüan by stating that Yao pressed him to tell Mao that the situation in Peking then was critical, much like that at the August 1970 Lushan Conference when Lin Piao attempted a coup. Yao did not deny his part, but emphasized that it was Chiang Ch'ing who organized the plan to frame Chou and Teng. Two other witnesses, Mao's niece Wang Hai-jung (a vice-minister of foreign affairs) and Nancy Tang (Mao's favorite English translator), testified that Chiang Ch'ing had asked them to defame Chou and Teng before Mao, but they had refused.

The chief justice accepted the prosecution's argument that the evidence was "sufficient" and "conclusive" that the Chiang Ch'ing Group (now replacing the term the "Gang of Four") framed Chou and Teng to create favorable conditions for themselves to usurp party leadership and state power.

The Group was also accused of illegally prosecuting three-quarters of a

5. Full text of indictment in *A Great Trial in Chinese History* (Peking, 1981), 18-26, 149-98.

million people and killing 34,375 of them during the decade of 1966-76. To prove the crimes, the prosecution projected on a large screen grisly pictures of the badly bruised corpse of a former minister of coal mining and played chilling tape recordings of screaming, wailing, and moaning from intellectuals who refused to cooperate with Chiang Ch'ing and were abused in her private torture chamber.[6]

Ch'en Po-ta, assumed dead since his purge in 1970, appeared feeble and old. He admitted having plotted with Chiang Ch'ing and K'ang Sheng in July 1967 the purge of Liu Shao-ch'i and his death in prison. Ch'en further confessed to ordering the purges of Teng Hsiao-p'ing (then party general secretary), Tao Chu (Canton party leader), and Lu Ting-i (party Propaganda Department head). In late 1967 Ch'en even had pressed false charges against Chu Teh, the co-founder of the Red Army and chairman of the National People's Congress. In all, Ch'en was charged with wrongful persecution of 84,000 people and the deaths of 2,950 during the Cultural Revolution.

Under repeated questioning, Chiang Ch'ing broke down and admitted that she wrote letters to a group in charge of the persecution of Liu Shao-ch'i, instructing that Liu be hounded to death. She also admitted having said Liu "should be cut into a thousand pieces."[7] This irrefutably incriminating testimony enabled the prosecution to score a major breakthrough in the trial.

In a final move to prove Chiang Ch'ing's guilt, the prosecution produced a list of Central Committee members prepared by her for purge during the Cultural Revolution: Liu Shao-ch'i, Teng Hsiao-p'ing, Marshal P'eng Te-huai, and Mayor P'eng Chen of Peking. History showed that all of them had been forced out of office, the prosecution asserted, and all on false charges brought by Chiang Ch'ing.

As regards the six defendants of the Lin Piao Group, all pleaded guilty to the charge that they plotted the murder of Mao while he was touring the country in September 1971, on orders from Lin. The oldest of them, Ch'en Po-ta, expressed the sentiments of all six when he said that he had nothing to say in his own defense, but that he asked for mercy from the party.

After 27 days of trial and many recesses extending nearly two months, the court concluded its work on December 29, 1980, without announcing a verdict. Eight of the ten defendants admitted their guilt as charged, but

6. The coal minister was Chang Lin-chih (Zhang Linzhi). Eleven professors and acquaintances of Liu Shao-ch'i were tortured and three of them died. *A Great Trial*, 43-45, 56-57.
7. *Ibid.*, 39.

Chang Ch'un-ch'iao consistently refused to cooperate and Chiang Ch'ing remained unrepentant to the end.

Prosecutor Chiang Wen asked that Chiang Ch'ing be given a severe penalty (though not necessarily death) in view of her "particularly serious, particularly wicked" counterrevolutionary activities. He came close to condemning Mao when he said: "The people of all nationalities throughout the country understand that Chairman Mao was responsible, so far as his leadership was concerned, for their plight during the Cultural Revolution, and that he was also responsible for failing to see through the Lin Piao and Chiang Sh'ing counter-revolutionary cliques." Nonetheless, the prosecutor was quick to add, Mao made great contributions toward overthrowing imperialism, feudalism, and bureaucratic-capitalism, and was responsible for the founding of the People's Republic and for pioneering the socialist cause in China. Echoing the views of Teng, the prosecutor said that Mao's achievements were primary and his mistakes secondary.[8]

Still feisty, unrepentant, and haughty, Chiang Ch'ing shouted in the court: "Fine. Go ahead! You can't kill Mao—he is already dead—but you can kill me. Still I regret nothing. I was right!" Declaring it would be more glorious to have her head chopped off than to yield to her accusers, she told the court: "I dare you to sentence me to death in front of one million people in Tien-an-men Square."

Ultimately, just as the trial was political, so the verdict had to reflect the views of the present leadership; and a consensus was not reached for three weeks. On January 25, 1981, the Special Court announced the death sentence for Chiang Ch'ing and Chang Ch'un-ch'iao, but with a 2-year suspended execution. Wang Hung-wen was given life imprisonment. Yao Wen-yüan received 20 years and Ch'en Po-ta, 18 years imprisonment. The other five generals were sentenced to 16 to 18 years in jail.

Looking at the excruciatingly long time that it took the leadership to reach their verdict, one can only be reminded of the still powerful influence of the patriarch Mao. In life he sponsored the rise of his wife, and in death he shielded her from immediate execution. Chiang Ch'ing seemed to know that the present leadership dared not repudiate Mao and the Ninth and Tenth Party Congresses without severely discrediting the party itself. Indeed, her main line of defense was never directly refuted by the prosecution. But her role in the persecution and deaths of 34,375 people and her maintenance of a private torture chamber were unforgivable crimes, for which even a death sentence seemed too charitable. Yet the verdict, like the trial itself, was politically motivated and would be politi-

8. *Ibid.*, 105.

cally decided. The specter of Mao inhibited the present leadership from killing her.

In a sense, the trial may be viewed as an indirect trial of Mao, with Chiang Ch'ing serving as his surrogate. To probe still deeper, one may say that the trial was an indictment of the entire system which allowed Mao to overpower the Central Committee and allowed his wife's group to drive the country to the brink of anarchy and economic disaster. The net effect of the trial was a further erosion of the image of Mao and the effectiveness of the system he created.

A not unexpected repercussion of the trial was its adverse effect on the future of Chairman Hua. As minister of public security during the heyday of the Gang, he signed warrants authorizing arrests, prison sentences, and even deaths for enemies of the Gang, for which some said he should be held responsible. During the trial, dialogues occasionally went beyond the prescribed script and implicated him in the "Criticize Teng, Antirightist Deviationist" Campaign and in the suppression of the Tien-an-men Square Incident as well. The Tengists were quick to exploit the situation by accusing him of practicing "personality cultism" after the smashing of the Gang and of mismanaging modernization programs by stressing revolutionary zeal over solid economic planning. The Tengists saw Hua's leadership as a continuation of "wrong" Maoist policy which might provide a future rallying point for disaffected ultraleftists to oppose the course of the present leadership. Great pressure was exerted on him to resign.

Reportedly, Hua offered to resign as party chairman at the November 1980 meeting of the Politburo, to be effective at the next plenum of the party Central Committee in June 1981. Though still chairman until then, Hua had been reduced to a figurehead. The real power of the party fell into the hands of General Secretary Hu Yao-pang, and Teng Hsiao-p'ing became the de facto chairman of the Military Commission. While no one denied Hua's key role in smashing the Gang, Chinese newspapers increasingly reported that credit for such an exploit should not go to individuals but to the rules of historical development and to the will of the people. With the phasing out of Hua, the Maoist age came closer to an end.

Assessments of Mao

Few issues in Chinese politics had caused such deep dissension as the assessment of Mao. The pragmatic leaders, the victims of the Cultural Revolution and the Gang of Four, and a considerable number of youths favored a candid and critical assessment of Mao's achievements and failings. They regarded his legacy as unfit for the new mission of modernization and cited his failure to lift China from poverty as proof of the inadequacy of his approach. On the other hand, a large number of party and military leaders and cadres who owed their status to Mao's revolution, particularly the "old-timers" and those who rose during the Cultural Revolution, considered such an assessment tantamount to a defamation of the man who led the revolution to success, founded the People's Republic, and pioneered the socialist cause in China. Sympathetic to this latter view were many others who grew up under the influence of ceaseless indoctrination that Mao saved China from imperialism, feudalism, and capitalism, and gave the people a new life. These people could not easily renounce their habitual reverence of the Leader.

In the vortex of such conflicting views, consensus was difficult to reach; the closest thing to it was Vice-Chairman Teng's statement that Mao's achievements were primary and his mistakes secondary. The party adopted the principle of separating Mao's thought from Mao's leadership qualities. Mao's thought was said to represent the sum total of the Chinese revolutionary experience: all the contributions made by all the participants in that revolution. Hence it did not reflect Mao's thinking alone but was the legacy of all Chinese, and they had to continue to treasure it. On the other hand, Mao as a leader made many contributions as well as some very serious mistakes. To assess his achievements and errors impartially was to seek truth from facts and to learn from past experiences.

THE PARTY ASSESSMENT. The long-awaited party assessment of Mao finally came at the Sixth Plenary Session of the Eleventh Central Committee held between June 27 and 29, 1981. In a leadership reshuffle, the plenum accepted Hua Kuo-feng's resignation, naming Hu Yao-pang chairman of the Central Committee, and Teng Hsiao-p'ing chairman of the Military Commission, thereby breaking the tradition of one man holding both posts. Another personnel change was the appointment of Premier Chao Tzu-yang as a vice-chairman of the party, forming a "collective triumvirate" with Hu and Teng. Hua Kuo-feng became the most junior of the six vice-chairmen, a powerless but nevertheless respectable position.[9]

The plenum adopted a 35,000-word "Resolution on Certain Questions in the History of Our Party Since the Founding of the People's Republic," which detailed the party's accomplishments in the past 60 years, especially since 1949. However, a major point of the document was the Cultural Revolution and Mao's role in it. The party's stand was unequivocally critical: "The 'Great Cultural Revolution' from May 1966 to Octo-

9. "Communiqué of the Sixth Plenary Session of the 11th Central Committee of CPC," adopted June 29, 1981, *Beijing Review*, No. 27:6-9 (July 6, 1981).

ber 1976 caused the most devastating setback and heavy losses to the party, the state, and the people in the history of the People's Republic, and this 'Great Cultural Revolution' was initiated and led by Comrade Mao Tse-tung."

The movement, said the party, was neither in conformity with the thought of Mao nor with Marxism-Leninism or the realities of China. Indeed, the assertion that the "Great Cultural Revolution" was a struggle against revisionism and capitalism was, in retrospect, groundless. False accusations were prevalent: the "capitalist roaders" who were knocked down were in fact leaders from different levels of the party and the government who were the core force of socialist construction; the so-called "Liu-Teng Bourgeois Headquarters" never really existed. The persecution of "reactionary intellectual authorities" sacrificed countless talented and accomplished scholars. Moreover, the Cultural Revolution was conducted in the name of the masses but was actually divorced from both the party and the masses. Pointedly, the document stated: "Comrade Mao's leftist mistakes and personal leadership actually replaced the collective leadership of the party center and he became the object of fervent personal worship." The prominence of Mao's errors made a critical assessment absolutely necessary, for "to overlook mistakes or to whitewash them is not only impermissible but is in itself a mistake." The verdict was:

> Practice has shown that the "Great Cultural Revolution" did not in fact institute a revolution or social progress in any sense, nor could it possibly have done so. It was we and not the enemy who were thrown into disorder by it. Therefore, from beginning to end, it did not turn "great disorder under heaven" into "great order under heaven." . . . History has shown that the "Great Cultural Revolution," initiated by a leader laboring under a misconception and capitalized on by counter-revolutionary cliques, led to domestic turmoil and brought catastrophe to the party, the state, and the whole people. . . . Comrade Mao . . . far from making a correct analysis of many problems . . . confused right and wrong and the people with the enemy. . . . Herein lies his tragedy.

The party document also assessed Hua Kuo-feng; and while it gave him due recognition for his constructive role in the smashing of the Gang and certain economic works, it criticized him for leftist thinking. Hua was branded a "Whateverist"—supporting *whatever* decisions Mao made and implementing *whatever* instructions he issued—thereby perpetuating leftist errors. Hua took part in the "Anti-Teng" campaign, and after becoming party chairman blocked moves to correct the wronged cases, including the rehabilitation of victimized cadres and the Tien-an-men Square Incident.

He even practiced "personality cult." Worse, at the August 1977 Eleventh Central Committee meetings he obstructed attempts to assess the Cultural Revolution critically and instead used his influence to affirm it; later he was also responsible for proceeding with hasty, leftist economic policies. The document concluded: "Obviously, for him to lead the party in correcting the leftist errors, and particularly in reestablishing the fine tradition of the party, is an impossibility."[10]

A terse and crisp assessment of Mao was made by Chairman Hu in his first major address celebrating the 60th anniversary of the founding of the party (July 1, 1981). Mao's greatest contribution, Hu stated, was his early rejection of the "childish sickness" of worshipping foreign (Soviet) experience in the 1920s and 1930s. He creatively integrated Marxist universal principles with concrete Chinese revolutionary conditions to form a new synthesis that suited the Chinese situation. The Thought of Mao was the "crystallization of the collective wisdom of the party and a record of the victories of the great struggles of the Chinese people," and its creativeness had enriched the storehouse of Marxism. As such, Hu said, it was, is, and shall be the guiding principle of the party. Having complimented Mao, Hu then made the official criticism:

> Comrade Mao, like many great historical figures of the past, was not free from shortcomings and mistakes. The principal shortcoming occurred during his later years when, due to long and ardent support by the party and by all the people, he became smug and increasingly and seriously lost contact with realities and the masses. He separated himself from the collective leadership of the party, often rejecting or even suppressing the correct views of others. Mistakes thus became inevitable. A long period of comprehensive, serious mistakes led to the outbreak of the "Great Cultural Revolution," which brought the most severe misfortune to the party and the people. Of course, we must admit that neither before nor after the outbreak of the "Great Cultural Revolution" was the party able to prevent and turn Comrade Mao from his mistakes. On the contrary, it accepted and approved some of his erroneous proposals. We who are long-time comrades-in-arms with Comrade Mao, his long-time followers and students, must realize deeply our own responsibility and resolutely accept the necessary lessons.
>
> Nonetheless, though Comrade Mao committed serious errors in his late years, it is clear that from the perspective of his entire life his contributions to the Chinese revolution far outweigh his mistakes.

10. Chinese text of the "Resolution" in the *People's Daily*, July 1, 1981. Tr. mine. An English version, somewhat less literal than mine, may be found in *Beijing Review*, ibid., 10-39.

. . . He was both a party founder and the principal creator of the glorious People's Liberation Army. After the establishment of the People's Republic under the leadership of the party center and Comrad Mao, China was able to stand on its feet and pioneer the socialist cause. Even when he was making serious mistakes during his last years, Comrade Mao still vigilantly guarded the independence and security of the motherland, correctly assessed new developments in world politics and led the party and the people to resist all the pressures of hegemonism, opening a new direction for our foreign relations. During the long period of struggle, all party members absorbed wisdom and strength from Comrade Mao and his thought. They nurtured the successive generations of our leaders and cadres, and educated our people of different nationalities. Comrade Mao was a great Marxist, a great proletarian revolutionary, theorist, strategist, and the greatest national hero in Chinese history. He made immense contributions to the liberation of all oppressed peoples of the world and to human progress. His great contributions are immortal![11]

It is noteworthy that in the entire speech Hu never once referred to Mao as chairman but only as comrade. The "Great Cultural Revolution" was mentioned in quotes, implying his refusal to recognize its legality.

A HISTORIAN'S VIEW. In an objective assessment of Mao, historians who seek truth from facts will be among the first to recognize Mao's greatness as a revolutionary leader, founder of the People's Republic, and pioneer of the socialist cause in China. But they would be remiss if they overlooked his various policy blunders and their consequences. First and foremost was his rejection of any population control. Experts, including Peking University President Ma Yin-ch'u,[12] warned of the serious economic and social consequences of a population explosion, but Mao argued that population problems existed only in capitalist societies. The Soviet Union had no population control and did not suffer any negative consequences—why should China be different? Mao dismissed Malthusian population theories in the naïve belief that more people could do more work—more work meant more production and faster economic development. The result was an uncontrolled increase in population from 500 million in the early 1950s to over a billion today, while arable land, rather than increasing, actually decreased due to national disasters, increased industrial use of land, and the removal of trees for fuel by the poor.

The task of feeding, clothing, and providing shelter and employment

11. Chinese text of Hu's speech in the *People's Daily*, July 2, 1981. Tr. mine.
12. And others such as Ch'en Ta and Wu Ching-ch'ao.

for a billion people was a gigantic burden which no other country on earth faced. It drained much of the national resources which otherwise could be used for economic development. With 80 percent of the huge population based in the countryside, improvement of agricultural production was basic to China's socialist construction. Yet Mao followed the Soviet model of investing heavily in heavy industry and lightly in agriculture, resulting in extremely low agricultural productivity. Mao did not heed the Marxist dictum that the foundation of a society lay in the agricultural laborer's productive rate exceeding the individual needs of the laborer, creating a surplus to support the other sectors of the state. For thirty years, China's agricultural sector was neglected and semi-independent, necessitating the importation of food to meet domestic needs, and thereby consuming a considerable amount of scarce foreign exchange reserves. In 1978-79 per farm capita production amounted to only $50 a year—pitifully below any surplus that might support economic growth and improve the standard of living. Unless agricultural conditions were vastly improved and birth control was strictly enforced, China's march toward modernization would, at best, be slow and labored. In retrospect, Mao's population and agricultural policies had created the most serious obstacle to rapid modernization.

The second major policy blunder was the enforced isolation of the country. Except for the 1950s when Sino-Soviet cooperation was in full swing, China had been virtually cut off from the outside world for twenty years. Under the ideology of self-reliance, Chinese science, technology, arts, education, and other aspects of culture were deprived of the benefits of developments in other countries. It was exactly during the decades of the sixties and seventies that phenomenal progress was made in the West and Japan, while China preoccupied itself with civil strife and class struggle. The cost of this isolation to China was practically incalculable.

The third policy blunder was "leftist blind actionism" and "adventurism" in economic development. At a central work conference in December 1980, party Vice-Chairman Ch'en Yün, an economist, pointedly declared: "Since the founding of the People's Republic, the main mistake in economic development was 'leftism.' The situation before 1957 was relatively good, but after 1958 'leftist' mistakes became increasingly serious. It was a principal mistake . . . and the main source of that mistake was leftist leadership thinking."[13] An obvious manifestation of this "wrong thinking" was an overfondness for quick results, in total disregard of objective economic realities, which resulted in "taking fantasy as truth, working stub-

---

13. "Leadership Thinking in Rectifying Economic Work: On the Leftist Mistakes in Economic Construction," by a special commentator of the *People's Daily*. Chinese text in the *People's Daily*, Apr. 9, 1981. Tr. mine.

bornly according to self-will, and carrying out work today that might be possible for the future." Such "leftist adventurism" severely undermined the productive relations in the economic structure. Furthermore, the doctrine of uninterrupted revolution led to reckless "blind actionism" which set up unrealistic economic targets supported by a level of investment that far exceeded the country's ability to pay. What followed was an unending flow of falsified figures to deceive the leadership, and the twin evils of adventurism and blind actionism together drove the people to the brink of bankruptcy.[14]

Mao's fourth major mistake was his assumption of an unquestionable superiority within the party, destroying party democracy and collective leadership, and opening the way for "one-man rule." As the revolutionary leader and founding father of the People's Republic, Mao endowed himself with the status of a patriarch (*chia-chang* 家長), tolerating no opinion except his own (*i-yen t'ang* 一言堂). His actions reflected the feudalistic notion that he who conquered the country controlled it as a family possession (*chia t'ien-hsia* 家天下). Thus, official documents frequently started with the phrase "the Chairman, and the party center . . . ," suggesting one man towering above the party, mocking the collective leadership affirmed in the Eighth Party Congress. Sycophants such as secret service head K'ang Sheng whetted Mao's appetite by asserting that a party history mentioning the contributions of other leaders belittled Mao and created a rival center. Thus, the party history became a chronicle of the Leader's continuous feuds with others until, one by one, he had knocked them all down.[15] Like the emperors of the past, Mao was a patriarch, Helmsman, and even god-hero, who could do no wrong. He acted with total impunity in "designating" his successor and sponsoring the rise of his wife far beyond her worth. It appears the Actonian dictum "power corrupts and absolute power corrupts absolutely" holds true even in a dictatorship of the proletariat!

In addition to granting him absolute control, Mao's political style became an example for party secretaries in the provinces and districts to follow; they behaved like small patriarchs and smaller patriarchs in their respective jurisdictions. Nothing could be done without their approval, establishing a highly bureaucratic and privileged class throughout the country.[16]

How could the party allow all this to happen? The Chinese themselves

14. *Ibid.*
15. *Ibid.*, Sept. 18, 1980, 5.
16. "On the Necessity of Reforming the Leadership System," by a commentator of the *Red Flag* magazine. Chinese text in *Hongqi* (Red Flag), No. 17:2-4, 1980.

were hard put to find a proper explanation but finally came up with two interpretations. The first stated that in China, as in other communist countries, the leader of the revolutionary party was empowered with great discretionary authority and freedom of action during the seizure of power. Once success was achieved, the concentration of power had a tendency to continue; and due to the obviously great contributions of the leader, his followers readily accepted his exalted status. His status eventually became institutionalized, and he received lifelong tenure as *the* Leader, as well as credit for the fruits of others' labor.

The broader official explanation concerned the profound impact of China's feudal past on the thought and action of all. The vestiges of the distinction between high and low, of the rank and grade system, and the role of the family head could be seen everywhere. Farmers and small producers were unaccustomed to controlling their own fates, relying instead on the graces of the emperor as "savior," giving in return their loyalty and gratitude. Thus, there was a powerful social precedent for the high concentration of authority in one man. Even the party itself reflected this feudal influence, permitting the emergence of a situation in which no one dared criticize the patriarch. Consequently, collective leadership and democratic centralism became meaningless: in the former, one was "more equal" than others, and in the latter, centralism prevailed over democracy.[17]

These explanations were certainly valid, but they omitted one key element: it was Mao's firm control of the army, the secret police, the security apparatus, the 8341 unit, and the network of intelligence and investigatory agencies, which made opposition to him virtually impossible. Those who dared to criticize him risked their futures and even their lives.

In conclusion, historians would agree that Mao was extremely successful as a revolutionary but disappointingly erratic as a nation builder. His great achievements before 1957 were a source of inspiration to others, but his serious mistakes thereafter must serve as a lesson to all.

A New Leadership and a New Order

With the delicate assessment of Mao finally out of the way and the party's guilt recognized, a heavy psychological burden was lifted. The new power structure put Hu, Teng, and Chao firmly in control of the party, the military, and the government, and cemented their plans for China's future. They were committed to a revolution of modernization. Lest there be any misunderstanding, Hu impressed upon the nation the following six points

17. *Ibid.*, 5-8.

in his speech commemorating the 60th anniversary of the founding of the party:[18]

1. All party members must dedicate themselves to modern construction of Chinese socialism regardless of personal sacrifice, and serve the people with all their hearts and minds.

2. Under the new historical conditions, we must advance Marxism and the Thought of Mao Tse-tung. [Hu reaffirmed the importance of the four basic principles: the socialist line, the proletarian dictatorship, the leadership of the Communist party, and Marxism-Leninism and the Thought of Mao.]

3. We must further strengthen the democratic life of the party and tighten the party discipline. . . . We must forbid any form of individual worship. . . . All important issues must be decided after collective discussions by appropriate party committees. No one man should have the final say. All members of the committees involved must abide by such decisions. At all levels of party organization, we must implement collective leadership . . . with emphasis on quality and efficiency. [To enhance party democracy,] any member has the right to criticize party leaders at party meetings, even including the top leaders at the center, with impunity. [But no one is allowed to create his own] independent kingdom.

4. We must regularly dust ourselves off in order to insure revolutionary youthfulness permanently within the political framework of the government. [Very pointedly Hu admitted] in the past excessive struggles resulted in a counterproductive situation in which no one dared to make self-criticism or offer criticism. We must rectify this unhealthy style of behavior.

5. We must select young and vigorous cadres of character and knowledge for different levels of leadership.

6. We must persist in supporting internationalism and share the breathings and the lives of the proletarian class and people all over the world. . . . In dealing with strong and rich countries we must preserve our national dignity and independence, never permitting any cowering or toadying action and thought. We must be determined to unite all people including those on Taiwan in the sacred struggle for the return of Taiwan to the motherland.

It is clear that Mao's political work style and approach to economic development would not be followed: there would be no more personal worship, no suppression of free expression in party meetings, and no penalty for criticizing the leaders. However, all cadres were to be subordinated to orders from above to carry out economic construction, without feigning

18. Full text in the *People's Daily*, July 2, 1981. Tr. mine.

compliance while secretly resisting implementation. Also rejected were Maoist ideas of class struggle, disdain for intellectuals and foreign contacts, and opposition to limited private enterprise.

To ensure the success of socialist modernization, which was "a great revolution in itself," Hu called for intraparty and party-citizen unity as well as international exchanges in economics, culture, and science and technology, in order to develop a "prosperous, strong, highly democratic, and highly cultured modern socialist power," which would ultimately lead China to the communist utopia.

Hu's speech, together with the communiqué and resolution of the Sixth Plenum, was a crowning testimony to the victory of the pragmatists. In the spirit of unity, stability, conciliation, pragmatism, democracy, and realistic economic development, a new order was born under new historical conditions. With this, the Maoist era had come to an end.

Chinese Communism: A Thirty-five-Year Review

On October 1, 1949, standing atop the Gate of Heavenly Peace to proclaim the establishment of the People's Republic, Mao shouted triumphantly: "The Chinese people have stood up!" What a heroic voice, what an auspicious beginning! Foreign imperialism and domestic opposition had been swept away, and the country was unified in a way unknown since the mid-19th century. China was a blank canvas for the artist Mao; and his revolutionary romanticism, vision, idealism, and egalitarianism had caught the imagination of millions inside and outside China. The charismatic leader's articulation of a national purpose and the promise of the future reinforced the desire of 500 million people to rebuild their country. The galaxy of talents surrounding Mao and contributing to the success of the revolution seemed to ensure China's goals of domestic security, international respectability, and eventual emergence as a world power.

That these goals had been met at least in part was obvious. The greatest accomplishments had been the unification of China (except Taiwan) under one central government, the attainment of the status of a major participant in world affairs, the elimination of the curse of landlordism, the laying of a foundation for industrialization, the improvement of public health, the selective development of science and technology (especially in nuclear power and rocketry), the improvement of literacy, and significant archeological finds that could result in new interpretations of ancient Chinese history. The provision of subsistence-level food, housing, clothing, and employment for over a billion people answered a challenge no other country on earth had ever had to meet. Finally, statistics showed considerable

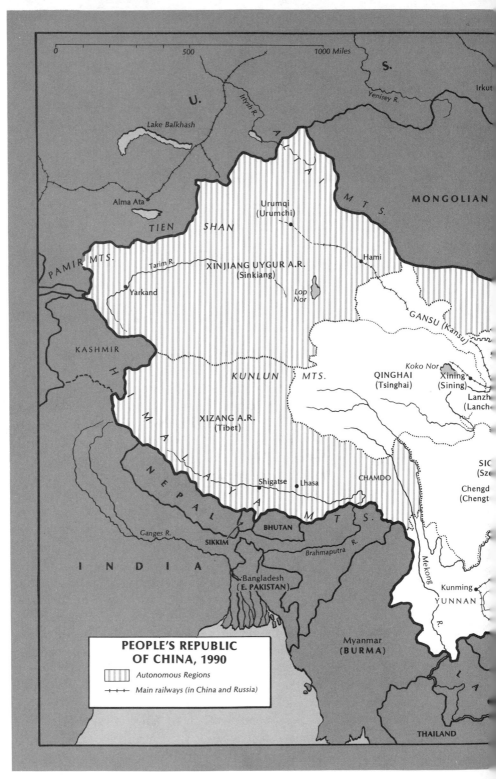

PEOPLE'S REPUBLIC
OF CHINA, 1990

||||| Autonomous Regions

++++ Main railways (in China and Russia)

834

S.

Lake
Baikal
Nerchinsk

Komsomolsk

R.

Amur R.

Khabarovsk

Manchouli

Ulaan Baatar
(Ulan Bator)

REPUBLIC

HEILONGJIANG (Heilungkiang)

Qiqihar
(Tsitsihar)

Sungari R.

Harbin

Ussuri R.

(Inner Mongolia)

GREATER KHINGAN MTS.

JILIN
(Kirin)

Changchun

Shenyang
(Mukden)

Vladivostok

Hohhot
(Huhehot)

MONGGOL A.R.

Baotou
(Paotow)

GREAT WALL

Beijing
(Peking)

Tangshan

LIAONING

Port
Arthur

Dairen

NORTH KOREA

Pyongyang

Seoul

HEBEI
(Hopeh)

Taiyuan

Tianjian
(Tientsin)

Yantai
(Chefoo)

SOUTH KOREA

SHANXI
(Shansi)

Jinan
(Tsinan)

SHANDONG
(Shantung)

Qingdao
(Tsingtao)

SHAANXI (Shensi)

Yellow
R.

Kaifeng

Grand
Canal

JIANGSU
(Kiangsu)

J A P A N

Xian
(Sian)

HENAN
(Honan)

Nanjing
(Nanking)

HUBEI
(Hupeh)

Wuhan

Hangzhou
(Hangchow)

Shanghai

Chongqing
(Chungking)

Yangtze R.

Nanchang

ZHEJIANG
(Chekiang)

Ningbo
(Ningpo)

Changsha

HUNAN

JIANGXI
(Kiangsi)

Wenzhou
(Wenchow)

RYUKYU IS.

OKINAWA

Guiyang
(Kweiyang)

FUJIAN
(Fukien)

Fuzhou
(Foochow)

HOU
how)

Kweilin

(Kwangtung)

Xiamen
(Amoy)

Taipei

TAIWAN (FORMOSA)
Nationalist China

Pacific

ANGXI A.R.
(Kwangsi)

West R.

GUANGDONG

Guangzhou
(Canton)

Swatow

PESCADORES

Ocean

Nanning

Hong Kong (Br.)
Macao (Port.)

IAM

HAINAN

PHILIPPINE
IS.

South China Sea

Vaughn Gray

increases in total industrial and agricultural productions and in social services.

Yet, recent Chinese leaders had openly acknowledged that despite selective progress, the country remained in a state of dire poverty and widespread scarcity (*i-ch'iung erh-pai*). The physical complexion of the country and the livelihood of the people had not substantially changed in twenty years;[19] the gap between China and advanced countries was probably wider now than in 1949 due to phenomenal progress in science, technology, and economic development in other countries. What was it that kept China from making greater progress in the three decades following liberation?

The chief deterrents within China had been political instability and the destruction of the principle of democracy within and without the party. Except for seven years (1949-56) of revolutionary momentum and the euphoria of building a new order, China's recent history had been plagued with such upheaval and strife that the country had been nearly destroyed. Political turmoil had resulted in the loss of much able talent, interruptions in economic development, and devastation of intellectual and artistic creativity.

In considering the source of Chinese political instability, the Eighth Party Congress of 1956 was generally viewed as the Rubicon in political development. This Congress, like the previous one, stressed party democracy and free discussion of issues. It affirmed the collective leadership of Mao, Liu, Chou, and Chu Teh, and the period from 1956 to early 1957 was regarded as a "springtime" in party history.

Into this springtime Mao introduced the Hundred Flowers Campaign, and intellectuals criticized sharply certain party policies. Mao responded as though Chinese communism itself were endangered by the criticism, and launched the Antirightist Campaign which adversely affected as many as one million people. Mao overthrew the decisions of the Eighth Congress, proclaiming that "the decisions of the Eighth Party Congress referring to the major contradiction between the productive forces of advanced and backward socialism is incorrect," and that capitalist-proletarian class conflict and capitalist-socialist line struggle remained the principal contradictions in Chinese society. From then on Mao enlarged the scope of class struggle, which caused ceaseless turmoil. Rejecting collective leadership, the concept of patriarchal rule gained ascendancy and the party center became "The Hall of One Voice" (*I-yen t'ang*). When Defense Minister P'eng Te-huai expressed his views on the Great Leap in 1959, he was dis-

19. According to Hu ch'iao-mu, president of the Chinese Academy of Social Sciences. See his article, "Observe Economic Laws, Speed Up the Four Modernizations," *Peking Review*, 47:18-19 (Nov. 24, 1978).

missed and disgraced as a "rightist opportunist." Thereafter, no one dared speak out. Party democracy was shattered.[20]

Once Mao's "omnipotence" was demonstrated, opportunists and intriguers crowded around him, gaining power by controlling access to him. Their power struggles reverberated throughout the nation in intensified class struggle and in the Cultural Revolution. The dominance of K'ang Sheng, Lin Piao, and the Gang of Four turned the proletarian dictatorship into a fascist one, with the added features of feudalism and revisionism.[21]

Political instability and the disappearance of party democracy inevitably affected economic development and the people's lives. On July 1, 1979, an editorial in the *People's Daily* commented:

> In the past thirty years whenever party democracy was relatively sufficient and democratic centralism relatively healthy, party leadership in economic work was in tune with reality. When problems arose, they were discovered and corrected easily, bringing about rapid socialist economic development. Whenever there was a lack of democracy in the party, nobody dared to speak out or speak the truth. Blind obedience was prevalent and the party's economic policies frequently lost touch with reality and objective laws. Socialist economic development then slowed, stagnated, or even retrogressed.[22]

Chinese statistics show that economic development was marked by three periods of growth (1949-57, 1963-65, 1977-88) and two periods of decline (1958-62, 1966-76). During Mao's 27-year rule, only 1952-57 were years of genuine growth, while 1949-52 represented a recovery from the civil war and 1963-65 a recovery from the Great Leap. The following statistics summarize development in the past thirty-five years.

China's erratic development demonstrates that revolutionary leaders who were skillful in political struggle were not necessarily knowledgeable in economic matters. Mao in particular would not heed the advice of economic experts or act according to economic laws and the realities of the country. In the fifties he adopted the Soviet model for development and emphasized heavy industry over agriculture and light industry, when the concrete situation in China suggested the reverse as more logical. When the Soviet model proved ill-suited, he precipitously resorted to the Commune and the Great Leap. What followed for the next two decades was the familiar saga of "leftist adventurism," which caused waste of time, energy, capital, and valuable talents. Consequently, Chinese per capita in-

20. Lu Chung-chien, "San-shih-nien ti chiao-hsün (The lessons of thirty years), *Cheng-Ming* Magazine, Hong Kong, No. 24:8, 11 (Oct. 1, 1979).
21. *Ibid.*, 14.
22. *People's Daily*, July 1, 1979. Tr. mine.

Periods of Recovery and Growth

	1949-52	1953-57	1963-65	1977	1978-86
Industry	36%	19.2%	7.9%	14.1%	134.3%
Agriculture	14%	4.5%	11.1%		67.2%
National Income	n.a.	n.a.	14.5%		

Periods of Decline

	1958-62	1967	1968	1974	1976
Industry	+3.8%	−13.8%	−5%	+0.3%	+1.3%
Agriculture	−4.3%		−2.5%		
National Income	−3.9%				

Figures from the *Journal of Philosophy and Social Sciences*, Nanking University, No. 3: 1-8, 1979; CIA, *China-Economic Policy and Performance in 1987* (Washington, D.C., 1988).

come ($300) ranked last among the socialist countries, and productivity lagged far behind that of Hong Kong and South Korea.[23]

Yet, in spite of political instability and erratic economic performance, substantial progress was registered in agricultural and industrial productions. The former averaged 2 to 3 percent annual growth rate, and the latter, a 9 to 10 percent rate of growth. Between 1952 and 1987, grains grew from 163.9 million metric tons to 402 million tons; coal, from 66.4 million tons to 920 million tons; steel, from 1.35 million tons to 56.22; crude oil, from 440,000 tons to 134 million tons; and electric power, from 7.3 billion kwh to 496 billion kwh.[24] However, many of the benefits of increased productions were compromised by the population explosion from 570 million to 1.08 billion, resulting in an extremely low standard of living for most Chinese.

The lessons to be learned from the past thirty-five years are many. First and foremost, there must be political and social stability to enable the government to carry out orderly reform and development. Second, population control must be strictly enforced so as to achieve a zero growth rate. Third, international cooperation must be strengthened in all areas including science, technology, education, and the arts. Fourth, war must be avoided if possible, for it is the biggest waster of financial and human resources. Fifth, the political system must be reformed to prevent the recurrence of

23. Lu Chung-chien, 6.
24. Xue Muqiao (ed.), *Almanac of China's Economy, 1985/1986* (Hong Kong, 1986), 26; State Statistical Bureau, Peking, Feb. 23, 1988, *Beijing Review*, March 7-13, 1988.

patriarchal rule and to ensure democracy within and without the party. Bureaucratism, lifelong tenure of cadres, and special privileges should be reduced (if not eliminated), with the institution of a benevolent retirement system. Sixth, economic development must be neither "leftist" nor "rightist" but "centrist," based on realities and economic laws. Last but not least, the party must take note of the Actonian dictum that power corrupts and absolute power corrupts absolutely. It must rid itself immediately of the widespread corruption and special privileges for the few and stamp out the rampant inflation. Unless these serious problems are effectively addressed, the party's credibility before the people would be vastly compromised. (See Chapter 41 for "Chinese Communism at 45.")

Further Reading

A Great Trial in Chinese History (Beijing, 1981).

"Ai-kuo chu-i shih chien-she she-hui chu-i ti chu-ta ching-shen li-liang" 愛國主義是建設社會主義的巨大精神力量 (Patriotism is a great spiritual power in building socialism), *Renmin Ribao* (People's Daily), Mar. 19, 1981.

Bonavia, David, *Verdict in Peking: The Trial of the Gang of Four* (London, 1984).

"Ch'uan-li pu-neng kuo-fen chi-chung yü ko-jen" 權力不能過份集中於個人 (Power should not be overly concentrated in one man), *Hongqi* (Red Flag), 17:5-8 (1980). Written by a special commentator of the journal believed to be one close to Vice-Chairman Teng Hsiao-p'ing.

"Communiqué of the Sixth Plenary Session of the 11th Central Committee of CPC," adopted on June 29, 1981, *Beijing Review*, No. 27:6-8 (July 6, 1981).

Hsiung, James C. (ed.), *Symposium: The Trial of the "Gang of Four" and Its Implication in China* (Baltimore, 1981). Occasional Papers/Reprint Series in Contemporary Asian Studies, University of Maryland, School of Law.

Hu Yaobang (Hu Yao-pang), "Speech Commemorating the 60th Anniversary of the Founding of the Chinese Communist Party," Chinese text in the *People's Daily*, July 2, 1981; English tr. in *Beijing Review*, No. 28:9-24 (July 13, 1981).

Huang, Kecheng, "How to Assess Chairman Mao and Mao Zedong Thought," *Beijing Review*, 17:15-23 (Apr. 27, 1981).

Johnson, Chalmers, "The Failure of Socialism in China," *Issues & Studies*, XV:7: 22-33 (July 1979).

Kallgren, Joyce K. (ed.), *The People's Republic of China after Thirty Years: An Overview* (Berkeley, 1979).

Li, Hung-lin 李洪林, "K'o-hsüeh ho mi-hsin" 科學和迷信 (Science and superstition), *Renmin Ribao* (People's Daily), Peking, Oct. 2, 1978.

——— (Li Honglin), "Chinese Communist Party Is Capable of Correcting Its Mistakes," *Beijing Review*, No. 25:17-20 (June 22, 1981).

Li, Victor H., *Law without Lawyers: A Comparative View in the United States and China* (Boulder, 1978).

"Ling-tao chih-tu pi-hsü kai-ke" 領導制度必須改革 (The leadership system must be reformed), *Hongqi* (Red Flag), 17:2-4 (1980). Written by a commentator of the journal.

Liu, Kwang-ching, "World View and Peasant Rebellion: Reflections on Post-Mao Historiography," *The Journal of Asian Studies,* XL:2:295-326 (Feb. 1981).

Lu, Chung-chien 盧中堅, "San-shih-nien ti chiao-hsün" 三十年的教訓 (The lessons of thirty years), *Cheng-ming* Magazine, 爭鳴, Hong Kong, No. 24:5-15 (Oct. 1, 1979).

Lu, Shih 魯實, "'Mao-hsüan' wu-chuan ying-tang ch'ung-shen ch'ung-pien," 「毛選」五卷應當重審重編 (Vol. 5 of Mao's *Selected Works* should be re-examined and re-edited), *Cheng-ming* Magazine, Hong Kong, No. 24:16-17 (Oct. 1, 1979).

Morath, Inge, and Arthur Miller, *Chinese Encounters* (New York, 1979).

Pye, Lucian W., *Mao Tse-tung: The Man in the Leader* (New York, 1976).

"Resolution on Certain Questions in the History of Our Party Since the Founding of the People's Republic of China." Adopted June 27, 1981, at the Sixth Plenary Session of the 11th Central Committee of Chinese Communist Party. Chinese text in *People's Daily,* July 1, 1981; English translation in *Beijing Review,* No. 27:10-39 (July 6, 1981).

Shao, Yü-ming 邵玉銘, "Shih-lun Chung-Kung cheng-ch'üan tsai Chung-kuo chin-tai-shih shang ti kung-kuo" 試論中共政權在中國近代史上的功過 (An appraisal of the achievements and failures of the Chinese Communist regime in modern Chinese history), *Hai-wai hsüeh-jen* 海外學人 (Overseas Scholars), Taipei, 99:6-20 (Oct. 1980).

"Tuan-cheng chin-chi kung-tso ti chih-tao ssu-hsiang: Lun chin-chi chien-she chung ti tso-ch'ing ts'o-wu" 端正經濟工作的指導思想：論經濟建設中的左傾錯誤 (The leadership's thinking in rectifying economic work: On the leftist mistakes in economic construction), *Renmin Ribao* (People's Daily), Peking, Apr. 9, 1981.

Wilson, Dick (ed.), *Mao Tse-tung in the Scales of History* (Cambridge, Eng., 1977).

Witke, Roxane, *Comrade Chiang Ch'ing* (Boston, 1977).

Yahuda, Michael, "Political Generations in China," *The China Quarterly,* 80:793-805 (Dec. 1979).

"Yao kung-k'ai-ti k'o-hsüeh-ti p'ing-Mao" 要公開地、科學地評毛 (Mao should be openly and scientifically assessed) *Cheng-ming* Magazine, Hong Kong, No. 24:4, editorial (Oct. 1, 1979).

37

Building Socialism with Chinese Characteristics

The party conference of December 1978 (Third Plenum, Eleventh Central Committee) was a major landmark in the political and economic life of the post-Mao era. It signaled the rise of Teng Hsiao-p'ing as the paramount leader and adopted the key decisions of accelerating economic development and opening the door to the outside world. Teng became the architect of a new socialist transformation that promised to lift China out of her poverty and developmental stagnation.

The Vision of Teng Hsiao-p'ing

Initially, Teng had no master plan. He had only the pragmatic sense that, in order for any transformation to be successful, socialist construction in China must have Chinese characteristics and that Marxism-Leninism must be integrated with Chinese realities. In this, he was not unlike Mao, who early recognized that a successful Communist revolution in China depended on the same integration. History will take note of Mao's revolution and Teng's construction as two of the most powerful events in China— and to a degree in the world—during the second half of the 20th century. Each will receive proper recognition.

As the year following the conference passed, Teng gradually evolved a clearer vision of his plans for the future of China. In December 1979, when the visiting Japanese Prime Minister, Masayoshi Ohira, asked: "What is the aim of your Four Modernizations?" Teng readily replied that it was to quadruple the then current gross national product (GNP) of $250 bil-

lion to $1 trillion by the end of the century, with a per capita GNP of
$1,000. Later, he clarified his statement by taking into consideration the
inevitable population increase from one billion to 1.2 billion and lowered
the goal of a per capita GNP to $800 by A.D. 2000, while leaving the gross
GNP goal unchanged at $1 trillion. Once this target was reached, China
would have a solid foundation from which further gains could be made. It
could then join the ranks of the more advanced nations within 30 to 50
years. The figure of $1 trillion by A.D. 2000 quickly caught on and became
a national fixation.[1]

Reaching the goal, of course, would require the dedication of the entire
nation; accelerated economic growth; and the absorption of foreign capital,
science and technology, and managerial skills. Hence, it was essential to
adopt the dual policy of economic reform and opening up the country to
the outside world. Since 80 percent of the people lived in the countryside,
invigorating the rural economy and raising farm income and the peasants'
standard of living became the first order of business. Successful rural reforms would be followed by industrial reforms in the urban areas. Meanwhile, a long-range open-door policy was launched to increase foreign
trade; to promote tourism; and to absorb foreign capital, technology, and
managerial skill. It was stressed that the open-door policy was necessary
for China to advance: the closed-door policy from the mid-Ming period to
the Opium War (1840) and the unfortunate period of 1958-76 resulted in
years of ignorance and backwardness.[2]

Teng assured his countrymen that their fear that the Open Door represented a capitalistic erosion of socialism was unfounded. The mainstays of
the Chinese economy would remain socialistic: China would maintain the
socialist principle of distribution, and the state still owned the means of
production and all the basic economic structures. Influx of foreign capital
could not undermine the socialist economic foundation as joint ventures
with foreigners would be at least 50 percent Chinese. To be sure, the open-
door policy would have some negative effects, but it would not lead to a
capitalistic revival. Even with a per capita GNP of several thousand dollars, there should be no fear of the rise of a new capitalist class. Teng
asked, "What's wrong with increasing the wealth of the country and the
people?"[3]

Teng was realistic enough to know that different regions of the country
were endowed with different natural and human resources, so that no two

1. Teng Hsiao-p'ing, *Building Socialism with Chinese Characteristics* (Peking, Foreign
   Language Press, 1985), 35-40, 49-52, 58-59, 70-73.
2. *Ibid.*, 61.
3. *Ibid.*, 62.

areas could develop at the same pace. He would permit some regions and people to get rich first as examples for others to emulate. The city of Soochow, 70 miles from Shanghai, was a source of inspiration, for it had already reached the level of affluence represented by a per capita GNP of $800. Curious about the quality of life there—which might give an inkling of what China would be like in A.D. 2000—Teng made a visit in 1983. He found the local people well clothed, well fed, living in more spacious quarters than other places (on average 20 square meters per person), enjoying television, and willing to invest in local education. The crime rate was low, and the local inhabitants exuded a spirit of well-being and confidence. Their life-style was characterized by a deep love for their native locality and a conspicuous lack of desire to move to large cities such as Peking and Shanghai.[4]

Teng became more and more confident that his dream could be realized. On October 1, 1984, on the occasion of the 35th anniversary of the founding of the People's Republic, he confidently announced to the nation that the annual economic growth rates of 7.9 percent during the period 1979-83 and of 14.2 percent in 1984 surpassed the 7.2 percent needed to quadruple the GNP to $1 trillion by the year A.D. 2000. If the growth rate continued, China could reach the projected target. The World Bank also seemed to agree with this.[5] Teng's pragmatic strategy was "one step at a time; watch out and keep the momentum going."

Agricultural Reform

Traditionally, agriculture was the foundation of the Chinese state and economy. Therefore, it was considered essential to institute radical reforms in agriculture first. For the 20 years between 1957 and 1978, agriculture had been in a sorry state, with the annual growth of grain production only 2.6 percent; and of cotton, 2.1 percent.[6] China had had to import large quantities of grain to feed its growing population. The rural economy was listless, if not lifeless. The standard of living on farms had not improved for two decades, and a strong incentive to work was almost nonexistent.

4. *Ibid.*, 53.
5. *China: Long-Term Development Issues and Options* (Washington, D.C., The World Bank, 1985). The World Bank believes the goal of quadrupling the GNP is possible if China (1) invests efficiently in building its infrastructure at a rate of 30 percent of national income, (2) makes reasonable improvement in the use of energy and raw materials, and (3) keeps its population to no more than 1.2 billion by A.D. 2000.
6. Nicholas R. Lardy, "Overview, Agricultural Reform and the Rural Economy," *China's Economy Looks Toward the Year 2000*, Joint Economic Committee, Congress of the United States (Washington, D.C., 1986), Vol. 1, *The Four Modernizations*, 325, 331.

Everybody knew that the most serious obstacle to the re-invigoration of the rural economy was the commune system. Since it was the kingpin of the Maoist rural economic structure, nobody dared criticize or tamper with this sacrosanct institution as long as Mao lived. Now it was recognized that nothing less than a fundamental reform could inject new life into the stultified rural economy, rekindle enthusiasm for work, release the vast potential of the peasant masses, and improve their standard of living.

The party conference of December 1978 adopted the drastic decisions of using greater material incentives and loosening the control mechanisms that had heretofore constrained growth in the rural sector. In the months that followed, discussions between local and central government leaders led to the adoption of what was described as the "Responsibility System," or *pao-kan tao-hu,* meaning literally, "full responsibility to the household." Under this system, land remained public, but each household received a plot for cultivation and negotiated a contract with the commune production team or economic cooperative. The contract specified quantities of crops to be planted and the quota of output to be handed to the production team or cooperative as payment for the use of the land. This payment also covered such common expenses as irrigation fees, health care, and welfare. Each household had full control of its labor resources and could either keep or sell in the free market the products that exceeded the contracted quota. The farming household assumed full responsibility for the entire process of production—from the selection of seeds, choice of fertilizer, labor allocation, work schedule, and preparation of soil, all the way to the final product.

The Responsibility System, which began in 1979, gradually spread through the provinces in 1980-81, and the process accelerated during the 1982-83 period, so that by 1984 some 98 percent of farm households came under it. A plot of land was initially assigned to each household for a season or a year. But later, in 1984, the assignments were extended to 15 years to encourage long-term planning and investment in the land. The longer contracts were in consideration of such agricultural concerns as the intensity of farming, choice of crops—especially slow-maturing fruit trees—and development of soil fertility.[7] More recently, the land contract was made inheritable, as was the house on the land, to encourage further long-range investment. However, cancellation of contracts could occur in cases of nonfulfillment of the original terms.

The government also encouraged rural workers to specialize in crops,

7. Frederick W. Crook, "The Reform of the Commune System and the Rise of the Township-Collective-Household System," *China's Economy Looks Toward the Year 2000,* Vol. 1, 362-63.

livestock, forestry, fisheries, or other diverse sidelines. This was in contrast to Mao's heavy emphasis on grain production. Gradually, "specialized households" (*chuan-yeh hu*) emerged, which did not till the land but engaged exclusively in noncrop production. Between the specialized households and the ordinary farming households were the "key households" (*chung-tien hu*), which tilled the land but were primarily occupied with noncrop activities such as fishery and animal husbandry. These two types of households amounted to some 24 million by October 1984, or 13 percent of total rural households.[8]

The expansion of the *Pao-kan* Responsibility System increasingly superseded the functions of the commune until the latter was all but extinct. Today only a few models are left as historic landmarks or as showpieces for foreign visitors and students of Chinese social and economic history. In 1984 a further significant restructuring came when the individual household was allowed to transfer the contracted land to another household with the approval of the local cooperative unit. This was instituted to allow for times when a household might be beset with sickness, death, or other problems that would prevent it from utilizing its parcel of land. In 1987, the Thirteenth Party Congress further liberalized the sale of right to land utilization by one household to another. Theoretically, it is not inconceivable that one household could have acquired the right to cultivate the contracted land of two or three or more neighbors. Some critics pointed to this possibility as a case of incipient capitalism, but such occurrences were rare. In any event, land remained public and chances for renewed capitalism seemed small.

As a result of the agricultural reforms, both yield and productivity rose sharply. In 1987, rice and wheat yields had risen 50 percent over those obtained under the commune system. More importantly, the farmer spent only an average of 60 days a year on the crops, compared with 250 to 300 days a year in the field in the days of the farm collectives. The time saved was spent on sideline activities aimed at profit. Cash income quadrupled and the standard of living vastly improved. This newfound prosperity was soon reflected in new brick houses; new televisions and furniture; and new, more colorful clothes for the participating households. In Szechwan and many other provinces, the contracted quota accounted for approximately one-sixth of the total output, and although most plots were less than one acre in size, there was enough food raised for each household. The farmers led an ownerlike life, and quite a few of them earned incomes in excess of 10,000 yuan annually (*wan-yuan hu*).[9]

8. Crook, 370.
9. *Los Angeles Times*, Nov. 25, 1987.

Thus, the dismantling of the commune was not abrupt but took place over a period of five years. Now a new type of township-collective-household rural structure emerged, which assumed some of the former functions of the commune, but, with a clear-cut division of labor. The township concerned itself with government and administrative affairs, the party committee did party work, and the economic collectives performed such functions as signing the responsibility contracts with the individual households. Not infrequently, the former commune production brigades and teams became new economic cooperatives, and many former commune departments were transformed into "village and township enterprises" engaged in production, processing, transportation, marketing, and service industries.[10]

The results of the reforms were nothing less than spectacular. The growth rate in annual grain production rose from 2.1 percent during the period 1957-78 to 4.9 percent during the period 1979-84, with record harvests of 407 million metric tons in 1984, compared with the previous 305 million metric tons. Per capita grain production also surpassed the high 302 kilograms achieved in 1957 and even the pre-Liberation production peaks. Total production of crops and livestock increased 49 percent between 1978 and 1984.[11]

Major Farm Products (in million tons)

	1952	1957	1965	1978	1980	1984	1987
Grain	163.42	195.05	194.53	304.77	320.56	407.31	402.41
Cotton	1.30	1.64	2.09	2.16	2.07	6.25	4.19
Oil-bearing Crops	4.19	4.19	3.62	5.21	7.69	11.91	15.25
Sugarcane	7.11	10.39	13.39	21.11	22.80	39.51	46.85

Sources: Xue Muqiao (ed.), *Almanac of China's Economy, 1985/86* (Hong Kong, 1986), p. 19; State Statistical Bureau figures, Feb. 23, 1988, *Beijing Review*, March 7-13, 1988.

The remarkable advance in agricultural output changed China from being a net importer to an exporter of grains, soybeans, and raw cotton. China achieved a trade surplus of $4 billion in agricultural products between 1980 and 1984, the largest gain in 35 years.[12] Per capita farm income rose from Y134 in 1978 to Y310 in 1983 and Y463 in 1987. Success came not only through hard work and good planning, but also from the higher

10. Crook, 364-65, 368-69.
11. Frederick M. Surls, "China's Agriculture in the Eighties," *China's Economy Looks Toward the Year 2000*, Vol. I, 338.
12. Lardy, 327.

procurement prices the government paid for farm products (a dramatic 50 percent increase between 1978 and 1983), as well as noncrop sideline income from livestock, as well as fishery, and forestry enterprises. The success was all the more remarkable in light of a reduction of some 50 percent in state agricultural investment between 1978-79 and 1981-82. However, the slack was partially taken up by an increase in credit offered by the People's Bank and by private investments in machine tools, tractors, and housing, with housing totaling Y15.7 billion in 1982 and Y21.4 billion in 1983.[13]

The dismantling of the commune resulted in a progressive neglect of large projects formerly serviced by the commune, such as the mechanized pumping of the irrigation system and the use of heavy tractors for preparation of the land. Social services, health care, and primary education also suffered. Moreover, the state incurred a heavy new burden because it paid higher prices for farm products but could not raise commodity prices in the city for fear of inflation and public anger. State subsidies to cereals and oils grew from Y4 billion in 1974 to Y20 billion in 1983.[14]

There were other agricultural problems as well. First, there were limits to what hard work and material incentive could achieve; beyond a certain point, saturation was reached and no amount of hard work or willpower would make a difference. Greater government investment in agriculture was needed to improve productive power, but government finances were tight, and the amount budgeted for agricultural investment actually decreased from 13.3 percent of the national expenditure in 1978 to 6.8 percent in 1983 and to 5.6 percent in 1985. Secondly, decline in collectivization resulted in deferred maintenance of irrigation, reduction of mechanization, and greater use of low-grade chemical fertilizers. The ability of the farm units to deal with natural calamities had been vastly reduced. Thirdly, low prices for grains resulted in a low profit margin compared with other production activities. In 1985 a farmer in the relatively well-off Feng District of Kiangsu Province earned Y650 annually, compared with Y2,375 for an animal husbandry or fishery worker, Y4,199 for an industrial worker, Y4,033 for a construction worker, and Y4,762 for a communications worker. On the average, a nonagricultural worker earned 4.1 times more than a grain producer.[15] It is little wonder that a number of

---

13. *Ibid.*, 328-330. Procurement prices increased 22.1 percent in 1979, 7.1 percent in 1980, 5.9 percent in 1981 and 2.2 percent in 1982. See Nai-Ruenn Chen and Jeffrey Lee, *China's Economy and Foreign Trade, 1981-85* (U.S. Dept. of Commerce, 1984), 6.
14. Lardy, 333.
15. *Ching-chi Hsüeh Chou-pao* (Economic Weekly), Shanghai, July 20, 1986. Article on food problems.

grain producers turned to other lines of work for support or turned farm work into a mere sideline activity. Fourthly, loss of productive farm land continued owing to national appropriation of such land for industrial uses; for new farm housing; and for the development of timber, livestock, and fisheries. In 1985 alone the loss amounted to 15 million *mou* (2.5 million acres) of land. Fifthly, such perennial problems as high illiteracy; lack of agricultural technicians (1 in 4,000 farm households); and periodical flooding, drought, and fire persisted.

A much more fundamental challenge was the burgeoning population explosion, which threatened to consume most, if not all, of the increased agricultural and industrial production, thereby neutralizing the benefits of the reforms. To check the population from getting out of control, the government initiated a "one-child-per-family" policy, supported by material rewards for those who observed it (i.e. job security, promotion, favorable housing and school assignments) and by penalties for those who did not (demotion, monetary fines, food rationing, etc.). Given vast publicity during 1982, the policy was a success in the cities but less so in the countryside. The farmers still preferred male offspring as potential helpers and heirs and would often resort to female infanticide in order to win a second chance for a male issue. More recently, economic affluence on the farm led many farmers to defy the governmental orders by deliberately having a second or third son while paying the fines willingly. In China today, children from one-child families have often become so spoiled by doting parents and grandparents that they behave like "little emperors." These children, usually the center of a family's affection, attention, and social life, are increasingly becoming more steeped in egoism and individualism than in Marxist-Maoist ideology. It will be of great interest to follow the development and maturation of this new generation of Communist brats.

In spite of all these problems, grain production in 1986 reached 390 million tons, an increase of 10 million tons over 1985; and average farm income reached Y425 in 1986, a 7 percent increase over the previous year. In 1987 grain production rose further to 402.41 million tons, an increase of 2.8 percent.[16]

All in all, the first five years of agricultural reform released such vast hidden potential in the agricultural sector that the government was encouraged to tackle industrial reform in the urban areas as well. Here the problems were much more complicated.

16. State Statistical Bureau, Peoples Republic of China, *Statistics for 1987: Socio-Economic Development*, issued Feb. 23, 1988. In *Beijing Review*, March 7-13, 1988.

Industrial Reform

Chinese industrial growth had averaged a respectable 9.8 percent annually from 1952 to 1983.[17] But efficiency, productivity, and work incentive were all hampered by numerous "irrational practices," as Western economists would call them. The industrial structure built by Mao in the early 1950s was modeled after the Soviet system, which was characterized by central planning and emphasis on the development of heavy industry. As the owner, operator, and employer, the state planned, directed, and funded all public enterprises. The state provided land, plants, equipment, basic materiel, working capital, managers, and everything else throughout the entire process of production. It also set the prices of the finished goods regardless of their cost and quality. There was no recognition of the "law of value" or the principle of supply and demand. State enterprises were required to remit to the central government all their profits and depreciation funds.

Under this system, enterprises received state support regardless of their performance records, and the workers received their standard wages regardless of the quality of their work. A saying went, "Every enterprise eats from the Big Pot of the state, and every worker eats from the Big Pot of the enterprise." The system worked at first because of the momentum of revolution, patriotism, and personal dedication to the building of a new socialist society. But as time went on, it became clear that the merits of an enterprise or of a worker were immaterial. The reward would remain the same in any case: the plant would receive the same allotted funding; and the worker, the same low pay. The socialist boast of full employment virtually guaranteed lifelong job security, and dismissal of indolent workers was well nigh impossible. Attempts by a manager to dismiss a worker would likely result in the disgrace of the manager for unsympathetic leadership rather than the dismissal of the worker. Similarly, penalties for inefficiently run state enterprises or debt-ridden plants were rare or unheard of. In 1979-80 roughly 25 to 30 percent of state enterprises operated at a loss.[18] Pricing structure was even more "irrational": the state set the prices of all commodities without regard to production cost or quality. Not infrequently, an article cost more to produce than it could be sold for, and a

17. Robert Michael Field, "China, The Changing Structure of Industry," in *China's Economy Looks Toward the Year 2000,* Selected Papers submitted to the Joint Economic Committee, Congress of the United States (Washington, D.C., 1986), Vol. I, *The Four Modernizations,* 505. (Hereafter to be cited as *China's Economy*).

18. Nai-Ruenn Chen and Jeffrey Lee, *China's Economy and Foreign Trade, 1981-85,* U.S. Department of Commerce (Washington, D.C., 1984), 13. (Hereafter Chen and Lee).

low-quality product would sell for more than a similar item of higher quality. Irrational as this system was, the country had become accustomed to it over the previous 30 years. Reform of any part of it would disrupt the balance that had existed in the vast interlocking network of planning, management, production, marketing, and pricing. Millions of cadres were involved in the process, and any change in any part of it would adversely affect their lives. The worst fear was price decontrol, which raised the dreaded specter of inflation. The central government wanted to avoid any action that would arouse public anger or unrest. It took one step at a time, tested the reaction, evaluated the results, and then either proceeded or took a step back.

THE FIRST PHASE. The spirit of the industrial reform during the period 1978-84 was to rekindle work enthusiasm, to unleash the full potential of the workers, to "enliven" the industrial structure, and to raise the living standard. The method used was none other than material incentive—the most disdained of values in the Maoist revolutionary days.

Beginning in the period 1978-79, various profit retention schemes were experimented with in Szechwan and other selected areas. When improvements were achieved, they were extended throughout the country. The heart of the reform was the institution of an Industrial Responsibility System whereby a state enterprise signed a "profit and loss contract" (*yin-k'ui pao-kan*) with its supervisory body, agreeing to remit a quota of profit to the state but retain a share of the "basic profits" above the quota. By 1980 some 6,600 state enterprises had come under this system. Profits so retained could be used for bonuses, employee welfare benefits, and further industrial innovations.[19] It was decreed that more work would yield more pay and that different types of work (skilled vs. unskilled, intellectual vs. manual) should command different renumerations. Work enthusiasm returned overnight.

In the period 1981-82, the profit retention system was refined to allow a larger share of profit above the quota for the enterprises and also partial retention of budgetary savings through reduced losses. The retention rates averaged 10 percent for high-profit industries, 30 percent for low-profit ones, and 20 percent for all others. By the end of 1982 all industrial enterprises had come under the Responsibility System. They were made responsible for all their economic decisions, as well as their return of a profit or loss. Plant managers could hire and fire employees, determine wages and bonuses, and set prices within a state-approved price range. But the man-

19. Barry Naughton, "Finance and Planning Reforms in Industry," *China's Economy*, I, 608.

agers themselves were no longer given lifelong tenure. Effective January 1, 1985, they were appointed for four-year terms, renewable up to three times.[20]

One immediate repercussion of the Responsibility System was the vast reduction of funds the state received from the enterprises with the corresponding increase in funds kept by the enterprises and the localities, which were used for capital construction without central control or coordination, to the tune of Y42 billion by the end of 1982.[21] The state was dealt the double blow of budgetary deficits and loss of control over local investments. A construction boom led to shortages of building materials and inflation.

On June 1, 1983, the government substituted a new income tax for profit remission (*i-shui tai-li*). Large and medium-sized enterprises were required to pay 55 percent of their profit as tax, and small enterprises paid according to an eight-grade progressive tax schedule, thus severing the direct relationship between the state enterprises and the government business bureaus.[22] In addition, three different kinds of taxes were to be phased in gradually and their relative share of the total profit was as follows: (1) product tax, 40 percent of profit; (2) income tax, 33 percent; and (3) adjustment tax or a surtax for the better developed coastal areas, 12 percent. The amount of profit now retained by the state enterprises amounted to some 15 percent. In addition, there were two other exactions imposed on a local level: a capital user's fee and a municipal tax for the use of land, roads, housing, and urban services.[23]

The introduction of income tax in a Communist system was an epochal event. Formerly, the plants, as public properties, paid no rent on land, little or no interest on working capital, and little or no amortization on fixed capital investments provided by the state. They retained high profits even though many operated inefficiently or in the red. Now industrial profits became a source of tax revenue, and the income taxes had the temporary effect of adjusting price distortions—a kind of substitute for price reforms.[24]

These new measures had indeed brought improvement to the industrial sector in the form of a higher standard of living, new business bonuses, and a construction boom; but there was little evidence of improvement in the efficiency of the enterprises. The performance of Chinese industry had not become more effective as first expected.[25] In fact, in 1982 some 30

20. Christine Wong, "The Second Phase of Economic Reform in China," *Current History*, September 1985, 261, 278.
21. Chen and Lee, 13.
22. Field, 532.
23. Naughton, 612.
24. *Ibid.*, 611.
25. *Ibid.*, 608-9.

percent of enterprises still operated in the red with a loss of Y4 billion, or 4 percent of the state budget revenue. Some 42,000 industrial enterprises were consolidated or amalgamated between 1983 and 1985.[26]

Perhaps the most visible result of the economic reforms was the mushrooming of private businesses and free markets in both rural and urban areas. Private businesses grew in number from 100,000 in 1978 to 5.8 million in 1983 and 17 million by 1985, with some making impressive profits in the capitalist fashion.[27] The service industry, considered a tertiary sector, also made great strides. Its share in the GNP grew from 18.7 percent in 1980 to 21.3 percent in 1985, and as of 1985, it employed 73.68 million people.[28] Rural free markets numbered some 40,000 by 1985, and urban free markets totaled some 3,000. Together they accounted for 6.6 percent of total retail sales in 1978, 9.5 percent in 1979, 10.2 percent in 1980, and 11.4 percent in 1981. These free markets and private businesses constituted a lively sector in the vast sea of state-owned enterprises.[29]

The most sensitive issue involved in the urban economic reform was the introduction of a realistic pricing system that would ultimately obviate the need for government subsidies for consumer goods. Gingerly the government lifted price controls on selected articles to minimize the effect on the market. Between 1979 and 1982, prices of coal, iron ore, cigarettes, and liquor increased whereas those of machinery and tires decreased. In 1983 price changes affected 100,000 items with a total value of Y40 billion, including increases in the prices of chemical products by 20-50 percent, in railway freight transport by 20 percent, and in light industry consumer durables such as electric fans and color television sets by 8-17 percent. People complained of inflation, which outpaced both wage increases and cost-of-living adjustments. Inflation was officially put at 4 percent in 1979, 6 percent in 1980, 2.4 percent in 1981, and 1.9 percent in 1982,[30] though the unofficial estimates ranged from 15 to 20 percent or more annually. In October 1984, when the party announced its accelerated program of urban reform, people flocked to the banks to withdraw money to purchase merchandise suspected to be in short supply in an effort to beat price increases. Confusion, fear, and disbelief were prevalent. Having experienced the hardships caused by the hyperinflationary period of 1945-49, the Chinese dreaded any sign of its recurrence.

26. Chen and Lee, 14.
27. Sung Ting-ming, "Review of Eight Years of Reform," *Beijing Review*, Dec. 22, 1986, 15.
28. Jung-hsia Li, "Tertiary Industry Takes Off in China," *Beijing Review*, Feb. 9, 1987, 18-19.
29. Chen and Lee, 15.
30. *Ibid.*, 8.

The urban reforms of 1979-84 were not intended to create a free market system. They were only a patchwork, designed to amend the inefficient old structure with some economic realism and market mechanisms in hopes of breathing life into an otherwise ossified body. In the process, the government backed away from the bureaucratic command economy characterized by central planning and directives (*chi-lin-hsing chi-hua*) to a position of planning through guidance (*chi-tao-hsing chi-hua*). By 1984, only 30 to 40 percent of industrial production could be attributed to central planning measures, 20 percent to the market economy, and 40 to 50 percent to locally planned or guidance-planned output.

THE SECOND PHASE. Encouraged by the success of agricultural reform and progress made in the industrial sector, the party, on October 20, 1984, passed a new "Resolution on the Reform of the Economic System" to accelerate the pace of urban reform. This was a curious and interesting document in that it was not a blueprint for reform but an optimistic statement of intent and principles for the guidance of the 44 million party members. It exuded confidence of future success, coming as it did in the wake of a string of good news: record grain production (407 million tons), unprecedented foreign reserves (US $20 billion), and exceptional performance by Chinese athletes at the Summer Olympics in Los Angeles (32 gold, silver, and bronze medals). Moreover, the gross industrial and agricultural output surpassed the grand psychological mark of Y1 trillion for the first time in Chinese history.[31]

There was yet another favorable development that bolstered China's new confidence. This was the successful negotiations with Britain for the return of Hong Kong. For more than two years the two countries had been discussing the future of three pieces of territory that China had ceded and leased to Britain in the 19th century under the unequal treaties: (1) the island of Hong Kong, ceded by the Treaty of Nanking in 1842; (2) the southern part of the Kowloon peninsula and Stonecutters Island, ceded by the convention of Peking in 1860; and (3) the New Territories and 235 nearby islands which comprise 92 percent of the land area of Hong Kong, leased in 1898 to Britain for 99 years until June 30, 1997.

In view of the impending expiration of the lease, both Britain and China desired an amiable negotiated settlement. In this, the British hoped to retain some administrative role beyond 1997 so as to ensure Hong

---

31. Chao Tzu-yang, "Tang-ch'ien ti ching-chi hsin-shih ho ching-chi t'i-chih kai-ke" (On the economic conditions before us and the economic institutional reforms), Report before the Sixth National People's Congress, March 27, 1985 (Hong Kong, 1985).

Kong's continued stability and prosperity. The Chinese, on the other hand, insisted on the complete restoration of sovereignty over all three territories. After British Prime Minister Margaret Thatcher's visit to Peking in September 1982, substantial progress was made in accommodating the Chinese wish. On September 26, 1984, an agreement on the future of Hong Kong was reached, under which Hong Kong would become a Special Administrative Region of China after June 30, 1997, but would still retain a high degree of autonomy in its legal, educational, and, most importantly, its economic and financial structures, including its free enterprise system.[32] China pledged not to interfere with Hong Kong's socioeconomic systems for 50 years beyond 1997, in effect, creating a "One Country, Two Systems" arrangement, which Teng Hsiao-p'ing also wished to someday apply to Taiwan.

The initialing of the agreement in October 1984 and its formal signing in December by Margaret Thatcher and Premier Chao Tzu-yang marked an epoch-making triumph for China. It signaled an end to the last vestiges of foreign imperialism in China. This triumph was all the more satisfying because the British Prime Minister twice came to China in connection with the settlement, and, for the first time in history, the British Crown was scheduled to visit China.

It was in this state of euphoria that the second stage of urban reform was launched. What was wrong with the existing economy? Premier Chao stated that it was stultified because the government and economic enterprises were not treated as separate entities and that the former controlled the latter too tightly. Disregard of proper interaction between commercial production, the law of value, and market forces caused imbalances that had to be redressed. In distribution there was too much emphasis on "averageism" (*p'ing-chün chu-i*), resulting in everyone eating from the same "Big Pot" of the state and nobody wanting to work hard. The enterprises and the workers had lost their initiative and creativity, lapsing into a general state of paralysis. On top of this malaise was the "leftist" tendency, in force since 1957, to deprecate any effort to develop a commodity economy as a revival of capitalism. Chao admitted that it would require a bold liberation of thought to correct such ossified thinking: "The basic function of socialism is to develop the productive power of the society, ceaselessly to increase its wealth, and to meet the increasing material and cultural needs

---

32. For details of the agreement, see *A Draft Agreement between the Government of the United Kingdom of Great Britain and Northern Ireland and the Government of the People's Republic of China on the Future of Hong Kong.* Hong Kong, Sept. 26, 1984. It should be noted that Portugal also reached an agreement with China in Oct. 1986 on the return of Macao on Dec. 20, 1999.

of the people. Socialism wants to end poverty: pauperism is not socialism."[33]

Chao's method of injecting life into the economy was to loosen state control of the large and medium-sized enterprises. Public ownership need not be equated with direct state control: ownership and management were two separate functions. Within the framework of state governance, Chao felt, enterprises should be granted enough autonomy to make their own decisions about supplies, sales, capital utilization, hiring and firing, salaries, wages, and bonuses and about the prices of the finished products as well. An enterprise should be made to function legally as an individual person, responsible for its own profit and loss.

> We must break loose from the traditional concept that a planned economy and a commodity economy stood opposed to each other and recognize clearly that, under socialism, the planned economy must spontaneously rely on and utilize the law of value to build a planned commodity economy on the basis of public ownership. The full development of a commodity economy is an indispensable step in the development of a socalist economy. It is a necessity in the modernization of our economy. The difference between a socialist and a capitalist economy does not lie in the existence of a commodity economy or in the functioning of the law of value, but in the ownership system, in the existence or not of an exploiting class, and in whether the workers are the masters of the house.[34]

Chao considered the gradual lifting of price control to be the heart of urban economic reform, for it would enable the state to withdraw its subsidies and allow prices to float according to the law of value and the market forces of supply and demand. But fluctuation in prices had to be kept within limits; and people's salaries and wages, adjusted according to inflation. Profit in an economic enterprise should come from better management, not from individually determined price increases, which could only distort market conditions. To fight the debilitating effects of "averageism," Chao reaffirmed the principle of reward based on work: "Those who labor more shall receive more and those who labor less shall receive less." The much abused practice of equal pay for all was a chief stumbling block in increasing the productive power of society.

Chao declared that to equate "common wealth" with "equal wealth for all at the same speed" was not only impossible, but would lead to common poverty. Certain regions, enterprises, and people should be allowed to grow

33. *Chung-kung Chung-yang kuan-yu ching-chi t'i-chih kai-ke ti chu-ting* (Third Plenum, Twelfth Central Committee, Chinese Communist Party, Resolution on the Reform of the Economic System, Oct. 20, 1984). (Hong Kong, 1984), 4-6, 17. Tr. mine.
34. *Ibid.*, 9-10. Tr. mine.

rich first so as to exert a ripple effect on others. But China would not allow exploitation by a small number of people to plunge the majority into poverty.[35]

Finally, Chao approved the continued growth of private enterprise to supplement the public ownership system and also the leasing or contracting of small and medium-sized state enterprises to private operation to enrich the variety of economic life. Such developments would not jeopardize the socialist foundation, but were seen as necessary to the progress of socialism.

Even as the resolution was being passed, the economy was charging ahead. The gross output value of agriculture and industry increased at an annual rate of 10 percent between 1978 and 1986, and the national income grew at a rate of 8.7 percent annually. Particularly impressive were the sharp increases in capital construction outside the state budget, from 16.7 percent of total investment in 1978 to 57 percent in 1984. The construction boom was evident everywhere. Capital investment rose 25 percent in 1982, 23.8 percent in 1984, and 42.8 percent in 1985. The level of investment was the highest since the disastrous Great Leap of 1958-60.

However, this pace was clearly unsustainable, creating with it shortages of construction materials, waste, confusion, and inflation. With such high capital investment, industrial growth naturally was fast, at 14 percent in 1984, and 18 percent in 1985; but only the most strenuous effort kept the 1986 rate to 9.2 percent. In 1987, the rate rose again to 16.5 percent, achieving a total industrial output value of almost Y1.4 trillion.[36]

The overheated economy created many adverse effects on the long-term interests of the state. The following were among the most obvious:

1. State budgetary deficits between 1978 and 1985 came to Y100 billion, largely owing to excessive capital investment and large subsidies.
2. Large trade deficits amounted to US$ 28 billion in 1985 and 1986.
3. High inflation rates of 12.5 percent in 1985, 7 percent in 1986, and 8 percent in 1987 prevailed although unofficial figures put the rates between 15 and 20 percent annually.
4. Widespread economic crime and corruption, especially among the children and relatives of high cadres who exploited their special positions to practice favoritism and other business irregularities.
5. Fast growth took place in durable consumer goods in 1986 such as

35. *Ibid.,* 17-18. Tr. mine.
36. Chu-yuan Cheng, "China's Economy at the Crossroads," *Current History,* Sept. 1987, 272; State Statistical Bureau, Peking, Feb. 23, 1988. *Beijing Review,* March 7-13, 1988.

washing machines (9 million units), electric fans (33 million), and refrigerators (2.85 million).

6. An energy bottleneck was caused by the slow growth of primary energy supplies (2.9 percent in 1986), which could not sustain the demands of the rapidly expanding industrial growth (9.2 percent), forcing many plants to operate only four days a week.

7. There was a rapid development of village industry to 820,000 units in 1985, with a total value of Y137.5 billion, accounting for 15.7 percent of total industrial output. But the quality of rural industrial products was frequently low.

8. Cultivated land was lost at the rate of 20 million *mou* or 3.29 million acres annually owing to increased industrial use, new housing, and an annual population growth of 14 million.[37]

Facing the serious problems of rapid industrialization, conservative leaders clashed with the pragmatists over the direction, speed, and scope of economic reform, and the former attacked the open-door policy as a source of foreign "spiritual pollution" of Chinese life. The clash resulted in the retrenchment policy of January 1986, which highlighted four key concepts: (1) to *consolidate* the gains already made and to solidify the foundation of reforms, (2) to *digest* the changes made necessary by price reforms and wage adjustments and to tackle the problems of reform in line with the

Output of Major Industries

	1952	1957	1965	1978	1981	1984	1987
Coal (100 mill. tons)	0.66	1.31	2.36	6.18	6.22	7.89	9.20
Crude Oil (1 mill. tons)	.14	1.46	11.31	104.05	101.22	114.61	134.00
Natural Gas (100 mill. cubic meters)	0.08	0.7	11.00	137.30	127.40	124.30	140.15
Electricity (bill. kwh.)	7.3	19.3	67.6	256.6	309.3	377.0	496.0
Rolled Steel— Final Products (mill. tons)	1.06	4.15	8.81	22.08	26.70	33.72	43.91
Steel (mill. tons)	1.35	5.35	12.23	31.78	35.60	43.47	56.02
Pig Iron (mill. tons)	1.93	5.94	10.77	34.79	34.17	40.01	54.33

Sources: *Almanac.* 26; State Statistical Bureau, Feb. 23, 1988, *Beijing Review*, March 7-13, 1988; *Monthly Bulletin of Statistics*, China, March 1988.

37. Chu-yuan Cheng, 272-73.

financial and physical capacities of each unit, (3) to *supplement* and amend the imperfect and unhealthy links in the chain of reform so as to improve coordination, and (4) to *improve* the macroeconomic controls so as to achieve a better balance between supply and demand.

The retrenchment policy, however, could not stop the momentum of economic expansion. In 1987 the agricultural output value climbed 4.7 percent to Y444.7 billion, and the industrial output grew 16.5 percent to Y1,378 billion—both new records. But the growth was uneven, as the energy and transportation sectors continued to lag behind, with coal production registering only a 2.9 percent increase and crude oil a 2.6 percent increase. The energy bottleneck would constrain the economic growth for years to come and inhibit a balanced development. Nevertheless, the general trend toward continued growth would persist, if only haphazardly.

The Open-Door Policy

During the first decade of the People's Republic (1949-59), China maintained diplomatic and commercial relations only with the Soviet Union and the Eastern European satellite states. There was no trade between China and the United States. After the Sino-Soviet split in 1960, China became extremely isolated in the international community, simultaneously facing both the Soviet Union and the United States as potential enemies. It was not until after the visit of President Richard Nixon to China in 1972 that limited commercial relations began. In 1972 American-Chinese trade amounted to only $92 million, but it rapidly grew to $1,189 million in 1978, $5,478 million in 1981, $8 billion in 1986, and $13.5 billion in 1988, amounting to approximately 10 percent of China's total foreign trade.[38]

The rapid growth of foreign trade after 1978 was the result of the new open-door policy adopted by the party in December 1978 (Third Plenum, Eleventh Central Committee). It was a complete reversal of the Maoist policy of seclusion that had been in force for the 20 years between 1958 and 1978. Teng Hsiao-p'ing and his pragmatic followers realized that China could not develop in isolation and that she must import foreign science, technology, capital, and management skills in order for her modernization to succeed.

Japan, Hong Kong, the United States, and West Germany were China's

38. U.S. figures do not agree with Chinese data because the U.S. used the "country of origin" method in its calculation whereas the Chinese used their customs figures, which often include insurance and freight fees. In 1987 the Chinese figure for the total U.S.-China trade came to $8.5 billion, compared with the U.S. figure of $10.5 billion.

largest trading partners. In 1983, Japan's share came to $9,764 million; Hong Kong's, $8,341 million; the United States', $4,425 million; and West Germany's, $1,743 million. The nature of the trade goods China imported and exported had changed substantially. Initially, Chinese imports consisted largely of raw materials such as agricultural products (primarily grains), synthetic fibers, lumber, and chemicals. But later, as China became almost self-sufficient in agricultural production, the focus of importation shifted to industrial machinery, finished manufactured goods, technology, office equipment, commercial aircraft, and services.

Japan enjoyed a special relationship with China because of her geographical proximity and a certain degree of cultural affinity. This gave Japan a deeper understanding of both Chinese psychology and her immediate economic needs, much more so than other foreign traders might have had. Japan's technological and financial successes in the world markets enabled her to offer concessional loans, grants, and preferential tariffs under the Generalized System of Preferences (GSP). Hong Kong also had a special place in China's opening world trade: it served as a link between China and the outside world. The United States and West Germany did not share these advantages, but the Chinese traditionally had great respect for American and German aircraft, machinery, and scientific products.

A prime objective of the foreign trade program was to generate sufficient foreign exchange to help finance modernization. To improve their competitive edge, the Chinese diversified their products, raised quality levels, devalued the yuan, and eagerly learned international business practices. In purchasing, they adhered strictly to the three criteria of good prices, good quality, and financing arrangements on concessionary terms. They wanted transfer of the latest technology; but they would also accept less sophisticated plants at bargain prices, as in the cases of the purchase of an older steel mill, a semiconductor line, and a textile mill in the United States during the 1982 recession.[39]

Through strict control of foreign currency, export expansion, and import restraint, China steadily built up a foreign currency reserve. When China encountered protectionism in the West against her textiles, she opened new markets in the Middle East, Latin America, Eastern Europe, and the Soviet Union, although the volume of trade in these areas remained limited. In early 1981 China crossed the line from being a debtor nation to being a creditor nation. By the end of 1983 her foreign currency reserve reached an unprecedented $20 billion, the tenth largest in the world and ahead of Britain's $18.2 billion. Part of the foreign earnings

39. Helen Louise Noyes, "United States-China Trade," *China's Economy*, II, 343.

were used for domestic infrastructural projects in the long neglected areas of energy, transportation, communication, and light industries.[40]

One of the persistent difficulties with China's foreign trade system was the irrationality of her domestic pricing system, which made necessary government subsidies for both imports and exports. Domestic prices in China were out of line with world prices. Though it was profitable for China to export primary goods and to import manufactured goods, it was highly unprofitable to do the opposite at the official exchange rate of Y2.8 to the dollar. Before 1981, Chinese exports generally created losses and imports gains—a sure sign of overvaluation of the yuan, creating the need for government subsidy for exports. To counteract these pricing distortions, China devalued her currency to Y3.7 to the dollar in 1984, but it resulted in increased subsidies for many imports, again producing a loss.[41]

An unexpected source of foreign trade profit came from arms sales to the states of the Middle East through North Korea and Egypt, rising from almost nothing in 1980 to $1.5 billion in 1983 and $5 billion in 1987. Most famous among these arms were the Silkworm short-range missiles sold to Iran, and the intermediate range (2,000 miles) CSS-2 missiles sold to Saudi Arabia. Arms were also sold to Iraq, Jordan, Egypt, Syria, and Israel.[42] Tourism provided another source of income, which came to $1.84 billion in 1987, up 20.3 percent from 1986.[43]

Except for the retrenchment years 1981-83, when importation of capital goods declined, Chinese foreign trade grew annually by 20 percent in imports and 10 to 15 percent in exports.[44]

To attract foreign capital and investments, China adopted a number of measures to improve the investment climate. The following were some of the more salient steps taken:

> 1. Opening four "Special Economic Zones" in 1979 (with preferential treatment) in Shenzhen, Chu-hai (on the opposite banks of the Pearl River estuary close to Canton), Swato (on the northern Kuangtung coast), and Amoy (on the southern Fukien coast). These were not "Export Processing Zones" (EPZ) as in Taiwan, but "laboratories" for transforming the Chinese economy.[45]

40. John L. Davie, "China's International Trade and Finance," *China's Economy Looks Toward the Year 2000*, II, 311-12, 323.
41. *Ibid.*, 319.
42. *Los Angeles Times*, May 4, 1988.
43. State Statistical Bureau, Beijing, Feb. 23, 1988.
44. Davie, 318.
45. Victor C. Falkenheim, "China's Special Economic Zones," *China's Economy*, II, 348-50; Y. C. Jao and C. K. Leung (eds.), *China's Special Economic Zones: Policies, Problems, and Prospects* (Hong Kong, 1986).

2. Opening fourteen coastal sites and Hainan Island in 1984 to foreign investment, with preferential terms on taxes and import duties.
3. Hosting international conferences to advertise projects that needed foreign advice, capital, equipment, management, and marketing.
4. Permission given to local authorities to arrange foreign investments without central government approval, which resulted in a burgeoning of imports of foreign supplies—steel, nonferrous metals, lumber, plastics, with a substantial outflow of foreign reserve.
5. Passing laws and regulations on taxation, liability, patent protection, and foreign trademarks.
6. Clarification of arbitration procedures, labor compensation, and repatriation of foreign profit from China.[46]

China's success in attracting foreign investments was very limited. By the end of 1983, there were only 188 "equity" and 1,047 "contractual" joint Chinese foreign ventures, with $6.6 billion pledged but only $2.3 billion paid-in. Three of the largest American-Chinese ventures were the Great Wall Hotel ($11 million), the Jiangguo Hotel ($11 million), and American Motor's Beijing Jeep Corporation ($16 million). Atlantic Richfield took part in exploring and drilling gas fields near Hainan Island at a cost of $250-300 million.[47]

Foreigners found the Chinese environment unconducive to investment. Endless negotiations and long bureaucratic delays strained patience, and business and residential facilities were substandard. Many foreign firms had to set up offices in hotels and paid high rents and fees for Chinese services even though their Chinese employees received only a fraction of the foreign wages paid while the lion's share went to the government's business bureaus. Many foreign firms left out of frustration and lack of prospects for profit.

SINO-JAPANESE TRADE. As China's leading trading partner, Japan was the largest supplier of modern plants and equipment and also of financial and technical assistance. In addition, Japan provided a market for Chinese exports of oil and other labor-intensive products. The needs of the two countries complemented each other, and the Chinese were successful in keeping the trade deficit small with the strategy that what Japan could sell to China must be coordinated with what she could buy from China. The Japanese helped China develop exports needed to generate foreign exchange for the purchase of Japanese goods and also offered low-interest concessional loans or grants. In 1983, Sino-Japanese trade reached $10 bil-

46. Noyes, 340.
47. Davie, 324-25; Noyes, 341.

lion, with $4.9 billion in Japanese exports to China and $5.1 billion in Chinese exports to Japan. The two-way trade represented 22 percent of China's total foreign trade and 3 percent of Japan's.[48] In 1983 China's largest exports were crude oil (40.9 percent of total exports)—representing 5.2 percent of Japan's total petroleum imports—and coal, 4.4 percent of Japan's coal imports. Other Chinese exports included light manufactured goods, agricultural products, meats, fish, shellfish, antiques, art work, and firearms.

Japanese exports to China were mostly metals and metallic articles (iron and steel), which accounted for 49.6 percent of Japan's total exports in 1983. Heavy machinery and mechanical apparatus represented 28.5 percent; chemical goods, 11 percent; and textiles, 5.8 percent.

Some of the reasons for the Japanese success in China included the following: (1) geographical proximity, which cut transportation costs—Tokyo is 5,000 miles closer to Shanghai than San Francisco; (2) cultural affinity, which enabled the Japanese to gain a deeper understanding of Chinese psychology and taste, adapt their products to suit the Chinese life-style,

Sino-Japanese Trade, 1979-92
[Millions U.S. dollars]

	Chinese exports	Chinese imports	Chinese trade balance
1979	$2,954	$3,698	−744
1980	4,323	5,078	−755
1981	5,291	5,097	+194
1982	5,352	3,510	+1,842
1983	5,087	4,912	+175
1984	5,155	8,057	−2,902
1985	6,091	15,178	−9,087
1986	5,079	12,461	−7,384
1987	6,392	10,087	−3,695
1988	8,046	11,062	−3,016
1989	8,395	10,534	−2,139
1990	9,210	7,656	+1,554
1991	10,265	10,079	+186
1992	17,000 (est.)	12,000 (est.)	+5,000 (est.)

Source: International Monetary Fund, *Direction of Trade Statistics Yearbook, 1992* (Washington, D.C.).

48. Dick K. Nanto and Hong Nack Kim, "Sino-Japanese Economic Relations," *China's Economy*, II, 454, 466.

China's Merchandise Exports to Japan by Commodity, 1980-83
[*in thousands of dollars*]

Commodity	1980	1981	1982	1983	1983 share (percent)
Animal products	297,108	316.311	272.910	262,467	5.2
Fish shellfish	181,979	188,042	138,042	131,314	2.6
Vegetable products	321,623	407,377	377,977	484,236	9.5
Mineral products	2,514,233	3,060,980	3,212,072	2,926,877	57.5
Coal	116,519	188,676	212,536	212,958	4.2
Crude oil	1,949,172	2,332,960	2,340,918	2,080,959	40.9
Textiles and textile articles	682,967	691,504	722,582	806,577	15.9
Silk/silk fabrics	171,611	116,587	153,262	158,487	3.1
Cotton/cotton fabrics	92,180	115,865	118,248	140,121	2.8
Garments	230,704	242,748	263,896	270,895	5.3
Others	507,443	815,628	766,876	607,200	11.9
Total	4,323,374	5,291,800	5,352,417	5,087,356	100.0

Japan's Merchandise Exports to China by Commodity, 1980-83
[*in thousands of dollars and percent*]

Commodity	1980	1981	1982	1983	1983 share (percent)
Chemical goods	575,416	559,599	512,139	539,674	11.0
Chemical fertilizers	244,476	213,120	84,712	17,509	0.4
Metals and articles thereof	1,686,655	1,255,421	1,355,788	2,434,133	49.6
Iron and steel and articles thereof	1,618,233	1,197,407	1,292,616	2,253,334	45.9
Machinery and mechanized apparatus	2,154,309	2,440,450	1,007,491	1,399,656	28.5
General machinery	1,164,226	1,440,696	399,967	545,107	11.1
Electrical machinery	422,428	554,861	203,868	264,502	5.4
Transport machinery	426,746	225,294	309,836	320,580	6.5
Scientific optical and precision apparatus	140,909	219,599	163,820	269,466	5.5
Textiles and textile articles	403,900	599,233	368,220	286,567	5.8
Man-made fibers	156,127	201,815	115,869	81,977	1.7
Others	258,055	242,486	197,817	252,304	5.2
Total exports	5,078,335	5,097,189	3,510,825	4,912,334	100.0

Source: U.S. Congress, *China's Economy Looks Toward the Year* 2000 (Washington, D.C., 1986), II, 460-63.

and lobby the Chinese bureaucracy effectively with lavish promotional gifts and free trips to Japan; (3) Japanese businesspersons offered competitive prices and sales training, and the Japanese government offered preferential tariff rates under the Generalized System of Preferences (GSP) in addition to the Most-favored-nation treatment (MFN); (4) Tokyo offered concessional loans and credit in yen, at attractive, low interest rates; credit of Y300 billion ($1.5 billion) at a 3 percent interest rate for the period 1979-83 and Y470 billion ($2.1 billion) in 1984 at 3.5 percent interest to finance key development projects such as railways, ports, telephone equipment, and hydroelectric power stations; (5) between 1978 and 1982 the yen weakened against both the dollar and the Chinese yuan, making Japanese products highly competitive in price; after 1982 the yen gained in value against the dollar and the yuan, but the attractive concessional financial arrangements still stood Japanese products in good stead; and (6) the basic policy of helping China to develop resources to generate the needed foreign exchange to buy Japanese goods.[49]

Yet, in spite of these advantages, the Japanese did not score an unqualified success. The Chinese often complained of "second quality" and even shoddy Japanese products, nonconformity with the original orders, and the dubious business practice of sponsoring great "show-how" but withholding the genuine "know-how."

SINO-AMERICAN TRADE. The Americans did not have the same advantages as the Japanese, nor could they offer concessionary loans and credit. But they had a deep reservoir of goodwill among the Chinese, who preferred American aircraft, computers, electronics, telecommunication equipment, and oil-drilling tools for their reputed quality and durability. The Americans imported considerable quantities of Chinese textiles, apparel, and oil products and an assortment of hand tools, housewares, pharmaceuticals, furniture, antiques, and art. In 1981, the Chinese-American trade temporarily peaked at $5.5 billion, which represented 14 percent of China's foreign trade and 3 percent of America's. Every year between 1972 and 1983, China sustained a trade deficit with the United States, resulting in a cumulative shortfall of $7.7 billion.[50]

As China progressed in her modernization, she would need more state-of-the-art technology, as well as fertilizers, chemicals, and timber products.

49. *Ibid.*, 465-66.
50. Noyes, 335-37.

Sino-American Trade, 1972-92
[*Millions U.S. dollars*]

	Chinese exports	Chinese imports	Chinese surplus or deficit
1972	32	60	−28
1973	61	741	−680
1974	112	819	−707
1975	156	304	−148
1976	202	135	+67
1977	203	171	+32
1978	324	865	−541
1979	594	1,724	−1,130
1980	1,059	3,755	−2,696
1981	1,875	3,603	−1,728
1982	2,275	2,912	−637
1983	2,244	2,173	+71
1984	3,065	3,004	+61
1985	3,865	3,856	+6
1986	4,771	3,106	+1,665
1987	6,293	3,497	+2,796
1988	8,511	5,021	+3,490
1989	11,990	5,755	+6,235
1990	15,237	4,806	+10,431
1991	18,969	6,278	+12,691
1992	25,728	7,418	+18,309
1993	31,530	8,760	+22,770

Sources: These 1972-82 figures derived from *China's Economy Looks Toward the Year 2000* (Washington, D.C., 1986), vol. 2, 329-30. The 1983-92 figures derived from *U.S. Foreign Trade Highlights*, U.S. Department of Commerce, International Trade Administration, 1989 and June 1993.

Her 300,000 state industrial enterprises could use American advice on upgrading, management techniques, and infrastructure development. There was a great potential in the service industry sector, and the United States could offer advice and guidance in tree cultivation, insecticidal chemistry, water conservancy, food preservation, coal extraction, cargo handling, birth

Highlights of China's Exports to the United States
[*Million U.S. dollars, FAS*]

Category	1983 Value	1983 Share (percent)	First half, 1984 Value	First half, 1984 Share (percent)
Total	2,244	100.0	1,482	100.0
Manufactures	1,026	45.7	689	46.5
Apparel and accessories	774	34.5	499	33.7
Wicker, basketware	58	2.6	35	2.4
Footwear	34	1.5	24	1.6
Fuels	430	19.2	263	17.7
Gasoline	309	13.8	149	10.1
Crude oil	79	3.5	62	4.2
Intermediate manufactures	390	17.4	272	18.4
Textile yarn and fabrics	241	10.7	184	12.4
Chemicals, related products	131	5.8	79	5.3
Fireworks	29	1.3	19	1.3
Pharmaceuticals	25	1.1	13	0.9
Food	112	5.0	78	5.3
Canned vegetables	34	1.5	30	2.0
Tea	10	0.4	8	0.5
Crude materials	97	4.3	55	3.7
Barium sulfate and carbonate	26	1.2	14	0.9
Down and feathers	8	0.3	6	0.4
Machinery, transport equipment	42	1.9	29	2.0
Miscellaneous	10	0.5	12	0.8
Beverages and tobacco	4	0.2	2	0.1
Beer	2	0.1	1	0.1
Animal/Vegetable fats, oils	2	0.1	2	0.1

control, integrated circuits—and any number of developmental activities. The possibilities were limitless.[51]

Irritation and frustration within Sino-American trade was inevitable on both sides. The Americans were critical of the seemingly endless delays in negotiations, lack of a bilateral investment treaty, and the difficulty of repatriating funds from China. The Chinese, on the other hand, were unhappy with American protectionism against Chinese textile imports and limitations on American hi-tech exports. As a Communist state, China came under the U.S. Export-Import Bank provision restricting loans ex-

51. Davie, 324, Noyes, 342, 344.

Highlights of China's Imports from the United States
[Million U.S. dollars, FOB]

Category	1983 Value	1983 Share (percent)	First half, 1984 Value	First half, 1984 Share (percent)
Total	2,173	100.0	1,162	100.0
Machinery, transport eqpt	586	27.0	284	24.4
Aircraft and parts	235	10.8	49	4.2
Construction, mining eqpt	52	2.4	45	3.9
Office eqpt	52	2.4	34	2.9
Food	541	24.9	284	24.4
Grain	536	24.7	283	24.4
Chemicals	354	16.3	265	22.8
Fertilizers	168	7.7	119	10.2
Plastics	92	4.2	73	6.3
Crude materials	300	13.8	173	14.9
Conifer logs	228	10.5	129	11.1
Manufactures	220	10.1	61	5.2
Aluminum	87	4.0	3	0.3
Paper	41	1.9	22	1.9
Miscellaneous	172	7.9	94	8.1
Electrical meters, controls	92	4.2	46	4.0

Source: U.S. Congress, *China's Economy Looks Toward the Year 2000* (Washington, D.C., 1986), II, 336-40.

ceeding $50 million, and the Generalized System of Preferences (GSP) excluded China from duty-free treatment. Also, the United States' rule on merchandise's "country-of-origin" made the Chinese feel discriminated against, and the modest American investments in China caused disappointment. Irritations aside, the Sino-American trade was expected to continue to grow and reach new heights in the years to come.

Future Prospect of Growth

The economic and technological benefits of the reforms and opening were obvious. There was growing prosperity in the countryside and substantial improvement in the farmer's standard of living. Urban life had become more colorful, open, and relaxed, and commercial and scientific exchanges with foreign countries grew by leaps and bounds. The successful leadership of Teng Hsiao-p'ing was compared by foreign observers to that of Colbert of France, Frederick the Great of Prussia, and the leaders of the

early Meiji period in Japan.[52] Others forecast rosily that barring a cata-
strophic occurrence such as the outbreak of war with the Soviet Union or a
major upheaval on the scale of the Cultural Revolution, China's GNP,
growing at a feasible annual rate of 8 percent, could surpass those of Italy
and England before the year 2000 and soar past those of West Germany
and France by A.D. 2020.[53]

The forecast was plausible, statistically. Ten years into the reforms,
China's economic indicators continued to skyrocket and showed no sign of
slackening, but major catastrophes of the types mentioned could not be
ruled out entirely. Destabilizing influences that could prove highly detri-

GDP Projections of China, India, and Certain Western European
States, 1980-2020

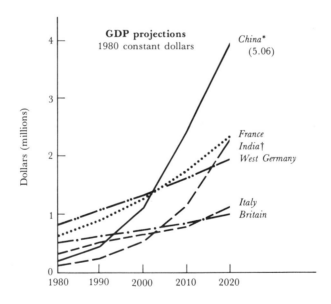

Source: The Economist, Dec. 21, 1985, 69.

52. Paul Kennedy, *The Rise and Fall of the Great Powers: Economic Change and Military Conflict from 1500 to 2000* (New York, 1987), 448.
53. *Ibid.*, 455.

mental in the long run and that eluded statistics were already lurking in the background. Foremost among them were these: (1) the progressive erosion of faith in Marxism-Leninism-Maoism and in the leadership of the ruling party; (2) widespread corruption in the party and the government, which threatened a breakdown of public morality; and (3) raging inflation in the wake of price decontrol, causing massive discontent.

As the decade of reform and opening drew to a close, there were ominous signs of ideological confusion, economic imbalance, social unrest, and moral degradation. Serious challenges, as well as new opportunities, waited to test the leadership.

Further Reading

A Draft Agreement between the Government of the United Kingdom of Great Britain and Northern Ireland and the Government of the People's Republic of China on the Future of Hong Kong. Hong Kong, Sept. 26, 1984.

Bannister, Judith, *China's Changing Population* (Stanford, 1987).

Burns, John P., *Policy Conflicts in Post-Mao China* (Armonk, N.Y., 1986).

Barnett, A. Doak, and Ralph N. Clough (eds.), *Modernizing China: Post-Mao Reform and Development* (Boulder, 1985).

Chao, Tzu-yang, "Tang-ch'ien ti ching-chi hsin-shih ho ching-chi t'i-chih kai-ke" (On the economic conditions before us and the economic institutional reforms), Report before the Sixth National People's Congress, March 27, 1985 (Hong Kong, 1985).

Chen, Nai-Ruenn, and Jeffrey Lee, *China's Economy and Foreign Trade* (Washington, D.C., 1984), Dept. of Commerce report.

China: Economic Policy and Performance in 1987 (Washington, D.C., 1988). A report by the Central Intelligence Agency presented to the Subcommittee on National Security Economics of the Joint Economic Committee, U.S. Congress.

China: Long-Term Development Issues and Options (Washington, D.C., 1985). The World Bank report.

"China and India: Two Billion People Discover the Joys of the Market," *The Economist*, Dec. 21, 1985, 65-70.

China's Economy Looks Toward the Year 2000, Joint Economic Committee, U.S. Congress (Washington, D.C., 1986), 2 vols.

Ching, Frank, *Hong Kong and China: For Better or for Worse* (New York, 1985).

Chiu, Hungdah, Y. C. Jao, and Yuan-li Wu (eds.), *The Future of Hong Kong: Toward 1997 and Beyond* (New York, 1987).

Chow, Gregory C., *The Chinese Economy* (Hong Kong, 1986).

Cremer, R. (ed.), *Macau, City of Commerce and Culture* (Hong Kong, 1987).

Croll, Elisabeth, *The Family Rice Bowl* (Geneva and London, 1983).

Griffin, Keith (ed.), *Institutional Reform and Economic Development in the Chinese Countryside* (Hong Kong, 1986).

Hinton, William, *Shenfan: The Continuing Revolution in a Chinese Village* (New York, 1984).

Ho, Samuel P. S., and Ralph W. Huenemann, *China's Open Door Policy: The Quest for Foreign Technology and Capital* (Vancouver, 1984).

Ishikawa, Shigeru, "Sino-Japanese Economic Cooperation," *The China Quarterly*, 109:1-21 (March 1987).

Jao, Y. C., and C. K. Leung (eds.), *China's Special Economic Zones: Problems and Prospects* (New York, 1986).

Joffe, Ellis, *The Chinese Army After Mao* (Cambridge, Mass., 1987).

Kelley, Ian, *Hong Kong: A Political-Geographic Analysis* (Honolulu, 1986).

Kirby, Richard J. R., *Urbanization in China: Town and Country in a Developing Economy, 1949-2000 A.D.* (New York, 1985).

Lardy, Nicholas R., *Agriculture in China's Modern Economic Development* (New York, 1983).

Lee, Ching Hua (Patrick), *Deng Xiaoping: The Marxist Road to the Forbidden City* (1985).

Leeming, Frank, *Rural China Today* (New York, 1985).

Madsen, Richard, *Morality and Power in a Chinese Village* (Berkeley, 1986).

Mathur, Ike, and Chen Jai-sheng, *Strategies for Joint Ventures in the People's Republic of China* (New York, 1987).

Parish, William L. (ed.), *Chinese Rural Development: The Great Transformation* (Armonk, N.Y., 1985).

Perkins, Dwight, and Shahid Yusuf, *Rural Development in China* (Baltimore, 1984).

Perkins, Dwight H., *China: Asia's Next Economic Giant* (Seattle, 1986).

Perry, Elizabeth J., and Christine Wong (eds.), *The Political Economy of Reform in Post-Mao China* (Cambridge, Mass., 1985).

Rabushka, Alvin, *Forecasting Political Events: Future of Hong Kong* (New Haven, 1986).

Saith, Ashwani (ed.), *The Re-emergence of the Chinese Peasantry: Aspects of Rural Decollectivisation* (London, 1987).

Shaw, Yu-ming (ed.), *Mainland China: Politics, Economics, and Reform* (Boulder, 1985).

Starr, John B. (ed.), *The Future of U.S.-China Relations* (New York, 1984).

Stavis, Benedict, *The Politics of Agricultural Mechanization in China* (Ithaca, N.Y., 1978).

Stephan, John J., and V. P. Chichkanov (eds.), *Soviet-American Horizons on the Pacific* (Honolulu, 1986).

Tam, On Kit, *China's Agricultural Modernization: The Socialist Mechanization Scheme* (Dover, N.H., 1985).

Teng, Hsiao-p'ing, *Building Socialism with Chinese Characteristics* (Peking, 1985).

The Chinese People's Liberation Army 60 Years On: Transition towards a New Era, four articles, *The Chinese Quarterly*, 112:541-630 (Dec. 1987).

Tsao, James T. H., *China's Development Strategies and Foreign Trade* (Lexington, Mass., 1987).

Tsou, Tang, *The Cultural Revolution and Post-Mao Reforms: A Historical Perspective* (Chicago, 1986).

Wesley-Smith, Peter, *Unequal Treaty, 1898-1997: China, Great Britain, and Hong Kong's New Territories* (New York, 1984).

Wik, Philip, *How to Do Business with the People's Republic of China* (Englewood Cliffs, N.J., 1984).

Youngson, A. J. (ed.), *China and Hong Kong: An Economic Nexus* (New York, 1985).

Zweig, David. "Prosperity and Conflict in Post-Mao Rural China," *The China Quarterly*, 105:1-18 (March 1986).

38

China in Transition, 1986-88:
The Cultural Impact
of the Open-Door Policy

When the policy of opening China's doors to the outside world was first adopted in December 1978, the central leadership had hoped to import foreign science and technology without importing foreign culture and values. But ideas travel across boundaries like the wind; they cannot be stopped by decree. The cultural impact of the open-door policy went far beyond anything the CCP leadership had imagined, causing the conservative ideologues to decry the invasion of Western "bourgeois liberalism" as foreign "spiritual pollution." From the standpoint of ideological orthodoxy, the fears and accusations were understandable, but the fact remained that Westernization was a global phenomenon that could not be stopped.

In China's case, after 30 years of isolation from the West, the doors were suddenly flung open: foreign ideas, news, films, plays, music, literature, and popular culture swept in like a windstorm. In the 10 years that followed, 60,000 students and visiting scholars, as well as tens of thousands of officials and delegates, went abroad to study and visit, creating international exchange between China and the outside world such as had not been seen for decades.

The workings of Western democracy and the freedom of its people made a deep impression on Chinese visitors to the West. Particularly enlightening was seeing the functioning of institutional checks and balances; of the division of power in government; of judicial supremacy and the rule of law; political pluralism; and the freedoms of speech, of assembly, and of the press. Many Chinese came to believe that their country needed political democracy as the Fifth Modernization, without which a true modern transformation would not be possible. University students, in particular, felt a social responsibility to be the vanguard of such change.

Not only the West, but Japan and the Four Little Dragons of Asia—Hong Kong, Singapore, Taiwan, and South Korea—were sources of inspiration. Any comparison between them and China would lead to the inevitable questions as to whether Marxism-Maoism was the proper ideology for China's modernization and whether the Chinese Communist Party was the most effective agent to lead China into the 21st century. Doubts persisted on both counts among a number of social critics, writers, journalists, artists, professors, scientists, and college students, although most of them chose to remain silent.

The party itself was torn between the need for liberalization and the urgency of maintaining some level of orthodoxy. Ideology was in a state of flux. Many party members wondered whether a Communist system should adopt the capitalist devices of material incentives and market mechanisms based on supply and demand and the law of value, capitalist-style management, and private enterprise. Others asked whether the Four Cardinal Principles—the socialist line, the proletarian dictatorship, the leadership of the Communist Party, and Marxism-Leninism and the Thought of Mao—should not be downplayed to defuse both the demand for political pluralism and Taiwan's thrust as an alternative model of development.

China, from 1986 to 1988, was experiencing the growing pains of rapid economic development and the agony of changing values. Society was buffeted by student unrest, ideological confusion, a leadership crisis, widespread corruption, high inflation, and a loss of a clear sense of direction. If there were such a thing as a "misery index," the period of 1986-88 would have had the highest readings of any period since the end of the Cultural Revolution. Yet, within the depths of the chaos and turbulence, some promise of regeneration and a better future were dimly discernible.

Student Demonstrations

In December 1986 gigantic student demonstrations broke out in 15 major cities in China, which left the party and the government in deep disarray. One hundred thousand students from 150 colleges and universities marched in the streets to demand the freedoms of speech, assembly, and of the press, as well as democratic elections. Their message was clear: the Chinese youths wanted political liberalization.

Student activism was nothing new in modern Chinese history; many precedents existed for the incidents of December 1986. The May Fourth Movement of 1919 was the most famous of these. More recently, in September 1985, students had protested Japan's new economic aggression and the Chinese government's ingratiating attitude toward Japan. This protest had been timed to coincide with the 54th anniversary of the Japanese invasion of Manchuria on September 18, 1931. The protest also served as a vehicle for the students to air other grievances, such as rising prices, economic crimes, bureaucratic irregularities, and nepotism and favoritism for the children of high cadres. A demonstration planned by students of Peking and Tsinghua Universities on December 9, 1985, had been aborted owing to government intervention. The discontent, however, remained.

The first protest of December 1986 broke out in Ho-fei, in Anhwei province, early in the month. The students of the Chinese University of Science and Technology, which is located within the city's electoral district, protested the local party secretaries' designation of eight prospective delegates to the National People's Congress without consulting the university. On December 1, 1986, a big poster appeared on the campus calling for a general boycott of the "faked" election scheduled for December 8. At a student-faculty meeting the following day, the renowned astrophysicist and Vice-President of the university, Fang Li-chih, called for decisive action to achieve a breakthrough in the struggle for political democracy. He repeated his much touted saying that democracy could not be bestowed from above but must be won from below—what was bestowed could be withdrawn, but what was won could not.

The students of the Chinese University of Science and Technology were joined in their protest by students from two other institutions of higher learning in the area.[1] In all, 3,000 students marched on the municipal government on December 5, demanding democratic elections, freedom of the press and of assembly, and immunity from persecution and insisting that the media be allowed to report their protest. They further demanded the formation of a Democratic Alliance of All Students of Higher Learning. The students aimed at breaking the party's control of the press and of the students. The local party authorities reluctantly agreed to assign four delegates to the university precinct and postpone the election to December 29.[2] The students continued to press for election reform, chanting: "We want democracy. We want liberty. We want freedom of the press. No democracy, no modernization."[3]

In Shanghai, there was a huge sympathetic response. On December 19

1. Anhwei University and Ho-fei Institute of Technology.
2. Professor Fang Li-chih was elected with 3,503 votes; Professor Weng Yuan-kai, 2,406 votes; student leader Sa Ma, 2,164 votes; the fourth is unknown.
3. *Shih-yüeh P'ing-lun* (October Review), Hong Kong, 14:4-5:6-8, special issue (1987).

some 30,000 students,[4] joined by an estimated 100,000 workers, marched on the municipal government. This marked the high tide of the protest. In Peking, 4,000 students marched on Tien-an-men Square and burned bundles of the party newspaper, the *Peking Daily*. One of the protesters, watching the rising smoke, was heard to say elatedly: "Gone with the wind!"[5]

These demonstrators made it clear that, although they sought democratization, they were true supporters of Teng Hsiao-p'ing's policy of economic reforms and an Open Door to the outside world. This prompted a sarcastic comment from the *Peking Daily* that they waved the Red Flag to attack the Red Flag.

Caught off guard, the party was unable to be decisive on the issue. The conservatives pressed for forceful repression, but General Secretary Hu Yao-pang took the enlightened attitude that youthful idealism should not be blunted but guided toward constructive goals. The Vice-Minister of the State Education Commission,[6] himself a demonstrator in his youth, said that most of the demonstrators, who constituted less than 2 percent of China's 2 million college population, were patriotic and well-meaning, but that they were misguided by Western liberalism. He said, "God allows young people to make mistakes. When we were younger, we basically did the same thing. . . . Our policy toward them is to educate them and to advise them and to give them a proper orientation." But he warned that the students must not violate the Four Cardinal Principles lest the country's hard-won stability and unity be compromised.[7]

Other authorities were likewise moderate in their reactions. The Mayor of Shanghai[8] acknowledged the student action to be "just, legal, and patriotic" and permissible under the constitution. He promised that there would be no retribution.[9] Police and public security agents conducted themselves with restraint. They took pictures of the demonstrators but avoided inflaming them lest the demonstration get out of control. Party newspapers urged the students to cherish China's material progress and the improvement in the quality of life made in recent years, including their educational privileges. A famed sociology professor[10] asked the stu-

---

4. Mostly from T'ung-chi and Chiao-t'ung Universities and, to a lesser extent, from Fu-tan University.
5. *Los Angeles Times*, Jan. 6, 1987. The protester was proud of knowing the title of the famed American movie.
6. Ho Tung-ch'ang, a Central Committee member.
7. *Los Angeles Times*, Dec. 31, 1986.
8. Chiang Tze-min.
9. *Los Angeles Times*, Dec. 22, 1986.
10. Professor Fei Hsiao-t'ung.

dents to air their grievances through the existing channels and offer their criticisms in the interest of stability and unity lest a good cause turn sour. The party authorities did not question the students' motives but feared that their actions might create traffic problems or disrupt economic production!

The more conservative leaders blamed Western bourgeois liberalism as the source of trouble. A small number of students were arrested, and most were soon released. However, a few remained in custody for months.[11] Time was on the government's side. Final examinations soon preoccupied the students, followed by the winter vacation. On the whole, the official position toward the students was initially lenient, but toughened later. In an effort to tighten further central control of China's student population, the government reinstated military training and political indoctrination on campus and revived the unpopular policy of sending students to "grass roots" farms and factories for a year before being sent to their job assignments.[12] But the government's policy toward the workers had been very strict from the beginning: they were strictly prohibited from joining the demonstrations. The authorities were determined to keep a Polish-style union strike from happening in China.

General Secretary Hu's handling of the unrest raised the ire of the hard-line conservatives. They were impatient with his liberal approach and became increasingly critical of his leadership. A large group of them in the party, the army, and the government formed a grand alliance to denounce him before the paramount leader Teng Hsiao-p'ing. They blamed the Voice of America and Taiwan's Voice of Free China as outside instigators of the turmoil, but even more emphatically they attacked Western bourgeois liberalism as the source of discontent and disharmony. This conservative group held a number of secret meetings in the second half of December, and before the end of the month they had won Teng over. Decisions were made to clamp down on the student agitation, to deal with Hu sternly, and to fight Western liberalism. The *People's Daily* editorialized, "It's time to wake up. Bourgeois liberalism is indeed an ideological trend. It is in the process of poisoning our youth, imperiling our social stability and unity, interfering with our reforms and opening-up policy,

11. Among them was one Yang Wei, 32, holder of an M.A. in molecular biology from the University of Arizona, where he studied from 1983 to 1986. Arrested in January 1986 in Shanghai, he was not tried until December. He was sentenced to two years of imprisonment for "counterrevolutionary" activities, including participation in a prodemocracy demonstration in Shanghai, and association with the Chinese Alliance for Democracy and its publication, *China Spring*. Headquartered in New York, the Alliance and its publication are highly critical of the lack of democracy in China.

12. *Christian Science Monitor*, Jan. 6, 1988.

and obstructing modernization's move forward. Can we afford to do nothing about it?" As the New Year approached, all party members were admonished to raise their vigilance against the spread of "bourgeois liberalization" in China.[13] The fate of General Secretary Hu hung in a precarious balance.

BOURGEOIS LIBERALISM. The student unrest was not an isolated and sudden outburst of youthful enthusiasm, but a movement several years in the making. Ever since the Third Plenum of the Eleventh Party Congress in December 1978, when the policy of modernization and opening up was adopted, foreign news, contacts, and travel had swamped the Chinese public with cultural and political concepts once considered alien and inconceivable. A new vista was opened to the Chinese as to the nature of the modern world. Ideas such as human rights, democracy, free elections, free speech, free assembly, free press, division of power, and "loyal opposition" captured the Chinese imagination and won their deepest appreciation. Particularly admired was the idea that the governing bodies could be supervised by the governed through a watchful press, the right to dissent, and political pluralism, all of which could serve as checks on the excesses of the government. These ideas, to the Chinese, symbolized the true character of a democracy dedicated to the fullest development of the human potential.

Several international events contributed to this state of mind among the reading public: the overthrow of the Marcos regime in the Philippines by "people's power" in February 1986; the expulsion of the Haitian dictator, Jean-Claude Duvalier, by a popular revolt; the political liberalization of Taiwan and, with it, the creation of the opposition Democratic Progressive Party; and the student strike in France against the adoption of an elitist educational policy.

Domestically, there were powerful stirrings in intellectual and cultural circles. Ever since the Third Plenum of December 1978, when the central focus of the party shifted from class struggle to economic development, intellectuals had been given higher status because modernization demanded knowledge and talent. Teng Hsiao-p'ing proclaimed the intellectuals to be part of the proletariat, deserving both respect and good treatment from the state—no longer the "stinking No. 9" at the bottom of the social heap as under Mao. Emboldened by their new status, independent-minded writers, artists, scientists, and thinkers began to speak their minds on pressing social and political issues.

13. *People's Daily*, Jan. 6, 1987. The word "liberalization" rather than "liberalism" is used to denote an unfolding process rather than a fact—a subtle difference.

These dissenters frequently touched a raw nerve in the party, which jealously guarded its right to rule and to be obeyed unquestioningly. Dissent was equated with defiance, disloyalty, and a challenge to the authority of the party. In 1979, when the "Democracy Wall" movement sprang up, the government condemned its leader[14] to a 15-year prison sentence. In 1983, when it perceived Western ideas, theories, and practices as invasive of the Chinese public consciousness, the party instituted an "Anti-Spiritual Pollution" movement to counter the inroads of "decadent" values.

When President Ronald Reagan visited China in April 1984, he was imbued with a mission to extol the virtues of American democracy, faith in God, freedom, individual initiative, free enterprise, and the spirit of progress. In a speech delivered at the Great Hall of the People before 600 Chinese social leaders, he declared:

> From our roots we have drawn tremendous power from two great forces—faith and freedom. America was founded by people who sought freedom to worship God and to trust in Him to guide them in their daily lives with wisdom, strength, goodness, and compassion.
>
> Our passion for freedom led to the American Revolution, the first great uprising for human rights and independence against colonial rule. We knew each of us could not enjoy liberty for ourselves unless we were willing to share it with everyone else. And we knew our freedom could not truly be safe unless all of us were protected by a body of laws that treated us equally.
>
> George Washington told us we would be bound together in a sacred brotherhood of free men. Abraham Lincoln defined the heart of American democracy when he said, "No man is good enough to govern another man without that other's consent. . . ." These great principles have nourished the soul of America, and they have been enriched by values such as the dignity of work, the friendship of neighbors, and the warmth of family.
>
> "Trust the people"—these three words are not only the heart and soul of American history, but the most powerful force for human progress in the world today . . . the societies that have made the most spectacular progress in the shortest period of time are not the most rigidly organized nor even the richest in natural resources. No, it's where people have been allowed to create, compete, and build, where they've been permitted to think for themselves, make economic decisions, and benefit from their own risks that societies have become the most prosperous, progressive, dynamic, and free.[15]

14. Wei Ching-sheng.
15. U.S. Dept. of State, Bureau of Public Affairs, Washington, D.C., "President Reagan, A Historic Opportunity for the U.S. and China," April 27, 1984.

These were strong and agitating messages for an atheistic and authoritarian society. Obviously taking offense, the Chinese leadership decided to censor what it did not want the people to hear during later media coverage. Chinese television audiences did not hear the statements on faith and freedom, nor the exhortation that "free people build free markets that ignite development for everyone." Presidential Deputy Press Secretary Larry Speakes stated that while the decision to censor was "an internal matter for the Chinese people to decide," he regretted the omission of "key passages dealing with the President's view of values that Americans cherish . . . which would have given the Chinese people a better understanding of our country and its people."[16]

Reagan repeated the same themes with special emphasis on human rights in a televised speech at Fu-tan University in Shanghai: "We believe in the dignity of each man, woman, and child. Our entire system is founded on an appreciation of the special genius of each individual—and of his special right to make his own decisions and lead his own life."[17] The Chinese allowed no translation of the text during the telecast, but there were enough people who could understand Reagan. The full texts of both speeches quickly appeared in the free markets and were snapped up instantly. Reagan got his message across, and the Chinese audiences secretly admired him for his audacity and "guts."[18]

The Chinese youths were strongly affected by their new exposure to Western political ideas, both abroad and through Reagan's visit. They came to feel that any political system that deviated from the ideals of the Western democracies inhibited the full development of individuals and must be condemned as backward, authoritarian, and out of touch with the modern era.

With the urban economic reforms in 1984, there were increasing instances of crime, corruption, inflation, and favoritism toward the children of the high cadres under the guise of promoting third-tier leaders. Seeing themselves as the conscience of the younger generation, the students felt a social responsibility to protest this degradation of morals. They wanted human rights; democracy; the rule of law; freedom of speech, assembly and the press; and prohibition of arbitrary arrests and ad hoc extrajudiciary proceedings.

Not only the students but also the general cultural, artistic and intellectual communities came under the impact of Western liberalization as well. Writers and artists demanded greater freedom in creative endeavours,

16. *Los Angeles Times*, April 28, 1984.
17. *The Christian Science Monitor*, May 1, 1984.
18. *Sino Express, New York*, May 6, 1984.

and public-minded intellectuals longingly discussed such subjects as free elections, the three-way division of power, the right to dissent, and the multiparty system. A writer in Shanghai wrote: "Ideology should be free, emancipated, rich and pluralistic. . . . I have established for myself four basic ideological principles—to shake off ideological bonds, to free myself from following others blindly and from superstition, to think for myself, and to help my own ideology eventually jump from the realm of necessity to the realm of freedom."[19]

Fang Li-chih, the noted professor of astrophysics, was a tireless proponent of human rights and democracy. Dubbed the Chinese "Sakharov," he made frequent speaking tours and appeared at universities and before research institutes. His favorite topics were the rule of law, free elections, free speech, a free press, and political pluralism. Imbued with the conviction that scholars had a social responsibility to speak out on the vital issues of the day, he declared that he feared nobody and would accept the consequences of his actions. A research university, he stated, should honor science, democracy, creativity, and independence as the Four Basic Creeds. He advocated these in place of the Four Cardinal Principles, which he disparagingly compared with superstition, autocracy, conservatism, and dependency. He partook fully of the spirit and ideas of the May Fourth Movement—science, democracy, and complete Westernization. He considered Marxism to be absolescent and held it responsible for keeping China backward and for sanctioning the party's unrestrained and inviolable exercise of power. The party's rule, he pointed out, was based on military conquest, not merit, and its overall record in the past 35 years could only be considered a failure. Fang urged the students to fight for their rights, but he never asked them to take to the streets.[20]

Liu Pin-yen, an investigative reporter for the *People's Daily,* and Wang Jo-wang, a writer, were devastating in their exposé of the darker side of bureaucratism and official domination of literature and of the press. They particularly took exception to Mao's 1942 talk in Yenan, which stated that literature and journalism should serve politics.

A vice-chairman of the Chinese Writers Association, Liu Pin-yen ridiculed the Four Cardinal Principles and questioned the logic of insisting on socialism in China. He asked: If the socialism of the period 1953-76 under Mao was correct, what would the socialism of the period 1979-86 under

19. Sha Yeh-hsin, quoted in *Los Angeles Times,* Jan. 7, 1987.
20. Gist of Fang's speeches delivered before Chiao-t'ung University, Shanghai, Nov. 15, 1986; T'ung-chi University, Nov. 18, 1986; Anhwei Economic and Cultural Research Center, Sept. 27, 1986.

Teng be? The two were so different. How could they both be correct? With contempt he asked:

> What is the essence of extreme leftism? It is mutual despite, mutual destruction, and mutual cruelty. It makes human inhuman. It makes a free man unfree. It turns a person of independent personality into a submissive tool. It turns man into beast. In the process, the conscience is lost, and the sense of self-reproof disappears. In their place, there is mutual hostility and hatred, and mutual suspicion and cruelty. Finally, fear fills the air—fear of brutal power, fear of leaders, and fear of authorities. . . . Hence, I must submit that the essence of leftism is inhumanism.

Liu chided his countrymen for delusion with four fantasies: (1) that socialism is perfect, (2) that the Communist party is infallible, (3) that Marxism-Leninism is eternal truth, and (4) that there is a deep chasm between socialism and capitalism and that nothing in the capitalist societies that took place in the past several hundred years should be given proper recognition. For these misconceptions, Liu stated, the country paid dearly in the form of wanton sacrifices and unnecessary losses. And all for what?[21]

The other critic, Wang Jo-wang, a council member in the Chinese Writers Association, took on Teng himself in November 1986. Teng had agreed to an interview with the CBS journalist Mike Wallace and stated that Chinese socialism would permit some to get rich first but would prevent the rise of polarity between the very rich and the very poor. In any event, Teng believed that it would be difficult for anyone to become a millionaire under the Chinese socialist system and that there should be no concern for the rise of a new capitalistic class. The statement sounded innocuous enough, but Wang took issue with several points in this line of thinking. Polarity in a developing socialist society was unavoidable, Wang argued. The opposite of polarity would be egalitarian or averageism ("Everybody eats from the same Big Pot."), which China had practiced for 30 years and which had brought on nothing but abject Communist poverty. To avoid polarity was to remove incentives for greater production and profit, Wang argued. To predict the impossibility of millionaires was to preset a limit to one's expectations and upward mobility. In a country as large as China, Wang suggested, even 3,000 or 5,000 millionaires would

---

21. Party internal materials for use by district party secretaries and regiment commanders and above: critiques of Fang Li-chih and Liu Pin-yen, reprinted in *Chiu-shih Nientai* (The Decade of the Nineties), Hong Kong, June 1987, 37-40. Tr. by author, with minor editing.

not be too many and being products of Chinese socialism, they would be different from the capitalists of Marxist definition.[22]

It was not just intellectuals like Fang, Liu, and Wang who showed the influence of Western liberalism. In society at large, there was a craving for anything Western. Thousands would stand in line for hours to see a Picasso exhibit, a performance by The Royal Ballet, or an Arthur Miller play; but few would visit Mao's Revolutionary Military Museum. Almost anything foreign was attractive: political thought, social theories, futurology, novels, plays, art, fashion, and even such mundane things as Coca-Cola, Maxwell House Coffee, and Kentucky Fried Chicken. The most prized wedding gift of the time was a set of the *Encyclopaedia Britannica* in Chinese. The chairman of the translation and editorial board explained the public enthusiasm:

> For more than three decades, we treated Western culture as taboo and abandoned everything from the West. As a result, our knowledge of the West remained a blank. The works of Milton, Shaw, Rousseau, Balzac, Boccaccio, and Goethe, the music of Bach and Mozart, and the dramas of Shakespeare and Ibsen were considered "bourgeois" and their publication and performance were forbidden. How many Chinese are there who don't even know about the stories in the Bible?[23]

There is no denying that Western influence was pervasive and growing stronger by the day.

In the party, reaction to the new trend was mixed. General Secretary Hu Yao-pang took a laissez-faire attitude toward it in the belief that the information revolution of the world had brought countries close to each other and that it was in China's interest to experiment with new ideas and to create a new image for Chinese communism. His liberal approach pained the conservative elders who considered Western liberalism to be a disturbing influence, a source of decadence and spiritual pollution in Chinese life. They were distressed by the growing popularity of disco music, rock 'n' roll, blue jeans, sexy movies, and foreign publications that suggested freedom and democracy. They concluded that the inroads of bourgeois liberalism prompted a loss of regard for the party, a lack of respect for the leaders, a questioning of the legitimacy of Communist rule, and a

22. Wang Jo-wang's criticism of Teng's views appeared under the title, "Liang-chi fen-hua chih wo-chien—yu Teng Hsiao-p'ing t'ung-chih shang-ch'üeh" (My views on the polarity of the two extremes—A discussion with Comrade Teng Hsiao-p'ing), *Workers Daily,* Shenzhen Economic Zone, Nov. 15, 1986, reprinted in *The Decade of the Nineties,* Hong Kong, Feb. 1987, 58-59.
23. Shang Jung-kuang, "Bridging Ocean-Wide Chasm," *Beijing Review,* Jan. 18-24, 1988, 31. His name is Liu Tsun-ch'i, 76 years old.

blind worship of Western values. If bourgeois liberalism were not stopped in time, they insisted, there would be no place for Communist orthodoxy. Hu, the supposed ideological leader of the party, was too soft to lead the fight. Only the hard-line ideologues could defend the purity of the Faith. Thus, the Long March generation septuagenarians and octogenarians assigned themselves the sacred duty of remaining in power to fight Western liberalism.

The clash between the orthodox ideologues and progressive reformers could only be resolved by the paramount leader, Teng. He was a curious mixture of economic progressivism and political conservatism, endowed with a gift for playing a balancing act as political necessity dictated. In a system where the rule of man superseded the rule of law, he was the supreme arbiter. In his mind, economic reforms and an open-door policy were but means by which to borrow foreign technology, capital, and managerial skills. These were seen as tools with which to strengthen the Communist rule, but never as steps to move the country toward a Western-style democracy. He ridiculed the three-way division of power as three governments in one country and considered political pluralism to be totally out of order. Only his Four Cardinal Principles, announced in March 1979, could ensure stability and order and provide the necessary environment for modernization. The country could not tolerate any disruption, such as the student turmoil, or any disturbing influence, such as Western liberalism. In short, he was interested in Western science, but not Western values.[24]

In September 1986, at the Sixth Plenum of the Twelfth Congress, with Teng's acquiescence, the conservative ideologues outmaneuvered the progressives by blocking the discussion of political reforms. Instead, they passed a much-revised (eight times) resolution on the "Spiritual Construction of Socialism," which highlighted a provision titled "Anti-Bourgeois Liberalization." A shattering blow was dealt to the cause of democratization. It was against this background that the student demonstrations erupted in December 1986.

DISMISSAL OF HU YAO-PANG. A major result of the student unrest was the ouster of General Secretary Hu, who was too liberal and outspoken for his own good. He had survived for five years (1981-86) in that post, primarily owing to the support of Teng, who had groomed him to be his successor and who once confidently remarked that, with Hu in charge of the party and Chao at the helm of the government, even if heaven should come crashing

24. Central Committee Documents, Nos. 2 and 4, 1987, reprinted in *Ch'ao-liu Yüeh-k'an* (Tide Monthly), Hong Kong, April 15, 1987, 14-17.

down, he would have no fear. But now the conservatives in the party, the government, and the military descended on Teng to demand the resignation of Hu on grounds of his ineptness in dealing with the students and in stopping the inroads of Western bourgeois liberalism. P'eng Chen, chairman of the Standing Committee of the People's Congress, three other iedologues, and two powerful military leaders headed the move to oust Hu.[25] A shrewd politician, Teng knew the futility of fighting this powerful coalition; besides, his own confidence in Hu had faltered. After several secret meetings in the second half of December 1986, Teng painfully accepted the decision that Hu had to go. Teng sacrificed Hu to keep the support of the conservative leaders.

After the New Year, Hu was not seen in public; his fall was imminent. Meanwhile, intensive preparations were made for the convocation of a special enlarged Politburo, which opened on January 16, 1987. At the end of the day a formal announcement was made that Hu had resigned after making a "self-criticism of his mistakes on major issues of political principles in violation of the party's principle of collective leadership."

A glimpse of the intraparty squabbling at the highest councils may be seen from the eight party documents distributed for internal circulation during the early months of 1987. The scrutiny of Hu fell under six headings, which revealed policy differences as well as personal vendettas:

1. Hu practiced factionalism by favoring the promotion of members of the Communist Youth Corps, of which he was the former head.
2. Hu spoke carelessly on diplomatic occasions: during a visit to Tokyo in 1983 he declared that if the United States intervened in the Taiwan question, China might consider canceling the mutual visits of Premier Chao Tzu-yang and President Ronald Reagan.
3. Hu did not carry out party rectification effectively.
4. Hu advocated an inordinately fast pace of economic reform creating economic imbalances and loss of control of the situation.
5. Hu favored the rule of man over the rule of law in administering party work.
6. Hu did not observe organization discipline but revealed state secrets to foreigners and journalists.

The enlarged Politburo meeting summarized Hu's mistakes as follows:

1. He violated the collective leadership of the party.

25. P'eng and three other powerful ideologues, Teng Li-chün, Hu Ch'iao-mu, and Po I-po, all Politburo members, were sometimes dubbed the new Gang of Four. The two military leaders were General Yang Te-chih, Army Chief-of-Staff, and General Yang Shang-k'un, Vice-Chairman, the Military Commission.

2. He committed mistakes in major political principles.
3. He repeatedly and arbitrarily interfered with the work of government.
4. He repeatedly ignored the advice of Teng Hsiao-p'ing.
5. He made arbitrary decisions on important diplomatic issues.
6. He showed a tendency toward complete Westernization in his political style of work.

Behind these formal charges was the deep resentment of Hu's constant plea for the retirement of the party elders. Reportedly, he had planned to announce his plans to retire Central Committee members at age 60 and Politburo members at age 72 at the Thirteenth Party Congress scheduled for October 1987. To be sure, the principle of retiring the aged leaders was accepted by the party and Teng, but the actual implementation was left to the General Secretary, and it proved extremely sensitive. With no official retirement system, position meant power, political clout, and economic privileges. The Long March generation elders were particularly sensitive to their indispensability. They would do anything to avoid retirement, which was equated with worthlessness. Teng, too, may have been uncomfortable with Hu's talk of retirement, especially since Hu suggested that Teng retire first to set an example for the other elders. To be sure, Teng had repeatedly protested that he wanted to retire but others would not let him. Hu might have naively taken Teng's words literally and irritated him, leading Teng to question Hu's political astuteness. The student upheaval and Teng's displeasure provided the older leaders with a golden opportunity to get rid of Hu and to keep themselves in power.

The military had no use for Hu either, for he had no close relationship with the army and had assigned a low priority to benefits for the military in the Four Modernizations. Besides, Hu had advocated a cut in the armed forces totaling one million and was an outspoken critic of the Vietnam War. Since September 1986, the military had predicted and worked toward his removal.

The hard-line elders were determined to oust Hu before the convocation of the Thirteenth Congress so as to deny him a role in the selection of the delegates and in the preparation of the agenda. With the support of the military, their combined strength was formadible; and Teng, a political conservative-centrist at heart, bowed to their demands, perhaps in the interest of preserving party unity and orthodoxy.

When Hu made his self-criticism and offered to resign, 21 of the 40 participants spoke to criticize him, including some of his closest former allies. Hu wept but agreed "never to regret and reverse" the verdict. The case was closed with Hu's retaining his seat in the Politburo and the Central Committee. There was no official disgrace; in fact, Hu's popularity soared.

Premier Chao was appointed Acting General Secretary, concurrently with his other posts. The reformist cause had been dealt a severe setback, and conservative forces were on the rise.

ANTI-BOURGEOIS LIBERALIZATION. P'eng Chen and Teng Li-chün now unleashed their "Anti-Bourgeois Liberalization" campaign with a vengeance. A Media and Publications Office was created under the State Council in January 1987 to monitor the news media and publication of books, magazines, and newspapers. Three leading critics of the party were summarily dismissed from the party: the astrophysicist Fang, the journalist Liu, and the writer Wang. However, no physical abuse was visited on any of them. Fang was transferred from the prestigious position as Vice-President of the Chinese University of Science and Technology to a far less visible post as a researcher in a Peking observatory. He was allowed to continue his research and attend scholarly meetings in China—and occasionally abroad—but was deprived of a public forum to speak from. His fame may have protected him from greater penalty, as was the case with the other two critics. Several other cultural and intellectual figures received chastisements of varying severity, but no one was subjected to the harsh treatment of the Cultural Revolution period.[26]

In response, a thousand Chinese students and scholars studying in the United States jointly dispatched a letter of protest to the party and the State Council on January 19. The letter expressed concern over the dismissal of Hu, the expulsion of Fang and others from party membership, and the rising tide of leftism reminiscent of "the cruel excesses of the Cultural Revolution." The students reaffirmed their support of reform, democracy, and the rule of law; and they stated their opposition to regression.

Teng lent his support to the Anti-Bourgeois Liberalization campaign but did not allow it to be developed at the expense of stability and unity. The campaign was to be limited to the party, the government, the army, and urban enterprises. The countryside, the other political parties, and independent intellectuals were off limits. Also prohibited was the use of any expressions or terminology reminiscent of the Cultural Revolution.[27]

26. They included Su Shao-chih, philosopher and reformist theoretician and head of the Marxist Institute; Wang Jo-shui, a former deputy editor of the *People's Daily;* Wu Tsu-kuang, a playwright; and a number of other writers. They were asked to resign from the party.
27. Central Committee Documents, Nos. 2 and 4, 1987.

From January to May 1987, the rising tide of conservatism penetrated deep into ideological, cultural, literary, and journalistic circles with harsh rhetoric and vicious innuendo. The Anti-Bourgeois Liberalization campaign threatened to envelop the major cities, and the reformers looked disorganized and powerless. The hard-liners called for the revival of the Maoist three-way coalition among the old, the middle-aged, and the young. They promoted the old virtues of frugality, hard work, plain living, and devotion to the state as an antidote to Western influence. The Maoist slogan "Learn from Lei Feng," a fictitious model worker, was revived. Cautious people started to store away their Western clothes in favor of Mao jackets.

But the majority of people did not want to return to the old days. Once exposed to some freedom and creature comforts such as television and refrigerators, they could not bear the thought of going back to the Spartan life of Maoist times. Even the children and grandchildren of the hard-liners took exception to their elders' high-flown moralistic preaching, and there was a national desire to avoid the reappearance of another Cultural Revolution. By May the high tide of Anti-Bourgeois Liberalization was spent. The party elders painfully discovered that they were not moving in the main stream of society. Their pronouncements fell on deaf ears; most people simply counted on their passing.

The party was indeed in a dilemma. It had enjoyed a monopoly of power since 1949, and its right to rule had never been questioned before. Now the younger generation clamored for democratization and yearned for a freer life such as that found in other modern societies. Marxism and Maoism had lost appeal, and the party itself was experiencing a crisis of confidence. Not to grant economic reforms and greater political relaxation would further alienate the people and drive the party farther from the realities of the time. But permitting democracy could lead to social disharmony and ultimately to the demise of communism in China. The top leadership strove to find a middle course that would make it neither unfaithful to its ideology nor guilty of "orthodox Marxist sectarianism." It groped for a theory that would permit market mechanisms, importation of foreign capital and technology, and borrowing of capitalist management skills—all within a flexible framework that could be viewed as both Marxist and Chinese. Such a formula, if found, could justify an extension of both economic reform and the open-door policy, strengthen the position of the reformers, and deflate the zeal of the conservatives.

Premier Chao, an economic technocrat at heart, temporized with the hard-liners to avoid a split. Only once did he confront them at an enlarged Politburo meeting on May 13, 1987, in an attempt to curb the excesses of leftism.[28] Teng was conspicuously quiet throughout the spring, nursing his wounds over the loss of Hu. In the end he realized that what was at issue was not who should resign, but rather the future of economic reforms and of the open-door policy. If the conservative ascendency were not checked, modernization itself would stall. The luxury of silence was over; Teng was compelled to declare his position.

On four occasions in May and June, before foreign visitors, Teng vigorously affirmed the need to curb leftism and expand the scope of economic reforms and the open-door policy.[29] Newspapers and journals, which had been awaiting a signal, quickly echoed this line. This was a triumph for antileftism and a deflation of the Anti-Bourgeois Liberalization campaign. The reformers had regained a measure of coherence and started to plan for an offensive at the forthcoming enlarged Politburo meetings at the summer resort of Pei-tai-ho in July.

At Pei-tai-ho, intense political jockeying took place among the various power brokers, and an agreement was reached on four principal issues: (1) a political report on the future course of development to be delivered at the Thirteenth Party Congress scheduled for October 1987; (2) retirement of the elders and selection of the future leadership; (3) an official statement on the nature of the present stage of socialist construction; and (4) political restructuring. The deliberations were kept secret until their final ratification by the Thirteenth Congress.

The Thirteenth Party Congress

The much-awaited Congress was held in the Great Hall of the People in Peking from October 25 to November 1, 1987, and attended by 1,936 delegates representing 46 million party members. For the first time, 200 foreign journalists, including some from Taiwan, were invited to view the opening and closing ceremonies. The Congress was significant in several ways. First and foremost, it reaffirmed the correctness of the policy of reforms and the Open Door that was adopted at the Third Plenum of the Eleventh Congress held in December 1978, and it made economic development the central task of the party. Secondly, it achieved a rejuvenation of the leadership by the voluntary retirement of the Long March generation of elders and secured their replacement with younger and better-educated technocrats. Thirdly, it adopted a new theoretical framework for

28. Chao criticized three leading conservative ideologues, Teng Li-chün, Hu Ch'iao-mu, and Po I-po.
29. The foreign visitors included the Canadian Prime Minister Brian Mulroney, the First Vice-Premier of Singapore, and Japanese and Yugoslavian delegations.

the market-oriented reforms previously thought un-Marxist by the conservatives. Fourthly, it defined the scope of political restructuring so that administrative efficiency could be improved.

Acting General Secretary Chao Tzu-yang opened the Congress with a glowing political report on the major accomplishments of the nine years since December 1978. The gross national product (GNP) and the average income of rural and urban residents had doubled, and the overwhelming majority of the one billion strong population was adequately fed and clothed. According to Chao, building "socialism with Chinese characteristics" under Teng ranked as one of the "two major historic leaps" in the 60-year history of the Chinese revolution, the other being the success of Mao's New Democratic Revolution of 1949. By implication, Teng was placed on a par with Mao as one of the two leading contributors to the enrichment of Marxism-Leninism in China. Many now viewed the current modernization as a Second Revolution or a new Long March.[30]

PERSONNEL CHANGE. Teng personally orchestrated the retirement of more than 90 party elders who were critics of the market-oriented reforms, including P'eng Chen, 85, chairman of the Standing Committee of the National People's Congress; Ch'en Yun, 82, leading party economist and central planner; and Hu Ch'iao-mu, 75, and Teng Li-ch'ün, 72, two orthodox ideologues and harsh critics of bourgeois liberalism; and Li Hsien-nien, 78, president of the People's Republic. Teng himself retired from all positions in the party but kept the chairmanship of the Military Commission through a special party constitutional amendment that allowed him to hold the post without being a Central Committee member. It was also arranged that General Secretary Chao would become the first vice-chairman of the Military Commission; and General Yang Shang-k'un, the permanent vice-chairman. These arrangements could have been a hint that the military would accept Chao as the potential successor to Teng.

The new 285-member Central Committee consisted of 175 regulars and 110 alternates. Some 150 aged leaders (43 percent) of the previous 348-man Central Committee failed to win reelection. Interestingly, Hua Kuo-feng, successor to Mao in 1976, kept his membership. Chao Tzu-yang was confirmed as General Secretary by an overwhelming vote. The average age of the new ruling body was 55.2 years, down from the 59.1 of its predecessor. Eighty-seven of the full and alternate members were new, and 209 (73 percent) of all Central Committee members were college-educated.[31]

---

30. *Beijing Review*, Nov. 2-8, 1987, 10, 12, 18-19; *U.S. News and World Report*, Oct. 12, 1987, 41.
31. *Beijing Review*, Nov. 16-22, 1987, 6.

The Politburo, with 17 regulars and 1 alternate, was packed with younger supporters of reform. Nine of the 20 previous members retired, and former General Secretary Hu Yao-pang retained his seat; so did his close associates, Vice-Premiers Wan Li and T'ien Chi-yüan. The average age was 63, 7 years younger than that of the previous body. In the new Standing Committee of the Politburo, the average age was 64, 13 years younger than that of its predecessor.

A balance of opinion was maintained in the all-powerful Standing Committee, with the election of the following five members: Chao Tzu-yang, 68; Li P'eng, 59; Hu Ch'i-li, 58; Yao I-lin, 70; and Ch'iao Shih, 63. It is generally accepted that Chao and Hu were ardent supporters of mar-ket-oriented economic development and that Li and Yao were inclined to-ward the Soviet-style central planning, with Ch'iao Shih, a security spe-cialist, in the middle. Actually, all of them claimed to be supporters of reform; they differed only in style, method, pace, and scope. Economic de-velopment had become such a national passion that few politicians dared to profess anything else.

Other important party appointments included Ch'en Yun as chairman of the 200-member Central Advisory Commission of Discipline Inspection and Hu Ch'i-li as the leader of a 4-member Secretariat.

With most of the conservative critics out of the way, the reformers gained a mandate to go full steam ahead. Yet they knew that the elders who retired did not relinquish their influence. Their willingness to retire probably was made with the understanding that their favorite choice, Li P'eng, would be appointed to the Politburo Standing Committee as well as to the future premiership. Indeed, barely three weeks after the close of the Congress, Li was named Acting Premier, and later, at the National People's Congress in March 1988, was confirmed as premier.

Li P'eng, the son of an early Communist martyr,[32] grew up in the household of the late premier Chou En-lai. Li joined the party at 17 and was sent to the Soviet Union in 1948 to study electrical engineering. Returning home 6 years later, he rose steadily as an expert in electric power and energy resources. Close to the party elders Ch'en Yun, P'eng Chen, and Teng Ying-ch'ao (Chou's wife), Li enjoyed a broad spectrum of support among the conservative wing of the party, the bureaucrat-followers of Chou in the government, and the Soviet-trained Chinese of the 1950s, who were well-placed in many different walks of life. Sensi-tive of his educational background, Li made a point of disclaiming any preference for the Soviet economic system or central planning. Immedi-

32. Li Shih-hsün.

ately after his election to the Politburo Standing Committee, he announced on November 2, 1987: "The allegation that I am in favor of a centrally planned economy is a complete misunderstanding. The economic system [of China] must be restructured." Upon his appointment as Acting Premier, he announced his support for the policy of economic reform and opening China to the outside world "while continuing to maintain the political stability and unity, and pursue, as always, the country's independent foreign policy."[33] In the political report delivered by General Secretary Chao at the Congress, the statement "the state regulates the market and the market guides (economic) enterprises" could have been a contribution of Li's.[34]

THE PRIMARY STAGE OF SOCIALISM. The constant charge of the conservative hard-liners and the lingering doubts among many others that market-oriented mechanisms were basically un-Marxist were irritating thorns in the side of the progressive movement. Yet market forces had to be recognized to make economic development work, and contact with the outside world was essential to modernization. The most urgent task for the reformers was to develop a theoretical framework to justify their work as being neither capitalistic nor un-Marxist, but as highly necessary and permissible within the limits of socialism. After groping in the dark for a long time, Chinese social scientists came up with the new concept that China was in the primary stage of socialism during which market forces, capitalistic techniques and management skills, and a mixed economy characterized by a multi-ownership system were all acceptable. Sanctioned by this theoretical support, Chao stated confidently in his political report: "Reform is the only process through which China can be revitalized. It is a process which is irreversible and which accords with the will of the people and the general trend of events."

Chao advanced the thesis that China did not go through the proper stage of capitalism because of her previously backward productive forces and underdeveloped commodity economy. To insist on China's going from capitalism to socialism was to be mechanistic and to commit the mistake of the political Right. But to believe that she could skip the primary stage of socialism and proceed straight to socialism was to be utopian and to commit the mistake of the political Left. "During this [primary] stage we shall accomplish industrialization and the commercialization, socialization and modernization of production, which many other countries have achieved under capitalist conditions," concluded Chao.

33. *Los Angeles Times,* Nov. 25, 1987.
34. *Beijing Review,* Nov. 2-8, 1987, 12.

Building socialism with Chinese characteristics, Chao proclaimed, was an experiment that could not possibly have been foreseen by 19th-century European theorists.

> We are not in the situation envisaged by the founders of Marxism, in which socialism is built on the basis of highly developed capitalism, nor are we in exactly the same situation as other socialist countries. So we cannot blindly follow what the books say, nor can we mechanically imitate the examples of other countries [Russia?]. Rather, proceeding from China's actual conditions and integrating the basic principles of Marxism with those conditions, we must find a way to build socialism with Chinese characteristics through practice.

Chao urged his fellow citizens to find ways to allow a multi-ownership system so as to avoid rigidity in economic structure, to expand the commodity economy, to raise labor productivity, and to achieve the Four Modernizations.

The theory was further developed along the following lines. The central task during this primary stage was to end poverty and backwardness. It was no longer class struggle, though that still existed as a contradiction. To ensure the policies of economic development and the Open Door, China had to have stability and unity through the application of the Four Cardinal Principles. The primary stage of socialism could last as long as 100 years from the 1950s, when the private means of production went through a socialist transformation, to the mid-21st century, when the socialist modernization would have been largely completed. This long process was divided into three phases: first, to double the GNP of 1980 and solve the problems of food and clothing for the people—a goal largely achieved; secondly, to double the GNP again by the year A.D. 2000 so as to provide a relatively comfortable standard of living; and thirdly, to reach a level of affluence enjoyed by most medium-developed countries by the middle of the next century. The expansion of science, technology, and education were the keys to this success.[35]

The gnawing questions that plagued the country were these: "Is the reform making China capitalistic, and is a certain new measure socialistic or capitalistic by nature?" Many Chinese and foreigners alike tended to associate "market mechanisms" with capitalist systems, and "central planning" with the Soviet or socialist system. The Chinese leadership was now

35. Chao Tzu-yang," Advance along the Road of Socialism with Chinese Characteristics"—a report delivered at the Thirteenth National Congress of the Communist Party of China on October 25, 1987, *Beijing Review*, Nov. 9-15, 1987, 23-49.

reconciled to the idea that market mechanisms and central planning were both "neutral means and methods that do not determine the basic economic system of a society."[36] Hence, adoption of capitalist techniques and management skills, a mixed economy characterized by a multi-ownership system, and increased grassroots participation in political affairs were all permissible during the primary stage. A current saying went thus: "Whatever promotes economic development is good; whatever hinders it is no good." Such ideas almost sound like American pragmatism.

POLITICAL RESTRUCTURING. Political restructuring did not signify that a Western style reform in which a democratic system complete with free elections, a free press, a three-way division of power, and alternating control of government by different parties would evolve. Rather, it simply meant improvement in administrative efficiency, simplification of unwieldy bureaucratic structures, and elimination of overstaffing. A key feature was the separation of the party from the day-to-day operation of the government and economic enterprises. Government administration of economic enterprises would be replaced by indirect control. Bureaucracy would be streamlined from top to bottom, and a meritocratic civil service system installed.

According to the party leadership, China had to maintain its unique style of government with the distinct character of a socialist democracy. People's congresses at different levels, democratic centralism, and "multiparty" cooperation would continue while grass-roots participation in government increased. With the development of an efficient legal system, people's rights would be protected from arbitrary official violations and extralegal procedures. In this way, a social democracy could be built.

AN ASSESSMENT. The Thirteenth Congress was remarkable for several reasons. It firmly launched China onto the road of accelerated economic development and greater opening to the outside world. In no other Communist state had any ruling group voluntarily relinquished power in favor of a younger leadership. Even more remarkably, Chinese leaders had found that the traditional Communist system was unworkable unless it was adulterated with market mechanisms. The Soviet leader Mikhail Gorbachev seemed to have made the same discovery, but China was ahead of the Soviet Union in breaking away from the bondage of orthodox Marxism. China had forged an important new ideological tool—the development of

36. David Holly, "New Leaders, Reforms to be Weighed at Chinese Party Congress," *Los Angeles Times*, Oct. 24, 1987.

a new theory to fit reality rather than bending "reality to theory."[37] Teng had turned economic reform into an irreversible commitment that enjoyed the vast support of the people. Meanwhile, Gorbachev was still "grinding slowly uphill in low gear," tackling nearly the entire Soviet bureaucracy. Knowledgeable sources placed greater odds on China's economic success than on Russia's.[38]

But, remarkable as it was, the Thirteenth Congress left many questions unanswered. First of all, its work represented no clear victory for the progressive reformers, but rather a compromise among disparate groups within the party. The conservative elders had retired but had not relinquished their influence and could still have used it to block more drastic liberalization.[39] Many sensitive issues such as price decontrol, inflation, leasing state enterprises to private operations, bankruptcy law, and transfer of the land utilization right (a euphemism for private ownership) could still prove explosive. In particular, the Weng-chou model of private ownership, a pet project of Chao's, was certain to touch off controversy.

Secondly, Teng had arranged for Chao to be the First Vice-Chairman of the Military Commission, but there was no assurance that Chao could succeed him as the commander-in-chief or the paramount leader. In the past, all designated heirs had fallen before coming to power: Liu Shao-ch'i in 1966, Lin Piao in 1972, and Hua Kuo-feng in 1978. Instability has been inherent in the leadership succession in any Communist state. Common sense tempts one to ask, If Chao was protected by Teng during his (Teng's) lifetime, what would happen to him after Teng was gone? It was necessary at this point that the rule of law replace the rule of man if stability was to be achieved.

Thirdly, accelerated economic development and greater opening to foreign influences would inevitably revive the old questions of "spiritual pollution," "bourgeois liberalization," and the perennial issue of "Chinese essence vs. foreign value." The Thirteenth Congress skirted these issues.

Fourthly, separation of party functions from the government and economic enterprises would affect the vested interests of millions. Its implementation could be excruciatingly slow and difficult.

Fifthly, the supremacy of the Four Cardinal Principles prohibited any rule other than the Communist and any freedom beyond what was per-

---

37. Henry A. Kissinger, "China Now Changing Rules and Ruling Party," *Los Angeles Times*, Oct. 25, 1987.

38. Joseph C. Harsch, "A New Look," *The Christian Science Monitor*, Nov. 5, 1987, and "Fortunes Shift for Leaders of World's Three Powers," *Ibid*. Nov. 6, 1987.

39. Adi Ignatius, "China's Party Meeting Unlikely to Settle Success Issue," *The Wall Street Journal*, Oct. 23, 1987.

mitted by the party. A modicum of dissent might be allowed to a select few—such as a famous astrophysicist, a journalist, a writer, or an artist—for purposes of window dressing and nothing more, but the extent of tolerance would be tightly controlled and strictly limited.

All in all, the Congress was a qualified success insofar as it represented a concensus among the disparate leaders to move the country forward economically. But the search for the higher goals of democracy, pluralism, and human rights would necessitate another Long March.

The Coastal Development Plan

In the period 1987-88 China evolved a new economic strategy for the coastal regions by which the development of these areas was to be accelerated and closely tied to international markets. The Coastal Development Plan had two basic components: (1) the importation of raw materials from abroad and (2) the exportation of processed finished goods to world markets to earn foreign exchange. The capital raised would be used to purchase high technology and modern equipment, which were meant to help finance China's heavy industry. These, in turn, would help the development of agriculture. The coastal economy would be made heavily export-oriented, with both ends of the production process—the supply of raw materials and the marketing of the processed goods—deeply involved in the world economy. The importation of foreign materials was seen as necessary at first, because China's inland provinces could not provide them; but as conditions improved, some of the raw materials could come from Western China. In the long run, the benefit would "trickle down" thoroughout the entire country.

The Coastal Development Plan followed logically from the adoption of the open-door policy in December 1978. From that point, the country continued to move in the direction of greater opening and more international contact, as evidenced by the creation of Four Special Economic Zones in 1979; the opening of 14 seaports; and the more recent opening of the Yangtze Delta, the Pearl River Delta, and the Southern Fujian Triangle, as well as the designation of Hainan island as the fifth economic zone, in 1985. There was a realistic acceptance of the fact that no two regions of China could develop at the same speed and that China should accelerate the development of the coastal areas now to maximize the benefits of the export trade.

The theoretical basis of this plan was advanced in June 1987 by a 34-year-old economic planner, who vigorously expounded the advantage of integrating China's coastal economy with international markets in a continuing cycle of importation of raw materials for processing and exportation of finished goods for foreign exchange to finance China's modernization.[40] The plan called for three stages of development. During the first stage of 5 to 7 years, the coastal economy would be made export-oriented, especially in the areas of textiles, foodstuffs, small electric appliances and light industrial products. Meanwhile, efforts would be made to improve communication with the inland provinces, but investment in heavy industry would have to wait until exports had earned sufficient foreign capital. In the second stage, again, of 5 to 7 years, inland products would begin to enter the international markets and greatly enlarge the foreign exchange earning capacity of the labor-intensive industries. In the third stage, between 1996 and 2000, there would be a substantial increase in the export of sophisticated, technology-intensive industrial products and a corresponding decrease in the export of labor-intensive goods. More surplus labor would flow into high-tech products, accelerating the speed and quality of economic growth. During the first stage, some 60 million farm laborers could be absorbed into the export-oriented activities; and during the second and third, 120 million. By the end of the century, China would be able to export $150 billion worth of goods yearly, which would require an annual export growth rate of 12 percent. This would not be excessively high compared with Japan's 17 percent, Brazil's 16 percent, and South Korea's 40 percent, during their peak years of economic development.

The Coastal Development Plan took into consideration the peculiar, dichotomized economic structure of China that had existed since the Liberation in 1949. Under Soviet influence, heavy industry received favored treatment at the expense of light industry, and agriculture was kept at a low level of technology, with vast amounts of farm labor tied to the land. In this dichotomy, contradictions were inevitable: greater industrialization would release more labor, but developing a higher technology would obviate the need for labor. Also, labor-intensive industries tended to enhance consumption, but development of these sophisticated industries required capital. Where to find it? Through the global market!

The coastal areas of China were ideally suited for labor-intensive, export-oriented industries because of the abundance of intelligent, diligent, and relatively well-trained, but inexpensive labor. These areas were also endowed with considerable scientific, technical information and telecom-

40. Wang Chien, "Hsüan-tse cheng-chüeh ti ch'ang-chi'i fa-chan chan-lüeh—kuan-yü 'Kuo-chi ta hsün-huan' ching-chi fa-chan chan-lüeh ti kou-hsiang" (Correctly select a long-term development strategy: the concept of a "Grand International Cycle" economic development plan), *Ching-chi Jih-pao* (Economic Daily), Peking, Jan. 23, 1988, 3. Ideas in this seminal article provided most of the information of this section.

munication facilities, which offered a suitable environment for combined labor-intensive and knowledge-intensive activities. Indeed, rural industries were already booming in the eastern and southeastern coast: the Pearl (Chu-chiang) River Delta near Canton, the Yangtze (Ch'ang-jiang) River Delta, the Southern Fukien Triangle, and the Shantung and Liaotung Peninsulas—in short, the municipalities of Canton, Shanghai, and Tientsin; the provinces of Kwangtung, Fukien, Chekiang, Kiangsu, and Shantung; and the southern tip of Liaotung Peninsula (Lu-shun and Darien). The rural industries were self-reliant and highly efficient, responsible for their own profits and losses. In 1987, they could boast of 85 million well-trained laborers producing Y450 billion worth of goods, outpacing the value of production of the agricultural sector. This vast, inexpensive labor force, if turned export-oriented, would place China in a highly favorable and competitive position for at least 20 years relative to the Four Little Dragons of Asia—South Korea, Taiwan, Hong Kong, and Singapore.[41] Furthermore, the currencies of Japan, South Korea, and Taiwan were rising in value whereas the American dollar and the Chinese yuan were depreciating, making Chinese products even more attractive in global markets. The China coast was ready to "process foreign raw materials," "to accept foreign orders according to specifications," and "to assemble foreign parts"; and the government was ready to "subsidize foreign traders with Chinese products" (*San-lai i-pu*).

The Coastal Development Plan appealed to the reform-minded General Secretary Chao Tzu-yang and was deliberated on in the high councils of state. After the Thirteenth Congress, Chao made two inspection trips to the coastal areas of Shanghai, Kiangsu, Chekiang, and Fukien provinces in November 1987 and January 1988, and he was thoroughly convinced of the feasibility of the plan. His optimistic report on "The Strategic Problems of Coastal Economic Development" won the full support of Teng Hsiao-p'ing.[42] Teng endorsed the plan with the remark, dated January 23, 1988: "Completely approved. It is imperative that you go ahead boldly, speedily, and not miss this key opportunity!"[43] On February 6, 1988, the Politburo formally approved the plan, and coastal authorities were instructed to work for its success and to welcome foreign investment, joint ventures, and foreign management of Chinese enterprises in their respective jurisdictions.

41. The average wage of a Chinese worker in 1987 was one-fifth of that of a comparable worker in Taiwan and South Korea, and one-eighth of those in Hong Kong and Singapore. *Ta Kung Pao,* Hong Kong, Feb. 12, 1988, 1.
42. A summary of the report appears in English under the title, "Chao on Coastal Areas' Development Strategy," *Beijing Review,* Feb. 8-14, 1988, 18-23.
43. *Ta Kung Pao,* Hong Kong, March 21, 1988, 2.

The coastal development strategy must be viewed as an ingenious masterstroke of the reformers, who used it to link China with the world economy and, at the same time, to block any attempt by the conservatives to move the economy back to central planning and economic isolation. Yet the plan also made China heavily dependent on the international economy, subjected her to the mercy of volatile and fluctuating foreign markets, and forced China to forfeit control of her economic fate. It was essential that Chinese products be of high quality to compete with the exports from the other Pacific nations and to secure a foothold in the global markets. To win foreign orders and contracts and to set up networks of distribution in foreign countries, it would be necessary for China to train large numbers of sales and trade representatives who should be well versed in foreign languages, market conditions, and international business practices. Moreover, for the project to be successful, time was of the essence. If it had been put into practice 20 or 30 years before, China would have taken a proud place alongside Japan and the Four Little Dragons of the Pacific Rim, but now China would have to face much stiffer resistance from the West and Japan owing to rising protectionism and what seemed to be a looming threat of worldwide economic slowdown or recession in the early 1990s. In the final analysis, the adoption of the coastal development strategy confirmed the unwelcome truth that communism by itself could not transform China into a modern state; it needed capitalist help.

Society in Flux: Rising Inflation and Falling Ethics

China's fast economic growth was accompanied by a high inflation rate, officially put at 12.5 percent in 1985, 7 percent in 1986, and 8 percent in 1987, but unofficially estimated at 20 to 30 percent a year. The ravages of inflation were felt by all salaried people and affected their outlook, lifestyle, behavior, and ultimately their social ethics. Without a doubt, of all the reform measures, price decontrol and wage adjustment were the most painful and least successful. The government was determined to carry out pricing reforms because the old system of subsidies on food, cooking fuel, housing, and virtually everything else ignored the law of value and drained the state budget. It insisted that price decontrol was the key to the success of the economic reforms and that momentary discomfort was to be preferred to future pains. But for the people, decontrol led to price rises that always outpaced wage adjustments. In May and June of 1988, control on four foods—pork, eggs, vegetables, and sugar—was lifted, and prices shot up 30-60 percent overnight in the big cities. The government subsidy of

Y10 to each worker, Y8 to each college student, and Y7 to each middle school student, was hardly sufficient to offset inflation, even though the government argued that fast economic growth in the past ten years should have mitigated the pains of deregulation.[44]

Inflation heightened corruption in government, nepotism among the children of high cadres, backdoor deals, and black-marketeering. An adviser to the government admitted: "Nowadays it is almost impossible to do anything without bribing officials."[45] The profit motive transcended all other considerations. Frequently, an article carried four prices: a state-set price, a market price, a negotiated price, and a foreigner's price. Airline and train tickets were not only hard to get, but they carried different prices for different travelers. Universities were asked to profit from sideline activities to subsidize education. Knowledgeable persons worried that China had become a country of traders (*ch'uan-min chieh-shang*). A pun became current in society: *hsiang-ch'ien k'an*, which homonymously means "looking forward" or "looking to money." Practice of one-upmanship, fraud, and economic crimes was rampant; and gouging and overcharging were common in hotels, restaurants, and state offices.

The hardships common in the 1980s seemed to reduce people's feelings of self-worth and make them grumpy and sullen. Life became characterized by indifference, callousness, and rudeness. Psychologists suggest that such an attitude reflected utter frustration and hopelessness rather than enjoyment of being rude. Inflation added yet another burden to the weary souls of those already struggling with economic problems. Watching their life savings steadily eroded by rising prices, people dreaded the return of the hyperinflation of the period 1945-49 and desperately sought ways to beat it.

Between 1983 and 1988, most salaried city dwellers saw their purchasing power cut by 100 percent or more; and their standard of living, drastically lowered. A dinner in a good restaurant cost five times more in 1988 than in 1983. A college graduate who made Y55 a month in 1956 could support a family of four, but in 1988 with a higher salary of Y133 he could not; his daily wages could only buy him two watermelons! In terms of gold, his 1987 income was only 49 percent of the 1979 value and 15.7 percent of the 1956 value.

In 1987 an average worker's family spent 35 to 45 percent of its income on food and another 25 to 35 percent on other necessities.[46] There was lit-

---

44. *Beijing Review*, May 23-29, 1988, 10; Jupne 20-26, 1988, 7-9.
45. *Newsweek*, Hong Kong, June 6, 1988, 25-26.
46. Feng Ching, "The Life of Ordinary Chinese People," *Beijing Review*, July 4-10, 1988, 21-26.

tle left to meet all the inevitable exigencies of life. Frustration gave rise to anxiety, selfishness, resentment, and rudeness. Public ethics sank to the point of disintegration. Confusion in values and chaos in management were everywhere.

Decontrol provided the occasion for price increases, but two other fundamental causes of inflation should not be overlooked. One was the rise in the money supply at a rate of 20 percent annually between 1984 and 1988, resulting in a rapid increase in the issuance of paper money. The other was the large amounts of surplus cash in bank deposits (Y307.5 billion) and liquid cash—some Y420 billion by the end of 1987. Frequently, people rushed to the banks to withdraw money in order to buy needed goods before a new round of price increases. When too much money chased after too few goods, the proverbial inflation resulted.

Public discontent and protest against price rises led to many stormy sessions in the Politburo at the summer resort of Pei-tai-ho in July 1988. General Secretary Chao continued to espouse price reform as the heart of his economic restructuring, while Premier Li P'eng and Vice Premier Yao I-lin argued for a slowdown if not suspension of decontrol on grounds that wage increases had not kept pace with price increases. In the end, concern over public discontent prevailed. The government took measures to cool the economy by controlling credit, tightening the money supply, and curbing construction and capital outlay. In late August, it announced that there would be no further price decontrol for the rest of 1988 and 1989. People sighed with relief.

One group, however, seemed to thrive in these unsettled times. They were the emerging individual entrepreneurs and private businesspersons (*Ko-t'i-hu*), who mostly engaged in handicrafts, light manufacturing, home appliance repair and sales, and transport and consumer services. They achieved success through hard work, vision, judgment, risk-taking, and good management—much as anywhere else. Numbering 225,000 in June 1988 and employing 3.6 million people, these private enterprises were condoned by the state as legitimate on grounds that they were conducive to the development of productive forces. Many of these individuals became millionaires. A farmer in Shen-yang, Liaoning Province, organized a transport team and made Y1 million in 1987. Another entrepreneur invested Y1 million in a steel rolling mill and maintained 100 workers on her payroll. A 31-year-old owner of a one-hour Kodak film processing shop in Foochow controlled assets of $539,000 and lived in a $108,000 air-conditioned house. A wood-carving "king" employed 3,000 workers and amassed $20 million. A motorcycle helmet maker in Peking did so well that a foreign firm offered him a joint venture and paid him $82,000 a year in salary.[47] On a lesser scale, many individual entrepreneurs of little education—owners of small appliance shops, tea farmers, owner-drivers of taxis—were making Y30,000 to Y100,000 a year, 10 to 30 times the salaries of professors and surgeons. Not surprisingly, some of the most money-hungry elements turned to crime and engaged in the robbing of ancient tombs in hopes of finding instant wealth.[48]

Had money-mania driven China to capitalism? Most Chinese theorists thought not because the total output of the private sector hardly amounted to 1 percent of the national industrial production value.[49] Nonetheless, the spirit of enterprise was gripping the nation, and it was not surprising that the Chinese translation of Lee Iacocca's autobiography, which extols the virtue of individual initiative, was a best-seller in 1987-88.[50]

In this climate, the primacy of ideology was extenuated, leading to a blurring of direction and a rise in self-doubt. Reverence for Marxism and Maoism yielded to more pragmatic assessments. Chao Tzu-yang regarded Mao's last two decades as "twenty lost years," and Teng Hsiao-p'ing, in June 1988, advised the visiting President of Mozambique "not to practice socialism."[51] One Chinese theorist asked: "Who knows what Marxism is, anyways? Today, we live in a technological world beyond the imagination of Marx."[52] Another stated that Marx saw only the early stages of capitalism and Lenin experienced little "real-life socialism in his lifetime." Still others believed that capitalism and socialism should borrow from each other for mutual benefit, and China's turn to a "socialist commodity economy," which, in practice, meant adoption of some of the mechanisms of capitalism, was entirely justified during the primary stage of socialism.[53]

China in 1988 was in a state of flux. The Maoist order was largely gone, but a substitute was not yet born. With a 17 percent economic growth rate and 26 percent inflation, China was experiencing the growing pains of a developing nation. In those unsettled and unsettling times, economic euphoria, ideological confusion, falling morality, and widespread corruption formed a vortex of paradoxes, from which a new order was struggling to emerge.

47. *The Christian Science Monitor*, June 9, 1988; Oct. 4, 1988. *Asiaweek*, Hong Kong, July 1, 1988, 19; *U.S. News and World Report*, Sept. 8, 1986.
48. *China Daily*, Beijing, June 28, 1988.
49. *Beijing Review*, July 18-24, 1988, 12-13.
50. *U.S. News and World Report*, Feb. 8, 1988, 30.
51. President Joaquim Chissano. *Asiaweek*, Hong Kong, July 1, 1988, 18.
52. *U.S. News and World Report*, Feb. 8, 1988, 30.
53. *Asiaweek*, July 1, 1988, 20-21.

Further Reading

Amnesty International, *China: Violations of Human Rights* (London, 1984).

Benton, Gregor (ed.), *Wild Lillies, Poisonous Weeds: Dissident Voices from People's China* (Dover, N.H., 1982).

Brzezinski, Zbigniew, *The Grand Failure: The Birth and Death of Communism in the Twentieth Century* (New York, 1989).

Ch'en, Hsi-yuan, "Ts'ung kuo-chi ta hsun-huan tao chia-k'uai yen-hai ching-chi fa-chan" (From the "Grand International Cycle" to acceleration of coastal development), *Chung-Kung Yen-chiu* (Chinese Communist Studies), Taipei, 22:7:88-94 (July 1988).

Cohen, Roberta, "People's Republic of China: The Human Rights Exception," *Human Rights Quarterly*, 9:447-549 (1987).

Chao, Tzu-yang, "Advance Along the Road of Socialism with Chinese Characteristics"—a report delivered to the Thirteenth National Congress of the Communist Party of China, Oct. 25, 1987, *Beijing Review*, Nov. 9-15, 1987, 9-15.

——, "Chao on Coastal Areas' Development Strategy," *Beijing Review*, Feb. 8-14, 1988, 18-23.

Chu, David S. K., *Sociology and Society in Contemporary China, 1979-1983* (Armonk, N.Y., 1984).

Duke, Michael S., *Blooming and Contending: Chinese Literature in the Post-Mao Era* (Bloomington, 1985).

Edwards, R. Randle, Louis Henkin, and Andrew J. Nathan, *Human Rights in Contemporary China* (New York, 1986).

Goldman, Merle, *China's Intellectuals: Advise and Dissent* (Cambridge, Mass., 1981).

——, Timothy Cheek, and Carol Lee Hamrin (eds.), *China's Intellectuals and the State: In Search of a New Relationship* (Cambridge, Mass., 1987).

Goodman, David S. G., *Beijing's Street Voices: The Poetry and Politics of China's Democracy Movement* (New York, 1981).

Hamrin, Carol Lee, *China and the Future: Decision Making, Economic Planning, and Foreign Policy* (Boulder, 1987).

——, and Timothy Cheek (eds.), *China's Establishment Intellectuals* (Armonk, N.Y., 1986).

Hayhoe, Ruth, and Marianne Bastid, *China's Education and the Industrial World: Studies in Cultural Transfer* (Armonk, N.Y., 1987).

Kallgren, Joyce K., and Denis Fred Simar (eds.), *Educational Exchanges: Essays on the Sino-American Experience* (Berkeley, 1987).

Kissinger, Henry A., "China Now Changing Rules and Ruling Party," *Los Angeles Times*, Oct. 25, 1987.

Kinkley, Jeffrey C. (ed.), *After Mao: Chinese Literature and Society, 1978-1981* (Cambridge, Mass., 1985).

Lampton, David M., Joyce A. Modancy, and Kristen M. Williams, *A Relationship Restored: Trends in U.S.-China Educational Exchanges, 1978-1984* (Washington, D.C., 1986).

Link, Perry (ed.), *Roses and Thorns: The Second Blooming of the Hundred Flowers in Chinese Fiction, 1979-80* (Berkeley, 1984).

———, *Stubborn Weeds: Popular and Controversial Chinese Literature After the Cultural Revolution* (Bloomington, 1983).

Liu, Pin-yen, "Ti-erh-chung chung-ch'eng," (The second kind of loyalty), *Cheng Ming* magazine, Hong Kong, 10:48-61 (October 1985).

Louie, Kam, *Inheriting Tradition: Interpretation of the Classical Philosophers in Communist China, 1949-1966* (New York, 1986).

Mosher, Steven W., *Journey to the Forbidden City* (New York, 1985).

Nathan, Andrew J., *Chinese Democracy* (New York, 1985).

———, R. Randle Edwards, and Louis Henkin, *Human Rights in Contemporary China* (New York, 1986).

Orleans, Leo A., *Chinese Students in America: Policies, Issues, and Numbers* (Washington, D.C., 1988).

Rozman, Gilbert, *The Chinese Debate About Soviet Socialism, 1978-1985* (Princeton, 1987).

Sa, Kung-ch'iang, "P'ing Chung-Kung yen-hai ti-ch'ü ching-chi fa-chan hsin chan-lüeh" (On Communist China's new coastal economic development strategy), *Chung-kuo ta-lu yen-chiu* (Studies on Mainland China), *Taipei*, 30:11 (May 1988).

Schram, Stuart R., "China After the 13th Congress," *The China Quarterly*, No. 114:177-197 (June 1988).

Seymour, James D., *China's Satellite Parties* (Armonk, N.Y., 1987).

——— (ed.), *The Fifth Modernization: China's Human Rights Movement, 1978-1979* (Stanford, 1980).

Shapiro, Judith, and Liang Heng, *Cold Winds, Warm Winds: Intellectual Life in China Today* (Middletown, Conn., 1986).

Siu, Helen F., and Zelda Stern (eds.), *Mao's Harvest: Voices from China's New Generation* (New York, 1983).

Sullivan, Lawrence R., "Assault on the Reforms: Conservative Criticism of Political and Economic Liberalization in China, 1985-86," *The China Quarterly*, 114: 198-222 (June 1988).

Tung, Constantine, and Collin MacKerras (eds.), *Drama in the People's Republic of China* (Albany, N.Y., 1987).

Wang, Chien, "Hsüan-tse cheng-ch'üeh ti ch'ang-ch'i fa-chan chan-lüeh—kuan-yü 'Kuo-chi ta hsün-huan' ching-chi fa-chan chan-lüeh ti kou-hsiang" (Correctly select a long-term development strategy: the concept of a "Grand International Cycle" economic development plan), *Ching-chi jih pao* (Economic Daily), Peking, Jan. 23, 1988.

Wortzel, Larry M., *Class in China: Stratification in a Classless Society* (New York, 1987).

39

Taiwan's "Economic Miracle" and the Prospect for Unification with Mainland China

Walking down the main streets of Taipei, one witnesses an unending flow of motorcycles, buses, and cars, with hotels, modern apartments, and high-rise office buildings on each side. Inside the offices is the controlled chaos of successful enterprise; the electronic chorus of elevators, air conditioners, typewriters, and computers is punctuated by the ringing of telephones and the raised voices of those making overseas calls. Farmers, a world away from Taipei's bustle, proudly show the fruits of their labor; nicely clothed and fed, with modest but comfortable homes, they appear content with their lives.

This is modern Taiwan, transformed from an agricultural society to an industrial power within a generation. The title "Little Japan" evokes a mixed reaction of open displeasure and secret pride. In East Asia, Taiwan is indeed second only to Japan in terms of industrialization, foreign trade, and quality of life. Taiwan's success is its most important weapon in the struggle for survival, security, and international ties. A close look at Taiwan's accomplishments will help explain its current position and future course.

By 1988, Taiwan had enjoyed peace, stability, and sustained economic growth for more than 35 years. The rate of growth averaged 7.3 percent per annum in the 1950s, 9.1 percent in the 1960s, and nearly 10 percent in the 1970s. According to *Euromoney*,[1] during the decade of 1974-84, Taiwan had the second highest growth rate in the world, next only to

---

1. A European financial journal.

Singapore. The speed of development accelerated even more during the 1980s. In 1980 Taiwan registered a GNP of $40.3 billion, a per capita GNP of $2,100, and a foreign exchange reserve of $7.4 billion. In 1984, the 10.52 percent growth generated a GNP of $57.5 billion, a per capita GNP of $3,046, foreign trade amounting to $52 billion,[2] and a foreign currency reserve of $16 billion. In 1992 the surging economy generated a GNP of $211 billion, a per capita GNP of $10,215, a foreign trade of $139 billion, and a foreign currency reserve of $82 billion, the highest in the world. Amazingly, with only 20 million people, Taiwan held about 10 percent of the world's reserves.[3] Most of Taiwan's reserve was derived from trade surpluses with the United States—$13.6 billion in 1986, $19 billion in 1987, $14.1 billion in 1988, and $13 billion in 1989. If the economic growth rate could maintain a steady 6.5 percent per annum until A.D. 2000, the per capital income would reach $15,000 by then; and the foreign trade, $290 billion, making Taiwan a developed country and one of the 10 largest trading nations on earth.[4]

It is noteworthy that wealth had not been concentrated in a few hands but was shared by a majority of people, in fulfillment of the ancient ideal of an "equitable distribution of wealth" (*ch'ün-fu* 均富). The income ratio of the highest and lowest 20 percent of wage earners in 1952 was 15 : 1; but in 1964 only 5.33 : 1; and by 1987, 4.69 : 1, less of a gap than in the United States.[5] Televisions, refrigerators, washing machines, and telephones had become common; and unemployment was 1.69 percent in 1988. The average lifespan for a man was 71, and for a woman, 76; and the daily intake of calories was 2,845 and of protein, 80 grams, both exceeding international standards. Inflation has been kept low, at only 1.1 percent in 1988. Almost unique in the simultaneous attainment of rapid economic growth, price stability, and the equitable distribution of wealth, Taiwan can be said today to enjoy the highest standard of living in Chinese history.

Causes of Taiwan's Economic Success

ECONOMIC STRATEGY. Taiwan's economic strategy has given priority to agriculture, light industry, and heavy industry in descending order. From

2. Thirty billion dollars in exports against $22 billion in imports.
3. Figures of the International Monetary Fund as reported in *Los Angeles Times,* Jan. 4, and March 7, 1988; *The Free China Journal,* Taipei, Feb. 19 and 23, 1993.
4. *The Free China Journal,* Taipei, March 24, 1986.
5. *The Christian Science Monitor,* editorial, July 1980. In the United States, 9 to 1; in Mexico, 20 to 1. *The Free China Journal,* Taipei, August 11, 1988.

1949 to 1960, the thrust of policy was toward development of agriculture and light industry, beginning with a three-stage land reform program and followed by labor and technological innovations that accelerated farm production. Measures were also taken to expand the physical and social infrastructure, to stabilize prices, to reform the foreign exchange system, and to develop light, import-replacement industries.

Small and resource-poor, Taiwan had no choice but to depend on trade. Recognizing this, the economic planners emphasized industrialization and exports in the 1960s. Measures were adopted to increase the production of durable consumer goods (light industry); encourage labor-intensive, export-oriented assembly industries; and diversify agricultural products for export. Electronics, synthetic fiber, and plastics industries grew rapidly and moved into the world market. With labor costs low and quality control high, Taiwan's products competed successfully in foreign markets.

In the 1970s the emphasis was shifted to developing sophisticated and heavy industry as well as to expanding the infrastructure. In 1973 Ten Major Projects (seven of them infrastructure-related) were launched at a cost of $7 billion: (1) the North-South Freeway known as the Sun Yat-sen Memorial Express Way, (2) the international airport at Taoyuan outside Taipei named after Chiang Kai-shek, (3) electrification of the west coast railway trunk line, (4) the northern coastal railway, (5) the Taichung Harbor, (6) the Suao Harbor expansion, (7) the nuclear power plant at Chin-shan near the northern tip of the island with two generating units, (8) the modern steel mill called the China Steel Corporation, (9) the giant China Ship Building Corporation, and (10) a petrochemical complex. The last three were located at or near Kaohsiung, which was transformed into a special municipality in 1979 on a par with Taipei.

With the completion of these ten projects in 1979, Taiwan took on the appearance of a "rich developing nation." The projects injected a large dosage of capital into the economy and relieved the recession of 1974-75. Moreover, a large number of economic planners, engineers, and technicians working on these projects gained valuable experience; and numerous workers received training in the process, giving all a new confidence in their ability to build a modern society with their own hands.

To utilize fully these new skills and to modernize the island further, the government immediately launched the Twelve New Projects with emphasis on technology- and capital-intensive industries. These included expanding steel mills; adding nuclear power plants; constructing new cross-island highways; completing Taichung Harbor and the round-the-island railway system; extending the freeway; improving regional irrigation and drainage; building major sea dikes; increasing farm mechanization; and constructing

General Secretary Chao Tzu-yang.

Premier Li P'eng.

President Lee Teng-hui, Republic of
China on Taiwan, 1988.

Premier Lien Chan, ROC.

new towns, cultural centers, and housing. With the successful completion of these projects at a cost of $5.75 billion, Taiwan became one of the Newly Developed Countries.

In 1985 the government launched the Fourteen Key Projects to enlarge the economic base at an estimated cost of $20 billion. These projects consisted largely of infrastructural construction and improvement of existing structures. They included the third-phase expansion of the China Steel Corporation; the construction of new railways; an underground rapid transit system in metropolitan Taipei; modernization of telecommunication facilities; development of four national parks, additional flood control projects, dikes, and garbage disposal sites; exploitation of oil and water resources; and preservation of the natural ecology. Projected completion dates were scheduled for 1990-91.

MEANS OF MODERNIZATION. Modernization required capital, qualified personnel, and scientific management. Although government and domestic private investment had constituted a major source of capital, foreign and overseas Chinese (ethnic Chinese of Singapore, Hong Kong, the United States, and so forth) investments had been heavy owing to Taiwan's favorable investment conditions. In 1960 the Statute for Encouragement of Investment offered a deferrable five-year income tax exemption for capital-intensive and high-technology industries. It set a maximum income tax liability of 25 percent after the tax holiday and offered other privileges such as exemptions from export taxes, customs duties, business taxes, and so on. In 1979 the statute was revised to offer tax credit further to attract foreign capital, and the substantial investments from abroad continued to grow.

The creation of three Export Processing Zones—two in Kaohsiung area and one near Taichung—in the 1960s also encouraged foreign investment by simplifying customs procedures and export regulations. By 1974 a total of $156,755,000 had been invested in 291 Export Zone projects, and exports totaled $511,322,000 with a favorable balance of over $200 million.

Introduction of new production techniques from abroad had also played an important role in rapid economic development. Under the Statute on Technical Cooperation, a total of 837 cases of technical cooperation under private agreements was made between 1952 and 1974. Japan led the way with 615 cases, and the United States followed with 151 while Europe accounted for 57 cases, and other countries, 14.[6]

The qualified personnel that encouraged investment and modernization

6. *A Review of Public Administration: The Republic of China,* compiled by the Administrative Research and Evaluation Commission, Executive Yüan, 1975. (Taipei, 1975), 79-80, 96.

had been provided by Taiwan's excellent educational system. A nine-year free education was available to all. By the period 1977-78, 99.6 percent of the elementary school age group were attending primary schools; 50.9 percent of those between 15 and 17 years of age were attending senior high school; and 25.2 percent of those between 18 and 21 were attending the 101 colleges and universities. In addition, Taiwan had sent a large number of students to the United States for advanced studies.[7] Thus, there was no shortage of trained personnel on the island.

Abundant incentives and privileges as well as trained personnel and cheap labor prompted foreign and overseas Chinese investments to grow quickly from a few million dollars each year in the 1950s to $213 million in 1978. In 25 years (1952-78) those investments totaled $1.92 billion— 31 percent ($595 million) from overseas Chinese, 30 percent ($586 million) from the United States, 17 percent ($321 million) from Japan, and 12 percent ($227 million) from Europe. Favorite investment targets were electronic and electrical products ($633 million) and chemicals ($291 million), followed by services, machinery and instruments, metal products, and textiles.

The United States' recognition of the People's Republic of China on January 1, 1979, had little impact on economic growth and foreign investments in Taiwan. The 1979 GNP registered a 20 percent increase; foreign trade, 31 percent; and foreign investments, 50 percent. More remarkably, informal trade with mainland China via Hong Kong and Japan reached $100 million in 1979 and $2.7 billion in 1988.[8] Taiwan's foreign trade in 1987 was the second largest in Asia and fourteenth in the world.

Rapid industrialization had caused inevitable environmental and social problems including traffic, pollution, industrial waste, inflation, juvenile delinquency, and the migration of the rural inhabitants to urban centers, swelling the cities and causing a steady decline of the farm population.[9] A serious social and economic problem was the high birthrate, which stood at 2.064 percent in 1980, creating a population density of 564 people per square kilometer in 1990, and 10,160 in Taipei—higher than New York's

7. In the period 1979-80, there were 17,560 students from Taiwan enrolled in United States colleges and universities, next only to Iran's 51,310. Other large foreign student groups included these: Nigeria, 16,360; Canada, 15,130; Japan, 12,260; and Hong Kong, 9,900. See *The Chronicle of Higher Education,* May 11, 1981, 14. In 1985, Taiwan sent 22,590 students to study in U.S. colleges and universities, comprising the largest number of foreign students.

8. U.S. Department of State, "Review of Relations with Taiwan," Current Policy No. 190, June 11, 1980 (Washington, D.C.). *The Free China Journal,* March 30, 1989.

9. Farm population fell from 51.9 percent of the total in 1953 to 21.6 percent in 1986. *The Free China Journal,* May 12, 1986.

9,050 and Tokyo's 5,388.[10] Years of government effort at birth control finally brought the rate down to the desired 1.241 percent in 1990. In 1993 there were about 20.79 million people living in Taiwan.

SOCIAL CHANGE. Under the impact of rapid industrialization, profound changes took place in social structure and family relationships. The most significant of these changes was in the rapid rise of the middle class within the last 20 years, accounting for 50 percent of the population by 1985. They were a fluid heterogeneous group of small businesspersons, professionals, technical and managerial personnel, who had received a good education and possessed considerable wealth. Prominent among them were lawyers, engineers, architects, doctors, pharmacists, accountants, public functionaries, and business managers. Vocal and public-spirited, they participated in politics and spoke out forcefully on public issues. They enjoyed life through spending rather than saving as was the old Chinese tradition. Having reached a preferred status in life, they were fearful of radical social and political change. Everything considered, they had no use for communism.

Family relations also underwent significant and far-reaching changes. Parental authority declined sharply in matters relating to children's marriages and career selections. Children chose their own mates or met them through introductions by friends or matchmakers: parental consent was sought later but generally only as a formality. Parents also lost the traditional power to oblige children to follow in the family line of work. Universal education and plentiful openings in industry made the young more independent and assertive than ever before. They made their way in society through hard work rather than family influence; 74 percent of those surveyed showed little interest in family inheritance as a means of gaining social recognition. Those who lived at home did not automatically surrender their paychecks to parents as in the past, and those who lived away sent token remittances. The values of the traditional society rapidly gave way to new ones created by rapid industrialization; greater individual freedom and faster social mobility were the hallmarks of change.[11]

Rapid accumulation of wealth had its adverse effects on economic behavior and social ethics. Years of trade surpluses and the unusually high savings rate of 34 percent flooded society with idle cash looking for outlets. Speculation in stocks, bonds, commodities, and real estate became rampant,

---

10. *The Free China Journal*, April 4, 1991.
11. Lai Tse-han, "She-hui pien-ch'ien chung ti chia-t'ing chih-tu" (The family system in a changing society), *Overseas Scholars,* Jan. 31, 1985, 11-15.

driving prices ever higher. In the process, numerous economic crimes and fraudulent schemes were perpetrated, injuring the moral fiber of society. Other signs of nouveau riche behavior included conspicuous spending, lavish entertainments, and a near obsession with sponsoring beauty contests.[12]

In a society where the old values were rapidly changing, there was a need for identity and reassurance. An American political caricaturist, Ranan R. Lurie, was invited to draw a cartoon character of the typical Chinese on Taiwan. In December 1985, the celebrated "Cousin Lee"—a young man in a martial arts (*kung-fu*) costume with a strong chin, large ears, bushy eyebrows, and a shock of black hair and exuding determination, strength, and vitality—was born. The government authorities were pleased with what they saw as a positive image of Taiwan.

Chiang Ching-kuo deserves much credit for his leadership in economic development and in weathering the storm of American rapprochement with Peking. An efficient administrator surrounded by economic, scientific, and managerial experts, he was keenly interested in economic affairs and the livelihood of the people. As premier in 1973 he initiated the Ten Major Modernization Projects, followed by Twelve more large ones. At Chiang Kai-shek's death in 1975, Vice-President C. K. Yen succeeded as president, and the younger Chiang was elected chairman of the Nationalist Party. Yen finished his term in May 1978, and Chiang Ching-kuo, 68, was elected president for a six-year term by a nearly unanimous vote in the National Assembly.

Chiang Ching-kuo maintained an extreme filial piety to his father, but the two were quite different in training, outlook, temperament, and lifestyle. The elder Chiang was more formal, stern, distant, and militarily oriented; the younger Chiang was more personable, approachable, and economically oriented. He visited farmers, soldiers, and hospital patients and mixed with intellectuals, artists, writers, and baseball players with equal ease. In this way he acted in the ancient tradition of "loving the people" (*ch'in-min*). He was well liked by both Taiwanese and mainlanders on Taiwan, for he symbolized unity and economic development as well as improvement of the people's livelihood.

Chiang's administration was characterized by political innovation, economic development, social stability and welfare, and military preparedness. He believed that the age of individual heroism was past and urged all to

12. In 1988 the "Miss Republic of China" contest was revived for the first time in 24 years, with the first prize of $350,000; the "Miss Wonderland" contest took place in May 1988, with a $20,000 first prize; and the 37th "Miss Universe" contest was held in Taiwan with a $250,000 first prize. The Taiwanese sponsors spent $4.2 million to stage the last pageant. *The Free China Journal,* April 11, May 9 and 16, 1988.

contribute their utmost in building a promising future. He was especially mindful of cultivating able young talent and of promoting native Taiwanese into the mainstream of government. His vice-president (Shieh Tung-ming), the governor of Taiwan, the mayor of Taipei, several cabinet ministers, and a large number of the members of the Legislative Yüan as well as of the representative assemblies were Taiwanese. Chiang's administration exhibited an increasing liberalism, but he steadfastly refused to yield the Nationalist claim to jurisdiction over all of China or to negotiate with Peking (unless it gave up communism); nor would he tolerate the Taiwanese Independence Movement, negotiations with the Soviet Union, the propagation of communism on Taiwan, or attacks on the Nationalist party and the Three People's Principles.

THE LEGACY OF CHIANG CHING-KUO (1910-88). President Chiang was elected to a second term in 1984, with Dr. Lee Teng-hui, a Taiwan-born agricultural economist, as vice-president. Though he continued to promote major infrastructural construction (i.e. Fourteen Key Projects) to strengthen the economic base of Taiwan, Chiang, who was aging and feeling the effects of diabetes, became increasingly occupied with the larger and more fundamental issues of the future of the Kuomintang and Taiwan.

By the period 1986-87, international events and radical social changes at home convinced him that only through political liberalization and strengthening of the rule of law could Taiwan evolve into a true democracy characterized by economic prosperity, political maturity, and social stability. The new order he envisioned would blend attributes of Western democracy with the Chinese political and cultural heritage and create a distinct polity that would serve as an alternative to the way of life on Mainland China.

Chiang's thinking was doubtlessly influenced by major developments abroad and in Taiwan in the years immediately preceding. The tide of democracy had swept through several countries in Asia. In the Philippines, the people's power had overthrown the Marcos regime in 1986; and in South Korea, there had been continuous protest against the authoritarian regime of President Chun Doo-Hwan. Students on Mainland China had demonstrated during 1985-86 to demand greater freedom and democratization. Chiang wanted to spare Taiwan the pain of similar occurrences by taking the initiative and moving toward liberalization.

Rapidly changing social conditions at home drew his attention to the rising aspirations of the new middle class. Economically affluent and well educated, they were socially active and politically conscious. They viewed public participation as part of modern citizenship and did not hesitate to

voice their opinions on such issues as human rights, economic crimes, a
free press, political pluralism, and air pollution and traffic jams. The role
of the Kuomintang came under increasing scrutiny. Originally founded as
a revolutionary party to overthrow the Ch'ing Dynasty, the foreign imperialists, and the warlords of China, it became the dominant force that led
the government during the period of Political Tutelage (1928-48), with
its director-general ruling supreme. His word was law, and rule by decree
became common. But now times had changed; the rule of man would
have to be replaced by the rule of law. The political structure would have
to evolve toward a constitutional democracy. It was time for democracy on
Taiwan to become institutionalized and independent of the human factor.

Within the KMT itself, internal democratization and reorganization
seemed necessary in order to break the Old Guard's grip on the party.
Democratic elections of party officials for fixed terms had to be held; and
key features of the platform, approved by elected representatives at a national convention, much as in any democratic society. Public sentiment
and liberal elements within the KMT favored a gradual transformation of
the KMT from a revolutionary party to a regular party—ready and willing
to renounce power and assume the position of the Opposition if and when
it failed to win popular support.[13]

Taking these sentiments to heart, President Chiang was persuaded that
the time had come for the people, the government, and the party to assume
the greater responsibilities of democratic politics. He took the initiative of
laying the groundwork for a constitutional democracy, characterized by the
rule of law, peaceful change, social stability, and a renovated KMT. He
knew that only he, the last supreme, charismatic leader, could effect these
fundamental changes and set in motion a process that would not be unlike
mountain-climbing—very hard and time-consuming, but exhilarating nonetheless. Herein lay the true contributions of Chiang Ching-kuo. He voluntarily gave up his family control of the party and loosened the party monopoly of political power, thus paving the way for the growth of law and
democracy.

To assist him in implementing his master plan, Chiang, in March 1986,
appointed a 12-member commission from the KMT Central Committee to
study and report on 6 fundamental issues: (1) the lifting of the Emergency Decree that activated martial law, (2) the legalization of the formation of new political parties, (3) the strengthening of local autonomy,

13. Views of Dr. Ma Ying-jeou, Deputy Secretary-General, KMT. *Free China Review*,
    editorial, March 1988; Alexander Ya-li Lu, "Democratic Values Win Another Round,"
    Ibid., October 1987, 8-9; Wei Tsai, "Transformations in the Body Politics," *Ibid.*,
    14-15.

(4) the implementation of a genuine parliamentary system, (5) reform within the KMT, and (6) falling ethics and rising crime. Out of these deliberations came the decisions to lift the 38-year-old martial law (July 15, 1987), to recognize the legitimacy of "loyal opposition" by granting permission to form new political parties, and to allow Taiwan residents to visit relatives on Mainland China (November 2, 1987).

Actually, political parties had existed "illegally" before this time; now, on September 28, 1986, 6 of them filed for formal status, the largest being the Democratic Progressive Party (DPP). In the December 1986 elections, the DPP won 11 seats in the National Assembly with 18.9 percent of the popular vote and 12 seats in the Legislative Yüan with 22.17 percent of the popular vote[14]—an impressive show for an opposition party.

Another major contribution of Chiang's was his persistent introduction of new blood into his party and government. Most of the new members were well educated and relatively young—holders of advanced academic degrees abroad, predominantly from famous American universities.[15] They contributed greatly to the more liberal atmosphere in Taiwan. Indeed, the third and fourth generation KMT leaders were quite a different breed from their predecessors. Many of them were Taiwan-born, indicating a trend to "return political power to the local people."

After the New Year in 1988, Chiang's health failed rapidly. Death came on January 13. The sense of loss was overwhelming, for he had been a beloved leader. Within hours, Vice-President Lee was sworn in as the new president in strict accordance with the constitutional procedures. The transition was smooth, peaceful, and swift; there was no succession crisis.

THE ERA OF LEE TENG-HUI. President Lee, 65 at the time of his accession to the presidency, was a scholarly statesman who held a doctorate in agricultural economics from Cornell University. A devout Christian with broad vision and a sterling character, he was an experienced administrator, having been mayor of Taipei (1978-81), governor of Taiwan (1981-84), and vice-president from 1984 until 1988. He symbolized a new generation of leaders who came from the grass-roots segment of the populace, without the benefits of influential family connections, but who rose to the top through dedication, managerial skill, and political common sense. He

14. *The Free China Journal*, Taipei, Feb. 21, 1987; Peter Chang, "Party Politics Redefined," an interview with Dr. Ma Ying-jeoh, Deputy Secretary-General, KMT, *Free China Review*, Oct. 1987, 16-17.
15. To name a few: Lien Chan, Ph.D., University of Chicago; Frederick Chien, Ph.D., Yale University; Ma Ying-Jeou, S.J.D., Harvard Law School; Wei Yung, Ph.D., Stanford; James Soong, Ph.D., Georgetown; and Shaw Yu-ming, Ph.D., Chicago.

pledged to further the unfinished work of his predecessor, especially in the three areas of the Taiwanization (*pen-t'u hua*) of the Nationalist leadership, the greater opening to mainland China, and the advancement of constitutional democracy. His statesmanship was marked by a high degree of pragmatism.

At the Thirteenth Congress of the KMT in July 1988, Lee was confirmed as chairman of the party, and many Taiwan-born members rose to leadership positions. For the first time, a majority (16) of the 31-member Central Standing Committee were Taiwanese. In the cabinet reshuffle that followed, 8 of the 15 new appointees were Taiwanese, 13 of whom held advanced academic degrees from abroad while 2 others held professional degrees from Taiwan.[16] The cabinet was doubtless one of the best-educated in the world.

The Congress adopted several measures with respect to people-to-people contacts between Taiwan and the mainland, providing for the following:

1. Indirect trade with, and investments in, mainland China to enable the industrial sector of Taiwan to secure raw materials from, and relocate outdated labor-intensive production in, China.
2. Freedom for Taiwanese journalists to go to the mainland on assignment.
3. Permission for mainland intellectuals living abroad and scholars who eschewed Communism and who fought for academic freedom to visit Taiwan.
4. Permission for mainland residents to visit sick relatives or attend funerals on Taiwan.
5. Cultural and sports exchanges.
6. A special government office to take charge of relations with the mainland and to make sure that Taiwan's security and social stability were not compromised.[17]

Nothing was closer to Lee's heart than the advancement of constitutional democracy, and his commitment to political pluralism and party politics was unwavering. He saw clearly the great economic and social changes—the affluence, higher literacy rate, growing middle class, and increasing number of civic groups and labor unions—that had generated a "participatory political culture" and turned Taiwan into a "demanding civic society."

The opposition groups, known as the *tang-wai* ("outside the Nationalist

---

16. *The Free China Journal*, July 25, 1988. Ten of them held the doctorate—and one, an M.A.—from American universities; one each, a doctorate from a West German and Japanese university; and two others, an M.A. and an LL.B. from Taiwanese universities.
17. *Los Angeles Times*, July 13, 1988; *The Free China Journal*, July 18, 1988.

Party"), especially the Democratic Progressive Party (DPP), clamored for greater political participation and demanded the removal of all obstacles to a true constitutional democracy. In particular, they attacked the unfair representations in the three central representative bodies: the National Assembly, the Legislative Yuan, and the Control Yuan, whose members, largely elected on the mainland in 1947, were allowed to keep their seats without reelection through constitutional interpretations by the Council of Grand Justices in the early 1950s. For 40 years they had represented the interests and opinions of neither the constituencies on Taiwan nor of those on the mainland. Although since 1969 the government had held supplementary elections to increase the local representations in these bodies, still 900 of the 1,000 deputies in the National Assembly and 200 of the 300 members in the Legislative Yuan in the 1980s were old incumbents elected on the mainland. Too old to attend the regular sessions for debate and discussion, they had the sole function of appearing at the last minute for the final vote on a bill. Thus, the *Tang-wai* groups ridiculed them as a "voting machine" and argued that as long as they remained in power, there could be no real possibility of shaking the Nationalist control; for the National Assembly, which served as an Electoral College, elected the president and amended the constitution, and the president appointed the governor and the mayors of the two key cities of Taipei and Kaohsiung.[18]

In February 1989 President Lee proposed phased retirements of the old incumbents and phased elections of the new members. His plans called for a reduction of the total memberships of the three bodies to 579, with the National Assembly allotted 375 deputies; the Legislative Yuan, 150; and the Control Yuan, 54—all to be phased in over a period. For instance, the National Assembly quota for the new members would be increased from 84 to 230 in 1992 and to 375 in 1998; the Legislative Yuan quota, from 98 to 130 in December 1989 and to 150 in 1992; and the Control Yuan quota of 54 to be filled all at once in 1992.[19]

In June 1991 the Council of Grand Justices ruled that all mainland-elected members of the three bodies had to retire by December 31, 1991. On December 21, 1991, the celebrated first general election since 1949 took place. It was a historic moment. The Nationalist Party campaigned on a platform of constitutional reform and eventual reunification with the

---

18. Yangsun Chou and Andrew Nathan, "Democratizing Transition in Taiwan," Occasional Papers/Reprints Series in Contemporary Asian Studies, No. 3, 1987 (80):21-22. School of Law, University of Maryland.

19. Wu Wen-cheng and Chen I-hsiu, "Entering the Age of Party Politics," *Free China Review*, April 1989, 52-57; Chen Wen-tsung and Richard R. Vuylsteke, "Democratization in the ROC," *Free China Review*, March 1989, 46-56.

mainland. The Democratic Progressive Party advocated Taiwan independence, substitution of the title "Republic of Taiwan" for "Republic of China," and membership in the United Nations. Such a bold program was a calculated risk on the part of the DPP, and it backfired. Peking had repeatedly warned that it would attack if Taiwan declared independence, and the Nationalist leadership also condemned the Taiwan Independence Movements as illegal. Many voters apparently took these warnings to heart and opted for the milder course advocated by the Nationalists. The KMT won 71 percent of the popular vote and 254 of the 325 seats in the National Assembly whereas the DPP polled 24 percent of the popular vote and 66 seats—much less than the 30 percent of popular vote it had won in 1989. The splinter parties won five seats.

Including the 78 members elected in 1986, the Nationalists were assured of 79 percent of the seats in the National Assembly, enough to meet the 75 percent margin needed to pass constitutional revisions in the following year. Additionally, by a special formula of proportional allotment of "nationwide" and "Chinese Overseas" seats, the Nationalists were assigned 60 seats in the former category and 15 in the latter, whereas the DPP received 20 and 5 respectively. With these elections, Taiwan-born members of the three representative bodies became the majorities even though many of them were offspring of mainlanders.

In the December 19, 1992, election of the Legislative Yuan, the DPP emerged stronger and with greater credibility than ever before, polling an impressive 31 percent of the popular vote compared to the Nationalists' 53 percent. Although the Nationalists still controlled the majority with 103 seats—80 regional, 19 nationwide, and 4 Chinese Overseas seats— the overall Nationalist representation shrank from 74 percent to 63 percent whereas the DPP grew from 14.4 percent to 31.1 percent. Thus, Taiwan seemed moving toward an institutionalized bipartisan politcial system in which the DPP functioned as a "Loyal Opposition." In view of the fact that between 1980 and 1989 the Nationalists lost 14.4 percent of the popular vote and the DPP gained 14.2 percent of the vote in all elections, it is possible that in 10 to 15 years the DPP could become the dominant party.[20]

The elections of a new National Assembly and a new Legislative Yuan marked a giant step forward in Taiwan's march toward constitutional democracy. Much, however, remained to be done. The governor of Taiwan and the mayors of Taipei and Kaohsiung were still not popularly elected;

20. Zhou Xiaomeng, "On the Road Toward Democracy: Party Politics in Taiwan," Papers of the Center for Modern China, Vol. 3, No. 10 (Oct. 1992): 14; *The New York Times*, Dec. 22, 1991; *Los Angeles Times*, Dec. 21, 1991.

neither was the President, who was elected by the National Assembly. For a true constitutional democracy to function, these key posts would have to be popularly elected. Furthermore, the three major television networks had to be liberated from official control, and new networks and channels added for popular use. The democratization of Taiwan could inspire democratic aspirations on the mainland and challenge the government there to move toward political liberalization.

Though proud of its "economic miracle" and political democratization, Taiwan takes an even greater pride in its role as preserver of the Chinese cultural heritage. The Palace Museum holds more than 300,000 invaluable Chinese paintings, calligraphy, porcelain, jade and bronze pieces, and other art objects from the previous imperial collections. The new National Theater and the Concert Hall in the Chiang Kai-shek Memorial Park, completed in October 1987, greatly enrich the cultural life of Taiwan with their programs of Chinese operas and Western symphonic and chamber music. Taiwanese openly proclaim that though their island is small, it is economically dynamic and culturally great and has the potential of becoming "a great country of tomorrow."

Taiwan is determined to create a model of modern development based on free enterprise and the Three People's Principles as an alternative to the system on the mainland, where the democratization of Taiwan is closely observed, and hopes that Peking will ultimately recognize it as the way to political liberalization.

The Prospect for Reunification

The reunification of China is the common desire of both Nationalist and Communist Chinese. Taiwan had openly expressed concern for the welfare of the people on the mainland, and not a few mainlanders harbored secret admiration for the economic success of Taiwan. Peking no longer talked of liberating Taiwan but of its reunification with the motherland. Taiwan no longer talked of reconquering the mainland. Instead, the professed goal was the unifying of all China under the political philosophy of the Three People's Principles of Dr. Sun Yat-sen.

Peking considered the reunification a top priority of the 1980s. To this end, it offered a reunification plan based on three principles. Taiwan was to (1) give up all claims to being the legal government of all China; (2) retain its present economic and social system and its standard of living; and (3) retain a degree of autonomy, including the maintenance of an army. To begin the reconciliation, on New Year's Day 1979, Peking proposed postal, commercial, and air-and-shipping relations with Taiwan, and on

National Day, October 1, 1981, advanced a nine-point proposal suggesting talks between the two ruling parties. In effect, Peking asked the Nationalists, in the larger interest of reunification, to forfeit their de facto independence and their claim to represent all China, to cease to be a separate political entity in international politics, and to accept the status of an autonomous special region of China on a par with the other provinces.

President Chiang Ching-kuo adamantly rejected the peace overture as a "smiling diplomacy" of the United Front variety, designed to disarm Taiwan's vigilance. Cooperation would be possible, said Chiang, only if Peking gave up Communism in favor of free enterprise and the Three People's Principles. Until then, there would be "no contact, no negotiations, and no compromise." The Nationalists pointed out that experience of the past two collaborations during 1924–27 and 1937–41 had shown that cooperation with the Communists was futile. The current situation was made worse by Peking's insistence on the Four Cardinal Principles—the socialist line, the proletarian dictatorship, the leadership of the Communist Party, and Marxism-Leninism and the Thought of Mao—which ruled out any meaningful participation by the Nationalists in a coalition government. Taiwan also rejected Peking's call for the "Three Links" (*San-t'ung*)—mail, trade, and air-and-shipping services—and the "Four Exchanges" (*Ssu-liu*)—mutual visits of relatives, tourists, and academic, cultural, and sports groups.

After China's successful negotiations with Britain in 1984 for the return of Hong Kong in 1997, Teng Hsiao-p'ing proudly announced his formula of "One country, two systems," as the way to reunification. Just as China would allow Hong Kong to keep its social and economic systems for 50 years beyond 1997, China would also permit Taiwan to keep its political, economic, and military systems after its unification with the mainland provided that Taiwan accepted the status of a Special Administrative Region under the central government in Peking.

Taiwan found such a formula repugnant and the idea of placing it and Hong Kong in the same category unrealistic. The two were, the Taiwanese argued, in very different positions vis-à-vis China. First of all, Britain had no intention of fighting over Hong Kong and, therefore, retained no bargaining power. Second, Hong Kong had no army of its own and depended on the mainland for its food and water supply. The people of Taiwan, in contrast, were politically active, self-sufficient in food supplies, and desirous of retaining their way of life. Taiwan maintained a highly trained, modern military force, ever ready and able to defend the island against any invader. Furthermore, Taiwan was "protected" by the United States Taiwan Relations Act of 1979, under which the United States supplied defensive weapons to Taiwan. Any invasion of the island by the mainland could be construed as detrimental to the "security of the Western Pacific and of grave concern to the United States."

Above all, Taiwan's leaders equated the Three People's Principles with the Lincolnian ideals of "government of the people, by the people, and for the people." They wanted the people of China to see Taiwan as an alternative to Communism and accept the Three People's Principles as the basis for reunification. Hence, the idea of "One country, two systems" was totally unacceptable to Taiwan.

Friends of Taiwan and enlightened elements on the island had long urged a more flexible policy toward the mainland, describing the "Three No's Policy" as negative, self-limiting, and lacking in initiative. President Lee accepted political realism as the cornerstone of his administration and championed a flexible and pragmatic diplomacy. At the Thirteenth Nationalist Party Congress in July 1988, he asked all members to "strive with greater determination, pragmatism, flexibility and vision to develop a foreign policy based primarily on substantive relations." Lee was a practical-minded statesman who was always more concerned with substance than words. Visiting Singapore in March 1989, he was described by the local press as "the President from Taiwan" rather than as "the President of the Republic of China" on Taiwan. He chose not to protest, explaining: "It is unnecessary for us to care too much about the name. If we keep being bothered by these minor problems, there is no way to break out" of Taiwan's isolation. In sending his finance minister, Dr. Shirley Kuo, to the Asian Development Bank meetings in Peking in May 1989 under the designation of "Taipei, China," he explained on June 3: "The ultimate goal of the foreign policy of the Republic of China is to safeguard the integrity of the nation's sovereignty. We should have the courage to face the reality that we are unable for the time being to exercise effective jurisdiction on the mainland. Only in that way will we not inflate ourselves and entrap ourselves, and be able to come up with pragmatic plans appropriate to the changing times and environment."[21]

Lee's "pragmatic diplomacy" led the Republic of China on Taiwan to renounce its claim to being the legal government of all China. No longer did it declare, "There is only one China and the Republic of China on Taiwan is China." It realistically acknowledged that the Republic of China had jurisdiction only over Taiwan, the Penghu (Pascadores), Kinmen, Matsu, Pratas, and Spratly Islands and their surrounding islets. Thus, there was the "Taiwan Area" under the Nationalist control and the "Mainland Area" under the Communist control. With the termination in April 1991 of the "Period of National Mobilization for Suppression of the Com-

21. Fredrick F. Chien, "A View from Taipei," *Foreign Affairs* (Winter 1991-92): 97-98.

munist Rebellion," which had been in effect for 40 years, Taiwan no longer considered Peking "an illegal rebel." Taiwan's mainland policy thus drastically changed from one of confrontation and Cold War rhetoric to one of peaceful competition and cultivation of mutual trust.

Lee rejected Peking's "one country, two systems" formula, believing it would relegate Taiwan to a provincial status. However, he would negotiate with Peking if it agreed to (1) implement a political democracy and a free economic system, (2) regard Taiwan as a political entity of equal status, (3) renounce the use of force in the Taiwan Straits, and (4) refrain from blocking or isolating Taiwan in international organizations and activities.[22] Peking would accept none of these conditions, attacking Lee's "pragmatic diplomacy" as a plot to create "one China, one Taiwan" or "two Chinas." Taiwan suggested that a more realistic description of the situation on the two sides of the Straits should be: "one country, two areas, two political entities."

Because the reunification process would be long and arduous and could not be hurried, on February 23, 1991, Taiwan adopted "Guidelines for National Unification," which set no timetable for achieving a united China that is "democratic, free, and equitably prosperous"—conditions that clearly Peking would not accept. The document accepted the "one China" principle and insisted on peaceful unification and proper respect for the rights and interests of the people on Taiwan. It proposed three stages to allow both sides to acclimatize themselves to a unification framework.

The first stage encouraged unofficial exchanges and reciprocity and people-to-people contact. The second stage aimed at building mutual trust and cooperation by opening official communication channels. Taiwan would assist the mainland in developing its southeastern coastal area in order to narrow the gap in the standards of living, and both sides should take part in international organizations and exchange official visits. During the third stage, a joint consultative council for national unification should be established to work on the grand task of national unification on the basis of political democracy, economic freedom, social justice, and nationalization of the armed forces. Taiwan hoped that its "economic miracle" and political democratization could not only be a source of inspiration and aspiration for the mainland but also stimulate the peaceful change on the continent toward the eventual unification.[23]

President Lee believed unification possible in six to ten years, but public

---

22. Lee Teng-hui, "Opening a New Era for the Chinese People," Inaugural address by the eighth-term President of the Republic of China, May 20, 1990, in *Creating the Future towards a New Era for the Chinese People* (Taipei, 1992), 8.

23. Jason C. Hu, "Building Democracy for Unification," an address to the Los Angeles World's Affairs Council, Oct. 26, 1992; Lee, 126-27.

opinion on Taiwan was less optimistic. Only 10 percent of the population were strong supporters of unification, 5 to 12 percent favored Taiwan independence, and the silent majority in between wanted "one China, but not now." Another survey in April 1992 showed that 76.5 percent of Taiwan's population were in no hurry for unification and 7.8 percent wanted to scrap the idea altogether.[24] With Taiwan's growing democracy and a per capita GNP of U.S. $10,215 in 1992, compared with the Leninist dictatorship on the mainland and a per capita GNP of $350, the gap was too great to be bridged. There was the lurking fear that unification would enable the Big Fish to swallow up the Little Fish. Both the Nationalists and the DPP regarded unification as an issue to be "handled" but not "resolved" anytime soon.[25]

Until unification is peacefully achieved, Taiwan would continue to climb to a higher plateau of constitutional democracy, to charge ahead with a six-year, U.S. $303 billion infrastructural development program that would double the GNP and raise the per capita income from the current $10,215 to $14,000 by 1996, and to improve the quality of life and the environment. Many believed that Taiwan's political democratization and economic affluence would facilitate the ultimate unification. Meanwhile, it jealously guarded its own way of life and the status quo while methodically working through the three-stage process of unification.

Peking ridiculed Taiwan's claim to equal status, pointing out that the Nationalist government had lost control of the mainland in 1949 and had thus forfeited its right to represent China. The government at Peking was recognized by 155 countries of the world, compared with Taiwan's 29 (in 1993). With only 1/266 of mainland's territory and 1/55 of its population, how could Taiwan claim to be a separate but equal political entity? Insisting that Taiwan was an autonomous province of China, Peking warned of military invasion if Taiwan declared independence or colluded with foreign powers to split up China or put off unification "too long."[26] Li Jui-huan, a member of the Politburo Standing Committee, declared in November 1992 that Peking would attack Taiwan to stop its independence even at the price of suspending its own economic development.[27]

However, Peking welcomed people-to-people, economic, trade, and cultural exchanges, and on July 6, 1988, promulgated "Stipulation on Encouraging Taiwan Corporations' Investment in the Mainland," which

---

24. Lee, 101; *The Free China Journal*, April 10, 1992.
25. Zhou Xiaomeng, 7.
26. Statement of Chien Wei-chang, Vice chairman of Peking's Political Consultative Council, Dec. 9, 1992.
27. Cited in *The Free China Journal*, Nov. 6, 1992.

offered preferential terms on loans, tariffs, taxes, and other favors to Taiwan investors. Obviously, Peking welcomed Taiwan investments, capital, and technical know-how to finance and accelerate its economic development. Between 1987 and 1992 some five million Taiwan residents visited the mainland and spent an average of U.S.$4,000 apiece on gifts, travel, hotel, and the like, injecting an estimated U.S.$20 billion into the mainland coffers. Approximately 10,000 Taiwan corporations relocated their production facilities to the mainland because of cheap labor and abundance of raw materials and land, investing U.S.$10 billion.[28] Taiwan's indirect trade with the mainland via the entrepôt of Hong Kong zoomed to U.S. $7.4 billion in 1992 and possibly $9.3 billion in 1993.

There was a growing feeling of economic interdependence on both sides. Taiwan's investment accounted for U.S.$8.6 billion of the GNP on the mainland, and U.S.$7.2 of its per capita income.[29] On the other hand, the indirect exports to the mainland helped maintain Taiwan's 7.3 percent economic growth rate in 1991 and 6.9 percent in 1992, when most of the world was in a recession.[30] Taiwan enjoyed a continuous trade surplus with the mainland—U.S.$1.76 billion in 1988, $2.30 billion in 1989, $2.51 billion in 1990, $3.54 billion in 1991, and $5.16 billion in 1992.[31] There was hope on Taiwan that the trade surplus and the vast China market could propel Taiwan's economy to a "Second Miracle" and that mutual economic benefits could help resolve difficult problems that could not be resolved by political means.

Along with growing economic interdependence, scholarly, scientific, journalistic, athletic, and cultural exchanges were on the rise. The more memorable events included the visit to Taiwan of the Central Ballet of China in October 1992, the delegation of nine Peking economists in November 1992, and the two-week engagement of the famous Peking opera star, Mei Pao-chu, son of the great Mei Lan-fang, in April 1993. In the opposite direction, Taiwan's Academia Sinica president, Wu Ta-you, visited Peking in June 1992.

Then came the much-awaited, semiofficial meeting in Singapore on April 28–29, 1993, between Taiwan's C. F. Koo (Ku Chen-fu), chairman of the Straits Exchange Foundation, and Peking's Wang Tao-han (Wang Daohan), chairman of the Association for Relations Across the Taiwan Straits and former mayor of Shanghai. Billed as a nonpolitical meeting, the discussions focused on economic matters such as protection of Taiwan's

28. Central Bank of China figures, Taipei, cited in *The Free China Journal*, Nov. 27, 1992.
29. *The Free China Journal*, Oct. 30, 1992.
30. *Ibid.*, July 9, 1993.
31. Hong Kong government statistics.

investment on the mainland, intellectual property rights, joint efforts to fight piracy, reciprocal repatriation of illegal refugees and criminals, settlement of fishing disputes, compensation for lost registered mail, and joint development of natural resources. The historic meeting produced four documents and a decision to meet four times a year in the future. The Koo-Wang meeting augured well for a closer relationship between the two sides of the Straits after 40 years of hostility. That it happened at all testified to the pragmatic statesmanship on both sides.[32] Genuine unification, however, remained a distant prospect.

A hundred years ago leaders of the Self-strengthening Movement sought "wealth and power" as the key to China's survival in the modern world. During the May Fourth Movement (1919), survival dictated national independence and unity, science and democracy, liberty and the improvement of the people's livelihood. Later, the demand for freedom of thought and for the liberation of the creative spirit of individuals appeared.[33] Constitutional democracy was the dream. Any government that could fulfill these aspirations would have the support of all Chinese.

In its 4,000 years of recorded history, China has been divided and reunited countless times. If history is any guide and if politics is the art of the possible, then one need not lose heart over the present difficulties. The genius of the Chinese people will find a way to make all Chinese one again.

Further Reading

Bader, William, and Jeffery T. Bergner (eds.), *The Taiwan Relations Act: A Decade of Implementation* (Indianapolis, 1989).

Brzezinski, Zbigniew, *Power and Principle: Memoirs of the National Security Advisor, 1977-1981* (New York, 1983).

Carnegie Council on Ethics and International Affairs, *Republic of China in International Perspective* (New York, 1992).

Chang, King-yuh, *A Framework for China's Unification* (Taipei, 1987).

——— (ed.), *ROC-US Relations under the Taiwan Relations Act: Practice and Prospects* (Taipei, 1988).

Chang, Maria, "Taiwan's Mainland Policy and the Unification of China." Asian Studies Center, Claremont Institute of California, 1990.

Chen, Wen-tsung, and Richard R. Vuylsteke, "Democratization in the ROC," *Free China Review*, Taipei (March 1989): 46-56.

Chiu, Hungdah, "Chinese Communist Policy Toward Taiwan and the Prospect of Unification." Reprint Series in Contemporary Asian Studies, No. 5—1991 (106). School of Law, University of Maryland.

32. *The New York Times*, April 28, 1993; *Los Angeles Times*, April 28 and 29, 1993.

33. Shao Yü-ming, "Shih lun Chung-Kung cheng-ch'uan tsai Chung-kuo chin-tai-shih shang ti kung-kuo" (An appraisal of the achievements and failures of the Chinese Communist regime in modern Chinese history), *Hai-wai hsüeh-jen*, 99:8 (Oct. 1980).

Chou, Yangsun, and Andrew J. Nathan, "Democratizing Transition in Taiwan." Reprint Series in Contemporary Asian Studies, No. 3—1987 (80). School of Law, University of Maryland.

Clough, Ralph, *Island China* (Cambridge, Mass., 1978).

Cohen, Marc J., *Taiwan at the Crossroad* (Washington, D.C., 1988).

Cohen, Myron L., *House United, House Divided: The Chinese Family in Taiwan* (New York, 1976).

Copper, John F., *A Quiet Revolution: Political Development in the Republic of China* (Lanham, Md., 1988).

————, *Taiwan: Nation-state or Province* (Boulder, 1989).

————, "Taiwan's Recent Elections: Fulfilling the Democratic Promise." Reprint Series in Contemporary Asian Studies, 1990. School of Law, University of Maryland.

————, *China Diplomacy: The Washington-Taipei-Beijing Triangle* (Boulder, 1992).

Chu, Yungdeh Richard (ed.), *China in Perspectives: Prospects of China's Reunification* (Hong Kong, 1987).

Damrosch, Lori Fischer, *The Taiwan Relations Act After Ten Years* (Baltimore, 1990).

Faurot, Jeannette I. (ed.), *Chinese Fiction from Taiwan* (Bloomington, 1980).

Gold, Thomas B., *State and Society in the Taiwan Miracle* (Armonk, N.Y., 1986).

Gregor, A. James, and Maria Hsia Chang, "The Taiwan Independence Movement: The Failure of Political Persuasion," *Political Communication and Persuasion*, 2:4 (1985): 363-91.

Ho, Samuel P. S., *Economic Development of Taiwan, 1860-1970* (New Haven, 1978).

Jing, Wei, "Overstretched: Taiwan's 'Elastic Diplomacy,'" *Beijing Review*, April 3-9, 1989, 7.

Knapp, Ronald G. (ed.), *China's Island Frontier: Studies in the Historical Geography of Taiwan* (Honolulu, 1981).

Kuo, Shirley W. Y., *The Taiwan Economy in Transition* (Boulder, 1983).

Lasater, Martin L., *Policy in Evolution: The U. S. Role in China's Reunification* (Boulder, 1989).

Lee, Teng-hui, *Creating the Future: Towards a New Era for the Chinese People* (Taipei, 1992).

Li, Cheng, and Lynn White, "Elite Transformation and Modern Change in Mainland China and Taiwan: Empirical Data and the Theory of Technocracy," *The China Quarterly*, March 1990, 1-35.

Li, Dahong, "Mainland-Taiwan Economic Relations on the Rise," *Beijing Review*, April 3-9, 1989, 24-27.

Li, Jiaquan, "More on Reunification of Taiwan with the Mainland," *Beijing Review*, Jan. 16-22, 1989, 26-30.

————, "Taiwan's New Mainland Policy Raises Concern," *Beijing Review*, May 22-28, 1989, 23-25.

Long, Simon, *Taiwan: China's Last Frontier* (London, 1991).

————, "Taiwan's National Assembly Elections," *The China Quarterly*, March 1992, 216-28.

Myers, Ramon H. (ed.), *A Unique Relationship: The United States and the Republic of China Under the Taiwan Relations Act* (Stanford, 1989).

——— (ed.), *Two Societies in Opposition: The Republic of China and the People's Republic of China after Forty Years* (Stanford, 1991).

Peng, Huai-en, *Tai-wan cheng-chih pien-ch'ien ssu-shih nien* [Forty years of political change in Taiwan] (Taipei, 1987).

Political Trends in Taiwan since the Death of Chiang Ching-kuo. Committee on Foreign Affairs, House of Representatives, Washington, D.C., 1989.

Schive, Chi, *The Foreign Fact: The Multinational Corporations' Contribution to the Economic Modernization of the Republic of China* (Stanford, 1990).

Senese, Donald J., and Diane D. Pikcunas, *Can the Two Chinas Become One?* (Washington, D.C., 1989).

Silin, Robert H., *Leadership and Values: The Organization of Large-Scale Taiwanese Enterprises* (Cambridge, Mass., 1976).

Sutter, Robert G., *Taiwan: Entering the 21st Century* (Lanham, Md., 1988).

Tien, Hung-mao, *The Great Transition: Political and Social Change in the Republic of China* (Stanford, 1987).

Ts'ai Ling, and Ramon H. Myers, "Surviving the Rough-and-Tumble of Presidential Politics in an Emerging Democracy: The 1990 Elections in the Republic of China on Taiwan," *The China Quarterly*, March 1992, 123-48.

Tsai, Wen-hui, "In Making China Modernized: Comparative Modernization between Mainland and Taiwan." Reprint Series in Contemporary Asian Studies, No. 4—1993 (117). School of Law, University of Maryland.

Vogel, Ezra F., *The Four Dragons: The Spread of Industrialization in East Asia* (Cambridge, Mass., 1991).

Wu, Wen-cheng, and Chen I-hsin, "Entering the Age of Party Politics," *Free China Review*, April 1989, 52 57.

Wu, Yuan-li, *Becoming An Industrialized Nation: ROC Development on Taiwan* (New York, 1985).

40

The Violent Crackdown at
T'ien-an-men Square, June 3-4, 1989

Nineteen eighty-nine, the Year of the Snake, opened with ominous signs of an impending explosion in China. The country faced rising inflation, falling ethics, widespread corruption, official profiteering, a widening gap in income between the privileged few and the great masses, and an increasing loss of faith in communism. To be sure, these phenomena existed in the preceding years (see chapter 38), but the underlying discontent was brought to the surface and now threatened to come to a head. Frustration and unrest were rampant in many parts of the country.

The forces of democracy and liberalization ceaselessly clashed with those of repression and authoritarianism, both openly and secretly, but tension was always present. This clash took on added significance in light of two major trends in international politics: the rising tide of freedom and democracy, and the retreat of communism in Poland, Hungary, and the three Baltic states (Lithuania, Latvia, and Estonia). Even in the Soviet Union, *glasnost* and *perestroika* were changing the face of communism. Instability, unrest, and compromise seemed to characterize the Communist world, and some political scientists predicted the demise of communism as a shaping force of history.[1]

The year was also significant because of the many anniversaries of historical importance: the fortieth anniversary of the founding of the People's Republic of China (October 1, 1949), the seventieth anniversary of the May Fourth Movement (1919), and the bicentennial anniversary of

1. Zbigniew Brzezinski, *The Grand Failure: The Birth and Death of Communism in the Twentieth Century* (New York, 1989).

926

the French Revolution (July 14, 1789), which championed liberty, equality, and fraternity. On any of these occasions pro-democracy demonstrations could erupt, warned university students, whose quest for political liberalization had been cut short two years earlier. Feeling ill at ease, the government called for redoubled vigilance on the part of party members and the armed forces, urging them to guard against disturbances and to prevent a Polish-style Solidarity movement from happening in China.

The Gathering Storm

The year had barely begun when noted astrophysicist Fang Li-chih sent a letter to Teng Hsiao-p'ing on January 6 calling for the granting of a general amnesty either on the occasion of the seventieth anniversary of the May Fourth Movement or the fortieth anniversary of the founding of the People's Republic. Fang specifically asked for the release of Wei Ching-sheng, an electrical worker sentenced in 1979 to fifteen years in prison for promoting democracy as a "Fifth Modernization" and warning that Teng could become another Mao. Fang's letter won the endorsement of fifty-one famous Chinese scholars abroad and thirty-nine leading intellectuals in China. Incensed, Teng and the government refused to respond.

The democratic cause suffered a setback during President George Bush's visit to China in late February 1989. Following President Ronald Reagan's precedent of hosting a group of Soviet dissidents at the American Embassy in Moscow the previous summer, Bush invited four leading Chinese liberals to a state dinner on Sunday, February 26. The guests included Fang Li-chih, his wife Li Shu-hsien, and three others,[2] all of whom were active supporters of General Secretary Chao Tzu-yang's reforms and sharp critics of the conservative leadership. Most Chinese regarded the invitation as an American statement of support for democracy and liberalization in China. President Yang Shang-k'un and Premier Li P'eng threatened to boycott the dinner if Fang and his wife attended. However, they did accept the compromise situation whereby President Bush would not move from table to table to exchange toasts with Fang and his wife. Yet even this proved too much of an affront. At the last minute the Chinese leaders changed their minds and sent police to keep Fang and his wife from attending the banquet. Bush was unaware of the incident, assuming all through dinner that they were among the guests.

The government action outraged Chinese students and intellectuals. They waited for an opportunity to vent their wrath and renew the pro-

2. Su Shao-chih, a Marxist theoretician; Wu Tsu-kuang, a playwright, and Yen Chia-ch'i, a political scientist.

democracy demonstrations cut short two years earlier. The death of former General Secretary Hu Yao-pang on April 15 provided just such an opportunity. Dismissed in January 1987 for his lenient attitude toward the students, Hu had become for many a symbol of openness and political liberalization. The students wished to honor his memory with an elaborate commemorative service, and they also planned to use the occasion to insist on the clearing of his name and to push forward demands for freedom of speech, assembly, and the press, as well as strong anti-corruption measures. Posters appeared at Peking University eulogizing Hu and lampooning the conservative leaders: "A good man has died, but many bad ones are still living," and "A man of sincerity has passed away, but hypocrites are still around."

The party refused to clear Hu's name; to do so would have been an admission of guilt by all those involved in his dismissal, including Teng and other hard-liners. Thousands of students marched in the streets and staged a sit-in at T'ien-an-men Square, chanting "Long live democracy! Long live freedom! Down with corruption!" On April 22, three students ridiculed the government by kneeling at the steps of the Great Hall of the People, with one of them raising above his head a large scroll listing their demands and shouting: "This is how petitions were presented to emperors. What era is this? We still have to use this method, which means we have no freedom." The students boycotted their classes and continued the demonstrations in the Square for six weeks, drawing increasing support from fellow students in the provinces, as well as from local and provincial workers, intellectuals, journalists, professors, researchers, musicians, actors, ordinary citizens, and even some members of the party and the armed forces. By mid-May the ranks of the pro-democracy protest had grown to over a million. Smaller demonstrations also broke out in twenty-three other cities. It had become a tidal wave of protest that cut across classes, creating a stirring spectacle for television viewers both at home and abroad. The crowning insult to the leadership came on May 30, when students of the Central Art Institute erected on the Square a thirty-foot statue of the Goddess of Democracy, loosely modeled upon the Statue of Liberty. Positioned in the northern sector of the Square, opposite the entrance to the Imperial Palace, the statue stared defiantly at the giant portrait of Mao.

The Party Split

General Secretary Chao Tzu-yang, like his predecessor Hu Yao-pang, displayed surprising tolerance of the demonstrations and sympathy with the student's motives. As a promoter of reform and modernization, he had hoped to guide the party toward greater openness and a gradual transformation in the direction of political liberalization, though not necessarily a capitalistic democracy. He came into sharp conflict with Premier Li P'eng and other veteran hard-liners such as Ch'en Yun, P'eng Chen, and President Yang Shang-k'un. Influenced by his advisors at the Chinese Academy of Social Sciences and his four "think tanks,"[3] Chao may have wanted to draw on the rising student power to strengthen his position vis à vis the conservatives and to advance the cause of reform. On the other hand, the conservatives, who had won the first round of the contest two years earlier when they persuaded Teng to dismiss Hu for mishandling the student demonstrations, now schemed to implicate Chao as a secret patron of the demonstrations and to oust him from the post of general secretary. Thus, the student demonstrators had unwittingly become pawns in the seething political struggle within the party.

The hard-liners saw in student demonstrations a rare opportunity to crush the democracy movement and derail economic reform. A deftly precipitated clash with the students and Chao would enable them to kill two birds with one stone. Ch'en Yun, the economist, and Yao I-lin, the vice-premier, masterminded a plot whereby Premier Li P'eng was to be absolutely stern and unyielding with regard to student demands, thereby goading students into greater belligerence. Meanwhile, Chao was to be attacked as a secret supporter of the demonstrators—a traitor within the party. The growing insolence of the students and Chao's sympathy for them would drive Teng into a rage and cause him to react violently. He could then be manipulated to smash the students and Chao in one fell swoop, just as he had been persuaded to oust Hu two years earlier. Thus, the key to success was to infuriate the demonstrators and Teng in order to escalate the confrontation to the point where a military crackdown could be justified.[4]

On April 24, a Peking municipal party secretary,[5] at the behest of the conservative elders, submitted a "war report" to the Central Committee. The report stated that, based on student posters, slogans, and secret reports compiled by state security agencies, it appeared that the demonstrations had been two years in the making, with the avowed purpose of negating

---

3. Central Political Structure Reform Institute; China Economic Structure Reform Institute; State Council Agricultural Research and Development Institute; and China Trust International Relations Institute.

4. These views of Ch'en I-tzu, former director of the China Economic Structure Reform Institute, and Su Shao-chih, former director of the Institute of Marxism-Leninism-and the Thought of Mao, Peking, were reported in the *Central Daily News,* Taipei, Sept. 10, 1989, and in *Ming Pao,* Hong Kong, Sept. 6, 1989.

5. Li Hsi-min.

the socialist cause and overthrowing the Communist leadership. The Polit-buro Standing Committee and President Yang Shang-k'un agreed that the demonstrations were an "organized, planned, and premeditated anti-party, anti-socialist activity." On the following day Teng declared that the dem-onstrations constituted a "conspiracy" or "turmoil" (*tung luan*) that must be suppressed. At his direction, two party writers[6] formerly connected with the Cultural Revolution and the 1986 "Anti-Bourgeois Liberalization" campaign, wrote an editorial for the April 26 issue of *People's Daily* enti-tled "Clearly Raise the Banner of Opposition to the Turmoil." The essence of the article was that the government was locked in a grand political struggle against a "turmoil" that had as its target the destruction of party leadership and the socialist system. Capitulation to student demands, it warned, would turn a promising country into a hopeless, turbulent one. The contents of the editorial were wired to Chao, who was visiting North Korea, and received his endorsement in principle.

It was apparent that the hard-liners had won Teng over to their cause, just as they had two years earlier, and that the party would take a tougher stand against the demonstrators. Li P'eng steadfastly refused to meet with student leaders, thereby losing a chance to defuse the issue and negotiate an early settlement. When he finally did consent to meet with them, on May 18, he sternly lectured student leaders Wang Tan and Wu-er K'ai-hsi, leaving no room for a dialogue or a meeting of minds. Meanwhile, in Shanghai the party boss, Chiang Tze-min, dismissed Ch'in Pun-li, the editor-in-chief of the *World Economic Herald,* for his pro-democracy stand.

When Chao returned home on April 29, he was urged by his advisors[7] to refute the April 26 editorial and to fight for an affirmation of the student action as spontaneous, patriotic, and in conformity with the government's own anti-corruption policy. He tried to reverse the April 26 editorial, but to no avail. Then, on May 4, in a speech to the Asia Development Bank, which was meeting in Peking, Chao implied that there was division in the party, but that he was certain that the government would follow a "cool, reasoned, disciplined, and orderly way of resolving problems in a demo-cratic and lawful manner." The hard-liners were incensed. They charged that Chao had revealed leadership dissension in a statement that had not been cleared by the party in advance. Li P'eng declared that Chao's speech represented his personal view and that the party view could only be repre-sented by Teng. Li increasingly succeeded in pitting Teng against Chao.

6. Lu Jen and Hsü Wei-ch'eng.
7. Pao T'ung, director of the Central Political Structure Reform Institute, and Tu Jun-sheng, director of the Rural Research Center.

A further controversy arose on May 16 when Chao told visiting Soviet General Secretary Mikhail S. Gorbachev that since the Thirteenth Party Congress, in November 1987, all important decisions had to be deferred to Teng for approval, implying that Teng, not Chao, was responsible for all key decisions as well as mistakes. Teng and the hard-liners were angry that Chao had disclosed a "state secret" to a foreign guest.

During Gorbachev's three-day visit, the Chinese leadership was thoroughly humiliated by the million occupants of T'ien-an-men Square. The official reception of the honored guest had to be shifted from the Square to the airport. Gorbachev's tour of the Forbidden City was canceled because of the large crowds assembled at its entrance, and plans for him and his wife, Raisa, to lay a wreath at the Hero's Monument in the center of the Square also had to be canceled. His press conference was transferred from the Great Hall of the People to the Tiao-yü-t'ai State Guest House, where he was lodged. Except for a brief moment, his famous walkabouts had to be omitted for safety's sake. These last-minute changes created an impression that the Chinese leadership was weak, indecisive, and incapable of controlling the situation. Teng was all the more determined to teach the young people a stern lesson once the guest had departed. As patriarch of the extended Chinese family, Teng might have felt that the students were ungrateful for all the benefits his economic reforms had brought them, including the improved quality of higher education and student life. The price of insubordination would be unrelenting retribution.

The hard-liners insisted that the root cause of student unrest lay within the party itself. It was the mismanagement of the economy by Chao and his predecessor, Hu, that led to high inflation, economic imbalance, and confusion. It was also their permissiveness toward the students and their halfhearted sponsorship in 1983 of the "Anti-Spiritual Pollution" campaign (which lasted only twenty days) and the "Anti-Bourgeois Liberalization" campaign in 1986 that led to the current trouble. Further evidence pointed to Chao's aiding and abetting the demonstrators through his open sympathy with their cause and leakage of vital information to them. Frequently the students learned of the Politburo's decisions before anyone else. The patronage by Chao and his liberal advisors boosted their sagging morale and emboldened them to be more unyielding. Two of Chao's advisors[8] were charged with having made a "most furious and vicious" May 17 Declaration, in which they bluntly stated: "Because the autocrat controls the unlimited power, the government has lost its own obligation and

8. Yen Chia-ch'i, a political scientist and former director of the Political Science Institute of the Chinese Academy of Social Sciences, and Pao Tsun-hsin, a philosopher and an associate research fellow at the Historical Research Institute.

normal human feelings. . . . Despite the death of the Ch'ing Dynasty 76 years ago, there is still an emperor in China, though without such a title, a senile and fatuous autocrat . . . Gerontocratic politics must end and the autocrat must resign." The inference was clear: Teng was that "decrepit autocrat."[9] Some demonstrators burned Teng's effigy, shouting "Teng Step Down," "Li P'eng Step Down," and "Long Live Chao Tzu-yang."

This development further convinced the hard-liners that Chao and his advisors had conspired to unleash student power to split the party. By advocating a different course of action from Teng's toward the demonstrators, he was, in fact, setting up an alternative headquarters within the party, with a second voice that confused both the party and the people. In addition to this grave sin, Chao was also guilty, as charged earlier, of attempting to reject the April 26 editorial, of making an unauthorized statement to the Asian Development Bank, of disclosing a state secret to Gorbachev, and of permitting his two sons to engage in commercial specu-lation and profiteering. Teng, who had handpicked Chao as his successor, now called him a "traitor." The split was complete; only punishment re-mained to be meted out.

A "war council" at the highest level, including only the seven or eight most senior party elders who had made great contributions to the party and the state, convened on or about May 17.[10] Their purpose was to decide what to do about the demonstrations and Chao. They reduced the first issue to the choice of whether or not the party should retreat in the face of the student threat, which daily grew more serious as the ranks of the protestors increased and won a wide spectrum of popular support. To retreat meant acceptance of the students' demand for democracy and free-dom; not to retreat meant forcible repression of the unrest.

The gerontocratic leaders were in a state of despair. The spectacle of a million demonstrators marching, singing, shouting, and waving huge ban-ners day after day in the Square left them feeling impotent. Furthermore, demonstrations had erupted in twenty-three other cities, raising the specter of an uncontrollable mass uprising. The elders concluded that, in the final analysis, the situation was tantamount to a war between communism and democracy. To give in would lead to the downfall of the leadership, the

9. Peking Mayor Ch'en Hsi-tung's speech before the Eighth Session of the Seventh Na-tional People's Congress, held June 30, 1989, entitled, "Report on Checking the Turmoil and Quelling the Counter-Revolutionary Rebellion," *Beijing Review,* July 17–23, 1989, 12–13.

10. These eight were Teng Hsiao-p'ing, former president Li Hsien-nien, President Yang Shang-k'un, Vice-President Wang Chen, economist Ch'en Yun, former chairman of the Standing Committee of the National People's Congress P'eng Chen, Teng Ying-ts'ao (Madame Chou En-lai), and perhaps Po I-po, an economic affairs expert.

overthrow of the socialist order, and ultimately to a capitalist restoration of a bourgeois government, just as John Foster Dulles would have wished. They reasoned that the students' desire for change was insatiable; if one gave them an inch, they would take a mile. Where would it all end? They feared that the students would not quit until they had overthrown the Communist party and the Four Cardinal Principles (the proletarian dictatorship, the socialist line, the leadership of the Communist party and Marxism-Leninism and the Thought of Mao). The veteran economist Ch'en Yun, aged eighty-five, delivered an emotional speech, perhaps with the intention of striking fear into the hearts of all concerned: "We seized power and established the People's Republic after decades of struggle and fighting, in which hundreds of thousands of our revolutionary heroes lost their lives. Are we to give it all up just to satisfy the students?" The council voted unanimously against retreat and for immediate dismissal of Chao as general secretary of the party.[11]

Clearly, the elders wished to cling to their political power and economic privileges at all costs; to do so they had to defend the socialist order that made possible their special status. Killing the demonstrators was of no concern because they were "anti-party counter-revolutionaries" who deserved to be eliminated. Even the threat of a loss of tourism, foreign investment, trade, credit, and loans paled when viewed from the larger perspective of the survival of the Communist leadership.

Teng declared that he did not fear foreign and domestic public opinion or bloodshed in any confrontation. He journeyed to Wuhan on May 19 to convene an enlarged meeting of the Military Commission, with the purpose of enlisting support for his policy of repression and possibly setting up "a second headquarters" in case the situation in Peking deteriorated. Reportedly, preparations were also made for a secret flight abroad if all else failed. In Wuhan, Teng won support from the Navy, the Air Force, and virtually all the army regional commanders. Thus fortified, on May 20 he ordered Premier Li to announce the imposition of martial law in Peking. Shortly thereafter, Teng also won the endorsement of his policy from nearly all the provinces and special municipalities. The stage was set for a military crackdown.

The Mind-set of the Gerontocracy

The psychology of the party elders played a key role in their decisions. As products of the Long March generation, with a lifelong experience of

11. Speech by President Yang Shang-k'un before an Enlarged Emergency Meeting of the Military Commission, May 24, 1989.

sabotage, class struggle, civil war, foreign war, and endless mass campaigns of one sort or another, they had become highly sensitive to the question of safety, security, and, above all else, survival. Obsessed with power, they embraced the one-dimensional view that power was life, and life without power was not worth living. They believed that in any struggle one could not afford to be kind and humane. To survive, one had to be hard, brutal, and even heartless. He who struck first gained the advantage. Teng espoused this view unabashedly when he declared: "Whoever wins the battle for state power gets to occupy the throne. This is the way it was in the past, and the way it still is in China as well as abroad."[12] Since their lives, positions, and privileges all depended on the survival of the socialist order, the party elders would go to any length to defend it.

Faced with widespread criticism of their advanced age, poor health, declining mental faculties, and alienation from mainstream society, the conservative leadership was all the more anxious to prove its capacity for vital decision-making and forceful action. Yet their sense of insecurity and fear of change remained. In this confused and paradoxical state, they tended to overreact and to be impulsive and precipitous in their judgment.

The crackdown policy stemmed from a misreading, whether intentional or otherwise, of the students' objectives. They asked to work with the government on anti-corruption measures and to discuss with its leaders the prospects for democracy, freedom of speech, of assembly, and of the press—the basic rights found in most civilized modern states. When the government refused to listen to their concerns, the frustrated students shouted for the dismissal of Li, Teng, and Yang, but they had neither the means nor the power to remove them or to overthrow the government. They lacked a platform of goals, a program of action, or an experienced, charismatic leader who could articulate their aspirations and unite the various splinter groups. Although the students in the Square and their supporters abroad declared that the government was so corrupt and irresponsible that it had lost its moral authority to rule, such a view lacked the force of a real threat. In opting to suppress the demonstrations by military force, the leadership lost a valuable opportunity to respond wisely and responsibly to the popular will and channel the tremendous potential of the people toward the constructive ends it had hoped to achieve.

After declaring marial law, the government leaders went into hiding and were not seen until after the crackdown. Chao, however, made his last public appearance on May 19, when he visited the hungry strikers in the Square and apologized for his tardy arrival. Immediately afterward, he was

---

12. Cited in Fox Butterfield, "Deng Is Said to Link Fear to Safety of Party," *The New York Times,* June 17, 1989, p. A4.

ousted from his job. On June 24, Chiang Tze-min was appointed as the new general secretary.

The Massacre

From the declaration of martial law on May 20 to the bloody crackdown on June 4, a full two weeks intervened, long enough for Teng and Yang to develop an elaborate, well-planned military operation. President Yang, a professional soldier trained in the Soviet Union during the Stalinist era, was a powerful military figure, second in command to Teng in the military hierarchy. Many of Yang's family members were placed in key army positions, so that they formed what the Chinese dubbed the "Yang family of generals." Yang himself was Executive Vice Chairman and Secretary General of the party's Military Commission; his half-brother, Yang Pai-ping, was Director of the Army's General Political Department; his son-in-law, Ch'ih Hao-t'ien, was the Army's Chief-of-staff; and his nephew, Yang Chien-hua, was commander of the 27th Army, which was to inflict most of the killing on June 3–4. Teng himself could handily draw from his own experience in military operations and political repression. In late 1948 he participated in the direction of the Huai-Hai (Hsuchow) Battle during the Chinese civil war and used tanks and artillery to annihilate the Nationalist opponents. In 1957, during the Anti-rightist Movement, he was party general secretary in charge of persecuting more than half a million intellectuals, winning high praise from Mao for his thoroughness.

Teng and Yang mobilized 300,000 troops from ten armies[13] across the country, as well as an armored division, a parachute division, and other special units. They were transferred to the outskirts of Peking not to fight the students but to forestall any possible palace revolution or coup d'état by Chao and his military supporters. Of course, the sheer size of the military force could also coerce students into submission and pressure the Politburo and the People's National Congress into supporting the policy of repression.

During the two-week lull, the students who had "occupied" T'ien-an-men Square for six weeks grew weary and exhausted. Many local students returned home or to their schools for rest and recuperation, but students from the provinces continued to pour in. Having traveled long distances, they were not willing to quit so soon. The local citizenry showed solidarity by offering them food, shelter, and other necessities. Many civilians helped students build barricades at key intersections to block the advancing

13. The 12th, 20th, 24th, 27th, 28th, 38th, 54th, 63rd, 64th, and 65th armies.

troops. When the first contingent of troops arrived, they carried no weapons and were friendly, also accepting food and drink from the citizens. There was a general belief that the troops would not shoot their own people. Still, martial law had been declared and no one could be certain of the outcome. Few suspected that the friendliness of the first contingent of troops could have been a disarming ruse. Some professors quietly urged the students to disperse, since they had made their point, but the students would not leave. They set June 20, the date marking the opening of the emergency session of the National People's Congress, as the time of evacuation. With each passing day, fear of an impending military attack alternated with a sense of relief that no violence had occurred. But in their hearts everybody knew that bloodshed was unavoidable.

On the evening of June 3 ominous signs of an impending crisis became apparent. A government television announcer gravely warned the citizens of Peking to stay away from T'ien-an-men Square because the People's Liberation Army would take any measures necessary to restore order. Students at Peking University again took this warning as an empty threat. To show their solidarity, they headed straight for the Square.

T'ien-an-men Square itself was filled with an impending sense of doom. At 4 p.m. on Saturday, June 3, an anonymous phone call was made to the student leaders' command post warning that the army was about to attack. The Student Association asked everyone to leave in order to avoid bloodshed, but 40,000 to 50,000 students and 100,000 other citizens vowed to stay and die, if necessary, in the cause of democracy and freedom. They still believed that the troops would not fire on their own unarmed people.

At 10 p.m. Premier Li ordered the troops to move at top speed to the Square, shoot all demonstrators without compunction, and clear the Square by dawn. Tanks, armored vehicles, and soldiers with automatic weapons struck from three directions in strict accordance with prearranged plans. One column attacked from the Revolutionary Military Museum located four miles from the Square, moving along the western section of the Avenue of Eternal Peace toward the Square, shooting and killing everyone in sight. Another column attacked from the eastern section of the avenue, and a third descended from the north, all converging at the Square. Much of the killing occurred before the troops and tanks reached the Square itself.

Before midnight two armored vehicles sped into the Square with loudspeakers blaring a shrill warning of "notification." In the early hours of June 4, thirty-five heavy tanks charged into the main encampment, crushing those students who were still inside. At 4 a.m. the lights in the Square were turned off and loudspeakers again ordered the remaining demonstrators to clear out. Four hunger strikers, including the Taiwanese pop singer Hou Te-chien, next secured from certain army officers safe passage for the remaining students, but before the latter could be notified the soldiers started to attack the Heroes' Monument area. At 4:40 a.m. a barrage of red flares burst overhead, signaling another onslaught. Soldiers and military police stormed out of the Great Hall of the People, their automatic weapons firing dum dum bullets, using electric cattle prods, rubber truncheons, and other types of special weapons, while tanks and armored vehicles rode roughshod through the crowds of terror-stricken demonstrators. Eleven students—two from Peking University and nine from Tsinghua University—linked hands in a symbolic gesture to protect the Goddess of Democracy; they were mowed down together with the statue. By 6 a.m. those who could had already escaped, while the dead or maimed were scattered all over the blood-soaked killing field. The soldiers hurriedly bulldozed the bodies into large piles for burning on the spot or packed them in plastic bags for cremation outside the city at Pa-pao-shan, where no registry was permitted to be divulged. The carnage was over in seven hours.[14]

An accurate accounting of the casualties is impossible. Western sources estimated that there were 3,000 dead and 10,000 or more wounded, though a later report by *The New York Times* revised the death toll to 400 to 800.[15] On June 16 Chinese government spokesman Yuan Mu told NBC anchorman Tom Brokaw that "in the whole process of clearing the square, there was no casualty [*sic*]. No one was shot down or crushed under the wheels of armored vehicles. The reports abroad that there was a bloodbath and that many people were crushed were incorrect." However, the government admitted that twenty-three students had been killed accidentally outside the Square, while 5,000 soldiers were wounded and 150 of them had died.[16]

One cannot question the right of any government to defend itself, nor can one expect it to surrender supinely when threatened with the danger of extinction. The question here, however, was not survival or extinction but meeting with student leaders to discuss anti-corruption measures and

14. See two moving eyewitness accounts of what happened at T'ien-an-men Square June 3–4, 1989: one by a twenty-year-old student at Tsing-hua University (*The San Francisco Examiner*, June 11, 1989, and also *The New York Times*, June 12, 1989), the other by a twenty-three-year-old student leader, Chai Ling (*The Free China Journal*, Taipei, June 15, 1989).

15. *The New York Times*, June 12, 13, 21, p. A6; *The Chronicle of Higher Education*, June 14, 1989.

16. *Beijing Review*, June 12–25, 1989, 9, July 3–9, 1989, 15–16.

political liberalization. Essentially, the issue was whether the government judged the challenge correctly and honestly and devised counter-measures appropriate to the occasion. The answer must be "no." In most civilized nations crowd control in massive public demonstrations involves the use of nonlethal implements—water cannons, tear gas, and even riot police armed with batons and shields—but not tanks and guns. In the last analysis, the threat to the Chinese leadership in May-June 1989 was largely fabricated, ultimately giving the government an excuse to kill the peaceful demonstrators as "anti-party counter-revolutionaries." If the octogenarians had been more patient and less impulsive, the demonstration would have exhausted itself in two or three weeks, since the students had already declared their intention to quit by June 20. All the bloodshed could easily have been avoided.

There is a tendency among many commentators to compare the violent events of 1989 with the Cultural Revolution, but there are fundamental differences between the two. First and foremost, the Cultural Revolution was directed from the top down by Mao, who used the students as Red Guards to crush his political enemies. The latter, like Mao, were veteran revolutionaries fully cognizant of the rules and dangers of intra-party struggles. It was a struggle between the hard-line rivals themselves. On the other hand, the demonstrations of 1989 were a spontaneous and patriotic movement started by young and idealistic students from the bottom up. The government used tanks and guns to crush peaceful protesters. The contest, if one can call it that, was brutal and utterly one-sided.

Rewriting History

Students of totalitarian systems realize that lies are routinely perpetrated during power struggles and liquidations of political enemies. The greater the violence, the greater the need for rewriting history through distortion. Stalin liquidated his Bolshevik associates on the pretext that they colluded with Leon Trotsky to oust him. Hitler's Minister of Propaganda, Joseph Goebbels, the inventor of the Big Lie theory, claimed that the bigger the lie, the greater the likelihood that people would believe it. In China's recent past, when Chiang Ch'ing (Mao's wife) lost her bid to be Mao's successor, she was charged by the party, among other crimes, of being a secret agent of the Nationalist Party!

A catastrophe as ugly as the T'ien-an-men massacre required the most blatant and implausible lies as a cover-up. Shortly after the crackdown, Teng told a group of military leaders that it was necessary to "create public opinion on a grand scale and make the people understand what really happened."[17] The party propaganda machine then churned out stories claiming that a massacre never occurred, and that reports by foreign journalists were based on misinformation and misunderstanding of Chinese realities. What the heroic People's Liberation Army had suppressed were not student demonstrations but a "counter-revolutionary rebellion" fomented by a small group of ruffians, hooligans, bandits, and certain "good but naive" students, who received financial support from reactionary forces in the United States, Britain, Hong Kong, and Taiwan.[18] Some misguided party leaders[19] secretly lent them support and encouragement in the hope of using them to precipitate a coup d'état to topple the government. These class enemies rightly deserved to be smashed, and if they escaped the initial crackdown they must be picked up "like rats" in follow-up sweeps. Each and every one of them had to be brought to justice, with no show of mercy; only then could the manhunt be thorough enough to prevent any potential resurgence of protest.

The mass arrests began almost immediately after the bloody crackdown. An all-points bulletin was issued to ferret out 21 student leaders, and citizens were urged to inform on them. By July 17, some 4,600 arrests had been made, and 29 of the prisoners were given a quick trial and then shot in the back of the head. The scene was reminiscent of the "White Terror" perpetrated after the 1917 Bolshevik Revolution and during the Cultural Revolution (1966–76).

China in June 1989 was cerily Orwellian. Big Brother was everywhere. Tales abounded of a mother turning in her son and a sister informing on her brother. People walked through the streets but did not talk; they whispered among themselves and occasionally winked at passing foreigners. Laughter, vivacity, and openness disappeared from daily life. The government papers carried story after story about what had happened in T'ien-an-men Square, but most urban dwellers did not buy the Big Lie, though a fair number of the rural population, cut off from urban life, probably accepted the official version. Hong Kong and other foreign sources provided much information to people in the cities, and the Voice of America did its share of broadcasting information. In an age of long distance calls, fax machines, computer networks, and satellite transmission of televised images, it was impossible for the Chinese government to seal

17. *The New York Times,* June 17, 1989, p. A4; see also Harrison E. Salisbury, "China's Peasants Get the Bad News," *The New York Times,* June 19, 1989, p. A15.
18. The Voice of America was particularly vilified for misreporting the China situation and for misinforming the Chinese public as to what happened. Two of its bureau chiefs in Peking, Alan W. Pessin and Mark W. Hopkins, were expelled from China.
19. Chao Tzu-yang and his liberal advisors?

off the country completely. In the end, the micro-chip that helped create the information revolution may have defeated the obsolescent totalitarian control of information.[20] Lu Hsün, the famous writer, wrote the following in 1926: "Lies written in ink can never disguise facts written in blood. . . . This is not the conclusion of an incident, but a new beginning."[21]

The repercussions of the crackdown and the subsequent mass arrests were staggering. The image of the Chinese government as an increasingly responsible member of the international community with a potent stabilizing influence on world affairs was shattered beyond repair. Virtual universal condemnation followed the crackdown, along with economic and military sanctions, as well as a tacit ostracism of China from important diplomatic meetings. Relations with Hong Kong and Taiwan suffered severely. A million Hong Kong residents demonstrated to protest the T'ien-an-men Square massacre and to register their distrust of Peking's pledge that it would honor the capitalistic way of life in Hong Kong for fifty years beyond 1997. As for reunification with Taiwan, the prospect became more remote than ever. Most tragic of all, the spirit of the Chinese people had been mortally wounded, and the painful experience of the crackdown would not be forgotten or forgiven for a long time. Many believed that the government's moral authority to rule (*t'ien-ming* or the Mandate of Heaven) was beginning to slip away, and the question of its legitimacy was very much on the people's minds.

Further Reading

Amnesty International, *People's Republic of China: Preliminary Findings on Killings of Unarmed Civilians, Arbitrary Arrests and Summary Executions Since June 3, 1989* (New York, August 1989).

Brzezinski, Zbigniew, *The Grand Failure: The Birth and Death of Communism in Twentieth Century* (New York, 1989).

Cheng, Chu-yuan, *Behind the Tiananmen Massacre: Social, Political, and Economic Ferment in China* (Boulder, 1990).

Fang, Lizhi, *Bringing Down the Great Wall: Writings on Science, Culture, and Democracy in China* (New York, 1991). Ed. by James H. Williams and tr. by him and others.

Gargan, Edward A., "Beijing Diary: Eyewitness at Tian An Men Square," *Los Angeles Times Magazine*, July 16, 1989, 6-20, 37-39.

Gold, Thomas B., "The Class of '89," *California Monthly*, Sept. 1989, 18.

Hayhoe, Ruth, "China's Universities Since Tiananmen: A Critical Assessment," *The China Quarterly*, June 1993, 291-309.

Hicks, George (ed.), *The Broken Mirror: China After Tiananmen* (Essex, Eng., 1990).

Kristof, Nicholas D., "China Update: How the Hardliners Won," *The New York Times Magazine*, Nov. 12, 1989, 39-41, 66-71.

———, "Escape from Tiananmen Square: A Chinese Odyssey," *The New York Times Magazine*, May 5, 1991, 28-31, 49-51.

Lin, Nan, *The Struggle for Tiananmen: Anatomy of the 1989 Mass Movement* (Westport, Conn., 1992).

Liu, Banyan, with Ruan Ming and Xu Gang, *Tell the World: What Happened in China and Why* (New York, 1989). Tr. by Henry L. Epstein.

Lukin, Alexander, "The Initial Soviet Reaction to the Events in China in 1989 and the Prospects for Sino-Soviet Relations," *The China Quarterly*, March 1991, 119-36.

Mann, Jim, "China's Lost Generation: After Tian An Men, the Best and Brightest Say They Can't Go Home Again," *Los Angeles Times Magazine*, March 25, 1990, 10-19, 38-39.

Oksenberg, Michel, Lawrence R. Sullivan, and Marc Lambert (eds.), *Beijing Spring, 1989: Confrontation and Conflict, the Basic Documents* (Armonk, N.Y., 1990).

Rosen, Stanley, "The Effect of Post-4 June Re-education Campaigns on Chinese Students," *The China Quarterly*, June 1993, 310-34.

Saich, Tony, "The Rise and Fall of the Beijing People's Movement," *The Australian Journal of Chinese Affairs*, 24: (July 1990): 181-208.

Salisbury, Harrison E., "China's Peasants Get the Bad News," *The New York Times*, June 19, 1989, A15.

Simmie, Scott, and Bob Nixon, *Tiananmen Square* (Seattle, 1989).

Tsao, Hsingyuan, "The Birth of the Goddess of Democracy," *California Monthly*, Sept. 1989, 16-17.

Walder, Andrew G., "The Political Sociology of the Beijing Upheaval of 1989," *Problems of Communism*, 38:5 (Sept.-Oct. 1989): 30-40.

———, "Workers, Managers, and the State: The Reform Era and the Political Crisis of 1989," *The China Quarterly*, Sept. 1992, 467-92.

Whyte, Martin King, "The Social Sources of the Student Demonstrations," *The Chinese Intellectual*, 23 (Spring 1991): 16-22. Chinese translation by Lo Wen-fu.

Williams, Daniel, "China's Underground Presses Seized as Crackdown on Media Continues," *Los Angeles Times*, June 20, 1989, Part I, 10.

Yang, Winston L. Y., and Marsha L. Wagner (eds.), *Tiananmen: China's Struggle for Democracy*. Occasional Papers/Reprints Series in Contemporary Asian Studies, No. 2-1990 (97). School of Law, University of Maryland.

41

The Chinese Model of Development: Quasi-Capitalism in a Political Dictatorship

Barely six months after the T'ien-an-men crackdown in June 1989, Eastern Europe was engulfed in popular protests against Communist rule, precipitating its largely unforeseen downfall. By the end of 1991, the Soviet Union itself had disintegrated. In light of these cataclysmic events, what would be the future of China?

Teng Hsiao-p'ing's Stratagem

The T'ien-an-men crackdown brought China not only universal condemnation and severe international economic and military sanctions but diplomatic ostracism as well. Teng Hsiao-p'ing concluded that the next three to five years would be most critical to the survival of Chinese Communism and that China should strive to achieve domestic stability and international noninvolvement. At this point in history, Teng believed that "action is not as good as inaction" and that debate on political and economic policies, dismissal of key personnel, and changes in political direction would work to China's detriment. To the West, he announced; "You practice your capitalism; we practice our socialism. The well water shall not invade the river water."

Haunted by the parallels between the T'ien-an-men crackdown and the Rumanian turmoil that led to the violent deaths of President Nicolae Ceausescu and his wife, the Chinese leadership severely criticized Mikhail S. Gorbachev for subverting socialism from within. His refusal to intervene in Eastern Europe was attacked as an abdication of his leadership in the Warsaw Pact. His report to the Enlarged Soviet Central Committee held during February 5–7, 1990 (which renounced the party's monopolistic leadership; permitted a multiparty, "presidential" system; and endorsed a "humane, democratic socialism") was condemned as a betrayal of Marxism-Leninism. To the Chinese, Gorbachev was seen as rejecting class struggle and practicing Western style parliamentary democracy, thus repudiating the very foundations of Communism.[1]

Would the fate of Eastern Europe and the Soviet Union be repeated in China? At first, Peking had confidence in China's uniqueness, trusting in Chinese Communism's strict adherence to Marxist-Leninism-Maoism and its powerful mass base. But by the end of 1991, Soviet Communism, the oldest and most authentic home-grown variety, collapsed. Buffeted now by the winds of change from both the West and the Soviet Union, Peking, on the one hand, detested Gorbachev's "humane, democratic socialism" because it blurred the boundaries of "true socialism" and confused the ideology; and, on the other hand, Peking was certain that the West, having victimized the Soviet Union, would resort to the same sinister scheme of "peaceful evolution" to subvert China. To forestall this fate, China determined to tighten its ideological control and military preparedness against Western ideas, culture, and values while putting on a benign and peaceful front. Peking's double-barreled posture was known as *nei-ching wai-sung* (tight internal control, relaxed external appearance).

The Chinese leadership had come to some hardheaded decisions. First, it would never voluntarily give up its monopoly of power as the Soviet leaders had foolishly done. In this they were confident of success because the democratic forces had been mostly crushed; and, except for the largely subdued, if still defiant, Tibetans, Chinese nationalities did not seek independence. Above all, China, unlike the Soviet Union, enjoyed the long tradition of a unified state. Second, economic development must continue, and the people's livelihood must improve. Third, the government media should regularly play up the difficult life in the Soviet Union by reporting vividly the food shortages, runaway inflation, black market, political squabbles, economic dislocation, and the contempt and hatred heaped upon the former Communist leaders. Presenting the specter of chaos (*luan*) if the Soviet experience were repeated in China, the authorities promised economic prosperity and political stability as long as they were not opposed.

1. Chinese Communist Central Committee Document: A Critique of the Soviet Central Committee Second Plenary Session. Confidential; original in Chinese; reprinted in *China Spring*, May 1990, 64-65.

Peaceful Evolution (Ho-p'ing yen-pien)

The Chinese leadership subscribed to the orthodox Marxist view that the clash between capitalism and socialism was inevitable and that the West never tired of plotting against Communism. The real aim of Western intervention in the Korean War, which broke out in June 1950 barely nine months after the establishment of the Chinese People's Republic, Peking claimed, was to destabilize and overthrow the new Chinese government. Failing that, the West would resort to the peaceful means of infiltrating and sabotaging socialist countries, hoping that over a period of time it would gradually disarm the vigilance of the socialist youths and party cadres and destroy Communism from within. The architect of this stratagem was thought to be the American Secretary of State John Foster Dulles, who in the 1950s urged the opening of all channels and points of contact with Communist states, transmitting by whatever means possible Western ideas, values, religions, literature, art, and way of life. Such psychological warfare was believed to be effective when socialist productive power and standard of living were lower than those of the West, making socialist youth easy prey to Western temptations.[2]

From the Chinese perspective, the collapse of Soviet Communism left China with no choice but to serve as an "Iron-and-steel Great Wall of Socialism" to fight the conspiracy of "peaceful evolution." "Just as in the past only socialism could save China, now only China can save socialism," declared the Chinese leadership. Clearly, they refused to see the global trend toward democracy, economic freedom, free trade, and human rights. They did not want to admit that the Four Little Dragons of Asia—Hong Kong, Singapore, Taiwan, and South Korea—had achieved economic prosperity through peaceful evolution. Certainly, they did not want to face the unsettling fact that the social benefits in the United States and Britain were substantially greater than those of the Soviet Union, Cuba, Eastern Europe, and China, where all were provided with a subsistence level of existence ("iron rice bowls") that encouraged laziness, laxity, and a low standard of living.[3]

All this seemed irrelevant to the conservative leadership of China. Captives of the "Yenan psychology," they continued to insist that "foreign

2. Liu Hsueh-min, "Ho-p'ing yen-pien chiu Chung-kuo" (Peaceful evolution saves China), *T'an So* (The Quest), June 1991, 35-36.

3. Liu Tsung-hsiang, "Yeh hsing-jen ti k'ou-hsiao: ho-p'ing yen-pien chiu Chung-kuo" (A night traveler's whistle: peaceful evolution saves China), *T'an So* (The Quest) Nov. 1991, 54; Ch'ien Chia-chu, "Ho-p'ing yen-pien yu ho-p'ing yen-pien ti wen-t'i" (Peaceful evolution and questions about peaceful evolution), *T'an So* (The Quest) April 1992, 13.

hostile forces never gave up the intention of exterminating us." They distrusted American intentions in spite of assurances of the American Secretary of State James A. Baker that "the American agenda for China is for all, Chinese and Americans, to see. We want to protect human rights and advance liberty. We want to counter the threat of nuclear and missile proliferation. We want free and fair trade that benefits both countries and the region . . . Our ideals and values must be an essential part of our engagement with China. We will fight against political repression and religious persecution."[4] They also turned increasingly critical of Teng and his policy of opening, which they insisted let in poisonous bourgeois ideas that corrupted the minds of the youth and led to the student demonstrations of 1989. Teng Li-ch'un (Deng Liqun), the die-hard conservative ideologue under Ch'en Yun, hinted: "Within the party there are capitalist roaders, but I don't mean Teng Hsiao-p'ing!"[5] Indeed, Teng's reform and his promotion of a "socialist market economy" could be viewed as a modified form of peaceful evolution. Even President Bush stated that he knew of no means whereby a country could import the world's goods and services but stop foreign ideas at its borders.

The collapse of the Soviet Union was a sobering experience for Teng. More pragmatist than ideologue, he wanted to borrow elements of capitalism to save Communism, and he openly proclaimed that without his reform and opening China would share the fate of Eastern Europe and the Soviet Union. Intimating—without openly admitting—that peaceful evolution had saved China, Teng was ready to launch a new initiative to move the country back to the track of reform and opening. At 87, he saw such a strategy as his last chance to establish a place in history. To the booming South he went with this message: accelerated economic development and political stability were to be the order of the day. Now was not the time to push for ideological purity and world revolution.

Teng's Southern Tour

In January 1992 Teng Hsiao-p'ing made a highly publicized tour of the South, where economic growth in the past decade had been the greatest, with Kwangtung scoring an amazing 21 percent in 1991, perhaps the fastest in the world. Accompanied by his family, Teng visited the economic showplace of Shenzhen, the Pearl River Delta, and Canton, giving his "enthusiastic endorsement of the brave new world of skyscrapers and ad-

---

4. James A. Baker, III, "America in Asia: Emerging Architecture for a Pacific Community," *Foreign Affairs*, Winter 1991/92, 15-16.
5. Cited in *T'an So* (The Quest) April 1992, 19.

vertising, investment incentives."[6] With a radiant face, he showed the world that there was enough strength left in him to carry out a new round of reform and opening that would push the country toward quasi-capitalism—a new economic revolution.

Teng's speech was often blurred, earthy, and biting; it sometimes required interpretation by his daughter, Teng Jung (Deng Rong). But the salient points could be summarized as follows:

1. Economic development is the central task of government. Reform and opening are its two pillars.
2. Stability is essential to economic development.
3. The superiority of socialism can be demonstrated by improving the national economy.
4. Experiments with stock markets and more foreign capital, foreign advanced technology, foreign experience, and joint ventures should be welcomed.[7]
5. Capitalism in Hong Kong can continue for 100 years after the island's return to China in 1997.
6. Kwangtung province should aspire to become the "Fifth Little Dragon of Asia" in 20 years—after Hong Kong, Singapore, Taiwan, and South Korea—and China, a superdragon.[8]

Teng's call for a new Great Leap in reform and opening put him on a collision course with his arch rival, Ch'en Yun. Whereas Teng insisted on economic development as the central task of government, Ch'en advocated ideological tightening and resistance to peaceful evolution. On March 10, 1992, the enlarged Politburo resolved that the country should guard against turning "right" (i.e., into capitalism) but that even more importantly it should guard against turning "left" (to radical Communism). The position of the Politburo was a tactful endorsement of Teng's line.

But it was not a total victory for him. Teng viewed his critics as parochial in outlook and deficient in the understanding of world affairs. On January 15, 1992, he spoke in Shanghai:

> There are party members whose brains are muddled and yet they consider themselves masters of Marxism. They criticized, attacked, and obstructed the current conditions, objecting to this and that while

6. Richard Holme, "China: Tiger's Gilded Cage," *New Statesman and Society*, England, Jan. 22, 1993, 19.
7. Originally *San-tzu ch'i-yeh*, referring to three types of foreign investments: (1) sole foreign investment; (2) Chinese-foreign joint investment; and (3) Chinese-foreign joint enterprise.
8. CCP Central Committee, Document No. 2, Feb. 28, 1992. Complete text reprinted in *Cheng-ming* Magazine, Hong Kong, April 1992, 23-27.

mistaking minor currents for the mainstream. Is it scientific Marxist pragmatism? I believe that the capitalist society is changing and the world is changing, but our comrades are not and they keep their doctrinal orthodoxy. Could this promote social progress? If we had not undertaken reform and other measures in the past decade, our situation would have been worse than that of the Soviet Union.[9]

Statistics of economic progress supported Teng. During the decade of the 1980s, the GNP grew by 8.9 percent annually, which was three times as fast as the world average, reaching Ch$1,700 billion in 1990. The village and township rural enterprises grew by leaps and bounds, numbering 18.88 million and producing Ch$950 billion in 1990, or 58.5 percent of the total agricultural production. People's livelihood was materially improved.

Internationally, the Asia-Pacific Region was burgeoning with economic activity and seemed poised for a gigantic advance in the next decade or two. Teng was determined to participate in this advance and not let slip the golden opportunity that history seemed to be offering China. He believed that he was moving in the mainstream of world economy whereas his rival, Ch'en, was stuck in a muddy little creek.[10]

The Fourteenth Party Congress

The Fourteenth Party Congress, convened in Peking during the week of October 12, 1992, was attended by 1,991 delegates. Its main mission was to approve economic reform as the central task of the party, to elect a new leadership, and to permit a moderate loosening of cultural and ideological controls.

General Secretary Chiang Tse-min (Jiang Zemin) delivered the political report calling for the acceleration of economic reform and opening and the construction of a "socialist market economy." "Reform," he said, "is also a revolution, a revolution whose goal is to liberate productive forces. It is the only way to modernize China. If we cling to outmoded ideas and remain content with the status quo, we shall accomplish nothing . . . Poverty is not socialism." Yet no definition of "socialist market economy" was given until January 1993, when the State Council interpreted it to mean a multifaceted economic structure in which public ownership, individual household industry, private enterprise, and foreign investment all competed on an equal basis. In practice, the new structure could mean moving away

9. Wang Wei-hsin, "Teng Hsiao-p'ing lu-hsien ti kuan-ch'a" (An observation of the Teng Hsiao-p'ing line), *T'an So* (The Quest), May 1992, 41.
10. *Ibid.*, 41-42.

from central planning and accepting capitalistic practices based on a market economy, a gradual freeing of price control, private ownership of properties and businesses, stock markets, foreign investment and joint ventures, and restructuring of the public sector of the economy.

The political report also reaffirmed the party dictatorship, the ideology of Marxism-Leninism-Maoism, and the need to suppress political turmoil. It called for vigilance against peaceful evolution and declared that "the goal of this reform is to build a socialist democracy suited to the Chinese conditions and absolutely not a Western, multiparty, parliamentary system."

Although the report seemed contradictory in calling for a market economy in a political dictatorship, Teng saw no inconsistency in promoting a semicapitalist economy in an authoritarian political order. In the late nineteenth century, Meiji Japan and Bismarckian Germany had done the same splendidly; and in more recent times Singapore, Taiwan, and South Korea had achieved economic miracles under authoritarian rule.[11] The Singaporean model of Premier Lee Kuan Yew struck a particularly responsive cord in Teng. Singapore was clean, orderly, rich, and strictly controlled by the government; there were no street demonstrations, pornography, drugs, chewing of gum in public, or jabbering about human rights. Teng remarked: "We should learn from their experience, and we should do a better job than they do."[12] In his mind, China could become a superdragon of Asia through the accelerated development of the South and the coastal areas. Such a possibility seemed more realizable under an effective authoritarian government, which could provide the necessary stability for the economic takeoff than under a weak democratic government torn by dissension and undermined by instability.

Teng's conviction that political stability was essential to economic development became stronger than ever after the collapse of the Soviet Union. The Chinese model of promoting quasi-capitalism in a political dictatorship won the praise of many Russian traditionalists and ex-Communists, who were frustrated by the shortages and disorder in their country. "The Chinese people feed themselves, and 1.2 billion are living a life of which we can only dream," remarked Arkady Volsky, a longtime aid to Andropov and Brezhnev.[13] In Eastern Europe there was a nostalgia for the certainty of life under the old order and also a yearning for a "third way" between

11. Maurice Meisner, "What Beijing Leaders Know that Critics Won't See: Repression Can Be Profitable," *Los Angeles Times*, Oct. 25, 1992, M2.

12. Nicholas D. Kristof, "China Sees Singapore As a Model of Progress," *The New York Times*, August 9, 1992: see also Richard Holme, 19-20.

13. William Safire, "Vision in Collision," *The New York Times*, Dec. 23, 1992, A25.

Jiang Zemin (Chiang Tse-min), President of the People's Republic of China and General Secretary of the Chinese Communist Party.

Zhu Rongji (Chu Jung-chi), Vice Premier and Economic Chief.

Zou Jiahua, Vice Premier and Minister in charge of the State Planning Commission.

Qiao Shi (Ch'iao Shih), Chairman, Standing Committee of the National People's Congress.

Communist central planning and Western democratic freedom and market economy.[14]

PERSONNEL SHUFFLE. Teng set about to entrench a new generation of leaders committed to his twin policies of economic reform and political dictatorship. Forty-seven percent of the new Central Committee of 189 regulars and 130 alternates were elected for the first time and many of them were technocrats with college degrees (84 percent). Their average age was 56. These technocrats knew that revolutionary prestige was a thing of the past and that only economic progress could justify their continued stay in power and the party's legitimacy.

The die-hard conservatives fared poorly in the election, and three of their prominent members failed to win seats on the Central Committee.[15] Their removal from their media, propaganda, and cultural posts augured a freer atmosphere in the arts, culture, and intellectual life. The "Crown Prince Group," children of high party officials, also did poorly; Ch'en Yuan, son of Ch'en Yun, failed to make the Central Committee. Many of them left politics for business.

The Politburo of 20 members also saw a radical reshuffle. Two new reformers joined the rank—Deputy Premier Tsou Chia-hua (Zou Jiahua) and Foreign Minister Ch'ien Ch'i-chen (Qian Qichen). Together with the two continuing reformer members—T'ien Chi-yun (Tian Jiyun), an agricultural expert, and Li T'ieh-ying (Li Tieying), the education minister, they formed a "liberal bloc." Five other new members came from the provinces. The Politburo seemed more liberal and dedicated to promoting market-oriented reforms, including the freeing of prices, expansion of stock markets, and restructuring of state enterprises so that they could compete in the marketplace.

The Politburo Standing Committee, the real seat of power, consisted of four continuing members and three new ones. The former included General Secretary Chiang Tse-min, 66; Premier Li P'eng, 63; Ch'iao Shih, 68; and the reformer-ideologue Li Jui-huan (Li Ruihuan), 58. The latter included Deputy Premier Chu Jung-chi (Zhu Rongji), 64, the economic chief and a protégé of Teng; General Liu Hua-ch'ing (Liu Huaqing), 76, a vice-chairman of the Military Commission; and Hu Ching-t'ao (Hu Jingtao), 49, a former party secretary in Tibet. All seven were also given

---

14. Roger Cohen, "An Empty Feeling Is Infecting Eastern Europe," *The New York Times*, March 21, 1993, E3.
15. Kao Ti (Gao Di) editor-in-chief of the *People's Daily*; Wang Jen-chih (Wang Renzhi), director of the Propaganda Department of the party; and Ho Ching-chih (He Jingzhi), the acting minister of culture.

important government posts at the Eighth National People's Congress in March 1993.[16]

The conspicuous losers were the "Yang Brothers," who had become too powerful for the comfort of Teng. President Yang, 85, was retired on grounds of old age, and his half brother, Yang Pai-ping (Yang Baibing), 73, was nominally "promoted" to the Politburo while losing his positions in the Central Military Commission and Central Committee Secretariat.

On the whole, the reformers gained a degree of ascendancy in the power shuffle, but the continuation of the Chiang-Li (Jiang-Li) Leadership suggested a lack of breakthrough in new directions.

In historical perspective, Teng's strategy might have saved Communism in China from collapse for the time being, but it also swept away many of the economic underpinnings of Communism, affecting profoundly the fundamental nature of the economy and society. What was left of Communism was an empty shell, propped up by the military and the secret police.

Chinese Communism at 45: Quasi-capitalism in a Political Dictatorship

So fundamental and far-reaching have been the social and economic changes in China in the last decade that if Chairman Mao were to return for a visit, he would be stunned beyond belief by what he would see. Communist ideology is largely ignored as irrelevant. The entrepreneurial spirit fills the air, and the new religion is money. Life along the coast is throbbing with the activities of the stock markets, privatization of homes and state enterprises, foreign investments, joint ventures, imports and exports, and conspicuous consumption. At the same time, social and cultural controls have markedly loosened, and the information revolution makes it all but impossible for the government to control the spread of ideas; moreover, the migration of tens of millions of farm labor to the cities challenges the party control of population movement and residential permits. Yet although the social and economic underpinnings of Communism are crumbling, the political structure remains unchanged. This emergence of quasi-capitalism in a Leninist dictatorship is unprecedented.

THE IDEOLOGICAL CONTRADICTION. Teng was sheepish about introducing a market economy, which was never a part of the classic Marxist economic

16. Chiang Tse-min, President of the People's Republic; Li P'eng, Premier; Chu Jung-chi, First Deputy Premier; Ch'iao Shih, Chairman, The National People's Congress; and Li Jui-huan, Chairman, People's Political Consultative Council.

structure and which in Maoist times had been condemned as "capitalist enemy No. 1," targeted for destruction. From the beginnings of his reforms he was specious about the true nature of market economy, couching it in socialist terminology to make it more palatable to his followers. Calling it "a planned commodity economy on the basis of public ownership system," he declared that "the planned economy should be the mainstream and market regulation the supplement" and called for the "internal unification" of the two. It was not until 1992 that he broke loose from all inhibitions and openly proclaimed the building of a "socialist market economy" as the central task of government. His proclamation was tantamount to a repudiation of the central dogmas of Marxist economic theory: public ownership and central planning. To soften the "apostasy," the word *socialist* was added to *market economy*. To his enemies, the phrase seemed an inglorious oxymoron.[17]

THE PARTY'S NEW MISSION: MONEY AND POWER. The collapse of Communism in the Soviet Union and the economic boom in China led many Chinese to believe that revolutionary prestige was a thing of the past and that wealth was the new source of power, status, and the good life. This new attitude, infecting the entire society from the highest party officials to the common men and women in the streets, caused many children and grandchildren of party leaders to choose business over politics. Their plunge into commercial activities is described, somewhat pejoratively, as *hsia-hai*, or "plunging into the sea."

Meanwhile, the party itself had quietly undergone a subtle transformation from an ascetic, revolutionary organization to an elitist political instrument, no longer committed to world revolution, international brotherhood, national liberation, and the Communist utopia, but to the perpetuation of its dictatorial power and the enrichment of its members and their families. It parceled out state enterprises at bargain prices to its preferred members, who overnight became board chairmen, presidents, vice presidents, stock holders, and managers of valuable enterprises in control of sizable portfolios of financial assets and properties at home and substantial investments and real estate abroad.

As the middle- and lower-echelon party members likewise delved into business and commercial activities, so did everybody else in Chinese society. Even the political dissidents jumped on board to raise funds for their future fight for democracy. The result of this feverish pursuit of wealth

17. Leng Heng-mei, "Shih-ssu-ta wei ta-lu mai-hsia-liao tung-luan ti ho-keng" (The Fourteenth Party Congress sowed the seed of future disorder on Mainland China), *T'an So* (The Quest), Dec., 1992, 35-36.

has been an explosion of national energy that generated an unprecedented tidal wave of commercialism. All of a sudden, the once poverty-stricken Communist state had turned into a gigantic economic machine in open bid for wealth, respectability, and acceptance in the capitalist world—the very objectives that the Marxists had vowed to destroy![18]

PRIVATIZATION OF ENTERPRISE. With one-third of the state enterprises operating in the red, the government was anxious to end its heavy subsidies. Several plans were considered. One was to privatize unprofitable businesses, contracting private individuals as operators and allowing them to keep the profit above a predetermined sum. A second way was to auction off medium-sized state enterprises with a net worth of Ch$10 million or less to party members in good standing who were able to pay a 10 to 15 percent down payment, had three years of experience running a profitable business, and could pay off the purchase price in five years. A third way was to lease the businesses to private investors and split the profit fifty-fifty with the state. Most of these investors turned out to be favored party members who were the incumbent managers or key members of the management teams. Their investment allowed many to accumulate considerable wealth. A few years earlier a family with Ch$10,000 a year income was considered rich. By 1993, only millionaires were admired, and Kwangtung province had 40,000 of them. In all China, there are reportedly five million millionaires.[19]

HOME OWNERSHIP. Since the early 1950s living quarters had always been assigned through work units, which were controlled by the party. Now the government began regarding apartments and houses as commercial commodities to be sold or rented at market prices. In Peking, Shanghai, and several other large cities, the government offered the current occupants favorable terms for the purchase of their two-bedroom apartments at U.S. $3,100, with a third down payment and a monthly payment of $12 under a 15-year interest-free mortgage. By then the apartment was expected to be worth ten times more.

Initially, the program did not seem that appealing to the public: most Chinese are frugal, and they were used to their heavily subsidized low rent of $2 a month. But in the South and along the coast there were more

18. Lao Pei, "Ts'ung pien-ko ti chueh-tu k'an Chung-kuo ta-lu ti wei-lai" (The future of Mainland China from the standpoint of reform), *T'an So* (The Quest), Jan. 1993, 76-77.

19. Li Hsiao-ming, "Chung-kuo nan-fang ti tzu-pen chu-i je" (The capitalistic fever in southern China), *China Spring*, March 15, 1993, 9-10; Sheryl WuDunn, "As China Leaps Ahead, the Poor Slip Behind," *The New York Times*, May 23, 1993.

affluent citizens who were potential home owners. A Western consulting firm (McKinsey) in Hong Kong estimated that 5 percent of the Chinese population, that is, 50 million, had picked up Western patterns of consumption.[20]

The real estate boom in the South and along the coast had been fueled by Hong Kong and Taiwan money, which was quick to take advantage of the new regulations that permitted lease of land for 70 years by foreigners and private investors. Land values in Canton-Shenzhen-Pearl River Delta shot up from Ch$50 per square meter to Ch$1,000, and 30 percent of the choicest vacant land was snapped up by Hong Kong investors. A medium-priced apartment of 100 square meters in Shenzhen sold for almost as much as a comparable condominium in Los Angeles, Ch$500,000 or U.S.$95,000. Indeed, Shenzhen had become an outskirt of Hong Kong; and Fukien province, a playground of Taiwan investors, who poured in U.S.$1 billion for land and real estate acquisitions. Fukien, some quipped, had become a new colony of Taiwan. Land speculation in Hainan Island was even more rampant.

THE STOCK MARKETS. Although stock markets are denigrated as symbols of decadent Western capitalism in most Communist countries, in China today they are a novel attraction. The Shanghai Stock Market was opened in December 1989; that in Shenzhen, in July 1991. Two types of stocks were offered: Class A for Chinese investors and Class B for foreigners. Class A stocks have performed very well, rising sometimes 20 percent a day and boasting an astronomical price-earnings ratio of 146, compared with 20 or 30 in the New York Stock Exchange. Class B stocks, however, have been stagnating, perhaps because of the small capitalization of the two markets, which, at U.S.$1.1 billion, cannot handle large foreign orders. Foreigners prefer to deal through the Hong Kong Stock Market, where the capitalization is U.S.$250 billion.

Most Chinese naively believe that the stock market can only go up because the government wants the experiment to succeed. With an average saving rate of 38 percent, the Chinese had accumulated Ch$1,000 billion by the end of 1991. When the shares of the Shenzhen Development Bank shot up 80 times in a short while, all wanted to climb aboard and get rich quick. In August 1992 a million potential investors stood in line outside

20. Nicholas D. Kristof, "A New Class of Chinese Is Emerging: Home Owners," *The New York Times*, Nov. 22, 1992; Richard Holme, "China: Tiger's Gilded Cage," *New Statesman and Society*, England, Jan. 22, 1993, 19; Yang Man-k'e, "Yu-fang chieh-chi chueh-ch'i ta-lu" (Home-owning class arises on Mainland China), *China Spring*, March, 1993, 47-48; Sheryl WuDunn, "China Sells Off Public Land to the Well-Connected," *The New York Times*, May 8, 1993.

the Shenzhen Stock Market to wait for the application forms to buy stocks. When they heard that half of these forms had been snapped up by corrupt officials in collusion with the stock market managers, they rioted in the streets. This dramatic demonstration of capitalist impulse in a Communist country made for a good story on Wall Street.

As of May 1993 there were two million stockholders in China, and each week the number grows by 50,000. The Shanghai market rose 167 percent in 1992, and the volume of trade grew prodigiously. In time it could surpass the Hong Kong and Tokyo markets to be the largest in Asia. Duncan P. F. Mount, managing director of CEF Investment Management Ltd. in Hong Kong, observed: "China is the world's fastest-growing economy. There are about 300,000 incorporated companies, of which about 70 are listed. In time, it's obviously going to be Asia's biggest stock market, probably bigger than Japan." William A. Schreyer, 65, chairman of Merrill Lynch & Company, opened a branch in Shanghai in April 1993 and remarked: "The modernization of China's economy, I think, is one of the most exciting endeavors in this century, and we want to be there and be part of it." He added a personal note: "If I were 25 years younger, I'd study Chinese and ask Merrill to post me there."

The stock market experiment is significant for China in transition. It confirms the economic move toward quasi-capitalism, forces the state enterprises to be more alert to profitability and efficiency, and establishes entrepreneurial interest as an alternative value system of power and prestige. Above all, it creates a new group of stockholders and investors who rapidly form the core of an urban middle class in an emerging civil society.[21]

ENTREPRENEURS' PARADISE. It has been said that whereas 75 years of Communism in the Soviet Union stunted the entrepreneurial spirit, 45 years of Communism in China whetted the appetite for businesses and commercialism.[22] Certainly, in no other Communist state has there been so blatant a worship of money and the things it buys. With an annual economic growth of 9.5 percent since 1980, China is much like Japan, South Korea, and Taiwan at a comparable stage of development. Indeed, a 9 percent growth rate will quintuple the GNP in 19 years; China's goal of quadrupling the 1980 GNP by A.D. 2000 can be achieved ahead of schedule.[23]

21. Nicholas D. Kristof, "Don't Joke About This Stock Market," *The New York Times*, May 9, 1993.
22. Nicholas D. Kristof, "Entrepreneurial Energy Sets Off A Chinese Boom," *The New York Times*, Feb. 14, 1993.
23. *Ibid.*

China today is a paradise for freewheeling entrepreneurs. A little-known small town has become the button-making capital of the world by producing 12 billion buttons in 1992. One owner of a factory enjoyed sales of U.S.$500,000; and another, U.S.$200,000. Public display of credit cards, gold bracelets, diamond watches, and jewelry is a new way of life. Even the local party secretary, who rides in a chauffeured Audi, sees a new role for himself: "My most important job is building up the economy. People here say, 'If you push the economy along, you are a good leader. Otherwise, you are not.' "[24]

Success stories abound. Two brothers in their twenties founded the Sky Dragon Charter Airline Company and netted U.S.$2 million in 1992. A woman shop owner sells imported shoes from Europe for U.S.$200 a pair and has no lack of customers. Wenchow (Wenzhou), a city of "Dickensian capitalism," has privatized 90 percent of its economy. One owner of a shoe factory employs 100 workers and produces 500 pairs a day, grossing U.S.$1 million in 1992. He lives in a "palatial" house with marble floor, complete with air-conditioning and a satellite dish on the roof for his Panasonic television and Kenwood stereo system. Such a lifestyle was unheard of in ascetic Maoist days.[25]

An even greater success story revolves around the 52-year-old Mu Ch'i-chung (Mu Qizhong), who borrowed U.S.$55 in 1979 to form the Land Economic Group. Today he has offices in five countries engaged in tourism, investment, manufacturing, and barter trade. His current occupation is to barter 600,000 pairs of socks and shoes, canned food, and thermos to fill 500 trains for four Russian 164-seat TU-154M passenger jets, which he sells to a Chinese airline for U.S.$75 million, netting $25 million in profit. He is one of 21 million entrepreneurs in China, of whom 100,000 own or operate large private businesses.[26]

FOREIGN INVESTMENT. The booming economy has turned from 60 million to 300 million Chinese into conspicuous consumers.[27] To capture this "legendary" China market, foreign companies all wish to get in on the ground floor. American corporations represented there read like a *Fortune* 500 list: GM, Ford, Chrysler, A T & T, GE, IBM, Digital Equipment,

24. Nicholas D. Kristof, "Poor Chinese Town Bets Its Shirt on Making Buttons and, Bingo!" *The New York Times*, Jan. 18, 1993.
25. Nicholas D. Kristof, "Backed by China, Go-Getters Get Rich," *The New York Times*, Jan. 27, 1993.
26. Nicholas D. Kristof, "A Tycoon Named Mu: Product of Old China Leading the New," *The New York Times*, Aug. 30, 1992.
27. According to Jerome A. Cohen, formerly associate dean of Harvard Law School and currently representing American business interests in China.

Motorola, Procter & Gamble, Avon Cosmetics, Campbell Soup, Big Mac, Kentucky Fried Chicken, Coca Cola, Heinz baby food, Pabst beer, Nike, Lux soap, Merrill Lynch, and so on.[28] Charles D. Brown, Nike manager in Canton, said: "I don't think we are in a dream stage. This is definitely an economic boom. It could be one of the largest ever."[29] With a total investment in China of U.S.$100 billion in 1993, foreign investors like the low Chinese wages of $2 to $4 a day and the high quality of the work force, a good percentage of whom are computer-literate. China's foreign trade has been growing rapidly, registering U.S.$165.63 billion in 1992 and surpassing that of Taiwan for the first time. In 1992 China enjoyed a U.S.$18.3 billion trade surplus with the United States and a foreign currency reserve in excess of U.S.$40 billion.

RURAL INDUSTRY. With the introduction of the "Responsibility System" in the countryside in the early 1980s, agricultural production rose rapidly. Owned by individual households or village or township collectives, rural industries sprang up overnight to process agricultural products and reap handsome profits. Numbering 18.88 million in 1988 with an average output of Ch$50,000, they expanded rapidly at a rate of 27.9 percent annually from 1979 to 1988, and the value of their output rose from 23.5 percent of the national agricultural production in 1980 to 58.1 percent in 1988. In 1990 they exported U.S.$12.5 billion of goods, about one-fifth of China's total export. Rural industries have thus become the most vibrant sector of the economy and have generated a powerful momentum toward market economy and privatization of enterprises. With 18.88 million of them spreading over the best cropping areas near the major urban centers, they have changed the landscape of China. They are engaged in a variety of activities: industrial production, 69.73 percent; construction, 12.73 percent; commerce, 8.24 percent; transportation, 7.28 percent; and agricultural production, 1.78 percent. They have absorbed a large portion of the migrant farm labor (See next section).

The rural industrial boom created many problems, however, the foremost of which was damage to the environment. Located outside the city limits, rural industries paid scant attention to city ordinances and government regulations on pollution control, chemical discharge, soil erosion,

---

28. *The New York Times*, June 15, 1992; Feb. 24, 1993. However, Levi Strauss & Co., the giant clothing maker of San Francisco, decided to withdraw from China because of its poor human rights record. *Los Angeles Times*, May 4, 1993.
29. Sheryl WuDunn, "Booming China Is Dream Market for West," *The New York Times*, Feb. 15, 1993; *Los Angeles Times*, October 18, 1993.

waste management, and ecology. As a result, air pollution in many Chinese cities is seven or eight times worse than that in New York City, and lung diseases and cancer deaths are on the rise.[30] By A.D. 2000 rural industries will account for 50 percent of the national industrial output, and state and private enterprises will each contribute 25 percent. The Chinese economy by then can hardly be called socialistic.

MIGRATION OF FARM LABOR. The economic boom has caused a historic migration of farm workers to the urban areas. The *China Population Journal* reported that by 1990 some 50 million migrant workers had left rural areas for the cities. In Shanghai the migrants accounted for 1.83 million or 26 percent of all residents; in Peking, 1.15 million, or 22 percent; and in Canton, 880,000 or 33 percent.[31] Although, historically, the appearance of transients symbolized unrest, disorder, and perhaps the end of a dynasty, the migrants of today are a better omen. Mostly young males (93.1 percent) of 25 years or less (70.40 percent), with only a small percentage above 35 (7.50 percent), they are rather well educated, with 67.3 percent having a junior high background and 15.4 percent from senior high. They are ambitious and full of personal initiative and are anxious to seek a better life in a more exciting setting.[32]

The massive migration also signals the breakdown of the party control of population movement and residential permits. Since the beginnings of the People's Republic, the government controlled people's lives by controlling the assignment of their work and their living quarters through their work units. Without a residential permit, one could not receive housing, medical care, food coupons, or permission to travel. Now, however, the migrant workers have challenged the government orders by moving to urban areas on their own, risking punishment and the loss of the perquisites. Although they live in crowded quarters, receive no food and medical subsidies, and accept low-paying menial jobs, after a while their lives improve because of their hard work and personal initiative. Gradually, they win social acceptance and contribute to the economic growth of their adopted land.

30. Da Ning, "Chung-kuo ta-lu hsiang-chen ch'i-yeh nung-yeh ho nung-yeh huan-ching wen-ti" (China's rural industries and their effects on agriculture and agricultural environment), Papers of the Center for Modern China, No. 16 (Feb. 1992): 2-7; Sheryl WuDunn, "Chinese Suffer from Rising Pollution as Byproduct of the Industrial Boom," *The New York Times*, Feb. 28, 1993. *See also* David Zweig, "Internationalizing China's Countryside: The Political Economy of Exports from Rural Industry," *The China Quarterly*, Dec. 1991, 717-18.
31. Ch'eng Chi, "Tang-ch'ien Chung-kuo ti liu-min wen-t'i (The problem of migrant workers in Mainland China today), *T'an So* (The Quest), Jan. 1993, 40-41.
32. According to a survey of 500 farm migrants in Peking. *Ibid.*, 42-43.

THE GREATER CHINA ECONOMIC SPHERE. The burgeoning trade, investment, and tourism linking coastal China, Hong Kong, and Taiwan in recent years gives credence to the idea of a Greater China Economic Sphere. Although still in its embryonic stage, this triangular regional market in time could emerge to resemble the European Economic Community and the North American Free Trade Agreement.[33] The complementary nature of the three economies is obvious in the increase of trade between Taiwan and China via Hong Kong from 1987 to 1992.

Indirect Taiwan-China Trade via Hong Kong
[Millions of U.S. dollars]

	Taiwan exports to China	China exports to Taiwan	Taiwan surplus
1987	$1,226.53	$ 288.94	$ 937.59
1988	2,242.22	478.69	1,763.53
1989	2,896.49	586.90	2,309.59
1990	3,278.25	765.36	2,512.89
1991	4,667.15	1,125.85	3,541.30
1992	6,287.93	1,118.97	5,168.96

Source: Hong Kong Government, Bureau of Statistics.

With China's vast endowment of land, natural resources, labor supply, and production facilities; Taiwan's enormous capital and technological sophistication; and Hong Kong's entrepreneurial and marketing expertise, economic interdependence is growing. Some 12,000 Taiwan corporations have relocated their production plants to the mainland and invested U.S. $11 billion by 1992. At the same time, China has invested U.S.$11 billion in Hong Kong trade, real estate, and financial enterprises; and the Bank of China in Hong Kong is the second largest in the city. Some 80 percent of Hong Kong manufacturers have established branches in China, investing HK$30 billion (U.S.$4 billion) and employing 3 million workers. With the expected admission of China and Taiwan into GATT (General Agreement on Tariffs and Trade, soon to be renamed World Trade Organization) in 1995, trading between the two and Hong Kong will accelerate and tariffs be reduced, giving new impetus to greater market integration and the emergence of a "Bamboo Network."[34]

33. James A. Baker, III, "America in Asia: Emerging Architecture for a Pacific Community," *Foreign Affairs*, Winter 1991-92, 16; Allen Pun, "Galbraith Supports China Sphere Idea," *The Free China Journal*, Nov. 17, 1992, Taipei.

34. Murrey Weidenbaum, "Rising Chinese Economy Creates Prime Opportunity for U.S. Investors," *Los Angeles Times*, June 6, 1993; *The Free China Journal*, Oct. 15, 1993.

The emergence of a Greater China Economic Community is not universally welcomed, however. Japan, fearing competition and perhaps out of spite as well, does not relish such a development; and political resistance on both sides of the Taiwan Straits is also considerable. The conservative Communist leaders fear the influx of bourgeois ideas of freedom and democracy that will inevitably accompany greater opening and the exacerbation of the already wide economic gap between the North and the South and between the coastal areas and the inland provinces. At the same time, some leaders on Taiwan are concerned that closer economic ties with the mainland will strengthen Communist rule and increase Taiwan's dependency, making it more vulnerable to absorption by the mainland.[35] Nevertheless, the benefits of closer economic ties are hard to dispute, given the vast China market and Taiwan's lucrative trade surplus with the mainland (U.S.$3.54 billion in 1991; U.S.$5.16 billion in 1992); thus, in spite of political hesitancy, economic realities are generating momentum for the gradual emergence of a Greater China Economic Market.

The social and economic change in the last decade has been so drastic that some view it as a "White Revolution,"[36] sometimes euphemistically called "socialism with Chinese characteristics." In the scale of history, Teng's economic revolution may turn out to be more significant than Mao's political revolution.

Sino-American Relations: The Most-Favored Nation Status

The Chinese view of the post-Cold War world was incisive and realistic, postulating that although hostilities between the two superpowers may have ended, nations would continue to compete, cooperate, compromise, and fight according to national interests. To survive, any state had as its most fundamental consideration absolute control of its natural resources and economic and military power.

Two powerful forces were apparent in the post-Cold War era: intensified economic competition and a new political equilibrium dominated by no one single state. Although the United States was generally considered the only remaining superpower, its strength was compromised by its large

---

35. Li Tzu-hsiang, "Kung-ch'uang Ta-Chung-hua chin-chi kung-t'ung-ch'uan," (Together we create the Greater China Economic Sphere), *T'an So* (The Quest), March 1992, 35-38.
36. Li Hsiao-ming, "Chung-kuo nan-fang ti tzu-pen chu-i je" (The capitalistic fever in southern China), *China Spring*, March 15, 1993, 10.

national debt, debilitating budgetary deficit, and serious domestic problems. It was no longer the undisputed leader of a Western alliance against the nonexistent Soviet Union, but one of the many contending states, albeit a very powerful one. Although its hegemonistic ambitions could not be ignored, it was a superpower in decline.

China, on the other hand, was a developing country on the rise, with growing economic and military power. In international affairs it assumed a low posture, eschewing any intention of succeeding the Soviet Union as the new center of Communism or confronting the United States on military, economic, and ideological issues. Instead, China concentrated its energies on domestic economic development, military modernization, and intensified political control. Although China lost some strategic advantage in the triangular relations with the United States and Russia, it remained a vitally important state in geopolitical terms, being in control of a large territory, a vast population, a nuclear capability, rich natural resources, a veto power in the United Nations, and rising economic and military strength. With its growing economy and political stability, China had become an increasingly attractive place for foreign investments and international trade, earnings from which could be used to finance its modernization programs. Hence the most-favored-nation status in the United States was far more significant than a trading benefit. It was the money tree that substantively fueled China's economic development. The post-Cold War world offered China a golden chance to grow and to become rich and powerful—the twin goals that had eluded the country for more than a century. It was a historic opportunity not to be squandered.[37]

BUSH'S POSITION ON CHINA AND MFN. China expected the United States to view favorably its policy of economic reform and opening and rapid integration into the world market and to regard its political stability as contributing to world peace. In July 1991, President Bush announced his policy of "constructive engagement" in China, based on the premise that the United States did not intend to isolate China but wished to increase contact with its government and people. Through these contacts Bush believed that gentle persuasion could be applied to urge the Chinese to

37. Wang Yang-ming, "Ts'ung 'chien-she-hsing chiao-feng' yu 'T'ien-ch'i san-ma' k'an Chung-Mei kuan-hsi ti hsin tsou-hsiang" (New trends in Sino-American relations from the standpoint of "constructive engagement" and "T'ien-ch'i horse race"), *T'an So* (The Quest), Sept. 1992, 74-77; Nicholas D. Kristof, "As China Looks at World Order, It Detects New Struggles Emerging," *The New York Times*, April 21, 1992; Gen. John Galvin, "America's Asian Challenge," *U.S. News & World Report*, Oct. 19, 1992.

improve their human rights record, release political prisoners, participate in international arms control, prevent nuclear proliferation and arms sales, observe international trade practice, honor intellectual copyrights, bar products of prison labor for export, and open the Chinese domestic markets.[38]

China was receptive to Bush's approach to a degree, for it was anxious to end the international isolation imposed after the T'ien-an-men crackdown. Knowing China's need for trade and high technology, Teng Hsiao-p'ing chose not to confront the United States but to cooperate with it selectively as in the cases of Cambodia and the Gulf War. China also released political prisoners at critical times of its own choosing to show that it did not act under foreign pressure. That China piled up trade surpluses with the United States of U.S.$12 billion in 1991, U.S.$18.3 billion in 1992, and U.S.$22.7 billion in 1993, and built up a foreign currency reserve in excess of U.S.$40 billion by the end of 1991, showed how valuable the MFN status was. Peking would leave no stone unturned to fight for its renewal.

Actually, the most-favored-nation treatment is not really a special privilege. About 160 of America's trading partners—including Libya, Syria, and until recently Iraq—were granted the status; but China, as a Communist state, was not given MFN until 1979–80 when President Carter normalized relations with China and signed a series of diplomatic and trade agreements granting mutual MFN status. The Jackson Vanik Amendment of 1974 required the President to certify each year to Congress that any Communist country, in order to receive the MFN status, must allow the freedom of emigration. Since the Chinese quotas for immigration, including those from Hong Kong and Taiwan, had always been filled, the President had no difficulty certifying for China. But the T'ien-an-men crackdown, the human rights abuses, the arms sales, and the ballooning trade surpluses led many to demand a reexamination of China's MFN status.

China did not always enjoy trade surplus with the United States. From 1972 to 1982, it had a trade deficit almost annually with the United States and accumulated a total loss of U.S.$8.196 billion. The trade was more or less balanced between 1983 and 1985, but then it turned rapidly in China's favor.[39]

38. James A. Baker, III, "America in Asia: Emerging Architecture for a Pacific Community," *Foreign Affairs*, Winter 1991-92, 15-16.
39. U.S. Department of Commerce figures, International Trade Administration, Office of Trade and Economic Analysis, 1989 and June 1993; also *Los Angeles Times*, May 4, 1991.

U.S. Trade Deficits with China

1983	−U.S.$	.071 billion
1984	− $	.061
1985	− $	.006
1986	− $	1.665
1987	− $	2.796
1988	− $	3.490
1989	− $	6.235
1990	−	$10.431
1991	−	$12.691
1992	−	$18.309
1993	−	$22.770

The galloping trade deficit alarmed many, especially members of Congress. There were suggestions to cut the deficit by revoking China's MFN status, which would increase the import dues on Chinese goods from 8 percent to 40 percent on average, or by attaching conditions to its annual renewal. The conditions most mentioned were these: improvement in China's human rights record; elimination of exports made by prison labor; reduction in arms sales and nuclear aid to Pakistan, Iran, and Algeria; cessation of crackdown in Tibet; accessibility to the domestic Chinese markets; and intellectual copyrights, especially in computer software and pharmaceuticals. What was once a routine renewal became an annual hot debate in American politics, with the Administration favoring unconditional renewal and Congress insisting on conditional extension.

Congresswoman Nancy Pelosi of California was particularly vocal on using trade to force Chinese concessions, and she had the support of Senate majority leader George J. Mitchell, Senator Jesse A. Helms, and House majority leader Richard A. Gephardt. Of Peking's leadership, Pelosi said: "They may not love capitalism, they may not love human rights, but they love money."[40] Under her sponsorship, the House passed a bill in November 1991, 409 to 21, which required China, if it was to receive the MFN status for 1992, to release all T'ien-an-men prisoners numbering about 1,000; stop selling long-range missiles to Syria and Iran; show "progress" in granting free speech, a free press, and free religion in China as well as in Tibet; give "assurance" not to sell nuclear technology; and end export of prison labor products to the United States.

In February 1992 the Senate voted 59 to 39 in favor of conditional renewal of China's MFN status but fell short of the two-thirds majority needed to override the promised presidential veto. Bush was able to extend

40. *The New York Times*, May 13, 1991.

China's status unconditionally for 1992, as he had done in the previous three years.

THE CHINESE ARGUMENT. The Chinese campaigned vigorously for the unconditional renewal of their MFN status. From the high councils in Peking to its ambassador in Washington D.C., Chu Ch'i-chen (Zhu Qizhen); to the American public relations and law firms retained by China to lobby the Administration and Congress, four carefully crafted themes were followed: (1) the paramount value of the Sino-American geopolitical relationship; (2) the recognition of the Chinese concept of human rights, which emphasized meeting the material needs of its people rather than permitting Western-style civil liberties; (3) the likely possibility of a German or Japanese take-over of the American position in China if MFN were revoked; and (4) the unfairness of determining the American trade deficit based on "country of origin" because such a standard included Chinese exports that passed through Hong Kong although American goods shipped to China through a third country were not included in the United States exports to China.[41] Between 1981 and 1990, Chinese export to the United States accounted for 8 percent of the total export, but the Chinese import from the United States grew to 10-13 percent of the total import, indicating a continuous increase in American export to China.[42]

The Chinese complaint about the "country of origin" method of computation was well taken. Since 1987 many Hong Kong and Taiwan manufacturers had moved their production plants to mainland China because of its low wages and cheap raw materials. When the finished products were transshipped via Hong Kong to the United States, they were classified as Chinese exports. Thus, the American trade deficit with China grew dramatically whereas those with Hong Kong and Taiwan declined correspondingly. The total American trade deficit with all three—the Greater China Region—remained relatively stable, ranging between U.S.$22 to U.S.$29 billion annually.[43]

In a conciliatory move, Peking showed a willingness to discuss the human rights issue with foreign representatives. Richard Schifter, assistant secretary of state in charge of human rights, was invited to China; Ambassador James R. Lilley was given permission to visit Tibet. When former president Jimmy Carter was in China, the Chinese listened politely

41. Keith Bradsher, "A Hard Line From China on Trade," *The New York Times*, May 13, 1991.
42. Shih Liu-tzu, "Tsui-hui-kuo tai-yu ti chueh-li" (The contest about the most-favored-nation treatment). *China Spring*, June 1991, 45.
43. U.S. Department of Commerce figures, cited in Nicholas D. Kristof, "China Is Making Asia's Goods, and the U.S. Is Buying," *The New York Times*, March 12, 1993.

	China	Hong Kong	Taiwan	Greater China Region (all 3)
1987	−U.S.$ 2.7 billion	−6.5	−19.0	−28.2
1988	− $ 3.4	−5.1	−14.1	−22.6
1989	− $ 6.2	−3.4	−13.0	−22.6
1990	− $10.4	−2.8	−11.2	−24.4
1991	− $12.7	−1.1	− 9.8	−23.6
1992	− $18.3	−0.7	− 9.4	−28.4

to his plea that they pardon all political dissidents. Undersecretary of State Robert Kimmit went to China in May 1991 to open discussions on human rights, prevention of the spread of nuclear technology and missiles, and bilateral trade, as well as the Middle East and Cambodia.

In November 1991, Secretary of State James A. Baker III came to Peking and raised the same issues. In response, the Chinese gave some information requested by the United States on China's 800 political prisoners but balked at the American suggestion that Peking free some of them or allow the Red Cross to visit them in prison. China, however, did agree to grant exit visas to some dissidents who had no criminal charges pending against them and sometime after Baker's visit allowed several of them to go abroad.[44]

On the question of arms sales and nuclear nonproliferation, Peking made no firm commitment beyond a promise to "observe" the Missile Technology Control Regime of 1987 in exchange for Washington's lifting of the ban on high-speed computers and satellite parts sales to China and also to approve the nonproliferation treaty on nuclear weapons before the year's end. On the basis of these promises, Washington lifted the ban in December 1991 on sales of high-speed computers and satellite parts, despite the absence of proof that Peking would comply with either the missile agreement or the Nuclear Nonproliferation Treaty. In fact, American intelligence continued to detect Chinese shipments of M-9 and M-11 ballistic missile components to Pakistan and Syria and nuclear arms to Algeria. Moreover, diplomatic sources indicated that after the Bush administration agreed to sell 150 F-16 fighter jets to Taiwan in September 1992, the Chinese attitude changed dramatically. They considered the American action a violation of the 1982 protocol worked out during the Reagan administration in which the United States pledged to reduce gradually arms sales to Taiwan. Peking retaliated by exporting missile parts to Pakistan. Thus, the arms trade, too lucrative for China to give up, remained a source of friction between Washington and Peking.

---

44. Including dissident journalist Tai Ch'ing (Dai Qing), Hou Hsiao-t'ien (Hou Xiaotian), Wang Yu-ts'ai (Wang You-cai), and Han Tung-fang (Dan Dongfang).

Chinese students, scholars, and dissidents in the United States generally favored using economic pressure to force Peking to improve human rights; but they did not want to hurt Sino-American trade or the Chinese entrepreneurs, reformers, and workers connected with the export trade. Revocation of the MFN status, they feared, would play into the hands of the die-hard conservatives, who opposed reform and opening.[45]

CLINTON'S DECISION. President Bush, ambassadorial liaison officer in Peking during 1975–76, considered himself an expert on China, but his repeated unconditional extension of China's MFN status came under criticism by many as "shameful." During the presidential election of 1992, Democratic candidate Bill Clinton attacked him for "coddling" Chinese leaders and ignoring their human rights abuses.[46] Clinton denounced "tyrants from Baghdad to Beijing" and proposed to use trade to force Chinese concessions on human rights, arm sales, unfair trade practice, religious persecution, and the Tibetan crackdown. The nervous Chinese leadership decided not to go to a trade war with the United States. Chinese toys accounted for 30 percent of the total U.S. imports in that category; shoes, 40 percent; and textiles and clothes, 26 percent. The situation was too good to be spoiled by human rights and intellectual copyrights issues.[47]

A confidential Foreign Ministry document warned that any setback in trade would hurt China more than the United States. The export industries in the South and along the coast, especially new products created by Hong Kong and Taiwan investments expressly for export to the United States, would be ruined. Direct consequences to China could be these: (1) a loss of a million jobs in coastal areas where large farm migration had taken place, causing social and economic unrest; (2) a loss of low-interest credits and loans from the World Bank; and (3) the discouragement of foreign investment and trade. China could ill afford such setbacks at a critical time of "economic takeoff," and any deterioration in the Sino-American relationship would deal a severe blow to the basic policy of reform and opening. The report recommended that China should publicly protest American interference in Chinese domestic affairs but secretly regard political prisoners as "hostages" to be exchanged for greater American economic benefits. "One step in political concession will lead to two steps in economic gains," stated the document. It would be a win-win game plan.

45. Li Xianlu and Lu Mai, "Renew China's Trade Status," *The New York Times*, July 23, 1991.
46. *The New York Times*, Nov. 20, 1992.
47. Ho Tsin, "Chung-Kung tui jen-ch'uan wen-t'i hui tso-ch'u hui-ying ma?" (Will the Chinese Communists respond to the question of human rights?), *T'an So* (The Quest); Feb. 1993, 16-18.

Teng agreed and decreed that economic pragmatism take precedence over all other concerns because the central task of government was economic development. Indeed, economic gains were best politics. "Hostages for trade"[48] became an accepted practice, and a set of guidelines was adopted giving priority to short-term political prisoners with a record of good behavior and excluding violent prisoners. Neither reduction of prison terms nor admission of wrongful arrests was to be made, nor was the release to be effected under direct foreign pressure. Instead, releases were to be made on grounds of medical treatment to show "revolutionary humanism."

Thus, the release of political prisoners was a calculated political gamble rather than an acceptance of the international standards on human rights. At best, it could be interpreted as Chinese acknowledgment of human rights as a legitimate international concern.[49] As a result, a string of prominent political prisoners were released at critical times to win American goodwill. Pao Tsun-hsin (Bao Zunxin), 55, a philosopher and a leader in the 1989 prodemocracy demonstrations who had been sentenced to a 5-year term was released early in November 1992 for medical treatment abroad; Wang Tan, the student leader in the 1989 demonstrations, was released in February 1993, 4½ months ahead of a 4-year sentence; and Hsü Wen-li (Xu Wenli), 49, a prominent democracy seeker in the early 1980s, was released in May 1993 after serving 12 years of a 15-year sentence. Others were allowed to go abroad. Finally, in an all-out effort to win the Olympic site for Peking in A.D. 2000, the Chinese government released its most famous political prisoner, Wei Ching-sheng (Wei Jingsheng), in mid-September 1993, a half a year ahead of the completion of his 15-year sentence.

China sent purchasing missions to America on a shopping spree to show its sincere desire to narrow the trade gap. In the first half of 1993, they signed an U.S.$800 million agreement with Boeing to purchase 21 jetliners with an option to buy or lease 215 more that were worth U.S.$9 billion in the next few years. China ordered 4,600 vehicles each from General Motors, Ford, and Chrysler, worth U.S.$160 million. It purchased U.S.$200 million of oil drilling and exploration equipment from companies in Texas, Louisiana, and Washington, and U.S. $800 million of satellites from Hughes Aerospace. There were other purchases of grains, phosphate

48. This phrase reminds one of Colonel Oliver North's "guns for hostages" formula during the Irangate.
49. Lu Min-sheng, "Tui-Mei chin-chi le-so: Chung-Kung yung ta-lu ti cheng-chih yen-chi jen-shih tang jou-p'iao" (Economic blackmail of the United States: Chinese Communists used political dissidents as human hostages), *T'an So* (The Quest), Feb. 1993, 77-80.

fertilizers, electric machinery, and telephone and scientific equipment. American businessmen estimated that the China trade created 150,000 jobs in the United States and the future of the Chinese markets seemed limitless. AT&T expected to install half of the 15 million telephone lines in China each year for several years. "We are talking about a business in China that is as big as, if not bigger than, that in the United States," said its vice president, Randall L. Tobias. He wanted to use telecommunication to open up China and make it less repressive.[50]

After his election, Clinton moderated his stance on China. He acknowledged Bush's gains in reducing China's human rights and trade violations and disavowed any intention of isolating China. As he pondered the best course to pursue, he came under intense pressure from 298 large corporations and 37 trade groups connected with the China trade, including General Motors and the Chamber of Commerce. On May 12, 1993, they jointly addressed him not to attach human rights or any conditions to the MFN issue lest American access to the Chinese markets be lost. They stressed that their very presence in China was conducive to promoting fair trade, reform, opening, and democratic ways.[51]

Foreign pleadings were also in evidence. Governor Christopher Patten of Hong Kong, though locked in a fight with China over the future democracy of the island, came to urge Clinton to extend China's MFN status because it could be an effective, peaceful weapon to influence changes in China, and it would also save 150,000 jobs in Hong Kong. The Tibetan spiritual leader, the Dalai Lama, visited the White House to ask for protection of Tibetan religious and cultural heritage in the face of massive Chinese immigration.

Clinton strove to develop a single strategy that would cover trade deficit, arms sales, human rights, prison labor, and Tibetan persecution. He knew that if he did not come up with a satisfactory solution, Congress would legislate a China bill that might be too tough for the Chinese to accept. On May 28, 1993, President Clinton signed an executive order to extend China's MFN status for a year; but he made it very clear that its renewal in mid-1994 would be contingent on "overall, significant progress" in China's human rights record, including releasing or accounting for its political prisoners. Secretary of State Warren Christopher was directed to recommend the following year whether there should be another extension after July 3, 1994, on the basis of China's performance in the following areas:

50. Calvin Sims, "China Steps Up Spending to Keep U.S. Trade Status," *The New York Times*, May 7, 1993.
51. *The New York Times*, May 7, 13, 14, 1993.

1. Releasing or accounting for political prisoners.
2. Ensuring humane treatment of all prisoners and access to prisons by international human rights organizations such as the Red Cross.
3. Protecting Tibet's distinctive religious and cultural heritage.
4. Permitting international radio and television broadcasts into China.
5. Prohibiting prison labor products for exports to America.
6. Ending discrimination against American businesses.
7. Adhering to international treaties governing the transfer of missile technologies and nuclear materials.[52]

China accused the United States of politicizing the trade issue and violating the principles underlying American-Chinese diplomatic relations. Peking insisted that human rights was a domestic concern that permitted no foreign interference. At the Asia-Pacific Economic Cooperation (APEC) forum held in Seattle in November 1993, Chinese President Chiang Tse-min (Jiang Zemin) flatly refused to open a dialogue with President Clinton on human rights and missile sales. Instead, he delivered a 15-minute monologue on the importance of noninterference in the internal affairs of other countries. The feeling was prevalent among Chinese officials that America's own record of human rights was spotty and its position hypocritical. Finance Minister Liu Tsung-li (Liu Zhongli) sarcastically remarked: "The United States maintains a triple standard. For their own human rights problems they shut their eyes. For some other countries' human rights questions they open one eye and shut the other. And for China, they open both eyes and stare."[53] Most Asian states agreed, and some secretly applauded China for standing up to the United States and resisting its pressure to impose Western values on China.

Within the Clinton administration different views arose as to how best to promote American interests in China. The Treasury and Commerce departments had been bombarded with petitions from American corporations concerning the need to maximize trade and investment opportunities in the booming China market. While accepting the importance of human rights, Treasury Secretary Lloyd Bentsen and Commerce Secretary Ronald Brown argued that business investment, trade, missile sales, military co-

---

52. On August 25, 1993, the United States imposed limited sanctions on China for selling components for 24 M-11 surface-to-surface missiles to Pakistan. Washington banned U.S.$1 billion worth of American high technology equipment sales to China in the next two years. The M-11 has a range of 190 miles and a payload of 1,100 pounds, and the M-9 has a range of 300 miles (500 kilometers), which are beyond the 185 miles (300 kilometers) and the 1,100 pounds (500 kilograms) limits specified in the 1987 Missile Technology Control Regime. China did not sign the missile agreement but promised in November 1991 to observe it. The United States was required by its export-control laws to impose sanctions on any country that violated the missile agreement.

53. *The New York Times*, March 20, 1994.

operation, and regional security should also receive due attention in the formulation of national policy. China's economy, already the third largest in the world, could equal that of the United States by the year 2015. Its planned $575 billion infrastructure investment beween now and the end of the century could result in numerous contracts for American corporations and countless jobs for American workers. Robert E. Rubin, chairman of the National Economic Council, and W. Bowman Cutter, Deputy Assistant to the President for Economic Policy, went so far as to say that the annual threat of canceling MFN was outdated and counterproductive, and punishing China by revoking MFN would hurt the United States as much as China.

On the other hand, a softer and more subtle approach—increasing contact through trade and investment—could initiate an evolutionary process leading to greater openness and acceptance of Western business practices and ideas, gradual liberalization, and eventual democracy. It would benefit Chinese reformers and reduce suspicion among the conservatives. One need only witness the proliferation of satellite dishes, fax machines, cellular phones, color television sets, stock markets, and foreign newspapers, banks, and financial services to realize how far China had come since the Maoist days. Indeed, the export of goods was as important as the export of ideas since economic progress was bound to lead to political liberalization.[54] The United States' China policy should therefore be based on economic pragmatism, trade, business investment, security considerations, and the disassociation of trade from human rights.

Clinton saw the logic of this argument, but he was bound by his own restrictive executive order. To regain the diplomatic initiative, he launched a campaign of "intensive engagement" according to which a succession of high-ranking officials would be sent to Peking to persuade the Chinese to improve their record on human rights, trade, and weapons sales, and to cultivate better overall relations between the two countries. First to go, in January 1994, were Agriculture Secretary Mike Espy and Treasury Secretary Lloyd Bentsen; both were well received. Then, in March, came Secretary of State Warren Christopher. What a controversial visit that proved to be!

An outspoken critic of China's human rights abuses, Christopher came with a message urging an improvement in China's record so as to create favorable conditions for Clinton to renew MFN. A week before his arrival,

54. Thomas L. Friedman, "New Tack on China," *The New York Times,* Jan. 23, 1994; Stephen Robert, "In China, Let Free Markets Aid Liberty," *The New York Times,* April 24, 1994; Bill Bradley, "Trade, the Real Engine of Democracy," *The New York Times,* May 25, 1994.

John Shattuck, Assistant Secretary of State for Human Rights, had come to prepare for the visit. Much to the dismay of Chinese officials, on February 27 he met with the arch-dissident Wei Ching-sheng (Wei Jing-sheng) for 90 minutes. Wei called on Clinton to keep economic pressure on the Chinese government and force it to release political prisoners. Meanwhile, seven leading intellectuals appealed to President Chiang (Jiang) for an end to repression and the release of all political prisoners. The government reacted by arresting Wei and a dozen or so prodemocracy leaders. It was a deliberate insult to the visiting American secretary of state.

At the official meetings Premier Li P'eng (Li Peng) lectured Christopher, insisting that China would never yield to American pressure and would never accept its concept of human rights. If President Clinton revoked MFN, Li warned, American business interests in China would suffer. Foreign Minister Ch'ien Ch'i-chen (Qian Qichen) accused Shattuck of violating Chinese law by meeting with "a criminal on parole." He disputed the American claim that China had a $22.7 billion trade surplus with the United States, stating that a large percentage of the exports in question involved Hong Kong, Taiwan, and Singapore investments in China, as well as American-Chinese joint ventures. If Washington revoked MFN, China could redirect these exports to Asian markets and suffer no more than a 10 percent loss, which would not slow its 9 percent annual economic growth. Being largely self-sufficient, China could live quite well without MFN, just as it did prior to Nixon's visit in 1972.

American corporations sided with the Chinese government. Two hundred American business leaders in Peking chided Christopher for the "misguided" policy of linking trade with human rights and placing American businesses at a competitive disadvantage with respect to Japan and the European countries. William Warrick, an AT&T executive, warned that the revocation of MFN would drive American businesses out of the huge Chinese market. Christopher meekly defended himself by stating that the United States was not asking for the transformation of the Chinese political system but only its adherence to the Universal Declaration of Human Rights. In reality, however, the human rights issue went to the very heart of the Chinese political system.

THE CHINESE STRATEGY. Chinese belligerency was puzzling to Americans but explainable in terms of Chinese politics. At a time when the health of the paramount leader, Teng (Deng), was rapidly failing (he was 89 at the time) politicians in Peking outdid each other in protesting their allegiance to the Communist system. A tough stance against the United States could be an asset in the forthcoming struggle for succession. This political climate foredoomed Christopher's visit, the more so because he came at a time when the National People's Congress was in session.

Any analysis of the Chinese attitude toward human rights first assumes an understanding of the nature of the Communist system. The Leninist dictatorship rested on the coercive instruments of secret police, army, and party monopoly of power. Thought control and labor camps were a way of life in which human rights had no part. Even in today's more open China, the government continues to view political dissent and democratic stirrings as counterrevolutionary threats. Political pluralism and power-sharing were never in the Marxist vocabulary. Intellectual agitation was regarded as a prelude to political turmoil and social chaos, which could disrupt economic growth and bring down the existing order. When survival of the Communist system, real or imagined, was at stake, the loss of MFN paled into insignificance.

Shattuck's meeting with Wei Ching-sheng (Wei Jingsheng) was viewed by the Chinese authorities as collusion with a criminal and encouragement of the dissident movement. It enhanced Wei's international reputation and emboldened the dissidents. Upon learning of Wei's message to his friends that they could expect "a great event" during Christopher's visit, the authorities instinctively feared that Wei would seek a meeting with Christopher and that the dissidents would capitalize on the occasion to mount a major demonstration. The officials struck first by detaining the leading dissidents or sending them away from the capital, prohibiting all intellectuals from meeting with foreign envoys. They did this in accordance with Teng's (Deng's) standing order that they should never hesitate to crack down on dissidents nor worry about foreign reaction.

The government also felt vulnerable with respect to the rising tide of social and economic tensions generated by the 100 million "floating" population of surplus labor; the 26 percent urban inflation rate, which pushed up vegetable prices by 54 percent, grains by 40 percent, and meat by 30 percent; and the discontent of peasants, who received IOUs rather than cash from the government for their grain. Worker unrest and rising unemployment led to 6,000 strikes and 200 riots in 1993. In fact, the government feared organized labor as much as it did political dissidents and would go to any length to maintain "social stability."

Against this background of fear and insecurity, American intervention in the Chinese human rights issue was perceived as a disturbing influence and an opening salvo in a larger plot to destabilize China. Previously the party had waged "anti-spiritual pollution" and "anti-bourgeois liberalization" campaigns to combat foreign influence; now it fought American interference in Chinese domestic affairs with a vengeance, humiliating Christopher by regarding him as Clinton's surrogate.

Second, with hundreds of American corporations vying for investment opportunities in China and eyeing infrastructure projects worth $575 billion, and with Chinese imports growing at a 9 percent rate annually, Peking was sure that the United States would not risk being left out of this lucraive market. It gambled that in the end American economic interests would triumph over human rights considerations, leaving Clinton no choice but to renew China's MFN status. If not, the revocation would hurt the United States more than China.

Third, the knowledge that the United Nations would need China's cooperation in sanctioning North Korea for refusing to permit inspections of its nuclear plants made Peking confident that the United States would pay the price of extending MFN. Besides, no European or Asian states linked trade to human rights, so why should the United States? On a trip to Peking in March 1994, Japanese Prime Minister Morihiro Hosokawa remarked that Western concepts of human rights should not be applied blindly to all countries.

Last, Clinton was caught in the impossible situation of confronting China on human rights while hoping to improve economic and military ties. Such a major contradiction, according to the Marxist view, should be fully exploited. Employing a "divide and rule" tactic, China courted American corporations and let them lead the charge against their own government. The Chinese also subtly cultivated the goodwill of the Treasury and Commerce departments while taking a tough stance toward the State Department, always pitting business interests against human rights. They dared Clinton to cancel MFN, convinced that he would not do so since he could not afford to offend corporate America, whose support he would need in future elections. The Chinese understood American politics well and succeeded in driving Washington into disarray.

THE U.S. POLICY REVERSAL. When President Clinton issued his executive order in May 1993, it was generally praised as a brilliant stroke of statesmanship, but in reality he was merely postponing the difficult decision on China for a year and allowing himself little flexibility of action. When the Chinese saw through the hollowness of his threat, Clinton faced the painful decision of either canceling MFN or disassociating trade from human rights, fully aware that the former would cause disastrous economic consequences to both countries, and that the latter would seriously undermine his credibility. Meanwhile, his critics—among them human rights organizations—argued that if the United States would simply stand firm on its principles, China would be forced to comply since it needed the American market to finance its modernization and offset its $12.8 billion overall trade deficit.

Strangely, no one in the Clinton administration and few in Congress actually wanted to cancel MFN. The pressing question was how to find a face-saving way of salvaging Clinton's China policy. Clearly, Washington needed China's cooperation. Sensing victory, Premier Li P'eng declared that China would make "great efforts" to improve relations with the United States. On April 23 and May 14, 1994, Peking released two leading activists involved in the 1989 prodemocracy demonstrations—Wang Chun-t'ao (Wang Juntao), aged 35, and Ch'en Tse-ming (Chen Zeming), aged 41—for medical treatment abroad. China reached an agreement with the Swiss Red Cross permitting selective prison inspection. In April 1994 a 200-member Chinese purchasing delegation signed $11 billion worth of contracts with American corporations. Finally, on May 18 Peking agreed to receive American technicians to discuss an end to the jamming of Voice of America broadcasts.

With China making only limited progress in the area of human rights, Clinton could revoke MFN completely, partially, or selectively. The mood in Congress was for disassociating trade from human rights, but Senate Majority leader George J. Mitchell (D-Me.) and Representative Nancy Pelosi (D-Ca.) pushed for "targeted sanctions," according to which the United States would cancel preferential tariffs on Chinese imports made by state enterprises and military industries but not on imports from the private sector. However, administering such a procedure would be a nightmare, and Peking categorically rejected any conditional renewal of MFN.

On May 24 Christopher reported to the President that China had met the two mandatory requirements of ending the export of prison-labor products and allowing certain dissidents and their families to emigrate. However, China had not made "overall significant progress" in the five other areas: adhering to the Universal Declaration of Human Rights; releasing and accounting for political prisoners; permitting the International Red Cross to inspect prisons; ending the jamming of international radio and television broadcasts into China; and preserving Tibet's religious and cultural heritage.

Clinton agonized over the MFN decision. In a defensive and almost anguished tone, on May 26 he announced that while China "continues to commit very serious human rights abuses," the United States economic and strategic interests led him to renew China's MFN status, with the token proviso that there would be a ban on imports of Chinese-made guns and ammunition—estimated at $200 million a year, with Chinese imports totaling $31 billion! With this announcement Clinton officially disassociated trade from human rights. "That linkage has been constructive during the past year," said the President, "but I believe, based on our aggressive contacts with the Chinese in the past few months, that we have reached the end of the usefulness of that policy, and it is time to take a new path toward the achievement of our constant objectives. We need to place our relationship into a larger and more productive framework."

The stunning policy reversal signified a tacit admission that Clinton's executive order was misguided and that "Bush was right." Furthermore, the message that the United States could be defied with impunity reverberated worldwide. Clinton's credibility had sustained a major blow. Senator Mitchell and Congresswoman Pelosi threatened to introduce legislation to reverse the President's decision and to impose greater sanctions against China. But many members of Congress as well as foreign policy experts considered the President's decision sound, if politically embarrassing. His decision would stand unless rejected by both houses.

The driving force behind Clinton's action was his concern for overall American-Chinese relations, American business interests in China, and American jobs connected with the China trade. Furthermore, he found compelling reasons for his decision in the geopolitical need for China's cooperation in dealing with North Korea, the future of China's reform movement, and the conviction that the dynamics of change would lead China to adopt a steady course that would ultimately result in a freer and more open society.

Viewed from a broader perspective, the President's action reflected the spirit of a new age in which economic interests dominate foreign policy decisions, and profits, not principles, are the prime concern. It signaled the end of the T'ien-an-men era's brand of diplomacy and the beginning of a period of rising international commerce and greater scientific, educational, and cultural exchanges.

China, of course, was pleased with Clinton's decision and predicted improved relations with the United States. Most other Asian and European countries were relieved that the annual fight over MFN was finally over.

Teng's Place in History

As architect of the policy of reform and opening, Teng Hsiao-p'ing is responsible for the economic boom that is sweeping over China. Doubtless, the economic transformation of China is a major event in the history of the late twentieth century, with a far-reaching impact on Asia and the world at large. As such, an assessment of Teng's place in history seems in order.

China's paramount leader since 1979, Teng is an "indestructible" revolutionary of the Long March generation. Making his historical debut in France in 1920 as a 16-year-old work-study student from Szechuan, he returned home a committed Communist. He was present at the historic Tsunyi Conference in January 1935 as a *Red Star* editor, supported Mao Tse-tung, and rose quickly thereafter. Mao, on a trip to Moscow in 1957, pointed Teng out to Khrushchev: "See that little man there? He's highly intelligent and has a great future ahead of him."[55] Mao's words were prophetic. Shortly after Mao's death, Teng ascended rapidly to the pinnacle of power and became the second "core" leader of Chinese Communism.

Ironically, however, Teng's greatest claim to a place in history was his courage in reversing the Maoist course in 1979. In condemning the Cultural Revolution as a "decade of great catastrophe," he ushered in a new age in which his "Economics in Command" replaced Mao's "Politics in Command." Promoting two new slogans—"Seek truth from facts" and "Practice is the sole criterion of truth"—Teng demystified Mao as a demi-God and denied the sanctity of his Thought, which was now to be subject to the scrutiny of facts, practice, and, above all, truth. Having thus brought Mao down from his high pedestal, Teng single-handedly initiated the bold venture of economic reform and opening to the outside world. It was his way to alleviate the shortcomings and limitations of Communism, which obviously had failed to lift China from poverty and backwardness for 30 years.

Although Teng borrowed Western science, technology, and managerial skill to strengthen the Communist rule, he never intended to move China toward a Western-style parliamentary democracy. To him, the three-way division of power was like having three governments in one country, each vying for a greater voice in national affairs and contributing to divisiveness and dissension. Political pluralism was anathema because it encouraged squabbles, compromises, mutual attrition, and ultimate weakness. In Teng's view, China could not afford disruptions such as student demonstrations or disturbing influences like Western bourgeois liberalism. Only a strong executive—one who could demand absolute obedience and unquestioning loyalty—ensured national stability, the essential condition for the modernization that would bring China ultimate wealth and power.

Yet although Teng found no contradiction between economic progressivism and his conservative politics, the two were intrinsically incompatible and could not coexist except in temporary expediency. In the end, economic

55. Nikita Khrushchev, *Khrushchev Remembers: The Last Testament* (Boston, 1974), 253. Tr. and ed. by Strobe Talbott.

prosperity and social affluence would dilute the revolutionary ethics and erode the power base of the proletarian dictatorship. Thus, in adopting quasi-capitalism to save Communism, Teng may have sown the seeds of its ultimate demise and unwittingly set in motion, however imperceptibly at first, a process for eventual political liberalization.

Teng's lasting regret had to be his agonizing decision to crack down on the T'ien-an-men student demonstrations in 1989. That General Secretary Chao Tzu-yang, who counseled against the use of force, was not charged with treason pointed to Teng's post-factum realization that he had over-reacted. Like many—if not most—politicians, Teng was incapable of admitting his mistakes, yet his promotion of economic prosperity and improved living conditions for the people was perhaps a quiet effort partially to atone for the bloody massacre that horrified not only China but also the world.

Doubtless, in the scale of history, Teng was less weighty than Mao, but his economic revolution could well have a greater impact on China and the world than Mao's political revolution. The two, in fact, complemented each other. Mao broke down the Old Order; Teng laid the foundations of a new one, based on economic prosperity and the adoption of Western science, technology, and managerial skill. Yet because of the persistence of the Leninist dictatorship, Teng's New Order could only be transitory. Until political liberalization—the Fifth Modernization—is achieved, China's transformation remains incomplete.

China in the Year 2000

What will China be like in 10 to 15 years? Lawrence H. Summers, formerly senior economist of the World Bank and currently undersecretary of the Treasury for international affairs in the Clinton Administration, spoke of China's future in historical perspective:

> It may well be that when the history of the late 20th century is written 100 years from now, the most significant event will be the revolutionary change in China, which will soon be Communist only in a rhetorical sense . . . For more than a century, the United States has been the world's largest economy. The only nation with a chance of surpassing it in the next generation in absolute scale is China.[56]

China's economic progress seems irreversible regardless of the change in leadership. On the basis of purchasing power, the International Monetary Fund puts China's economy as the third largest in the world—U.S.$1.66 trillion in GDP in 1991—after the United States and Japan.[57] By A.D.

56. Cited in Kristof, "Entrepreneurial . . ." *The New York Times*, Feb. 14, 1993.

World of Money: 2 Ways of Counting

A new study has dramatically altered the lineup of the world's largest economies, giving more weight to China and other developing countries. The study measures the value of each nation's gross domestic product in terms of the purchasing power of its own currency at home, rather than the currency's value on international exchanges.

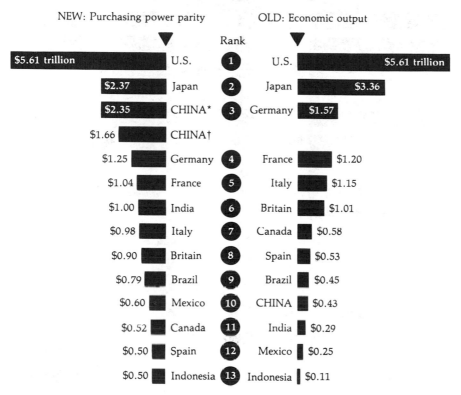

NEW: Purchasing power parity   OLD: Economic output

		Rank		
$5.61 trillion	U.S.	1	U.S.	$5.61 trillion
$2.37	Japan	2	Japan	$3.36
$2.35	CHINA*	3	Germany	$1.57
$1.66	CHINA†			
$1.25	Germany	4	France	$1.20
$1.04	France	5	Italy	$1.15
$1.00	India	6	Britain	$1.01
$0.98	Italy	7	Canada	$0.58
$0.90	Britain	8	Spain	$0.53
$0.79	Brazil	9	Brazil	$0.45
$0.60	Mexico	10	CHINA	$0.43
$0.52	Canada	11	India	$0.29
$0.50	Spain	12	Mexico	$0.25
$0.50	Indonesia	13	Indonesia	$0.11

*World Bank estimate
†I.M.F. estimate

Figures exclude the republics
of the former Soviet Union.

*Sources: Organization for Economic Cooperation and Development.
World Bank, International Monetary Fund.*

2020 the Chinese economy could be the largest on earth. Politically, the government is searching for a unifying philosophy to reinforce the bankrupt Communism and seems to incline to a mixture of traditional Confucianism, nationalism, patriotism, and a touch of Great Chinaism that subtly reflects the Middle Kingdom concept. At the same time it keeps up the economic momentum by sponsoring a new version of the comprador-styled, profit-oriented quasi-capitalism that is well integrated into the world economy.

57. Steven Greenhouse, "New Tally of World's Economies Catapults China into Third Place," *The New York Times,* May 20, 1993.

Measuring Per Capita Income

Per capita income in each country according to the two different measure. Countries are ranked according to the new system.

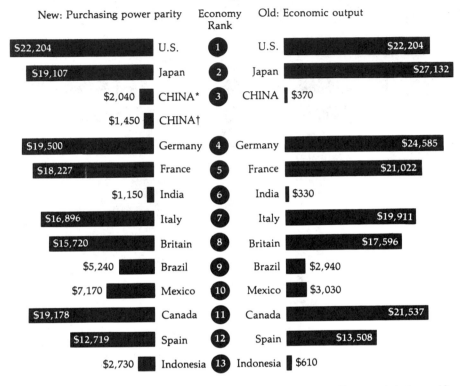

New: Purchasing power parity		Economy Rank	Old: Economic output	
$22,204	U.S.	1	U.S.	$22,204
$19,107	Japan	2	Japan	$27,132
$2,040	CHINA*	3	CHINA	$370
$1,450	CHINA†			
$19,500	Germany	4	Germany	$24,585
$18,227	France	5	France	$21,022
$1,150	India	6	India	$330
$16,896	Italy	7	Italy	$19,911
$15,720	Britain	8	Britain	$17,596
$5,240	Brazil	9	Brazil	$2,940
$7,170	Mexico	10	Mexico	$3,030
$19,178	Canada	11	Canada	$21,537
$12,719	Spain	12	Spain	$13,508
$2,730	Indonesia	13	Indonesia	$610

*World Bank estimate
† I.M.F. estimate

Figures exclude the republics
of the former Soviet Union.

*Sources: Organization for Economic Cooperation and Development.
World Bank, International Monetary Fund.*

Investment in the military will continue to grow—13 percent each year between 1990 and 1992—to make China a major player in international politics.[58] Control of the South China Sea and extension of influence into the Indian Ocean are part of Chinese strategic thinking.

58. Nicholas D. Kristof, "China, the Conglomerate, Seeks a New Unifying Principle," *The New York Times*, Feb. 21, 1993. His views are somewhat different from mine. See also views of Professor Liu Shu-hsien of the Chinese University of Hong Kong, cited in Chin Chung, "Chung-kuo hsu-yao tsung-chiao ching-shen" (China needs the spirit of religion), *K'ai-fang*, Hong Kong, April 1992, 49-50.

What is the vision of the people? Most of them have little respect for the government as it now stands. Yet, abhorring chaos and disorder (*luan*), they want no repetition of the Cultural Revolution nor any part of the chaos, confusion, and shortages accompanying the collapse of Communism in the former Soviet Union. They do not advocate a violent, bloody overthrow of the government. Rather, they favor a peaceful transformation of the Communist system from within.[59] The passing of the current leaders is only a matter of time. The next generation of leaders may be nominal Communists, but it is certain that they will be pragmatic technocrats whose chief concerns will be economic progress and the raising of living standards.

Increasingly, Chinese leadership will come under the pressure of the new middle class in an emerging civil society. The powerful new rich—the entrepreneurs, the financiers, the bankers, the investors, and the directors of corporations and international trading firms—will demand greater political participation and the rule of law, as well as a voice in legislation and budgetary decisions. Step-by-step, the government will be forced to accommodate their demands and permit the freedom of speech, assembly, and the press, as well as freedom of religion, creative and artistic expression, and popular elections. With this scenario of peaceful evolution, a Chinese-style democracy may finally emerge.[60]

59. Shen Tong, "The Next Generation," *The New York Times*, Sept. 2, 1992.
60. Yang Mai-k'e, "Ta-lu ti hsien-tai-hua sao-tung" (The modernization agitation on the Mainland), *T'an So* (The Quest) Nov. 1992, 43-45. *See also* a different view in Nicholas D. Kristof, "Chinese Communism's Secret Aim: Capitalism," *The New York Times*, Oct. 19, 1992.

Further Reading

Baker, James A., III, "America in Asia: Emerging Architecture for a Pacific Community," *Foreign Affairs*, Winter 1991-92, 1-18.

Barme, Geremie, "Traveling Heavy: The Intellectual Baggage of the Chinese Diaspora," *The Chinese Intellectual*, No. 26 (Winter 1992): 3-20.

Borthwick, Mark, *Pacific Century: The Emergence of Modern Pacific Asia* (Boulder, 1992).

Bradsher, Keith, "A Hard Line from China on Trade," *The New York Times*, May 13, 1991.

Brugger, Bill, and David Kelly, *Chinese Marxism in the Post-Mao Era* (Stanford, 1990).

Brzezinski, Zbigniew, *Out of Control: Global Turmoil on the Eve of the 21st Century* (New York, 1993).

Catton, Chris, "Great Leap Backward," *New Statesman & Society*, England, Jan. 8, 1993, 28-30.

Chai, Joseph C. H., "Consumption and Living Standards in China," *The China Quarterly*, Sept. 1992, 721-49.

Ch'eng, Chi, "Tang-ch'ien Chung-kuo ti liu-min wen-t'i" (The problem of migrant workers in Mainland China today), *T'an So* (The Quest), Jan. 1993, 40-44.

Ch'i, Hsi-sheng, *Politics of Disillusionment: The Chinese Communist Party under Deng Xiaoping* (Armonk, N.Y., 1991).

Chien, Frederick F., "A View from Taipei," *Foreign Affairs, Winter* 1991-92, 92-103.

Chin, Chung, "Chung-kuo hsü-yao tsung-chiao ching-shen," (China needs the spirit of religion), *K'ai-fang*, Hong Kong, April 1992, 49-50.

China: Between Plan and Market (Washington, D.C., 1990). The World Bank.

"China in Transformation," *Daedalus*, Spring 1993.

China's Economic Dilemmas in the 1990s: The Problems of Reforms, Modernization, and Interdependence (Washington, D.C., 1991). 2 vols. Study Papers submitted to the Joint Economic Committee, Congress of the United States.

Ch'ien, Chia-chü, "Ho-p'ing yen-pien yu ho-p'ing yen-pien ti wen-t'i" (Peaceful evolution and questions about the peaceful evolution), *T'an So* (The Quest), April 1992, 11-16.

Chinese Communist Central Committee: *A Critique of the Soviet Central Committee Second Plenary Session*. Confidential. Complete text reprinted in *China Spring*, May 1990, 64-65.

———, Document 2, Feb. 28, 1992. Complete text reprinted in *Cheng-meng* magazine, Hong Kong, April 1992, 23-27.

Cohen, Roger, "An Empty Feeling Is Affecting Eastern Europe," *The New York Times*, March 21, 1993, E3.

Copper, John F., and Ta-ling Lee, *Tiananmen Aftermath: Human Rights in the People's Republic of China, 1990*. Occasional Papers/ Reprint Series, in Contemporary Asian Studies, No. 4—1992. School of Law, University of Maryland.

Da, Ning, "Chung-kuo ta-lu hsiang-chen ch'i-yeh nung-yeh ho nung-yeh huan-ching wen-t'i" (China's rural industries and their effect on agriculture and agricultural environment). Paper of the Center for Modern China, No. 16, Feb. 1992, 1-10.

"Deng Xiaoping: An Assessment," *The China Quarterly*, Sept. 1993, 409-572.

Galvin, Gen. John, "America's Asian Challenge," *U.S. News & World Report*, Oct. 19, 1992.

Garver, John W., "China's Push Through the South China Sea: The Interaction of Bureaucratic and National Interests," *The China Quarterly*, Dec. 1992, 999-1028.

———, "The Chinese Communist Party and the Collapse of Soviet Communism," *The China Quarterly*, March 1993, 1-26.

Greenhouse, Steven, "New Tally of World's Economies Catapults China Into Third Place," *The New York Times*, May 20, 1993.

Harding, Harry, *A Fragile Relationship: The United States and China Since 1972* (Washington, D.C., 1992).

———, *China's Second Revolution: Reform After Mao* (Washington, D.C., 1987).

Ho, Tsin, "Chung-Kung tui jen-ch'üan wen-t'i hui tso-ch'ü hui-ying ma?" (Will the Chinese Communists respond to the question of human rights?), *T'an So* (The Quest), Feb. 1993, 16-18.

Holley, David, "Beijing Wants Open Market but Closed Society," *Los Angeles Times*, May 9, 1992.

——, "The New Religion Is Money," *Los Angeles Times*, Jan. 15, 1993.

Holme, Richard, "China: Tiger's Gilded Cage," *New Statesman & Society*, England, Jan. 22, 1993, 19-20.

Hsu, John C., *China's Foreign Trade Reforms: Impact on Growth and Stability* (Cambridge, England, 1989).

Kallgren, Joyce K., *Building a Nation-State: China After Forty Years* (Berkeley, 1990).

Kristof, Nicholas D., "China Sees Singapore as a Model of Progress," *The New York Times*, Aug. 9, 1992.

——, "Looking Beyond Deng, Leaders Find Chinese Looking Beyond Them," *The New York Times*, Oct. 11, 1992.

——, "A New Class of Chinese Is Emerging: Home Owners," *The New York Times*, Nov. 22, 1992.

——, "Entrepreneurial Energy Sets Off a Chinese Boom," *The New York Times*, Feb. 14, 1993.

——, "Don't Joke About This Stock Market," *The New York Times*, May 9, 1993.

——, "Chinese Communism's Secret Aim: Capitalism," *The New York Times*, Oct. 19, 1992.

——, "Poor Chinese Town Bets Its Shirt on Making Buttons and, Bingo," *The New York Times*, Jan. 18, 1993.

——, "Backed by China, Go-Getters Get Rich," *The New York Times*, Jan. 27, 1993.

——, "A Tycoon Named Mu: Product of Old China Leading the New," *The New York Times*, August 30, 1992.

——, "As China Looks at World Order, It Detects New Struggles Emerging," *The New York Times*, April 21, 1992.

——, "China Is Making Asia's Goods, and the U.S. Is Buying," *The New York Times*, March 12, 1993.

——, "China: the Conglomerate, Seeks a New Unifying Principle," *The New York Times*, Feb. 21, 1993.

Kueh, Y. Y., "Foreign Investment and Economic Change in China," *The China Quarterly*, Sept. 1992, 637-90.

Lao, Pei, "Ts'ung pien-ko ti chueh-tu k'an Chung-kuo ta-lu ti wei-lai" (The future of Mainland China from the standpoint of reform), *T'an So* (The Quest), Jan. 1993, 76-77.

Lardy, Nicholas R., *Foreign Trade and Economic Reforms in China, 1978-1990* (Cambridge, England, 1992).

——, "Chinese Foreign Trade," *The China Quarterly*, Sept. 1992, 691-720.

Leng, Heng-mei, "Shih-ssu-ta wei ta-lu mai-hsia-liao tung-luan ti ho-keng" (The Fourteenth Congress sowed the seed of future disorder in Mainland China), *T'an So* (The Quest), Dec. 1992, 35-39.

Li, Cheng, and Lynn White, "Elite Transformation and Modern Change in Mainland China and Taiwan: Empirical Data and the Theory of Technocracy," *The China Quarterly*, March 1990, 1-35.

Li, Hsiao-ming, "Chung-kuo nan-fang ti tzu-pen chu-i je" (The capitalistic fever in southern China), *China Spring*, Oct. 1992, 6-10.

Li, Tzu-hsiang, "Kung-chuang Ta-Chung-hua chin chi kung-t'ung-ch'uan" (To-gether we create the Greater China Economic Common Sphere), *T'an So* (The Quest), March 1992, 35-38.

Li, Xianlu, and Lu Mai, "Renew China's Trade Status," *The New York Times*, July 23, 1991.

Liu, Hsüeh-min, "Ho-p'ing yen-pien chiu Chung-kuo" (Peaceful evolution saves China), *T'an So* (The Quest), June 1991, 35-38.

Liu, Yia-ling, "Reform from Below: The Private Economy and Local Politics in the Rural Industrialization of Wenzhou," *The China Quarterly*, June 1992, 293-316.

Lord, Winston, "China and America: Beyond the Big Chill," *Foreign Affairs*, Fall 1989, 1-26.

Lu, Min-sheng, "Tui-Mei chin-chi le-so: Chung-Kung yung ta-lu ti cheng-chih yen-chi jen-shih tang jou-p'iao" (Economic blackmail of the United States: Chinese Communists used Mainland's political dissidents as human hostages), *T'an So* (The Quest), Feb. 1993, 77-80.

McCord, William, *The Dawn of the Pacific Century: Implications for Three Worlds of Development* (New Brunswick, 1991).

Meisner, Maurice, "What Beijing Leaders Know That Critics Won't See: Repression Can Be Profitable," *Los Angeles Times*, Oct. 25, 1992, M2.

Nathan, Andrew, *China's Crisis: Dilemmas of Reform and Prospects for Democracy* (New York, 1990).

Rocca, Jean-Louis, "Corruption and Its Shadow: An Anthropological View of Cor-ruption in China," *The China Quarterly*, June 1992, 402-16. Tr. from French.

Safire, William, "Vision in Collision," *The New York Times*, Dec. 23, 1992, A25.

Saich, Tony, "The Fourteenth Party Congress: A Programme for Authoritarian Rule," *The China Quarterly*, Dec. 1992, 1136-1160.

Salisbury, Harrison E., *The New Emperors: China in the Era of Mao and Deng* (Boston, 1992).

Shen, Tong, "The Next Generation," *The New York Times*, Sept. 2, 1992.

Solinger, Dorothy J., *China's Transition from Socialism? Statist Legacies and Market Reforms, 1980-1990* (New York, 1992).

Tsai, Wen-hui, "New Authoritarianism, Neo-Conservatism, and Anti-Peaceful Evo-lution: Mainland China's Resistance to Political Modernization," *Issues & Studies*, Taipei, Vol. 28, No. 2, 1992, 1-22.

"The People's Republic After 40 Years," *The China Quarterly*, No. 119, Sept. 1989, pp. 420-630.

Tucker, Nancy Bernkopf, "China and America, 1941-1991," *Foreign Affairs*, Winter 1991-92, 75-92.

Vogel, Ezra, *One Step Ahead in China: Guangdong Under Reform* (Cambridge, Mass., 1989).

———, *The Four Dragons: The Spread of Industrialization in East Asia* (Cam-bridge, Mass., 1991).

Wang, Wei-hsin, "Teng Hsiao-p'ing lu-hsien ti kuan-ch'a" (An observation of the Teng Hsiao-p'ing line), *T'an So* (The Quest), May 1992, 40-42.

Weidenbaum, Murrey, "Rising Chinese Economy Creates Prime Opportunity for U.S. Investors," *Los Angeles Times*, June 6, 1993.

WuDunn, Sheryl, "Economy Is a Pawn In China's Power Game," *The New York Times*, July 5, 1992.

————, "Booming China Is Dream Market for West," *The New York Times*, Feb. 15, 1993.

————, "Chinese Suffer from Rising Pollution As Byproduct of the Industrial Boom," *The New York Times*, Feb. 28, 1993.

————, "As China Leaps Ahead, The Poor Slip Behind," *The New York Times*, May 23, 1993.

Yang, Mai-k'e, "Ta-lu ti hsien-tai-hua sao-tung" (The modernization agitation on the Mainland), *T'an So* (The Quest), Nov. 1992, 43-45.

————, "Yu-fang chieh-chi chueh-ch'i ta-lu," (A home-owning class arises on Mainland China), *China Spring*, March 1993, 47-48.

Zheng, Decheng, "Ya-T'ai ti-ch'ü chiu-shih nien-tai ti chin-chi ch'ü-shih chi ch'i tui Chung-kuo ti cheng-ts'e han-i" (The economic trends in the Asian-Pacific Region to 2000: Implications to P. R. China), *Papers of the Center for Modern China*, Vol. 3, No. 6, June 1992, 1-26.

Zweig, David, "Internationalizing China's Countryside: The Political Economy of Exports from Rural Industry," *The China Quarterly*, Dec. 1991, 716-41.

Appendix I

Chinese Personal Names, Places, and Terms Mentioned in Part VII
in Both Wade-Giles and Pinyin Systems

Wade-Giles	Pinyin	Wade-Giles	Pinyin
Ah Chia	Ah-Jia	Ch'ün-yen-t'ang	Qun yan tang
		Chung-fa	Zhongfa
Canton	Guangzhou	Chung-nan hai	Zhongnanhai
Ch'ai Tse-min	Chai Zemin		
Chang Ai-p'ing	Zhang Aiping	Fan Yi	Fang Yi
Chang Ching-fu	Zhang Jingfu	Fang Li-chih	Fang Lizhi
Chang Ch'un-ch'iao	Zhang Chunqiao	Fei Hsiao-t'ung	Fei Xiaotong
Chao Ts'ang-pi	Zhao Cangbi		
Chao Tzu-yang	Zhao Ziyang	Han Tung-fang	Han Dongfang
Ch'en Hsi-lien	Chen Xilian	Hai Jui	Hai Rui
Ch'en Po-ta	Chen Boda	*Ho-p'ing yen-pien*	*Heping Yanbian*
Ch'en Yuan	Chen Yuan	Ho Ching-chih	He Jingzhi
Ch'en Yün	Chen Yun	Ho Lung	He Long
Ch'en Yung-kuei	Chen Yonggui	Hou Hsiao-t'ien	Hou Xiaotian
Ch'eng-tu	Chengdu	*hsia-fang*	*xiafang*
Chi-tung	Jidong	Hsieh Fu-chih	Xie Fuzhi
Chi Teng-k'uei	Ji Dengkui	Hsu Wen-li	Xu Wenli
chia-chang-chih	*jia zhang zhi*	Hu Ch'i-li	Hu Qili
chia-t'ien hsia	*jia tian xia*	Hu Ching-t'ao	Hu Jingtao
Chiang Ch'ing	Jiang Qing	Huang Chen	Huang Zhen
Chiang Hua	Jiang Hua	Huang Huo-ch'ing	Huang Huoqing
Chiang Tse-min	Jiang Zemin	Hsü Hsiang-ch'ien	Xu Xiangqian
Chiang Wen	Jiang Wen	Hsü Shih-yu	Xu Shiyou
Ch'iao Shih	Qiao Shi	Hu Ch'iao-mu	Hu Qiaomu
Ch'ien Ch'i-chen	Qian Qichen	Hu Yoa-pang	Hu Yaobang
Chou Chia-hua	Zhou Jiahua	Hua Kuo-feng	Hua Guofeng
Chou En-lai	Zhou Enlai	Huang Hua	Huang Hua
Chu Ch'i-chen	Zhu Qizhen	Huang Huo-ch'ing	Huang Huoqing
chu-fan	*zhu fan*	Huang K'e-ch'eng	Huang Kecheng
Chu Jung-chi	Zhu Rongji		
Chu Teh	Zhu De	*I-ch'iung erh-pai*	*Yi qiong er bai*
Ch'ü Ch'iu-pai	Qu Qiubai	I-yen-t'ang	Yi yan tang

985

Wade-Giles	Pinyin	Wade-Giles	Pinyin
K'ang Sheng	Kang Sheng	Ta-ch'ing	Daqing
Kao Ti	Gao Di	Tai Ch'ing	Dai Qing
Kuang-ming Jih-pao	*Guangming ribao*	T'ang-shan	Tangshan
kung-nung-ping	gong nong bing	T'ao Chu	Tao Zhu
		Teng Jung	Deng Rong
Li Hsien-nien	Li Xiannian	Teng Li-ch'ün	Deng Liqun
Li Jui-huan	Li Ruihuan	Teng Hsiao-p'ing	Deng Xiaoping
Li P'eng	Li Dazhao	*t'i-yung*	*tiyong*
Li Ta-chao	Li Peng	Tiao-yü-t'ai	Diaoyu tai
Li T'ieh-ying	Li Tieying	T'ien Chi-yun	Tian Jiyun
Liang Hsiao	Liang Xiao	*T'ien-tsai jen-huo*	Tianzai renhuo
Liao Mo-sha	Liao Mosha	Tientsin	Tianjin
Lin Piao	Lin Biao	Ting Sheng	Ding Sheng
Liu Hua-ch'ing	Liu Huaqing	Tsing hua	Qinghua
Liu Pin-yen	Liu Binyan	Tsunyi	Zunyi
Liu Shao-ch'i	Liu Shaoqi		
Lu Ting-i	Lu Dingyi	Wang Chen	Wang Zhen
luan	*luan*	Wang Hai-jung	Wang Hairong
		Wang Hung-wen	Wang Hongwen
Mao Tse-tung	Mao Zedong	Wang Jen-chung	Wang Renzhong
Mao Yüan-hsin	Mao Yuanxin	Wang Jen-chih	Wang Renzhi
Mu Ch'i-chung	Mu Qizhong	Wang Jo-wang	Wang Ruowang
		Wang Tan	Wang Dan
Nanking	Nanjing	Wang Tung-hsing	Wang Dongxing
Nei-ching wai-sung	*Neijing waisong*	Wang Yu-ts'ai	Wang Youcai
		Wei Ching-sheng	Wei Jingsheng
Pao-shan	Baoshan	Wei Kuo-ch'ing	Wei Guoqing
Pao Tsun-hsin	Bao Zunxin	Wen-chow	Wenzhou
Peking	Beijing	Wu Te	Wu De
P'eng Chen	Peng Zhen	Wu Tse-t'ien	Wu Zetian
P'eng Ch'ung	Peng Chong		
P'eng Te-huai	Peng Dehuai	Yang Ching-jen	Yang Jingren
p'ing-fan	*ping fan*	Yang Pai-ping	Yang Baibing
Po I-po	Bo Ibo	Yang Shang-k'un	Yang Shangkun
		Yang Te-chih	Yang Dezhi
Shen-yang	Shenyang	Yao I-lin	Yao Yilin
Su Yü	Su Yu	Yao Wen-yüan	Yao Wenyuan
Su Chen-hua	Su Zhenhua	Yeh Chien-ying	Ye Jianying
Szechwan	Sichuan	Yü Ch'iu-li	Yu Qiuli
		Yü Hui-yung	Yu Huiyong
Ta-chai	Dazhai		

Appendix II

Hanyu Pinyin/Wade-Giles Conversion Table

Pinyin	Wade-Giles	Pinyin	Wade-Giles	Pinyin	Wade-Giles
a	a	chao	ch'ao	ding	ting
ai	ai	che	ch'e	diu	tiu
an	an	chen	ch'en	dong	tung
ang	ang	cheng	ch'eng	dou	tou
ao	ao	chi	ch'ih	du	tu
		chong	ch'ung	duan	tuan
ba	pa	chou	ch'ou	dui	tui
bai	pai	chu	ch'u	dun	tun
ban	pan	chuai	ch'uai	duo	to
bang	pang	chuan	ch'uan		
bao	pao	chuang	ch'uang	e	e, o
bei	pei	chui	ch'ui	ei	ei
ben	pen	chun	ch'un	en	en
beng	peng	chuo	ch'o	eng	eng
bi	pi	ci	tz'u	er	erh
bian	pien	cong	ts'ung		
biao	piao	cou	ts'ou	fa	fa
bie	pieh	cu	ts'u	fan	fan
bin	pin	cuan	ts'uan	fang	fang
bing	ping	cui	ts'ui	fei	fei
bo	po	cun	ts'un	fen	fen
bu	pu	cuo	ts'o	feng	feng
				fo	fo
ca	ts'a	da	ta	fou	fou
cai	ts'ai	dai	tai	fu	fu
can	ts'an	dan	tan		
cang	ts'ang	dang	tang	ga	ka
cao	ts'ao	dao	tao	gai	kai
ce	ts'e	de	te	gan	kan
cen	ts'en	dei	tei	gang	kang
ceng	ts'eng	deng	teng	gao	kao
cha	ch'a	di	ti	ge	ke, ko
chai	ch'ai	dian	tien	gei	kei
chan	ch'an	diao	tiao	gen	ken
chang	ch'ang	die	tieh	geng	keng

From Endymion Wilkinson, *The History of Imperial China: A Research Guide,* Harvard Asian Monographs, No. 49 Cambridge, Mass.: Harvard University Press, 1975.

Pinyin	Wade-Giles	Pinyin	Wade-Giles	Pinyin	Wade-Giles
gong	kung	keng	k'eng	mo	mo
gou	kou	kong	k'ung	mou	mou
gu	ku	kou	k'ou	mu	mu
gua	kua	ku	k'u		
guai	kuai	kua	k'ua	na	na
guan	kuan	kuai	k'uai	nai	nai
guang	kuang	kuan	k'uan	nan	nan
gui	kuei	kuang	k'uang	nang	nang
gun	kun	kui	k'uei	nao	nao
guo	kuo	kun	k'un	ne	ne
		kuo	k'uo	nei	nei
ha	ha			nen	nen
hai	hai	la	la	neng	neng
han	han	lai	lai	ni	ni
hang	hang	lan	lan	nian	nien
hao	hao	lang	lang	niang	niang
he	he, ho	lao	lao	niao	niao
hei	hei	le	le	nie	nieh
hen	hen	lei	lei	nin	nin
heng	heng	leng	leng	ning	ning
hong	hung	li	li	niu	niu
hou	hou	lia	lia	nong	nung
hu	hu	lian	lien	nou	nou
hua	hua	liang	liang	nu	nu
huai	huai	liao	liao	nuan	nuan
huan	huan	lie	lieh	nuo	no
huang	huang	lin	lin	nü	nü
hui	hui	ling	ling	në	nüeh
hun	hun	liu	liu		
huo	huo	long	lung	o	o
		lou	lou	ou	ou
ji	chi	lu	lu		
jia	chia	luan	luan	pa	p'a
jian	chien	lun	lun	pai	p'ai
jiang	chiang	luo	lo	pan	p'an
jiao	chiao	lü	lü	pang	p'ang
jie	chieh	lüe	lüeh	pao	p'ao
jin	chin			pei	p'ei
jing	ching	ma	ma	pen	p'en
jiong	chiung	mai	mai	peng	p'eng
jiu	chiu	man	man	pi	p'i
ju	chü	mang	mang	pian	p'ien
juan	chüan	mao	mao	piao	p'iao
jue	chüeh	mei	mei	pie	p'ieh
jun	chün	men	men	pin	p'in
		meng	meng	ping	p'ing
ka	k'a	mi	mi	po	p'o
kai	k'ai	mian	mien	pou	p'ou
kan	k'an	miao	miao	pu	p'u
kang	k'ang	mie	mieh		
kao	k'ao	min	min	qi	ch'i
ke	k'e, k'o	ming	ming	qia	ch'ia
ken	k'en	miu	miu	qian	ch'ien
				qiang	ch'iang

Pinyin	Wade-Giles	Pinyin	Wade-Giles	Pinyin	Wade-Giles
qiao	ch'iao	si	szu	ya	ya
qie	ch'ieh	song	sung	yan	yen
qin	ch'in	sou	sou	yang	yang
qing	ch'ing	su	su	yao	yao
qiong	ch'iung	suan	suan	ye	yeh
qiu	ch'iu	sui	sui	yi	i
qu	ch'ü	sun	sun	yin	yin
quan	ch'üan	suo	so	ying	ying
que	ch'üeh			yong	yung
qun	ch'ün	ta	t'a	you	yu
		tai	t'ai	yu	yü
ran	jan	tan	t'an	yuan	yüan
rang	jang	tang	t'ang	yue	yüeh
rao	jao	tao	t'ao	yun	yün
re	je	te	t'e		
ren	jen	teng	t'eng	za	tsa
reng	jeng	ti	t'i	zai	tsai
ri	jih	tian	t'ien	zan	tsan
rong	jung	tiao	t'iao	zang	tsang
rou	jou	tie	t'ieh	zao	tsao
ru	ju	ting	t'ing	ze	tse
ruan	juan	tong	t'ung	zei	tsei
rui	jui	tou	t'ou	zen	tsen
run	jun	tu	t'u	zeng	tseng
ruo	jo	tuan	t'uan	zha	cha
		tui	t'ui	zhai	chai
sa	sa	tun	t'un	zhan	chan
sai	sai	tuo	t'o	zhang	chang
san	san			zhao	chao
sang	sang	wa	wa	zhe	che
sao	sao	wai	wai	zhei	chei
se	se	wan	wan	zhen	chen
sen	sen	wang	wang	zheng	cheng
seng	seng	wei	wei	zhi	chih
sha	sha	wen	wen	zhong	chung
shai	shai	weng	weng	zhou	chou
shan	shan	wo	wo	zhu	chu
shang	shang	wu	wu	zhua	chua
shao	shao			zhuai	chuai
she	she	xi	hsi	zhuan	chuan
shei	shei	xia	hsia	zhuang	chuang
shen	shen	xian	hsien	zhui	chui
sheng	sheng	xiang	hsiang	zhun	chun
shi	shih	xiao	hsiao	zhuo	cho
shou	shou	xie	hsieh	zi	tzu
shu	shu	xin	hsin	zong	tsung
shua	shua	xing	hsing	zou	tsou
shuai	shuai	xiong	hsiung	zu	tsu
shuan	shuan	xiu	hsiu	zuan	tsuan
shuang	shuang	xu	hsü	zui	tsui
shui	shui	xuan	hsüan	zun	tsun
shun	shun	xue	hsüeh	zuo	tso
shuo	shuo	xun	hsün		

Illustration Credits

Section 1 (following page 122)

1. From *Ch'ing-tai Ti-hou Hsiang* (Pictures of Ch'ing Emperors and Empresses), Palace Museum, Peking, 1931. 2. The Metropolitan Museum of Art, Rogers Fund, 1942. 3. The Metropolitan Museum of Art, Rogers Fund, 1942. 4. The Metropolitan Museum of Art, Rogers Fund, 1942. 5. From Sarah Pike Conger, *Letters from China*, 1909. 6. From John Nieuhof, *An Embassy from the East Indian Company of the United Provinces to the Grand Tartar*, London, 1669. Rare Book Division, New York Public Library. 7. The Palace Museum, Peking. 8. *China Pictorial*, Number 5, 1980. 9. The Palace Museum, Peking. 10. Author's picture, 1981. 11. Peking Slides Studio, Peking. 12. Private collection. 13. Private collection. 14. Private collection. 15. From Yen Kung-cho, *Ch'ing-tai Hsüeh-che Hsiang-chuan* (Pictorial Biographies of Ch'ing Scholars), Shanghai, 1930. 16. From Nieuhof, *An Embassy*. 17. From Nieuhof, *An Embassy*.

Section 2 (following page 220)

1. From Stanley F. Wright, *Hart and the Chinese Customs*, Belfast, 1950. 2. Radio Times Hulton Picture Library. 3. From Henry McAleavy, *The Modern History of China*, New York, 1967. 4. Radio Times Hulton Picture Library. 5. Courtesy of the Boston Museum of Fine Arts. 6. Radio Times Hulton Picture Library. 7. Radio Times Hulton Picture Library. 8. From Alexander Michie, *The Englishman in China During the Victorian Era*, Edinburgh and London, 1800. 9. The Bettmann Archive. 10. Radio Times Hulton Picture Library.

Section 3 (following page 294)

1. The Bettmann Archive. 2. The Bettmann Archive. 3. From Yen Kung-cho, *Ch'ing tai*. 4. Radio Times Hulton Picture Library. 5. From Hsiang-hsiang Wu, *Hsiang-hsiang Tseng-shih Wen-hsien*, Taipei, 1965. 6. From Gideon Chen, *Tso Tsung-t'ang*, Peking, 1938. 7. From Wright, *Hart and the Chinese Customs*. 8. Radio Times Hulton Picture Library. 9. Radio Times Hulton Picture Library. 10. From William A. P. Martin, *A Cycle of Cathay*, New York and Chicago, 1896. 11. From Thomas E. La Fargue, *China's First Hundred*, Pullman, Washington, 1942. 12. From La Fargue, *China's First Hundred*.

13. Charles Phelps Cushing. 14. The Freer Gallery of Art. 15. The Metropolitan Museum of Art, gift of Franklin Jasper Walls, 1948.

Section 4 (following page 418)

1. From Alexander H. Smith. *China in Convulsion*, Edinburgh and London, 1901. 2. The Metropolitan Museum of Art, anonymous gift, 1942. 3. The Bettmann Archive. 4. Charles Phelps Cushing. 5. From Timothy Richard, *Forty-five Years in China*, New York, 1916. 6. From Richard, *Forty-five Years*. 7. Radio Times Hulton Picture Library. 8. Radio Times Hulton Picture Library. 9. Radio Times Hulton Picture Library. 10. Compliments of Huang Hsing's son-in-law, Professor Chün-tu Hsüeh of the University of Maryland. 11. Li Wen-ta, *Jen-lung-jen: Pu-yi hua-chuan* (Hong Kong, 1987).

Section 5 (following page 492)

1. Radio Times Hulton Picture Library. 2. Eastfoto. 3. Courtesy of the Academia Sinica. 4. *China Pictorial*, Number 10, 1976. 5. Shanghai People's Publishing House. 6. From Ōkubo Yasushi, *Chūgoku Kyōsantō shi*, Tokyo, 1971. 7. From *Li Ta-chao chuan*, Peking, People's Publishing House, 1980. 8. *China Pictorial*, Number 8, 1977. 9. Shensi People's Publishing House. 10. Shensi People's Publishing House.

Section 6 (following page 644)

1. *China Pictorial*, Number 8, 1977. 2. Author's picture. 3. *China Pictorial*, Number 5, 1980. 4. *Chou En-lai hsüan-chi*, Peking, People's Publishing House, 1980. 5. *China Pictorial*, Number 2, 1980. 6. *China Pictorial*, Number 6, 1967. 7. *China Pictorial*, Number 6, 1977. 8. The First Tractor Plant, Loyang, China.

Section 7 (following page 706)

1. Wide World Photos. 2. *China Pictorial*, Supplement to Number 11, 1972. 3. *China Pictorial*, Number 4, 1972. 4. *China Pictorial*, Special Issue Number 11, 1973. 5. *China Pictorial*, Special Issue Number 11, 1973. 6. *China Pictorial*, Number 11, 1973. 7. Museum of Pottery Figures of Warriors and Horses, Sian, Shensi province. 8. Museum of Pottery Figures of Warriors and Horses, Sian, Shensi province. 9. Museum of Pottery Figures of Warriors and Horses, Sian, Shensi province. 10. *China Pictorial*, Number 6, 1977. 11. *China Pictorial*, Number 6, 1977. 12. *Sankei Shimbun*, Aug. 15, 1974.

Section 8 (following page 784)

1. China Pictorial, Number 12, 1976. 2. *China Pictorial*, Number 1, 1977. 3. Author's picture. 4. Author's picture. 5. *China Pictorial*, Number 11, 1977. 6. *China Pictorial*, Number 10, 1977. 7. *China Pictorial*, Number 12, 1977. 8. *Cheng-ming* Magazine, *Hong Kong*, Number 40, 1981–82. 9. *China Pictorial*, Number 11, 1977. 10. *A Great Trial in Chinese History*, New World Press, 1981. 11. *A Great Trial in Chinese History*, New World Press, 1981. 12. *Taipei Pictorial*, Number 158, 1981.

Section 9 (following page 906)

1. *Beijing Review*, Volume 30, Number 46, Nov. 16–22, 1987. 2. *Beijing Review*, Volume 30, Number 49, Dec. 7–13, 1987. 3. *Overseas Scholars*, Taipei, Number 186, 1988. 4. *Sinorama*, Jan. 1986.

Section 10 (following page 948)

3. *Free China Review*, May 1993. 4. *Beijing Review*, Oct. 9-15, 1989. 5. Reuters/Bettmann. 6. *Beijing Review*, April 19-25, 1993. 7. *Beijing Review*, April 5-11, 1993. 8. Copyright © 1993 by *The New York Times*, May 20, 1993. Reprinted by permission. 9. Copyright © 1993 by *The New York Times*, May 20, 1993. Reprinted by permission.

Index

Abahai, 23–25, 441

Academia Sinica, on Taiwan, 756

Age of Discovery, 3, 6–7; spirit of adventurism, 92

Age of Enlightenment, 105

Agricultural Reform (1979–), 843; replacement of commune by the "Responsibility System," 844; farmer's life improved, 845; the 10,000-yüan family, 845; increased production, 846; new problems, 847; spectacular results, 848; record grain production of 407 million tons in 1984, 853

Aigun, 216–18

Aksu, 320

Albazin, 3, 108, 110–12, 117

Alcock, Rutherford, and Alcock Convention, 297, 299–300

Allen, Young J., influence of on Chinese reform thought, 358

American Revolution, influence of on China, 453

Amherst Mission, 163–66

Amoy: Ming "loyalist" opposition at, 27–28, 189; foreign trade at, 94, 139–40; opening of, 190, 200

Amur Region in Sino-Russian relations, 33–34, 108, 110–13, 216–18, 347

An Te-hai, 264, 307

An-Fu Club, 484

Ancient Texts (Ku-wen), 363–64

Anglo-Japanese Alliance, 533

Anhwei Clique, 484

Anking, 244, 279

Annam: Ch'ing suzerainty over, 40–41, 130, 132; Chinese trade with, 134; in Sino-French relations, 325–30

"Anti-Ch'ing Revive Ming" (Fan-Ch'ing fu-Ming), 127–30, 231, 241, 453

"Anti-Bourgeois Liberalization" campaign (1986): led by P'eng Chen and Teng Li-chün, 886; supported by Teng for a while, 886; height of campaign, 887; party's dilemma, 887; Premier Chao's position in, 887–88; Teng's position in, 886–88; ended, 888

Anti-Comintern Pact, 582, 585

Anti-rightist Campaign, 663

"Antiquity-Doubters" (I-ku p'ai), 510

Araki, Sadao, 580–81

Arrow War, 206–7; Tientsin settlement, 208–11; Shanghai Tariff Conference, 211–12; Taku repulse, 213–14; Convention of Peking, 215; Russian advance, 216–18

Association of God Worshippers (Pai Shang-ti hui), 228

Autumn Harvest Uprising, 554

Ayüki, 112–13

Baikov, Feodor, 108

Baker, James A., III: clarifies American agenda for China, 945; visits Peking in Nov. 1991, 964

Bandung Conference, 662, 664; "Five Principles of Peaceful Co-existence," 730

Banner system: under Nurhaci, 22; under

993